Lecture Notes in Computer Science 4416

Commenced Publication in 1973
Founding and Former Series Editors:
Gerhard Goos, Juris Hartmanis, and Jan van Leeuwen

Alberto Bemporad Antonio Bicchi
Giorgio Buttazzo (Eds.)

Hybrid Systems: Computation and Control

10th International Conference, HSCC 2007
Pisa, Italy, April 3-5, 2007
Proceedings

Springer

Volume Editors

Alberto Bemporad
Università di Siena
Dip. Ingegneria dell'Informazione
Via Roma, 56, 53100 Siena, Italy
E-mail: bemporad@unisi.it

Antonio Bicchi
Università di Pisa
Centro "E. Piaggio", Facoltà di Ingegneria
Via Diotisalvi 2, 56100 Pisa, Italy
E-mail: bicchi@ing.unipi.it

Giorgio Buttazzo
RETIS Lab., Area CNR, Scuola Superiore S. Anna
Via Moruzzi 1, 56124 Pisa, Italy
E-mail: buttazzo@sssup.it

Library of Congress Control Number: 2007922928

CR Subject Classification (1998): C.3, C.1.3, F.3, D.2, F.1.2, J.2, I.6

LNCS Sublibrary: SL 1 – Theoretical Computer Science and General Issues

ISSN 0302-9743
ISBN-10 3-540-71492-8 Springer Berlin Heidelberg New York
ISBN-13 978-3-540-71492-7 Springer Berlin Heidelberg New York

Springer is a part of Springer Science+Business Media

springer.com

Printed in Germany

Typesetting: Camera-ready by author, data conversion by Scientific Publishing Services, Chennai, India
Printed on acid-free paper SPIN: 12038794 06/3142 5 4 3 2 1 0

Preface

This volume contains the proceedings of the 10th International Conference on Hybrid Systems: Computation and Control (HSCC 2007) held in Pisa, Italy, April 3-5, 2007. The conference, tenth in a series of successful annual meetings, was dedicated to research in embedded reactive systems involving the interplay between symbolic/switching and continuous dynamical behaviors, and attracted academic as well as industrial researchers to exchange information on the latest developments of theoretical advancements and applications for the design, analysis, control, optimization, and implementation of hybrid systems. This year, the scope was broadened to include, along with traditional HSCC topics, areas of research where the convergence of information and control is currently hottest. Embedded and real-time control and control of/over communication networks are two such areas.

The previous workshops in the HSCC series were held in Berkeley, USA (1998), Nijmegen, The Netherlands (1999), Pittsburgh, USA (2000), Rome, Italy (2001), Palo Alto, USA (2002), Prague, Czech Republic (2003), Philadelphia, USA (2004), Zurich, Switzerland (2005), and Santa Barbara, USA (2006). We were honored to co-chair this prestigious event, which has marked the history of the convergence of computation and control science and engineering; and even more so, as this was the tenth anniversary of HSCC.

The program consisted of 3 keynote speeches, 44 regular papers, and 39 short papers selected from 167 regular submissions. The program covered topics such as tools for analysis and verification, control and optimization, modeling, and engineering applications. More details about the conference, the program, and other activities are available on the conference Web site http://hscc07.dii.unisi.it.

We would like to thank the Program Committee members and reviewers for an excellent job of evaluating the submissions and participating in the online Program Committee discussions. Special thanks go to Ed Brinksma (Embedded Systems Institute, Eindhoven, The Netherlands), Shankar Sastry (University of California, Berkeley, USA), and John A. Stankovic (University of Virginia, USA) for their participation as keynote speakers. We are also grateful to the HSCC Steering Committee for helpful guidance and support. We would like to express our gratitude to HYCON and ARTIST, Networks of Excellence of the Sixth Framework Programme of the European Commission, for sponsoring the event.

January 2007

Alberto Bemporad
Antonio Bicchi
Giorgio Buttazzo

Organization

HSCC 2007 was organized by the University of Pisa, the University of Siena, and the Scuola Superiore Sant'Anna of Pisa in cooperation with ACM/SIGBED.

General Chairs

Alberto Bemporad (University of Siena, Italy)
Antonio Bicchi (University of Pisa, Italy)
Giorgio Buttazzo (Scuola Superiore Sant'Anna, Italy)

Steering Committee

Rajeev Alur (University of Pennsylvania, USA)
Bruce Krogh (Carnegie Mellon University, USA)
Oded Maler (VERIMAG, France)
Manfred Morari (ETH Zurich, Switzerland)
George Pappas (University of Pennsylvania, USA)
John Rushby (SRI Int.l)

Program Committee

Tarek Abdalzaher (University of Illinois Urbana Champaign, USA)
Karl-Erik Arzen (Lund University, Sweden)
Calin Belta (Boston University, USA)
Michael Branicky (Case Western Reserve University, USA)
Jennifer Davoren (University of Melbourne, Australia)
Luca de Alfaro (University of California, Santa Cruz, USA)
Bart De Schutter (Delft University, The Netherlands)
Magnus Egerstedt (Georgia Inst. of Technology, USA)
Petru Eles (Linkoping University, Sweden)
Emilio Frazzoli (University of California, Los Angeles, USA)
Antoine Girard (University of Grenoble, France)
Alessandro Giua (Universitá di Cagliari, Italy)
Radu Grosu (S.U. N.Y. at Stony Brook, USA)
Maurice Heemels (Technical University of Eindhoven, The Netherlands)
Joao Hespanha (University of California, Santa Barbara, USA)
Franjo Ivancic (NEC Labs, Princeton, USA)
Karl-Henrik Johansson (Royal Inst. of Technology, Sweden)
Daniel Liberzon (University of Illlinois Urbana Champaign, USA)
Pau Marti (University of Politecnica Catalunya, Spain)
Ian Mitchell (University of British Columbia, Canada)

Joel Ouaknine (Oxford University, UK)
Luigi Palopoli (Università di Trento, Italy)
Benedetto Piccoli (CNR-IAC Rome, Italy)
Maria Prandini (Politecnico di Milano, Italy)
Jean F. Raskin (Université Libre de Bruxelles, Belgium)
Olaf Stursberg (Technical University of Munich, Germany)
Paulo Tabuada (Notre Dame University, USA)

Referees

E. Aaron
T. Abdalzaher
A. Alessandri
K.-E. Ärzen
E. Asarin
V. Azhmyakov
G. Batt
C. Belta
L. Benvenuti
M. Bernadsky
H.P. Bieker
Y. Boers
F. Borrelli
U. Boscain
O.r Bournez
P. Bouyer
M. Branicky
T. Brihaye
L. Busoniu
M.P. Cabasino
S. Cadambi
P. Caines
K. Camlibel
F. Cassez
F. Ceragioli
D. Chatterjee
F. Chevalier
P. Collins
D. Corona
J. Costa
J. Daafouz
T. Dang
J. Davoren
L. de Alfaro
B. de Jager
C. De Persis
B. De Schutter
M. De Wulf
M. Di Bernardo
S. Di Gennaro
A. Donze
L. Doyen
M. Egerstedt
P. Eles
E. Farcot
A. Fehnker
G. Ferrari Trecate
E. Frazzoli
G. Frehse
L. Fribourg
M. Garavello
A. Girard
E. Girejko
A. Giua
R. Grosu
A. Halasz
M. Heemels
J. Hespanha
F. Ivancic
R. Izmailov
K.-H. Johansson
A.A. Julius
E. Klavins
M. Kloetzer
S. Kremer
F. Laroussinie
M. Lazar
C. Le Guernic
M. Lemmon
D. Liberzon
S. Loizou
G. Lowe
J. Lunze
N. Markey
P. Marti
P. Mason
R. Mayr
I. Mitchell
J.M. Moehlis
B. Munsky
I. Necoara
J. Ouaknine
L. Palopoli
S. Paoletti
A. Paoli
M. Pavone
B. Piccoli
A. Pisano
S. Pogromsky
G. Pola
M. Prandini
M. Rabi
J.F. Raskin
J. Riehl
M. Roetteler
A. Rondepierre
S. Sankaranarayanan
C. Seatzu
R. Sepulchre
Y. Sharon
S. Simic
A. Singh
S. Smolka
O. Stursberg
S. Susca

P. Tabuada
D.C. Tarraf
M. Theobald
E. Trelat
P. Valigi
J. van Ast
B. van Beek
N. van der Wouw
M. Velasco
T. Villa
L. Vu
Y. Wardi
S. Weiland
E. Witrant
J. Worrell
K. Zadarnowska

Sponsoring Institutions

University of Pisa
HYCON, a Network of Excellence of the Sixth Framework Programme
ARTIST2, a Network of Excellence of the Sixth Framework Programme
EECI, the European Embedded Control Institute

Table of Contents

Hybrid Systems: Computation and Control

Keynote Speeches

Regular Papers

Short Papers

Networked Embedded Systems: From Sensor Webs to Cyber-Physical Systems

Shankar Sastry

University of California at Berkeley - Berkeley, CA, USA

Abstract. There has been a great deal of excitement in recent years concerning the evolution of sensor webs of smart dust. There has been a very substantive active world wide in this area and in particular at Berkeley there have now been over six generation of "motes" for these sensor webs, and at least three new start ups have arisen to commercialize these developments. I will survey these developments and where they have brought us in a very important new class of computing involving an integration of communication and computing. I will describe how the technology push is matched by the applications pull on numerous different applications.

Throughout the talk, I will highlight the efforts of my group and that of my colleagues especially Culler, Pister, Wagner and Brewer in "closing the loop" around these networked embedded systems. We believe that this closing the loop brings into sharp focus the real time constraints and issues inherent in the use of networked embedded systems. Further, the most important new directions in sensor webs involve this new direction beyond simply sensing and monitoring the physical environment and infrastructure. In particular, I will describe the range of methods and algorithms needed to track multiple targets in a sensor web and to be able to pursue them. The culmination of this project was a 557 node demonstration that we conducted at the Richmond Field Station in August 2005. Some areas of future development in sensor networks involve the use of high bandwidth sensors (such as camera motes) and mobile sensor webs. I will give a preview of some of the most exciting opportunities in this regard.

Finally, with our increase dependency on computing and communication to instrument physical infrastructures, such as electric power, water, gas, etc. we find that they are not high confidence: that is they are complex systems which may not be correct by construction, or fault tolerant and are vulnerable to information attack of networked embedded systems. Such systems are being referred to as high confidence cyber physical systems. To address this grand challenge societal problem of builing high confidence cyber physical systems, I will give a snap shot of the kinds of techniques with, privacy and policy work, in the area of secure network embedded systems.

A. Bemporad, A. Bicchi, and G. Buttazzo (Eds.): HSCC 2007, LNCS 4416, p. 1, 2007.

Control Challenges in Wireless Sensor Networks

John A. Stankovic

Department of Computer Science University of Virginia, USA

Abstract. Wireless sensor networks (WSN) composed of large numbers of small devices that self-organize are being investigated for a wide variety of applications. Applications, such as military surveillance and large scale assisted living facilities are key examples of applications that can benefit from WSN. Current research for WSN is widespread. However, many of the proposed solutions are developed with simplifying assumptions about wireless communication and the environment, even though the realities of wireless communication and environmental sensing are well known. Many of the solutions are evaluated only by simulation. In this talk I describe a fully implemented system, called VigilNet, consisting of a suite of more than 30 synthesized protocols (40,000 lines of code). The system supports a power aware surveillance, tracking and classification application running on 203 XSM motes and evaluated in a realistic, large-area environment. Technical details and evaluations are presented for several of the key services. In developing such systems various types of control challenges occur. I will also discuss a number of these challenges, possible solutions and open challenges.

A. Bemporad, A. Bicchi, and G. Buttazzo (Eds.): HSCC 2007, LNCS 4416, p. 2, 2007.

The Challenges of Embedded Systems Engineering

Ed Brinksma

Embedded Systems Institute, Eindhoven, The Netherlands

Abstract. Embedded system technology has become an important, if not dominating component in the realization of all sorts of high-tech products, machines, and infrastructures. The temptation to create systems with new, powerful, intelligent features has turned embedded software into an essential high-tech ingredient that, exploiting the hardware capabilities afforded by Moore's law, is subject to exponential growth. As has been pointed out by many authors before, the complexity of the embedded software is not just a product of its growing size, but also results from the required relation between the software and its physical environment, both in terms of its execution on physical platforms and in its interaction with the system environment. This combination of digital control and physical phenomena makes it plausible that hybrid modelling and hybrid systems theory have a role to play in the design of embedded systems. Last year Henzinger and Sifakis (The Embedded Systems Design Challenge) suggested that we need the development of more physically informed models of computation combining analytical and constructive elements. This would provide a more fundamental basis for embedded systems design, as well as provide a much needed paradigmatic change for computer science.

Being in principle sympathetic to the proposed programme from a scientific point of view, in this talk we want to examine what are the most pressing problems from an engineering point of view, in particular from the overall system perspective. The software complexity of high-tech systems is often related to the system integration, and not to the embedded software of individual components. The interpretation of the required integral functionality usually includes engineering disciplines beyond those related to hardware, software, and control, with particular methods, models, and tools. It remains to be seen whether deep (i.e. at the semantic level) integration of relevant models and methods will ultimately outperform more loosely coupled coalitions of specialized approaches that are closer to the cultures of the contributing disciplines. Another practically dominant concern often is the sheer size of the collective system software. Alternative approaches to design, therefore, must scale up and support the management of large quantities of design software (programs, models, specifications, etc.). In our presentation we will draw on a number of big industry-as-laboratory projects carried out by the Embedded Systems Institute.

A. Bemporad, A. Bicchi, and G. Buttazzo (Eds.): HSCC 2007, LNCS 4416, p. 3, 2007.

Computational Approaches to Reachability Analysis of Stochastic Hybrid Systems

Alessandro Abate[1], Saurabh Amin[1], Maria Prandini[2], John Lygeros[3], and Shankar Sastry[1]

[1] University of California, at Berkeley - Berkeley, CA, USA
{aabate,saurabh,sastry}@eecs.berkeley.edu
[2] Politecnico di Milano - Milano, Italy
prandini@elet.polimi.it
[3] ETH Zurich - Zurich, Switzerland
lygeros@control.ee.ethz.ch

Abstract. This work investigates some of the computational issues involved in the solution of probabilistic reachability problems for discrete-time, controlled stochastic hybrid systems. It is first argued that, under rather weak continuity assumptions on the stochastic kernels that characterize the dynamics of the system, the numerical solution of a discretized version of the probabilistic reachability problem is guaranteed to converge to the optimal one, as the discretization level decreases. With reference to a benchmark problem, it is then discussed how some of the structural properties of the hybrid system under study can be exploited to solve the probabilistic reachability problem more efficiently. Possible techniques that can increase the scale-up potential of the proposed numerical approximation scheme are suggested.

1 Introduction

This paper addresses the problem of determining the control policy that maximizes the probability that a stochastic system will remain within a safe set over some look-ahead time horizon (finite-time probabilistic reachability problem). We focus on the discrete time controlled stochastic hybrid system (DTSHS) model introduced in [1], and consider the case when the control input to be applied at a certain time is selected based only on the value of the state at that same time (Markov policy). Following the approach in [1,2], the stochastic reachability problem of interest can be formulated as a finite-horizon optimal control problem with a multiplicative cost function to be maximized. This optimal control problem, in turn, can be solved by dynamic programming (DP). This requires to introduce a cost-to-go function and to determine the value of the control input maximizing the cost-to-go function along the reference time horizon for all values of the state within the safe set. Since an analytic solution to the DP equation is generally hard to find, the computational aspects of the problem are of key importance to its actual implementation. This is the main motivation of the present work.

A. Bemporad, A. Bicchi, and G. Buttazzo (Eds.): HSCC 2007, LNCS 4416, pp. 4–17, 2007.

There are two approaches to the problem: the first is to resort to a numerical approximation scheme relying on the discretization of the continuous state and control input spaces (gridding approach). Alternatively, one can introduce a family of finitely parameterized functions, and then look for the cost-to-go function within that family (neuro-dynamic programming approach [9]).

Here, we study a gridding procedure for the numerical solution to the DP equation of the stochastic reachability problem. We assess the convergence of the numerical solution to the actual solution as the grid size goes to zero, and derive explicit bounds on the level of approximation introduced for a given small but nonzero grid size. The study is inspired by the reference work [3], discussing discretization procedures for the numerical solution to DP in the additive cost case and for stochastic –non hybrid– systems: we extend this approach to a hybrid system setting with multiplicative cost and general disturbances. A numerical approximation scheme was proposed in [6] for estimating the probability of remaining within a safe set for a certain class of autonomous, continuous time stochastic hybrid systems. The discretization process in that case involved gridding the system both in time and in space. Convergence of the estimate to the true probability as the grid size goes to zero was proven, but no bounds were provided for assessing the quality of the estimates derived for a small but nonzero grid size.

Furthermore, we reinterpret some ideas proposed in [4] and [5] within the hybrid systems framework to suggest that structural properties of the problem, such as its decentralized nature, may be exploited to obtain a more compact state representation and efficient implementation of the computations involved in the solution to DP. This feature may partly mitigate the curse of dimensionality that affects DP as well as other approaches proposed in the literature to address the reachability problem [8].

The rest of the paper is organized as follows. We first briefly recall the DTSHS model of [1] in Section 2 and describe the DP algorithm to solve the probabilistic reachability problem in Section 3. Section 4 proposes a numerical approximation scheme for solving the DP algorithm. Based on some regularity assumptions on the stochastic kernels that characterize the system dynamics, convergence of the numerical scheme and explicit bounds assessing the quality of the approximated solution to DP are shown in Section 5. Section 6 illustrates the convergence properties and scaling issues of the proposed numerical approximation scheme with reference to a multi-room heating benchmark. Possible extensions regarding efficient representations and computation of solutions are finally outlined in Section 7.

2 Stochastic Hybrid System Model

In this section we briefly recall the discrete time stochastic hybrid system (DTSHS) model first introduced in [1].

Definition 1. *A discrete time stochastic hybrid system (DTSHS) is a tuple* $\mathcal{H} = (\mathcal{Q}, n, \mathcal{A}, T_x, T_q, R)$, *where*

- $\mathcal{Q} := \{q_1, q_2, \dots, q_m\}$, *for some* $m \in \mathbb{N}$, *represents the discrete state space;*
- $n : \mathcal{Q} \to \mathbb{N}$ *assigns to each discrete state value* $q \in \mathcal{Q}$ *the dimension of the continuous state space* $\mathbb{R}^{n(q)}$. *The hybrid state space is then given by* $\mathcal{S} := \cup_{q\in\mathcal{Q}}\{q\} \times \mathbb{R}^{n(q)}$*;*
- $\mathcal{A}$ *is a compact Borel space representing the control space;*
- $T_x : \mathcal{B}(\mathbb{R}^{n(\cdot)}) \times \mathcal{S} \times \mathcal{A} \to [0,1]$ *is a Borel-measurable stochastic kernel on* $\mathbb{R}^{n(\cdot)}$ *given* $\mathcal{S} \times \mathcal{A}$*, which assigns to each* $s = (q,x) \in \mathcal{S}$ *and* $a \in \mathcal{A}$ *a probability measure on the Borel space* $(\mathbb{R}^{n(q)}, \mathcal{B}(\mathbb{R}^{n(q)}))$*:* $T_x(dx|(q,x),a)$
- $T_q : \mathcal{Q} \times \mathcal{S} \times \mathcal{A} \to [0,1]$ *is a discrete stochastic kernel on* $\mathcal{Q}$ *given* $\mathcal{S} \times \mathcal{A}$*, which assigns to each* $s \in \mathcal{S}$ *and* $a \in \mathcal{A}$*, a probability distribution over* $\mathcal{Q}$*:* $T_q(q|(q,x),a)$*;*
- $R : \mathcal{B}(\mathbb{R}^{n(\cdot)}) \times \mathcal{S} \times \mathcal{A} \times \mathcal{Q} \to [0,1]$ *is a Borel-measurable stochastic kernel on* $\mathbb{R}^{n(\cdot)}$ *given* $\mathcal{S} \times \mathcal{A} \times \mathcal{Q}$*, that assigns to each* $s = (q,x) \in \mathcal{S}$*,* $a \in \mathcal{A}$*, and* $q' \in \mathcal{Q}$*, a probability measure on the Borel space* $(\mathbb{R}^{n(q')}, \mathcal{B}(\mathbb{R}^{n(q')}))$*:* $R(dx|(q,x),a,q')$. □

The system initialization at time $k = 0$ is specified through some probability measure $\pi : \mathcal{B}(\mathcal{S}) \to [0,1]$ on the Borel space $(\mathcal{S}, \mathcal{B}(\mathcal{S}))$, where $\mathcal{B}(\mathcal{S})$ is the σ-field generated by the subsets of $\mathcal{S}$ of the form $\cup_q\{q\} \times B_q$, with B_q denoting a Borel set in $\mathbb{R}^{n(q)}$. With reference to the time horizon $[0,N]$, we next define the notion of Markov policy.

Definition 2. *Consider a DTSHS* $\mathcal{H} = (\mathcal{Q}, n, \mathcal{A}, T_x, T_q, R)$. *A Markov policy for* $\mathcal{H}$ *is a sequence* $\mu = (\mu_0, \mu_1, \dots, \mu_{N-1})$ *of universally measurable maps* $\mu_k : \mathcal{S} \to \mathcal{A}$, $k = 0, 1, \dots, N-1$. *We denote the set of Markov policies as* $\mathcal{M}_m$. □

For conciseness sake, we can introduce the Borel-measurable stochastic kernel $T_s : \mathcal{B}(\mathcal{S}) \times \mathcal{S} \times \mathcal{A} \to [0,1]$ on $\mathcal{S}$ given $\mathcal{S} \times \mathcal{A}$, which assigns to each $s = (q,x)$, $s' = (q',x') \in \mathcal{S}$, $a \in \mathcal{A}$ a probability measure on the Borel space $(\mathcal{S}, \mathcal{B}(\mathcal{S}))$ as follows:

$$T_s(ds'|s,a) = \begin{cases} T_x(dx'|(q,x),a)T_q(q'|s,a), & \text{if } q' = q \\ R(dx'|(q,x),a,q')T_q(q'|s,a), & \text{if } q' \neq q. \end{cases} \tag{1}$$

Definition 3. *An execution for a DTSHS* $\mathcal{H} = (\mathcal{Q}, n, \mathcal{A}, T_x, T_q, R)$ *associated with a policy* $\mu = (\mu_0, \mu_1, \dots, \mu_{N-1}) \in \mathcal{M}_m$ *and an initial distribution* π *is a stochastic process* $\{\mathbf{s}(k), k \in [0,N]\}$ *with values in* $\mathcal{S}$ *whose sample paths are obtained according to the following algorithm:*

extract from $\mathcal{S}$ a value s_0 for $\mathbf{s}(0)$ according to π;

for $k = 0$ to $N-1$

  set $a_k = \mu_k(s_k)$;

  extract from $\mathcal{S}$ a value s_{k+1} for $\mathbf{s}(k+1)$ according to $T_s(\cdot\,|s_k, a_k)$;

end □

A DTSHS $\mathcal{H}$ can then be described as a controlled Markov process with state space $\mathcal{S}$, control space $\mathcal{A}$, and controlled transition probability function $T_s : \mathcal{B}(\mathcal{S}) \times \mathcal{S} \times \mathcal{A} \to [0,1]$ defined in (1). Thus, the execution $\{\mathbf{s}(k), k \in [0,N]\}$ associated with $\mu \in \mathcal{M}_m$ and π is a time inhomogeneous stochastic process defined on the canonical sample space $\Omega = \mathcal{S}^{N+1}$, endowed with its product topology $\mathcal{B}(\Omega)$, with probability measure P_π^μ uniquely defined by the initial probability measure π on $(\mathcal{S}, \mathcal{B}(\mathcal{S}))$ and one-step transition kernels $T_s^{\mu_k}(ds'|s) := T_s(ds'|s, \mu_k(s))$, $k = 0, 1, \ldots, N-1$. When π is concentrated at $\{s\}$, $s \in \mathcal{S}$, that is $\pi(ds) = \delta_s(ds)$, we shall write simply P_s^μ.

3 Probabilistic Reachability Problem

Given a stochastic hybrid system $\mathcal{H}$, a Borel compact set $D \in \mathcal{B}(\mathcal{S})$, and a Markov policy $\mu \in \mathcal{M}_m$, let

$$p_\pi^\mu(D) := P_\pi^\mu(\mathbf{s}(k) \in D \text{ for all } k \in [0,N])$$

denote the probability that the execution of $\mathcal{H}$ associated with policy μ and with the initial state distribution π will stay within set D over the time horizon $[0,N]$. If π is concentrated at $\{s\}$, $s \in \mathcal{S}$, we use the notation $p_s^\mu(D)$. If set D represents a safe set for $\mathcal{H}$, by computing $p_s^\mu(D)$, we shall evaluate the safety level for system $\mathcal{H}$ when it starts from $s \in D$ and is subject to policy μ. The objective is to determine the Markov policy that maximizes the probability $p_\pi^\mu(D)$.

Let $\mathbf{1}_C : \mathcal{S} \to \{0,1\}$ denote the indicator function of a set $C \subseteq \mathcal{S}$: $\mathbf{1}_C(s) = 1$, if $s \in C$, and 0, if $s \notin C$. Observe that

$$\prod_{k=0}^{N} \mathbf{1}_D(s_k) = \begin{cases} 1, & \text{if } s_k \in D \text{ for all } k \in [0,N] \\ 0, & \text{otherwise,} \end{cases}$$

where $s_k \in \mathcal{S}$, $k \in [0,N]$. Then,

$$p_\pi^\mu(D) = P_\pi^\mu\left(\prod_{k=0}^{N} \mathbf{1}_D(\mathbf{s}(k)) = 1\right) = E_\pi^\mu\left[\prod_{k=0}^{N} \mathbf{1}_D(\mathbf{s}(k))\right]. \tag{2}$$

One can then introduce functions $V_k^\mu : \mathcal{S} \to [0,1]$, $k = 0, 1, \ldots, N$, associated with a Markov policy μ:

$$V_k^\mu(s) := \mathbf{1}_D(s) \int_{\mathcal{S}^{N-k}} \prod_{l=k+1}^{N} \mathbf{1}_D(s_l) \prod_{h=k+1}^{N-1} T_s(ds_{h+1}|s_h, \mu_h(s_h)) T_s(ds_{k+1}|s, \mu_k(s)),$$

$s \in \mathcal{S}$, where T_s is the controlled transition function of the embedded controlled Markov process, and $\int_{\mathcal{S}^0}(\ldots) = 1$. These functions are known as *cost-to-go functions* because they satisfy $V_k^\mu(s) = E_\pi^\mu[\prod_{h=k}^{N} \mathbf{1}_D(\mathbf{s}(h))|\mathbf{s}(k) = s]$ for any $s \in \mathcal{S}$ within the support of the distribution of $\mathbf{s}(k)$. Thus, $V_k^\mu(s)$ returns the value of the probability of remaining within D over the (residual) time horizon $[k, N]$ starting from s at time k, under policy $\mu \in \mathcal{M}_m$ applied from π.

For any policy $\mu \in \mathcal{M}_m$, the cost-to-go functions $V_k^\mu : \mathcal{S} \to [0,1]$, $k = 0, 1 \ldots, N$, can be computed by the backward recursion:

$$V_k^\mu(s) = \mathbf{1}_D(s) \int_{\mathcal{S}} V_{k+1}^\mu(s_{k+1}) T_s(ds_{k+1}|s, \mu_k(s)),\ s \in \mathcal{S}, \tag{3}$$

initialized with $V_N^\mu(s) = \mathbf{1}_D(s)$, $s \in \mathcal{S}$, [1].

From equation (2) we have that

$$p_\pi^\mu(D) = \int_{\mathcal{S}} E_\pi^\mu[\prod_{k=0}^{N} \mathbf{1}_D(\mathbf{s}(k))|\, s(0) = s]\pi(ds) = \int_{\mathcal{S}} V_0^\mu(s)\pi(ds). \tag{4}$$

Moreover, given that $\mu \in \mathcal{M}_m$ and that the execution associated with a Markov policy is a Markov process, it is easily seen that $E_\pi^\mu[\prod_{k=0}^{N} \mathbf{1}_D(\mathbf{s}(k))|\, s(0) = s] = p_s^\mu(D)$; hence, $p_\pi^\mu(D) = \int_{\mathcal{S}} \mathcal{P}_s^\mu(D)\pi(ds)$.

Definition 4. *Let $\mathcal{H} = (\mathcal{Q}, n, \mathcal{A}, T_x, T_q, R)$ be a DTSHS and $D \in \mathcal{B}(\mathcal{S})$ a safe set. A Markov policy μ^* is maximally safe if $p_s^{\mu^*}(D) = \sup_{\mu \in \mathcal{M}_m} p_s^\mu(D)$, $\forall s \in D$.* □

In view of (4), a maximally safe Markov policy in fact maximizes $\mathcal{P}_\pi^\mu(D)$ for any initial state distribution π. The following theorem was shown in [1]:

Theorem 1. *Define functions $V_k^* : \mathcal{S} \to [0,1]$, $k = 0, 1, \ldots, N$, by the following dynamic programming algorithm:*

$$V_k^*(s) = \sup_{a \in \mathcal{A}} \mathbf{1}_D(s) \int_{\mathcal{S}} V_{k+1}^*(s_{k+1}) T_s(ds_{k+1}|s, a),\ s \in \mathcal{S}, \tag{5}$$

initialized with $V_N^*(s) = \mathbf{1}_D(s)$, $s \in \mathcal{S}$.

Then, $V_0^*(s) = \sup_{\mu \in \mathcal{M}_m} \mathcal{P}_s^\mu(D)$ for all $s \in \mathcal{S}$. Moreover, if $U_k(s, \lambda) = \{a \in \mathcal{A} | \mathbf{1}_D(s) \int_{\mathcal{S}} V_{k+1}^*(s_{k+1}) T_s(ds_{k+1}|s, a) \geq \lambda\}$ is compact for all $s \in \mathcal{S}$, $\lambda \in \mathbb{R}$, $k \in [0, N-1]$, then there exists a maximally safe policy $\mu^* = (\mu_0^*, \ldots, \mu_{N-1}^*)$, with $\mu_k^* : \mathcal{S} \to \mathcal{A}$, $k \in [0, N-1]$, given by

$$\mu_k^*(s) = \arg\sup_{a \in \mathcal{A}} \mathbf{1}_D(s) \int_{\mathcal{S}} V_{k+1}^*(s_{k+1}) T_s(ds_{k+1}|s, a),\ \forall s \in \mathcal{S}, \tag{6}$$

and $V_k^{\mu^*}(s) = V_k^*(s)$, $s \in \mathcal{S}$, $k = 0, 1, \ldots, N$.

In the sequel, we consider the case when

Assumption 1. *The control space $\mathcal{A}$ is a finite set.* □

Under this assumption, the compactness condition in Theorem 1 for the existence of a maximally safe Markov policy is not required. The results illustrated next can be extended to the case when $\mathcal{A}$ is a compact uncountable set in an Euclidean space following a similar line of reasoning.

4 Numerical Approximation Scheme

In this section we describe a numerical scheme for determining an approximately maximally safe policy based on the characterization of a maximally safe policy given in Theorem 1.

For the purpose of numerical approximation, it is important to note that the DP algorithm (5) as well as the optimal argument in equation (6) can be restricted to the compact set D of the state space as follows:

$$V_k^*(s) = \max_{a \in \mathcal{A}} \int_D V_{k+1}^*(s_{k+1}) T_s(ds_{k+1}|s,a),\ s \in D, \tag{7}$$

initialized with $V_N^*(s) = 1$, $s \in D$, and

$$\mu_k^*(s) = \arg\max_{a \in \mathcal{A}} \int_D V_{k+1}^*(s_{k+1}) T_s(ds_{k+1}|s,a),\ \forall s \in D. \tag{8}$$

This is quite intuitive, since for values of the state outside D the cost-to-go function is identically zero for any μ and the optimal policy $\mu^* : \mathcal{S} \to \mathcal{A}$ can be set arbitrarily. Thus, we just have to consider the values for the state within the compact set D. The advantage of restricting the state space to the compact set D is that we can adopt a *finite* discretization in the numerical approximation scheme for solving the dynamic programming algorithm and determining the optimal policy μ^*. Moreover, under suitable regularity conditions on the transition kernels defining the DTSHS, the optimal cost-to-go functions can be shown to be Lipschitz continuous over D in the continuous state component. This property (valid only within D, given the discontinuity when passing from a safe state within D to an unsafe state outside D) is used for determining bounds to the numerical approximated solution.

4.1 Discretization Procedure

State discretization. As discussed before, we can restrict computations to the compact safe set D. Thus we only need to discretize D. The set $D \subset \mathcal{S}$ is given by $D = \cup_{q \in \mathcal{Q}} \{q\} \times X^q$. The size of the continuous state space within D is measured by $\lambda := \max_{q \in \mathcal{Q}} \mathcal{L}(X_q)$, where $\mathcal{L}(X_q)$ denotes the Lebesgue measure of the set $X_q \subset \mathbb{R}^{n(q)}$. For simplicity, we assume that the compact set X^q is not empty, for all $q \in \mathcal{Q}$. Let us introduce a partition of cardinality m_q of the set $X^q \subset \mathbb{R}^{n(q)}$, $q \in \mathcal{Q}$: $X^q = \cup_{i=1}^{m_q} X_i^q$, where X_i^q, $i = 1, \dots, m_q$, are pairwise disjoint Borel sets $X_i^q \in \mathcal{B}(\mathbb{R}^{n(q)})$, $X_i^q \cap X_j^q = \emptyset$, $\forall i \neq j$. For any q and i, pick a hybrid state value $v_i^q \in \{q\} \times X_i^q$. The set of all discrete values for the hybrid state is $\mathcal{G} := \{v_i^q, i = 1, \dots, m_q,\ q \in \mathcal{Q}\}$. Notice that the compactness assumption on D ensures the finiteness of the cardinality of $\mathcal{G}$. Denote with d_i^q the diameter of the set X_i^q, $d_i^q = \sup\{\|x - x'\| : x, x' \in X_i^q\}$. Then, $\Delta := \max_{i=1,\dots,m_q, q \in \mathcal{Q}} d_i^q$ represents the *grid size parameter*.

Note that, differently from [3] where the system dynamics is described through a difference nonlinear equation affected by a stochastic disturbance taking value in a finite set, we do not have any disturbance input appearing explicitly. The definition of the dynamics of the system via stochastic kernels incorporates both the disturbance effect and the deterministic contribution to the system evolution (see the example in Section 6). As a consequence, by discretizing the state space, we implicitly define a discretization of the disturbance space.

Dynamic Programming approximation. With reference to the finite state grid $\mathcal{G}$, we introduce a discretized version of the dynamic programming algorithm (7). For $k = 0, 1, \ldots, N-1$, compute the approximated optimal cost-to-go functions as follows

$$\begin{aligned} \hat{V}_k^*(v_i^q) &= \max_{a \in \mathcal{A}} \int_D \hat{V}_{k+1}^*(s) T_s(ds|v_i^q, a), \quad \text{if } v_i^q \in \mathcal{G} \\ \hat{V}_k^*(s) &= \hat{V}_k^*(v_i^q), \text{ if } s \in \{q\} \times X_i^q, \text{ for some } i \in \{1, \ldots, m_q\}, \, q \in \mathcal{Q}, \end{aligned} \tag{9}$$

with $\hat{V}_N^*(s) = 1$, $s \in D$.

Note that due to the piecewise constant approximation of the optimal cost-to-go function and to the definition of T_s in Eqn. (1), the integral in equation (9) can be rewritten as

$$\begin{aligned} \hat{V}_k^*(v_i^q) = \max_{a \in \mathcal{A}} \Big\{ &\sum_{j=1,\ldots,m_q} \hat{V}_{k+1}^*(v_j^q) T_q(q|v_i^q, a) \int_{X_j^q} T_x(dx|v_i^q, a) \\ &+ \sum_{\substack{j=1,\ldots,m_{\bar{q}}, \\ \bar{q} \neq q \in \mathcal{Q}}} \hat{V}_{k+1}^*(v_j^{\bar{q}}) T_q(\bar{q}|v_i^q, a) \int_{X_j^{\bar{q}}} R(dx|v_i^q, a, \bar{q}) \Big\}, \end{aligned}$$

which explicitly shows that (9) consists of a computation on the finite grid $\mathcal{G}$.

Based on the approximated optimal cost-to-go $\hat{V}_k^*$, we define a Markov policy $\hat{\mu}^* = (\hat{\mu}_0^*, \ldots, \hat{\mu}_{N-1}^*)$, $\hat{\mu}_k^* : \mathcal{S} \to \mathcal{A}$, $k \in [0, N-1]$, as follows:

$$\begin{aligned} \hat{\mu}_k^*(v_i^q) &= \arg\max_{a \in \mathcal{A}} \int_D \hat{V}_{k+1}^*(s) T_s(ds|v_i^q, a), \quad \text{if } v_i^q \in \mathcal{G}, \\ \hat{\mu}_k^*(s) &= \hat{\mu}^*(v_i^q), \text{ if } s \in \{q\} \times X_i^q, \text{ for some } i \in \{1, \ldots, m_q\}, \, q \in \mathcal{Q}. \end{aligned} \tag{10}$$

As for any other policy, $\hat{\mu}^*$ can be arbitrarily selected outside D.

The performance of such policy $\hat{\mu}^*$ is given by the corresponding values for the cost-to-go functions $V_k^{\hat{\mu}^*}$, $k = 0, 1, \ldots, N$, that can be computed by the recursion in (3). In particular, $V_0^{\hat{\mu}^*}(s)$, $s \in D$, provides the value of the probability that the system will remain within D in the time horizon $[0, N]$ starting from $s \in D$ under policy $\hat{\mu}^*$.

In the following section, we shall show that, under proper assumptions, the performance of policy $\hat{\mu}^*$ tends to the one of a maximally safe policy, as the grid size parameter Δ goes to zero.

5 Convergence Analysis

We suppose that the stochastic kernels T_x and R on the continuous component of the hybrid state in Definition 1 of the DTSHS admit density t_x and r. We further assume that t_x and r as well as the stochastic kernel T_q satisfy the following Lipschitz condition.

Assumption 2

1. $|T_q(\bar{q}|s,a) - T_q(\bar{q}|s',a)| \leq k_1\|x-x'\|$, *for all* $s=(q,x), s'=(q,x') \in D$, $a \in \mathcal{A}$, *and* $\bar{q} \in \mathcal{Q}$,
2. $|t_x(\bar{x}|s,a) - t_x(\bar{x}|s',a)| \leq k_2\|x-x'\|$, *for all* $s=(q,x), s'=(q,x') \in D$, $a \in \mathcal{A}$, *and* $(q,\bar{x}) \in D$,
3. $|r(\bar{x}|s,a,\bar{q}) - r(\bar{x}|s',a,\bar{q})| \leq k_3\|x-x'\|$, *for all* $s=(q,x), s'=(q,x') \in D$, $a \in \mathcal{A}$, $(\bar{q},\bar{x}) \in D$, *and* $\bar{q} \neq q$,

where k_1, k_2 *and* k_3 *are suitable Lipschitz constants.* □

Based on this assumption, we can prove that the optimal cost-to-go functions satisfy some Lipschitz condition over D. This property will be fundamental in proving the convergence result. Due to space limitations, proofs are omitted.

Theorem 2. *Under Assumption 2 the optimal cost-to-go functions satisfy the following Lipschitz condition over* D*:*

$$|V_k^*(s) - V_k^*(s')| \leq \mathcal{K}\|x-x'\|, \forall\, s=(q,x),\ s'=(q,x') \in D, \tag{11}$$

for any $k \in [0,N]$*. The constant* $\mathcal{K}$ *is given by* $\mathcal{K} = mk_1 + \lambda\big(k_2+(m-1)k_3\big)$.

Based on Theorem 2, the following convergence result can be proven.

Theorem 3. *Under Assumption 2, there exist positive constants* γ_k, $k=0,\dots,N$, *such that the solutions* $\hat{V}_k^*$ *to the approximated dynamic programming equations* (9) *and the cost-to-go functions of the corresponding Markov policy* $\hat{\mu}^*$ *defined in* (10) *satisfy:*

$$\begin{aligned}\big|V_k^*(s) - \hat{V}_k^*(s)\big| &\leq \gamma_k \Delta,\ \ s \in D,\\ \big|V_k^*(s) - V_k^{\hat{\mu}^*}(s)\big| &\leq \nu_k \Delta,\ \ s \in D.\end{aligned}$$

where $\gamma_k = \gamma_{k+1} + \mathcal{K}$, $k=1,2\dots,N-1$, *initialized with* $\gamma_N = 0$, $\nu_k = \gamma_k + \gamma_{k+1} + \mathcal{K} + \nu_{k+1}$, $k=1,2\dots,N-1$, *initialized with* $\nu_N = 0$, *and* $\mathcal{K} = mk_1 + \lambda\big(k_2+(m-1)k_3\big)$.

From this theorem it follows that the quality of the approximation by the numerical procedure described in equations (9) and (10) improves as the grid size parameter Δ decreases. The rate of convergence is linear in Δ with a constant that depends on the Lipschitz constants k_1, k_2, and k_3 in Assumption 2 through the $\mathcal{K}$ constant defined in Theorem 2. This is not surprising because we are using a piecewise constant approximation of the optimal cost-to-go function and we expect that the optimal cost-to-go function is smoother as k_1, k_2, and k_3 are smaller. As the time horizon grows, the approximation error propagates. This is taken into account by the constants γ_k and ν_k in Theorem 3 that grow linearly as k decreases from N to 0, where N is the length of the time-horizon.

6 Computational Study

In this section we present the results of a computational study for a multi-room heating benchmark inspired by [1,7]. We numerically analyze the convergence of quantities computed by the discretization scheme proposed in Section 4.1. We also propose possible improvements in the implementation when the underlying structure of the DTSHS can be exploited to implement the DP algorithm in a computationally efficient manner.

The benchmark in [7] deals with the problem of regulating temperature in a house with n rooms via m heaters. In this report we focus on $m = 1$, the single heater case. The state of the system can be described as a hybrid state with discrete state component described by the position and status of the heater. The continuous state component can be described by the average temperature in each of the rooms. Let Δt be the time step and N be the total number of time intervals. Let $\mathbf{x}_i(k)$ denote the average temperature in room i at time k, x_a the ambient temperature and h_i a boolean vector of size equal to the number of rooms with components equal to 1 if the heater is present and in the "on" status in the corresponding room, and 0 otherwise. The average temperature in room i is governed by the following linear stochastic difference equation:

$$\mathbf{x}_i(k+1) = \mathbf{x}_i(k) + \Big(b_i(x_a - \mathbf{x}_i(k)) + \sum_{i \neq j} a_{i,j}(\mathbf{x}_j(k) - \mathbf{x}_i(k)) + c_i h_i\Big)\Delta t + \mathbf{n}_i(k) \quad (12)$$

where, $a_{i,j}, b_i, c_i$ are constants and $\{\mathbf{n}_i(k), k = 0, \ldots, N\}$ is a sequence of i.i.d Gaussian random variables with zero mean and variance ν^2. For $i \neq j$, $E[\mathbf{n}_i \mathbf{n}_j^T] = \mathbf{0}$. The heater is controlled by a thermostat that is prone to delay and failures in switching the heater between one room to another and between the "on" and "off" status: the effect of these control actions on the discrete state transitions is specified by a finite-state, finite-action, controlled Markov chain which is independent of the continuous state, that is, $T_q : \mathcal{Q} \times \mathcal{Q} \times \mathcal{A} \to [0,1]$. One can easily check that the number of possible discrete states is $n+1$ and the maximum number of available control actions is $n(n+1)+1$. We define the compact safe set to be

$$D = \cup_{q \in \{1,\ldots,(n+1)\}} \cup_{i \in \{1,\ldots,n\}} \{q\} \times \{i\} \times [x_{li}^q, x_{ui}^q],$$

where x_{ui}^q and x_{ui}^q specify the lower and upper limits for the desired temperature in room i for discrete state q. For simplicity, these are assumed to be independent of i and q. We now describe the discretization procedure as follows: we adopt a uniform partitioning of the set $[x_{li}^q, x_{ui}^q]$ into m disjoint intervals each of size $\varkappa = (x_{ui}^q - x_{li}^q)/m$. Therefore, $[x_{li}^q, x_{ui}^q] = [x_{li}^q, x_{li}^q + \varkappa) \cup \ldots \cup [x_{li}^q + (m-1)\varkappa, x_{ui}^q]$. The value of the temperature in room i for the discrete state q is defined by

$e^q_{r_i i} = x^q_{li} + (r_i - 1)\varkappa$, where $r_i \in \{1, \dots, m\}$. Define $r = [r_1, \dots, r_n]^T$. We pick $v^q_r = (q, [e^q_{r_1 1}, \dots, e^q_{r_n n}]^T)$ as hybrid state value. Thus, the set of all discrete values for the hybrid state is $\mathcal{G} = \{v^q_r, r = [r_1, \dots, r_n]^T; r_i = 1, \dots, m; i = 1, \dots, n; q = 1, \dots, (n+1)\}$. Let $\mathcal{N}(\cdot; \eta, \sigma^2)$ denote the probability measure over $(\mathbb{R}, \mathcal{B}(\mathbb{R}))$ associated with a Gaussian density function with mean η and variance σ^2. Then the stochastic kernel $T_s(ds'|v^q_r, a)$ that is used in the discretized dynamic programming equations (9) can be defined as follows:

$$T_s(ds'|v^q_r, a) = T_x(dx'|v^q_r, a) T_q(q'|q, a),$$

for $v^q_r \in \mathcal{G}$, $a \in \mathcal{A}$, and $s' \in \mathcal{S}$. Here, $T_x(\cdot|v^q_r, a) = \mathcal{N}(\cdot; \mu^q_r, \nu^2 I_n)$, I_n being the identity matrix of size n, $\mu^q_r = [\mu^q_{r_1}, \dots, \mu^q_{r_n}]^T$ and $\mu^q_{r_i} = e^q_{r_i} + \Big(b_i(x_a - e^q_{r_i}) + \sum_{i \neq j} a_{i,j}(e^q_{r_j} - e^q_{r_i}) + c_i h_i\Big)\Delta t$. It is easy to check that $T_x(dx'|v^q_r, a)$ and $T_q(q'|q, a)$ satisfy the Assumption 2.

6.1 Convergence Properties

We first analyze the convergence properties of the discretization scheme for the case when $n = 2$ (two rooms). The number of modes is 3 and maximum number of allowable control actions is 7, as shown in Figure 1(a). The computations are performed for the safe set $D = \cup_{q \in \{1,2,3\}} \cup_{i \in \{1,2\}} \{q\} \times \{i\} \times [17.5, 22]^o\text{C}$. The size of time interval is $\Delta t = 1/15$ and the number of intervals is $N = 60$. The parameters values in equation (12) are: $x_a = 6$, $b_1 = b_2 = 0.25, a_{12} = a_{21} = 0.33, c_1 = 12, c_2 = 14$ and $\nu^2 = 0.9$. All the parameters should be interpreted in appropriate units. For each control action by the thermostat that elicits a transition between two different modes of the heater, the transition happens with probability 0.8. The remaining 0.2 probability is divided evenly between the "do nothing" transition that models the delay and the transition to the third, non-recommended mode that models a faulty behavior.

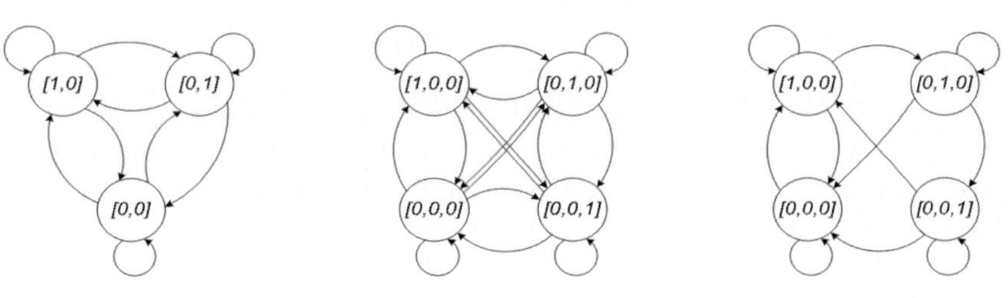

Fig. 1. (a) Maximum available control actions for $n = 2$. (b) Maximum available control actions for $n = 3$. (c) Reduced number of available control actions for $n = 3$. The discrete states are assigned numbers clockwise starting from the top-left state.

The computations of the solutions $\hat{V}^*_0$ to the approximated DP equations in (9) were performed for four discretization levels: $m \in \{9, 18, 36, 45\}$. Inspired by [1], we define the *approximately maximal probabilistic safe sets* $\hat{S}^*(\epsilon)$ with

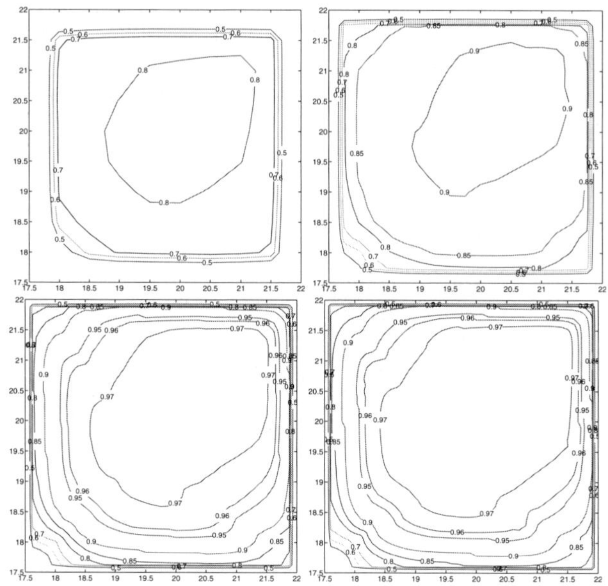

Fig. 2. Maximal probabilistic safe sets corresponding to safety levels: $0.5, 0.6, 0.7, 0.8, 0.85, 0.9, 0.95, 0.96$ and 0.97 for the case $n = 2$ and initial discrete state "off". In going from left-to-right and top-to-bottom, the plots correspond to discretization levels of $9, 18, 36$ and 45 respectively.

safety level $(1-\epsilon)$ as $\hat{S}^*(\epsilon) = \{s \in \mathcal{S} : \hat{V}_0^*(s) \geq (1-\epsilon)\}$. Figure 2 shows the approximately maximal safe sets when the initial discrete state is "off", and corresponding to different safety levels. As expected, the maximal safe sets get smaller as the required safety level increases. Furthermore, as the discretization level decreases, the maximal safe sets tend to graphically converge: this visually confirms the numerical convergence of the proposed discretization scheme.

The optimal control actions for the case when the initial discrete state is "off" are plotted in Figure 3 for the four discretization levels and $k = 1$. The optimal actions at finer resolution were obtained from that of coarser resolution by nearest neighbor interpolation. It can be noticed that the regions of optimal recommended actions become more well-formed and again visually converge as the discretization step decreases.

6.2 Scaling to Higher Dimensions

We now present the results from the three-room, one heater benchmark case. For this case, the number of continuous states is $n = 3$, the number of discrete states is 4 and maximum number of allowable actions is 13, as shown in Figure 1(b). The safe set is specified to be $D = \cup_{q\in\{1,2,3,4\}} \cup_{i\in\{1,2,3\}} \{q\} \times \{i\} \times [17.5, 22]^o$C. The size of time interval is $\Delta t = 1/15$ and the number of intervals is $N = 60$. The parameters values in equation (12) are: $a_{12} = a_{21} = 0.80$, $a_{13} = a_{31} = 0.60$,

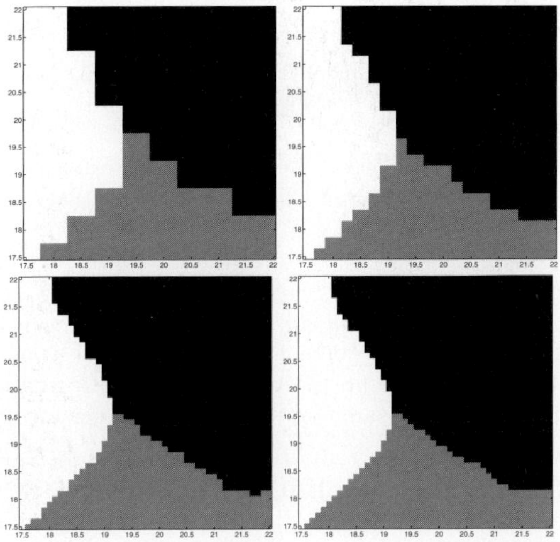

Fig. 3. Maximally safe actions for the case $n = 2$, initial discrete state "off" and $k = 1$. In going from left-to-right and top-to-bottom, the plots correspond to discretization levels of $9, 18, 36$ and 45 respectively. The colors *black*, *white* and *grey* respectively stand for "do nothing", "switch heater to room 1" and "switch heater to room 2" actions.

Fig. 4. Maximal probabilistic safe sets corresponding to a safety level of 0.95 for the case when $n = 3$ and initial discrete state is "off'. Available control actions are shown in Figure 1(b). In going from left to right and top to bottom, the plots correspond to $k = 60, 55, 50, 40, 20$ and 1.

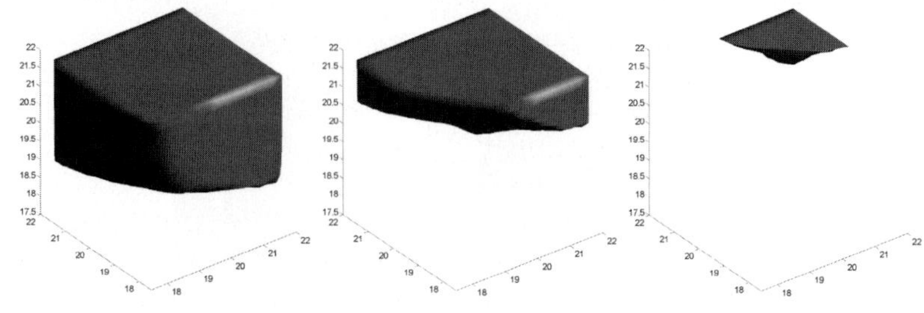

Fig. 5. Maximal probabilistic safe sets corresponding to the safety level 0.95 for the case $n = 3$ and initial discrete state "off". The reduced set of available control actions is shown in Figure 1(c). In going from left-to-right, the plots correspond to $k = 55, 50$ and 45. The safe set for $k = 60$ is same as the corresponding safe set in Figure 4.

$a_{23} = a_{32} = 0.70$, $x_a = 6$, $b = [0.30, 0.20, 0.30]^T$, $c = [12.00, 14.00, 12.00]^T$ and $\nu^2 = 0.33$. Similar to the two-room case, the effect of control actions is described by a controlled Markov chain. The exact details are omitted due to space limitations. The computation of the DP algorithm was performed for the discretization level $m = 18$. Figure 4 shows the maximal safe sets corresponding to the safety level $1 - \epsilon = 0.95$, at different times. As expected, as the number of steps-to-go increases, the size of the safe sets also decreases. It is of interest to compare the effect of number of available control actions on the size of the maximal safe set. In order to study this, we performed the DP computations for the three-room, one heater example for the reduced set of actions shown in Figure 1(c). The resulting maximal safe sets corresponding to safety level $1 - \epsilon = 0.95$ are shown in Figure 5. We observe that the maximal safe set is becomes very small and eventually decreases to the empty set as the number of steps-to-go increases.

We finally notice an important structural property of the benchmark, namely the conditional independence of the continuous stochastic kernel: $T_x(\bar{x}|v_r^q, a) = T_x(\bar{x}_1|v_r^q, a) \times \ldots \times T_x(\bar{x}_n|v_r^q, a)$. This enables us to efficiently compute the state transition probabilities.

7 Possible Extensions and Future Work

The above discretization schemes can be directly extended to the case of uncountable, but compact control space (see Assumption 1) in a similar way. We shall include the details of this in a future work.

Even in the presence of the conditional independence property of the continuous transitions kernel, increasing the problem size further will make the computation of approximate cost-to-go value functions prohibitively expensive. This motivates the study of more efficient approaches to solve the DP algorithm in hybrid state space; the literature suggests some methods to attack this problem. One technique exploits some decentralization in the structure of the state-space in order to distribute the computations: the HS structure naturally yields itself to this distributed approach according to the topology of the underlying graph consisting of the modes and the edges of the HS. In [4], an approach to asynchronously perform in parallel the computations with proven convergence is suggested. A second more recent approach is to solve large-scale Markov Decision Processes (MDPs) by approximating the optimal value function by a linear combination of basis functions and finding the associated optimal weights by linear programming [5]. The authors are currently investigating and experimenting these methods that leverage on the structure of the DP to achieve computationally attractive performances for the proposed schemes, to be further tested on the benchmark [7] and compared to other approaches in the literature.

References

1. Amin, S., Abate, A., Prandini, M., Lygeros, J., Sastry, S.: *Reachability analysis for controlled discrete time stochastic hybrid systems.* In J. Hespanha and A. Tiwari, eds.: Hybrid Systems: Computation and Control. Lecture Notes in Computer Science 3927. Springer Verlag, pp. 49-63 (April 2006).
2. Abate, A., Amin, S., Prandini, M., Lygeros, J., Sastry, S.: *Probabilistic Reachability for Safety and Regulation of Controlled Discrete-Time Stochastic Hybrid Systems.* In the Proceedings of the 45th IEEE Conference on Decision and Control, San Diego, CA, pp. 258-263 (December 2006).
3. Bertsekas, D.: *Convergence of Discretization Procedures in Dynamic Programming.* IEEE Transactions on Automatic Control, vol. 20, 3, pp. 415-419 (June 1975).
4. Bertsekas, D.: *Distributed Dynamic Programming.* IEEE Transactions on Automatic Control, vol. 27, 3, pp. 610-616 (June 1982).
5. Kveton, B., Hauskrecht, M., Guestrin, C.,: *Solving factored MDPs with Hybrid State and Action Variables.* Journal of Artificial Intelligence Research, vol. 27, pp. 1-49 (September 2006).
6. Prandini, M., and Hu, J.: *Stochastic reachability: Theoretical foundations and numerical approximation.* In C. Cassandras and J. Lygeros, eds.: Stochastic Hybrid Systems. Taylor & Francis Group/CRC Press (2006).
7. Fehnker, A., Ivancic F.: *Benchmarks for Hybrid Systems Verifications.* In R. Alur and G.J. Pappas, eds.: Hybrid Systems: Computation and Control. Lecture Notes in Computer Science 2993. Springer Verlag, pp. 326-341 (April 2004).
8. Mitchell, I., Templeton J.: *A Toolbox of Hamilton-Jacobi Solvers for Analysis of Nondeterministic Continuous and Hybrid Systems.* In M. Morari and R. Tiele, eds.: Hybrid Systems: Computation and Control. Lecture Notes in Computer Science 3414. Springer Verlag, pp. 480-494 (March 2005).
9. Bertsekas, D.P., Tsitsiklis, J.N.: *Neuro-Dynamic Programming.* Athena Scientific, Belmont, MA (1996).

Groupoids in Control Systems and the Reachability Problem for a Class of Quantized Control Systems with Nonabelian Symmetries

Alessandro Arsie and Emilio Frazzoli

Laboratory for Information and Decision Systems,
Massachusetts Institute of Technology, Cambridge, MA
{arsie,frazzoli}@mit.edu

Abstract. The aim of this paper is twofold. On one hand we present an approach to the general problem of nonlinear control in the framework of (differentiable) groupoids, which, in our opinion deserves further investigation. On the other hand, using recently-developed algebraic tools, we show that for a control system whose state space is a semisimple Lie group (like $SO(3)$), it is possible to reach a dense subset of the state space using just two properly chosen discrete controls, and this property is robust with respect to the choice of controls.

1 Introduction

Traditionally, system theory focused separately on discrete event systems or controlled dynamic systems with a continuous state space. On the other hand, the actual system configuration in many real-life applications involves a closed interaction of such components. For instance a physical plant, modeled in terms of a differential controlled system, is interfaced to the controller via encoding of its outputs with a finite set of symbols (an alphabet) through an A/D converter. Usually, in this framework also the control actions decided by the controller and fed-back to the plant are symbols encoded in a (possibly) different alphabet. More specifically, quantized control systems (QCSs) provide a unified framework to deal with several real-world systems, ranging from plants where the hardware implementations admits information transfer with a finite bandwidth to applications implementing switching actuators.

A complete analysis of the reachable sets for linear QCSs appears for the first time in [1] using deep results in number theory. This line of research has been further extended to include the analysis of reachable sets and the synthesis of steering paths (motion planning) for some classes of nonlinear systems, such as systems which are feedback equivalent to chained form systems (see [2]), systems that are feedback equivalent to the strictly triangular form (see [3]) and systems consisting of a polyhedral body rolling on a planar surface (see [4]).

A somehow different perspective has been championed in [5, 6, 7] and more recently in [8]. In these works the focus is on the purposeful introduction of control quantization in the design of control systems through the implementation

A. Bemporad, A. Bicchi, and G. Buttazzo (Eds.): HSCC 2007, LNCS 4416, pp. 18–31, 2007.

of a Motion Description Language. In particular, in [7] through the introduction of motion primitives as equivalence classes of suitable trajectories and the composition of *compatible* maneuver motion primitives and trim primitives solve the motion planning for a small helicopter through a Maneuver Automaton. A feature of these works is the use of group-theoretic methods, resulting in a group-theoretic formulation of the QCSs [9]. In this paper, we present a different theoretical framework for quite general control systems, through the use of groupoids, which provide a wide generalization of previous set-ups. In particular groupoids enables one to preserve information concerning the action of a group on a set, even when the action is restricted to a non-invariant subset of the original set, thus destroying the original action. In this contribution we will just give a first insight on the possible use of groupoids in control theory, and we think that further study should be reserved to this area.

Another common aspect of the aforementioned papers on QCSs is the necessity of having *Abelian* symmetries in order to analyze the structure of reachable sets and to plan the motion among initial and final configurations. Indeed, generally speaking, those analyses are based on a *global* splitting of the variables of a control system in *base variables* and *fiber variables*, thus relying on the structure of a global fibration. In the systems analyzed so far it turns out that the reachable sets have either the structure of a lattice or are dense in the configuration space, depending on the fact that some resonance conditions, involving the control quanta, are satisfied or not. For instance, in [3] it is proved that for driftless nonholonomic systems that are feedback equivalent to the strictly triangular form a control quanta set $\Delta = \{\delta_1, \ldots, \delta_N\}$ is *exhaustive* in $\mathbb{R}^n$ (namely the reachable set is a dense subset of the state manifold $\mathbb{R}^n$) iff $N \geq n+1$, the first n elements of Δ (up to a permutation) are linearly independent and moreover the following condition holds:

$$\frac{\delta_N \cdot \delta_i}{\|\delta_i\|} \in \mathbb{R}\backslash\mathbb{Q}, \qquad i = 1, \ldots, n. \tag{1}$$

From a practical point of view, conditions like (1) can not be exactly checked or met in real-world applications. Moreover, the conditions expressed in (1) are unstable with respect to arbitrarily small perturbation or disturbance. So the structure of the reachable sets is not robust to small errors in the set Δ.

The main contribution of this paper is twofold. First of all, we sketch a framework for the use of groupoids in control modeling. Secondly, we show that there is a very special class of nonlinear driftless QCSs, possessing nonabelian symmetries, for which the reachable sets are dense in the state space and they are robust with respect to uncertainties concerning control quanta. These examples arise as control systems whose state spaces are semisimple Lie groups.

2 A Groupoid-Theoretic Point of View

The extension from groups to groupoids starts with the desire to describe reversible processes which may traverse a number of states. Here are two equivalent formal definitions:

Definition 1 (Short). *A groupoid $\mathcal{G}$ is a small category (i.e. a category whose class of objects is a set), where each morphism is an isomorphism.*

Definition 2 (Long). *A groupoid $\mathcal{G}$ is quintuple (G_0, G_1, s, t, i), where G_0 is the set of objects, G_1 is the set of arrows, $s : G_1 \to G_0$ is the* source map *(it identifies the source of the arrow to which is applied), $t : G_1 \to G_0$ is the* target map *(it identifies the target of the arrow to which is applied) and $i : G_0 \to G_1$ is the* identity map *associating to each object p the identity arrow $i(p) = 1_p : p \to p$ (this implies in particular that the source and target maps are both onto). Moreover, a partially defined composition of arrows is enforced: namely for those arrows $x, y \in G_1$ such that $t(x) = s(y)$, there is a composition $x \star y \in G_1$ (which is simply the arrow derived by the compatible composition of x and y) such that $s(x \star y) = s(x)$ and $t(x \star y) = t(y)$. Finally there are axioms for a left and right identity for each arrow (if $x \in G_1$, then $i(s(x))$ is a left identity, that is $i(s(x)) \star x = x$, and $i(t(x))$ is a right identity, that is $x \star i(t(x)) = x$), an* inverse *x^{-1} for each arrow x such that $s(x^{-1}) = t(x)$, $t(x^{-1}) = s(x)$, and the usual associativity axiom for composition of compatible arrows.*

We say that G_1 is a groupoid over G_0.

Before recasting a wide class of control systems in this framework let us give some meaningful examples of groupoids.

Example 1. Consider an (left) action of a group G on a set X, given by $\Phi : G \times X \to X$. We can think of this action as a groupoid, as it is immediate to see. Indeed, consider a groupoid (G_0, G_1) where the triples $(\Phi(g, x), g, x) \in G_1$ for some $g \in G$ and $x \in X$, while $G_0 = X$ and the source map is given by $s : (\Phi(g, x), g, x) \mapsto x$ and the target map is given by $t : (\Phi(g, x), g, x) \mapsto \Phi(g, x)$. What is actually remarkable about groupoids is that we can restrict the action of G to *any* subset of X. Indeed, suppose that M is a subset of X which *is not invariant* under the action of G. So in principle, there could be a point $y_1 \in M$, such that $\Phi(g, y) \notin M$ for any $g \in G$, while for another point $y_2 \in M$, the set of $g \in G$ such that $\Phi(g, y_2) \in M$ is a just a subgroup of G. In any case, we can define a new groupoid $(G_0, G_1)_M$, where $(G_1)_M = \{(\Phi(g, x), g, x), x \in M, g \in G, \Phi(g, x) \in M\}$, $(G_0)_M = M$ and the target and source maps are the obvious ones. So even if there is no action of G on M (or even there is no action of any nontrivial subgroup of G on M), the groupoid $(G_0, G_1)_M$ takes into account the maximal amount of symmetry still remaining in M.

Example 2. If B is any set, the Cartesian product $B \times B$ is a groupoid over B with $s((x, y)) = x$, $t((x, y)) = y$ and composition $(x, y) \star (y, z) = (x, z)$. The (left and right) identities are (x, x) and inverses are constructed as $(x, y)^{-1} = (y, x)$. Observe that a subgroupoid of $B \times B$, that is a set closed under product and inversion and containing all the identity elements is just an *equivalence relation* on B.

Until now we have thought of groupoids as a construction generalizing groups. On the other hand, the previous example suggest that we can consider groupoids as generalized equivalence relations as well. From this point of view, a groupoid over a set B (set of states) tells us not only which elements of B are equivalent to one another, but it also parametrizes the *different ways* in which two elements can be equivalent.

Consider now time-invariant, (possibly) non-linear control systems of the form

$$x^+ = f(x, u), \tag{2}$$

where x belongs to a state space $\mathcal{X}$, x^+ indicates either time derivative (in the case of continous systems), or the evolution of x under the dynamic prescribed by f for a certain choice of the control u. In this framework u can be either a measurable function $u : [0, T] \to U \subset \mathbb{R}^m$, a control law $u : V \subset X \to U \subset \mathbb{R}^m$ or an admissible input word formed by symbols in U (where this time U is a finite or countable alphabet). The only requirement is that for any state x, there exists a "control" u such that either x does not evolve or evolves according to $x^+ = f(x, 0)$, and if there is a "control" u steering the system from configuration x_1 to configuration x_2, there must exist a "control" u^{-1} steering the system from configuration x_2 to configuration x_1. We call such systems *symmetric*. Moreover, by abuse of language, we call the evolution according to $x^+ = f(x, 0)$ evolution along *relative equilibria*. We have the following

Theorem 1. *Any symmetric control system is a groupoid on the state space or on the space of relative equilibria of the system.*

Proof. Suppose that the symmetric control system under consideration is such that for any state x there exists a control u under which x does not evolve (as in the case of driftless, affine-in-control systems). Then we consider a groupoid $\mathcal{G}$, where $G_0 := X$, and G_1 is the set of admissible controls (an element of G_1 is an arrow between two states of X and is obtained specifying completely the control u and the initial state (or the final state)). The source and target maps are well defined on G_1 and also is well defined the injective map $i : G_0 \to G_1$. The axioms for a groupoid are readily checked using the assumption that the system is symmetric (observe that we do not need the system to be controllable). On the other hand, consider now a symmetric control system for which there always exists a control u such that any configuration x evolve according to a relative equilibrium. For instance, we can consider in this class non-linear, affine-in-control systems of the form

$$x^+ = g(x) + f(x)u, \tag{3}$$

where we identify a relative equilibrium as the evolution obtained setting $u = 0$. In this example the different relative equilibria correspond to the orbits induced by the dynamic $x^+ = g(x)$ and control actions are simply viewed as arrows connecting different orbits. Also in this case it is immediate to see that such a symmetric control system is a groupoid on the space of relative equilibria. In this case G_0 is identified with the space of relative equilibria and G_1 is the set of arrows among relative equilibria, obtained specifying suitable control actions. ■

Remark 1. Quite often, in abstract terms, control systems are introduced using *state-transition maps.* A common adopted definition (especially for quantized control systems) is like the following. A control system is a quintuple $(X, \mathcal{T}, U, \Omega, \mathcal{A})$, where X denotes as usual the configuration space, $\mathcal{T}$ is an ordered time set, U is a set of admissible input symbols, possibly depending on the configuration, Ω is a set of admissible input words formed by symbols in U and $\mathcal{A} : \mathcal{T} \times \Omega \times X \rightarrow X$ is a state-transition map. Notice however that these state-transition maps are not *uniformly defined* over X, in the sense that the set Ω over which $\mathcal{A}$ is defined depends in general on the current state $x \in X$. With the use of groupoids, this problem is readily solved.

Now we recast in this framework some examples, also in order to clarify the meaning of the Theorem 1.

Example 3. Consider the following differentiable control system:

$$\dot{x} = e_1 + u, \tag{4}$$

where $x \in \mathbb{R}^2$, $e_1 = (1, 0)^T$ and u is a measurable function with values in $\mathbb{R}^2$, such that $\|u\| < 1$. The vector field e_1 is never vanishing, so it identifies a foliation (which in this case is also a fibration) on $\mathbb{R}^2$, whose leaves are just the orbits of the vector field. In this simple system we can think of these orbits as relative equilibria, generated by the Lie group $\mathbb{R}$ acting by time translations. This system is not small time locally controllable, but it is symmetric, in that using suitable control functions u, if it is possible to move from one relative equilibrium e_1 to another e_2, then it is possible also to move back from e_2 to e_1. Fixing a certain relative equilibrium e_1 it is possible to consider the set of all control functions steering the system from e_1 to e_1. This is set is clearly a group and its structure strongly depends on the type of control actions we enable. As long as we consider measurable open-loop control actions, any relative equilibrium can be reached from any other; on the other hand, if we restrict to constant and discrete control actions, such as for instance $U = \{(0, 0), (-1/2, 0), (0, -1, 2), (1/2, 0), (0, 1/2)\}$, then the set of relative equilibria reachable from a starting position is not discrete. Finally notice that U does not be symmetric for the system to be symmetric (for instance, in this specific example $U = \{(0, 0), (-1/4, 0), (0, -1, 2), (\pi/5, 0), (0, 1/2)\}$ will do the same job).

A more general example is the following:

Example 4. Consider the following differentiable control system:

$$\dot{x} = f(x) + g(x)u, \tag{5}$$

where $x \in M^n$, a complete differentiable manifold, $f(x)$ is a C^∞-vector field, $g(x)$ is a $n \times m$ matrix whose entries are C^∞-functions and $u \in U \subset \mathbb{R}^m$. Let us recall, that a complete manifold is a manifold for which all C^∞-vector fields are complete, and a vector field is complete if all its integral curves can be infinitely continued backward and forward.

If $f(x)$ is never vanishing on M^n it induces a foliation, which is not in general a fibration (so it is not possible to identify globally on M^n base and fiber variables). It is not always possible to select non vanishing smooth vector fields on a differentiable manifold; for instance on $\mathbb{S}^2$ there is no such a field (a fact known as the hairy ball theorem). Once again we identify relative equilibria with the orbits of $f(x)$. A sufficient condition for the system to be symmetric is that along any orbit of $f(x)$, the dimension of $< g(x)u > + < f(x) >)$ remains constant, where $< g(x)u >$ is the vector subspace spanned in T_xM^n as u varies in U. Notice that also in this case this condition does not imply that the system is small time locally controllable. Moreover, the condition can be weakened considering a stratification of M^n dictated by $g(x)$.

A quantized control system example is the following:

Example 5. In [2] and in [4] the authors provide a deep study of an hybrid system consisting on a polyhedron rolling on a plane. Possible actions on this system are restricted to be rotations about one of the edges of the face lying on the plane, by exactly that amount that bring an adjacent face to the plane. The state-space X is clearly given by $\mathbb{R}^2 \times \mathbb{S}^1 \times F$, where F is simply the set of faces of the polyhedron. For any given configuration $((x, y), \theta, f_i)$, the set of admissible control actions Ω_i is the set of all sequences of adjacent faces beginning with a face adjacent to f_i. If we consider the set of all sequences of adjacent faces beginning and ending with a face adjacent to f_i, we get a subset $\Omega_{i,i}$ of Ω_i, which is indeed a group. It is immediate to see that the set Ω of all compatible sequences of faces (namely just sequence of adjacent faces) is a ! groupoid over X. All the groups $\Omega_{i,i}$ as f_i varies in F are conjugate one to the other (even though not in a canonical way, since it is necessary to select a sequence of compatible faces to move from f_i to f_j). Suppose now that some edges on the polyhedron are "fragile", so that it is not possible to rotate the polyhedron along those edges. As long as it is possible to move the configuration from one face f_i to any other face f_j, the automorphism groups $\{\Omega_{i,i}\}_{i=1,\dots,\#(F)}$ are still conjugate. On the other hand, if the fragile edges form closed loop on the polyhedron, than the automorphism groups $\{\Omega_{i,i}\}_{i=1,\dots,\#(F)}$ are no more conjugate one to another. Despite of this the set of all admissible control actions Ω is still a groupoid on X and it grasps in an intrinsic way the overall behavior of the system.

Finally let us recast in our framework the hybrid control structure introduced in [7]:

Example 6. The main problem faced in this paper is to develop an approach for the efficient solution of motion planning problems for time-invariant dynamical control systems which posses symmetries. More specifically, motion plans are described in terms of concatenations of well-defined motion primitives (a sort of "elementary" motions), according to some compatibility relations.Given a time-invariant control system $\mathcal{S}$, we say that the Lie group G is a symmetry group for $\mathcal{S}$ if there is a (left) action of G on the state space of $\mathcal{S}$, such that it commutes with the evolution of $\mathcal{S}$. (This evolution can be thought as an action of $\mathbb{R}$ on the state space of $\mathcal{S}$, provided the corresponding vector fields are complete and certain choices of input curves which may lead to finite explosion times are avoided. However even in the worst cases, namely those involving finite explosion times, the evolution of $\mathcal{S}$ can be modeled as a groupoid). A motion primitive $[\pi]$ is the class of trajectories equivalent to a trajectory π, where two trajectories are declared equivalent if there is a time translation and an action by an element $g \in G$, such that the two trajectories can be identified (for more detailed see [7]). Indicating with $\mathcal{P}(\mathcal{S}, G)$ the set of all motion primitives for a time-invariant control system $\mathcal{S}$, with symmetry group G, it is easy to see that on $\mathcal{P}(\mathcal{S}, G)$ there is a partially defined binary operation, given by the composition of *compatible* motion primitives. More specifically, two primitives $[\pi_1]$ and $[\pi_2]$ are compatible if there exists trajectories $\alpha_1 \in [\pi_1]$ and $\alpha_2 \in [\pi_2]$ such that the final state of α_1 coincides with the initial state of α_2. Moreover, the *concatenation* of compatible primitives gives rise to another motion primitive. Inside $\mathcal{P}(\mathcal{S}, G)$, two different classes of motion primitives are identified. One is called the class of *trim primitives* $\mathcal{T}(\mathcal{S}, G)$, which is identified with steady-state motions, also known as relative equilibria or trim trajectories in the aeronautical community. Trim primitives are generate by exponentiation of the elements of the Lie algebra of G and corresponds to finite flows along left-invariant vector fields, keeping the controls constant. The other class $\mathcal{M}(\mathcal{S}, G)$ of motion primitives is called a *maneuver primitive*, which is a nontrivial primitive compatible, on the right and on the left, with trim primitives. Thus, a maneuver is nothing else than an equivalence class of trajectories connecting two steady-state conditions.

The proposed method for motion planning is based on the choice of a finite set of maneuvers $\Sigma \subset \mathcal{M}(\mathcal{S}, G)$ and the generation of complex trajectories thorough the concatenation of maneuvers in Σ. In [7] a Finite State Machine is used to select all strings in Σ^* (the free monoid generated by Σ), which are legal sequences, namely which correspond to feasible trajectories. This Finite State Machine (called Maneuver Automaton) can be represented as a directed graph in which vertices represent relative equilibria (i.e. trim primitives) and edges represent maneuvers. In general, this graph can not be considered as a groupoid, due to the fact that it is a directed graph, so it may well happen that while there is a way to go from steady-state $[\pi_1]$ to steady-state $[\pi_2]$, there is no way to go back. However, in applications it is readily seen that this is not the case. For instance, in [7] the authors provide an application of the proposed motion planning methodology to a realistic model of a small helicopter, namely an X-Cell.60 SE. The Maneuver Automaton discussed in the example turns out to be a groupoid, in that for any maneuver connecting in one direction two trim primitives, there is always a maneuver (or sequence of maneuvers) going in the opposite direction. Moreover, there can be in principle more than one maneuver (beside the "empty" maneuver) connecting a trim primitive with itself. This is for instance the case of a "loop" maneuver which connects the forward level flight with itself.

In [11] a set of maneuvers are developed to train astronauts and reduce their adaptation time in a microgravity environment. Each maneuver produces a net rotation around a specified axis, modifying the moment of inertia of the body and using internal torques produced by the muscles. Therefore each maneuver can be represented as an element of $SO(3)$. However, the transition from one maneuver to another is not always obtained through a rotation, but sometimes through a re-displacement of the elements of the body (for instance move both arms overhead). In this problem, therefore, we are forced to interpret this control system as a groupoid, where the group $SO(3)$ provides the possible control actions (the arrows of the groupoid) and the state space is the space of all possible configurations of the human body in a microgravity environment. In any given configuration of the body, the elements of $SO(3)$ which can be generated from that configuration depends on the configuration itself. Thus this system can not be represented as a principal bundle. Using group-theoretic methods in dealing with control systems it is necessary to consider a group homomorphism from the group of symmetries of the system under consideration to the automorphisms (or diffeomorphisms) group of the state space. The groupoid approach bypasses these considerations and enables to consider *local group of symmetries* for the system under consideration even when there is no global group of symmetry acting on the total state-space as in the case just described.

3 A Class of QCSs with Semisimple Symmetries

In this section we present a class of discrete nonlinear driftless control system and we analyze the structure of the reachable sets under quantization. Such systems arise as dynamical systems on semisimple Lie groups (i.e. Lie groups whose corresponding Lie algebras have no nonzero Abelian ideals). For example, the special linear group $SL(n)$ and the special orthogonal group $SO(n)$ are semisimple, whereas triangular groups of matrices are not. We propose the following:

Definition 3. *Given a semisimple Lie group G, a nonlinear driftless control system described by the following equation:*

$$x^+ = u \circ x \tag{6}$$

where $\circ$ is the group multiplication in G, $x \in G$ and $u \in U \subset G$ is a finite set is called a semisimple QCS*.*

We are interested in analyzing the structure of the reachable sets of semisimple QCS. To this aim, we use the following definition from [2] and adapted to semisimple QCSs:

Definition 4 (Approachability). *Given a Riemannian metric on G, a configuration x_f (final configuration) is* approachable *from a configuration x_0 (initial configuration) if for every $\epsilon > 0$, there exists an element $u \in < U >$, such that $d(u \circ x_0, x_f) < \epsilon$, where $< U >$ is the group generated by the elements of U and d is the distance associated to the chosen Riemannian metric on G. Let us call $R(x_0)$ the the set of reachable configurations starting from x_0. We say that the system is* locally approachable *from x_0 if the closure of the reachable set $R(x_0)$*

contains a neighborhood of x_0*, it is* approachable from x_0 *if* $R(x_0)$ *is dense in* G*. Finally we say that the system is* approachable *if the closure of* $R(x)$ *coincide with* G *for any* $x \in G$*.*

For semisimple QCSs we have the following

Theorem 2. *For any control system described by equation* (6) *there exists an* open *neighborhood* W *of the identity* e *in* G *and a closed subvariety* $R \subset W \times W$*, such that, if* $(g_1, g_2) \in (W \times W) \backslash R$ *and if the discrete control set is given by* $U = \{e, g_1, g_2, g_1^{-1}, g_2^{-1}\}$*, then the corresponding semisimple QCS is approachable.*

Proof: Let G be a connected real semisimple Lie group. It is well known that its real semisimple Lie algebra $\mathfrak{g}$ is generated by 2 elements. Indeed, indicating with $< x, y >$ the Lie subalgebra generated by elements x and y, it turns out that $S = \{(x, y) \in \mathfrak{g} \times \mathfrak{g} \colon < X, Y > \neq \mathfrak{g}\}$ is an algebraic subvariety in $\mathfrak{g} \times \mathfrak{g}$ (see for instance [12], VIII, 3, Exercise 8). Obviously the point $(0, 0) \in \mathfrak{g} \times \mathfrak{g}$ does not belong to the subvariety S. If we restrict S to a suitable product neighborhood of $(0, 0)$ and then take exponentiation, we end up with a so called *exponential algebraic subvariety* R which is contained in $W \times W$, where W is a suitable identity neighborhood in G. By construction, $(e, e) \notin R$. In [13] it is proved that no matter how you choose (g_1, g_2) in $(W \times W) \backslash R$, the subgroup generated by taking arbitrary words on $(g_1, g_2, g_1^{-1}, g_2^{-1})$ is dense in G. Consider the discrete control system described by (6), and assume that the starting configuration is a given $x_0 \in G$. Using the quantized control belonging to $U = \{e, g_1, g_2, g_1^{-1}, g_2^{-1}\}$, the reachable sets is just given by $< g_1, g_2 > \circ x_0$, where $< g_1, g_2 >$ is the subgroup generated by g_1 and g_2. Since by assumption $(g_1, g_2) \in (W \times W) \backslash R$, then $< g_1, g_2 > := H$, where H is a dense subgroup of G. Therefore the reachable set is $H \circ x_0$ and thus is still dense in G, which is also the state space. ∎

Remark 2. Theorem 2 provides us with a large class of control systems which are approachable with just five control actions (the five elements of U), independently on the dimension of the state space (recall that the dimension of $SL(n)$ is $n^2 - 1$, whereas the dimension of $SO(n)$ is $\frac{n(n-1)}{2}$). Moreover, the approachability result stated in the previous theorem is *robust* with respect to small perturbations of control actions g_1 and g_2, since, as long as they belong to the *open* set $(W \times W) \backslash R$ the approachability property still holds. In a sense, compared to the result previously analyzed in the literature, a special class of systems exhibiting nonabelian symmetries (the class presented here) may perform better as far as robustness is concerned.

In the proof of Theorem 2 it is not necessary to specify how the open neighborhood $Z := (W \times W) \backslash R$ must be chosen. Anyway, if we want to have even a remote possibility to apply concretely this result, we need a way to have a further insight on how Z is constructed. We will deal with this problem in the next section, where we will focus our attention on the case of $SO(3)$.

Remark 3. We want to underline that in this section we just studied the reachability problem for *a class* of QCSs with nonabelian symmetries. It is obviously not true that a general QCS possessing nonabelian symmetries can be reduced to a system having a Lie group as a space state. Therefore, the case presented here is just a very special subclass inside the class of QCSs possessing nonabelian symmetries.

4 The Case of $SO(3)$

In this section, we work out in the detail the construction outlined in the previous paragraph for the special case of $SO(3)$. First of all, we derive a concrete expression for the algebraic subvariety S. This is the content of the following

Theorem 3. *Let A_1 and A_2 two elements of $so(3)$, the Lie algebra of $SO(3)$. Identifying $so(3)$ with the 3-dimensional real vector space of 3×3 traceless antisymmetric matrices, we can assume that*

$$A_1 = \begin{bmatrix} 0 & x_1 & y_1 \\ -x_1 & 0 & z_1 \\ -y_1 & -z_1 & 0 \end{bmatrix} \quad A_2 = \begin{bmatrix} 0 & x_2 & y_2 \\ -x_2 & 0 & z_2 \\ -y_2 & -z_2 & 0 \end{bmatrix},$$

where $(x_1, y_1, z_1) \times (x_2, y_2, z_2) \in \mathbb{R}^3 \times \mathbb{R}^3 \cong \mathbb{R}^6$. Then A_1 and A_2 generate the Lie algebra $so(3)$ if and only if $(x_1, y_1, z_1, x_2, y_2, z_2) \notin S$, where S is the hypersurface of $\mathbb{R}^6$ defined by the following equation:

$$S = \{(x_1, y_1, z_1, x_2, y_2, z_2) \in \mathbb{R}^6 \colon z_1^2(x_2^2 + y_2^2) + z_2 y_2(x_1^2 + y_1^2) -$$

$$-y_1 y_2 z_1^2 - x_1 x_2 z_1 z_2 - y_2 z_1(x_1 x_2 + y_1 y_2) + (y_1 x_2 - x_1 y_2)^2 = 0\} \tag{7}$$

Proof: The characterization of $so(3)$ in terms of traceless antisymmetric matrices is well-known and we will not deal with it. It is also clear that any two elements A_1 and A_2 of $so(3)$ can be expressed in the form given above. If $A_3 := [A_1, A_2] \in \mathrm{Span}\{A_1, A_2\}$, where $\mathrm{Span}\{A_1, A_2\}$ is the real vector space generated over A_1 and A_2, then $A_3 = \lambda A_1 + \mu A_2$ for some real λ and μ.Therefore, if this is the case, then A_1 and A_2 can not generate the 3-dimensional Lie algebra $so(3)$, since, iterating the Lie bracket (the commutator in this case), one gets: $[A_3, A_1] = -\mu A_3$ and $[A_3, A_2] = \lambda A_1$. On the other hand, if $A_3 \notin \mathrm{Span}\{A_1, A_2\}$, then A_1, A_2 and A_3 generate the Lie algebra as it is readily seen. So A_1 and A_2 *fail* to generate $so(3)$ if and only if

$$[A_1, A_2] = \lambda A_1 + \mu A_2. \tag{8}$$

Using the coordinate form of A_1 and A_2 we can rewrite equation (8) as the following set of equations:

$$y_2 z_1 - y_1 z_2 = \lambda x_1 + \mu x_2, \tag{9}$$

$$x_1 z_2 - x_2 z_1 = \lambda y_1 + \mu y_2, \tag{10}$$

$$x_2 y_1 - x_1 y_2 = \lambda z_1 + \mu z_2. \tag{11}$$

We can interpret equations (10), (11) and (11) as defining an algebraic subvariety of the product $\mathbb{R}^6 \times \mathbb{R}^2$ (where, $(\lambda, \mu) \in \mathbb{R}^2$). As such it describes an algebraic family $\mathcal{F}$ of subvarieties of $\mathbb{R}^6$, parametrized by $\mathbb{R}^2$, The locus S of the pairs (A_1, A_2) which fail to generate $so(3)$, is simply given by the projection of $\mathcal{F} \subset \mathbb{R}^6 \times \mathbb{R}^2$ to the first factor (namely $\mathbb{R}^6$). Projecting $\mathcal{F}$ to $\mathbb{R}^6$ is equivalent to eliminate the variables λ and μ from equations (10), (11), and (11). A straightforward but lengthy calculation shows that S is the hypersurface defined the quartic homogeneous polynomial (7). ■

In order to be able to apply Theorem 2 in this specific case, we need to understand how to construct the open neighborhood W and the subvariety $R \subset W \times W$. As far as the subvariety R, this is immediate. Indeed, the pair $(g_1, g_2) \in R$, if and only if $g_1 = \exp(A_1)$ and $g_2 = \exp(A_2)$, where $(A_1, A_2) \in \mathfrak{g}$. On the other hand, characterizing the open neighborhood W is more difficult.

First of all, we are going to think to the elements of $SO(3)$ and $so(3)$ as matrices; moreover we are going to use the Hilbert-Schmidt norm, instead of the operator norm.

Definition 5. *Given a $n \times n$ matrix X with entries in $\mathbb{C}$, its Hilbert-Schmidt norm is*

$$\|X\| = \left(\sum_{k,l=1}^{n} |X_k l|^2 \right)^{\frac{1}{2}}$$

Since we are working with linear operators on finite dimensional vector spaces, the two norms induce the same topology.

Due to the fact that we want to use the Lie algebra to control the motion in the Lie group, it is natural to ask to what extent $\exp : \mathfrak{g} \to G$ is a diffeomorphism. Generally speaking the exponential function establishes a local isomorphism from a neighborhood U of the origin in the Lie algebra to a neighborhood V of the identity in the Lie group. First of all, let us recall when the logarithm (i.e. the inverse of exp) is defined.

Theorem 4. *The function*

$$\log(A) := \sum_{m=1}^{\infty} (-1)^{m+1} \frac{(A-I)^m}{m}$$

is defined and continous on the set of all $n \times n$ matrices A with $\|A - I\| < 1$, where I is the identity matrix. Moreover, for all A with $\|A - I\| < 1$,

$$e^{\log(A)} = A$$

and for all X, with $\|X\| < \log(2)$, we have

$$\|e^X - I\| < 1 \text{ and } \log(e^X) = X.$$

Proof: See [14], Theorem 2.7. ■

In the light of Theorem 4, if we choose two elements A_1 and A_2 in $so(3)$, such that $\|A_1\| < \log(2)$ and $\|A_2\| < \log(2)$ and $(A_1, A_2) \notin S$, then the corresponding elements $g_1 = \exp(A_1)$ and $g_2 = \exp(A_2)$ generate a dense subgroup in $SO(3)$. Let us spell out these conditions with the following:

Proposition 1. *Let A_1 and A_2 two elements of $so(3)$, such that $(A_1, A_2) \notin S$. Using the representations for A_1 and A_2 introduced in Theorem 3, we have that if*

$$x_1^2 + y_1^2 + z_1^2 < 0.045 \quad \text{and} \quad x_2^2 + y_2^2 + z_2^2 < 0.045,$$

then the subgroup $< g_1, g_2 >$ is dense in $SO(3)$, where $g_1 \exp(A_1)$ and $g_2 = \exp(A_2)$.

Proof: This is nothing more than a rephrasing of the conditions $\|A_1\| < \log(2)$ and $\|A_2\| < \log(2)$. Observing that $\log(2) \approx 0.3$ and the fact that $\|A_1\| < \log(2)$ is equivalent to

$$2(x_1^2 + y_1^2 + z_1^2) < (0.3)^2.$$

■

Example 7. If we take as A_1 and A_2 the following matrices:

$$A_1 = \begin{bmatrix} 0 & 0.2 & 0 \\ -0.2 & 0 & 0 \\ 0 & 0 & 0 \end{bmatrix} \quad A_2 = \begin{bmatrix} 0 & 0 & 0.2 \\ 0 & 0 & 0 \\ -0.2 & 0 & 0 \end{bmatrix},$$

then it is immediate to see that $(A_1, A_2) \notin S$ and moreover they satisfy the numerical bounds of Proposition 1. Therefore, using $\exp(A_1)$ and $\exp(A_2)$, which just describe "small" rotations around the x-axis and the y-axis respectively, we can reach in principle a configuration which is arbitrarily close to a desired configuration in $SO(3)$.

Up to now, we have not touched the problem of planning a quantized control path, from a starting configuration to a desired configuration. Indeed, the complete solution of the synthesis problem is definitely beyond current techniques, even in the special case of $SO(3)$. Here we sketch one possible approach, which turns out to have remarkable links with the "lattice" framework developed by other authors such as [2]. In the following methodology, instead of quantizing directly the possible displacement in the group G, we quantize inputs through elements of the Lie algebra $\mathfrak{g}$. This idea is exploited in the following:

Theorem 5 (Synthesis). *Let G be a connected, n-dimensional real semisimple Lie group and let $\mathfrak{g}$ be the corresponding Lie algebra. Let $X_1, \ldots, X_n, X_{n+1}$ be nonzero elements of $\mathfrak{g}$, linearly independent n by n. Identifying $\mathfrak{g}$ with $\mathbb{R}^n$, suppose moreover that*

$$\frac{(X_{n+1})^T X_i}{\|X_i\|} \in \mathbb{R} \backslash \mathbb{Q}, i = 1, \ldots, n. \tag{12}$$

Then any final configuration g on G is approachable through

$$\exp(Z_1) \circ exp(Z_2) \circ \cdots \circ \exp(Z_k), \tag{13}$$

where Z_i belongs to the subgroup generated by $X_1, \ldots, X_n, X_{n+1}$ for any i.

Proof: Consider an arbitrary final configuration $g \in G$. Since G is connected, there exists a (smooth) path $\gamma : [0,1] \to G$, such that $\gamma(0) = I$ (the initial configuration), and $\gamma(1) = g$. In general, the exponential mapping is not surjective on G, so we can not find in general an element $X \in \mathfrak{g}$ such that $\exp(X) = g$. On the other hand, the exponential mapping establishes a local diffeomorphism between a neighborhood of the origin U in $\mathfrak{g}$ and a neighborhood of the identity V in G. Now, since $[0,1]$ is compact, one can choose a sufficiently fine subdivision of the interval $[0,1]$, $0 = t_0 < t_1 < \cdots < t_k = 1$, such that $\gamma(t_i) = g_i$ (where g_i is a certain element of G, with $g_k = g$ and $g_0 = I$). If the subdivision $t_0, \dots, t_k$ is sufficiently small, then $B_{i-1} := g_{i-1}^{-1} \circ g_i \in V$. Moreover, it is readily seen that $g = g_k = B_0 \circ B_1 \circ \cdots \circ B_{k-1}$. Since each $B_i \in V$, then there exists a $Y_i \in \mathfrak{g}$, such that $\exp(Y_i) = B_i$. Due to condition (12), it is proved in [2] that the subgroup $\mathfrak{h}$ generated by $X_1, \dots, X_{n+1}$ is dense in the group $\mathfrak{g}$ (where the group structure on $\mathfrak{g}$ is the one induced by being a vector space). However this does not imply that $Y_i \in \mathfrak{h}$. In any case, $\mathfrak{h} \cap U$ is dense in U and therefore, $\exp(\mathfrak{h} \cap U)$ is dense in V. Therefore, instead of considering of considering the path γ, we consider a perturbation of the path $\tilde{\gamma} : [0,1] \to G$, such that $\tilde{\gamma}(0) = I$, $\tilde{\gamma}(1)$ is ϵ-close to g and such that $\tilde{\gamma}(t_i) = \tilde{B}_{i-1} \in \exp(\mathfrak{h} \cap U)$. Such a perturbation can always be found due to the density of $\exp(\mathfrak{h} \cap U)$ in V. ■

The main drawback of the previous synthesis methodology is due to the fact, already observed at the beginning of this note, that conditions (12) are not robust with respect to any perturbation, no matter how small it is. For instance, in the case of $SO(3)$ the elements of $so(3)$ can be identified with angular velocities with respect to given axis. The previous Theorem provides a way of reaching a configuration in $SO(3)$ which is arbitrarily close to any assigned configuration, quantizing the angular velocities (and the rotation axes). If there is a *resonance* among the angular velocities (that is to say (12) is false), then we lose the approachability property of this QCS.

5 Conclusions

In this paper we have presented a methodology to deal with control systems of general type through the use of groupoids. We have not attempted to study directly a control system on a Lie groupoid, but in analogy with what is under development in the field of integrable dynamical systems, it is likely that such a generalization provides interesting applications also in the case of control systems. We believe that this line of research deserves further insight and this is one of our aim for future investigations. Moreover, at the best of our knowledge we presented a first analysis of a very special class of quantized non-linear control systems having nonabelian symmetries, that is QCSs having a semi-simple Lie group as a state space. Certainly this is just a first insight, and lots of work has to be done in this area before a knowledge comparable to the case of systems with abelian symmetries is available. Anyway, we have seen that the class of semisimple QCSs presents some peculiar aspects, such as the robustness of the approachability property with respect to perturbation of quantization and the small number of quantized controls needed to achieve approachability, number which is independent on the dimension of the state space. Both these properties should be compared to the typical behavior of QCSs with abelian symmetries, where the approachability is typically not robust and where the number of control quanta increases with the dimension of the state space. An interesting challenge is to integrate the groupoid approach to control systems, with the reachability analysis so far developed in those contexts where both issues are relevant such as the problem of computational modeling of astronaut orientation sketched in [11].

References

1. Y. Chitour and B. Piccoli. Controllability for discrete systems with a finite control set. *Math. Control Signals Syst.*, 14(2), 2001.
2. A. Bicchi, A. Marigo, and B. Piccoli. On the reachability of quantized control systems. *IEEE Trans. on Automatic Control*, 47(4), April 2002.
3. A. Marigo and A. Bicchi. Steering driftless nonholonomic systems by control quanta. In *Proc. IEEE Conf. on Decision and Control*, 1998.
4. A. Bicchi, Y. Chitour, and A. Marigo. Reachability and steering of rolling polyhedra: a case study in discrete holonomy. *IEEE Transactions on Automatic Control.*, 49(5), May 2004.
5. R. W. Brockett. Formal languages for motion description and map making. In R. W. Brockett, editor, *Robotics*, volume 41 of *Proc. of Symposia in Applied Mathematics*, pages 181–193. American Mathematical Society, Providence, RI, 1990.
6. V. Manikonda, P. S. Krishnaprasad, and J. Hendler. Languages, behaviors, hybrid architectures and motion control. In J. Bailleul and J. C. Willems, editors, *Mathematical Control Theory*, pages 199–226. Springer Verlag, New York, NY, 1998.
7. E. Frazzoli, M. A. Dahleh, and E. Feron. Maneuver-based motion planning for nonlinear systems with symmetries. *IEEE Trans. on Robotics*, 21(6):1077–1091, December 2005.
8. A. Fagiolini, L. Greco, A. Bicchi, B. Piccoli, and A. Marigo. Symbolic control for underactuated differentially flat systems. In *Proc. IEEE Int. Conf. on Robotics and Automation*, May 2006.
9. A. Bicchi, A. Marigo, and B. Piccoli. A group-theoretic characterization of quantized control systems. In *Proc. IEEE Conf. Dec. Contr.*, 2002.
10. L.C. Li. Coboundary dynamical poisson groupoids and integrable systems. *Int. Math. Res. Notices*, (51), 2003.
11. L. Stirling. The computational modeling of astronaut orientation. PhD proposal, MIT.
12. N. Bourbaki. *Groupes et algébres de Lie*, volume Chapitres 7 et 8. Hermann, 1975.
13. E. Breuillard and T. Gelander. On dense free subgoups of lie groups. *Journal of Algebra*, (261), 2003.
14. Brian C. Hall. *Lie Groups, Lie Algebras and Representations.* Graduate Texts in Mathematics. Springer Verlag, 2004.

Minimum Time for a Hybrid System with Thermostatic Switchings

Fabio Bagagiolo

Dipartimento di Matematica, Università di Trento, Via Sommarive 14, loc. Povo, 38050 Trento, Italy
bagagiol@science.unitn.it

Abstract. In this paper we study a minimum time problem for a hybrid system subject to thermostatic switchings. We apply the Dynamic Programming method and the viscosity solution theory of Hamilton-Jacobi equations. We regard the problem as a suitable coupling of two minimum-time/exit-time problems. Under some controllability conditions, we prove that the minimum time function is the unique bounded below continuous function which solves a system of two Hamilton-Jacobi equations coupled via the boundary conditions.

1 Introduction

In this paper we study a minimum time problem for a hybrid system in $\mathbb{R}^n$ whose evolution y is subject to a switching parameter which may take the values, 1 and -1. In particular, the switching rule is subject to the evolution of an assigned component of the state y, and it is governed by a so-called thermostatic (or relay-type) hysteresis input-output relationship. That is, the switching between the two values occurs when such fixed component of y reaches (or better, gets over) some fixed thresholds, see Figure 1 for an example.

We are interested in applying the Dynamic Programming method to such kinds of problem, and then study the corresponding Hamilton-Jacobi-Bellman equation in the framework of the viscosity solutions theory. In this paper, under some controllability hypotheses, we prove the continuity of the minimum time function, we derive a Hamilton-Jacobi problem satisfied by the minimum time function in the viscosity sense, and prove that the latter is indeed the unique viscosity solutions. Such a Hamilton-Jacobi problem is given by a system of two Hamilton-Jacobi equations coupled by some part of the boundary conditions, which are also expressed in the viscosity sense. Our method consists in interpreting the problem as a coupling of two optimal control problems which present feature of minimum-time as well as exit-time problems.

Optimal control problems for systems with thermostatic behavior are of course important for applications. Many mechanical, physical, economical, biological systems have such a kind evolution (see for instance [12] for motivations from magnetism, [15] for a biological motivation, [14] from economics). Moreover, the thermostat is a simple example of hysteresis operator, which is also fundamental

A. Bemporad, A. Bicchi, and G. Buttazzo (Eds.): HSCC 2007, LNCS 4416, pp. 32–45, 2007.

to construct the so-called Preisach model of hysteresis, which is one of the most interesting and versatile analytical description of hysteresis phenomena (it is a superposition of a continuum of thermostats).

The dynamic programming method for the optimal control of systems with hysteresis feature is our main motivation. The present author has already studied, via dynamic programming method and viscosity solutions theory, some optimal control problems with hysteresis (see [2], [3], [4], [5]) (which lead to different Hamilton-Jacobi equations with respect to present one). Some possible directions of future investigations on the minimum time problem are: the case of a "discrete" Preisach operator, i.e. a finite sum of thermostats, see [3] for a one-dimensional infinite horizon problem in this framework; the case of the "true" Preisach operator, i.e. a superposition of a continuum of thermostats, see [4] for a one-dimensional finite horizon problem in this framework; the case of laking controllability conditions on the switching points, which leads to discontinuous solutions of Hamilton-Jacobi equations, see [5] for a multidimensional finite horizon problem in this framework. We also recall the works of Belbas and Mayergoyz on some applications of the Dynamic Programming method to optimal control problems with hysteresis (see for instance [8] and [9]).

Systems with thermostatic (or relay-type) switchings are also good examples of hybrid systems where the switchings are mandatory, when the state reaches some particular interdict zones (which is probably one of the most difficult behavior to treat in the framework of Dynamic Programming method and viscosity solutions). The application of the Dynamic Programming method to hybrid optimal control problems was first outlined by Branicky, Borkar, and Mitter in [11]. Bensoussan and Menaldi in [10] were the first to apply the viscosity solutions theory to a hybrid control problem, and they proved uniqueness of the value function as continuous viscosity solutions of the corresponding Hamilton-Jacobi problem. However, in the study in the Hamilton-Jacobi problem, they suppose that the system has no mandatory switchings. The present author, as already outlined, in [3] and [5], applied the viscosity solution theory to hybrid problems with thermostats, where the switchings are mandatory (although there are only mandatory switchings). In particular in [5] the case where the value function is discontinuous is addressed. Recently, other works on this subject have appeared. Dharmatti and Ramaswamy in [13] studied a problem with continuous value function, whereas Zhang and James in [18] studied a problem with discontinuous value function.

Finally we recall that the mathematical theory of hysteresis operators may be found in Visintin [17], and the theory of viscosity solutions for Hamilton-Jacobi-Bellman equations in Bardi-Capuzzo Dolcetta [7].

The present paper is organized as follows. In Section 2 we describe the delayed relay switching rule and introduce the minimum time problem; in Section 3 we prove the continuity of the minimum time function; in section 4 we prove that the minimum time function is the unique viscosity solution of the associated Hamilton-Jacobi problem; in the Appendix we give some results for an optimal control problem (without switchings) which is a combination of minimum time and exit-time features (which is, at least as formulation, rather new). In the Appendix. we also give the definitions of viscosity solutions and of boundary conditions in the viscosity sense.

2 Models and Problems

2.1 The Delayed Relay Switching Rule

For more details on the argument of this subsection see Visintin [17]. Let us fix two thresholds $\rho_1, \rho_2 \in \mathbb{R}$ with $\rho_1 < \rho_2$, and write $\rho := (\rho_1, \rho_2)$. For every continuous input function $g : [0, +\infty[\to \mathbb{R}$, and for every initial state $w_0 \in \{1, -1\}$, we define the following discontinuous output function $z(\cdot) := h_\rho[g, w_0](\cdot) : [0, +\infty[\to \{1, -1\}$ by

$$z(0) := \begin{cases} 1 \ \text{ if } g(0) > \rho_2, \\ -1 \text{ if } g(0) < \rho_1 \\ w_0 \text{ if } \rho_1 \le g(0) \le \rho_2, \end{cases}$$

and, for $t > 0$, we define $X(t) := \left\{\tau \in [0, t] \middle| g(\tau) < \rho_1 \text{ or } g(\tau) > \rho_2\right\}$, and

$$\begin{cases} z(t) = w_0 \text{ if } X(t) = \emptyset, \\ z(t) = 1 \quad \text{if } X(t) \neq \emptyset, \text{ and } g(\sup X(t)) \ge \rho_2, \\ z(t) = -1 \text{ if } X(t) \neq \emptyset, \text{ and } g(\sup X(t)) \le \rho_1. \end{cases}$$

For instance, if $z(t) = 1$ (which of course implies that $g(t) \ge \rho_1$), then z will remain constantly equal to 1 until g will possibly get over (downward) the threshold ρ_1, and after that time, z will be switched on -1 until g will possibly get over (upward) the threshold ρ_2. For example, if $z(t) = -1$ and, in the time interval $[t, t']$, g strictly increases from a value $g_1 < \rho_2$ to a value $g_2 > \rho_2$, passing on the threshold ρ_2 at the time t'', then the output is $z \equiv -1$ in $[t, t'']$, $z \equiv 1$ in $]t'', t']$. If instead g, after reaching the threshold ρ_2 at the time t'', changes the monotonicity and stays below or equal to ρ_2, then $z \equiv -1$ in $[t, t']$.

2.2 The Delayed Controlled Dynamical System

Let us consider a set of constant controls $A \subset \mathbb{R}^m$, for some m, a function

$$f : \mathbb{R}^n \times \{-1, 1\} \times A \to \mathbb{R}^n,$$

a fixed unit vector $S \in \mathbb{R}^n$, and $\rho = (\rho_1, \rho_2)$ a couple of thresholds for a delayed relay h_ρ. Let us also define the set of measurable controls

$$\mathcal{A} := \left\{\alpha : [0, +\infty[\to A \middle| \alpha \text{ is measurable}\right\}.$$

Then, we consider the following dynamical system (the dot "·" between vectors is the usual scalar product)

$$\begin{cases} y'(t) = f(y(t), z(t), \alpha(t)), \ t > 0, \\ z(t) = h_\rho[y(\cdot) \cdot S, w](t), \quad t \ge 0, \\ y(0) = x, \end{cases} \tag{1}$$

where $\alpha \in \mathcal{A}$, the initial state is the couple (x, w), and it is admissible if $w = 1$ and $x \cdot S \geq \rho_1$, or $w = -1$ and $x \cdot S \leq \rho_2$. Hence the solution $(y(\cdot), z(\cdot))$ of (1) (if it exists) can be view as a trajectory starting from (x, w) and evolving in the subset of $\mathbb{R}^n \times \mathbb{R}$

$$\overline{\mathcal{H}} := \overline{\mathcal{H}}_1 \cup \overline{\mathcal{H}}_{-1},$$

where,

$$\begin{aligned} \overline{\mathcal{H}}_1 &:= \left\{ (x,1) \in \mathbb{R}^n \times \{1\} \middle| x \cdot S \geq \rho_1 \right\}, \\ \overline{\mathcal{H}}_{-1} &:= \left\{ (x,-1) \in \mathbb{R}^n \times \{-1\} \middle| x \cdot S \leq \rho_2 \right\}, \end{aligned} \tag{2}$$

with a rule for switching from one connected component $\overline{\mathcal{H}}_w$ to the other, given by the switching rule of the delayed relay. See Figure 1. In the sequel, for any $w \in \{-1, 1\}$, we will denote by $\mathcal{H}_w$ the set, defined as in (2), but with the strict inequality for $x \cdot S$; in some sense it is the "interior" of $\overline{\mathcal{H}}_w$. Moreover, with some abuse of notations, we will denote by $\partial\overline{\mathcal{H}}_w$ the set, defined as in (2), but with just the equality only for $x \cdot S$; in some sense it is the "boundary" of $\overline{\mathcal{H}}_w$ (it is one of the two "switching boundaries").

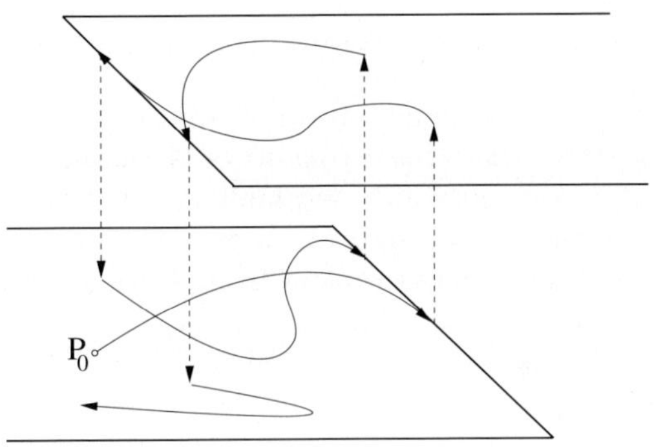

Fig. 1. Delayed switching evolution, starting from P_0

Some problems arise in defining a solution of (1): it even may not exist (see Alt [1] and also Bagagiolo [3] for a discussion on such a problem). We give the following definition of solution. For instance, let (x, w) be an initial state. We denote by $\tilde{y}(\cdot)$ the solution (if it exists) defined in $[0, +\infty[$ of the system

$$\begin{cases} \tilde{y}'(t) = f(\tilde{y}(t), w, \alpha(t)) \ t > 0, \\ \tilde{y}(0) = x. \end{cases}$$

Hence, we let the relay switch exactly when $h_\rho[\tilde{y}(\cdot), w]$ should switch by its switching rule. If $\tilde{t}$ is the switching time, we define, as solution of (1) in $[0, \tilde{t}]$, $(y(\cdot), z(\cdot)) = (\tilde{y}(\cdot), w)$. For $t > \tilde{t}$, we consider the solution in $\mathbb{R}^n$ $\tilde{y}_1$ starting from $\tilde{y}(\tilde{t})$ with dynamics $f(\tilde{y}_1, -w, \alpha)$. Let t_1 be the (possible) switching time for $h_\rho[y_1(\cdot - \tilde{t}), -1]$. Then we define, as solution of (1) in $]\tilde{t}, \tilde{t}+t_1]$, $(y(\cdot), z(\cdot)) = (y_1(\cdot), -1)$. We go on in this way. Since $\rho_1 < \rho_2$, then, any possible switching time is larger than the previous one plus an independent quantity $\delta > 0$ (recall that f is bounded and hence the velocity of $y \cdot S$ is bounded). Hence, we guess that such a construction of solution is possible for all the time, and moreover it is a good definition. This is Proposition 1, whose proof is now easy. We suppose that

$$\begin{cases} A \text{ is compact}, \quad f \text{is continuous and bounded} \\ \exists L > 0 \text{ s.t. } |f(x_1, w, a) - f(x_2, w, a) \leq L|x_1 - x_2| \\ \qquad \forall x_1, x_2 \in \mathbb{R}^n, \ w \in \{-1, 1\}, \ a \in A. \end{cases} \tag{3}$$

Proposition 1. *Under the hypothesis (3), for every initial state $(x, w) \in \overline{\mathcal{H}}$, and for every measurable control $\alpha \in \mathcal{A}$, there exists a unique solution (in the sense given above) $(y(\cdot), z(\cdot)) \in C^0([0, +\infty[; \mathbb{R}^n) \times L^\infty(0, +\infty; \mathbb{R})$ of the system (1). We will denote such a solution by $\big(y_{(x,w)}(\cdot; \alpha), z_{(x,w)}(\cdot; \alpha)\big)$.*

Remark 1. Note that, in general, for the solution as above, it is not true that $z(t) = h_\rho[y(\cdot) \cdot S](t)$ for all $t \geq 0$. Indeed, a trajectory switches when, if it does not switch, $y \cdot S$ is going to cross the threshold. But it may happens that, as the trajectory switches, the new dynamics $f(\cdot, -w, \cdot)$ is such that $y \cdot S$ does not cross the threshold. Hence, the glued trajectory has never crossed the threshold, and so, for the true switching delayed relay rule, there should not be any switching. A discussion more detailed on such definition of solution, is reported in Bagagiolo [3]. Here, we only say that, if a "true solution" of (1) exists, then it must coincide we the one above. Moreover, as will be explained in the sequel, such a definition seems useful for transforming an hybrid optimal control problem with such a kind of switching, in an exit-time problem from a closed set, which is, in some sense, more stable than other exit-time problems.

2.3 The Minimum Time Problem

Let $\mathcal{T} \subset \overline{\mathcal{H}}$ be fixed. The minimum time problem is to reach the target $\mathcal{T}$ as quickly as possible. Hence, for every initial state $(x, w) \in \overline{\mathcal{H}}$, and for every control $\alpha \in \mathcal{A}$, we define the reaching time for the corresponding trajectory of (1)

$$t_{(x,w)}(\alpha) = \inf \left\{ t \geq 0 \middle| \big(y_{(x,w)}(t; \alpha), z_{(x,w)}(t; \alpha)\big) \in \mathcal{T} \right\},$$

with the convention $\inf \emptyset = +\infty$. The minimum time function is then defined in $\overline{\mathcal{H}}$ as

$$T(x, w) = \inf_{\alpha \in \mathcal{A}} t_{(x,w)}(\alpha).$$

We consider the following hypotheses

$$\begin{array}{c} \mathcal{T} \text{ is the closure of its interior, has compact boundary } \partial\mathcal{T} \\ \text{which is also a } C^2 \text{ manifold,} \\ \forall (x, w) \in \partial\mathcal{T}, \text{ denoting by } n(x, w) \text{ the outer normal to } \partial\mathcal{T}, \\ \inf_{a \in A} f(x, w, a) \cdot n(x, w) < 0. \end{array} \tag{4}$$

Except for the compactness of $\partial\mathcal{T}$, the other regularity hypotheses on $\mathcal{T}$ may be changed, for instance in order to take account of a single point. Roughly speaking we need of the usual hypotheses of controllability on $\partial\mathcal{T}$ in order to have continuity of the minimum time function.

We also suppose the following controllability properties hold on the switching boundaries: for every $(x, w) \in \partial\overline{\mathcal{H}}_w$ there exist $a_1, a_2 \in A$ such that

$$f(x, w, a_1) \cdot S < -c < 0 < c < f(x, w, a_2) \cdot S, \tag{5}$$

with $c > 0$ independent from (x, w).

Let us define the controllable set

$$\mathcal{R} = \left\{(x, w) \middle| \exists\alpha \in \mathcal{A}, t_{(x,w)}(\alpha) < +\infty\right\} \tag{6}$$

Finally note that, both $\mathcal{T}$ and $\mathcal{R}$, can be split in to the disjoint union of two sets $\mathcal{T} = \mathcal{T}_1 \cup \mathcal{T}_{-1}$, $\mathcal{R} = \mathcal{R}_1 \cup \mathcal{R}_{-1}$, where, for $w \in \{1, -1\}$

$$\mathcal{T}_w := \mathcal{T} \cap \overline{\mathcal{H}}_w, \quad \mathcal{R}_w = \mathcal{R} \cap \overline{\mathcal{H}}_w.$$

3 Continuity

Proposition 2. *Let (3), (4), (5) hold. Then T is continuous in $\mathcal{R}$.*

We first recall the following lemma. For the proof, which is quite standard, see for instance Soner [16] (see also Bagagiolo-Bardi [6])

Lemma 1. *With the same hypotheses as in Proposition 2, let $w \in \{-1, 1\}$ and $K \subseteq \overline{\mathcal{H}}_w$ compact be fixed. Let us also fix $t \geq 0$. Then there exists a radius $r > 0$, and a constant $C > 0$ (both depending on t and K) such that, whenever for some $(x, w) \in K$ and for some $\alpha \in \mathcal{A}$ we have (here $B((x, w), r)$ is the ball of radius r around (x, w)),*

$$\left(y_{(x,w)}(\tau; \alpha), z_{(x,w)}(\tau; \alpha)\right) \in \overline{\mathcal{H}}_w \ \forall\tau \in [0, t],$$

then, for every $(\xi, w) \in B((x, w), r) \cap \overline{\mathcal{H}}_w$, there exists a control $\overline{\alpha}$ such that

$$\left(y_{(\xi,w)}(\tau; \overline{\alpha}), z_{(\xi,w)}(\tau; \overline{\alpha})\right) \in \overline{\mathcal{H}}_w, \quad \forall\tau \in [0, t],$$

and

$$|y_{(x,w)}(t; \alpha) - y_{(\xi,w)}(t; \overline{\alpha})| \leq C|x - \xi|. \tag{7}$$

Proof. (Of Proposition 2.) Let us sketch the proof that T is continuous in $\mathcal{R}_1$ (the other case is similarly treated). First of all note that $\mathcal{R}_1$ may be empty, and, in such a case there is nothing to prove. Hence, let us suppose that it is not empty (note that, by the hypotheses made, at least one of $\mathcal{R}_1$ and $\mathcal{R}_{-1}$ is not empty). Let $(x, 1), (\xi, 1)$ be two points of $\mathcal{R}_1$, and, for every $\varepsilon > 0$, let us take a control $\alpha_\varepsilon \in \mathcal{A}$ such that

$$T(x, 1) \geq t_{(x,1)}(\alpha_\varepsilon) - \varepsilon.$$

Let r_ε be the number of switchings of the trajectory $\big(y_{(x,1)}(\cdot;\alpha_\varepsilon), z_{(x,1)}(\cdot;\alpha_\varepsilon)\big)$, in the time interval $[0, t_{(x,1)}(\alpha_\varepsilon)]$, and note that, by the fact that the switchings are delayed and that f is bounded, there exists $N > 0$ such that

$$r_\varepsilon \leq N, \ \forall \varepsilon > 0. \tag{8}$$

Let U be a bounded open neighborhood of $(x,1)$ in $\overline{\mathcal{H}}_1$.

Let us first suppose that $r_\varepsilon = 0$. Then the trajectory starting from $(x,1)$ with control α_ε does not switch up to the time $t_{(x,1)}(\alpha_\varepsilon)$. Hence, for a suitable ball $B \subseteq U$ around $(x,1)$, and a suitable constant $C > 0$, applying Lemma 1, for every point $(\xi,1) \in B \cap \overline{\mathcal{H}}_1$, we obtain a control $\overline{\alpha}_\varepsilon$ such that the corresponding trajectory starting from $(\xi,1)$ does not switch, and (7) holds. Hence, since

$$\big(y_{(x,1)}(t_{(x,1)}(\alpha_\varepsilon);\alpha_\varepsilon), z_{(x,1)}(t_{(x,1)}(\alpha_\varepsilon);\alpha_\varepsilon)\big) \in \partial\mathcal{T},$$

by the controllability hypothesis on the boundary of the target, if $|x-\xi|$ is small enough, we obtain that

$$t_{(\xi,1)}(\overline{\alpha}_\varepsilon) = t_{(x,1)}(\alpha_\varepsilon) + O(|x-\xi|). \tag{9}$$

Now, let us suppose that $r_\varepsilon = 1$, and let t_1 be the switching instant. Again, for a suitable ball $B_1 \subseteq U$ around $(x,1)$, and for a suitable constant $C_1 > 0$, for every point $(\xi,1) \in B_1 \cap \overline{\mathcal{H}}_1$, we obtain a control $\overline{\alpha}_\varepsilon^1$ such that

$$\big(y_{(\xi,1)}(\tau;\overline{\alpha}_\varepsilon^1), z_{(\xi,1)}(\tau;\overline{\alpha}_\varepsilon^1)\big) \in \overline{\mathcal{H}}_1 \ \forall \tau \in [0,t_1],$$

and (7) holds with C_1 as constant. Since $\big(y_{(x,1)}(t_1;\alpha_\varepsilon), z_{(x,1)}(t_1;\alpha_\varepsilon)\big) \in \partial\mathcal{H}_w$, by the controllability hypotheses (refeq:switchingcondition), if $|x-\xi|$ is small enough, we can use a suitable control in order to make the trajectory starting from $(\xi,1)$ switch in a lap of time of order $O(|x-\xi|)$. Then we have two new starting points on $\overline{\mathcal{H}}_{-1}$ which are $\big(y_{(x,1)}(t_1;\alpha_\varepsilon), -1\big)$ and, say, $(\xi_1,-1)$ with $|y_{(x,1)}(t_1;\alpha_\varepsilon) - \xi_1| = O(|x-\xi|)$. Since in the remaining time interval $[t_1, t_{(x,1)}(\alpha_\varepsilon)]$, the trajectory starting from $(x,1)$ does not switch anymore, we eventually construct a control $\overline{\alpha}_\varepsilon$ for which (9) still holds, for $|x-\xi|$ sufficiently small.

Finally, repeating the previous steps r_ε times, for $r_\varepsilon > 1$, we obtain, for $|x-\xi|$ sufficiently small, let say less than $\mu > 0$, the same relation as in (9). In particular μ depends only on $\overline{U}$ and on N.

Recalling (8), we can say that, for every r_ε, we can use the same infinitesimal error-function O in (9).

Hence, if $(x,1),(\xi,1) \in U$ and $|x-\xi| \leq \mu$, then, supposing $T(x,1) \leq T(\xi;1)$ we get, for the arbitrariness of $\varepsilon > 0$,

$$0 \leq T(\xi,1) - T(x,1) \leq O(|x-\xi|).$$

Otherwise, if $T(x,1) > T(\xi,1)$, we exchange the role of x and ξ and note that all the previous estimates remain unchanged, with the same constant and error-functions O. In particular, the number of switchings r_ε cannot increase, since every switching requires a lap of time $\delta > 0$, and hence from (9), we would get an absurd. Hence we obtain

$$0 \le T(x,1) - T(\xi,1) \le O(|x-\xi|),$$

and we conclude. □

Remark 2. By the proof of Proposition 2 we also get the fact that $\mathcal{R}$ is open in $\overline{\mathcal{H}}$ (for the induced topology). Indeed, if $(x,w) \in \mathcal{R}_w$, then we have shown the existence of a ball B around it such that $B \cap \overline{\mathcal{H}}_w \subset \mathcal{R}_w$.

4 DPP and HJB

In this section we want to study a Hamilton-Jacobi-Bellman (HJB) problem for the minimum time function, which arises by applying the Dynamic Programming Principle (DPP). Let $(x,w) \in \mathcal{R} \setminus \mathcal{T}$, and a control $\alpha \in \mathcal{A}$ be fixed. Considering the corresponding trajectory, we define the *first exit time* from $\overline{\mathcal{H}}_w$ as

$$\tau_x^w(\alpha) := \inf \left\{ t \ge 0 \middle| y_{(x,w)}(t;\alpha) \notin \overline{\mathcal{H}}_w \right\},$$

which is nothing but the *first switching time* (for the trajectory). If $a, b \in [0,+\infty]$, then we are going to use the following convention: $\chi_{\{a \le b\}} = 1$ if $a \le b$, $\chi_{\{a \le b\}} = 0$ otherwise (similarly for $\chi_{\{a<b\}}$).

Proposition 3. *For every $(x,w) \in \overline{\mathcal{H}}$, we have*

$$\begin{aligned} T(x,w) = \inf_{\alpha \in \mathcal{A}} \Big(& \min(t_{(x,w)}(\alpha), \tau_x^w(\alpha)) \\ & + \chi_{\{t_{(x,w)}(\alpha) > \tau_x^w(\alpha)\}} T(y_{(x,w)}(\tau_x^w(\alpha);\alpha), -w)) \Big) . \end{aligned} \tag{10}$$

Proof. It can be easily proved by using the following lemma and dynamic programming techniques.. □

Lemma 2. *For every $(x,w) \in \partial\overline{\mathcal{H}}_w$, and for every $\alpha \in \mathcal{A}$ such that $\tau_x^w(\alpha) = 0$, we have*

$$T(x,w) \le T(x,-w), \quad t_{(x,w)}(\alpha) \ge T(x,-w). \tag{11}$$

Proof. Let us prove the first inequality in (11). By the controllability hypothesis (5), there exists a control $\overline{\alpha} \in \mathcal{A}$ such that $\tau_x^w(\overline{\alpha}) = 0$ (i.e. there is an immediate switching). For every $\varepsilon > 0$, we have

$$T(x,w) \le \varepsilon + T\left(y_{(x,w)}(\varepsilon;\overline{\alpha}), -w\right),$$

and, letting $\varepsilon \to 0^+$, we conclude by the continuity of T.

To prove the second inequality, for every $\varepsilon > 0$, we have

$$t_{(x,w)}(\alpha) = \varepsilon + t_{(y_{(x,w)}(\varepsilon;\alpha),-w)}(\alpha(\cdot+\varepsilon)) \ge \varepsilon + T\left(y_{(x,w)}(\varepsilon;\alpha), -w\right),$$

and, again, we conclude by the continuity of T. □

Remark 3. Note that, in (10), we can replace $t_{(x,w)}(\alpha)$ with the following instant

$$t_x^w(\alpha) := \inf\left\{t \geq 0 \middle| \left(y_{(x,w)}(t;\alpha), w\right) \in \mathcal{T}_w\right\}.$$

Hence, Proposition 3 says that, for every $w \in \{1, -1\}$ fixed, we can regard our problem in $\overline{\mathcal{H}}_w$ as the problem of minimizing the reaching time of $\mathcal{T}_w$, subject to stopping the process and paying the time elapsed plus an exit cost if we exit from $\overline{\mathcal{H}}_w$ before reaching $\mathcal{T}_w$. In particular, the exit cost is given by $T(\cdot, -w)$, i.e. our minimum time function evaluated on the point where we "switch down" after exit from $\overline{\mathcal{H}}_w$.

In the sequel, by $\partial_{\overline{\mathcal{H}}_w}\mathcal{R}$ we will denote the boundary of $\mathcal{R}$ with respect to the induced topology in $\overline{\mathcal{H}}_w$, and by $int_{\overline{\mathcal{H}}_w}\mathcal{R}$ the interior of $\mathcal{R}$, with respect to the same topology. In particular note that such interior may intersect the boundary $\partial\overline{\mathcal{H}}_w$ of $\overline{\mathcal{H}}_w$.

Using Remark 3 and Proposition 5 in the Appendix, we can say that the minimum function $T : \mathcal{R} \to [0, +\infty[$ solves the following problem in the unknown u:

$$\begin{cases} \text{for every } w \in \{-1, 1\},\ u \text{ is a viscosity solution of} \\ \begin{cases} \sup_{a\in A}\{-\nabla u(x,w)\cdot f(x,w,a)\} = 1 & \text{in } (\mathcal{R}_w \cap \mathcal{H}_w) \setminus \mathcal{T}_w, \\ u = 0 & \text{on } \partial\mathcal{T}_w, \\ u(x,w) \to +\infty & \text{as } (x,w) \to \partial_{\overline{\mathcal{H}}_w}\mathcal{R}_w, \\ u(\cdot,w) = u(\cdot,-w) & \text{on } \left(int_{\overline{\mathcal{H}}_w}\mathcal{R}_w \cap \partial\overline{\mathcal{H}}_w\right) \setminus \mathcal{T}_w. \end{cases} \end{cases} \tag{12}$$

In particular, in (12), the last boundary condition has to be understood in the *viscosity sense*. Now, we prove that T is indeed the unique solution of (12).

Theorem 1. *The minimum time function is the unique bounded below continuous function from $\mathcal{R}$ to $\mathbb{R}$ which solves the problem (12).*

Proof. First of all, for every $w \in \{1, -1\}$, let us denote by $(12)_w$ the Hamilton-Jacobi boundary problem in $\overline{\mathcal{H}}_w$ which appear in (12). Note that, even if we have a uniqueness result for each single problem $(12)_w$, we cannot immediate conclude that we have uniqueness for the problem (12), since the boundary conditions are intrinsic to the problem: they are part of the solutions.

For every $w \in \{1, -1\}$ we define the set

$$S^w := \left\{(x,w) \in \overline{\mathcal{H}}_w \middle| (x,-w) \in \partial\overline{\mathcal{H}}_{-w}\right\},$$

and note that it is exactly the set where we arrive when we exit from $\overline{\mathcal{H}}_{-w}$.

Let $u : \mathcal{R} \to \mathbb{R}$ be a bounded below continuous solution of (12). For every $w \in \{1, -1\}$ we extend ("by continuity") u from $(S^w \cap \mathcal{R}) \setminus \mathcal{T}$ to $S^w \setminus \mathcal{T}$ by setting $u = +\infty$. From the uniqueness result Theorem 2 of the Appendix, we know that $u(\cdot, w)$ is the value function of the minimum/exit time problem in $\overline{\mathcal{H}}_w$, given by reaching the target $\mathcal{T}_w$ or exit from $\overline{\mathcal{H}}_w$ paying the spent time plus the cost $u(\cdot, -w)$. So, it possibly differs from the problem solved by the minimum time T on $\overline{\mathcal{H}}_w$, only for the exit cost. We are going to prove that, for every $w \in \{1, -1\}$ the two problems $(12)_w$ solved respectively by T and u have the same exit cost, and hence the thesis will be proved.

For every $\delta > 0$ let us define the set

$$\mathcal{R}(\delta) = \left\{ (x, w) \in \overline{\mathcal{H}} \middle| T(x, w) \leq \delta \right\}.$$

Note that, since the dynamics f is bounded, and since the switchings are delayed, there exists $\delta > 0$ such that for every starting point $(x, -w) \in S^{-w}$ we need a time strictly larger than δ in order to exit from $\overline{\mathcal{H}}_{-w}$, whichever is the control we are using.

We first prove that $u \geq 0$. To this end we prove that $u \geq 0$ on $S^w \cup S^{-w}$, from which the claim follows by the interpretation of u as value function. Let us suppose that there exists $(x, w) \in S^w$ such that $u(x, w) < 0$. This means that there are trajectories starting from (x, w) which reach points of the boundary $(x', w) \in \partial\overline{\mathcal{H}}_w$ where the cost is $u(x', -w) < -\delta$. But, the same argumentation shows that there should exist trajectories starting from $(x', -w)$ which reach points of the boundary $(x'', -w) \in \partial\overline{\mathcal{H}}_{-w}$ where the cost is $u(x'', w) < -2\delta$. Iterating such procedure, we obtain a contradiction to the fact that u is bounded below.

For the point of the (possibly empty) set $S^{-w} \cap \mathcal{R}(\delta)$ the value of u does not depend on the exit cost, since, from those points, it is not convenient to reach the boundary and pay the exit cost because it needs a time larger than δ and the exit cost is nonnegative. Hence we have $u = T$ on $S^w \cap \mathcal{R}(\delta)$, Now, let us consider the point of the (possibly empty) set $S^w \cap \mathcal{R}(2\delta)$. For every such a point, the value of u is equal to T if $T(x) \leq \delta$ (since we do not switch), otherwise it may be conditioned by the value of the exit cost $u(\cdot, -w)$ on the points of $\partial\overline{\mathcal{H}}_w$. But for reaching such boundary points (x, w) from $S^w \cap \mathcal{R}(2\delta)$ we spent a time larger than δ and hence we certainly have $u(x, -w) = T(x, -w)$ since $(x, -w) \in S^{-w} \cap \mathcal{R}(\delta)$. Again, iterating such a process, we obtain that $u = T$ on $S^w \cup S^{-w}$, and we conclude. □

Remark 4. Arguing as in the Remark 5, we can say that the couple $(\mathcal{R}, T)$ is the unique couple given by an open set $\mathcal{O}$ in $\overline{\mathcal{H}}$ (for the induced topology) which contains $\mathcal{T}$, and by a bounded below continuous function $u : \mathcal{O} \to \mathbb{R}$ which solves the corresponding problem (12), where the set $\mathcal{R} = \mathcal{R}_1 \cup \mathcal{R}_{-1}$ is replaced by $\mathcal{O} = \mathcal{O}_1 \cup \mathcal{O}_{-1}$.

References

1. H. W. Alt: *On the thermostat problem*, Control Cybernet., 14 (1985), 171-193.
2. F. Bagagiolo: *Dynamic programming for some optimal control problems with hysteresis*, NoDEA Nonlinear Differential Equations Appl., 9 (2002), 149–174.
3. F. Bagagiolo: *An infinite horizon optimal control problem for some switching systems*, Discrete Contin. Dyn. Syst. Ser. B, 1 (2001), no. 4, 443-462.
4. F. Bagagiolo: *Viscosity solutions for an optimal control problem with Preisach hysteresis nonlinearities*, ESAIM Control Optim. Calc. Var., 10 (2004), 271–294.

5. F. Bagagiolo: *Optimal control of finite horizon type for a multidimensional delayed switching system*, Discrete Contin. Dyn. Syst. Ser. B, 5 (2005), no. 2, 239–264.
6. F. Bagagiolo, M. Bardi: *Singular perturbation of a finite horizon problem with state-space constraints*, SIAM J. Control Optim., 36 (1998), 2040–2060.
7. M. Bardi, I. Capuzzo Dolcetta: *Optimal Control and Viscosity Solutions of Hamilton-Jacobi Bellman Equations*, Birkhäuser, Boston, 1997.
8. S. A. Belbas, I. D. Mayergoyz: *Optimal control of dynamical systems with Preisach hysteresis*, Internat. J. Non.Linear Mech., 37 (2002), 1351–1361.
9. S. A. Belbas, I. D. Mayergoyz: *Hadamard-like derivatives in Preisach modeling and control*, Physica B-Condensed Matter, 372 (2006), 87-90.
10. A. Bensoussan, J. L. Menaldi: *Hybrid control and dynamic programming*, Dynam. Contin. Discrete and Impuls. Systems, 3 (1997), 395–492.
11. M. S. Branicky, V. S. Borkar, S. K. Mitter: *A unified framework for hybrid control: Model and optimal control theory*, IEEE Trans. Automat. Control., 43 (1998), 31–45.
12. E. Della Torre: *Magnetic Hysteresis*, IEEE Press, New York, 1999.
13. S. Dharmatti, M. Ramaswamy, *Hybrid control system and viscosity solutions*, SIAM J. Control Optim., 44 (2005), 1259–1288.
14. M. Göcke: *Various concepts of hysteresis applied in economics*, J. Economic Surveys, 16 (2002), 167–188.
15. S. M. Lenhart, T. I. Seidman, J. Yong: *Optimal control of a bioreactor with modal switching*, Math. Models Methods Appl. Sci., 11 (2001), 933–949.
16. H. M. Soner: *Optimal control with state-space constraints I*, SIAM J. Control Optim., 31 (1993), 132–146.
17. A. Visintin: *Differential Models of Hysteresis*, Springer-Verlag, Heidelberg, 1994.
18. H. Zhang, M. R. James: *Optimal control of hybrid systems and a system of quasi-variational inequalities*, SIAM J. Control Optim., 45 (2006), 722–761.

Appendix: On a Minimum/Exit Time Problem

For many results concerning the theory of viscosity solutions, adopted in this section, we refer the reader to the book Bardi-Capuzzo Dolcetta [7].

Let $\Omega \subset \mathbb{R}^n$ be an open set, $\mathcal{T} \subseteq \overline{\Omega}$ be a closed set satisfying (4), and $\psi : \partial\Omega \to [0, +\infty]$ be a continuous function (for the usual topology on $[0, +\infty]$). Using the same notations as before for controls, considering a bounded continuous function $f : \mathbb{R}^n \times A \to \mathbb{R}^n$ satisfying the analogous of (3) (neglecting w), and the analogous on $\partial\Omega$ of (5) (again neglecting w), we consider the optimal control problem of minimizing the time spent for reaching $\mathcal{T}$ or the time spent plus a cost for exit from $\overline{\Omega}$, subject to the controlled (non switching) dynamical system

$$\begin{cases} y'(t) = f(y(t), \alpha(t)), \, t > 0, \\ y(0) = x, \end{cases} \tag{13}$$

We denote by $y_x(\cdot;\alpha)$ the solution of (13). We define the following instants

$$t_x(\alpha) := \inf \left\{ t \geq 0 \middle| y_x(t;\alpha) \in \mathcal{T} \right\}, \text{ \textit{reaching time}},$$
$$\tau_x(\alpha) := \inf \left\{ t \geq 0 \middle| y_x(t;\alpha) \notin \overline{\Omega} \right\} \text{ \textit{exit time}},$$

and define the value function

$$V(x) := \inf_{\alpha\in\mathcal{A}} \left(\min(t_x(\alpha), \tau_x(\alpha)) + \chi_{\{t_x(\alpha)>\tau_x(\alpha)\}}\psi(y_x(\tau_x(\alpha);\alpha))\right)$$

Let us define the set

$$\mathcal{R} := \left\{x \in \overline{\Omega}\middle| V(x) < +\infty\right\}.$$

Proposition 4. *With all the hypotheses made before, the set $\mathcal{R}$ is open in $\overline{\Omega}$ for the induced topology, and the value function V is continuous in $\mathcal{R}$.*

Proof. It is similar to that of Proposition 2. □

For every $x \in \mathcal{R}$, and for every $\alpha \in \mathcal{A}$ we define

$$\epsilon_x(\alpha) := \min(t_x(\alpha), \tau_x(\alpha)) = \inf\left\{t \geq 0\middle| y_x(t;\alpha) \notin \overline{\Omega} \setminus \mathcal{T}\right\}.$$

We have the following Dynamic Programming Principle: for every $x \in \mathcal{R}$ and for every $t \geq 0$

$$V(x) = \inf_{a\in A} \left\{\min(t, \epsilon_x(\alpha)) + V(y_x(\min(t, \epsilon_x(\alpha)); \alpha))\right\}. \tag{14}$$

Proposition 5. *Let all the hypotheses of Proposition 4). Then the value function V is a viscosity solution of the following problem in the unknown u:*

$$\begin{cases} \sup\limits_{a\in A} \{-\nabla u(x) \cdot f(x,a)\} = 1, & \text{in } (\mathcal{R}\cap\Omega)\setminus\mathcal{T}, \\ u = 0, & \text{on } \partial\mathcal{T}, \\ u(x) \to +\infty, & \text{as } x \to \partial_{\overline{\Omega}}\mathcal{R}, \\ u(x) = \psi(x), & \text{on } (int_{\overline{\Omega}}\mathcal{R}\cap\partial\Omega)\setminus\mathcal{T}. \end{cases} \tag{15}$$

In particular, the last boundary condition of (15), has to be understood in the viscosity sense.

A continuous function $u : \mathcal{R} \to \mathbb{R}$, satisfying the second boundary condition (the limit one) is a viscosity solution of (15) if it is a subsolution and a supersolution. Being a subsolution (respectively: a supersolution) means that $u \leq 0$ (respectively $u \geq 0$) on $\partial\mathcal{T}$, and moreover for every C^1 test function $\varphi : \overline{\Omega} \to \mathbb{R}$, and for every $x \in \mathcal{R}\setminus\mathcal{T}$ such that $u-\varphi$ has a local maximum in x (respectively: a local minimum) with respect to $\overline{\Omega}$, the following holds

$$\begin{cases} \sup\limits_{a\in A}\{-\nabla\varphi(x)\cdot f(x,a)\} \leq 1, \\ \text{if either } x \in (\mathcal{R}\cap\Omega)\setminus\mathcal{T} \text{ or } x \in (int_{\overline{\Omega}}\mathcal{R}\cap\partial\Omega)\setminus\mathcal{T} \text{ and } u(x) > \psi(x); \\ \Big[\text{respectively: } \sup\limits_{a\in A}\{-\nabla\varphi(x)\cdot f(x,a)\} \geq 1, \\ \text{if either } x \in (\mathcal{R}\cap\Omega)\setminus\mathcal{T} \text{ or } x \in (int_{\overline{\Omega}}\mathcal{R}\cap\partial\Omega)\setminus\mathcal{T} \text{ and } u(x) < \psi(x)\Big]. \end{cases}$$

Proof. The proof is almost standard, we only check the last boundary condition. Note that, by the controllability hypothesis on $\partial\Omega$, we may only have $V \leq \psi$ on $\partial_{\overline{\Omega}}\mathcal{R} \cap \partial\Omega$. Let us fix $x \in int_{\overline{\Omega}}\mathcal{R} \cap \partial\Omega$, we have only to analyze the case $V(x) < \psi(x)$. Since $x \in \mathcal{R} \setminus \mathcal{T}$, then there exists $\delta > 0$ such that, for every t small, there is a minimizing sequence of controls α_n for (14) with $\epsilon_x(\alpha_n) \geq \delta$. Hence, for every test function $\varphi \in C^1(\overline{\Omega})$ such that $V - \varphi$ has a local minimum in x with respect to $\overline{\Omega}$, by the usual technique we get

$$\sup_{a \in A} \{-\nabla\varphi(x) \cdot f(x,a)\} \geq 1.$$

□

We now suppose the following "internal cone condition" in $\overline{\Omega}$: there exists a constant $c > 0$ and a uniformly continuous function $\eta : \overline{\Omega} \to \mathbb{R}^n$ such that, for every $x \in \overline{\Omega}$ ($B(x + s\eta(x), cs)$ is the ball around $x + s\eta(x)$ with radius cs)

$$B(x + s\eta(x), cs) \subseteq \Omega, \ \forall 0 < s \leq c. \tag{16}$$

Theorem 2. *Let the hypotheses Proposition 4 and (16) hold. Then the value function is the unique continuous bounded below function $u : \mathcal{R} \to \mathbb{R}$ which is a viscosity solution of the problem (15).*

Proof. We sketch a standard technique. We first introduce the so-called Kruzkov transformation. Let $u : \mathcal{R} \to \mathbb{R}$ be a continuous bounded below function satisfying the first two boundary conditions in (15). Then, we define the bounded continuous function $\tilde{u} : \overline{\Omega} \to \mathbb{R}$ as follows

$$\tilde{u}(x) := \begin{cases} 1 - e^{-u(x)} & \text{if } x \in \mathcal{R}, \\ 1 & \text{if } x \in \overline{\Omega} \setminus \mathcal{R}. \end{cases}$$

We also denote by $\tilde{\psi} : \partial\Omega \to \mathbb{R}$ the Kruzkov transform of the boundary datum ψ

$$\tilde{\psi}(x) := \begin{cases} 1 - e^{-\psi(x)} & \text{if } \psi(x) \in \mathbb{R}, \\ 1 & \text{if } \psi(x) = +\infty. \end{cases}$$

If u is a continuous bounded below viscosity solutions of (15), then $\tilde{u} : \overline{\Omega} \to \mathbb{R}$ is a continuous bounded viscosity solution of the problem in the unknown v

$$\begin{cases} v + \sup_{a \in A} \{-\nabla v \cdot f(x,a)\} = 1 \text{ in } \Omega \setminus \mathcal{T}, \\ v = 0 & \text{on } \partial\mathcal{T}, \\ v = \tilde{\psi} & \text{on } \partial\Omega \setminus \mathcal{T}, \end{cases} \tag{17}$$

where the last boundary condition has to be understood in the viscosity sense.

Hence, our thesis will come from a uniqueness result for problem (17). Such a problem is a little bit different from the usual ones found in the literature, since it is a minimum/exit time problem. In particular, the target $\mathcal{T}$ and the exit boundary $\partial\Omega$ may intersect. Since, the open set on which the Hamilton-Jacobi equation must be verified is $\Omega \setminus \mathcal{T}$, it is not in general true that a internal cone condition, similar to (16) should hold for the closure of such a set. It depends on how $\mathcal{T}$ and $\partial\Omega$ intersect. And the internal cone condition is in general necessary for the uniqueness result where the boundary conditions are in the viscosity sense. However, in our case, the condition on $\partial\mathcal{T}$ is not in the viscosity sense, it is just a classical boundary condition. Moreover, since $\partial\mathcal{T}$ is compact by hypothesis, we can say that for every open (in $\overline{\Omega}$) neighborhood $\mathcal{U}$ of $\partial\mathcal{T}$, a sort of (uniformly) internal cone condition holds for the set $\overline{\Omega}\setminus\mathcal{U}$, in the following sense: there exists a constant $c' > 0$ (depending on $\mathcal{U}$), such that, for every $x \in \overline{\Omega}\setminus\mathcal{U}$, we have

$$B(x + s\eta(x), c's) \subseteq \Omega\setminus\mathcal{T} \ \ \forall 0 < s \leq c'. \tag{18}$$

As usual, the uniqueness is proved by a comparison result between a continuous subsolution v_1 and a continuous supersolution v_2 (i.e. by proving that $v_1 \leq v_2$ in $\overline{\Omega}\setminus\mathcal{T}$). But, on $\partial\mathcal{T}$, they must satisfy $v_1 \leq 0 \leq v_2$, and then, for every $\delta > 0$ we found a neighborhood $\mathcal{U}_\delta$ of $\partial\mathcal{T}$ such that $v_1 - v_2 \leq \delta$ in $\mathcal{U}_\delta$. This permits to use the standard double variables/penalization technique. Indeed, one usually suppose by absurd that there exists $\tilde{x}$ such that $v_1(\tilde{x}) - v_2(\tilde{x}) = \delta > 0$. In our case, we certainly have $\tilde{x} \in \overline{\Omega}\setminus\mathcal{U}_\delta$. Hence, the usual machinery, and especially (18) may be used.

In the end, we get that there exists a unique bounded continuous viscosity solution of (17). □

Remark 5. Let us suppose that $\mathcal{R}' \subseteq \overline{\Omega}$ is another open set in $\overline{\Omega}$ (for the induced topology), which contains $\mathcal{T}$, and that $u' : \mathcal{R}' \to \mathbb{R}$ is a continuous bounded below viscosity solutions of the corresponding problem (15), where $\mathcal{R}$ is replaced by $\mathcal{R}'$. Then, the Kruzkov transformation applied to u' leads to the same problem (17). Hence, by uniqueness, we can say that the couple $(\mathcal{R}, T)$ is the unique couple given by an open set in $\overline{\Omega}$ containing $\mathcal{T}$, and by a bounded below continuous function on such an open set, which solves the corresponding problem (15).

Complexity Reduction for the Design of Interacting Controllers*

A. Balluchi[1], E. Mazzi[1,2], and A.L. Sangiovanni Vincentelli[1,3]

[1] PARADES, Via di S.Pantaleo, 66, 00186 Roma, Italy
Tel.: +39 06 68807923; Fax: +39 06 68807926
{balluchi,emazzi,alberto}@parades.rm.cnr.it.
[2] Centro Interdipartimentale di Ricerca "E. Piaggio", Università di Pisa,
Via Diotisalvi 2, 56126 Pisa, Italy
Tel.: +39 050 2217050; Fax: +39 050 2217051
[3] Department of Electrical Engineering and Computer Sciences,
University of California at Berkeley, Berkeley, CA 94720, USA
Tel.: +1 510 6421792; Fax: +1 510 6435052
alberto@eecs.berkeley.edu.

Abstract. The complexity of embedded controllers in important industrial domains such as automotive and avionics is growing constantly making error-free system integration almost impossible. We address the complexity issues posed by the analysis and design of interacting controllers introducing approximation techniques that are shown to be effective on a substantial industrial test case: the control system for common-rail fuel-injection developed by Magneti Marelli Powertrain.

1 Introduction

This paper is motivated by the analysis of the development flow for embedded control systems in the automotive industry reported in [1]. The design flow for ECUs adopted by Tier–1 automotive companies such as Bosch and Magneti-Marelli is captured by the so–called *V-diagram*, representing a synthesis flow and an integration and testing flow [1,2]. The synthesis flow is articulated in the following steps: system specification, functional deployment, control system, and HW/SW components. In functional deployment, system specifications are mapped onto a control-system architecture: design requirements are derived for each control algorithm, sensor and actuator, so that the system specifications are met. Next, each control algorithm is designed independently, according to the specification defined in the previous step.

Today, functional deployment is largely guided by the experience of system engineers. In fact, while there are many established techniques to carry out the design of single control algorithms, a methodology is still lacking for the design of a control system architecture and the management of interactions between control algorithms. Management of control algorithm interactions is by no means trivial, due to:

– *Complexity of control system architectures:* Often, ECUs may have more than one hundred I/O signals, they may execute more than four hundred control algorithms,

* Research partially supported by the Network of Excellence HYCON, E.U. grant IST-511368.

A. Bemporad, A. Bicchi, and G. Buttazzo (Eds.): HSCC 2007, LNCS 4416, pp. 46–60, 2007.

with nested control loops, and may share with the other ECUs more than one hundred signals.

- *Synchronization issues:* The control system is multi-rate, with heterogeneous time–domains, and frequency and phase drifts between fixed sampling–time actions and event driven actions. To make things even more complicated, proper synchronization has to be achieved between algorithms implemented on different ECUs connected by a communication network.
- *Hybrid behavior:* Control algorithms are often characterized by several operation modes, encoding different regions of operation as well as different phases of the life of the product. "Inside" each mode, the system is characterized by hybrid dynamics where the evolution of the system occurs in discrete and continuous domains.

Given its conceptual and practical complexity, effective management of the control algorithm interactions can be achieved only by formulating the design and verification problem at a level of abstraction that is high enough to allow to analyze the properties of interest in a quantitative way.

We present in this paper a systematic procedure to reduce the complexity of the problem by abstracting the behavior of the system while maintaining accuracy when needed of the approximations used. We leverage model-reduction techniques as well as finite-state machine abstractions to simplify the analysis and synthesis problem. These techniques are combined in a flow that is motivated by automotive applications but could be extended easily to other application domains. While the derivation of robust demonstrable bounds on the approximation procedure presented here is still missing, the effectiveness of the approach is demonstrated on an industrial design problem, common-rail fuel injection, made available by Magneti Marelli Powertrain. The rest of the paper is organized as follows. In Section 2, we briefly review fundamental results on model reduction for continuous systems and abstractions of Finite State Machines (FSM). In Section 3, the proposed approach is illustrated, while details and an application to a case study in common rail control are given in Sections 4 and 5. Some concluding remarks are given in Section 6.

2 Background

The complexity reduction approach we propose is based on results on model reduction for continuous systems and abstractions of FSMs that we briefly review in this section.

2.1 Model Reduction for Continuous Systems

Model-reduction simplifies system analysis and controller design by reducing the dimensions of the state space of a continuous time system while maintaining accuracy within a given bound. Model reduction has close connections to approximation theory and system identification. The literature dealing with this important domain is very rich. Numerical analysts, ODE experts, automatic control researchers and the Electronic Design Automation (EDA) community have all investigated a variety of techniques.

In particular, EDA researchers developed a number of very effective numerical techniques that have been implemented in industrial tools in the context of simulation of

VLSI circuits with parasitics. Physical modeling of IC components results in high dimensional state models (up to ten thousands of states), for which low dimension approximations have been obtained [3]. Reductions of up to three orders of magnitude were obtained. Algorithms, such as PRIMA, PVL and Arnoldi schemes, generate guaranteed-passive models for systems with special internal structure, using numerically stable and efficient Krylov-subspace iterations [4].

Our approach is closer to model reduction techniques well-established in the control community (e.g., *Hankel-Norm approximation* [5]; *Truncated Fourier* and *Taylor expansion* [6]). In particular, we borrow from the following two techniques:

Proper Orthogonal Decomposition (POD), also called *Karhunen-Love expansion*, can be applied to both linear and nonlinear ordinary differential equations, as well as to partial differential equations. The reduced order model is obtained by identification using trajectory profiles given by simulations of the original model. Given a state space model, snapshots $x(t_i)$ of state profiles over a time horizon $[t_0, t_N]$ are collected in a matrix X. Then, a reduced sub-space where most of the snapshots $x(t_i)$ lie is determined. If the approximation error is to be minimized in a least-square sense, then the problem can be expressed using the Frobenius norm as in singular value decomposition [7]. After inspection of the singular value of the matrix X, a suitable state subspace is chosen. The reduced order model is obtained by simple truncation of "small states", so-called *Galerkin projection*. Notice that the POD method *does not guarantee that the reduced model is a close approximation of the original model*. In some cases one can prove that stability of equilibrium points is preserved between the original and the projected model [8].

Balanced truncation was first proposed in [9]. A common physical interpretation of balanced truncation is given in the state space: "small states" are removed by truncation of the original state–space model. This interpretation has close connections to POD, but it does not capture the connection between balanced truncation and Singular Value Decomposition (SVD) methods (see [7]). For linear time-invariant systems, if the Lyapunov observability and controllability differential inequalities have any solution, i.e., the observability and controllability Gramians are greater than 0, for all t, then there exists a transformation matrix $T(t)$ that obtains a *balanced realization* of the system. When the system is transformed into a balanced form, the reduced model is obtained by state truncation. Balanced truncation for nonlinear systems is discussed in [10] and [11]. Time-varying balanced truncation based on both SVD and on input-output data is presented in [12]. Interesting results on balanced truncation of linear time-varying systems in both discrete–time and continuous–time are reported in [13]. Both upper and lower error bounds for the truncated models can be easily derived using time-varying Lyapunov equations or inequalities. In [14] it is shown how to construct *truncated balanced realization*-like methods that generate guaranteed passive reduced models for state-space systems with arbitrary internal structure.

2.2 FSM Abstraction

FSM abstraction is defined on the basis of the notion of refinement (see [15] and [16]). An FSM S_1 *refines* an FSM S_2 if: (i) inputs(S_1) = inputs(S_2); (ii) outputs(S_1) = outputs (S_2); (iii) behaviors$(S_1) \subseteq$ behaviors(S_2). A behavior of an FSM is a pair (x, y) where

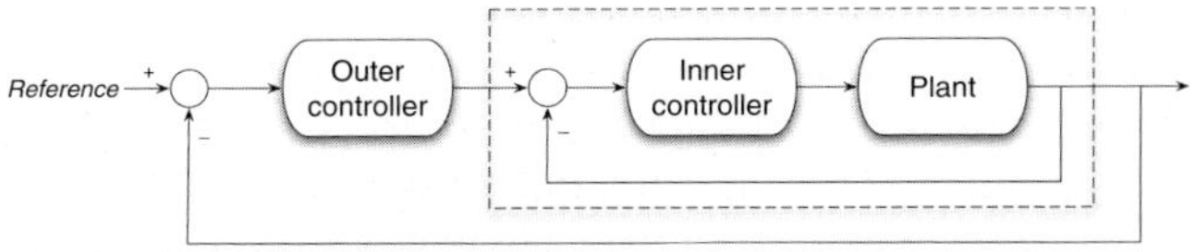

Fig. 1. Two interacting controllers connected in a cascade architecture

x is an input sequence and y is a corresponding output sequence. If S_1 refines S_2, then S_1 is a more detailed description (a *refinement*) of S_2, while S_2 is an *abstraction* of S_1. S_1 and S_2 are said to be *equivalent* if: (i) inputs(S_1) = inputs(S_2); (ii) outputs(S_1) = outputs(S_2); (iii) behaviors(S_1) = behaviors(S_2). An equivalent and minimal FSM to a given FSM can be obtained using the notion of indistinguishable states [17]. Let I_a be a generic sequence of inputs $i_1,\ldots,i_k$ and let U_{a,s_j} be the corresponding sequence outputs $u_1,\ldots,u_k$, starting from the initial state s_j. Let $\lambda()$ denote the input–output function, i.e. $U_{a,s_j} = \lambda(s_j, I_a)$. States s_i and s_j are indistinguishable, i.e. $s_i \sim s_i$, if $U_{a,s_i} = \lambda(s_i, I_a) = \lambda(s_j, I_a) = U_{a,s_j}$ for all I_a. Since $s_i \sim s_i$, $s_i \sim s_j \leftrightarrow s_j \sim s_i$ and $s_i \sim s_i \wedge s_j \sim s_k \rightarrow s_i \sim s_k$, then indistinguishability is an equivalence relation. States can be collected in equivalence classes and the corresponding equivalent quotient system defines the minimal equivalent FSM.

3 Complexity Reduction Techniques for Hybrid Control Systems

We consider control algorithms connected in a cascade structure as in Figure 1 (e.g. a control system consisting of an outer–loop heat pump controller and an inner–loop compressor controller). The objective is to analyze the interaction between the inner loop controller and the outer loop controller. Analyzing the interactions using the complete hybrid model of the plant, the inner controller and the outer controller is often too complex and not even necessary. This is especially true when there are several nested control loops. Hence, we propose an approach based on complexity reduction of the inner–closed loop system behavior. The proposed approach can be cast in the framework of *approximate simulation* of hybrid systems which was defined in [18]. The main new contribution of the present work is the formulation of a procedure for complexity reduction based on which approximated models of hybrid systems can be obtained.

The proposed complexity reduction procedure consists of three main steps that borrow from the literature presented in Section 2:

***Step 1*: Model reduction of the continuous dynamics associated to locations of the complete hybrid system**

This step is a direct application of model reduction techniques. In [19], continuous dynamics model reduction was proposed as an approach to simplify reachability analysis. While this step is useful, it will not lead to an effective simplification of the discrete behavior of the inner–loop hybrid system. To do so, we will follow two paths: (i) transform part of the hybrid system into a purely continuous time system; (ii) find classes of equivalence in the graph of locations by equating discrete states that have "similar" continuous dynamics "inside". We can apply the first strategy, when switching between a set of locations occurs sufficiently fast so that a mean–value continuous behavior can closely approximate the switching behavior of the hybrid system. Simple examples of this behavior are: bang-bang and sliding mode control, cascade control with switching in the inner loop faster than in outer loop. In all of these cases, mean–value modeling can be applied to abstract the discrete behavior of the inner control loop. However, this transformation is a heuristic and there is no guarantee that the simplified inner–loop yields an overall behavior of the system that is not too distant from the one of the complete system. Hence, we need to verify this condition before we actually accept the simplification. This reasoning yields the following step:

***Step 2*: Abstraction of independent cyclic paths**

- ***2.a*:** Identification of fast–switching independent cyclic paths;
- ***2.b*:** Mean–value modeling of the continuous behavior on cyclic paths;
- ***2.c*:** Validation of the abstract model.

Once, the validation has been carried out, the resulting hybrid system can be minimized using state abstraction methods:

***Step 3:* Equivalent minimal realization of the discrete behavior**

- ***3.a*:** Hybrid system discrete representation;
- ***3.b*:** Reduction of the discrete representation.

Since Step 1 is a straightforward application of model reduction techniques, we will not analyze it in further details, while we will take a closer look at Step 2 and 3 in the following sections.

4 Abstraction of Independent Cyclic Paths

Identification of fast–switching independent cyclic paths. This step can be easily carried out exploiting the insight of the control system designer or by inspection since in some cases, the hybrid model is the composition of several models, one of which is a hybrid automaton represented by only one cycle.

Mean–value modeling of the continuous behavior on cyclic paths. Once a cyclic path is selected, a model of the mean continuous behavior of the system performing the cycle at high frequency has to be obtained. Variable–structure control theory provides established methods to solve this problem in case the system switches across a sliding surface. The Filippov's continuation method [20] and the method of the equivalent control [21,22] can be used for this purpose. Notice that the equivalent dynamics depends on the sliding surface only and not the dynamics between which the system switches. Consequently, the mean–value model of a hybrid system may depend on the guard conditions only and not on the continuous dynamics associated to the locations. Alternatively, the model can be obtained by identification on simulation traces of the hybrid system.

Validation of the abstract model. Identification techniques inspired us to use a set of metrics to evaluate the "distance" between the approximate model and the complete one. In particular, we make use of:

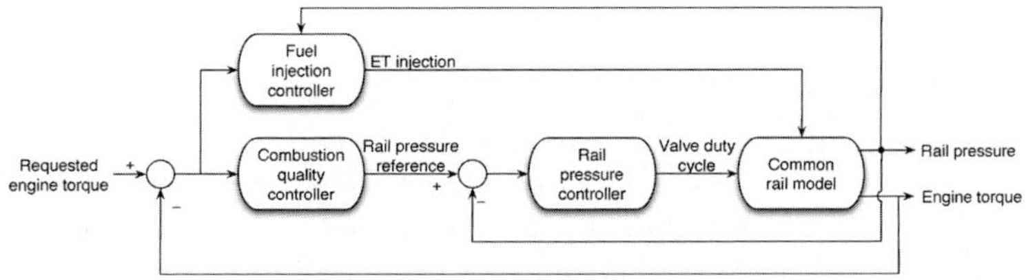

Fig. 2. Fuel system control architecture

Sum of Square Error. Also called sum square of residuals, it is a measure of the total deviation of the fit $\hat{y}_i$ to the system response y_i (best fit $SSE = 0$):

$$SSE = \sum_{i=1}^{n} (y_i - \hat{y}_i)^2 , \qquad SSE_{weighted} = \sum_{i=1}^{n} w_i (y_i - \hat{y}_i)^2 . \tag{1}$$

R-Square. It measures how successful the fit is in explaining the variation of the data. R^2 is the square of the correlation between the response values and the predicted response values. It is also called the square of the multiple correlation coefficient or the coefficient of multiple determination. R^2 is defined as follows:

$$R^2 = \frac{SSR}{SST} = 1 - \frac{SSE}{SST} \quad \text{with } SSR = \sum_{i=1}^{n} w_i (\hat{y}_i - \bar{y})^2 , \; SST = \sum_{i=1}^{n} w_i (y_i - \bar{y})^2 , \tag{2}$$

where SSR is the sum of squares of the regression and SST is the total sum of squares (also sum of squares about the mean). R^2 assumes values lower than 1 (with 1 indicating best fit). When R^2 is negative, it cannot be interpreted as the square of a correlation.
Percentage relative error. It is the local relative error in percent

$$PRE_{max} = 100 \max_{i=1,n} \left[\left(\frac{|y_i - \hat{y}_i|}{y_i} \right) \right] , \qquad PRE_{mean} = 100 \frac{1}{n} \sum_{i=1}^{n} \left[\left(\frac{|y_i - \hat{y}_i|}{y_i} \right) \right] . \tag{3}$$

4.1 A Case Study: Common Rail Pressure Controllers

In this section, an application of ***Step 2*** to the design of a controller for common rail pressure regulation is presented. Figure 2 reports the fuel system control architecture.

Common rail model. The case study is based on a hybrid model of the common–rail injection system marketed by Magneti Marelli Powertrain, referred to as $\mathcal{H}_{IMV}$, proposed in [23]. This model describes accurately the interacting discrete and continuous behaviors of the injection system components. The components of the model are represented in Figure 3. The rail pressure p is the controlled output. Its evolution depends on the balance between the HP pump flow q^P, the injector flow q^{INJ} and the DRV flow rate q^{DRV}. The IMV and DRV duty cycles u_{IMV}, $u_{DRV} \in [0,1]$ are the control inputs that modulate the pump inlet flow q^M and the DRV flow rate q^{DRV} respectively. The injector fuel flow q^{INJ} is considered as a disturbance to be compensated. It depends on:

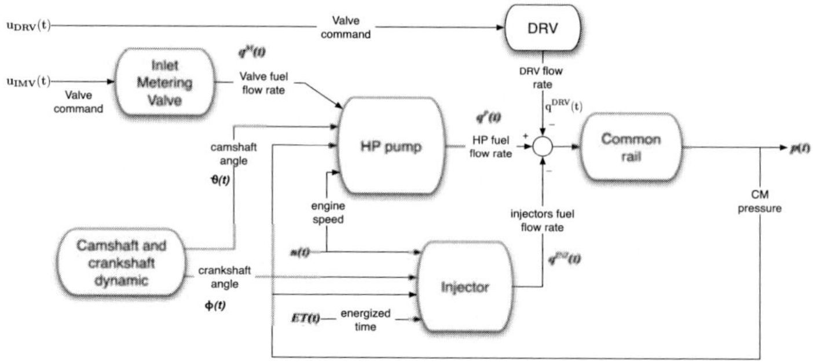

Fig. 3. Hybrid model $\mathscr{H}_{IMV}$ of the common rail

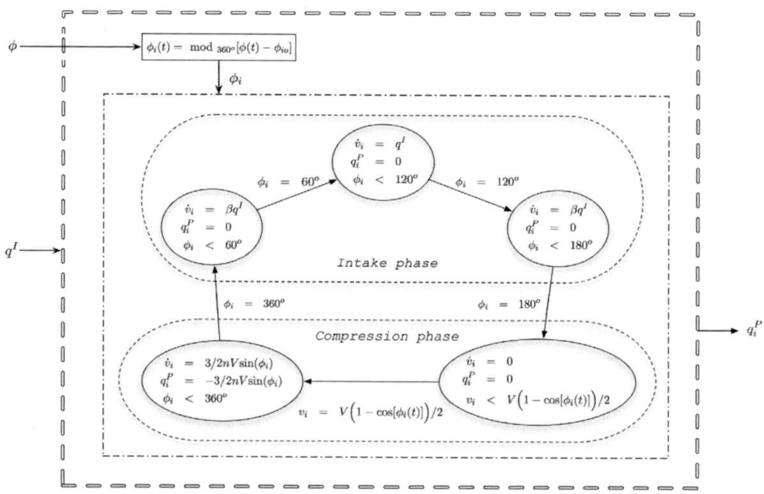

Fig. 4. Hybrid model of the i-th ram of the HP pump

the injector opening times ET, the engine speed n, and the rail pressure p. Finally, the evolutions of the HP pump and the injectors are synchronized by the camshaft angle ϕ and crankshaft angle θ, respectively.

The HP pump. The model of the HP pump is the composition of the hybrid models of the three identical rams that composed the pump. The hybrid model of the i-th ram is shown in Figure 4. Its evolution is determined by the ram angle $\phi_i \in [0,360]^o$ given by $\phi_i(t) = \mod_{360^o}[\phi(t) - \phi_{io}]$, with ϕ_{1o} the ram phase. The model has two macro–modes: intake and compression. The pumping cycle starts with the intake phase, which starts at $\phi_i = 0^o$ and ends when ϕ_i approaches 180°. Since the three rams are mounted with a relative phase of 120°, then the intake phases of the rams partially overlap. Intake overlapping results in different supplying fuel flows to the rams. The compression mode starts

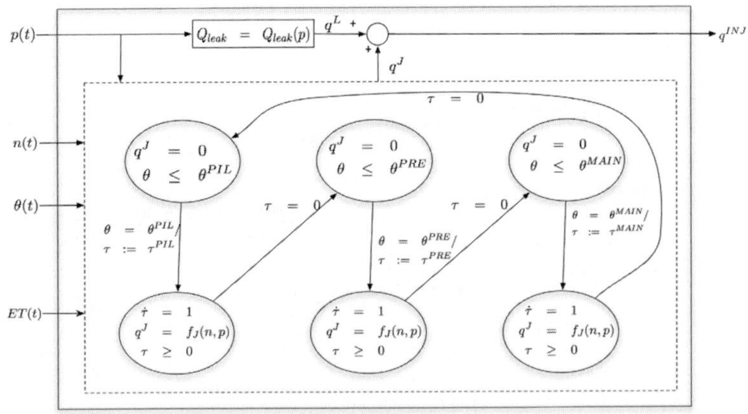

Fig. 5. Hybrid model of the injectors

at $\phi_i = 180^o$ and ends when ϕ_i approaches 360°. It consists of two modes: C_1, modeling fuel condensation in the ram, and C_2, modeling fuel delivery to the rail.

The injectors. The common rail supplies four injectors, one for each cylinder of the engine. In multi–jet engines, each injection phase is composed by a sequence of 3 to 5 distinct injections. Having the engine four cylinders, the frequency of injection sequences is twice the engine speed. Depending on the engine operating condition, the engine torque controller implemented in the engine control unit defines the amount of fuel to be injected and, consequently, the durations $ET = (\tau^{PIL}, \tau^{PRE}, \tau^{MAIN})$ and phases $(\theta^{PIL}, \theta^{PRE}, \theta^{MAIN})$ of each fuel injection.

The rail. The dynamics of the rail pressure is obtained by considering the balance between the HP pump flow, the injector flows and the DRV actuator flow:

$$\dot{p}(t) = \frac{K_{Bulk}(p)}{V_{rail}} \left[q^P(t) - q^{INJ}(t) - q^{DRV}(t) \right] . \tag{4}$$

Model simplification. The common rail model $\mathscr{H}_{IMV}$ presented in the previous paragraph is the composition of several hybrid models, among which the HP pump model and the injector model, shown in Figures 4 and 5 respectively, have discrete behaviors described by independent cyclic paths. Furthermore, in the executions of the hybrid model $\mathscr{H}_{IMV}$, such cycles are performed at high frequency. Hence, according to ***Step 2.a***, independent cyclic paths are identified and the corresponding simplifications of $\mathscr{H}_{IMV}$ can be investigated to verify whether they are accurate enough for controller design. ***Step 2.b*** is performed by computing mean–value signals in time obtaining the following two abstracted models:

– $\mathscr{H}^P_{CM}$: obtained by replacing in $\mathscr{H}_{CM}$ the HP pump hybrid model in Figure 4 with:

$$Q_{HP}(s) = \eta(n,p)\frac{1}{1+\tau s}Q^M(s) \qquad \text{where } \eta(\cdot) \text{ is HP pump efficiency.} \tag{5}$$

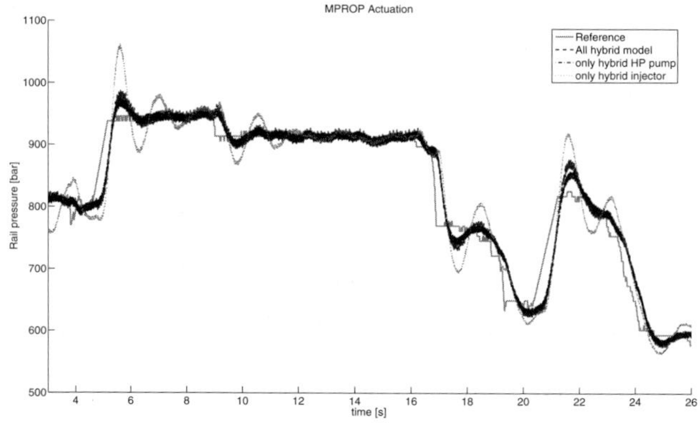

Fig. 6. Profiles of the rail pressure evolution obtained with the abstracted models $\mathscr{H}^P_{IMV}$ (red) and $\mathscr{H}^I_{IMV}$ (black) and the original model $\mathscr{H}_{IMV}$ (blue)

– $\mathscr{H}^I_{CM}$: obtained by replacing in $\mathscr{H}_{CM}$ the injector hybrid model in Figure 5 with:

$$Q_{inj} = \sum_{i=\{\text{PIL, PRE, MAIN}\}} \frac{n}{30} Q^i_{inj}(p, ET) \tag{6}$$

where $Q^i_{inj}(\cdot)$ is a piecewise affine function modeling injector flow rate.

In model $\mathscr{H}^P_{IMV}$ the HP pump is considered as a continuous system that provides a continuous flow rate of fuel to the rail, which corresponds to the mean–value flow of the original hybrid HP pump model. In model $\mathscr{H}^I_{IMV}$, the discrete behavior of fuel injections is abstracted away and injectors are modeled as continuous valves that deliver mean–value fuel flow rates to the combustion chamber.

In the next step, ***Step 2.c***, we verify whether the abstractions $\mathscr{H}^P_{IMV}$ and $\mathscr{H}^I_{IMV}$ of $\mathscr{H}_{IMV}$ are accurate enough for controller design. Due to the unstable behavior of pressure dynamics, models $\mathscr{H}^P_{IMV}$ and $\mathscr{H}^I_{IMV}$ have to be evaluated in closed–loop. A simple PID controller is used to stabilize the system. In Figure 6, profiles of the rail pressure evolution obtained with abstracted models $\mathscr{H}^P_{IMV}$ and $\mathscr{H}^I_{IMV}$ are compared to that produced by the original hybrid model $\mathscr{H}_{IMV}$. Simulation results show that while model $\mathscr{H}^I_{IMV}$ with injector abstraction reproduces closely the evolution of the original model $\mathscr{H}_{IMV}$, the behavior of the model $\mathscr{H}^P_{IMV}$ with pump abstraction is not satisfactory. Model $\mathscr{H}^P_{IMV}$ correctly reproduces the steady–state behavior but not the transient one. In addition, model $\mathscr{H}^P_{IMV}$ does not exhibit the high frequency rail pressure ripple that is present in the original model $\mathscr{H}_{IMV}$, while the ripple is precisely reproduced by model $\mathscr{H}^P_{IMV}$. This qualitative analysis is confirmed by the computation of metrics. Index R^2 in (2) evaluates to 96,95% for $\mathscr{H}^I_{CM}$, and 69,46% for $\mathscr{H}^P_{CM}$. Since a standard figure for good model fitting using R^2 is 90%, then $\mathscr{H}^I_{CM}$ is a good abstraction while $\mathscr{H}^P_{CM}$ is not. When the rail pressure is controlled using the *DRV* actuator, index R^2 does not give any information since it is negative. The quality of our models $\mathscr{H}^I_{DRV}$ and $\mathscr{H}^P_{DRV}$ are evaluated using the relative error metrics (3) and results similar to the previous case are obtained. For $\mathscr{H}^I_{DRV}$, we have $RE_{max} = 0.62\%$ and $RE_{mean} = 0.23\%$; for the $\mathscr{H}^P_{DRV}$, we have $RE_{max} = 14.31\%$ and $RE_{mean} = 2.03\%$. In conclusion, for rail pressure controller design, the discrete behavior of the injectors can be approximated away, modeling fuel injections as a continuous phenomenon, while the discrete behavior of the HP pump has to be represented because it affects significantly the closed–loop behavior.

5 Equivalent Minimal Realization of the Discrete Behavior

In ***Step 3***, an equivalent minimal realization of the discrete behavior of the hybrid system is computed.

Hybrid system discrete representation. First, the hybrid system model is projected to the discrete domain, by abstracting away continuous dynamics. A new FSM, referred to as $\mathscr{F}$, is defined as follows:

- states and transitions of $\mathscr{F}$ correspond to the hybrid system locations and transitions, respectively;
- To each transition of $\mathscr{F}$ we associate
  - an output label referring to the continuous time dynamics of the entering location in the hybrid system;
  - an input label referring to the guard condition of the corresponding transition in the hybrid system.

Reduction of the discrete representation. In this step, a minimal equivalent realization of the FSM $\mathscr{F}$ is computed. The well-known *Paull-Unger recursive rule* (see [24,25]) states that two states s_i and s_j are equivalent if for any input i_a, the corresponding outputs $\lambda(\ )$ are equal and the next states $\delta(\ )$ are equivalent, that is: $s_i \sim s_i \leftrightarrow \forall i_a\, \lambda(s_i,i_a)=\lambda(s_j,i_a)$ and $\delta(s_i,i_a)\sim\delta(s_j,i_a)$. Given that input and state sets are finite, the *Paull-Unger recursive rule* is guaranteed to terminate. Indistinguishability relations are computed by the *Table of Implications*, obtained as the Cartesian product of the set of states by itself. Due to symmetry and reflexiveness, only the lower triangular part of the table is filled in. Each entry of the table contains either:

- the symbol $\nsim$: if states are not equivalent;
- the symbol $\sim$: if states are equivalent;
- a pair of states whose equivalence implies the equivalence of the states corresponding to the entry.

By solving the implications defined in the third type of entries, the *Table of Implications* is refined until all entries are assigned to either $\nsim$ or $\sim$. The refined table defines the equivalence classes and hence a minimal FSM equivalent to $\mathscr{F}$. The approximate simulation relation proposed in [18] imposes conditions weaker than equivalence, namely the observations of the system and its approximation have to be close one another. In our approach, equivalence is imposed since closeness of continuous behaviors is analyzed in ***Step 1*** and ***2***.

5.1 A Case Study: The Combustion Quality Controller

We present in this case study an application of ***Step 3*** for the reduction of the model of the common–rail system controlled by the pressure controller discussed in the previous

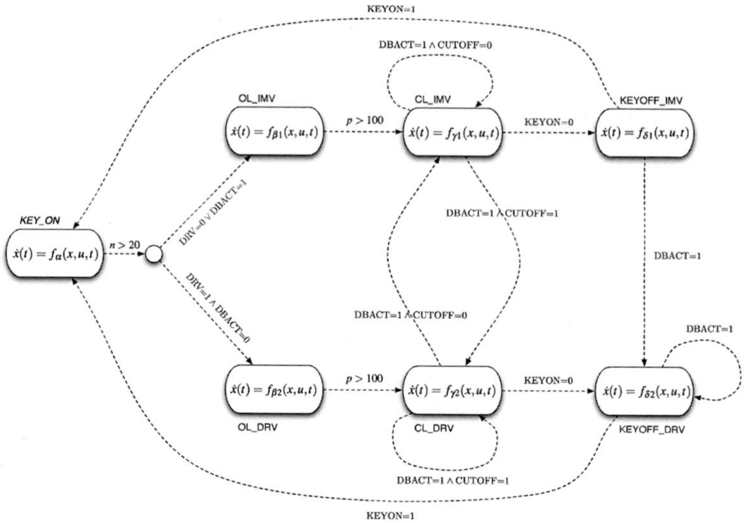

Fig. 7. Hybrid model of the combustion quality controller

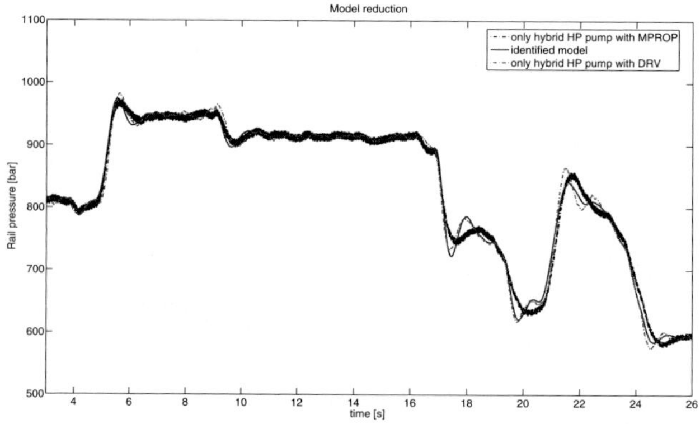

Fig. 8. Rail pressure profiles simulated with the detailed hybrid model using the IMV actuator (black line) and the DRV actuator (red-dot line), compared to the common identified continuous model

case study. As described in Figure 2, the fuel system control architecture contains in the outer control loop the Combustion Quality Controller (*CQC*), which defines the rail pressure reference based on the engine conditions and the requested engine torque. The model of the discrete behavior of the *CQC* is shown in Figure 7. Locations are related to the different engine modes (i.e. cranking, cut-off and power-off) and different actuators to be used for rail pressure control: either the IMV (a valve mounted on the HP pump) or the DRV (a valve mounted on the rail). Notice that *CQC* is designed to work properly in engine equipped with IMV only, DRV only, or both. In the last case, the *CQC* selects on–line the type of actuator to be used to minimize fuel consumption, engine noise and

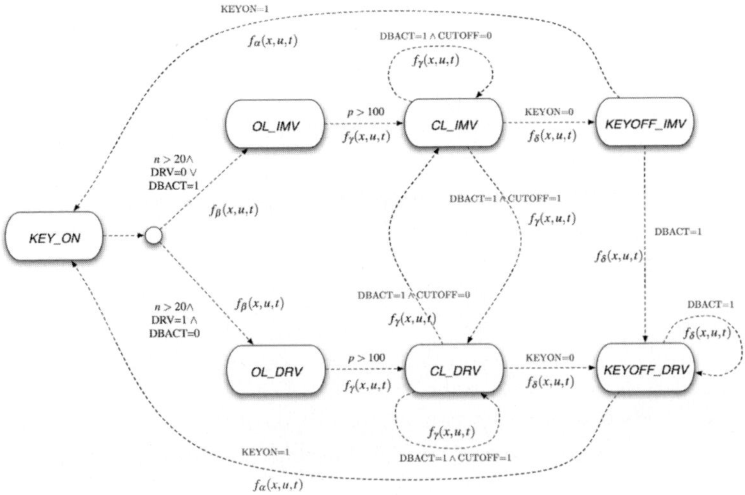

Fig. 9. FSM extracted from the CQC hybrid model according to ***Step 3.a***

tailpipe emissions. In each location, *CQC* defines the target rail pressure to be tracked. The initial mode is *KEY-ON*. In this mode, parameter initialization is performed. As soon as cranking starts, the *CQC* switches to the next mode, which depends on the actuator configuration: $DBACT = 1$ denotes both IMV and DRV are present; $DBACT = 0$ denotes only one of them is present. In modes *OL-IMV* and *OL-DRV*, the rail pressure controller acts in open–loop, until the fuel pressure overcomes 100 bar. Then, the *CQC* moves to nominal operation modes, either to *CL-IMV* or *CL-DRV* depending of the chosen actuation. During fuel cut-off, the *CQC* always switches to *CL-DRV*, if DRV is present. For a proper power–off of the engine, the *CQC* switches either to mode *Keyoff-IMV* or to mode *Keyoff-DRV*. At the next key–on the *CQC* switches back to the *KEY-ON* mode.

In this case study we apply ***Step 3*** of the proposed technique to obtain a simplified version of the common rail closed–loop model.

First, according to ***Step 1*** for each *CQC* location, the hybrid model of the common rail in closed–loop with the pressure controller is reduced to a simple continuous model. The Proper Orthogonal Decomposition method described in Section 2 is applied using input-output data obtained by simulations on profiles extracted by ECE cycle. The reduced continuous systems are of the type

$$G(s) = \frac{K(1+T_z s)}{(1+(2zT_w)s+T_w^2 s^2)(1+T_p s)} \,. \tag{7}$$

In Figure 8, the evolutions of the fuel injection models with DRV and IMV actuators are compared with the identified ones. For a same location, the identified models using the IMV actuator ($f_{\beta 1}$) and the DRV actuator ($f_{\beta 2}$) are quite close, meaning that the inner pressure controller achieves good compensation of the different dynamics of the

Table 1. Table of Implications

	keyon	OL–IMV	OL–DRV	CL–IMV	CL–DRV	Keyoff–IMV
OL–IMV	×					
OL–DRV	×	CL–IMV/ CL–DRV				
CL–IMV	×	×	×			
CL–DRV	×	×	×	Keyoff–IMV/ Keyoff–DRV		
Keyoff–IMV	×	×	×	×	×	
Keyoff–DRV	×	×	×	×	×	∼

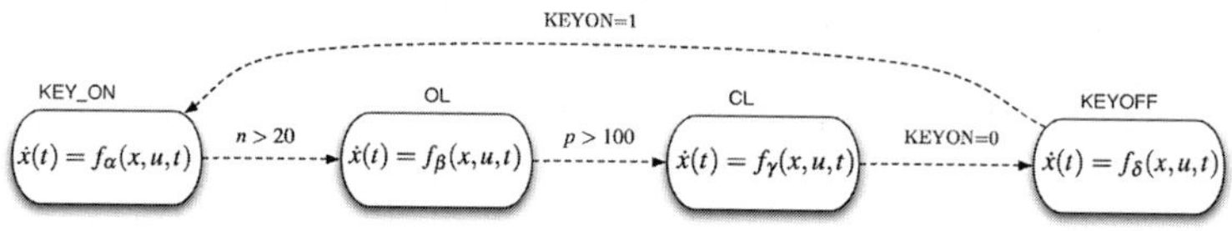

Fig. 10. Reduced model of the injection system

actuators. By evaluating the R^2 index (2), on the identified models (7), we obtain good fitting (greater than 90%): 91,25% for *IMV* and 92,3% for *DRV*. Hence, the IMV closed–loop dynamics $f_{\beta 1}$ and the DRV closed–loop dynamics $f_{\beta 2}$ can be replaced by a unique reduced dynamics f_β. The same results holds for the other modes, and dynamics $f_{\gamma 1}$, $f_{\gamma 2}$ are replaced by f_γ; while dynamics $f_{\delta 1}$, $f_{\delta 2}$ are replaced by f_δ.

Figure 9 reports the FSM extracted from the hybrid model according to ***Step 3.a***. The *Table of Implications* of the FSM, computed according to ***Step 3.b***, is reported in Table 1. The following equivalent classes can be defined:

$$\begin{aligned} \textbf{\textit{OL}} &= \{\textit{OL-IMV, OL-DRV}\} \\ \textbf{\textit{CL}} &= \{\textit{CL-IMV, CL-DRV}\} \\ \textbf{Keyoff} &= \{\textit{Keyoff-IMV, Keyoff-DRV}\}\ . \end{aligned}$$

Therefore, the original hybrid model that includes the common rail, the pressure controller and the *CQC*, can be reduced to the hybrid system reported in Figure 10 with the four locations: {*Keyon*, ***OL***, ***CL***, ***Keyoff***}.

6 Conclusion

A methodology for reducing hybrid system model complexity was presented. The proposed methodology is based on three main steps: reduction of continuous dynamics associated to locations and abstraction of independent cyclic paths, and equivalent minimal realization of the discrete behavior. Evaluation of abstraction quality is based on statistic metrics defined in identification theory. The proposed methodology was illustrated in an interesting case study in automotive applications: the common-rail fuel-injection system for Diesel engines developed by Magneti Marelli Powertrain. Future work involves the generalization of the approach to other connection architecture and the theoretical analysis of provable bounds for the proposed procedure.

References

1. Balluchi, A., Benvenuti, L., Ferrari, A., Sangiovanni-Vincentelli, A.: Hybrid systems in automotive electronics design. International Journal of Control **79, Special Issue on "Advanced design methodologies in automotive control"** (2006) 375–394
2. Isermann, R.: Mechatronic systems — innovative products with embedded control. In: Proc. of the 16th IFAC World Congress, Prague, CZ (2005) Best Applications Paper Prize.
3. Daniel, L.: Simulation and Modeling Techniques for Signal Integrity and Electromagnetic Interference on High Frequency Electronic Systems. PhD thesis, Engineering - Electrical Engineering and Computer Sciences, University of California at Berkeley (2003)
4. Daniel, L., Philips, J.: Model order reduction for strictly passive and causal distributed systems. In: IEEE/ACM 39th Design Automation Conference, New Orleans (2002)
5. Dewilde, P., van der Veen, A.: Time-varying systems and computations. 3rd edn. Kluwer Academic Publisher (1998)
6. Zadeh, L.: Frequency analysis of variable networks. Proceedings of the Institute of Radio Engineers **38** (1950) 291–299
7. Golub, G., Loan, F.V.: Matric computation. 3rd edn. Johns Hopkins University Press (1996)
8. Prajna, S.: Pod model reduction with stability guarantee. In: In Proceedings of 42nd IEEE Conference on Decision and Control, Maui, Hawaii (2003) 5254–5258
9. Moore, B.: Principal component analysis in linear systems: controllability, observability, and model reduction. IEEE Transactions on Automatic Control **26**(1) (1981) 17–32
10. Lall, S., Marsden, J., Glavaski, S.: A subspace approach to balanced truncation for model reduction of nonlinear control systems (2002)
11. Hahn, J., Edgar, T.F., Marquardt, W.: Controllability and observability covariance matrices for the analysis and order reduction of stable nonlinear systems. Journal of Process Control **13** (2003) 115–127
12. Shokoohi, S., Silverman, L.M.: Identification and model reduction of time-varying discrete-time systems. Automatica **23**(4) (1987) 509–521
13. Sandeberg, H., Rantzer, A.: Balanced truncation of linear time-varying systems. IEEE Transactions on Automatic Control **49**(2) (2004) 217–229
14. Philips, J., Daniel, L., Silveira, L.: Guaranteed passive balancing transformations for model order reduction. IEEE Transactions on Computer-Aided Design of Integrated Circuits ans Systems **22**(8) (2003)
15. Buttazzo, G.: Sistemi in tempo reale. Pitagora Editrice (2001)
16. Astrom, K., Wittenmark, B.: Computer-Controlled Systems: Theory and Design. Prentice Hall (1999)
17. Kohavi, Z.: Switching and Finite Automata Theory. McGraw-Hill (2002)
18. Girard, A., Julius, A.A., Pappas, G.J.: Approximate simulation relations for hybrid systems. In: 2nd IFAC Conference on Analysis and Design of Hybrid Systems, Alghero, Sardinia, Italy (2006)
19. Han, Z., Krogh, B.: Reachability analysis of hybrid control systems using reachability analysis of hybrid control systems using reduced-order models. In: American Control Conference, Boston, Massachusetts (2004) 1183–1190
20. Filippov, A.: Differential Equations with Discontinuous Right-Hand Sides. Kluwer, The Netherlands (1988)
21. Utkin, V.: Variable structure systems with sliding modes: a survey. IEEE Trans. on Automat. Contr. **22** (1977) 212–222
22. DeCarlo, R., Zak, S., Matthews, G.: Variable structure control of nonlinear multivariable systems: a tutorial. Proceedings of the IEEE **76**(3) (1988) 212–232

23. Balluchi, A., Bicchi, A., Mazzi, E., , Sangiovanni-Vincentelli, A.L., Serra, G.: Hybrid modelling and control of the common rail injection system. In Hespanha, J., Tiwari, A., eds.: Hybrid Systems: Computation and Control, HSCC2006. Volume 3927., Springer-Verlag, Berlin Heidelberg (2006) 79–92
24. Micheli, G.D.: Synthesis and Optimization of Digital Circuits. McGraw-Hill (2001)
25. Katz, R.: Contemporary Logic Design. The Benjamin/Cummings Publishing Company (1994) chapter: 1-10.

Model Checking Genetic Regulatory Networks with Parameter Uncertainty

Grégory Batt[1], Calin Belta[1], and Ron Weiss[2]

[1] Center for Information and Systems Engineering and Center for BioDynamics, Boston University, Brookline, MA, USA
[2] Department of Electrical Engineering and Department of Molecular Biology, Princeton University, Princeton, NJ, USA
batt@bu.edu, cbelta@bu.edu, rweiss@princeton.edu

Abstract. The lack of precise numerical information for the values of biological parameters severely limits the development and analysis of models of genetic regulatory networks. To deal with this problem, we propose a method for the analysis of genetic regulatory networks with parameter uncertainty. We consider models based on piecewise-multiaffine differential equations, dynamical properties expressed in temporal logic, and intervals for the values of uncertain parameters. The problem is then either to guarantee that the system satisfies the expected properties for every possible parameter value - the corresponding parameter set is then called valid - or to find valid subsets of a given parameter set. The proposed method uses discrete abstractions and model checking, and allows for efficient search of the parameter space. This approach has been implemented in a tool for robust verification of gene networks (RoVerGeNe) and applied to the tuning of a synthetic network build in *E. coli*.

1 Introduction

Numerous cellular processes are controlled at the molecular level by networks of interactions between genes, proteins and small molecules, called *genetic regulatory networks*. Understanding how the cellular behavior emerges from these networks of interactions is a central problem in systems and synthetic biology [1,2]. Arguably, the most widely-used modeling frameworks for the analysis of the dynamics of these networks are based on differential equations [3]. With few exceptions [4], it is generally assumed that the numerical values of state variables and model parameters are precisely known. However, given the current limitations of experimental measurement techniques, and the fact that parameter values themselves vary with the ever-fluctuating extra- and intracellular environments, the results obtained by these techniques may be of limited validity.

In this work, we present a method for the analysis of genetic regulatory networks with *parameter uncertainty*. We consider gene network models based on piecewise-multiaffine (PMA) differential equations, dynamical properties expressed in temporal logic (LTL), and intervals for the values of uncertain

A. Bemporad, A. Bicchi, and G. Buttazzo (Eds.): HSCC 2007, LNCS 4416, pp. 61–75, 2007.

parameters. The problem is then either to *guarantee* that the system satisfies the expected properties for *every* possible parameter value - the corresponding parameter set is then called *valid* - or to *find* valid subsets of a given parameter set.

In the proposed approach, we use a partition of the *state space* induced by the piecewise nature of the models and specific properties of multiaffine functions [5] to define an equivalence relation on parameters. *Discrete abstractions* [6] are used to transpose the problem defined on (infinite) continuous state and parameter spaces into a problem defined on (finite) discrete spaces. Algorithmic analysis by *model-checking* [7] is then possible. Conservative approximations are used that guarantee that the parameter sets returned by the procedure are valid. However, not all valid parameter sets are guaranteed to be found. This approach has been implemented in a tool for Robust Verification of Gene Networks (RoVerGeNe) and applied to the analysis of the tuning of a synthetic gene network, build in the bacterium *E. coli.* This case study demonstrate the practical applicability and biological relevance of the proposed approach.

This paper is organized as follows. Section 2 introduces preliminary notions. PMA models are presented in Section 3, and the proposed approach is detailed in Section 4. The application to the tuning of a network is presented in Section 5. The final section discusses the results in the context of related work.

2 Preliminaries

All the notions and notations presented here are described at length in [8]. We consider Kripke structures $T = (S, \rightarrow, \Pi, \models)$ defined over sets of atomic propositions Π, and simply called *transition systems* [7]. S is a (finite or infinite) set of states, $\rightarrow \subseteq S \times S$, a total transition relation, and $\models \subseteq S \times \Pi$, a satisfaction relation. An *execution* of T is an infinite sequence $e = (s_0, s_1, s_2, \ldots)$ such that for every $i \geq 0$, $s_i \in S$ and $(s_i, s_{i+1}) \in \rightarrow$. We use the syntax and the semantics of *Linear Temporal Logic* (LTL) formulas defined over executions of Kripke structures given in [7]. We refer to [6] for the usual notions of *simulation* between transition systems and of *quotient transition systems.* T_1 simulates T_2 is denoted $T_2 \preceq T_1$, and we recall that simulation relations weakly preserve LTL [7].

Polytopes are bounded intersections of finitely-many open or closed halfspaces. A polytope P is *hyperrectangular* if $P = P_1 \times \ldots \times P_n$ with $P_i = \{x_i \in \mathbb{R} \mid x = (x_1, \ldots, x_n) \in P\}$, $i \in \{1, \ldots, n\}$. The definitions of the closure, vertices, faces and facets of a polytope are recalled in [8]. $\overline{P}$ and $\mathcal{V}_P$ denote respectively the closure and the set of vertices of a polytope P. A function $f : \mathbb{R}^n \rightarrow \mathbb{R}^m$ is multiaffine if it is a polynomial with the property that the degree of f in any of its variable is at most 1. Theorem 1 is proven in [5].

Theorem 1. *Let $f : \mathbb{R}^n \rightarrow \mathbb{R}^m$ be a multiaffine function and P be a hyperrectangular polytope in $\mathbb{R}^n$, $n, m \in \mathbb{N}$. Then, for every $x \in P$, $f(x)$ is a convex combination of the values of f at the vertices of P.*

3 Uncertain PMA Models of Genetic Regulatory Networks

3.1 PMA Systems and LTL Specifications

In this section, we present a formalism for modeling gene networks. The notations and terminology are adapted from [9]. We consider a gene network consisting of n genes. The state of the network is given by the vector $x = (x_1, \ldots, x_n)$, where x_i is the concentration of the protein encoded by gene i. The state space $\mathcal{X}$ is a hyperrectangular subset of $\mathbb{R}^n$: $\mathcal{X} = \prod_{i=1}^{n} [0, max_{x_i}]$, where max_{x_i} denotes a maximal concentration of the protein encoded by gene i. Some parameters may be *uncertain*: $p = (p_1, \ldots, p_m)$ is the vector of uncertain parameters, with values in the parameter space $\mathcal{P} = \prod_{j=1}^{m} [min_{p_j}, max_{p_j}]$, where min_{p_j} and max_{p_j} denote a minimal and a maximal value for p_j.

The dynamics of the network is given by the differential equations

$$\dot{x}_i = f_i(x,p) = \sum_{j \in P_i} \kappa_i^j \, r_i^j(x) - \sum_{j \in D_i} \gamma_i^j \, r_i^j(x) \, x_i, \quad i \in \{1, \ldots, n\}, \tag{1}$$

where P_i and D_i are sets of indices, $\kappa_i^j > 0$ and $\gamma_i^j > 0$ are (*possibly uncertain*) *production* and *degradation rate parameters*, and $r_i^j : \mathcal{X} \to [0,1]$ are continuous, PMA functions, called *regulation functions*. PMA functions arise from products of ramp functions r^+ and r^- (Figure 1(a)) used for representing complex gene regulations or protein degradations (Figure 5(b) Eq. 4, and [8]). With the additional assumption that r_i^j does not depend on x_i for $j \in D_i$,[1] it holds that $f = (f_1, \ldots, f_n) : \mathcal{X} \times \mathcal{P} \to \mathbb{R}^n$ is a (non-smooth) *continuous* function of x and p, a *piecewise-multiaffine* function of x and an *affine* function of p. Note that production and degradation rate parameters may be uncertain, but regulation functions (with their threshold parameters) must be known precisely. Each component of the vector p of uncertain parameters is either a production or a degradation rate parameter. Finally, Equation (1) is easily extended to account for constant inputs u by considering u as a new variable satisfying $\dot{u} = 0$.

A number of dynamical properties of gene networks can be specified in temporal logic by LTL formulas over atomic propositions of type $x_i < \lambda$ or $x_i > \lambda$, where $\lambda \in \mathbb{R}_{\geq 0}$ is a constant. We denote by Π the set of all such atomic propositions. A *PMA system* Σ is then defined by a piecewise-multiaffine function f defined as above and a set of atomic propositions Π: $\Sigma = (f, \Pi)$.

Consider the cross-inhibition network represented in Figure 1(b). This system can be represented by the PMA differential equations given in Figure 1(c). For example, the first equation states that protein A synthesis is inhibited by protein B (r^- function) and that its degradation is not regulated. Parameter values are given in Figure 1(d). Synthesis parameters are unknown: $(\kappa_a, \kappa_b) \in \mathcal{P} = [0,40] \times [0,20]$. For illustrating our purpose, we also consider $p_1 \in \mathcal{P}$ with $p_1 = (36, 17)$. This network is known to be *bistable*: it has two stable equilibrium

[1] This assumption requires that a protein does not regulate its own degradation. In practice, this assumption is generally satisfied.

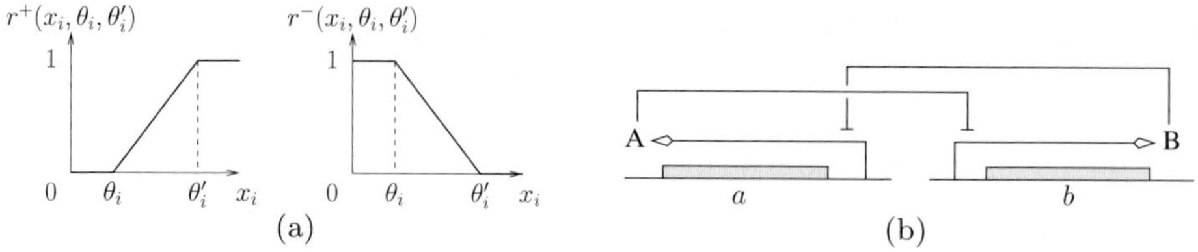

$$\dot{x}_a = \kappa_a r^-(x_b, \theta_b^1, \theta_b^2) - \gamma_a x_a$$
$$\dot{x}_b = \kappa_b r^-(x_a, \theta_a^1, \theta_a^2) - \gamma_b x_b$$

(c)

$$\begin{aligned}\phi_1 = \ & (x_a < \theta_a^1 \wedge x_b > \theta_b^2 \rightarrow \mathrm{G}\ (x_a < \theta_a^1 \wedge x_b > \theta_b^2)) \\ & \wedge (x_b < \theta_b^1 \wedge x_a > \theta_a^2 \rightarrow \mathrm{G}\ (x_b < \theta_b^1 \wedge x_a > \theta_a^2))\end{aligned}$$

(e)

	κ_i	γ_i	θ_i^1	θ_i^2
x_a	[0, 40]	2	8	12
x_b	[0, 20]	1	8	12

(d)

Fig. 1. (a) Ramp functions r^+ and r^-. θ_i and θ'_i are threshold parameters. (b) Gene network comprising two genes, a and b, coding for two repressor proteins, A and B. Each protein represses the expression of the other gene, forming a cross-inhibition network. (c) PMA model of the network in (b). Because of its simplicity, this model is actually piecewise-affine. (d) Known and uncertain parameter values. (e) Bistability property expressed in LTL.

states, corresponding to protein A and B concentrations being respectively high and low, or low and high. This property can be expressed in LTL by the property ϕ_1 (Figure 1(e)). For example, the first part of the property expresses that if the concentrations of protein A and B are respectively low ($x_a < \theta_a^1$) and high ($x_b > \theta_b^2$), then the system will always (G) remain in such a state. We refer the reader to [10] for a discussion of the use of invariants to express stability in biology.

PMA models of gene networks were proposed in [11] (see [12] for a related, piecewise-continuous formalism). The models considered here are also related to the piecewise-affine (PA) models proposed in [13] (see also [9]). However, contrary to the step functions used in PA models, ramp functions capture the graded response of gene expression to continuous changes in effector concentrations.

3.2 Embedding Transition Systems

The specific form of the PMA functions f suggests a division of the state space $\mathcal{X}$ into hyperrectangular regions (Figure 2(a) for our example). For every $i \in \{1, \ldots, n\}$, let $\Lambda_i = \{\lambda_i^j\}_{j \in \{1,\ldots,l_i\}}$ be the ordered set of all threshold constants in f, and of all atomic proposition constants in Π, associated with gene i, together with 0 and max_{x_i}. The cardinality of Λ_i is l_i. Then, we define $\mathcal{R}$ as the following set of n-dimensional hyperrectangular polytopes $R \subseteq \mathcal{X}$, simply called *rectangles*:

$$\mathcal{R} = \{R_c \mid c = (c_1, \ldots, c_n) \text{ and } \forall i \in \{1, \ldots, n\} : c_i \in \{1, \ldots, l_i - 1\}\},$$

where

$$R_c = \{x \in \mathcal{X} \mid \forall i \in \{1, \ldots, n\} : \lambda_i^{c_i} < x_i < \lambda_i^{c_i+1}\}.$$

c is the *coordinate* of the rectangle R_c. The union of all rectangles in $\mathcal{X}$ is denoted by $\mathcal{X}_\mathcal{R}$: $\mathcal{X}_\mathcal{R} = \cup_{R \in \mathcal{R}} R$. Note that $\mathcal{X}_\mathcal{R} \neq \mathcal{X}$. Notably, threshold hyperplanes are not included in $\mathcal{X}_{\mathcal{R}}$. Two rectangles R and R', are said *adjacent*, denoted $R \rightleftharpoons R'$, if they share a facet. $coord : \mathcal{R} \to \prod_{i=1}^{n}\{1,\ldots,l_i-1\}$ maps every rectangle $R \in \mathcal{R}$ to its coordinate, and $rect : \mathcal{X}_{\mathcal{R}} \to \mathcal{R}$ maps every point x in $\mathcal{X}_{\mathcal{R}}$ to the rectangle R such that $x \in R$. For the cross-inhibition network, the set $\mathcal{R} = \{R_{11},\ldots,R_{33}\}$ of all rectangles is represented in Figure 2(b). R_{11} and R_{21} are adjacent, whereas R_{11} and R_{22} are not.

Formally, the semantics of a PMA system Σ is defined by means of an embedding transition system.

Definition 1 (Embedding transition system). *Let $p \in \mathcal{P}$. The embedding transition system associated with the PMA system $\Sigma = (f, \Pi)$ is $T_{\mathcal{X}}(p) = (\mathcal{X}_{\mathcal{R}}, \to_{\mathcal{X},p}, \Pi, \models_{\mathcal{X}})$ defined such that:*

- *$\to_{\mathcal{X},p} \subseteq \mathcal{X}_{\mathcal{R}} \times \mathcal{X}_{\mathcal{R}}$ is the transition relation defined by $(x, x') \in \to_{\mathcal{X},p}$ iff there exists a solution ξ of (1) and $\tau \in \mathbb{R}_{>0}$ such that $\xi(0) = x$, $\xi(\tau) = x'$, $\forall t \in [0,\tau]$, $\xi(t) \in \overline{rect}(x) \cup \overline{rect}(x')$, and either $rect(x) = rect(x')$ or $rect(x) \rightleftharpoons rect(x')$,*
- *$\models_{\mathcal{X}} \subseteq \mathcal{X}_{\mathcal{R}} \times \Pi$ is the satisfaction relation defined by $(x,\pi) \in \models_{\mathcal{X}}$ iff $x = (x_1,\ldots,x_n)$ satisfies the proposition π (of type $x_i < \lambda$ or $x_i > \lambda$) with the usual semantics.*

Remark. Not all solution trajectories of (1) are represented by executions of the embedding. First, due to our restricted notion of adjacency ($\rightleftharpoons$), solution trajectories of (1) that go from a rectangle to another by passing through a face of low ($< n-1$) dimension are not represented in the embedding. Second, the dynamics of the system in $\mathcal{X} \setminus \mathcal{X}_{\mathcal{R}}$ (including the threshold hyperplanes) is not described by the embedding. However, since the vector field is continuous everywhere, trajectories originating in full-dimensional rectangles can not "disappear" in a facet by sliding along the supporting hyperplane. Consequently, the embedding describes *almost* all solution trajectories of (1), which is satisfying for all practical purposes.

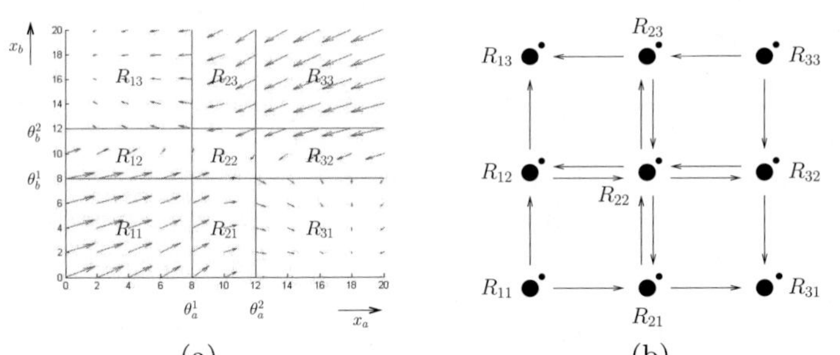

(a) (b)

Fig. 2. (a) Continuous dynamics in the state space of the cross-inhibition network for parameter $p_1 = (\kappa_a, \kappa_b) = (36, 17)$. (b) Discrete abstraction of the dynamics in (a). Dots denote self transitions.

A PMA system Σ satisfies an LTL formula ϕ for a given parameter $p \in \mathcal{P}$ if $T_{\mathcal{X}}(p) \models \phi$, that is, if every execution of $T_{\mathcal{X}}(p)$ satisfies ϕ. Then, valid parameter sets are defined as follows.

Definition 2. *Let Σ be a PMA system and ϕ an LTL formula. A parameter set $P \subseteq \mathcal{P}$ is* valid *for ϕ iff Σ satisfies ϕ for almost all $p \in P$.*

Again, the use of *almost all* is motivated by the fact that this criteria is sufficient for all practical purposes. Finally, we consider the following problems.

Problem. *Let Σ be a PMA system, $\mathcal{P}$ an hyperrectangular parameter space, and ϕ an LTL formula.*
1. Robustness analysis: *Check whether $\mathcal{P}$ is valid for ϕ.*
2. Synthesis: *Find a set $P \subseteq \mathcal{P}$ such that P is valid for ϕ.*

4 Analysis of PMA Systems with Parameter Uncertainty

4.1 Discrete Abstraction

We use discrete abstractions [6] to obtain finite transition systems preserving dynamical properties of $T_{\mathcal{X}}(p)$ and amenable to algorithmic verification [7]. Let $\sim_{\mathcal{R}} \subseteq \mathcal{X}_{\mathcal{R}} \times \mathcal{X}_{\mathcal{R}}$ be the (proposition-preserving) equivalence relation defined by the surjective map *rect*: $x \sim_{\mathcal{R}} x'$ iff $rect(x) = rect(x')$. $\mathcal{R}$ is the set of equivalence classes. Then, the discrete abstraction of $T_{\mathcal{X}}(p)$ is the *quotient* of $T_{\mathcal{X}}(p)$ given the equivalence relation $\sim_{\mathcal{R}}$.

Definition 3. *Let $p \in \mathcal{P}$. The discrete abstraction of $T_{\mathcal{X}}(p)$ is $T_{\mathcal{R}}(p) = (\mathcal{R}, \rightarrow_{\mathcal{R},p}, \Pi, \models_{\mathcal{R}})$, the quotient of $T_{\mathcal{X}}(p)$ given the equivalence relation $\sim_{\mathcal{R}}$.*

For our example network, $T_{\mathcal{R}}(p_1)$ is represented in Figure 2(b). By definition of quotient transition systems, $T_{\mathcal{R}}(p)$ *simulates* $T_{\mathcal{X}}(p)$.

$$\text{For every } p \in \mathcal{P},\ T_{\mathcal{X}}(p) \preceq T_{\mathcal{R}}(p). \tag{2}$$

In words, the discrete transition system $T_{\mathcal{R}}(p)$ is a *conservative approximation* of the continuous dynamics of the PMA system described by $T_{\mathcal{X}}(p)$. Because simulation relations *weakly preserve* LTL, we have for any LTL formula ϕ: if $T_{\mathcal{R}}(p) \models \phi$ then $T_{\mathcal{X}}(p) \models \phi$. The converse does not necessarily hold.

By exploiting specific properties of multiaffine functions defined over hyperrectangular polytopes [5], we provide the following characterization.

Proposition 1. *Let $p \in \mathcal{P}$. $T_{\mathcal{R}}(p) = (\mathcal{R}, \rightarrow_{\mathcal{R},p}, \Pi, \models_{\mathcal{R}})$, where*

- *$\rightarrow_{\mathcal{R},p} \subseteq \mathcal{R} \times \mathcal{R}$ is such that $(R, R') \in \rightarrow_{\mathcal{R},p}$ iff $R = R'$, or $R \mathrel{\smile\!\!\!\frown} R'$ and there exists $v \in \mathcal{V}_R \cap \mathcal{V}_{R'}$ such that*
$$f_i(v, p)(c'_i - c_i) > 0,$$
with $c = coord(R)$, $c' = coord(R')$ and $i \in \{1, \ldots, n\}$ such that $c_i \neq c'_i$.
- *$\models_{\mathcal{R}} \subseteq \mathcal{R} \times \Pi$ is such that $(R, \pi) \in \models_{\mathcal{R}}$ iff for every $x \in R$, $(x, \pi) \in \models_{\mathcal{X}}$.*

Proof. Let $R, R' \in \mathcal{R}$. By Definition 1 and 3, it is clear that if neither $R = R'$ nor $R \diamond R'$, there can not exist a transition from R to R'. If $R = R'$, then since it exists a solution of (1) that remains in R on $[0, \tau]$ for some $\tau > 0$, there exists a (self) transition from R to R' (Definition 1 and 3). The last case is when $R \diamond R'$. Then, let $c = coord(R)$, $c' = coord(R')$ and $i \in \{1, \ldots, n\}$ such that $c_i \neq c'_i$ and let F be the facet shared by R and R'. We assume without loss of generality that $c'_i - c_i = 1$, the other case $(= -1)$ being symmetrical.
$\Rightarrow$ (by contradiction): Suppose that for every $v \in \mathcal{V}_R \cap \mathcal{V}_{R'} = \mathcal{V}_F$, $f_i(v, p) \leq 0$. Using Theorem 1, it holds that for every $x \in F$, $f_i(x, p) \leq 0$. Consequently, no solution can enter R' from R and $(R, R') \notin \rightarrow_{\mathcal{R},p}$.
$\Leftarrow$: Assume that there exists $v \in \mathcal{V}_F$ such that $f_i(v, p) > 0$. By continuity of f, there exists a ball $B_{v,\epsilon}$ of center v and radius ϵ such that $\forall x \in B_{v,\epsilon}$, $f_i(x, p) > 0$. In particular, there exist $x_f \in F$, $x_f \neq v$, such that $f_i(x_f, p) > 0$. Then, there exists a solution entering R' from R without leaving $\overline{R} \cup \overline{R'}$, and by Definition 1 and 3, $(R, R') \in \rightarrow_{\mathcal{R},p}$.

The characterization of $\models_{\mathcal{R}}$ follows immediately from the fact that the equivalence relation $\sim_{\mathcal{R}}$ preserves the atomic propositions in Π. □

Informally, Proposition 1 simply states that there is a transition between two adjacent rectangles if and only if there exists at least one common vertex at which the direction of the vector field $(f_i(v, p))$ is in agreement with the relative position of the two rectangles $(c'_i - c_i)$. Similar rules have been proposed in [14]. Consider the two rectangles R_{11} and R_{21} in Figure 2(a). They share two vertices: $v_1 = (\theta_a^1, 0)$ and $v_2 = (\theta_a^1, \theta_b^1)$. From Proposition 1, there is a transition from R_{11} to R_{21}, because $f_a(v_1, p_1) > 0$, and there is no transition from R_{21} to R_{11}, because neither $f_a(v_1, p_1) < 0$ nor $f_a(v_2, p_1) < 0$ (check with Figure 2(b)).

For known parameters, Proposition 1 provides a means to compute the relation $\rightarrow_{\mathcal{R},p}$ by evaluating f at all the vertices. The computation of the set of states $\mathcal{R}$ and of the relation $\models_{\mathcal{R}}$ are trivial. So $T_{\mathcal{R}}(p)$ can be computed and one can use model checking for testing whether $T_{\mathcal{R}}(p) \models \phi$. If the abstract system $T_{\mathcal{R}}(p)$ satisfies ϕ, then so does the original system $T_{\mathcal{X}}(p)$ (Property (2)), and p is valid for ϕ. Conversely, if $T_{\mathcal{R}}(p)$ does not satisfy ϕ, no conclusion on the validity of p can be obtained. If some parameters are unknown, we will use Proposition 1 to define an equivalence relation on parameters.

4.2 Parameter Equivalence Classes

Consider a vertex $v \in \mathcal{V}_R$, $R \in \mathcal{R}$. Because f is an affine function of p, $f_i(v, p)$ is an affine expression in p: $f_i(v, p) = a^T p + b$, with $a \in \mathbb{R}^m$ and $b \in \mathbb{R}$. Let Ψ be the set of all such non-constant $(a \neq 0)$ affine expressions:

$$\Psi = \{f_i(v, p) = a_{i,v}^T p + b_{i,v} \mid i \in \{1, \ldots, n\}, v \in \mathcal{V}_R, R \in \mathcal{R} \text{ and } a_{i,v} \neq 0\}.$$

After having removed repeated elements, we denote by n_Ψ the cardinality of Ψ and order the elements in Ψ: $\Psi = \{\psi_1, \ldots, \psi_{n_\Psi}\}$. For our example network, with uncertain parameters κ_a and κ_b, out of the 32 affine expressions only 4 different non-constant expressions exist: $n_\Psi = 4$ (Figure 3(a)).

$$\Psi = \{\psi_1, \psi_2, \psi_3, \psi_4\}, \text{ with}$$
$$\psi_1(p) = \kappa_a - 24,$$
$$\psi_2(p) = \kappa_b - 12,$$
$$\psi_3(p) = \kappa_b - 8,$$
$$\psi_4(p) = \kappa_a - 16.$$

(a)

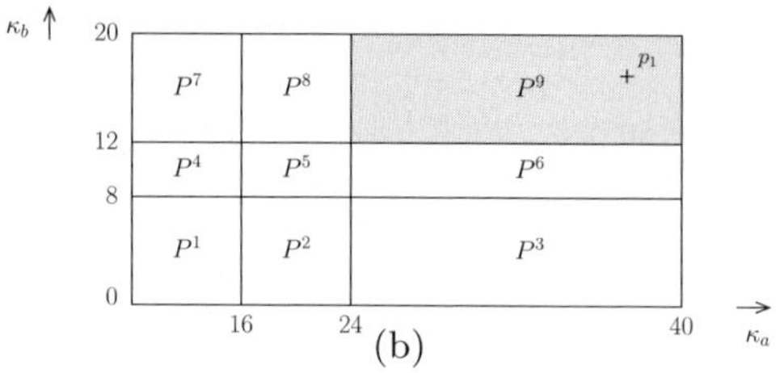

(b)

Fig. 3. (a) Set of affine expressions for the cross-inhibition network with unknown parameters κ_a and κ_b. (b) Parameter space in the dimensions of κ_a and κ_b. $p_1 = (36, 17)$ is represented. The shaded region is the set of all valid parameters for property ϕ_1.

The affine predicates $\psi_i(p) = 0$, $\psi_i \in \Psi$, divide the *parameter space* into polyhedral regions (Figure 3(b))[2]. These regions can be represented by a Boolean encoding. Let $\mathcal{B}^l$ be the set of Boolean numbers of length l: $\mathcal{B}^l = \{0, 1\}^l$. We denote by ϵ the Boolean of length 0. Then, to every Boolean $b \in \mathcal{B}^l$, $l \in \{0, \ldots, n_\Psi\}$, we associate the parameter set P_b such that $P_\epsilon = \mathcal{P}$ and, if $b \neq \epsilon$,

$$P_b = \{p \in \mathcal{P} \mid \forall i \in \{1, \ldots, l\} : \psi_i(p) < 0, \text{ if } b_i = 0, \text{ and } \psi_i(p) > 0, \text{ if } b_i = 1\}.$$

The sets P_b are subsets of $\mathcal{P}$ obtained by adding constraints of type $\psi_i(p) < 0$ or $\psi_i(p) > 0$, with $\psi_i \in \Psi$. If b is a prefix of b', then $P_{b'} \subseteq P_b$. The hierarchy between the sets P_b induced by the set-inclusion partial-order is represented in Figure 4 for the cross-inhibition network (see [15,16] for similar ideas in the context of predicate abstraction).

We say that two parameters p and p' are *equivalent* if their associated discrete transition systems $T_\mathcal{R}(p)$ and $T_\mathcal{R}(p')$ are isomorphic. A similar definition is used in [17,9]. Naturally, a PMA system satisfies the same LTL properties for two equivalent parameters.

Definition 4. *Let $\sim_\mathcal{P} \subseteq \mathcal{P} \times \mathcal{P}$ be the equivalence relation defined by $p \sim_\mathcal{P} p'$ iff $T_\mathcal{R}(p) = T_\mathcal{R}(p')$.*

Proposition 2. *Let $b_\Psi \in \mathcal{B}^{n_\Psi}$. For every $p, p' \in P_{b_\Psi}$, $p \sim_\mathcal{P} p'$.*

Proof. Let $b_\Psi \in \mathcal{B}^{n_\Psi}$ and $p, p' \in P_{b_\Psi}$. Then, $\forall i \in \{1, \ldots, n\}$, $R \in \mathcal{R}$ and $v \in \mathcal{V}_R$, $f_i(v, p) \# 0$ iff $f_i(v, p') \# 0$, with $\# \in \{<, >\}$. So, by Proposition 1, $T_\mathcal{R}(p) = T_\mathcal{R}(p')$ and $p \sim_\mathcal{P} p'$.

The above proposition states that the set of all predicates $\psi_i(p) = 0$, $\psi_i \in \Psi$, divide the parameter space in equivalence classes. Consequently, with $b_\Psi \in \mathcal{B}^{n_\Psi}$, if for some $p \in P_{b_\Psi}$, $T_\mathcal{R}(p) \models \phi$, then using Propositions 2 and Property (2), it holds that for all $p \in P_{b_\Psi}$, $T_\mathcal{X}(p) \models \phi$: P_{b_Ψ} is a valid parameter set. Since we can compute $T_\mathcal{R}(p)$ for any given p (Proposition 1), solutions to Problem 1 and 2 can be obtained by testing for every equivalence class $P_{b_\Psi} \subseteq \mathcal{P}$ whether

[2] Note that, in general, the partition of the parameter space is not hyperrectangular.

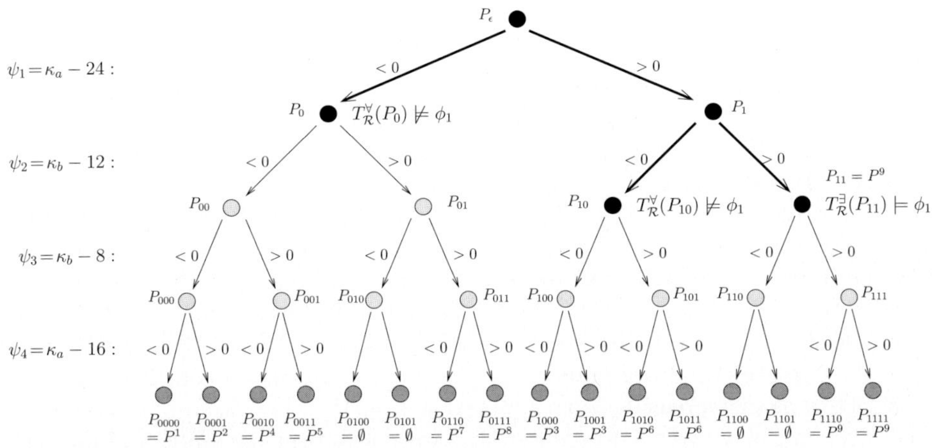

Fig. 4. Hierarchy between the parameter sets P_b, represented as a binary tree. Arrows indicate set inclusion: $P \to P'$ means $P' \subseteq P$. Leaves (dark gray) correspond to parameter equivalence classes. $P^1, \ldots, P^9$ refer to regions in Figure 3. The fragment of the tree actually computed during hierarchical parameter space exploration for the analysis of property ϕ_1 is emphasized. Model checking results used for backtracking are shown at the nodes where the recursive search stops.

$T_{\mathcal{R}}(p) \models \phi$ for some (randomly chosen) $p \in P_{b_\Psi}$. Note however that if $T_{\mathcal{R}}(p) \not\models \phi$, no conclusion can be obtained on P_{b_Ψ}. On our example network, only two equivalence classes, P_{1110} and P_{1111}, both corresponding to P^9, are found to be valid for the bistability property ϕ_1 (Figure 4 and 3). However, this naive approach is impractical since the number of equivalence classes (*i.e.* the leaves of the tree in Figure 4) increases exponentially with the number of affine predicates, the latter increasing exponentially with the number of variables and uncertain parameters. A more efficient approach is proposed in the next section.

4.3 Hierarchical Parameter Space Exploration

Our goal is to describe the behavior of the network for *sets* of parameters $P \subseteq \mathcal{P}$. To do so, we introduce two transition systems, $T^{\exists}_{\mathcal{R}}(P)$ and $T^{\forall}_{\mathcal{R}}(P)$.

Definition 5. *Let* $P \subseteq \mathcal{P}$. *Then* $T^{\exists}_{\mathcal{R}}(P) = (\mathcal{R}, \to^{\exists}_{\mathcal{R},P}, \Pi, \models_{\mathcal{R}})$ *and* $T^{\forall}_{\mathcal{R}}(P) = (\mathcal{R}, \to^{\forall}_{\mathcal{R},P}, \Pi, \models_{\mathcal{R}})$, *where*
- $(R, R') \in \to^{\exists}_{\mathcal{R},P}$ *iff* $\exists p \in P$ *such that* $(R, R') \in \to_{\mathcal{R},p}$ *in* $T_{\mathcal{R}}(p)$, *and*
- $(R, R') \in \to^{\forall}_{\mathcal{R},P}$ *iff* $\forall p \in P$, $(R, R') \in \to_{\mathcal{R},p}$ *in* $T_{\mathcal{R}}(p)$.

In words, $T^{\exists}_{\mathcal{R}}(P)$ contains all the transitions present in at least one transition system $T_{\mathcal{R}}(p)$ and $T^{\forall}_{\mathcal{R}}(P)$ contains only the transitions present in all the transition systems $T_{\mathcal{R}}(p)$. For every $p \in P$, $T^{\exists}_{\mathcal{R}}(P)$ simulates $T_{\mathcal{R}}(p)$, which simulates $T^{\forall}_{\mathcal{R}}(P)$. This follows immediately from the definition of simulation between transition systems, using the fact that $\to^{\forall}_{\mathcal{R},P} \subseteq \to_{\mathcal{R},p} \subseteq \to^{\exists}_{\mathcal{R},P}$. Informally, $T^{\exists}_{\mathcal{R}}(P)$ and

$T_{\mathcal{R}}^{\forall}(P)$ can be respectively considered as over- and under-approximations of the possible behaviors of $T_{\mathcal{R}}(p)$, when p varies.

Proposition 3. *For every* $p \in P$, $T_{\mathcal{R}}^{\forall}(P) \preceq T_{\mathcal{R}}(p) \preceq T_{\mathcal{R}}^{\exists}(P)$.

Using Proposition 3 and Property (2), it holds that for *any* $P \in \mathcal{P}$, if $T_{\mathcal{R}}^{\exists}(P) \models \phi$ then $\forall p \in P$, $T_{\mathcal{X}}(p) \models \phi$: P is a valid parameter set. Alternatively, using Proposition 3, it also holds that if $T_{\mathcal{R}}^{\forall}(P) \not\models \phi$, then $\forall p \in P$, $T_{\mathcal{R}}(p) \not\models \phi$: no valid parameter can be found in P using our approach, either because P contains no valid parameter, or because the discrete abstraction is overly conservative. Otherwise ($T_{\mathcal{R}}^{\exists}(P) \not\models \phi$ and $T_{\mathcal{R}}^{\forall}(P) \models \phi$), it is worth inspecting subsets of P, that may contain valid parameter sets. Accordingly, we propose an algorithm that explores $\mathcal{P}$ in a hierarchical manner by considering parameter sets P_b associated with Booleans of increasing length, starting from P_ϵ. This amounts to explore recursively a binary tree, represented in Figure 4 for our example, and for each node, to compute P_b, $T_{\mathcal{R}}^{\exists}(P_b)$ and $T_{\mathcal{R}}^{\forall}(P_b)$, and test whether $T_{\mathcal{R}}^{\exists}(P_b) \models \phi$ and whether $T_{\mathcal{R}}^{\forall}(P_b) \models \phi$. As explained above, the recursive search can be stopped if either $T_{\mathcal{R}}^{\exists}(P_b) \models \phi$ or $T_{\mathcal{R}}^{\forall}(P_b) \not\models \phi$. For the leaves (*i.e.* the equivalence classes), $T_{\mathcal{R}}^{\exists}(P_b) = T_{\mathcal{R}}^{\forall}(P_b)$ and the search necessarily terminates. A more detailed description of the algorithm can be found in [8]. Note that in general, $T_{\mathcal{R}}^{\forall}(P)$ does *not* provide information on the *original* system $T_{\mathcal{X}}(p)$, since no relation exists between $T_{\mathcal{R}}^{\forall}(P)$ and $T_{\mathcal{X}}(p)$. Nevertheless, it makes it possible to identify (potentially large) regions of the parameter space in which no valid parameter can be found. Consequently, it plays a key role when exploring large parameter spaces where only small regions are valid sets. The fragment of the tree actually computed for the analysis of property ϕ_1 is represented in Figure 4. The same result is obtained as previously (P^9 is a valid parameter set), but in much fewer tests.

We have not yet explained how $T_{\mathcal{R}}^{\forall}(P)$ and $T_{\mathcal{R}}^{\exists}(P)$ can be computed.

Proposition 4 (Computation of $T_{\mathcal{R}}^{\exists}(P)$ and $T_{\mathcal{R}}^{\forall}(P)$). *Let* $P \subseteq \mathcal{P}$.
– $(R, R') \in \rightarrow_{\mathcal{R},P}^{\exists}$ *iff either* $R = R'$, *or* $R \mathrel{\substack{\leftarrow\\\rightarrow}} R'$ *and* $P \cap g(R, R') \neq \emptyset$,
– $(R, R') \in \rightarrow_{\mathcal{R},P}^{\forall}$ *iff either* $R = R'$, *or* $R \mathrel{\substack{\leftarrow\\\rightarrow}} R'$ *and* $P \subseteq g(R, R')$,
where $g(R, R') = \{p \in \mathcal{P} \mid \exists v \in \mathcal{V}_R \cap \mathcal{V}_{R'}$ *such that* $f_i(v, p)(c'_i - c_i) > 0\}$, *with* $c = coord(R)$, $c' = coord(R')$ *and* $i \in \{1, \ldots, n\}$ *such that* $c_i \neq c'_i$.

Proof. Let $P \subseteq \mathcal{P}$ and $R, R' \in \mathcal{R}$ be such that $R \mathrel{\substack{\leftarrow\\\rightarrow}} R'$ (the other cases being trivial). From Proposition 1, it is easy to see that $g(R, R')$ is the set of parameters $p \in \mathcal{P}$ for which there is a transition from R to R' in $T_{\mathcal{R}}(p)$. Then, the result follows from the definition of the transition relations $\rightarrow_{\mathcal{R},P}^{\exists}$ and $\rightarrow_{\mathcal{R},P}^{\forall}$ (Definition 5).

Given that $f_i(v, p)$ is an affine expression in p, the sets $g(R, R')$ correspond to unions of polytopes in $\mathcal{P}$. Consequently, for polyhedral sets P, the computation of the transition systems $T_{\mathcal{R}}^{\exists}(P)$ and $T_{\mathcal{R}}^{\forall}(P)$ using Proposition 4 simply amounts to compute intersections and inclusions of unions of polytopes, which are standard polyhedral operations efficiently implemented in toolboxes. This method has been implemented in a freely-available tool for Robust Verification of Gene Networks

(RoVerGeNe, see `http://iasi.bu.edu/~batt/rovergene/rovergene.htm`). It is written in Matlab on top of several other tools (MPT, MatlabBGL, NuSMV). Because the efficiency of the computations may significantly depend on the order in which the affine predicates $\psi_i(p) = 0$, $\psi_i \in \Psi$, are considered during the search, we implemented a simple heuristic that orders first the predicates splitting the parameter space the more evenly (*i.e.* yielding two polytopes of similar volumes). Additionally, RoVerGeNe supports an extension of the method presented here, dealing with problems specifically encountered when verifying liveness properties, and described in [8,18].

5 Tuning of a Transcriptional Cascade

The method presented in the previous section is applied to the analysis of the steady-state input/output (I/O) behavior of a synthetic transcriptional cascade build and analyzed in [19] (Figure 5(a)). We have developed a PMA model of this system, represented in Figure 5(b). Parameter values were estimated based on experimental data available in [19].

The cascade is ultrasensitive: the steady-state I/O behavior is such that the output (EYFP) undergoes a dramatic change for a moderate change of the input (aTc) in a transition region. The cascade is expected to present at least a 1000-fold increase of the output value for a two-fold increase of the input value. Using F Gp ("eventually, p will be always true") to express that property p holds at equilibrium, the specifications in Figure 5(c) can be translated in LTL as follows.

$$\begin{array}{lrl} \phi_2 = & u_{aTc} < 100 \rightarrow & \mathrm{FG}(x_{eyfp} > 2.5\,10^2 \wedge x_{eyfp} < 5\,10^2) \\ \wedge & 100 < u_{aTc} < 200 \rightarrow & \mathrm{FG}(x_{eyfp} > 2.5\,10^2 \wedge x_{eyfp} < 10^6) \\ \wedge & u_{aTc} > 200 \rightarrow & \mathrm{FG}(x_{eyfp} > 5\,10^5 \wedge x_{eyfp} < 10^6). \end{array}$$

The actual network does not meet its specifications. So, we tried to tune it by finding valid parameter sets for property ϕ_2 (Problem 2). Using RoVerGeNe, we found a valid set, P_1, by tuning three production rate parameters:

$$P_1 : 1832.43 < \kappa_{lacI} < 3350.62,\ 393.46 < \kappa_{cI} \ \text{ and } 6495.42 < \kappa_{eyfp} < 12995.42$$

In order to evaluate the significance of these constraints, we computed by numerical simulation the steady-state I/O behavior of the system for different parameters in P_1, notably using extreme values (Figure 5(c)). This clearly reveals that relevant constraints on the parameters have been identified by our method.

With a partition of the state space having 1500 rectangles, 18 affine predicates on parameters were found, defining $> 200\,000$ equivalence classes. The computation lasted < 2 hours (PC, 3.4 GHz processor, 1 Gb RAM) and only 350 different parameter sets were analyzed. This computational time can be considered as very reasonable, given the difficulty of the problem: we systematically explore a 3-dimensional parameter space, testing a non-trivial dynamical property for any initial condition in a 5-dimensional (1 input and 4 state variables) state-space. As explained elsewhere [8,18], we have also been able to assess the robustness of the network with 11 uncertain parameters.

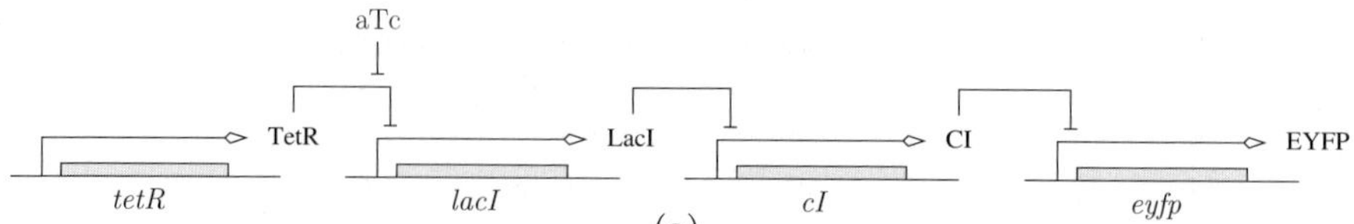

(a)

$$\dot{x}_{tetR} = \kappa_{tetR} - \gamma_{tetR}\, x_{tetR}, \tag{3}$$

$$\dot{x}_{lacI} = \kappa^0_{lacI} + \kappa_{lacI}(1 - r^+(x_{tetR}, \theta^1_{tetR}, \theta^2_{tetR})\, r^-(u_{aTc}, \theta^1_{aTc}, \theta^2_{aTc})) - \gamma_{lacI}\, x_{lacI}, \tag{4}$$

$$\dot{x}_{cI} = \kappa^0_{cI} + \kappa_{cI}\, r^-(x_{lacI}, \theta^1_{lacI}, \theta^2_{lacI}) - \gamma_{cI}\, x_{cI}, \tag{5}$$

$$\dot{x}_{eyfp} = \kappa^0_{eyfp} + \kappa_{eyfp}\, r^-(x_{cI}, \theta^1_{cI}, \theta^2_{cI}) - \gamma_{eyfp}\, x_{eyfp}, \tag{6}$$

$(\theta^1_{aTc}, \theta^2_{aTc}) = (80, 4000)$; $(\kappa_{tetR}, \gamma_{tetR}, \theta^1_{tetR}, \theta^2_{tetR}) = (260, 0.013, 4500, 5500)$; $(\kappa^0_{lacI}, \kappa_{lacI}, \gamma_{lacI}, \theta^1_{lacI}, \theta^2_{lacI}) = (2.40, 875.6, 0.013, 500, 4500)$; $(\kappa^0_{cI}, \kappa_{cI}, \gamma_{cI}, \theta^1_{cI}, \theta^2_{cI}) = (3.90, 386, 0.013, 600, 23000)$; $(\kappa^0_{eyfp}, \kappa_{eyfp}, \gamma_{eyfp}) = (4.58, 4048, 0.013)$

(b)

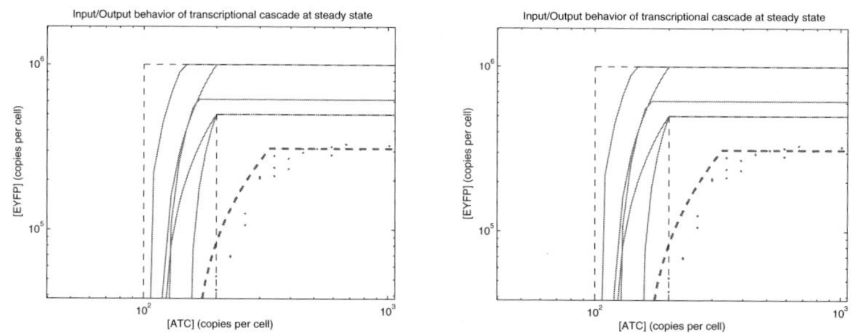

(c) (d)

Fig. 5. (a) Synthetic transcriptional cascade made of four genes. *tetR* inhibits *lacI*, *lacI* inhibits *cI*, and *cI* inhibits *eyfp*. The input aTc relieves the inhibition of *lacI* by TetR. The fluorescence of the protein EYFP is the output. (b) PMA model. Equation (4) states that *lacI* is repressed when the protein TetR is present and aTc absent. (c) I/O response of the cascade at steady state (zoomed in (d)). Measured (red dots), predicted (red thick dashed line) and expected (region delimited by black dashed lines) behaviors of the actual network. Predicted (magenta solid lines) behaviors for different parameters in the set P_1.

6 Discussion

We have presented a method for the analysis of genetic regulatory networks with parameter uncertainty. Given a PMA model, a property expressed in LTL over rectangular predicates and a polyhedral parameter set, the proposed approach can be used to test whether the property is satisfied for every parameter in the parameter set -the set is then called valid-, or to find valid subsets of the given parameter set. To do so, we use a discrete abstraction $T_{\mathcal{R}}(p)$ of an embedding continuous transition system $T_{\mathcal{X}}(p)$ to define an equivalence relation onbreak parameters p, in the sense that two equivalent parameters are associated to the same discrete abstraction. Then we define discrete transition systems, $T_{\mathcal{R}}^{\exists}(P)$ and $T_{\mathcal{R}}^{\forall}(P)$, that over- and under-approximate $T_{\mathcal{R}}(p)$ with parameter p in a set P, and show how they can be used to search the parameter space efficiently. The proposed approach is conservative: if a parameter set is found, it is guaranteed to be valid. However, not all valid parameter sets are guaranteed to be found. The method is implemented in a publicly-available tool called RoVerGeNe, and its practical applicability and biological relevance is demonstrated on the tuning of a synthetic network build in *E. coli.* Network tuning is a central problem in synthetic biology, since most initial attempts at constructing gene networks do not result in a system exhibiting the desired behavior [2].

Other approaches have been proposed for the verification of continuous or hybrid systems with parameter uncertainties. In most approaches, unknown parameters are represented as symbolic constants, and symbolic operations are used to manipulate sets of states and compute (approximations of) sets of predecessors or successors [17,20,21,22,23]. A major limitation is that the computational techniques supporting these symbolic operations currently apply only to systems having rather simple continuous dynamics, such as timed automaton [20,21], linear hybrid automaton [22], piecewise-affine systems [17], or affine hybrid automaton [23]. Alternatively, numerical approaches have been proposed in which parameter uncertainties are captured by means of differential inclusions (*e.g.* $\dot{x} \in \mathrm{hull}(\{f(x,p) \mid p \in P\})$) [24]. For large parameter sets, these approaches can be very conservative. In this paper, we propose an approach which is successively symbolic (synthesis of parameter constraints) and numerical (computation of transition systems). The results of the first step are used to refine the parameter set considered in the second step, in order to limit (though not eliminate) overconservatism, while preserving efficiency.

In the field of systems biology, several approaches use formal verification to analyze uncertain models, often with a focus on parameter identification. In [25,26], solution trajectories are computed by numerical simulation for parameter values chosen in specified intervals. Model checking is used to select trajectories satisfying the expected properties. This approach applies to very general classes of models, but can not provide guaranties for dense sets of parameters. Alternatively, exhaustive search or symbolic computations have been used to obtain constraints on parameters of discrete models having finite parameter spaces [26,27], or of piecewise-affine models having dense parameter spaces [23]. However, these models do not capture complex genetic regulations with graded responses, as we do in the transcriptional cascade example.

Motivated by applications in synthetic biology, we view two directions for further work. A first improvement would be to deal also with uncertain *threshold* parameters. A second desirable extension would be to allow for the verification of the frequently-encountered properties involving timing constraints.

Acknowledgements. We would like to thank Boyan Yordanov for contributions to model development and acknowledge financial support by NSF 0432070.

References

1. Kitano, H.: Systems biology: A brief overview. Science **295**(5560) (2002) 1662–1664
2. Andrianantoandro, E., Basu, S., Karig, D., Weiss, R.: Synthetic biology: New engineering rules for an emerging discipline. Mol. Syst. Biol. (2006)
3. Szallasi, Z., Stelling, J., Periwal, V., eds.: System Modeling in Cellular Biology: From Concepts to Nuts and Bolts. MIT Press (2006)
4. de Jong, H., Ropers, D.: Qualitative approaches to the analysis of genetic regulatory networks. In [3] (2006) 125–148
5. Belta, C., Habets, L.C.G.J.M.: Controlling a class of nonlinear systems on rectangles. IEEE Trans. Aut. Control **51**(11) (2006) 1749–1759
6. Alur, R., Henzinger, T.A., Lafferriere, G., Pappas, G.J.: Discrete abstractions of hybrid systems. Proc. IEEE **88**(7) (2000) 971–984
7. Clarke, E.M., Grumberg, O., Peled, D.A.: Model Checking. MIT Press (1999)
8. Batt, G., Belta, C.: Model checking genetic regulatory networks with applications to synthetic biology. CISE Tech. Rep. 2006-IR-0030, Boston University (2006)
9. de Jong, H., Gouzé, J.-L., Hernandez, C., Page, M., Sari, T., Geiselmann, J.: Qualitative simulation of genetic regulatory networks using piecewise-linear models. Bull. Math. Biol. **66**(2) (2004) 301–340
10. Abate, A., Tiwari, A.: Box invariance of hybrid and switched systems. In: Proc. ADHS'06. (2006)
11. Belta, C., Habets, L.C.G.J.M., Kumar, V.: Control of multi-affine systems on rectangles with applications to hybrid biomolecular networks. In: Proc. CDC'02. (2002)
12. Mestl, T., Plahte, E., Omholt, S.: A mathematical framework for describing and analysing gene regulatory networks. J. Theor. Biol. **176** (1995) 291–300
13. Glass, L., Kauffman, S.: The logical analysis of continuous non-linear biochemical control networks. J. Theor. Biol. **39**(1) (1973) 103–129
14. Kloetzer, M., Belta, C.: Reachability analysis of multi-affine systems. In Hespanha, J., Tiwari, A., eds.: Proc. HSCC'06. LNCS 3927, Springer (2006) 348–362
15. Alur, R., Dang, T., Ivancic, F.: Progress on reachability analysis of hybrid systems using predicate abstraction. In Pnueli, A., Maler, O., eds.: Proc. HSCC'03. LNCS 2623, Springer (2003) 4–19
16. Koutsoukos, X., Antsaklis, P.J.: Safety and reachability of piecewise linear hybrid dynamical systems based on discrete abstractions. J. Discrete Event Dynamic Systems **13**(3) (2003) 203–243
17. Batt, G., Ropers, D., de Jong, H., Geiselmann, J., Mateescu, R., Page, M., Schneider, D.: Validation of qualitative models of genetic regulatory networks by model checking : Analysis of the nutritional stress response in *E. coli*. Bioinformatics **21**(Suppl.1) (2005) i19–i28
18. Batt, G., Belta, C., Weiss, R.: Model checking liveness properties of genetic regulatory networks. To appear in Grumberg, O., Huth, M., eds.: Proc. TACAS'07. LNCS, Springer (2007)
19. Hooshangi, S., Thiberge, S., Weiss, R.: Ultrasensitivity and noise propagation in a synthetic transcriptional cascade. Proc. Natl. Acad. Sci. USA **102**(10) (2005) 3581–3586
20. Annichini, A., Asarin, E., Bouajjani, A.: Symbolic techniques for parametric reasoning about counter and clock systems. In Emerson, E., Sistla, A., eds.: Proc. CAV'00. LNCS 1855, Springer (2000) 419–434

21. Wang, F.: Symbolic parametric safety analysis of linear hybrid systems with BDD-like data-structures. IEEE Trans. Softw. Eng. **31**(1) (2005) 38–51
22. Henzinger, T., Ho, P.-H., Wong-Toi, H.: HYTECH: A model checker for hybrid systems. Software Tools Technology Transfer **1**(1-2) (1997) 110–122
23. Ghosh, R., Tomlin, C.J: Symbolic reachable set computation of piecewise affine hybrid automata and its application to biological modelling: Delta-Notch protein signalling. IEE Proc. Syst. Biol. **1**(1) (2004) 170–183
24. Lin, H., Antsaklis, P.J.: Robust regulation of polytopic uncertain linear hybrid systems with networked control system applications. In Antsaklis, P., Liu, D., eds.: Stability and Control of Dynamical Systems with Applications. Birkhauser (2003)
25. Antoniotti, M., Piazza, C., Policriti, A., Simeoni, M., Mishra, B.: Taming the complexity of biochemical models through bisimulation and collapsing: Theory and practice. Theor. Comput. Sci. **325**(1) (2004) 45–67
26. Calzone, L., Chabrier-Rivier, N., Fages, F., Soliman, S.: Machine learning biochemical networks from temporal logic properties. In Priami, C., Plotkin, G., eds: Trans. Comput. Syst. Biol. VI. LNBI 4220, Springer (2006) 68–94
27. Bernot, G., Comet, J.-P., Richard, A., Guespin, J.: Application of formal methods to biological regulatory networks: Extending Thomas' asynchronous logical approach with temporal logic. J. Theor. Biol. **229**(3) (2004) 339–347

MARCO: A Reachability Algorithm for Multi-affine Systems with Applications to Biological Systems

Spring Berman, Ádám Halász, and Vijay Kumar

GRASP Laboratory, School of Engineering and Applied Science,
University of Pennsylvania, Philadelphia PA 19104, USA
{spring,halasz,kumar}@grasp.upenn.edu

Abstract. We present a new algorithm for the reachability analysis of multi-affine hybrid systems. In our previous work on reachability analysis and that of our collaborators [1,2,3], we exploited the convexity of multi-affine functions and the fact that the vector field in modes with rectangular invariants is uniquely determined by its values at the rectangle vertices. In this paper, we explicitly calculate conical overapproximations of the reachable set in the invariant of each mode. We describe our Multi-Affine Reachability analysis using Conical Overapproximations, MARCO, and show that it yields results that are superior to those obtained by existing methods for multi-affine hybrid systems. Finally, we demonstrate the application of MARCO to the analysis of an ant house hunting model that incorporates quorum sensing [4] and the analysis of bi-stability of the lactose induction system regulated by glucose and lactose [5].

1 Introduction

Multi-affine equations are often used to model systems in the areas of molecular biology and population biology. Bio-molecular networks can be modeled by multi-affine rate equations that describe chemical reactions among species [1]. In population biology, the Volterra-Lotka predator-prey equations constitute a familiar example of a multi-affine model [6]. The spread of information or disease within a population can also be described using multi-affine equations, as in the example of honeybee recruitment to a nest site [4]. As a result, multi-affine models have potential applications in understanding biology, synthesizing new biomolecular circuits, and designing bio-inspired controllers for networked robotic systems.

We are particularly interested in *multi-affine hybrid systems* because of their relevance in biology. These systems consist of discrete modes that are each characterized by multi-affine continuous dynamics. While biological systems exhibit smooth behavior on some level, it is often convenient to develop hybrid abstractions to describe switches that are associated with such phenomena as gene regulation or quorum-sensing. For example, a gene may be turned on (or off) when a threshold concentration of regulatory biomolecular species is exceeded,

A. Bemporad, A. Bicchi, and G. Buttazzo (Eds.): HSCC 2007, LNCS 4416, pp. 76–89, 2007.

which results in a change in the dynamics of the network [7]. Similarly, ants emigrating from a nest in search of alternative nests change their behavior when they detect a quorum at any of the candidate nests [4]. Thus, we have found it useful to use hybrid system abstractions to describe biological networks [1,8].

In order to accurately approximate the global behavior of a set of hybrid system trajectories, or to verify that they do not enter an undesirable region, it is productive to consider *reachability analysis*, a well-known symbolic analysis technique [9,10,11]. A typical reachability problem is to determine whether a certain region of the state space can be reached by a system, starting from a given set of initial conditions. The reachability problem is decidable when the continuous dynamics are constant (timed and multirate automata), take values in a constant interval (rectangular automata) [10], or fall into certain classes of linear systems [12]. If the dynamics are not of these types, an overapproximation of the reachable set can be computed in one of two ways. One option is to pursue a discrete abstraction of the hybrid system via an *indirect method.* Alternatively, the reach set can be directly calculated the on the state space via a *direct method.*

In the indirect method, one generally partitions the continuous state space of the system into a finite number of sets and explores how states in one set may reach states in another set. Sets are usually convex regions of the state space; the exact representation of a set depends on a particular method. In this paper, sets are represented by hyper-rectangles in the n-dimensional state space. The multi-affine reachability algorithm developed in [1,2,3] is referred to here as the MAR1 algorithm. It exploits the convexity of multi-affine functions on hyper-rectangles in a manner similar to [13], which describes a technique for controlling affine systems on general polytopes. Once a state is inside a hyper-rectangle, the algorithm considers the entire hyper-rectangle to be reachable. Because of this, the algorithm computes *conservative approximations* of the reach set. While this approximation is guaranteed to include all reachable states, it can be overly conservative and in many simple cases (for example, constant vector fields along the diagonals of the hyper-rectangles) yield little insight into the actual behavior of the system.

In this paper, we present a new direct reachability analysis algorithm for multi-affine hybrid systems, MARCO (Multi-Affine Reachability via Conical Overapproximations), which attempts to overcome some of the shortcomings of the MAR1 algorithm. We consider the problem of computing less conservative reachable sets without sacrificing accuracy. As in the MAR1 method, the algorithm performs a computationally inexpensive reachability analysis within each mode by exploiting the convexity property of multi-affine vector fields on rectangles. However, we determine a better conical approximation for the reachable set, thus providing a finer level of granularity for the reachable set without incurring a significantly higher penalty for computations. As before, a higher degree of precision for the entire reachable set can be achieved by increasing the resolution of the rectangular partitions.

Our technique for overapproximating the reachable set within a mode is similar in spirit to that used in HyTech [10] and PHAVer [14], which are tools for the verification of linear hybrid automata. This class of automata has piecewise constant bounds on the derivatives of the continuous state variables. HyTech and PHAVer overapproximate affine continuous dynamics by linear formulas over the derivatives. PHAVer also has the ability to partition reachable modes recursively along user-defined hyperplanes.

Due to the simplicity of its reachability operations, the MARCO algorithm is suitable for multi-affine hybrid systems with many modes, such as a system that closely approximates a hybrid automaton with nonlinear dynamics. Thus, in principle, it is more readily applicable to such systems than existing reachability algorithms that use direct techniques for nonlinear hybrid systems, such as MATISSE [15] and CheckMate [11].

2 Theory

We define a *hyper-rectangular multi-affine switched system (HMS)* as the seven-tuple $H = (X, X_0, \Omega, I, F, T, A)$. $X \subset \mathbb{R}^n$ is the continuous space of state variables x, and $X_0 \subset X$ is a set of initial states, and Ω is a set of discrete modes. I maps the modes to subsets of X such that when the system is at mode $\omega \in \Omega$, $x \in I(\omega)$, the location invariant of ω. The location invariants are n-dimensional hyper-rectangles, which are defined as follows. For each dimension $j = 1, ..., n$, we specify a strictly monotonically increasing sequence of values, $\{x_0^{(j)}, x_1^{(j)}, \cdots, x_{D_j}^{(j)}\}$. A mode ω is labeled by an n-dimensional coordinate vector $\omega = (k_1, \cdots, k_n)$, where $k_j \in \{1, ..., D_j\}$. Then $I(\omega)$ is the hyper-rectangle $[x_{k_1-1}^{(1)}, x_{k_1}^{(1)}] \times [x_{k_2-1}^{(2)}, x_{k_2}^{(2)}] \cdots \times [x_{k_n-1}^{(n)}, x_{k_n}^{(n)}]$. F is a map that assigns a continuous, autonomous vector field to each mode ω, $\dot{x} = f_\omega(x) \in \mathbb{R}^n$. f_ω is a multi-affine function of x.

Definition 1 (Multi-affine function). *A multi-affine function $f : \mathbb{R}^n \rightarrow \mathbb{R}^n$ has the following form:*

$$f(x) = \sum_{j=0}^{2^n-1} c_j x_1^{i_1(j)} x_2^{i_2(j)} ... x_n^{i_n(j)} \;; \qquad c_j \in \mathbb{R}^n, \tag{1}$$

where $x = (x_1, ..., x_n)$ and the concatenation $i_1(j)i_2(j)...i_n(j)$, where $\{i_1(j), ..., i_n(j)\} \in \{0,1\}^n$, is a binary representation of the integer j.

Henceforth, Θ_j will denote the concatenation $i_1(j)i_2(j)...i_n(j)$.

Proposition 1 ([1]). *Let $f_\omega : I(\omega) \rightarrow \mathbb{R}^n$ be a multi-affine function and let $x \in I(\omega)$. Then $f_\omega(x)$ is a convex combination of the values of f_ω at the 2^n vertices of $I(\omega)$.*

T is a finite set of transitions between modes, each defined by a three-tuple $(\omega, \omega', g_{\omega,\omega'})$, in which $\omega, \omega' \in \Omega$ and $g_{\omega,\omega'} \subset \partial I(\omega)$ is a guard set. The transition from ω to ω' is enabled when $x \in g_{\omega,\omega'}$. Each guard $g_{\omega,\omega'}$ of mode ω corresponds to a facet that $I(\omega)$ shares with $I(\omega')$. We denote this shared facet by $H(\omega, \omega')$.

A is a finite set of symbols that label the transitions. We now define the trajectories, footprints, and reachable sets of an HMS.

Definition 2 (Mode trajectory [16]). *A trajectory $(\omega, \tau, x_\omega(t))$ associated with mode $\omega \in \Omega$ consists of a nonnegative time τ and a continuous and piecewise differentiable function $x_\omega : [0, \tau] \to \mathbb{R}^n$ such that $x_\omega(t) \in I(\omega)$ and $\dot{x}_\omega(t) = f_\omega(x_\omega(t))$ for all $t \in (0, \tau)$.*

Definition 3 (Trajectory of an HMS [16]). *A trajectory of an HMS starting from $x_{\omega_0}(0) \in X_0 \subset I(\Omega_0)$, where $\Omega_0 \subset \Omega$, is defined as an infinite sequence of mode trajectories,*

$$(\omega_0, \tau_0, x_{\omega_0}(t)) \xrightarrow{a_0} (\omega_1, \tau_1, x_{\omega_1}(t)) \xrightarrow{a_1} (\omega_2, \tau_2, x_{\omega_2}(t)) \xrightarrow{a_2} \cdots \qquad (2)$$

such that at the event times $t_{\omega_j} = \sum_{i=0}^{j} \tau_i$, $x_{\omega_j}(t_{\omega_j}) \in H(\omega_j, \omega_{j+1})$. Since the HMS is defined to be a switched system, $x_{\omega_j}(t_{\omega_j}) = x_{\omega_{j+1}}(0)$. The jth transition is labeled by $a_j \in A$.

The ordered set of modes in equation (2) after a finite number of transitions is represented by a *filiation sequence* of length $d \in \mathbb{N}$, $s = \{\omega_0, \omega_1, \cdots, \omega_{d-1}\}$. We define a concatenation operation similar to that which is used for strings: $s * \{\sigma\} = \{\omega_0, \cdots, \omega_{d-1}, \sigma\}$. In the following definitions, ϕ_s designates an HMS trajectory whose first d modes comprise sequence s, given some $x_{\omega_0} \in X_0$.

Definition 4 (Footprint). *A footprint of degree d and filiation sequence s, $X_{s,\omega_d}^{(d)} \subset H(\omega_{d-1}, \omega_d)$, is the set consisting of $x_{\omega_{d-1}}(t_{\omega_{d-1}})$ from each ϕ_s.*

Definition 5 (Forward reachable set of a mode). *The forward reachable set of mode ω_d from a set B, where $B = X_0$ if $d = 0$ and $B = X_{s,\omega_d}^{(d)}$ if $d > 0$, is $X_{r,\omega_d}(B) \subset I(\omega_d)$. It consists of the union of states*

$$x_{\omega_{d-1}}(t_{\omega_{d-1}}) = x_{\omega_d}(0) \ \cup \ \{x_{\omega_d}(t) \mid t \in (0, \tau_d)\} \ \cup \ x_{\omega_d}(t_{\omega_d}) = x_{\omega_{d+1}}(0) \qquad (3)$$

from each ϕ_s for which $\omega_d \in s$.

Definition 6 (Forward reachable set of an HMS). *The forward reachable set X_r from an initial set X_0 of an HMS is the set of all continuous states $x_\omega(t)$ associated with each ϕ_s.*

Definition 7 (Time-elapse cone). *The time-elapse cone C_ω for mode $\omega = (k_1, \cdots, k_n)$ is the cone generated by nonnegative linear combinations of the velocity vectors at the vertices of $I(\omega)$:*

$$C_\omega = \{ \sum_{j=0}^{2^n-1} \lambda_{\Theta_j} f_\omega(x^{(1)}_{k_0+(k_1-k_0)i_1(j)}, \cdots, x^{(n)}_{k_{n-1}+(k_n-k_{n-1})i_n(j)}) \mid \lambda_{\Theta_j} \geq 0 \} \, . \qquad (4)$$

Proposition 2. *Let $x_\omega(t)$ be defined as in Definition 2. The displacement vector $\Delta x_\omega(t) = x_\omega(t) - x_\omega(0)$, $t \in [0,\tau]$, is contained in the convex hull of the set of velocities at the vertices of $I(\omega)$, scaled by the elapsed time t. That is, $(\exists)$ $\{\Lambda_{\Theta_j}\}$ where $\Lambda_{\Theta_j} \in [0,1]$, $j = 0, ..., 2^n - 1$, and $\sum_{j=0}^{2^n-1} \Lambda_{\Theta_j} = 1$, such that:*

$$\Delta x_\omega(t) = t \sum_{j=0}^{2^n-1} \Lambda_{\Theta_j} f_\omega(x^{(1)}_{k_0+(k_1-k_0)i_1(j)}, \cdots, x^{(n)}_{k_{n-1}+(k_n-k_{n-1})i_n(j)}) \ . \quad (5)$$

Proof. The solution to $\dot{x}_\omega(t) = f_\omega(x_\omega(t))$, $t \in [0,\tau]$, is $x_\omega(t) = x_\omega(0) + \int_0^t f_\omega(x_\omega(s))ds$. From Proposition 1, for $s \in [0,\tau]$, $(\exists)$ $\{\lambda_{\Theta_j}(s)\}$ where $\lambda_{\Theta_j}(s) \in [0,1]$, $j = 0, ..., 2^n - 1$, and $\sum_{j=0}^{2^n-1} \lambda_{\Theta_j}(s) = 1$, such that:

$$f_\omega(x_\omega(s)) = \sum_{j=0}^{2^n-1} \lambda_{\Theta_j}(s) f_\omega(x^{(1)}_{k_0+(k_1-k_0)i_1(j)}, \cdots, x^{(n)}_{k_{n-1}+(k_n-k_{n-1})i_n(j)}) \quad (6)$$

The existence of $\{\lambda_{\Theta_j}(s)\}$ is guaranteed but it is not unique; we can choose one set. The displacement vector $\Delta x_\omega(t) = x_\omega(t) - x_\omega(0)$ at t is:

$$\begin{aligned}\Delta x_\omega(t) &= \int_0^t \sum_{j=0}^{2^n-1} \lambda_{\Theta_j}(s) f_\omega(x^{(1)}_{k_0+(k_1-k_0)i_1(j)}, \cdots, x^{(n)}_{k_{n-1}+(k_n-k_{n-1})i_n(j)})ds \\ &= \sum_{j=0}^{2^n-1} f_\omega(x^{(1)}_{k_0+(k_1-k_0)i_1(j)}, \cdots, x^{(n)}_{k_{n-1}+(k_n-k_{n-1})i_n(j)}) \int_0^t \lambda_{\Theta_j}(s)ds \quad (7)\end{aligned}$$

Define Λ_{Θ_j} as the integrated quantity divided by t:

$$0 \leq \Lambda_{\Theta_j} = \frac{1}{t}\int_0^t \lambda_{\Theta_j}(s)ds \ \leq 1 \quad (8)$$

$$\sum_{j=0}^{2^n-1} \Lambda_{\Theta_j} = \sum_{j=0}^{2^n-1} \frac{1}{t}\int_0^t \lambda_{\Theta_j}(s)ds = \frac{1}{t}\int_0^t \sum_{j=0}^{2^n-1} \lambda_{\Theta_j}(s)ds = 1 \ . \quad (9)$$

□

Corollary 1. *The set of continuous states $x_\omega(t)$, $t \in [0,\tau]$, in a trajectory of mode ω is a subset of $x_\omega(0) \oplus C_\omega$, the Minkowski sum of $x_\omega(0)$ and the time-elapse cone.*

The following definitions specify the core steps of the MARCO reachability algorithm. The proofs demonstrate that the reachable set computed by the algorithm contains the exact reachable set X_r.

Definition 8 (Overapproximated reach set of a mode). *Consider a mode ω and a set $B \subset I(\omega)$. The overapproximated reach set in mode ω with initial set B is defined as:*

$$R_\omega(B) \;=\; (B \oplus C_\omega) \cap I(\omega) \;\; . \tag{10}$$

Proposition 3. $X_{r,\omega}(B) \subset R_\omega(B)$.

Proof. From Definition 5, $X_{r,\omega}(B)$ is the set of all states $x_\omega(t)$ in the trajectory of mode ω such that $x_\omega(0) \in B$, which by Corollary 1 is a subset of $x_\omega(0) \oplus C_\omega$:

$$X_{r,\omega}(B) = \{x_\omega(t) \mid x_\omega(0) \in B, t \in [0,\tau]\} \subset \{x_\omega(0) \oplus C_\omega \mid x_\omega(0) \in B\} = B \oplus C_\omega$$

Since $X_{r,\omega}(B) \subset I(\omega)$ by definition, $X_{r,\omega}(B) \subset (B \oplus C_\omega) \cap I(\omega) = R_\omega(B)$. □

Definition 9 (Overapproximated footprint). *An overapproximated footprint of degree d and filiation sequence s, $F^{(d)}_{s,\omega_d} \subset H(\omega_{d-1}, \omega_d)$ is generated as follows.*

$$\begin{aligned} F^{(1)}_{\{\omega_0\},\omega} &= (X_0 \oplus C_{\omega_0}) \cap H(\omega_0, \omega) \\ F^{(d+1)}_{s*\{\omega_d\},\omega} &= (F^{(d)}_{s,\omega_d} \oplus C_{\omega_d}) \cap H(\omega_d, \omega) \end{aligned} \tag{11}$$

The footprints and their corresponding overapproximated reach sets form a tree structure, which in practical implementations is organized as a linked list. The sequence s distinguishes among repeated passages through the same mode during the reachability calculation.

Proposition 4 (Validity of overapproximation). *The set of states $x_\omega(t)$ in the first d mode trajectories of an HMS trajectory ϕ_s with $x_{\omega_0}(0) \in X_0$ is contained in the union of $R_{\omega_0}(X_0)$ with $R_{\omega_j}(F^{(j)}_{\{\omega_0,\ldots,\omega_{j-1}\},\omega_j})$, $j = 1, \ldots, d-1$.*

Proof. By Proposition 3, $x_{\omega_0}(t) \in R_{\omega_0}(X_0)$ for $x_{\omega_0}(0) \in X_0$, $t \in [0, \tau_0]$. Therefore, by Definition 8, $x_{\omega_0}(\tau_0) \in (X_0 \oplus C_{\omega_0}) \cap I(\omega_0)$. Also, $x_{\omega_0}(\tau_0) = x_{\omega_0}(t_{\omega_0}) \in H(\omega_0, \omega_1) \subset I(\omega_0)$. Thus, by Definition 9, $x_{\omega_0}(\tau_0) = x_{\omega_1}(0) \in F^{(1)}_{\{\omega_0\},\omega_1}$. By Proposition 3 again, $x_{\omega_1}(t) \in R_{\omega_1}(F^{(1)}_{\{\omega_0\},\omega_1})$ for $t \in [0, \tau_1]$. The same set inclusions may be defined for the remaining modes in s. □

There are two possible termination conditions for the algorithm.

Proposition 5 (Termination condition 1). *If $R_{\omega_d}(F^{(d)}_{s,\omega_d})$ is a subset of R_{ω_d}, the union of the reach sets previously computed for mode ω_d, then all states $x_\omega(t)$ in HMS trajectories with $x_{\omega_0}(0) \in F^{(d)}_{s,\omega_d}$ are contained in R_{ω_d} and all reach sets evolving from R_{ω_d}.*

Since the reach set might grow by very small amounts for a long time, a second heuristic condition may be applied to ensure termination within a reasonable amount of time. Each iteration of the algorithm generates a new set of conical overapproximations and footprints; let $V(R_i)$ be the volume of the newly computed reach set at iteration i and $V(S)$ be the volume of the state space.

Proposition 6 (Termination condition 2). *For a small constant ζ, stop if $V(R_i) < V(R_{i-1})$ and $V(R_i) < \zeta\, V(S)$.*

3 Implementation

The MARCO algorithm is written in Matlab and uses the Multi-Parametric Toolbox (MPT) for polyhedral operations. Figure 1 illustrates its steps for a two-dimensional state space, and Figure 2 gives an outline of the algorithm.

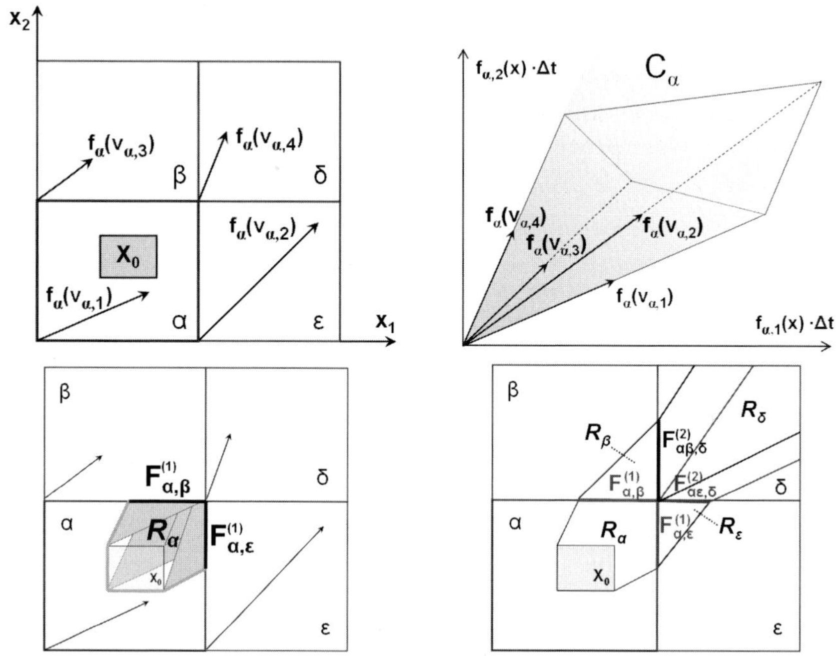

Fig. 1. Illustration of the MARCO algorithm. (a) (upper left) Initial set X_0 and velocities at vertices of mode α; (b) (upper right) definition of the time-elapse cone C_α; (c) (lower left) computation of reachable set R_α and footprints $F^{(1)}_{\alpha,\beta}$ and $F^{(1)}_{\alpha,\epsilon}$ of adjacent modes; (d) (lower right) computation of R_β, R_δ, and R_ϵ.

The user inputs the specifications of the hybrid system H. First, the set Ω of reachable modes is initialized with the modes $\Omega_0 \subset \Omega$ that contain the initial set X_0. These modes are identified as members of generation 0. In Figure 1a, $\Omega_0 = \alpha$. The portion of X_0 that intersects the mode invariant $I(\omega)$ for $\omega \in \Omega_0$ is the first incoming footprint of mode ω. For each mode in generation 0, a time-elapse cone C_ω is found according to Definition 7. Figure 1a-b shows the creation of the cone C_α from the velocities at the vertices of mode α. The cone is scaled to extend past the mode boundaries. C_ω is added to the mode footprint via a Minkowski sum and then bounded by the facets of the mode to produce the overapproximated mode reachable set, $R_\omega(X_0 \cap I(\omega))$ (Figure 1c).

Input: System dimension n, mode dividers, vertices of initial set I_0, dynamical parameters
Output: $R = \{R_{\omega_0}, ..., R_{\omega_N}\}$, $\omega_i \in \Omega$

$\Omega := \{\omega_i \mid I(\omega_i) \cap X_0 \neq \emptyset\}$
for all $\omega_i \in \Omega$: Generation$(\omega_i) = 0$; $R_{\omega_i} = \emptyset$
G $= -1$
do
  G = G + 1
  for all $\{\omega_i \mid \text{Generation}(\omega_i) = \text{G}\}$
    $R^{gen}_{\omega_i} = \emptyset$
    Calculate velocities at vertices of ω_i
    Create time-elapse cone C_{ω_i}
    Combine overlapping footprints of ω_i
    for all footprints $F^{(G)}_{s,\omega_i}$:
      $R_{\omega_i}(F^{(G)}_{s,\omega_i}) = (F^{(G)}_{s,\omega_i} \oplus C_{\omega_i}) \cap I(\omega_i)$
      for all $\{\omega_j \mid F^{(G+1)}_{s*\{\omega_i\},\omega_j} = (F^{(G)}_{s,\omega_i} \oplus C_{\omega_i}) \cap H(\omega_i, \omega_j) \neq \emptyset\}$
        if $\omega_j \notin \Omega$
          $\Omega = \Omega * \{\omega_j\}$
          $R_{\omega_j} = \emptyset$
        Generation$(\omega_j) = \text{G} + 1$
      end
      $R^{gen}_{\omega_i} = R^{gen}_{\omega_i} * \{R_{\omega_i}(F^{(G)}_{s,\omega_i})\}$
    end
    if $R^{gen}_{\omega_i} \not\subset R_{\omega_i}$
      $R_{\omega_i} = R_{\omega_i} * \{R^{gen}_{\omega_i}\}$
  end
until $R^{gen}_{\omega_i} \subset R_{\omega_i} \ \forall \ \{\omega_i \mid Generation(\omega_i) = G\}$

Fig. 2. MARCO reachability algorithm

Next, each adjacent mode ω' with a facet that has a nonempty intersection with $R_\omega(X_0 \cap I(\omega))$ is added to Ω if it is not already in the list, and the intersection is designated as the overapproximated incoming footprint of that mode, $F^{(1)}_{\{\omega\},\omega'}$. These modes are identified as members of the next generation. In Figure 1d, the footprints are $F^{(1)}_{\alpha,\beta}$ and $F^{(1)}_{\alpha,\epsilon}$, and modes β and ϵ are in generation 1.

The algorithm repeats the reach set overapproximation and footprint identification for modes in each consecutive generation. Note that a mode ω may have multiple footprints, as does mode δ in Figure 1d. Each footprint generates a reach set, and the concatenation of these sets is the total reach set within the mode. The algorithm terminates according to Proposition 5, Proposition 6, or when there are no new modes in the current generation, which occurs when the reach set hits the boundary of the state space X, as in Figure 1d. The algorithm returns the total reach set, stored as polyhedral subsets of mode invariants, that is attained from X_0.

4 Examples

Our first set of examples illustrates the improvement of MARCO over the method in [1]. Figures 3 and 4 display reachable sets computed by MARCO and by a Matlab implementation of the MAR1 algorithm. The MARCO reach sets are shown in dark gray, while the MAR1 sets consist of light gray boxes in the 2D examples and transparent boxes in the 3D and 4D examples. In each example, both algorithms used the same state space boundaries and mode partition. All examples were run on a standard 2 GHz. laptop.

In Figure 3a, the dynamics in each mode consist of the constant vector field $\dot{x}_1 = 1$, $\dot{x}_2 = 0.5$, and the initial set is the box in the lower left corner. The reach set computed by MARCO is the exact reachable set. However, the MAR1 algorithm predicts that all modes are reached.

Figure 3b displays a vector field whose integral curves are spirals with a steady state at the origin. The dynamics are given by

$$\dot{x}_1 = -x_1 + ax_2 \qquad \dot{x}_2 = -ax_1 - x_2 \ , \tag{12}$$

where $a = 2$. The initial set is the box containing the steady state. The MARCO algorithm terminates and returns a conservative but finite reach set around the equilibrium point; it essentially recognizes the presence of the steady state. The MAR1 method considers the entire space to be reached, due to the velocity components pointing out of the center mode.

Figure 3c shows the computation of the reach set for a three-dimensional vector field with integral curves that are helical spirals. The results are similar to those of Figure 3b.

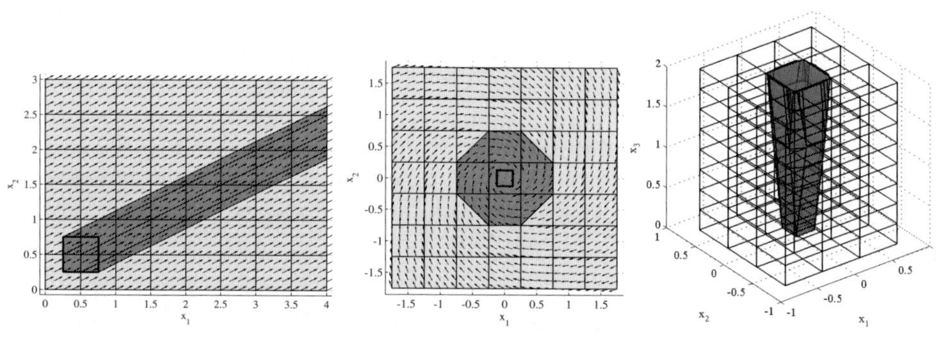

Fig. 3. Reachable sets for (a) 2D constant field; (b) 2D linear field; (c) 3D linear field

As another example, consider the bistable vector field,

$$\dot{x}_1 = f(x_2) - x_1 \qquad \dot{x}_2 = x_1 - x_2 \ , \tag{13}$$

where $f(x_2)$ is a piecewise-linear approximation, $P + Qx_2$, of a sigmoid-shaped function. P and Q for a mode depend on the particular x_1 interval that contains the average x_1 coordinate of the mode. In Figure 4a, the initial set is located at a

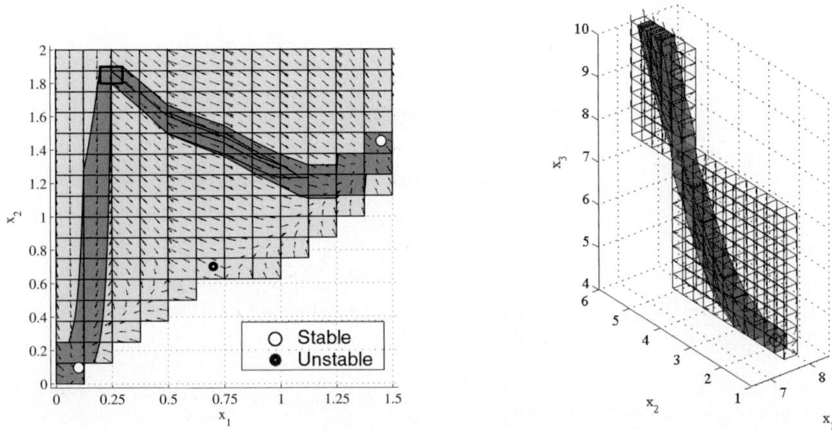

Fig. 4. Reachable sets for (a) 2D affine field; (b) 4D multi-affine field

place where the vector field diverges. The MARCO reach set correctly approaches and terminates at the two steady states while avoiding the unstable steady state. The MAR1 reach set is much more conservative.

Figure 4b illustrates a four-dimensional multi-affine system with 24 equilibria. In particular, the equilibrium $x_{e1} = (10.5, 7.5, 1.5, 4.5)$ is stable and the equilibrium $x_{e2} = (10.5, 7.5, 4.5, 7.5)$ is unstable. The initial set for the reachability computation is a box surrounding x_{e2}. Figure 4b shows a projection of the reach set onto the x_2, x_3, x_4 dimensions. The reach set diverges at x_{e2}: one branch terminates at x_{e1}, while the other runs into the state space boundary. Again, the MAR1 reach set fails to attain the precision of the MARCO set under the same mode partition.

Table 1 compares the performance of the two algorithms in terms of the computation time and volume fraction of the state space reached for each example. Notes that although MAR1 is faster on all examples, its overly conservative predictions of the reach set cannot be refined with iterative partitioning.

Table 1. Comparison of computation times and reachable set precision

Vector field	Time (sec)		Reached vol./State space vol.	
	MARCO	MAR1	MARCO	MAR1
2D constant	4.17	0.42	0.255	1.000
2D linear	2.83	0.42	0.329	1.000
3D linear	4.78	0.78	0.078	1.000
2D affine	7.27	0.94	0.266	0.714
4D multi-affine	130.31	2.53	0.022	0.061

5 Applications

5.1 Ant House Hunting Model

We consider a portion of the model of ant house hunting presented in [4]. This model, constructed from experimental observations of *Temnothorax albipennis* ants, predicts the behavior of a colony that is faced with a choice between two new nest sites, labeled 1 and 2, following the destruction of its original nest, site 0. The state variables represent the number of ants in different roles: naive ants, X; assessors of site i, Z_i; and recruiters to site i, Y_i. The method of recruitment used by Y_i ants varies depending on whether they have reached a quorum T. The model equations are as follows [4]:

$$\begin{aligned} \dot{X} &= -(\mu_1 + \mu_2)X - \lambda_1 Y_1 \theta(X)\theta(T - Y_1) - \lambda_2 Y_2 \theta(X)\theta(T - Y_2) \\ \dot{Y_1} &= k_1 Z_1 - \rho_{12} Y_1 \qquad \dot{Y_2} = k_2 Z_2 + \rho_{12} Y_1 \\ \dot{Z_1} &= \mu_1 X + \lambda_1 Y_1 \theta(X)\theta(T - Y_1) - \rho_{12} Z_1 - k_1 Z_1 \\ \dot{Z_2} &= \mu_2 X + \lambda_2 Y_2 \theta(X)\theta(T - Y_2) + \rho_{12} Z_1 - k_2 Z_2 \\ &\theta(X) = 1 \text{ when } X > 0, \ 0 \text{ otherwise} \end{aligned} \tag{14}$$

We performed reachability analysis to determine whether a quorum of recruiters at site 1 will ever be reached for a certain value of k_1, which reflects the quality of site 1. We reduce the model to a four-dimensional affine system by using the ant conservation constraint $X + Y_1 + Y_2 + Z_1 + Z_2 = N$ to eliminate X. This conservation constraint forms a boundary of the state space, along with nonnegativity constraints and the hyperplane corresponding to the quorum. The initial set is the four-dimensional unit cube to approximate the biologically realistic situation in which all ants start as naive. We set $N = 52$ and $T = 10$, according to the values in [4].

Figure 5a shows the new reach set volume per iteration of the algorithm as a fraction of the total state space volume. The algorithm was set to terminate according to Proposition 6 with $\zeta = 0.05$.

Figure 5b shows the projection of the reach set onto the Y_1, Y_2 dimensions. The curved black lines are the solutions of the continuous model starting at the vertices of the initial set. From comparison with these solutions, the reachable set correctly predicts that site 1 will never achieve a quorum of 10. The large reach set projection to the right of $Y_1 = 4$ resulted from defining some relatively large modes and from covering footprints with bounding boxes to reduce polyhedral complexity.

5.2 Inducibility in the *lac* Operon Control Network

The *lac* operon and its control network is an important example of bistability in a genetic network. We apply reachability analysis to a model of this system, due to Santillán and Mackey [5]. The system of equations is nonlinear and we replace it with a piecewise approximation. A diagram of the model network is given in Figure 6a. We follow closely the model by [5], referring the reader there

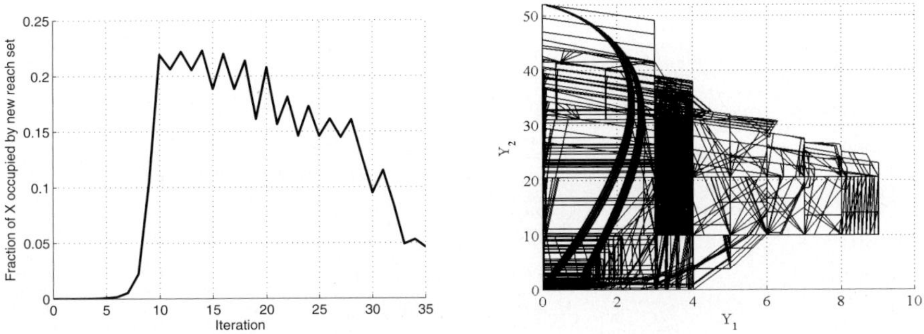

Fig. 5. (a) Increase in reachable set volume at each iteration divided by state space volume as a function of the number of iterations. (b) Projection of 4D reachable set for $k_1 = 0.0025$ [run time = 9251 sec].

for further details. The model variables $\{x_1, x_2, x_3, x_4, x_5, x_6\}$, are respectively the concentrations of: β-galactosidase mRNA, permease mRNA, β-galactosidase, permease, total allolactose, and total cAMP:

$$\begin{aligned}
\dot{x}_1 &= k_m x_4 \eta(x_6, x_5) - (\mu + \xi_M) x_1 \qquad \dot{x}_2 = k_m x_4 \eta(x_6, x_5) - (\mu + \xi_M) x_2 \\
\dot{x}_3 &= \frac{1}{4}\kappa_B x_1 - (\mu + \xi_B) x_3 \qquad \dot{x}_4 = \kappa_P x_2 - (\mu + \xi_P) x_4 \\
\dot{x}_5 &= \frac{1}{2}\phi_{L_1} \frac{L_e}{\Phi_{L_1} + L_e} \frac{\Phi_{G_1}}{\Phi_{G_1} + G_e} \frac{x_4 x_5}{x_5 + \Phi_{L_1}/2} - \frac{1}{2}\phi_{L_2} \frac{x_3 x_5}{x_5 + \Phi_{L_2}/2} \\
\dot{x}_6 &= \phi_C \frac{\Phi_C}{\Phi_C + G_e} - \xi_C \omega(x_6) - \mu x_6
\end{aligned} \tag{15}$$

The concentrations of external glucose G_e and lactose L_e are inputs to the system. The remaining symbols are constants taken from [5]. The system (15) is bistable for some combinations of the external inputs; the system has two equilibria, an induced state with high concentrations of β-galactosidase and high lactose metabolism, and an uninduced state with very little enzyme. One can cause an uninduced population to induce by exposing it temporarily to a high lactose/low glucose environment, steering the system trajectory around the bistable region. Thus, induction can be framed as a reachability problem: what is the region in the $L_e - G_e$ plane where high enzyme concentration states are reachable from an uninduced state?

In this example we use the methodology discussed in detail in [3]. Our procedure involves the piecewise multi-affine approximation of the equations of motion (15), including a two variable piecewise approximation of the function $\eta(\cdot, \cdot)$. These are preliminary results obtained with an implementation in gnu C, on a linux workstation with four Pentium Xeon processors. We perform our reachability calculations using a grid of 4105728 hyper-rectangles in the model space,

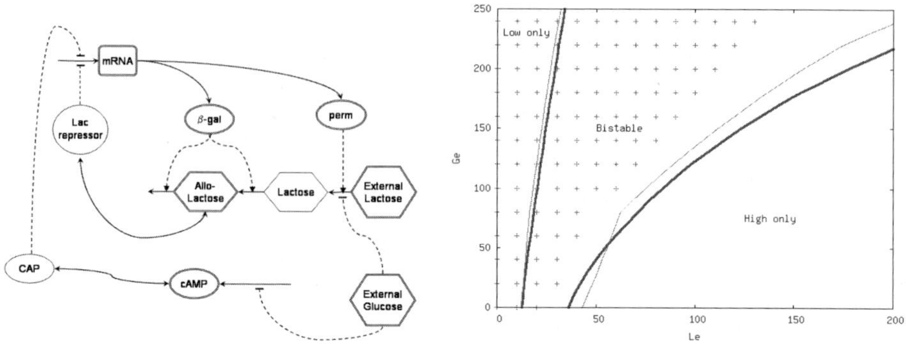

Fig. 6. (a) Diagram of the lactose-glucose network. External glucose and external lactose are external inputs. (b) Bistability region in the $L_e - G_e$ plane, in the exact model and the piecewise approximation. Superimposed are points where reachability forbids upward switching. All units are μM.

for various values of the external inputs. In each calculation we evaluate the set of all hyper-rectangles reachable from the mode with the lowest values of the concentrations of all substances. Figure 6b summarizes the reachability results. Points signify values for which induction is not possible according to reachability. The plot also shows the boundary of the bi-stable region. The non-inducible points are inside the bistability region, as expected intuitively.

6 Conclusion

Multi-affine hybrid systems arise naturally in biological systems. The main contribution of this paper is the development of MARCO, a reachability analysis algorithm that can be used to calculate reachable sets for biological systems. We have shown that MARCO overapproximates reachable sets and yet provides results that are quantitatively better than the MAR1 algorithm.

There are several directions for future work. First, it is necessary to use information on the volume of the time-elapsed cones and footprints to adaptively regulate the growth of reachable sets to improve efficiency and to ensure automatic termination of the algorithm. Second, we are working on many computational techniques to speed up the performance of the algorithm. Finally, our ultimate goal is to be able to use reachability analysis as a tool for bio-inspired synthesis of controllers for collective behaviors.

Acknowledgements. We gratefully acknowledge the support of NSF grants CCR02-05336 and IIS-0427313, and ARO Grants W911NF-05-1-0219 and W911NF-04-1-0148.

References

1. Belta, C., Habets, L., Kumar, V.: Control of multi-affine systems on rectangles with applications to hybrid biomolecular networks. In: 41st IEEE Conference on Decision and Control, Las Vegas, NV (2002)
2. Kloetzer, M., Belta, C.: Reachability analysis of multi-affine systems. In Hespanha, J.P., Tiwari, A., eds.: Proc. 9th Int'l Workshop on Hybrid Systems: Computation and Control (HSCC). Volume 3927 of LNCS. (2006) 348–362
3. Halász, A.M., Kumar, V., Imielinski, M., Belta, C., Sokolsky, O., Pathak, S., Rubin, H.: Analysis of lactose metabolism in E.coli using reachability analysis of hybrid systems. (Accepted to IET Systems Biology 2006; available at http://www.seas.upenn.edu/~spring)
4. Franks, N., Pratt, S., Mallon, E., Britton, N., Sumpter, D.: Information flow, opinion polling and collective intelligence in house-hunting social insects. Phil Trans Roy Soc London **B**(357) (2002) 1567–1584
5. Santillan, M., Mackey, M.C.: Influence of catabolite repression and inducer exclusion on the bistable behavior of the lac operon. Biophys. J. **86** (2004) 1282–1292
6. Sastry, S.: Nonlinear systems: analysis, stability, and control. Springer-Verlag, New York (1999)
7. Ptashne, M., Gann, A.: Genes and Signals. Cold Spring Harbor Laboratory Press, New York (2002)
8. Berman, S., Halász, A., Kumar, V., Pratt, S.: Algorithms for the analysis and synthesis of a bio-inspired swarm robotic system. In: Swarm Robotics SAB 2006 International Workshop, Rome, Italy (2006) to appear in LNCS.
9. Dang, T., Maler, O.: Reachability analysis via face lifting. In Henzinger, T., Sastry, S., eds.: Hybrid Systems : Computation and Control. Volume 1386 of Lecture Notes in Computer Science. Springer Verlag, Berlin (1998) 96–109
10. Henzinger, T., Ho, P., Wong-Toi, H.: HyTech: A model checker for hybrid systems. Software Tools for Technology Transfer **1**(1) (1998) 110–122
11. Silva, B., Richeson, K., Krogh, B., Chutinan, A.: Modeling and verifying hybrid dynamic systems using CheckMate. In: Proc. 4th Conference on Automation of Mixed Processes. (2000) 323–328
12. Lafferriere, G., Pappas, G., Yovine, S.: A new class of decidable hybrid systems. In: Hybrid Systems: Computation and Control (HSCC). Volume 1569 of LNCS. (1999) 137–151
13. Habets, L., van Schuppen, J.: A control problem for affine dynamical systems on a full-dimensional polytope. Automatica **40** (2004) 21–35
14. Frehse, G.: PHAVer: Algorithmic verification of hybrid systems past HyTech. In Morari, M., Thiele, L., eds.: Proc. 5th Int'l Workshop on Hybrid Systems: Computation and Control (HSCC). Volume 3414 of LNCS. (2005) 258–273
15. Girard, A., Pappas, G.: Approximate bisimulations for nonlinear dynamical systems. In: Proc. 44th IEEE Conf. on Decision and Control, Seville, Spain (2005)
16. van der Schaft, A.J., Schumacher, J.M.: An introduction to hybrid dynamical systems. Springer-Verlag, Berlin (2000)

Symbolic Analysis for GSMP Models with One Stateful Clock*

Mikhail Bernadsky and Rajeev Alur

Department of Computer and Information Science
University of Pennsylvania
{mbernads,alur}@cis.upenn.edu

Abstract. We consider the problem of verifying reachability properties of stochastic real-time systems modeled as generalized semi-Markov processes (GSMPs). The standard simulation-based techniques for GSMPs are not adequate for solving verification problems, and existing symbolic techniques either require memoryless distributions for firing times, or approximate the problem using discrete time or bounded horizon. In this paper, we present a symbolic solution for the case where firing times are random variables over a rich class of distributions, but only one event is allowed to retain its firing time when a discrete change occurs. The solution allows us to compute the probability that such a GSMP satisfies a property of the form "can the system reach a target, while staying within a set of safe states". We report on illustrative examples and their analysis using our procedure.

1 Introduction

Engineering of complex systems such as hardware devices, communication protocols, multimedia systems and networks requires accurate reliability modeling and performance evaluation at many stages of development [10,8]. For such systems, it is often the case that the event occurrence times, which determine the evolution of a system, interactions between components and between a system and its environment can be described by probabilistic assumptions. This observation has led to extensive research on *probabilistic model checking* of probabilistic and stochastic models. The goal of probabilistic model checking is to check algorithmically that a model of a system satisfies a probabilistic correctness property, for example, "every message is delivered within 1ms with probability 0.9."

Recently, results were obtained on model checking of discrete and continuous time Markov chains (DTMCs and CTMCs), with specifications written in temporal logics such as PCTL and CSL [5, 11]. While CTMCs can be used as building blocks to approximate distributions with unbounded support [7], approximation of distributions whose support is bounded, for instance, uniform or beta distributions, is problematic. It may be a serious restriction in modeling of real-time systems with mutually exclusive events. To circumvent this

* This research was supported by the US National Science Foundation via grants CCR-0410662 and CSR-EHS 0509143.

A. Bemporad, A. Bicchi, and G. Buttazzo (Eds.): HSCC 2007, LNCS 4416, pp. 90–103, 2007.

restriction, non-Markovian formalisms were proposed, but they either require that non-exponential distributions are deterministic [12], or that at any given moment there is at most one active event with a general distribution [6].

Our goal is to develop algorithms for the probabilistic model checking problem for systems modeled as *Generalized Semi-Markov Processes* (GSMPs) [9, 14, 8]. In our model of *finite-state* GSMPs, the system can be in one of the finitely many states, and can have a finite number of scheduled events. When the event with the least remaining firing time happens, the state is updated probabilistically, and new events can be scheduled at times chosen randomly according to the specified distributions. In [2], the authors show how to check *qualitative* probabilistic properties, that is, whether a GSMP satisfies a property with probability 0 or 1, and this analysis is based on the so-called region graph introduced for analysis of non-probabilistic real-time systems modeled using timed automata [3]. In a recent paper, we showed that if we are given a bound on the number of events, then exact symbolic analysis for verifying quantitative probabilistic properties of GSMPs is possible [1]. None of these techniques, however, suggest a general method for symbolic analysis of GSMPs.

In this paper, we present a symbolic analysis technique for the class of GSMPs where firing times are random variables over a rich class of distributions, but only one event is allowed to retain its firing time when a discrete change occurs. We call this class of processes 1GSMPs. In particular, we focus on model checking of *until* properties: given a set of destination and safe locations, we wish to compute the probability that an execution of the 1GSMP will reach a destination location while staying within the set of safe locations.

In our solution, we first derive a system of integral equations of harmonic functions such that each function is associated with a region of clock values in a particular location, and gives the probability of satisfying the until property as a function of the firing time of the stateful clock upon entry. This step can be easily generalized to work for all GSMPs.

In [13] the integral equations of similar structure were proposed for a related problem for semi-Markov processes. The authors cited [8], noting that solving these equations either by using numerical methods for integral equations or by applying Laplace transforms is not practical and works only for small models. In this paper, we describe a novel method that directly transforms the system of the integral equations into a system of ordinary differential equations. Each integral term in an integral equation is converted into a sum of differentials of newly introduced unknown functions (we call such functions 'auxiliary') and GSMP density functions. The original unknown functions and the auxiliary functions are linked by differential equations. The algorithm that constructs such equations uses the characterization of the density functions as solutions of linear homogeneous ordinary differential equations. The resulting system can then be solved to compute the desired probability for any given initial state.

We illustrate the proposed modeling and analysis techniques using classical examples of component failures and of a $GI/G/1$ queue [4].

2 Generalized Semi-Markov Processes

Let $\mathbb{N}$ be the set of all natural numbers, $\mathbb{N}_0$ be $\mathbb{N} \cup \{0\}$, $\mathbb{R}$ be the set of reals, and $\mathbb{R}_+$ be the set of all non-negative reals.

We start by reviewing some facts from the theory of differential equations and the probability theory. The solutions to the class of differential equations that we will describe form a class of expressions that we will use later to define a class of density functions, which, in turn, will be used in the definition of GSMP.

2.1 Preliminaries

Linear homogeneous ordinary differential equation with constant coefficients (LHODE) of order n is an equation of the form

$$a_n \frac{\mathrm{d}^n}{\mathrm{d}x^n} y(x) + a_{n-1} \frac{\mathrm{d}^{n-1}}{\mathrm{d}x^{n-1}} y(x) + \cdots + a_1 \frac{\mathrm{d}}{\mathrm{d}x} y(x) + a_0 y(x) = 0,$$

where $a_0, \ldots, a_n \in \mathbb{R}$.

Characteristic equation for this LHODE is $a_n\lambda^n + a_{n-1}\lambda^{n-1} + \cdots + a_1\lambda + a_0 = 0$. This equation has exactly n (complex, possibly repeated) roots and they determine, up to constants, all solutions of the LHODE.

Specifically, if there are $\lambda_1 = \cdots = \lambda_k$, $k \geq 1$ real repeated roots, then LHODE has a solution $y(x) = e^{\lambda_1 x}(c_1 + \cdots + c_k x^{k-1})$, where $c_1, \ldots, c_k \in \mathbb{R}$ are arbitrary constants. If $\lambda_1 = \cdots = \lambda_k$ are complex repeated roots equal to $\alpha + \beta i$, $\alpha, \beta \in \mathbb{R}$ and $\beta \neq 0$, then the equation should also have k repeated conjugate roots $\bar{\lambda}_1 = \cdots = \bar{\lambda}_k$ equal to $\alpha - \beta i$. All these $2k$ roots correspond to a solution $y(x) = e^{\alpha x}(c_1 + \cdots + c_k x^{k-1}) \sin \beta x + e^{\alpha x}(d_1 + \cdots + d_k x^{k-1}) \cos \beta x$, where $c_1, \ldots, c_k, d_1, \ldots, d_k \in \mathbb{R}$ are arbitrary constants. Summing solutions for such groups of roots we obtain the general solution for the LHODE.

We say that $expr(x)$ is a *DESOL expression* iff it is a sum of terms, such that each term is either of the form $cx^m e^{\mu x} \sin(\alpha x)$ or of the form $cx^m e^{\mu x} \cos(\alpha x)$, where $c, \mu, \alpha \in \mathbb{R}$ and $m \in \mathbb{N}_0$. For every DESOL expression it is possible to construct an LHODE that has this expression as its solution. This 'encoding' of the DESOL expressions with LHODEs will be used later in Section 4 to convert a system of integral equations into a system of differential equations.

Let $Expr(x)$ be the set of all DESOL expressions. Consider a partition $R_a = \cup_{i=1}^{a}\{(i-1, i]\}$ of $(0, a]$, which consists of a unit intervals. The constant a is the *width* of R_a. Let $Int(x)$ be a function defined for all $x \in (0, a]$, such that if $x \in (i-1, i]$, then $Int(x) = i$. We say that a function $f(x)$ is a *piecewise DESOL function*, with finite support on R_a, if there exists a map $M_f \colon \{1, \ldots, a\} \to Expr(x)$, such that for all $x \in (0, a]$, $f(x) = M_f(Int(x))(x')$, where $x' = x - Int(x) + 1$. Thus to compute $f(x)$, we determine the interval of R_a to which x belongs, find DESOL expression for this interval and then evaluate this expression at x'. Notice, that $x' \in (0, 1]$, so every expression is evaluated only in that interval. This leads to simplifications in our algorithms. By $f^j(x)$, $1 \leq j \leq a$ we will denote the expression of $f(x)$ that corresponds to the interval $(j-1, j]$.

In a GSMP the time between scheduling an event and its occurrence (or firing time) is modeled as a positive random variable. A random variable X is characterized by its *cumulative distribution function* (cdf) $distr(x) = \Pr(X < x)$, and if $distr(x)$ is continuous then also by a *probability density function* (pdf) $dens(x)$, defined by the equation $distr(x) = \int_0^x dens(y)\,dy$.

An unidimensional random variable X has a *DESOL distribution* of width a, if there exists a piecewise DESOL function $dens(x) \geq 0$ on R_a, such that for all $t \in \mathbb{R}_+$, $\Pr(X < t) = \int_0^t dens(y)\,\mathrm{d}y$ [1].

2.2 Modeling Stochastic Processes

A finite-state generalized semi-Markov process (GSMP) with the firing distributions of width a is a tuple $G = (Q, \Sigma, E, init, distr, next)$ where:

- Q is a finite set of locations;
- Σ is a finite set of events;
- $E\colon Q \to 2^{\Sigma}$ assigns to each location $q \in Q$ a set of events that are *active* in q. A location q is *absorbing* iff $E(q) = \emptyset$;
- $init\colon Q \to [0,1]$ is a probability measure on Q, which for each location $q \in Q$ gives the probability that q is the initial location of G;
- $distr\colon \Sigma \to (\mathbb{R}_+ \to [0,1])$ assigns to each event its *firing time distribution*, which is a DESOL distribution of width a. For a cdf $distr(e)$, $dens_e$ denotes the corresponding pdf.
- $next\colon Q \times \Sigma \to 2^{\Sigma} \times (Q \to [0,1])$ defines transitions between locations of G. This function takes as its arguments a source location q and an active event e of q and returns a set of events $E^{q,e}_{\text{reset}}$ and a probability measure $P^{q,e}_{\text{next}}$ on Q. For each location q', $P^{q,e}_{\text{next}}(q')$ gives the probability that a run of G will move from q to q' if e fires, and $E^{q,e}_{\text{reset}}$ is the set of events that are reset when the transition occurs. We require that $\sum_{q' \in Q} P^{q,e}_{\text{next}}(q') = 1$, $E^{q,e}_{\text{reset}} \subseteq E(q)$, and $E^{q,e}_{\text{reset}} \subseteq E(q'')$, for every location q'', such that $P^{q,e}_{\text{next}}(q'') > 0$ [2].

Notice that since we use random variables with density functions that do not have mass points and are discontinuous only at a finite number of points, we do not consider the possibility that several events would fire at the same time.

It is convenient to think that a clock is assigned to each event e. The clock, denoted t_e, shows the time remaining until the next occurrence of e. Upon scheduling/resetting of e we update its clock to a new value chosen independently at random according to $distr(e)$. All clocks of the current active events run down with the same rate equal to 1.

Let us say that $\nu\colon \Sigma \to \mathbb{R}_+$ is a clock valuation (or simply valuation) if ν maps events to the values of their clocks. If an event is not active in the current location we assume that its value is undefined.

[1] Assume that $dens(y)$ is 0 for $y \in (a, +\infty)$.

[2] Adding a possibility that some events can be reset in a transition from one location to another does not make our model more powerful, but it is useful for modeling, and we add it as a "syntactic sugar".

A *configuration* of the GSMP G is a pair (q, ν), where $q \in Q$ and ν is a clock valuation. Given a configuration $s = (q, \nu)$, let $t^*(s) = \min\{\nu(e), e \in E(q)\}$ be the time until the next transition, and $e^*(s) = \arg\min(\nu(e), e \in E(q))$ be the event that causes the transition. For any $t \leq t^*(s)$ we denote by $\nu - t$ the valuation ν' such that for all $e \in E(q)$, $\nu'(e) = \nu(e) - t$. We say that $s \xrightarrow{t} s'$ is a *timed transition* between configurations $s = (q, \nu)$ and $s' = (q, \nu')$ if $\nu' = \nu - t$. If $t^*(s) = 0$, and e^* is such that $\nu(e^*) = 0$, then $s \xrightarrow{\mu} s'$ denotes a *discrete transition* between configurations $s = (q, \nu)$ and $s' = (q', \nu')$, where q' is chosen according to the probability measure $\mu = P_{\text{next}}^{q,e^*}$, and the valuation ν' is constructed as follows:

- if an event $e \in E_{\text{inherited}}(q, e^*, q')$, where $E_{\text{inherited}}(q, e^*, q') = E(q') \cap [E(q) \setminus \{e^*\} \setminus E_{\text{reset}}]$ is the set of events that were active in q and continue to be active in q', excluding e^* and excluding the events that were reset, then $\nu'(e) = \nu(e)$;
- if $e \in E_{\text{new}}(q, e^*, q')$, where $E_{\text{new}}(q, e^*, q') = E(q') \setminus E_{\text{inherited}}(q, e^*, q')$, then $\nu'(e)$ is chosen independently at random according to *distr*(e) (i.e. the events in $E_{\text{new}}(q, E^*, q')$ are (re-)scheduled);
- if $e \in E_{\text{cancelled}}(q, e^*, q')$, where $E_{\text{cancelled}}(q, e^*, q') = E(q) \setminus E(q')$ is the set of canceled events that were active in q but no longer active in q', then $\nu'(e)$ is undefined.

A *run* σ of G is a sequence of alternating timed and discrete transitions:

$$\sigma = s_0 \xrightarrow{t^*(s_0)} s_0' \xrightarrow{\mu_0} s_1 \xrightarrow{t^*(s_1)} s_1' \xrightarrow{\mu_1} s_2 \xrightarrow{t^*(s_2)} s_2' \xrightarrow{\mu_2} \ldots$$

The run σ starts at the initial configuration $s_0 = (q_0, \nu_0)$, q_0 is the initial location, which was chosen according to *init*, and ν_0 is the initial valuation of the events in $E(q_0)$, scheduled according to the corresponding firing time distributions. A run can have a finite or infinite number of transitions; a run that has reached an absorbing location will stay in that location forever.

We say that a GSMP is *normalized* iff (i) for every q_{pred}, e^* and q, such that $P_{\text{next}}^{q_{\text{pred}},e^*}(q) > 0$, $E_{\text{inherited}}$ and E_{new} do not depend on q_{pred} and e^*, i.e. $E_{\text{inherited}}(q_{\text{pred}}, e^*, q) = E_{\text{inherited}}(q)$ and $E_{\text{new}}(q_{\text{pred}}, e^*, q) = E_{\text{new}}(q)$. And (ii) if q is an initial location, i.e. $init(q) > 0$, then $E_{\text{inherited}}(q_{\text{pred}}, e^*, q) = \emptyset$.

Condition (i) states that the partition of a location's events into sets of inherited and new events is the same for every visit to that location. Condition (ii) requires that the set of inherited events of every initial location be empty.

We modify the definition of GSMP, and say that a normalized GSMP is a tuple $G_{\text{norm}} = (Q, \Sigma, E_{\text{inherited}}, E_{\text{new}}, init, distr, next)$ to emphasize that the partition into sets of inherited and new events is fixed for every location.

Given a GSMP G, we can determine for every location all possible event partitions and then for every found partition, by creating a clone of that location and connecting it to the corresponding (clones of) predecessor and successor locations, create a normalized GSMP G_{norm}.

G and G_{norm} are equivalent in the sense that the answers to the questions that we consider in this paper are the same for G and G_{norm}.

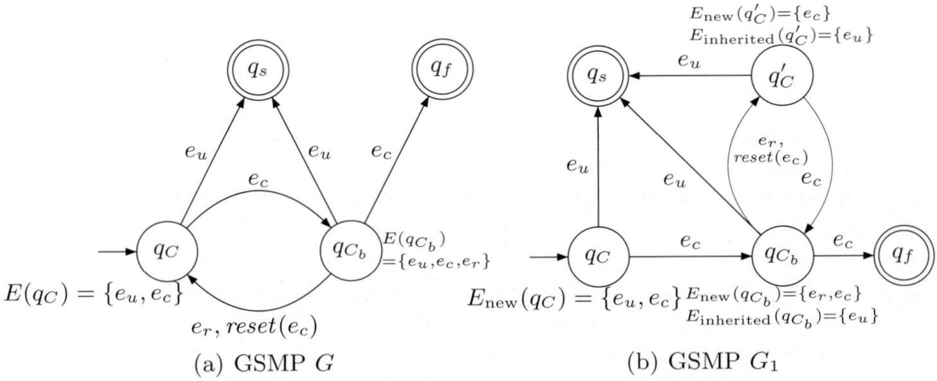

Fig. 1. Sample GSMP and its normalized version

A normalized GSMP is called *1GSMP* if for every location q, $E_{\text{inherited}}(q)$ consists of at most one event.

In 1GSMP at most one event can retain the value of its clock upon a transition. The other active events are either reset or canceled in the transition target location. In the sequel, we will be interested only in 1GSMPs.

For a clock of an active event, we say that it is *new* if it is the clock of a new event, and we call it *inherited* otherwise. We know the distributions of values of all new clocks upon reaching the current location q, but the distribution of the inherited clock is unknown and depends on the path to q.

A *history* π of length n of a run σ is a sequence of locations and transitions between them marked by the events that have fired:

$$q_0 \xrightarrow{e_1^*} q_1 \xrightarrow{e_2^*} \dots \xrightarrow{e_n^*} q_n.$$

2.3 Computing Probabilities of Until Properties

Suppose that we are given a 1GSMP G with firing time distributions of size a. The locations of G are partitioned into three disjoint sets: Q_s, Q_u and Q_d, which are called the set of *safe locations*, the set of *unsafe locations*, and the set of *destination locations*, respectively. We require that from every $q_s \in Q_s$, a location in $Q_u \cup Q_d$ is reachable with probability one. This property can be checked by a method presented in [2].

Let $\Pi^n_{\text{until}} \subseteq \Pi$ be a set of histories of length less than or equal to n, such that only the last location of every history in Π^n_{until} is in Q_u, while the other locations are in Q_s. Let $\Pi_{\text{until}} = \cup_{n \geq 0} \Pi^n_{\text{until}}$.

We consider the following *model-checking problem*:

– *What is the probability P_{until} that a run σ of G has a history $\pi \in \Pi_{\text{until}}$?*

In addition to solving this problem, our approach will enable us to determine the probability of reaching an unsafe location before any of the destination locations from any location q, if we specify the valuation of the event in $E_{\text{inherited}}(q)$.

2.4 Illustrative Example

We illustrate the given definitions by an example that we will use throughout the paper. Consider a system that crucially depends on a component C. To ensure an uninterrupted service, the system has a back-up component C_b. While C is active, C_b is in the standby mode, but if C fails, C_b is activated immediately. We are interested to know the probability of the system failure.

The system can be modeled as a GSMP G depicted in Figure 1(a). Each location is marked with its active events. Each transition is marked with the event that causes this transition, and $reset(e)$ indicates that e was reset.

Location q_C is the initial location, and it models the configuration in which C is operating and C_b is in the standby mode. Location q_{C_b} represents the configuration when C has failed and C_b is active. There are two absorbing locations q_s and q_f, the former is the destination location, it models the successful completion of the task, the latter is the unsafe location of G, and it models the state of the system when both components have failed.

Two events e_u and e_c are active in q_C. The event e_u, which is scheduled only upon the initial visit to q_C, models the time interval the system should be up. Firing of this event indicates that the system has completed its task successfully and has reached q_s; e_u is scheduled using a random variable whose density function is $dens_{e_u}(x) = \frac{1}{1-e^{-1}}e^{-x}$ on $(0,1]$ and 0 otherwise[3]. On $(0,1]$ this density function is a solution of the differential equation $\frac{\mathrm{d}}{\mathrm{d}x}dens_{e_u}(x) + dens_{e_u}(x) = 0$. The second active event of q_C is e_c, it is scheduled every time a run reaches q_C, and it models a crash event of C. If it fires a run of G moves to q_{C_b}. The event e_c is scheduled using a random variable with the beta density function $dens_{e_c}(x) = \frac{1}{2}x$ on $(0,1]$, the LHODE for $dens_{e_c}(x)$ is $\frac{\mathrm{d}^2}{\mathrm{d}x^2}dens_{e_c}(x) = 0$.

Every time location q_{C_b} is reached, two events e_r and e_c are scheduled. The firing time of e_r is determined by a random variable with the uniform density function $dens_{e_r}(x) = 1$ on $(0,1]$ and 0 otherwise (its LHODE is $\frac{\mathrm{d}}{\mathrm{d}x}dens_{e_r}(x) = 0$), and firing of this event indicates that C was replaced and the run returns to q_C. But if in q_{C_b} the event e_c fires before e_r, it means that C_b had failed before C was replaced, the system failed and the run moves to the location q_f.

G is not a normalized GSMP — upon the first visit to q_C both e_u and e_c are scheduled, upon every subsequent visit e_c is reset, but the clock of e_u keeps its value. The normalized GSMP G_1, constructed from G, is shown in Figure 1(b). Location q_C of G was split into two locations q_C and q'_C, and, as in G, q_C is the only initial location of 1GSMP G_1.

---

[3] The densities in this example are chosen to make our approach clear and not to model accurately a real system.

3 System of Integral Equations for Harmonic Functions

To solve the model-checking problem we will use a method similar to the "first step analysis" for the Markov chains. For every location we introduce a set of harmonic functions. If a location q does not have an inherited clock, then its set consists of one constant function that we denote by H_q, otherwise it consists of a functions $H_q^1(t), \dots, H_q^a(t)$.

Each $H_q^i(t)$ is defined on $(0, 1]$. The interpretation of the functions is the following — if q is reached at the moment when the value of the inherited clock t satisfies $i - 1 < t \leq i$, then the probability to reach an unsafe location before any of the destination locations is $H_q^i(t - (i-1))$ [4]. If there is no inherited clock in q, then all clocks are rescheduled (or reset) upon reaching q and the constant H_q is the desired probability.

Our method constructs integral equations that capture dependencies between harmonic functions of a location and all its immediate successor locations. This is a general method and it works not only for 1GSMPs but also for all normalized GSMPs, however restriction to 1GSMPs allows us a transformation from the system of constructed integral equations to a system of ordinary differential equations.

Before describing our algorithm, we need to introduce a class of partitions of clock valuations that we will use. The same class of partitions was used in [1] for solving bounded model-checking problem.

3.1 Diagonal Mesh Partitions

For a set of variables $t_1, \dots, t_n$, an n-dimensional *diagonal mesh partition* $R_a(t_1, \dots, t_n)$ of width $a \in \mathbb{N}$ is a partition of $\mathbb{R}_+^n$ into regions such that each region is described by:

- *mesh constraints:* for each variable t, by a constraint of the form $b-1 < t \leq b$, where $b \in \mathbb{N}$ and $1 \leq b \leq a$;
- *diagonal constraints:* for every pair of distinct variables t and t', by an ordering on the fractional parts of the variables, i.e. by a constraint of the form $(t - \lfloor t \rfloor) \sim (t' - \lfloor t' \rfloor)$, where $\sim \in \{<, >\}$, and $\lfloor s \rfloor$ denotes the largest integer not greater than s.

For a region r and a variable t, let $Int_r(t)$ be the function that returns the mesh constraint constant of t in r.

Given a region r we consider a (total) *region ordering* $\prec_r$ of fractional parts of $t_1, \dots, t_n$, i.e. $t_i \prec_r t_j$ iff $(t_i - \lfloor t_i \rfloor) < (t_j - \lfloor t_j \rfloor)$. Thus, each region r in R_a can be described by the order $\prec_r$ and the unit intervals of all variables.

4 The reason for all harmonic functions to be defined on the same interval $(0, 1]$ will become clear later when we present the algorithm that constructs integral equations.

3.2 Algorithm

Suppose we are given a 1GSMP $G = (Q, \Sigma, E_{\text{inherited}}, E_{\text{new}}, init, distr, next)$, Q_d is the set of destination locations, and Q_u is the set of unsafe locations. We assume that all locations in $Q_d \cup Q_u$ are absorbing. We describe the algorithm in two steps. We start by discussing the main loop, and then we describe construction of right-hand sides of the integral equations.

In the main loop, the algorithm goes over all locations of G. For each destination location q_d it outputs equation $H_{q_d} = 0$, which states that in a destination location, the probability to reach an unsafe location before any of destination locations is zero. For each unsafe location q_u, the algorithm outputs $H_{q_u} = 1$. If q is neither a destination nor an unsafe location, then, in case q has an inherited event $e_q^{\text{inherited}}$, the algorithm constructs a equations for each of the functions $H_q^1, \ldots, H_q^a$ of the same argument $t_{e_q^{\text{inherited}}}$. In case q does not have an inherited event, a single equation for H_q is constructed.

Algorithm 1 returns the right-hand side for the equation that defines the harmonic function $H_q^i(t_{e_q^{\text{inhereted}}})$, where $e_q^{\text{inhereted}}$ is the only inherited event of q (the algorithm for H_q, such that q has only new events is similar).

Let us assume that the number of active events in q is n. For every non-constant function $H_p^j(t)$ let $\tilde{H}_p^j(t) = H_p^j(1-t)$.

At line 2 we have the loop that goes through all regions in the diagonal mesh partition $R_a(t_{e_1}, \ldots, t_{e_n})$ for which $t_{e_q^{\text{inhereted}}} \in (i-1, i]$. The restriction is required because we are constructing RHS for $H_q^i(t_{e_q^{\text{inhereted}}})$, which implies that $t_{e_q^{\text{inhereted}}} \in (i-1, i]$. At line 3 we determine the clock that should fire, i.e. the clock which is minimal in respect to $\prec_r$ among all clocks that have the minimal value returned by Int_r. At line 4 we ensure that we consider every transition that may be caused by firing of e_r^* along with its probability. At lines 8 – 10 we create a product of all new clock densities, each enters with its own variable. At line 11 we check if the target location has an inherited clock. If it does then at line 15 or 18 we determine to which interval this inherited clock belongs. This is uniquely determined by the region r. At line 22 we construct the integrand.

From line 24 to line 36 we have the loop that goes over all active event clocks of q. For each new clock we integrate over all possible values that this clock can have in r. The limits of integration are constructed in such a way that they respect $\prec_r$. For example, suppose that in the ordering $t_{e_1} \prec_r t_{e_2} \prec_r t_{e_3} \prec_r t_{e_4} \prec_r t_{e_5} \prec_r t_{e_6} \prec_r \cdots \prec_r t_{e_n}$, t_{e_3} is the inherited clock of q and the others are new clocks. Then we know that t_{e_1} can have values between 0 and t_{e_2}, t_{e_2} between 0 and t_{e_3}, t_{e_4} between t_{e_3} and t_{e_5}, t_{e_5} between t_{e_3} and t_{e_6} and so on. The last clock t_{e_n} can have values between t_{e_3} and 1. This idea is captured in the algorithm.

At line 37 we add the constructed integral to RHS, and at line 40 we output the entire constructed expression.

Let us return to our sample GSMP G_1. For q_{C_b}, for example, the algorithm generates the following integral equation:

Algorithm 1. Generate RHS(q, i)

$RHS \leftarrow 0$
for all $r\colon (r \in R_a(t_{e_1}, \ldots, t_{e_n}), e_i \in E(q)) \wedge Int_r(t_{e_q^{\text{inherited}}}) = i$ **do**
  $e_r^* \leftarrow$ the firing clock of r
  for all $tran \in next(q, e_r^*)$ **do**
    $q_{\text{target}} \leftarrow$ the target location of the transition $tran$
    $Prob \leftarrow$ the probability of the transition $tran$, which is $P_{\text{next}}^{q,e_r^*}(q_{\text{target}})$
    $DensProduct \leftarrow 1$
    for all $e \in E_{\text{new}}(q)$ **do**
      $DensProduct \leftarrow DensProduct * dens_e^{Int_r(t_e)}(t_e)$
    end for
    if $E_{\text{inherited}}(q_{\text{target}}) = \emptyset$ **then**
      $HarmonicFunction \leftarrow H_{q_{\text{target}}}$
    else
      if $t_{e_r^*} \prec_r t_{e_{q_{\text{target}}}^{\text{inherited}}}$ **then**
        $index \leftarrow Int_r(t_{e_{q_{\text{target}}}^{\text{inherited}}}) - Int_r(t_{e_r^*}) + 1$
        $HarmonicFunction \leftarrow H_{q_{\text{target}}}^{index}(t_{e_{q_{\text{target}}}^{\text{inherited}}} - t_{e_r^*})$
      else
        $index \leftarrow Int_r(t_{e_{q_{\text{target}}}^{\text{inherited}}}) - Int_r(t_{e_r^*})$
        $HarmonicFunction \leftarrow \tilde{H}_{q_{\text{target}}}^{index}(t_{e_r^*} - t_{e_{q_{\text{target}}}^{\text{inherited}}})$
      end if
    end if
    $Integrand = HarmonicFunction * DensProduct$
    $LowerLimit \leftarrow 0$
    for $i = 1$ to n **do**
      $e_{\text{cur}} \leftarrow e_i$, where t_{e_i} is the ith element in $t_{e_1} \prec_r t_{e_2} \prec_r \cdots \prec_r t_{e_n}$
      if $e_{cur} \in E_{\text{inherited}}(q)$ **then**
        $LowerLimit \leftarrow t_{e_{\text{cur}}}$
      else
        if $i < n$ **then**
          $UpperLimit \leftarrow t_{e_{i+1}}$, where $t_{e_{i+1}}$ is the $(i+1)^{\text{th}}$ element in $t_{e_1} \prec_r t_{e_2} \prec_r \cdots \prec_r t_{e_n}$
        else
          $UpperLimit \leftarrow 1$
        end if
        $Integrand = \int_{LowerLimit}^{UpperLimit} Integrand \, \mathrm{d}t_{e_{\text{cur}}}$
      end if
    end for
    $RHS \leftarrow RHS + Prob * Integrand$
  end for
end for
return RHS

$$H^1_{q_{C_b}}(t_{e_u}) = \int_{t_{e_u}}^{1} \int_{0}^{t_{e_u}} dens_{e_c}(t_{e_c})\, dens_{e_r}(t_{e_r})\, \mathrm{d}t_{e_c}\, \mathrm{d}t_{e_r}$$
$$+ \int_{0}^{t_{e_u}} \int_{0}^{t_{e_r}} dens_{e_c}(t_{e_c})\, dens_{e_r}(t_{e_r})\, \mathrm{d}t_{e_c}\, \mathrm{d}t_{e_r}$$
$$+ \int_{t_{e_u}}^{1} \int_{0}^{t_{e_u}} H^1_{q'_C}(t_{e_u} - t_{e_r})\, dens_{e_c}(t_{e_c})\, dens_{e_r}(t_{e_r})\, \mathrm{d}t_{e_r}\, \mathrm{d}t_{e_c}$$
$$+ \int_{0}^{t_{e_u}} \int_{0}^{t_{e_c}} H^1_{q'_C}(t_{e_u} - t_{e_r})\, dens_{e_c}(t_{e_c})\, dens_{e_r}(t_{e_r})\, \mathrm{d}t_{e_r}\, \mathrm{d}t_{e_c}$$

The first term is for the region r_1 with the ordering $t_{e_c} \prec_{r_1} t_{e_u} \prec_{r_1} t_{e_r}$, the second term is for r_2, $t_{e_c} \prec_{r_2} t_{e_r} \prec_{r_2} t_{e_u}$, the third term is for r_3, $t_{e_r} \prec_{r_3} t_{e_u} \prec_{r_3} t_{e_c}$, and the last term is for r_4, $t_{e_r} \prec_{r_4} t_{e_c} \prec_{r_4} t_{e_u}$. The first two terms do not include harmonic functions because $H_{q_f} = 1$. There are no terms for the orderings that start with t_{e_u} because $H_{q_s} = 0$.

4 System of Differential Equations

In this section we show how to convert the system of integral equations constructed in the previous section into a system of differential equations such that the solution of the former system is a solution of the latter. Compared to the existing methods (e.g. [15]), our method converts every integral term individually and can handle equations with terms that contain multiple integrals.

Due to the lack of space, we give intuition only for the main step. The other steps use the fact that the class of DESOL expressions is closed under addition, multiplication, integration and differentiation [16]. The resulting system will consist of equations obtained after converting the integral equations, equations that define auxiliary functions and additional equations described in Section 4.1.

We show how to convert, for example, a term $T = \int_a^b H(t - t')dens(t)\,\mathrm{d}t$, where the integration limits a and b can be 0, 1 or clock variables, $H(t - t')$ is a harmonic function and $dens(t)$ is a function, which is a solution of a LHODE:

$$\sum_{l=1}^{n} a_l \frac{\mathrm{d}^l dens(t)}{\mathrm{d}t^l} + a_0 dens(t) = 0, \tag{1}$$

For $H(t - t')$, we introduce an auxiliary function $A(t - t')$. The differential equation that links $A(t - t')$ to $H(t - t')$ is constructed from (1):

$$H(t - t') = \sum_{l=1}^{n} (-1)^l a_l \frac{\mathrm{d}^l A(t - t')}{\mathrm{d}t^l} + a_0 A(t - t').$$

Replacing $H(t - t')$ in T with this equation, we will obtain that

$$T = \sum_{l=1}^{n} (-1)^l a_l \int_a^b \frac{\mathrm{d}^l A(t - t')}{\mathrm{d}t^l} dens(t)\,\mathrm{d}t + a_0 \int_a^b A(t - t')dens(t)\,\mathrm{d}t. \tag{2}$$

Consider the summand T_n for $l = n$ in the equation above. Let us apply the integration by parts method to it:

$$T_n = (-1)^n a_n \int_a^b \frac{\mathrm{d}^n A(t-t')}{\mathrm{d}t^n} dens(t)\,\mathrm{d}t = (-1)^n a_n \int_a^b dens(t)\,\mathrm{d}\left(\frac{\mathrm{d}^{n-1} A(t-t')}{\mathrm{d}t^{n-1}}\right)$$

$$= (-1)^n a_n \left(dens(t) \left.\frac{\mathrm{d}^{n-1} A(t-t')}{\mathrm{d}t^{n-1}}\right|_{t=a}^{t=b} - \int_a^b \frac{\mathrm{d}^{n-1} A(t-t')}{\mathrm{d}t^{n-1}} \frac{\mathrm{d}\,dens(t)}{\mathrm{d}t}\,\mathrm{d}t \right).$$

Next we apply the integration by parts method $n-1$ more times:

$$\begin{aligned} T_n &= (-1)^n a_n \left(dens(b) \left.\frac{\mathrm{d}^{n-1} A(t-t')}{\mathrm{d}t^{n-1}}\right|_{t=b} - dens(a) \left.\frac{\mathrm{d}^{n-1} A(t-t')}{\mathrm{d}t^{n-1}}\right|_{t=a} \right) \\ &+ (-1)^n a_n \sum_{k=2}^{n-1} (-1)^{n-k} \left(\left.\frac{\mathrm{d}^{n-k} dens(t)}{\mathrm{d}t^{n-k}} \frac{\mathrm{d}^{k-1} A(t-t')}{\mathrm{d}t^{k-1}}\right|_{t=a}^{t=b} \right) \\ &- a_n \left(\left.\frac{\mathrm{d}^{n-1} dens(t)}{\mathrm{d}t^{n-1}}\right|_{t=b} A(b-t') - \left.\frac{\mathrm{d}^{n-1} dens(t)}{\mathrm{d}t^{n-1}}\right|_{t=a} A(a-t') \right) \\ &+ a_n \int_a^b A(t-t') \frac{\mathrm{d}^n dens(t)}{\mathrm{d}t^n}\,\mathrm{d}t = TD_n + a_n \int_a^b A(t-t') \frac{\mathrm{d}^n dens(t)}{\mathrm{d}t^n}\,\mathrm{d}t, \end{aligned} \tag{3}$$

where TD_n is a sum of differentials. Now we apply the same conversion to the remaining summands of (2). Consider the sum of all converted summands:

$$\begin{aligned} T &= \sum_{l=1}^{n} TD_l + \sum_{l=1}^{n} a_l \int_a^b A(t-t') \frac{\mathrm{d}^l dens(t)}{\mathrm{d}t^l}\,\mathrm{d}t + a_0 \int_a^b A(t-t') dens(t)\,\mathrm{d}t \\ &= \sum_{l=1}^{n} TD_l + \int_a^b A(t-t') \left(\sum_{l=1}^{n} a_l \frac{\mathrm{d}^l dens(t)}{\mathrm{d}t^l} + a_0 dens(t) \right) \mathrm{d}t = \sum_{l=1}^{n} TD_l. \end{aligned}$$

The last step follows from (1).

Let us return to our example. Consider the integral equation for q'_C:

$$H^1_{q'_C}(t_{e_u}) = \int_0^{t_{e_u}} H^1_{q_{C_b}}(t_{e_u} - t_{e_c})\, dens_{e_c}(t_{e_c})\,\mathrm{d}t_{e_c}.$$

Conversion gives $H^1_{q'_C}(t_{e_u}) = C_1\, dens_{e_c}(t_{e_u}) - C_2 \frac{\mathrm{d}}{\mathrm{d}t} dens_{e_c}(t) + 2\,A(t_{e_u})$, where $C_1 = \frac{\mathrm{d}}{\mathrm{d}t} A(0)$ and $C_2 = A(0)$ are constants that we can choose to be zero, and $H^1_{q_{C_b}}(t) = \frac{\mathrm{d}^2}{\mathrm{d}t^2} A(t)$.

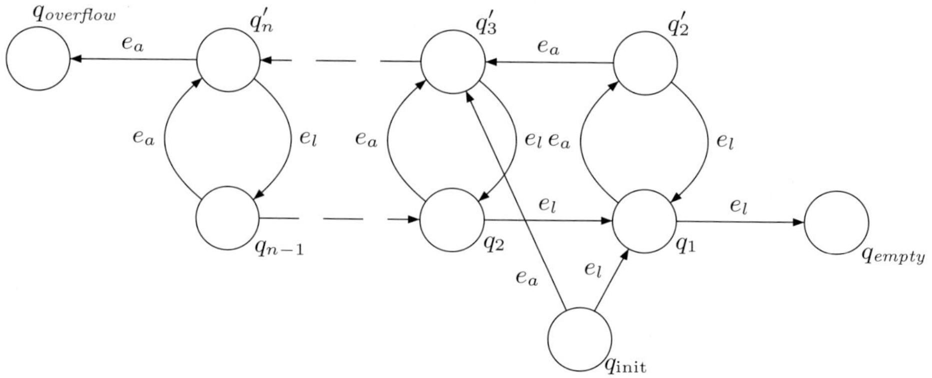

Fig. 2. Queue

4.1 Additional Equations

Consider a product $\left.\frac{\mathrm{d}^{n-k} dens(t)}{\mathrm{d}t^{n-k}} \frac{\mathrm{d}^{k-1} A(t-t')}{\mathrm{d}t^{k-1}}\right|_{t=b}$ from (3), and let us assume that $b = 1$, then using the calculus chain rule:

$$\left.\frac{\mathrm{d}^{n-k} dens(t)}{\mathrm{d}t^{n-k}} \frac{\mathrm{d}^{k-1} A(t-t')}{\mathrm{d}t^{k-1}}\right|_{t=1} = (-1)^{k-1} \left.\frac{\mathrm{d}^{n-k} dens(t)}{\mathrm{d}t^{n-k}} \frac{\mathrm{d}^{k-1} A(t-t')}{\mathrm{d}t'^{k-1}}\right|_{t=1}$$

$$= (-1)^{k-1} C_{dens} \frac{\mathrm{d}^{k-1} A(1-t')}{\mathrm{d}t'^{k-1}} = (-1)^{k-1} C_{dens} \frac{\mathrm{d}^{k-1} \tilde{A}(t')}{\mathrm{d}t'^{k-1}},$$

where $C_{dens} = \left.\frac{\mathrm{d}^{n-k} dens(t)}{\mathrm{d}t^{n-k}}\right|_{t=1}$ and $\tilde{A}(t) = A(1-t)$.

Recall that our system may also include unknowns of the form $\tilde{H}(t) = H(1-t)$. To ensure that the system can be solved, we should add additional equations for $\tilde{H}(t)$ and $\tilde{A}(t)$. For $\tilde{H}(t)$ we take the equation with $H(t)$ on its left-hand side and do a change of variable from t to $t' = 1 - t$. For $\tilde{A}(t)$, we do the same for the equation that defines $A(t)$. We add the equations for all such functions to our system and then solve it.

5 Implementation

To demonstrate our tool, we consider an example motivated by research on power-aware devices. Suppose that a device processes requests. Unprocessed requests can be stored in a queue of a finite length n. To save power it is preferable to accumulate in the queue as many requests as possible and then process them in one batch until the queue is empty. We know the distribution of time intervals between two successive requests, and the distribution of time to process a request. These distributions are not exponential, so we are dealing with a $GI/G/1$ queue.

Suppose that we are given the number of requests k in the queue at the moment when the device starts batch processing. We want to know what is the probability that the queue overflows before it gets empty.

The 1GSMP for our example (when k is set to 2) is shown in Figure 2. A numeric index of a location is also the number of requests in the queue while a run is in that location. The firings of events e_a and e_l indicate a new request arrival and request completion, respectively. The density for e_a is t_{e_a} on $(0, 1]$ and $t_{e_a} - 1$ on $(1, 2]$. The density for e_l is 2/3 on $(0, 1]$ and 1/3 on $(1, 2]$.

Experiments were conducted on a Windows XP computer with a Pentium D processor running at 2.80 GHz with 2 GB of RAM.

Parameters		**Results**	
n	k	P_{overflow}	Running time
16	1	7.5401361×10^{-8}	5 min. 4 sec.
16	8	0.00010769	5 min. 2 sec.
16	16	0.53083234	5 min. 9 sec.
32	1	$7.4714055 \times 10^{-16}$	47 min. 56 sec.
32	16	1.8367065×10^{-8}	41 min. 38 sec.
32	32	0.53083236	42 min. 46 sec.

References

1. R. Alur and M. Bernadsky. Bounded model checking for GSMP models of stochastic real-time systems. In *Proc. of HSCC'06*, LNCS 3927, pages 19–33. Springer, 2006.
2. R. Alur, C. Courcoubetis, and D.L. Dill. Model-checking for probabilistic real-time systems. In *Proc. of ICALP'91*, LNCS 510, pages 115–136. Springer-Verlag, 1991.
3. R. Alur and D.L. Dill. A theory of timed automata. *TCS*, 126:183–235, 1994.
4. S. Asmussen. *Applied Probability and Queues.* Springer, 2006.
5. A. Aziz, K. Sanwal, V. Singhal, and R.K. Brayton. Model-checking continuous-time Markov chains. *ACM Trans. on Computational Logic*, 1(1):162–170, 2000.
6. H. Choi, V.G. Kulkarni, and K. Trivedi. Markov regenerative stochastic Petri nets. Performance Evaluation, 20:337–357, 1994.
7. A. Cumani. Esp — A package for the evaluation of stochastic Petri nets with phase-type distributed transition times. In *Proc. of Int. Workshop on Timed Petri Nets*, pages 144–151, Torino (Italy), 1985. IEEE Computer Society Press no. 674.
8. R. German. *Performance analysis of communication systems: Modeling with non-Markovian stochastic Petri nets.* J. Wiley & Sons, 2000.
9. P.W. Glynn. A GSMP formalism for discrete event systems. In *Proc. of the IEEE*, 77(1):14–23, 1988.
10. B. Haverkort. *Performance of computer-communication systems: A model-based approach.* Wiley & Sons, 1998.
11. M.Z. Kwiatkowska. Model checking for probability and time: from theory to practice. In *Proc. of the 18th IEEE Symposium on Logic in CS*, pages 351–360, 2003.
12. C. Lindemann and A. Thümmler. Numerical Analysis of Generalized Semi-Markov Processes. Research Report No. 722, Dept. of CS, University of Dortmund, 1999.
13. G. López, H. Hermanns and J.-P. Katoen. Beyond Memoryless Distributions: Model Checking Semi-Markov Chains. In *Proc. of PAPM-PROBMIV*, LNCS 2165, pages 57–70, 2001.
14. G.S. Shedler. *Regenerative stochastic simulation.* Academic Press, 1993.
15. H. Ye and R. Corless. Solving linear integral equations in Maple. In *Proc. of the Int. Symposium on Symbolic and Algebraic Computation*, pages 95–102, 1992.
16. D. Zeilberger. Holonomic Systems Approach To Special Functions. *J. Computational and Applied Math*, 32:321–368, 1990.

Robust, Optimal Predictive Control of Jump Markov Linear Systems Using Particles

Lars Blackmore[1], Askar Bektassov[2], Masahiro Ono[1], and Brian C. Williams[1]

[1] Massachusetts Institute of Technology, Cambridge, MA 02139
larsb@mit.edu, hiro_ono@mit.edu, williams@mit.edu
[2] Università degli Studi Roma Tre, Rome, Italy 00146
askar.bektassov@gmail.com

Abstract. Hybrid discrete-continuous models, such as Jump Markov Linear Systems, are convenient tools for representing many real-world systems; in the case of fault detection, discrete jumps in the continuous dynamics are used to model system failures. Stochastic uncertainty in hybrid systems arises in both the continuous dynamics, in the form of uncertain state estimation, disturbances or uncertain modeling, and in the discrete dynamics, which are themselves stochastic.

In this paper we present a novel method for optimal predictive control of Jump Markov Linear Systems that is robust to both continuous and discrete uncertainty. The approach extends our previous 'particle control' approach, which approximates the predicted distribution of the system state using a finite number of particles. Here, we present a *weighted* particle control approach, which uses importance weighting to ensure that low probability events such as failures are considered. We demonstrate the method with a car braking scenario.

1 Introduction

Hybrid discrete-continuous models,such as Jump Markov Linear Systems(JMLS), are convenient tools for representing many real-world systems[1,2]. In the case of fault detection and fault-tolerant control, discrete jumps in the continuous dynamics are used to model component failures[3]. Stochastic uncertainty in hybrid systems arises in both the continuous dynamics, in the form of uncertain state estimation, disturbances or uncertain modeling, and in the discrete dynamics, which are themselves stochastic.

Control of stochastic systems has received a great deal of attention in recent years, see [4] for a survey. Much work has been done in the area of feedback control for JMLS, see [5] for a survey. By contrast, *predictive* optimal stochastic control takes into account probabilistic uncertainty in dynamic systems and aims to control the predicted distribution of the system state in some optimal manner. In the case of stochastic linear dynamic systems, recent work has developed tractable algorithms for optimal, *robust* predictive control[6,7,8,9,10]. These methods are robust in the sense that they ensure the system state leaves a given feasible region with probability at most δ. This *chance-constrained*

A. Bemporad, A. Bicchi, and G. Buttazzo (Eds.): HSCC 2007, LNCS 4416, pp. 104–117, 2007.

formulation is a powerful one, as it enables the user to specify a desired level of conservatism, which can be traded against performance.

Chance-constrained optimal stochastic control of JMLS has a number of important applications. In the case of fault-tolerant control, we would like to be able to control a system in an optimal manner while taking into account both continuous disturbances and the possibility of system failures, such that task failure is below a certain threshold. For example, in controlling an autonomous ground vehicle we would like to ensure that, despite having brakes that may fail, collision with obstacles or other vehicles happens with low probability. Recent work developed a Model Predictive Control approach for JMLS which imposes constraints on the mean and covariance of the system state[11]. These are not the same as chance constraints even when all forms of uncertainty are Gaussian since the state distribution in JMLS is multimodal.

In this paper we develop a tractable algorithm for chance-constrained optimal predictive control of JMLS that extends our previous work on particle-based control of continuous systems[12]. The key idea behind this approach is to approximate all probability distributions using a finite number of samples, or 'particles'. In doing so, we approximate the stochastic predictive control problem as a deterministic one, with the property that as the number of particles tends to infinity, the approximation tends to the original stochastic problem. The resulting optimization problem can be solved efficiently using Mixed Integer Linear Programming (MILP). The approach generalizes previous work by [17] by handling stochastic uncertainty with general distributions, in both the continuous and discrete dynamics.

In this paper we present first a straightforward extension of the particle control method to JMLS. This extension uses particles to represent uncertainty in the discrete mode sequences as well as the continuous variables. An empirical validation with a ground vehicle braking scenario shows that the method is effective, but is prone to neglect low-probability events such as failures. We therefore develop a new *weighted* particle control approach that overcomes these difficulties by drawing on the idea of importance weighting from particle filtering[13,14,15,16]. The key idea is that by sampling mode sequences from a proposal distribution, and representing the discrepancy between the proposal distribution and the true distribution by an analytic weight, an increase in sampling efficiency can be achieved. The resulting optimization can be solved efficiently and to global optimality using MILP. We demonstrate empirically that a dramatic improvement in performance is achieved by employing the weighted particle control approach.

2 Problem Statement

In this paper we are concerned with the following stochastic control problem:

Design a finite, optimal sequence of control inputs $\mathbf{u}_{0:T-1}$*, taking into account probabilistic uncertainty, which ensures that the continuous state trajectory* $\mathbf{x}_{c,1:T}$ *of a JMLS leaves a defined feasible region* F *with probability at most* δ*, and satisfies constraints on the expected system state.*

We consider four sources of stochastic uncertainty; initial state uncertainty; system model uncertainty; disturbances, modeled as stochastic processes; and random mode transitions. These transitions can model component failures, for example. We assume that the p.d.f.s of the uncertainty mentioned here are known at least approximately, but we make no assumptions about the form the distributions take. We assume a cost function that is piecewise linear in the control inputs; previous work has shown that minimum control effort and minimum time problems can be posed using such functions[18]. Finally, we assume that the feasible region F is a polytope, and that the control inputs $\mathbf{u}_t$ are subject to interval constraints.

We define a Jump Markov Linear System as a system with hybrid discrete-continuous state $\mathbf{x} = \langle \mathbf{x}_c, x_d \rangle$. The discrete state x_d is a Markov chain that can take one of M values and evolves according to:

$$p(x_{d,t+1} = j | x_{d,t} = i) = T_{ij}. \tag{1}$$

The continuous state $\mathbf{x}_c$ evolves according to:

$$\mathbf{x}_{c,t+1} = A(x_{d,t})\mathbf{x}_{c,t} + B(x_{d,t})\mathbf{u}_t + \nu_t. \tag{2}$$

The initial hybrid discrete-continuous state is random, with a known distribution $p(\mathbf{x}_{c,0}, x_{d,0})$. The variable ν_t is a random disturbance process distributed according to $p(\nu_t | x_{d,t})$, which we assume independent from the initial state. Modeling errors can be modeled as an additional stochastic disturbance. For notational simplicity we assume a single disturbance process.

The key idea behind solving this stochastic control problem is to approximate all distributions using samples, or particles, and then solve the resulting deterministic problem. In Section 3 we review some results relating to sampling from random variables. In Section 4 we review the chance-constrained particle control approach introduced in [12] for systems with continuous state. We then extend this approach to JMLS in Section 5 and show that the resulting problem can be solved using MILP. In Section 6 we introduce a novel weighted particle control method that gives a dramatic increase in performance by using a proposal distribution to sample from discrete mode sequences. In Section 7 we provide empirical results.

3 Sampling from Random Variables

Previous work has shown that approximating the probability distribution of a random variable using samples drawn from that distribution, or particles, can lead to tractable algorithms for estimation and control[19]. Here we review some properties of samples drawn from random variables.

Suppose that we have a multivariate random variable X with p.d.f. $p(\mathbf{x})$. We draw N independent, identically distributed random samples $\mathbf{x}^{(1)}, \cdots, \mathbf{x}^{(N)}$ from this distribution. Often, we would like to calculate an expectation involving this random variable:

$$E_X[f(X)] = \int_X f(\mathbf{x})p(\mathbf{x})d\mathbf{x} \tag{3}$$

In many cases this integral cannot be evaluated in closed form. Instead it can be approximated using the sample mean:

$$\hat{E}_X[f(X)] = \frac{1}{N}\sum_{i=1}^{N} f(\mathbf{x}^{(i)}). \tag{4}$$

From the strong law of large numbers, the sample mean converges to the true expectation as N tends to infinity. This can be used to approximate the probability of a certain event, such as the event $f(\mathbf{x}) \in A$. This is given exactly by:

$$P_A = \int_{f(\mathbf{x})\in A} p(\mathbf{x})d\mathbf{x} = E_X[g(\mathbf{x})] \quad \text{where} \quad g(\mathbf{x}) = \begin{cases} 1 & f(\mathbf{x}) \in A \\ 0 & f(\mathbf{x}) \notin A. \end{cases} \tag{5}$$

We can therefore approximate P_A as:

$$\hat{P}_A = \frac{1}{N}\sum_{i=1}^{N} g(\mathbf{x}^{(i)}) \qquad \text{where} \qquad \hat{P}_A \longrightarrow P_A \text{ as } N \longrightarrow \infty. \tag{6}$$

Note that $\sum_{i=1}^{N} g(\mathbf{x}^{(i)})$ is simply the number of particles for which $f(\mathbf{x}^{(i)}) \in A$. Assuming that evaluating $f(\cdot)$, and checking whether a given value is in A, are both straightforward, calculating $\hat{P}_A$ is also; we simply need to count how many of the propagated particles, $f(\mathbf{x}^{(i)})$ fall within A. By contrast, evaluating P_A as in (5) requires a finite integral over an arbitrary probability distribution, where even calculating the bounds on the integral may be intractable. Hence the particle-based approximation is extremely useful, especially given the convergence property in (6). In Section 4 we use this property to approximate the stochastic control problem defined in Section 2.

3.1 Importance Weighting

In certain situations, drawing samples from the distribution $p(\mathbf{x})$ may be intractable. In such cases, previous work proposed sampling from an alternative *proposal* distribution and using *importance sampling* to correct for the discrepancy between the desired distribution and the proposal distribution[19]. We review relevant results here.

The proposal distribution $q(\mathbf{x})$ is chosen so that sampling from $q(\mathbf{x})$ is easy, and so that $p(\mathbf{x}) > 0$ implies $q(\mathbf{x}) > 0$. We draw N independent, identically distributed random samples $\mathbf{x}^{(1)}, \cdots, \mathbf{x}^{(N)}$ from $q(\mathbf{x})$. To each sample we assign an importance weight w_i, where $w_i = p(\mathbf{x}^{(\mathbf{i})})/q(\mathbf{x}^{(\mathbf{i})})$. In order to approximate the expectation of the function $f(\cdot)$ we now use the *weighted* sample mean:

$$\hat{E}_X[f(X)] = \frac{1}{N}\sum_{i=1}^{N} w_i f(\mathbf{x}^{(i)}). \tag{7}$$

From the strong law of large numbers, we have the convergence property as N tends to infinity:

$$\hat{E}_X[f(X)] \longrightarrow E_X[f(X)]. \quad (8)$$

In order to approximate the probability of the event $f(\mathbf{x}) \in A$ we use the *weighted* number of propagated particles that fall within A:

$$\hat{P}_A = \frac{1}{N}\sum_{i=1}^{N} w_i g(\mathbf{x}^{(i)}), \quad (9)$$

where $g(\cdot)$ is as defined in (5). As in (6) we have the convergence property $\hat{P}_A \longrightarrow P_A$ as $N \rightarrow \infty$.

4 Review of Particle Control Approach

In this section we review the chance-constrained particle control approach introduced in [12] for robust control of systems with continous state.

The key observation behind the method is that, by approximating all probabilistic distributions using particles, an intractable stochastic optimization problem can be approximated as a tractable deterministic optimization problem. By solving this deterministic problem we obtain an approximate solution to the original stochastic problem, with the additional property that as the number of particles used tends to infinity, the approximation becomes exact.

The method is outlined as follows:

1. Generate N samples from the joint distribution of initial state and disturbances.
2. Express the distribution of the future state trajectories approximately as a set of N *analytic* particles, where each particle $\mathbf{x}_{1:T}^{(i)}$ corresponds to the state trajectory given a particular set of samples. Each particle *depends explicitly on the control inputs* $\mathbf{u}_{0:T-1}$, which are yet to be generated.
3. Approximate the chance constraints in terms of the generated particles; the probability of $\mathbf{x}_{1:T}$ falling outside of the feasible region is approximated as the *fraction* of particles $\mathbf{x}_{1:T}^{(i)}$ that do so.
4. Approximate the cost function in terms of particles.
5. Solve the deterministic constrained optimization problem for control inputs $\mathbf{u}_{0:T-1}$.

The method is illustrated in Fig. 1. The general particle control problem results in a deterministic optimization problem that is intractable, except for very small problems. However in [12] we showed that for a polytopic feasible region F, piecewise linear cost function h and linear system dynamics $\mathbf{x}_{t+1} = A\mathbf{x}_t + B\mathbf{u}_t$, the deterministic optimization can be solved to global optimality in an efficient manner using MILP. This relies on the fact that each particle $\mathbf{x}_{1:T}^{(i)}$ is a *linear* function of the control input sequence $\mathbf{u}_{0:T-1}$.This is also true for time-varying linear systems. In Section 5 we use this to extend the method to JMLS.

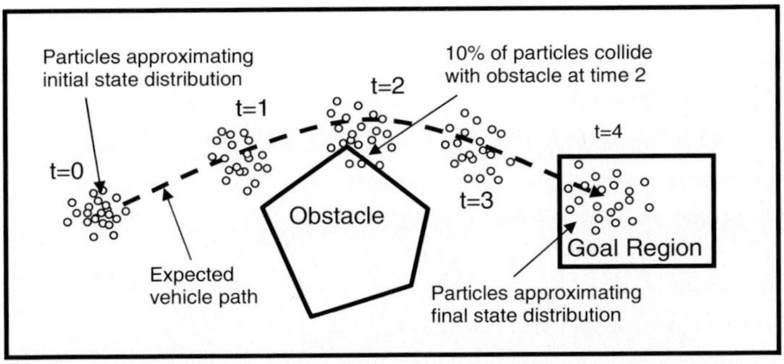

Fig. 1. Illustration of chance constrained particle control method for continuous systems. For this vehicle path planning scenario, the feasible region is defined so that the plan is successful if the vehicle avoids the obstacles at all time steps and is in the goal region at the final time step. The objective is to find the optimal sequence of control inputs so that the plan is successful with probability at least 0.9. The particle control method approximates this so that at most 10% of the particles fail.

5 Straightforward Extension of Particle Control to JMLS

For JMLS, we approximate the stochastic control problem by sampling from *discrete mode sequences* as well as disturbances. Given a discrete mode sequence and samples for all of the disturbance variables, the future system state trajectory is a known deterministic function of the control inputs. Hence each particle provides a sample of the future state trajectory corresponding to a sample of the discrete mode sequence and disturbances.

Note that the mode sequence is independent of the control inputs and the continuous state in JMLS, and hence:

$$p(\mathbf{x}_{c,1:T}, x_{d,1:T}|\mathbf{u}) = p(\mathbf{x}_{c,1:T}|x_{d,1:T}, \mathbf{u})p(x_{d,1:T}). \tag{10}$$

We therefore first generate samples of the mode sequence $x_{d,1:T}$, and for each sample $x^{(i)}_{d,1:T}$, we generate samples of the disturbance variables. While there are M^T different mode sequences, sampling from $p(x_{d,1:T})$ is straightforward due to the Markov property. The algorithm is described in full in Table 1. From the results in Section 3 we have convergence of the approximated problem to the original stochastic problem as the number of particles tends to infinity.

5.1 MILP Solution of JMLS Particle Control

We now show that the approximated problem can be solved efficiently using MILP. For a given particle, the mode at each time step in the horizon is known, as are the disturbances at each time step. From the definition of JMLS in Section 1 we obtain the following expression for each particle:

Table 1. Straightforward Particle Control Approach for JMLS

1) Generate N samples of the initial discrete mode $\{x_{d,0}^{(1)},\dots,x_{d,0}^{(N)}\}$ according to the distribution $p(x_{d,0})$.
2) For each sample $x_{d,0}^{(i)}$, generate a sample of the initial continuous state $\{\mathbf{x}_{c,0}^{(1)},\dots,\mathbf{x}_{c,0}^{(N)}\}$ according to $p(\mathbf{x}_{c,0}|x_{d,0}^{(i)})$.
3) For each sample $x_{d,0}^{(i)}$ generate a sample of the discrete mode sequence $x_{d,1:T}^{(i)}$ according to $p(x_{d,1:T}|x_{d,0})$.
4) For each sample $x_{d,0:T}^{(i)}$ generate a sample of the disturbances $\{\nu_0^{(i)},\dots,\nu_{T-1}^{(i)}\}$ from the distribution $p(\nu_0,\dots,\nu_{T-1}|x_{d,0:T}^{(i)})$.
5) Express the distribution of the future state trajectories approximately as a set of N particles, where each particle $\mathbf{x}_{c,1:T}^{(i)}$ corresponds to the continuous state trajectory given a particular set of samples $\{\mathbf{x}_0^{(i)},x_{d,1:T}^{(i)},\nu_0^{(i)},\cdots,\nu_{T-1}^{(i)}\}$. Each particle depends explicitly on the control inputs $\mathbf{u}_0,\cdots,\mathbf{u}_{T-1}$, which are yet to be generated.
6) Approximate the expected state constraints and chance constraints in terms of the generated particles.

$$E[\mathbf{x}_{1:T}^{(i)}]\approx\frac{1}{N}\sum_{i=1}^{N}\mathbf{x}_{1:T}^{(i)}=\mathbf{x}_{1:T}^{equality}\qquad p(\mathbf{x}_{c,1:T}\notin F)\approx\frac{1}{N}\sum_{i=1}^{N}g(\mathbf{x}_{1:T}^{(i)})\le\delta,\quad(11)$$

where $g(\cdot)$ is as defined in (5).
7) Approximate the cost function in terms of particles.
8) Solve deterministic constrained optimization problem for inputs $\mathbf{u}_{0:T-1}$.

$$\mathbf{x}_{c,t}^{(i)}=\sum_{j=0}^{t-1}\Big(\prod_{l=1}^{t-j-1}A(x_{d,l}^{(i)})\Big)\Big(B(x_{d,j}^{(i)})\mathbf{u}_j+\nu_j^{(i)}\Big)+\Big(\prod_{l=1}^{t}A(x_{d,l}^{(i)})\Big)\mathbf{x}_{c,0}^{(i)}.\qquad(12)$$

Note that this is a linear function of the control inputs, and that $\mathbf{x}_{c,0}^{(i)}$, $\nu_j^{(i)}$ and $x_{d,l}^{(i)}$ are all known values. Hence each particle $\mathbf{x}_{c,1:T}^{(i)}$ is linear in the control inputs.

In accordance with (11), we need to constrain the number of particles that fall outside of the feasible region. In the same manner as described in [12], we define a set of N binary variables $z_1,\cdots,z_N$, where $z_i\in\{0,1\}$. These binary variables are defined so that $z_i=0$ implies that particle i falls inside the feasible region. We then constrain the sum of these binary variables:

$$\frac{1}{N}\sum_{i=1}^{N}z_i\le\delta.\qquad(13)$$

This constraint ensures that the fraction of particles falling outside of the feasible region is at most δ. In [12] we showed how to impose constraints such that $z_i=0\implies\mathbf{x}_{1:T}^{(i)}\in F$ for convex and non-convex polygonal feasible regions. We do not repeat this here, but we do note that the linearity of (12) and piecewise linearity of the cost function h ensures that the encoding results in a MILP, which can be solved efficiently to global optimality. We have therefore introduced a new method for robust optimal control of JMLS, where the probability distributions of uncertain variables can take an arbitrary form.

Table 2. Weighted Particle Control Approach for JMLS

1) Generate N samples of the initial discrete mode $\{x_{d,0}^{(1)}, \dots, x_{d,0}^{(N)}\}$ according to the distribution $p(x_{d,0})$.
2) For each sample $x_{d,0}^{(i)}$, generate a sample of the initial continuous state $\{\mathbf{x}_{c,0}^{(1)}, \dots, \mathbf{x}_{c,0}^{(N)}\}$ according to $p(\mathbf{x}_{c,0}|x_{d,0}^{(i)})$.
3) For each sample $x_{d,0}^{(i)}$ generate a sample of the discrete mode sequence $x_{d,1:T}^{(i)}$ *according to the proposal distribution* $q(x_{d,1:T})$.
4) For each sample $x_{d,0:T}^{(i)}$ generate a sample of the disturbances $\{\nu_0^{(i)}, \dots, \nu_{T-1}^{(i)}\}$ from the distribution $p(\nu_0, \dots, \nu_{T-1}|x_{d,0:T}^{(i)})$.
5) For each sample $x_{d,0:T}^{(i)}$ calculate $p(x_{d,0:T}^{(i)})$ and assign weight w_i as in (16).
6) Express the distribution of the future state trajectories approximately as a set of N particles, where each particle $\mathbf{x}_{c,1:T}^{(i)}$ corresponds to the continuous state trajectory given a particular set of samples $\{\mathbf{x}_0^{(i)}, x_{d,1:T}^{(i)}, \nu_0^{(i)}, \cdots, \nu_{T-1}^{(i)}\}$.
7) Approximate the chance constraints using the *weighted fraction* of particles outside of the feasible region:
$$p(\mathbf{x}_{1:T} \notin F) \approx \frac{1}{N} \sum_{i=1}^{N} w_i g(\mathbf{x}_{1:T}^{(i)}) \leq \delta. \tag{14}$$
8) Approximate the expected state constraints using the *weighted* sample mean approximation, for example:
$$E[\mathbf{x}_{1:T}] = \mathbf{x}_{1:T}^{equality} \quad \text{becomes} \quad \frac{1}{N} \sum_{i=1}^{N} w_i \mathbf{x}_{1:T}^{(i)} = \mathbf{x}_{1:T}^{equality}. \tag{15}$$
9) Approximate the cost function in terms of weighted particles.
10) Solve the deterministic constrained optimization problem for inputs $\mathbf{u}_{0:T-1}$.

6 Weighted Particle Control for JMLS

We now extend the method described in Section 5 to deal more efficiently with low probability mode transitions such as failures. The key idea behind the extension is to sample mode sequences from a proposal distribution designed to ensure that low probability events such as failures are more likely to be taken into consideration. Drawing on the idea of importance weighting in particle filtering[19], the discrepancy between the actual distribution over mode sequences and the proposal distribution is represented using an analytical weighting. In doing so, we retain the convergence property that the approximate problem converges to the original stochastic problem as the number of particles tends to infinity. The algorithm is described in Table 2.

We now show how to calculate the weights w_i. For the approximated problem to converge to the original stochastic problem as the number of particles tends to infinity, weights must be assigned according to[19]:

$$w_i = \frac{p(\mathbf{x}_{c,1:T}^{(i)}, x_{d,1:T}^{(i)}|\mathbf{u}_{0:T-1})}{q(\mathbf{x}_{c,1:T}^{(i)}, x_{d,1:T}^{(i)}|\mathbf{u}_{0:T-1})}. \tag{16}$$

Since we sample the disturbances from their true distributions, the joint proposal $q(\mathbf{x}_{c,1:T}, x_{d,1:T})$ can be written in terms of the proposal over mode sequences to give:

$$w_i = \frac{p(\mathbf{x}_{c,1:T}^{(i)}|x_{d,1:T}^{(i)}, \mathbf{u}_{0:T-1})p(x_{d,1:T}^{(i)})}{p(\mathbf{x}_{c,1:T}^{(i)}|x_{d,1:T}^{(i)}, \mathbf{u}_{0:T-1})q(x_{d,1:T}^{(i)})} = \frac{p(x_{d,1:T}^{(i)})}{q(x_{d,1:T}^{(i)})}. \tag{17}$$

Since calculating both the true probability of a given mode sequence and its probability according to the proposal distribution is straightforward, calculating the weight to assign to a sampled mode sequence is also.

We now show that the weighted particle control problem for JMLS can be solved using MILP. The key insight is that, since the weights do not depend on the control inputs $\mathbf{u}_{0:T-1}$, incorporating weighted particles does not affect the form of the optimization problem.

The weighted particle control problem can be formulated in exactly the same manner as the unweighted approach described in Section 5, except for the approximate chance constraint and the approximate cost function. We now must constrain the *weighted* fraction of particles that fall outside of the feasible region. Defining again binary variables z_i such that $z_i = 0 \implies \mathbf{x}_{c,1:T}^{(i)} \in F$, we constrain the *weighted* sum of the binary variables:

$$\frac{1}{N}\sum_{i=1}^{N} w_i z_i \leq \delta. \tag{18}$$

The weights w_i do not depend on the control inputs, as shown in (17). Hence (18) is a linear constraint on the binary variables z_i. The expected cost is now approximated using the *weighted* sample mean as follows:

$$E[h] \approx \hat{h} = \frac{1}{N}\sum_{i=1}^{N} w_i h(\mathbf{u}_0, \cdots, \mathbf{u}_{T-1}, \mathbf{x}_{c,1:T}^{(i)}). \tag{19}$$

As the number of particles tends to infinity, we have the convergence result $\hat{h} \longrightarrow E[h]$. Furthermore, since the weights w_i do not depend on the control inputs, the approximate value $\hat{h}$ is piecewise-linear in the control inputs assuming a piecewise-linear cost function h. Similarly, the expected state is approximated using the *weighted* sample mean. This weighted sample mean is a linear function of the control inputs, hence expected state constraints such as (15) are linear.

In summary, therefore, the weighted particle control problem for JMLS can be posed as a MILP. It now remains to choose a proposal distribution $q(x_{d,1:T}^{(i)})$.

6.1 Choosing a Proposal Distribution

The convergence of the approximate problem to the original deterministic problem applies for any choice of the proposal distribution $q(x_{d,1:T})$ subject to the constraint that $q(x_{d,1:T}) > 0$ wherever $p(x_{d,1:T}) > 0$. However for a finite number of particles the performance of the weighted particle control approach is affected greatly by the choice of $q(x_{d,1:T})$. As in particle filtering, the appropriate choice of $q(x_{d,1:T})$ depends on the application, and a great deal of work has focussed on developing proposal distributions for specific applications, for example [14,20]. We now introduce a proposal distribution designed to improve the performance of the particle control approach for JMLS when dealing with low-probability transitions such as faults.

Consider first a proposal distribution equal to the true mode sequence distribution:

$$q(x_{d,1:T}) = p(x_{d,1:T}). \tag{20}$$

In a JMLS with low-probability transitions such as faults, there is a high probability that no fault transitions will be sampled if this proposal is used.

Next consider a proposal equal to the pseudo-uniform distribution $q(x_{d,1:T}) = U(x_{d,1:T})$, where $U(\cdot)$ assigns an equal probability to each mode sequence for which $p(x_{d,1:t}) > 0$. More precisely:

$$U(x_{d,1:T}) = \begin{cases} 1/n_p & p(x_{d,1:T}) > 0 \\ 0 & p(x_{d,1:T}) = 0, \end{cases} \tag{21}$$

where n_p is the number of mode sequences for which $p(x_{d,1:T}) > 0$. Using this proposal ensures that sequences involving faults are sampled with the same likelihood as the mode sequence without failures, which in reality has much higher probability. This means that the control algorithm is more likely to take into account the sequences involving faults in the control design. The drawback in using this proposal is that there is a significant likelihood that the nominal mode sequence is not sampled. If this occurs, the deterministic optimization will typically be infeasible, since achieving most control tasks requires nominal operation of the system components with non-zero probability.

We therefore choose a proposal distribution $q^*(x_{d,1:T})$ that increases the probability of sampling failure sequences, while ensuring that the nominal mode sequence is sampled at least once with a probability λ:

$$q^*(x_{d,1:T}) = \begin{cases} P_{nom} & x_{d,1:T} = x_{d,1:T}^{nom} \\ \frac{1-P_{nom}}{n_p-1} & x_{d,1:T} \neq x_{d,1:T}^{nom} \end{cases} \quad \text{where } P_{nom} = 1-(1-\lambda)^{1/N}. \tag{22}$$

The proposal distribution $q^*(x_{d,1:T})$ therefore ensures a minimum probability of sampling the nominal mode sequence and shares the remaining probability space evenly among the remaining mode sequences.[1]

In Section 7 we give an empirical analysis that shows that using this proposal distribution the weighted particle control algorithm significantly outperforms straightforward particle control for JMLS when there are low-probability transitions such as failures.

---

[1] For simplicity of exposition, $q^*(x_{d,1:T})$ described here assumes a single nominal mode sequence. The extension to multiple nominal mode sequences is straightforward.

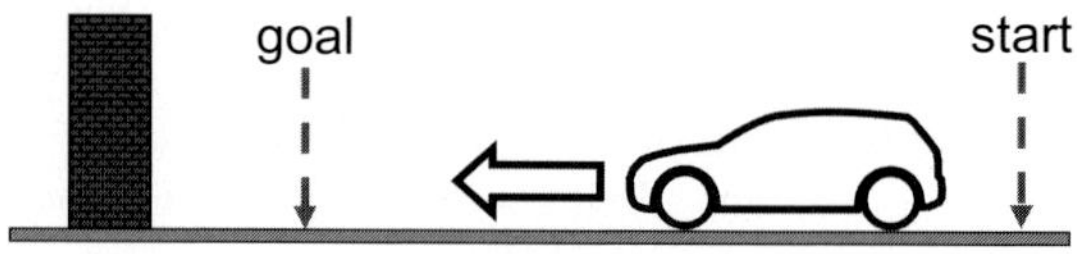

Fig. 2. Illustration of ground vehicle brake failure scenario. The expected vehicle position must arrive at the goal in the minimum possible time, while avoiding collision with the wall.

7 Results

In order to illustrate the new particle control approach for JMLS we use a simple ground vehicle braking example. In this example the system to be controlled is a ground vehicle that can accelerate and brake along a one-dimensional track. The brakes however, can be in one of two modes; mode 1 = *ok* and mode 2 = *faulty*. In the *ok* mode, accelerations and decelerations can be applied to the vehicle, however when the brakes are in the *faulty* mode, decelerations cannot be applied. The continuous system state $\mathbf{x}_c$ is comprised of the position along the track y and the velocity $\dot{y}$. The continuous state evolves according to:

$$\dot{\mathbf{x}}_c = \begin{bmatrix} \dot{y} \\ \ddot{y} \end{bmatrix} = \begin{bmatrix} 0 & 1 \\ 0 & -b_{fric} \end{bmatrix} \begin{bmatrix} y \\ \dot{y} \end{bmatrix} + B(x_{d,t}) \begin{bmatrix} u_{power} \\ u_{brake} \end{bmatrix} + \nu_t, \tag{23}$$

where the control inputs u_{power} and u_{brake} are both constrained to be greater than or equal to zero (in other words neither negative power nor negative braking can be applied). The term b_{fric} represents a damping term due to friction. Random disturbances ν_t act on the vehicle. The matrix $B(x_{d,t})$ is defined as follows:

$$B(x_{d,t}) = \begin{cases} \begin{bmatrix} 1 & -1 \end{bmatrix} & x_{d,t} = ok \\ \begin{bmatrix} 1 & 0 \end{bmatrix} & x_{d,t} = faulty. \end{cases} \tag{24}$$

The discrete state evolves according to the transition matrix:

$$T = \begin{bmatrix} 0.999 & 0.001 \\ 0.0 & 1.0 \end{bmatrix}. \tag{25}$$

We consider the problem where the car is initially at rest and must travel to the goal and stop, as illustrated in Fig. 2. Task failure is defined as collision with the wall.

Fig. 3 compares two typical solutions generated by the weighted particle control approach for a maximum probability of failure of 0.01 and 10^{-6} respectively. The more conservative solution takes $9s$, while the aggressive one takes only $6s$. We now demonstrate that the weighted particle control approach enables the controller to take into account the low probability brake failures. Fig. 4 compares two typical solutions generated with and without weighting respectively.

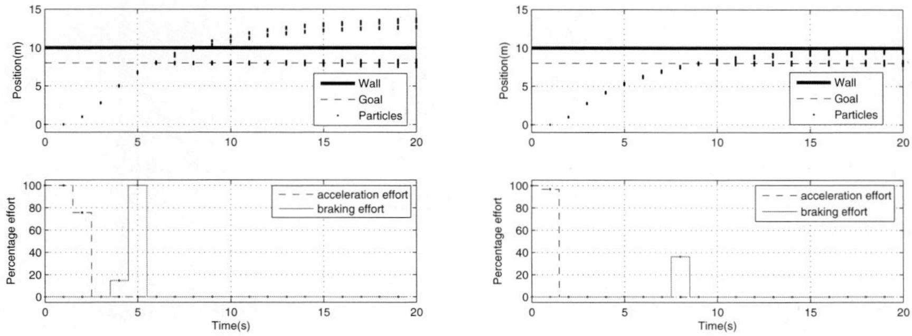

Fig. 3. Two typical solutions with 100 particles. Left: Maximum probability of task failure set to 0.01. The vehicle arrives at the goal within $6s$, but will collide with the wall if a brake failure occurs at or before $5s$. This particular solution gives a true probability of task failure of approximately 0.006. Right: Maximum probability of task failure set to 10^{-6}. The vehicle travels more slowly and arrives later than with the more aggressive solution. In the case of brake failure, however, friction brings the vehicle to rest before collision with the wall. This solution is therefore robust to brake failure, giving a probability of task failure of approximately 1.0×10^{-6}.

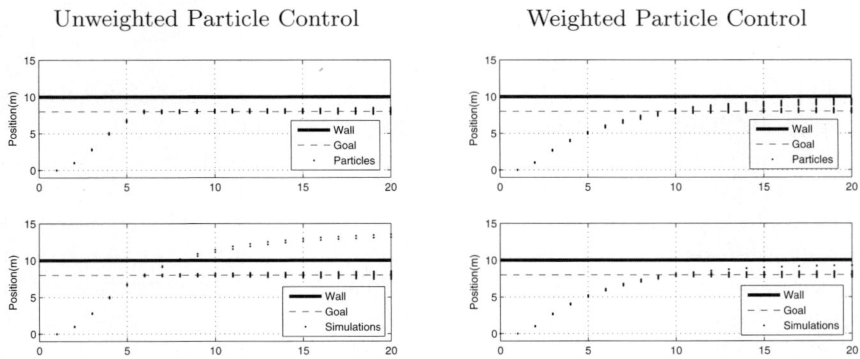

Fig. 4. Typical solutions with and without weighting for $\delta = 10^{-6}$ and 100 particles. The top row shows the particles used for planning, while the bottom row shows Monte-Carlo simulations of the true state trajectory. **Left:** *Without* weighting, no particles have sampled the brake failure so the controller plans aggressively. In reality, there is a probability of approximately 0.0050 that a brake failure occurs at or before $t = 5s$, causing the vehicle to collide with the wall. **Right:** *With* weighting, many particles have sampled brake failures, hence the controller plans taking brake failures into account. The controller is less aggressive, giving a collision probability of approximately 1.0×10^{-6}.

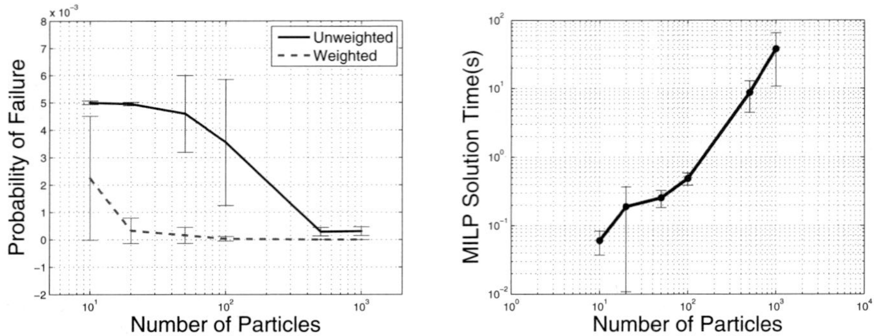

Fig. 5. Left: True probability of failure against number of particles for weighted method and unweighted method. The desired probability of failure was 10^{-6}. The weighted approach achieves a true probability of failure dramatically closer to the desired value than the unweighted approach. With a very small particle set, the effect of weighting is diminished since the probability of sampling the nominal sequence must be high in order to satisfy constraints on the probability of a feasible solution. **Right:** MILP solution time for weighted particle control with ground vehicle scenario using ILOG CPLEX 9.0 on Intel Pentium 4 2.8GHz machine with 1GB RAM. The specified maximum probability of task failure was 0.01.

In the unweighted case, the algorithm did not sample any of the failure transitions and so has generated an inappropriately aggressive control policy that does not take into account the possibility of brake failure. By increasing the probability of sampling failure transitions, the weighted algorithm by contrast has taken into account brake failure, generating an appropriately conservative plan.

Fig. 5 compares the weighted particle control approach against the unweighted particle control approach in terms of the true probability of task failure. In this example the desired probability of task failure was 10^{-6}. The weighted approach achieves a true probability of failure dramatically closer to the desired value than the unweighted approach. Notice also that for larger particle sets the unweighted case approaches the weighted one, except that the variance is much greater in the unweighted case. This is because on the rare occasion that brake failure transitions are sampled, the solution is very different from the average case. This variance is particularly undesirable for control. Fig. 5 also shows the solution time as a function of the number of weighted particles used.Solutions were found in seconds even for relatively large particle sets.

8 Conclusion

In this paper we have presented a novel approach to optimal stochastic control for Jump Markov Linear Systems that takes into account probabilistic uncertainty due to disturbances, uncertain state estimation, modeling error and stochastic mode transitions. The new weighted particle control method is robust in ensuring that the probability of task failure is less than a defined threshold δ. By approximating the original stochastic problem as a deterministic one using a number of importance-weighted particles, the approach is able to handle arbitrary probability distributions. Furthermore the approximation error tends to zero as the number of particles tends to infinity. Importance weighting is used in conjunction with sampling from a proposal distribution to make sure that the method takes into account low probability events such as component failures.

References

1. Hofbaur, M.W., Williams, B.C.: Hybrid estimation of complex systems. In: IEEE Trans. on Systems, Man, and Cybernetics - Part B: Cybernetics. (2004)
2. Oh, S.M., Rehg, J.M., Balch, T., Dallaert, F.: Learning and inference in parametric switching linear dynamic systems. In: Proc. IEEE Int. Conf. Comp. Vision. (2005)
3. Blackmore, L., Rajamanoharan, S., Williams, B.C.: Active estimation for switching linear dynamic systems. In: Proc. of the CDC. (2006)
4. Pham, H.: On some recent aspects of stochastic control and their applications. Probability Surveys **2** (2005) 506–549
5. Costa, O.L.V., Fragoso, M.D., Marques, R.P.: Discrete-Time Markovian Jump Linear Systems. Springer-Verlag (2005)
6. Schwarm, A., Nikolaou, M.: Chance-constrained model predictive control. AlChE Journal **45** (1999) 1743–1752
7. Li, P., Wendt, M., Wozny, G.: A probabilistically constrained model predictive controller. Automatica **38** (2002) 1171–1176
8. Batina, I.: Model Predictive Control for Stochastic Systems by Randomized Algorithms. PhD thesis, TU Eindhoven, The Netherlands (2004)
9. Hessem, D.H.V.: Stochastic Inequality Constrained Closed-loop Model Predictive Control. PhD thesis, Technische Universiteit Delft, Delft, The Netherlands (2004)
10. Blackmore, L., Li, H.X., Williams, B.C.: A probabilistic approach to optimal robust path planning with obstacles. In: Proceedings of the ACC. (2006)
11. Vargas, A.N., Furloni, W., do Val, J.B.R.: Constrained MPC of jump linear systems with noise and non-observed markov state. In: Proc. ACC. (2006)
12. Blackmore, L.: A probabilistic particle control approach to optimal, robust predictive control. In: Proc. of the AIAA GNC Conference. (2006)
13. Metropolis, N., Ulam, S.: The Monte Carlo method. American Statistical Association **44** (1949) 335–341
14. Morales-Menendez, R., de Freitas, N., Poole, D.: Real-time monitoring of complex industrial processes with particle filters. In: Proceedings of NIPS. (2002)
15. Doucet, A., de Freitas, N., Murphy, K., Russell, S.: Rao-Blackwellised particle filtering for dynamic Bayesian networks. In: Proceedings of UAI. (2000)
16. Gordon, N.J., Salmond, D.J., Smith, A.F.M.: Novel approach to nonlinear/non-Gaussian Bayesian state estimation. IEE Proceedings - F **140**(2) (1993) 107–113
17. Bemporad, A., Cairano, S.D.: Optimal control of discrete hybrid stochastic automata. In Morari, M., Thiele, L., eds.: Lecture Notes in Computer Science 3414. Springer, Berlin (2005) 151–167
18. Schouwenaars, T., Moor, B.D., Feron, E., How, J.: Mixed integer programming for multi-vehicle path planning. In: Proc. European Control Conference. (2001)
19. Doucet, A., de Freitas, N., Gordon, N.J.: Sequential Monte Carlo Methods in Practice. Springer Verlag (2001)
20. de Freitas, N.: Rao-Blackwellised particle filtering for fault diagnosis. IEEE Aero. (2002)

Optimal Switching of 1-DOF Oscillating Systems

Paolo Bolzern[1], Patrizio Colaneri[1], and José Claudio Geromel[2]

[1] Politecnico di Milano, Dipartimento di Elettronica e Informazione, Piazza Leonardo da Vinci 32, 20133 Milano, Italy
bolzern[colaneri]@elet.polimi.it
http://www.elet.polimi.it
[2] School of Electrical and Computer Engineering
UNICAMP, CP 6101, 13083 - 970, Campinas, SP, Brazil
geromel@dsce.fee.unicamp.br
http://www.dt.fee.unicamp.br/geromel/

Abstract. This paper considers the class of hybrid linear second-order oscillating systems, in which two parameters are free to be assigned in a finite set of values. The control task is to decide, at any time instant, the value of the parameters as a function of the system state vector, in order to minimize a quadratic functional over an infinite horizon. The problem lends itself to cope with a variety of important applications, in diverse engineering fields. In the paper a numerical algorithm to compute the optimal switching rule is presented. Then the algorithm is applied to a simplified model of a vehicle suspension system with the aim of minimizing the chassis acceleration (comfort-oriented control).

1 Introduction

Switched systems have received a great deal of attention in the last decades. Roughly speaking, the design problem relies on the determination of a switching rule $\sigma(\cdot)$ which selects, at each instant of time, the actual mode of the underlying dynamical system, among M available ones. This paper considers the class of the so-called *switched linear systems*. For this class the control problem has an inherent feedback nature, as the control signal has to be determined from the available measurements in order to improve some specified performance.

The stability analysis of continuous time switched linear systems has been addressed by many authors, [1]- [5]. In reference [2] the interested reader can find an thorough discussion on a collection of results on uniform stability of switched systems. However, less attention has been devoted to the design of stabilizing feedback control laws. The reader is referred to [6]-[9] for a rather complete review on stability of continuous time switched linear systems, where special attention is given to the case of switching between two linear systems.

On the other hand, the optimal control problem for switched systems has been also investigated in the last years both from the theoretical and numerical viewpoints. Most of the available literature studied necessary and/or sufficient optimality conditions with the introduction of new versions of the maximum principle, see e.g. [10], [11], [12], [13]. The problem was also investigated in

A. Bemporad, A. Bicchi, and G. Buttazzo (Eds.): HSCC 2007, LNCS 4416, pp. 118–130, 2007.

[14] for the case of two subsystems. In [15], the problem of optimal control of autonomous switched systems was studied for a quadratic cost functional on an infinite horizon and a fixed number of switches. In this setting, the optimal control law can be computed by a discretization of the unitary semi-sphere. In later works, the same procedure was extended to the case where an infinite number of switches are allowed, [16], [17].

In this paper we approach the optimal control problem for 1-DOF (one degree of freedom) oscillating systems with switching stiffness and damping parameters. The method is based on the underlying Hamilton-Jacobi equation, [18]. The Lyapunov function is determined by a discretization of the unit circle and a thorough sensitivity analysis with respect to the parameter variation is worked out. A realistic practical application of a switched linear system is included. The problem consists in the design of a switching control strategy for semi-active suspensions in road vehicles. This is an important problem in the automotive field, as witnessed by numerous studies in the very recent literature, see e.g. [19], [20]. The optimal method discussed in the present paper is applied to a simplified model of a semiactive suspension system, obtained by assuming an infinite tire stiffness. The results are then compared with those obtained by the classical Sky-Hook approach introduced in [21]. The design methodology proves its effectiveness in all realistic simulations, also for values of the stiffness of the tire in the medium range.

The paper is organized as follows. In the next section the class of systems to be considered is introduced and the optimal control problem is formulated. An algorithm that provides the solution to this problem is discussed in Section 3. In Section 4 the attention focuses on a oscillating system with a fixed stiffness coefficient and a damping parameter assuming two different values. At the end of the section the case of a switching stiffness parameter is also considered. In Section 5 a simplified model of a suspension system is introduced and the optimal switching strategy is worked out. The results are compared with those obtained by the Sky-Hook approach and passive suspensions via extensive simulations. Finally, Section 6 concludes the paper.

2 The Switching Oscillating System

Our analysis focuses on a second order system of the form

$$\ddot{y}(t) = -\alpha_i \dot{y}(t) - \beta_j y(t) + w(t) \tag{1}$$

where $y(t) \in \mathbb{R}$, $i \in \Omega_\alpha = \{1, 2, \cdots, n_\alpha\}$, $j \in \Omega_\beta = \{1, 2, \cdots, n_\beta\}$, and the values of α_i and β_j are known parameters. In mechanical systems α_i can be interpreted as the damping coefficient and β_j as the stiffness coefficient. The input $w(t)$ is a scalar disturbance to be specified later.

The above model lends itself to describe a large variety of physical systems, whose coefficients may be switched within a finite set in order to improve some given performance. We say that the system is operating in the (i, j) mode when the underlying parameters take the values (α_i, β_j). Let $\sigma(t) \in \Omega_\alpha \times \Omega_\beta$ represent the switching signal. As $\sigma(t)$ changes, the evolution of the system is switched from one mode to another. Notice that the positiveness of α_i and β_j is a necessary and sufficient condition for the stability of the single (i,j) mode. However, in general, even if all modes are stable, there might exist a switching signal that makes the resulting time-varying system unstable, [8].

Let us now introduce the performance variable (scalar or 2-dimensional vector)

$$z(t) = \gamma_j y(t) + \delta_i \dot{y}(t)$$

and the performance index

$$J = \int_0^\infty z(t)'z(t)dt \tag{2}$$

The (vector) coefficients γ_j, $j = 1, 2, \cdots, n_\beta$ and δ_i, $i = 1, 2, \cdots, n_\alpha$, may depend on the switching signal $\sigma(t)$ in order to weight differently the contribution of the individual modes in the performance index.

Our aim is at finding a state-feedback strategy $\sigma = u(y, \dot{y})$ that minimizes J when $w(\cdot) = 0$ and the initial state $(y(0), \dot{y}(0))$ is given, albeit arbitrary. Notice that this problem admits a solution whenever the switched system is stabilizable, see [8]. This occurs for instance when a single (i,j) mode is stable. The problem generalizes to switched system the classical linear quadratic optimal control theory. It is interesting to stress that the solution to this problem also provides the optimal switching strategy in the case when the initial state is zero and $w(t)$ is an impulsive signal. Indeed, the latter situation reduces to the former by taking an initial state $y(0) = 0$ and $\dot{y}(0) = 1$. In addition the optimal strategy minimizes the variance of $z(t)$ when $w(t)$ is a white noise process.

3 Computation of the Optimal Switching

The optimal control problem for the switched system can be solved by a suitable adaptation of the Hamilton-Jacobi equation, see e.g. [18]. To compact the notation we are well advised to rewrite the system in state-space form

$$\dot{x}(t) = A_{\sigma(t)} x(t) + Bw(t) \tag{3}$$
$$z(t) = E_{\sigma(t)} x(t) \tag{4}$$

where

$$x = \begin{bmatrix} y \\ \dot{y} \end{bmatrix}, \quad A_\sigma = \begin{bmatrix} 0 & 1 \\ -\beta_j & -\alpha_i \end{bmatrix}, \quad B = \begin{bmatrix} 0 \\ 1 \end{bmatrix}, \quad E_\sigma = \begin{bmatrix} \gamma_j \ \delta_i \end{bmatrix}$$

The solution to the optimal control problem exists if it is possible to compute a continuous, piecewise differentiable and positive definite function $V(y, \dot{y}) = V(x)$ satisfying

$$0 = \min_\sigma \left(\frac{\partial V}{\partial x} A_\sigma x + x' E_\sigma' E_\sigma x \right) \tag{5}$$

The optimal switching rule is then given by

$$\sigma = u(y, \dot{y}) = u(x) = \arg\min_{\sigma} \left(\frac{\partial V}{\partial x} A_\sigma x + x' E'_\sigma E_\sigma x \right) \tag{6}$$

and $V(x(0))$ represents the optimal value of the performance index when $x(0)$ is the initial state. It is obvious that a sufficient condition for the existence of the optimal solution is the existence of a stabilizing switching rule. For instance, this condition is guaranteed when one of the modes is already stable or when there exists a stable convex combination of the $M = n_\alpha n_\beta$ modes, see e.g. [8].

The solution to equation (5) can be found through an iterative numerical procedure. It is expedient to perform a change of coordinates from the phase plane $(y, \dot{y})$ to the polar coordinates (ρ, θ). To this purpose we write

$$x = \begin{bmatrix} \rho cos(\theta) \\ \rho sin(\theta) \end{bmatrix}, \quad W(\rho,\theta) = V(x), \quad \frac{\partial V}{\partial x} = \begin{bmatrix} \frac{\partial W}{\partial \rho} & \frac{\partial W}{\partial \theta} \end{bmatrix} \begin{bmatrix} cos(\theta) & sin(\theta) \\ -\rho^{-1} sin(\theta) & \rho^{-1} cos(\theta) \end{bmatrix}$$

Notice now that the optimal switching rule is invariant with respect to a scaling of the norm of $x(0)$ and a change of sign. Consequently, for each real number ϵ and each initial state $x(0) \in \mathbb{R}^2$, we have $V(\epsilon x(0)) = \epsilon^2 V(x(0))$. This reflects in simple constraints for $W(\rho, \theta)$, namely $W(\rho,\theta) = \rho^2 \bar{W}(\theta)$ and $\bar{W}(\theta - \pi) = \bar{W}(\theta)$. By using the polar coordinates and recalling the definitions of A_σ and E_σ, equation (5) can be equivalently rewritten as

$$0 = \min_{\sigma} H(\theta, \sigma) \tag{7}$$

where

$$\begin{aligned} H(\theta,\sigma) = {} & 2sin(\theta)\left((1-\beta_j)cos(\theta) - \alpha_i sin(\theta)\right)\bar{W} \\ & - \left(sin(\theta)^2 + \beta_j cos(\theta)^2 + \alpha_i sin(\theta)cos(\theta)\right)\frac{d\bar{W}}{d\theta} \\ & + \left(\gamma_j cos(\theta) + \delta_i sin(\theta)\right)'\left(\gamma_j cos(\theta) + \delta_i sin(\theta)\right) \end{aligned} \tag{8}$$

As obvious, the role of ρ becomes immaterial and the only unknown is the function $\bar{W}(\theta)$. This means that the switching surfaces are straight line in the phase plane. Moreover, being $H(\theta + \pi, \sigma) = H(\theta, \sigma)$, such surfaces turn out to be symmetric with respect to the origin and the modes activation regions are cones, as already known, see e.g. [16].

The problem is then to find a solution $\bar{W}^o(\theta)$, $\theta \in [0, \pi)$, and the optimal switching strategy σ as a function of θ, namely

$$\sigma^o = u^o(\theta) = \arg\min_{\sigma} H(\theta, \sigma) \tag{9}$$

We have devised a simple discretization algorithm to work out the solution. Precisely, consider a discretization of the upper unit semicircle $\theta = k\Delta\theta$, $\Delta\theta = \frac{\pi}{N}$, $k = 0, 1, \cdots, N-1$ and take the symmetric approximation of the derivative, i.e.

$$\frac{d\bar{W}}{d\theta} \simeq \frac{\bar{W}(\theta + \Delta\theta) - \bar{W}(\theta - \Delta\theta)}{2\Delta\theta}, \quad \bar{W}(-\Delta\theta) = \bar{W}((N-1)\Delta\theta), \quad \bar{W}(\pi) = \bar{W}(0)$$

Now letting

$$s = \begin{bmatrix} \sigma(0) \\ \sigma(\Delta\theta) \\ \sigma(2\Delta\theta) \\ \vdots \\ \sigma((N-1)\Delta\theta) \end{bmatrix}, \; v = \begin{bmatrix} \bar{W}(0) \\ \bar{W}(\Delta\theta) \\ \bar{W}(2\Delta\theta) \\ \vdots \\ \bar{W}((N-1)\Delta\theta) \end{bmatrix}, \; h(s) = \begin{bmatrix} H(0,\sigma(0)) \\ H(\Delta\theta,\sigma(\Delta\theta)) \\ H(2\Delta\theta,\sigma(2\Delta\theta)) \\ \vdots \\ H((N-1)\Delta\theta,\sigma(N-1)\Delta\theta) \end{bmatrix}$$

we can rewrite (8) as

$$h(s) = L(s)v + m(s) \tag{10}$$

where the N^2 square matrix $L(s)$ and the vector $m(s)$ can be easily deduced from (8). Notice that $L(s)$ is a tridiagonal matrix except for the first and last rows. The algorithm starts with an initial vector $v_{(0)}$, for instance a vector with identical positive entries, or the one obtained from the Lyapunov function of a stable mode. Then, the core of the algorithm is based on equations (7), (9) and (10). The main iteration step is to compute

$$s_{(i)} = \begin{bmatrix} \sigma_{(i)}(0) \\ \sigma_{(i)}(\Delta\theta) \\ \sigma_{(i)}(2\Delta\theta) \\ \vdots \\ \sigma_{(i)}((N-1)\Delta\theta) \end{bmatrix}$$

and $v_{(i+1)}$ in the following way

$$\begin{aligned} s_{(i)} &= \arg\min_{s} \left(L(s)v_{(i)} + m(s)\right) \\ v_{(i+1)} &= -L(s_{(i)})^{-1}m(s_{(i)})\eta + (1-\eta)v_{(i)} \end{aligned}$$

where the above minimization of the vector $L(s)v_{(i)} + m(s)$ is considered elementwise and $\eta \in (0,1]$ is a parameter controlling the smoothness of the solution. The algorithm ends when $\|v_{(i^\star+1)} - v_{(i^\star)}\|$ is smaller than a given tolerance. The entries of $s_{(i^\star)}$ yield the optimal control strategy in the θ grid points. Finally, the optimal value of the performance index is $J^o = \rho(0)^2\bar{W}(\theta(0))$. This last value, in the grid points, can be found by taking the appropriate entry of vector $v_{(i^\star)}$. The convergence analysis of the algorithm as well as its computational complexity are worth of further investigation. However, the algorithm was tested in many examples and convergence was always observed when at least one mode was stable.

4 A Special Case

This section is mainly devoted to discuss the special situation of equation (1) when the stiffness parameter β_j is fixed, i.e. $\Omega_\beta = \{1\}$, $\beta_1 = \beta > 0$, and the damping parameter α_i may switch between two values, i.e. $\Omega_\alpha = \{1,2\}$, $\alpha_1 =$

$\alpha_{min} \geq 0$, $\alpha_2 = \alpha_{max} > \alpha_{min}$. For simplicity we set $\alpha_{min} = 0$. We assume that the performance index is the integral of $\ddot{y}(t)^2$, so that $\delta_i = \alpha_i$ and $\gamma_j = \beta$. In mechanical systems this corresponds to minimizing the integral of the squared acceleration. The case when also the parameter β_j can switch is briefly discussed at the end of the section.

The algorithm presented in the previous section has been run for different values of β and α_{max} and $N = 500$. In all outcomes the optimal switching surfaces have the shape drawn in Figure 1. As can be noticed, one commutation

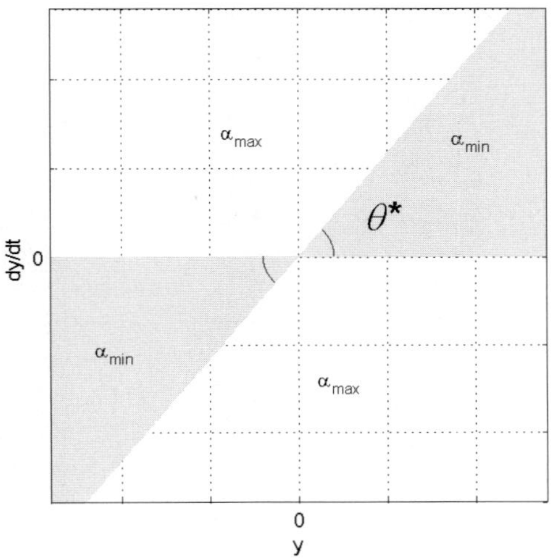

Fig. 1. Shape of the switching surfaces

occurs when the velocity $\dot{y}$ changes its sign, whereas the second commutation is triggered by the crossing of a straight line with angle $\theta^\star(\alpha_{max}, \beta)$. Therefore, the optimal strategy suggests that a null damping coefficient is more effective when y and $\dot{y}$ have the same sign and the ratio $\dot{y}/y$ is below a given threshold, namely $tan(\theta^\star)$. Figure 2 shows the value (in degrees) of $\theta^\star(\alpha_{max}, \beta)$ as a function of α_{max} for different values of β. In order to illustrate the role of the switching rule, in Figure 3 the phase portrait of the optimal switched system is plotted for the particular choice $\alpha_{max} = 1$, $\beta = 1$.

Finally, we have computed the performance index corresponding to the particular initial condition $\theta(0) = \pi/2$ and $\rho(0) = 1$. In Figure 4 the optimal performance index J^o is plotted against α_{max} for different values of β. The dashed curves correspond to the L_2 performance associated with the constant damping coefficient α_{max}. It is apparent that the switched damping improves significantly on the constant specially for high values of α_{max}.

The transient behavior of $\ddot{y}(t)$ is plotted in Figure 5 in the case $\alpha_{max} = 1$. The solid curve corresponds to the optimal switching (OS), while the dashed curve is obtained with constant damping α_{max}. The advantage of commuting to $\alpha_{min} = 0$ at appropriate time-instants is apparent.

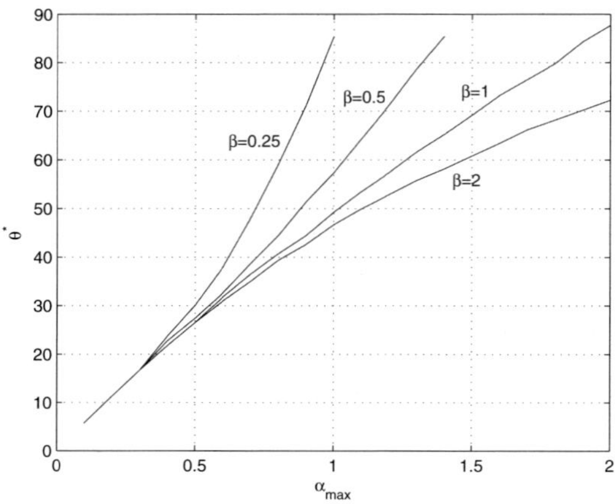

Fig. 2. $\theta^\star(\alpha_{max}, \beta)$ as a function of α_{max} for different values of β

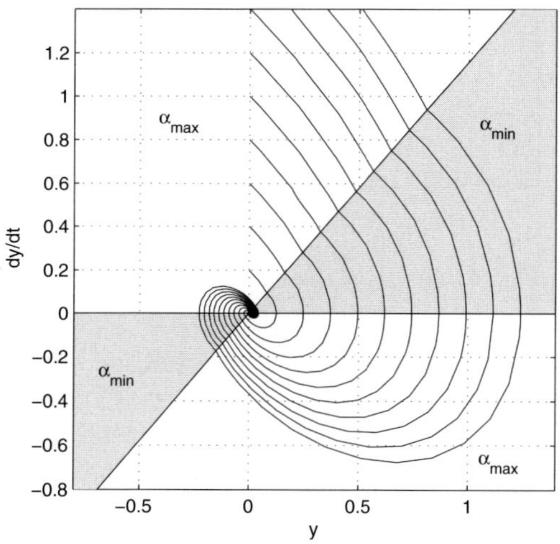

Fig. 3. Phase portrait of the optimal switched system for $\alpha_{max} = 1$ and $\beta = 1$

To enlighten the potentiality of the algorithm, we have considered the same optimization problem by allowing, in addition, for a switching stiffness parameter, namely $\Omega_\beta = \{1\ ,2\}$, $\beta_1 = \beta_{min} > 0$, $\beta_2 = \beta_{max} > \beta_{min}$. For the sake of conciseness, we report the results only for the case $\alpha_{max} = 1$, $\beta_{max} = 1$, $\beta_{min} = 0.5$. In Figure 6 the resulting optimal switching surfaces are shown. This more complicated switching rule obviously gives a better performance. For instance, the performance index associated with $\theta(0) = \pi/2$ and $\rho(0) = 1$ is $J^o = 0.664$, that is lower than the corresponding points in Figure 4 (curves $\beta = 1$ and $\beta = 0.5$).

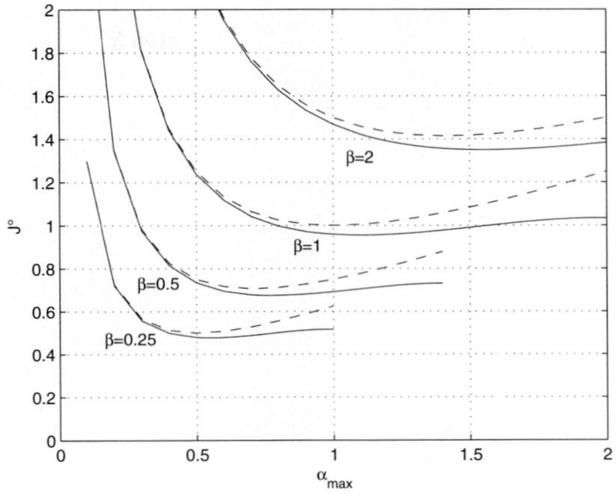

Fig. 4. Optimal performance index with $\theta(0) = \pi/2$ and $\rho(0) = 1$

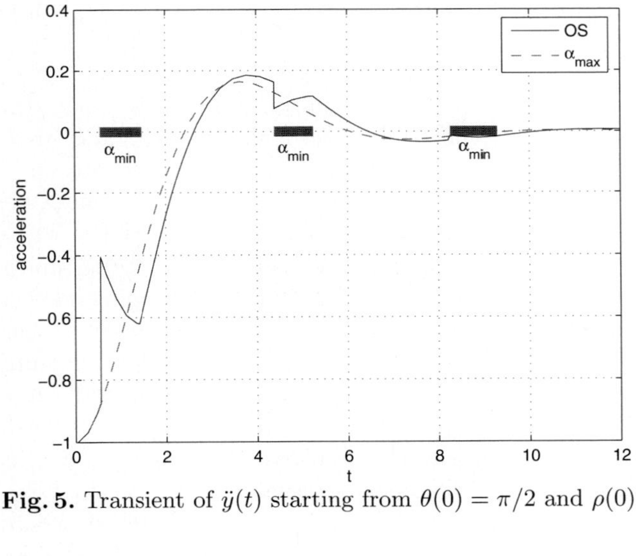

Fig. 5. Transient of $\ddot{y}(t)$ starting from $\theta(0) = \pi/2$ and $\rho(0) = 1$

5 An Application

This section discusses a practical application of the optimal switching control design presented before. Precisely, we consider the problem of comfort-oriented control of a semi-active suspension system in road vehicles. Our aim is to compare the achievable performance with the one provided by the classical switching rule based on the so-called two-state Sky-Hook (SH) approach, [21]. The model is as follows:

$$M\ddot{\xi}(t) = -c(t)(\dot{\xi}(t) - \dot{\xi}_t(t)) - k(\xi(t) - \xi_t(t)) + k\Delta_s - Mg$$

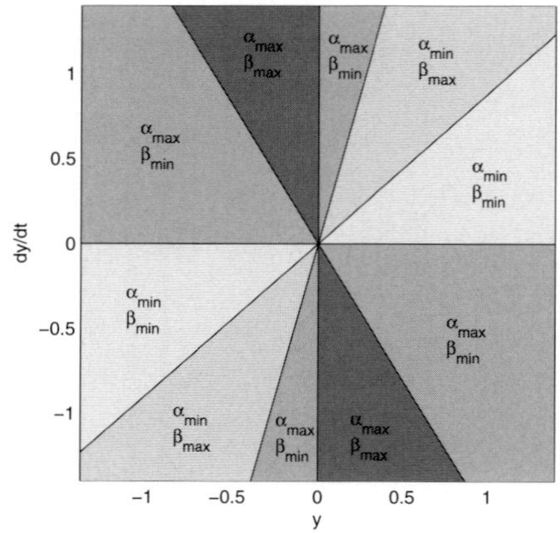

Fig. 6. Optimal switching surfaces with both switching damping and switching stiffness

$$m\ddot{\xi}_t(t) = c(t)(\dot{\xi}(t) - \dot{\xi}_t(t)) + k(\xi(t) - \xi_t(t)) - k_t(\xi_t(t) - \xi_r(t)) - k\Delta_s + k_t\Delta_t - mg$$
$$\dot{c}(t) = -\eta c(t) + \eta c_{in}(t)$$

where $\xi(t)$, $\xi_t(t)$ and $\xi_r(t)$ are the vertical position of the body, the unsprung mass and the road profile, respectively. The coefficients M and m are the quarter-car body mass and the unsprung mass (tire, wheel, brake, etc...), respectively. The parameters η, k and k_t are the bandwidth of the active shock absorber, the stiffness of the suspension spring and of the tire, respectively. The coefficients Δ_s and Δ_t are the length of the unloaded suspension spring and of the tire. Finally, $c(t)$ and $c_{in}(t)$ are the actual and requested damping coefficients of the passive shock-absorber. In order to simplify the computations we assume that η is large enough so that $c(t) \sim c_{in}(t)$. Moreover we consider a genuine switching strategy, so that $c(t) = c_i$ can assume only two values, namely $c_1 = c_{min} \geq 0$ and $c_2 = c_{max} > c_1$, to be specified later on.

The control objective consists in minimizing the chassis vertical acceleration $\ddot{\xi}(t)$ by a suitable choice of the control variable $c(t) \in \{c_{min},\ c_{max}\}$. In the classical two-state SH approach [21], the system is switched according to the sign of $\dot{\xi}(t)(\dot{\xi}(t) - \dot{\xi}_t(t))$. In order to fit this example in the framework of the present paper, let us take the variations $\delta\xi(t)$ and $\delta\xi_t(t)$ of $\xi(t)$ and $\xi_t(t)$ around an equilibrium point associated with zero road profile, arriving to the system

$$M\delta\ddot{\xi}(t) = -c_i(\delta\dot{\xi}(t) - \delta\dot{\xi}_t(t)) - k(\delta\xi(t) - \delta\xi_t(t)) \tag{11}$$
$$m\delta\ddot{\xi}_t(t) = c_i(\delta\dot{\xi}(t) - \delta\dot{\xi}_t(t)) + k(\delta\xi(t) - \delta\xi_t(t)) - k_t(\delta\xi_t(t) - \xi_r(t)) \tag{12}$$

Notice that this is a 2-DOF system. In order to apply the optimal switching control design previously discussed, we make the (realistic) assumption that k_t is sufficiently high so that the displacement of the tire can be approximated by the road profile, i.e. $\delta\xi_t(t) \simeq \xi_r(t)$. Consequently, letting $y(t) = \delta\xi(t) - \xi_r(t)$, the approximated model can be written as

$$\ddot{y}(t) = -\frac{c_i}{M}\dot{y}(t) - \frac{k}{M}y(t) + \ddot{\xi}_r(t)$$

Thus, we have recovered equation (1) with $\alpha_i = c_i/M$, $\beta_j = \beta = k/M$ and $w(t) = \ddot{\xi}_r(t)$. Moreover, to improve comfort, it is advisable to minimize the integral of $\ddot{y}(t)^2$. The situation is exactly the one discussed in Section 4, and, consequently, the optimal switching surfaces are those qualitatively depicted in Figure 1. The following parameters have been selected, see [19]: $M = 400kg$, $m = 50kg$, $k = 2.0 \times 10^4 N/m$, $k_t = 2.5 \times 10^5 N/m$, $c_1 = c_{min} = 3.0 \times 10^2 Ns/m$ and $c_2 = c_{max} = 3.9 \times 10^3 Ns/m$. The optimal switching angle has been computed on the basis of α_{max} and β through the numerical algorithm of Section 3 with $N = 500$ grid points. It turns out $\theta^\star = 86.6^o$.

Two sets of simulations have been carried out, by applying both the Sky-Hook (SH) and the optimal switching (OS) control laws to the 2-DOF system (11), (12). The first set of simulations refers to the response to a unit impulse on the

Table 1. Performance of the different control strategies under an impulsive or a white noise disturbance

	OS	SH	PS_1	PS_2
$\int_0^\infty \ddot{y}(t)^2 dt$ for $\ddot{\xi}_r = \delta(t)$	7.446	8.288	26.548	8.307
$\frac{\int_0^{20} \ddot{y}(t)^2 dt}{\int_0^{20} \ddot{\xi}_r(t)^2 dt}$ for $\ddot{\xi}_r \sim WN$	0.623	0.787	3.558	0.719

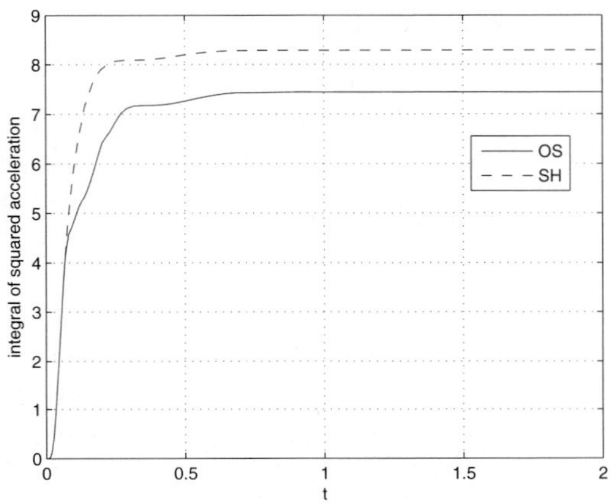

Fig. 7. Time history of the integral of $\ddot{y}(t)^2$ due to an impulse of $\ddot{\xi}_r(t)$

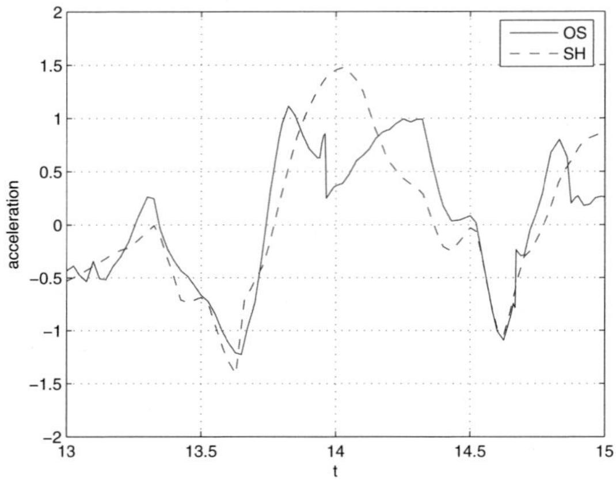

Fig. 8. Chassis acceleration during a short interval under a random road profile

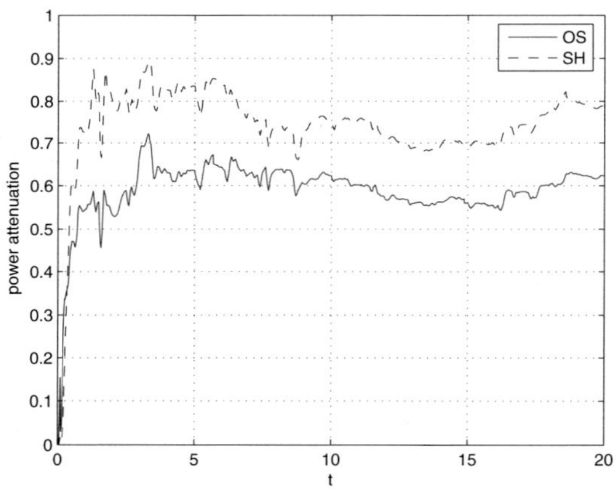

Fig. 9. Power attenuation under a random road profile

road acceleration $w(t)$, namely a ramp on the road profile. The first row of Table 1 reports the integral of the squared chassis acceleration obtained with different control strategies. The notation PS_1 and PS_2 refers to a passive suspension with fixed damping coefficient equal to c_{min} and c_{max}, respectively. As apparent from Table 1, the algorithm OS outperforms all other strategies.

Figure 7 shows the integral of the square of the chassis acceleration against time. It can be seen that OS is capable of lowering the acceleration in the transient better than SH, even if its design is based on a simplified 1-DOF model.

In the second set of simulations the road profile $\xi_r(t)$ has been generated as the double integral of a sample realization of a white noise process with power $\chi^2 = 0.1$. The performance of the four algorithms above has been measured as the power attenuation on the chassis acceleration, namely the ratio

$$\Theta_T = \frac{\int_0^T \ddot{y}(t)^2 dt}{\int_0^T \ddot{\xi}_r(t)^2 dt}$$

This value, for $T = 20$ *sec.*, is reported in the second row of Table 1. Figure 8 shows the behavior of the acceleration. The plot has been restricted to an interval of 2 seconds, in order to better represent the effects of the commutations. The OS strategy outperforms SH at the price of faster switching commutation and shorter dwell intervals.

Finally the power attenuation Θ_T as a function of T is plotted in Figure 9.

6 Conclusion

In this paper we have developed an optimal switching control law for 1-DOF oscillating systems. For such systems, two parameters can be assigned in finite sets of values in order to minimize a quadratic cost. The special case of 2 different values is thoroughly discussed and the practical application of comfort-oriented control of a suspension system has been also presented. In the authors' feeling, the methodology developed in this work can be successfully applied to diverse engineering problems where a quadratic objective has to be minimized. In particular further studies will be carried out for the control of suspension systems where, in addition to the damping coefficient, also the stiffness coefficient can assume different values depending on the measurements (reactive control).

References

1. Branicky, M. S., "Multiple Lyapunov functions and other analysis tools for switched and hybrid systems". *IEEE Trans. Automat. Contr.*, vol. 43, pp. 475–482, 1998.
2. Hespanha, J. P.,"Uniform stability of switched linear systems : extensions of LaSalle's principle" *IEEE Trans. Automat. Contr.*, vol. 49, pp. 470–482, 2004.
3. Hockerman-Frommer, J., Kulkarni, S. R., and Ramadge, P. J., "Controller switching based on output predictions errors", *IEEE Trans. Automat. Contr.*, vol. 43, pp. 596–607, 1998.
4. Johansson, M., and Rantzer, A., "Computation of piecewise quadratic Lyapunov functions for hybrid systems", *IEEE Trans. Automat. Contr.*, vol. 43, pp. 555–559, 1998.
5. Ye, H., Michel, A. N., and Hou, L., "Stability theory for hybrid dynamical systems", *IEEE Trans. Automat. Contr.*, vol. 43, pp. 461–474, 1998.
6. DeCarlo, R. A., Branicky, M. S., Pettersson S., and Lennartson, B., "Perspectives and results on the stability and stabilizability of hybrid systems", *Proceedings of the IEEE*, vol. 88, no. 7, pp. 1069–1082, 2000.
7. Liberzon, D., and Morse, A. S., "Basic problems in stability and design of switched systems", *IEEE Control Systems Magazine*, vol. 19, pp. 59-70, 1999.
8. Liberzon, D., *Switching in Systems and Control*, Birkhauser, 2003.
9. Geromel, J. C., and Colaneri, P., "Stabilization of continuous-time switched linear systems", *SIAM J. Control Optim.*, to appear, 2006.

10. Sussmann, H. ., "New theories of set-valued differentials and new versuions of the maximum principle of optimal control theory", in *Nonlinear Control in the year 2000*, A. Isidori, F. Lamnabhi-Lagarrige, W. Respondek Eds, Vol2, pp- 487-526, Springer verlag, 2001.
11. Egerstedt M., P. Ogren, O. Shakernia, J. Ligeros, "Toward optimal control of switched linear systems", Proc. 39th IEEE CDC, pp. 587-592, 2000.
12. P. Riedinger, C. Iung, and F. Kratz, "An optimal control approach for hybrid systems", *European J. of Contr.*, vol. 49, no. 5, pp. 449-458, 2003.
13. Xuping Xu and Panos J.Antsaklis, "Results and Perspectives on Computational Methods for Optimal Control of Switched Systems", *Sixth International Workshop on Hybrid Systems Computation and Control (HSCC)*, Prague, Czech Republic, 2003.
14. S. Bengea and R. DeCarlo, "Optimal control of switching system", *Automatica*, vol. 41, no. 2, pp. 11.27, 2005.
15. Giua, A., Seatzu, C. and Van der Mee, C.M., "Optimal control of autonomous linear systems switched with a preassigned finite sequence", in Proc. of the *ISIC*,pp. 144-149, 2001.
16. Bemporad, A., Corona, D., Giua, A. and Seatzu, C., "Optimal state feedback quadratic regulation of linear hybrid automata", in Proc. of the *IFAC ADHS*, pp. 407-412, 2003.
17. Seatzu C., D. Corona, A. Giua, A. bemporad, "Optimal control of continuous time switched affine systems", *IEEE Trans. on Automatic Control*, Vol. 51, N.5, pp. 726-741, 2006.
18. Athans, M. and Falb, P., "Optimal Control: An Introduction to the Theory and its Applications", New York: McGraw-Hill, 1966.
19. Savaresi, S. M., Silani, E., and Bittanti, S., "Acceleration-Driven-Damper (ADD): An optimal control algorithm for confort-oriented semiactive suspensions", *Trans. of the ASME*, vol. 127, pp. 218-228, 2006.
20. Geromel, J.C., Colaneri P. and Bolzern, P.,"Dynamic output feedback control of switched linear systems", submitted, 2006.
21. Williams R.A., "Automotive active suspensions, part I: Basic principles", *IMechE*, 211, pp. 415-426, 1997.

Feedback Scheduling for Pipelines of Tasks*

Tommaso Cucinotta[1] and Luigi Palopoli[2]

[1] Scuola Superiore S. Anna - Pisa (Italy)
author_secondname@sssup.it
[2] University of Trento - Trento (Italy)
author_secondname@dit.unitn.it

Abstract. The problem analysed in this paper is how to effectively share a pool of resources amongst software applications consisting of pipelines of communicating tasks. The goal is to guarantee that specified Quality of Service (QoS) requirements are met. To this end, we advocate the use of a scheduling mechanism able to reserve fraction of the different resources to the competing tasks. Our work is focused on a feedback controlled adaptation of these fractions based on measurements of the QoS experienced by the application.

1 Introduction

In this paper, we consider embedded software applications consisting of multiple tasks, which run on different and networked computing nodes. Significant examples include (but are not limited to) MPEG streaming, video-surveillance and Voice-over-IP. The particular problem we deal with is how to effectively *share* resources without compromising the real-time behaviour of the applications. In fact, while resource sharing is a cost effective and flexible solution enabled by modern operating systems and middleware infrastructures, it also introduces non-deterministic scheduling delays affecting the Quality of Service of the application.

In the past few years, researchers have been confronted with the problem of constructing a real-time software infrastructure matching the temporal guarantees of a dedicated solution with the efficiency of resource sharing. A large body of results stemming from this activity has focused on predictable scheduling mechanisms [1,2,3,4]. The idea is to reserve a certain fraction of the resources to the competing tasks guaranteeing that this allocation will be respected in time (within a specified granularity). In the Resource Kernels project [3], the resource reservation approach has been successfully applied to different types of resources (including disk and network).

More recently, this technique has been complemented with adaptive mechanisms able to dynamically track the resource requirements of each task: the idea of *feedback scheduling*. In this framework, the parameters of the schedulers are

* This work has been supported by the FP6/2005/IST/5-034026 European Commission Project named FRESCOR.

A. Bemporad, A. Bicchi, and G. Buttazzo (Eds.): HSCC 2007, LNCS 4416, pp. 131–144, 2007.

used as "actuators" to adjust the QoS measured by appropriate *sensors* within acceptable bounds. In particular, Stankovic et al. [5] propose a similar mechanism using the deadline of an EDF scheduler as an *actuation* mechanism. In [6,7,8], the use of a better suited scheduler enables the design of the feedback law based on a precise dynamic model of the "plant" to be controlled, thus making for a better founded application of control theory to this problem.

The results cited above do not offer a general solution to the problem of QoS management, since do not support real-time applications using multiple resources. In general, the need for different types of resources may generate undesired effects of unexpected harshness, unless the interaction of different allocation mechanisms is adequately accounted for. In particular, Rajikumar et al. [9] developed a framework (called QRAM) that decides the fraction of different resources to be allocated to the applications by solving an optimisation (NP-hard) problem. However, the resulting allocation is static and it does not allow the scheduler to accommodate applications with dynamically varying or scarcely known resource requirements. What we need is an appropriate combination of feedback control and multiple resource management.

This paper offers a first contribution in this direction. We consider applications consisting of pipelines of tasks which communicate by means of intermediate buffers, where tasks use resources of different type. For each resource, we use a reservation-based scheduler [3] that allows one to reserve a specified fraction (bandwidth) of the resource to each task using it. This technique allows to define a *dynamic model* that describes the temporal evolution of the system. In particular, we introduce a performance metric (called scheduling error) that is a good indicator for both the QoS offered by the application and its efficiency in utilising resources. These issues are discussed in Section 2. Based on this model, we define a control strategy that dynamically changes the bandwidth of each task. This technique is called *adaptive reservations* and has been proposed in previous work for a single resource [10,6]. The central contribution of this paper is to extend it to pipeline of tasks. To this end, we use a decentralised control algorithm, in which each stage of the pipeline is associated to a local *task controller* while a global supervisor enforces consistency on the total allocated bandwidth for each resource. Each task controller is based on a combination of a predictor and of a controller that counters the fluctuations of the resource requirements. Contrary to previous work on the control of pipelined applications [11], we can prove practical stability of our algorithm, based on the dynamic model of the plant. The control design is shown in Section 3. Finally in Section 4, we show simulations proving the effectiveness of the approach (an implementation of the technique in the Linux kernel is under way).

2 Task Model and Scheduling

We consider a set of applications $\mathcal{A}^{(1)}, \ldots, \mathcal{A}^{(L)}$ sharing a pool of resources $\mathcal{R} = \{\mathcal{R}^1, \ldots, \mathcal{R}^R\}$. Resources can be of potentially different kind (e.g., CPU, disks, network links etc). An application $\mathcal{A}^{(i)}$ consists of a pipeline of $n^{(i)}$ tasks $\mathcal{A}^{(i)} = (\tau^{(i)[1]}, \ldots, \tau^{(i)[n^{(i)}]})$, which are pairwise connected by uni-directional buffers. For notational convenience, we will henceforth omit the (i) superscript whenever the discussion refers to a single application. Each resource is allocated by a scheduler that operates using scheduling parameters decided for the different tasks.

2.1 The Task Model

Consider an application $\mathcal{A}$ associated to a pipeline $\tau^{[1]}, \ldots, \tau^{[n]}$. Task $\tau^{[1]}$ acquires its input from a data source, which produces data blocks (tokens) at a regular rate. Tokens are processed in sequence by each task in the pipeline and arrive to $\tau^{[n]}$ which sends the result to an output device (e.g. a screen if we are dealing with a video stream). The first task in the pipeline is periodically activated at the production period T of the data source (*time-triggered* activation). The remaining tasks of the pipeline are activated as soon as a new element is pushed into the input buffer by the task in the previous stage of the pipeline (*event-triggered* activation).

When task $\tau^{[j]}$ receives the kth token, it instantiates a job $J_k^{[j]}$, which cannot start before the termination of $J_{k-1}^{[j]}$ (e.g. the decoding of a frame in a MPEG stream cannot start if the previous frame is still being decoded). The job consumes one input token and produces one output token, which is placed on the output buffer. When a job is terminated, in absence of a new token to process, the task is blocked on a read operation. On the contrary, write operations are assumed to be non-blocking. As discussed next, our adaptive scheduling solution allows us to respect this semantic with a finite number of buffers. We denote by $s_k^{[j]}$ the start time of $J_k^{[j]}$ and by $f_k^{[j]}$ its finishing time.

As a simplifying assumption, we require that each task $\tau^{(i)[j]}$ uses *only one* resource, denoted by $r^{(i)[j]}$. The resource requirement of the kth job of task $\tau^{[j]}$ is denoted by $c_k^{[j]}$. Since we are interested in real-time applications, job $J_k^{(i)[j]}$is associated a deadline $d_k^{(i)[j]}$, which is a *soft* execution constraint, i.e., occasional failures are tolerated provided that the problem be kept in check. Since the activation pattern of the pipeline is periodic, it is reasonable to consider periodically spaced out deadlines: $d_{k+1}^{[j]} = d_k^{[j]} + T$. Concerning the initial values of the absolute deadlines $\left\{d_1^{[j]}\right\}$, we set $d_1^{[j]} = d_1^{[j-1]} + T$, resulting into $d_k^{[j]} \triangleq (k + j - 1)T$.

Example. Figure 1 shows an example of the activation pattern of the jobs. The bottom line reports the execution of the first task in the pipeline, whose job activations are periodically activated. For the second task in the pipeline (top line), the job $J_k^{[2]}$ cannot start until job $J_k^{[1]}$ finishes and produces the kth token. On the contrary, $J_{k+1}^{[2]}$ starts right after $J_k^{[2]}$ finishes because $J_{k+1}^{[1]}$ has already terminated by that time and the $(k+1)^{th}$ token is already available.

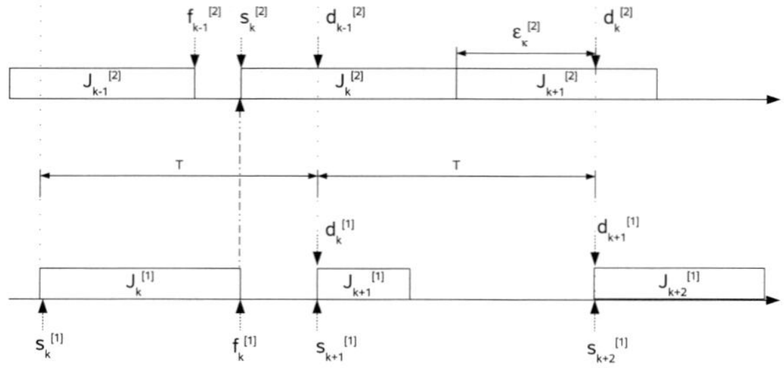

Fig. 1. Example showing the activation pattern of the jobs and the notation

2.2 The Scheduler

The scheduler plays the role of a "plant" to be controlled. Therefore, we need an algorithm exposing *sensors* and *actuators*. Moreover, a sound design for a feedback scheduler has to be founded upon a realistic dynamic model relating the QoS evolution to the control choices.

To attain these goals, we *restricted* our choice to scheduling algorithms that closely approximate a *fluid allocation* of the resource (see [12]). In simple terms, it means that each task $\tau^{(i)[j]}$ executes as if using a dedicated resource whose speed is a fraction (which we call *bandwidth*) of the actual resource $r^{[j]}$.

Actuators. In our framework, the bandwidth $b_k^{(i)[j]}$ can be set for each job of $\tau^{(i)[j]}$. Therefore, it can be used as an *actuator* to control the evolution of the QoS. However, the sum of the bandwidths reserved for a resource $\mathcal{R}^r$ must never exceed its total capacity C^r:

$$\forall r \in \mathcal{R}, \forall t \qquad \sum_{i,j,k\,:\,r^{(i)[j]}=r \,\wedge\, s_k^{(i)[j]} \leq t < f_k^{(i)[j]}} b_k^{(i)[j]} \leq C^r, \tag{1}$$

Sensors. We introduce the scheduling error $\epsilon_k^{(i)[j]} \triangleq f_k^{(i)[j]} - d_k^{(i)[j]}$ as a metric to quantify the violation of a deadline, which is measured by *appropriate* sensors inside the operating system. This quantity is a good indicator for both the QoS experienced by the task and its efficiency in utilising resources. Indeed, if task $\tau^{(i)[i]}$ produces an output to the end-users, a large positive scheduling error corresponds to an increased latency for the output of the output. In a real-time application, this degradation has to be coped with by by either dropping tokens or increasing the size of the buffers (which is an expensive). On the other hand, a negative value for the scheduling error is associated to an excessive allocation of the resource, as the job would have completed on-time with a smaller resource allocation as well. Therefore, it is required that the value of the scheduling error be kept as near as possible to zero.

Dynamic model. The evolution of the pipeline can be described by a discrete event model. Consider the kth token produced by the data source, which

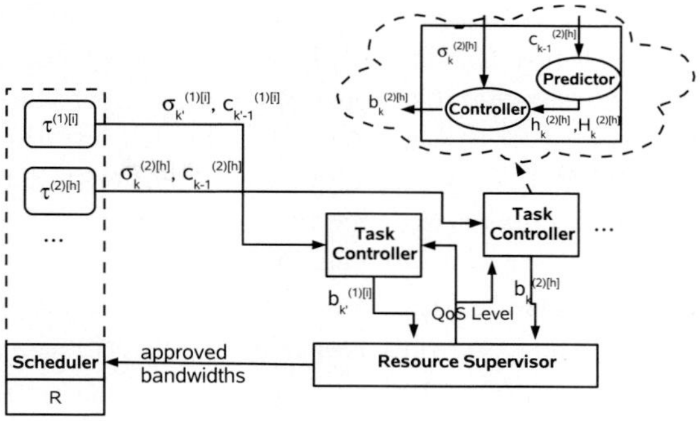

Fig. 2. The control Scheme adopted in this paper. We zoom in one task controller to show its internal structure.

determines the subsequent activations of the kth jobs for all the tasks in the application pipeline $J_k^{[1]}, \ldots, J_k^{[n]}$. The state of the pipeline can be described by a vector of state variables $\boldsymbol{\epsilon}_k$, where each component is the scheduling error $\epsilon_k^{[i]}$ that the ith element of the pipeline experiences when it processes the token.

Due to lack of space, in this context we omit a technical discussion on the derivation of the dynamic model (see [13] for details). For our purposes it is sufficient to say that the evolution of $\epsilon_k^{[i]}$ is given by:

$$\epsilon_k^{[j]} = \sigma_k^{[j]} + \frac{c_k^{[j]}}{b_k^{[j]}} - T, \text{ with } \sigma_k^{[j]} = \begin{cases} \max\left\{\epsilon_k^{[j-1]}, \epsilon_{k-1}^{[j]}\right\} & j \geq 2, k \geq 2 \\ \epsilon_k^{[j-1]} & j \geq 2, k = 1 \\ \max\left\{\epsilon_k^{[j-1]}, 0\right\} & j = 1, k \geq 2 \\ 0 & j = 1, k = 1 \end{cases}, \quad (2)$$

where, for notational convenience, the symbol $\sigma_k^{[j]} = s_k^{[j]} - d_{k-1}^{[j]}$ has been introduced to denote the start time of $J_k^{[j]}$ relative to the "ideal" value $d_{k-1}^{[j]}$.

3 Control Design

The system described in this paper is comprised of different software applications that evolve independently and asynchronously from each other. Moreover, referring to a single application, the different components of its state vectors are asynchronously collected at different times (in the state vector $\boldsymbol{\epsilon_k}$, index k refers to a token and not to a time instant). These considerations dictate a decentralised control scheme, as shown in Figure 2, in which each resource controller consists of a supervisor and of a collection of task controllers.

There is a strong separation of concerns between the task controllers and the supervisor. Roughly speaking, the role of task controllers is to track the resource requirements of each task to maintain or recover an equilibrium condition where the QoS level is regarded as acceptable. On the contrary, the role of the supervisor is to ensure that consistency Condition (1) is never violated. To this end, the supervisor is allowed to change the working conditions of the task controllers (i.e. the QoS level of the task) to lower their bandwidth requests. An important remark regarding both the task controllers and the supervisor is that, since the entire machinery is activated at every scheduling decision, the control algorithms are bound to take a few dozens of numeric operations.

Control decisions are taken at the *start time of a new job* $J_k^{[j]}$ according to the following scheme: 1) the task controller acquires information about the state of the task (in particular $\sigma_k^{[j]}$ and the computation time of the previous job $c_{k-1}^{[j]}$); 2) it computes the new bandwidth $b_k^{[j]}$ to be used for $J_k^{[j]}$ and submits the request to the supervisor; 3) the supervisor grants the request if it does not violate Condition (1), otherwise it changes the working mode of some of the task controllers to reduce the cumulative required bandwidth. For the sake of brevity, in this paper we will restrict the focus only to the design of the task controllers and discuss their properties. For what concerns the supervisor, we will simply offer some insight into how a controlled QoS degradation (leading to diminished bandwidth requirements) can actually be obtained. For a complete description of the supervisor the reader is referred [13].

3.1 Task Controllers

A task controller operates on a single task $\tau^{[j]}$ of the application $\mathcal{A}$, but the coordinated action of the different task controllers for $\mathcal{A}$ aims at attaining stability properties for the entire application. To this regard, a perfect allocation could be one where the each task experiences a null scheduling error and it receives, at each step, a bandwidth equal to $b_k^{[j]} = \frac{c_k^{[j]}}{T}$. This goal is not attainable, since it would entail predictive knowledge of $c_k^{[j]}$. A more realistic situation is one where the task controller uses a predictor able to produce, at the beginning of each job, a range $P_k^{[j]} = [h_k^{[j]}, H_k^{[j]}]$ such that $c_k^{[j]} \in P_k^{[j]}$(see Figure 2). The design of the predictor is largely application dependent (see [6,14]) and is out of the scope of the present paper. Because of the resource constraints, it is important to quantify a saturation level for the control laws, which is associated to the maximum bandwidth reserved to the tasks.

Considering this control scheme, a natural notion of practical stability is the following (see [15]):

Definition 1. *Consider the system defined by Equation* (2) *and let* $\mathcal{H}, \mathcal{G}$ *be two sets such that* $\mathcal{H} \subseteq \mathcal{G} \subseteq \mathbb{R}^n$. *Let* $\varepsilon_\mathbf{k} \triangleq [\varepsilon_k^{[1]} \ldots \varepsilon_k^{[n]}]$ *and* $\mathbf{P}_k \triangleq P_k^{[1]} \times \cdots \times P_k^{[n]}$. *The system is said* $(\mathcal{H}, \mathcal{G})$*-stabilisable in* M *steps iff there exists a control such that:*

1. $\mathcal{H}$ *is a Robust Controlled Invariant Set (RCIS):* $\forall k_0 : \boldsymbol{\epsilon}_{k_0} \in \mathcal{H} \wedge \mathbf{c_k} \in \mathbf{P}_k \forall k > k_0$ *implies* $\boldsymbol{\epsilon}_k \in \mathcal{H} \forall k > k_0$*;*
2. $\mathcal{H}$ *is robustly attractive from* $\mathcal{G}$ *in* M *steps:* $\forall k_0 : \boldsymbol{\epsilon}_{k_0} \in \mathcal{G} \wedge \mathbf{c_k} \in \mathbf{P}_k \forall k \in [k_0, \ldots k_0 + M]$ *implies* $\boldsymbol{\epsilon}_k \in \mathcal{G} \forall k \in [k_0 + 1, \ldots k_0 + M] \wedge \boldsymbol{\epsilon}_{k_0+M} \in \mathcal{H}$.

The first property in the definition above refers to the equilibrium condition: we require that the state evolves in a small set countering the possible fluctuations of $c_k^{[j]}$ within $P_k^{[j]}$. As far as the geometry of the $\mathcal{H}$ set is concerned, we are interested in a hypercube where the state of each task is constrained in the interval $\mathcal{I} = [-e, E]$ with $e, E \in \mathbb{R}^+$, thus $\mathcal{H} = \mathcal{I}^n = \mathcal{I} \times \ldots \times \mathcal{I}$. E quantifies the maximum delay a task can suffer, e quantifies the efficiency in utilising resources. Moreover, it can be shown [13] that the number of buffer elements required for intertask communications between two stages of the pipeline can be bounded by $\left\lfloor \frac{e+E}{T} \right\rfloor + 1$; therefore, with this number of elements, it never happens that a task finds a full buffer in its write operation.

The second property in Definition 1 relates to the evolution of the state when it is initially outside of $\mathcal{I}^n$. This situation occurs as a result of a perturbation such as a temporary system overload which prevents the task controller to use all the bandwidth it needs. After a perturbation of the equilibrium has terminated, k_0 is the first job entering the first task of the pipeline. Generally, the scheduling errors experienced by the task instances $k_0 - 1$ on all the pipeline stages $\left\{ J_{k_0-1}^{[j]} \right\}_{j=1,2,\ldots}$ deviate from $\mathcal{I} = [-e, E]$ and are in the set $\mathcal{J} =[-e, L]$, with $e, L \in \mathbb{R}^+$ and $L \geq E$. In this case, L quantifies the maximum delay that tasks in the pipeline will suffer. The definition requires that we are able to reduce the state from $\mathcal{G} = \mathcal{J}^n$ to $\mathcal{H} = \mathcal{I}^n$ in at most M steps. The requirements of a fixed number of steps (as opposed to an asymptotic definition) makes the property of practical interest for system design.

In the sequel, we will separately look at control schemes that attain Robustly Controlled Invariance and Robust attractivity.

Maintaining the equilibrium. The following offers necessary and sufficient conditions for the existence of a RCIS, and describes a family of control laws attaining it.

Theorem 1. *A control law attaining robust controlled invariance of $\mathcal{I}^N$ with $\mathcal{I} =[-e, E]$ exists iff the following conditions hold*

$$\tilde{H} \leq T \wedge \begin{cases} e + \alpha^{[1]} E \geq T \left(1 - \alpha^{[1]}\right) \\ e + E \geq T \left(\frac{1-\alpha}{\alpha}\right), \end{cases} \tag{3}$$

where $\tilde{H} \triangleq \max_j \left\{ \frac{\sup_k \{H_k^{[j]}\}}{B^{[j]}} \right\}$, $\alpha^{[1]} \triangleq \inf_k \left\{ \alpha_k^{[1]} \right\}$, $\alpha \triangleq \min_{j \geq 2} \left\{ \inf_k \left\{ \alpha_k^{[j]} \right\} \right\}$, *with* $\alpha_k^{[j]} \triangleq \frac{h_k^{[j]}}{H_k^{[j]}}$. *Furthermore, the control laws meeting the goal requirements have to be chosen in the range* $b_k^{[j]} \in [\mathcal{B}_L^{[j]}(\sigma_k^{[j]}), \mathcal{B}_H^{[j]}(\sigma_k^{[j]})]$, *where:*

$$\mathcal{B}_L^{[j]}(\sigma) = \frac{H_k^{[j]}}{T + E - \sigma_k^{[j]}}, \; \mathcal{B}_H^{[j]}(\sigma) = \min \left\{ \frac{h_k^{[j]}}{T - e - \sigma_k^{[j]}}, B^{[j]} \right\} \tag{4}$$

Proof. See Appendix.

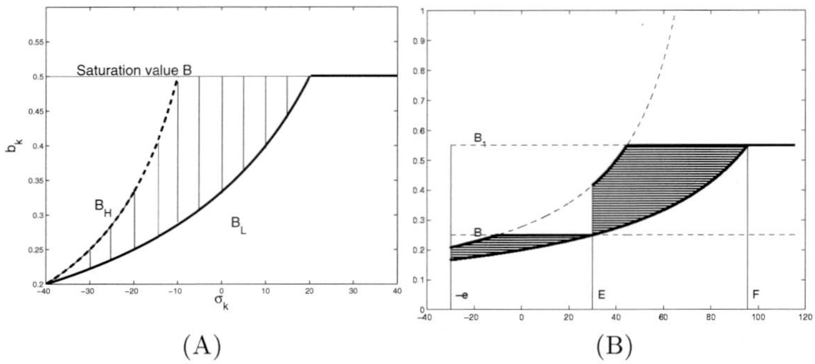

Fig. 3. (A) Family of control laws ensuring controlled invariance; (B) Family of control laws used to restore the equilibrium

The following remarks are very useful in the design of the supervisor.

Remark 1. Condition (4) identifies a family of controllers attaining the control goal (See Figure 3.(A)). The choice of one element of this family depends on the different trade-offs sought in the application. For instance, the lower bound $\mathcal{B}_L^{[j]}$ is clearly the most conservative in terms of used bandwidth. On the contrary, picking a control value closer to the upper bound $\mathcal{B}_H^{[j]}$ is the most robust choice for the QoS against possible un-modelled effects. Moreover, it is easy to show that robust controlled invariance of $\mathcal{I}$ is preserved if we change the value of the bandwidth during the job, as long as its value always belongs to the $[\mathcal{B}_L^{[j]}(\sigma), \mathcal{B}_H^{[j]}(\sigma)]$ range. Therefore, the supervisor is allowed to change the bandwidth allocated to the task up to $\mathcal{B}_L^{[j]}$ without compromising the invariance of $\mathcal{I}^N$.

It is possible for a task to specify different values for the upper-bound E of the RCIS, which correspond to different QoS levels. In a normal situation, the task aims at the maximum QoS level, but it can switch to a degraded QoS (i.e., a larger value of E) if an overload occurs. Indeed, by doing so, its minimum bandwidth requirement $\mathcal{B}_L^{[j]}(\sigma)$ decreases. This type of degradation is another degree of freedom exposed to the supervisor to manage overload situations.

Restoring the equilibrium. A control law that attains $(\mathcal{J}^n,\mathcal{I}^n)$−attractivity in M steps can be built using, for each task controller j, two saturation levels $B'^{[j]}$ and $B^{[j]}$ with $B'^{[j]} > B^{[j]}$. $B'^{[j]}$is used to recover the equilibrium and $B^{[j]}$ to maintain it. Let $F_k^{[j]}$ be a number such that $\frac{H_k^{[j]}}{T+E-F_k^{[j]}} = B'^{[j]}$. A control strategy is built as follows:

$$\begin{aligned} b_k^{[j]} &= B'^{[j]} \text{ if } \sigma_k^{[j]} \geq F_k^{[j]}, \\ b_k^{[j]} &\in [\mathcal{B}_L^{[j]}(\sigma), \min B'^{[j]}, \frac{h_k^{[j]}}{T-e-\sigma_k^{[j]}}], \text{ if } E \leq \sigma_k^{[j]} \leq F_k^{[j]}, \\ b_k^{[j]} &\in [\mathcal{B}_L^{[j]}(\sigma_k^{[i]}), \mathcal{B}_H^{[j]}(\sigma_k^{[i]})], \text{if } -e \leq \sigma_k^{[j]} \leq E. \end{aligned} \tag{5}$$

The rational is very simple. When we are far off from the target equilibrium, we use the maximum available bandwidth B' to quickly reduce the scheduling

Table 1. Experimental probability for a scheduling error in the target RCIS

Predictor	$E[\alpha_k^{[1]}]$	$\Pr\{\epsilon_k^{[1]} \in \mathcal{I}\}$	$E[b_k^{[1]}]$
Fix. 0.15/0.70	-	15.71%	=
Fix. 0.17/0.75	-	23.41%	=
MMA[3,3]	0.767	72.42%	16.9%
MMA[12,3]	0.796	86.31%	15.4%

Predictor	$E[\alpha_k^{[2]}]$	$\Pr\{\epsilon_k^{[2]} \in \mathcal{I}\}$	$E[b_k^{[2]}]$
Fix. 0.15/0.70	-	25.34%	=
Fix. 0.17/0.75	-	57.22%	=
MMA[3,3]	0.900	89.78%	73.7%
MMA[12,3]	0.916	89.31%	73.5%

error. Then we reach a zone ($\sigma_k \in [E, F]$) where we can reach the equilibrium in one step. Finally, when we are inside the target we can switch to the control laws that allow us to maintain the equilibrium (which requires a lower saturation value). The family of control laws constructed in this way are those in the striped area delimited by the tick lines in Figure 3.(B). The effectiveness of this control policy is shown in the next Theorem.

Theorem 2. *Under the assumption of Theorem 1, the family of control laws in Equation* (5) *attains* $(\mathcal{J}^n, \mathcal{I}^n)$*−attractivity (with* $\mathcal{J} = [-e, L]$ *and* $\mathcal{I} = [-e, E]$*) in* $M+1$ *steps for the system in Equation* (2) *if* $\hat{H} \leq \frac{T+E-L+MT}{M+1}$*, where* $\hat{H} = \max_j \left\{ \frac{\sup_{k>k_0}\{H_k^{[j]}\}}{B'^{[j]}} \right\} \leq \tilde{H}$.

Proof. See Appendix.

Remark 2. As one would expect, the saturation level required for attractivity is higher than the one required for mere controlled invariance. Indeed, the two values coincide if $M \to \infty$.

4 Experimental Results

We applied the control techniques shown so far to an MPEG-2 decoder, whose behaviour has been simulated by using execution traces measured from a real application running on Linux. The application consists of a pipeline of two tasks. The first task loads the frames from the disk; therefore its resource requirements are proportional to the frame size. The second task decodes the buffered frames; the resource requirements are given, in this case, by the decoding times measured from the real application.

We ran the simulations experiment considering different scenarios. In the first scenario, we used a fixed bandwidth allocation, in which the bandwidth chosen for the two stages were slightly above the mean value of the resource requirements. In the second scenario, we considered a fixed allocation with a greater bandwidth value for the two stages. In the third and in the fourth scenario, we used our control scheme with different predictors, based on multiple moving averages (MMA(3,3), and MMA(12,3), meaning respectively 3 and 12 independent averages of 3 samples). The use of these predictors is motivated by the periodic coding scheme used for the considered MPEG2 stream [6]. The mean time required to decode the frame was $\mu_{c^{[2]}} = 26.85ms$. The mean time required to load

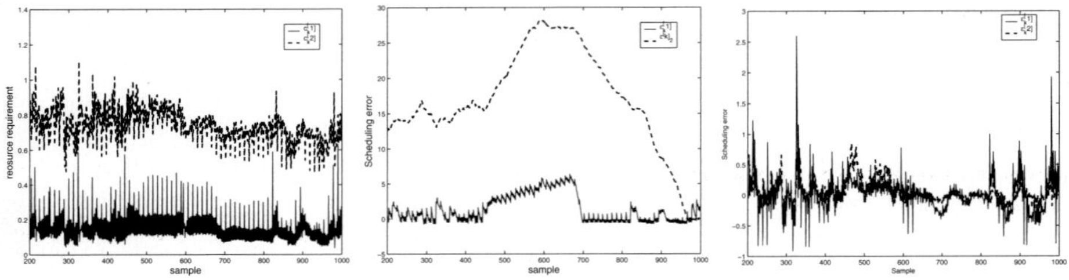

(a) (b) (c)

Fig. 4. Segments of execution traces. (a) Resource requirements, (b) Evolution of the scheduling error in the case of fixed bandwidth (0.17/0.75). (c) Evolution in case of feedback control with MA[12,3] predictor.

a frame for the disk and its standard deviation were respectively $\mu_{c[1]} = 4.941ms$. For both tasks we specified a desired RCIS $[-e, E] = [-16ms, 16ms]$. The application has a period of $T = 40ms$ (corresponding to 25 frames-per-second stream).

Figure 4.(a) shows the temporal evolution of the resource requirements in the two stages for 800 frames (as a percentage of the task period). In this segment, the $c_k^{[2]}$ value is often above the average. Moreover, $c_k^{[1]}$ displays several peaks above the average. As a result, if we look at the scheduling error for fixed bandwidth (Figure 4.(b)), we can experience large delays on the second stage (which suffers for both the peaks of $c_k^{[1]}$ and of $c_k^{[2]}$). The feedback scheme, although based on a predictor that occasionally fails, is able to compensate this effect. If we want to evaluate the system performance on the entire stream, it is useful to compute the experimental probability of having a scheduling error inside the target RCIS. This information is reported in Table 1, where the first column reports the mean value of the $\alpha_k^{[i]}$ parameter (i.e. *mean* $\frac{h_k^{[i]}}{H_k^{[i]}}$) produced by the predictors for the two stages. This is related to the quality of the prediction and to the variability of the trace (a lower value corresponds to a higher variability). The last column reports the average bandwidth allocated by the controllers. The improvement achieved by using feedback control is evident. The last column of the table also highlights that the average bandwidth allocated by the controller is very close to the average value of the computation requirements.

5 Conclusions and Future Work

In this paper, we have shown the application of a feedback controller to the problem of dynamical allocation of resources to a time-sensitive application consisting of a pipeline of tasks. The use of a scheduling mechanism that approximates a fluid partitioning of the resources enables the definition of precise dynamic model for the system, which can be used in the design of a the feedback controller providing guarantees on its closed loop performance. Most of the future work will be concentrated in evaluating different strategies for the supervisor and on the application of stochastic control techniques to the design of the feedback scheduler.

References

1. Stoica, I., Abdel-Wahab, H., Jeffay, K., Baruah, S.K., Gehrke, J.E., Plaxton, C.G.: A proportional share resource allocation algorithm for real-time, time-shared systems. In: Proceedings of the IEEE Real-Time Systems Symposium. (1996)
2. Mercer, C.W., Rajkumar, R., Tokuda, H.: Applying hard real-time technology to multimedia systems. In: Workshop on the Role of Real-Time in Multimedia/Interactive Computing System. (1993)
3. Rajkumar, R., Juvva, K., Molano, A., Oikawa, S.: Resource kernels: A resource-centric approach to real-time and multimedia systems. In: Proceedings of the SPIE/ACM Conference on Multimedia Computing and Networking. (1998)
4. Abeni, L., Buttazzo, G.: Integrating multimedia applications in hard real-time systems. In: Proceedings of the IEEE Real-Time Systems Symposium, Madrid, Spain (1998)
5. C. Lu, J. Stankovic, G.T., Son, S.: Feedback control real-time scheduling: Framework, modeling and algorithms. Ppecial issue of RT Systems Journal on Control-Theoretic Approaches to Real-Time Computing **23**(1/2) (2002)
6. Abeni, L., Cucinotta, T., Lipari, G., Marzario, L., Palopoli, L.: Adaptive reservations in a linux based environment. In: Proceeding of the Real-Time Application Symposium (RTAS 04), Toronto (Canada), IEEE (2004)
7. Goel, A., Walpole, J., Shor, M.: Real-rate scheduling. In: Proc.of rtas04. (2004) 434
8. Eide, E., Stack, T., Regehr, J., Lepreau, J.: Dynamic cpu management for real-time, middleware-based systems. In: Proc. of 10th IEEE Real-Time and Embedded Technology and Applications Symposium, Toronto, Canada (2004)
9. Rajkumar, R., Lee, C., Lehoczky, J.P., Siewiorek, D.P.: Practical solutions for QoS-based resource allocation. In: RTSS. (1998) 296–306
10. Palopoli, L., Cucinotta, T., Bicchi, A.: Quality of service control in soft real-time applications. In: Proc. of the IEEE 2003 conference on decision and control (CDC02), Maui, Hawai, USA (2003)
11. Steere, D., Shor, M.H., Goel, A., Walpole, J., P, C.: Control and modeling issues in computer operating systems: Resource management for real-rate computer applications. In: Proceedings of 39th IEEE Conference on Decision and Control (CDC00). (2000)
12. Abeni, L., Palopoli, L., Lipari, G., Walpole, J.: Analysis of a reservation-based feedback scheduler. In: Proc. of the Real-Time Systems Symposium, Austin, Texas (2002)
13. Luigi Palopoli, T.C.: Feeback scheduling for pipelines of tasks. Technical report, Uiversity of Trento (2006)
14. Calafiore, G., Camp, M.: Interval predictors for unknown dynamical systems: an assessment of reliability. In: 41st IEEE Conference on Decision and Control (cdc02), 2002 (2002)
15. Blanchini, F.: Set invariance in control. Automatica (1999)

A Proofs of the Stability Theorems

Robust controlled invariance. In order to show Theorem 1, we need some preliminary Lemmas.

Lemma 1. *Consider the job* $J_k^{[j]}$ *of task* $\tau^{[j]}$ *with* $j \geq 2$. *A control law exists guaranteeing that* $\epsilon_k^{[j]} \in \mathcal{I}$, $\forall \sigma_k^{[j]} \in \mathcal{I}$, $\forall c_k^{[j]} \in [h_k^{[j]}, H_k^{[j]}]$ *if and only if* $\frac{H_k^{[j]}}{T} \leq B^{[j]} \wedge e + E \geq T\left(\frac{1-\alpha_k^{[j]}}{\alpha_k^{[j]}}\right)$. *Moreover, the bandwidth guaranteeing such property has to be chosen in the range identified by Equation* (4).

Proof. For the sake of brevity, the proof is given only for the case $e + E < T$. Consider job $J_k^{[j]}$, with $k \geq 2$, of $\tau^{[j]}$, with $j \geq 2$. Under the stated hypotheses $\sigma_k^{[j]} \in [-e, E]$, in view of Equation (2), we can have $\epsilon_k^{[j]} \in [-e, E]$ iff $T - e - \sigma_k^{[j]} \leq \frac{c_k^{[j]}}{b_k^{[j]}} \leq T + E - \sigma_k^{[j]}$. Since $e + E < T$, it is possible to re-write the condition as: $\frac{c_k^{[j]}}{T+E-\sigma_k^{[j]}} \leq b_k^{[j]} \leq \frac{c_k^{[j]}}{T-e-\sigma_k^{[j]}}$. As $c_k^{[j]}$ is not known, and it may take any value in the range $[h_k^{[j]}, H_k^{[j]}]$, the only possibility for the controller not to violate last equation is the choice of a bandwidth value belonging to the intersection of all the ranges corresponding to any possible value of $c_k^{[j]}$, i.e. $\frac{H_k^{[j]}}{T+E-\sigma_k^{[j]}} \leq b_k^{[j]} \leq \frac{h_k^{[j]}}{T-e-\sigma_k^{[j]}}$. Such a choice exists if and only if $e + \alpha_k^{[j]} E \geq (1 - \alpha_k^{[j]})(T - \sigma_k^{[j]})$. The latter condition must hold for any possible value of the start time $\sigma_k^{[j]} \in [-e, E]$, leading to $e + E \geq T\left(\frac{1-\alpha_k^{[j]}}{\alpha_k^{[j]}}\right)$. Furthermore, the chosen control value must be legal, i.e. $b_k^{[j]} \leq B^{[j]}$. This is possible if and only if $\frac{H_k^{[j]}}{T+E-\sigma_k^{[j]}} \leq B^{[j]}$. This must hold for each possible value of $\sigma_k^{[i]} \leq E$, which leads to the existence condition for the controller: $\frac{H_k^{[j]}}{B^{[j]}} \leq T$.

With simular arguments, we can prove the following for the first stage.

Lemma 2. *Consider job* $J_k^{[1]}$ *of task* $\tau_k^{[1]}$ *with* $k \geq 2$. *A control law exists guaranteeing that* $\epsilon_k^{[1]} \in \mathcal{I}$, $\forall \sigma_k^{[j]} \in [0, E]$, $\forall c_k^{[1]} \in [h_k^{[1]}, H_k^{[1]}]$, *if and only if* $\frac{H_k^{[1]}}{T} \leq B^{[1]}$ *and* $e + \alpha_k^{[1]} E \geq T\left(1 - \alpha_k^{[1]}\right)$.

Proof. (of Theorem 1). First we focus on sufficiency of the theorem condition, which may be proved by proving the following statement $S(k)$ inductively on $k > k_0$:

$$\begin{cases} \max_{k \in]k_0, k]} \max_j \left\{\frac{H_k^{[j]}}{B^{[j]}}\right\} \leq T \\ e + \alpha_k^{[1]} E \geq T\left(1 - \alpha_k^{[1]}\right) \\ e + E \geq T\left(\frac{1-\alpha_{k_0}}{\alpha_{k_0}}\right) \; j \geq 2 \end{cases} \text{implies} \begin{cases} \varepsilon_{\mathbf{k_0}} \in \mathcal{H} \\ \mathbf{c_h} \in \mathbf{P}_h \forall h \in]k_0, k] \end{cases} \implies \varepsilon_{\mathbf{k+1}} \in \mathcal{H}.$$

Pick a value for k_0 and consider the base inductive case $k = k_0$. Assume $\varepsilon_{\mathbf{k_0}} \in \mathcal{H} \wedge \mathbf{c_{k+1}} \in \mathbf{P}_{k+1}$. For Lemma 2, $\frac{H_{k_0}^{[1]}}{T} \leq B^{[1]}$ and $e + \alpha_{k_0}^{[1]} E \geq T\left(1 - \alpha_{k_0}^{[1]}\right)$

guarantees that $\varepsilon^{[1]}_{k_0+1} \in \mathcal{I}$. Furthermore, for Lemma 1, $\frac{H^{[j]}_{k_0}}{T} \leq B^{[j]}$ and $e + E \geq T\left(\frac{1-\alpha^{[j]}_{k_0}}{\alpha^{[j]}_{k_0}}\right)$ guarantees that $\varepsilon^{[j]}_{k_0+1} \in [-e, E]$. Therefore, $\varepsilon_{\mathbf{k_0+1}} \in \mathcal{H}$ is derived from the intersection of all these conditions, i.e. the left-hand side of $S(k_0)$.

For a generic k, assume $S(k-1)$ holds, $\varepsilon_{\mathbf{k_0}} \in \mathcal{H}$ and $\mathbf{c_h} \in \mathbf{P}_h \forall h \in]k_0, k]$. Then, under the conditions of the left-hand side of $S(k-1)$, we have: $\varepsilon_{\mathbf{k}} \in \mathcal{H}$. For the two cited lemmas, $\varepsilon_{\mathbf{k+1}} \in \mathcal{H}$ is thus derived from the intersection of the conditions $\frac{H^{[1]}_k}{T} \leq B^{[1]}$, $e+\alpha^{[1]}_k E \geq T\left(1-\alpha^{[1]}_k\right)$, $\frac{H^{[j]}_k}{T} \leq B^{[j]}$ and $e+E \geq T\left(\frac{1-\alpha^{[j]}_k}{\alpha^{[j]}_k}\right)$, plus the ones in the left-hand side of $S(k-1)$. These constitute exactly the left-hand side of $S(k)$. Therefore, $S(k)$ is true for any $k > k_0$. Sufficiency of the theorem condition is obtained by observing that it is obtained as $\lim_{k\to+\infty} S(k)$.

Concerning necessity of the theorem condition, we have to observe that in the definition of RCIS (1) the stated property is required to hold for each k_0. This means that, if we consider only a single evolution step of the state vector from k_0 to k_0+1, Lemma 2 and 1 require $\frac{H^{[1]}_{k_0}}{T} \leq B^{[1]}$, $e + \alpha^{[1]}_{k_0} E \geq T\left(1-\alpha^{[1]}_{k_0}\right)$, $\frac{H^{[j]}_{k_0}}{T} \leq B^{[j]}$ and $e+E \geq T\left(\frac{1-\alpha^{[j]}_{k_0}}{\alpha^{[j]}_{k_0}}\right)$ as necessary conditions to guarantee that $\varepsilon_{\mathbf{k_0+1}} \in \mathcal{H}$. Therefore, by considering any possible value for k_0, we obtain that all of these conditions (at varying k_0) must hold true, leading to the theorem proof.

$(\mathcal{H}, \mathcal{G})$ **attractivity.** We recall that $F^{[j]}_k$ has been defined as a number such that $\frac{H^{[j]}_k}{T+E-F^{[j]}_k} = B'^{[j]}$. Before showing Theorem 2, we need the following.

Lemma 3. *Under the assumptions of Theorem 2, any control law chosen in the family in Equation* (5) *guarantees that: 1) if* $\sigma^{[j]}_k \geq -e$*, then* $\epsilon^{[j]}_k \geq -e$ *, 2) if* $\sigma^{[j]}_k \in [-e, E]$*, then* $\epsilon^{[j]}_k \in [-e, E]$*, 3) if* $\sigma^{[j]}_k \in [E, F^{[j]}_k]$*, then* $\epsilon^{[j]}_k \in [-e, E]$.

Proof. The Lemma immediately derives from the way the family of controllers is constructed and from the evolution of the system, condensed in Equation (2). For instance, to violate the first claim, we should choose a value for $b^{[1]}_k \geq \frac{h^{[1]}_k}{T-e-\sigma^{[j]}_k}$, which is never done as shown in Figure 3.B. Similar arguments can be used for the other claims.

Proof. (of Theorem 2) Consider k_0 as defined above. As a preliminary step, we prove the following property:

$$\forall j, \forall n \in [1, M+1], -e \leq \epsilon^{[j]}_{k_0+n} \leq \max\left\{L + n(\hat{H} - T), E\right\}. \qquad (6)$$

The proof is by induction on the stages of the pipeline. Let's focus on the first stage. In this case, $\sigma^{[1]}_{k_0+n} = \max\{0, \epsilon^{[1]}_{k_0+n}\}$. In view of the first property in Lemma 3 and of the evolution of the system in Equation (2), $\epsilon^{[1]}_{k_0+n} \geq -e$ is easily verified by induction on n. As far as the upper bound below E is concerned, we observe that, if for any $k \in [k_0, k_{0+n}]$ $\epsilon_k^{[1]} \leq F_k^{[1]}$, the claim is an immediate result of the second and third claims in Lemma 3. On the contrary, assuming that for all $k \in [k_0, k_{0+n}]$ $\epsilon_k^{[1]} > F_k^{[1]}$, we use the saturation value $B'^{[1]}$ as a control value; therefore we can write $\epsilon_{k_0+n}^{[1]} \leq \epsilon_{k_0}^{[1]} + \frac{\sum_{h=k_0+1}^{k_0+n} H_h^{[1]}}{B'^{[1]}} - nT \leq L + \frac{\hat{H} n B'^{[1]}}{B'^{[1]}} - nT = L + n(\hat{H} - T)$.

Now, let's consider stage j of the pipeline assuming that the property holds for stage $j-1$. In this case, $\sigma_k^{[j]} = \max\{\epsilon_k^{[j-1]}, \epsilon_{k-1}^{[j]}\}$.By inductive hypotheses, $\epsilon_k^{[j-1]} \geq -e$, which leads to $\sigma_k^{[j]} \geq -e$. Hence, in view of the first property of Lemma 3, we have $\epsilon_k^{[j]} \geq -e$.

As far as the upper bound is concerned, the dynamic evolution is described in Equation (2) and it can lead to one of the following cases:

$$\epsilon_{k_0+n}^{[j]} = \begin{cases} \epsilon_{k_0}^{[j]} + \sum_{h=k_0+1}^{k_0+n} \frac{c_h^{[j]}}{B'^{[j]}} - nT & \text{if } \epsilon_{k-1}^{[j]} > \epsilon_k^{[j-1]}\, \forall k \in [k_0+1, , k_0+n] \\ \epsilon_{k_0+1}^{[j-1]} + \sum_{h=k_0+1}^{k_0+n} \frac{c_h^{[j]}}{B'^{[j]}} - nT & \text{if } \epsilon_{k-1}^{[j]} > \epsilon_k^{[j-1]}\, \forall k \in [k_0+2, k_0+n] \wedge\, \epsilon_{k_0+1}^{[j-1]} > \epsilon_{k_0}^{[j]} \\ \ldots & \\ \epsilon_{k_0+n}^{[j-1]} + \frac{c_{k_0+n}^{[j]}}{B'^{[j]}} - T & \text{if } \epsilon_{k_0+n}^{[j-1]} > \epsilon_{k_0+n-1}^{[j]} \end{cases}$$

The first case is dealt with exactly as shown for the previous stage of the pipeline and it leads to Equation (6). Consider the generic case $\epsilon_{k_0+n}^{[j]} = \epsilon_{k_0+m}^{[j-1]} + \sum_{h=k_0+m}^{k_0+n} \frac{c_h^{[j]}}{B'^{[j]}} - (n-m+1)T$, with $m \geq 1$. The application of the inductive hypothesis $\epsilon_{k_0+m}^{[j-1]} \leq \max\left\{L + m(\hat{H} - T), E\right\} \equiv \rho(m)$ leads us to $\epsilon_{k_0+n}^{[j]} \leq \rho + \sum_{h=k_0+m}^{k_0+n} \frac{c_h^{[j]}}{B'^{[j]}} - (n-m+1)T \leq L + (n+1)(\hat{H} - T) \leq L + n(\hat{H} - T)$.

If $\rho = L + m(\hat{H} - T)$, then we can write $\epsilon_{k_0+n}^{[j]} \leq L + (n+1)(\hat{H} - T) \leq L + n(\hat{H} - T)$. If $\rho = E$, then we can write $\epsilon_{k_0+n}^{[j]} \leq E + (n-m+1)(\hat{H} - T) \leq E$ (since $\hat{H} \leq \tilde{H} \leq T$ by hypotheses). This terminates the proof of property in Equation (6).

As a consequence of Lemma 3, the claim of the theorem is proved if: 1) $\sigma_k^{[j]} \geq -e$ for $k = k_0 + 1, \ldots, k_0 + M + 1$ and 2) $\sigma_{k_0+M+1}^{[j]} \leq F_{k_0+M+1}^{[j]}$. Condition 1) is part of Equation (6). Because of Equation (6), we can also write (considering the only relevant case with $L + M(\hat{H} - T) > E$) : $\sigma_{k_0+M+1}^{[j]} = \max\{\varepsilon_{k_0+M+1}^{[j-1]}, \varepsilon_{k_0+M}^{[j]}\} \leq L + M(\hat{H} - T)$. On the other hand, $F_k^{[j]} = T + E - \frac{H_k^{[j]}}{B'^{[j]}} \geq T + E - \hat{H}$, thus condition 2) is satisfied if:$L + M(\hat{H} - T) \leq T + E - \hat{H}$, which can be written as $\hat{H} \leq \frac{T+E-L+MT}{M+1}$, which completes the proof.

On Simulations and Bisimulations of General Flow Systems*

J.M. Davoren[1] and Paulo Tabuada[2]

[1] Department of Electrical & Electronic Engineering
The University of Melbourne, VIC 3010 Australia
davoren@unimelb.edu.au
[2] Department of Electrical Engineering
The University of California at Los Angeles, CA 90095 USA
tabuada@ee.ucla.edu

Abstract. We introduce a notion of bisimulation equivalence between general flow systems, which include discrete, continuous and hybrid systems, and compare it with similar notions in the literature. The interest in the proposed notion is based on our main result, that the temporal logic $\mathbf{GFL}^\star$ – an extension to general flows of the well-known computation tree logic $\mathbf{CTL}^\star$ – is semantically preserved by this equivalence.

1 Introduction

There is growing interest in the study of simulation and bisimulation relationships within general classes of dynamical and control systems including hybrid systems [1,2,3,4,5,6]. A core motivation is the potential for using these relationships, whenever they preserve significant structural and behavioral properties, as a means to reduce complexity in the analysis and design of systems.

In this paper, we define a notion of bisimulation equivalence that is sufficient to preserve the semantics of the general flow logic $\mathbf{GFL}^\star$, introduced in [7], and thus to preserve all system properties expressible in that logic. $\mathbf{GFL}^\star$ extends the discrete-time semantics of the well-known temporal logic $\mathbf{CTL}^\star$ to the class of *general flow systems* [7], which offer a unified treatment of discrete-time transition systems, continuous-time differential inclusions, hybrid-time systems such as hybrid automata and stochastic hybrid systems, as well as more complex systems requiring higher-dimensional time lines; e.g., a "meta-hybrid automaton" as a finite state machine with a hybrid automaton at each discrete state [8].

The framework of general flow systems is given in general set-theoretic terms, and builds on the notion of a *time line* as a suitably structured linear order, and of *finite paths* as functions from bounded and finite-duration subsets of a time line into some value space. A general flow system Φ over a state space X associates with each initial state $x \in X$ the set $\Phi(x)$ of all possible finite paths or trajectories starting from x, and satisfies a generalized version of the *semigroup property* or "Axiom of State" from *Behavioural Systems theory* [9]. The system

* First author partially supported by Australian Research Council grant DP0208553; second author partially supported by NSF CAREER award 0446716.

A. Bemporad, A. Bicchi, and G. Buttazzo (Eds.): HSCC 2007, LNCS 4416, pp. 145–158, 2007.

class is essentially Aubin's model of an *Evolutionary System* [10], generalized from discrete or continuous time to arbitrary time lines, and "deconstructed", so that the basic objects are finite, bounded duration paths, rather than functions from the whole time line. In moving to the hybrid time line, this perspective simply fails to transfer: if we take the limit of any infinite sequence of longer and longer finite hybrid trajectories, then we do not end up with a function defined on the whole hybrid time line, but rather with a function whose time domain is the union of a finite or infinite sequence of disjoint intervals, separated by infinite gaps in the time line. Revising the treatment in [7,11], we develop a theory of *maximal extensions* of finite paths by taking the limit of infinite sequences of longer and longer finite paths, where those sequences are indexed by transfinite ordinals, up to the ordinal length of the underlying time line (so in the case of continuous or hybrid time, up to the ordinal of the cardinality of the continuum). The payoff from this apparent transfinite generosity is the crucial equivalence, expressed in Theorem 1, between the finitary system property of being deadlock-free or non-blocking, and the infinitary property of being maximally extendible, in the sense that every finite path of the system has a maximal extension.

The maximal extension property is essential for the semantics of the logic **GFL***, which straight-forwardly generalize the semantics of the logic **CTL*** with respect to ω-length execution sequences of non-blocking transition systems or state machines, with the singular and crucial exception of the *next-time* operator. To cover general time lines, we introduce a generalized *next-times* operator that behaves as the discrete successor if there is one, and otherwise, in the presence of a dense sub-interval of time, has the meaning "*immediately after now*". This operator is also a key ingredient in our new notions of simulation and bisimulation, and in our semantic preservation theorem for the logic **GFL***.

The body of the paper is organized as follows. Section 2 consists of preliminary theory of time lines and paths. Section 3 briefly reviews general flow systems and develops the theory of maximal extensions. In Section 4, we give three, progressively stronger, concepts of simulation and bisimulation between general flow systems, allowing *differing* time lines in the systems compared for the first two of these. The new notion we call *p-simulation* is strictly intermediate between two other concepts of simulation found in the current literature [1,2,3,4,5,6], and we illustrate the differences with some simple examples. Section 5 reviews the syntax and semantics of Full General Flow Logic **GFL*** and discusses its expressibility. The main result is in Section 6, where we establish that our notion of p-bisimulation preserves the semantics of **GFL***.

2 Preliminaries: Time Lines and Paths

We use relations/set-valued maps $r : X \rightsquigarrow Y$, with $r(x) \subseteq Y$ for $x \in X$, and let $[X \rightsquigarrow Y]$ denote the set of all such maps, so $[X \rightsquigarrow Y] = 2^{X\times Y}$. A map $r : X \rightsquigarrow Y$ has a converse $r^{-1} : Y \rightsquigarrow X$; domain $\mathrm{dom}(r) := \{x \in X \mid r(x) \neq \varnothing\}$; range $\mathrm{ran}(r) := \mathrm{dom}(r^{-1}) \subseteq Y$; and r is *total on* X if $\mathrm{dom}(r) = X$. Writing $r : X \to Y$ means r is a single-valued function total on X, with values $r(x) = y$, and $[X \to Y]$ is the set of all such functions. For partial functions, writing $r : X \dashrightarrow Y$ means that on $\mathrm{dom}(r) \subseteq X$, r is single-valued; we write $r(x) = y$ when $x \in \mathrm{dom}(r)$ with value y, and $r(x) = \text{UNDEF}$ when $x \notin \mathrm{dom}(r)$, and $[X \dashrightarrow Y]$ is the set of all such maps. So $[X \to Y] \subseteq [X \dashrightarrow Y] \subseteq [X \rightsquigarrow Y]$.

Let $(L, <, 0)$ be a *linear order* with least element 0 and no largest element. We will call L a (future) *time line* if the following three conditions are satisfied:

(i) L is Dedekind-complete (sup's and inf's exist for non-empty bounded subsets);
(ii) there exists a *linearly ordered abelian group* $(\overline{L}, <, +, 0)$ such that $(L, <, +, 0)$ is a linearly ordered sub-semigroup of $\overline{L}$, and $L \subseteq \{l \in \overline{L} \mid l \geqslant 0\}$;
(iii) L is equipped with an extended metric function $d_L : (L \times L) \to \mathbb{R}_0^{+\infty}$ together with a continuous order-preserving total function (a *fibering map*) $p : L \to M$ into a countable linear order $(M, <_M)$ such that,
(a) for each $m \in M$, the *fibre* $p^{-1}(m) \subseteq L$ is a metric space under d_L;
(b) for all $m, m' \in M$, $a \in p^{-1}(m)$, $b \in p^{-1}(m')$: $d_L(a, b) < \infty$ iff $m = m'$;
(c) for all $a, b, c \in L$, $a \leqslant c$, $d_L(a, c) < \infty$: $d_L(a, c) = d_L(a, b) + d_L(b, c)$ iff $a \leqslant b \leqslant c$;
(d) for all $a, b, c \in L$, $\quad d_L(b, c) = d_L(a + b, a + c)$.

From the group $\overline{L}$, a time line L has a family of order-isomorphisms $\{\sigma^{+a}\}_{a \in L}$ such that $\sigma^{+0} = \mathrm{id}_L$ and for each $a \in L$, the *right a-shift* $\sigma^{+a} : L \to L$ is given by $\sigma^{+a}(l) := l + a$, and with inverse $\sigma^{-a} := (\sigma^{+a})^{-1} : [a, \infty) \to L$ the *left a-shift*. A subset $T \subseteq L$ will be called $<$*-unbounded* if for all $a \in L$, there exists $t \in T$ such that $t > a$, and it will be called $<$*-bounded* otherwise. For any subset $T \subseteq L$, define the set's *total duration* $\mathrm{dur}(T) \in \mathbb{R}_0^{+\infty}$ as follows:

$$\mathrm{dur}(T) := \sum_{m \in M} \sup \{ \, d_L(t, t') \mid t \in T \cap p^{-1}(m) \ \wedge \ t' \in T \cap p^{-1}(m) \, \}$$

where for $S \subseteq \mathbb{R}_0^{+}$, we take $\sup(S) = 0$ if $S = \varnothing$. A subset $T \subseteq L$ will be called *duration-bounded* if $\mathrm{dur}(T) < \infty$, and *duration-unbounded* otherwise.

Basic examples are the discrete time line $\mathbb{N}$, and the dense continuum time line $\mathbb{R}_0^{+} := [0, \infty)$, whose linearly ordered abelian groups under addition are $\mathbb{Z}$ and $\mathbb{R}$ respectively. For $L = \mathbb{N}$ and $L = \mathbb{R}_0^{+}$, the group operation also gives a suitable metric: take $d_L(a, b) := \max\{a - b, b - a\}$, and take $p : L \to \{0\}$ constant, so there is only one fibre, $p^{-1}(0) = L$. For these standard time lines, the metric is finite everywhere and for all subsets $T \subseteq L$, we have the following equivalences: T is $<$-bounded iff T is duration-bounded iff $T \subseteq [0, b]$ for some $b \in L$.

The hybrid time set $L = \mathbb{H} := \mathbb{N} \times \mathbb{R}_0^{+}$ is linearly ordered *lexicographically*: $(i, t) <_{\text{lex}} (j, s)$ iff $i < j$ or else both $i = j$ and $t < s$. The least element is $\mathbf{0} := (0, 0)$ and the ordering is Dedekind-complete. The linear order $\mathbb{H}$ is the non-negative *quarter* of the abelian group $\mathbb{Z} \times \mathbb{R}$, defined by: $(i, t) + (j, s) := (i + j, t + s)$; with the lexicographic ordering, $\mathbb{Z} \times \mathbb{R}$ is a linearly ordered abelian group. For the extended metric on $L = \mathbb{H}$, the fibering map $p_{\mathbb{H}} : \mathbb{H} \to \mathbb{N}$ is simply $p_{\mathbb{H}}(i, t) := i$ for all $(i, t) \in \mathbb{H}$, and for each $i \in \mathbb{N}$, the fibre under $p_{\mathbb{H}}$ is $p_{\mathbb{H}}^{-1}(i) = \{i\} \times \mathbb{R}_0^{+}$. We only assign a finite distance between time positions $a = (k, r)$ and $b = (i, t)$ with discrete time coordinates $k = i$ the same; define $d_{\mathbb{H}} : (\mathbb{H} \times \mathbb{H}) \to \mathbb{R}_0^{+\infty}$ such that, for all $a = (k, r)$ and $b = (i, t)$ in $\mathbb{H}$, $d_{\mathbb{H}}(a, b) := d_{\mathbb{R}}(r, t)$ if $k = i$, and $d_{\mathbb{H}}(a, b) := \infty$ if $k \neq i$. Thus $(\mathbb{H}, <_{\text{lex}}, \mathbf{0}, +, d_{\mathbb{H}}, p_{\mathbb{H}})$ is a time line. In $L = \mathbb{H}$, the two concepts of boundedness are not equivalent: the interval $T = \{42\} \times [0, \infty)$ is $<$-bounded but duration-unbounded, while the time domain of a *Zeno* infinite hybrid trajectory is duration-bounded but $<$-unbounded.

For any linear order $(L, <)$, and for any subset $T \subseteq L$, the T-*successor* partial function $\mathrm{succ}_T : T \dashrightarrow T$ is defined by:

$$\forall a, b \in T, \qquad \mathrm{succ}_T(a) = b \quad \Leftrightarrow \quad [\; a < b \;\wedge\; (\forall t \in T)\, t \leqslant a \;\vee\; b \leqslant t \;].$$

For example, for $T = L = \mathbb{N}$, we have $\mathrm{dom}(\mathrm{succ}_L) = L$, which means L is *everywhere discrete*, while for $T = L = \mathbb{R}_0^+$, the map succ_L is defined nowhere, which means L is *everywhere dense*. If $T \subset \mathbb{H}$ is the domain of a hybrid trajectory, then the partial function succ_T will be defined only at the switching times in T. For the purpose of formulating a new concept of bisimulation later in the paper, as well as for giving a semantics to a "*next-times*" operator in temporal logic, we have the need for a *progress* operator acting on initial subsets T of time lines.

Definition 1. (***Progress operator on initial subsets of time lines***)
For any time line L, and any initial subset $T \subseteq L$ with $0 \in T$, define:

$$\mathrm{Pro}(T) := \{\, t \in T \mid t > 0 \;\wedge\; (\forall s \in \mathrm{ran}(\mathrm{succ}_T))\; t \leqslant s \,\}$$

Hence if $0 \in \mathrm{dom}(\mathrm{succ}_T)$ then $\mathrm{Pro}(T) = \{\mathrm{succ}_T(0)\}$; if $0 \notin \mathrm{dom}(\mathrm{succ}_T)$ but $\mathrm{ran}(\mathrm{succ}_T) \neq \varnothing$ then $\mathrm{Pro}(T) = (0, s_T]$ where $s_T := \min(\mathrm{ran}(\mathrm{succ}_T))$, while if T is everywhere dense, so $\mathrm{ran}(\mathrm{succ}_T) = \varnothing$, then $\mathrm{Pro}(T) = T - \{0\}$.

For the usual time lines $L = \mathbb{N}$ and $L = \mathbb{R}_0^+$, the basic form of a time domain for a *path* is a closed bounded interval $[0, b]$, which in $\mathbb{N}$ evaluates to $\{0, 1, \ldots, b\}$. For the hybrid time line $L = \mathbb{H}$, finite hybrid trajectories are typically functions taking values in a space $X \subseteq Q \times \mathbb{R}^n$, with Q a finite set, and we can represent their time domains as disjoint unions of the form:

$$T \;=\; \bigcup_{i<N} \{i\} \times [s_i, s_i + \Delta_i] \;=\; \bigcup_{i<N} [\,(i, s_i),\, (i, s_{i+1})\,] \tag{1}$$

where $s_0 := 0$ and $s_{i+1} := s_i + \Delta_i$ and $(\Delta_0, \Delta_1, \ldots, \Delta_{N-1})$ is a finite sequence of *interval durations* $\Delta_i \in \mathbb{R}_0^+$ for $i < N$, and $(s_1, \ldots, s_{N-1}, s_N)$ is the corresponding sequence of *switching times*. Along a hybrid trajectory, for $i < N - 1$, we have $\mathrm{succ}_T(i, s_{i+1}) = (i + 1, s_{i+1})$, but relative to the underlying ordering $\mathbb{H}$, there is a distinct **gap** between these two time positions, in the form of the interval $\{i\} \times (s_{i+1}, \infty)$ followed by the interval $\{i + 1\} \times [0, s_{i+1})$.

Definition 2. (***Bounded time domains and paths***)
Given a time line L, define a bounded time domain *in L to be a subset $T \subset L$ such that $T = \bigcup_{n<N} [a_n, b_n]$ with $N \in \mathbb{N}^+$, $a_0 = 0$, $b_{N-1} = b_T := \max(T)$, and $a_n \leqslant b_n < a_{n+1} \leqslant b_{n+1}$ for all $n < N - 1$, and $d(a_n, b_n) < \infty$ for all $n < N$. Let $\mathsf{BT}(L) \subset 2^L$ be the set of all bounded time domains in L, and let $\mathsf{BI}(L) := \{T \in \mathsf{BT}(L) \mid (\exists b \in L)\, T = [0, b]\,\}$ be the interval time domains. Over any set $X \neq \varnothing$, define:*

$$\mathsf{Path}(L, X) := \{\, \gamma : L \dashrightarrow X \mid \mathrm{dom}(\gamma) \in \mathsf{BT}(L) \,\}$$
$$\mathsf{IPath}(L, X) := \{\, \gamma : L \dashrightarrow X \mid \mathrm{dom}(\gamma) \in \mathsf{BI}(L) \,\}$$

For $\gamma \in \mathsf{Path}(L,X)$*, define* $b_\gamma := \max(\mathrm{dom}(\gamma))$*, so that* $\gamma(0) \in X$ *is* γ*'s start-point, and* $\gamma(b_\gamma) \in X$ *is* γ*'s end-point. The* total duration *of* γ *is* $\mathrm{dur}(\gamma) := \mathrm{dur}(\mathrm{dom}(\gamma)) < \infty$ *(so* $\mathrm{dom}(\gamma)$ *is both duration-bounded and* $<$*-bounded). Let* $\epsilon \in [L \dashrightarrow X]$ *denote the empty path; for any* $\mathcal{P} \subseteq \mathsf{Path}(L,X)$*, let* $\mathcal{P}_\epsilon := \mathcal{P} \cup \{\epsilon\}$.

Given a time line L, the set $\mathsf{BT}(L)$ is *partially ordered* via the linear ordering on L: for $T, T' \in \mathsf{BT}(L)$, we say T' is an *ordered extension* of T, and (re-using notation) we write $T < T'$, iff $T \subset T'$ and $t < t'$ for all $t \in T$ and all $t' \in T' - T$. Likewise, the path set $\mathsf{Path}_\epsilon(L,X)$ is partially ordered: $\gamma < \gamma'$ iff $\gamma \subset \gamma'$ and $\mathrm{dom}(\gamma) < \mathrm{dom}(\gamma')$, in which case we say the path γ' is a (proper) *extension* of γ. For any set of paths $\mathcal{P} \subseteq \mathsf{Path}_\epsilon(L,X)$, $\mathcal{P}$ is $<$*-unbounded* (or *extension-unbounded*) if for all $\gamma \in \mathcal{P}$, there exists $\gamma' \in \mathcal{P}$ such that $\gamma < \gamma'$.

We use the following operations; for $\gamma, \gamma' \in \mathsf{Path}_\epsilon(L,X)$, $t \in \mathrm{dom}(\gamma)$, $x \in X$:

- the *trivial path* $\theta_x : [0,0] \to X$ given by $\theta_x(0) = x$.
- *restriction* or *prefix* ending at t: $\gamma|_t \in \mathsf{Path}_\epsilon(L,X)$ where $(\gamma|_t)(l) := \gamma(l)$ for all $l \in \mathrm{dom}(\gamma|_t) := [0,t] \cap \mathrm{dom}(\gamma)$.
- *translation* or *suffix* starting at t: ${}_t|\gamma \in \mathsf{Path}_\epsilon(L,X)$ where $({}_t|\gamma)(l) := \gamma(l+t)$ for all $l \in \mathrm{dom}({}_t|\gamma) := \sigma^{-t}([t, b_\gamma] \cap \mathrm{dom}(\gamma))$.
- *point-concatenation* at $x \in X$: $\gamma *_x \gamma' \in \mathsf{Path}_\epsilon(L,X)$ where, for all $l \in L$:

$$(\gamma *_x \gamma')(l) := \begin{cases} \gamma(l) & \text{if } l \in \mathrm{dom}(\gamma) \ \wedge \ \gamma'(0) = \gamma(b_\gamma) = x \\ \gamma'(l - b_\gamma) & \text{if } l \in \sigma^{+b_\gamma}(\mathrm{dom}(\gamma')) \ \wedge \ \gamma'(0) = \gamma(b_\gamma) = x \\ \text{UNDEF} & \text{otherwise} \end{cases}$$

  and $\mathrm{dom}(\gamma *_x \gamma') = \mathrm{dom}(\gamma) \cup \sigma^{+b_\gamma}(\mathrm{dom}(\gamma'))$ [1].

The path extension ordering and point-concatenation are related as follows:

$$\gamma < \gamma' \quad \text{iff} \quad \gamma' = \gamma *_x \gamma'' \text{ for some } \gamma'' \in \mathsf{Path}(L,X) \text{ and } x \in X \text{ with } \gamma'' \neq \theta_x \quad (2)$$

3 General Flow Systems, and Their Infinitary Extensions

Introduced in [7], and further developed in [11], the class of general flow systems generalizes to arbitrary time lines Aubin's model of an *Evolutionary System* [10].

Definition 3. (***General flow systems and finitary properties***)
Let L be a time line, and let $X \neq \varnothing$ *be an arbitrary value space. A* general flow system *over* X *with time line* L *is a map* $\Phi\colon X \rightsquigarrow \mathsf{Path}(L,X)$ *satisfying, for all* $x \in \mathrm{dom}(\Phi)$*, for all* $\gamma \in \Phi(x)$*, and for all* $t \in \mathrm{dom}(\gamma)$*:*

(**GF0**) initialization*:* $\gamma(0) = x$*;*

(**GF1**) time-invariance *or* suffix-closure*:* ${}_t|\gamma \in \Phi(\gamma(t))$*;*

(**GF2**) point-concatenation*:* $\gamma|_t *_y \gamma' \in \Phi(x)$ *for all* $\gamma' \in \Phi(y)$ *with* $y = \gamma(t)$*.*

[1] For the discrete time line $L = \mathbb{N}$, the interval path set $\mathrm{IPath}_\epsilon(\mathbb{N}, X) = X^*$ is the set of all finite *words* or sequences over X. The usual operation of *word-concatenation* from automata theory equips X^* as a total monoid with identity ϵ; word-concatenation can be readily defined in terms of point-concatenation using length-2 connecting words formed from the end-value of the first word and the start-value of the second.

• Φ *is* deadlock-free *if* $\Phi(x) \neq \{\theta_x\}$*, for every* $x \in \mathrm{dom}(\Phi)$*;*
• Φ *is* $<$-unbounded *if the path set* $\Phi(x)$ *is* $<$*-unbounded, for every* $x \in \mathrm{dom}(\Phi)$*;*
• Φ *is* deterministic *if* $\Phi(x)$ *is linearly-ordered by* $<$*, for every* $x \in \mathrm{dom}(\Phi)$*;*
• Φ *is* point-controllable [9] *if for all* $x', x'' \in \mathrm{dom}(\Phi)$*, there exists* $\gamma \in \Phi(x')$ *and* $t \in \mathrm{dom}(\gamma)$ *such that* $\gamma(t) = x''$*;*
• Φ *is* path-controllable [9] *if for all* $x, x', x'' \in \mathrm{dom}(\Phi)$ *and for all* $\gamma' \in \Phi(x)$*, if* $x' = \gamma'(b_{\gamma'})$*, then for all* $\gamma'' \in \Phi(x'')$*, there exists* $\gamma \in \Phi(x')$ *and* $t \in \mathrm{dom}(\gamma)$ *such that* $(\gamma' *_{x'} \gamma|_t *_{x''} \gamma'') \in \Phi(x)$*.*

The following results are readily established [7,11]: Φ is point-controllable iff Φ is path-controllable; Φ is deadlock-free iff Φ is $<$-unbounded. In terms of *Behavioural Systems theory* [9], condition **GF1** corresponds to the *time invariance* property, while condition **GF2** corresponds to the "Axiom of State" principle.

Example 1. Let L be any time line, and consider the map $\Phi_L : L \rightsquigarrow \mathsf{Path}(L, L)$ given by $\Phi_L(a) := \{\gamma \in \mathsf{Path}(L, L) \mid (\exists s \in L)\ \gamma = (\sigma^{+a})|_s\}$. Then Φ_L is an interval-path, deterministic and deadlock-free general flow system over L, but it is not point-controllable (being only uni-directional since L is a semigroup).

Further examples of general flow systems include automata and state transition systems over $L = \mathbb{N}$, differential equations and inclusions over $L = \mathbb{R}_0^+$, and hybrid automata and impulse differential inclusions over hybrid time $L = \mathbb{H}$ [7], as well as stochastic hybrid and continuous time systems.

In order to directly represent hybrid trajectories, and their constituent parts, we choose to take as our primitive objects paths of finite duration, with a start-point and an end-point. However, we still have many reasons to "go to infinity" by finding "maximal extensions" of finite duration paths, including formalizing asymptotic properties of systems such as stability, as well as the eventuality and *until* type properties expressible in temporal logics [7], and also comparing and utilizing work on existing system-theoretic models, e.g. *Evolutionary Systems* [10]; *Behavioural Systems* [9]; and various hybrid system classes [12,1,2,3,6].

A general flow Φ is deadlock-free iff it is $<$-unbounded, so for each $x \in dom(\Phi)$ and finite path $\gamma \in \Phi(x)$, we can recursively construct an ω-length extending sequence of paths $\{\gamma_n\}_{n<\omega}$ starting from $\gamma_0 = \gamma$ with $\gamma_n \in \Phi(x)$ and $\gamma_n < \gamma_{n+1}$ for all $n < \omega$. Motivated by this fact, we view "maximal extensions" or "completions" of paths as infinitary objects, arising as limits of infinite sequences of finite paths. Revising earlier work in [7,11], we now work with ν-length infinite sequences of paths $\{\gamma_n\}_{n<\nu}$, for all limit ordinals $\nu \leq \kappa$, where $\kappa = |L|$, the cardinality of the time line L and the initial limit ordinal of that cardinality. The crucial pay-off from this transfinite generosity is Theorem 1.

Definition 4. (***κ-extension of path sets***)
For a time line L*, let* $\kappa = |L|$*, and let* $\mathrm{LO}(\kappa)$ *be the set of all limit ordinals* $\nu \leq \kappa$ *with* $\nu \neq 0$*. For any path set* $\mathcal{P} \subseteq \mathsf{Path}_\epsilon(L, X)$*, define the* κ-extension *of* $\mathcal{P}$*:*

$$\begin{aligned} \mathsf{Ext}(\mathcal{P}) := \{\, \beta \in [L \dashrightarrow X] \mid\ & (\exists \nu \in \mathrm{LO}(\kappa))\,(\exists \overline{\gamma} \in [\nu \to \mathsf{Path}(L, X)])\,(\forall n < \nu) \\ & \gamma_n := \overline{\gamma}(n) \ \wedge\ \gamma_n \in \mathcal{P} \ \wedge\ (\forall n' < \nu)\,(n < n' \ \Rightarrow\ \gamma_n < \gamma_{n'}) \\ & \wedge\ \beta = \textstyle\bigcup_{m<\nu} \gamma_m \,\} \end{aligned}$$

Define $\mathsf{EPath}(L, X) := \mathsf{Ext}\,(\,\mathsf{Path}_\epsilon(L, X)\,)$, $\mathsf{EIPath}(L, X) := \mathsf{Ext}\,(\,\mathsf{IPath}_\epsilon(L, X)\,)$.

The κ-extension $\mathsf{Ext}(\mathcal{P})$ contains all the partial functions $\beta : L \dashrightarrow X$ that can arise as the limit of a ν-length chain of paths in $\mathcal{P}$, for some limit ordinal $\nu \in \mathrm{LO}(\kappa)$. The *total duration* of a limit path $\alpha \in \mathsf{EPath}(L, X)$ is defined $\mathrm{dur}(\alpha) := \mathrm{dur}(\mathrm{dom}(\alpha))$, where $\mathrm{dur}(T)$ is as defined in Section 2 for any $T \subseteq L$. The path extension ordering $<$ on bounded paths induced by the linear order on L can also be lifted to limit paths, and if $\alpha < \alpha'$ then we must have $\mathrm{dur}(\alpha) < \infty$.

For reasoning about the *asymptotic* behaviour of a path set or general flow, the Ext operation will not quite do, as the set of limit paths $\mathsf{Ext}(\mathcal{P})$ also includes limit paths α that are *too short* to be of *maximal* extension or duration, as witnessed by there being some actual, finite-duration path $\gamma \in \mathcal{P}$ that properly extends α. For general flows Φ, we want to additionally require the limit paths in $\mathsf{M}\Phi(x)$ to be not only maximal w.r.t. the extension partial ordering, but also collectively *complete* in their representation of Φ, in that for every finite path $\gamma \in \Phi(x)$, there is at least one limit path $\alpha \in \mathsf{M}\Phi(x)$ properly extending γ.

Definition 5. (***Maximal extension & infinitary properties of path sets***) *Let L be a time line and let $X \neq \varnothing$ be any value space.*
For any path set $\mathcal{P} \subseteq \mathsf{Path}_\epsilon(L, X)$, define the maximal extension *of $\mathcal{P}$ to be the limit path set $\mathsf{M}(\mathcal{P})$, with $\mathsf{M}(\mathcal{P}) \subseteq \mathsf{Ext}(\mathcal{P}) \subseteq \mathsf{EPath}(L, X)$ defined by:*

$$\mathsf{M}(\mathcal{P}) := \{ \alpha \in \mathsf{Ext}(\mathcal{P}) \mid (\forall \gamma \in \mathcal{P})\ \alpha \not< \gamma \}$$

A path set $\mathcal{P} \subseteq \mathsf{Path}_\epsilon(L, X)$ will be called maximally extendible *if for all $\gamma \in \mathcal{P}$, there exists $\alpha \in \mathsf{M}(\mathcal{P})$ such that $\gamma < \alpha$.*
Given a general flow system $\Phi\colon X \rightsquigarrow \mathsf{Path}(L, X)$, define the maximal extension *of Φ to be the map $\mathsf{M}\Phi\colon X \rightsquigarrow \mathsf{EPath}(L, X)$ given by $(\mathsf{M}\Phi)(x) := \mathsf{M}(\Phi(x))$ for all $x \in \mathrm{dom}(\mathsf{M}\Phi) := \mathrm{dom}(\Phi)$. A general flow system Φ will be called* maximally extendible *if for all $x \in \mathrm{dom}(\Phi)$, the path set $\Phi(x)$ is maximally extendible.*

Theorem 1. [Assume the Axiom of Choice.] *For any set $\mathcal{P} \subseteq \mathsf{Path}_\epsilon(L, X)$,*
$\mathcal{P}$ is maximally extendible iff $\mathcal{P}$ is $<$-unbounded.
Hence for any general flow system $\Phi\colon X \rightsquigarrow \mathsf{Path}(L, X)$,
Φ is maximally extendible iff Φ is $<$-unbounded iff Φ is deadlock-free;
if Φ is deadlock-free, then: Φ is deterministic iff $\mathsf{M}\Phi$ is a partial function.

4 Bisimulation Relations Between General Flow Systems

The most basic notion of simulation and bisimulation is reachability-preserving but not time-preserving or path-matching. This is what is known as "*time-abstract*" simulation and bisimulation for the case of *transition systems* (including transition system representations of hybrid and continuous systems [1]), which are general flow systems over time $L = \mathbb{N}$.

Definition 6. *Given time lines L_1 and L_2, possibly different, and general flows $\Phi_1\colon X_1 \rightsquigarrow \mathsf{Path}(L_1, X_1)$, $\Phi_2\colon X_2 \rightsquigarrow \mathsf{Path}(L_2, X_2)$, a relation $R : X_1 \rightsquigarrow X_2$ is a* reachability simulation (*or* r-simulation) *of Φ_1 by Φ_2 if $\mathrm{dom}(\Phi_1) \subseteq \mathrm{dom}(R)$ and for all $x_1, x_1' \in X_1$ and for all $x_2 \in X_2$ such that $(x_1, x_2) \in R$,*
if *there exists $\gamma_1 \in \Phi_1(x_1)$ and $t_1 \in \mathrm{dom}(\gamma_1)$ such that $t_1 > 0$ and $x_1' = \gamma_1(t_1)$,*

then *there exists* $x_2' \in X_2$ *and* $\gamma_2 \in \Phi_2(x_2)$ *and a time point* $t_2 \in \mathrm{dom}(\gamma_2)$ *such that* $t_2 > 0$ *and* $x_2' = \gamma_2(t_2)$ *and* $(x_1', x_2') \in R$.
A map $R : X_1 \rightsquigarrow X_2$ *is a* reachability bisimulation (*or* r-bisimulation) *between* Φ_1 *and* Φ_2 *if both* R *and* R^{-1} *are r-simulations.*

For general flows Φ_i, $i = 1, 2$, let $Q_i : X_i \rightsquigarrow X_i$ be the (strict) *reachability relation* of the system: for all $x, x' \in X_i$, define $(x, x') \in Q_i$ iff $x' = \gamma(t)$ for some $\gamma \in \Phi_i(x)$ and $t \in \mathrm{dom}(\gamma)$ with $t > 0$. So $\mathrm{dom}(Q_i) \subseteq \mathrm{dom}(\Phi_i)$, and for sets $A \subseteq X_i$, the Φ_i-reachable region from A is the $\exists$-post-image of relation Q_i applied to A; that is, $\mathrm{Reach}_i(A) := Q_i^{\exists}(A) = \{x' \in X_i \mid (\exists x \in A)\,(x, x') \in Q_i\}$. This is at the heart of any transition system representation of hybrid or continuous dynamical systems. For any map/relation $R : X_1 \rightsquigarrow X_2$, the $\exists$-pre-image operator $R^{-\exists}$ is given by $R^{-1}(B) := \{x_1 \in X_1 \mid (\exists x_2 \in B)\,(x_1, x_2) \in R\}$ for sets $B \subseteq X_2$. The following results are straight-forward, and motivate our choice of name "reachability simulation" (alternatively, "time abstract simulation").

Proposition 1. *Given time lines* L_1 *and* L_2*, possibly different, and general flows* $\Phi_1 : X_1 \rightsquigarrow \mathsf{Path}(L_1, X_1)$ *and* $\Phi_2 : X_2 \rightsquigarrow \mathsf{Path}(L_2, X_2)$*, and a map* $R : X_1 \rightsquigarrow X_2$*, suppose that* $\mathrm{dom}(\Phi_1) \subseteq \mathrm{dom}(R)$*. Then the following are equivalent:*
(1.) *R is an r-simulation of Φ_1 by Φ_2;*
(2.) $R^{-1} \circ Q_1 \subseteq Q_2 \circ R^{-1}$*;*
(3.) $\mathrm{Reach}_1(R^{-\exists}(B)) \subseteq R^{-\exists}(\mathrm{Reach}_2(B))$ *for all* $B \subseteq X_2$.
If R *is an r-bisimulation and* $Q_i^{-1} = Q_i$ *for* $i = 1, 2$ (*e.g. if both flows* Φ_i *are point-controllable*)*, then* $R^{\exists}(\mathrm{Reach}_1(A)) = \mathrm{Reach}_2(R^{\exists}(A))$ *for all* $A \subseteq X_1$.

Proposition 2. *Given* L_1 *and* L_2*, and general flows* $\Phi_1 : X_1 \rightsquigarrow \mathsf{Path}(L_1, X_1)$ *and* $\Phi_2 : X_2 \rightsquigarrow \mathsf{Path}(L_2, X_2)$*, suppose that* $R : X_1 \rightsquigarrow X_2$ *is an r-simulation of* Φ_1 *by* Φ_2*, and* $\mathrm{dom}(\Phi_2) \subseteq \mathrm{ran}(R)$*. If* Φ_1 *is deadlock-free, then* Φ_2 *is deadlock-free.*

Next, we introduce a slightly stronger notion of simulation and bisimulation which requires some "matching" of time points along paths, but not an exact matching, thus relating systems defined over different time lines.

Definition 7. *Given time lines* L_1 *and* L_2*, possibly different, and general flows* $\Phi_1 : X_1 \rightsquigarrow \mathsf{Path}(L_1, X_1)$*,* $\Phi_2 : X_2 \rightsquigarrow \mathsf{Path}(L_2, X_2)$*, a relation* $R : X_1 \rightsquigarrow X_2$ *is a* progress simulation (*or* p-simulation) *of* Φ_1 *by* Φ_2 *if* $\mathrm{dom}(\Phi_1) \subseteq \mathrm{dom}(R)$ *and for all* $x_1, x_1' \in X_1$ *and for all* $x_2 \in X_2$ *such that* $(x_1, x_2) \in R$,
if *there exists* $\gamma_1 \in \Phi_1(x_1)$ *and* $t_1 \in \mathrm{Pro}(\mathrm{dom}(\gamma_1))$ *such that* $x_1' = \gamma_1(t_1)$,
then *there exists* $x_2' \in X_2$ *and* $\gamma_2 \in \Phi_2(x_2)$ *and* $t_2 \in \mathrm{Pro}(\mathrm{dom}(\gamma_2))$ *such that* $x_2' = \gamma_2(t_2)$ *and* $(x_1', x_2') \in R$*, and for all intermediate times* $s_2 \in (0, t_2] \cap \mathrm{dom}(\gamma_2)$*, there exists* $s_1 \in (0, t_1] \cap \mathrm{dom}(\gamma_1)$ *such that* $(\gamma_1(s_1), \gamma_2(s_2)) \in R$.
Map $R : X_1 \rightsquigarrow X_2$ *is a* progress bisimulation (p-bisimulation) *between* Φ_1 *and* Φ_2 *if both* R *and* R^{-1} *are p-simulations.*

As we show in the main result, Theorem 2 below, this notion of p-bisimulation is strong enough to yield a semantic-presevation bisimulation theorem for the logic $\mathbf{GFL}^{\star}$ of general flow systems, yet is still flexible enough to allow that the time lines of the two general flows be different. Note that, in the case that both time lines are discrete, with $L_1 = L_2 = \mathbb{N}$, and both general flows are determined by a (one-step) transition relation on the state space, then a p-simulation is a (standard) simulation relation in the original sense of Milner [13].

Definition 8. *Given general flow systems* $\Phi_1 \colon X_1 \rightsquigarrow \mathsf{Path}(L, X_1)$ *and* $\Phi_2 \colon X_2 \rightsquigarrow \mathsf{Path}(L, X_2)$ *over the same time line* L, *a relation* $R : X_1 \rightsquigarrow X_2$ *is a* timed simulation (t-simulation) *of* Φ_1 *by* Φ_2 *if* $\mathrm{dom}(\Phi_1) \subseteq \mathrm{dom}(R)$, *and for all* $x_1, x_1' \in X_1$, *and* $x_2 \in X_2$ *such that* $(x_1, x_2) \in R$, *and for all times* $t > 0$,
if *there exists* $\gamma_1 \in \Phi_1(x_1)$ *such that* $x_1' = \gamma_1(t)$,
then *there exists* $x_2' \in X_2$ *and* $\gamma_2 \in \Phi_2(x_2)$ *such that* $x_2' = \gamma_2(t)$ *and* $\mathrm{dom}(\gamma_2) = \mathrm{dom}(\gamma_1)$ *and* $(\gamma_1(s), \gamma_2(s)) \in R$ *for all* $s \in \mathrm{dom}(\gamma_2) \cap [0, t]$.
A relation $R : X_1 \rightsquigarrow X_2$ *is a* timed bisimulation (*or* t-bisimulation) *between* Φ_1 *and* Φ_2 *if both* R *and* R^{-1} *are t-simulations.*

It follows directly from the definitions that when $L_1 = L_2$ and R is a t-simulation, R is also a p-simulation, and for any L_1 and L_2, if R is a p-simulation, then R is an r-simulation. Other notions of simulation and bisimulation have been recently investigated in the context of continuous [5,6] and hybrid systems [1,2]. All of them require an exact matching between the time parameterizing trajectories and thus are t-bisimulations between the general flow systems defined by the corresponding continuous or hybrid systems. Notions of bisimulation not requiring exact time matching were implicitly considered in [4]. Although [4] is based on the standard Milner notion of bisimulation between transition systems, different embeddings of linear control systems into transition systems resulted in different notions of bisimulation: t-bisimulation and r-bisimulation. The notions of simulation and bisimulation developed for hybrid I/O automata in [3] come out as intermediate between the r-simulations and p-simulations here.

We conclude this section with a brief discussion of some examples. We deliberately choose systems with simple deterministic dynamics so as to keep the focus on illustrating the various simulation relationships.

Example 2. Consider a discrete-time deterministic system with general flow map $\Phi_1 : X_1 \rightsquigarrow \mathsf{Path}(\mathbb{N}, X_1)$ over state space $X_1 := \{q_1, q_2, q_3, q_4\}$ generated by the transition function $\delta : X_1 \to X_1$ with $\delta(q_k) := q_{k+1}$ for $k = 1, 2, 3$ and $\delta(q_4) = q_1$. Next, consider a continuous-time deterministic system with general flow map $\Phi_2 : X_2 \rightsquigarrow \mathsf{Path}(\mathbb{R}_0^+, X_2)$ over the state space $X_2 := \mathbb{R}^2 - \{(0, 0)\}$ given by the differential equation: $\dot{x}_1 = x_2$ and $\dot{x}_2 = -x_1$. So $\Phi_2(x_1, x_2) = \{\gamma : [0, b] \to X_2 \mid b \geq 0 \wedge (\forall t \in \mathrm{dom}(\gamma))\, \gamma(t) = (x_1 \cos(t) + x_2 \sin(t), x_2 \cos(t) - x_1 \sin(t))\}$, and the paths correspond to circular motion in clockwise direction, with radius of the circle $r = \sqrt{x_1^2 + x_2^2}$. Then consider the relation $R : X_1 \rightsquigarrow X_2$ given by:

$$R(q_1) = \{(x_1, x_2) \in X_2 \mid x_1 \leqslant 0 \wedge x_2 > 0\},\ R(q_2) = \{(x_1, x_2) \in X_2 \mid x_1 > 0 \wedge x_2 \geqslant 0\}$$
$$R(q_3) = \{(x_1, x_2) \in X_2 \mid x_1 \geqslant 0 \wedge x_2 < 0\},\ R(q_4) = \{(x_1, x_2) \in X_2 \mid x_1 < 0 \wedge x_2 \leqslant 0\}$$

It is clear that $\mathrm{dom}(\Phi_1) = X_1 = \mathrm{dom}(R)$ and $\mathrm{dom}(\Phi_2) = X_2 = \mathrm{ran}(R)$, and it is readily established that R is an r-simulation of the discrete system Φ_1 by the continuous system Φ_2, but it is not a p-simulation. To see why p-similarity fails, consider $(q_2, (\frac{1}{2}, \frac{1}{2})) \in R$ and note that, in Φ_1, along the unique discrete path

$\gamma_1 \in \Phi_1(q_2)$ with $\mathrm{dom}(\gamma_1) = \{0, 1, 2, 3, 4\}$, we have $q_3 = \gamma_1(1)$ with time $t_1 = 1 \in \mathrm{Pro}(\mathrm{dom}(\gamma_1))$, and this is matched in reachability terms by the unique continuous path $\gamma_2 \in \Phi_2(\frac{1}{2}, \frac{1}{2})$ with $\mathrm{dom}(\gamma_2) = [0, 2\pi]$, at time $t_2 = \frac{\pi}{2} \in \mathrm{Pro}(\mathrm{dom}(\gamma_2))$ and state $(\frac{1}{2}, -\frac{1}{2}) = \gamma_2(\frac{\pi}{2})$, since $(q_3, (\frac{1}{2}, -\frac{1}{2})) \in R$. However, if we pick the intermediate time point $s_2 = \frac{\pi}{4} \in (0, t_2] \cap \mathrm{dom}(\gamma_2) = (0, \frac{\pi}{2}]$, and the state $(\frac{1}{\sqrt{2}}, 0) = \gamma_2(\frac{\pi}{4})$, then we *cannot* find any matching time point $s_1 \in (0, t_1] \cap \mathrm{dom}(\gamma_1) = \{1\}$ such that $\gamma_1(s_1) = q_2$ – because $\gamma_1(1) = q_3$, and thus we cannot satisfy $(\gamma_1(s_1), \gamma_2(s_2)) = (\gamma_1(s_1), (\frac{1}{\sqrt{2}}, 0)) \in R$.

We can, however, easily construct a variant map $\hat{R} : X_1 \rightsquigarrow X_2$ such that $\hat{R}$ is a p-simulation of Φ_1 by Φ_2. Let $\hat{R} : X_1 \rightsquigarrow X_2$ be given by:

$$\hat{R}(q_1) = \{(x_1, x_2) \in X_2 \mid x_1 = 0 \wedge x_2 > 0\},\ \hat{R}(q_2) = \{(x_1, x_2) \in X_2 \mid x_1 > 0 \wedge x_2 = 0\}$$
$$\hat{R}(q_3) = \{(x_1, x_2) \in X_2 \mid x_1 = 0 \wedge x_2 < 0\},\ \hat{R}(q_4) = \{(x_1, x_2) \in X_2 \mid x_1 < 0 \wedge x_2 = 0\}$$

While we have chosen for illustrative purposes to simulate a discrete-time system by a continuous-time system, the process of temporal sampling and spatial quantization would go the other way, simulating continuous by discrete.

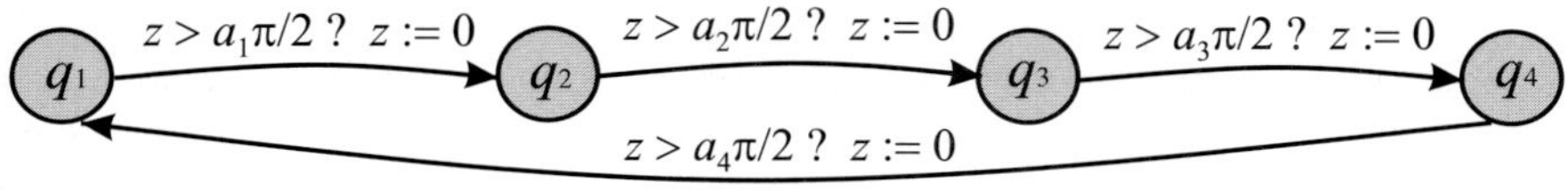

Example 3. Consider the hybrid system defined by the timed automaton H over state space $X_3 := \bigcup_{k \in K} \{q_k\} \times [0, (a_k + 1)\frac{\pi}{2}]$ represented in the figure above, where z is the (sole) clock variable and for $k \in K = \{1, 2, 3, 4\}$, $a_k > 0$ are fixed real constants; let $\Phi_3 : X_3 \rightsquigarrow \mathsf{Path}(\mathbb{H}, X_3)$ be its general flow. Then consider the relation $S : X_3 \rightsquigarrow X_2$ defined for all $z \in \mathbb{R}_0^+$ by:

$$\begin{aligned}
S(q_1, z) &= \{(x_1, x_2) \in X_2 \mid x_1 \leq 0 \wedge x_2 > 0 \wedge z = a_1 \tfrac{\pi}{2} \arctan(\tfrac{x_1}{x_2})\} \\
S(q_2, z) &= \{(x_1, x_2) \in X_2 \mid x_1 > 0 \wedge x_2 \geq 0 \wedge z = a_2 \tfrac{\pi}{2} \arctan(\tfrac{-x_2}{x_1})\} \\
S(q_3, z) &= \{(x_1, x_2) \in X_2 \mid x_1 \geq 0 \wedge x_2 < 0 \wedge z = a_3 \tfrac{\pi}{2} \arctan(\tfrac{-x_2}{x_1})\} \\
S(q_4, z) &= \{(x_1, x_2) \in X_2 \mid x_1 < 0 \wedge x_2 \leq 0 \wedge z = a_4 \tfrac{\pi}{2} \arctan(\tfrac{x_1}{x_2})\}
\end{aligned}$$

Then S is a p-bisimulation between the hybrid system Φ_3 and the continuous-time system Φ_2, but it cannot be a t-bisimulation since the time-lines are different. However, we can give a continuous-time model Φ_3' of the timed automaton H such that, if $a_1 = a_2 = a_3 = a_4 = 1$, S is a t-bisimulation between Φ_3' and Φ_2.

5 Full General Flow Logic GFL*

The logic *Full General Flow Logic*, **GFL***, first introduced in [7], extends to general flow models the semantics of *Full Computation Tree Logic*, **CTL***, introduced by Emerson and Halpern in 1983 [14,15] for formalizing reasoning about executions of concurrent systems (hardware or software) in discrete time.

The syntax here is a labelled variant of that of **CTL***, the labelling allowing for semantic models consisting of a family of deadlock-free general flow systems.

A *signature* is a pair $\Sigma = (\text{Sys}, \text{Prp})$, where Sys is a countable set of system labels, and Prp is a countable set of atomic propositions. The temporal logic language $\mathcal{F}(\Sigma)$ consists of the set of all formulas φ generated by the grammar:

$$\varphi ::= p \mid \neg\varphi \mid \varphi_1 \vee \varphi_2 \mid \varphi_1 \, \mathbf{U}_a \, \varphi_2 \mid \mathbf{X}_a\varphi \mid \forall_a \varphi$$

for atomic propositions $p \in \text{Prp}$, and system labels $a \in \text{Sys}$. Define logical constants *true*, $\top \stackrel{\text{def}}{=} p \vee \neg p$, for any $p \in \text{Prp}$, and *false*, $\bot \stackrel{\text{def}}{=} \neg\top$. The other propositional (Boolean) connectives are defined in a standard way, and the path quantifiers $\forall_a$ have classical negation duals $\exists_a$, as follows:

$$\varphi_1 \wedge \varphi_2 \stackrel{\text{def}}{=} \neg(\neg\varphi_1 \vee \neg\varphi_2) \qquad \varphi_1 \rightarrow \varphi_2 \stackrel{\text{def}}{=} \neg\varphi_1 \vee \varphi_2$$
$$\varphi_1 \leftrightarrow \varphi_2 \stackrel{\text{def}}{=} (\varphi_1 \rightarrow \varphi_2) \wedge (\varphi_2 \rightarrow \varphi_1) \qquad \exists_a\varphi \stackrel{\text{def}}{=} \neg\forall_a\neg\varphi$$

The *next-times* temporal operators, $\mathbf{X}_a$, for $a \in \text{Sys}$, will be given their semantics using the progress operator (Definition 1). The formula $\mathbf{X}_a\varphi$, read "at *next* times, φ, along $\mathsf{M}\Phi_a$-paths", will hold along any maximal limit path $\eta \in \text{ran}(\mathsf{M}\Phi_a)$ if for some time $t \in \text{Pro}(\text{dom}(\eta))$, φ holds at all time points $s \in (0, t]$. This means that if 0 has a discrete sucessor within $\text{dom}(\eta)$, then φ will hold at that (unique) time, while if 0 does not have a discrete sucessor within $\text{dom}(\eta)$, then φ will hold "immediately after now", throughout the left-open interval $(0, t]$. With this construction, we recover the standard meaning of *next* in discrete time, the same at that for **CTL***, while if 0 is followed by a dense interval within $\text{dom}(\eta)$, then we gain a rather useful notion of "dense next".

In earlier work on the logic **GFL***, in [7], we worked with the strictest version of the *until* operator, and used a well-known method to *define* a discrete "next-time" operator as well as a separate dense "immediately after now" operator both in terms of this strictest *until*. Here, we take both *until* and *next* as syntactic and semantic primitives, as is standard in the presentation of **CTL*** [14,15], but give new semantics for *next* to allow for denseness in the time domains of paths. This is better for the formulation and proof of preservation of semantics by suitable bisimulations, but still gives a logic equivalent in expressive power to the original.

Definition 9. *A* general flow logic model *of signature* $\Sigma = (\text{Sys}, \text{Prp})$ *is a structure* $\mathfrak{M} = (X, \mathcal{L}, \mathcal{S}, \mathcal{P})$, *where:*
- $X \neq \varnothing$ *is the state space, of arbitrary non-zero cardinality;*
- $\mathcal{L}$ *is a function mapping each symbol* $a \in \text{Sys}$ *to a time line* $L_a := \mathcal{L}(a)$*;*
- $\mathcal{S}$ *is a function mapping each symbol* $a \in \text{Sys}$ *to a deadlock-free general flow system* $\Phi_a := \mathcal{S}(a) : X \rightsquigarrow \mathsf{Path}(L_a, X)$ *over the space* X*, with time line* L_a*;*
- $\mathcal{P} : \text{Prp} \rightsquigarrow X$ *maps each* $p \in \text{Prp}$ *to a set* $\mathcal{P}(p) \subseteq X$ *of states.*

The maximal path space *of a model* $\mathfrak{M}$ *is* $\mathsf{MPath}(\mathfrak{M}) := \bigcup_{a \in \text{Sys}} \text{ran}(\mathsf{M}\Phi_a)$.

Let $\mathbb{GF}(\Sigma)$ denote the class of all general flow logic models of signature Σ, and for the case of a single time line L, let $\mathbb{GF}(L, \Sigma)$ denote the subclass of all logic models $\mathfrak{M}$ such that $\mathcal{L}(a) = L$ for all $a \in \mathrm{Sys}$. For the further special case where $|\mathrm{Sys}| = 1$ and Prp is countably infinite, let $\mathbb{TR}(\mathbb{N})$ denote the subclass of all discrete time logic models $\mathfrak{M}$ with one general flow $\Phi_S : X \rightsquigarrow \mathsf{IPath}(\mathbb{N}, X)$ from a total transition relation $S : X \rightsquigarrow X$ [14,15]).

Definition 10. *For $\varphi \in \mathcal{F}(\Sigma)$ and maximal limit path $\eta \in \mathsf{MPath}(\mathfrak{M})$, the relation "$\varphi$* is satisfied along path *η* in model *$\mathfrak{M}$", written $\mathfrak{M}, \eta \models \varphi$, is defined by induction on the structure of formulas, with $p \in \mathrm{Prp}$ and $a \in \mathrm{Sys}$:*

$$
\begin{array}{lll}
\mathfrak{M}, \eta \models p & \text{iff} & \eta(0) \in \mathcal{P}(p) \\
\mathfrak{M}, \eta \models \neg\varphi & \text{iff} & \mathfrak{M}, \eta \not\models \varphi \\
\mathfrak{M}, \eta \models \varphi_1 \vee \varphi_2 & \text{iff} & \mathfrak{M}, \eta \models \varphi_1 \ \text{or} \ \mathfrak{M}, \eta \models \varphi_2 \\
\mathfrak{M}, \eta \models \varphi_1 \,\mathbf{U}_a\, \varphi_2 & \text{iff} & \eta \in \mathrm{ran}(\mathsf{M}\Phi_a) \ \text{and} \ \exists t \in \mathrm{dom}(\eta),\ t \geqslant 0 \ \text{such that} \\
& & \mathfrak{M}, {}_t|\eta \models \varphi_2 \ \text{and} \ \forall s \in [0,t) \cap \mathrm{dom}(\eta),\ \mathfrak{M}, {}_s|\eta \models \varphi_1 \\
\mathfrak{M}, \eta \models \mathbf{X}_a \varphi & \text{iff} & \eta \in \mathrm{ran}(\mathsf{M}\Phi_a) \ \text{and} \ \exists t \in \mathrm{Pro}(\mathrm{dom}(\eta)) \ \text{such that} \\
& & \forall s \in (0,t] \cap \mathrm{dom}(\eta),\ \mathfrak{M}, {}_s|\eta \models \varphi \\
\mathfrak{M}, \eta \models \forall_a \varphi & \text{iff} & \forall \eta' \in \mathsf{M}\Phi_a(\eta(0)),\ \mathfrak{M}, \eta' \models \varphi
\end{array}
$$

For formulas $\varphi \in \mathcal{F}(\Sigma)$, the maximal path denotation set *$[\![\varphi]\!]^{\mathfrak{M}} \subseteq \mathsf{MPath}(\mathfrak{M})$, and the* state denotation set *$[\![\varphi]\!]^{\mathfrak{M}}_{\mathsf{st}} \subseteq X$, are defined by:*

$$[\![\varphi]\!]^{\mathfrak{M}} := \{\eta \in \mathsf{MPath}(\mathfrak{M}) \mid \mathfrak{M}, \eta \models \varphi\}$$

$$[\![\varphi]\!]^{\mathfrak{M}}_{\mathsf{st}} := \{x \in X \mid \exists \eta \in \mathsf{MPath}(\mathfrak{M}) : \mathfrak{M}, \eta \models \varphi \ \textit{and} \ x = \eta(0)\}$$

For a logic model $\mathfrak{M} \in \mathbb{GF}(\Sigma)$, state x in the state space of $\mathfrak{M}$, class of logic models $C \subseteq \mathbb{GF}(\Sigma)$, and for formulas $\varphi \in \mathcal{F}(\Sigma)$, we say:
- *φ is* satisfied *in $\mathfrak{M}$ at state x, if $x \in [\![\varphi]\!]^{\mathfrak{M}}_{\mathsf{st}}$, and* satisfiable *in $\mathfrak{M}$, if $[\![\varphi]\!]^{\mathfrak{M}}_{\mathsf{st}} \neq \varnothing$;*
- *φ is* true *in $\mathfrak{M}$, written $\mathfrak{M} \models \varphi$, if $\mathfrak{M}, \eta \models \varphi$ for every $\eta \in \mathsf{MPath}(\mathfrak{M})$;*
- *φ is C-*valid*, written $\models_C \varphi$, if $\mathfrak{M} \models \varphi$ for every $\mathfrak{M} \in C$.*

Define $\mathbf{Valid}(C) := \{\psi \in \mathcal{F}(\Sigma) \mid \models_C \psi\}$ *to be the set of all C-valid formulas, and define* $\mathbf{GFL}^\star(L, \Sigma) := \mathbf{Valid}(\mathbb{GF}(L, \Sigma))$*, for any given time line L.*

It is immediate that when restricting to discrete-time systems in $\mathbb{TR}(\mathbb{N})$, we have $\mathbf{Valid}(\mathbb{TR}(\mathbb{N})) = \mathbf{CTL}^\star$ [14,15]. For each system label $a \in \mathrm{Sys}$, the one-place operators for *eventually* $\Diamond_a$ and *always* $\Box_a$ are definable in the standard way from the two-place $\mathbf{U}_a$ plus $\top$, and the range of $\mathsf{M}\Phi_a$ is also definable:

$\Diamond_a \varphi \stackrel{\text{def}}{=} \top \,\mathbf{U}_a\, \varphi$	paths in $\mathrm{ran}(\mathsf{M}\Phi_a)$ along which φ is *eventually* true
$\Box_a \varphi \stackrel{\text{def}}{=} \neg(\top \,\mathbf{U}_a\, (\neg\varphi))$	paths in $\mathrm{ran}(\mathsf{M}\Phi_a)$ along which φ is *always* true, plus all the limit paths in $\mathsf{MPath}(\mathfrak{M}) - \mathrm{ran}(\mathsf{M}\Phi_a)$
$\mathsf{Beh}_a \stackrel{\text{def}}{=} \mathbf{X}_a \top$	set of all paths in $\mathrm{ran}(\mathsf{M}\Phi_a)$ = the *behaviour* of Φ_a

As is the case for $\mathbf{CTL}^\star$ [14], properties expressible in $\mathbf{GFL}^\star$ include safety, invariance, eventuality and "return infinitely often" fairness-type properties. Other properties of interest that can be expressed include:

• *Safety with event sequence behaviour* [16]: suppose the finite family of regions $\{S_1, \ldots, S_k\}$ forms a cover of the designated *Safe* portion of the state space X, and $K = \{1, \ldots, M\}$ and $next : K \leadsto K$ is a total map describing the permitted sequence orderings of traversal through the regions. The requirement is that every maximal Φ_a trajectory that enters a region S_k remains in S_k until it enters into $S_{k'} - S_k$ for some $k' \in next(k)$. This can by expressed by:

$$\mathfrak{M} \models ((\bigvee_{k\in K} \mathsf{S}_k) \leftrightarrow \mathsf{Safe}) \wedge \bigwedge_{k\in K} \forall_a(\mathsf{S}_k \rightarrow \bigvee_{k'\in next(k)} (\mathsf{S}_k\, \mathbf{U}_a(\mathsf{S}_{k'} \wedge \neg \mathsf{S}_k)))$$

• Aubin's notion of *viability with target* [10,12] is a "weak until" concept which is expressible in the logic, as is the dual notion of *invariance with target*. The set of maximal Φ_a trajectories that are *viable* within state set $K = [\![\,\mathsf{K}\,]\!]^{\mathfrak{M}}_{\mathsf{st}}$ until capturing target $C = [\![\,\mathsf{C}\,]\!]^{\mathfrak{M}}_{\mathsf{st}}$ can be defined in the logic as follows:

$$\mathsf{K}\,\mathbf{V}_a\,\mathsf{C} \stackrel{\text{def}}{=} \mathsf{Beh}_a \wedge (\Box_a \mathsf{K} \vee \mathsf{K}\,\mathbf{U}_a(\mathsf{K} \wedge \mathsf{C}))$$

• Fix $\Sigma = (\mathrm{Sys}, \mathrm{Prp})$ with $\alpha \in \mathrm{Sys}$ and Prp having at least two distinct symbols. Given an interval-path deadlock-free general flow Φ, Φ is *deterministic* iff $\mathfrak{M} \models \exists_\alpha \varphi \rightarrow \forall_\alpha \varphi$ for every formula $\varphi \in \mathcal{F}(\Sigma)$ and every model $\mathfrak{M} = (X, \mathcal{L}, \mathcal{S}, \mathcal{P})$ of signature Σ over the space X such that $\mathcal{L}(\alpha) = L$ and $\mathcal{S}(\alpha) = \Phi$.

• The properties of *point-controllability* (and hence of *path-controllability*) for deadlock-free general flows are expressible in the logic by an inference rule which is valid in every model which includes that flow.

6 Bisimulation Theorem for the Logic GFL*

In this section, we announce that the notion of p-bisimulation, intermediate between "time-abstract" reachability-preserving and exact time- and path-matching, is adequate for preservation of the semantics of **GFL***.

Definition 11. *Fix a signature* $\Sigma = (\mathrm{Sys}, \mathrm{Prp})$, *and for* $i = 1, 2$, *let* $\mathfrak{M}_i = (X_i, \mathcal{L}_i, \mathcal{S}_i, \mathcal{P}_i)$ *be two logic models of signature* Σ, *and for each system label* $a \in \mathrm{Sys}$ *and* $i = 1, 2$, *let* $L_{ia} := \mathcal{L}_i(a)$ *be the time line in the model* $\mathfrak{M}_i$ *for the (deadlock-free) general flow system* $\Phi_{ia} := \mathcal{S}_i(a)\colon X_i \leadsto \mathsf{Path}(L_{ia}, X_i)$.
A relation $R : X_1 \leadsto X_2$ *is a* p-simulation *of model* $\mathfrak{M}_1$ *by model* $\mathfrak{M}_2$ *if:*
(i) *for each system label* $a \in \mathrm{Sys}$, *relation* R *is a p-simulation of* Φ_{1a} *by* Φ_{2a}; *and*
(ii) *for each atomic proposition* $p \in \mathrm{Prp}$, *and for all* $x_1 \in X_1$ *and* $x_2 \in X_2$,
if $x_1\, R\, x_2$ *and* $x_1 \in \mathcal{P}_1(p)$, *then* $x_2 \in \mathcal{P}_2(p)$.
A relation $R : X_1 \leadsto X_2$ *is a* p-bisimulation *between model* $\mathfrak{M}_1$ *and model* $\mathfrak{M}_2$ *if* R *is a p-simulation of* $\mathfrak{M}_1$ *by* $\mathfrak{M}_2$, *and* R^{-1} *is a p-simulation of* $\mathfrak{M}_2$ *by* $\mathfrak{M}_1$.

Recall from the definition of p-bisimulations for general flow systems that if R is a p-bisimulation between $\mathfrak{M}_1$ and $\mathfrak{M}_2$, then we have $\mathrm{dom}(\Phi_{1a}) \subseteq \mathrm{dom}(R)$ and $\mathrm{dom}(\Phi_{2a}) \subseteq \mathrm{ran}(R)$ for each system label $a \in \mathrm{Sys}$.

Theorem 2. [Semantic preservation of **GFL*** for p-bisimulations]
Fix a signature $\Sigma = (\mathrm{Sys}, \mathrm{Prp})$*, and for* $i = 1, 2$*, let* $\mathfrak{M}_i = (X_i, \mathcal{L}_i, \mathcal{S}_i, \mathcal{P}_i)$ *be two logic models of signature* Σ*, and suppose* $B : X_1 \rightsquigarrow X_2$ *is a p-bisimulation between* $\mathfrak{M}_1$ *and* $\mathfrak{M}_2$*. Then for all* $x_1 \in X_1$ *and* $x_2 \in X_2$*,* if $x_1 \, B \, x_2$, then *for all* $\varphi \in \mathcal{F}(\Sigma)$*,* $\left[x_1 \in [\![\varphi]\!]_{\mathsf{st}}^{\mathfrak{M}_1} \Leftrightarrow x_2 \in [\![\varphi]\!]_{\mathsf{st}}^{\mathfrak{M}_2} \right]$.

Corollary 1. *If* $B : X_1 \rightsquigarrow X_2$ *is a p-bisimulation between* $\mathfrak{M}_1$ *and* $\mathfrak{M}_2$*, and both* B *and* B^{-1} *are total maps (on* X_1 *and* X_2*, respectively), then for all formulas* $\varphi \in \mathcal{F}(\Sigma)$*,* $\mathfrak{M}_1 \models \varphi$ *iff* $\mathfrak{M}_2 \models \varphi$*.*

A journal-length paper covering this material, with detailed proofs and more examples, is available from the authors.

References

1. Haghverdi, E., Tabuada, P., Pappas, G.: Bisimulation relations for dynamical, control and hybrid systems. Theoretical Computer Science **342** (2005) 229–261
2. Julius, A.: On Interconnection and Equivalences of Continuous and Discrete Systems: A Behavioural Perspective. The University of Twente (2005) PhD thesis.
3. Lynch, N., Segala, R., Vaandrager, F.: Hybrid I/O Automata. Information and Computation **185** (2003) 105–157
4. Pappas, G.: Bisimilar linear systems. Automatica **39** (2003) 2035–2047
5. Tabuada, P., Pappas, G.: Bisimilar control affine systems. Systems and Control Letters **52** (2004) 49–58
6. van der Schaft, A.: Equivalence of dynamical systems by bisimulation. IEEE Trans. Automatic Control **49** (2004) 2160–2172
7. Davoren, J., Coulthard, V., Markey, N., Moor, T.: Non-deterministic temporal logics for general flow systems. In: Hybrid Systems: Computation and Control (HSCC'04). LNCS 2993, Springer-Verlag (2004) 280–295
8. Davoren, J.: On hybrid systems and the modal mu-calculus. In: Hybrid Systems V. LNCS 1567, Springer-Verlag (1999) 38–69
9. Willems, J.: Models for dynamics. Dynamics Reported **2** (1989) 171–269
10. Aubin, J.P., Dordan, O.: Dynamical qualitative analysis of evolutionary systems. In: Hybrid Systems: Computation and Control (HSCC'02). LNCS 2289, Springer-Verlag (2002) 62–75
11. Davoren, J., Moor, T.: Non-deterministic reactive systems, from hybrid systems and behavioural systems perspectives. In: Proc. 2nd IFAC Conference on Analysis and Design of Hybrid Systems (ADHS'06), IFAC (2006) 409–416
12. Aubin, J.P., Lygeros, J., Quincampoix, M., Sastry, S., Seube, N.: Impulse differential inclusions: A viability approach to hybrid systems. IEEE Trans. on Automatic Control **47** (2002) 2–20
13. Milner, R.: Communication and Concurrency. Prentice Hall (1989)
14. Emerson, E.: Temporal and modal logic. In van Leeuwen, J., ed.: Handbook of Theoretical Computer Science. Elsevier Science (1990) 997–1072
15. Emerson, E., Halpern, J.: "Sometimes" and "Not Never" revisited: on branching versus linear time. Journal of the ACM **33** (1986) 151–178
16. Moor, T., Davoren, J.: Robust controller synthesis for hybrid systems using modal logic. In: Hybrid Systems: Computation and Control (HSCC'01). LNCS 2034, Springer-Verlag (2001) 433–446

A Partial Order Approach to Discrete Dynamic Feedback in a Class of Hybrid Systems

Domitilla Del Vecchio

University of Michigan, Department of Electrical Engineering and Computer Science,
1301 Beal Avenue, Ann Arbor, MI 48109
ddv@umich.edu
http://www.eecs.umich.edu/~ddv

Abstract. We consider the dynamic feedback problem in a class of hybrid systems modeled as (infinite) state deterministic transition systems, in which the continuous variables are available for measurement. The contribution of the present paper is twofold. First, a novel framework for performing dynamic feedback is proposed which relies on partial orders on the sets of inputs and of discrete states. Within this framework, a state estimator updates a lower and an upper bound of the set of current states. A controller then uses such upper and lower bounds to compute the upper and lower bounds of the set of inputs that maintain the current state in a desired set. Second, we show that under dynamic controllability assumptions, the conditions that allow to apply the developed algorithms can always be verified. Therefore, the partial order approach to dynamic feedback is general. A multi-robot system is presented to show the computational advantages in a system in which the size of the state set can be so large as to render enumeration and exhaustive techniques inapplicable.

1 Introduction

Controller design problems under language specification have been extensively studied for discrete systems in the computer science literature (see [10] for an overview). A control perspective in the context of discrete event systems was given by [7]. The approach has been extended to specific classes of hybrid systems such as timed automata [1] and rectangular automata [11]. These works are mainly concerned with state feedback. An output map is considered in the literature of viability theory for hybrid systems (see for example [2] and [5]), in which static output feedback is usually performed. In this paper, we consider the dynamic control problem for systems with continuous and discrete variables in the case in which the continuous variables are measured. This simplified scenario has practical interest in multi-robot systems in which the continuous variables represent the position and the velocity of a robot, while the discrete variables regulate the internal communication and coordination protocol. This work thus relates also to the computer science literature addressing control under partial observation of automata and of discrete event systems. In [7], the control problem of discrete event systems under language specifications is considered. The proposed control algorithms with full observation have polynomial complexity in the number of states. In the case of partial observations, the control problem becomes NP complete at worse. In a practical system, the number of states can be exponential in the number of constituent processes,

A. Bemporad, A. Bicchi, and G. Buttazzo (Eds.): HSCC 2007, LNCS 4416, pp. 159–173, 2007.

and therefore these control methodologies are prohibitive. Caines and Wang [6], consider the problem of steering the state of a partially observed automata to a final desired state. A dynamic programming methodology is proposed, which leads to a complexity of the control computation that is polynomial in the size of the state set, of the input set, and of the output set. Modular synthesis and special structures on the process are suggested (by [8], for example) in order to reduce computation.

In this work, we exploit a partial order structure on the set of inputs and of states to construct a feedback system that updates the lower and upper bound of the set of possible current system states and gives as output the lower and upper bounds of the set of inputs that satisfy the system specifications. This can be achieved under suitable order preserving assumptions of the system dynamics with respect to the state and to the input. We then show that if the system is controllable by dynamic output feedback one can always find partial orders on which the assumptions needed for the construction of the proposed controller are verified. We finally show how these assumptions can be relaxed. A multi-robot example is proposed, which shows how to apply the proposed methodology in an attack-defense scenario. This paper is organized as follows. In section 2, we introduce the system model. In section 3, we introduce the control problem on a partial order. In section 4, we give a solution to the problem and in Section 5 we show that the proposed construction is possible if the system is controllable. In section 6, a relaxed version of the main theorem is proposed and a multi-robot example is illustrated. An appendix contains notions on partial order theory and the proofs.

2 Deterministic Transition Systems

Definition 1. A *deterministic transition system* is a tuple $\Sigma = (Q, \mathcal{I}, \mathcal{Y}, F, g)$ in which Q is a set of states, $\mathcal{Y}$ is a set of outputs, $\mathcal{I}$ is a set of inputs, $F : Q \times \mathcal{I} \longrightarrow Q$ is a transition function, and $g : Q \longrightarrow \mathcal{Y}$ is an output function.

An *execution* of Σ is any sequence $\sigma = \{s(k)\}_{k\in\mathbb{N}}$ such that $s(0) \in Q$ and $s(k+1) = F(s(k), u(k))$ for $u(k) \in \mathcal{I}$ for all $k \in \mathbb{N}$. The set of all executions of Σ is denoted $\mathcal{E}(\Sigma)$. The output sequence $g(\sigma)$ is also denoted $\{y(k)\}_{k\in\mathbb{N}}$ with $y(k) = g(s(k))$. Given a system execution σ, $s(k) = \sigma(k)(s)$ denotes the value of the state at step k along such an execution. Let $S \subseteq Q$ be a subset of the state set. We would like to design a control algorithm that based on the output sequence $\{y(k)\}_{k\in\mathbb{N}}$ of Σ determines control inputs that guarantee that $\sigma(k)(s) \in S$ for all k. The initial set, denoted $X_0 \subseteq Q$ is the set in which the initial condition of the system Σ is constrained to lie, that is, $s(0) \in X_0$. The next definition proposes a concept of dynamic output feedback analogous to the one proposed by [9].

Definition 2. The system Σ is said to be *controllable by dynamic output feedback* with respect to set S and initial set $X_0 \subseteq S$ if there exist a feedback system $\Sigma_f = (\mathcal{P}(Q), \mathcal{Y}, \mathcal{P}(\mathcal{I}), H_2, H_1)$ such that for all executions $\sigma \in \mathcal{E}(\Sigma)$ with output sequence $\{y(k)\}_{k\in\mathbb{N}}$ if $X(k+1) = H_2(X(k), y(k))$, $u(k) \in H_1(X(k), y(k))$, with $X(0) = X_0$, then (i) $\sigma(k)(s) \in X(k)$ and (ii) $X(k) \subseteq S$ for all k.

In this definition, $X(k)$ is the set of all possible states compatible with the system dynamics and with the output sequence, while H_2 is the update function of a state estimator.

The function H_1 for all set of possible states X, determines the set of inputs that map such a set inside S. Let $y \in \mathcal{Y}$, we denote the set of all possible states compatible with such an output by $O_y(\Sigma) = \{s \in Q \mid g(s) = y\}$. We refer to this set as an *output set*.

Proposition 1. *Let $X_0 \subseteq S$. System Σ is controllable by dynamic output feedback with respect to set S and initial set $X_0 \subseteq S$ if and only if $\{u \in \mathcal{I} \mid F(O_y(\Sigma) \cap S, u) \subseteq S\} \neq \emptyset$.*

The theorems that will be proven rely on the condition that a system is controllable by dynamic output feedback with respect to a set S. This proposition allows to replace such controllability condition by $\{u \in \mathcal{I} \mid F(O_y(\Sigma) \cap S, u) \subseteq S\} \neq \emptyset$ for all $y \in \mathcal{Y}$. In this paper, we do not focus on the problem of checking whether the condition of Proposition 1 is verified in a given system, but we focus on how to construct a dynamic feedback controller when such a condition is verified. For completeness, a system Σ is said to be *controllable by static output feedback* if for all $y \in \mathcal{Y}$ the set $\{u \in \mathcal{I} \mid F(O_y(\Sigma), u) \subseteq S\}$ is not empty. A system that is controllable by dynamic output feedback is not necessarily controllable by static output feedback. In fact, in the static output feedback no memory is needed in the controller. This memory is instead used in the dynamic output feedback case, in which a state estimator on-line restricts at each step the set of all possible current system states. We next specialize the structure of system Σ to explicitly model the evolution of continuous and discrete variables.

3 Problem Setup

Given a deterministic transition system $\Sigma = (Q, \mathcal{I}, \mathcal{Y}, F, g)$, we specialize it to the case $Q = \mathcal{A} \times \mathcal{Z}$, in which $\mathcal{A}$ is a discrete set of variables denoted $\alpha \in \mathcal{A}$, $\mathcal{Z}$ is a set of continuous variables denoted $z \in \mathcal{Z}$, and $\mathcal{I}$ is a discrete set of inputs denoted $u \in \mathcal{I}$. The transition function is the pair $F = (f, h)$, in which $f : \mathcal{A} \times \mathcal{Z} \times \mathcal{I} \longrightarrow \mathcal{A}$ and $h : \mathcal{A} \times \mathcal{Z} \longrightarrow \mathcal{Z}$. The set of outputs is defined as $\mathcal{Y} = \mathcal{Z} \times \mathcal{Z}$ and the output function is $g : \mathcal{A} \times \mathcal{Z} \longrightarrow \mathcal{Y}$. For the remainder of this paper, we denote by $\Sigma = (\mathcal{A} \times \mathcal{Z}, \mathcal{I}, \mathcal{Y}, (f, h), g)$ the system represented by the following difference equations

$$\alpha(k+1) = f(\alpha(k), z(k), u(k)), \qquad z(k+1) = h(\alpha(k), z(k)) \tag{1}$$
$$(y_1(k), y_2(k)) = (z(k), h(\alpha(k), z(k))).$$

Any execution of the system Σ is of the form $\sigma = \{\alpha(k), z(k)\}_{k \in \mathbb{N}}$ and the output sequence is given by $\{y(k)\}_{k \in \mathbb{N}} = \{y_1(k), y_2(k)\}_{k \in \mathbb{N}}$. Given any execution σ of the system, we will denote the values of z and α at step k along such an execution by $\sigma(k)(z)$ and $\sigma(k)(\alpha)$, respectively. Given the measured variables z, we consider the problem of determining the input u such that the discrete state α is kept inside a set $S \subseteq \mathcal{A}$. If $\mathcal{A}$ and $\mathcal{I}$ are finite and discrete, in order to compute the set of inputs that map a set $X \subseteq \mathcal{A}$ inside S, we can compute $f(\alpha, z, u)$ for all $u \in \mathcal{I}$ and all $\alpha \in X$ and check whether it is contained in S. Assuming the size of X and the size of S of the order of the size of $\mathcal{A}$, this requires a number of computations of order $|\mathcal{I}||\mathcal{A}|^2$. If $\mathcal{A}$ is given by the product of a number of sets (as it is in the multi-agent systems that we consider) this approach is not practical as the number of computations is exponential in the number of agents. We thus propose an alternative procedure, whose idea can be explained in the following simple example.

Assume $\alpha \in \mathbb{N}$, $X = [2, 11]$, $S = [1, 10]$, $u \in \mathbb{Z}$, and that $f(\alpha, z, u) = f(\alpha, u) = \alpha + u$. For computing the set of inputs in $\mathbb{Z}$ such that $f(X, u) \subset S$, it is enough to compute the set of $u \in \mathbb{Z}$ such that $f(2, u) \geq 1$ and the set of $u \in \mathbb{Z}$ such that $f(11, u) \leq 10$, and then intersect these two sets. These two sets are intervals in $\mathbb{Z}$: $[-1, \infty)$ and $(-\infty, -1]$, respectively. The intersection of these two sets gives the answer $u = \{-1\}$. This simplification is due to the fact that the spaces $\mathcal{A}$ and $\mathcal{I}$ are equipped with an order (total in this case), while the function f preserves such orders. This argument will be formalized in a general framework in this paper by using partial order theory. We next state the problem of determining a feedback system Σ_f that updates the lower and upper bounds of the set of possible current states and gives as output the lower and upper bounds of the set of allowed inputs.

Problem 1. (Dynamic Output Feedback on a Lattice) Given system $\Sigma = (\mathcal{A}\times\mathcal{Z}, \mathcal{I}, \mathcal{Y}, (f, h), g)$ with initial set $X_0 \subseteq S$, find a deterministic transition system $\Sigma_f = (\chi\times\chi, \mathcal{Y}, \tilde{\mathcal{I}}\times\tilde{\mathcal{I}}, (H_{21}, H_{22}), (H_{11}, H_{12}))$ with $H_{21} : \chi\times\chi\times\mathcal{Y} \to \chi$, $H_{22} : \chi\times\chi\times\mathcal{Y} \to \chi$, $H_{11} : \chi\times\chi\times\mathcal{Y} \to \tilde{\mathcal{I}}$, $H_{12} : \chi\times\chi\times\mathcal{Y} \to \tilde{\mathcal{I}}$, $(\chi, \leq)$ and $(\tilde{\mathcal{I}}, \leq)$ lattices, with $\mathcal{A} \subseteq \chi$ and $\mathcal{I} \subseteq \tilde{\mathcal{I}}$, such that if $u(k) \in [H_{11}(L(k), U(k), y(k)), H_{12}(L(k), U(k), y(k))] \cap \mathcal{I}$, $L(k+1) = H_{21}(L(k), U(k), y(k))$, $U(k+1) = H_{22}(L(k), U(k), y(k))$, $L(0), U(0) \in \chi$, and $\{y(k)\}_{k\geq 0} = g(\sigma)$, we have (i) $\sigma(k)(\alpha) \in [L(k), U(k)] \cap \mathcal{A}$ and (ii) $[L(k), U(k)] \cap \mathcal{A} \subseteq S$.

The variables $L(k)$ and $U(k)$ are the lower and the upper bounds in a partial order $(\chi, \leq)$ of the set of possible current states. The functions H_{11} and H_{12} determine the lower and upper bounds of the set of inputs that map the set $[L(k), U(k)] \cap \mathcal{A}$ inside S. In the next section, we determine the form of the functions $H_{11}, H_{12}, H_{21}, H_{22}$ that solve this problem.

4 Problem Solution

To solve Problem 1, we need to re-define the original system on the partial orders.

Definition 3. Consider the system $\Sigma = (\mathcal{A}\times\mathcal{Z}, \mathcal{I}, \mathcal{Y}, (f, h), g)$. An extension of Σ on partial orders $(\chi, \leq)$ and $(\tilde{\mathcal{I}}, \leq)$ with $\mathcal{A} \subseteq \chi$ and $\mathcal{I} \subseteq \tilde{\mathcal{I}}$ is given by a new system $\tilde{\Sigma} = (\chi\times\mathcal{Z}, \tilde{\mathcal{I}}, \mathcal{Y}, (\tilde{f}, \tilde{h}), \tilde{g})$, in which

(i) $(\tilde{\mathcal{I}}, \leq) = \bigcup_x (\tilde{\mathcal{I}}(x), \leq)$, where for all $x \in \chi$, $(\tilde{\mathcal{I}}(x), \leq)$ is a sublattice of $(\tilde{\mathcal{I}}, \leq)$ with $\mathcal{I} \subseteq \tilde{\mathcal{I}}(x)$ and with the sublattices $(\tilde{\mathcal{I}}(x), \leq)$ for all $x \in \chi$ compatible partial orders;
(ii) $\tilde{f}|_{\mathcal{A}\times\mathcal{Z}\times\mathcal{I}} = f$, $\tilde{h}|_{\mathcal{A}\times\mathcal{Z}} = h$, and $\tilde{g}|_{\mathcal{A}\times\mathcal{Z}} = g$.

Item (i) requires to have input set extensions allowed at different states in χ, which all contain the inputs in $\mathcal{I}$. Item (ii) requires that the extended system is equal to the original system when restricted to the original sets $\mathcal{A}$ and $\mathcal{I}$. In the sequel, we will denote by $\tilde{\Sigma}|_{\mathcal{I}}$ the system $\tilde{\Sigma}$ in which the input set is restricted to $\mathcal{I}$.

Definition 4. The pair $(\tilde{\Sigma}, (\chi, \leq))$ is said to be *output interval compatible* if

(i) for all $y \in \mathcal{Y}$, $O_y(\tilde{\Sigma})$ is an interval lattice, that is, $O_y(\tilde{\Sigma}) = [\bigwedge O_y(\tilde{\Sigma}), \bigvee O_y(\tilde{\Sigma})]$;
(ii) $\tilde{f} : ([\bigwedge O_y(\tilde{\Sigma}), \bigvee O_y(\tilde{\Sigma})], z, u) \to [\tilde{f}(\bigwedge O_y(\tilde{\Sigma}), z, u), \tilde{f}(\bigvee O_y(\tilde{\Sigma}), z, u)]$ is an order isomorphism for all $(z, u) \in \mathcal{Z}\times\mathcal{I}$.

If the pair $(\tilde{\Sigma}, (\chi, \leq))$ is output interval compatible, we can use the result of [4], in which a state estimator on a partial order that updates a lower bound L and an upper bound U of the set of all possible current states is given by the update laws

$$L(k+1) = \tilde{f}(L(k) \vee \bigwedge O_{y(k)}(\tilde{\Sigma}), z(k), u(k)) \tag{2}$$
$$U(k+1) = \tilde{f}(U(k) \wedge \bigvee O_{y(k)}(\tilde{\Sigma}), z(k), u(k)). \tag{3}$$

These update laws are such that $\sigma(k)(\alpha) \in [L(k), U(k)] \cap \mathcal{A}$. As a consequence, the functions H_{21} and H_{22} that solve item (i) of Problem 1 are given by equations (2) and (3), respectively. One contribution of this work is to determine also the functions H_{11} and H_{12} of Problem 1, which determine the dynamic feedback law. In order to proceed, we give the following definition.

Definition 5. The pair $(\tilde{\Sigma}, (\tilde{\mathcal{I}}, \leq))$ is *input interval compatible* if for all $x \in \chi$ and $z \in \mathcal{Z}$

(i) $\tilde{f} : (x, z, \tilde{\mathcal{I}}(x)) \rightarrow [\tilde{f}(x, z, \bigwedge\tilde{\mathcal{I}}(x)), \tilde{f}(x, z, \bigvee\tilde{\mathcal{I}}(x))]$ is order preserving and onto;
(ii) $\tilde{f}$: $(x, z, \tilde{\mathcal{I}}(x)) \rightarrow [\tilde{f}(x, z, \bigwedge\tilde{\mathcal{I}}(x)), \tilde{f}(x, z, \bigvee\tilde{\mathcal{I}}(x)]$ is either $\vee$-preserving or $\tilde{f}(x, z, \bigvee\tilde{\mathcal{I}}(x)) = \bigvee\tilde{S}$, and it is either $\wedge$-preserving or $\tilde{f}(x, z, \bigwedge\tilde{\mathcal{I}}(x)) = \bigwedge\tilde{S}$.

This definition establishes that $\tilde{f}$ preserves the order in the second argument. The $\vee$ ($\wedge$) preserving properties guarantee that the set of inputs that is mapped to the same point through $\tilde{f}$ is a lattice. The following theorem gives the expressions of the functions H_{11} and H_{12}. A pictorial interpretation of H_{11} and H_{12} is given in Figure 1. Denote $\tilde{f}^{-1}_{x,z}(w) := \{u \in \tilde{\mathcal{I}}(x) \mid f(x, z, u) = w\}$.

Theorem 1. *Let system $\Sigma = (\mathcal{A} \times \mathcal{Z}, \mathcal{I}, \mathcal{Y}, (f, h), g)$ be controllable by dynamic output feedback with respect to $S \subseteq \mathcal{A}$ and initial set $X_0 \subseteq S$. Let $(\chi, \leq)$ and $(\tilde{\mathcal{I}}, \leq)$ be such that $\mathcal{A} \subseteq \chi$ and $\mathcal{I} \subseteq \tilde{\mathcal{I}}$. Let $\tilde{S} \subseteq \chi$ be an interval lattice such that $\tilde{S} \cap \mathcal{A} = S$. Assume that the extension $\tilde{\Sigma} = (\chi \times \mathcal{Z}, \tilde{\mathcal{I}}, \mathcal{Y}, (\tilde{f}, \tilde{h}), \tilde{g})$ is such that*

(i) $\tilde{\Sigma}|_{\mathcal{I}}$ is controllable by dynamic output feedback with respect to $\tilde{S}$;
(ii) the pair $(\tilde{\Sigma}, (\chi, \leq))$ is output interval compatible;
(iii) the pair $(\tilde{\Sigma}, (\tilde{\mathcal{I}}, \leq))$ is input interval compatible.

Then, a solution to Problem 1, Σ_f, is given by functions H_{21} and H_{22} given by expressions (2) and (3), respectively, with $L(0) = \bigwedge\tilde{S}$, $U(0) = \bigvee\tilde{S}$, and

$$\begin{aligned} H_{11}(L(k), U(k), y(k)) &= \bigwedge\tilde{f}^{-1}_{L'(k),z(k)}\left(\tilde{f}(L'(k), z(k), \bigwedge\tilde{\mathcal{I}}(L'(k))) \vee \bigwedge\tilde{S}\right) \\ &\quad \vee \bigwedge\tilde{f}^{-1}_{U'(k),z(k)}\left(\tilde{f}(U'(k), z(k), \bigwedge\tilde{\mathcal{I}}(U'(k))) \vee \bigwedge\tilde{S}\right) \\ H_{12}(L(k), U(k), y(k)) &= \bigvee\tilde{f}^{-1}_{L'(k),z(k)}\left(\tilde{f}(L'(k), z(k), \bigvee\tilde{\mathcal{I}}(L'(k))) \wedge \bigvee\tilde{S}\right) \\ &\quad \wedge \bigvee\tilde{f}^{-1}_{U'(k),z(k)}\left(\tilde{f}(U'(k), z(k), \bigvee\tilde{\mathcal{I}}(U'(k))) \wedge \bigvee\tilde{S}\right), \end{aligned} \tag{4}$$

in which $L'(k) = L(k) \vee \bigwedge O_{y(k)}(\tilde{\Sigma})$, $U'(k) = U(k) \wedge \bigvee O_{y(k)}(\tilde{\Sigma})$.

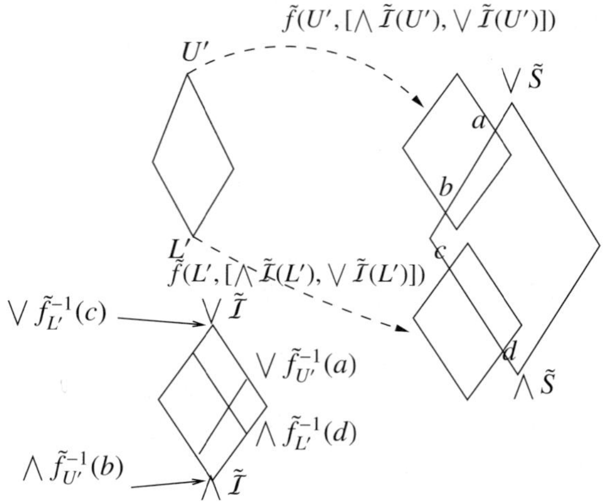

Fig. 1. Abstraction of Hasse diagrams to rhombi. In the picture, $a = \tilde{f}(U', \bigvee\tilde{\mathcal{I}}(U')) \wedge \bigvee\tilde{S}$, $b = \tilde{f}(U', \bigwedge\tilde{\mathcal{I}}(U')) \vee \bigwedge\tilde{S}$, $c = \tilde{f}(L', \bigvee\tilde{\mathcal{I}}(L')) \wedge \bigvee\tilde{S}$, $d = \tilde{f}(L', \bigwedge\tilde{\mathcal{I}}(L')) \vee \bigwedge\tilde{S}$, $H_{11} = \bigvee\tilde{f}_{U'}^{-1}(a)$, and $H_{12} = \bigwedge\tilde{f}_{L'}^{-1}(d)$. The dependencies on z and on k have been omitted.

5 Generality of the Partial Order Approach

We next show that if the system Σ is controllable by dynamic output feedback, the assumptions of Theorem 1 can be verified by suitable choices of $(\chi, \leq)$, $(\tilde{\mathcal{I}}, \leq)$, and $\tilde{\Sigma}$.

Theorem 2. *If system Σ is controllable by dynamic output feedback with respect to $S \subseteq \mathcal{A}$ and initial set $X_0 \subseteq S$, then there are partial orders $(\chi, \leq)$ and $(\tilde{\mathcal{I}}, \leq)$, an interval lattice $\tilde{S} \subseteq \chi$ with $\tilde{S} \cap \mathcal{A} = S$, and an extension $\tilde{\Sigma}$, such that $\tilde{\Sigma}|_{\mathcal{I}}$ is controllable by dynamic output feedback with respect to $\tilde{S}$, $(\tilde{\Sigma}, (\chi, \leq))$ is output interval compatible, and $(\tilde{\Sigma}, (\tilde{\mathcal{I}}, \leq))$ is input interval compatible.*

The assumption that Σ is controllable by dynamic output feedback with respect to S is needed to show that $\tilde{\Sigma}|_{\mathcal{I}}$ is also controllable by dynamic output feedback with respect to the interval lattice $\tilde{S}$. Such assumption is not needed to show output and input interval compatibility of $(\tilde{\Sigma}, (\chi, \leq))$ and of $(\tilde{\Sigma}, (\tilde{\mathcal{I}}, \leq))$, respectively. In case Σ is not controllable by dynamic output feedback with respect to S, $\tilde{\Sigma}|_{\mathcal{I}}$ will also not be controllable by dynamic output feedback with respect to any interval lattice $\tilde{S}$ such that $\tilde{S} \cap \mathcal{A} = S$ and for any choice of partial orders. This implies that $[H_{11}, H_{12}] \cap \mathcal{I}$ might be empty.

Example 1. The proof of Theorem 2 is constructive (see Appendix). We illustrate in this example how to construct the extended input partial order on a finite state/finite input system. Let $\mathcal{A} = \{\alpha_1, \alpha_2, \alpha_3\}$, $\mathcal{I} = \{u_1, u_2, u_3\}$, and let the update function F be given in the table of Figure 2. According to the proof of Theorem 2, we have $(\chi, \leq) = (\mathcal{P}(\mathcal{A}), \subseteq)$ and for all $x \in \chi$, we have $\tilde{\mathcal{I}}(x) = \mathcal{P}(\mathcal{I}) \cup I_x$, in which I_x contains "silent inputs" introduced to satisfy the onto property of item (i) of Definition 5. We start with $x = \alpha_1 \vee \alpha_2 \vee \alpha_3$. By computing $\tilde{f}(x, \tilde{u})$ (with $\tilde{f}(x, \tilde{u}) = f(x, \tilde{u})$ where $x \subseteq \mathcal{A}$ and $\tilde{u} \subseteq \mathcal{I}$ in the righthand side) for all $\tilde{u} \in \mathcal{P}(\mathcal{I})$, we note that $\tilde{f}(x, \tilde{u}) = \alpha_1 \vee \alpha_2 \vee \alpha_3$ for

$\tilde{u} \in \{u_1 \vee u_2 \vee u_3, u_1 \vee u_2, u_2, u_2 \vee u_3\}$, $\tilde{f}(x, \tilde{u}) = \alpha_1 \vee \alpha_2$ for $\tilde{u} \in \{u_1, u_2 \vee u_3\}$, $\tilde{f}(x, u_3) = \alpha_1$. As a consequence, the elements in χ that are less than $\tilde{f}(x, u_1 \vee u_2 \vee u_3)$ (where $u_1 \vee u_2 \vee u_3 = \vee \tilde{\mathcal{I}}(x)$) for which there is not an input in $\mathcal{P}(\mathcal{I})$ that map x to them are given by $\{\alpha_1 \vee \alpha_2, \alpha_2, \alpha_2 \vee \alpha_3, \alpha_3\}$. Thus, the set of silent inputs is $I_x = \{\epsilon_1, \epsilon_2, \epsilon_3, \epsilon_4\}$ such that $\tilde{f}(x, \epsilon_1) = \alpha_1 \vee \alpha_2$, $\tilde{f}(x, \epsilon_2) = \alpha_2$, $\tilde{f}(x, \epsilon_3) = \alpha_2 \vee \alpha_3$, $\tilde{f}(x, \epsilon_4) = \alpha_3$. We then establish the order among the elements in $I_x \cup \mathcal{P}(\mathcal{I})$ by following the procedure outlined in item 3 of the proof of Theorem 2 to guarantee the $\vee$-preserving property of item (ii) of Definition 5. Since $\tilde{f}(x, \epsilon_1) \prec \alpha_1 \vee \alpha_2 \vee \alpha_3$ and $\sup_{\tilde{u} \in \mathcal{P}(\mathcal{I})}\{\tilde{u} \mid \tilde{f}(x, \tilde{u}) = \alpha_1 \vee \alpha_2 \vee \alpha_3\} = u_1 \vee u_2 \vee u_3$, we set $\epsilon_1 \prec u_1 \vee u_2 \vee u_3$. Also, $\tilde{f}(x, \epsilon_1) \succ \alpha_1$ and $\sup_{\tilde{u} \in \mathcal{P}(\mathcal{I})}\{\tilde{u} \mid \tilde{f}(x, \tilde{u}) = \alpha_1\} = u_3$. Then, we set $\epsilon_1 \succ u_3$. Finally, $\epsilon_1 \succ \epsilon_2$ because $\tilde{f}(x, \epsilon_1) \succ \tilde{f}(x, \epsilon_2)$. Proceeding in a similar way for all of the other silent inputs, we obtain the additional relations: $\epsilon_2 \prec \epsilon_3$, $\epsilon_4 \prec \epsilon_3$, $\epsilon_4 \prec u_1 \vee u_3$, $\epsilon_3 \prec u_1 \vee u_2 \vee u_3$. The resulting extended input partial order $\tilde{\mathcal{I}}(x)$ is shown in the left plot of Figure 2. For $x = \alpha_2 \vee \alpha_3$, the resulting $\tilde{\mathcal{I}}(x)$ is shown in the right plot of Figure 2.The reader can verify that when $x = \alpha_i$ for some i, $I_x = \emptyset$.

	α_1	α_2	α_3
u_1	α_1	α_3	α_1
u_2	α_2	α_1	α_3
u_3	α_1	α_1	α_1

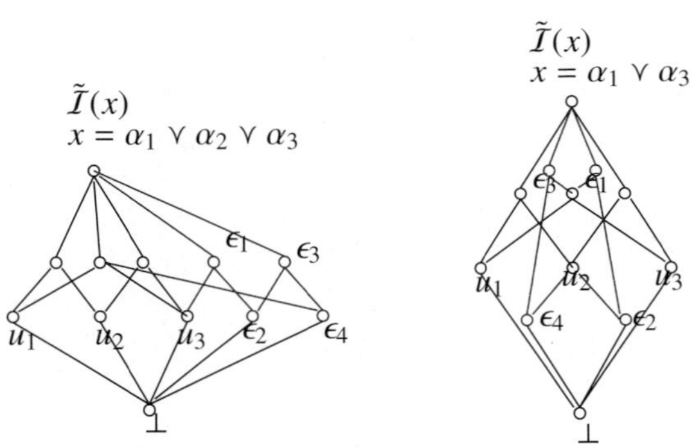

Fig. 2. Example 1. The table represents the update function $F(\alpha, u)$. The pictures at the center and at the right represent the extended input sets for $x = \alpha_1 \vee \alpha_2 \vee \alpha_3$ and $x = \alpha_1 \vee \alpha_3$ with the associated partial orders, respectively. The blue elements are the silent inputs I_x.

Computational considerations. For a finite state-finite input system, the sizes of $\tilde{\mathcal{I}}$ and of χ are related to the computational load of the proposed algorithms as these partial order structures need to be computed and stored in memory. The size of these partial orders does not affect computation in those systems in which the partial orders have algebraic properties as we will see in Example 2 of the next section. The amounts of computation c needed for computing such partial orders can be estimated to be $c \leq K \sum_{i=1}^{|\mathcal{A}|^2} |X_i||\mathcal{I}||S|$, in which X_i are the sets on which the estimator evolves, $|\mathcal{A}|^2$ is the number of such sets, and $K > 0$. This amount of computation is comparable to the one obtained by using enumeration and exhaustive techniques.

In this section, we have shown that the partial order approach to dynamic feedback is general and that the worst case computation is proportional to the one of exhaustive searches. The partial orders constructed in the proof of Theorem 2 and in Example 1 are not unique and have mainly a theoretical relevance as they are impractical for implementation in systems with a large number of states and inputs. Thus, we propose in the next section a relaxed version of Theorem 1.

6 Relaxations and Application to a Multi-robot Example

Consider the case in which partial orders $(\chi, \leq)$ and $(\tilde{\mathcal{I}}, \leq)$ have been chosen and the assumptions of Theorem 1 do not all hold. Some possible relaxations of the basic assumptions of Theorem 1 are as follows:

(R1) the set $\tilde{S} \subseteq \chi$ such that $\tilde{S} \cap \mathcal{A} = S$ is given by $\tilde{S} = \bigcup_{i=1}^{M} \tilde{S}^i$, in which $\tilde{S}^i$ are intervals and $\tilde{S} \cap O_y(\tilde{\Sigma})$ is an interval;

(R2) $\tilde{f} : \chi \times \mathcal{Z} \times \mathcal{I} \to \chi$ is a piece-wise order isomorphism, that is, for all interval $[L, U] \subseteq \chi$, we have that there are disjoint intervals $[L^j, U^j]$ with $\bigcup_j [L^j, U^j] = [L, U]$ such that $\tilde{f}([L^j, U^j], z, u) \to [\tilde{f}(L^j, z, u), \tilde{f}(U^j, z, u)]$ is an order isomorphism for all j and any $u \in \mathcal{I}$;

(R3) for all interval $[L, U] \subseteq \tilde{S} \cap O_y(\tilde{\Sigma})$ there are a function $\tilde{f}' : \chi \times \tilde{\mathcal{I}} \to \chi$ with $\tilde{f}' : (x, [\bigwedge\tilde{\mathcal{I}}, \bigwedge\tilde{\mathcal{I}}]) \to [\tilde{f}'(x, \bigwedge\tilde{\mathcal{I}}), \tilde{f}'(x, \bigvee\tilde{\mathcal{I}})]$ an order isomorphism for all $x \in \chi$ and an order preserving map in the first argument, constants $L^* \leq U^* \in \chi$, and constants $L^S \leq U^S \in \chi$ such that $\{u \in \mathcal{I} \mid \tilde{f}([L, U], z, u) \subseteq \tilde{S}\} \supseteq \mathcal{I} \cap \{\tilde{u} \in \tilde{\mathcal{I}} \mid \tilde{f}'([L^*, U^*], \tilde{u}) \subseteq [L^S, U^S]\}$, with the righthand set not empty.

It is always possible to determine a set $\tilde{S}$ that is a union of intervals and any function can always be broken into order isomorphisms. The expressions of the functions H_{12} and H_{11} as given in formulas (4) stay the same, but one should substitute $\tilde{f}'$ in place of $\tilde{f}$, L^* and U^* in place of L' and U', and L^S and U^S in place of $\bigwedge\tilde{S}$ and $\bigvee\tilde{S}$. Due to the piecewise isomorphic nature of the function $\tilde{f}$, the update laws (2-3) become: $L(k+1) = \bigwedge_{\bar{L}^j \leq \bar{U}^j} \bar{L}^j$, $\bar{L}^j = \tilde{f}(L^j(k), z(k), u(k)) \vee \bigwedge O_{y(k+1)}(\tilde{\Sigma})$ and $U(k+1) = \bigvee_{\bar{L}^j \leq \bar{U}^j} \bar{U}^j$, $\bar{U}^j = \tilde{f}(U^j(k), z(k), u(k)) \vee \bigvee O_{y(k+1)}(\tilde{\Sigma})$, in which L^j, U^j establish the intervals where $\tilde{f}$ is an order isomorphism in the first argument, and $L(0) = \bigwedge O_{y(0)}(\tilde{\Sigma})$, $U(0) = \bigvee O_{y(0)}(\tilde{\Sigma})$.

Example 2. We consider a version of the "capture the flag" game for robots called "RoboFlag Drill" already considered in [4], in which now the attackers can use their estimates of the assignments of the opponents to decide the next action to take. Briefly, some number of robots with positions $(z_i, 0) \in \mathbb{R}^2$, which we refer to as blue robots, must defend their zone $\{(x, y) \in \mathbb{R}^2 \mid y \leq 0\}$ from an equal number of incoming robots, which we refer to as red robots. The positions of the red robots are $(x_i, y_i) \in \mathbb{R}^2$. The red robots move toward the blue defensive zone. The blue robots are assigned each to a red robot and they coordinate to intercept the red robots. In this work, we allow the red robots to swap their horizontal location with a nearby red robot as appropriate. Let N represent the number of robots in each team. The RoboFlag Drill system can be specified by the rules $y_i(k+1) = y_i(k) - \delta$ if $y_i(k) \geq \delta$,

$$z_i(k+1) = z_i(k) + \delta \text{ if } z_i(k) < x_{\alpha_i(k)}, \quad z_i(k+1) = z_i(k) - \delta \text{ if } z_i(k) > x_{\alpha_i(k)} \tag{5}$$

$$(\alpha_i(k+1), \alpha_{i+1}(k+1)) = (\alpha_{i+1}(k), \alpha_i(k)) \text{ if } x_{\alpha_i(k)} \geq z_{i+1}(k) \wedge x_{\alpha_{i+1}(k)} \leq z_{i+1}(k) \tag{6}$$

$$(x_i(k+1), x_{i+1}(k+1)) = (x_{i+1}(k), x_i(k)) \text{ if } swap_{i,i+1}(k). \tag{7}$$

The variable α_i is the red robot that blue robot i is required to intercept. Equation (6) establishes that two blue robots trade their assignments only when the current assignments cause them to go toward each other. Rule (7) allows two adjacent red robots to swap their horizontal position. If the red robots never swap horizontal position, the assignments of the blue robot reaches an equilibrium value in which no more conflicts among the assignments of the blue robots are present (the attackers have all been intercepted). In this work, we want to solve the following problem: *Given measurements* $z(k)$ *determine control inputs* $swap_{i,i+1}(k)$ *such that there are always at least two pairs of blue robots with conflicting assignments*. To formalize this problem, we translate the rules (5-6-7) to the form $\Sigma = (\mathcal{A} \times \mathcal{Z}, \mathcal{I}, \mathcal{Y}, (f,h), g)$. Thus, let $\{1, ..., N\}$ be the locations at which the red robots can reside, that is, the location denotes the order along the x direction at which the red robots are displaced. With abuse of notation, let x_i denote the x coordinate of location i and α_i the location to which blue robot i is assigned. We assume $x_i \leq z_i \leq x_{i+1}$ for all i and for all time. We set $\mathcal{A} = \text{perm}(N)$, $\mathcal{Z} = \mathbb{R}^N$, $\mathcal{I} = \{u \in \{-1, 0, 1\}^N \mid u_i = 1 \Leftrightarrow u_{i+1} = -1, u_N \neq 1, u_1 \neq -1\}$ ($u_i = 1$ iff $swap_{i,i+1}$ is true), $\mathcal{Y} = \mathbb{R}^N \times \mathbb{R}^N$. The functions are defined as follows: $f(\alpha, z, u) = G(F(\alpha, z), u)$, in which $F(\alpha, z)$ is represented by relations (6) and $G(\beta, u) = \beta'$, with $u_j = 1 \Rightarrow$ [(if $\beta_i = j \Rightarrow \beta'_i = j + 1$) and (if $\beta_i = j + 1 \Rightarrow \beta'_i = j$)]. The function $h(\alpha, z)$ is represented by relations (5). Let the entropy of the blue robots be defined by $E = \frac{1}{2}\sum_{i=1}^{N} |\alpha_i - i|$. In the absence of input to the system (i.e. $u(k) = 0$ for all k), E converges to zero. We define the set S as $S = \{\alpha \mid E \geq 2\}$, which can be computed and is given by $S = \{\alpha \mid \exists\, i, j,$ with $j > i + 1$ such that $\alpha_i \neq i$ and $\alpha_j \neq j\}$. If $\alpha \in S$, there are at least two pairs of blue robots with conflicting assignments.

To apply the dynamic control algorithm using the relaxations (R1-R2-R3), we need to determine the partial orders $(\chi, \leq)$ and $(\tilde{\mathcal{I}}, \leq)$, the set $\tilde{S}$ satisfying item (R1), the extended function $\tilde{f}$, and finally the function $\tilde{f}'$ with the intervals $[L^*, U^*]$ and $[L^S, U^S]$ as given in item (R3). Set $(\chi, \leq) = (\mathbb{N}^N, \leq)$ with order established component-wise. Given any set $X \subseteq \mathbb{N}^N$, we denote $[X]_j$ the projection along j of such set. Then, a set $\tilde{S} \subseteq \chi$ is given by $\tilde{S} = \bigcup_i \tilde{S}^i$, in which $\tilde{S}^i$ for each i are intervals of four types: (a) there are $l < j$ such that $[S^i]_l = [l + 1, N]$ and $[S^i]_j = [j + 1, N]$; (b) there are $l < j$ such that $[S^i]_l = [1, l - 1]$ and $[S^i]_j = [j + 1, N]$; (c) there are $l < j$ with $j > l + 1$ such that $[S^i]_l = [l + 1, N]$ and $[S^i]_j = [1, j - 1]$; (d) there are $l < j$ such that $[S^i]_l = [1, l - 1]$ and $[S^i]_j = [1, j - 1]$. This can be checked by recalling that $\alpha \in \bigcup_i \tilde{S}_i$ if $\alpha \in \tilde{S}^i$ for some i. Define the extension $\tilde{F} : \chi \times \mathcal{Z} \to \chi$ as F with now $\alpha \in \mathbb{N}^N$. Clearly, $\tilde{F}|_{\mathcal{A}\times\mathcal{Z}} = F$. Also, we define $\tilde{h} : \chi \times \mathcal{Z} \to \mathcal{Z}$ as h with $\alpha \in \mathbb{N}^N$, for which $\tilde{h}|_{\mathcal{A}\times\mathcal{Z}} = h$. One can check that the output set is an interval and that the function $\tilde{F}$ is an order isomorphism on the output set. The function $\tilde{G} : \chi \times \mathcal{I} \to \chi$ is defined as G in which the first argument belongs to $\mathbb{N}^N$. Then $\tilde{f} = \tilde{G} \circ \tilde{F}$, in which one can check that $\tilde{f}_{\mathcal{A}\times\mathcal{Z}\times\mathcal{I}} = f$. We thus have defined the extended system $\tilde{\Sigma} = (\chi \times \mathcal{Z}, \mathcal{Y}, \mathcal{I}, (\tilde{f}, \tilde{h}), \tilde{g})$. For the input set, we consider $\tilde{\mathcal{I}} = \{-1, 0, 1\}^N$ with order established componentwise. It is easy to show that the extended system $\tilde{\Sigma}|_{\mathcal{I}} = (\chi \times \mathcal{Z}, \mathcal{I}, \mathcal{Y}, (\tilde{f}, \tilde{h}), \tilde{g})$ satisfies the dynamic controllability condition with respect to $\tilde{S}$ if $N > 4$. We are left to determine the function $\tilde{f}'$ with the intervals $[L^*, U^*]$ and $[L^S, U^S]$ as given in item (R3). For all $x = (x_1, ..., x_N) \in \chi$ and $\tilde{u} = (\tilde{u}_1, ..., \tilde{u}_N) \in \tilde{\mathcal{I}}$, we set $\tilde{f}'(x, \tilde{u}) = (\tilde{f}'_1(x_1, \tilde{u}_1), ..., \tilde{f}'_N(x_N, \tilde{u}_N))$, in which $\tilde{f}'(x_i, \tilde{u}_i) := x_i + \tilde{u}_i$. The following algorithm, computes the sets $[L^*, U^*]$ and $[L^S, U^S]$ component-wise for all intervals $[L, U] \subseteq \tilde{S} \cap O_y(\tilde{\Sigma})$. Let $P' = \tilde{F}([L, U], z)$

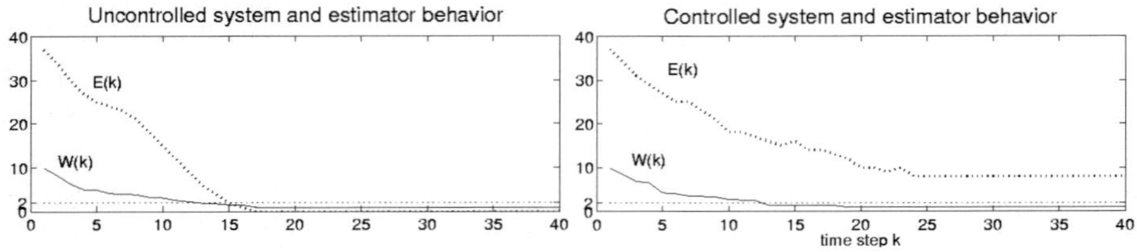

Fig. 3. Convergence plots of the estimator and of the entropy with $N = 15$ and $\alpha(0) = (4, 8, 9, 2, 13, 15, 6, 5, 12, 10, 1, 14, 3, 7, 11)$. For the controlled system $E > 2$ always. From the uncontrolled system plot, one realizes that a strategy that estimates the state first and then computes the controller once the estimator has converged does not work.

Algorithm

Initialize $flag_i = 0$, $L_i^* = U_i^* = i$, $L_i^S = 1$, and $U_i^S = N$ for all i
For $i = 1 : N$
  If $\min(P_i') = i$ and $flag_{i-1} = 0 \Longrightarrow L_i^S = 1,\ U_i^S = i - 1$ and $flag_i = 1$
End
For $i = 2 : N$
  If $\max(P_{i-1}') = i - 1$ and $flag_{i-1} = 0 \Longrightarrow L_{i-1}^S = i,\ U_{i-1}^S = N$ and $flag_i = 1$
End
For $i = 1 : N$
  If $\min(P_i') \geq i + 1$ and $flag_{i+1} = 0 \Longrightarrow L_i^* = U_i^* = i + 1,\ L_i^S = i + 1,\ U_i^S = N$
  If $\max(P_i') \leq i - 1$ and $flag_i = 0 \Longrightarrow L_i^* = U_i^* = i - 1,\ L_i^S = 1\ U_i^S = i - 1$
End.

The idea behind this algorithm is as follows. Say that $[P']_i = [i, N]$ and that we want to remove i from it by swapping red robot i with red robot $i-1$. This can be done by asking that $i + \tilde{u}_i \in [1, i - 1]$, which gives $\tilde{u}_i \leq -1$. Finally, note that the function $\tilde{f}$ is a composition of a function $\tilde{F}$, which is an order isomorphism, and a function $\tilde{G}$, which is a piecewise order isomorphism. To see this, let $\tilde{u}_i = -1$ and $P_i' = [i, N]$ for example, then we can re-write $[i, N] = [i, i] \cup [i + 1, N]$ so that $\tilde{G}_i : ([i, i], -1) \to [\tilde{G}_i(i, -1), \tilde{G}_i(i, 1)] = [i - 1, i - 1]$ and $\tilde{G}_i : ([i + 1, N], -1) \to [\tilde{G}_i(i + 1, -1), \tilde{G}_i(N, -1)] = [i + 1, N]$ are order isomorphisms. Figure 3 shows the behavior of the estimator error (given by $W(k) = 1/N \sum_{i=1}^{N} |m_i(k)|$, in which $m_i(k)$ is the coordinate set $[L_i(k), U_i(k)]$ minus all the singletons that occur at other coordinates) and of the entropy $E(k) = 1/2 \sum_{i=1}^{N} |\alpha_i(k) - i|$. In this example, the computation requirement for the implementation of the dynamic controller is proportional to N (number of variables to control and estimate). If we had not used any structure, we would have had a number of computations at least of the order of $(N!)^2$ as the size of the output set and of the set S are both of the order of $N!$. Note also that this simplification is not due to the fact that the dynamics decouples as it is heavily coupled between the robots.

7 Conclusions

We have proposed a partial order approach to dynamic feedback for the discrete variables of a hybrid system, which relies on partial order theory to compute only suitable lower and upper bounds to determine the dynamic controller. We have shown that such an approach is general as it can be applied to any system that is controllable by dynamic output feedback. The worst case computation load of the proposed approach does not exceed the one of exhaustive searches under partial observations. The main computational advantage is obtained when one can choose suitable partial orders in which the computation of joins and meets is efficiently performed. A multi-robot example showed this point. The next step is to consider the dynamic feedback problem also for the continuous variables and establish system structures that allow efficient choices of partial orders. As we mentioned, this work was not concerned with analysis problems: these are left to our future work, in which we would like to determine efficient computations of escape tubes and controlled invariance kernels by the computation of suitable lower and upper bounds, only. Finally, we plan to consider in our future work uncertainty in the system dynamics by modeling it as nondeterminism. On the application side, we will extend these results to the design of safety controllers under partial observation in the context of intelligent transportation systems.

References

1. E. Asarin, O. Maler, and A. Pnueli. Symbolic controller design for discrete and timed systems. *Hybrid Systems: Computation and Control,* Lecture Notes in Computer Science, volume 999, P. Antsakilis, W. Kohn, A. Nerode, and S. Sastry, Eds. Springer Verlag, pages 1–20, 1995.
2. J. Aubin, J. Lygeros, M. Quincampoix, S. Sastry, and N. Seube. Impulse differential inclusions: A viability approach to hybrid systems. *IEEE Transactions on Automatic Control*, 47(1):2–20, 2002.
3. B. A. Davey and H. A. Priesteley. *Introduction to Lattices and Order.* Cambridge University Press, 2002.
4. D. DelVecchio, R. M. Murray, and E. Klavins. Discrete state estimators for systems on a lattice. *Automatica*, 42(2):271–285, 2006.
5. A. Deshpande and P. Varaiya. Viable control of hybrid systems. In *Hybrid Systems II,* Lecture Notes in Computer Science, volume 999, P. Antsaklis, W. Kohn, A. Nerode, and S. Sastry, Eds. Springer Verlag, pages 128–147, 1995.
6. P. E.Caines and S. Wang. Classical and logic based regulator design and its complexity for partially observed automata. In *Conf. on Decision and Control*, pages 132–137, 1989.
7. P. J. Ramadge and W. M. Wonham. The control of discrete event systems. *Proceedings of the IEEE*, 77(1):81–98, 1989.
8. K. Rohloff and S. Lafortune. On the synthesis of safe control policies in decentralized control of discrete event systems. *IEEE Transactions on Automatic Control*, 48(6):1064–1068, 2003.
9. E. D. Sontag. *Mathematical Control Theory*. Springer, 1998.
10. W. Thomas. On the synthesis of strategies in infinite games. In *Proceedings of the STACS 95,* E. W. Mayr and C. Puech, Eds. Springer Verlag, pages 1–13, 1995.
11. H. Wong-Toi. The synthesis of controllers for linear hybrid automata. In *Conf. on Decision and Control*, pages 4607–4613, 1997.

Appendix I. Partial Order Theory

In this section, we introduce the main notation and definitions about partial orders that will be used in this work. For a complete overview, the reader is referred to [3]. A partial order is a set χ with a partial order relation "$\leq$", and we denote it by the pair $(\chi, \leq)$. for all $x, w \in \chi$, the $\sup\{x, w\}$ is the smallest element that is larger than both x and w. In a similar way, the $\inf\{x, w\}$ is the largest element that is smaller than both x and w. We define the *join* "$\vee$" and the *meet* "$\wedge$" of two elements x and w in χ as $x \vee w = \sup\{x, w\}$ and $x \wedge w = \inf\{x, w\}$. If $S \subseteq \chi$, $\bigvee S = \sup\ S$ and $\bigwedge S = \inf\ S$. If $x \wedge w \in \chi$ and $x \vee w \in \chi$ for all $x, w \in \chi$, then $(\chi, \leq)$ is a *lattice*. Let $(\chi, \leq)$ be a lattice and let $S \subseteq \chi$ be a non-empty subset of χ. Then, $(S, \leq)$ is a *sublattice* of χ if $a, b \in S$ implies that $a \vee b \in S$ and $a \wedge b \in S$. Any interval sublattice of $(\chi, \leq)$ is given by $[L, U] = \{w \in \chi \mid L \leq w \leq U\}$ for $L, U \in \chi$. That is, this special sublattice can be represented by only two elements. for all set S, we denote by $\mathcal{P}(S)$ the set of all subsets of S. On $\mathcal{P}(S)$, it is possible to establish a partial order relation determined by the inclusion relation. Therefore, $(\mathcal{P}(S), \leq)$ with "$\leq$" established by the inclusion relation is a lattice. Let $(\chi, \leq)$ be a partial order and let $x, w \in \chi$. We will use the notation $x \prec w$ to say that $x \leq w$ and there is not an element that is larger than x and smaller than w; we will use the notation $x \succ w$ to say that $x \geq w$ and there is not an element that is larger than x and smaller than w. for all $w, x \in \chi$, we denote $w \parallel x$ if they are not related by the order relation. Let $(P, \leq_P)$ and $(Q, \leq_Q)$ be two partial orders and let $X = P \cap Q$. They are said to be *compatible partial orders* if for all pair $x_1, x_2 \in X$ we have that $x_1 \leq_P x_2$ if and only if $x_1 \leq_Q x_2$. Let $(P, \leq_P)$ and $(Q, \leq_Q)$ be two compatible partial orders. Then, the union of the two partial orders, denoted $(P, \leq_P) \cup (Q, \leq_Q)$ is the new partial order $(R, \leq)$, in which $R = P \cup Q$ and for all $x_1, x_2 \in R$ we have that $x_1 \leq x_2$ if and only if $x_1 \leq_Q x_2$ or $x_1 \leq_P x_2$. In the sequel, when we will have two compatible partial orders, we will omit the subscript of "$\leq$" as there will be no ambiguity on the partial order relation between any two elements. We now consider maps on partial orders. Let $(P, \leq)$ and $(Q, \leq)$ be partially ordered sets. A map $f : P \to Q$ is (i) an *order preserving map* if $x \leq w \implies f(x) \leq f(w)$; (ii) an *order embedding* if $x \leq w \iff f(x) \leq f(w)$; (iii) an *order isomorphism* if it is order embedding and it maps P *onto* Q. The map $f : P \to Q$ is said to be $\vee$-preserving if for all $x, w \in P$, we have that $f(x \vee w) = f(x) \vee f(w)$. It is said to be $\wedge$-preserving if for all $x, w \in P$, we have that $f(x \wedge w) = f(x) \wedge f(w)$. One can show that if f is order preserving, then for all $x, y \in P$, we have that $f(x) \vee f(w) \leq f(x \vee w)$ and $f(x) \wedge f(w) \geq f(x \wedge w)$.

Appendix II. Proof of Theorems and Propositions

Proof. (Proof of Proposition 1) ($\Leftarrow$)Choose functions H_1 and H_2 as $H_2(X(k), y(k)) = F(X(k) \cap O_{y(k)}(\Sigma), u(k))$, $u(k) \in H_1(X(k), y(k))$ with $H_1(X(k), y(k)) = \{u \in \mathcal{I} \mid F(X(k) \cap O_{y(k)}(\Sigma), u) \subset S\}$ and $X(0) = X_0$. We show that the set $H_1(X(k), y(k))$ is not empty for all k and that properties (i) and (ii) of Definition 2 are satisfied. We proceed by induction argument on the step k. (Base case) By assumption, $X(0) \subseteq S$ and $s(0) \in X(0)$. As a consequence, $\{u \in \mathcal{I} \mid F(X(0) \cap O_{y(0)}(\Sigma), u) \subseteq S\}$ is not empty. (Induction step) Assume $X(k) \subseteq S$ and $s(k) \in X(k)$, then $H_1(X(k), y(k)) = \{u \in \mathcal{I} \mid F(X(k) \cap O_{y(k)}(\Sigma), u) \subseteq S\} \neq \emptyset$ because $\{u \in \mathcal{I} \mid F(X(k) \cap O_{y(k)}(\Sigma), u) \subseteq S\} \supseteq \{u \in \mathcal{I} \mid F(S \cap O_{y(k)}(\Sigma), u) \subseteq S\}$ and the latter set is nonempty by assumption. Thus, if $u(k) \in H_1(X(k), y(k))$ we have by construction that $X(k+1) \subseteq S$. Also, since $s(k) \in X(k)$ and $s(k) \in O_{y(k)}(\Sigma)$, we have that $s(k+1) \in X(k+1)$.

($\Rightarrow$) Assume that $\{u \in \mathcal{I} \mid F(O_y(\Sigma) \cap S, u) \subseteq S\} = \emptyset$ for some y. Let $s(0) \in S$ be such that $y(0) = g(s(0))$ and $\{u \in \mathcal{I} \mid F(O_{y(0)}(\Sigma) \cap S, u) \subseteq S\} = \emptyset$. Thus $\{u \in \mathcal{I} \mid F(X_0 \cap O_{y(0)}(\Sigma), u) \subseteq S\} = \emptyset$. Assume that the system is controllable by dynamic output feedback with respect to $X_0 \subseteq S$. Then, there are functions $H_1 : \mathcal{P}(Q) \times \mathcal{Y} \longrightarrow \mathcal{P}(\mathcal{I})$ and $H_2 : \mathcal{P}(Q) \times \mathcal{Y} \times \mathcal{I} \longrightarrow \mathcal{P}(Q)$ such that $X(1) = H_2(X_0, y(0)) \subseteq S$, $s(1) \in X(1)$ and $u(0) \in H_1(X_0, y(0))$. For guaranteeing $s(1) \in X(1)$ with $s(1) = F(s(0), u(0))$ and $s(0) \in X_0 \cap O_{y(0)}(\Sigma)$, we need that $F(X_0 \cap O_{y(0)}(\Sigma), u(0)) \subseteq X(1)$. However, $F(X_0 \cap O_{y(0)}(\Sigma), u(0)) \nsubseteq S$ and $X(1) \subseteq S$. This leads to a contradiction.

Proof. (Proof of Theorem 1) The dependencies on z are neglected. Equations (2-3) imply that $\alpha(k) \in [L(k), U(k)] \cap \mathcal{A}$. Thus, property (i) of Problem 1 is true. We next show that

(a) $\{u \in \mathcal{I} \mid \tilde{f}([L'(k), U'(k)], u) \subseteq [\bigwedge\tilde{S}, \bigvee\tilde{S}]\} = \mathcal{I} \cap [H_{11}(L(k), U(k), y(k)), H_{12}(L(k), U(k), y(k))]$;

(b) $[H_{11}(L(k), U(k), y(k)), H_{12}(L(k), U(k), y(k))] \cap \mathcal{I}$ is not empty.

Proof of (a). Since $[L'(k), U'(k)] \subseteq O_{y(k)}(\tilde{\Sigma})$, the function $\tilde{f}$ preserves the ordering in the first argument. As a consequence, we have that $\{u \in \mathcal{I} \mid \tilde{f}([L'(k), U'(k)], u) \subseteq [\bigwedge\tilde{S}, \bigvee\tilde{S}]\} = \{u \in \mathcal{I} \mid \bigwedge\tilde{S} \leq \tilde{f}(L'(k), u) \leq \bigvee\tilde{S}\} \cap \{u \in \mathcal{I} \mid \bigwedge\tilde{S} \leq \tilde{f}(U'(k), u) \leq \bigvee\tilde{S}\}$. Also, we have that $\{u \in \mathcal{I} \mid \bigwedge\tilde{S} \leq \tilde{f}(L'(k), u) \leq \bigvee\tilde{S}\} = \mathcal{I} \cap \{u \in \tilde{\mathcal{I}}(L'(k)) \mid \bigwedge\tilde{S} \leq \tilde{f}(L'(k), u) \leq \bigvee\tilde{S}\}$ and that $\{u \in \mathcal{I} \mid \bigwedge\tilde{S} \leq \tilde{f}(U'(k), u) \leq \bigvee\tilde{S}\} = \mathcal{I} \cap \{u \in \tilde{\mathcal{I}}(U'(k)) \mid \bigwedge\tilde{S} \leq \tilde{f}(U'(k), u) \leq \bigvee\tilde{S}\}$. As a consequence, we have that

$$\begin{aligned} &\{u \in \mathcal{I} \mid \tilde{f}([L'(k), U'(k)], u) \subseteq [\bigwedge\tilde{S}, \bigvee\tilde{S}]\} = \\ &\mathcal{I} \cap \{u \in \tilde{\mathcal{I}}(L'(k)) \mid \bigwedge\tilde{S} \leq \tilde{f}(L'(k), u) \leq \bigvee\tilde{S}\} \cap \\ &\{u \in \tilde{\mathcal{I}}(U'(k)) \mid \bigwedge\tilde{S} \leq \tilde{f}(U'(k), u) \leq \bigvee\tilde{S}\}. \end{aligned} \tag{8}$$

One can readily verify that $\{u \in \tilde{\mathcal{I}}(L'(k)) \mid \bigwedge\tilde{S} \leq \tilde{f}(L'(k), u) \leq \bigvee\tilde{S}\} = \tilde{f}^{-1}_{L'(k)}((\tilde{f}(L'(k), \tilde{\mathcal{I}}(L'(k))) \cap [\bigwedge\tilde{S}, \bigvee\tilde{S}])$, which derives from the definition of $\tilde{f}^{-1}_{L'(k)}$. By the onto property in item (i) of Definition 5, we also have that $\tilde{f}(L'(k), \tilde{\mathcal{I}}(L'(k))) = [\tilde{f}(L'(k), \bigwedge\tilde{\mathcal{I}}(L'(k))), \tilde{f}(L'(k), \bigvee\tilde{\mathcal{I}}(L'(k)))]$. As a consequence, we obtain that $\{u \in \tilde{\mathcal{I}}(L'(k)) \mid \bigwedge\tilde{S} \leq \tilde{f}(L'(k), u) \leq \bigvee\tilde{S}\} = \tilde{f}^{-1}_{L'(k)}([\tilde{f}(L'(k), \bigwedge\tilde{\mathcal{I}}(L'(k))) \vee \bigwedge\tilde{S}, \tilde{f}(L'(k), \bigvee\tilde{\mathcal{I}}(L'(k))) \wedge \bigvee\tilde{S}])$. We are thus left to show that $\tilde{f}^{-1}_{L'(k)}([\tilde{f}(L'(k), \bigwedge\tilde{\mathcal{I}}(L'(k))) \vee \bigwedge\tilde{S}, \tilde{f}(L'(k), \bigvee\tilde{\mathcal{I}}(L'(k))) \wedge \bigvee\tilde{S}]) = [\bigwedge\tilde{f}^{-1}_{L'(k)}(\tilde{f}(L'(k), \bigwedge\tilde{\mathcal{I}}(L'(k))) \vee \bigwedge\tilde{S}), \bigvee\tilde{f}^{-1}_{L'(k)}(\tilde{f}(L'(k), \bigvee\tilde{\mathcal{I}}(L'(k))) \wedge \bigvee\tilde{S})]$. To show this, we show that any element of the first set belongs to the second and *viceversa*. Any element of the second set is also an element of the first set due to the order preserving property of $\tilde{f}$ in the second argument as established in item (i) of Definition 5. Assume now that u is in the first set, then $\tilde{f}(L'(k), \bigwedge\tilde{\mathcal{I}}(L'(k))) \vee \bigwedge\tilde{S} \leq \tilde{f}(L'(k), u) \leq \tilde{f}(L'(k), \bigvee\tilde{\mathcal{I}}(L'(k))) \wedge \bigvee\tilde{S}$. We next show that $w \leq u$ in which $w = \bigwedge\tilde{f}^{-1}_{L'(k)}\left(\tilde{f}(L'(k), \bigwedge\tilde{\mathcal{I}}(L'(k))) \vee \bigwedge\tilde{S}\right)$. If $\tilde{f}(L'(k), \bigwedge\tilde{\mathcal{I}}(L'(k))) = \bigwedge\tilde{S}$, we have that $\bigwedge\tilde{f}^{-1}_{L'(k)}\left(\tilde{f}(L'(k), \bigwedge\tilde{\mathcal{I}}(L'(k))) \vee \bigwedge\tilde{S}\right) = \bigwedge\tilde{\mathcal{I}}(L'(k))$ and therefore we have that $\bigwedge\tilde{f}^{-1}_{L'(k)}(\tilde{f}(L'(k), \bigwedge\tilde{\mathcal{I}}(L'(k))) \vee \bigwedge\tilde{S}) \leq u$. If instead $\tilde{f}(L'(k), \bigwedge\tilde{\mathcal{I}}(L'(k))) \neq \bigwedge\tilde{S}$, by item (ii) of Definition 5, we have that $\tilde{f}$ is $\wedge$-preserving in the second argument. Since $w \leq \tilde{f}(L'(k), u)$, it must be that either $\bigwedge\tilde{f}^{-1}(w) \leq u$ or $\bigwedge\tilde{f}^{-1}(w) \parallel u$ by the order preserving property of $\tilde{f}$ in the second argument. Let us show that $\bigwedge\tilde{f}^{-1}(w) \parallel u$ is not possible. By the $\wedge$-preserving property, we have that $\tilde{f}(\bigwedge\tilde{f}^{-1}_{L'(k)}(w) \wedge u) = w \wedge$

$\tilde{f}(L'(k), u)$. Since $w \leq \tilde{f}(L'(k), u)$, we have that $w \wedge \tilde{f}(L'(k), u) = w$, which in turn implies that $\tilde{f}(L'(k), \bigwedge \tilde{f}^{-1}(w) \wedge u) = w$. By the definition of $\bigwedge \tilde{f}^{-1}_{L'(k)}(w)$, it follows that we must have $\bigwedge \tilde{f}^{-1}_{L'(k)}(w) \leq u$. One can proceed in a similar way to show that $u \leq \bigvee \tilde{f}^{-1}_{L'(k)}\left(\tilde{f}(L'(k), \bigvee \tilde{I}(L'(k))) \wedge \bigvee \tilde{S}\right)$. As a consequence, we have concluded that

$$\begin{aligned}\{u \in \tilde{I}(L'(k)) \mid \bigwedge \tilde{S} \leq \tilde{f}(L'(k), u) \leq \bigvee \tilde{S}\} = \\ [\bigwedge \tilde{f}^{-1}_{L'(k)}\left(\tilde{f}(L'(k), \bigwedge \tilde{I}(L'(k))) \vee \bigwedge \tilde{S}\right), \bigvee \tilde{f}^{-1}_{L'(k)}\left(\tilde{f}(L'(k), \bigvee \tilde{I}(L'(k))) \wedge \bigvee \tilde{S}\right)].\end{aligned} \tag{9}$$

Similar reasonings can be used to show that equation (9) holds for $U'(k)$. Equations (8), (9) and (9) with $U'(k)$ in place of $L'(k)$ prove (a). Given (a), to show (b) one can show that $\{u \in \mathcal{I} \mid \tilde{f}([L'(k), U'(k)], u) \subseteq [\bigwedge \tilde{S}, \bigvee \tilde{S}]\}$ is not empty. This is true if $[L'(k), U'(k)] \subseteq O_y(\tilde{\Sigma}) \cap \tilde{S}$ as by assumption $\tilde{\Sigma}|_{\mathcal{I}}$ is controllable by dynamic output feedback with respect to $\tilde{S}$. We can show that $[L'(k), U'(k)] \subseteq O_{y(k)}(\tilde{\Sigma}) \cap \tilde{S}$ by induction on the step k. In fact, $[L'(0), U'(0)] \subseteq O_y(\tilde{\Sigma}) \cap \tilde{S}$ as $L(0) = \bigwedge \tilde{S}$ and $U(0) = \bigvee \tilde{S}$. Assume that $[L'(k), U'(k)] \subseteq O_{y(k)}(\tilde{\Sigma}) \cap \tilde{S}$, let us show that also $[L'(k+1), U'(k+1)] \subseteq O_{y(k+1)}(\tilde{\Sigma}) \cap \tilde{S}$. Since $[L'(k), U'(k)] \subseteq O_{y(k)}(\tilde{\Sigma}) \cap \tilde{S}$, we have that $\mathcal{I} \cap [H_{11}(L(k), U(k), y(k)), H_{12}(L(k), U(k), y(k))]$ is not empty. We thus can take $u(k) \in \mathcal{I} \cap [H_{11}(L(k), U(k), y(k)), H_{12}(L(k), U(k), y(k))]$ and apply it to the system. By construction of $H_{11}(L(k), U(k), y(k))$ and $H_{12}(L(k), U(k), y(k))$, we have that $[L(k+1), U(k+1)] \subseteq \tilde{S}$. Thus, $[L'(k+1), U'(k+1)] \subseteq \tilde{S} \cap O_{y(k+1)}(\tilde{\Sigma})$. Therefore, (b) is shown.

Proof. (Proof of Theorem 2) We determine a system extension $\tilde{\Sigma}$ and we show that the properties of Definition 5 are satisfied.

1. Define $\chi = \mathcal{P}(\mathcal{A})$ and $(\chi, \leq) = (\mathcal{P}(\mathcal{A}), \subseteq)$. The bottom element is $\perp_\chi = \emptyset$. Let $x \in \chi$ be given by $x = \alpha_1 \vee ... \vee \alpha_n$ with $\alpha_i \in \mathcal{A}$, for all $u \in \mathcal{I}$ we define the function $\tilde{f} : \chi \times \mathcal{I} \to \chi$ as $\tilde{f}(x, u) = f(\alpha_1, u) \vee ... \vee f(\alpha_n, u)$ for $u \in \mathcal{I}$. Output interval compatibility of the pair $(\tilde{\Sigma}, (\chi, \leq))$ follows immediately.

2. for all $x \in \chi$, the extended input set $\tilde{I}(x)$ is defined as $\mathcal{P}(\mathcal{I}) \bigcup I_x$, in which the order among the elements in $\mathcal{P}(\mathcal{I})$ is established according to inclusion relation, and the sets I_x for all x are called the sets of *silent inputs* and are defined as follows. for all $\tilde{u} \in \mathcal{P}(\mathcal{I})$, we have $\tilde{u} = u_1 \vee ... \vee u_p$ for some $u_i \in \mathcal{I}$. Then, we define $\tilde{f}(x, \tilde{u}) = \tilde{f}(x, u_1) \vee ... \vee \tilde{f}(x, u_p)$. Let us initialize $I_x = \emptyset$ and let $\mathcal{I} = \{u_1, ..., u_m\}$. for all $w \in \chi$ such that $w \leq \tilde{f}(x, u_1 \vee ... \vee u_m)$, if there is not a $\tilde{u} \in \mathcal{P}(\mathcal{I})$ such that $\tilde{f}(x, \tilde{u}) = w$, define a silent input ϵ such that $\tilde{f}(x, \epsilon) = w$. Thus, we add such silent input to I_x, that is, $I_x = I_x \cup \epsilon$.

3. We next establish the order among the silent inputs and the inputs in $\mathcal{P}(\mathcal{I})$. for all $\epsilon \in I_x$, let $w = \tilde{f}(x, \epsilon)$. By construction, $\tilde{f}(x, \epsilon) \leq \tilde{f}(x, u_1 \vee ... \vee u_m)$. Let $\{w_1, ..., w_k\}$ be the set of elements with $w_i \leq \tilde{f}(x, u_1 \vee ... \vee u_m)$ such that either $w_i \prec \tilde{f}(x, \epsilon)$ or $w_i \succ \tilde{f}(x, \epsilon_i)$. Let $\tilde{u}_i \in \tilde{I}(x)$ be such that $\tilde{f}(x, \tilde{u}_i) = w_i$. If $\tilde{u}_i \in I_x$ then set $\tilde{u}_i \succ \epsilon$ if and only if $w_i \succ \tilde{f}(x, \epsilon_i)$ and $\tilde{u}_i \prec \epsilon$ if and only if $w_i \prec \tilde{f}(x, \epsilon_i)$. If $\tilde{u}_i \in \mathcal{P}(\mathcal{I})$, there may be several such inputs so that $\tilde{f}(x, \tilde{u}_i) = w_i$. Let $\tilde{u}_i$ be the greatest of such inputs, that is, $\tilde{u}_i = \sup_{(\mathcal{P}(\mathcal{I}), \leq)}\{\tilde{u} \mid \tilde{f}(x, \tilde{u}) = w_i\}$. (By the way $\tilde{f}$ has been defined on elements of $\mathcal{P}(\mathcal{I})$ it follows that $\tilde{f}(x, \tilde{u}_i) = w_i$.) Then, we set $\tilde{u}_i \succ \epsilon$ if and only if $w_i \succ \tilde{f}(x, \epsilon_i)$ and $\tilde{u}_i \prec \epsilon$ if and only if $w_i \prec \tilde{f}(x, \epsilon_i)$. Let $\perp_{\mathcal{I}} = \bigwedge \tilde{I}(x)$ such that every element that does not have a lower element is strictly greater than it. We also define $\tilde{f}(x, \perp_{\mathcal{I}}) = \perp_\chi$. Note that by construction, the top element of $\tilde{I}(x)$ is given by $u_1 \vee ... \vee u_m = \bigvee \tilde{I}(x)$. By construction, $(\tilde{I}(x), \leq)$ are compatible partial orders.

From item 2., it follows that $\tilde{f} : (x, \tilde{I}(x)) \to [\tilde{f}(x, \wedge\tilde{I}(x)), \tilde{f}(x, \vee\tilde{I}(x))]$ is onto. To show that it is also order preserving, we show that for all $\tilde{u}_1 \leq \tilde{u}_2$ in $\tilde{I}(x)$ also $\tilde{f}(x, \tilde{u}_1) \leq \tilde{f}(x, \tilde{u}_2)$. Let $\tilde{u}_{1,1} \prec \tilde{u}_{1,2} \prec \ldots \prec \tilde{u}_{1,k}$ be the chain between $\tilde{u}_{1,1} = \tilde{u}_1$ and $\tilde{u}_{1,k} = \tilde{u}_2$. Consider any consecutive pair $\tilde{u}_{1,i} \prec \tilde{u}_{1,i+1}$. Then if $\tilde{u}_{1,i}, \tilde{u}_{1,i+1} \in \mathcal{P}(\mathcal{I})$, by the definition of $\tilde{f}$ on elements in $\mathcal{P}(\mathcal{I})$ given in item 1., we have $\tilde{f}(x, \tilde{u}_{1,i}) \leq \tilde{f}(x, \tilde{u}_{1,i+1})$. If either one of $\tilde{u}_{1,i}, \tilde{u}_{1,i+1}$ is in I_x (that is, it is a silent input), by the definition of the order in item 3., we have that $\tilde{u}_{1,i} \prec \tilde{u}_{1,i+1}$ if and only if $\tilde{f}(x, \tilde{u}_{1,i}) \prec \tilde{f}(x, \tilde{u}_{1,i+1})$. Since, this holds for all consecutive pair $(\tilde{u}_{1,i}, \tilde{u}_{1,i+1})$, we thus have that $\tilde{f}(x, \tilde{u}_1) \leq \tilde{f}(x, \tilde{u}_2)$.

To show property (ii) of Definition 5, note that $\tilde{S} = [\perp_{\mathcal{X}}, \vee\tilde{S}]$ for $\vee\tilde{S} = \mathcal{P}(S) \in \mathcal{X}$. As a consequence, we have that $\tilde{f}(x, \wedge\mathcal{I}(x)) = \wedge S$. Thus, we are left to show that for all $x \in \mathcal{X}$, $\tilde{f}(x, \cdot)$ is $\vee$-preserving in the second argument when the second argument is ranging in $\tilde{I}(x)$. Let $\tilde{u}_1, \tilde{u}_2 \in \tilde{I}(x)$, we need to show that $\tilde{f}(x, \tilde{u}_1 \vee \tilde{u}_2) = \tilde{f}(x, \tilde{u}_1) \vee \tilde{f}(x, \tilde{u}_2)$ for all $x \in \mathcal{X}$. By the order preserving property of $\tilde{f}$ in the second argument, we already know that $\tilde{f}(x, \tilde{u}_1) \vee \tilde{f}(x, \tilde{u}_2) \leq \tilde{f}(x, \tilde{u}_1 \vee \tilde{u}_2)$. Let us denote $\tilde{f}(x, \tilde{u}_1) \vee \tilde{f}(x, \tilde{u}_2) = a$ and let us in fact show that $a = \tilde{f}(x, \tilde{u}_1 \vee \tilde{u}_2)$. Let $\tilde{u}$ be such that $\tilde{f}(x, \tilde{u}) = a$. If $\tilde{u} \in \mathcal{P}(\mathcal{I})$, then let it be the largest such $\tilde{u}$. Consider the two chains $w_1 \prec w_2 \prec \ldots \prec w_{k_1}$ and $v_1 \prec v_2 \prec \ldots \prec v_{k_2}$, in which $w_1 = \tilde{f}(x, \tilde{u}_1)$, $v_{k_2} = w_{k_1} = a$, and $\tilde{f}(x, \tilde{u}_2) = v_1$. for all two consecutive elements on such chains $w_i \prec w_{i+1}$, there are $\tilde{u}_{1,i}, \tilde{u}_{1,i+1} \in \tilde{I}(x)$ such that $\tilde{f}(x, \tilde{u}_{1,i}) = w_i$ and $\tilde{f}(x, \tilde{u}_{1,i+1}) = w_{i+1}$. If $\tilde{u}_{1,i}, \tilde{u}_{1,i+1}$ are both in $\mathcal{P}(\mathcal{I})$ then $w_i \leq w_{i+1}$. Also, if $\tilde{u}_{1,i} \in \mathcal{P}(\mathcal{I})$, we assume it is the largest input such that $\tilde{f}(x, \tilde{u}_{1,i}) = w_i$. If one or both of the inputs $\tilde{u}_{1,i}, \tilde{u}_{1,i+1}$ is a silent input, by item 3., we have that $\tilde{u}_{1,i} \prec \tilde{u}_{1,i+1}$. Since this is true for all $i \in \{1, \ldots, k-1\}$, we finally obtain that $\tilde{u}_1 \leq \tilde{u}$. Repeating this process for the chain $v_1 \prec v_2 \prec \ldots \prec v_{k_2}$, one also obtains that $\tilde{u}_2 \leq \tilde{u}$. Since $\tilde{u}_1$ and $\tilde{u}_2$ cannot have two different joins, it must be that either $\tilde{u} \leq \tilde{u}_1 \vee \tilde{u}_2$ or $\tilde{u}_1 \vee \tilde{u}_2 \leq \tilde{u}$. By the order preserving property of $\tilde{f}$, we have that $\tilde{u}_1 \vee \tilde{u}_2 \leq \tilde{u}$ implies $\tilde{f}(x, \tilde{u}_1 \vee \tilde{u}_2) \leq \tilde{f}(x, \tilde{u}) = a$. But, we assumed that $a < \tilde{f}(x, \tilde{u}_1 \vee \tilde{u}_2)$, as a consequence it must be that $\tilde{u} \leq \tilde{u}_1 \vee \tilde{u}_2$. However, by definition $\tilde{u}_1 \vee \tilde{u}_2$ is the smallest element that is larger than both $\tilde{u}_1$ and $\tilde{u}_2$. This in turn implies $\tilde{u} = \tilde{u}_1 \vee \tilde{u}_2$ and therefore $a = \tilde{f}(x, \tilde{u}_1 \vee \tilde{u}_2)$. Finally, we set $(\tilde{I}, \leq)$ as the union of the lattices $(\tilde{I}(x), \leq)$ constructed above. This union is well defined as all of the partial orders $(\tilde{I}(x), \leq)$ are compatible by construction. Thus, one can add a bottom and a top element for $\tilde{I}$ to make $(\tilde{I}, \leq)$ a lattice. To conclude the proof, we need to show that $\{u \in \mathcal{I} \mid \tilde{f}([\perp, x], u) \subseteq [\perp, \vee\tilde{S}]\}$ is not empty for $[\perp, x] \subseteq \tilde{S} \cap O_y(\tilde{\Sigma})$. Note that $\{u \in \mathcal{I} \mid \tilde{f}([\perp, x], u) \subseteq [\perp, \vee\tilde{S}]\} = \{u \in \mathcal{I} \mid \tilde{f}(x, u) \leq \vee\tilde{S}\}$. The latter set is also equal to $\{u \in \mathcal{I} \mid f(x, x) \subseteq S\}$. This is not empty as $x \subseteq S \cap O_y(\Sigma)$ and Σ is controllable by dynamic output feedback with respect to S. Thus, $\tilde{\Sigma}|_{\mathcal{I}}$ is controllable by dynamic output feedback with respect to $\tilde{S}$.

Systematic Simulation Using Sensitivity Analysis

Alexandre Donzé and Oded Maler

VERIMAG
2, Avenue de Vignate
38610 Gières, France
Alexandre.Donze@imag.fr

Abstract. In this paper we propose a new technique for verification by simulation of continuous and hybrid dynamical systems with uncertain initial conditions. We provide an algorithmic methodology that can, in most cases, verify that the system avoids a set of bad states by conducting a *finite* number of simulation runs starting from a finite subset of the set of possible initial conditions. The novelty of our approach consists in the use of *sensitivity analysis*, developed and implemented in the context of numerical integration, to efficiently characterize the coverage of sampling trajectories.

1 Introduction

Numerical simulation is a commonly-used method for predicting or validating the behavior of complex dynamical systems. It is often the case that due to incomplete knowledge of the initial conditions or the presence of external disturbances, the system in question may have an infinite and non-countable number of trajectories, only a finite subset of which can be covered by simulation. Two major directions for attacking this coverage problem have been reported in the literature. The first approach, see e.g. [ACH+95,DM98,CK98,ADG03,Gir05] consists of an adaptation of discrete verification techniques to the continuous context via reachability computation, namely computing by geometric means an over-approximation of the set of states reached by all trajectories. The other complementary approach attempts to find conditions under which a finite number of well-chosen trajectories will suffice to prove correctness and cover in some sense all the trajectories of the system [KKMS03,BF04,BCLM05,BCLM06,GP06].This paper is concerned with the second approach.

The main contribution of the paper is an algorithm whose input consists of an arbitrary dynamical system, an initial set $\mathcal{X}_0$, a set of "bad" states $\mathcal{F}$, a time interval $[0, T]$ and some $\delta > 0$. The algorithm picks a point in $\mathcal{X}_0$ and simulates the trajectory of the system. If a bad state occurs in the trajectory the algorithm stops and declares the system "unsafe"; otherwise, a systematic refinement operator is used to extend the sampling of the initial points from which new trajectories are numerically simulated and so on. The algorithm is guaranteed to terminate in a finite number of steps, either by finding a bad trajectory and declaring the system unsafe, by declaring the system "safe" when the sampled

A. Bemporad, A. Bicchi, and G. Buttazzo (Eds.): HSCC 2007, LNCS 4416, pp. 174–189, 2007.

trajectories provably cover the reachable set, or returning an "uncertain" answer when sampled trajectories are less than δ-close to $\mathcal{F}$.

The novelty of this paper w.r.t. similar works is the notion of coverage used, which is based on the concept of an *expansion function* which characterizes how neighboring trajectories are getting closer to or further from one another as time goes by. We show that this notion can be effectively approximated[1] using the concept of *sensitivity function* implemented in standard numerical integrators. The rest of the paper is organized as follows. In Section 2 we define samples, their dispersion and expansion functions and use them to introduce an abstract algorithm for safety verification. The algorithm is then concretized and implemented using a numerical solver, a grid-based sample refinement scheme (Section 3) and sensitivity to approximate expansion (Section 4). In Section 5 we demonstrate the behavior of our implementation of the algorithm on a linear time-varying system and on two nonlinear analog circuits, the tunnel diode oscillator and the voltage controlled oscillator. The extension of sensitivity analysis to hybrid systems is the topic of Section 6, which is followed by discussions of future steps in simulation-based verification.

2 Verification by Simulation

We consider a dynamical system of the form

$$\dot{\mathbf{x}} = f(t, \mathbf{x}), \ \mathbf{x}(0) = \mathbf{x}_0 \in \mathcal{X}_0, \tag{1}$$

where $\mathbf{x}$ is a vector in $\mathbb{R}^n$, $\mathcal{X}_0$ is compact and f is assumed of class C^1. The problem is known to have a unique solution noted $\xi_{\mathbf{x}_0}(t)$. We say that the system is *safe* if all trajectories starting from any $\mathbf{x}_0 \in \mathcal{X}_0$ do not intersect a set $\mathcal{F}$ of bad states. We consider a bounded time horizon $[0, T]$. Let $\text{Reach}_{\leq t}(\mathcal{X}_0)$ (resp. $\text{Reach}_{=t}(\mathcal{X}_0)$) be the set reachable from $\mathcal{X}_0$ in less than (resp. exactly) t units of time. A usual way to prove that the system is safe is to prove emptiness of the intersection of the reachable set $\text{Reach}_{\leq T}(\mathcal{X}_0)$ with the set $\mathcal{F}$. In this section, we introduce the concept of *expansion function* which allows to characterize how a finite set of trajectories covers the reachable set and use such a set of trajectories trying to refute intersection with the bad set $\mathcal{F}$.

2.1 Preliminaries

We use a metric d and extend it to distance from points to set using $d(\mathbf{x}, \mathcal{X}) = \inf_{\mathbf{y} \in \mathcal{X}}\big(d(\mathbf{x}, \mathbf{y})\big)$ When used, the notation $\|\cdot\|$ denotes the infinity norm. It extends to matrix with the usual definition: $\|\mathbf{A}\| = \sup_{\|\mathbf{x}\|=1} \|\mathbf{A}\mathbf{x}\|$. A ball $\mathcal{B}_\delta(\mathbf{x})$ is the set of points $\mathbf{x}'$ satisfying $d(\mathbf{x}, \mathbf{x}') \leq \delta$. Given a set $\mathcal{X}$, a cover for $\mathcal{X}$ is a set of sets $\{\mathcal{X}_1, \dots \mathcal{X}_k\}$ such that $\mathcal{X} \subset \bigcup_{i=1}^k \mathcal{X}_i$. A ball cover is a cover consisting of balls. We extend the notation $\mathcal{B}_\delta$ to sets and trajectories as follows

[1] For linear time varying systems, as we show, these two notions coincide.

$$\mathcal{B}_\delta(\mathcal{S}) = \bigcup_{\mathbf{x}\in\mathcal{S}} \mathcal{B}_\delta(\mathbf{x}) \text{ and } \mathcal{B}_\delta(\xi_\mathbf{x}) = \bigcup_{t\in[0,T]} \mathcal{B}_{\delta(t)}(\xi_\mathbf{x}(t))$$

A *sampling* of $\mathcal{X}$ is a set $\mathcal{S} = \{\mathbf{x}_1, \ldots, \mathbf{x}_k\}$ of points in $\mathcal{X}$. The intuitive notion of the "coverage" of $\mathcal{X}$ by $\mathcal{S}$ is formalized by

Definition 1 (Dispersion). *The dispersion $\alpha_\mathcal{X}(\mathcal{S})$ is the smallest radius ϵ such that the union of all ϵ radius closed balls with their center in $\mathcal{S}$ covers $\mathcal{X}$.*

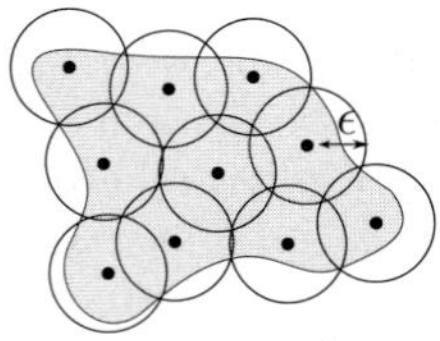

$$\alpha_\mathcal{X}(\mathcal{S}) = \min_{\epsilon>0} \{\epsilon \mid \mathcal{X} \subset \mathcal{B}_\epsilon(\mathcal{S})\} \tag{2}$$

We now define the process of *refining* a sampling, which simply consists in finding a new sampling with a strictly smaller dispersion.

Definition 2 (Refinement). *Let $\mathcal{S}$ and $\mathcal{S}'$ be samplings of $\mathcal{X}$. We say that $\mathcal{S}'$ refines $\mathcal{S}$ if and only $\alpha_\mathcal{X}(\mathcal{S}') < \alpha_\mathcal{X}(\mathcal{S})$.*

A refining sampling can be constructed from the set to refine (e.g. by adding sufficiently many points) or be found independently. In both cases, we can assume that it is obtained through a *refinement operator* which we define next.

Definition 3 (Refinement operators). *A refinement operator $\rho : 2^\mathcal{X} \mapsto 2^\mathcal{X}$ maps a sampling $\mathcal{S}$ to another sampling $\mathcal{S}' = \rho_\mathcal{X}(\mathcal{S})$ such that $\mathcal{S}$ refines $\mathcal{S}'$. A refinement operator is* complete *if $\forall \mathcal{S}$,*

$$\lim_{k\to\infty} \alpha_\mathcal{X}\big(\rho_\mathcal{X}^{(k)}(\mathcal{S})\big) = 0$$

where $\rho_\mathcal{X}^{(k)}(\mathcal{S})$ is the result of k application of $\rho_\mathcal{X}$ to $\mathcal{S}$.

In other terms, a refinement operator is complete if a sampling of $\mathcal{X}$ which has been infinitely refined is *dense* in $\mathcal{X}$. Until we define one in section 3, we assume the existence of a complete refinement operator ρ.

2.2 Expansion Function

The intuitive idea is to draw "tubes" around trajectories so that the union of these tubes will provide an over-approximation of the reachable set. The expansion function then simply maps time t to the radius of the tube at t, given an initial state $\mathbf{x}_0$ and an initial radius ϵ.

Definition 4 (Expansion function). *Given $\mathbf{x}_0 \in \mathcal{X}_0$, and $\epsilon > 0$, the* expansion function *of $\xi_{\mathbf{x}_0}$, denoted by $\mathcal{E}_{\mathbf{x}_0,\epsilon} : \mathbb{R}^+ \mapsto \mathbb{R}^+$ maps t to the smallest non-negative number δ such that all trajectories with initial state in $\mathcal{B}_\epsilon(\mathbf{x}_0)$ reach a point in $\mathcal{B}_\delta(\xi_{\mathbf{x}_0}(t))$ at time t:*

$$\mathcal{E}_{\mathbf{x}_0,\epsilon}(t) = \sup_{d(\mathbf{x}_0,\mathbf{x})\le\epsilon} d\big(\xi_{\mathbf{x}_0}(t), \xi_\mathbf{x}(t)\big) \tag{3}$$

Clearly, a first property of the expansion functions is that it approaches 0 as ϵ tends toward 0:

$$\forall t > 0, \lim_{\epsilon \to 0} \mathcal{E}_{\mathbf{x},\epsilon}(t) = 0 \tag{4}$$

This results directly from the continuity of $\xi_{\mathbf{x}}(t)$ w.r.t. $\mathbf{x}$.

The expansion function value $\mathcal{E}_{\mathbf{x}_0,\epsilon}(t)$ gives the radius of the ball which over-approximate tightly the reachable set from the ball $\mathcal{B}_\epsilon(\mathbf{x}_0)$ at time t. Obviously, if we take several such balls so that the initial set $\mathcal{X}_0$ is covered, we obtain a corresponding cover of $\text{Reach}_{=t}(\mathcal{X}_0)$. This is stated in the following

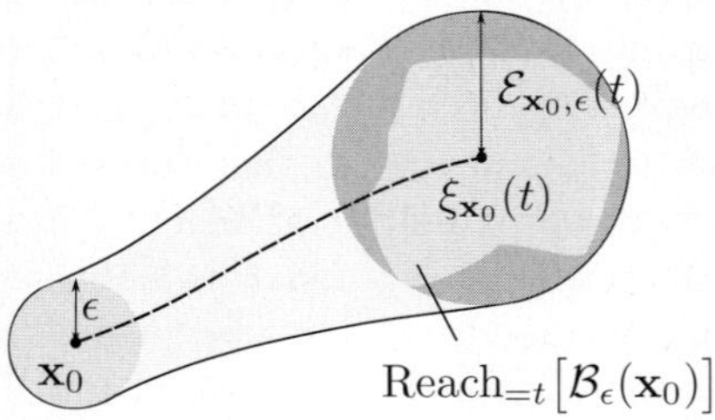

Proposition 1. *Let* $\mathcal{S} = \{\mathbf{x}_1, \ldots, \mathbf{x}_k\}$*be a sampling of* $\mathcal{X}_0$ *such that* $\bigcup_{i=1}^k \mathcal{B}_{\epsilon_i}(\mathbf{x}_i)$ *is a ball cover of* $\mathcal{X}_0$ *for some* $\{\epsilon_1, \ldots, \epsilon_k\}$. *Let* $t > 0$ *and for each* $1 \le i \le k$, *let* $\delta_i = \mathcal{E}_{\mathbf{x}_i,\epsilon_i}(t)$. *Then* $\bigcup_{i=1}^k \mathcal{B}_{\delta_i}(\xi_{\mathbf{x}_i}(t))$ *is a ball cover of* $\text{Reach}_{=t}(\mathcal{X}_0)$.

Proof. By definition, the ball cover of $\mathcal{X}_0$ contains $\mathcal{X}_0$, and each $\mathcal{B}_{\delta_i}(\xi_{\mathbf{x}_i}(t))$ contains $\text{Reach}_{=t}(\mathcal{B}_{\epsilon_i}(\mathbf{x}_i))$, and the rest follows from the commutativity of the dynamics with set union and containment. □

In particular, if $\mathcal{S}$ is a sampling of $\mathcal{X}_0$ with dispersion ϵ then we are in the case where $\epsilon_i = \epsilon$ for all $1 < i < k$ and since the result is true for all $t \in [0, T]$, we have the following

Corollary 1. *Let* $\mathcal{S} = \{\mathbf{x}_1, \mathbf{x}_2, \ldots, \mathbf{x}_k\}$ *be a sampling of* $\mathcal{X}_0$ *with dispersion* $\alpha_{\mathcal{X}_0}(\mathcal{S}) = \epsilon$. *Let* $\delta > 0$ *be an upper bound for* $\mathcal{E}_{\mathbf{x}_i,\epsilon}(t)$ *for all* $1 < i < k$ *and* $t \in [0, T]$, *then the following inclusions hold*

$$\text{Reach}_{[0,T]}(\mathcal{X}_0) \subseteq \bigcup_{\mathbf{x} \in \mathcal{S}} \mathcal{B}_{\mathcal{E}_{\mathbf{x},\epsilon}}(\xi_{\mathbf{x}}) \subseteq \bigcup_{\mathbf{x} \in \mathcal{S}} \mathcal{B}_\delta(\xi_{\mathbf{x}}) \subseteq \mathcal{B}_\delta\big(\text{Reach}_{[0,T]}(\mathcal{X}_0)\big) \tag{5}$$

Proof. The first inclusion is a direct application of the proposition. The second results from the fact that δ is an upper-bound and the third inclusion is due to the fact that $\forall (\mathbf{x}_i, t) \in \mathcal{S} \times [0, T]$, $\xi_{\mathbf{x}_i}(t) \in \text{Reach}_{[0,T]}(\mathcal{X}_0)$. □

In other terms, if we bloat the sampling trajectories starting from $\mathcal{S}$ with a radius δ, which is an upper bound for expansion functions of these trajectories, then we get an over-approximation of the reachable set which is between the exact reachable set and the reachable set bloated with δ. Because of (4), it is clear that δ, and then the over-approximation error, decreases when ϵ gets smaller.

The second corollary of proposition 1 underlies our verification strategy.

Corollary 2. *Let* $\mathcal{S} = \{\mathbf{x}_1, \ldots, \mathbf{x}_k\}$ *be a sampling of* $\mathcal{X}$ *such that* $\bigcup_{i=1}^k \mathcal{B}_{\epsilon_i}(\mathbf{x}_i)$ *is a ball cover of* $\mathcal{X}_0$. *For* $t \in [0, T]$ *and* $1 \le i \le k$, *let* $\delta_i(t) = \mathcal{E}_{\mathbf{x}_i,\epsilon_i}(t)$. *If for all* $t \in [0, T]$,

$$\mathcal{B}_{\delta_i(t)}(\xi_{\mathbf{x}_i}(t)) \cap \mathcal{F} = \emptyset,$$

then for all trajectory $\xi_{\mathbf{x}}$ starting from $\mathbf{x} \in \mathcal{X}_0$, the intersection of $\xi_{\mathbf{x}}$ and the bad set $\mathcal{F}$ is empty and thus the system is safe.

2.3 A Verification Algorithm

In theory, then, from previous proposition and corollaries, a unique trajectory could be sufficient to verify the system: take one point $\mathbf{x}_0$, find ϵ such that $\mathcal{X}_0 \subset \mathcal{B}_\epsilon(\mathbf{x}_0)$, then check for all $t \in [0,T]$ that $\mathcal{B}_{\delta(t)}(\xi_{\mathbf{x}_i}(t)) \cap \mathcal{F} = \emptyset$, where $\delta(t) = \mathcal{E}_{\mathbf{x}_0,\epsilon}(t)$. If this is the case, then the system is safe. Obviously, the opposite case does not mean that the system is unsafe since $\mathcal{B}_{\delta(t)}(\xi_{\mathbf{x}_i}(t))$ is actually an over-approximation of $\text{Reach}_{=t}(\mathcal{X}_0)$, rather it indicates that the distance between the reachable set and $\mathcal{F}$ is less than $\delta(t)$. If this indication is not sufficient, then more trajectories have to be simulated until a sufficiently dense sampling of $\mathcal{X}_0$ is found. This is what Algorithm 1 does.

Algorithm 1. Safety Verification. The algorithm takes $\delta > 0$ as input.

$\mathcal{U} \leftarrow \emptyset$, $\mathcal{S}_0 \leftarrow \{\mathbf{x}_0\}$ with $\mathbf{x}_0 \in \mathcal{X}_0$, $k \leftarrow 0$
loop
  /* For $\mathcal{S}_k$, check each trajectory and compute an upper bound δ_k for $\mathcal{E}$ */
  $\epsilon_k \leftarrow \alpha_{\mathcal{X}_0}(\mathcal{S}_k, \mathcal{X}_0)$, $\delta_k \leftarrow 0$
  for all $\mathbf{x} \in \mathcal{S}_k$ **do**
    if $\xi_{\mathbf{x}} \cap \mathcal{F} \neq \emptyset$ **then**
      return (`unsafe`, $\{\mathbf{x}\}$)
    else if $\mathcal{B}_{\mathcal{E}_{\mathbf{x},\epsilon_k}}(\xi_x) \cap \mathcal{F} \neq \emptyset$ **then**
      $\mathcal{U} \leftarrow \mathcal{U} \cup \{\mathbf{x}\}$
    end if
    $\delta_k \leftarrow \max\left(\sup_{t\in[0,T]} (\mathcal{E}_{\mathbf{x},\epsilon_k}(t)), \delta_k \right)$
  end for
  /* Stop either if no uncertain trajectory was found (safe) or if the upper bound is smaller than precision δ. Else refine the sampling and loop */
  if $\mathcal{U} = \emptyset$ **then**
    return `safe`
  else if $\delta_k < \delta$ **then**
    return (`uncertain`, $\mathcal{U}$)
  else
    $\mathcal{S}_{k+1} \leftarrow \rho_{\mathcal{X}_0}(\mathcal{S}_k)$, $\mathcal{U} \leftarrow \emptyset$, $k \leftarrow k+1$ /* Refine the sample */
  end if
end loop

Theorem 1. *Under assumptions mentioned above, algorithm 1 terminates and its output satisfies:*

- *it is* `safe` *only if the system is safe.*
- *it is* (`unsafe`, $\{\mathbf{x}\}$) only if the system is unsafe and $\{\mathbf{x}\}$ is a counter-example, i.e.: ξ_x intersects $\mathcal{F}$.
- it is (`uncertain`, $\mathcal{U}$) only if all the points in $\mathcal{U}$ induce uncertain trajectories: $\forall\, \mathbf{x} \in \mathcal{U},\ d(\xi_{\mathbf{x}}, \mathcal{F}) \leq \delta$.

Proof. For `unsafe`, the result is obvious from the algorithm. For `safe` and `uncertain`, the algorithm terminates because ρ is complete, $\lim_{k\to 0} \epsilon_k = 0$ and $\lim_{\epsilon_k \to 0} \delta_k = 0$. Consequently for some k, $\delta_k < \delta$. Now, if $\mathcal{U}$ was found empty, at or before iteration k, this means that corollary 2 applies which proves that the system is safe, while if $\mathcal{U}$ is still not empty at iteration k i.eif the algorithm has returned (`unsafe`, $\mathcal{U}$), then $\mathcal{U}$ contains states $\mathbf{x}$ for which

$$d(\xi_{\mathbf{x}}, \mathcal{F}) \leq \sup_{t\in[0,T]} (\mathcal{E}_{\mathbf{x},\epsilon_k}(t)) \leq \delta_k \leq \delta.$$

Note that moreover, the inclusions (5) are true for $\mathcal{S}_k$. □

The algorithm requires some $\delta > 0$ as input to guarantee termination. In fact, the problematic case is when the distance between the reachable set and the bad set is exactly 0. In this case, there is no way to get an answer other than `uncertain`. On the other hand, we can state the following theorem:

Theorem 2. *If $d\big(Reach_{\leq T}(\mathcal{X}_0), \mathcal{F}\big) > 0$, then there exist a $\delta > 0$ for which algorithm 1 returns* `safe`.

Proof. This is true for any $\delta < d\big(\text{Reach}_{\leq T}(\mathcal{X}_0), \mathcal{F}\big)$. Indeed, since for some k, the inclusions (5) in corollary 1 are true for $\mathcal{S}_k$ then

$$\mathcal{B}_\delta\big(\text{Reach}_{[0,T]}(\mathcal{X}_0)\big) \cup \mathcal{F} = \emptyset \Rightarrow \mathcal{B}_\delta\Big(\bigcup_{\mathbf{x}\in\mathcal{S}} \xi_{\mathbf{x}}\Big) \cup \mathcal{F} = \emptyset$$

so $\mathcal{U}$ is empty at the end of the **for** loop and the algorithm will return `safe`. □

3 An Efficient Sampling Strategy

In this section, we focus on the sampling strategy, that is, on the sample refinement operator.

3.1 Local Refinement

First, we can remark that when Algorithm 1 ends with the `uncertain` answer, one choice is to try a smaller δ. In that case, it is not necessary to restart the algorithm from scratch; the set $\mathcal{U}$ can be used. Indeed, it is clear that any trajectory starting from the ball $\mathcal{B}_{\epsilon_k}(\mathbf{x})$, where $\mathbf{x} \in \mathcal{S}_k$ has not been inserted into the uncertain set $\mathcal{U}$, is safe. Thus the set $B_{\epsilon_k}\big(\mathcal{S}_k \setminus \mathcal{U}\big)$ is safe and it is enough to verify the set $\mathcal{B}_{\epsilon_k}\big(\mathcal{U}\big)$. In fact, this observation is also relevant at each iteration of the **for** loop inside Algorithm 1. Instead of refining *globally* $\mathcal{S}_k$ by the instruction $\mathcal{S}_{k+1} \leftarrow \rho_{\mathcal{X}_0}(\mathcal{S}_k)$, $\mathcal{S}_k$ could be refined *locally* only around uncertain states. This can be done simply by replacing $\mathcal{X}_0$ with $B_{\epsilon_k}(\mathcal{U})$ and refine this new initial set. Line 19 thus becomes:

$\mathcal{X}_0 \leftarrow \mathcal{B}_{\epsilon_k}(\mathcal{U})$
$\mathcal{S}_{k+1} \leftarrow \rho_{\mathcal{X}_0}(\mathcal{U}), \mathcal{U} \leftarrow \emptyset, k \leftarrow k+1$

Now we proceed to describe the particular sampling method that we use in the implementation of the algorithm.

3.2 Hierarchical Grid Sampling

We saw that algorithm 1 terminates as soon as δ_k gets sufficiently small, which requires also the dispersion ϵ_k to be sufficiently small. Then we need that the convergence of sequence $(\epsilon_k)_{k\in\mathbb{N}}$ towards 0 to be as fast as possible. This requires that for a given number of points, our sampling strategy tries to minimize its dispersion. In this section, we assume that with an appropriate change in variables, sampling $\mathcal{X}_0$ is equivalent to sampling the unit hypercube $[0,1]^n$. In case we use the L_∞ metrics, i.e. $d(\mathbf{x},\mathbf{y}) = \max_i(|x_i - yi|)$, the solution of the problem of minimizing dispersion of N points in $[0,1]^n$ is known (see e.g. [LaV06]): the minimum possible dispersion is $\frac{1}{2\lfloor N^{1/n}\rfloor}$ and is obtained by placing the points at the center of smaller hypercubes of size $\frac{1}{\lfloor N^{1/n}\rfloor}$, partitioning the unit hypercube. Note that there are $(\lfloor N^{1/n}\rfloor)^n$ such hypercubes, which may be less than N. In this case, the remaining points can placed anywhere without affecting the dispersion. Obtained grids are referred to as *Sukharev grids*. The picture on the right gives an example of such a grid for $n = 2$ and $N = 49$.

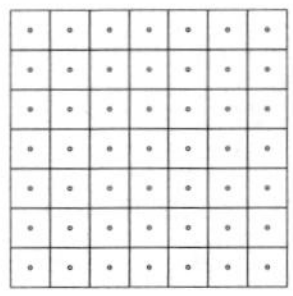

Sukharev grids have optimal dispersion but for a fixed number of points while in our verification algorithm, we do not know in advance how many points in the initial set we will have to use. In fact, we need to implement the refinement operator to be used throughout Algorithm 1, starting from the singleton sampling set $\mathcal{S}_0$. An elegant way to do it is to superpose *hierarchically* Sukharev grids, the refinement process being defined simply in a recursive fashion. Let $\mathcal{X}$ be a hypercube of size $\frac{1}{2^l}$ (we say that such a cube is part of the grid of resolution l) and $\mathcal{S}$ be a sampling of $\mathcal{X}$, then:

- if $\mathcal{S} = \emptyset$ then $\rho_{\mathcal{X}}(\mathcal{S}) = \{\mathbf{x}\}$, where $\mathbf{x}$ is the center of the hypercube $\mathcal{X}$;
- if $\mathcal{S} \neq \emptyset$ then $\rho_{\mathcal{X}}(\mathcal{S}) = \mathcal{S} \cup \bigcup_{i=1}^{2^n} \rho_{\mathcal{X}_i}(\mathcal{S}_i)$ where the sets $\mathcal{X}_i$ are the 2^n hypercubes of size $\frac{1}{2^{l+1}}$ partitioning $\mathcal{X}$ and the sets $\mathcal{S}_i$ contain the points of $\mathcal{S}$ that are inside $\mathcal{X}_i$.

On figure 1, we show the effect of three iterations in dimension 2 and 3 along with an example of successive local refinements. From this definition, it is clear that the operator ρ is a refinement operator and that it is complete since we have:

$$\alpha_{\mathcal{X}}(\rho_{\mathcal{X}}(\mathcal{S})) = \frac{1}{2}\alpha_{\mathcal{X}}(\mathcal{S}).$$

In [LYL04], a simple procedure is described for choosing the order in which the partitioning cubes are processed so that the mutual distance between two consecutive cubes is maximized. This is an interesting feature from the point of view of fast falsification since it means that every two consecutive simulation runs will be far from each other and that for any $k \in \mathbb{N}$, the initial states of the first k trajectories will constitute a good coverage of the initial set $\mathcal{X}_0$.

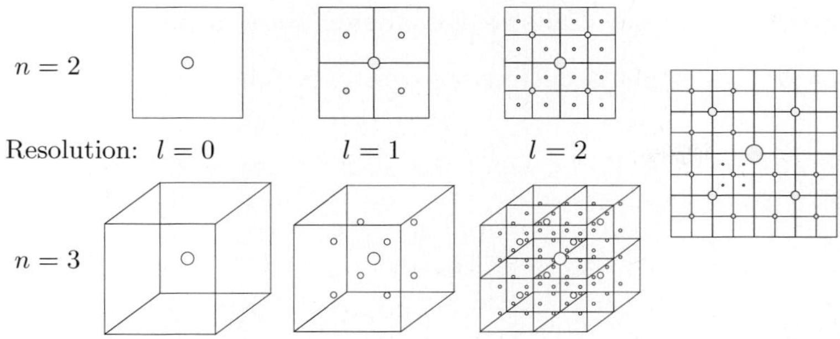

Fig. 1. Refinements for $n = 2$ and $n = 3$ dimensions for resolutions from $l = 0$ to $l = 2$. On the right, local refinements until resolution 3.

4 Implementation Using Sensitivity Analysis

4.1 Sensitivity Analysis Theory

Recall that we consider dynamics of the general form: $\dot{\mathbf{x}} = f(t, \mathbf{x}),\ \mathbf{x}(0) \in \mathcal{X}_0$. As a function of $\mathbf{x}_0$, the flow $\xi_{\mathbf{x}_0}$ is differentiable w.r.t $\mathbf{x}_0$. Thus the *sensitivity to initial conditions* at time t is well defined by:

$$\mathbf{s}_{\mathbf{x}_0}(t) \triangleq \frac{\partial \xi_{\mathbf{x}_0}}{\partial \mathbf{x}_0}(t). \tag{6}$$

where $\mathbf{s}_{\mathbf{x}_0}(t)$ is a square matrix of order n. To compute the sensitivity matrix, we first apply the chain rule to get the derivative of $\mathbf{s}_{\mathbf{x}_0}$ w.r.t. time:

$$\frac{\partial}{\partial t}\frac{\partial \xi_{\mathbf{x}_0}}{\partial \mathbf{x}_0}(t) = \frac{\partial}{\partial \mathbf{x}_0} f\big(t, \xi_{\mathbf{x}_0}(t)\big) = D_f(\xi_{\mathbf{x}_0}(t))\,\frac{\partial \xi_{\mathbf{x}_0}}{\partial \mathbf{x}_0}(t)$$

which gives the following *sensitivity equation*:

$$\dot{\mathbf{s}}_{\mathbf{x}_0}(t) = D_{f,\mathbf{x}_0}(t)\ \mathbf{s}_{\mathbf{x}_0}(t) \tag{7}$$

where $D_{f,\mathbf{x}_0}$ is the Jacobian matrix of f along trajectory $\xi_{\mathbf{x}_0}$. Hence, this equation is a linear time-varying ordinary differential equation (ODE). Note that this is a *matrix* differential equation but it can be viewed as a system of n ODEs of order n. The ij^{th} element of $\mathbf{s}_{\mathbf{x}_0}(t)$ basically represents the influence of variations in the ith coordinate x_0^i of $\mathbf{x}_0$ on the jth coordinate $x^j(t)$ of $\xi_{\mathbf{x}_0}(t)$. Then it is clear that the initial value $\mathbf{s}_{\mathbf{x}_0}(0)$ of $\mathbf{s}_{\mathbf{x}_0}$ must be the identity matrix, $\mathbf{I}_n$. Efficient solvers exist that implement the computation of sensitivity functions (in our implementations, we use the tool suite described in [SH05]).

A particularly interesting case is when the dynamics is linear time-varying, i.e. when $f(t, \mathbf{x}) = A(t)\ \mathbf{x}$. Indeed, in this case, we know that the Jacobian matrix of f is just the matrix A which means that sensitivity matrix $\mathbf{s}_{\mathbf{x}_0}(t)$ share *the same dynamics* as the flow $\xi_{\mathbf{x}_0}$. In fact, the columns of sensitivity matrix are solutions of the system dynamics equation where initial conditions are the canonical vectors of $\mathbb{R}^n$.

4.2 Sensitivity Functions and Expansion Functions

The following important result relates sensitivity functions to expansion functions:

Theorem 3. *Let* $\mathbf{x}_0 \in \mathcal{X}_0$, $t \in [0, T]$ *and assume that* f *is* C^2. *Then there exists a real* $M > 0$ *such that* $\forall \epsilon > 0$:

$$|\mathcal{E}_{\mathbf{x}_0,\epsilon}(t) - \|\mathbf{s}_{\mathbf{x}_0}(t)\|\ \epsilon\ | \leq M\epsilon^2 \tag{8}$$

Proof. Since f is C^2, the flow $\xi_{\mathbf{x}_0}$ is also C^2 w.r.t. $\mathbf{x}_0$ ([HS74]). Let $\mathbf{x} \in \mathcal{X}_0$. Then the Taylor expansion of $\xi_{\mathbf{x}_0}(t)$ around $\mathbf{x}_0$ shows that there exist a bounded function φ_t such that:

$$\begin{aligned} \xi_{\mathbf{x}}(t) &= \xi_{\mathbf{x}_0}(t) + \frac{\partial \xi_{\mathbf{x}_0}}{\partial \mathbf{x}_0}(t)\ (\mathbf{x} - \mathbf{x}_0) + \|\mathbf{x} - \mathbf{x}_0\|^2\ \varphi_t(\mathbf{x} - \mathbf{x}_0) \\ \Leftrightarrow \xi_{\mathbf{x}}(t) - \xi_{\mathbf{x}_0}(t) &= \mathbf{s}_{\mathbf{x}_0}(t)\ (\mathbf{x} - \mathbf{x}_0) + \|\mathbf{x} - \mathbf{x}_0\|^2\ \varphi_t(\mathbf{x} - \mathbf{x}_0) \end{aligned} \tag{9}$$

Equation (9) implies that $\forall \mathbf{x} \in \mathcal{B}_\epsilon(\mathbf{x}_0)$,

$$\|\xi_{\mathbf{x}}(t) - \xi_{\mathbf{x}_0}(t)\| \leq \|\mathbf{s}_{\mathbf{x}_0}(t)\| \|\mathbf{x} - \mathbf{x}_0\| + \|\mathbf{x} - \mathbf{x}_0\|^2\ \|\varphi_t(\mathbf{x} - \mathbf{x}_0)\| \leq \|\mathbf{s}_{\mathbf{x}_0}(t)\|\epsilon + \epsilon^2\ M$$

which implies in turn that

$$\mathcal{E}_{\mathbf{x}_0,\epsilon} - \|\mathbf{s}_{\mathbf{x}_0}(t)\|\epsilon \leq M\epsilon^2 \tag{10}$$

On the other hand, 9 can be rewritten as

$$\begin{aligned} \mathbf{s}_{\mathbf{x}_0}(t)\ (\mathbf{x}_0 - \mathbf{x}) &= \xi_{\mathbf{x}}(t) - \xi_{\mathbf{x}_0}(t) - \|\mathbf{x} - \mathbf{x}_0\|^2\ \varphi_t(\mathbf{x} - \mathbf{x}_0) \\ \Rightarrow \|\mathbf{s}_{\mathbf{x}_0}(t)\ (\mathbf{x}_0 - \mathbf{x})\| &\leq \|\xi_{\mathbf{x}}(t) - \xi_{\mathbf{x}_0}(t)\| + \|\mathbf{x} - \mathbf{x}_0\|^2\ \|\varphi_t(\mathbf{x} - \mathbf{x}_0)\| \\ &\leq \mathcal{E}_{\mathbf{x}_0,\epsilon}(t) + \epsilon^2 M \end{aligned} \tag{11}$$

From the definition of matrix norm, we know that we can find a unit vector $\mathbf{y}$ such that $\|\mathbf{s}_{\mathbf{x}_0}(t)\| = \|\mathbf{s}_{\mathbf{x}_0}(t)\ \mathbf{y}\|$. The inequality (11) is true for all $\mathbf{x} \in \mathcal{B}_\epsilon(\mathbf{x}_0)$ so in particular for $\mathbf{x} = \mathbf{x}_0 + \epsilon\mathbf{y}$ in which case

$$\|\mathbf{s}_{\mathbf{x}_0}(t)\ (\mathbf{x}_0 - \mathbf{x})\| = \|\mathbf{s}_{\mathbf{x}_0}(t)\ (\epsilon\mathbf{y})\| = \|\mathbf{s}_{\mathbf{x}_0}(t)\|\epsilon.$$

If we substitute in the right hand side of (11) and subtract $\mathcal{E}_{\mathbf{x}_0,\epsilon}(t)$, we get:

$$\|\mathbf{s}_{\mathbf{x}_0}(t)\|\epsilon - \mathcal{E}_{\mathbf{x}_0,\epsilon}(t) \leq M\epsilon^2 \tag{12}$$

The conjunction of inequalities (10) and (12) proves the result. □

When the dynamics of the system is affine, i.e. when $f(t, \mathbf{x}) = A(t)\mathbf{x} + b(t)$, where $A(t)$ and $b(t)$ are time varying matrices of appropriate dimensions, then expansion function can be computed exactly.

Theorem 4. *Let* $\mathbf{x}_0 \in \mathcal{X}_0$, $t \in [0, T]$ *and assume that* f *is affine. Then* $\forall \epsilon > 0$:

$$\mathcal{E}_{\mathbf{x}_0,\epsilon}(t) = \|\mathbf{s}_{\mathbf{x}_0}(t)\|\epsilon \tag{13}$$

Proof. This follows immediately from the fact that if f is affine, φ_t in equation (9) is null. Indeed, following the remark at the end of the previous subsection, we know from (7) that the lines of matrix $\mathbf{s}_{\mathbf{x}_0}(t)$ are solutions of the homogeneous system $\dot{\mathbf{x}} = A(t)\mathbf{x}$. Since this is a linear system, the vector $\mathbf{s}_{\mathbf{x}_0}(t)\ (\mathbf{x} - \mathbf{x}_0)$ is also solution of this system. Then $\xi_{\mathbf{x}_0}(t) + \mathbf{s}_{\mathbf{x}_0}(t)\ (\mathbf{x} - \mathbf{x}_0)$ is solution of the full system $\dot{\mathbf{x}} = A(t)\mathbf{x} + b(t)$. Furthermore, as $\mathbf{s}_{\mathbf{x}_0}(0)$ is the identity matrix,

$$\xi_{\mathbf{x}_0}(0) + \mathbf{s}_{\mathbf{x}_0}(0)\ (\mathbf{x} - \mathbf{x}_0) = \mathbf{x}_0 + (\mathbf{x} - \mathbf{x}_0) = \mathbf{x}.$$

In other words, $\xi_{\mathbf{x}_0} + \mathbf{s}\ (\mathbf{x} - \mathbf{x}_0)$ and $\xi_{\mathbf{x}}$ are two trajectories of the system with the same initial conditions so by uniqueness, they are equal. Then clearly,

$$\begin{aligned} \xi_{\mathbf{x}}(t) - \xi_{\mathbf{x}_0}(t) &= \mathbf{s}_{\mathbf{x}_0}(t)\ (\mathbf{x}_0 - \mathbf{x}) \\ \Rightarrow \sup_{\mathbf{x} \in \mathcal{B}_\epsilon(\mathbf{x}_0)} \|\xi_{\mathbf{x}}(t) - \xi_{\mathbf{x}_0}(t)\| &= \sup_{\mathbf{x} \in \mathcal{B}_\epsilon(\mathbf{x}_0)} \mathbf{s}_{\mathbf{x}_0}(t)\ (\mathbf{x}_0 - \mathbf{x}) \\ \Leftrightarrow \mathcal{E}_{\mathbf{x}_0,\epsilon}(t) &= \|\mathbf{s}_{\mathbf{x}_0}(t)\|\epsilon. \end{aligned}$$

□

From what precedes, then, we can approximate $\mathcal{E}_{\mathbf{x}_0,\epsilon}(t)$ with the quantity $\|\mathbf{s}_{\mathbf{x}_0}\|\epsilon$ and use it to implement Algorithm 1. In the case of affine systems, the implementation is exact and Theorem 1 applies to the concrete implantation. In the general case, when f may be nonlinear, we know that the error is quadratic with respect to ϵ. In order to take this error into account, we can force the algorithm to guarantee that the initial set is sampled with a sufficiently small dispersion ϵ. The new algorithm then takes an additional input parameter $\epsilon > 0$, refine globally the initial sampling until the dispersion ϵ is reached (while checking only for unsafe trajectories), and then continues and terminates as Algorithm 1 with local refinements.

5 Examples

We have implemented the techniques described in the preceding sections on top of a numerical simulation tool that supports sensitivity analysis and have applied it to several examples.

5.1 A High Dimensional Affine Time-Varying System

We consider a system with affine dynamics of the form $\dot{\mathbf{x}} = \mathbf{A}(t)\mathbf{x} + \mathbf{b}(t)$ with $\mathbf{A}(t) = e^{-t}\mathbf{M} - \mathbf{I}_{50}$ and $\mathbf{b}(t) = \mathbf{b}_0 e^{-t}\sin t$ where $\mathbf{M}$ and $\mathbf{b}_0$ are respectively 50×50 and 50×1 matrices with random coefficients in $[0, 1]$. We used a 2-dimensional $\mathcal{X}_0 = [0.5, 1.5] \times [0.5, 1.5] \times \{1\}^{48}$. The bad set $\mathcal{F}$ is the half plane given by an inequality of the form $x_1 \leq d$. The figure below illustrates the behavior of the verification algorithm in different scenarios (projected on the three first coordinates). In all cases, a small number of trajectories was needed to obtain the answer.

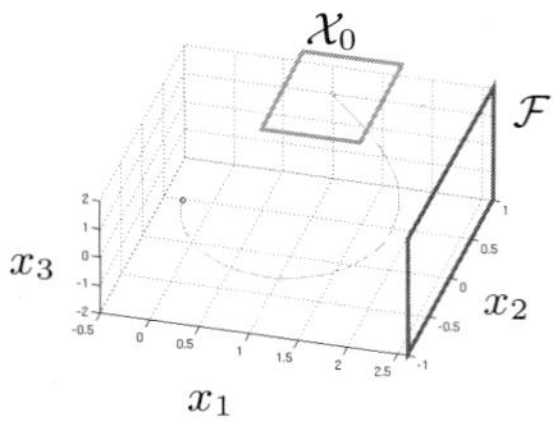

$d = 2.6$: One trajectory was enough to prove that the system is safe.

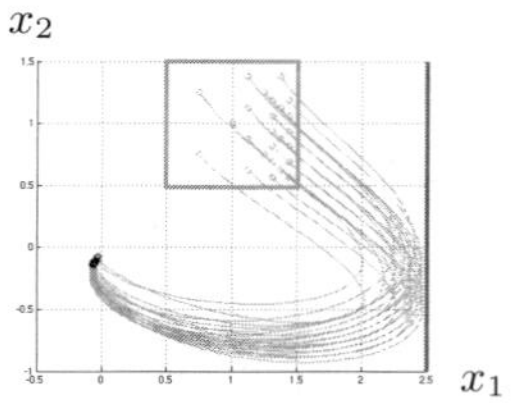

$d = 2.5$: The system is declared uncertain using $\delta = 0.1$ after 25 trajectories.

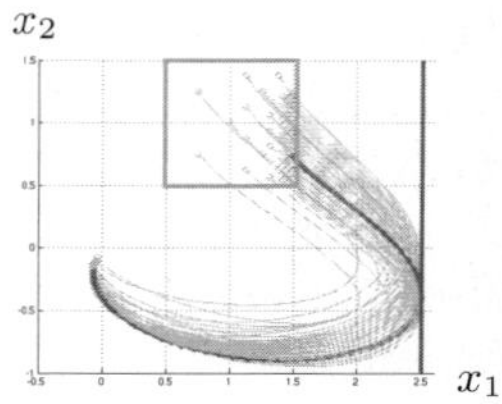

$d = 2.5$: The system was found unsafe with $\delta = 0.01$ after 63 trajectories.

5.2 Verifying the Invariant of Two Oscillator Circuits

For the following examples, our goal is to prove that a set is invariant for an unbounded horizon. To do this, the classical idea is to show that for a certain $T > 0$, the set $\text{Reach}_{=T}$ is contained in $\text{Reach}_{\leq t}$ with $t < T$ which implies that $\text{Reach}_{\leq T}$ is the reachable set for unbounded horizon. Our method is to use our verification algorithm slightly modified so that *every* trajectory is considered as uncertain. As previously, the algorithm stops when $\delta_k < \delta$. At this point, then, we can characterize the reachable sets thanks to inclusions (5) of Corollary 1.

We applied this idea to analyze the periodic steady state behavior of two analog oscillator circuits. The first one is a *tunnel-diode oscillator* (TDO) whose second-order nonlinear dynamics is given in Figure 2. The second circuit is a *voltage controlled oscillator* (VCO) circuit, the schema of which is given on the right. Its dynamics is governed by a third-order nonlinear equation. A fully-detailed model can be found in [FKR06].

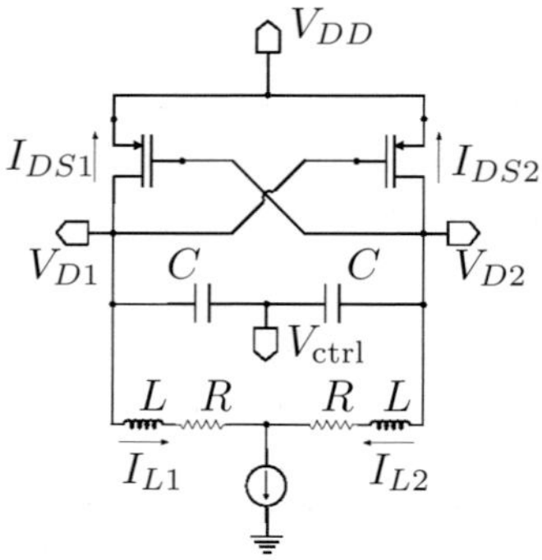

VCO schema

What makes this problem difficult for traditional tools performing reachability is that most often, the reachable set is computed step by step forward in time and each step increases the error of over-approximation. This error after one period may have become too large to prove the invariant property. This is particularly serious for the VCO for which the limit cycle is much less contractive than for the TDO. In [FKR06], this problem is addressed using forward-backward refinement. Our method, however, does not suffer from this problem. If we neglect the inherent error of the numerical solver used to compute the trajectories, the quality of the over-approximation that we get with the sensitivity function does not depend on the time we measure it. If the dynamics is neither really contractive nor diverging as for the VCO, this means that the norm of the sensitivity function will remain near 1 and then we know from Theorem 1 and Theorem 3

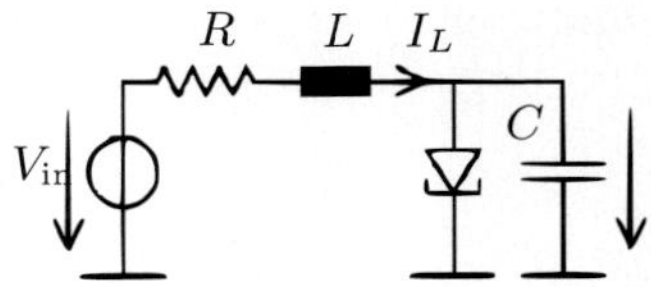

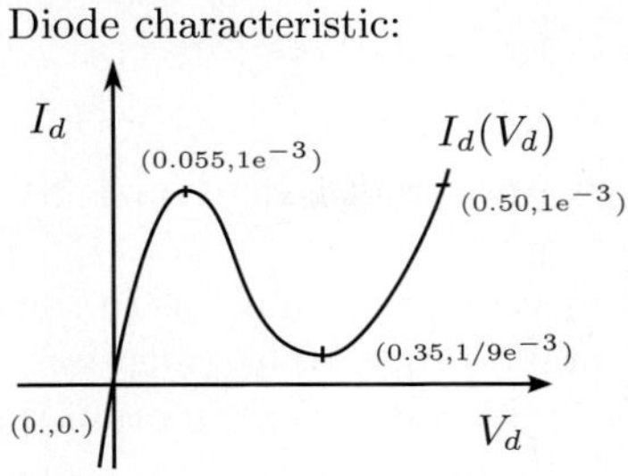

$$\dot{V}_d = \frac{1}{C}(-I_d(V_d) + I_L)$$
$$\dot{I}_L = \frac{1}{C}(-I_d(V_d) + I_L)$$

where $C = 1pF$, $L = 1\mu H$, $G = 5m\Omega^-1$, $V_{\text{in}} = 0.3V$.

Fig. 2. Tunnel Diode Oscillator Circuit

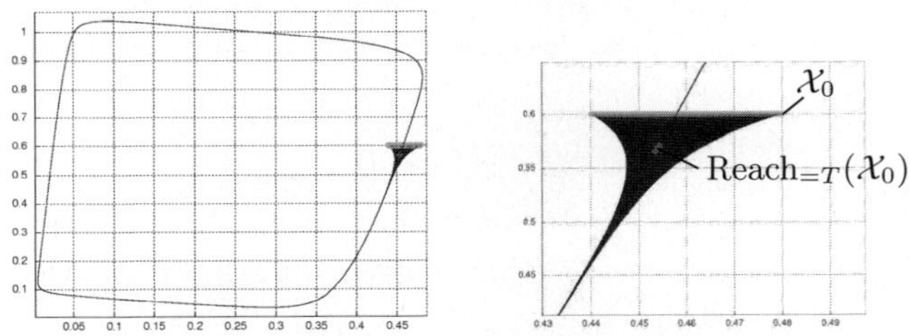

Reachable sets for the TDO.

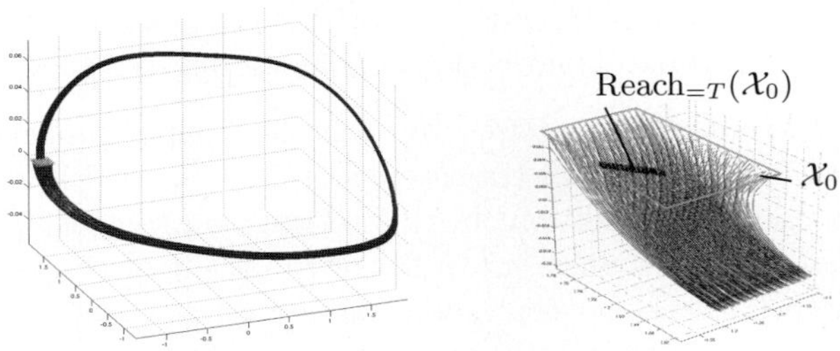

Reachable sets for the VCO.

Fig. 3. Verifying the invariant of two oscillator circuits

that if we sample the initial set with a precision ϵ then we will get an approximation of the reachable set at time T with a precision which is quadratic w.r.t. ϵ. We show some results we obtained in Figure 3. In both cases, we sampled the initial set with $\epsilon = 1 \times 10^{-3}$. Since ϵ^2 is then negligible before the size of the reachable sets $\text{Reach}_{=T}(\mathcal{X}_0)$ that we obtain, we can conclude that the sets are invariant. Since, we must note that with this method, we did not yet obtain a *formal* proof of the result because of the remaining indetermination of constant M in Theorem 3. To get this formal bound, a deeper analysis of the dynamics equations still needs to be done. Techniques and tools recently developed are related to this issue (see [SB06]).

6 Extension to Hybrid Systems

Extending sensitivity analysis to the hybrid case is not straightforward and even in the simplest case of a transition with state continuity, a discontinuity often appears in the sensitivity function that needs to be evaluated. The most general setting for sensitivity analysis includes hybrid systems for which dynamics in each mode is governed by differential algebraic equations [HP00,BL02] To simplify the presentation and get the intuition of what are the changes induced by the hybridicity of the dynamics, we restrict the study to the case of a unique transition between two modes. Let us assume that transitions are governed by crossings of an hyper-surface $\mathcal{G}$ implicitly defined by a continuous function g and that the flow is continuous. The dynamics of this system is described by:

$$\dot{\mathbf{x}} = \begin{cases} f_1(t, \mathbf{x}) \text{ if } g(\mathbf{x}) < 0 \\ f_2(t, \mathbf{x}) \text{ if } g(\mathbf{x}) \geq 0 \end{cases}, \mathbf{x}(0) \in \mathcal{X}_0 \tag{14}$$

We consider a trajectory $\xi_{\mathbf{x}_0}$ performing a first transition at time $\tau > 0$, i.e. such that $g(\xi_{\mathbf{x}_0}(t)) < 0 \ \forall t \in [0, \tau[$ and $g(\xi_{\mathbf{x}_0}(t)) = 0$. We make the following standard assumption:

Assumption 1. *At the crossing point* $\mathbf{x}$, $\langle \nabla_{\mathbf{x}} g(\mathbf{x}), f_1(\tau^-, \mathbf{x}) \rangle \neq 0$, *where* $\langle , \rangle$ *is the Euclidean cross product. Moreover, there exists a neighborhood* $\mathcal{N}$ *of* $\mathbf{x}_0$ *such that for all* $\mathbf{x} \in \mathcal{N}$, *this assumption is also true for the flow* $\xi_{\mathbf{x}}$.

Assumption 1 prevents the trajectory to cross the frontier tangentially and ensures that there exists a tube of trajectories around $\xi_{\mathbf{x}_0}$ which also crosses the frontier under the same condition. In this setting, we consider the most standard behavior of a hybrid system, i.e. it follows a continuous trajectory for some time, then switches to another continuous mode for again some time and so on. During a continuous evolution, we know how sensitivity evolves. The remaining question is about its continuity at transition times. We have the following

Proposition 2. *Under assumptions 1, the sensitivity function at time* τ *satisfies:*

$$\mathbf{s}(\tau^+) - \mathbf{s}(\tau^-) = \frac{d\tau}{d\mathbf{x}_0}\big(f_2(\tau, \xi_{\mathbf{x}_0}(\tau)) - f_1(\tau, \xi_{\mathbf{x}_0}(\tau))\big) \tag{15}$$

$$\textit{where } \frac{d\tau}{d\mathbf{x}_0} = \frac{\langle \nabla_{\mathbf{x}} g(\xi_{\mathbf{x}_0}(\tau)), \mathbf{s}_{\mathbf{x}_0}(\tau) \rangle}{\langle \nabla_{\mathbf{x}} g(\xi_{\mathbf{x}_0}(\tau)), f_1(\tau, \xi_{\mathbf{x}_0}) \rangle} \tag{16}$$

We omit the proof for brevity (it can be found in [HP00,BL02]). Rather, we provide a picture in Figure 4 giving an intuition of why a a discontinuity happens.

Proposition 2 provides a constructive formula to compute the values of the jumps. This means that sensitivity functions can be computed for hybrid

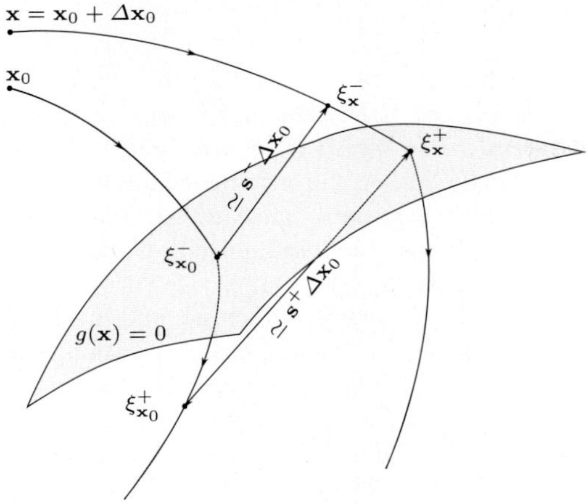

Fig. 4. Illustration of sensitivity discontinuity. The jump condition 15 results from the fact that between τ^- and τ^+ ($-$ and $+$ superscripts in the figure), the flows $\xi_{\mathbf{x}_0}$ and $\xi_{\mathbf{x}}$ evolve with different dynamics f_1 and f_2.

trajectories and thus Algorithm 1 can be implemented. The assumption made is reasonable in the sense that it is very likely that the set of points for which the frontier is crossed tangentially has a zero measure. Still, current work investigates the behavior of our algorithm around such points, along with the adaptation of Theorem 3 and 4 to the hybrid case.

7 Conclusion

We have developed a novel and general simulation-based method for proving safety of arbitrary continuous systems and demonstrated its effectiveness on several non trivial examples. This method can treat arbitrary nonlinear systems and can be particularly efficient for affine time-varying systems. As shown in Section 6, it can, under reasonable assumptions, be extended to hybrid systems as well. The use of the tunable tolerance parameter δ gives an elegant solution to the eternal tension between finite algorithmic termination and the potential infinite precision of the real numbers. In addition to its theoretical properties and efficiency, our method is more likely to be accepted by practitioners who already use simulation as a key validation tool.

Future work will extend the implementation to hybrid automata, look at the problem of unbounded horizon and, most importantly, deal with *non-autonomous* systems with *bounded inputs.* An appropriate sampling of the input space combined with the use of search heuristics (branch and bound, RRT etc.), and/or optimal control techniques should provide interesting results which could be incorporated naturally into the design process of complex control systems.

References

ACH$^+$95. R. Alur, C. Courcoubetis, N. Halbwachs, T.A. Henzinger, P.-H. Ho, X. Nicollin, A. Olivero, J. Sifakis, and S. Yovine. The algorithmic analysis of hybrid systems. *Theoretical Computer Science*, 138:3–34, 1995.

ADG03. E. Asarin, T. Dang, and A. Girard. Reachability analysis of nonlinear systems using conservative approximation. In Oded Maler and Amir Pnueli, editors, *Hybrid Systems: Computation and Control*, LNCS 2623, pages 20–35. Springer-Verlag, 2003.

BCLM05. M. S. Branicky, M. M. Curtiss, J. Levine, and S. Morgan. Sampling-based reachability algorithms for control and verification of complex systems. In *Proc. Thirteenth Yale Workshop on Adaptive and Learning Systems*, June 2005.

BCLM06. M.S. Branicky, M.M. Curtiss, J. Levine, and S. Morgan. Sampling-based planning, control and verification of hybrid systems. *Control Theory and Applications, IEE Proceedings*, 153:575–590, Sept. 2006.

BF04. A. Bhatia and E. Frazzoli. Incremental search methods for reachability analysis of continuous and hybrid systems. In R. Alur and G. J. Pappas, editors, *Hybrid Systems: Computation and Control*, volume 2993 of *LNCS*, pages 142–156. Springer, 2004.

BL02. P. I. Barton and C. Kun Lee. Modeling, simulation, sensitivity analysis, and optimization of hybrid systems. *ACM Trans. Model. Comput. Simul.*, 12(4):256–289, 2002.

CK98. A. Chutinan and B.H. Krogh. Computing polyhedral approximations to dynamic flow pipes. In *Proc. of the 37th Annual International Conference on Decision and Control, CDC'98*. IEEE, 1998.

DM98. T. Dang and O. Maler. Reachability analysis via face lifting. In T.A. Henzinger and S. Sastry, editors, *Hybrid Systems: Computation and Control*, LNCS 1386, pages 96–109. Springer-Verlag, 1998.

FKR06. G. Frehse, B. H. Krogh, and R. A. Rutenbar. Verifying analog oscillator circuits using forward/backward abstraction refinement. In *DATE 2006: Design, Automation and Test in Europe*, March 2006.

Gir05. A. Girard. Reachability of uncertain linear systems using zonotopes. In *Hybrid Systems : Computation and Control*, LNCS 3414, pages 291–305. Springer, 2005.

GP06. A. Girard and G. J. Pappas. Verification using simulation. In *Hybrid Systems : Computation and Control*, volume 3927 of *LNCS*, pages 272–286. Springer-Verlag, 2006.

HP00. I.A. Hiskens and M.A. Pai. Trajectory sensitivity analysis of hybrid systems. *IEEE Transactions on Circuits and Systems I: Fundamental Theory and Applications*, 47(2):204–220, February 2000.

HS74. M.W. Hirsch and S. Smale. *Differential Equations, Dynamical Systems and Linear Algebra*. Academic Press, 1974.

KKMS03. J. Kapinski, B. H. Krogh, O. Maler, and O. Stursberg. On systematic simulation of open continuous systems. In *HSCC*, pages 283–297, 2003.

LaV06. S. M. LaValle. *Planning Algorithms*, chapter 5. Cambridge University Press, Cambridge, U.K., 2006. Available at http://planning.cs.uiuc.edu/.

LYL04. S. R. Lindemann, A. Yershova, and S. M. LaValle. Incremental grid sampling strategies in robotics. In *Proceedings Workshop on Algorithmic Foundations of Robotics*, pages 297–312, 2004.

SB06. A. B. Singer and P. I. Barton. Bounding the solutions of parameter dependent nonlinear ordinary differential equations. *SIAM Journal on Scientific Computing*, 27(6):2167–2182, 2006.

SH05. R. Serban and A. C. Hindmarsh. Cvodes: the sensitivity-enabled ode solver in sundials. In *Proceedings of IDETC/CIE 2005*, Long Beach, CA., Sept. 2005.

Motion Programs for Puppet Choreography and Control

Magnus Egerstedt[1], Todd Murphey[2], and Jon Ludwig[3]

[1] Georgia Institute of Technology, Electrical and Computer Engineering
Atlanta, GA 30323, USA
magnus@ece.gatech.edu

[2] University of Colorado, Electrical and Computer Engineering
Boulder, CO 80309, USA
murphey@colorado.edu

[3] The Center for Puppetry Arts, Atlanta, GA 30309, USA
jonludwig@puppet.org

Abstract. This paper presents a motion description language (MDLp) for specifying and encoding autonomous puppetry plays in a manner that is faithful to the way puppetry choreography is currently formulated. In particular, MDLp is a formal language whose strings, when parsed by a dynamical system (the puppet) produces optimized, hybrid control laws corresponding to strings of motions, locations, and temporal durations for each agent. The paper is concerned with the development of this language as well as with an optimization engine for hybrid optimal control of MDLp strings, and with the generation of motion primitives within the "Imitate, Simplify, Exaggerate" puppetry paradigm.

1 Introduction

One of the main drivers behind the rapidly emerging abstraction-based approach to control and software design is the ability to specify the desired system behavior at a high-level of abstraction, without having to take the actual implementation details into account [9,21]. In other words, the key idea is to be able to give high-level specifications in some language such as linear temporal logic (LTL) [17,22], Computation and Control Languages (CCL) [16], maneuver automata [13], or Motion Description Languages (MDL) [4,10,14,20], and then be assured that the transitions from high-level specifications to actual control signals are achieved in a stable and correct manner.

In this paper we pursue this issue of abstraction-based control for a particular application, namely *autonomous puppetry*, which apart from being a conceptual oxymoron, presents a number of technical challenges. The ultimate objective is to be able to input high-level descriptions of desired puppet motions, denoted *plays*, and then go from such plays to actual control laws for implementing the plays on autonomous marionette puppets, as shown in Figure 1.

A. Bemporad, A. Bicchi, and G. Buttazzo (Eds.): HSCC 2007, LNCS 4416, pp. 190–202, 2007.

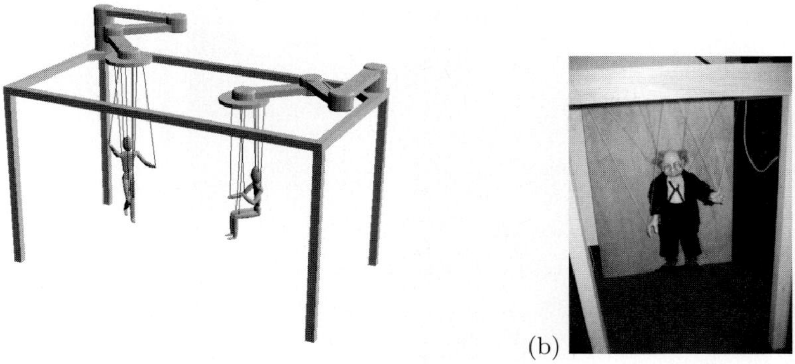

(a) (b)

Fig. 1. A *Mathematica* graphical representation of the mechanical system for autonomous puppetry control (a), together with one of the marionettes used (b)

The control strategy that we will employ will be hybrid, for three distinct reasons, namely:

1. Plays are naturally described as sequences of distinctive motions, which means that the controller must switch between different modes of operation;
2. As the objective is to mimic human (or animal) behaviors, puppeteers typically work with a set of established motion primitives, such as "walk", "run", "dance", "hop" [11]; and
3. The marionette platform is in itself hybrid in that the strings (actuation modalities) are in a number of different configurations during a play, including "free", "controlled", "locked", or "grouped" [15].

We will discuss these three hybrid aspects of the autonomous marionette project, and the framework that we propose in this paper for formalizing high-level specifications for puppetry is based Motion Description Languages. Specifically, a MDL is a string of pairs, each specifying what control law the system should be executing and an interrupt condition corresponding to the termination of this control law. The particular language that we propose is slightly more structured than the standard MDL (or MDLe, where "e" stands for "extended") and we will call this language MDLp, with "p" meaning "puppetry". In order for this language to be successful, it is important that it is expressive enough to be able to characterize actual puppet plays, and as such we draw inspiration from the way such plays are staged by professional puppeteers.

As an example, consider a part of an actual play, as shown in Figure 2. The play that this example comes from is the "Rainforest Adventures" - an original puppet play staged at the Center for Puppetry Arts in Atlanta during 2005 [7,19]. It shows how the basic building blocks for a formal language for puppet choreography can be derived from existing practices in puppeteering.

In fact, the standard way in which puppet plays are described is through four parameters, namely *temporal duration*, *agent*, *space*, and *motion* (when?,

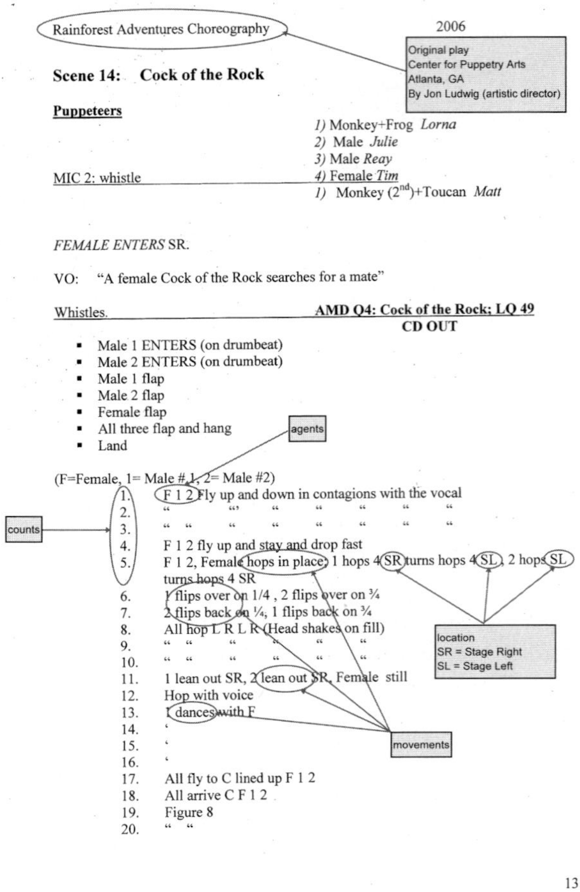

Fig. 2. Rainforest Adventures: This figure is an original puppet choreography sheet from the Center for Puppetry Arts in Atlanta [7]. It shows how the basic building blocks for a formal language for puppet choreography can be derived from existing practices in puppeteering.

who?, where?, and what?) [2,11]. Most plays are based on *counts* in that each puppet motion is supposed to happen at a particular count. (This becomes even more important if multiple puppets are acting simultaneously on stage or if the play is set to music). At each specified count, a motion is initiated and/or terminated. Following this standard practice, we, in the following sections, will propose a formal language for describing such puppet plays, and the outline of this paper is as follows: In Section 2, we recall the basic definitions of a Motion Description Language and show how these definitions can be augmented to form the MDLp, suitable for specifying puppet plays. We then, in Section 3, use the Calculus of Variations for parsing MDLp strings in an optimal way in order to produce effective control programs, deployable on the robot platform. The question of how to generate the motion primitives that constitute the building blocks of MDLp is the focus of Section 4. In fact, professional puppeteers use an expression, “Imitate, Simplify, Exaggerate”, to describe the basic steps in making a puppet perform a given behavior [11]. First, imitate the behavior that one observes, then simplify it down to its basic components, and finally exaggerate the resulting behavior to convey the correct level of animation and emotional content to the viewer, who is often quite distant from the stage. These three steps have formal mathematical counterparts, which is the focus of Section 4, followed by the conclusions, in Section 5.

2 Choreography

2.1 Motion Description Languages

As the complexity of many control systems increases, due both to the system complexity (e.g. manufacturing systems, [6]) and the complexity of the environment in which the system is embedded (e.g. autonomous robots [1,18]), multi-modal control has emerged as a useful design tool. The main idea is to define different modes of operation, e.g. with respect to a particular task, operating point, or data source. These modes are then combined according to some discrete switching logic and one attempt to formalize this notion is through the concept of a *Motion Description Language* (MDL) [4,10,14,20].

Each string in a MDL corresponds to a control program that can be operated on by the control system. Slightly different versions of MDLs have been proposed, but they all share the common feature that the individual atoms, concatenated together to form the control program, can be characterized by control-interrupt pairs. In other words, given a dynamical system

$$\dot{x} = f(x,u),\ x \in \mathbb{R}^N,\ u \in U$$
$$y = h(x),\ y \in Y,$$

together with a control program $(k_1,\xi_1),\ldots,(k_z,\xi_z)$, where $k_i : Y \to U$ and $\xi_i : Y \to \{0,1\}$, the system operates on this program as $\dot{x} = f(x, k_1(h(x)))$ until $\xi_1(h(x)) = 1$. At this point the next pair is read and $\dot{x} = f(x, k_2(h(x)))$ until $\xi_2(h(x)) = 1$, and so on. (Note that the interrupts can also be time-triggered, which can be incorporated by a simple augmentation of the state space.)

A number of results have been derived for such (and similar) systems, driven by strings of symbolic inputs. For example, in [3], the set of reachable states was characterized, while [12] investigated optimal control aspects of such systems. In [8,14,20], the connection between MDLs and robotics was investigated.

2.2 Puppet Dynamics

Before we can establish a suitable formalism for motion control of autonomous marionettes, an appropriate puppet model is needed. In fact, we let the puppet be modeled using well-known Euler-Lagrange methods for articulated mechanical systems [5]. Our puppet system is similar to a closed-chain of rigid bodies, but differs in that one of the linkages has no mass. This implies that one cannot apply a force to the link or use the inertia of the link. In fact, by including the parameters that define the string link (e.g., the location of the string end-points) in the configuration, we get a globally degenerate inertia matrix and, correspondingly, degenerate equations of motion. Associating a mass with the string parameters to avoid this difficulty introduces other problems in that a small mass yields stiff differential equations for the motion while a large mass will unrealistically affect the dynamics of the system.

The solution is to treat the string as a constraint that indirectly applies to the system inputs. Hence, we will treat the mechanical superstructure as a kinematic system while we treat the puppet itself as a dynamic system. Thus, we are modeling the puppet using a mixed dynamic-kinematic model.

The validity of such a partial kinematic reduction can be verified as follows (under the assumption that the mechanical system controlling the puppet is fully-actuated). The mechanical system, including both the puppet and the mechanism controlling the puppet, can be described using the constrained Euler-Lagrange equations with $q = [q_D, q_K, q_M]$, where q_M describes the configuration of the mechanism, q_K describes the configuration of the strings, and q_D describes the configuration of the puppet. The equations of motion can be written as: $\tilde{\nabla}_{\dot{q}}\dot{q} = u_i Y^i$ where $\tilde{\nabla}$ is the constrained affine connection, u^i are the m inputs, and Y_i are the associated input vector fields. (See [5] for a complete description of this formalism.) In this context, a system is *kinematically reducible* (i.e, all paths on the configuration manifold Q correspond to trajectories on TQ and vice-versa) if $\langle Y_i, Y_j \rangle \in span\{Y_k \mid 1 \leq k \leq m\}$ where $\langle X, Y \rangle = \tilde{\nabla}_X Y + \tilde{\nabla}_Y X$. Now, because the strings are massless, the inertia tensor is block diagonal in $[q_D, q_K]$. This allows us to address the constraints independently as constraints between q_D and q_M so that

$$\begin{aligned}\tilde{\nabla}_{\dot{q}_M}\dot{q}_M &= u_i Y^i \\ \tilde{\nabla}_{\dot{q}_D}\dot{q}_D &= 0.\end{aligned}$$

Because the mechanism controlling the puppet is *assumed* to be fully actuated, the first equation for the mechanism dynamics trivially satisfies the condition $\langle Y_i, Y_j \rangle \in span\{Y_k \mid 1 \leq k \leq m\}$ for kinematic reducibility. Hence, separating the kinematic reduction of the mechanism controlling the puppet from the (*not* kinematically reducible) dynamics of the puppet is a mechanically valid description of the system.

Lastly, the string constraints are actually inequality constraints. In fact, the strings can go slack. This can be included by monitoring the Lagrange multipliers enforcing the constraints and using projection operators to provide impulses that release the constraint when the string goes slack and enforce them again when the end of the string is reached.

2.3 MDLp

Now that we have a model of the puppet dynamics, we note that the general MDL outlined in Section 2.1 does not lend itself to be directly applicable to the scenario described in Figure 2. In fact, what we will do in this section is to augment the standard MDL formulation to include factors such as spatial location. For this, assume that the play starts at time t_0 and that it ends at time t_f. Moreover, let the temporal resolution (the length of each "count") be Δ, and assume that $(t_f - t_0)/\Delta = M$. Following this, the set of all times over which the play is specified is $\mathcal{T} = \{t_0, t_0 + \Delta, t_0 + 2\Delta, \ldots, t_0 + M\Delta\}$.

Moreover, assume that the stage is divided into N different sections (typically this number is 6, namely LowerLeft, LowerCenter, LowerRight, MiddleLeft, MiddleCenter, MiddleRight, UpperLeft, UpperCenter, UpperRight), whose planar center-of-gravity coordinates are given by $r_1, \ldots, r_N$, with the set of regions being given by $\mathcal{R} = \{r_1, \ldots, r_N\}$.

From the arguments in Section 2.2, we can assume that each puppet has a dynamics given by

$$\dot{x}^i = f^i(x^i, u^i),\ y^i = \pi^i(x^i),$$

where the superscript i denotes agent i, and the output $y^i \in \mathbb{R}^2$ is given by a projection π^i from X^i to the plane. Now, given that we have constructed a number of control laws κ^i_j, $j = 1 \ldots, C^i$, corresponding to different moves that puppet i can perform, with each control law being a function of x^i (state), t (time), and α_i (a parameter characterizing certain aspects of the motion such as speed, energy, or acceleration, as is the normal interpretation of the parametrization of biological motor programs), we can let the set of moves that puppet i can perform be given by $\mathcal{K}^i = \{\kappa^i_1, \ldots, \kappa^i_{C^i}\}$. In fact, we will often use the shorthand $f^i_j(x^i, t, \alpha)$ to denote the impact that control law κ^i_j has on puppet i through $f^i(x^i, \kappa^i_j(x^i, t, \alpha))$.

As already pointed out, each instruction in the puppet play language is a four-tuple designating *when*, *who*, *where*, and *what* the puppets should be doing. In other words, we let the motion alphabet associated with puppet i be given by $\mathcal{L}^i = \mathcal{T} \times \mathcal{T} \times \mathcal{R} \times \mathcal{K}^i$. Each element in $\mathcal{L}^i$ is thus given by (T_0, T_1, r, κ), where the interpretation is that the motion should take place during the time interval $T_1 - T_0$, largely in region r, while executing the control law κ.

For the reminder of this section, we will drop the explicit dependence on i, and assume that we are concerned with a given, individual puppet. We can then follow the standard notation in the formal language field and let $\mathcal{L}^\star$ denote the set of all finite-length concatenations of elements in $\mathcal{L}$ (including the empty string), and let puppet plays be given by words $\lambda \in \mathcal{L}^\star$. In particular, if we let $\lambda = (t_0, T_1, r_1, \kappa_1), (T_1, T_2, r_2, \kappa_2), \ldots, (T_{p-1}, T_p, r_p, \kappa_p)$, then the puppet operates on this string through

$$\dot{x} = \begin{cases} f_1(x, t, \alpha_1),\ t \in [t_0, T_1) \\ f_2(x, t, \alpha_2),\ t \in [T_1, T_2) \\ \vdots \\ f_p(x, t, \alpha_p),\ t \in [T_{p-1}, T_p]. \end{cases}$$

This seems fairly natural, but two essential parameters have been left out. First, the motion parameters $\alpha_1, \ldots, \alpha_p$ have not yet been specified. Moreover, the desired regions $r_1, \ldots, r_p$ have not been utilized in any way. In order to remedy this, we need to construct not just a *parser* for puppet plays, as given above, but also a *compiler* that selects the "best" parameters (as well as durations) for the different moves so that the play is executed as efficiently as possible.

3 A Puppet Play Compiler

Consider the following optimal control problem:

$$\min_{\tau,\alpha_1,\alpha_2} J(\tau,\alpha_1,\alpha_2) = \int_0^{t_f} L(x,t)dt + C_1(\alpha_1) + C_2(\alpha_2) + D(\tau) + \Psi_1(x(\tau)) + \Psi_2(x(\tau)),$$

where

$$\dot{x} = \begin{cases} f_1(x,t,\alpha_1), & t \in [0,\tau) \\ f_2(x,t,\alpha_2), & t \in [\tau,t_f] \end{cases}$$
$$x(0) = x_0.$$

This optimal control problem is the atomic problem involving how to execute the two-instruction play $(0,T,r_1,\kappa_1),(T,t_f,r_2,\kappa_2)$ under the following interpretations

$D(\tau) =$ function that penalizes deviations from$T$, e.g. $(\tau - T)^2$
$C_i(\alpha_i) =$ function that measures how much energy it takes to use parameter α_i
$\Psi_i(x(\tau \text{ or } T)) =$ function that ensures that the projection $\pi((x(\tau))$ is close to $r_1$and similarly for $\pi(x(T))$
$L(x,t) =$ function that may be used to ensure that a reference trajectory is followed.

By forming the Lagrangian

$$\tilde{J} = \int_0^{\tau} (L + \lambda_1(f_1 - \dot{x}))dt + \int_{\tau}^{T} (L + \lambda_2(f_2 - \dot{x}))dt + C_1 + C_2 + D + \Psi_1 + \Psi_2$$

and using a standard variational argument, with $\tau \to \tau + \epsilon\theta$, $\alpha_1 \to \alpha + \epsilon a_1$, $\alpha_2 \to \alpha_2 + \epsilon a_2$, we can obtain the corresponding variation in the trajectory as $x(t) \to x(t) + \epsilon\eta(t)$, with $\eta(0) = 0$.

By letting $\tilde{J}_\epsilon$ denote the Lagrangian from the variational system, we get that

$$\begin{aligned}
\lim_{\epsilon \to 0} \frac{\tilde{J}_\epsilon - \tilde{J}}{\epsilon} &= \int_0^{\tau} \left(\left(\frac{\partial L}{\partial x} + \lambda_1 \frac{\partial f_1}{\partial x} + \dot{\lambda}_1 \right) \eta + \lambda_1 \frac{\partial f_1}{\partial \alpha_1} a_1 \right) dt \\
&+ \int_{\tau}^{T} \left(\left(\frac{\partial L}{\partial x} + \lambda_2 \frac{\partial f_2}{\partial x} + \dot{\lambda}_2 \right) \eta + \lambda_2 \frac{\partial f_2}{\partial \alpha_2} a_2 \right) dt \\
&+ \left(-\lambda_1(\tau) + \lambda_2(\tau) + \frac{\partial \Psi_1}{\partial x} \right) \eta(\tau) + \left(-\lambda_2(T) + \frac{\partial \Psi_2}{\partial x} \right) \eta(T) \\
&+ \left(\lambda_1(\tau)(f_1(x(\tau)) - f_2(x(\tau)) + \frac{\partial D}{\partial \tau} + \frac{\partial \Psi_1}{\partial x} f_2(x(\tau)) \right) \theta \\
&+ \frac{\partial C_1}{\partial \alpha_1} a_1 + \frac{\partial C_2}{\partial \alpha_2} a_2.
\end{aligned}$$

This gives us the optimality conditions as

$$\begin{aligned}
\frac{\partial J}{\partial \tau} &= \lambda(\tau_-)f_1(x(\tau)) - \lambda(\tau_+)f_2(x(\tau)) + \frac{\partial D}{\partial \tau} \\
\frac{\partial J}{\partial \alpha_2} &= \xi(\tau_+) \\
\frac{\partial J}{\partial \alpha_1} &= \xi(0),
\end{aligned}$$

where the costates λ and ξ satisfy the following discontinuous (backwards) differential equations:

$$\begin{aligned}
\lambda(T) &= \frac{\partial \Psi_2}{\partial x}(x(T)) \\
\dot{\lambda} &= -\frac{\partial L}{\partial x} - \lambda \frac{\partial f_2}{\partial x}, \ t \in (\tau, T) \\
\lambda(\tau_-) &= \lambda(\tau_+) + \frac{\partial \Psi_1}{\partial x}(x(\tau)) \\
\dot{\lambda} &= -\frac{\partial L}{\partial x} - \lambda \frac{\partial f_1}{\partial x}, \ t \in [0, \tau) \\
\xi(T) &= \frac{\partial C_2}{\partial \alpha_2} \\
\dot{\xi} &= -\lambda \frac{\partial f_2}{\partial x}, \ t \in (\tau, T) \\
\xi(\tau_-) &= \frac{\partial C_1}{\partial \alpha_1} \\
\dot{\xi} &= \lambda \frac{\partial f_1}{\partial x}, \ t \in [0, \tau).
\end{aligned}$$

By a direct generalization to more than two modes, this construction allows us to produce a compiler that takes *plays* and outputs *strings of control modes with an optimized temporal duration and mode-parametrization*, as given in the algorithm below:

```
Algorithm
Given (t_0,T_1,r_1,κ_1),(T_1,T_2,r_2,κ_2),...,(T_{p-1},T_p,r_p,κ_p)
Set τ_i(0) = T_i, i = 1,...,p-1
Initialize α_i(0), i = 1,...,p
Optimization (k = 0)
 Repeat
  Compute x(t) forwards using α_i(k), τ_i(k)
  Compute λ(t),ξ(t) backwards (including jumps)
  Compute ∂J/∂τ_i, ∂J/∂α_i
  Gradient Descent
   Set τ_i(k+1) = τ_i(k) - γ(k) ∂J/∂τ_i (τ_i(k)), i = 1,...,p-1
   Set α_i(k+1) = α_i(k) - ℓ(k) ∂J/∂α_i (α_i(k)), i = 1,...,p
  Set k = k+1
 Until ||∇J|| ≤ δ
```

An example of this proposed approach is shown in Figure 3, in which an oscillator is switching between two modes - one slow (corresponding to walking) and one fast (corresponding to running). The frequencies and dynamics associated with the two modes are

$$\omega_1 = 5\frac{\alpha_1^2}{\alpha_1^2+10} + 5$$
$$\omega_1 = 10\frac{\alpha_2^2}{\alpha_2^2+10} + 10$$
$$\dot{x} = \begin{bmatrix} 0 & -\omega_{1,2} \\ \omega_{1,2} & 0 \end{bmatrix} x$$

and the cost is

$$J(\tau, \alpha_1, \alpha_2) = \int_0^3 0.1x(t)^T x(t)dt + (\tau - 2)^2 + 0.1\alpha_1^2 + 0.2\alpha_2^2 + (x(\tau) - [-1,0])^T(x(\tau) - [-1,0]^T) + 5x(3)^T x(3).$$

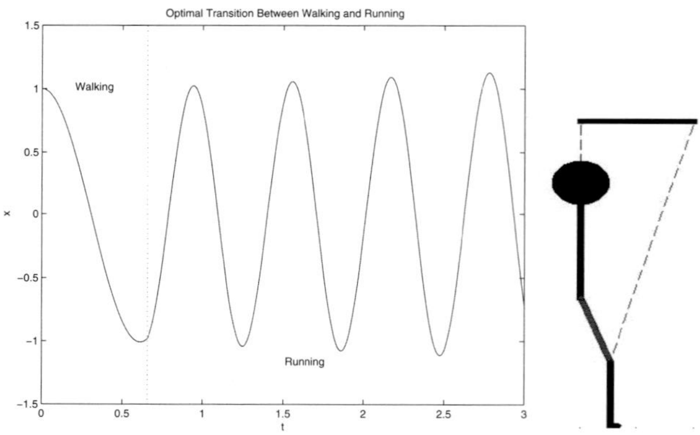

Fig. 3. A simplified locomotion model in which walking and running are defined through linear oscillators with different frequencies. The figure on the left depicts the waveform of the two gaits, and the figure on the right is a snapshot of the animation showing the result.

4 Motion and Caricature: *"Imitate, Simplify, Exaggerate"*

A system with limited expressive powers, such as a marionette, needs to be able to convey the proper emotions in such a way that a human audience understands what is being conveyed. As previously mentioned, puppeteers achieve this through the three steps: *Imitate*, *Simplify*, and *Exaggerate*. We will make these three steps formal in that human motion is being mimicked, after which the resulting motions are projected onto the constrained space over which the marionette operates (see Section 2.2), followed by a transformation of the resulting motion in such a way that the "energy" of the original (non-projected) motion is conserved.

4.1 Conservation of Energy

The method we propose for achieving this is by studying the way professional puppeteers actually control their puppets, as well as draw inspiration from human-like motions, since if the puppet is a human marionette, we would typically like to be able to execute human-like motions. However, since marionettes are constrained in such a way that they cannot be as expressive as human motions, we will first identify human motions (corresponding to the Imitate phase in puppetry) project human motion down onto the space of available puppet motions (the Simplify phase) and then exaggerate these motions in order to make them sufficiently expressive (the Exaggerate phase). Formally speaking, given a desired trajectory $z(t) \in Z$ that we would like the puppet (whose state is $x(t) \in X$, $\dim(X) \leq \dim(Z)$) to follow, we define a projection-like mapping $\rho : Z \to X$. The Simplify-phase thus consists of trying to minimize expressions like

$$\int_0^T L(x(t) - \rho(z(t)))dt,$$

subject to the puppet dynamics $\dot{x} = f(x, u)$, with u being the control input, and where T is the temporal duration of the movement.

Moreover, if TZ and TX are the tangent spaces associated with Z and X respectively, we define the energy conservation cost through the mapping $\phi : TX \to TZ$ in the following manner

$$\int_0^T \mathcal{E}(\phi(f(x(t), u(t)) - \dot{z}(t))dt,$$

and the combined Simplify-Exaggerate optimization problem becomes

$$\min_u \int_0^T (L(x - \rho(z)) + \mathcal{E}(\phi(f(x, u)) - \dot{z}))dt.$$

As an example, consider the situation when the puppet dynamics is given by the completely controllable linear control system

$$\dot{x} = Ax + Bu, \; x \in \mathbb{R}^n$$

with $z \in \mathbb{R}^m$, $m \geq n$. Moreover, let the projection ρ be given by a linear projection Pz and similarly let ϕ be given by a linear relation $E(Ax + Bu)$. The instantaneous cost thus becomes

$$\begin{aligned} L(x - \rho(z)) + \mathcal{E}(\phi(f(x, u)) - \dot{z}) &= \frac{1}{2}(x - Pz)^T Q(x - Pz) \\ &+ \frac{1}{2}(EAx + EBu - \dot{z})^T R(EAx + EBu - \dot{z}), \end{aligned}$$

given positive definite weight matrices Q, R.

Given free initial and final positions $x(0)$ and $x(T)$, this is a standard LQ-optimization problem that can be readily solved, and an example solution is shown in Figure 4. In that example,

$$\dot{z} = \begin{bmatrix} 0 & -1 & 0 \\ 1 & 0 & 0 \\ -0.1 & -0.4 & -1 \end{bmatrix} z, \; z(0) = \begin{bmatrix} 1 \\ 1 \\ 1 \end{bmatrix}$$
$$\dot{x} = \begin{bmatrix} 0 & -1.1 \\ 1.1 & -0.1 \end{bmatrix} x + \begin{bmatrix} 0 \\ 1 \end{bmatrix} u$$
$$Q = I_2, \; R = 0.01 I_3$$
$$P = \begin{bmatrix} 1 & 0 & 0 \\ 0 & 1 & 0 \end{bmatrix}, \; E = \begin{bmatrix} 2/3 & 0 \\ 0 & 2/3 \\ 1/3 & 1/3 \end{bmatrix}$$

4.2 Conservation of "Emotive" Energy

In the previous discussion, *energy* was defined in terms of motion, but one can easily picture a somewhat more esoteric yet perhaps more relevant notion of *emotive energy.* In other words, the puppet is asked to capture a particular emotive state through its motion. But, due to its constrained configuration space, a direct mapping from human emotions to puppet emotions is not feasible. (Our faces, for example, have many degrees of freedom, while the puppets' have significantly less.) The same approach as previously discussed would still apply, but

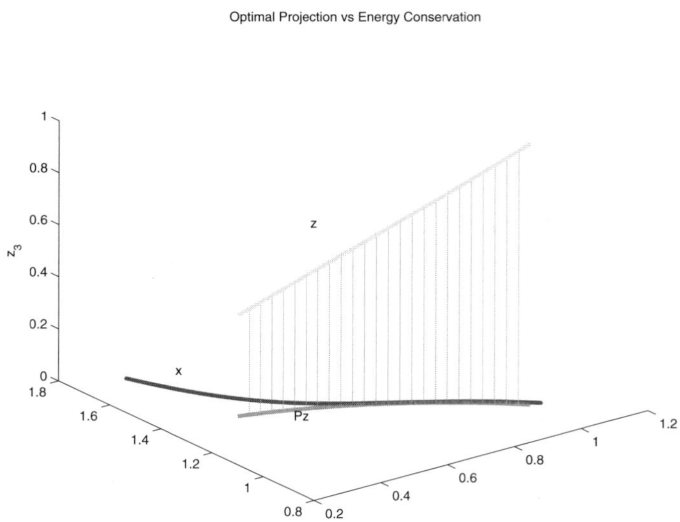

Fig. 4. This figure shows a simple example in which an optimal trade-off between tracking and energy-maintenance is achieved for linear systems. This method provides the basic building-block for the Simplify-Exaggerate phases of the construction of basic motion primitives.

Fig. 5. *Anger* representation: The *Mathematica* stick-figure on the left depicts a nominal allocation between body and face movements

with the difference that now *emotive energy measures* will have to be generated. As a simple example, consider the stick-figure drawing in Figure 5. There the idealized puppet is asked to express *anger* and by constraining either the facial expressions or the body language, the *emotive energy* is maintained by either exaggerating the body or the facial movements. In the middle figure, the stick-figure has to work really hard to move its face, while in the right figure it has to work hard to move its arms. Because of this, the middle figure has a more mild facial expression but more dramatic body expression and the figure on the right has a severe facial expression with almost no body expression. The middle and right figures have been generated automatically (in *Mathematica*) from the left figure by an optimization process similar to what was discussed in the previous paragraphs.

5 Conclusions

In this paper we presented the motion description language MDLp, which is a MDL that allows for slightly more structured instructions, making it useful for specifying and encoding autonomous puppetry plays in a manner that is faithful to standard puppetry choreography. In particular, MDLp is a formal language whose strings, when parsed by a dynamical system, produces optimized, hybrid control laws corresponding to strings of motions, locations, and temporal durations for each agent. The paper is concerned with the development of this language as well as with an optimization engine for hybrid optimal control of MDLp strings, and with the generation of motion primitives within the "Imitate, Simplify, Exaggerate" puppetry paradigm.

References

1. R.C. Arkin. *Behavior Based Robotics.* The MIT Press, Cambridge, MA, 1998.
2. B. Baird. *The Art of the Puppet.* Mcmillan Company, New York, 1965.
3. A. Bicchi, A. Marigo, and B. Piccoli. Encoding Steering Control with Symbols. *IEEE Conference on Decision and Control*, Maui, Hawaii, Dec. 2003.

4. R.W. Brockett. On the Computer Control of Movement. In the *Proceedings of the 1988 IEEE Conference on Robotics and Automation*, pp. 534–540, New York, April 1988.
5. F. Bullo and A.D. Lewis. *Geometric Control of Mechanical Systems.* Number 49 in Texts in Applied Mathematics. Springer-Verlag, 2004.
6. C.G. Cassandras and S. Lafortune. *Introduction to Discrete Event Systems.* Kluwer Academic Publishers, Norwell, MA, 1999.
7. Center for Puppetry Arts. http://www.puppet.org/.
8. M. Egerstedt. Motion Description Languages for Multi-Modal Control in Robotics. In *Control Problems in Robotics, Springer Tracts in Advanced Robotics* , (A. Bicchi, H. Cristensen and D. Prattichizzo Eds.), Springer-Verlag, pp. 75-90, Las Vegas, NV, Dec. 2002.
9. M. Egerstedt and C.F. Martin. Conflict Resolution for Autonomous Vehicles: A Case Study in Hierarchical Control Design. *International Journal of Hybrid Systems*, Vol. 2, No. 3, pp. 221-234, Sept. 2002.
10. M. Egerstedt and R.W. Brockett. Feedback Can Reduce the Specification Complexity of Motor Programs. *IEEE Transactions on Automatic Control*, Vol. 48, No. 2, pp. 213–223, Feb. 2003.
11. L. Engler and C. Fijan. *Making Puppets Come Alive.* Taplinger Publishing Company, New York, 1973.
12. E. Frazzoli. Explicit Solutions for Optimal Maneuver-Based Motion Planning. *IEEE Conference on Decision and Control*, Maui, Hawaii, Dec. 2003.
13. E. Frazzoli, M. A. Dahleh, and E. Feron. Maneuver-Based Motion Planning for Nonlinear Systems with Symmetries. *IEEE Transactions on Robotics*, Vol. 21, No. 6, pp. 1077–1091, Dec. 2005.
14. D. Hristu-Varsakelis, M. Egerstedt, and P.S. Krishnaprasad. On The Structural Complexity of the Motion Description Language MDLe. *IEEE Conference on Decision and Control*, Maui, Hawaii, Dec. 2003.
15. E. Johnson and T. Murphey. Dynamic Modeling and Motion Planning for Marionettes: Rigid Bodies Articulated by Massless Strings. Submitted to *ICRA*, 2007.
16. E. Klavins. A language for modeling and programming cooperative control systems. In *Proceedings of the International Conference on Robotics and Automation*, 2004.
17. M. Kloetzer and C. Belta. Hierarchical Abstractions for Robotic Swarms. *IEEE International Conference on Robotics and Automation*, Orlando, FL, 2006
18. D. Kortenkamp, R.P. Bonasso, and R. Murphy, Eds. *Artificial Intelligence and Mobile Robots.* The MIT Press, Cambridge, MA, 1998.
19. J. Ludwig. Rainforest adventures. http://www.puppet.org/perform/rainforest.shtml.
20. V. Manikonda, P.S. Krishnaprasad, and J. Hendler. Languages, Behaviors, Hybrid Architectures and Motion Control. In *Mathematical Control Theory*, Eds. Willems and Baillieul, pp. 199–226, Springer-Verlag, 1998.
21. G.J. Pappas, G. Laffierier, and S. Sastry. Hierarchically consistent control sytems. *IEEE Trans. Automatic Control*, 45(6):1144–1160, June 2000.
22. P. Tabuada and G. Pappas. Linear Time Logic Control of Discrete-Time Linear Systems. Accepted for publication in *IEEE Transactions on Automatic Control.*

Hierarchical Synthesis of Hybrid Controllers from Temporal Logic Specifications

Georgios E. Fainekos[1], Antoine Girard[2], and George J. Pappas[3]

[1] Department of Computer and Information Science, Univ. of Pennsylvania, USA
fainekos@cis.upenn.edu

[2] Université Joseph Fourier, Laboratoire de Modélisation et Calcul, Grenoble, France
antoine.girard@imag.fr

[3] Department of Electrical and Systems Engineering, Univ. of Pennsylvania, USA
pappasg@ee.upenn.edu

Abstract. In this paper, the problem of synthesizing a hybrid controller for a specification expressed as a temporal logic formula ϕ is considered. We propose a hierarchical approach which consists of three steps. First, the plant to be controlled is abstracted to a fully actuated system. Using the notion of approximate simulation relation, we design a continuous interface allowing the plant to track the trajectories of its abstraction with a guaranteed precision δ. The second step, which is also the main contribution of this paper, consists in deriving a more robust specification ϕ' from the temporal logic formula ϕ such that given a trajectory satisfying ϕ', any other trajectory remaining within distance δ satisfies ϕ. Third, we design a hybrid controller for the abstraction such that all its trajectories satisfy the robust specification ϕ'. Then, the trajectories of the plant satisfy the original specification. An application to the control of a second order model of a planar robot in a polygonal environment is considered.

1 Introduction

Modern engineering challenges involve controlling complex (possibly nonlinear and/or high order) systems to achieve complicated behaviors. An automated approach to synthesize controllers that are correct by design is very desirable since it can save much of the effort required for the verification of the controlled system. For that purpose, the use of a hierarchical approach is often necessary since trying to handle at once both the complexities of the dynamics and of the specification might lead to intractable computations. A hierarchical control system consists of (at least) two layers. The first layer consists of a coarse (and simple) model of the plant. A controller is designed so that this abstraction meets the specification of the problem. The control law is then refined to the second layer which consists of a detailed model of the plant. Architectures of hierarchical controllers based on the notion of simulation relations have been proposed in [1,2]. More recently, it has been claimed that approaches based on approximate simulation relations [3] would provide more robust control laws

A. Bemporad, A. Bicchi, and G. Buttazzo (Eds.): HSCC 2007, LNCS 4416, pp. 203–216, 2007.

while allowing to consider simpler discrete [4] or continuous [5] abstractions for control synthesis.

In this paper, we present such a hierarchical approach for the synthesis of hybrid controllers for specifications expressed as formulas ϕ in the propositional temporal logic over the reals which was introduced in [6]. Following [5], the system is abstracted to a fully actuated system. An interface is designed so that the system is able to track the trajectories of its abstraction with a given guaranteed precision δ. The control objective ϕ is then modified and replaced by a more robust specification ϕ'. The formula ϕ' is such that given a trajectory satisfying ϕ', any trajectory remaining within distance δ satisfies ϕ. This "robustification" procedure is the central step of our approach and constitutes also the main contribution of the paper. It then remains to design a controller for the abstraction such that all its trajectories satisfy the robust specification ϕ'. This is achieved by using one of the computational methods that have recently been developed for the synthesis of hybrid controllers from temporal logic specifications for fully actuated kinematic models [7] or for systems with affine dynamics with drift [8] or for general dynamical systems [9]. Finally, lifting the control law using the hierarchical control architecture, the controlled trajectories of the plant satisfy the original specification ϕ. Throughout the paper, an application to the control of a second order model of a planar robot in a polygonal environment is considered.

2 Problem Description

We consider a continuous time dynamical system

$$\Sigma : \begin{cases} \dot{x}(t) = f(x(t), u(t)), & x(t) \in \mathbb{R}^n,\ x(0) \in X_0 \subseteq \mathbb{R}^n,\ u(t) \in U \subseteq \mathbb{R}^p \\ y(t) = g(x(t)), & y(t) \in \mathbb{R}^k \end{cases} \tag{1}$$

where $x(t)$ is the state of the system, $u(t)$ is the control input and $y(t)$ is the observed output. The goal of this paper is to construct a hybrid controller that generates control inputs $u(t)$ for system Σ so that for the set of initial states X_0, the resulting output $y(t)$ satisfies a formula-specification ϕ in the propositional temporal logic over the positive real line $\mathbb{R}_+$ with the until connective [6]. Let us remark that we design state feedback controllers (i.e. the controller has full knowledge of the state $x(t)$). Thus, the observed output $y(t)$ is used only to specify the desired behavior of the plant.

For the high level planning problem, we consider the existence of a number of regions of interest to the user. Such regions could represent set invariants or sets that must be reached. Let $\Pi = \{\pi_0, \pi_1, \ldots, \pi_n\}$ be a finite set of symbols that label these areas. The denotation $[\![\cdot]\!]$ of each symbol in Π represents a subset of $\mathbb{R}^k$, i.e. for any $\pi \in \Pi$ it is $[\![\pi]\!] \subseteq \mathbb{R}^k$. Formally, $[\![\cdot]\!] : \Pi \to \mathcal{P}(\mathbb{R}^k)$, where $\mathcal{P}(\Gamma)$ denotes the powerset of a set Γ.

In order to make apparent the use of the propositional temporal logic for the composition of temporal specifications, we first give an informal description of the traditional and temporal operators. In this paper, we refer to this logic as RTL [10]. The formal syntax and semantics of RTL are presented in Section 4.

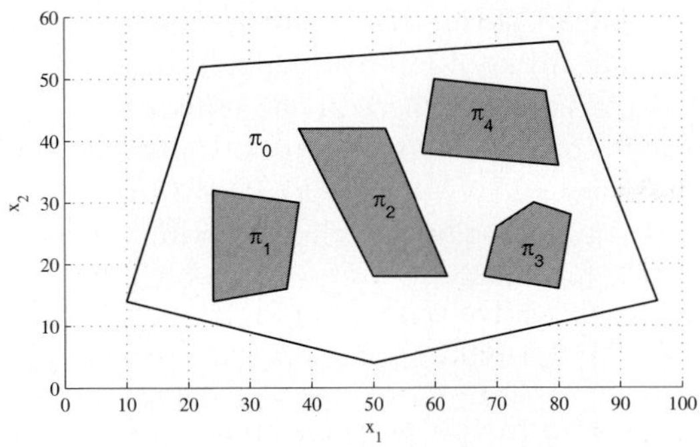

Fig. 1. The simple environment of Example 1. The four regions of interest $\pi_1, \pi_2, \pi_3, \pi_4$ appear in *gray* while the set labeled by π_0 in *white*.

RTL formulas are built over a set of atoms, the set Π in our case, using combinations of the traditional and temporal operators. Traditional logic operators are the *conjunction* ($\wedge$), *disjunction* ($\vee$), *negation* ($\neg$). Some of the temporal operators are *eventually* ($\Diamond$), *always* ($\Box$), *until* ($\mathcal{U}$) and *release* ($\mathcal{R}$). The Temporal Logic of the Reals can describe the usual properties of interest for control problems, i.e. *reachability* ($\Diamond\pi$) and *safety:* ($\Box\pi$ or $\Box\neg\pi$). Beyond the usual properties, RTL can capture sequences of events and certain infinite behaviours. For example:

- **Reachability while avoiding regions:** The formula $\neg(\pi_1 \vee \pi_2 \vee \cdots \vee \pi_n)\mathcal{U}\pi_{n+1}$ expresses the property that eventually π_{n+1} will be true, and until $[\![\pi_{n+1}]\!]$ is reached, we must avoid all unsafe sets $[\![\pi_i]\!]$, $i = 1, \ldots, n$.
- **Sequencing:** The requirement that we must visit $[\![\pi_1]\!]$, $[\![\pi_2]\!]$ and $[\![\pi_3]\!]$ in that order is naturally captured by the formula $\Diamond(\pi_1 \wedge \Diamond(\pi_2 \wedge \Diamond\pi_3))$.
- **Coverage:** Formula $\Diamond\pi_1 \wedge \Diamond\pi_2 \wedge \cdots \wedge \Diamond\pi_m$ reads as the system will eventually reach $[\![\pi_1]\!]$ and eventually $[\![\pi_2]\!]$ and ... eventually $[\![\pi_m]\!]$, requiring the system to eventually visit all regions of interest without any particular ordering.
- **Recurrence (Liveness):** The formula $\Box(\Diamond\pi_1 \wedge \Diamond\pi_2 \wedge \cdots \wedge \Diamond\pi_m)$ requires that the trajectory does whatever the coverage does and, in addition, will force the system to repeat the desired objective infinitely often.

More complicated specifications can be composed from the basic specifications using the logic operators. In order to better explain the different steps in our framework, we consider throughout this paper the following example.

Example 1 (Robot Motion Planning). A typical example of a system such as (1) is a robot which evolves in a planar environment. The state variable $x(t)$ models the internal dynamics of the robot whereas only its position $y(t)$ is observed. In this paper, we will consider a second order model of a planar robot:

$$\Sigma : \begin{cases} \dot{x}_1(t) = x_2(t), \ x_1(t) \in \mathbb{R}^2, \ x_1(0) \in X_{1,0} \\ \dot{x}_2(t) = u(t), \quad x_2(t) \in \mathbb{R}^2, \ x_2(0) = 0, \ \|u(t)\| \leq \mu \\ y(t) \ = x_1(t), \ y(t) \in \mathbb{R}^2 \end{cases}$$

where $\|\cdot\|$ is the Euclidean norm. The robot is moving in a convex polygonal environment π_0 with four areas of interest denoted by $\pi_1, \pi_2, \pi_3, \pi_4$ (see Fig. 1). Initially, the robot is placed somewhere in the region labeled by π_1 (i.e. $X_{1,0} = [\![\pi_1]\!]$) and its velocity is equal to zero. The robot must accomplish the following task : "Stay always in π_0 and visit area π_2, then area π_3, then area π_4 and, finally, return to and stay in region π_1 while avoiding areas π_2 and π_3," which is captured by the RTL formula:

$$\phi = \Box\pi_0 \wedge \Diamond(\pi_2 \wedge \Diamond(\pi_3 \wedge \Diamond(\pi_4 \wedge (\neg\pi_2 \wedge \neg\pi_3)\mathcal{U}\Box\pi_1))).$$

In this paper, for such spatio-temporal specifications, we provide a computational solution to the following problem.

Problem 1 (RTL Controller Synthesis). Given a system Σ, and an RTL formula ϕ, construct a hybrid controller H for Σ such that the observed trajectories of the closed-loop system satisfy the formula ϕ.

We propose a hierarchical synthesis approach which consists of three ingredients : tracking control using approximate simulation relations [5], robust satisfaction of RTL formulas and hybrid control for motion planning [7,9]. Firstly, Σ is abstracted to a first order fully actuated system:

$$\Sigma' : \dot{z}(t) = v(t),\ z(t) \in \mathbb{R}^k,\ z(0) \in Z_0 \subseteq \mathbb{R}^k,\ v(t) \in V \subseteq \mathbb{R}^k \tag{2}$$

where $Z_0 = g(X_0)$. Using the notion of approximate simulation relation, we evaluate the precision δ with which the system Σ is able to track the trajectories of the abstraction Σ' and design a continuous tracking controller that we call interface. Secondly, from the RTL formula ϕ and the precision δ, we derive a more robust formula ϕ' such that if a trajectory $z(t)$ satisfies ϕ', then any trajectory $y(t)$ remaining within distance δ from $z(t)$ satisfies the formula ϕ. Thirdly, we design a hybrid controller H' for the abstraction Σ', so that the trajectories of the closed loop system satisfy the formula ϕ'. Finally, by putting these three ingredients together, as shown in Fig. 2, we design a hybrid controller H solving Problem 1. In the following sections, we detail each step of our approach.

3 Tracking Control Using Approximate Simulation

In this section, we present a framework for tracking control with guaranteed error bounds. It allows to design an interface between the plant Σ and its abstraction Σ' so that Σ is able to track the trajectories of Σ' with a given precision. It is based on the notion of approximate simulation relation [3]. Whereas exact simulation relations requires the observations of two systems to be identical, approximate simulation relations allow them to be different provided their distance remains bounded by some parameter.

Definition 1 (Simulation Relation). *A relation $\mathcal{W} \subseteq \mathbb{R}^k \times \mathbb{R}^n$ is an approximate simulation relation of precision δ of Σ' by Σ if for all $(z_0, x_0) \in \mathcal{W}$,*

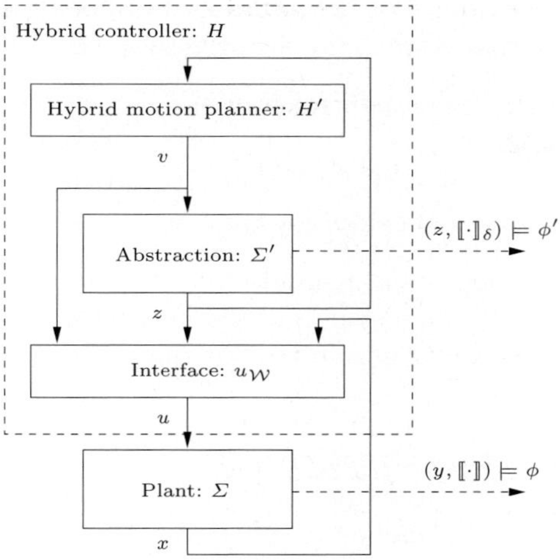

Fig. 2. Hierarchical architecture of the hybrid controller H

1. $\|z_0 - g(x_0)\| \leq \delta$
2. *For all state trajectories $z(t)$ of Σ' such that $z(0) = z_0$ there exists a state trajectory $x(t)$ of Σ such that $x(0) = x_0$ and satisfying*

$$\forall t \geq 0,\ (z(t), x(t)) \in \mathcal{W}.$$

Let us remark that for $\delta = 0$, we recover the notion of exact simulation relation as defined in [11,12]. An interface associated to the approximate simulation relation $\mathcal{W}$ allows to choose the input of Σ so that the states of Σ' and Σ remain in $\mathcal{W}$.

Definition 2 (Interface). *A continuous function $u_{\mathcal{W}} : V \times \mathcal{W} \to U$ is an interface associated to the approximate simulation relation $\mathcal{W}$, if for all $(z_0, x_0) \in \mathcal{W}$, for all trajectories $z(t)$ of Σ' associated to input $v(t)$ and such that $z(0) = z_0$, the trajectory of Σ given by*

$$\dot{x}(t) = f(x(t), u_{\mathcal{W}}(v(t), z(t), x(t))),\ x(0) = x_0$$

satisfies for all $t \geq 0$, $(z(t), x(t)) \in \mathcal{W}$.

Thus, by interconnecting Σ and Σ' through the interface $u_{\mathcal{W}}$ as shown on Fig. 2, Σ tracks the trajectories of the abstraction Σ' with precision δ.

Proposition 1 (Proof in [13]). *Let $x_0 \in X_0$, $z_0 = g(x_0) \in Z_0$ such that $(z_0, x_0) \in \mathcal{W}$, then for all trajectories $z(t)$ of Σ' associated to input $v(t)$ and initial state z_0, the observed trajectory $y(t)$ of Σ given by*

$$\begin{cases} \dot{x}(t) = f(x(t), u_{\mathcal{W}}(v(t), z(t), x(t))),\ x(0) = x_0 \\ y(t) = g(x(t)) \end{cases}$$

satisfies for all $t \geq 0$, $\|y(t) - z(t)\| \leq \delta$.

Let us remark that the choice of the initial state z_0 of the abstraction Σ' is not independent of the initial state x_0 of the system Σ ($z_0 = g(x_0)$).

Remark 1. Usual hierarchical control approaches assume that the plant Σ is simulated by its abstraction Σ'. In this paper, the contrary is assumed. The abstraction Σ' is (approximately) simulated by the plant Σ: the approximate simulation relation is used as a tool for tracking controller design.

The construction of approximate simulation relations can be done effectively using a simulation function [3], that is a positive function bounding the distance between the observations and non-increasing under the parallel evolution of the systems.

Definition 3 (Simulation Function). *Let $\mathcal{V} : \mathbb{R}^k \times \mathbb{R}^n \to \mathbb{R}_+$ be a continuous and piecewise differentiable function. Let $u_{\mathcal{V}} : V \times \mathbb{R}^k \times \mathbb{R}^n \to \mathbb{R}^p$ be a continuous function. $\mathcal{V}$ is a simulation function of Σ' by Σ, and $u_{\mathcal{V}}$ is an associated interface if for all $(z, x) \in \mathbb{R}^k \times \mathbb{R}^n$,*

$$\mathcal{V}(z, x) \geq \|z - g(x)\|^2, \tag{3}$$

$$\sup_{v \in V} \left(\frac{\partial \mathcal{V}(z, x)}{\partial z} v + \frac{\partial \mathcal{V}(z, x)}{\partial x} f(x, u_{\mathcal{V}}(v, z, x)) \right) \leq 0 \tag{4}$$

Then, approximate simulation relations can be defined as level sets of the simulation function.

Theorem 1 (Proof in [13]). *Let the relation $\mathcal{W} \subseteq \mathbb{R}^k \times \mathbb{R}^n$ be given by*

$$\mathcal{W} = \left\{(z, x) \mid \mathcal{V}(z, x) \leq \delta^2\right\}.$$

If for all $v \in V$, for all $(z, x) \in \mathcal{W}$, $u_{\mathcal{V}}(v, z, x) \in U$, then $\mathcal{W}$ is an approximate simulation relation of precision δ of Σ' by Σ and $u_{\mathcal{W}} : V \times \mathcal{W} \to U$ given by $u_{\mathcal{W}}(v, z, x) = u_{\mathcal{V}}(v, z, x)$ is an associated interface.

Example 2. Let us go back to our example. The system Σ modelling the dynamics of the robot is abstracted to a system Σ' such as (2) where the set of inputs is $V = \{v \in \mathbb{R}^2 | \ \|v\| \leq \nu\}$ and the set of initial states is $Z_0 = X_{1,0}$. Let $\alpha > 0$, we define the following functions

$$\mathcal{V}(z, x) = \max\left(Q(z, x), 4\nu^2\right) \text{ where } Q(z, x) = \|x_1 - z\|^2 + \alpha\|x_1 - z + 2x_2\|^2,$$
$$u_{\mathcal{V}}(v, z, x) = \tfrac{v}{2} + \tfrac{-1-\alpha}{4\alpha}(x_1 - z) - x_2.$$

First, let us remark that equation (3) clearly holds. If $Q(z, x) \leq 4\nu^2$, then it is clear that equation (4) holds. If $Q(z, x) \geq 4\nu^2$, then

$$\frac{\partial \mathcal{V}}{\partial z} v + \frac{\partial \mathcal{V}}{\partial x_1} x_2 + \frac{\partial \mathcal{V}}{\partial x_2} u_{\mathcal{V}} = 2(x_1 - z) \cdot (x_2 - v) + 2\alpha(x_1 - z + 2x_2) \cdot (x_2 - v + 2u_{\mathcal{V}}).$$

After the substitution of the expression of $u_{\mathcal{V}}$ and simplification, we arrive to

$$\frac{\partial \mathcal{V}}{\partial z}v+\frac{\partial \mathcal{V}}{\partial x_1}x_2+\frac{\partial \mathcal{V}}{\partial x_2}u_{\mathcal{V}}=-Q(z,x)-2(x_1-z)\cdot v\leq -Q(z,x)+2\nu\|x_1-z\|$$

because $\|v\|\leq \nu$. Since $\|x_1-z\|^2\leq Q(z,x)$, we have

$$\frac{\partial \mathcal{V}}{\partial z}v+\frac{\partial \mathcal{V}}{\partial x_1}x_2+\frac{\partial \mathcal{V}}{\partial x_2}u_{\mathcal{V}}\leq -Q(z,x)+2\nu\sqrt{Q(z,x)}\leq \sqrt{Q(z,x)}(2\nu-\sqrt{Q(z,x)}).$$

Since $Q(z,x)\geq 4\nu^2$, equation (4) holds and $\mathcal{V}$ is a simulation function of Σ' by Σ, and $u_{\mathcal{V}}$ is an associated interface.

Let us define $\mathcal{W}=\{(z,x)|\ \mathcal{V}(z,x)\leq 4\nu^2\}$. Let $v\in V$, $(z,x)\in\mathcal{W}$, let us remark that

$$\begin{aligned}\|u_{\mathcal{V}}(v,z,x)\| &= \left\|\frac{v}{2}+\frac{-1+\alpha}{4\alpha}(x_1-z)-\frac{1}{2}(x_1-z+2x_2)\right\| \\ &\leq \frac{\nu}{2}+\frac{|-1+\alpha|}{4\alpha}\sqrt{\mathcal{V}(z,x)}+\frac{1}{2}\sqrt{\frac{\mathcal{V}(z,x)}{\alpha}} \\ &\leq \frac{\nu}{2}\left(1+|1-1/\alpha|+2/\sqrt{\alpha}\right).\end{aligned}$$

We assume that the velocity bound ν of the abstraction Σ' has been chosen small enough so that $\frac{\nu}{2}\left(1+|1-1/\alpha|+2/\sqrt{\alpha}\right)\leq\mu$. Then, Theorem 1 applies and $\mathcal{W}$ is an approximate simulation relation of precision 2ν of Σ' by Σ and an associated interface is given by

$$u_{\mathcal{W}}(v,z,x)=\frac{v}{2}+\frac{-1-\alpha}{4\alpha}(x_1-z)-x_2.$$

Let the initial state of the abstraction Σ' be chosen so that $z(0)=x_1(0)$, then from Proposition 1, by interconnecting Σ' and Σ through the interface $u_{\mathcal{W}}$, the observed trajectories of system Σ tracks the trajectories of Σ' with precision 2ν.

4 RTL as a Controller Specification Language

Temporal logics are useful for reasoning about the occurrence of events with respect to some time model. In this paper, we advocate the applicability of the propositional temporal logic over the reals (RTL) [6,10] as a natural formalism for a controller specification language. RTL has the same temporal connectives as the Linear Temporal Logic (LTL) [14], but now the underlying time line is the positive real line instead of the natural numbers. In this section, after a brief presentation of RTL, we show how we can derive from an RTL specification ϕ a more robust specification ϕ' such that given a trajectory satisfying ϕ', any other trajectory remaining within distance δ satisfies ϕ. The goal of this "robustification" is then to design a controller for the abstraction Σ' so that it satisfies ϕ'. Then, refining this control law to Σ using the interface presented in the previous section, we can guarantee that the plant satisfies the original specification ϕ.

We first introduce the syntax of RTL formulas in Negation Normal Form (NNF). In NNF, we push the negations inside the subformulas such that the only allowed negation operators appear in front of atoms.

Definition 4 (RTL Syntax in NNF). *For $\pi \in \Pi$, the set Φ_Π of all well formed RTL formulas over Π in NNF is constructed using the grammar*

$$\phi ::= \pi \mid \neg\pi \mid \phi \vee \phi \mid \phi \wedge \phi \mid \phi \mathcal{U} \phi \mid \phi \mathcal{R} \phi$$

As usual, the boolean constants $\top$ (true) and $\bot$ (false) are defined as $\top = \pi \vee \neg\pi$ and $\top = \pi \wedge \neg\pi$ respectively.

Formally, the semantics of RTL formulas is defined over continuous time boolean signals. Here, we instantiate the definitions of the semantics over abstractions of the output trajectories of the system Σ with respect to Π. Let $(y, [\![\cdot]\!]) \models \phi$ to denote the satisfaction of the RTL formula ϕ over the output trajectory $y(t)$ starting at time $t = 0$ with respect to the atom mapping $[\![\cdot]\!]$. If all the output trajectories $y(t)$ of system Σ driven by a controller H and associated to an initial state in X_0 are such that $(y, [\![\cdot]\!]) \models \phi$, then we write $([\Sigma, H], [\![\cdot]\!]) \models \phi$ and we say that $[\Sigma, H]$ satisfies ϕ. In the following, given any function f from $\mathbb{R}_+$ to some normed space A, we define $f|_t$ for $t \in \mathbb{R}_+$ to be the t time shift of f with definition $f|_t(s) = f(t + s)$ for $s \in \mathbb{R}_+$.

Definition 5 (RTL Semantics). *Let $y(t)$ be a function from $\mathbb{R}_+$ to $\mathbb{R}^k$ and Π be the set of atoms. For $t, s \in \mathbb{R}_+$, the semantics of any formula $\phi \in \Phi_\Pi$ can be recursively defined as*

- $(y, [\![\cdot]\!]) \models \pi$ *iff* $y(0) \in [\![\pi]\!]$.
- $(y, [\![\cdot]\!]) \models \neg\pi$ *iff* $y(0) \notin [\![\pi]\!]$.
- $(y, [\![\cdot]\!]) \models \phi_1 \vee \phi_2$ *if* $(y, [\![\cdot]\!]) \models \phi_1$ *or* $(y, [\![\cdot]\!]) \models \phi_2$.
- $(y, [\![\cdot]\!]) \models \phi_1 \wedge \phi_2$ *if* $(y, [\![\cdot]\!]) \models \phi_1$ *and* $(y, [\![\cdot]\!]) \models \phi_2$.
- $(y, [\![\cdot]\!]) \models \phi_1 \mathcal{U} \phi_2$ *if there exists* $t \geq 0$ *such that* $(y|_t, [\![\cdot]\!]) \models \phi_2$ *and for all* s *with* $0 \leq s \leq t$ *we have* $(y|_s, [\![\cdot]\!]) \models \phi_1$.
- $(y, [\![\cdot]\!]) \models \phi_1 \mathcal{R} \phi_2$ *if for all* $t \geq 0$ *it is* $(y|_t, [\![\cdot]\!]) \models \phi_2$ *or there exists some* s *such that* $0 \leq s \leq t$ *and* $(y|_s, [\![\cdot]\!]) \models \phi_1$.

Note that due to the definition of the negation operator, the duality property of the logic holds and, thus, by using RTL in NNF we do not loose in expressive power. The path formula $\phi_1 \mathcal{U} \phi_2$ intuitively expresses the property that over the trajectory $y(t)$, ϕ_1 is true until ϕ_2 becomes true. Note that here the semantics of until require that ϕ_1 holds when ϕ_2 becomes true. Thus, this is a less expressive version of until than the strict until [10]. Intuitively, the release operator $\phi_1 \mathcal{R} \phi_2$ states that ϕ_2 should always hold, a requirement which is released when ϕ_1 becomes true. Furthermore, we can also derive additional temporal operators such as *eventually* $\Diamond\phi = \top \mathcal{U} \phi$ and *always* $\Box\phi = \bot \mathcal{R} \phi$. The formula $\Diamond\phi$ indicates that over the trajectory $y(t)$ the subformula ϕ becomes eventually true, whereas $\Box\phi$ indicates that ϕ is always true over $y(t)$.

Initially, an RTL formula ϕ is provided as a controller specification for the concrete system Σ. Since we design hybrid control laws for the subsystem Σ', we need to translate the initial RTL specification for Σ to a modified RTL specification ϕ' for Σ'. For this purpose, we introduce a new set of atoms and a new atom's mapping. First, similar to [15], we introduce the notion of δ-contraction for sets in order to define our notion of robustness.

Definition 6 (Contraction, Expansion). *Given a radius $\delta \in \mathbb{R}_+ \cup \{+\infty\}$ and a point α in a normed space A, the δ-ball centered at α is defined as $B_\delta(\alpha) = \{\beta \in A \mid \|\alpha - \beta\| \leq \delta\}$. If $\Gamma \subseteq A$, then $C_\delta(\Gamma) = \{\alpha \in A \mid B_\delta(\alpha) \subseteq \Gamma\}$ is the δ-contraction*[1] *and $B_\delta(\Gamma) = \{a \in A \mid B_\delta(a) \cap \Gamma \neq \emptyset\}$ is the δ-expansion.*

Consider now a new set of atomic propositions Ξ_Π such that $\Xi_\Pi = \{\xi_\alpha \mid \alpha = \pi \text{ or } \alpha = \neg\pi \text{ for } \pi \in \Pi\}$. For a given $\delta \in \mathbb{R}_+$, we define a new map $[\![\cdot]\!]_\delta : \Xi_\Pi \to \mathcal{P}(\mathbb{R}^k)$ based on the map $[\![\cdot]\!]$ as follows:

$$\forall \xi \in \Xi_\Pi, \quad [\![\xi]\!]_\delta =: \begin{cases} C_\delta(\overline{[\![\pi]\!]}) & \text{if } \xi = \xi_{\neg\pi} \\ C_\delta([\![\pi]\!]) & \text{if } \xi = \xi_\pi \end{cases}$$

Here, $\overline{\Gamma}$ denotes the complement of a set Γ. For clarity of the presentation, we define a translation algorithm $\mathbf{rob} : \Phi_\Pi \to \Phi_{\Xi_\Pi}$ which takes as input an RTL formula ϕ in NNF and it returns a formula $\mathbf{rob}(\phi)$ where the occurrences of atomic propositions of π and $\neg\pi$ have been replaced by the members ξ_π and $\xi_{\neg\pi}$ of Ξ_Π respectively.

The following theorem is the connecting link between the specifications satisfied by the abstraction Σ' and its concrete system Σ. Informally, it states that given $\delta \geq 0$ if we δ-expand the sets that must be avoided and δ-contract the sets that must be reached, then we will obtain a δ-robust specification. The latter implies that if a trajectory satisfies the δ-robust specification, then any other trajectory that remains δ-close to the initial one will also satisfy the same non-robust initial specification.

Theorem 2 (Proof in [13]). *Consider a formula $\phi \in \Phi_\Pi$ which is built on a set of atoms Π, a map $[\![\cdot]\!] : \Pi \to \mathcal{P}(\mathbb{R}^k)$, and a number $\delta \in \mathbb{R}_+$, then for all functions $y(t)$ and $z(t)$ from $\mathbb{R}_+$ to $\mathbb{R}^k$ such that for all $t \geq 0$, $\|z(t) - y(t)\| \leq \delta$, the following holds $(z, [\![\cdot]\!]_\delta) \models \mathbf{rob}(\phi) \implies (y, [\![\cdot]\!]) \models \phi$.*

Two remarks are in order here.

Remark 2. When $(z, [\![\cdot]\!]_\delta) \not\models \mathbf{rob}(\phi)$ we cannot conclude that $(y, [\![\cdot]\!]) \not\models \phi$. The only conclusion we can make in this case is that $z(t)$ is not a δ-robust trajectory in the sense of [16]. Potentially, we can gain more information if we define 3-valued semantics [17] for the satisfaction relation of $(z, [\![\cdot]\!]_\delta) \models \mathbf{rob}(\phi)$, but for the scope of this paper Theorem 2 is sufficient.

Remark 3. Theorem 2 does not particularly refer to the output trajectories of systems Σ and Σ'. If we consider both functions z and y to be trajectories of Σ', then Theorem 2 classifies which trajectories of Σ' are δ-robust.

[1] In cases when the δ-contraction of a set must be a polyhedral set, we under-approximate the δ-contraction by a δ-offset (see Sect. 6).

5 From RTL to Hybrid Controllers

It then remains to design a hybrid controller H' for the simpler system Σ' such that its trajectories satisfy an RTL specification ϕ', i.e. $([\Sigma', H'], [\![\cdot]\!]_\delta) \models \phi'$. In previous work, we have proposed two different solutions to this problem [7,9].

The first methodology [7] comprises the following steps. We first create an observation preserving partition of the state space of the system. By observation preserving we mean that all the states that belong to a set in the partition satisfy the same set of labels $\xi \in \Xi_\Pi$. The sets in the partition do not have to be convex and a particular set of labels can be mapped to more than one set in the partition. Then, we can abstract the continuous state space to a Finite State Machine (FSM) where each state corresponds to one set in the partition. Under the assumption of finite variability [18], i.e. within a finite interval of time there can exist only a finite number of changes in the atomic propositions, we can solve the RTL planning problem using Linear Temporal Logic (LTL) planning techniques [19]. The later consists of converting the LTL specification into a Büchi automaton, taking its product with the FSM and, then, finding a path on the product automaton that would satisfy the LTL formula [7]. The other important ingredient of the approach is the existence of simple feedback controllers [20,21] which are defined over polygons. These controllers satisfy the property that if a trajectory in their domain reaches a facet of the polygon, then every other trajectory in the domain of operation does the same. Using the discrete path on the FSM, we can guide the composition of the local feedback controllers and derive a hybrid automaton that generates trajectories that satisfy the initial RTL specification by construction. Note though that in our final construction, we have to make sure that no Zeno behavior occurs in order to satisfy the finite variability assumption.

On the other hand, in [9], we present some preliminary results on the design of hybrid controllers from RTL specifications by directly operating on RTL formulas. The method consists of 3 main steps. First, the RTL formula is converted to an abstract hybrid automaton where the dynamics in each discrete location remain undefined. Then, from each discrete location we extract controller constraints in the form of triplets $c = \{Init, Inv, Goal\}$, where $Init$ is a set of initial conditions, Inv is an invariant set and $Goal$ is a goal set. In the final step, we design controllers that satisfy the constraints c and derive a hybrid automaton whose trajectories satisfy the RTL specification. The main advantages of the new approach involve the possibility of using different control design methodologies for each discrete location of the hybrid automaton and the lack of mandatory partitioning of the environment. Currently, the theory solves the problem for a fragment of RTL, but we conjecture that the recent results by Maler et al. [22] can provide the basis for a solution to the complete RTL.

Both approaches [7,9] constitute valid solutions to the controller design problem that we consider in this paper. Nonetheless in the next section, we present some implementation results for our simple example using the framework presented in [9]. The choice of [9] over the approach in [7] was made for the following reasons. First, because the only modification required to the computational framework of [9] is just an implementation of the C_δ operator. Second, because approaches like [7,8] require the finest partition with respect to the set of atoms Ξ_Π. This implies that for each π in Π, we must create in advance two sets : a δ-contraction $C_\delta([\![\pi]\!])$ and a 2δ-annulus $B_\delta([\![\pi]\!]) \setminus C_\delta([\![\pi]\!])$, and then, take all the appropriate intersections with the contraction and annulus sets of the rest of the atoms. Finally, with the methodology proposed in [9] we can potentially design just one local feedback controller for each part of the specification. On the contrary, the methods in [7,8] would require a larger number of controllers even for the simple cases.

Given the hybrid controller H', the following theorem, which is immediate from Proposition 1 and Theorem 2, states the main result of the paper.

Theorem 3. *Let $\mathcal{W}$ be an approximate simulation relation of precision δ between Σ' and Σ and $u_\mathcal{W}$ be the associated interface. Consider a formula $\phi \in \Phi_\Pi$ and define $\phi' = \mathbf{rob}(\phi)$. Let H' be a controller for Σ' and H the associated controller for Σ obtained by interconnection of the elements as shown on Fig. 2. Then, $([\Sigma', H'], [\![\cdot]\!]_\delta) \models \phi'$ implies $([\Sigma, H], [\![\cdot]\!]) \models \phi$.*

6 Implementation and Simulations

In this section, we demonstrate the applicability of our framework by presenting some numerical results. We have implemented the algorithms presented in [9] in MATLAB using the polytope library of the Multi-Parametric Toolbox (MPT) [23]. In the following, we just provide an informal high-level description of our toolbox. The user inputs to the toolbox are : an RTL formula ϕ and the associated map $[\![\cdot]\!]$, the initial conditions x_0 and the maximum acceleration μ of the system Σ as well as the parameter α as described in Example 2 and an integration step ds. Currently, the set valued function $[\![\cdot]\!]$ is restricted to map only to convex or concave polyhedral sets. This is not a fundamental restriction and it was made only for simplifying the implementation of the operator C_δ.

Using the algorithms in [9], we first derive the controller specifications for the design of a hybrid controller H' such that $([\Sigma', H'], [\![\cdot]\!]_\delta) \models \mathbf{rob}(\phi)$. Note that the δ-contraction of a polyhedral set is not always a polyhedral set. In order to maintain a polyhedral description for all the sets, we under-approximate the δ-contraction by the inward δ-offset. Informally, the δ-offset of a polyhedral set is the inward δ-displacement of its facets along the corresponding normal directions. Since the δ-offset is an under-approximation of the δ-contraction, Theorem 2 still holds.

For the generation of the local feedback controllers such that they satisfy a controller specification triplet c (see Sect. 5), we use again a hierarchical approach. If the set Inv is convex, then we can use just one potential field controller [21] that converges inside $Goal$. If on the other hand the set invariant Inv is not convex, then we perform a convex decomposition on the set Inv and we apply the path planning methodology of Conner et al. [21]. By hierarchically composing the resulting controllers, in the sense that a controller that satisfies a triplet c can itself be a hybrid automaton, we obtain the hybrid controller H' for Σ'.

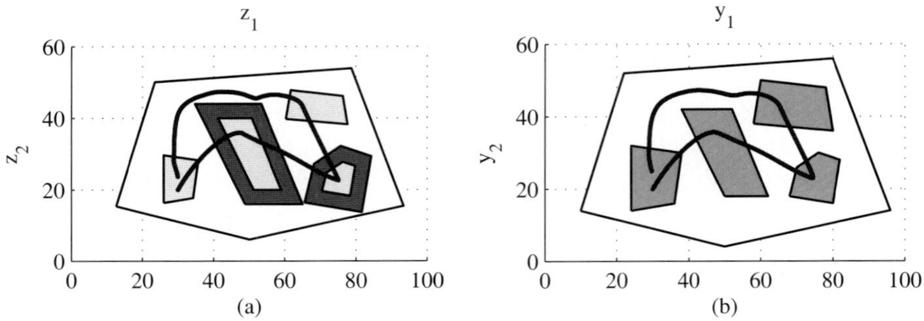

Fig. 3. Simulation results for $\nu = 1$. (a) A trajectory of the kinematic model - The *light gray* polygons denote the 2-contraction of the regions while the *dark gray* polygons the 2-expansion. (b) The corresponding trajectory of the dynamic model.

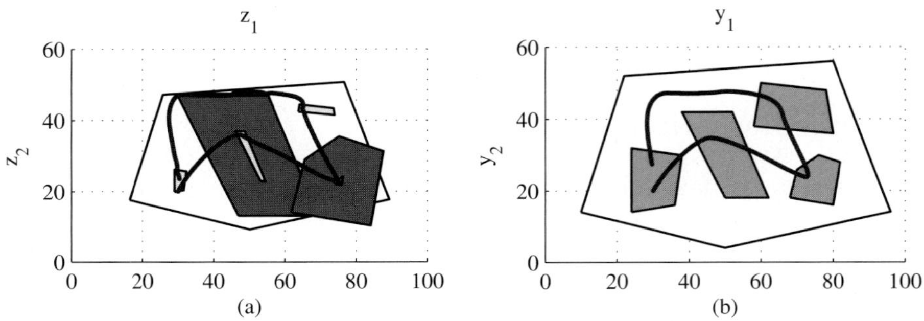

Fig. 4. Simulation results for $\nu = 2.5$. (a) Kinematic model. (b) Dynamic model.

Finally, the hybrid controller H for Σ is obtained from H' by composing with the interface $u_{\mathcal{W}}$ (see Fig. 2).

In the following, we present some numerical results for the robot motion planning example that we consider in this paper. In all the simulations, the parameter α is set to 100 and the initial conditions to $x_0 = [30\ 20\ 0\ 0]^T$. The first simulation (Fig. 3) shows the resulting trajectories for Σ' and Σ for $\nu = 1$ (i.e. the velocity bound for the kinematic model Σ'). It is easy to see that the resulting trajectory of Σ (Fig. 3.b) satisfies the RTL formula ϕ. The total running time for this example on a Pentium 4 at 2.4GHz with 768MB of RAM is 9 sec (including the simulation and the plotting of the graphs in Fig. 3).

Next, in Fig. 4 we present simulation results for $\nu = 2.5$. Note that there barely exists a 2ν-robust trajectory for the kinematic model with respect to the RTL specification. Nevertheless, the trajectory of the dynamic model does satisfy the formula. Also, in Fig. 5.a you can observe that the distance between the two trajectories $z(t)$ and $y(t)$ is always bounded by the precision $2\nu = 5$ of the approximate simulation relation and that this bound is tight. Finally, when we increase ν to 3, then there does not exist a 6-robust trajectory

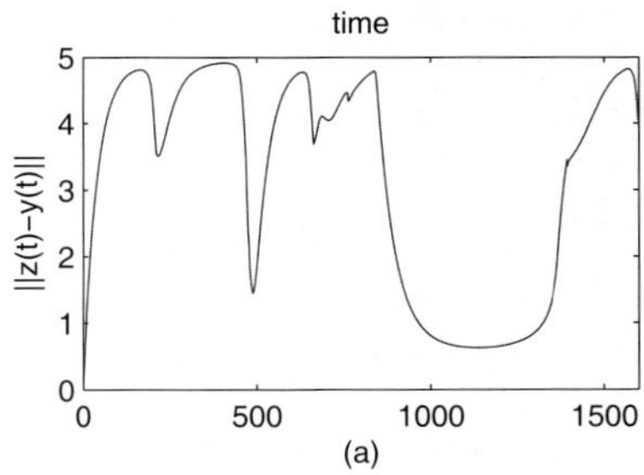

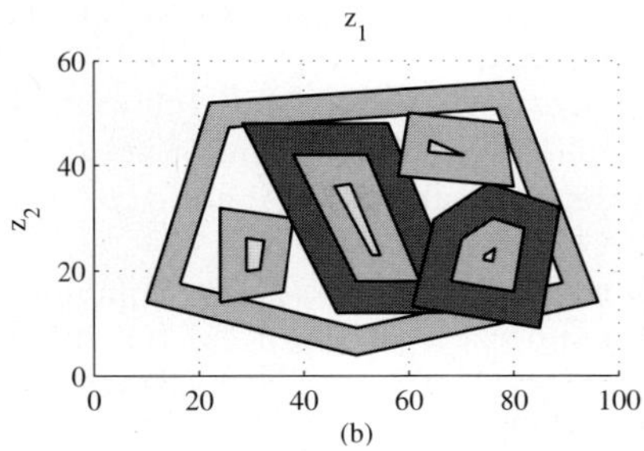

Fig. 5. (a) Distance between trajectories $z(t)$ and $y(t)$ when $\nu = 2.5$. (b) The modified environment when $\nu = 3$. *Medium gray* : original regions, *dark gray* : 6-expanded regions, *light gray* : 6-contracted regions, *white* : 6-contracted workspace π_0.

for the system Σ' (Fig. 5.b). Notice that the modified environment is partitioned into two disconnected components (the *white* workspace is separated by the *dark gray* regions).

7 Conclusions

We have presented a new approach to the hybrid controller synthesis problem from temporal logic specifications. Our proposed hierarchical framework is based on the notion of approximate simulation relations and a new definition of robustness for temporal logic formulas. The hierarchical synthesis approach comprises three basic steps : (i) tracking control using approximate simulation relations [5], (ii) robust satisfaction of RTL formulas and (iii) hybrid control for motion planning [7,9]. We strongly believe that a hierarchical approach can provide a viable solution to a large class of control problems. Future work will concentrate on developing interfaces for other types of systems (i.e. for the unicycle) and applications of the framework to control problems.

Acknowledgments. The authors would like to thank David Conner for providing them with his implementation of the potential field controllers and the reviewers for many useful remarks. This research is partially supported by NSF EHS 0311123, NSF ITR 0121431, and the Army Research Office MURI Grant DAAD 19-02-01-0383.

References

1. Pappas, G.J., Lafferriere, G., Sastry, S.: Hierarchically consistent control systems. IEEE Trans. on Auto. Cont. **45**(6) (2000) 1144–1160
2. Tabuada, P., Pappas, G.J.: Hierarchical trajectory generation for a class of nonlinear systems. Automatica **41**(4) (2005) 701–708
3. Girard, A., Pappas, G.J.: Approximation metrics for discrete and continuous systems. IEEE Trans. Auto. Cont. (2005) Accepted.

4. Tabuada, P.: An approximate simulation approach to symbolic control. (2006) Submitted.
5. Girard, A., Pappas, G.J.: Hierarchical control using approximate simulation relations. In: 45th IEEE Conference on Decision and Control. (2006)
6. Kamp, H.: Tense Logic and the Theory of Linear Order. PhD thesis, University of California (1968)
7. Fainekos, G.E., Kress-Gazit, H., Pappas, G.J.: Hybrid controllers for path planning: A temporal logic approach. In: 44th IEEE CDC and ECC. (2005)
8. Kloetzer, M., Belta, C.: A fully automated framework for control of linear systems from LTL specifications. In: Hybrid Systems: Computation and Control. Volume 3927 of LNCS. Springer (2006) 333–347
9. Fainekos, G.E., Loizou, S.G., Pappas, G.J.: Translating temporal logic to controller specifications. In: 45th IEEE Conference on Decision and Control. (2006)
10. Reynolds, M.: Continuous temporal models. In: the 14th Australian Joint Conference on Artificial Intelligence. Volume 2256 of LNCS., Springer (2001) 414–425
11. Pappas, G.J.: Bisimilar linear systems. Automatica **39**(12) (2003) 2035–2047
12. van der Schaft, A.: Equivalence of dynamical systems by bisimulation. IEEE Trans. Auto. Cont. **49** (2004) 2160–2172
13. Fainekos, G.E., Girard, A., Pappas, G.J.: Hierarchical synthesis of hybrid controllers from temporal logic specifications. Technical report, Dept. of CIS, Univ. of Pennsylvania (2006)
14. Emerson, E.A.: Temporal and modal logic. In van Leeuwen, J., ed.: Handbook of Theoretical Computer Science: Formal Models and Semantics. Volume B., North-Holland Pub. Co./MIT Press (1990) 995–1072
15. Moor, T., Davoren, J.M.: Robust controller synthesis for hybrid systems using modal logic. In: Proceedings of the 4th HSCC. Volume 2034 of LNCS., Springer (2001) 433–446
16. Fainekos, G.E., Pappas, G.J.: Robustness of temporal logic specifications. In: Proceedings of FATES/RV. Volume 4262 of LNCS., Springer (2006) 178–192
17. Fainekos, G.E.: An introduction to multi-valued model checking. Technical Report MS-CIS-05-16, Dept. of CIS, Univ. of Pennsylvania (2005)
18. Barringer, H., Kuiper, R., Pnueli, A.: A really abstract concurrent model and its temporal logic. In: POPL '86: Proceedings of the 13th ACM SIGACT-SIGPLAN symposium on Principles of programming languages, New York, NY, USA, ACM Press (1986) 173–183
19. Giacomo, G.D., Vardi, M.Y.: Automata-theoretic approach to planning for temporally extended goals. In: Proceedings of the 5th ECP. Volume 1809 of LNCS., Springer (1999) 226–238
20. Belta, C., Isler, V., Pappas, G.J.: Discrete abstractions for robot motion planning and control in polygonal environments. IEEE Trans. on Robotics **21**(5) (2005) 864–874
21. Conner, D.C., Rizzi, A., Choset, H.: Composition of local potential functions for global robot control and navigation. In: Proceedings of 2003 IEEE/RSJ International Conference on Intelligent Robots and Systems. (2003) 3546–3551
22. Maler, O., Nickovic, D., Pnueli, A.: From MITL to timed automata. In: Proceedings of FORMATS. Volume 4202 of LNCS., Springer (2006) 274–289
23. Kvasnica, M., Grieder, P., Baotić, M.: Multi-Parametric Toolbox (MPT) (2004) http://control.ee.ethz.ch/ mpt/.

Coupling from the Past in Hybrid Models for File Sharing Peer to Peer Systems

Bruno Gaujal[1] and Florence Perronnin[2]

[1] INRIA and Lab. ID-IMAG (CNRS, INPG, INRIA, UJF) 51, Av. J. Kunztmann, Montbonnot, France
Bruno.Gaujal@imag.fr

[2] UJF and Lab. ID-IMAG (CNRS, INPG, INRIA, UJF) 51, Av. J. Kunztmann, Montbonnot, France
Florence.Perronnin@imag.fr

Abstract. In this paper we show how file sharing peer to peer systems can be modeled by hybrid systems with a continuous part corresponding to a fluid limit of files and a discrete part corresponding to customers. Then we show that this hybrid system is amenable to perfect simulations (*i.e.* simulations providing samples of the system states which distributions have no bias from the asymptotic distribution of the system). An experimental study is carried to show the respective influence that the different parameters (such as time-to-live, rate of requests, connection time) play on the behavior of large peer to peer systems, and also to show the effectiveness of this approach for numerical solutions of stochastic hybrid systems.

1 Introduction

Hybrid systems are very useful to model discrete systems with several time and space scales. In that case, one typically uses *fluid limits* for the parts of the system with fastest and largest scales. These models have been introduced in various domains under the form of fluid queues [1], continuous Petri nets [2], or timed automata [3]. In this paper, we will consider one such example, namely peer to peer systems, where two types of dynamics are superimposed. The slow dynamics concerns the customers, who join and leave the system. The fast dynamics concerns the files and their transfers between the customers. A natural model for file sharing peer to peer systems mixes a discrete stochastic system to model the behavior of the customers and a deterministic differential equation for the mean behavior of the files, seen as a fluid quantity.

The analysis of such stochastic hybrid systems is often difficult on a mathematical as well as on a numerical point of view and such large hybrid systems are often considered computationally untractable. Simulation approaches are efficient alternatives to estimate their behavior by providing samples distributed according to their asymptotic distribution. However, simulation has several drawbacks. First, simulations do not make any sense unless the system has ergodicity properties, which are sometimes difficult to check. Second, even under

A. Bemporad, A. Bicchi, and G. Buttazzo (Eds.): HSCC 2007, LNCS 4416, pp. 217–230, 2007.

the right ergodicity conditions, classical simulation techniques only provide approximations of the asymptotic behavior. The longer the simulation the more accurate the result, but it is usually hard or impossible to be more precise than this general statement. Recently, Propp and Wilson have used backward coupling techniques ([4]) to design a simulation algorithm to get *perfect samplings* (*i.e.* whose distributions are not approximations but the exact asymptotic distributions) of *discrete time, finite* Markov chains.

In this paper we show how their idea can be adapted to the *infinite* and *continuous* case at hand, by using *regeneration points.* We use this property to design a perfect simulation algorithm of our peer to peer model, by using additional monotonicity properties. Finally, we carry an experimental study of the performances of the model based on this sampling technique. A similar approach has been used in [5] for decoupled hybrid systems (where the discrete part does not depend on the continuous one). In that case, the coupling of the perfect simulation occurs in a very controlled manner because the system is uniformly contracting. Here however, the interplay between the discrete and the continuous parts are more intricate so that the coupling time might have a larger variance.

Actually, the goal of this paper is two-fold. First, we propose a new hybrid model for file sharing P2P systems, combining the effects of popularity and age decays of files on the customers behavior. Second, we show how this kind of stochastic hybrid systems can be solved numerically by using a new and powerful simulation method based on coupling from the past properties. In practice, this approach has proved to be very fast and we have been able to treat very large cases (with state spaces of size larger than 10^7) within one or two minutes over a standard PC.

2 A Hybrid Model of Peer to Peer File Sharing Systems

Downloading popular multimedia content from the Internet can take a long time due to bandwidth bottlenecks and to Web server overload. The central idea of peer to peer (P2P) systems is to leverage the downloaders' own (often unused) resources to provide a globally better service. In the context of file download, the resource is upload bandwidth and the service is a faster diffusion of popular files. For instance, a popular file $\mathcal{F}$ downloaded by a user A may also be of interest to a user B which is "closer" to A than to the origin Web server hosting file $\mathcal{F}$. Then if B downloads file $\mathcal{F}$ from A instead of the origin server, the benefit is threefold : B downloads the file at a high rate on the local network; B doesn't contact the origin server, which reduces the load on this server and finally, bandwidth is saved on the wide-area network since the data is transferred locally.

An important aspect of P2P systems is that clients (downloaders) are also servers (uploaders). For this reason, these systems are said "self-scaling" since the resources increase with the demand.

Among P2P systems, one of the most popular applications is file sharing. The most famous systems such as KaZaA or Gnutella [6,7] belong to this category. They can mainly be described by the copies of files that all the users make available for download (typically, copies of files previously downloaded). These systems may be intricate and are extremely difficult to model in full details. Here are the main simplifications used in this paper. First, the peers are assumed to be statistically homogeneous and we only consider two variables, $N(t)$, the number of customers (or nodes) connected to the system at time t and $x(t)$, the number of copies of all the files which are available globally in the system. We will see below that $x(t)$ can also be viewed as the popularity index of the system. The second assumption is that downloads always succeed and download times are neglected since files can be split into fragments of "unit" size for download.

2.1 Modelling with a Hybrid System

The variable $N(t)$ being discrete, it only changes values at discrete times. Thus $N(t)$ is driven by *jump instants*, which forms a point process $T_0 = 0$ (time origin), $T_1, \dots T_n$. Let $(\tau_n)_{n\in\mathbb{N}}$ be the sequence of inter jump times: $\tau_n \stackrel{\text{def}}{=} T_n - T_{n-1}$. At each time T_n, each customer has the opportunity to see the global state of the system $(N(T)n), x(T_n))$. Based on this information, each customer decides either to join, or to leave or even to remain as is. Therefore, the total number of connected customers $N(T_n)$ is a continuous time Markov chain, which infinitesimal generator is given in Figure 1. This infinitesimal generator describes

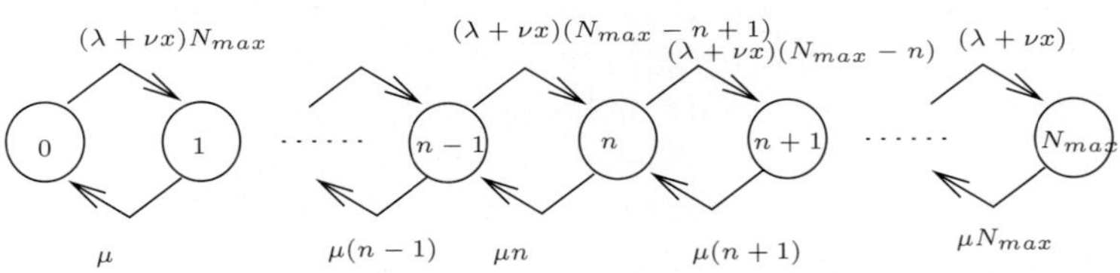

Fig. 1. The infinitesimal generator for the Markov process N

the fact that each customer leaves the system with a constant rate μ. Also, each customer joins the system with a rate proportional to the popularity of the system (which is proportional to x), plus a "blind" rate λ, independent of x. The behavior of $N(T_n)$ can be written under a constructive form as

$$N(T_n) = \varphi(N(T_{n-1}), x(T_{n-1}), \xi_n, \tau_n), \tag{1}$$

where $\{\xi_n\}_{n\in\mathbb{N}}$ is a random process of *innovations*, uniformly distributed over $[0, 1]$, and φ describes the dynamics of N at jump instants:

1. If $\xi > 1 - \lambda' + \nu' x(T_{n-1})(N_{max} - N(T_{n-1}))$, then the next event is a *customer arrival:* $N(T_n) = N(T_{n-1}) + 1$.
2. If $\xi < \mu' N(T_{n-1})$, then the next event is a *customer departure:* $N(T_n) = N(T_{n-1}) - 1$.
3. Otherwise, this is a *null event, i.e.* no customer arrives nor leaves and the system is left unchanged at time T_n: $N(T_n) = N(T_{n-1})$.

As for the continuous part, $x(t)$ is governed by a discrete process at jump instants, T_n. Upon a customer departure, it takes away all the documents it was responsible for. Since our model is symmetric over all customers and all files, the number of lost documents is uniform over all customers so that it corresponds to a proportional fraction of the total fluid. Therefore if the event at T_n is a departure then $x(T_n) = x(T_n^-)\frac{N(T_n^-)-1}{N(T_n^-)}$. This fraction of files lost may also be a random variable with mean $\frac{N(T_n^-)-1}{N(T_n^-)}$ to account for user heterogeneity as shown in Section 3.8. When a node joins the system, it does not bring exogenous files with him upon its arrival. This assumption can easily be relaxed as shown in Section 3.8. The increase of the number of files will come from the future downloads of the newcomer: $x(T_n) = x(T_n^-)$. Upon a null event, x is also left unchanged, $x(T_n) = x(T_n^-)$. We denote this evolution by

$$x(T_n) = h(N(T_n^-), x(T_n^-), \xi_n), \tag{2}$$

As for the behavior of $x(t)$ between jump times, it is given by a deterministic differential equation, In $[T_n, T_{n+1})$,

$$\frac{dx}{dt} = f(N(T_n), x, t) = \sigma N(T_n)e^{-\alpha x} + \sigma N(T_n)x\frac{N(T_n)C - x(t)}{N(T_n)C} - \theta x. \tag{3}$$

Here is a brief account on the dynamics of x between jumps. The first term $\sigma N e^{-\alpha x(t)}$ corresponds to the rate at which new files are introduced in the system. First, σ is the rate at which each customer requests files. As for $e^{-\alpha x(t)}$, it corresponds to the probability that the requested file is not yet present in the system (called the miss probability in the following) so that a download from outside is needed. The form of this term has been derived experimentally as follows.

The number of requests for each file typically follows a Zipf law with parameter β. Therefore, the number of copies per file also follows a Zipf law, since each request results in the creation of a new copy. In our model, we only know x, the total number of *copies* while the miss probability depends on the average number of distinct *files*, $d(x)$. Actually, the miss probability is uniform over all missing files (new files do not have any popularity yet) and is equal to $1-d(x)/C$. Computing $d(x)$ exactly is intricate and we could not find a close-form formula for it. However, we found empirically that $d(x)$ is very close to $C(1-exp(-\alpha x))$ as seen in Figure 2. Parameter α was found to be equal to $e^{-\beta}/C$. Finally, we can assume that new files are introduced in the system with rate $\sigma N(t)e^{-\alpha x(t)}$.

The second term $\sigma N(T_n)x(t)\frac{N(T_n)C-x(t)}{N(T_n)C}$ corresponds to the increase of the number of copies linearly in the popularity ($x(t)$), provided that the copy is not yet present locally (the probability of local absence being $\frac{N(T_n)C-x(t)}{N(T_n)C}$).

The last term $-\theta x(t)$ corresponds to the decrease of the number of copies due to obsolescence (each copy is removed from the system by the system administrator at rate θ).

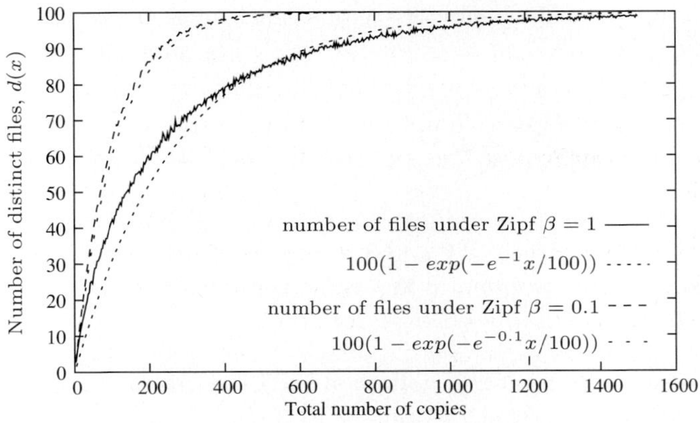

Fig. 2. Matching of the exponential with $d(x)$ under Zipf laws for popularity, with at most $C = 100$ different files and using Zipf parameters $\beta = 0.1$ and $\beta = 1$ respectively

This differential equation is Lipschitz everywhere, therefore, it admits a unique solution, once an initial condition x_0 is given, denoted $F(x_0, t)$. No closed form of the solution is known, so that one needs to use a numerical method for integration. This equation is prone to numerical instability because the second derivative of $x(t)$ is large when t goes to infinity. In our programs, we have used a finely tuned ad-hoc Runge-Kutta integrator with adaptative step sizes [8]. In any case, the step size is always larger than $\varepsilon/(T||f'||)$ where $||f'|| = \sup_{N,x} \frac{df(N,x)}{dx}$, for integrating with precision ε over an interval of length T. No further details will be given in this technical issue.

3 Coupling from the Past

Given the stochastic hybrid model for file sharing systems, we now show how to compute its steady state behavior.

3.1 Embedded Markov Chain

In this part we only study the system at its jump instants. The behavior at arbitrary instants can be easily derived as shown in Section 3.7.

Lemma 1. *The embedded sequence at jump times , $S_n \stackrel{\text{def}}{=} (N(T_n), x(T_n))$ is a homogeneous continuous time Markov chain over a continuous domain, $\mathcal{D} \subset \mathbb{N} \times \mathbb{R}$.*

Proof (sketch). The state of the process at time T_n only depends on the state at time T_{n-1}, the innovation ξ_n and the n-th inter-arrival of the jump process $\tau_n = T_n - T_{n-1}$, which value only depends on the state at time T_{n-1}. This means that $(N(T_n), x(T_n))$ is a homogeneous Markov chain over the domain of all reachable states, $\mathcal{D} \subset \mathbb{N} \times \mathbb{R}$.

In the following we will denote this global construction of this peer to peer Markov chain (P2P MC) as $S_n = \Phi(S_{n-1}, \xi_n)$. For simplicity, we also denote $\Phi(S, \xi_1, \ldots, \xi_n) \stackrel{\text{def}}{=} \Phi(\cdots \Phi(\Phi(S, \xi_1), \xi_2), \cdots)$.

The Markov chain S_n can be uniformized into a discrete time Markov chain (useful for simulation purposes) using the constant $\Lambda = N_{max}(\lambda + \nu C N_{max} + \mu)$.

Lemma 2. *The chain S_n is uniformly ergodic over $\mathcal{D}$: there exists a non-trivial measure φ over $\mathcal{D}$, some $m > 1$ and $0 \leq \beta \leq 1$ such that $\forall x \in \mathcal{D}$, $P^m(x, \cdot) \geq \beta\varphi(\cdot)$. Moreover, it has an atom in $(0,0)$ (i.e. (0,0) can be reached with a positive probability, starting from any other state).*

Proof. The discrete part, $N(T_n)$ follows a birth and death process. Therefore, $\mathbb{P}(N(T_n) = 0 | N(0) = N_0, x = x_0) \geq \mathbb{P}(N(T_n) = 0 | N(0) = N_{max}, x_0 = CN_{max}) \geq \mu^n N_{max}!/\Lambda^n$ if $n \geq N_{max}$. Again, $m = N_{max}$, $\varphi = \mathbb{I}_{(0,0)}$ and $\beta = N_{max}!(\mu/\Lambda)^{N_{max}}$ verify the definition of uniform ergodicity. The fact that $\varphi = \mathbb{I}_{(0,0)}$ means that $(0,0)$ is an atom.

Lemma 2 implies directly that S_n admits a unique stationary distribution (using general results for continuous Markov chains, see for example, [9]). Let $P^n(s, A)$ denote the transition law of n steps of the chain S_n (This is the probability $\mathbb{P}(S_n \in A | S_0 = s)$). The stationary measure Π of S satisfies $\Pi(A) = \int_{\mathcal{D}} P^1(s, A)\Pi(ds)$, for all measurable set A in $\mathcal{D}$.

3.2 Computing the Stationary State

If Π can be computed explicitly, there are many ways to draw samples from it. However, in most cases, analytical or even numerical computations of Π are impossible to obtain, either because the domain $\mathcal{D}$ is huge (in finite cases) or because the structure of the transition kernel $P^1(s, A)$ is too complex.

Without analytical or numerical knowledge of Π the most popular method for sampling from Π is simulation. The classical Monte-Carlo simulation consists in choosing an arbitrary initial value $S_0 = s_0$ in $\mathcal{D}$ and to use the constructive equations given in Equations (1), (2),(3) to generate $S_1, \ldots, S_n$ by using a random number generator for ξ. This technique works asymptotically because the sequence of samples converges in law, in the sup-norm, to the stationary distribution: $\lim_{n\to\infty} \sup_A |P^n(s_0, A) - \Pi(A)| = 0$. However, for a given finite n, the gap with the exact distribution depends on the convergence rate to the stationary distribution which is unknown in general. Here, we will show how to compute a sample in finite time which distribution is *exactly* Π (hence the name perfect), using a backward coupling technique. This technique was proposed for the first time in [4] for Markov chains over *finite* state spaces. The main idea is to run several simulations in parallel *starting in the past* from all possible initial states. If all the trajectories coincide at time 0 (meaning that they have all coupled at some point in the past) then the simulation stops and outputs the common value of all the trajectories at time 0, which happens to be a perfect sample of the chain.

Obviously, simulating trajectories from all initial states cannot be done for continuous state spaces. However, it was shown in [10] that backward coupling can also be defined for continuous state Markov chains. Here, we do not need the general theorems in [10] and we show how the backward coupling idea can be applied for the P2P MC using directly its uniform ergodicity.

Theorem 1. *The vertical backward coupling time*

$$K \stackrel{\text{def}}{=} \min\{n \geq 0 : \Phi(s, \xi_{-n}, \ldots, \xi_{-1}, \xi_0) = \Phi(r, \xi_{-n}, \ldots, \xi_{-1}, \xi_0), \forall r, s \in \mathcal{D}\},$$

is a well defined random variable. Furthermore, for all $s \in \mathcal{D}$, the random variable $\Phi(s, \xi_{-K}, \ldots, \xi_{-1}, \xi_0)$ is distributed according to measure Π (denoted $\Phi(s, \xi_{-K}, \ldots, \xi_{-1}, \xi_0) \sim \Pi$).

Proof. Let us consider the sub-sequence S_{mn} that makes m steps of the transition Kernel where $m = N_{max}$. As shown in Lemma 2, with probability $p_c > \mu^{N_{max}} N_{max}!/\Lambda^{N_{max}}$, then $S_{mn} = (0, 0)$. Therefore, K is a finite random variable, by Borel-Cantelli. Now, if s_∞ is any state distributed according to the stationary measure Π, then $\Phi(s_\infty, \xi_{-n}, \ldots, \xi_{-1}, \xi_0)$ is also stationary for all $n > 0$, by definition of Π. By choosing $n = K$, we get for an arbitrary state s, $\Phi(s, \xi_{-K}, \ldots, \xi_{-1}, \xi_0) = \Phi(s_\infty, \xi_{-K}, \ldots, \xi_{-1}, \xi_0) \sim \Pi$.

From Theorem 1, one may construct an algorithm providing a sample of the chain which has the stationary distribution Π: simulate backward in time starting from an arbitrary state and using steps of size m, until $\xi_{-n} < p_c$, where S_{-n+1} is set to $(0, 0)$. This algorithm has two major drawbacks: the kernel P^m is very difficult to compute and the run may last for long (because p_c is very small). They might make it impossible to use in practice. One way to deal with these two problems is to use structural properties of the P2P chain S_n, such as monotonicity.

3.3 Monotonicity Issues

The final class $\mathcal{D}$ admits the natural component-wise ordering: $(N, x) \leq (N', x')$ if $N \leq N'$ and $x \leq x'$. The chain S is said to be *monotone* if $(N, x) \leq (N', x')$ implies that $\Phi(N, x, \xi, \tau) \leq \Phi(N', x', \xi, \tau)$ for all ξ and all τ.

To test if the chain is monotone, one considers two chains S_1 and S_2 starting with ordered values, $N_1(0) \geq N_2(0)$ and $x_1(0) \geq x_2(0)$.

One must first consider the evolution between jumps. It should be clear that the differential equation (3) is monotone in N as well as in its initial condition. Therefore, if $N_1(0) \geq N_2(0)$ and $x_1(0) \geq x_2(0)$, then for all time $t \geq 0$, $x_1(t) \geq x_2(t)$.

As for the behavior of the chain at jump times, it is monotone as long as $\xi > 1 - \lambda' N_{max}$,However if $\xi < 1 - \lambda' + \nu' x(T_{n-1}) N_{max}$, the event may either be a departure or a null event for both chains. The following tricky situation can occur: $\mu' N_2(T_{n-1}) < \xi < \mu' N_1(T_{n-1})$. This corresponds to a departure for S_1 and a null event for S_2. Now, if $x_1(T_n^-)$ and $x_2(T_n^-)$ are too close, the following can happen: $x_1(T_n) = x_1(T_n^-) - x_1(T_n^-)/N_1(T_{n-1}) \leq x_2(T_n^-) = x_2(T_n)$. So that the chain is actually not monotone under such events.

3.4 Upper and Lower Envelopes

In order to deal with monotone systems, we introduce upper and lower envelopes of the trajectories of the Markov chain S, $S_1 = (N_1, x_1)$ and $S_2 = (N_2, x_2)$, respectively. The upper (resp. lower) envelope start at time $t = 0$ in state (N_{max}, C) (resp. $(0, 0)$). The upper (resp. lower) envelope evolve exactly as the Markov chain for all events which cannot cause a swap of the ordering between the two trajectories. Whenever a potential swapping event occurs, then, let $N_3 = \lfloor \xi/\mu' \rfloor$ be the largest value of N for which ξ is the null event. For N_3+1 and larger values of N, ξ would be a departure. We set $S_1(T_n) = (N_1(T_{n-1}) - 1, x_1)$ and $S_2(T_n) = (N_2(T_{n-1}), x_2(T_n^-)\frac{N_3}{N_3+1})$ so that the order remains unchanged between S_1 and S_2 Such an event can be seen as a "dummy" departure for S_2 and a "dummy" arrival in S_1. The construction of the envelopes $(S_1(T_n), S_2(T_n))$ given above can be written under the form of two new functions Γ_1, Γ_2 that describes the Markovian evolution of both envelopes at jump times: for $j = 1, 2$,

$$S_j(T_n) = \Gamma_j\Big(S_1(T_{n-1}), S_2(T_{n-1}), \xi_n, \tau_n\Big).$$

Note that by construction of Γ_1, Γ_2, the envelopes have been built such that $S_1(t)$ stays above $S_2(t)$ for all time $t \geq 0$ as soon as $S_1(0)$ is above $S_2(0)$. So the envelopes have a monotone behavior. Also note that by construction, for all initial state of the initial P2P MC, $S(0) = (N, x)$ and all time t: $S_1(t) \geq S(t) \geq S_2(t)$.

3.5 Perfect Simulation Algorithm for Peer to Peer Systems

The following theorem is the theoretical foundation of our perfect simulation algorithm for peer to peer systems.

Theorem 2. *The Markov chain $(S_1(T_n), S_2(T_n))$ hits the diagonal (i.e. states of the from $\big((N, x), (N, x)\big)$) in finite time a.s.. The hitting time $K'' =$*

$$\min\big\{n : \Gamma_1\big((N_{max}, C), (0, 0), \xi_{-K}, \ldots, \xi_0\big) = \Gamma_2\big((N_{max}, C), (0, 0), \xi_{-K}, \ldots, \xi_0\big)\big\}$$

is a vertical backward coupling time of the Markov chain S, so that for all initial state s, $\Phi(s, \xi_{-K''}, \ldots, \xi_{-1}, \xi_0) \sim \Pi$.

Proof. The first part of the proof is similar to the proof of the uniform ergodicity of chain S. Indeed, if a large number of departures occur, both envelopes will eventually reach $(0, 0)$. Since this happens with a positive probability, the Markov chain $(S_1(T_{n-1}), S_2(T_{n-1}))$ is uniformly ergodic and K'' is a finite random variable with finite expectation.

As for the second part of the proof, it simply uses the fact that $S_1(t) \geq S(t) \geq S_2(t)$ for all initial conditions for the chain S. Consider a stationary initial condition $S(-K'') \sim \Pi$. Then, $S(0) = \Phi(S(-K''), \xi_{-K''}, \ldots, \xi_{-1}, \xi_0) \sim \Pi$ by stationarity and $S_1(0) = S(0) = S_2(0)$ by definition of K''.

From Theorem 2, it is possible to derive an algorithm to compute both an upper bound on K'' and a sample from Π. Here is the outline of such an algorithm. Start at time $-k$ (at the beginning, $k = -1$) and simulate in parallel the lower and upper envelopes, starting respectively in states $(0, 0)$ and (N_{max}, C), using the same random variables, $(\tau_{-k}, \xi_{-k}), \cdots, (\tau_{-1}, \xi_{-1}), (\tau_0, \xi_0)$ for both of them. If at time 0, both envelopes reach the same state (N_0, x_0), then this means (using theorem 2) that the system has coupled, $K'' \leq k$ and $(N_0, x_0) \sim \Pi$. Otherwise, generate a new random innovation $(\tau_{-k-1}, \xi_{-k-1})$, and restart at time $-k-1$ with initial states $(0, 0)$ and (N_{max}, C).

A classical improvement is to double the number of steps backward at each iteration, so that the simulation time becomes linear in the total number of steps (κ) which is most twice the coupling time K''.

Another improvement is to stop when both envelopes are close enough. A stopping test on equality is theoretically possible since both envelopes couple in $(0, 0)$ with positive probability, and remain exactly equal from this point on. However, the probability that N ever reaches 0 is extremely small (less than 10^{-300} in the examples of Section 4). This means that the average vertical coupling time is huge. Testing for a small gap between both envelopes reduces the simulation time drastically and keep guarantees on the intervals on the measures as seen in the following. The complete algorithm is given as Algorithm 1.

Algorithm 1. Backward simulation for P2P MC

```
κ = 1;
repeat
  κ = 2κ;
  S1 := (N_max, C) ,S2 := (0,0) {Initialize the two envelopes at time −κ}
  for i = κ downto κ/2 + 1 do
    ξ[i] :=Random(Uniform over [0,1]); τ[i] :=Random(exponential with rate Λ)
  end for
  for i = κ downto 1 do
    S1 := Γ1(S1, S2, ξ[i], τ[i]) , S2 := Γ2(S1, S2, ξ[i], τ[i])
  end for
until S1.N = S2.N and S1.x − S2.x ≤ ε/3
return S1
```

In Figure 3, we display a perfect simulation of S_1, S_2. Note that several dummy events on S_1, S_2 are visible. They all correspond to discontinuous jumps of S_2 and cusps of S_1. Also note that the trajectories of x and N are a very high positive correlation: x and N both increase and decrease at the same time. This shows the effect of popularity: when the popularity is high, more customers get connected and they download more files, thus increasing the popularity. The total complexity of the program is $\kappa(c(\Gamma_1) + c(\Gamma_2)) + 2p$, where $c(\Gamma_i)$ is a constant corresponding to the time to compute Γ_i and p is the total number of steps needed to integrate the differential equation over the total simulated time, $T = \sum_{i=1}^{\kappa} \tau[i]$.

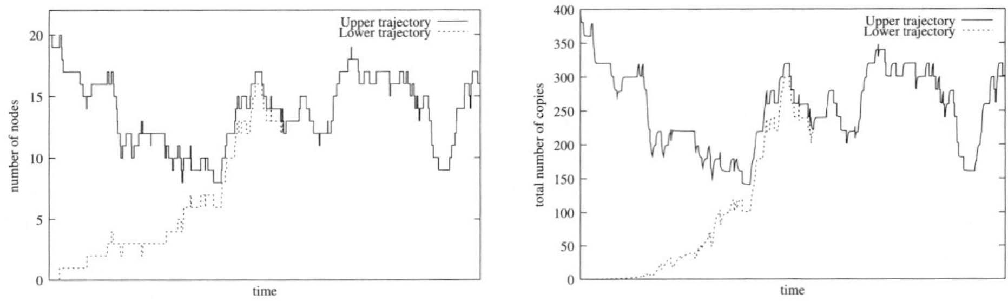

Fig. 3. Coupling of the trajectories of the two envelopes S_1, S_2 for N (left) and x (right) as generated by Algorithm 1

Since $\kappa \leq 2K''$ and $p \leq \frac{T^2}{\varepsilon} \sup_{N,x} \frac{df(N,x)}{dx}$, the complexity of the algorithm is given on average by the following lemma.

Lemma 3. *The average complexity of Algorithm 1, with precision ε is*

$$O\left(\mathbb{E}K'' + \frac{1}{\varepsilon}\left(\frac{\mathbb{E}K''}{\Lambda} + \frac{\mathbb{E}(K''^2)}{\Lambda^2}\right)\right). \tag{4}$$

Computing $\mathbb{E}K''$ is a difficult is out of the scope of this paper. We carried out several experiments to measure the number of backward steps $\kappa \leq 2K''$ in Figure 6.

3.6 Confidence Intervals

The outputs of the i-th run of the algorithm are numerical approximations of the couples $(S_1^i = (N_1^i, x_1^i), {S_2}^i = (N_2^i, x_2^i))$. The numerical integration of the differential equation yields a global error ϵ. The integration steps h are chosen small enough so that $\epsilon \leq \varepsilon/3$. Then, the outputs of the algorithm are such that the exact values verify $x_1^i - x_2^i \leq \varepsilon$.

Let E be an interval $E = [a, b] \subset [0, C]$ for which we want to compute $\pi_x(E)$ with confidence c. The central limit theorem gives the following confidence interval $I = \left[\hat{p_1} - \frac{\beta_c v}{2\sqrt{M}}, \hat{p_2} + \frac{\beta_c v}{2\sqrt{M}}\right]$, where β_c the c-percentile for the normal law and $v = \sqrt{\pi_x(E)(1 - \pi_x(E))} \leq 1/2$ and $\hat{p_1} = 1/M \sum_{i=1}^{M} \mathbf{1}\{x_1^i \in E \wedge x_2^i \in E\}$, $\hat{p_2} = 1/M \sum_{i=1}^{M} \mathbf{1}\{x_1^i \in E_\varepsilon \vee x_2^i \in E_\varepsilon\}$.

3.7 Stationary State at Arbitrary Instants

Algorithm 1 provides samples of the stationary distribution at jump instants. However, this distribution may significantly differ from the distribution of the system at arbitrary instants. From the PASTA [11] property, this latter distribution can be sampled simply by pursuing each trajectory during a random time independent of the system, distributed according to the jump process.

3.8 Adding Files at Jump Times

In this section we show how the model can be modified to take into account additional features of a P2P system. First, the fraction of files lost when a user leaves the system need not be a constant. It can be a random fraction η of the files present before the jump, with η following an arbitrary distribution with mean $\frac{N(T_n^-)-1}{N(T_n^-)}$. Also, some users may connect the system to inject new files in the system. These files may be *intrinsically* popular. This can easily be incorporated in the model by injecting a random number of copies δ when a node brings one of these files: upon a join event,

$$x(T_n) = x(T_n) + \begin{cases} 0 & \text{with probability } 1-p \\ \delta(x(T_n), N(T_n)) & \text{with probability } p \end{cases}$$

A possible model for δ is the following. Let $M = C - d(x)$ be the total number of missing files. When a file D is injected it is chosen according to the popularity Zipf distribution $P_M(D)$. The number of injected copies is then a fixed number proportional to the number of free places $NC-x$: $\delta = P(D)\omega(NC-x)$, where $0 < \omega < 1$ is a constant. This jump preserves the ordering between both trajectories and can be incorporated in the algorithm simply by modifying function h. This illustrates the flexibility and simplicity of the perfect simulation algorithm.

4 Numerical Experiments

The Algorithm 1 has been implemented in Java. All experiments reported here are carried out on a 2GHz Pentium 4 with 1GB of RAM. We have chosen realistic parameters for a rather large P2P model resembling a typical file sharing system. Here, $C = 1000$, $N_{max} = 1000$, $\mu = \lambda = 10^{-5}$ (an average customer stays connected/disconnected for 24 hours); $\nu = \lambda/(CN_{max})$, so that the popularity plays the same role as the exogenous process for the connection rate of customers; $\sigma = 10^{-4}$ (an average customer emits a request every 3 hours); $\theta = 10^{-4}$, (3 hour time-to-live for copies of files) $\alpha = e^{-1}/C$ (the Zipf-like popularity distribution

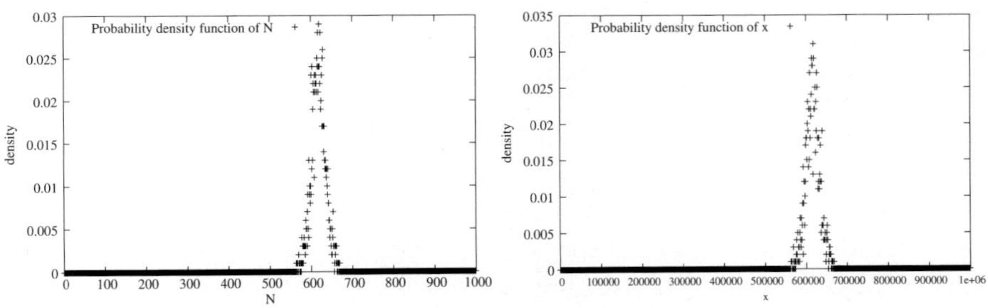

Fig. 4. The density of the stationary distribution of the number of customers (right) being connected and of the number of copies (left) in the hybrid model of a P2P system, as computed by perfect simulation

has parameter 1). The total computational time to get the asymptotic density of copies is about 2 hours, for a total of 10000 independent runs. Note that in Figure 4, the distributions of x and of N are centered on their average values and have rather small variances. In the following experiments, we will report only the average value of x or N (with confidence intervals).

Since θ is the only parameter that the system designer can control, we have computed the average value of the number of copies when θ varies. Figure 5 shows an interesting cut-off behavior: the number of files remains more or less constant until θ becomes very large ($\theta \approx 0.1$, meaning copies are removed after 10 seconds on average), where the number of copies drops to 0.

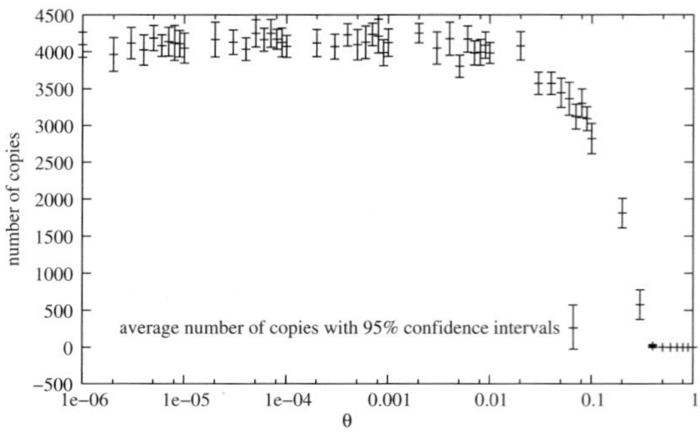

Fig. 5. Average number of copies as a function of θ

An important criterion for applicability of perfect simulation is the simulation time, which is random. The simulation time is directly related to the number of backward steps κ, which is doubled at each iteration in the algorithm. In order to evaluate experimentally how κ behaves when the state space of the system increases, we have run several simulation using the parameters $\lambda = \mu = \nu = \sigma = \theta = 1$ and we let the state space $N_{max} \times C$ grow from 0 to one million. As seen in Figure 6, the number of iterations κ of the simulation is almost deterministic and may only take two values as N and C grow. Also, κ is sub-linear in $N_{max} \times C$, which is quite good for perfect simulations.

5 Asymptotic Approximations

The solution x_N of the differential equation (3) admits an asymptote ℓ_N when t goes to infinity. The asymptote is the smallest solution of $f_N(x) = 0$. We consider the asymptotic *rate of convergence* of x_N to its asymptote, ℓ_N as

$$\gamma_N = \lim_{t\to\infty} \frac{f(x_N(t)) - f_N(\ell_N)}{x_N(t) - \ell_N} = \frac{df}{dx}(\ell_N) = -\alpha\sigma N e^{-\alpha\ell_N} + \sigma N - \frac{2\sigma\ell_N}{C} - \theta.$$

When the rate of convergence γ_N of $x_N(t)$ to its asymptotic value, is larger than the jump rate Λ, then in most cases, x will actually be very close to ℓ_N before

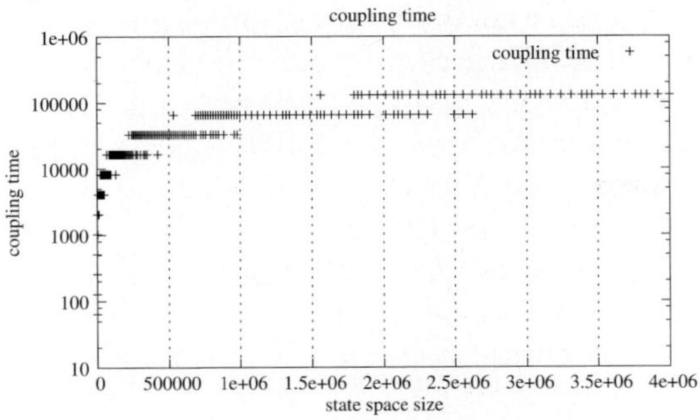

Fig. 6. Number of backward steps κ (in log scale for the number of iterations) as $N_{max} \times C$ vary

the next jump occurs. Therefore, one may disregard the transient behavior of x and let x jump directly from ℓ_{N_1} to ℓ_{N_2} whenever a jump from N_1 to N_2 occurs. This actually makes the system discrete since both x and N only take discrete values.

A visual illustration of this behavior is given in Figure 7 where two trajectories of the variable x are given with $\sigma = \Lambda$ and $\sigma = 10\Lambda$ respectively. In the second case, the rate of convergence of x to its asymptotic value is much larger than the jump rate and the trajectory of x looks like a staircase.

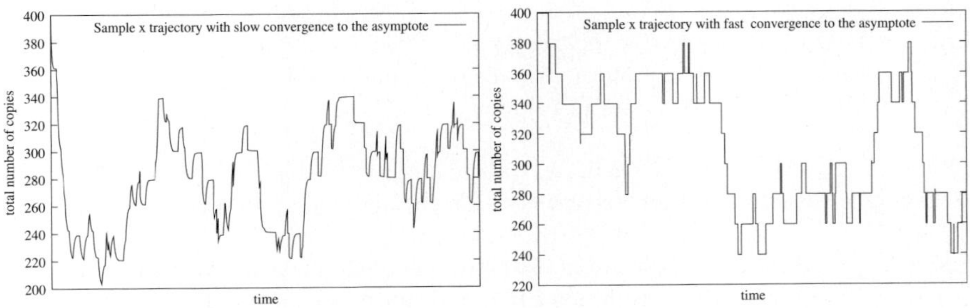

Fig. 7. Two behaviors of x with two rates of convergence

In that case, the stationary distribution Π of (N, x) can be approximated by

$$\Pi'(N = k, x = \ell_k) = \frac{1}{\Omega}\binom{N_{max}}{k}\frac{1}{\mu^k}\prod_{i=0}^{k-1}(\lambda + \nu\ell_i),$$

where the normalization constant Ω is such that all probabilities sum to one. Computing ℓ_k for each k is numerically easy using a Newton method. Here, the derivative of f at ℓ_k is large so that the computation can be done with a large precision quite fast. Then, generating samples according to the distribution Π' is rather simple and can be done using aliasing techniques [12]. We have compared this approximation with the exact samples (N_1, x_1) computed by the simulation algorithm presented in Section 3. Numerical evidence show that when $\gamma_N > \Lambda/10$, the approximation becomes very good and can be used instead of our perfect sampling method. If $\gamma_N < \Lambda/100$, then the approximation is not valid any longer.

6 Conclusion

In this paper we have presented a simulation study of a stochastic hybrid systems providing guaranteed samples of a complex peer to peer system modeled by hybrid equations.

Our simulations are based on backward coupling techniques that provide statistical guarantees on its samples. We believe this technique is well adapted to stochastic hybrid systems because they enable us to manipulate continuous variables in a coherent way and because they make numerical computations of the behavior of the system possible because the memory space required and the coupling time are small with respect to the size of the state space.

Acknowledgement. We thank Rémi Bertin who implemented Algorithm 1 and provided insights on monotonicity issues.

References

1. Dai, J.: On positive harris recurrence of multiclass queueing networks: a unified approach via fluid limit models. Annals of Applied Probability **5** (1995) 49–77
2. David, R., Alla, H.: Discrete, Continuous, and Hybrid Petri Nets. Springer-Verlag (2004)
3. Asarin, E., Maler, O., Pnueli, A.: Reachability analysis of dynamical systems having piecewise-constant derivatives. Theoretical Computer Science **138** (1995) 35–65
4. Propp, D., Wilson, J.: Exact sampling with coupled Markov chains and application to statistical mechanics. Random Structures and Algorithms **9**(1) (1996) 223–252
5. Gaujal, B., Perronnin, F., Bertin, R.: Perfect simulation of stochastic hybrid systems with an application to peer to peer systems. Technical report, INRIA (2006)
6. KaZaA: (http://www.kazaa.com/)
7. Gnutella: (http://www.gnutella.com/)
8. Press, W.H., Teukolsky, S.A., Vetterling, W.T., Flannery, B.P.: Numerical Recipes in C. Cambridge University Press (1992)
9. Meyn, S.P., Tweedy, R.L.: Markov chains and stochastic stability. Springer Verlag (1993)
10. Foss, S., Tweedy, R.: Perfect simulation and backward coupling. Stochastic Models **14** (1998) 187–204
11. Baccelli, F., Brémaud, P.: Elements of queueing theory. Springer-Verlag (1994)
12. Walker, A.: An efficient method for generating random variables with general distributions. ACM Trans. Math. Software (1974) 253–256

Approximately Bisimilar Finite Abstractions of Stable Linear Systems*

Antoine Girard

Université Joseph Fourier
Laboratoire de Modélisation et Calcul
B.P. 53, 38041 Grenoble, France
Antoine.Girard@imag.fr

Abstract. The use of bisimilar finite abstractions of continuous and hybrid systems, greatly simplifies complex computational tasks such as verification or control synthesis. Unfortunately, because of the strong requirements of bisimulation relations, such abstractions exist only for quite restrictive classes of systems. Recently, the notion of approximate bisimulation relations has been introduced, allowing the definition of less rigid relationships between systems. This relaxed notion should certainly allow us to build approximately bisimilar finite abstractions for more general classes of continuous and hybrid systems. In this paper, we show that for the class of stable discrete-time linear systems with constrained inputs, there exists an approximately bisimilar finite state system of any desired precision. We describe an effective procedure for the construction of this abstraction, based on compositional reasoning and samples of the set of initial states and inputs. Finally, we briefly show how our finite abstractions can be used for verification or control synthesis.

1 Introduction

Equivalence relationships for systems such as *bisimulation relations* [1,2] have been very useful to reduce the complexity of computational tasks such as verification or control synthesis for finite state systems. Early research on hybrid systems has focused on the characterization of continuous and hybrid dynamics with *bisimilar* finite abstractions. The first positive results on timed automata [3] were later extended to multirate hybrid automata [4] and hybrid systems with linear dynamics with a particular eigenstructure [5] (see [6] for a survey). More recently, the existence of bisimilar finite state systems has been shown for controllable discrete-time linear systems with unconstrained inputs [7]. The existence of such abstractions provides decidability results as well as computational procedures for verification or control synthesis for these classes of continuous and hybrid systems. Unfortunately, the class of hybrid dynamics admitting bisimilar finite abstractions is quite restrictive since even for very simple systems (*i.e.* three dimensional piecewise constant differential equations [8]), reachability verification is known to be undecidable.

* This work was partially supported by the ANR SETIN project VAL-AMS.

A. Bemporad, A. Bicchi, and G. Buttazzo (Eds.): HSCC 2007, LNCS 4416, pp. 231–244, 2007.

Recently, the notion of *approximate bisimulation relations* has been introduced in [9], allowing the definition of less rigid relationships between systems. As *exact* bisimulation relations require the observations of two systems to be (and remain) identical, approximate bisimulation relations allow the observations to be different provided the distance between them is (and remains) bounded by some parameter called *precision*. This relaxed assumption should certainly allow us to build approximately bisimilar finite abstractions for more general classes of continuous and hybrid systems.

In this paper, we show that for the class of stable discrete-time linear systems with constrained inputs, there exists an approximately bisimilar finite state system of any desired precision. We show that the linear system can be seen as the composition of two other linear systems, one autonomous and one with inputs but with an initial state set to zero. For each of these systems, we define an approximately bisimilar finite abstraction using a two step procedure. Firstly, by sampling the set of initial states or the set of inputs, we define discrete but infinite abstractions of the linear systems. Secondly, we show that this infinite state systems are approximately bisimilar to a finite state system. The composition of these systems provides us with the finite abstraction of original linear system. Our approach provides an effective way to compute this discrete abstraction. Finally, we briefly show how our finite abstractions can be used for verification or control synthesis.

Let us remark that the idea of sampling the sets of initial states and inputs to compute discrete abstractions for verification has already been proposed [10,11], though without further reduction to a finite state system. Contrary to these approaches, the abstractions we build are valid for an infinite time-horizon. In [12], a similar technique is proposed to build finite abstractions for stabilizable linear systems, however only a one-sided approximation result is provided making these abstractions suitable for control synthesis but not for verification.

2 Approximately Bisimilar Transition Systems

The notion of approximate bisimulation relation has been introduced in [9], in the framework of transition systems.

2.1 Transition Systems

Essentially, a transition system can be seen as an automaton, possibly with an infinite number of nodes and edges.

Definition 1. *A transition system (with observations) is a tuple* $T = (Q, \rightarrow, Q^0, \Pi, h)$ *that consists of:*

- *a (possibly infinite) set* Q *of states,*
- *a transition relation* $\rightarrow \subseteq Q \times Q$*,*
- *a (possibly infinite) set* $Q^0 \subseteq Q$ *of initial states,*
- *a (possibly infinite) set* Π *of observations,*
- *an observation map* $h : Q \rightarrow \Pi$*.*

If the set of states has a finite number of elements, we say that the transition system is finite. The transition $(q, q') \in\rightarrow$ is denoted $q \rightarrow q'$. A state trajectory of T is a finite sequence of transitions, $q^0 \rightarrow \cdots \rightarrow q^N$, where $q^0 \in Q^0$. Note that a transition system is possibly non-deterministic: for a given initial state there may be several state trajectories. An external or observed trajectory of T is a finite sequence of observations, $\pi^0 \dots \pi^N$ such that there exists a state trajectory of T which satisfies $h(q^i) = \pi^i$, for all $i \in \{0, \dots, N\}$. The language of T is denoted by $L(T)$ and consists of all external trajectories of T. The reachable set of T is the subset of Π defined by:

$$\text{Reach}(T) = \left\{\pi \in \Pi |\, \exists\, \pi^0 \dots \pi^N \in L(T),\ i \in \{0, \dots, N\},\ \text{such that } \pi^i = \pi\right\}.$$

In this paper, the transition systems we consider are observed over $\Pi = \mathbb{R}^p$.

We define a composition operator for transition systems that will be useful in the development of the paper.

Definition 2. *Let $T_1 = (Q_1, \rightarrow_1, Q_1^0, \Pi, h_1)$ and $T_2 = (Q_2, \rightarrow_2, Q_2^0, \Pi, h_2)$ be transition systems. The composition of T_1 and T_2, denoted by $T_1||T_2$, is the transition system $T = (Q, \rightarrow, Q^0, \Pi, h)$ defined by:*

- *the set of states $Q = Q_1 \times Q_2$,*
- *the transition relation $\rightarrow$ given by,*

$$(q_1, q_2) \rightarrow (q_1', q_2') \text{ if } q_1 \rightarrow_1 q_1' \text{ and } q_2 \rightarrow_2 q_2' \,,$$

- *the set of initial states $Q^0 = Q_1^0 \times Q_2^0$,*
- *the set of observations $\Pi = \mathbb{R}^p$,*
- *the observation map $h(q_1, q_2) = h_1(q_1) + h_2(q_2)$.*

Let us remark that the composition of two transition systems observed on $\mathbb{R}^p$ is also observed on $\mathbb{R}^p$. We can now introduce the notion of approximate bisimulation relation for transition systems as presented in [9].

2.2 Approximate Bisimulation Relations

The notion of exact bisimulation relation allows to characterize the observational equivalence of two transition systems [1,2]. The notion of approximate bisimulation relation is obtained from the exact one by relaxing the observational equivalence constraint. Instead of requiring that the observations of two systems are and remain the same we require that the distance between them is and remains bounded by some parameter called precision.

Definition 3. *Let $T_1 = (Q_1, \rightarrow_1, Q_1^0, \Pi, h_1)$ and $T_2 = (Q_2, \rightarrow_2, Q_2^0, \Pi, h_2)$ be transition systems. A relation $R \subseteq Q_1 \times Q_2$ is a δ-approximate bisimulation relation between T_1 and T_2, if for all $(q_1, q_2) \in R$:*

1. *$\|h_1(q_1) - h_2(q_2)\| \leq \delta$,*
2. *for all $q_1 \rightarrow_1 q_1'$, there exists $q_2 \rightarrow_2 q_2'$, such that $(q_1', q_2') \in R$,*
3. *for all $q_2 \rightarrow_2 q_2'$, there exists $q_1 \rightarrow_1 q_1'$, such that $(q_1', q_2') \in R$.*

Definition 4. *T_1 and T_2 are said to be approximately bisimilar with precision δ (denoted $T_1 \sim_\delta T_2$), if there exists R, a δ-approximate bisimulation relation between T_1 and T_2 such that:*

1. *for all $q_1 \in Q_1^0$, there exists $q_2 \in Q_2^0$, such that $(q_1, q_2) \in R$,*
2. *for all $q_2 \in Q_2^0$, there exists $q_1 \in Q_1^0$, such that $(q_1, q_2) \in R$.*

Remark 1. For $\delta = 0$, we recover the usual notion of exact bisimulation relation and of *exactly* bisimilar transition systems. Thus, $T_1 \sim_0 T_2$ will be denoted $T_1 \sim T_2$.

The following proposition states fundamental properties of approximate bisimulation relations. The proof is omitted here but can be found in [9].

Proposition 1. *Let T_1, T_2 and T_3 be transition systems, then:*

1. *for all $\delta \geq 0$, $T_1 \sim_\delta T_1$,*
2. *for all $\delta \geq 0$, $T_1 \sim_\delta T_2 \iff T_2 \sim_\delta T_1$,*
3. *for all $\delta \geq 0$, $\delta' \geq 0$, $T_1 \sim_\delta T_2$ and $T_2 \sim_{\delta'} T_3 \implies T_1 \sim_{\delta+\delta'} T_3$.*

Remark 2. Contrarily to the relation $\sim$, for $\delta > 0$, the relation $\sim_\delta$ is not an equivalence relation on the set of transition systems. However, the relation defined by $T_1 \equiv \mathbb{T}_2 \iff \exists \delta \geq 0,\ T_1 \sim_\delta T_2$ is an equivalence relation.

The following proposition shows that approximate bisimulation relations allow compositional reasoning.

Proposition 2. *Let T_1, T_2, S_1 and S_2 be transition systems, then:*

$$T_1 \sim_{\delta_1} S_1 \text{ and } T_2 \sim_{\delta_2} S_2 \implies T_1 || T_2 \sim_{\delta_1+\delta_2} S_1 || S_2 .$$

Proof. Let $T_1 = (Q_1, \rightarrow_1, Q_1^0, \Pi, h_1)$, $T_2 = (Q_2, \rightarrow_2, Q_2^0, \Pi, h_2)$, $S_1 = (P_1, \rightsquigarrow_1, P_1^0, \Pi, g_1)$ and $S_2 = (P_2, \rightsquigarrow_2, P_2^0, \Pi, g_2)$. $T_1 \sim_{\delta_1} S_1$ and $T_2 \sim_{\delta_2} S_2$, let R_1 and R_2 be the associated approximate bisimulation relations. Let $T_1||T_2 = (Q, \rightarrow, Q^0, \Pi, h)$ and $S_1||S_2 = (P, \rightsquigarrow, P^0, \Pi, g)$. Let us define the following relation $R \subseteq Q \times P$:

$$R = \{(q_1, q_2, p_1, p_2) \in Q \times P |\ (q_1, p_1) \in R_1 \text{ and } (q_2, p_2) \in R_2\} .$$

Let $(q_1, q_2, p_1, p_2) \in R$,

$$\begin{aligned} \|h(q_1, q_2) - g(p_1, p_2)\| &= \|h_1(q_1) + h_2(q_2) - g_1(p_1) - g_2(p_2)\| \\ &\leq \|h_1(q_1) - g_1(p_1)\| + \|h_2(q_2) - g_2(p_2)\| \leq \delta_1 + \delta_2 . \end{aligned}$$

Let $(q_1, q_2) \rightarrow (q_1', q_2')$, then $q_1 \rightarrow_1 q_1'$ and $q_2 \rightarrow_2 q_2'$. Since $(q_1, p_1) \in R_1$ and $(q_2, p_2) \in R_2$, there exist $p_1 \rightsquigarrow_1 p_1'$ and $p_2 \rightsquigarrow_2 p_2'$ such that $(q_1', p_1') \in R_1$ and $(q_2', p_2') \in R_2$. Then, $(p_1, p_2) \rightsquigarrow (p_1', p_2')$ and $(q_1', q_2', p_1', p_2') \in R$. Similarly, we can show that for all $(p_1, p_2) \rightsquigarrow (p_1', p_2')$ there exists $(q_1, q_2) \rightarrow (q_1', q_2')$ such that $(q_1', q_2', p_1', p_2') \in R$. Hence, R is a $\delta_1 + \delta_2$-approximate bisimulation relation between $T_1||T_2$ and $S_1||S_2$. Let $(q_1, q_2) \in Q^0$, then $q_1 \in Q_1^0$ and $q_2 \in Q_2^0$. There exist $p_1 \in P_1^0$ and $p_2 \in P_2^0$ such that $(q_1, p_1) \in R_1$ and $(q_2, p_2) \in R_2$. Then, $(p_1, p_2) \in P^0$ and $(q_1, q_2, p_1, p_2) \in R$. Similarly, we can show that for all $(p_1, p_2) \in P^0$ there exists $(q_1, q_2) \in Q^0$ such that $(q_1, q_2, p_1, p_2) \in R$. Therefore, we can conclude that $T_1||T_2 \sim_{\delta_1+\delta_2} S_1||S_2$. ■

2.3 Approximation Results

The precision of an approximate bisimulation relations allows to quantify how well two systems approximate each other. The following result shows that the distance between the external trajectories of approximately bisimilar systems ($T_1 \sim_\delta T_2$) is bounded by the precision δ.

Theorem 1. *Let T_1 and T_2 be transition systems such that $T_1 \sim_\delta T_2$, then for all $\pi_1^0 \dots \pi_1^N \in L(T_1)$, there exists $\pi_2^0 \dots \pi_2^N \in L(T_2)$, such that*

$$\text{for all } i \in \{0, \dots, N\},\ \|\pi_1^i - \pi_2^i\| \leq \delta$$

and conversely.

Proof. $T_1 \sim_\delta T_2$, let R be the associated δ-approximate bisimulation relation between T_1 and T_2. Let $\pi_1^0 \dots \pi_1^N \in L(T_1)$, let $q_1^0 \to_1 \dots \to_1 q_1^N$ be the associated state trajectory of T_1. Since $q_1^0 \in Q_1^0$, there exists $q_2^0 \in Q_2^0$ such that $(q_1^0, q_2^0) \in R$. Using the second property of Definition 3, we can show by induction that there exists $q_2^0 \to_2 \dots \to_2 q_2^N$ a state trajectory of T_2 such that for all $i \in \{0, \dots, N\}$, $(q_1^i, q_2^i) \in R$. Then, for all $i \in \{0, \dots, N\}$, $\|\pi_1^i - \pi_2^i\| = \|h_1(q_1^i) - h_2(q_2^i)\| \leq \delta$. ■

The previous result extends naturally to reachable sets.

Corollary 1. *Let T_1 and T_2 be transition systems such that $T_1 \sim_\delta T_2$, then*

$$d_H(\text{Reach}(T_1), \text{Reach}(T_2)) \leq \delta$$

where d_H is the Hausdorff distance[1].

Proof. Let $\pi_1 \in \text{Reach}(T_1)$, there exists $\pi_1^0 \dots \pi_1^N \in L(T_1)$ and $j \in \{0, \dots, N\}$, such that $\pi_1^j = \pi_1$. From Theorem 1, there exists $\pi_2^0 \dots \pi_2^N \in L(T_2)$ such that for all $i \in \{0, \dots, N\}$, $\|\pi_1^i - \pi_2^i\| \leq \delta$. Particularly, $\pi_2 = \pi_2^j \in \text{Reach}(T_2)$ and $\|\pi_2 - \pi_1\| \leq \delta$. Similarly, we can show that for all $\pi_2 \in \text{Reach}(T_2)$, there exists $\pi_1 \in \text{Reach}(T_1)$, such that $\|\pi_2 - \pi_1\| \leq \delta$. ■

3 Finite Abstractions of Stable Linear Systems

Let us consider the following discrete-time linear system:

$$\begin{cases} x_{k+1} = Ax_k + Bu_k,\ x_0 \in I,\ u_k \in U \\ y_k \quad\ = Cx_k \end{cases} \tag{1}$$

where A is a $n \times n$ matrix, B is a $n \times m$ matrix and C is a $p \times n$ matrix. The set of initial states I is a compact subset of $\mathbb{R}^n$ and the set of inputs U is a compact subset of $\mathbb{R}^m$ containing 0. We assume that the system is asymptotically stable (*i.e.* all the eigenvalues of A are strictly inside the unit circle in the complex plane).

[1] The Hausdorff distance between two subsets A, $B \subseteq \mathbb{R}^p$ is defined by $d_H(A, B) = \max\left(\sup_{a\in A} \inf_{b\in B} \|a - b\|, \sup_{b\in B} \inf_{a\in A} \|a - b\|\right)$.

Proposition 3. *There exists $\lambda \in (0,1)$ and a positive semi-definite symmetric $n \times n$ matrix M such that the following linear matrix inequalities hold:*

$$C^T C \le M \tag{2}$$
$$A^T M A \le \lambda^2 M \tag{3}$$

Proof. Let us search M under the form $M = N + C^T C$ where N is a positive semi-definite symmetric matrix. Then, equation (3) is equivalent to

$$\frac{1}{\lambda^2} A^T N A - N \le C^T C - \frac{1}{\lambda^2} A^T C^T C A = Q.$$

Q is symmetric and thus can be written as $Q = Q^+ - Q^-$ where Q^+ and Q^- are positive semi-definite symmetric matrices. Since all the eigenvalues of A are strictly inside the unit circle, for λ sufficiently close to 1, all the eigenvalues of A/λ are also strictly inside the unit circle. Then, the discrete-time Lyapunov equation $A^T N A/\lambda^2 - N = -Q^-$ has a unique solution which is positive semi-definite symmetric. Then, it is easy to see that $M = N + C^T C$ satisfies equation (2) and (3). ∎

We denote by $\|.\|_M$ the norm on $\mathbb{R}^n$ associated to M: $\|x\|_M = \sqrt{x^T M x}$. We define the radii of the sets of initial states and inputs for this norm.

$$r_I = \max_{x \in I} \|x\|_M \text{ and } r_U = \max_{u \in U} \|Bu\|_M.$$

The discrete-time linear system (1) can be seen as a transition system $\Sigma = (\mathbb{R}^n, \rightarrow, I, \mathbb{R}^p, h)$ where the transition relation $\rightarrow$ is given by

$$x \rightarrow x' \iff \exists u \in U,\ x' = Ax + Bu$$

and the observation map by $h(x) = Cx$. The dynamics of system (1) can be split into an autonomous dynamics and a controlled dynamics with the initial state set to zero. Thus, we define the following transition systems Σ_1 and Σ_2. $\Sigma_1 = (\mathbb{R}^n, \rightarrow_1, I, \mathbb{R}^p, h)$ where the transition relation is given by

$$x \rightarrow_1 x' \iff x' = Ax.$$

Note that the transition system Σ_1, which captures the autonomous dynamics, is deterministic. $\Sigma_2 = (\mathbb{R}^n, \rightarrow_2, \{0\}, \mathbb{R}^p, h)$, where the transition relation $\rightarrow_2 = \rightarrow$, holds for the controlled dynamics with zero initial state.

Proposition 4. *Σ and $\Sigma_1 || \Sigma_2$ are exactly bisimilar.*

Proof. Let us denote $\Sigma_1 || \Sigma_2 = (\mathbb{R}^{2n}, \twoheadrightarrow, I \times \{0\}, \mathbb{R}^p, g)$. We define the following relation $R \subseteq \mathbb{R}^n \times \mathbb{R}^{2n}$:

$$R = \left\{ (x, x_1, x_2) \in \mathbb{R}^n \times \mathbb{R}^{2n} |\ x = x_1 + x_2 \right\}.$$

Let $(x, x_1, x_2) \in R$, then

$$\|h(x) - g(x_1, x_2)\| = \|h(x) - h(x_1) - h(x_2)\| = \|C(x - x_1 - x_2)\| = 0.$$

Let $x \to x'$, there exists $u \in U$ such that $x' = Ax + Bu$. Then, we have $x_2 \to_2 x_2'$ where $x_2' = Ax_2 + Bu$ and $x_1 \to_1 x_1'$ where $x_1' = Ax_1$. Thus, $(x_1, x_2) \twoheadrightarrow (x_1', x_2')$ and

$$x' = Ax + Bu = A(x_1 + x_2) + Bu = Ax_1 + Ax_2 + Bu = x_1' + x_2'.$$

Then, $(x', x_1', x_2') \in R$. Similarly, it is easy to see that for all $(x_1, x_2) \twoheadrightarrow (x_1', x_2')$ there exists $x \to x'$ such that $(x', x_1', x_2') \in R$. Hence, R is an exact bisimulation relation between Σ and $\Sigma_1 || \Sigma_2$. For all $x \in I$, $(x, 0) \in I \times \{0\}$ satisfies $(x, x, 0) \in R$. Therefore, $\Sigma \sim \Sigma_1 || \Sigma_2$. ■

Remark 3. Proposition 4 is the translation in the framework of transition systems of the fundamental superposition principle in linear systems theory.

Thus, the linear system Σ can be seen as the composition of the autonomous dynamics Σ_1 and of the controlled dynamics Σ_2. In the following, we construct a finite state system S which is approximately bisimilar to Σ. The approach is the following. First, we construct finite state systems S_1 and S_2 that are approximately bisimilar respectively to Σ_1 and Σ_2. Then, from Proposition 2, we can obtain the abstraction S from the composition of S_1 and S_2.

3.1 Abstraction of the Autonomous Dynamics

The construction of a finite state system that is approximately bisimilar to Σ_1 is processed in two steps. First, by using a sample of the set of initial states I, we compute a system with a discrete but infinite set of states. Then, a finite abstraction is derived from this system by remarking that all its external trajectories converge to 0.

Lemma 1. *Let $\varepsilon_1 > 0$, there exists a finite set $I_\sigma = \{x_0, \dots, x_{J_1}\} \subseteq I$ such that*

$$\forall x \in I,\ \exists x_j \in I_\sigma,\ \|x - x_j\|_M \le \varepsilon_1\ .$$

Proof. Let α_M be the largest eigenvalue of the matrix M, then for all $(x, x') \in \mathbb{R}^n \times \mathbb{R}^n$, $\|x - x'\|_M \le \alpha_M \|x - x'\|$. Now, let us assume that for all finite set of points $\{x_0, \dots, x_r\} \subseteq I$, there exists $x \in I$, such that for all x_j, $\|x - x_j\| \ge \varepsilon_1/\alpha_M$. Then, starting from a point $x_0 \in I$, we can construct a sequence $\{x_j\}_{j \in \mathbb{N}}$ such that for all $j, j' \in \mathbb{N}$, $j \ne j'$, we have $\|x_j - x_{j'}\| \ge \varepsilon_1/\alpha_M$. Therefore, we cannot extract a converging subsequence of $\{x_j\}_{j \in \mathbb{N}}$ and I cannot be a compact set. Therefore, we proved by contradiction that there exists a finite set of points $I_\sigma = \{x_0, \dots, x_{J_1}\} \subseteq I$ such that for all $x \in I$, there exists $x_j \in I_\sigma$ satisfying the inequality $\|x - x_j\| \le \varepsilon_1/\alpha_M$. ■

Then, let us define the transition system $T_1 = (Q_1, \twoheadrightarrow_1, Q_1^0, \mathbb{R}^p, h_1)$ where the set of symbolic states is

$$Q_1 = \left\{ q_j^k |\ j \in \{0, \dots, J_1\},\ k \in \mathbb{N} \right\},$$

the transition relation $\twoheadrightarrow_1$ is deterministic, given by $q_j^k \twoheadrightarrow_1 q_j^{k+1}$, the set of initial states is $Q_1^0 = \left\{ q_j^0 |\ j \in \{0, \dots, J_1\} \right\}$, and the observation map $h_1(q_j^k) = CA^k x_j$.

Lemma 2. *The transition systems Σ_1 and T_1 are approximately bisimilar with precision ε_1.*

Proof. Let us define the relation $R \subseteq \mathbb{R}^n \times Q_1$:

$$R = \left\{(x, q_j^k)|\ \|x - A^k x_j\|_M \leq \varepsilon_1\right\}.$$

From equation (2), for all $(x, q_j^k) \in R$,

$$\|h(x) - h_1(q_j^k)\| = \|C(x - A^k x_j)\| \leq \|x - A^k x_j\|_M \leq \varepsilon_1.$$

Further, from equation (3),

$$\|Ax - A^{k+1} x_j\|_M = \|A(x - A^k x_j)\|_M \leq \lambda \|x - A^k x_j\|_M \leq \varepsilon_1.$$

Therefore, $(Ax, q_j^{k+1}) \in R$. Hence, R is an ε_1-approximate bisimulation relation between Σ_1 and T_1. Moreover, for all $x \in I$, there exists $x_j \in I_\sigma \subseteq I$ such that $\|x - x_j\|_M \leq \varepsilon_1$, (*i.e.* $(x, q_j^0) \in R$) and conversely. Thus, $\Sigma_1 \sim_{\varepsilon_1} T_1$. ■

Let us remark that as k goes to infinity, $h_1(q_j^k)$ converges to 0. Then, by replacing the states for large values of k (typically for k greater than a given parameter K_1) by an invariant state associated with the observation 0, we define the finite state transition system $S_1 = (P_1, \rightsquigarrow_1, P_1^0, \mathbb{R}^p, g_1)$ where the finite set of symbolic states is

$$P_1 = \left\{p_j^k|\ j \in \{0, \dots, J_1\},\ k \in \{0, \dots, K_1\}\right\} \cup \{p^\infty\},$$

and the transition relation $\rightsquigarrow_1$ is deterministic and given by

$$\forall j \in \{0, \dots, J_1\}, k \in \{0, \dots, K_1 - 1\},\ p_j^k \rightsquigarrow_1 p_j^{k+1},\ p_j^{K_1} \rightsquigarrow_1 p^\infty \text{ and } p^\infty \rightsquigarrow_1 p^\infty.$$

The set of initial states is $P_1^0 = \left\{p_j^0|\ j \in \{0, \dots, J_1\}\right\}$, and the observation map is given by

$$\forall j \in \{0, \dots, J_1\}, k \in \{0, \dots, K_1\},\ g_1(p_j^k) = CA^k x_j \text{ and } g_1(p^\infty) = 0.$$

Lemma 3. *The transition systems T_1 and S_1 are approximately bisimilar with precision $r_I \lambda^{K_1+1}$.*

Proof. We define the following relation $R \subseteq Q_1 \times P_1$:

$$R = \left\{(q_j^k, p_j^k)|\ j \in \{0, \dots, J_1\},\ k \in \{0, \dots, K_1\}\right\} \cup \left\{(q_j^k, p^\infty)|\ j \in \{0, \dots, J_1\},\ k > K_1\right\}.$$

Let $(q_j^k, p_j^k) \in R$, for $j \in \{0, \dots, J_1\}$, $k \in \{0, \dots, K_1\}$, then $\|h_1(q_j^k) - g_1(p_j^k)\| = 0$. If $k < K_1$, then $(q_j^{k+1}, p_j^{k+1}) \in R$. If $k = K_1$, then $(q_j^{K_1+1}, p^\infty) \in R$. Let $(q_j^k, p^\infty) \in R$, for $j \in \{0, \dots, J_1\}$, $k > K_1$, then from equations (2) and (3)

$$\|h_1(q_j^k) - g_1(p^\infty)\| = \|CA^k x_j\| \leq \|A^k x_j\|_M \leq \lambda^k \|x_j\|_M \leq \lambda^{K_1+1} r_I.$$

Further $(q_j^{k+1}, p^\infty) \in R$. Thus, R is a $r_I \lambda^{K_1+1}$-approximate bismulation relation between T_1 and S_1. Moreover, for all $q_j^0 \in Q_1^0$, there exists $p_j^0 \in P_1^0$ such that $(q_j^0, p_j^0) \in R$, and conversely. Therefore, $T_1 \sim_{r_I \lambda^{K_1+1}} S_1$. ■

From Lemmas 2 and 3 and from Proposition 1, the following result is straightforward.

Theorem 2. *The transition systems Σ_1 and S_1 are approximately bisimilar with precision $\varepsilon_1 + r_I \lambda^{K_1+1}$.*

Let us remark that by choosing appropriately the parameters ε_1 and K_1, any desired precision can be achieved by the finite abstraction S_1. The number of states of S_1 is $(J_1+1)(K_1+1)+1$.

3.2 Abstraction of the Controlled Dynamics

The construction of a finite abstraction that is approximately bisimilar to Σ_2 is also processed in two steps. First by using a sample of the set of inputs U, we compute an abstraction with an infinite number of states. Then, we show that this system is approximately bisimilar to a finite state system.

Lemma 4. *Let $\varepsilon_2 > 0$, there exists a finite list $U_\sigma = \{u_0, \dots, u_{J_2}\} \subseteq U$ such that $u_0 = 0$ and*

$$\forall u \in U,\ \exists u_j \in U_\sigma,\ \|Bu - Bu_j\|_M \le \varepsilon_2 \ .$$

The proof of this result is similar to that of Lemma 1. We define the alphabet $V = \{v_0, \dots, v_{J_2}\}$. We define the transition system $T_2 = (Q_2, \twoheadrightarrow_2, Q_2^0, \mathbb{R}^p, h_2)$ where the set of symbolic states is the set of infinite words on the alphabet V:

$$Q_2 = \{v_{j_0} v_{j_1} v_{j_2} \dots \mid \forall k \in \mathbb{N},\ j_k \in \{0, \dots, J_2\}\} \, .$$

The transition relation $\twoheadrightarrow_2$ is given by

$$\forall v_{j_0} v_{j_1} v_{j_2} \dots \in Q_2,\ \forall v_j \in V,\ v_{j_0} v_{j_1} v_{j_2} \dots \twoheadrightarrow_2 v_j v_{j_0} v_{j_1} v_{j_2} \dots \, .$$

The set of initial states consists of a single infinite word $Q_2^0 = \{v_0 v_0 v_0 \dots\}$. The observation map is given by

$$\forall v_{j_0} v_{j_1} v_{j_2} \dots \in Q_2,\ \ h_2(v_{j_0} v_{j_1} v_{j_2} \dots) = \sum_{k=0}^{k=\infty} C A^k B u_{j_k} .$$

Let us remark that since the linear system (1) is stable, the observation map is well defined for any infinite word.

Lemma 5. *The transition systems Σ_2 and T_2 are approximately bisimilar with precision $\varepsilon_2/(1-\lambda)$.*

Proof. Let us define the relation $R \subseteq \mathbb{R}^n \times Q_2$,

$$R = \left\{ (x, v_{j_0} v_{j_1} v_{j_2} \dots) \,\middle|\, \|x - \textstyle\sum_{k=0}^{k=\infty} A^k B u_{j_k}\|_M \le \varepsilon_2/(1-\lambda) \right\} .$$

From equation (2), for all $(x, v_{j_0} v_{j_1} v_{j_2} \dots) \in R$,

$$\begin{aligned} \|h(x) - h_2(v_{j_0} v_{j_1} v_{j_2} \dots)\| &= \|C(x - \textstyle\sum_{k=0}^{k=\infty} A^k B u_{j_k})\| \\ &\le \|x - \textstyle\sum_{k=0}^{k=\infty} A^k B u_{j_k}\|_M \le \varepsilon_2/(1-\lambda). \end{aligned}$$

Let $x \rightarrow_2 x'$, then there exists $u \in U$ such that $x' = Ax + Bu$. There exists $u_j \in U_\sigma$, such that $\|Bu - Bu_j\|_M \leq \varepsilon_2$. Then, $v_{j_0}v_{j_1}v_{j_2}\cdots \twoheadrightarrow_2 v_j v_{j_0}v_{j_1}v_{j_2}\ldots$ and from equation (3)

$$\begin{aligned}\|x' - Bu_j - \textstyle\sum_{k=0}^{k=\infty} A^{k+1}Bu_{j_k}\|_M &\leq \|A(x - \textstyle\sum_{k=0}^{k=\infty} A^k Bu_{j_k})\|_M + \|Bu - Bu_j\|_M \\ &\leq \lambda\|x - \textstyle\sum_{k=0}^{k=\infty} A^k Bu_{j_k}\|_M + \varepsilon_2 \leq \varepsilon_2/(1-\lambda).\end{aligned}$$

Thus, we have $(x', v_j v_{j_0}v_{j_1}v_{j_2}\ldots) \in R$. Similarly, it is easy to show that for all $v_{j_0}v_{j_1}v_{j_2}\cdots \twoheadrightarrow_2 v_j v_{j_0}v_{j_1}v_{j_2}\ldots$, there exists $x \rightarrow_2 x'$ such that we have $(x', v_j v_{j_0}v_{j_1}v_{j_2}\ldots) \in R$. Thus, R is an $\varepsilon_2/(1-\lambda)$-approximate bisimulation relation between Σ_2 and T_2. Moreover, it is clear that $(0, v_0v_0v_0\ldots) \in R$, then $\Sigma_2 \sim_{\varepsilon_2/(1-\lambda)} T_2$. ■

We now show that the system T_2 is approximately bisimilar to a finite system. Let us remark that for an infinite word $v_{j_0}v_{j_1}v_{j_2}\cdots \in Q_2$, the value of $h_2(v_{j_0}v_{j_1}v_{j_2}\ldots)$ does not depend much on v_{j_k} for large values of k (since the matrices A^k converge to 0). Then, by replacing an infinite word by its finite prefix of a given length, we can define a finite state transition system $S_2 = (P_2, \rightsquigarrow_2, P_2^0, \mathbb{R}^p, g_2)$ where the set of symbolic states is the set of words of length K_2+1 on the alphabet V:

$$P_2 = \left\{v_{i_0}\ldots v_{i_{K_2}} \mid \forall k \in \{0,\ldots,K_2\},\ i_k \in \{0,\ldots,J_2\}\right\}.$$

The transition relation $\rightsquigarrow_2$ is given by

$$\forall v_{i_0}\ldots v_{i_{K_2}} \in P_2,\ \forall v_i \in V,\ v_{i_0}\ldots v_{i_{K_2}} \rightsquigarrow_2 v_i v_{i_0}\ldots v_{i_{K_2-1}}.$$

The set of initial states consists of a single word of length K_2+1, $P_2^0 = \{v_0\ldots v_0\}$. The observation map is given by

$$\forall v_{i_0}\ldots v_{i_{K_2}}\cdots \in P_2,\ \ g_2(v_{i_0}\ldots v_{i_{K_2}}) = \sum_{k=0}^{k=K_2} CA^k Bu_{i_k}.$$

Lemma 6. *The transition systems T_2 and S_2 are approximately bisimilar with precision $r_U\lambda^{K_2+1}/(1-\lambda)$.*

Proof. We define the relation $R \subseteq Q_2 \times P_2$:

$$R = \left\{(v_{j_0}v_{j_1}v_{j_2}\ldots, v_{i_0}\ldots v_{i_{K_2}}) | \forall k \in \{0,\ldots K_2\}, j_k = i_k\right\}.$$

Let $(v_{j_0}v_{j_1}v_{j_2}\ldots, v_{i_0}\ldots v_{i_{K_2}}) \in R$, from equations (2) and (3)

$$\begin{aligned}\|h_2(v_{j_0}v_{j_1}v_{j_2}\ldots) - g_2(v_{i_0}\ldots v_{i_{K_2}})\| &= \|\textstyle\sum_{k=K_2+1}^{k=\infty} CA^k Bu_{j_k}\| \\ &\leq \|\textstyle\sum_{k=K_2+1}^{k=\infty} A^k Bu_{j_k}\|_M \\ &\leq \lambda^{K_2+1}\|\textstyle\sum_{k=0}^{k=\infty} A^k Bu_{j_{k-K_2-1}}\|_M \\ &\leq \lambda^{K_2+1}\textstyle\sum_{k=0}^{k=\infty} \lambda^k r_U = r_U\lambda^{K_2+1}/(1-\lambda).\end{aligned}$$

For all $v_{j_0}v_{j_1}v_{j_2}\cdots \twoheadrightarrow_2 v_j v_{j_0}v_{j_1}v_{j_2}\ldots$, we have $v_{i_0}\ldots v_{i_{K_2}} \rightsquigarrow_2 v_j v_{i_0}\ldots v_{i_{K_2-1}}$ and it is easy to see that $(v_j v_{j_0}v_{j_1}v_{j_2}\ldots, v_j v_{i_0}\ldots v_{i_{K_2-1}}) \in R$. Similarly, for all $v_{i_0} \dots v_{i_{K_2}} \rightsquigarrow_2 v_i v_{i_0} \dots v_{i_{K_2-1}}$, we have $v_{j_0} v_{j_1} v_{j_2} \dots \twoheadrightarrow_2 v_i v_{j_0} v_{j_1} v_{j_2} \dots$ and $(v_i v_{j_0} v_{j_1} v_{j_2} \dots, v_i v_{i_0} \dots v_{i_{K_2-1}}) \in R$. Hence, R is a $r_U \lambda^{K_2+1}/(1-\lambda)$-approximate bisimulation relation between T_2 and S_2. Moreover, $(v_0 v_0 v_0 \dots, v_0 \dots v_0) \in R$, then $T_2 \sim_{r_U \lambda^{K_2+1}/(1-\lambda)} S_2$. ∎

Intuitively, S_2 can be seen as an abstraction which keeps track of the last K_2+1 values of the input of Σ_2 (the larger K_2 the better the precision). From Lemmas 5 and 6 and from Proposition 1, the following result is straightforward.

Theorem 3. *The transition systems Σ_2 and S_2 are approximately bisimilar with precision $(\varepsilon_2 + r_U \lambda^{K_2+1})/(1-\lambda)$.*

Let us remark that by choosing appropriately the parameters ε_2 and K_2, any desired precision can be achieved by the finite abstraction S_2. The number of states of S_2 is $(J_2+1)^{(K_2+1)}$.

3.3 Abstraction of the Linear System

We now define the finite abstraction S of Σ by composition of S_1 and S_2: $S = S_1 || S_2$. Then, we can give the main result of the paper.

Theorem 4. *The transition system Σ associated with the stable linear system (1) and the finite state transition system S are approximately bisimilar with precision $(\varepsilon_1 + r_I \lambda^{K_1+1}) + (\varepsilon_2 + r_U \lambda^{K_2+1})/(1-\lambda)$.*

Proof. From Proposition 2 and Theorems 2 and 3, $\Sigma_1 || \Sigma_2$ and $S_1 || S_2$ are approximately bisimilar with precision $(\varepsilon_1 + r_I \lambda^{K_1+1}) + (\varepsilon_2 + r_U \lambda^{K_2+1})/(1-\lambda)$. Then, Theorem 4 follows from Propositions 1 and 4. ∎

Hence, for any stable linear systems of the form (1), there exists a finite abstraction that is approximately bisimilar. Moreover, by choosing appropriately ε_1, K_1, ε_2 and K_2, any desired precision can be achieved. Let us remark that the paper provides an effective way to compute these finite abstractions using samples of the sets of initial states and inputs.

Example 1. We consider the linear system of the form (1) given by the matrices

$$A = \begin{bmatrix} 0.4 & 0.2 \\ -0.2 & 0.4 \end{bmatrix}, \; B = \begin{bmatrix} 1 \\ 0 \end{bmatrix}, C = \begin{bmatrix} 1 & 0 \\ 0 & 1 \end{bmatrix},$$

the set of initial states $I = [4,6] \times [-1,1]$ and the set of inputs $U = [-1,1]$. We computed finite abstractions S_1 and S_2 of the autonomous and controlled dynamics. Figure 1 shows simple discrete abstractions of S_1 for $\varepsilon_1 = 1/2$ and $K_1 = 2$ (precision 0.85) and for $\varepsilon_1 = 1/5$ and $K_1 = 3$ (precision 0.44) and of S_2 for $\varepsilon_2 = 1/3$ and $K_2 = 1$ (precision 0.97) and for $\varepsilon_2 = 1/5$ and $K_2 = 2$ (precision 0.52). More precise abstractions for smaller ε_1 and ε_2 and larger K_1 and K_2 can be computed; however, the vizualisation becomes problematic.

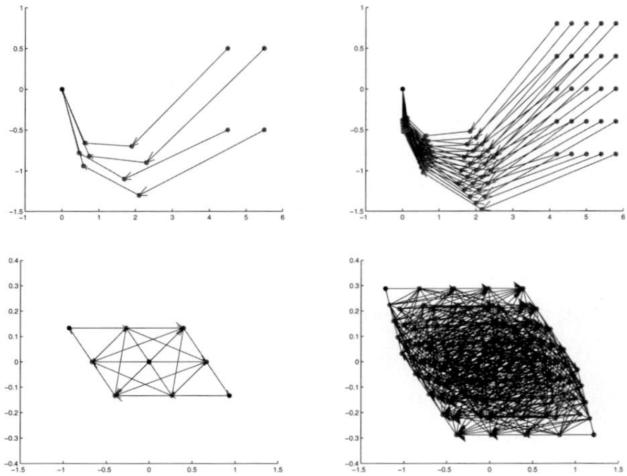

Fig. 1. Finite abstractions of the autonomous dynamics (top) and of the controlled dynamics (bottom)

4 Application to Verification and Control Synthesis

In this section, we show how the finite abstraction developed in the previous section can be used to solve verification or control synthesis problems. Let us define $\delta = (\varepsilon_1 + r_I \lambda^{K_1+1}) + (\varepsilon_2 + r_U \lambda^{K_2+1})/(1-\lambda)$ and let F be a subset of $\mathbb{R}^p$. We define the following sets:

$$\forall \mu \geq 0,\ I(F,\mu) = \{\pi \in F |\ \forall \pi' \in \mathbb{R}^p,\ \|\pi - \pi'\| \leq \mu \implies \pi' \in F\}\ .$$

$$\forall \mu \geq 0,\ O(F,\mu) = \{\pi \in \mathbb{R}^p |\ \exists \pi' \in F \text{ such that } \|\pi - \pi'\| \leq \mu\}\ .$$

$I(F,\mu)$ denotes the set of points of F whose distance to the boundary is greater than μ. $O(F,\mu)$ denotes the μ neighborhood of F.

4.1 Verification

We want to determine whether F is reachable by the linear system (1) (*i.e.* $\text{Reach}(\Sigma) \cap F \neq \emptyset$). The idea is to do the reachability computation on the finite abstraction S, which is a simple task since S is finite, and then to interpret the results for Σ using the following proposition.

Proposition 5. *The following assertions hold*

$$\text{Reach}(S) \cap I(F,\delta) \neq \emptyset \implies \text{Reach}(\Sigma) \cap F \neq \emptyset \tag{4}$$
$$\text{Reach}(S) \cap O(F,\delta) = \emptyset \implies \text{Reach}(\Sigma) \cap F = \emptyset \tag{5}$$
$$\text{Reach}(\Sigma) \cap I(F,2\delta) \neq \emptyset \implies \text{Reach}(S) \cap I(F,\delta) \neq \emptyset \tag{6}$$
$$\text{Reach}(\Sigma) \cap O(F,2\delta) = \emptyset \implies \text{Reach}(S) \cap O(F,\delta) = \emptyset \tag{7}$$

The proof of this result is not stated here but is straightforward from Corollary 1. Equations (4) and (5) allow us to determine whether Reach(Σ) intersects F. The interpretation of equation (6) is that if Reach(Σ) *sufficiently* intersects F, then the reachability analysis of the finite abstraction S will allow to conclude that Reach(Σ)$\cap F \neq \emptyset$. Similarly, equation (7) tells that if Reach(Σ) is *sufficiently* far from F, then the reachability analysis of S will allow to conclude that Reach(Σ)$\cap$ $F = \emptyset$. This means that the more robust Σ is with respect to the reachability property, the coarser the finite abstraction S we can use. Let us remark that, contrarily to most approximate reachability techniques, the approach based on approximately bisimilar abstractions allows to verify reachability properties on infinite time-intervals.

4.2 Control Synthesis

Let us assume that Reach(S) $\cap I(F,\delta) \neq \emptyset$, then from Proposition 5, we know that Reach(Σ) intersects F. We now want to determine an initial state and a sequence of inputs which drives the system Σ to an observation in F. For the sake of simplicity, we assume that $K_1 = K_2 = K$.

There exists an element $(p,w) \in P_1 \times P_2$ reachable by a state trajectory of S and such that $g_1(p) + g_2(w) \in I(F,\delta)$. We can show that there are only two possibilities:

1. $(p,w) = (p_j^k, v_{i_0} \dots v_{i_K})$ where for all $r \in \{k,\dots,K\}$, $v_{i_r} = v_0$. Then, a state trajectory of S leading to (p,w) is

$$(p_j^0, v_0 \dots v_0) \rightsquigarrow (p_j^1, v_{i_{k-1}} v_0 \dots v_0) \rightsquigarrow (p_j^2, v_{i_{k-2}} v_{i_{k-1}} v_0 \dots v_0) \rightsquigarrow \dots \quad (8)$$
$$\dots \rightsquigarrow (p_j^k, v_{i_0} \dots v_{i_{k-1}} v_0 \dots v_0).$$

2. $(p,w) = (p^\infty, v_{i_0} \dots v_{i_K})$. Then, a state trajectory of S leading to (p,w) is

$$(p_j^0, v_0 \dots v_0) \rightsquigarrow (p_j^1, v_{i_K} v_0 \dots v_0) \rightsquigarrow \dots \quad (9)$$
$$\dots \rightsquigarrow (p_j^K, v_{i_1} \dots v_{i_K} v_0) \rightsquigarrow (p^\infty, v_{i_0} v_{i_1} \dots v_{i_K}).$$

Thereof, we can choose an initial state and a sequence of inputs for the system Σ such that its associated external trajectory tracks with precision δ the external trajectory of the discrete abstraction S associated to the state trajectory given by equation (8) or (9).

5 Conclusion

In this paper, we showed that stable linear systems with constrained inputs admit approximately bisimilar finite abstractions of arbitrary precision. An effective way of computing these abstractions, based on compositional reasoning and samples of the set of initial states and inputs, is described. We showed how these abstractions can be used to simplify some computational tasks such as verification and control synthesis for reachability specifications.

Future work includes further reductions of these abstractions using approximation techniques for finite systems in order to obtain abstractions achieving a desired precision with a minimal number of states. We also intend to extend these results to continuous-time systems, as well as non-linear dynamical systems and to use our discrete abstractions to solve verification and control synthesis problems for more complex specifications such as those expressed in temporal logics using model checking or supervisory control for purely discrete systems.

Acknowledgements

The author would like to thank the members of the hybrid systems groups in Verimag and University of Pennsylvania for numerous valuable discussions on sampling based verification and approximate bisimulation relations that have tremendously inspired this work.

References

1. Clarke, E.M., Grumberg, O., Peled, D.A.: Model Checking. MIT Press (2000)
2. Milner, R.: Communication and Concurrency. Prentice-Hall (1989)
3. Alur, R., Dill, D.: A theory of timed automata. Theor. Comput. Sci. **126**(2) (1994) 183–235
4. Alur, R., Courcoubetis, C., Halbwachs, N., Henzinger, T., Ho, P.H., Nicollin, X., Olivero, A., Sifakis, J., Yovine, S.: The algorithmic analysis of hybrid systems. Theor. Comput. Sci. **138**(1) (1995) 3–34
5. Lafferriere, G., Pappas, G., Yovine, S.: A new class of decidable hybrid systems. In: HSCC'99. Volume 1569 of LNCS., Springer (1999) 137–151
6. Alur, R., Henzinger, T., Lafferriere, G., Pappas, G.: Discrete abstractions of hybrid systems. Proceedings of the IEEE **88** (2000) 971–984
7. Tabuada, P., Pappas, G.J.: Model checking LTL over controllable linear systems is decidable. In: HSCC'03. Volume 2623 of LNCS., Springer (2003) 436–451
8. Asarin, E., Maler, O., Pnueli, A.: Reachability analysis of dynamical systems having piecewise-constant derivatives. Theor. Comput. Sci. **138**(1) (1995) 35–65
9. Girard, A., Pappas, G.J.: Approximation metrics for discrete and continuous systems. IEEE Trans. Automatic Control (2005) To appear.
10. Kapinski, J., Krogh, B.H., Maler, O., Stursberg, O.: On systematic simulation of open continuous systems. In: HSCC'03. Volume 2623 of LNCS., Springer (2003) 283–297
11. Girard, A., Pappas, G.: Verification using simulation. In: HSCC'06. Volume 3927 of LNCS., Springer (2006) 272–286
12. Tabuada, P.: Symbolic control of linear systems based on symbolic subsystems. IEEE Trans. Automatic Control **51**(6) (2006) 1003–1013

Learning Cycle-Linear Hybrid Automata for Excitable Cells

R. Grosu[1], S. Mitra[2], P. Ye[1], E. Entcheva[3], I.V. Ramakrishnan[1], and S.A. Smolka[1]

[1] Department of Computer Science, Stony Brook University
[2] Computer Science and AI Lab, Massachusetts Institute of Technology
[3] Department of Biomedical Engineering, Stony Brook University

Abstract. We show how to automatically learn the class of Hybrid Automata called Cycle-Linear Hybrid Automata (CLHA) in order to model the behavior of excitable cells. Such cells, whose main purpose is to amplify and propagate an electrical signal known as the action potential (AP), serve as the "biologic transistors" of living organisms. The learning algorithm we propose comprises the following three phases: (1) Geometric analysis of the APs in the training set is used to identify, for each AP, the modes and switching logic of the corresponding Linear Hybrid Automata. (2) For each mode, the modified Prony's method is used to learn the coefficients of the associated linear flows. (3) The modified Prony's method is used again to learn the functions that adjust, on a per-cycle basis, the mode dynamics and switching logic of the Linear Hybrid Automata obtained in the first two phases. Our results show that the learned CLHA is able to successfully capture AP morphology and other important excitable-cell properties, such as refractoriness and restitution, up to a prescribed approximation error. Our approach is fully implemented in MATLAB and, to the best of our knowledge, provides the most accurate approximation model for ECs to date.

1 Introduction

Hybrid automata [2,19] are an increasingly popular modeling formalism for systems that exhibit both continuous and discrete behavior. Intuitively, Hybrid automata (HA) are extended finite-state automata whose discrete states correspond to the various modes of continuous dynamics a system may exhibit, and whose transitions express the switching logic between these modes.

Traditionally, HA have been used to model embedded systems, including automated highway systems, air traffic management, embedded automotive controllers, robotics, and real-time circuits. More recently, they have been used to model and analyze biological systems, such as cellular cycles and immune response [3], bio-molecular networks [1], gene-regulatory networks [7,17,23], protein-signaling pathways[11], and metabolic processes [4]. The hybrid-system metaphor has also been used to develop algorithms for large-scale simulation of

A. Bemporad, A. Bicchi, and G. Buttazzo (Eds.): HSCC 2007, LNCS 4416, pp. 245–258, 2007.

biological systems [15]. Biological systems are intrinsically hybrid in nature: biochemical concentrations may vary continuously, yet discrete transitions between distinct states are also possible.

Excitable cells (ECs) are a typical example of biological systems exhibiting hybrid behavior: transmembrane ion fluxes and voltages may vary continuously but the transition from the resting state to the excited state is generally considered an all-or-nothing discrete response. Traditionally, however, the preferred modeling approach for ECs uses large sets of coupled nonlinear differential equations. Although an invaluable asset for integrating genomics and proteomics data to reveal local interactions, such models are not typically amenable to control-theoretic techniques developed for linear systems, and render large-scale simulation impractical.

In previous work [25,26] we showed that it is possible to construct a conceptually simpler HA model for ECs that approximates with reasonable accuracy their electrical properties. We called these HA *Cycle-Linear HA* (CLHA) to highlight their cyclic structure and the fact that, while in each cycle they exhibit linear dynamics, the coefficients of the corresponding linear equations and mode-transition guards may vary in interesting ways from cycle to cycle.

The manual construction of CLHA, however, proved to be a tedious task, and the CLHA so derived were tied to a particular type of EC and to a particular species. Moreover, since recent advances in measuring *in vitro* the electrical activity of ECs have resulted in the availability of extensive data sets, it was natural to turn our attention to the following question: *Given a training set of electrical measurements of an EC, is it possible to automatically learn a CLHA that approximates the behavior of the EC up to a required error margin?*

In this paper we address this question, by presenting a methodology for automatically learning CLHA models for two types of ECs: the squid giant axon and the guinea pig ventricular cell. To the best of our knowledge, these are the most accurate approximation models (with per-AP-cycle linear dynamics), developed for these ECs to date, both in terms of electrical-signal morphology and typical excitable-cell characteristics such as refractoriness and restitution.

To simplify the process of obtaining training sets, we used *virtual* measurements obtained by applying the so-called *S1S2*-protocol to existing nonlinear models of these ECs. Extending the method outlined here to *in vitro* data obtained in the laboratory of the fourth author is a direction for future work.

The learning technique we have developed for CLHA is also of independent interest, as we learn all aspects of excitable-cell CLHA models up to a given error margin, including the number of modes; for each mode, the dimension of the state space and the coefficients of its linear time-invariant dynamics; and all aspects of the mode switching logic, including the jump conditions, thresholds and resets. To do so, we use the modified Prony's method to obtain an exponential fit for the continuous per-mode linear dynamics. Moreover, in learning the CLHA, we make no *a priori* assumptions about the dimension of the state space of the nonlinear system we are targetting, nor the degree of its input and output.

We also learn the functions that adjust a CLHA's mode dynamics and switching logic on a per-cycle basis. This aspect of our technique is critical in the case of excitable cells, which exhibit the following *restitution property*: the longer the recovery time for an EC, the longer in duration its subsequent *action potential*.

Organization. The rest of the paper is organized as follows. Section 2 defines Cycle Linear Hybrid Automata (CLHA). Section 3 describes the requisite biology of excitable cells. Our proposed methodology for learning a CLHA from a given training set is presented in Section 4. In this section, we also present our results for the training/testing set generated from the highly nonlinear Luo-Rudi dynamic model previously developed for Guinea Pig ventricular cells. Section 5 discusses related work. Section 6 contains our concluding remarks and directions for future research.

2 Hybrid Automata

We define *Cycle-Linear Hybrid Automata* as a restricted class of *Structured Hybrid Automata* [20]. SHA, which are derived from *Timed Input/Output Automata* [16], can model a general class of hybrid systems for which the input/output distinction intrinsic to the IOA methodology is ignored. We use a variable structure to specify states of an automaton. Let X be a set of variables. A *valuation* $\mathbf{x}$ for X is a function that associates with each $x \in X$ a value in its type. The set of all valuations of X is denoted by $val(X)$. For $\mathbf{x} \in val(X)$, let $\mathbf{x}.x$ denote the value of the variable x within the state $\mathbf{x}$. A *trajectory* $\tau : J \rightarrow val(X)$ specifies the values of all variables in X over a real time interval J. The *limit time* of a trajectory τ, written as $\tau.ltime$, is the supremum of the domain of τ. A *state model* for X is a collection F of differential and algebraic equations involving the continuous variables in X such that for every $\mathbf{x} \in val(X)$, there exists solution trajectory τ of F that starts from $\mathbf{x}$.

Definition 1. *A* Structured Hybrid Automaton *(SHA) with mode set $\mathcal{M}$ is a tuple $\mathcal{A} = (X, Q, \Theta, A, \mathcal{D}, P)$, where X is a set of* variables, *including a special discrete variable called* mode *of type $\mathcal{M}$; $Q \subseteq val(X)$ is the set of states; $\Theta \subseteq Q$ is a nonempty set of* start states*; A is a set of* actions, *$\mathcal{D} \subseteq Q \times A \times Q$ is a set of* discrete transitions*; and P is an indexed family F_i, $i \in \mathcal{M}$, of state models.*

As usual, we will specify the set of transitions in $\mathcal{D}$ corresponding to an action $a \in A$ by a guard predicate E_a and a reset map $R_a : X \rightarrow X$. A transition $(\mathbf{x}, a, \mathbf{x}') \in \mathcal{D}$ is called a *mode switch* if $\mathbf{x}.mode \neq \mathbf{x}'.mode$. The set $\mathcal{T}_\mathcal{A}$ of trajectories of an SHA $\mathcal{A}$ is defined as follows: a trajectory τ for X is in $\mathcal{T}_\mathcal{A}$ if the restriction of τ to the set of continuous variables in X satisfies the state model $F_{\tau(0).mode}$, and the restriction of τ to the discrete variables in X remain constant over the domain of τ. An *execution fragment* captures a particular run of $\mathcal{A}$; it is defined as an alternating sequence of actions and trajectories $\alpha = \tau_0 a_1 \tau_1 a_2 \ldots$, where each $\tau_i \in \mathcal{T}$, and if τ_i is not the last trajectory then $(\tau_i(ltime), a_{i+1}, \tau_{i+1}(0)) \in \mathcal{D}$.

The SHA model represents one end of the spectrum of hybrid-system models where one can specify systems with very general dynamics, and discrete mechanics. Linear and rectangular hybrid automata [2], at the other end of the spectrum, enable powerful analysis techniques by restricting the state models to linear differential equations. The Cycle-Linear Hybrid Automaton (CLHA) model was proposed in [26] for describing and analyzing highly nonlinear but periodic systems, and provides the mathematical basis for the remainder of this paper. Definition 2 gives a precise semantic definition of this model as restricted SHA. Informally, a CLHA captures a class of hybrid system where the state models, reset maps, and guards are all linear but with coefficients that are functions of a discrete state variable called *epoch*. The *epoch* variable is reset only when a particular mode is entered.

Definition 2. *A* Cycle-Linear Hybrid Automaton *(CLHA) with state space $Q \subseteq val(X)$, mode set $\mathcal{E} \times \mathcal{P}$, and snapshot map $S : Q \to \mathcal{E}$, is an SHA with mode set $\mathcal{M}$ satisfying:*

1. *Variable mode has type $\mathcal{M} = \mathcal{E} \times \mathcal{P}$, and its first component is a discrete variable of type $\mathcal{E}$ referred to as* epoch. *There is a unique $\zeta \in \mathcal{P}$ that is visited infinitely many times in any execution with an infinite number of mode switches.*
2. *For each $(\epsilon, p) \in \mathcal{M}$, $F_{\epsilon,p}$ is a linear state model. For each action $a \in A$, the guard E_a (reset map R_a) can be expressed as a linear predicate (resp. function) on X, with coefficients that are functions of epoch.*
3. *Suppose $(\mathbf{x}, a, \mathbf{x}')$ is a mode switch with $\mathbf{x}.mode = (\epsilon_1, p_1)$ and $\mathbf{x}'.mode = (\epsilon_2, p_2)$, for some $\epsilon_1, \epsilon_2 \in \mathcal{E}$, $p_1, p_2 \in \mathcal{P}$. If $p_2 = \zeta$ then $\epsilon_2 = S(\mathbf{x})$; otherwise, $\epsilon_2 = \epsilon_1$. A mode switch of the first type is called is called an* epoch transition.

3 Excitable Cells (ECs)

Action potentials. Excitable cells (ECs) can be viewed as active electrical circuits with nonlinear behavior, capable of amplifying and propagating electrical signals. ECs include neurons, cardiac cells, skeletal and smooth muscle cells. In cardiac cells for example, on each heart beat, an electrical control signal is generated by the sinoatrial node, the heart's internal pacemaking region. This signal is amplified and propagated as an electrical wave along a prescribed path, exciting cells in the main chambers of the heart (atria and ventricles) and assuring synchronous contractions. At the cellular level, the electrical signal is a change in the potential across the cell membrane caused by different ion currents flowing through the cell membrane. This electrical signal for each excitation event is known as an *action potential* (AP). A typical AP for Guinea-Pig ventricular cells is shown in Figure 1(a). Its morphology is usually defined as a sequence of six phases: *stimulated* (S), *upstroke* (U), *early repolarization* (E), *plateau* (P), *final repolarization* (F), and *resting* (R).

For non-pacemaking ECs, APs are externally triggered events: a cell fires an AP as an all-or-nothing response to a supra-threshold stimulus, and each

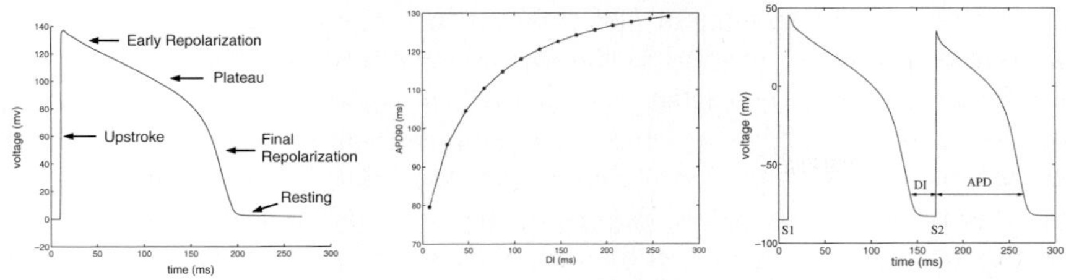

Fig. 1. (a) AP and its phases. (b) Restitution curve. (c) APD, DI and S1S2 protocol.

AP follows the same sequence of phases and maintains approximately the same magnitude regardless of the applied stimulus. After an initial step-like increase in the membrane potential, an AP lasts for a couple milliseconds to hundreds milliseconds in most mammals. During phases U, E, P and the first part of F, generally no re-excitation can occur. This portion of an AP is therefore known as the *absolute refractory period* (ARP). Starting with the second part of phase F, an altered secondary excitation event is possible if the stimulation strength or duration is raised. This portion of the AP is therefore known as the *relative refractory period* (RRP).

Restitution function. When an EC is subjected to repeated stimuli, two important time intervals can be identified: the *action potential duration* (APD), the period when the AP is above some prescribed percentage (e.g. 10%) of its maximum height, and the *diastolic interval* (DI), the period from the end of the APD to the end of the cycle, i.e. the end of phase R. Figure 1(c) illustrates the two intervals.

The function relating APD to DI as the cell is subjected to different stimulation frequencies is called the APD *restitution function.* As shown in Figure 1(b), the function is nonlinear and captures the phenomenon that a longer recovery time is followed by a longer APD. A physiological explanation of a cell's restitution is rooted in the ion-channel kinetics as a limiting factor in the cell's response to multiple stimuli over time. The sum of the APD and DI is called the *Basic Cycle Length* (BCL).

The *S1S2 protocol* is often used to determine the restitution function of an excitable cell. In this protocol, a cell is driven into a stable mode, in which a stable APD may be observed, by first subjecting it to a train of so-called S1 stimuli at a fixed BCL. Immediately thereafter, a single S2 stimulus, having a different (i.e. shorter) BCL is delivered. As such, one can associate a DI-APD pair with each running of the protocol, viz. the DI preceding the S2-induced APD. By repeating this procedure and varying the DIs before S2, one gradually constructs the graph of the restitution curve. Figure 1(c) illustrates the placement of the last S1 stimulus followed by the S2 stimulus.

Mathematical models of excitation. Modeling of the ionic processes that underlie cell excitation dates back to 1952, when Hodgkin and Huxley formulated their model of the squid giant axon [13]. Intuitively, the HH model is that of a nonlinear resistor-capacitor (RC) circuit with current sources, defining AP in terms of a stimulation current and three ionic currents: (fast) inward sodium, (slow) outward potassium, and a time-independent linear (leak) current. The ionic currents depend themselves on the AP via a gating mechanism (a time-varying conductance). The corresponding nonlinear system of equations is given below, where: V, m, n and h are continuous state variables; V is the AP, m, n and h are the ion-channel gates; $\overline{g}_{\text{Na}}$,$\overline{g}_{\text{K}}$,$\overline{g}_{\text{L}}$ are the constants which represent the maximum channel conductances for the sodium, potassium and leakage channel, respectively; E_{Na},E_{K},E_{L} are the constants for reversal potentials for these channels; m_∞, h_∞ and n_∞ are the ion-channel gates' steady-state values, and τ_m, τ_h and τ_n are their time-constant values; C is the constant cell capacitance and I_{st} is the stimulation current.

$$C\dot{V} = -\overline{g}_{\text{Na}}m^3h(V - E_{\text{Na}}) - \overline{g}_{\text{K}}n^4(V - E_{\text{K}}) - \overline{g}_{\text{L}}(V - E_{\text{L}}) + I_{\text{st}}$$

$$\begin{array}{lll}
\tau_m\,\dot{m} = m - m_\infty & \tau_m = 1/(\alpha_m + \beta_m) & m_\infty = \alpha_m/(\alpha_m + \beta_m) \\
\tau_h\,\dot{h} = h - h_\infty & \tau_h = 1/(\alpha_h + \beta_h) & h_\infty = \alpha_h/(\alpha_h + \beta_h) \\
\tau_n\,\dot{n} = n - n_\infty & \tau_n = 1/(\alpha_n + \beta_n) & n_\infty = \alpha_n/(\alpha_n + \beta_n)
\end{array}$$

$$\begin{array}{lll}
\alpha_m(V) = \frac{2.5-0.1V}{e^{2.5-0.1V}-1} & \alpha_h(V) = 0.07e^{-\frac{V}{20}} & \alpha_n(V) = \frac{0.1-0.01V}{e^{1-0.1V}-1} \\
\beta_m(V) = 4e^{-\frac{V}{18}} & \beta_h(V) = \frac{1}{e^{3-0.1V}+1} & \beta_n(V) = 0.125e^{-\frac{V}{80}}
\end{array}$$

The HH model with its 3 membrane currents, 4 state variables, and 12 fitted parameters laid the foundation for subsequent models of excitable cells of increasing complexity. All of these models use multiple continuous state variables (voltage, ion-channel gates, ion concentrations) to describe action potential in different cell types. One of the most popular cardiac-cell models is the dynamic Luo-Rudy model [18]. The LRd model uses 11 different membrane currents, more than 20 state variables and over 150 fitted parameters to describe the AP. Due to space constraints, the full structure of the LRd model is not listed here.

4 Learning the CLHA of an Excitable Cell

Given a training set of APs generated by applying the *S1S2*-protocol to an excitable cell of a particular species, our methodology for learning the CLHA that approximates the cell's behavior up to a given error margin consists of two phases. In the first phase, we obtain for each AP a linear Hybrid automaton (LHA) whose output is within the specified error bound. This involves identifying the segments of the APs that correspond to the modes of the LHA, deriving the flows for each mode, and the guards and reset maps for each transition. In each mode, we use the modified Prony's method (MPM) [21] to approximate the AP with a (normalized) linear dynamics, i.e., with a sum of exponentials.

In the second phase, we derive a CLHA that combines the behavior of all the LHAs and therefore captures all the APs in the training set. We exploit the fact that the coefficients defining the flows, guards, and reset maps of the CLHA are functions of the *epoch* variable which is updated during an *epoch transition*. We

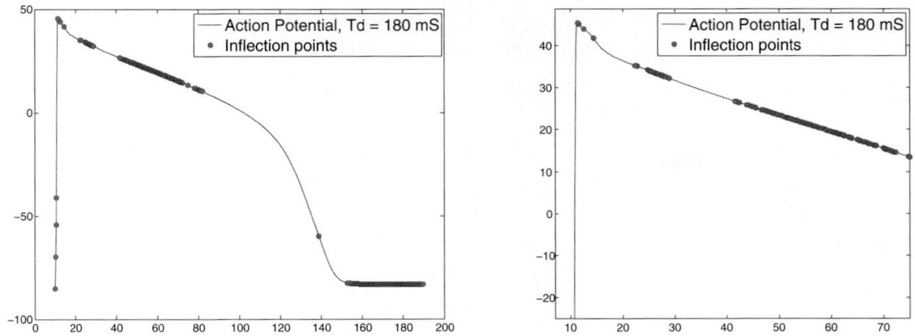

Fig. 2. Null/inflection points in the LRd APs

choose the variable to be a voltage-valued variable called v_0 and epoch transitions to be those that are brought about by the occurrence of a stimulus. In finding the snapshot map which sets the value of the epoch variable in the post-state of epoch transitions, we once again use MPM. Specifically, we estimate the voltage-dependent coefficients of the CLHA as an exponential regression of the constant coefficients in the LHAs obtained in the first phase.

Assumptions. Our goal is to derive a CLHA, the output of which is within ± 2 mv of the output of the Luo-Rudy model, under the following class of stimuli: each stimulation is a step of amplitude -80 μA/cm^2, duration 0.6 msecs, and BCL between 160 and 400 msecs. The set of 25 APs sampled every 0.2 msecs, corresponding to BCL 160 to 400 msecs, in 10 msecs intervals, serves as the *training set* for deriving the CLHA. The performance of the learned CLHA is evaluated on the *test set* consisting of APs with BCL from 165 to 405 msecs, in 10 msecs intervals, sampled at the same frequency.

Identifying Modes. To discover the points in the APs that correspond to mode transitions in the target LHAs, we computed the null points (zeros of the first-order derivative) and the inflection points (zeros of the second-order derivatives) of the APs. This approach worked very well for the HH model, and the sections of APs between successive null or inflection points were identified as the modes of the LHAs.

When directly applied to the LRd model, this approach yielded far too many modes. In particular, there exist trains of inflection points in the P and R phases of the APs (see Figure 2). This was somewhat surprising because the AP of these phases appears as rather smooth line segments corresponding to "stretched" inflection points. The higher-order nonlinearity of the LRd model seems to have dealt with such segments by generating trains of points whose tangent (first-order derivative) difference was smaller (in absolute value) than 10^{-5}. Based on this observation, we designed our own parameterized filter to eliminate such long sequences of closely-spaced inflexion points. The filter parameter enable us to increase or decrease the number of segments and thereby achieve the desired accuracy of the CLHA.

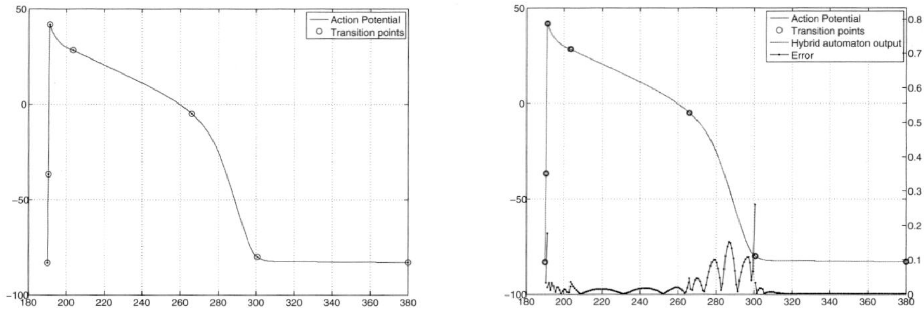

Fig. 3. (a) Inflection points after filtering. (b) Hybrid-automaton output.

Using the MPM described below, we were able to approximate each segment with two exponentials and the entire AP to within the desired accuracy. Since, however, this approach seemed to split each of the E and F phases in two, we decided to eliminate one inflection point in each. In doing so, we were not able to maintain the desired accuracy, unless we moved down the end-point of phase P and up the starting-point of phase R. The correctness of both transformations was confirmed by analyzing higher-resolution APs, where these points were indeed very close to their inferred position. The final seven points chosen are shown in Figure 3(a).

Using Modified Prony's Method to Obtain LHA. The null/inflection points partition the AP into sections defining six modes of the LHA: S, U, E, P, F and R. We denote the set of modes by $\mathcal{P}$. Since these modes are always visited in order, the voltages of the six inflection points define the guards (thresholds) for the corresponding transition edges. We denote the transition voltages by V_p, where $p \in \mathcal{P}$, is the mode in the post-state of the transition. For example, in the AP of Figure 3(a), the transition from U to E occurs at $V_{\mathsf{E}} = 45.32$ mv. To completely define the LHA, it remains to define the flows and the reset maps; for this we use the modified Prony's method [21].

The modified Prony's method is a technique for fitting exponential or sinusoidal functions to time-series data. For fixed n, MPM minimizes the $L2$ distance between time-series data and any function y that solves a differential equation with constant coefficients:

$$\sum_{i=1}^{n+1} c_i \frac{d^{i-1}y}{dt^i} = 0. \tag{1}$$

Depending on the coefficients c_i, the function estimating the solution of Equation 1 may be a complex or a real exponential, damped or undamped sinusoids. Furthermore, the input to the algorithm can be noisy periodic samples from the actual solution. Because of these attractive features, MPM has found many practical applications.

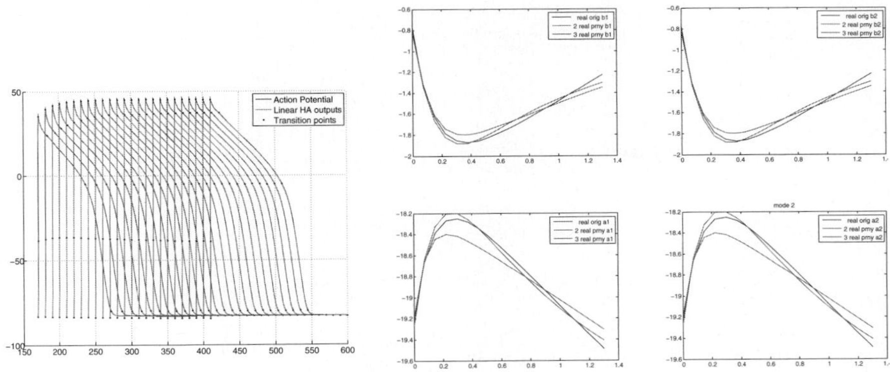

Fig. 4. (a) Original APs and superposed LHA outputs for training set. (b) Sums of 2 and 3 exponentials for estimating a_1, a_2, b_1 and b_2 for mode U.

Suppose the voltage in mode $p \in \mathcal{P}$ of the AP can be approximated as a sum of exponential functions:

$$v(t) = \sum_{i=1}^{n} a_{ip} e^{b_{ip} t} \tag{2}$$

Then, we can specify the flows in each mode as :

$$\forall\, i \in \{1, \ldots, n\}, \quad \dot{x}_i = b_{ip} x_i \text{ and } x_i(0) = a_{ip} \tag{3}$$
$$v = \sum_{i=1}^{n} x_i,$$

where the x_i's are the state variables. The initial condition on the state variables is set by the reset map of the transition from the previous mode. The accuracy of the above approximation is a function of n, that is, the number of state variables used. Using the MPM with $n = 2$, we obtained, approximations that were within the acceptable error bounds for all modes. The output of the resulting LHA, the original AP, and the error between the two, are plotted in Figure 3(b). We apply this procedure to obtain an LHA for each AP in the training set. The output of these automata, superimposed on the original APs, are shown in Figure 4(a).

Linear to Cycle-Linear HA. From the first phase, we obtain for each AP in the training set and for each mode $p \in \mathcal{P}$, the transition voltage V_p for the guards, and the coefficients b_{1p}, b_{2p}, and a_{1p}, a_{2p} corresponding to the the differential equations and initial values for the state variables x_1 and x_2. In other words, we obtain one linear hybrid automaton approximating each of the APs in the training set.

In the second phase, we combine these LHAs into a single CLHA by using the transition to mode S (stimulus arrival) as the epoch transition, setting the value of the epoch variable v_0. We call the value of v_0 the *epoch voltage*. For each mode, we find a function mapping v_0 of each LHA to transition voltages and coefficients; this function implicitly defines the snapshot map. We once again

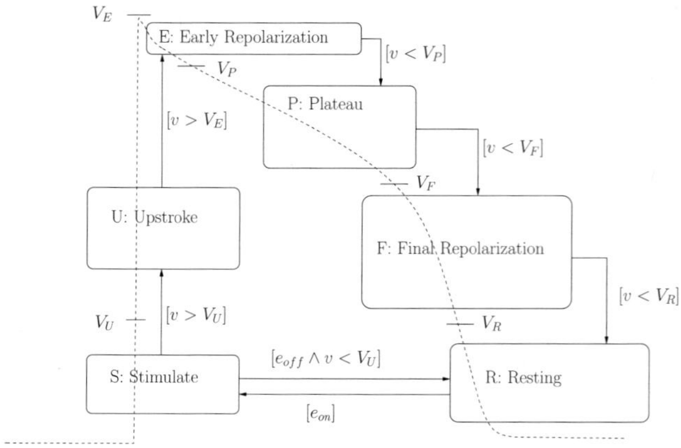

Fig. 5. Structure of the CLHA

use sums of two exponentials for these functions and obtain their coefficients by applying MPM. These functions are defined below, where $p \in \mathcal{P}$ and $i \in [1..2]$:

$$\begin{aligned} V_p(v_0) &= \vartheta_p \ e^{\theta_p v_0} + \vartheta'_p \ e^{\theta'_p v_0} \\ a_{ip}(v_0) &= \alpha_{ip} \ e^{\lambda_{ip} v_0} + \alpha'_{ip} \ e^{\lambda'_{ip} v_0} \\ b_{ip}(v_0) &= \beta_{ip} \ e^{\gamma_{ip} v_0} + \beta'_{ip} \ e^{\gamma'_{ip} v_0} \end{aligned}$$

Thus, a_{ip}, b_{ip} and V_p in the CLHA depend on the AP value stored in variable v_0 on the epoch transition between modes R and S. The way MPM approximates a_{1U}, a_{2U}, b_{1U} and b_{2U} with sums of two or three exponentials is shown in Figure 4(b). The structure of the CLHA thus obtained is given in Figure 5. For simplicity, the figure does not show the actions on the transitions and the flows within the modes.

While the above equations give the general pattern for the transition voltages and coefficients, a few observations are in place. First, by construction, V_F and V_R are constant in all LHAs and therefore no exponential fitting is necessary for the CLHA. Secondly, the a_i and b_i coefficients of modes F and R are up to a very small variation the same in all LHAs. Although we expressed them as functions in the CLHA, we are confident that using constants instead would have still satisfied the required accuracy. Thirdly, for the rest of the modes, the a_i and b_i obtained for the LHAs are complex values. We therefore separately fitted their real and imaginary parts. The constant coefficients ϑ_p, ϑ'_p, θ_p, θ'_p, α_{ip}, α'_{ip}, λ_{ip}, λ'_{ip}, β_{ip}, β'_{ip}, γ_{ip} and γ'_{ip} are complex too. Finally, due to space restrictions, we defer including a table with all voltage and coefficient values to the full version of the paper. We are happy, however, to provide them upon request.

Simulation results. We have implemented the above-described learning technique in MATLAB, and applied it to both the HH and LRd models. The accuracy of the resulting CLHA was analyzed on both the training and test sets. Due to space constraints, the results on the simpler HH model are omitted.

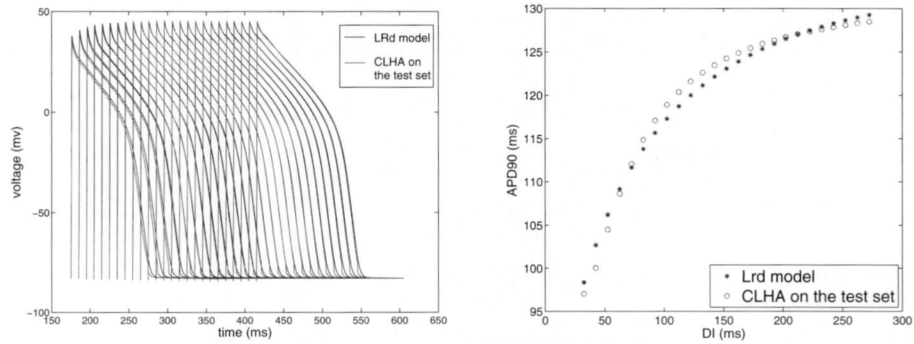

Fig. 6. (a) Comparison of AP. (b) Comparison of restitution curve.

The output of the CLHA on the LRd test set is shown superposed on the original APs in Figure 6(a). As can be observed, the morphology of the output, as well as the required accuracy, is maintained on this set. In Figure 6(b), the restitution curve obtained from the CLHA by running it on the v_0's specified in the test set is compared to the restitution curve obtained from the APs in the test set. Although not perfect, the results are very satisfactory. To our knowledge, these are the best results among the LRd-approximation models proposed so far.

5 Related Work

We have developed a learning/identification technique for cycle-linear hybrid automata (CLHA), and applied it to a classical, highly nonlinear model of ventricular cardiac myocytes. The technique of hybrid-automaton identification has been previously used in a number of communication and control applications, including interplanetary life-support systems [12], dynamic power management [8], autonomous systems and intelligent robots [14,10], and figure tracking [22]. To the best of our knowledge, our application of this technique in the area of systems biology, in general, and excitable cells, in particular, is the first of its kind.

Our approach to hybrid-automaton identification is further distinguished from prior work in the area by the novelty of the identification technique itself. Specific contributions in this regard include the following: (1) Our approach is applicable to continuous-time nonlinear systems that exhibit some level of periodicity and adaptation. Given such a system, the CLHA we learn are also continuous-time, specifically, linear time-invariant (LTI). In contrast, the techniques of [24,5] target discrete-time PWARX (piecewise-affine auto-regressive exogenous) models. Furthermore, in contrast to these approaches, when learning the CLHA for a system S, we make no *a priori* assumptions about the dimension of S's state space nor the degree of its input and output.

(2) Our technique learns all aspects of a hybrid automaton, including the number of modes; for each mode, the dimension of the state space and the coefficients of its LTI dynamics; and all aspects of the mode switching logic, including the jump conditions, thresholds and resets. To do so, we use a modified

Prony method to obtain an exponential fit for the continuous per-mode linear dynamics. Cf. [24], where polynomial fitting is used for the case of discrete-time PWARX systems.

(3) We also learn the functions that adjust a CLHA's mode dynamics and switching logic on a per-cycle basis. This aspect of our technique is critical in the case of excitable cells because of their *restitutional* nature (see Section 3). In this case, the coefficients of the mode dynamics and the voltage thresholds are functions of V_0, the cell's initial transmembrane voltage for the current cycle.

Other approximate models for cardiac-tissue excitability have been proposed in the literature, including the piecewise-linear model of Biktashev [6] and the nonlinear model of Fenton and Karma [9]. The CLHA models of excitable cells learned by our technique retain the simplicity of Biktashev's model without sacrificing the expressiveness of Fenton-Karma.

6 Conclusions

We have presented a method for automatically learning CLHA that approximate, up to a prescribed error margin, the complex, nonlinear processes of amplification and propagation of electrical signals (APs) in excitable cells. Our method, implemented in MATLAB, combines geometric analysis with exponential regression (using the modified Prony's method) to derive a CLHA that covers in a cycle-linear manner the input/output behavior of the original nonlinear system. Moreover, it provides, to the best of our knowledge, the most accurate approximation of extant nonlinear excitable-cell models, such as HH and LRd.

A source of complexity in the HH and LRd models is the coupling between state variables, which seemingly occurs continuously throughout an AP cycle. In contrast, the coupling between state variables in the CLHA model is markedly reduced: (i) the membrane voltage v_0 at the time a stimulus arrives determines the coefficients of the flows of the "gated voltages" x_1 and x_2 for a complete cycle; (ii) x_1 and x_2's flows determine the membrane voltage within a mode; and (iii) the transition voltages V_p, $p \in \mathcal{P}$, determine the mode-switching logic within a cycle. This decoupling of state variable within the CLHA model may provide additional insight into essential properties of ECs, such as refractoriness and restitution. The derivatives of x_1 and x_2 approximate in each mode of the CLHA the inward and the outward currents, respectively.

It was therefore no surprise that the AP phases that were most difficult to linearly approximate were upstroke (U) and early repolarization (E): it is in these phases when disparate time constants coexist. For smaller error margins, we will most likely require three exponentials in the MPM approximation of these phases, and therefore three state variables in the corresponding modes. The third variable presumably will distinguish between the K and Ca currents contributing to repolarization.

As future work, we plan to more carefully consider APs at higher stimulation frequencies, i.e. in the 130 to 160 msecs range; accurately capturing their behavior also seems to require a third state variable. We also plan to develop additional training sets based on protocols incorporating stimuli of varying shapes and intensity. Furthermore, we intend to study AP propagation within an array of CLHA. Our preliminary data [25], showed that a 400-x-400 cell array was able to produce spirals (arrhythmia-related phenomena). We expect that the increased accuracy of the learned CLHA will better match observed behavior.

Finally, we would like to better understand the class of nonlinear systems whose behavior CLHA can successfully approximate. Intuitively, they exhibit periodic, but nonetheless adaptive, behavior with respect to the input stimuli. For narrow ranges of the epoch voltage v_0, CLHA naturally provide a linear approximation, and well-established techniques for reachability, stability, observability and controllability analysis can be readily applied. It would be interesting, however, to investigate whether the additional structure provided by CLHA (its parametrization on v_0) can be exploited to extend the reach of such techniques.

References

1. R. Alur, C. Belta, F. Ivancic, V. Kumar, M. Mintz, G. Pappas, H. Rubin, and J. Schug. Hybrid modeling and simulation of biomolecular networks. In *Hybrid Systems: Computation and Control, HSCC 2001*, volume 2034 of *LNCS*, pages 1932, 2001.
2. R. Alur, C. Coucoubetis, N. Halbwachs, T.A. Henzinger, P.H. Ho, X. Nicolin, A. Olivero, J. Sifakis, and S. Yovine. The altorithmic analysis of hybrid systems. *Theoretical Computer Siences*, 138:334, 1995.
3. P. Barbano, M. Spivak, J. Feng, M. Antoniotti, and B. Misra. A coherent framework for multi-resolution analysis of biological networks with memory: Ras pathway, cell cycle and immune system. In *Proc. National Academy of Science*, volume 102, pages 62456250, 2005.
4. C. Belta, P. Finin, L. Habets, A. Halasz, M. Imielinski, V. Kumar, and H. Rubin. Understanding the bacterial stringent response using reachability analysis of hybrid systems. In *Hybrid Systems:Computation and Control, HSCC 2004*, volume 2993 of *LNCS*, pages 111126, 2004.
5. A. Bemporad, A. Garulli, S. Paoletti, and A. Vicino. A bounded-error approach to piecewise affine system identification. *IEEE Transactions on Automatic Control*, 50(10):14731634, 2005.
6. V. N. Biktashev. A simplified model of propagation and dissipation of excitation fronts. *International Journal of Bifurcation and Chaos*, 13:36053619, 2003.
7. H. de Jong, J.-L. Gouz, C. Hernandez, M. Page, T. Sari, and J. Geiselmann. Hybrid modeling and simulation of genetic regulatory networks: A qualitative approach. In *Hybrid Systems: Computation and Control, HSCC 2003*, volume 2623 of *LNCS*, pages 267282, 2003.
8. T. Erbes. Stochastic learning feedback hybrid automata for dynamic power management in embedded systems, 2004.
9. F. Fenton and A. Karma. Vortex dynamics in 3d continous myocardium with fiber rotation: Filament instability and fibrillation. CHAOS, 8:2047, 1998.
10. G.C. Rigatos, I.P Vlahavas, and C.D. Spyropoulos. Fuzzy stochastic automata for reactive learning and hybrid control. In *Methods and Applications of Artificial Intelligence*, pages 366377. LNCS, Vol. 2308, Springer-Verlag, 2002.
11. R. Ghosh and C.J. Tomlin. Symbolic reachable set computation of piecewise affine hybrid automata and its application to biological modeling: Delta-notch protein signaling. *IEEE Transactions on Systems Biology*, 1(1):170183, 2004.

12. M. M. Henry. Model-based estimation of probabilistic hybrid automata, 2002.
13. A. L. Hodgkin and A. F. Huxley. A quantitative description of membrane currents and its application to conduction and excitation in nerve. *J Physiol*, 117:500544, 1952.
14. M. Huber and R.A. Grupen. A hybrid architecture for learning robot control tasks. *Robotics Today*, 13(4), 2000.
15. K. Joshi, N. Neogi, and W. Sanders. Dynamic partitioning of large discrete event biological systems for hybrid simulation and analysis. In *Hybrid Systems: Computation and Control, HSCC 2004*, volume 2993 of *LNCS*, 2004.
16. D.K. Kaynar, N. Lynch, R. Segala, and F. Vaandrager. *The Theory of Timed I/O Automata.* Synthesis Lectures on Computer Science. Morgan Claypool, November 2005. Also available as Technical Report MIT-LCS-TR-917.
17. P. Lincoln and A. Tiwari. Symbolic systems biology: Hybrid modeling and analysis of biological networks. In *Hybrid Systems: Computation and Control, HSCC 2004*, volume 2993 of *LNCS*, pages 660672, 2004.
18. C. H. Luo and Y. Rudy. A dymanic model of the cardiac ventricular action potential: I. simulations of ionic currents and concentration changes. *Circ Res*, 74:1071 1096, 1994.
19. N. Lynch, R. Segala, and F. Vaandrager. Hybrid I/O automata.*Information and Computation*, 185(1):103157, 2003.
20. S. Mitra, D. Liberzon, and N. Lynch. Verifying average dwell time by solving optimization problems. In Ashish Tiwari and Joao P. Hespanha, editors,*Hybrid Systems: Computation and Control (HSCC 06)*, LNCS, Santa Barbara, CA, March 2006. Springer.
21. M.R. Osborne and G.K. Smyth. A modified Prony algorithm for exponential function fitting.*SIAM J. Sci. Comput.*, 16(1):119138, 1995.
22. V. Pavlovic, J.M. Rehg, T.-J. Cham, and K.P. Murphy. A dynamic bayesian network approach to figure tracking using learned dynamic models. In*Proc. of 7th IEEE Int. Conf. on Computer Vision*, 1999.
23. A. Singh and J. Hespanha. Models for generegulatory networks using polynomial stochastic hybrid systems. In*CDC05*, 2005.
24. R. Vidal, S. Soatto, Y. Ma, and S. Sastry. An algebraic geometric approach to the identification of a class of linear hybrid systems. In *Proc. of 42nd IEEE Conf. on Decision and Control*, 2003.
25. P. Ye, E. Entcheva, R. Grosu, and S.A. Smolka. Efficient modeling of excitable cells using hybrid automata. In CMSB05, *Computational Methods in Systems Biology Workshop*, Edinburgh, UK, April 2005.
26. P. Ye, E. Entcheva, S.A. Smolka, M.R. True, and R. Grosu. A cycle-linear approach to modeling action potentials. In *EMBC06, the IEEE International Conference of the Engineering in Medicine and Biology Society*, IEEE, New York City, NY, September 2006. IEEE.

Input-to-State Stability of Discontinuous Dynamical Systems with an Observer-Based Control Application*

W.P.M.H. Heemels, S. Weiland, and A. Lj. Juloski

Eindhoven University of Technology, P.O. Box 513,
5600 MB Eindhoven, The Netherlands
m.heemels@tue.nl

Abstract. In this paper we will extend the input-to-state stability (ISS) framework to continuous-time discontinuous dynamical systems adopting Filippov's solution concept and using non-smooth ISS Lyapunov functions. The main motivation for adopting non-smooth ISS Lyapunov functions is that "multiple Lyapunov functions" are commonly used in the stability theory for hybrid systems. We will show that the existence of a non-smooth (but Lipschitz continuous) ISS Lyapunov function for a discontinuous system implies ISS. Next, we will prove an ISS interconnection theorem for two discontinuous dynamical systems that both admit an ISS Lyapunov function. The interconnection will be shown to be globally asymptotically stable under a small gain condition. The developed ISS theory will be applied to observer-based controller design for a class of piecewise linear systems using an observer structure proposed by the authors. The LMI-based design of the state feedback and the observer can be performed separately.

1 Introduction

For plants in which the state variable is not available for feedback, one often resorts to an observer-based controller. Typically, such a controller consists of an observer that generates on the basis of inputs and outputs of the plant an estimate of the state variable. This estimate is substituted in a state feedback controller to generate the inputs to the plant (see Figure 1). The certainty equivalence principle is a rigorous justification for such a substitution. If the plant is linear (and detectable and stabilizable) one can separately design an observer with asymptotically stable estimation error dynamics and a state feedback controller that stabilizes the plant. The interconnection of the observer-based controller and the plant can be proven to be globally asymptotically stable (GAS). The linear case has been extended into the non-linear smooth direction by considering the concept of input-to-state stability (ISS), see e.g. [1,2,3,4,5]. The approach of designing the observer and the controller separately (as in the linear case), and

* This work is partially supported by European project grants SICONOS (IST2001-37172) and HYCON Network of Excellence, contract number FP6-IST-511368.

A. Bemporad, A. Bicchi, and G. Buttazzo (Eds.): HSCC 2007, LNCS 4416, pp. 259–272, 2007.

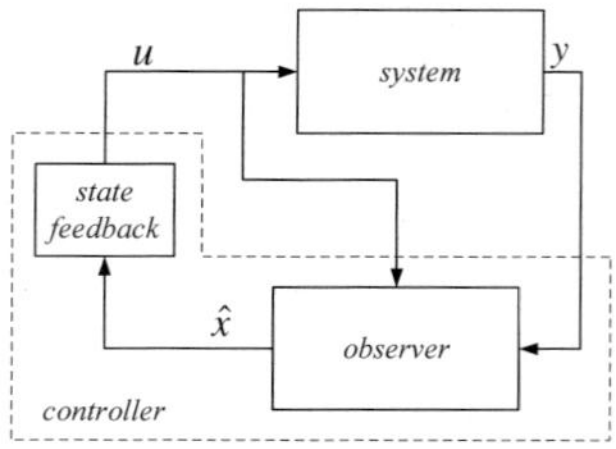

Fig. 1. Structure of an observer-based feedback controller

applying the ISS-interconnection approach to prove the stability of the closed loop system was used for *Lipschitz continuous* nonlinear systems in e.g. [6,7]. These results apply to continuous systems, and the designed observers result in GAS estimation error dynamics. However, there exist situations in which it is not possible to obtain GAS observers. One such situation of our interest are piecewise affine (PWA) systems, as we will see below. In that case one might still be able to design an observer that yields an estimation error dynamics that is ISS from state variable to estimation error. In case the plant-state feedback combination is ISS from estimation error to state variable, the closed-loop system can still be GAS, if a suitable small gain condition is satisfied. The (small gain) conditions for the stability of general interconnected *Lipschitz continuous* systems were presented in [3,5].

For continuous-time systems that are in general discontinuous, like PWA systems, the theory of [3,5] does not apply. The first reason that hampers the use of the results in [3,5] is that the system does not have a Lipschitz continuous vector field. The second reason is that the (ISS) Lyapunov functions for hybrid systems are generally non-smooth, while [1,2,3,4,5] require smooth Lyapunov functions. Indeed, for hybrid systems one typically uses multiple Lyapunov functions (see e.g. [8,9]). In particular for piecewise affine systems, piecewise quadratic Lyapunov functions (see e.g. [10,11]) are popular as they can be constructed from linear matrix inequalities (LMIs), which has clear computational advantages. This motivates the development of ISS theory for *continuous-time discontinuous* systems by using *non-smooth* Lyapunov functions.

Stability theory for continuous-time hybrid systems using non-smooth Lyapunov functions is well developed (see e.g. [8,9,10,11] and the references therein). Typically, the solution trajectories are considered in the traditional sense instead of a generalized sense like Filippov's definition [12]. An exception is formed by the work [13] in which stability theory is developed for discontinuous dynamical systems adopting non-smooth Lipschitz continuous Lyapunov functions. We will extend this work towards ISS and ISS interconnection theorems for Filippov's solution concept (including sliding modes). Recently, ISS of continuous-time hybrid systems is being researched [14,15,16]. However, the emphasis in [14,15,16] is on *smooth* Lyapunov functions, which might not have the computational advantages that non-smooth Lyapunov functions have. As such, the work in this paper extends [13] towards ISS and ISS interconnection theory, and extends the line of research in [14,15,16] towards using non-smooth ISS Lyapunov functions.

The developed general ISS theory is used to design an observer-based feedback controller for a class of piecewise linear (PWL) systems in which the currently active mode of the system is not known. The proposed output-feedback controller consists of a static state feedback and a switching state observer as proposed by the authors in [17]. Both the synthesis of the observer and state feedback will be based on LMIs. The interconnection of these ISS subsystems forms a typical situation that requires interconnection theory using non-smooth ISS Lyapunov functions as developed in this paper.

Notation. $\mathbb{R}_+$ denotes all nonnegative real numbers. For a set $\Omega \subseteq \mathbb{R}^n$, $\text{int}\Omega$ denotes its interior and $\text{co}\Omega$ its convex hull. A set $\Omega \subseteq \mathbb{R}^n$ is called a polyhedron, if it is the intersection of a finite number of open or closed half spaces. For a real-valued, differentiable function V, ∇V denotes its gradient. For a positive semi-definite matrix $A \in \mathbb{R}^n$, $\lambda_{min}(A)$ and $\lambda_{max}(A)$ will denote its minimal and maximal eigenvalue. In matrices we denote by $(*)$ at block position (i, j) the transposed matrix block at position (j, i), e.g. $\begin{bmatrix} A & B \\ (*) & C \end{bmatrix}$ means $\begin{bmatrix} A & B \\ B^\top & C \end{bmatrix}$, where $B^\top$ denotes the transposed matrix of B. The operator $\text{col}(\cdot, \cdot)$ stacks its arguments into a column vector, e.g. for $a \in \mathbb{R}^n$ and $b \in \mathbb{R}^m$ $\text{col}(a, b) = (a^\top, b^\top)^\top \in \mathbb{R}^{n+m}$. A function $u : \mathbb{R}_+ \mapsto \mathbb{R}^n$ is *piecewise continuous*, if on every bounded interval the function has only a finite number of points at which it is discontinuous. Without loss of generality we will assume that every piecewise continuous function u is right continuous, i.e. $\lim_{t\downarrow\tau} u(t) = u(\tau)$ for all $\tau \in \mathbb{R}_+$. With $|\cdot|$ we will denote the usual Euclidean norm for vectors in $\mathbb{R}^n$, and $\|\cdot\|$ denotes the L_∞ norm for time functions, i.e. $\|u\| = \sup_{t\in\mathbb{R}_+} |u(t)|$ for a time function $u : \mathbb{R}_+ \mapsto \mathbb{R}^n$. For two functions f and g we denote by $f \circ g$ their composition, i.e. $(f \circ g)(x) = f(g(x))$. A function $\gamma : \mathbb{R}_+ \to \mathbb{R}_+$ is of class $\mathcal{K}$ if it is continuous, strictly increasing and $\gamma(0) = 0$. It is of class $\mathcal{K}_\infty$ if, in addition, it is unbounded, i.e. $\gamma(s) \to \infty$ as $s \to \infty$. A function $\beta : \mathbb{R}_+ \times \mathbb{R}_+ \to \mathbb{R}_+$ is of class $\mathcal{KL}$ if, for each fixed $t \in \mathbb{R}_+$, the function $\beta(\cdot, t)$ is of class $\mathcal{K}$, and for each fixed $s \in \mathbb{R}_+$, the function $\beta(s, \cdot)$ is decreasing and tends to zero at infinity. A function $\gamma : \mathbb{R}_+ \to \mathbb{R}_+$ is called positive definite, if $\gamma(s) > 0$, when $s > 0$.

2 ISS for Discontinuous Dynamical Systems Using Non-smooth Lyapunov Functions

Consider the differential equation with discontinuous right-hand side of the form

$$\dot{x}(t) = f(x(t), u(t)) \tag{1}$$

with $x(t) \in \mathbb{R}^n$ and $u(t) \in \mathbb{R}^m$, the state and control input at time $t \in \mathbb{R}_+$, respectively. The vector field f is assumed to be a piecewise continuous function from $\mathbb{R}^n \times \mathbb{R}^m$ to $\mathbb{R}^n$ in the sense that

$$f(x, u) = f_i(x, u) \text{ when } \text{col}(x, u) \in \Omega_i, \; i = 1, 2, \ldots, N. \tag{2}$$

Here, $\Omega_1, \ldots, \Omega_N$ are closed subsets of $\mathbb{R}^n \times \mathbb{R}^m$ that form a partitioning of the space $\mathbb{R}^n \times \mathbb{R}^m$ in the sense that $\text{int}\Omega_i \cap \text{int}\Omega_j = \emptyset$, when $i \neq j$ and $\bigcup_{i=1}^N \Omega_i = \mathbb{R}^n \times \mathbb{R}^m$. Moreover, we assume that $\Omega_j \subseteq \text{cl}(\text{int}\Omega_j)$ for all $j = 1, \ldots, N$ implying that Ω_j is a not a subset of a lower dimensional manifold. The functions $f_i : \Omega_i \mapsto \mathbb{R}^n$ are locally Lipschitz continuous on their domains Ω_i (this means including the boundary). The class of piecewise affine systems forms a particular instance of piecewise continuous systems.

Since the system (1) has a discontinuous right-hand side, one has to use a generalized solution concept. The most commonly used solution concept in this context, is Filippov's convex definition [12, p. 50]. This replaces the differential equation (1) by a differential inclusion[1] of the form

$$\dot{x}(t) \in F(x(t), u(t)) \tag{3}$$

with

$$F(x,u) = \text{co}\{f_i(x,u) \mid i \in I(x,u)\} \text{ and } I(x,u) := \{i \in \{1, \ldots, N\} \mid \text{col}(x,u) \in \Omega_i\}. \tag{4}$$

$I(x,u)$ is an index set indicating the regions Ω_i to which the state-input vector $\text{col}(x,u)$ belongs.

Definition 1. *A function $x : [a,b] \mapsto \mathbb{R}^n$ is a Filippov solution to (1) for the piecewise continuous input function $u : [a,b] \mapsto \mathbb{R}^n$, if it is a* solution *to (3) for the input u, i.e. x is absolutely continuous*[2] *and satisfies $\dot{x}(t) \in F(x(t), u(t))$ for almost all $t \in [a,b]$.*

Under the conditions given here, it can be shown by using § 2.6 (page 69) in [12] that the mapping $(t,x) \mapsto F(x, u(t))$ defined as in (4) for any bounded piecewise continuous function u is upper semicontinuous in (t,x) on $I \times \mathbb{R}^n$, where I indicates any interval where u is continuous. As $F(x,u)$ as in (4) is bounded, convex and closed for any $\text{col}(x,u) \in \mathbb{R}^{n+m}$, local existence of solutions to (1) given an initial condition $x(t_0) = x_0$ and a piecewise continuous input function is guaranteed from Theorem 1, page 77 in [12]. To obtain also uniqueness, additional conditions have to be imposed on (1), see e.g. § 10 in [12].

2.1 ISS for Discontinuous Dynamical Systems

Given the possible non-uniqueness of Filippov solution trajectories, we define the concept of input-to-state stability (ISS) for (1) [1,3,5] as follows.

---

[1] Strictly speaking, we embedded the space of Filippov solutions in the solution space of the differential inclusion (3), which might be larger. Under conditions on the regions Ω_i and the input function u, the Filippov solution space and the solution space to (3) coincide, cf. § 2.6 in [12]. As the Filippov solution space is always included in the solution space of (3), properties of solutions to (3) hold for Filippov solutions as well.

[2] A function $x : [a,b] \mapsto \mathbb{R}^n$ is called absolutely continuous, if x is continuous and there exists a function $\dot{x}$ in $L_1[a,b]$, the set of integrable functions, such that $x(t) = x(a) + \int_a^t \dot{x}(\tau)d\tau$ for all $t \in [a,b]$. The function $\dot{x}$ is called the derivative of x on $[a,b]$. This implies that x is almost everywhere differentiable.

Definition 2. *The system* (1) *is said to be input-to-state stable (ISS) if there exists a function* β *of class* $\mathcal{KL}$ *and a function* γ *of class* $\mathcal{K}$ *such that for each initial condition* $x(0) = x_0$ *and each piecewise continuous bounded input function* u *defined on* $[0, \infty)$,

- *all corresponding Filippov solutions* x *of the system* (1) *exist on* $[0, \infty)$ *and,*
- *all corresponding Filippov solutions satisfy*

$$|x(t)| \leq \beta(|x_0|, t) + \gamma(\|u\|), \quad \forall t \geq 0. \tag{5}$$

In the study of hybrid systems often non-smooth or multiple Lyapunov functions are employed, see for instance [8,9,10,11]. As such, we will consider *continuous* Lyapunov functions that are composed of "multiple" Lyapunov functions V_j as

$$V(x) = V_j(x) \text{ when } x \in \Gamma_j,\ j = 1, \ldots, M, \tag{6}$$

where $\Gamma_1, \ldots, \Gamma_M$ are closed subsets of $\mathbb{R}^n$ that form a partitioning of the space $\mathbb{R}^n$, i.e. $\text{int}\Gamma_i \cap \text{int}\Gamma_j = \emptyset$, when $i \neq j$ and $\bigcup_{i=1}^{N} \Gamma_i = \mathbb{R}^n$. As before, we also assume $\Gamma_j \subseteq \text{cl}(\text{int}(\Gamma_j))$ for $j = 1, \ldots, M$. For each j we assume that V_j is a continuously differentiable function on some open domain containing Γ_j. Continuity of V implies that $V_i(x) = V_j(x)$ when $x \in \Gamma_i \cap \Gamma_j$. The continuity of the Lyapunov function is a typical condition used in the study of stability for piecewise affine systems in *continuous-time*, see e.g. [10]. Similarly, as in (4), we define the index set $J(x)$ as

$$J(x) := \{j \in \{1, \ldots, M\} \mid x \in \Gamma_j\}. \tag{7}$$

Definition 3. *A function* V *of the form* (6) *is said to be an ISS-Lyapunov function for the system* (1) *if:*

- V *is Lipschitz continuous,*
- *there exist functions* ψ_1, ψ_2 *of class* $\mathcal{K}_\infty$ *such that:*

$$\psi_1(|x|) \leq V(x) \leq \psi_2(|x|), \quad \forall x \in \mathbb{R}^n, \tag{8}$$

- *there exist functions* χ *of class* $\mathcal{K}$ *and* α *positive definite and continuous such that for all* $x \in \mathbb{R}^n$ *and* $u \in \mathbb{R}^m$ *the implication*

$$\{|x| \geq \chi(|u|)\} \Rightarrow \{\nabla V_j(x) f_i(x, u) \leq -\alpha(V(x)), \textit{for all } i \in I(x, u),\ j \in J(x)\} \tag{9}$$

holds, or stated differently, for all $x \in \mathbb{R}^n$ *and* $u \in \mathbb{R}^m$

$$\{|x| \geq \chi(|u|), \text{col}(x, u) \in \Omega_i \textit{ and } x \in \Gamma_j\} \Rightarrow \{\nabla V_j(x) f_i(x, u) \leq -\alpha(V(x))\}. \tag{10}$$

Definition 3 is similar to the one proposed in [1,3,5], the only difference being that here we use *non-smooth* Lyapunov functions and that it is used for systems (1) in which the vector field might be discontinuous.

We first derive conditions on the time derivative of an ISS Lyapunov V along Filippov solutions x of system (1) provided $\frac{dV}{dt}(x(t))$ and $\dot{x}(t)$ exist at time t. The complications are that a solution trajectory might go along a surface on which ∇V does not exist and that solutions are of Filippov type.

Theorem 1. *If there exists an ISS Lyapunov function V of the form (6) for system (1) in the sense of Definition 3, then*

$$\frac{d}{dt}V(x(t)) \leq -\alpha(V(x(t))) \tag{11}$$

at times t, where both $\frac{dV}{dt}(x(t))$ and $\dot{x}(t)$ exist and $|x(t)| \geq \chi(|u(t)|)$.

Proof. In [12, § 15] it is shown that if at time t both $\dot{x}(t)$ and $\frac{dV(x(t))}{dt}$ exist, then

$$\frac{d}{dt}V(x(t)) = \frac{d}{dh}V(x(t)+hy)\mid_{h=0} = \lim_{h\to 0}\frac{V(x(t)+hy)-V(x(t))}{h}, \tag{12}$$

where $\dot{x}(t) = y \in F(x(t), u(t))$. Hence, due to (4) y can be written as

$$y = \sum_{i\in I(x(t),u(t))} \alpha_i f_i(x(t), u(t)) \tag{13}$$

with $\alpha_i \geq 0$, $i \in I(x(t), u(t))$ and $\sum_{i\in I(x(t),u(t))}\alpha_i = 1$. As $x(t) \in \Gamma_j$ iff $j \in J(x(t))$ and V is continuous we have that $V(x(t)) = V_j(x(t))$ for all $j \in J(x(t))$. To evaluate the right-hand side of (12), we have to realize that $V(x(t) + hy) = V_j(x(t) + hy)$ for all $j \in J(x(t) + hy)$ in which the set $J(x(t) + hy)$ depends on h. Since $\frac{dV}{dt}(x(t))$ exists, this means that

$$\frac{d}{dt}V(x(t)) = \lim_{h\downarrow 0}\frac{V_j(x(t)+hy)-V_j(x(t))}{h},$$

for all $j \in \bar{J}(x(t), y)) := \bigcap_{h_0>0}\bigcup_{0<h<h_0} J(x(t) + hy)$. Due to closedness of Γ_j, it holds that $d(x(t)), \Gamma_j) := \inf_{z\in\Gamma_j}|z - x(t)| > 0$, when $j \notin J(x(t))$. Hence, for sufficiently small h, we have that $x(t)+hy \in \bigcup_{j\in J(x(t))}\Gamma_j$ and thus $J(x(t)+hy) \subseteq J(x(t))$ for sufficiently small h and thus $\bar{J}(x(t), y)) \subseteq J(x(t))$. Hence, we can conclude that for $x(t) \neq 0$ with $|x(t)| \geq \chi(|u(t)|)$ that

$$\begin{aligned}\frac{d}{dt}V(x(t)) &\leq \max_{j\in J(x(t))}\lim_{h\to 0}\frac{V_j(x(t)+hy)-V_j(x(t))}{h} = \max_{j\in J(x(t))}\nabla V_j(x(t))y \overset{(13)}{=} \\ &\overset{(13)}{=} \max_{j\in J(x(t))}\sum_{i\in I(x(t),u(t))}\alpha_i\nabla V_j(x(t))f_i(x(t),u(t)) \leq \\ &\leq \max_{i\in I(x(t),u(t)),j\in J(x(t))}\nabla V_j(x(t))f_i(x(t),u(t)) \leq -\alpha(V(x(t))).\end{aligned} \tag{14}$$

Using the above theorem we can now prove that the existence of an ISS Lyapunov function implies ISS of the system.

Theorem 2. *If there exists an ISS Lyapunov function V of the form (6) for system (1) in the sense of Definition 3, then system (1) is ISS.*

Proof. Consider initial condition $x(0) = x_0$ and let u be a piecewise continuous bounded input function. Let x denote a corresponding Filippov solution (might be non-unique) to (1). Define the set $S := \{x \mid V(x) \leq c\}$ with $c := \psi_2(\chi(\|u\|))$. Note that when $x(t) \notin S$, then $\psi_2(|x(t)|) \geq V(x(t)) > \psi_2(\chi(\|u\|))$, which implies $|x(t)| \geq \chi(|u(t)|)$ and thus $\nabla V_j(x(t)) f_i(x(t), u(t)) \leq -\alpha(V(x(t)))$ for all $i \in I(x(t), u(t))$ and all $j \in J(x(t))$. According to Theorem 1, inequality (11) holds for $x(t) \notin S$ (provided $\dot{x}(t)$ and $\frac{d}{dt}V(x(t))$ exist). We prove the following claim.

Claim: S is positively invariant, i.e. if there exists a t_0 such that $x(t_0) \in S$, then $x(t) \in S$ for all $t \geq t_0$.

Indeed, suppose this statement is not true. Due to closedness of S (continuity of V) there is an $\varepsilon > 0$ and time $\tilde{t} > t_0$ with $V(x(\tilde{t})) \geq c+\varepsilon$. Let $t^* := \inf\{t \geq t_0 \mid V(x(t)) \geq c+\varepsilon\}$ and $t_* := \sup\{t_0 \leq t \leq t^* \mid V(x(t)) \leq c\}$. Note that $t_0 \leq t_* < t^*$ and $V(x(t_*)) = c$ by continuity of V and x. Since $x(t) \notin S$ for $t \in (t_*, t^*)$ (11) holds if both $\dot{x}(t)$ and $\frac{dV(x(t))}{dt}$ exist. Since V is locally Lipschitz continuous and any solution to (3) is absolutely continuous, the composite function $t \mapsto V(x(t))$ is absolutely continuous and consequently, $t \mapsto V(x(t))$ is differentiable almost everywhere (a.e.) with respect to time t, and $\dot{x}(t)$ exists also a.e. Therefore,

$$V(x(t^*)) - V(x(t_*)) = \int_{t_*}^{t^*} \frac{dV(x(\tau))}{dt} d\tau \leq \int_{t_*}^{t^*} -\alpha(V(x(\tau)))d\tau \leq 0.$$

Hence, $V(x(t^*)) \leq V(x(t_*)) = c$, thereby contradicting that $V(x(t_*)) \geq c + \varepsilon$. This proves the claim.

Now let $t_1 = \inf\{t \geq 0 \mid x(t) \in S\} \leq \infty$ (note that t_1 might be infinity). Then it follows from the above reasoning and (8) that $\psi_1(|x(t)|) \leq V(x(t)) \leq c := \psi_2(\chi(\|u\|))$ for all $t \geq t_1$. Hence,

$$|x(t)| \leq \gamma(\|u\|) \text{ for all } t \geq t_1 \tag{15}$$

with $\gamma := \psi_1^{-1} \circ \psi_2 \circ \chi$ a $\mathcal{K}$-function. For $t < t_1$, $x(t) \notin S$ and consequently, (11) holds almost everywhere in $[0, t_1)$. This yields $\frac{d}{dt}V(x(t)) \leq -\alpha(V(x))$ a.e. in $[0, t_1)$. Lemma 4.4. in [4] now gives that there exists a $\mathcal{KL}$ function $\tilde{\beta}$ (only depending on α) such that $V(x(t)) \leq \tilde{\beta}(V(x_0), t))$ for $t \leq t_1$. Hence,

$$|x(t)| \leq \beta(x_0, t) \text{ for all } t \leq t_1, \tag{16}$$

where $\beta(r, t) := \psi_1^{-1}(\tilde{\beta}(\psi_2(r), t))$ is a $\mathcal{KL}$ function as well. Combining (15) and (16) yields (5) for this particular trajectory. Global existence of any trajectory can also be proven via Theorem 2 page 78 [12] by using the bound (5) (that shows that there cannot be "finite escape times.") As β and γ do not rely on the particular initial state nor on the input u, this proves ISS of the system. □

Remark 1. The proof follows similar lines as the proof of [2, Lemma 2.14] with the necessary adaptations for the non-smoothness of V and the discontinuity of the dynamics using Theorem 1.

2.2 Stability of Discontinuous Dynamical Systems

Consider the autonomous variant of the discontinuous system (1) given by

$$\dot{x}(t) = f(x(t)) = f_i(x(t)) \text{ when } x(t) \in \Omega_i \subseteq \mathbb{R}^n \tag{17}$$

with a similar generalization in terms of a differential inclusion

$$\dot{x}(t) \in F(x(t)). \tag{18}$$

We assume that 0 is an equilibrium of (17) (or equivalently (18)), which means that $f_i(0) = 0$ for all $i \in I(0)$ and thus $F(0) = \{0\}$.

Definition 4. *The system* (17) *is said to be globally asymptotically stable (GAS), if there exists a function β of class $\mathcal{KL}$ such that for each $x_0 \in \mathbb{R}^n$, all Filippov solutions x of the system* (1) *with initial condition $x(0) = x_0$ exist on $[0, \infty)$ and satisfy:*

$$|x(t)| \leq \beta(|x(0)|, t), \quad \forall t \geq 0. \tag{19}$$

As a corollary of Theorem 2 we obtain the following result.

Theorem 3. *Consider the discontinuous dynamical system* (17) *and a Lipschitz continuous function V of the form* (6). *Assume that*

– *there exist functions ψ_1, ψ_2 of class $\mathcal{K}_\infty$ such that:*

$$\psi_1(|x|) \leq V(x) \leq \psi_2(|x|), \quad \forall x \in \mathbb{R}^n$$

– *there exists a continuous positive definite function α such that*

$$\nabla V_j(x) f_i(x) \leq -\alpha(V(x)) \text{ when } x \in \Omega_i \cap \Gamma_j. \tag{20}$$

Then the discontinuous dynamical system (17) *is globally asymptotically stable.*

The above theorem is closely related to one of the main results (Theorem 3.1 to be precise) in [13]. In case of the particular form of V and f, the results coincide. Hence, in the particular context considered here one of the main result of [13] is recovered as a special case of our ISS result (Theorem 2).

2.3 An Interconnection Result

Consider the interconnected system (we dropped time t for shortness)

$$\dot{x}_a = f^a(x_a, x_b) = f^a_{i_a}(x_a, x_b) \text{ if } \mathrm{col}(x_a, x_b) \in \Omega^a_{i_a} \text{ for } i_a = 1, \ldots, N^a \tag{21a}$$
$$\dot{x}_b = f^b(x_a, x_b) = f^b_{i_b}(x_a, x_b) \text{ if } \mathrm{col}(x_a, x_b) \in \Omega^b_{i_b} \text{ for } i_b = 1, \ldots, N^b \tag{21b}$$

with partitionings $\{\Omega^a_1, \ldots, \Omega^a_{N^a}\}$ and $\{\Omega^b_1, \ldots, \Omega^b_{N^b}\}$, respectively, of $\mathbb{R}^{n_a+n_b}$ as in (1), where $x_a \in \mathbb{R}^{n_a}$ and $x_b \in \mathbb{R}^{n_b}$. $I^a(x_a, x_b)$ and $I^b(x_a, x_b)$ are defined similarly as in (4). The interconnected system in the combined state variable $x = \mathrm{col}(x_a, x_b)$ is given by

$$\dot{x} = f(x) = f_{(i_a, i_b)}(x) = \mathrm{col}(f^a_{i_a}(x_a, x_b), f^b_{i_b}(x_a, x_b)) \text{ when } x \in \Omega^a_{i_a} \cap \Omega^b_{i_b} \tag{22}$$

for each pair $(i_a, i_b) \in \{1, \ldots, N^a\} \times \{1, \ldots, N^b\}$. Hence, we have (at most) $N := N_a N_b$ regions for the interconnected system. We take $I(x)$ as given in (4) for the interconnected system in the form $I(x) = \{(i_a, i_b) \mid i_a \in I^a(x_a, x_b),\ i_b \in I^b(x_a, x_b)\}$.

Theorem 4. *Suppose that there exist ISS Lyapunov functions V^a and V^b of the form* (6) *for the systems* (21a) *and* (21b)*, respectively. Let* $(\psi_1^a, \psi_2^a, \chi^a, \alpha^a)$ *and* $(\psi_1^b, \psi_2^b, \chi^b, \alpha^b)$ *denote the bounding functions corresponding to V^a and V^b in the sense of Definition 3 with χ^a and χ^b $\mathcal{K}_\infty$-functions. Define*

$$\tilde{\chi}^a := \psi_2^a \circ \chi^a \circ [\psi_1^b]^{-1}, \qquad \tilde{\chi}^b := \psi_2^b \circ \chi^b \circ [\psi_1^a]^{-1}$$

and assume that the coupling condition

$$\tilde{\chi}^a \circ \tilde{\chi}^b(r) < r$$

holds for all $r > 0$. Then the interconnected system (21) is GAS.

Proof. The differential inclusion that replaces (21), when Filippov solutions are used, is given by $\dot{x} \in F(x)$ with $F(x) = F^a(x_a, x_b) \times F^b(x_a, x_b)$ and $F^a(x_a, x_b)$ the set (4) for (21a) and $F^b(x_a, x_b)$ the set (4) for (21b). Due to the coupling condition it holds that $\tilde{\chi}^b(r) < [\tilde{\chi}^a]^{-1}(r)$ for $r > 0$. According to Lemma A.1 in [3] there exists a $\mathcal{K}_\infty$-function σ, which is continuously differentiable and satisfies $\tilde{\chi}^b(r) < \sigma(r) < [\tilde{\chi}^a]^{-1}(r)$ for all $r > 0$ and $\sigma'(r) > 0$ for all $r > 0$ (thus the derivative σ' is positive definite and continuous).

Define the Lipschitz continuous function V similar as in [3] with $x = \mathrm{col}(x_a, x_b)$

$$V(x) = \max\{\sigma(V^a(x_a)), V^b(x_b)\}. \tag{23}$$

This function will be proven to be a Lyapunov function for (21). The function V is in the form (6) with a partitioning induced by the partitioning $\{\Gamma_1^a, \ldots, \Gamma_{M^a}^a\}$ of V^a, the partitioning $\{\Gamma_1^b, \ldots, \Gamma_{M^b}^b\}$ of V^b and the additional split up given by $\sigma(V^a(x_a)) \geq V^b(x_b)$ or $\sigma(V^a(x_a)) \leq V^b(x_b)$. Hence, $V(x) = V_{(j_a, j_b, 0)}(x) := \sigma(V_{j_a}^a(x_a))$, when $\sigma(V^a(x_a)) \geq V^b(x_b)$, $x_a \in \Gamma_{j_a}^a$, $x_b \in \Gamma_{j_b}^b$ and $V(x) = V_{(j_a, j_b, 1)}(x) := V_{j_b}^b(x_b)$, when $\sigma(V^a(x_a)) \leq V^b(x_b)$, $x_a \in \Gamma_{j_a}^a$, $x_b \in \Gamma_{j_b}^b$. Note that there is a slight abuse of notation as we characterize (7) using "indices" consisting of triples (j_a, j_b, p) with $p = 1$ related to $\sigma(V^a(x_a)) \leq V^b(x_b)$ and $p = 0$ related to $\sigma(V^a(x_a)) \geq V^b(x_b)$.

Note that

$$\max(\sigma(\psi_1^a(|x_a|)), \psi_1^b(|x_b|)) \leq V(x) \leq \max(\sigma(\psi_2^a(|x|)), \psi_2^b(|x|)).$$

As either $|x|^2 = |x_a|^2 + |x_b|^2 \leq 2|x_a|^2$ or $|x|^2 \leq 2|x_b|^2$, we obtain that

$$\begin{aligned} \max(\sigma(\psi_1^a(|x_a|)), \psi_1^b(|x_b|)) &\geq \frac{1}{2}[\sigma(\psi_1^a(|x_a|)) + \psi_1^b(|x_b|)] \geq \\ &\geq \frac{1}{2}\min[\sigma(\psi_1^a(\frac{1}{2}\sqrt{2}|x|)), \psi_1^b(\frac{1}{2}\sqrt{2}|x|)]. \end{aligned}$$

Since the minimum and maximum of two $\mathcal{K}_\infty$-functions are also $\mathcal{K}_\infty$-functions, we obtain that V is lower and uppper bounded by $\mathcal{K}_\infty$-functions. This proves that the first hypothesis of Theorem 3 is satisfied.

To check the second hypothesis of Theorem 3 let $x = \mathrm{col}(x_a, x_b) \in \Omega_{i_a}^a \cap \Omega_{i_b}^b$ and $x \in [\Gamma_{j_a}^a \times \Gamma_{j_b}^a]$.

Case 1: If $V^b(x_b) \leq \sigma(V^a(x_a))$, then $V(x) = \sigma(V^a(x_a)) = \sigma(V^a_{j_a}(x_a))$ and we have to verify (20) for "index" $(j_a, j_b, 0)$. The properties of σ and the definition of $\tilde{\chi}^a$ yield

$$\psi^b_1(|x_b|) \leq V^b(x_b) \leq \sigma(V^a(x_a)) < [\tilde{\chi}^a]^{-1}(V^a(x_a)) \leq \psi^b_1 \circ [\chi^a]^{-1}(|x_a|).$$

This implies that $|x_a| > \chi^a(|x_b|)$. Using (10) for subsystem (21a) gives

$$\nabla V_{(j_a,j_b,0)}(x_a, x_b) f_{(i_a,i_b)}(x) = \sigma'(V^a_{j_a}(x_a)) \nabla V^a_{j_a}(x_a) f^a_{i_a}(x_a, x_b) \leq$$
$$\leq -\sigma'(V^a_{j_a}(x_a))\alpha^a(V^a_{j_a}(x_a)) = -\tilde{\alpha}^a(V(x)), \tag{24}$$

if we set $\tilde{\alpha}^a(r) := \sigma'(\sigma^{-1}(r))\alpha^a(\sigma^{-1}(r))$, which is a positive definite and continuous function.

Case 2: If $V^b(x_b) \geq \sigma(V^a(x_a))$, then $V(x) = V^b(x_b) = V^b_{j_b}(x_b)$. Then using the properties of σ we obtain

$$\psi^b_2 \circ \chi^b(|x_a|) \leq \tilde{\chi}^b(V^a(x_a)) < \sigma(V^a(x_a)) \leq V^b(x_b) \leq \psi^b_2(|x_b|),$$

which implies that $|x_b| > \chi^b(|x_a|)$. Applying (10) for subsystem (21b) gives

$$\nabla V_{(j_a,j_b,1)}(x_a, x_b) f_{(i_a,i_b)}(x) = \nabla V^b_{j_b}(x_b) f^b_{i_b}(x_a, x_b) \leq -\alpha^b(V^b_{j_b}(x_b)) = -\alpha^b(V(x))). \tag{25}$$

Hence, V is a Lyapunov function for system (21) with $\alpha(s) = \min[\tilde{\alpha}^a(s), \alpha^b(s)]$ which is a positive definite and continuous function. Hence, GAS follows from Theorem 3.

3 Observer-Based Controllers for a Class of Pwl Systems

As an illustration of the above results, consider the bimodal PWL system

$$\dot{x}(t) = \begin{cases} A_1 x(t) + Bu(t), & \text{if } H^\top x(t) \leq 0 \\ A_2 x(t) + Bu(t), & \text{if } H^\top x(t) \geq 0 \end{cases} \tag{26a}$$

$$y(t) = Cx(t), \tag{26b}$$

where $x(t) \in \mathbb{R}^n$, $y(t) \in \mathbb{R}^p$ and $u(t) \in \mathbb{R}^m$ are the state, output and the input, respectively at time $t \in \mathbb{R}^+$ and $A_i \in \mathbb{R}^{n \times n}$, $i = 1, 2$, $B \in \mathbb{R}^{n \times m}$, $C \in \mathbb{R}^{p \times n}$ and $H \in \mathbb{R}^n$. The input $u : \mathbb{R}^+ \to \mathbb{R}^m$ is assumed to be a piecewise continuous function and solutions are considered in the Filippov sense.

We will design a stabilizing output-based controller for (26) consisting of an observer and a state feedback using the estimated state (cf. Figure 1).

3.1 Observer Design

As an observer for the system (26), we take a continuous-time bimodal system with a structure as proposed in [17]:

$$\dot{\hat{x}} = \begin{cases} A_1 \hat{x} + Bu + L_1(y - \hat{y}), & \text{if } H^\top \hat{x} + M^\top (y - \hat{y}) \leq 0 \\ A_2 \hat{x} + Bu + L_2(y - \hat{y}), & \text{if } H^\top \hat{x} + M^\top (y - \hat{y}) \geq 0 \end{cases} \tag{27a}$$

$$\hat{y} = C\hat{x}, \tag{27b}$$

where $\hat{x}(t) \in \mathbb{R}^n$ is the estimated state at time t and $L_1, L_2 \in \mathbb{R}^{n\times p}$ and $M \in \mathbb{R}^p$.

The dynamics of the state estimation error $e := x - \hat{x}$ is then described by

$$\dot{e} = f_{\mathrm{err}}(e,x) := \begin{cases} (A_1 - L_1C)e, & H^\top x \le 0,\ H^\top x + (M^\top C - H^\top)e \le 0 \\ (A_2 - L_2C)e + \Delta A x, & H^\top x \le 0,\ H^\top x + (M^\top C - H^\top)e \ge 0 \\ (A_1 - L_1C)e - \Delta A x, & H^\top x \ge 0,\ H^\top x + (M^\top C - H^\top)e \le 0 \\ (A_2 - L_2C)e, & H^\top x \ge 0,\ H^\top x + (M^\top C - H^\top)e \ge 0, \end{cases} \tag{28}$$

where x satisfies (26a) and $\Delta A := A_1 - A_2$. The error dynamics in the first and the fourth mode of (28) is described by an n-dimensional autonomous state equation, while in the two other modes the "external input signal" x is present in the right-hand side. The presence of x makes it generally not possible to obtain GAS error dynamics, although in some particular cases it can be the case (cf. [17]). However, ISS is still obtainable under a suitable condition.

Theorem 5. *The observer* (27) *yields estimation error dynamics* (28) *that is ISS with respect to the system state x as an external input, if there exist constants $\lambda \ge 0$, $\varepsilon_e \ge 0$ and $\mu_e > 0$ and a matrix $P_e = P_e^\top > 0$ such that the following matrix inequalities are satisfied:*

$$\begin{bmatrix} \begin{matrix}(A_i - L_iC)^\top P_e \\ +P_e(A_i - L_iC) + (\mu_e + 1)I\end{matrix} & (*) \\ \begin{matrix}(-1)^i \Delta A^\top P_e + \\ \frac{\lambda}{2} H(H^\top - M^\top C)\end{matrix} & -\lambda H H^\top - \varepsilon_e I \end{bmatrix} < 0,\ i = 1,2 \tag{29}$$

Furthermore, $V_e(e) = e^\top P_e e$ is an ISS-Lyapunov function for the error dynamics (28) *and Definition 3 is satisfied for $\psi_1(|e|) = \lambda_{min}(P_e)|e|^2$, $\psi_2(|e|) = \lambda_{max}(P_e)|e|^2$, $\chi(|x|) = \sqrt{\varepsilon_e}|x|$, and $\alpha(V_e(e)) = \frac{\mu_e}{\lambda_{max}(P_e)}V_e(e)$.*

Proof. Due to page limitations we will only sketch the proof here. For the quadratic Lyapunov function $V_e(e) = e^\top P_e e$, the conditions (10) can be reformulated in the matrix inequalities above by using the constraints $|e|^2 \ge \varepsilon_e |x|^2$ (i.e. $|e| \ge \chi(|x|)$) and the regional information in (28) via S-procedure relaxations. See [17] for more details.

Remark 2. The matrix inequalities in (29) are linear in $\{P_e, L_1^\top P_e, L_2^\top P_e, \lambda M, \lambda, \mu_e, \varepsilon_e\}$, and thus can be efficiently solved.

3.2 Controller Design

As a controller for the system (26) we propose the following controller:

$$u = K\hat{x}, \text{ where } K^\top \in \mathbb{R}^n. \tag{30}$$

By using $\hat{x} = x - e$ the system dynamics with the controller (30) becomes

$$\dot{x} = f_{\mathrm{sfc}}(x,e) = \begin{cases} (A_1 + BK)x - BKe, & H^\top x \le 0 \\ (A_2 + BK)x - BKe, & H^\top x \ge 0. \end{cases} \tag{31}$$

Theorem 6. *The closed-loop system* (31) *is ISS with respect to the estimation error* e *as an external input, if there exist positive constants* ε_x, $\mu_x > 0$ *and a matrix* $P_x = P_x^\top > 0$ *such that*

$$\begin{bmatrix} (A_i+BK)^\top P_x + \\ P_x(A_i+BK) + (\mu_x+1)P_x & -P_xBK \\ (*) & -\varepsilon_x P_x \end{bmatrix} < 0, \ i = 1,2 \tag{32}$$

Furthermore, the function $V_x(x) = x^\top P_x x$ *is an ISS-Lyapunov function for the system dynamics* (31) *and Definition 3 is satisfied for* $\psi_1(|x|) = \lambda_{min}(P_x)|x|^2$, $\psi_2(|x|) = \lambda_{max}(P_x)|x|^2$, $\chi(|e|) = \sqrt{\varepsilon_x \frac{\lambda_{max}(P_x)}{\lambda_{min}(P_x)}}|e|$, *and* $\alpha_x(V_x(x)) = \mu_x V_x(x)$.

Proof. Similar as proof of Theorem 5.

Remark 3. The matrix inequalities (32) are bilinear in the variables. However, by pre- and post-multiplying the whole matrix inequality by $\begin{pmatrix} P_x^{-1} & 0 \\ 0 & P_x^{-1} \end{pmatrix}$, one obtains

$$\begin{bmatrix} P_x^{-1}A_i^\top + P_x^{-1}K^\top B^\top + \\ A_iP_x^{-1} + BK^\top P_x^{-1} + (\mu_x+1)P_x^{-1} & -BKP_x^{-1} \\ (*) & -\varepsilon_x P_x^{-1} \end{bmatrix} < 0 \tag{33}$$

for $i = 1, 2$. Inequalities (33) are still bilinear in the variables. However, by fixing the values of ε_x and μ_x, the inequalities are linear in $\{P_x^{-1}, P_x^{-1}K^\top\}$.

3.3 Interconnection

Theorems 5 and 6 give the means to design the observer gains L_1, L_2, M and the feedback gain K, so that the system and the observer separately have quadratic ISS-Lyapunov functions. Of course, one could also have used a relaxation by adopting piecewise quadratic ISS Lyapunov functions as in [10] based upon the general theory of Theorem 2. However, the current choice of quadratic and thus smooth Lyapunov functions illustrates nicely that even in this case the Lyapunov function of the interconnection is still *non-smooth* (see (23) in the proof of Theorem 4). Together with the fact that we have discontinuous dynamics, the "smooth ISS theory" [1,2,3,4,5] does not apply directly and we have to resort to the developed theory in this paper. As a direct application of Theorem 4 we obtain the following sufficient conditions for GAS of the interconnection (28)-(31).

Theorem 7. *Consider the system* (26)*, the observer* (27) *and the controller* (30)*. Suppose that the observer is designed according to Theorem 5 and the state feedback according to Theorem 6. Then the closed-loop system is globally asymptotically stable if the following condition is satisfied:*

$$\frac{\lambda_{max}(P_e)}{\lambda_{min}(P_e)}\frac{\lambda_{max}(P_x)}{\lambda_{min}(P_x)}\varepsilon_e\varepsilon_x < 1 \tag{34}$$

4 Conclusions

The contribution of this paper is twofold. Firstly, we presented an ISS framework for differential equations with discontinuous right-hand sides using non-smooth (ISS) Lyapunov functions. Secondly, we applied this framework in the design of an observer-based controller for a class of piecewise linear systems.

The ISS framework introduced by Sontag was extended to continuous-time discontinuous dynamical systems and non-smooth ISS Lyapunov functions. The main motivation for the use of non-smooth ISS Lyapunov function was the use of "multiple Lyapunov functions" as is common in the stability theory for hybrid systems. We showed that the existence of a non-smooth (but Lipschitz continuous) ISS Lyapunov function for a discontinuous dynamical system adopting Filippov's solution concept implies ISS. As a special case, this provided also a stability result for discontinuous dynamical systems using non-smooth Lyapunov functions. Finally, we proved that the interconnection of two discontinuous dynamical systems, which both admit an ISS Lyapunov function, is globally asymptotically stable under a small gain condition.

The developed ISS theory was exploited for the output-based feedback controller design for a class of PWL systems. Via LMIs the design of the state feedback and the observer could be performed separately. A small gain condition had to be checked to verify the stability of the overall closed-loop system.

Several future research issues remain. Besides extending the ISS framework for discontinuous systems, also generalizations are possible for the observer-based controller design. We presented the case of a common quadratic ISS Lyapunov function for both the state feedback and the observer design. This result can be generalized to piecewise quadratic (ISS) Lyapunov functions to obtain relaxed conditions. However, it is of interest to investigate further extensions to include observers with state resets [18]. Also robustness of the observer-based controller design with respect to disturbances such as measurement noise and model mismatch will be investigated in future work.

References

1. Sontag, E.D.: The ISS philosophy as a unifying framework for stability-like behavior. In Isidori, A., Lamnabhi-Lagarrigue, F., Respondek, W., eds.: Nonlinear Control in the Year 2000. Lecture Notes in Control and Information Sciences. Springer-Verlag (2000) 443–468
2. Sontag, E., Wang, Y.: On characterisations of the input-to-state stability property. System and Control Letters (1995)
3. Jiang, Z., Mareels, I., Wang, Y.: A Lyapunov formulation of hte nonlinear small-gain theorem for interconnected ISS systems. Automatica **32** (1996) 1211–1215
4. Lin, Y., Sontag, E., Wang, Y.: A smooth converse Lyapunov theorem for robust stability. SIAM J. Control Optim. **34**(1) (1996) 124–160
5. Jiang, Z., Teel, A., Praly, L.: Small-gain theorem for ISS systems and applications. Mathematics of Control, Signals & Systems **7** (1994) 95–120
6. Arcak, M., Kokotović, P.: Observer based control of systems with slope-restricted nonlinearities. IEEE Transactions on Automatic Control **46**(7) (2001) 1146–1150

7. Arcak, M.: Certainty equivalence output feedback design with circle criterion observers. IEEE Transactions on Automatic Control **50** (2005)
8. Branicky, M.: Stability theory for hybrid dynamical systems. IEEE Transactions on Automatic Control **43**(4) (1998) 475–482
9. DeCarlo, R., Branicky, M., Pettersson, S., Lennartson, B.: Perspectives and results on the stability and stabilizability of hybrid systems. Proceedings of the IEEE (2000) 1069–1082
10. Johansson, M., Rantzer, A.: Computation of piecewise quadratic Lyapunov functions for hybrid systems. IEEE Transactions on Automatic Control **43**(4) (1998) 555–559
11. Pettersson, S., Lennartson, B.: LMI for stability and robustness of hybrid systems. (In: Proc. of the American Control Conference, Albuquerque, New Mexico, June 1997) 1714–1718
12. Filippov, A.: Differential Equations with Discontinuous Righthand Sides. Mathematics and its Applications. Kluwer, Dordrecht, The Netherlands (1988)
13. Shevitz, D., Paden, B.: Lyapunov theory for nonsmooth systems. IEEE Transactions on automatic control **39**(9) (1994) 1910–1914
14. Cai, C., Teel, A.: Results on input-to-state stability for hybrid systems. In: Proc. Conference Decision and Control. (2005) 5403–5408
15. Vu, L., Chatterjee, D., Liberzon, D.: ISS of switched systems and applications to switching adaptive control. In: 44th IEEE Conference on Decision and Control, Seville, Spain (2005) 120–125
16. Liberzon, D., Nesic, D.: Stability analysis of hybrid systems via small-gain theorems. (2006) 421–435
17. Juloski, A., Heemels, W., Weiland, S.: Observer design for a class of piecewise affine systems. In: Proc. of Conference on Decision and Control 2002, Las Vegas, USA (2002) 2606–2611
18. Pettersson, S.: Switched state jump observers for switched systems. In: Proceedings of the 16th IFAC World Congress, Prague, Czech Republic. (2005)

A Stochastic Framework for Hybrid System Identification with Application to Neurophysiological Systems

Nicolas Hudson and Joel Burdick

California Institute of Technology, Department of Mechanical Engineering
Pasadena, CA 91125, USA
hudson@caltech.edu, jwb@robotics.caltech.edu

Abstract. This paper adapts the Gibbs sampling method to the problem of hybrid system identification. We define a Generalized Linear Hiddenl Markov Model (GLHMM) that combines switching dynamics from Hidden Markov Models, with a Generalized Linear Model (GLM) to govern the continuous dynamics. This class of models, which includes conventional ARX models as a special case, is particularly well suited to this identification approach. Our use of GLMs is also driven by potential applications of this approach to the field of neural prosthetics, where neural Poisson-GLMs can model neural firing behavior. The paper gives a concrete algorithm for identification, and an example motivated by neuroprosthetic considerations.

1 Introduction

This paper develops a Gibbs Sampling based approach for the identification of a class of hybrid systems which we term Generalized Linear Hidden Markov Models (GLHMMs). This hybrid system model is based on probabilistic Markov switching, and continuous dynamics that are described by Generalized Linear Models (GLMs). GLMs can be viewed as a nonlinear extension to the Piece Wise Affine (PWA) systems often considered in hybrid system identification [1,2,3,4,5], and include these previously studied PWA systems as a special case.

The motivation for modeling dynamics via GLMs is application driven. Cortical neuroprotheses are being developed to restore motor function and provide communication channels for patients with spinal cord injuries and motor disorders, as well as patients with locked-in syndrome [6]. Prior work [7,8] has shown that for the purposes of supervisory control of a neural prosthetic, the evolution of the higher level planning signals found in the partial reach region (PRR) of the posterior parietal cortex can be approximately modeled as a finite state machine [8], with changes in neural activity corresponding to changes in a small set of discrete cognitive or behavioral states. In current laboratory settings where neural prosthetic activity is carried out, these discrete states can be inferred from experimental cues. However as neural prostheses are now transitioning to clinical devices, methods to simultaneously classify the discrete cognitive and

A. Bemporad, A. Bicchi, and G. Buttazzo (Eds.): HSCC 2007, LNCS 4416, pp. 273–286, 2007.

behavioral state of the brain in real time, as well as to identify representative neural dynamics, are needed. GLHMMs can model the dynamics of neural processes (which are based on point process models of neural behavior [9]), and our proposed identification method can potentially construct the needed models of cognitive switching behavior from recorded neural data sets. GLHMMs can also potentially model a wider class nonlinear physical systems and systems whose uncertainty properties are not well captured by Gaussian models.

This paper proposes a general Gibbs sampling approach to hybrid system identification. Several Maximum Likelihood (ML) algorithms for hybrid system identification already exist. Expectation maximization has been used in vision[10] and in speech recognition [11] to identify state space representations of PWA systems. Clustering approaches [5] have been developed for Piece-Wise Autoregressive Exogenous systems (PWARX). Gibbs sampling is a flexible method that has a number of potential advantages as compared to previously proposed methods. Unlike ML methods, Gibbs sampling utilizes prior information, and allows for models to be created based on the maximum a posteri (MAP) estimates, which is shown to be important in neural applications. Previous Bayesian methods incorporating prior information have been considered [3]. However, by allowing for asymptotically exact sampling of complex probability density functions, the Gibbs-based approach overcomes the approximations used in [3]. Given a sufficiently long sampling sequence, Gibbs sampling methods can also generally avoid getting stuck in local minima. Based on the discussion in Section 3.1, it is likely that Gibbs sampling based identification can be extended to other hybrid system models beside the one considered here.

2 Generalized Linear Hidden Markov Models

GLHMMs are based on two existing frameworks, Hidden Markov Models (HMMs), and Generalized Linear Models (GLMs). In this class of hybrid systems, HMMs model the switching between discrete system modes, while GLMs define the discrete time dynamics of the continuous state evolution. GLMs can be viewed as an nonlinear extension to AutoRegressive eXogenous (ARX) [12] systems frequently used in system identification. We are practically motivated to use GLMs because of the applications described in Sec. 5. To establish the definition and notation of GLHMMs, we first review standard concepts from HMMs and GLMs.

2.1 Hidden Markov Models

A Hidden Markov Model (HMM) is formed around a set of N unobservable discrete states, $\mathcal{S} = \{S_1, S_2, \ldots, S_N\}$, whose evolution is governed by a first order Markov process. Let discrete instants of time be indexed by $\{t_1, t_2, \ldots, t_k, \ldots, t_T\}$. At each t_k, let m_k denote the mode index, i.e., at t_k the system is in state S_{m_k}. There is an associated parameterized probability distribution which generates the observed system output at time t_k: $y_k \sim p\left(y_k|\theta_{m_k}\right)$, parameterized by $\theta_{m_k} \in \mathbb{R}^P$. The probability of switching between modes of the system is governed by a first order Markov chain with transition matrix $A = [a_{i,j}]$:

$$P(m_k = i|\, m_{k-1}, \ldots, m_1) = P(m_k = i|\, m_{k-1}) = a_{m_{k-1},i} \ . \tag{1}$$

Given an observed HMM output sequence, $Y = [y_1, \ldots, y_T]$, estimation of the parameters $\Theta = [\theta_1, \ldots, \theta_N]$ and the transition matrix A can be carried out using several methods, including the well known Baum-Welch method [13]. Because we mix HMMs with GLMs, we will pursue a second identification approach based on Gibbs sampling [14].

2.2 Generalized Linear Models

Generalized Linear Models (GLMs) [15], are an extension of linear regression that allows modeling of situations where observations of the system state are not normally distributed. In GLMs, a linear predictor is used to predict a function of the mean of the outcome (observed) variable, $\mu \in \mathbb{R}$:

$$g(\mu) = \eta = \beta^T x \ , \tag{2}$$

where $g(\cdot)$ is an smooth invertible *link function*, $x \in \mathbb{R}^m$ is a regressor vector of observed values or known inputs, and $\beta \in \mathbb{R}^m$ is a vector of parameters.

The observed output of a GLM, $\mathbf{y} = \{y_1, y_2, \ldots\}$, is distributed according to a distribution $f(\mu)$ whose mean μ is described by the inverse of the link function:

$$y \sim f(\mu), \text{ where } \mu = g^{-1}(\beta^T x) \ . \tag{3}$$

2.3 Generalized Linear Hidden Markov Models

With this background in mind, a SISO[1] Generalized Linear Hidden Markov Model (GLHMM) is defined as follows.

Definition 1. *A GLHMM is a system* $\mathcal{G} = \{\mathcal{S}, A, \mu, \mathcal{U}, \mathcal{Y}, \Sigma\}$, *where*[2]

1. $\mathcal{S} = \{S_1, \ldots, S_N\}$ *is the set of* N *discrete states, or modes of* $\mathcal{G}$. *The mode of the system a time* t_k *is denoted by* m_k.
2. *The evolution of the discrete states* $\mathcal{S}$ *is governed by a first order Markov process parameterized by a transition probability matrix* A, *as defined in* (1),
3. *The variable* $\mu \in \mathbb{R}$ *is a continuous state variable, and* μ_k *is the continous state at time* t_k,
4. $u_k \in \mathcal{U}$ *is a control input at* t_k. *For this paper, we assume* $u_k \in \mathbb{R}$,

[1] The GLHMM definition can be extended to a MISO system by increasing the size of the regressor, and to a MIMO system by creating separate models for each output.

[2] One can also define an initial distribution on the continuous states, $\pi(\mu_0)$, and an initial discrete distribution on the discrete states or modes, $\Pi(m_0)$. However, because this paper focuses on model identification, and not estimation, these initial conditions are not essential to our definition.

5. *$y_k \in \mathcal{Y}$ is the system output at t_k. Depending on Σ, $\mathcal{Y} = \mathbb{R}, \mathbb{R}^+, \mathbb{N}$.*
6. *Σ is a discrete time dynamical system taking the form of a GLM, defined as follows. A regressor vector x_k of previous inputs and outputs is defined as:*

$$x_k = \left[1, y_{k-1}, \ldots, y_{k-n_y}, u_{k-1}, \ldots, u_{k-n_u}\right]^T \quad . \tag{4}$$

The continuous state μ_k of the GLM is a nonlinear function of the regressor x_k and the discrete state m_k,

$$\mu_k = g^{-1}\left(h(x_k)\right) \quad . \tag{5}$$

The function $h(\cdot)$ is a switching function which relates the continuous dynamics to each discrete mode,

$$h(x_k) = \begin{cases} \theta_1^T x_k & \text{if } m_k = 1 \\ \vdots & \\ \theta_N^T x_k & \text{if } m_k = N \end{cases} \quad . \tag{6}$$

The parameters $\theta_i \in \mathbb{R}^m$, $i = 1, \ldots, N$ are associated with each discrete mode, and govern the dynamics within a mode. The output of the system, y_k, is distributed according to the distribution $f(\cdot)$,

$$y_k \sim f(\mu_k). \tag{7}$$

The distribution, $f(\cdot)$, and the link function, $g(\cdot)$, are smooth and are often constrained to be pairs of compatible functions, as described in Sec. 2.4. For convenience the following notation is adopted:

$$\Theta = [\theta_1, \ldots, \theta_N], \; M = [m_1, \ldots, m_T], \; Y = [y_1, \ldots, y_T], \; X = [x_1, \ldots, x_T]. \tag{8}$$

2.4 Relation of GLHMMs to Prior Work

HMMs are chosen as a switching criterion for this class of hybrid systems because they provide flexibility in the identification process. Several identification algorithms have been designed for Piece Wise ARX (PWARX) hybrid systems [1]. In PWARX systems, the guards consists of hyperplanes in the regressor space; the system changes its discrete state when a guard is crossed. However, most identification algorithms do not account for the guards explicitly in the identification process [1]. Instead, an initial process jointly identifies the parameter Θ and classifies the regressor x_k, and then a second process identifies the guards based on the classification of the regressors x_k. The advantage of this approach is that it allows powerful existing algorithms to form the guard estimates. Our HMM switching model does not currently allow for the addition of complicated Guards into the identification process. However, when data has been collected sequentially in time, it is more likely that sequential data points belong to the same mode. A Markov transition rule can be used to represent this correlation. After initial identification using our method, guard functions can be estimated in a second processing step, as is conventionally done. Recent work in nonstationary HMMs provide an extension where the probability of switching between modes is a function of the duration in the current mode [16]. This extension can be readily implemented in the current algorithm, and therefore provide a means of identifying timed automata.

GLMs are often identified via maximum likelihood methods [17]. To usefully identify GLMs using Gibbs sampling, it must be shown that the density function $p(\theta|x, y)$ is of a convenient form (e.g., log-concave). In practice, the distribution $f(\mu)$ is often constrained to the exponential family of distributions, with an associated compatible link function $g(\mu)$, as shown in Table 1. These compatible functional forms have been proven to yield log-concave likelihoods [18], which have efficient simulation techniques appropriate for Gibbs sampling. Gilks [19] describes an adaptive rejection sampling technique bounding a log-concave density function with upper and lower hulls, allowing single samples to be drawn from the distribution with only a few function evaluations.

Table 1. Log-Concave Likelihood Forms of $g()$ and $f()$

$f(\mu)$	$g(\mu)$
Normal	identity operator
Gamma	$g(\mu) = \log \mu$, or $g(\mu) = \mu^\gamma, (-1 \leq \gamma < 0)$
Poisson	$g(\mu) = \log \mu$ or $g(\mu) = \mu^\gamma, (-1 \leq \gamma < 0)$
Binomial	$g(\mu) = \text{logit}(\mu)$ or $g(\mu) = \Phi^{-1}(\mu)$, or $g(\mu) = log\,(-log(1-\mu))$

In this paper, the utility of the Normal and Poisson distributions are demonstrated. A normally distributed $f()$ with identity link $g(\mu) = \mu$ is equivalent to the standard regressor: $y_k = \theta^T x_k + \epsilon$, with $\epsilon \in \mathcal{N}\left(0, \sigma^2\right)$. If the regressor $x_k \in X$ is defined as (4) then the GLM is equivalent to an ARX system.

A Poisson-GLM with log-link function $g(\mu) = \log(\mu)$ models counting processes, which are relevant for applications in neuroscience (see Section 5). Here $y \sim f(\mu)$, where $\mu = e^{(\theta^T x)}$. A Poisson distribution is defined as $f(\mu) = \frac{e^{-\mu}\mu^y}{y!}$.

3 Identification of GLHMMs Using Gibbs Sampling

Markov Chain Monte Carlo (MCMC) sampling methods generate samples from a desired probability density function (see [20] for a review). In particular, we consider Gibbs sampling, a type of MCMC. The main advantage of this approach is the capacity to work with high dimensional and complex systems.

3.1 Gibbs Sampling for Hybrid System Identification

Gibbs sampling is a MCMC method for sampling from a potentially complicated joint pdf, $p\left(\phi_1, ..., \phi_n\right)$, where $\phi_1, \ldots, \phi_n$ are system states or parameters. Gibbs sampling can be usefully applied when the joint pdf $p(\phi_1, ..., \phi_n)$, has associated conditional pdfs,

$$p(\phi_1 | \phi_2, ..., \phi_n), \ldots, p(\phi_i | \phi_1, ..., \phi_{i-1}, \phi_{i+1}, ..., \phi_n), \ldots, p(\phi_n | \phi_1, ..., \phi_{n-1}), \tag{9}$$

that can be efficiently sampled from. A single step in the Gibbs sampling cycle requires one sample to be drawn sequentially from each of the conditional pdfs, using the most recent sampled value in subsequent conditional arguments. At the end of each step, a new sample $\hat{\phi} = \left[\hat{\phi}_1, \ldots, \hat{\phi}_n\right]$ has been drawn. As the Gibbs sampler iterates through many steps, the samples $\{\Phi\}$ tend asymptotically to the joint distribution [20]. In theory this property implies that the maximum of $p(\phi_1, ..., \phi_n)$ can always be found using Gibbs sampling, as opposed to EM methods where only a local maximum is guaranteed. In practice there are only a finite number of samples drawn, and multiple runs of Gibbs sampling are usually conducted to detect convergence.

Hybrid system identification is complex because it involves simultaneous parameter identification as well as classification. Gibbs samplers provide a unique form that can be exploited for this problem, as hybrid system identification can be viewed as equivalent to maximizing the joint posterior pdf,

$$p(\Theta, M | \text{ Observed Data}) \ , \tag{10}$$

where Θ represents the unknown system parameters, and M represents a vector which classifies all data points into their generative modes. In practice (10) is impossible to solve analytically. Gibbs Sampling allows (10) to be separated into the component parts of identification $p(\Theta | M, \text{Observed Data})$ and classification $p(M | \Theta, \text{Observed Data})$.

3.2 Gibbs Sampling for GLHMMs

Gibbs sampling draws samples from the joint distribution, $P(\Theta, A, M | X, Y)$. *Algorithm* 1 (described below) sequentially samples from the conditional regressor parameter density $p(\Theta | M, A, Y, X)$, the conditional Markov transition parameter density $p(A | \Theta, M, Y, X)$, and the conditional discrete state density $p(M | \Theta, A, Y, X)$.

After *algorithm* 1 is described below, assumptions underlying the construction of the algorithm are given in Sec. 3.3, while practical implementation issues are presented in Sec. 3.4.

Algorithm 1 (Gibbs Sampling for GLHMM). Draw z_{max} number of samples from the joint distribution $P(\Theta, A, M | X, Y)$ of a GLHMM given the data set $X = \{x_k\}, x_k \in \mathbb{R}^n$, $Y = \{y_k\}$, $k = 1, \ldots, T$, the number of discrete states N: $\mathcal{S} = \{S_1, \ldots, S_N\}$, the distribution $f(\cdot)$ and link function $g(\cdot)$.

1. Define parameterized conjugate prior distributions for Θ, and A, using prior information to select parameters:
   (a) set $p(\theta_i(j))$ as normal distributions for $i = 1, \ldots, N$ and $j = 1, \ldots, n$.

(b) each row of $A = \left[a_{(i,j)}\right]$ is assumed to be independent and the prior for A is defined as a set of Dirichlet distributions $\mathcal{D}(\cdot)$:

$$p\left(a_{(i,1:N)}\right) = \mathcal{D}\left(\alpha_1^i,\ \ldots,\ \alpha_N^i\right) \ , \tag{11}$$

where α_j^i are the parameters of the prior Dirichlet pdf (see [14]).

2. Initialize parameter samples: $\hat{A}^{(0)}$,$\hat{M}^{(0)}$,$\hat{\Theta}^{(0)}$. These initial samples can either be drawn from the priors or be set to an arbitrary initial guess.
3. **set** $z = 1$
4. Sample from $p\left(M \middle| \ \Theta, A, Y, X\right)$, the conditional discrete mode density. The discrete modes are sampled from sequentially:

   (a) set $\hat{m}_1^{(z)} = 1$

   (b) **for** $k = 2 : T - 1$, draw a sample $\hat{m}_k^{(z)}$ from the discrete distribution:

$$\hat{m}_k^{(z)} \sim P\left(m_k | y_k, \Theta, m_{k-1}, m_{k+1}\right) \tag{12}$$

$$= \frac{a_{\left(\hat{m}_{k-1}^{(z)}, m_k\right)} f\left(y_k | \theta_{m_k}, x_k\right) a_{\left(m_k, \hat{m}_{k+1}^{(z-1)}\right)}}{\sum_{j=1}^{s} a_{\left(\hat{m}_{k-1}^{(z)}, j\right)} f\left(y_k | \theta_j, x_k\right) a_{\left(j, \hat{m}_{k+1}^{(z-1)}\right)}} \ . \tag{13}$$

   end

   (c) draw sample $\hat{m}_T^{(z)}$ from the discrete distribution:

$$\hat{m}_k^{(z)} \sim P\left(m_T | y_T, \Theta, m_{T-1}\right) \tag{14}$$

$$= \frac{a_{\left(\hat{m}_{T-1}^{(z+1)}, m_T\right)} f\left(y_T | \theta_{m_T}, x_T\right)}{\sum_{j=1}^{s} a_{\left(\hat{m}_{T-1}^{(z)}, j\right)} f\left(y_T | \theta_j, x_T\right)} \ . \tag{15}$$

5. Sample from $p\left(A \middle| \ \Theta, M, Y, X\right)$. Each row of A is sampled independently: **for** $i = 1 : N$

$$\hat{a}_{(i,1:N)}^{(k)} \sim p\left(a_{i,1:N} | M\right) =$$
$$\mathcal{D}\left(\alpha_1^i + \sum_{k=2}^{T} \delta_{\left(\hat{m}_{k-1}^{(z)} = i\right)} \delta_{\left(\hat{m}_k^{(z)} = 1\right)}, \ldots, \alpha_N^i + \sum_{k=2}^{T} \delta_{\left(\hat{m}_{k-1}^{(z)} = i\right)} \delta_{\left(\hat{m}_k^{(z)} = N\right)}\right) \tag{16}$$

   end

6. Sample from $p\left(\Theta \middle| \ M, A, Y, X\right)$:

   (a) assign data into discrete modes using $\hat{M}^{(k)}$
   for $i = 1, \ldots, N$

$$\mathcal{X}^i = \left\{x_k \ : \ \hat{m}_k^{(z)} = i,\ k = 1, \ldots, T\right\} \tag{17}$$

$$\mathcal{Y}^i = \left\{y_k \ : \ \hat{m}_k^{(z)} = i,\ k = 1, \ldots, T\right\} \tag{18}$$

   end

(b) Conditioning on the sets $\mathcal{X}^i$,$\mathcal{Y}^i$, the distributions for the regressor parameters θ_i in each mode are sampled independently.
for $i = 1, \ldots, N$

$$\hat{\theta}_i^{(k)} \sim p\left(\theta_i \middle| \mathcal{Y}^i, \mathcal{X}^i\right) \tag{19}$$

$$\propto P\left(\mathcal{Y}^i, \mathcal{X}^i \middle| \theta_i\right) p\left(\theta_i\right) . \tag{20}$$

end
The regressor parameter density (19) is log concave and is sampled using adaptive rejection sampling [19].

7. **set** z=$z+1$, **if** $z > z_{max}$ stop, **else** goto step 4.

Recall that $a(i,j)$ are the elements of the Markov transition matrix A. The conditional distribution $f(y_k|\theta_{m_k}, x_k)$ refers to the distribution $f(\mu_k)$ where $\mu_k = \theta_{m_k}^T x_k$ from (7), and δ is a delta function.

3.3 Derivation of Algorithm 1: Gibbs Sampling for GLHMMs

Sampling from the conditional discrete mode density $p\left(M \middle| \Theta, A, Y, X\right)$ and Markov transition density $p\left(A \middle| \Theta, M, Y, X\right)$, in algorithm steps 4 and 5, follows directly from work on Gibbs sampling in HMMs [14].

The regressor parameter distribution $p\left(\Theta \middle| M, A, Y, X\right)$ in step 6 is independent of the parameter A, when the modes m_k of the data x_k, y_k are known:

$$p\left(\Theta \middle| M, A, Y, X\right) = p\left(\Theta \middle| M, Y, X\right) . \tag{21}$$

This reduction is based on the assumption that if the discrete variables m_k of a hybrid system are known exactly, then the only information pertaining to parameters θ_i, is the data generated by the mode S_i, defined by the sets $\mathcal{X}^i$ and $\mathcal{Y}^i$, as in (17), and prior information about the parameter. This assumption also implies that the parameters of each mode are independent given the discrete modes m_k:

$$p\left(\Theta \middle| M, Y, X\right) = \prod_{i=1}^{N} p\left(\theta_i \middle| \mathcal{Y}^i, \mathcal{X}^i\right) . \tag{22}$$

Hence drawing samples from each $p\left(\theta_i \middle| \mathcal{Y}^i, \mathcal{X}^i\right)$, $i = 1, \ldots, N$ is the same as drawing samples from $p\left(\Theta \middle| M, A, Y, X\right)$. Log concavity of the distributions $p\left(\theta_i \middle| \mathcal{Y}^i, \mathcal{X}^i\right)$, $i = 1, \ldots, N$ is shown in [18], for distributions $f(\cdot)$ and link functions $g(\cdot)$ shown in Table 1.

3.4 Practical Issues Associated with Algorithm 1

Sampling from discrete density functions: The densities (12) and (14) are discrete, have support in $\mathcal{S} = \{S_1, \ldots, S_N\}$, and can be formulated as the vector:

$$P_{m_k} = \left[P\left(m_k = 1 | y_k, \Theta, m_{k-1}, m_{k+1}\right), \ldots, P\left(m_k = N | y_k, \Theta, m_{k-1}, m_{k+1}\right)\right] , \tag{23}$$

where $\sum P_{m_k} = 1$. Drawing a sample from (12) and (14) is accomplished by generating a random number from a uniform distribution on $[0, 1]$, and then choosing $m_k = i$, where i is the first element of the cumulative sum of P_{m_k} that is greater than or equal to the random number.

Non-Markov switching behavior: If the modes at sequential time points k,$k+1$ are considered independent, then Markov behavior of the discrete state is no longer appropriate and (12)–(14) reduces to:

$$P(m_k|y_k, \Theta) = \frac{f(y_k|\theta_{m_k})}{\sum_{j=1}^{s} f(y_k|\theta_j)} \ . \tag{24}$$

Degenerate mode vector assignments: A known problem with Gibbs sampling is that during some sampling sequences there may be no data (x_k, y_k) attributed to a specific mode $S_i \Rightarrow \mathcal{X}^i$, $\mathcal{Y}^i = \varnothing$. Using proper priors mitigates this problem as the posteriors (22) are the same as the priors $p(\theta_i)$.

Recovering parameter estimates: Gibbs sampling exhibits a *burn-in period* where initial samples are not distributed as the joint distribution. These samples are removed [20], and parameter estimates are formulated using the remaining samples. Common estimates are the expectation $E[p(\phi)]$, which is the average $\frac{1}{Z}\sum_{z=1}^{Z} \hat{\phi}^{(z)}$, of the remaining samples $\hat{\phi}^{(z)}$ $z = 1, \ldots, Z$, and the mode of the samples, which approximates the Maximum A Posteri (MAP) estimate.

4 PWARX Identification

To demonstrate the flexibility of the GLHMM identification framework, a PWA systems is identified. To allow comparison of Gibbs sampling to existing methods of hybrid system identification, the intersecting hyperplanes PWARX system from the comparison paper on hybrid system identification [1] is considered:

The PWARX system is defined as $y_k = h(x_k) + e_k$, where h is:

$$h(x_k) = \begin{cases} [x_k\ 1]\,[0.5\ 0.5]^T & \text{if } x_k \in [-2.5, 0] \\ [x_k\ 1]\,[-1\ 2]^T & \text{if } x_k \in (0, 2.5] \end{cases} \ . \tag{25}$$

Data is generated by drawing the regressor from a uniform distribution $x_k \sim U[-2.5, 2.5]$, for $k = 1, \ldots, 100$. The noise, e_k, is gaussian with variance σ^2.

Algorithm 1 was used as following: Gaussian priors on all regressor parameters $p(\theta_i) = \mathcal{N}(0, 10^3)$ were used. As there is no Markov switching behavior, the algorithm was modified by using (24) instead of (12)–(14). The regressor parameter pdfs (19) were explicitly derived using Bayes theorem:

$$\begin{aligned} &p\left(\theta_i|\ \mathcal{X}^i, \mathcal{Y}^i\right) = \mathcal{N}(E(\theta_i), \Sigma_i) \text{ where} \\ &\quad \Sigma_i = \sigma^2(\mathcal{X}^{i\ T}\mathcal{X}^i + \alpha^2\mathrm{I})^{-1} \text{ and } \mathrm{E}(\theta_i) = (\mathcal{X}^{i\ T}\mathcal{X}^i + \alpha^2\mathrm{I})^{-1}\mathcal{X}^{iT}\mathcal{Y}_i \ . \end{aligned} \tag{26}$$

An example of the first 1500 samples drawn by the Gibbs sampler is shown in Fig. 1. The last 500 samples are used to estimate parameters, shown in Fig. 2. Linear programming was used to infer the position of the hyperplane guard.

Following a procedure[3] in [1], the effectiveness of the Gibbs sampling procedure is tested by running the algorithm on (25) with a range of noise intensities σ^2. It was found that Gibbs sampling was able to identify parameters to a level of precision between the algebraic approach [4] and the clustering-based procedure [5], when the regressor length and number of modes N was known exactly. This assumes that the burn in period is correctly identified in our Gibbs sampling procedure. Detecting the burn in period is explained in [20].

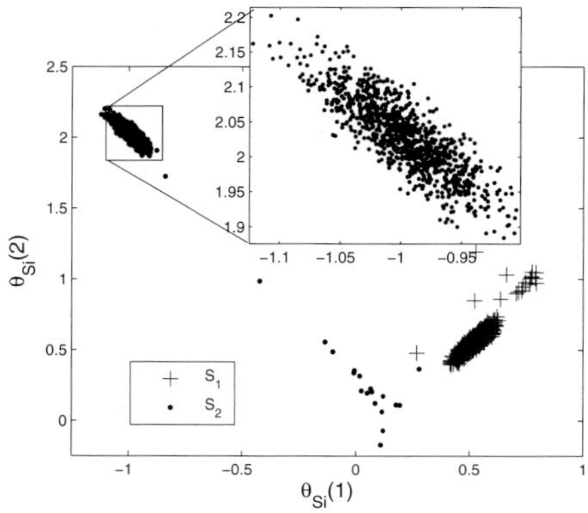

Fig. 1. Regressor parameter samples $\hat{\theta}_1$, $\hat{\theta}_2$ from Gibbs sampler

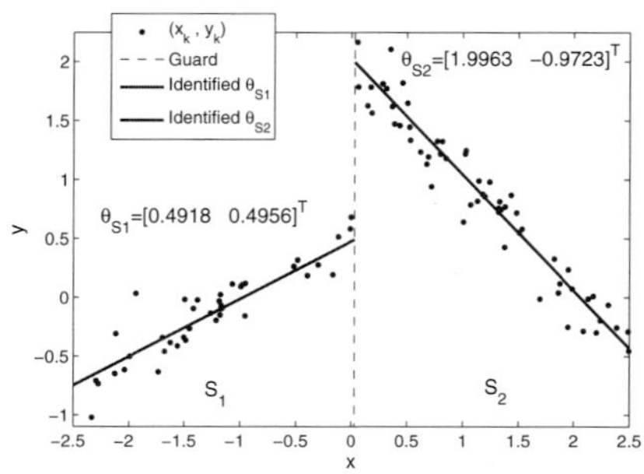

Fig. 2. Data points (x_k, y_k) from PWARX system and identified model parameters (solid lines)

5 Poisson GLM

Neurophysiological Background. Cortical neuroprostheses are being developed to restore motor function in individuals with high level spinal cord injuries or severe motor disorders (e.g. Lou Gehrig's disease). Neuroprostheses work by recording the spiking activity of multiple neurons in cortex, and decoding movement intent or movement plans from the neural activity to generate control signals that can be used to drive devices such as prosthetic arms or computer interfaces [6,7,8]. Future clinical neural prostheses will require a *supervisory decoder* whose job is to classify, in real time, the current cognitive or behavioral state of the brain. For example, the supervisory decoder must determine: (1) if the patient equipped with the neural prosthetic is awake; (2) if the patient wants to use the prosthetic; (3) if the brain is currently in the planning state; (4) if the patient actually wants to execute a planned movement; (5) if the patient wants to scrub or change the plan while it is being executed, etc. These different brain states and their associated neural dynamics can potentially be modeled as a hybrid system. Based on the current discrete state estimate, different algorithms for decoding movement plans or different neuroprosthetic system responses can be executed. For example, Shenoy et. al. [8] proposed that

[3] The metric $E[\Delta_\theta]$ from the comparison paper [1] is used.

PRR neural activity could be classified into three discrete states; a baseline state, a plan state and a reach state. They used neural recordings to develop a crude model and estimator in an ad-hoc fashion. For neural prosthetics to be used in real patients, systematic methods are needed to design the supervisory decoders based on hybrid system identification and estimation theory.

Neural activity is recorded using implanted electrodes which measure the extracellular components of action potentials, or spikes, from one or more neurons in the proximity of the each electrode. The continuous electrode waveforms are digitized and then spikes are detected in the considerably noisy signal. Using temporal alignment of the detected waveforms, principle component analysis, and clustering methods, the individual action potentials are then attributed to specific neurons. It is widely believed that only the arrival time of spike, and not its specific waveform shape, carries information. Thus, the end result of the initial electrode signal processing steps is a vector of spike arrival time points for each neuron. These vectors can contain considerable noise due to background electrical noise and misclassification. The arrival times are conventionally discretized into sufficiently small time bins (typically 1 msec) so that only one spike at most is assigned to each bin. Based on this conventional processing model, the dynamic firing behavior of neurons can be modeled using Poisson-GLMs [9].

5.1 Single Neuron Recording, a Simulated Example

A model of the recorded spiking activity from a single neuron present in a higher brain cortex is created. This neuron's spiking activity is dependent on the unobservable discrete state of the surrounding cortex. For this simple example, the cortex has two discrete states; S_1, an 'attention' state (i.e., the patient wants to actively use the neural prosthetic) and S_2, a 'baseline' state (i.e., sleep or disinterest in using the neural prosthetic). The transition of this cortical region between the 'attention' and 'baseline' states is assumed to follow Markov transition probabilities, and is illustrated in Fig 3. The simulations presented below directly simulate the number of spikes in sequential $0.01s$ time bins, and will simulate a $10s$ duration.

GLHMM Single Neuron Recording Model. The discrete modes are modeled by setting $m_1 = 1$ and evolving the discrete state m_k, $k = 1, \ldots, 1000$, using Markov transitions with parameters A:

$$P(m_{k+1} = j | m_k = i) = a_{ij}, \text{ where } [a_{ij}] = A = \begin{bmatrix} 0.9 & 0.1 \\ 0.1 & 0.9 \end{bmatrix} . \tag{27}$$

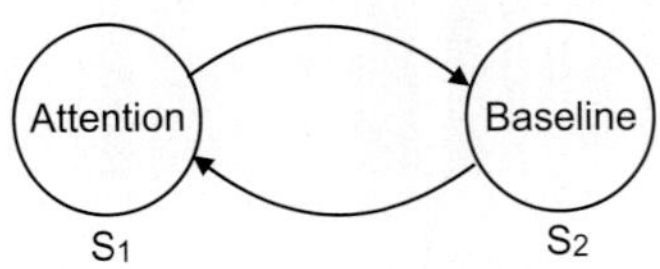

Fig. 3. Finite state machine representation of simulated neuron behavior

The neurons spiking activity in each mode is modeled with Poisson-GLMs. The firing rate, λ_k, in each mode S_i, is determined by two components: $\theta_i(1)$, the nominal firing rate of the mode, and $\theta_i(2)$, representing a change in rate depending on the spiking history. $\theta_i(2)$ can model refractory periods, a dwell period in spiking activity that is experienced immediately after spike firing.

$$\lambda_k = \begin{cases} e^{(\theta_1(1)+\theta_1(2)y_{k-1})} & \text{if } m_k = 1 \\ e^{(\theta_2(1)+\theta_2(2)y_{k-1})} & \text{if } m_k = 2 \end{cases} . \tag{28}$$

The following regressor parameters are used:

$$\theta_1 = \begin{bmatrix} -1 \\ -10 \end{bmatrix}, \quad \theta_2 = \begin{bmatrix} -2 \\ 0 \end{bmatrix} . \tag{29}$$

The parameters (29), correspond to a nominal firing rate of 36.78 Hz in the 'attention' state, and a nominal rate of 13.53 Hz in the 'baseline' state.

The number of spikes in the current time bin are generated from a Poisson distribution with rate λ_k:

$$y_k \sim \text{Poisson}\,(\lambda_k) \ . \tag{30}$$

An output sequence $y_k, k = 1, \ldots, 1000$ was generated from the single neuron model by using Poisson and discrete random number generators in Matlab. An example output sequence is plotted in Fig. 4. The associative generative modes m_k are also displayed. There were a total of 211 spike events over the simulated 10 second duration.

Algorithm 1 was run for 5000 iterations, the last 3000 of which were used for statistical analysis of parameters. Regressor parameter priors are set to dispersed normal distributions: $\theta_{i,j} = \mathcal{N}\left(0, 10^2\right)$ for $i \in 1, 2,\ j \in 1, 2$. Dirichlet priors are used for each row of A: $\left[a_{11}\ a_{12}\right] \sim \mathcal{D}\left(\left[\alpha_1\ \alpha_2\right]\right)$, $\left[a_{21}\ a_{22}\right] \sim \mathcal{D}\left(\left[\alpha_2\ \alpha_1\right]\right)$. Several different informative parameterizations were chosen that incorporate the assumption that sequential modes values m_k, m_{k+1} are more likely to belong to the same mode S_i:

$$\left[\alpha_1\ \alpha_2\right] = \left[90\ 10\right], \left[80\ 20\right], \left[70\ 30\right] \ . \tag{31}$$

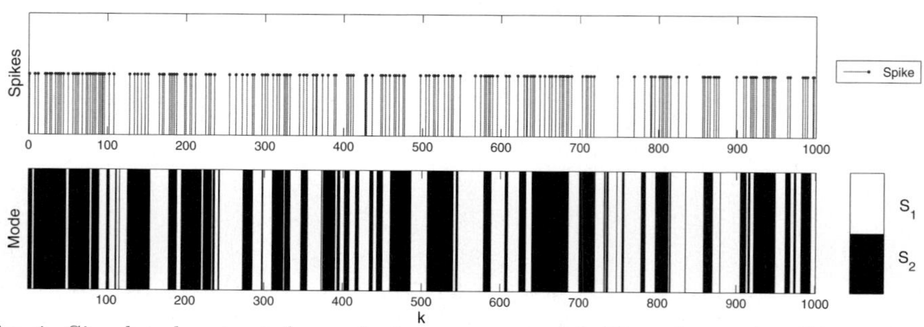

Fig. 4. Simulated output from single neuron model. The observed spiking activity is depicted in the top graph. The corresponding hidden discrete modes m_k, either 'attention', (black), or 'baseline', (white), are plotted directly below.

Table 2. Model parameter estimates. Expected value (E[·]) and the maximum a posteri (MAP) estimates are used, compared with actual parameter values (Model).

Parameter	Model	MAP	E[·]
A	$\begin{bmatrix} 0.9 & 0.1 \\ 0.1 & 0.9 \end{bmatrix}$	$\begin{bmatrix} 0.8980 & 0.1020 \\ 0.1017 & 0.8983 \end{bmatrix}$	$\begin{bmatrix} 0.8979 & 0.1021 \\ 0.0963 & 0.9037 \end{bmatrix}$
θ_1, θ_2	$\begin{bmatrix} -1 \\ -10 \end{bmatrix}, \begin{bmatrix} -2 \\ 0 \end{bmatrix}$	$\begin{bmatrix} -0.9643 \\ -2.7040 \end{bmatrix}, \begin{bmatrix} -2.0130 \\ 0.2014 \end{bmatrix}$	$\begin{bmatrix} -1.0031 \\ -25.0071 \end{bmatrix}, \begin{bmatrix} -2.1075 \\ -0.0218 \end{bmatrix}$

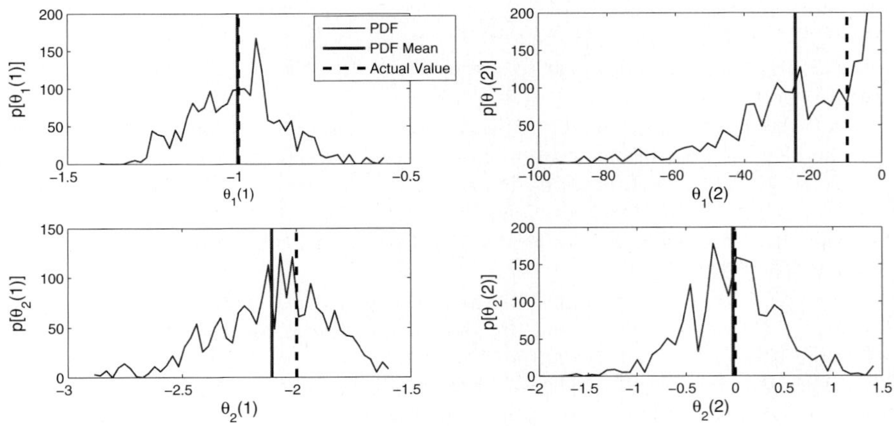

Fig. 5. Regressor parameter posterior densities and mean estimates

The solution was invariant when using different informative priors (31), and the key parameter estimates matched the model values (see Table 2).

Note the wide discrepancy between the MAP and Expectation estimates for the refractory parameter $\theta_1(2)$. The advantage of Gibbs sampling is that the posterior parameter densities can be analyzed, by generating a histogram of the samples, Fig. 5. The support of the posterior density for $\theta_1(2)$ is large, indicating that the parameter is largely unidentifiable from the generated data set. This problem arises because the refractery physics of spike firing dictate that no sequential outputs y_k and y_{k+1} in S_2 both contain spikes. Hence the only information that can be deduced by the algorithm from the data is the that refractory parameter $\theta_1(2)$ lowers the firing rate after a spike event has just occurred. The posterior densities thus allow the user to realize when a parameter is unidentifiable.

6 Conclusions

This paper demonstrated that Gibbs sampling can be an effective and flexible computational tool for hybrid system identification. The Generalized Linear Hidden Markov Models defined in this paper enable the creation of hybrid models representing neural activity in higher cortexes of the brain associated with arm reaching, as well as conventional PWARX systems. Their nonlinearities may also make them suitable models for many other physical hybrid systems.

References

1. Juloski, A., Ferrari-Trecate, G., Vidal, R., Paoletti, S., Niessen, J.: Comparison of four procedures for the identification of hybrid systems. Lecture Notes in Computer Science **3414** (2005) 354 – 369
2. Bemporad, A., Garulli, A., Paoletti, S., Vicino, A.: A greedy approach to identification of piecewise affine models. Lect. Notes in Comp. Sci. **2623** (2003) 97
3. Juloski, A., Weiland, S., Heemels, W.: A Bayesian approach to identification of hybrid systems. In: IEEE Conf. on Decision and Control. Volume 1., Nassau, Bahamas (2004) 13 – 19
4. Vidal, R., Soatto, S., Ma, Y., Sastry, S.: An algebraic geometric approach to the identification of a class of linear hybrid systems. In: IEEE Conference on Decision and Control. (2003)
5. Ferrari-Trecate, G., Muselli, M., Liberati, D., Morari, M.: A clustering technique for the identification of piecewise affine systems. Automatica **39** (2003) 205–217
6. Nicolelis, M.A.L.: Brainmachine interfaces to restore motor function and probe neural circuit. Nature Reviews Neuroscience **4** (2003) 417–422
7. Pesaran, B., Musallam, S., Andersen, R.: Cognitive neural prosthetics. Current Biology **16** (2006) 77–80
8. Shenoy, K., Meeker, D., Cao, S., Kureshi, S., Pesaran, B., Buneo, C., Batista, A., Mitra, P., Burdick, J., Andersen, R.: Neural prosthetic control signals from plan activity. Neuroreport **14**(4) (2003) 591–596
9. Truccolo, W., Eden, U.T., Fellows, M.R., Donoghue, J.P., Brown, E.N.: A point process framework for relating neural spiking activity to spiking history, neural ensemble, and extrinsic covariate effects. J Neurophysiol **93**(2) (2005) 1074–1089
10. Oh, S.M., Rehg, J.M., Dellaert, F.: Parameterized duration modeling for switching linear dynamic systems. IEEE Conf. on Computer Vision and Pattern Recog. (2006)
11. Rosti, A.V.: Linear Gaussian Models for Speech Recognition. PhD thesis, Wolfson College, University of Cambridge (2004)
12. Ljung, L.: System Identification, Theory for the User. 2nd edn. Prentice Hall (1999)
13. Rabiner, L.: A tutorial on hidden Markov models and selected applications in speech recognition. Proceedings of the IEEE **77** (1989) 257–286
14. Robert, C., Celeux, G., Diebolt, J.: Bayesian estimation of hidden Markov chains: A stochastic implementation. Statistics & Probability Letters **16** (1993) 77–83
15. Hoffmann, J.P.: Generalized Linear Models: an Applied Approach. Pearson Education (2004)
16. Djuric, P.M., Chun, J.H.: An MCMC sampling approach to estimation of nonstationary hidden Markov models. IEEE Trans. on Signal Processing **50**(5) (2002) 1113–1123
17. McCulloch, C.E.: Maximum likelihood algorithms for generalized linear mixed models. Journal of the American Statistical Association **92**(437) (1997) 162–170
18. Dellaportas, P., Smith, A.F.M.: Bayesian inference for generalized linear and proportional hazards models via Gibbs sampling. App. Stat. **42**(3) (1993) 443–459
19. Gilks, W.R., Wild, P.: Adaptive rejection sampling for Gibbs sampling. Applied Statistics **41**(2) (1992) 337–348
20. Gilks, W.R., Richardson, S., Spiegelhalter, D.J., eds.: Markov Chain Monte Carlo in Practice. Chapman & Hall (1996)

Reachability for Linear Hybrid Automata Using Iterative Relaxation Abstraction*

Sumit K. Jha[1], Bruce H. Krogh[2], James E. Weimer[2], and Edmund M. Clarke[1]

[1] Computer Science Department, Carnegie Mellon University
5000 Forbes Avenue, Pittsburgh, PA 15213,USA
{jha|emc}@cs.cmu.edu

[2] ECE Department, Carnegie Mellon University
5000 Forbes Avenue, Pittsburgh, PA15213, USA
{krogh|jweimer}@ece.cmu.edu

Abstract. This paper introduces *iterative relaxation abstraction* (IRA), a new method for reachability analysis of LHA that aims to improve scalability by combining the capabilities of current tools for analysis of low-dimensional LHA with the power of linear programming (LP) for large numbers of constraints and variables. IRA is inspired by the success of counterexample guided abstraction refinement (CEGAR) techniques in verification of discrete systems. On each iteration, a low-dimensional LHA called a *relaxation abstraction* is constructed using a subset of the continuous variables from the original LHA. Hybrid system reachability analysis then generates a regular language called the *discrete path abstraction* containing all possible counterexamples (paths to the bad locations) in the relaxation abstraction. If the discrete path abstraction is non-empty, a particular counterexample is selected and LP infeasibility analysis determines if the counterexample is spurious using the constraints along the path from the original high-dimensional LHA. If the counterexample is spurious, LP techniques identify an *irreducible infeasible subset* (IIS) of constraints from which the set of continuous variables is selected for the the construction of the next relaxation abstraction. IRA stops if the discrete path abstraction is empty or a legitimate counterexample is found. The effectiveness of the approach is illustrated with an example.

1 Introduction

Hybrid automata are a well studied formalism for representing and analyzing hybrid systems, that is, dynamic systems with both discrete and continuous state variables [1]. LHA are an important subclass of hybrid automata that can be

* This research was sponsored by the National Science Foundation under grant nos. CNS-0411152, CCF-0429120, CCR-0121547, and CCR-0098072, the US Army Research Office under grant no. DAAD19-01-1-0485, the Office of Naval Research under grant no. N00014-01-1-0796, the Defense Advanced Research Projects Agency under subcontract no. SA423679952, the General Motors Corporation, and the Semiconductor Research Corporation.

A. Bemporad, A. Bicchi, and G. Buttazzo (Eds.): HSCC 2007, LNCS 4416, pp. 287–300, 2007.

analyzed algorithmically and can asymptotically approximate hybrid automata with nonlinear continuous dynamics [2]. Tools for reachability analysis of LHA typically compute the sets of reachable states using polyhedra [3], but the sizes of the polyhedral representations can be exponential in the number of continuous variables of the LHA. Therefore, procedures for analysis of linear hybrid automata (LHA) do not scale well with the number of continuous state variables in the model. Although there has been considerable progress in the development of tools and algorithms for analyzing LHA [4,5], there is still a great need to develop new techniques that can handle high-dimensional LHA.

The approach to LHA reachability analysis proposed in this paper is inspired by the success of the *counterexample guided abstraction refinement* (CEGAR) technique for hardware and software verification [6,7,8]. In the CEGAR approach, in each iteration an abstraction of the original model (the concrete system) is constructed using only some of the state variables. The model with the smaller state space is then analyzed by a traditional model checking [9] algorithm. If this reduced model satisfies the given property, then the original system also satisfies the property and the algorithm terminates. Otherwise, the CEGAR loop picks a counterexample reported by the model checking algorithm, which determines a path in the location graph of the LHA. A decision procedure is then applied to the constraints along this path in the concrete system to determine if the counterexample is valid (the constraints can be satisfied by some run of the LHA) or spurious (the constraints cannot be satisfied) in the concrete system. The constraints used to check the feasibility of the counterexample involve all variables in the concrete system. If the constraints can be satisfied, the path corresponds to a true counterexample and the algorithm terminates. If the counterexample is spurious, a subset of variables is selected such that the constraints along the counterexample path are still infeasible by asking the decision procedure for an unsatisfiable core [10] or by using heuristics [11]. These variables are added to the set of variables used thus far and a new abstraction is constructed. On each iteration, the abstractions are therefore more refined and exclude any previously discovered spurious counterexamples.

The power of the above approach derives from its construction of abstractions with a small number variables for which model checking is feasible, while leveraging the power of decision procedures to deal with constraints involving many variables to test the feasibility of counterexamples in the original high-dimensional system. For LHA, we propose a similar approach in which full reachability analysis is performed on abstractions that have a small number of continuous variables. Linear programming (LP) is then applied as the decision procedure to check the validity of counterexamples using all of the variables in the original LHA. LP methods also find the variables to be used for constructing further abstractions. Linear programming for testing the feasibility of a path of a LHA was proposed in [12].

The following steps comprise this approach, which we call *iterative relaxation abstraction* (IRA). On each iteration, a low-dimensional LHA called a *relaxation abstraction* is first obtained by using a subset of the continuous variables from the original LHA. Hybrid system reachability analysis then generates a regular language called the *discrete path abstraction* containing all possible counterexamples (paths to the bad locations) in the relaxation abstraction. If the discrete path abstraction is non-empty, a particular counterexample is selected and linear programming (LP) determines if the counterexample is spurious using the constraints along the path from the original high-dimensional LHA. If the counterexample is spurious, infeasibility analysis of linear programs is applied to find an *irreducible infeasible subset* (IIS) of constraints [13] for the infeasible linear program corresponding to the spurious counterexample. The variables in the IIS are then used to construct the next relaxation abstraction. IRA stops if the discrete path abstraction is empty or a legitimate counterexample is found.

In the CEGAR loop for the analysis of discrete systems, the variables obtained from a spurious counterexample on each iteration are added to the set of variables used in previous iterations to construct a new abstraction. Such an approach would be counterproductive for LHA, however, as LHA reachability analysis does not scale well with increasing numbers of continuous variables. To avoid growth in the number of variables in the relaxation abstractions, only the variables in the current IIS are used on each iteration to construct the next relaxation abstraction, rather than adding these variables to the set of previously used variables. This assures that the number of variables in the LHA to which reachabililty analysis is performed is as small as possible. Counterexamples from previous iterations are excluded at each stage by intersecting the discrete path abstraction generated by the hybrid system analysis with the discrete path abstractions from previous iterations before checking for new counterexamples.

The paper is organized as follows. The next section introduces definitions and notation used throughout the paper. Section 3 defines the relaxation abstraction for LHA and a method for determining if a counterexample from a relaxation abstraction is also a counterexample for the original LHA. Section 4 presents the IRA procedure and Section 5 illustrates its application to a simple example. The performance of IRA is compared to the performance of PHAVer, a recently-developed LHA reachability analysis tool [5]. Section 6 summarizes the contributions of this paper and identifies directions for future work.

2 Preliminaries

2.1 Linear Constraints

We wish to apply a given set of constraints to different sets of variables. Therefore, we define a *linear constraint of order* m as a triple $l = (c, \sim, b)$ where $c = [c_1, \ldots, c_m] \in \mathbb{R}^m$, $\sim \in \{=, \geq, \leq\}$, and $b \in R$. L^m denotes the set of all linear constraints of order m. Given an ordered set of m variables $X = \{X_1, \ldots, X_m\}$ each ranging over the reals, l_X defines a (closed) *linear constraint over* X given by the expression $l_X : \sum_{i=1}^{m} c_i X_i \sim b$. For a given $x = [x_1, \ldots, x_m] \in \mathbb{R}^m$, $l_X(x)$ denotes the value of the expression l_X (TRUE or FALSE) for the valuation $X_1 = x_1, X_2 = x_2, \ldots, X_m = x_m$. Thus, l_X defines a predicate corresponding to a closed half-space in $\mathbb{R}^m$.

For $P \in \text{FIN}(L^m)$, where $\text{FIN}(A)$ denotes the set of finite subsets of a set A, $P_X \triangleq \bigwedge_{l \in P} l_X$, that is, P_X is the predicate over $\mathbb{R}^m$ defined by conjunction of the predicates determined by the linear constraints in P. The predicate P_X corresponds to a closed polyhedron in $\mathbb{R}^m$ defined by the intersection of the closed half-spaces determined by the linear constraints in P. We denote this polyhedron by $[\![P]\!]$ and write $P \Rightarrow P'$ to indicate that $[\![P]\!] \subseteq [\![P']\!]$.

Given a set of linear constraints $P \subset L^m$, the *support of* P, is defined as

$$support(P) = \{i \in \{1, \ldots, m\} | \; \exists \, l = (c, \sim, b) \in P \ni c_i \neq 0\}.$$

Given a second set of linear constraints $P' \subset L^m$, and a subset of indices $I \subset \{1, \ldots, m\}$, P' is said to be a *relaxation of* P *over* I, denoted $P' \sqsupseteq_I P$ if: (i) $P \Rightarrow P'$; and (ii) $support(P') \subseteq I$.

Example 1. If $P = \{([1\ 0\ 0], \geq, 0), ([0\ 1\ 0], \geq, 3), ([1\ 0\ 3], \geq, 8)\}$, $P' = \{([1\ 0\ 0], \geq, 0), ([0\ 1\ 0], \geq, 3)\}$, and $\mathcal{P}'' = \{([1\ 0\ 0], \geq, -1)\}$, then $P'' \sqsupseteq_{\{1\}} P' \sqsupseteq_{\{1,2\}} P$.

Relaxations of sets of linear constraints can be produced in many ways. For example, for an ordered set of m variables X, if X_I denotes the variables from X corresponding to the indices in a set $I \in \{1, \ldots, m\}$, applying the Fourier-Motzkin procedure [14] for existential quantifier elimination of the variables in $X - X_I$ from P_X produces the *projection* of P_X onto X_I. This corresponds to the tightest relaxation of P over I, which we denote by $proj_I(P)$. By "tightest" we mean that if P' is any relaxation of P over I, then $P' \sqsupseteq_I proj_I(P)$. A much looser relaxation of P is generated by simply eliminating the constraints involving variables not in X_I. We call this method of relaxation *localization* because of its similarity to the localization abstraction proposed by Kurshan for discrete systems [6]. Localization of the constraints in P to the variables with indices in I is given by

$$loc_I(P) = \{l \in P | support(l) \subseteq I\}.$$

A set of linear constraints $P \subset L^m$ is said to be *satisfiable* if there exists a valuation $x \in \mathbb{R}^m$ for a set X of m real-valued variables such that $P_X(x)$ is TRUE. We write $\text{SAT}(P)$ to indicate a set of linear constraints is satisfiable. If P is not satisfiable (in which case we write $\text{UNSAT}(P)$), we are interested in finding a minimal subset of constraints in P that is not satisfiable. This is known as an *irreducible infeasible subset* (IIS) of P, which is a subset $P' \subseteq P$ such that (i) $\text{UNSAT}(P')$ and (ii) for any $l \in P'$, $\text{SAT}(P' - \{l\})$ [13]. Although the problem of finding a minimal IIS (an IIS with the least number of constraints) is NP hard [15], several LP packages include functions implementing efficient heuristic procedures to compute IISs that are often minimal (e.g., MINOS (IIS)[16], CPLEX [17], IBM OSL [18], LINDO [19]).

2.2 Linear Hybrid Automata

Following [20], we define a *linear hybrid automaton* (LHA) [20] as a tuple $H = (G, n, \iota, \phi, \gamma, \rho)$, where

- $G = (Q, q_0, Q_{bad}, \Sigma, E)$ is the (labeled) *location graph* of H, where
  - Q is a finite set of *locations*;
  - $q_0 \in Q$ is the *initial* location;
  - $Q_{bad} \subset Q$ is the set of bad locations (the locations that should not be reachable);
  - Σ is a finite set of *labels*;
  - $E \subseteq (Q - Q_{bad}) \times \Sigma \times Q$ is finite set of (labeled) *transitions*, where no two outgoing transitions from a given location have the same label;
- n is the number of *continuous state variables*,
- $\iota : Q \longrightarrow \text{FIN}(L^n)$ identifies the *invariant* for each location.
- $\phi : Q \longrightarrow \text{FIN}(L^n)$ identifies the *flow constraints* for each location.
- $\gamma : E \longrightarrow \text{FIN}(L^n)$ identifies the *guard* for each transition.
- $\rho : E \longrightarrow \text{FIN}(L^{2n})$ identifies the *jump relation* for each transition.

Example 2. Figure 1 shows an LHA with continuous state variables $X = \{x_1, x_2\}$, discrete states $Q = \{q_0, q_1, q_2, q_3, q_4\}$, discrete state transition labels $\{a, b, c, d\}$, initial location q_0 (the unlabeled rectangle) with an invariant defining a unique initial continuous state $x(0) = \{.5, 0\}$, and $Q_{bad} = \{q_4\}$.

A *run* for $t \geq 0$ for an LHA H is a (possibly infinite) sequence of the form

$$q_0, x^0, \sigma_0, q_1, x^1, \sigma_1, q_2, x^2, \ldots,$$

where for all $k = 0, 1, \ldots$

- $x^k : [t_s^k, t_f^k] \rightarrow R^n$ denotes the continuous evolution of the continuous state variables for $t_s^k \leq t \leq t_f^k$, where $0 = t_s^0 \leq t_f^0 = t_s^1 \leq t_f^1 = t_s^2 \cdots$;
- $x^k(t) \in [\![\iota(q_k)]\!]$ (location invariants), where $t_s^k \leq t \leq t_f^k$;
- $\dot{x}^k \in [\![\phi(q_k)]\!]$ (flow constraints), where $t_s^k \leq t \leq t_f^k$;

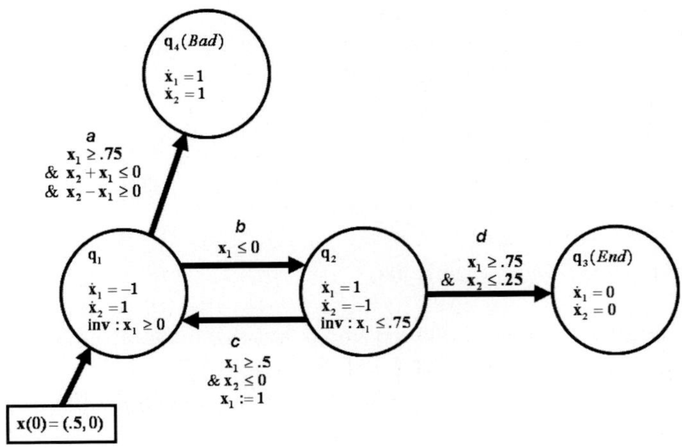

Fig. 1. An example LHA

– $(q_k, \sigma_{(k+1)}, q_{k+1}) \in E$ (transitions);
– $x^k(t_f^k) \in [\![\gamma((q_k, \sigma_{(k+1)}, q_{k+1}))]\!]$ (guards);
– $(x^k(t_f^k), x^k(t_s^{k+1})) \in [\![\rho((q_k, \sigma_k, q_{k+1}))]\!]$ (jump relation).

Projecting a run onto the transitions (i.e., eliminating the continuous state variable evolution) leads to a sequence of the form $\pi = q_0, \sigma_1, q_1, \sigma_2, q_2, \ldots$, which corresponds to a path in the location graph. Projecting a path onto the transition labels gives a sequence of labels, $\omega = \sigma_1, \sigma_2, \ldots$. Since the labels on the outgoing transitions from each location are distinct, the sequence ω corresponds to a unique path (π) in the location graph. A sequence of transition labels is said to be *feasible* if the path to which it corresponds could be generated by a run of the LHA; otherwise, the sequence is said to be *infeasible*. A path that leads to a state in Q_{bad} is called a *counterexample*.

We let $\mathcal{L}_{CE}(H)$ denote the set of all feasible sequences of transition labels generated by runs that lead to states in Q_{bad}. The definition of the transitions in the location graph precludes transitions from any state in Q_{bad}, therefore the sequences in $\mathcal{L}_{CE}(H)$ are all finite, that is, $\mathcal{L}_{CE}(H) \subseteq \Sigma^*$. $\mathcal{L}_{CE}(H)$ is not necessarily a regular language, however.

3 Relaxation Abstractions and Counterexample Analysis

In this section we first introduce a new class of abstractions for LHA based on relaxations of the linear constraint sets defining the invariants, flow constraints, guards, and jump relation for a given LHA. We then present a method using linear programming (LP) analysis for determining whether a counterexample for a relaxation abstraction is spurious for the original LHA.

Given an LHA $H = (G, n, \iota, \phi, \gamma, \rho)$ and an index set $I \subset \{1, \ldots, n\}$ with $|I| = n' < n$, a linear hybrid automaton $H' = (G', n', \iota', \phi', \gamma', \rho')$ is said to be a *relaxation of H over I*, denoted $H' \sqsupseteq_I H$, if

– $G' = G = (Q, q_0, Q_{bad}, \Sigma, E)$;
– for each $q \in Q$:
  - $\iota'(v) \sqsupseteq_I \iota(v)$ (invariants);
  - $\phi'(e) \sqsupseteq_I \phi(e)$ (flows);
– for each $e \in E$:
  - $\gamma'(e) \sqsupseteq_I \gamma(e)$ (guards);
  - $\rho'(e) \sqsupseteq_I \rho(e)$ (jump relations).

Example 3. Figure 2 shows a relaxed linear hybrid automaton derived from the LHA in Figure 1 over the index set $I = 1$. This relaxation is obtained by applying localization over I to each of the constraints in the original LHA.

Since the constraints defining a relaxation abstraction H' are constraints of the relation defining the original LHA H, it follows that any run for H is also a run for H'. Similarly, $\mathcal{L}_{CE}(H') \supseteq \mathcal{L}_{CE}(H)$.

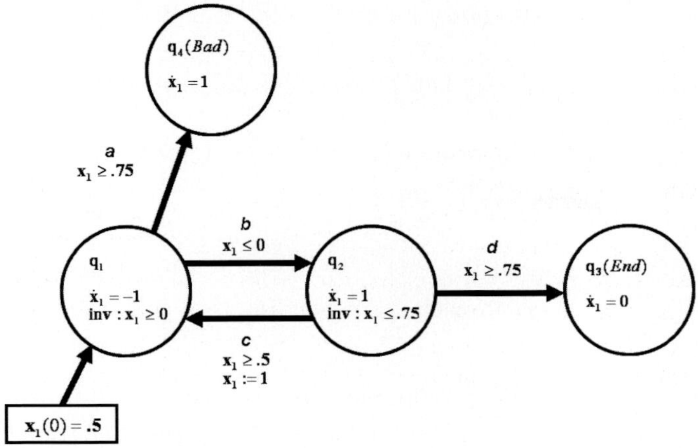

Fig. 2. A relaxation abstraction for the LHA in Fig. 1

Given an LHA H and an index set I, let H' be a relaxation of H over I. Given a counterexample for H', $ce = \sigma_1\sigma_2\ldots\sigma_K \in \mathcal{L}_{CE}(H')$, there is a *unique* corresponding path in the location graph G of H of the form $\pi_{ce} = q_0\sigma_1 q_1\sigma_2\ldots\sigma_K q_K$, where $q_K \in Q_{bad}$. To determine if there is a run for H corresponding to ce, we consider whether or not the constraints along this state-transition sequence are feasible as follows.

Given a feasible path $\pi = q_0\sigma_1 q_1\sigma_2\ldots\sigma_K q_K$, we introduce the following variables:

- X_s^0, corresponding to the initial continuous state in q_0;
- $X_s^1,\ldots,X_s^K$, corresponding to the values of the continuous states when the transitions occur *into* locations $q^1,\ldots,q^K$, respectively;
- $X_f^0,\ldots,X_f^{K-1}$, corresponding to the values of the continuous states when the transitions occur *out of* locations $q^0,\ldots,q^{K-1}$, respectively;
- $\Delta_0,\Delta_1,\ldots,\Delta_{K-1}$, corresponding to the amount of time the run spends in $q^0,\ldots,q^{K-1}$, respectively.

We let V_π denote the set of variables defined above for a path π. From the constraints in H we construct the following constraints over the variables in V_π that must be satisfied by a valid run for H. We denote this collection of linear constraints by $\mathcal{C}(H,\pi)$ or $\mathcal{C}_\pi$ depending on the context:

- $\iota(q_0)_{X_s^0}$: the initial continuous states must be in the invariant of q_0;
- for $k = 1,\ldots,K$, the kth transition in the path is $e_k = (q_{k-1},\sigma_k,q_k)$ and for each transition:
  - $\iota(q_{k-1})_{X_f^{k-1}}$: the continuous state before the transition must satisfy the invariant of q_{k-1};

- $\iota(q_k)_{X_s^k}$: the continuous state after the transition must satisfy the invariant of q_k; [1]
- $\gamma(e_k)_{X_f^{k-1}}$: the continuous state before the transition must satisfy the guard of e_k;
- $\rho(e_k)_{X_f^{k-1}, X_s^k}$: the continuous states before and after the transition must satisfy the jump relation for e_k;

– for $k = 0, \dots, (K-1)$:
  - $\hat{\phi}(q_k)_{(X_f^k, X_s^k, \Delta_k)}$, where $\hat{\phi}(q_k)$ is the set of linear constraints of the form $\hat{l} = ([c, -c, -b], \sim, 0)$, each corresponding to a constraint $l = (c, \sim, b) \in \phi(q_k)$.

The final constraint, which represents the flow constraint in each location, follows from the following derivation: for each $l = (c, \sim, b) \in \phi(q_k)$, $c\dot{x} \sim b$ for all $t_s \le t \le t_f$; this implies $c(x(t_f) - x(t_s)) = \int_{t_s}^{t_f} cx(\tau)d\tau \sim b(t_f - t_s)$; therefore, $cx(t_f) - cx(t_s) - b\Delta \sim 0$, where $\Delta = t_f - t_s$, which is the constraint $\hat{l}$.

As demonstrated in [12], a path is feasible if and only if the above linear constraints are feasible.

4 Iterative Relaxation Abstraction

We now present the IRA procedure to CEGAR-based reachability analysis of LHA. The following paragraphs describe the IRA steps shown in Fig. 3.

Step 1. Construct a relaxation abstraction over index set I_i of the LHA H, $H_i \sqsupseteq_{I_i} H$. Any relaxation method can be applied to the linear constraints in H.

Step 2. Compute the next discrete path abstraction $\mathcal{A}_{CE}^{i+1}$ (a regular language containing $\mathcal{L}_{CE}(H)$) as the intersection of the previous discrete path abstraction with $\widehat{\mathcal{L}}_{CE}(H_i)$, a regular language containing $\mathcal{L}_{CE}(H_i)$, and $(\Sigma^* - ce_i)$ to assure the previous counterexample is removed from the next discrete path abstraction.

Step 3. Choose a counterexample ce_{i+1} from $\mathcal{A}_{CE}^{i+1}$, or set $ce_{i+1} == null$ if $\mathcal{A}_{CE}^{i+1}$ is empty. This operation is performed by $Select\_CE(\mathcal{A}_{CE}^{i+1})$.

Step 4. $ce_{i+1} == null$ indicates that no bad states are reachable in the original linear hybrid automaton H since $\mathcal{A}_{CE}^{i+1}$ contains $\mathcal{L}_{CE}(H)$.

Step 5. Construct $C = \mathcal{C}(H, ce_{i+1})$, the set of linear constraints along the path in the location graph of H determined by ce_{i+1}.

Step 6. Apply LP to determine the feasibility of the constraints C. We know that SAT(C) if and only if ce_{i+1} is a valid counterexample in H [12].

---

[1] It is sufficient to check that the invariant holds at the beginning and end of the continuous state trajectory in each location because the invariants and flow constraints are convex [20].

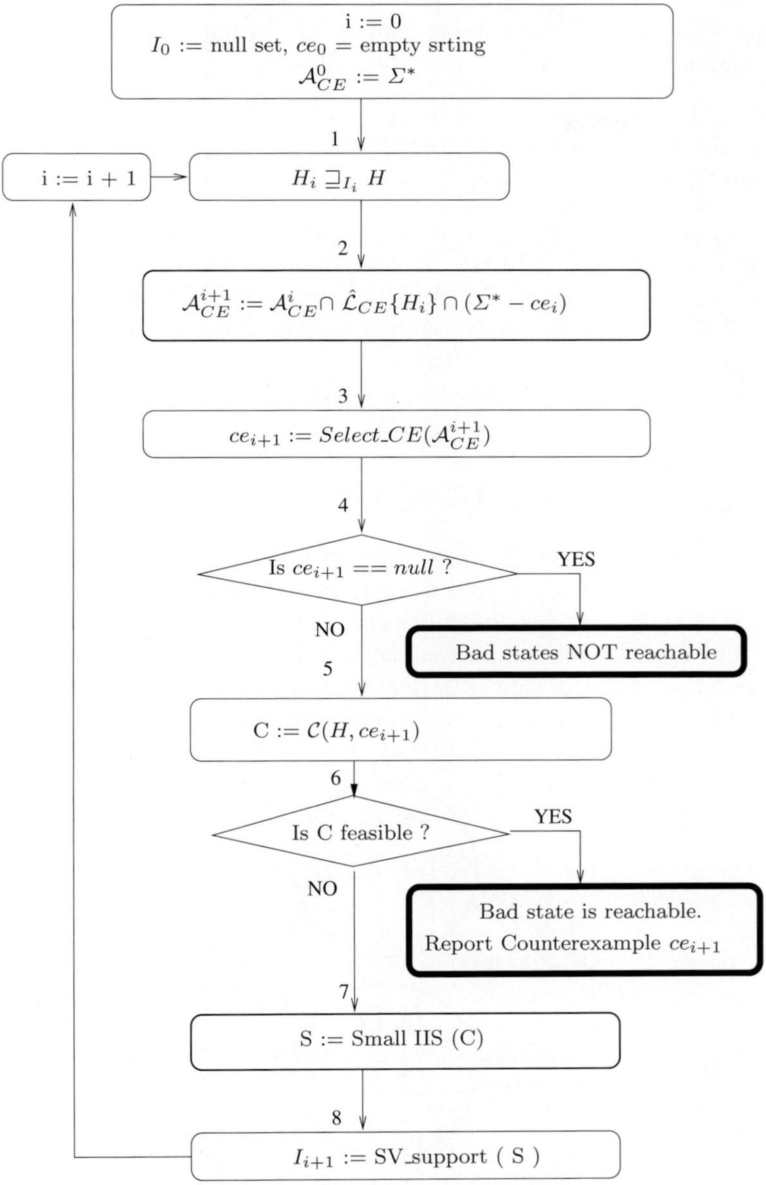

Fig. 3. The IRA procedure: Iterative relaxation abstraction reachability analysis for LHA

Step 7. If UNSAT(C), apply LP infeasibility analysis to find a minimal IIS for the constraints in C. (Heuristic procedures will actually find an IIS that may not be minimal.)

Step 8. Find the set of state variable indices corresponding to the variables with indices in $support(C)$. This is the operation represented by the function $SV\_support(C)$.

Correctness of the IRA procedure (in the sense that if it terminates, it is correct) is guaranteed since the LHA H^i and languages $\mathcal{A}^i$ are overapproximations of H and $\mathcal{L}_{CE}(H)$, respectively. Although termination cannot be guaranteed (because $\hat{\mathcal{L}}_{CE}$ (H_i) maybe an overapproximation of $\mathcal{L}_{CE}(H_i)$), the sequence of discrete path abstractions generated on by the iterations is monotonically decreasing in size and any counterexample that has been analyzed is eliminated in future iterations.

5 Implementation and Example

The IRA has been implemented using PHAVer [5] for LHA reachability analysis and the CPLEX [17] library for LP analysis. PHAVer builds overapproximate discrete abstractions of the linear hybrid automata represented by finite automata. The discrete abstractions in the IRA tool are stored and manipulated using the AT&T FSM library [21]. The IRA tool provides users the ability to write their own relaxation functions. We experimented with two versions of IRA,

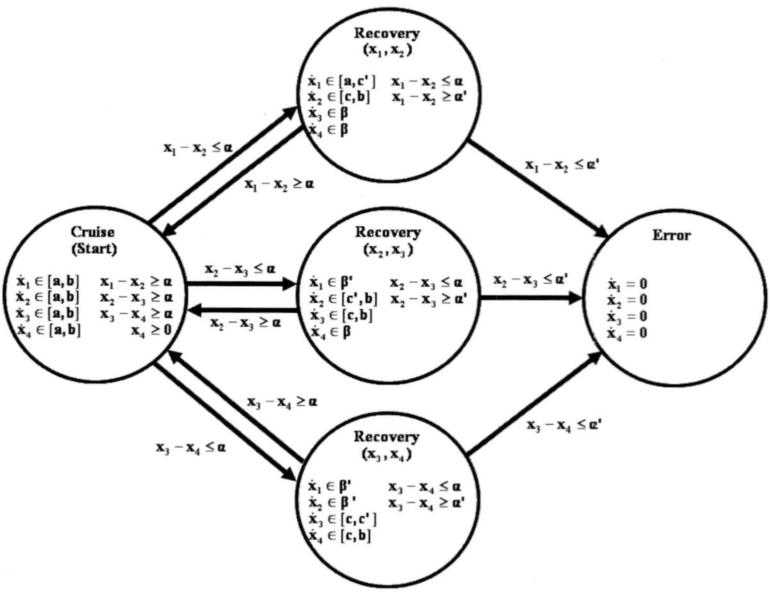

Fig. 4. A automated highway with 4 vehicles

Table 1. Comparison of analysis of Linear Hybrid Automata using PHAVer, IRA-Loc (localization relaxation) and IRA-FM (Fourier-Motzkin procedure)

	Time Taken in Seconds			Memory Used in KBs		
No of cars	**PHAVer**	**IRA-Loc**	**IRA-FM**	**PHAVer**	**IRA-Loc**	**IRA-FM**
6	0.26	1.34	61.05	5596	18344	452408
8	0.96	5.11	170.11	7532	20128	1000436
10	8.21	17.76	402.15	13904	22652	1876256
12	147.11	50.04	933.47	32800	26132	3155384
14	7007.51	123.73	1521.95	103408	30712	4194028
15	70090.06	181.74	2503.59	198520	33896	4193620
16	–	267.46	3519.51	–	36828	4194024
17	–	339.08	4741.75	–	40316	4194140
18	–	493.34	6384.94	–	44368	4194280
19	–	652.51	8485.49	–	49272	4194296

referred to as IRA-Localization and IRA-FM. IRA-Loc is the implementation that uses localization as the technique for building the relaxation. IRA-FM is another implementation which uses the Fourier Motzkin procedure for implementing quantifier elimination to build the relaxation.

As an example, we consider the model of a central arbiter for a automated highway and analyze the arbiter for safety properties, particularly for the specification that no two vehicles on the automated highway collide with each other. The electronic arbiter enforces speed limits on vehicles on the automated highway to achieve this purpose. The arbiter provides allowed ranges of velocities for each vehicle $[a, b]$. When two vehicles come within a distance α of each other, we call this a "possible" collision event. The arbiter asks the approaching car to slow down by reducing the upper bound to $[a, c']$ and asks the leading car to speed up by increasing the lower bound to $[c, b]$; it also requires that all other cars not involved in the possible collision slow down to a constant "recovery-mode" velocity β for cars behind the critical region and β' for cars in front of the critical region. When the distance between the two vehicles involved in the possible collision exceeds α, the arbiter model goes back to the dynamics of the cruise mode. The linear hybrid automata representing the case of four cars is shown in Fig. 4. The example is easily parameterized by varying the number of cars on the highway.

We ran this example on an AMD Opteron four-processor $x86\_64$ Linux machine running Fedora Core 5. The comparative results are shown in Table 1. In each case with n cars, both IRA versions verify in n-1 iterations that the bad states are not reachable. Consequently, the tighter Fourier Motzkin relaxation offer no advantage for this example. We expect, however, that tighter relaxations will be necessary to verify properties of more complex systems. Further empirical studies are currently being pursued. The plot of the log of the time taken vs. the

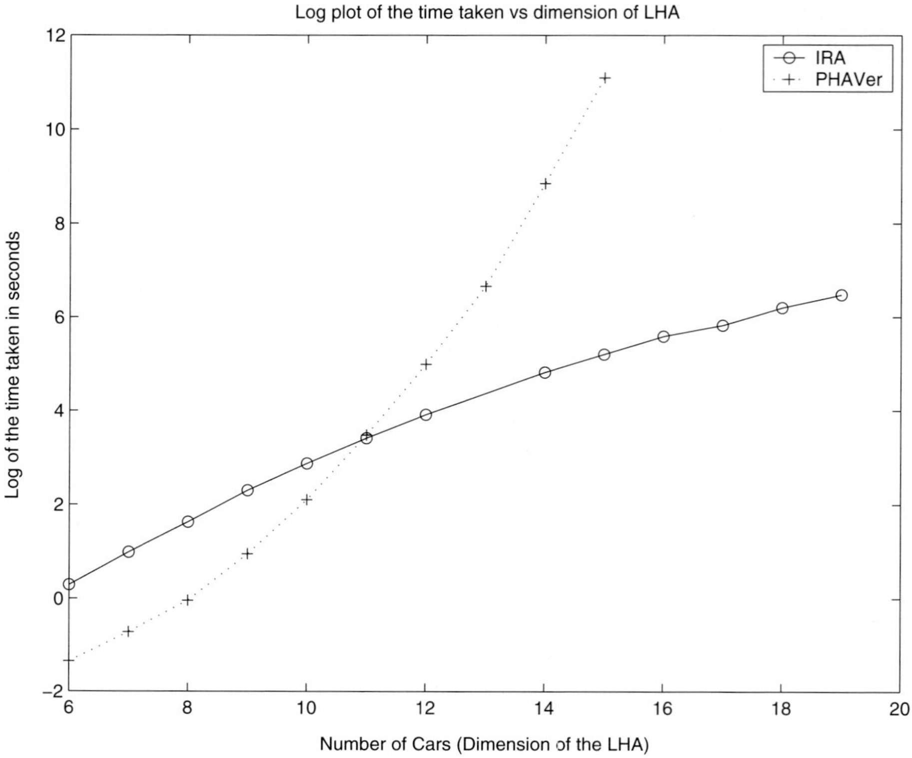

Fig. 5. Results: IRA-Loc vs. PHAVer

dimension of the LHA in Fig. 5 shows that PHAVer outperforms IRA-Loc for small dimensional systems. This happens as IRA-Loc spends time "reasoning" about the structure of the LHA and the possibility of reducing its dimension. For larger dimensions, IRA-Loc outperforms PHAVer significantly in both time and memory.

6 Discussion

IRA combines reachability analysis for low-dimensional LHA with the power of LP analysis for large numbers of variables. As proposed in [12], linear programming is used as an efficient counterexample validation algorithm. This idea of using linear programming as a counterexample validation is also used in [22]. As linear programming is in P [23], it is an efficient counterexample validation procedure for high dimensional LHA. Also, if a counterexample is found and validated, the reachability procedure can terminate immediately.

The IRA procedure uses linear constraint relaxation as a technique for generating abstractions of LHA. In contrast to previously proposed CEGAR techniques for hybrid system analysis in which abstractions are refined by splitting locations ([24],[25]), the relaxation abstraction retains the location graph as the original LHA.

We are working on extending the results in this paper to nonlinear hybrid automata. We are also currently evaluating the effectiveness of this procedure on a number of other benchmark problems.

Acknowldgments

The authors thank Xuandong Li and Prasad Sistla for several useful discussions, Goran Frehse for his help and guidance with PHAVer, and Alhad Arun Palkar for helping with the implementation.

References

1. Henzinger, T.: The Theory of Hybrid Automata. Lecture Notes in Computer Science (1996) 278
2. Alur, R., Courcoubetis, C., Halbwachs, N., Henzinger, T.A., Ho, P.H., Nicollin, X., Olivero, A., Sifakis, J., Yovine, S.: The algorithmic analysis of hybrid systems. Theoretical Computer Science **138**(1) (1995) 3–34
3. Henzinger, T.A., Ho, P.H., Wong-Toi, H.: HyTech: A model checker for hybrid systems. International Journal on Software Tools for Technology Transfer **1**(1–2) (1997) 110–122
4. Alur, R., Henzinger, T., Wong-Toi, H.: Symbolic analysis of hybrid systems. In: Proc. 37-th IEEE Conference on Decision and Control. (1997)
5. Frehse, G.: PHAVer: Algorithmic Verification of Hybrid Systems Past HyTech. [26] 258–273
6. Kurshan, R.: Computer-aided Verification of Coordinating Processes: The Automata Theoretic Approach. Princeton University Press, 1994. (1994)
7. Clarke, E.M., Grumberg, O., Jha, S., Lu, Y., Veith, H.: Counterexample-guided abstraction refinement. In: CAV '00: Proceedings of the 12th International Conference on Computer Aided Verification, London, UK, Springer-Verlag (2000) 154–169
8. Ball, T., Majumdar, R., Millstein, T.D., Rajamani, S.K.: Automatic Predicate Abstraction of C Programs. In: SIGPLAN Conference on Programming Language Design and Implementation. (2001) 203–213
9. E. M. Clarke, J., Grumberg, O., Peled, D.A.: Model Checking. MIT Press, Cambridge, MA, USA (1999)
10. Zhang, L., Malik, S.: Validating SAT Solvers Using an Independent Resolution-Based Checker: Practical Implementations and Other Applications. In: DATE, IEEE Computer Society (2003) 10880–10885
11. Chaki, S., Clarke, E., Groce, A., Strichman, O.: Predicate abstraction with minimum predicates. In: Proceedings of 12th Advanced Research Working Conference on Correct Hardware Design and Verification Methods (CHARME). (2003)
12. Li, X., Jha, S.K., Bu, L.: Towards an Efficient Path-Oriented Tool for Bounded Reachability analysis of Linear Hybrid Systems using Linear Programming. (2006)
13. Chinneck, J., Dravnieks, E.: Locating minimal infeasible constraint sets in linear programs. ORSA Journal on Computing **3** (1991) 157–168
14. Dantzig, G.B., Eaves, B.C.: Fourier-Motzkin elimination and Its Dual. J. Comb. Theory, Ser. A **14**(3) (1973) 288–297

15. Sankaran, J.K.: A note on resolving infeasibility in linear programs by constraint relaxation. Operations Research Letters **13** (1993) 1920
16. Chinneck, J.W.: MINOS(IIS): Infeasibility analysis using MINOS. Comput. Oper. Res. **21**(1) (1994) 1–9
17. ILOG: (http://www.ilog.com/products/cplex/product/simplex.cfm)
18. Hung, M.S., Rom, W.O., Waren, A.D.: Optimization with IBM OSL and Handbook for IBM OSL (1993)
19. Systems Inc., L.: (http://www.lindo.com/products/api/dllm.html)
20. Ho, P.H.: Automatic Analysis of Hybrid Systems, Ph.D. thesis, technical report CSD-TR95-1536, Cornell University, August 1995, 188 pages (1995)
21. Mohri, M., Pereira, F., Riley, M.: The design principles of a weighted finite-state transducer library. Theoretical Computer Science **231**(1) (2000) 17–32
22. Jiang, S.: Reachability analysis of Linear Hybrid Automata by using counterexample fragment based abstraction refinement. submitted (2006)
23. Karmarkar, N.: A new polynomial-time algorithm for linear programming. Combinatorica **4**(4) (1984) 373–395
24. Fehnker, A., Clarke, E.M., Jha, S.K., Krogh, B.H.: Refining Abstractions of Hybrid Systems Using Counterexample Fragments. [26] 242–257
25. Alur, R., Dang, T., Ivancic, F.: Counterexample-guided predicate abstraction of hybrid systems. Theor. Comput. Sci. **354**(2) (2006) 250–271
26. Morari, M., Thiele, L., eds.: Hybrid Systems: Computation and Control, 8th International Workshop, HSCC 2005, Zurich, Switzerland, March 9-11, 2005, Proceedings. In Morari, M., Thiele, L., eds.: HSCC. Volume 3414 of Lecture Notes in Computer Science., Springer (2005)

Sporadic Control of First-Order Linear Stochastic Systems

Erik Johannesson, Toivo Henningsson, and Anton Cervin

Department of Automatic Control, LTH
Lund University
Box 118, 221 00 Lund, Sweden
{erik,toivo,anton}@control.lth.se

Abstract. The standard approach in feedback control systems is to sample and control periodically. For some applications, such as networked control systems or severely energy-constrained systems, it could be advantageous to instead use event-based control schemes. Aperiodic control (that is, event-based control with no specified minimum inter-event time) of first-order stochastic systems has been investigated in previous work. In any real implementation, however, it is necessary to have a well-defined minimum inter-event time. In this paper, we explore two such sporadic control schemes for first-order linear stochastic systems and compare the achievable performance to both periodic and aperiodic control. The results indicate that sporadic control can give better performance than periodic control in terms of reduced process state variance and control action frequency.

1 Introduction

Digital feedback controllers are most often implemented using periodic sampling, computation, and actuation. This approach enables the control designer to utilize standard sampled-data system theory or to discretize a continuous-time controller assuming a fixed sampling rate and constant hold intervals [1].

For some applications, however, event-based control schemes may have an advantage over periodic schemes. In networked control applications [2], it could make sense to only transmit information when something significant has occurred in the system, in order to save bandwidth. In embedded applications [2], it may be essential to minimize the number of control actions in order to save energy. Also, in the application of inventory control it seems rational to replenish stock only when it is low rather than on a periodic basis, if there is a fixed transportation cost.

Event-based control as a technology is of course not new. It has been used for a long time in such diverse areas as engine control [3], robot path planning [4], and control of industrial processes [5]. Mostly, however, it has been applied in an ad-hoc way. This can be attributed to the lack of a comprehensive theory, which in turn can be explained by the mathematical difficulties involved.

From a control-theoretic point of view, event-based feedback control systems can be viewed as hybrid systems. In this paper, we consider first-order linear

A. Bemporad, A. Bicchi, and G. Buttazzo (Eds.): HSCC 2007, LNCS 4416, pp. 301–314, 2007.

stochastic systems, where an exogenous random disturbance (modelled as white noise) causes the process state to drift. The control law generates discrete events when the state crosses certain boundaries. Hence, our system falls into the category of stochastic hybrid systems as defined in [6].

Event-based control of first-order linear stochastic systems was previously studied in [7,8]. It was shown that, compared to periodic control, the output variance could be significantly reduced assuming the same mean time between events. The control was realized by applying an impulse action whenever the magnitude of the system state exceeded a certain threshold.

From a real-time systems point of view, however, tasks triggered by asynchronously generated events cannot be guaranteed service unless there is a well-defined minimum inter-arrival time. For the controller presented in [7,8] there was no such minimum inter-arrival time. In accordance with real-time systems terminology [9], we will refer to such a control policy as an *aperiodic* event-based control policy.

In this paper, we will explore the class of *sporadic* event-based controllers for first-order linear stochastic systems. Assuming a minimum inter-arrival time T between events, such a controller can be guaranteed not to consume more than a certain network bandwidth or CPU utilization. Two controllers within this class will be studied. For the first controller, it is assumed that the process state is measured continuously and that a control action can be taken at any point in time, but not more often than every T seconds. The second controller assumes that the process state is measured every T seconds and that the controller can then decide whether to apply a new control action or not.

The rest of this paper is organized as follows. The system and the control performance measure are presented in Section 2. Then, the two types of sporadic control are analyzed in Section 3. In Section 4, sporadic control is compared to periodic and aperiodic control. Finally, the conclusions are given in Section 5.

2 Problem Formulation

Consider the first-order system described by the linear stochastic differential equation

$$dx = ax\,dt + u\,dt + \sigma\,dw \tag{1}$$

where x is the state, u the control signal, w is a Wiener process with $\mathrm{E}(dw) = 0$, $\mathrm{E}(dw^2) = dt$, a is the pole of the system, and σ is the intensity of the process noise. The control signal u is zero except at events, when it is allowed to be a Dirac pulse of any magnitude. For simplicity we will often say that controllers *generate events* when they issue control actions.

The performance of the system is measured by a cost function with two terms. The state cost represents the stationary process variance and is given by

$$J_x = \lim_{t\to\infty} \frac{\int_0^t x^2\,ds}{t}$$

The control cost represents the average number of events per time unit and is given by

$$J_u = \lim_{t\to\infty} \frac{N_u(0,\, t)}{t}$$

where $N_u(0,\, t)$ is the number of control actions in the interval $(0,\, t)$. The total cost to be minimized is given by

$$J = J_x + \rho J_u \tag{2}$$

where $\rho \geq 0$ is a scalar representing the relative cost of control actions.

To reduce the number of free parameters we use coordinate scaling of x and t to fix two parameters ($\sigma = T = 1$) before performing the analysis. The original variables can then be retrieved from inverse scaling.

Let the transformed variables be described by

$$dt = T\,d\tau, \qquad dw = \sqrt{T}\,dv, \qquad x = \sigma\sqrt{T}x'$$

choosing dv so that $E(dw^2) = E(T\,dv^2) = T\,d\tau = dt$. The dynamics become

$$\begin{aligned} dx' &= \frac{1}{\sigma\sqrt{T}}\left(axT\,d\tau + uT\,d\tau + \sigma\sqrt{T}\,dv\right) \\ &= aTx'\,d\tau + \frac{\sqrt{T}}{\sigma}u\,d\tau + dv \\ &= a'x'\,d\tau + u'\,d\tau + dv \end{aligned}$$

where $a' = aT$ is the relevant measure of process speed. The original costs are retrieved as

$$J_x = \sigma^2 T J'_x, \qquad J_u = T^{-1}J'_u,$$

with $\rho' = \frac{1}{\sigma^2 T^2}\rho$ being the proper weighting after transformation.

3 Sporadic Control

3.1 General Observations

For the problem described above, we note that there are two properties that define a sporadic controller. Namely, under what conditions it generates events, and the magnitude of the control signal in the occurence of an event. It is easy to see that any sporadic controller that is optimal in the sense of the cost function given above must satisfy the following:

- At any event, u is chosen to bring x to the origin.
- When an event is permitted, the decision of whether to generate an event or not is a function of $|x|$.
- If the decision function is such that an event should be generated when $x = r$, there should also be an event whenever $|x| \geq r$.

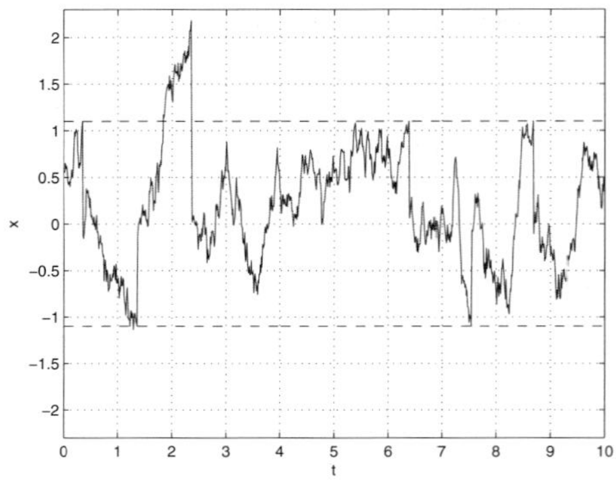

Fig. 1. A simulation example of sporadic control with continuous-time measurements, $a = 0$ and $r = 1.1$. When the state crosses the threshold $|x| = r$ and the controller is in the active mode, the state is reset. When the state passes the threshold at $t \approx 1.9$ the controller is in the inactive mode and has to wait until $t \approx 2.3$ before it can generate an event.

Thus, the only parameter left to specify the optimal controller is the threshold r, such that an event is triggered whenever $|x| \geq r$, if permitted. The threshold should be chosen to minimize J.

In the following subsections, we investigate two types of control schemes with different restrictions on when events are permitted. We refer to them as sporadic control with continuous- and discrete-time measurements, respectively.

3.2 Sporadic Control with Continuous-Time Measurements

It is assumed that the process state is measured continuously and that events may be generated at any time, with the restriction of a minimum inter-event time T. We say that the controller is in the *inactive state* when the time elapsed since the last event is less than T and in the *active state* otherwise. Thus, the optimal controller generates a new event at any point in time t when it is in the active state and $|x| \geq r$. A simulation example can be seen in Fig. 1.

To find the optimal threshold r, the closed-loop cost will be characterized as a function of r. Introduce the storage function $V(x)$ such that

$$E\Big(x^2\,dt + dV(x)\Big) = J\,dt \tag{3}$$

when the controller is in the active state and that

$$E\left(\int_{t_0}^{t_0+T} x^2\,dt + V\Big(x(t_0+T)\Big) - V\Big(x^-(t_0)\Big)\right) + \rho = JT \tag{4}$$

when there is an event at time t_0, and $x^-(t_0)$ is the state just prior to the event. The storage function can be seen as the state dependent part of the expected remaining cost. The function $V(x)$ will be an even function of x because of symmetry. Since only differences in $V(x)$ are of interest, we can arbitrarily let $V(0) = 0$.

We will derive an expression for J from the two last equations. Beginning from (3) and applying (1) (assuming $\sigma = 1$), it follows that

$$\begin{aligned} J\,dt &= x^2\,dt + V'(x)\,\mathrm{E}(dx) + \frac{1}{2}V''(x)\,\mathrm{E}(dx^2) \\ &= x^2\,dt + axV'(x)\,dt + \frac{1}{2}V''(x)\,dt \end{aligned}$$

so that

$$\frac{1}{2}V''(x) + axV'(x) + x^2 - J = 0.$$

To solve the above equation we note that the solution of

$$\frac{1}{2}f''(x) + axf'(x) = \kappa(x)$$

with $f(0) = f'(0) = 0$ is given by

$$f(x) = 2\int_0^x e^{-ay^2}\int_0^y e^{az^2}\kappa(z)\,dz\,dy. \tag{5}$$

In order for us to later solve for J, the storage function is written as

$$V(x) = JV_J(x) - V_c(x),$$

where V_J is found by inserting $\kappa(x) = 1$ in (5) and V_c by inserting $\kappa(x) = x^2$. This expression is valid for $|x| \le r$, beyond which $V(x)$ will be constant at $V(r)$.

To use (4) the following partial results are needed. The expected state cost during one period of inactive state is

$$\begin{aligned} J_T T &= \mathrm{E}\left(\int_{t_0}^{t_0+T} x^2\,dt \,\middle|\, x(t_0) = 0, u(t) = 0\right) \\ &= \begin{cases} \frac{e^{2aT} - 2aT - 1}{4a^2} & a \neq 0 \\ \frac{T^2}{2} & a = 0 \end{cases} \end{aligned} \tag{6}$$

After one period of inactive state x has a Gaussian distribution with zero mean and variance

$$\begin{aligned} V_T &= \mathrm{E}\left(x(t_0 + T)^2 \,|\, x(t_0) = 0, u(t) = 0\right) \\ &= \begin{cases} \frac{e^{2aT} - 1}{2a} & a \neq 0 \\ T & a = 0 \end{cases} \end{aligned} \tag{7}$$

Now let $\varphi(x)$ be the Gaussian probability density with zero mean and variance V_T. Then using the fact that $\int \varphi(x)\,dx = 1$ we can write (4) as

$$JT = J_T T - \underbrace{\int \varphi(x)\Big(V(r) - V(x)\Big)\,dx}_{\Delta V = J\Delta V_J - \Delta V_c} + \rho, \tag{8}$$

or splitting the integral into the part proportional to J and the constant part as

$$J\left(T + \Delta V_J\right) = J_T T + \Delta V_c + \rho.$$

Now we can solve for J as

$$J = \frac{J_T T + \Delta V_c + \rho}{T + \Delta V_J} = \frac{J_T T + \Delta V_c}{T + \Delta V_J} + \rho \frac{1}{T + \Delta V_J} = J_x + \rho J_u.$$

The optimal controller can be found by minimizing J as a function of r.

The given expressions involve integrals that are not straightforward to treat analytically. For the calculations presented here, all functions of x were approximated by piecewise polynomials on an interval of the real axis starting at zero. The terms and factors were constructed by spline interpolation with explicit endpoint derivatives, with knots spaced with constant x increment for low x and constant x^2 increment for high x to account for the growth of $e^{\pm ax^2}$. All operations—addition, subtraction, multiplication and integration—could then be done with no further approximation, yielding $J(r)$ as a piecewise rational function. The costs can be seen as a function of r in Fig. 2, and the optimal threshold as a function of ρ in Fig. 3.

The probability density $f(x)$ of the state in the active mode can be obtained from the diffusion equation

$$\frac{1}{2} f''(x) - axf'(x) - af(x) + \varphi(x)/\bar{T} = 0,$$

where $\bar{T}$ is the mean time between events and $f(x)$ is normalized so that $\int f(x)dx$ is the probability that the controller is in the active state.

The solution is

$$\frac{1}{2} f(x) = \frac{1}{\bar{T}} e^{ax^2} \int_{-r}^{x} e^{-ay^2} \int_{y}^{0} \varphi(z)\,dz\,dy.$$

3.3 Sporadic Control with Discrete-Time Measurements

We now assume that the process state is measured periodically with sample interval T and that an event may be generated at each sampling instant kT, for integer k. Thus, the optimal controller generates an event at time kT if $|x| \geq r$.

Sampling (1) with period T gives the state update equation

$$x(kT + T) = e^{aT} x(kT) + w(k)$$

where $\{w(k)\}_{k=1}^{\infty}$ is a sequence of independent Gaussian distributed random variables with $\mathrm{E}(w(k)) = 0$ and $\mathrm{E}(w(k)^2) = V_T$, as given by (7).

Assuming stationarity, the sampled equivalent of the state cost is

$$J_x = \frac{1}{T} \int_0^T \mathrm{E}\, x^2(t)\; dt = \frac{Q_1 \,\mathrm{E}\, x^2(kT)}{T} + J_T$$

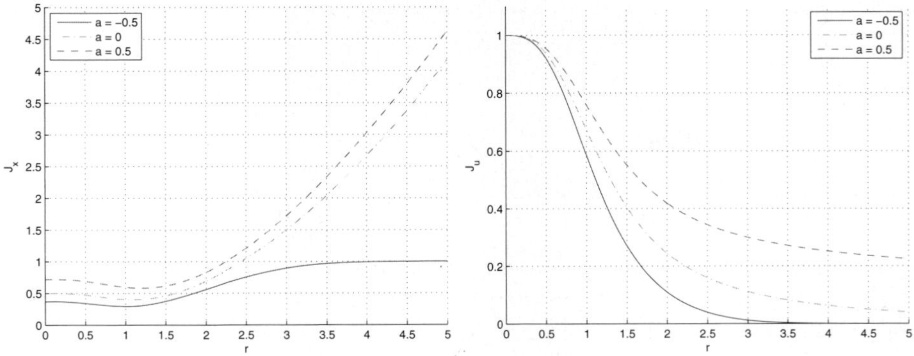

Fig. 2. Cost functions for sporadic control with continous-time measurements. Left: State cost J_x as a function of threshold r. Note the initial decrease of J_x as r increases. Right: Control cost J_u as a function of threshold r. Both functions are plotted for systems with $a = -0.5$, $a = 0$ and $a = 0.5$ respectively.

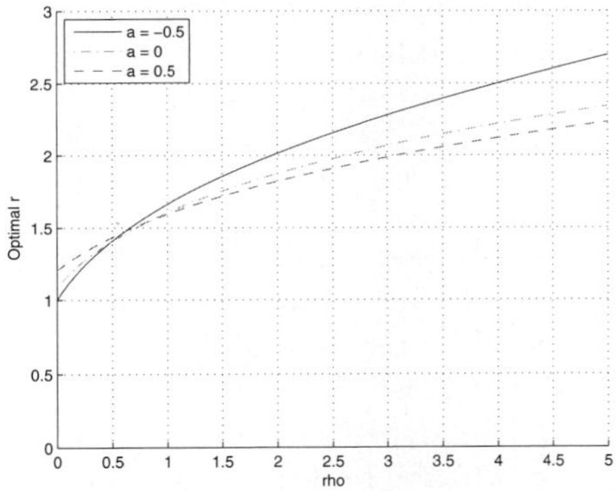

Fig. 3. Optimal threshold r as a function of relative cost of control actions ρ, with continous measurements, for systems with $a = -0.5$, $a = 0$ and $a = 0.5$ respectively

where J_T is given by (6), and

$$Q_1 = \begin{cases} \frac{e^{2aT}-1}{2a} & a \neq 0 \\ T & a = 0 \end{cases}$$

The control cost J_u is the probability of an event at each sample instant,

$$J_u = \frac{E[N_u(0,T)]}{T} = \frac{\text{Prob}(x(kT) \geq r)}{T}$$

Accordingly, the only things needed to calculate the loss function are the state variance and the probability of an event—both at time kT.

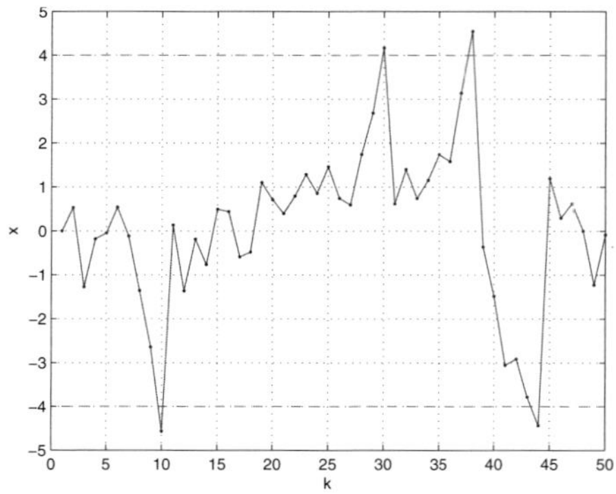

Fig. 4. A simulation example of sporadic control with discrete-time measurements, $a = 0$ and $r = 4$. The state is reset if it has passed the threshold, that is, if $|x| \geq r$ at a sampling instant. However it is never zero in the following sample due to the influence of noise. Note that the trajectory is interpolated between samples.

Since the controller brings x to zero whenever there is an event, the discrete-time closed-loop update equation becomes

$$x(kT + T) = \begin{cases} e^{aT}x(kT) + w(k) & x(kT) < r \\ w(k) & x(kT) \geq r \end{cases} \tag{9}$$

A simulation example can be seen in Fig. 4. Note that no event is generated if the state passes the threshold but returns before it is sampled.

The update equation (9) gives rise to a probability distribution of the state, which in turn determines the state variance as well as the event probability. Obviously there is a non-trivial dependence on the threshold parameter r.

Let $f(x, kT)$ denote the probability density function of the state at time kT, prior to a possible event. Then $f(x, kT + T)$ can be expressed as a mixture of two distributions, corresponding to the two separate cases of whether an event was generated at time kT or not:

$$\begin{aligned} &f(x, kT + T) = \\ &\text{Prob}(|x(kT)| < r) \left[\mathcal{S}_{aT}\left(f\left(x \,\middle|\, |x| < r, kT\right)\right) * \varphi(x)\right] \\ &+ \text{Prob}(|x(kT)| \geq r) \cdot \varphi(x) \end{aligned} \tag{10}$$

Here, $\varphi(x)$ is the Gaussian density function with zero mean and variance V_T, and $\mathcal{S}$ is a scaling operator defined as

$$\mathcal{S}_{aT}(g(x,t)) \stackrel{\text{def}}{=} \frac{1}{e^{aT}} g\left(\frac{x}{e^{aT}}, t\right) \tag{11}$$

Also, $f\left(x \,\middle|\, |x| < r, t\right)$ is the conditional probability density function of $x(t)$ given that $|x(t)| < r$. Since

$$f\left(x \,\middle|\, |x| < r, t\right) = \frac{f^r(x,t)}{\text{Prob}(|x(t)| < r)}$$

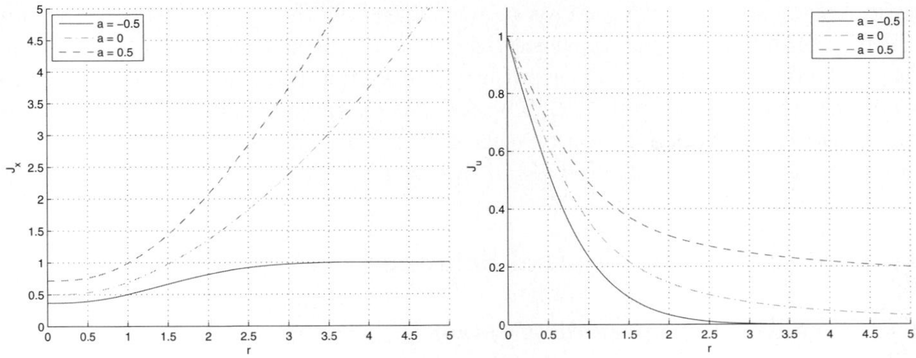

Fig. 5. Cost functions for sporadic control with discrete-time measurements. Left: State cost J_x as a function of threshold r. Right: Control cost J_u as a function of threshold r. Both functions are plotted for systems with $a = -0.5$, $a = 0$ and $a = 0.5$ respectively.

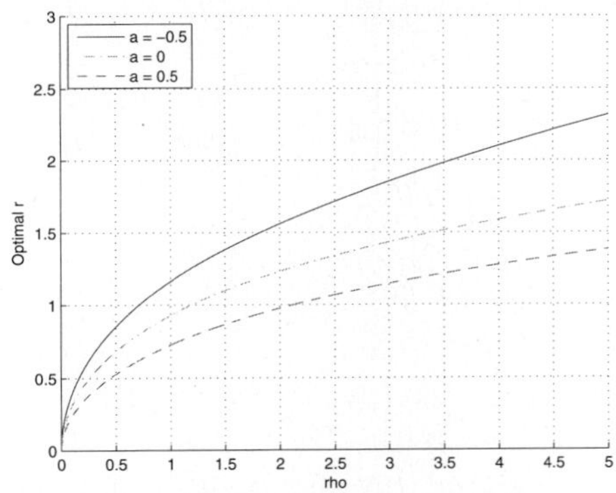

Fig. 6. Optimal threshold r as a function of relative cost of control actions ρ, with discrete measurements, for systems with $a = -0.5$, $a = 0$ and $a = 0.5$ respectively

where $f^r(x,t)$ is the truncation

$$f^r(x,t) = \begin{cases} f(x,t) & |x| < r \\ 0 & \text{otherwise} \end{cases}$$

we can now express (10) as

$$f(x, kT + T) = \mathcal{S}_{ah}\left(f^r(x, kT)\right) * \varphi(x) + \left(1 - \int_{-r}^{r} f(x, kT)\, dx\right) \cdot \varphi(x) \qquad (12)$$

Unfortunately the update equation (12) for the state distribution is very complicated and an analytical solution seems out of reach. However, an approximate solution can be found numerically if the distributions are discretized. One method is to start with a point distribution and iterate according to (12) until convergence. A better way is to recognize that the system (9) is a Markov process. If the state space is discretized it is a Markov chain, for which it is easy to find the stationary distribution. This distribution may then be used to approximate the solution to (12). The costs are easily calculated from the corresponding distributions and can be seen in Fig. 5 as a function of r. The optimal threshold as a function of ρ is plotted in Fig. 6.

4 Comparison of Control Schemes

4.1 Periodic and Aperiodic Control

Recall that an aperiodic controller sets the process state x to zero whenever $|x| \geq r$ using an impulse control action. We assume that periodic control is also implemented with impulse control action, such that x is periodically set to zero regardless of the measured value. The periodic sampling interval is restricted to be no shorter than for the sporadic schemes.

For the periodic controller, the cost functions are simply

$$\begin{aligned} J_x = J_T &= \begin{cases} \frac{e^{2aT}-2aT-1}{4a^2T} & a \neq 0 \\ \frac{T}{2} & a = 0 \end{cases} \\ J_u &= T^{-1} \end{aligned} \tag{13}$$

For the aperiodic controller, the cost functions can be found by letting $T \to 0$ in (8), yielding

$$J = \frac{V_c}{V_J} + \rho \frac{1}{V_J} = J_x + \rho J_u.$$

Optimal sampling intervals and thresholds are easily determined from these expressions.

4.2 Preliminaries

For the sporadic controllers, minimization of the loss function J for a given ρ implicitly determines an optimal threshold r. This in turn maps to an optimal average event frequency J_u. The same is true for aperiodic control. In periodic control, however, there is no threshold. Instead ρ determines the optimal sampling interval. Hence, it is possible to parameterize controllers from all four classes by average event frequency rather than threshold.

The four controllers are differentiated by the restrictions on when they can generate a control event. We should expect that a scheme that has fewer restrictions will be harder to implement but perform better according to the cost function J. The controllers can be ranked as follows, from less to more restrictive:

1. Aperiodic controller. Can generate an event at any time.
2. Sporadic controller with continous measurements.

3. Sporadic controller with discrete time measurements and periodic controller. A periodic controller can never emulate a sporadic one, and the sporadic one can not emulate the periodic unless it has an integer multiple of the periodic frequency. As long as the sporadic controller gets better with decreasing T, it will be better than the periodic one, however.

The question becomes, when does the investment in implementing a less restricted controller pay off?

In the limit when $J_u T \to 0$ and control is executed at a rate far below the maximum, the sporadic controllers should approach the aperiodic one since the inter-event time constraint is no longer in effect. When $\rho \to 0$, the sporadic control with discrete time measurements will approach periodic control since the sample rate is the only constraint.

Note that for asymptotically stable processes, that is, $a < 0$, J_x is bounded by the variance achieved without controller. It follows that also J is bounded. As the relative cost of control actions ρ increases, all controllers will generate events less often, so that $J_u \to 0$, and ultimately J_x will approach a maximum. The limit can be calculated from (13), where $J_x \to -1/2a$ as $T \to \infty$.

4.3 Comparison

The trade-off between state variance and average event frequency is made explicit in Fig. 7, where J_x and J_u are plotted against each other. The plots verify the ranking of the controllers. It is also seen that the sporadic controller with discrete time measurements always outperforms the periodic one, which is not so surprising since it has considerably more freedom.

An important insight given by Fig. 7 is what we consider the main advantage with event-based control: less events are needed to achieve the same state cost. Using a periodic controller, the variance increases quite rapidly when sampling less often. However, with sporadic control the average control action frequency can be decreased a lot more without the same variance increase. For example, when $a = 0.5$ the average number of control actions per time unit may be decreased by about 40 % with only a slight increase in variance, if using sporadic control with discrete-time measurements.

A surprising result is that for sporadic control with continuous-time measurements, the state variance can be made somewhat smaller by *decreasing* the average control frequency. This can also be seen in the left plot of Fig. 2, where J_x attains a minimum value for $r > 0$. Unlike the other controllers, it is thus not optimal to control as often as possible even if there is no control cost. The explanation is that when the controller issues an event, it is put in the inactive state and becomes unable to handle large state errors that may arise in the meantime. The optimal threshold is therefore such that the reduced variance due to the state being reset is balanced against the risk of large state errors arising while the controller is in the inactive mode. This phenomenon does not occur for sporadic controllers with discrete-time measurements since the ability of those controllers to generate events is independent of past actions.

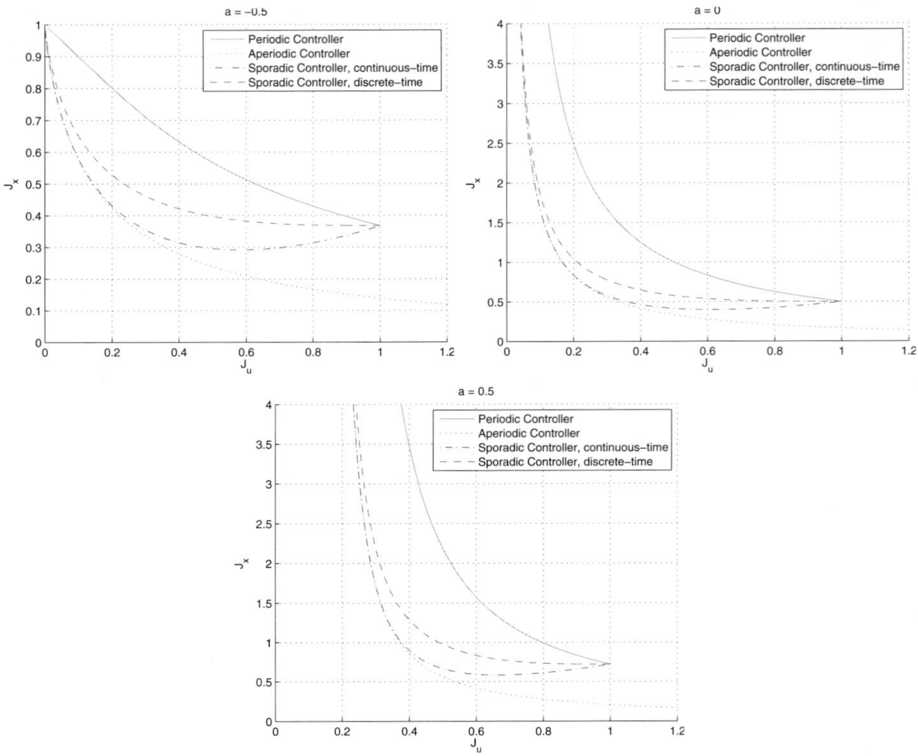

Fig. 7. Trade-off between process state variance J_x and average event frequency J_u for the four classes of controllers. Top left: $a = -0.5$. Top right: $a = 0$. Bottom: $a = 0.5$. Note the different vertical scales.

The minimum achievable costs are plotted in Fig. 8 as a function of ρ. The curves are nondecreasing and concave since starting from any point on the curve and fixing J_x and $J_u \geq 0$, the cost for any ρ is upper bounded by $J = J_x + \rho J_u$. When $a < 0$, the cost is upper bounded by $\frac{1}{-2a}$.

It is clear from the plots that the performance of the controllers agree with the ranking in Section 4.2. Note that the performance of sporadic control with continuous measurements is close to that of aperiodic control, except for small values of ρ. Except for the asymptotically stable process, the performance gap between the sporadic controllers appears to approach a constant value, making the relative difference small for large ρ.

When ρ is small, sporadic control with discrete-time measurements becomes periodic control. However, sporadic control with continuous-time measurements delivers a better result, since it can decrease J_x by choosing $J_u < T$. It is seen for the aperiodic controller that $J \to 0$ when $\rho \to 0$. However, this is of limited interest since the average event frequency approaches infinity.

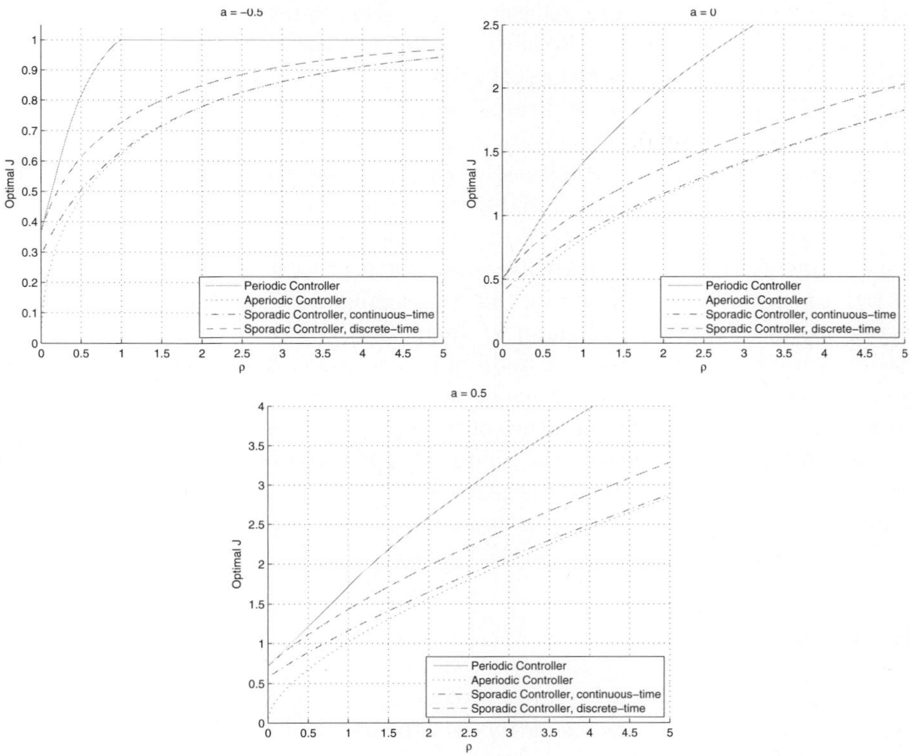

Fig. 8. Minimum achievable cost J as a function of ρ for the four classes of controllers. Top left: $a = -0.5$. Top right: $a = 0$. Bottom: $a = 0.5$. Note the different vertical scales.

5 Conclusions

In some applications there is a cost related to the execution of a control signal, regardless of the magnitude of that signal. If that cost is included in the performance objective of the controller, it will be meaningful to reduce the frequency of control actions. This may be accomplished with a periodic controller by increasing the sampling interval. However, the penalty in terms of increased process state variance is significant. Trying to make it better by not acting on small state errors naturally leads to the notion of event-based control.

In this paper, we show that sporadic control can provide a better trade-off and overall control performance, under the assumption of a fixed cost of control actions. It is noted that the average frequency of control events can be reduced with only a small increase in variance. Moreover, we show that sporadic control with continuous-time measurements can actually reduce both the average frequency of control events as well as the variance of the state.

Obviously, to implement sporadic control in situations where periodic control is used today would require certain changes. While sporadic control with discrete-time measurements would probably only require algorithmic changes, sporadic control with continuous-time measurements also requires some mechanism that can measure instantaneously if the threshold is passed. We doubt that the small relative difference is worth the extra effort.

Sporadic control has one additional parameter compared to periodic control, the threshold. This parameter should probably scale with the size of process disturbances. A mismatch in threshold will make the controller behave more like a periodic one if it is too small, and more like an aperiodic one if it is too large. In the second case we have to be aware that the threshold will work as a tolerable magnitude of error, and convergence may not proceed below the threshold.

Before sporadic control can be put to general use there has to be a further development of theory. The main problem that needs to be solved is how to design sporadic controllers for systems of higher order. How should the threshold be parameterized? Here, related work on optimal quantization [10] could possibly provide some ideas. How should the control signal be chosen if it is not possible to immediately set the state to zero? Controlling with Dirac impulses is unrealistic in most cases—it is merely used here to give a fair comparison between different control schemes. Further, if also measurements are non-periodic, how can output feedback then be applied?

References

1. Åström, K.J., Wittenmark, B.: Computer-Controlled Systems. Prentice Hall (1997)
2. Hristu-Varsakelis, D., Levine, W.S., eds.: Handbook of Networked and Embedded Control Systems. Birkhäuser (2005)
3. Hendricks, E., Jensen, M., Chevalier, A., Vesterholm, T.: Problems in event based engine control. In: Proc. American Control Conference. (1994) 1585–1587
4. Tarn, T.J., Xi, N., Bejczy, A.: Path-based approach to integrated planning and control for robotic systems. Automatica **32**(12) (1996) 1675–1687
5. Kwon, W.H., Kim, Y.H., Lee, S.J., Paek, K.N.: Event-based modeling and control for the burnthrough point in sintering processes. IEEE Transactions on Control Systems Technology **7**(1) (1999) 31–41
6. Hu, J., Lygeros, J., Sastry, S.: Towards a theory of stochastic hybrid systems. In: Proc. Hybrid Systems: Computation and Control. (2000)
7. Åström, K.J., Bernhardsson, B.: Comparison of periodic and event based sampling for first-order stochastic systems. In: Preprints 14th World Congress of IFAC. Volume J., Beijing, P.R. China (1999) 301–306
8. Åström, K.J., Bernhardsson, B.: Comparison of Riemann and Lebesque sampling for first order stochastic systems. In: Proceedings of the 41st IEEE Conference on Decision and Control. Volume 2. (2002) 2011–2016
9. Buttazzo, G.C.: Hard Real-Time Computing Systems: Predictable Scheduling Algorithms and Applications. Kluwer Academic Publishers (1997)
10. Elia, N., Mitter, S.K.: Stabilization of linear systems with limited information. IEEE Transactions on Automatic Control **46**(9) (2001) 1384–1403

Price-Based Optimal Control of Power Flow in Electrical Energy Transmission Networks

A. Jokic, M. Lazar, and P.P.J. van den Bosch

Dept. of Electrical Eng., Eindhoven Univ. of Technology,
P.O. Box 513, 5600 MB Eindhoven, The Netherlands
{a.jokic,m.lazar,p.p.j.v.d.bosch}@tue.nl

Abstract. This article presents a novel control scheme for achieving optimal power balancing and congestion control in electrical energy transmission networks via nodal prices. We develop an explicit controller that guarantees economically optimal steady-state operation while respecting all line flow constraints in steady-state. Due to these constraints, the resulting optimal control law has a piecewise affine structure. To optimize the dynamic response of the system and to satisfy line overload constraints during the transient period, the explicit optimal controller is complemented with a hybrid MPC controller. An example illustrates the effectiveness of the proposed hybrid control scheme.

1 Introduction

During the past decade there has been a tremendous amount of research devoted to a market-oriented approach for the electrical power system sector [1]. Electrical power systems have some unique properties, which make this a challenging task. For example, electrical energy cannot be efficiently stored in large amounts, which implies that production has to meet rapidly changing demands in real-time, making electricity a commodity with fast changing production costs. Furthermore, unlike other transportation systems, which assume a free choice among alternative paths between source and destination, the flow of power in electrical energy transmission networks is governed by physical laws and, for some fixed pattern of power injections, it can be influenced only to a certain degree. Therefore, physical and security limits on the maximal power flow in the lines of electrical energy transmission networks represent crucial system constraints, which cannot be neglected [2]. Due to the fast changing variable production costs, there is a general tendency in power markets towards increasing the speed with which the market price is updated. The usage of price as a real-time feedback signal for power balance control has been investigated in [3], however, with no line congestion limits considered. Analysis of real-time market dynamics was performed in [4], where the effects of congestion have been modeled by a static equality constraint representing the power flow in a congested line. A detailed overview of problems and current schemes for price-based control of power systems is presented in [2] and the references therein. The common characteristic

A. Bemporad, A. Bicchi, and G. Buttazzo (Eds.): HSCC 2007, LNCS 4416, pp. 315–328, 2007.

of virtually all existing approaches is that the congestion management problem is treated in a *static manner*, while for real-time control of power flow in the lines the automatic generation control (AGC) scheme [5] is utilized.

One of the main contributions of this paper is the development of an *explicit, price-based, dynamic* controller for real-time optimal power balancing and congestion management. The proposed explicit controller guarantees that, following any admissible change in the load, the power system will settle in the corresponding economically optimal steady-state point with all line flow constraints satisfied. Due to the inequality constraints representing the line flow limits, the optimal controller has a piecewise affine structure. Although steady-state optimal, the explicit controller does not guarantee that during the transients following load changes some of the power lines will not become overloaded to such an extent that it will threaten the safety of the system's operation. To solve this issue, we complement the explicit optimal controller with a hybrid model predictive controller (MPC), i.e. a MPC controller that uses a piecewise affine model for predictions. The MPC control action amends the explicit optimal control law so that constraints are met during the transients following load changes, and it converges to zero in steady-state. As a result, the response of the system is optimized and the constraints are satisfied in the transient period as well, while the advantageous steady-state properties of the explicit optimal controller are preserved. Consequently, the proposed hybrid control strategy achieves *both* optimal congestion management (in a *dynamic manner*) and *price-based AGC*.

The results of this paper add *the optimal power flow problem with congestion constraints* to a list of previously considered problems in electrical energy transmission systems, where the benefits of utilizing hybrid control schemes were already illustrated. The interested reader is referred to [6], [7], where hybrid MPC was utilized for efficient emergency voltage control, and to [8], where hybrid MPC was used for optimal control of co-generation power plants. Recently, in [9], hybrid MPC was employed for solving the problem of multi-period investments for maintenance and upgrade of electrical energy distribution networks.

1.1 Nomenclature

The field of real numbers is denoted by $\mathbb{R}$, while $\mathbb{R}^{m\times n}$ denotes m by n matrices with elements in $\mathbb{R}$. For a matrix $A \in \mathbb{R}^{m\times n}$, $[A]_{ij}$ denotes the element in the i-th row and j-th column of A. For a vector $x \in \mathbb{R}^n$, $[x]_i$ denotes the i-th element of x. $\ker A$ and $\operatorname{im} A$ denote the kernel and the image space of A, respectively. We use I_n and $\mathbf{1}_n$ to denote an identity matrix of dimension $n \times n$ and a column vector with n elements all being equal to 1, respectively. The operator $\operatorname{col}(\cdot,\dots,\cdot)$ stacks its operands into a column vector, and $\operatorname{diag}(\cdot,\dots,\cdot)$ denotes a square matrix with its operands on the main diagonal and zeros elsewhere. *All inequalities are interpreted elementwise.* We will use graph-theoretic terminology to represent power networks. With a slight abuse of notation we will often use the same symbol to denote a signal, i.e. a function of time, as well as possible values that the signal may take at any time instant.

2 Steady-State Optimal Controller Design

Consider a connected undirected graph $G = (V, E, A)$ as an abstraction of an electrical power network. $V = \{v_1, \ldots, v_n\}$ is the set of nodes, $E \subseteq V \times V$ is the set of undirected edges, and A is a weighted adjacency matrix. Undirected edges are denoted as $e_{ij} = (v_i, v_j)$, and the adjacency matrix $A \in \mathbb{R}^{n \times n}$ satisfies $[A]_{ij} \neq 0 \Leftrightarrow e_{ij} \in E$ and $[A]_{ij} = 0 \Leftrightarrow e_{ij} \notin E$. No self-connecting edges are allowed, i.e. $e_{ii} \notin E$. We associate the edges with the power lines of the electrical network and, for convenience, we set the weights in the adjacency matrix as follows: $[A]_{ij} = -\frac{1}{z_{ij}} = -b_{ij}$, where z_{ij} is the inductive reactance of a line, i.e. the imaginary part of the line impedance, and b_{ij} is the line susceptance, see [2] for details. Note that the matrix A has zeros on its main diagonal and $A = A^\top$. The set of neighbors of a node v_i is defined as $N_i \triangleq \{v_j \in V \mid (v_i, v_j) \in E\}$. Often we will use the index i to refer the node v_i. Define $I(N_i)$ as the set of indices corresponding to the neighbors of node i, i.e. $I(N_i) \triangleq \{j \mid v_j \in N_i\}$. We associate the nodes with the buses in the electrical energy transmission network.

2.1 Steady-State Optimal Control Problem

To define the steady-state optimization problem, with each node v_i we associate a singlet $\hat{p}_i$ and a quadruplet $(p_i, \underline{p}_i, \overline{p}_i, J_i)$, where $p_i, \underline{p}_i, \overline{p}_i, \hat{p}_i \in \mathbb{R}$, $\underline{p}_i < \overline{p}_i$ and $J_i : \mathbb{R} \to \mathbb{R}$ is a strictly convex, continuously differentiable function. The values p_i and $\hat{p}_i$ denote the reference values for node power injections into the network. Positive values correspond to a flow of power into the network (production), while negative values denote power extracted from the network (consumption). Both p_i and $\hat{p}_i$ can take positive as well as negative values, and the only difference is that, in contrast to $\hat{p}_i$, the value p_i has an associated objective function J_i and a constraint $\underline{p}_i \leq p_i \leq \overline{p}_i$. In the case of a positive p_i, the function J_i represents the variable costs of production, while for negative values of p_i, it denotes the negated benefit function of a consumer. We will refer to p_i as the power from a price-elastic producer/consumer (or simply, power from a price-elastic unit), and to $\hat{p}_i$ as the power from a price-inelastic producer/consumer (price-inelastic unit).

We use a "dc power flow" model [2] to determine the power flows in the network for given values of node power injections. This model is often used and, under certain reasonable assumptions, it is proven to be a relatively accurate approximation of a complex "ac power flow" model. In particular, convexity of the dc power flow model is a crucial property that we will exploit. With δ_i denoting a voltage phase angle at the node v_i, the power flow in a line $e_{ij} \in E$ is given by $p_{ij} = b_{ij}(\delta_i - \delta_j) = -p_{ji}$. If $p_{ij} > 0$, power in the line e_{ij} flows from node v_i to node v_j. The power balance in a node yields $p_i + \hat{p}_i = \sum_{j \in I(N_i)} p_{ij}$. With the abbreviations $p = \mathrm{col}(p_1, \ldots, p_n)$, $\hat{p} = \mathrm{col}(\hat{p}_1, \ldots, \hat{p}_n)$, $\delta = \mathrm{col}(\delta_1, \ldots, \delta_n)$ the overall network balance condition is $p + \hat{p} = B\delta$, where the matrix B is given by $B = A - \mathrm{diag}(A\mathbf{1}_n)$. We define the optimal power flow problem as follows.

Problem 1. **Optimal Power Flow (OPF) problem.**
For any constant value of $\hat{p}$,

$$\min_{p,\delta} J(p) \triangleq \min_{p,\delta} \sum_{i=1}^{n} J_i(p_i) \tag{1a}$$

subject to

$$p - B\delta + \hat{p} = 0, \tag{1b}$$

$$\underline{p} \leq p \leq \overline{p}, \tag{1c}$$

$$b_{ij}(\delta_i - \delta_j) \leq \overline{p}_{ij}, \ \forall (i, j \in I(N_i)), \tag{1d}$$

where $\underline{p} = \text{col}(\underline{p}_1, \ldots, \underline{p}_n)$, $\overline{p} = \text{col}(\overline{p}_1, \ldots, \overline{p}_n)$, and $\overline{p}_{ij} = \overline{p}_{ji}$ is the maximal allowed power flow in the line e_{ij}. □

We will refer to a vector p that solves the OPF problem as a *vector of optimal power injections.* For an appropriately defined matrix L and a suitably defined vector of power line limits $\overline{p}_L$, the set of constraints in (1d) can be written in a more compact form as follows:

$$L\delta \leq \overline{p}_L. \tag{2}$$

In a liberalized, market-oriented power system, different units are owned by separate parties and each of them acts autonomously to maximize its own benefit. In other words, when a price-elastic unit at node i receives the current price for electrical power, i.e. λ_i, it adjusts its production level p_i to be equal to $\tilde{p}_i$, where $\tilde{p}_i = \arg\min_{p_i} \left\{ J_i(p_i) - \lambda_i p_i \text{ subject to } \underline{p}_i \leq p_i \leq \overline{p}_i \right\}$. Since J_i is a strictly convex, continuously differentiable function, this relation defines a unique mapping from λ_i to $\tilde{p}_i$ for any $\lambda_i \in \mathbb{R}$. For convenience, we denote this mapping with $\Upsilon_i : \lambda_i \rightarrow p_i$, i.e.

$$\tilde{p}_i = \Upsilon_i(\lambda_i) \triangleq \arg \min_{p_i \in [\underline{p}_i, \overline{p}_i]} J_i(p_i) - \lambda_i p_i, \tag{3}$$

and define $\Upsilon(\lambda) \triangleq \text{col}(\Upsilon_1(\lambda_1), \ldots, \Upsilon_n(\lambda_n))$. The *operational goal in a liberalized power system* is to determine the nodal price λ_i for each node i in the network, in such a way that the total benefit of the system is maximized, while all constraints are fulfilled. Formally, we define the optimal nodal price problem as follows.

Problem 2. **Optimal Nodal Prices (ONP) problem.**
For any constant value of $\hat{p}$,

$$\min_{\lambda,\delta} \sum_{i=1}^{n} J_i(\Upsilon_i(\lambda_i)) \text{ subject to } \Upsilon(\lambda) - B\delta + \hat{p} = 0, \ L\delta \leq \overline{p}_L, \tag{4}$$

where $\lambda = \text{col}(\lambda_1, \ldots, \lambda_n)$ is a vector of nodal prices. □

We will refer to a vector λ that solves the ONP problem with the term *vector of optimal nodal prices.* The OPF and ONP problems are related through Lagrange duality, as it will be shown later in this section. The ONP problem is employed next to define the optimal steady-state control problem.

Consider a power network where each price-elastic unit is a dynamical system, and assign to each such unit an appropriate model G_i of its dynamics. We assume for simplicity that each model G_i is an *LTI system with respect to the input* $p_i = \Upsilon_i(\lambda_i)$, which is specified by its state-space realization, i.e.

$$G_i : \begin{cases} \dot{x}_i = A_i x_i + B_i p_i = A_i x_i + B_i \Upsilon_i(\lambda_i) \\ p_i^A = C_i x_i \end{cases}, \quad \forall i, \tag{5}$$

where the power reference signal p_i is the input, and the actual node power injection p_i^A is the output. We denote the actual power injection of a price-inelastic unit with $\hat{p}_i^A$. Note that (1b) is always fulfilled when p and $\hat{p}$ are replaced with $p^A = \text{col}(p_1^A, \ldots, p_n^A)$ and $\hat{p}^A = \text{col}(\hat{p}_1^A, \ldots, \hat{p}_n^A)$, since in that case (1b) represents the conservation law, i.e. $p^A - B\delta + \hat{p}^A = 0$. Achieving balance in *reference values* (1b), i.e. balance of the *desired* production and consumption, is a control problem. The production/consumption of price-inelastic units is an exogenous signal to the system and $\hat{p}^A = \hat{p}$, which yields:

$$p^A - B\delta + \hat{p} = 0. \tag{6}$$

The desired production/consumption of price-elastic units is a function of current nodal prices. Therefore, nodal prices can be effectively used as a feedback signal for power balance control. Each system G_i receives a price signal and, based on its benefit maximization objective (3), it maps the signal into a reference p_i. We will assume this mapping to be instantaneous, although the model can easily be extended with dynamics, time delays, threshold based rules, etc. *The complete dynamical model of the power system is described with the set of differential algebraic equations* (5)-(6). For a detailed presentation of power system modeling for real-time power balance control problems we refer to [5], Chapter 11, or [10], Chapter 12.

Note that the mapping $\Upsilon_i : \lambda_i \to p_i$ is linear only if J_i is a quadratic function and $\underline{p}_i = -\infty$, $\overline{p}_i = \infty$. In practice, however, it will always be a nonlinear mapping and therefore, the model (5)-(6) with the nodal prices λ_i as inputs, is nonlinear. Since for any strictly convex and continuously differentiable J_i the mapping $\Upsilon_i : \lambda_i \to p_i$ can be arbitrarily well approximated with a continuous piecewise affine mapping, one can always obtain a piecewise affine model that approximates (5) arbitrarily well. We will assume for the remainder of the article that each Υ_i is a piecewise affine function and thus, *the overall system model* (5)-(6) *is piecewise affine with respect to* λ_i *as inputs.*

To control the system, a measure of imbalance in (1b) has to be available. The network frequency serves that purpose. The system is in balance, in the sense of equality (1b), if the frequency is equal to its reference value, e.g. 50Hz in Europe. A change in any reference for power value causes the frequency to change. In steady-state, the frequency is equal for all nodes in the network and, if it is above its reference value, the total production in a network exceeds the total consumption. Finally, we are able to define the control problem.

Problem 3. **Optimal steady-state control problem.**
Design a feedback controller that has the network frequency and the line power flows as input, and the nodal prices as output (see Figure 1), such that the following objective is met: for any constant value of $\hat{p}$ such that the ONP problem is feasible, the state of the closed-loop system converges to a steady-state where the nodal prices are the *optimal nodal prices* as defined in Problem 2. □

2.2 Explicit Dynamic Optimal Controller

In this subsection we employ the relation between the solutions of the OPF and the ONP problems to obtain a solution to Problem 3. We start by presenting the following basic result from power system economics.

Proposition 1. *The optimal dual variable (Lagrange multiplier) associated with the power balance constraint* (1b) *in the Lagrange dual problem of OPF, is the vector of optimal nodal prices for the corresponding ONP problem.*

Proof. Consider some constant value $\hat{p}$ such that the OPF and ONP problems are feasible. The OPF problem is a convex problem which satisfies Slater's constraint qualification. This implies that strong duality holds and that first-order Karush-Kuhn-Tucker (KKT) conditions are *necessary and sufficient* conditions for optimality. For the OPF problem, the Lagrangian is given by $\mathcal{L}(p, \delta, \nu^+, \nu^-, \lambda, \mu) = J(p) - \lambda^\top(p - B\delta + \hat{p}) + (\nu^-)^\top(\underline{p} - p) + (\nu^+)^\top(p - \overline{p}) + \mu^\top(L\delta - \overline{p}_L)$ and the KKT conditions are given by:

$$p - B\delta + \hat{p} = 0, \tag{7a}$$

$$p - \overline{p} \leq 0, \quad \nu^+ \geq 0, \quad (\nu^+)^\top(p - \overline{p}) = 0, \tag{7b}$$

$$-p + \underline{p} \leq 0, \quad \nu^- \geq 0, \quad (\nu^-)^\top(-p + \underline{p}) = 0, \tag{7c}$$

$$L\delta - \overline{p}_L \leq 0, \quad \mu \geq 0, \quad \mu^\top(L\delta - \overline{p}_L) = 0, \tag{7d}$$

$$\nabla_p J(p) - \lambda + \nu^+ - \nu^- = 0, \tag{7e}$$

$$B\lambda + L^\top\mu = 0, \tag{7f}$$

where λ, ν^+, ν^- and μ are (*vector*) Lagrange multipliers. The multiplier λ is associated with the equality constraint (7a) and is therefore not sign restricted. At the optimum, the value of the objective function and the vector of optimal power injections in the OPF problem are equal to the value of the objective function and the vector of power injections in the ONP problem, i.e. $p = \Upsilon(\lambda)$. Assume that the optimal Lagrange multiplier λ from the OPF dual problem is taken to be the vector of nodal prices. In this case the conditions (7b),(7c) and (7e) correspond to the KKT conditions for the constrained minimization problem in (3) for all price-elastic units in the network. Therefore, for this particular λ, the vector of power injections $p = \Upsilon(\lambda)$ in the ONP problem corresponds to the vector of *optimal power injections* in the OPF problem, which concludes the proof. □

Remark 1. B is a singular matrix with rank deficiency one and with the kernel space spanned by the vector $\mathbf{1}_n$. Physically, this reflects the fact that only the relative voltage phase angles determine the power flow. Note also that $\mathbf{1}_n \notin \operatorname{im} B$, since $B = B^\top$ implies $\mathbf{1}_n^\top B = 0$, i.e. $\mathbf{1}_n$ is orthogonal to each column of B. Finally, analyzing the structure of L (see (1d) and (2)) one can easily observe that $\mathbf{1}_n^\top L^\top = 0$ and therefore, $\mathbf{1}_n \notin \operatorname{im} L^\top$. These properties of B and L will be used later in this section to prove Proposition 2.

Consider the OPF problem solution for some constant value $\hat{p}$ such that the problem is feasible. We denote the minimizers of OPF with $\tilde{p}$, $\tilde{\delta}$, and with $\tilde{\lambda}$ the value of the corresponding Lagrange multiplier. Strict convexity of each J_i implies that at the optimum $\tilde{p}$ is unique. On the other hand, due to singularity of B, if $\tilde{\delta}$ is a minimizer so is $\tilde{\delta}+\mathbf{1}_n c$ where $c \in \mathbb{R}$ is an arbitrary constant. However, note that for all minimizers the set of active constraints is uniquely determined. Furthermore, we denote with $\breve{\mu}$ and $\tilde{\mu}$ the Lagrange multipliers corresponding to inactive and active line power flow constraints, respectively. Analogously, we define $\breve{\nu}^+$, $\tilde{\nu}^+$ and $\breve{\nu}^-$, $\tilde{\nu}^-$. For inactive constraints $\operatorname{col}(\breve{\mu}, \breve{\nu}^+, \breve{\nu}^-) = 0$. The equality (7f) yields $B\tilde{\lambda} = -L^\top \tilde{\mu}$. This condition implies that $\tilde{\lambda} \in \ker B$ if no lines are congested. This further implies $\tilde{\lambda} = \mathbf{1}_n \lambda^\star$, $\lambda^\star \in \mathbb{R}$, i.e. at the optimum, there is one price in the network for all nodes. *In case at least one line in the system is congested, it follows that the optimal nodal prices will in general be different for each node in the system.*

Next, we present the explicit dynamic controller that solves Problem 3. We define $f_i = \frac{\dot{\delta}_i}{2\pi}$ as the network frequency [Hz] at node i, the abbreviation $f \triangleq \operatorname{col}(f_1, \ldots, f_n)$, and we use f_{ref} ($f_{\text{ref}} \in \mathbb{R}$) to denote the frequency reference value, e.g. $f_{\text{ref}} = 50\text{Hz}$. Let K_λ, K_f and K_p be diagonal matrices with positive elements on the diagonal, such that $K_f = \alpha K_\lambda, \alpha \in \mathbb{R}, \alpha > 0$, and let Γ denote a diagonal matrix of the same size as K_p. Consider the following explicit piecewise affine *dynamic* controller:

$$\begin{pmatrix} \dot{x}_\lambda \\ \dot{x}_\mu \end{pmatrix} = \begin{pmatrix} -K_\lambda B & -K_\lambda L^\top \\ 0 & 0 \end{pmatrix} \begin{pmatrix} x_\lambda \\ x_\mu \end{pmatrix} + \begin{pmatrix} -K_f & 0 \\ 0 & \Gamma \end{pmatrix} \begin{pmatrix} \Delta f \\ \Delta p_L \end{pmatrix}, \tag{8a}$$

$$\lambda = \begin{pmatrix} I_n & 0 \end{pmatrix} \begin{pmatrix} x_\lambda \\ x_\mu \end{pmatrix}, \tag{8b}$$

$$\begin{cases} [\Gamma]_{ii} = [K_p]_{ii} & \text{if} \quad [x_\mu]_i \geq 0 \;\&\; [\Delta p_L]_i \geq 0 \\ [\Gamma]_{ii} = [K_p]_{ii} & \text{if} \quad [x_\mu]_i > 0 \;\&\; [\Delta p_L]_i < 0 \\ [\Gamma]_{ii} = 0 & \text{if} \quad [x_\mu]_i = 0 \;\&\; [\Delta p_L]_i < 0, \end{cases} \tag{8c}$$

$$x_\mu(0) \geq 0, \tag{8d}$$

where x_λ and x_μ denote the controller states and the matrices K_λ, K_f and K_p represent the controller gains. In (8), the inputs $\Delta p_L = L\delta - \overline{p}_L$ and $\Delta f = f - \mathbf{1}_n f_{\text{ref}}$ denote the line power overflow and the frequency deviation, respectively, while the output λ denotes the vector of nodal prices. Note that, due to the initialization constraint (8d), it holds that $x_\mu(t) \geq 0$ for all $t \geq 0$.

Assumption 1 The closed-loop system resulting from the interconnection of the explicit dynamic controller (8) with the overall network model given by the differential algebraic equations (5)-(6) is globally asymptotically stable for any constant value of $\hat{p}$ (i.e. with respect to the corresponding steady-state) such that the ONP problem is feasible. □

Proposition 2. *Suppose that Assumption 1 holds. Then the explicit dynamic controller* (8) *solves the optimal steady-state control problem, as defined in Problem 3.*

Proof. To prove Proposition 2, it suffices to show that in steady-state, the vector of nodal prices λ in (8) coincides with the Lagrange multiplier λ in (7), and therefore, by Proposition 1, is a vector of optimal nodal prices. In steady-state, the frequency is equal for all nodes, i.e. $\Delta f = \mathbf{1}_n \Delta f^\star$, $\Delta f^\star \in \mathbb{R}$. Since in the power system dynamics (5)-(6) there is no direct feedthrough from the input λ to the outputs Δf and Δp_L, in *steady-state* the following conditions are satisfied: (i) for $\mu := x_\mu$ and from (8a) and (8b) it follows that $B\lambda + L^\top \mu + \mathbf{1}_n \alpha \Delta f^\star = 0$, which, together with $\mathbf{1}_n \notin \text{im}\begin{pmatrix} B & L^\top \end{pmatrix}$ (see Remark 1), implies that $\Delta f^\star = 0$ and $B\lambda + L^\top \mu = 0$, i.e. that the power balance constraint (7a) and the optimality condition (7f) are satisfied; (ii) from (8a) and (8c) it follows that (7d) is satisfied for $\mu := x_\mu$; (iii) (7b), (7c), (7e) are satisfied since they correspond to the KKT conditions for the optimization problem in (3), which concludes the proof. □

Remark 2. The property $\mathbf{1}_n \notin \text{im}\begin{pmatrix} B & L^\top \end{pmatrix}$, which ensures zero frequency deviation in steady-state ($\Delta f^\star = 0$), is not robust with respect to possible errors in the controller implementation, i.e. to perturbations of matrices B and L in (8a). To overcome this drawback and to robustly ensure $\Delta f^\star = 0$ in steady-state, the controller (8) can easily be modified to explicitly include integral action on the network frequency deviation. This modification is presented in Appendix A.

Remark 3. Due to the steady-state related complementarity conditions in (7d), there does not exist an LTI controller that solves Problem 3 and a hybrid controller, such as the piecewise affine controller (8), is a necessity. For fixed values of the gain matrices K_λ, K_f and K_p in (8), one can a posteriori check asymptotic stability of the resulting closed-loop system by searching for quadratic Lyapunov functions, and therefore, validate Assumption 1.

3 Hybrid MPC Control Scheme

Following a large disturbance acting on the system, for example, a relatively large, sudden change in $\hat{p}$, it is desirable that the closed-loop system response is such that the power lines overload Δp_L and the frequency deviations Δf are limited during the transient period, i.e.

$$\Delta p_L \leq \Delta p_L^{\max}, \quad -\Delta f^{\max} \leq \Delta f \leq \Delta f^{\max}, \tag{9}$$

at all times, where $\Delta p_L^{\max} > 0$, $\Delta f^{\max} > 0$ are some predefined values. Since the lines can be overloaded only for a short period of time, satisfying the limit

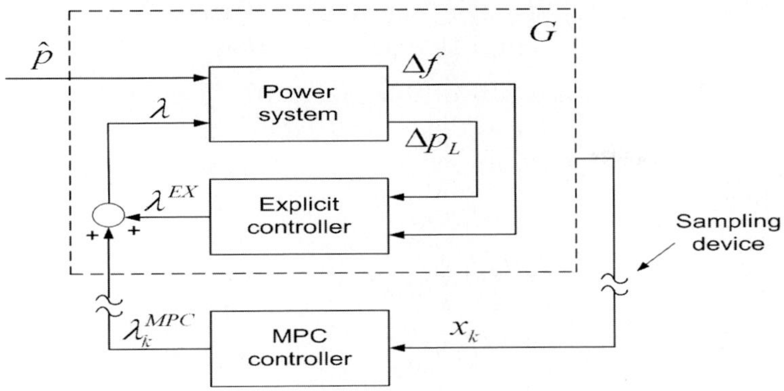

Fig. 1. Hybrid control scheme

constraints imposed on Δp_L and Δf, as well as fast convergence to the new steady-state, are crucial.

To guarantee the fulfillment of the constraints (9) during the transient period, following a change in $\hat{p}$, we design a hybrid MPC algorithm whose purpose is to complement the output λ^{EX} of the explicit optimal controller (8) such that inequalities (9) are satisfied at all times. The hybrid control scheme is depicted in Figure 1. Let G denote the continuous-time model of the overall network (5)-(6) in closed-loop with the explicit dynamic controller (8), as shown in Figure 1. To define the MPC algorithm, we first consider the following discrete-time approximation of G in closed-loop with the MPC controller:

$$G_D : \begin{cases} x_{k+1} = A_j x_k + B_j \lambda_k^{\text{MPC}} + a_j(\hat{p}_k) \\ y_k = C_j x_k = \text{col}(\Delta p_{Lk}, \Delta f_k, -\Delta f_k) \end{cases} \quad \text{if } H_j x_k \leq h_j, \quad k \geq 0, \tag{10}$$

where $A_j, B_j, a_j(\hat{p}_k), C_j, H_j, h_j$ are matrices and vectors of appropriate dimensions for all $j \in \mathcal{S} = \{1, \ldots, s\}$, with $\mathcal{S}$ a finite set of indices. The state vector x incorporates now both the states of the model describing the transmission network and the states of the explicit controller. As G is a piecewise affine system, an *equivalent* discrete-time piecewise affine counterpart cannot be obtained in general. However, one can obtain a discrete-time piecewise affine *approximation* of G by discretizing each continuous-time affine sub-system of G and taking into account physical insights when deriving the switching hyperplanes of the discrete-time model, see, for example, [6], [7].

The outputs of system (10) are given by Δp_{Lk} - the power flow deviation in the lines of the transmission network and Δf_k - the frequency deviation in the nodes of the transmission network ($-\Delta f_k$ is considered an output just to easily specify the constraints (9) in the MPC algorithm).

The MPC control action λ_k^{MPC} acts additively on the output of the explicit controller (8) present in the discrete-time dynamics G_D such that the constraints (9) are fulfilled at all times $k \geq 0$. In steady-state, λ_k^{MPC} converges to zero so that the network is controlled only by the output of the explicit controller, which then guarantees optimality and constraint satisfaction.

Problem 4. **MPC optimization problem.** Let $N \geq 1$ be given and let $x_k =: x_{0|k}$ denote the measured state at time $k \geq 0$. Let $\mathbf{c}_k := (c_{0|k}, \ldots, c_{N|k})$ denote a sequence of optimization variables and let Ψ be a positive definite and symmetric matrix. Minimize the cost:

$$J(\mathbf{c}_k) \triangleq \sum_{i=0}^{N-1} c_{i|k}^\top \Psi c_{i|k}, \tag{11}$$

subject to the constraints

$$\begin{cases} x_{i+1|k} = A_j x_{i|k} + B_j c_{i|k} + a_j(\hat{p}_{i|k}) \\ y_{i|k} = C_j x_{i|k} = \mathrm{col}(\Delta p_{i|k}, \Delta f_{i|k}, -\Delta f_{i|k}) \end{cases} \text{if } H_j x_{i|k} \leq h_j, \quad \forall i = 0, \ldots, N,$$

$$y_{N+1|k} = C_j x_{N+1|k} = \mathrm{col}(\Delta p_{N+1|k}, \Delta f_{N+1|k}, -\Delta f_{N+1|k}) \text{ if } H_j x_{N+1|k} \leq h_j, \tag{12a}$$

$$y_{i|k} \leq \mathrm{col}({\Delta p_L}^{\max}, \Delta f^{\max}, \Delta f^{\max}), \quad \forall i = 1, \ldots, N+1, \tag{12b}$$

$$c_{N|k} = 0. \tag{12c}$$

□

Let $\mathbf{c}_k^* := (c_{0|k}^*, \ldots, c_{N-1|k}^*, 0)$ denote an optimal sequence of variables obtained by solving Problem 4. Then, the MPC control law is defined as follows:

$$\lambda_k^{\mathrm{MPC}} := c_{0|k}^*; \quad k \geq 0. \tag{13}$$

The value of $\hat{p}_k$, which is employed in (12a), can be estimated from the measured values of the system state. With the current estimated value $\hat{p}_k$ available, to obtain the future outputs in (12a), we assume $\hat{p}_{i|k} = \hat{p}_k$ for all $i = 0, \ldots, N$.

Assumption 2 For any constant value of $\hat{p}_k$, the prediction horizon N is long enough such that the predicted state $x_{N|k}$ lies in a subset of the state-space that is invariant for the closed-loop system (10) with $\lambda_k^{\mathrm{MPC}} = 0$ for all $k \geq 0$, and where the constraints (12b) are satisfied. Furthermore, the global optimum is attained in Problem 4 for any x_k and all $k \geq 0$. □

Proposition 3. (i) *Suppose that Assumption 2 holds. If Problem 4 is feasible at time $k \geq 0$ for the measured state $x_k = x_{0|k}$, then Problem 4 remains feasible at time $k+1$ for state $x_{k+1} = A_j x_k + B_j \lambda_k^{MPC} + a_j(\hat{p}_k)$ if $H_j x_k \leq h_j$.*

(ii) *Suppose that Assumption 1 holds for G_D with $\lambda_k^{MPC} = 0$ for all $k \geq 0$ and for any constant value of $\hat{p}_k$ such that the ONP problem is feasible. Furthermore, suppose that Assumption 2 holds. Then the discrete-time closed-loop system* (10) *with the input λ_k^{MPC} defined as in* (13) *is asymptotically stable for any constant value of $\hat{p}_k$ such that the ONP problem is feasible.*

Fulfilment of the first part of Assumption 2 can be a priori guaranteed by adding a terminal[1] equality or inequality constraint to Problem 4, see, for example, [11].

[1] By this we mean a constraint on the predicted state $x_{N|k}$.

4 Illustrative Example

The proposed hybrid control scheme was implemented for a three nodes triangular network with a synchronous generator at each node. Each generator is modeled by a third order model, which is standardly used in AGC studies [5], [10]. The generator model is taken from [10], pp. 543-545. We have used the following parameter values: $\tau_g \in \{0.2, 0.25, 0.2\}$, $\tau_T \in \{0.5, 0.45, 0.5\}$, $H \in \{5, 5, 5\}$, $D \in \{0.7, 0.7, 0.8\}$, $R \in \{\frac{1}{22}, \frac{1}{24}, \frac{1}{20}\}$, where the first element in each set denotes the parameter value for a generator at the first node, etc., and the symbols are the ones from [10]. We use quadratic functions to represent the variable production costs J_i, with quadratic terms $\{1.2, 1.24, 1.4\}$ and linear terms $\{35.9, 36.1, 36\}$, respectively. For simplicity, no saturation limits were considered for the generators, i.e. $\underline{p}_i = -\infty$, $\overline{p}_i = \infty$, $\forall i$. Furthermore, we use the following reactance values for each line: $z_{12} = 0.1$, $z_{13} = 0.09$, $z_{23} = 0.13$. Appropriate, although not optimal, values for the gains of the explicit controller were chosen as $K_\lambda = \text{diag}(2, 2, 2)$, $K_f = \text{diag}(7, 7, 7)$. For the line connecting nodes 1 and 2, the *steady-state* power flow limit was set equal to 1.2, i.e. $\overline{p}_{12} = 1.2$, and the maximal allowed violation of this constraint in a transient period was set to 0.1, i.e. the corresponding element in $\Delta p_L^{\max}$ from (9) is $\Delta p_{L_{12}}^{\max} = 0.1$. For simplicity, no line flow limits were considered in the remaining lines, and the explicit controller is implemented to act only on Δp_{12} with gain $K_p = 2$. No frequency deviation constraints were imposed during transients.

The resulting closed-loop system G is a continuous-time piecewise affine model with 15 states and 3 affine sub-systems. The (robust) asymptotic stability of G for any values of $\hat{p}$ in the interval indicated below was established via a quadratic Lyapunov function, which validates Assumption 1. We obtained a discrete-time model G_D by discretizing each affine sub-system of G with a sampling period of $0.1s$ and keeping the same switching hyperplanes as the ones of the continuous-time model. The simulations showed that G_D is a good (i.e. for control purposes) approximation. The asymptotic stability test was successfully carried out for the discrete-time model G_D with $\lambda_k^{\text{MPC}} = 0$ for all $k \geq 0$ as well, which validates Assumption 1 for G_D. For the MPC controller, the following tuning parameters were used: $N = 12$, $\Psi = \text{diag}(0.1, 0.1, 0.1)$. For simplicity, we have used the following estimate for the value of $a_j(\hat{p}_k)$: $a_j(\hat{p}_k) = a_j(\hat{p}_{k-1}) = x_k - (A_j x_{k-1} + B_j \lambda_{k-1}^{\text{MPC}})$ if $H_j x_{k-1} \leq h_j$. A price-inelastic load was connected to each node. In the simulations, in nodes 1 and 3, the corresponding loads were set equal to constant values: $\hat{p}_1(t) = 0$, $\hat{p}_3(t) = 2$. The load at node 2 was taken to be a time varying signal given by $\hat{p}_2(t) = 3$ for $t < 5s$ and $\hat{p}_2(t) = 4$ for $t \geq 5s$. The system's response to the change in the load is presented in Figures 2 - 5. At time instant $t = 0$ the system is in a steady-state. In Figure 2, solid lines represent simulated trajectories of power flow in line e_{12}. One trajectory (labeled "Explicit") corresponds to the system in closed-loop with the the explicit optimal controller alone, while the other trajectory (labeled "Explicit + MPC") corresponds to the MPC closed-loop system described in Figure 1. In Figure 2, dashed lines represent the steady-state line flow limit $\overline{p}_{12} = 1.2$ and the transient line flow limit $\overline{p}_{12} + \Delta p_{L_{12}}^{\max} = 1.3$, respectively. Figure 3 presents

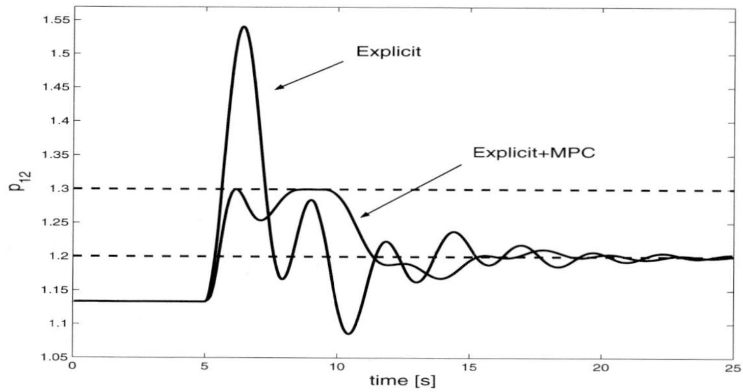

Fig. 2. Line power flow p_{12}. Dashed lines represent power flow restriction: 1.2 in steady-state; 1.3 in transient.

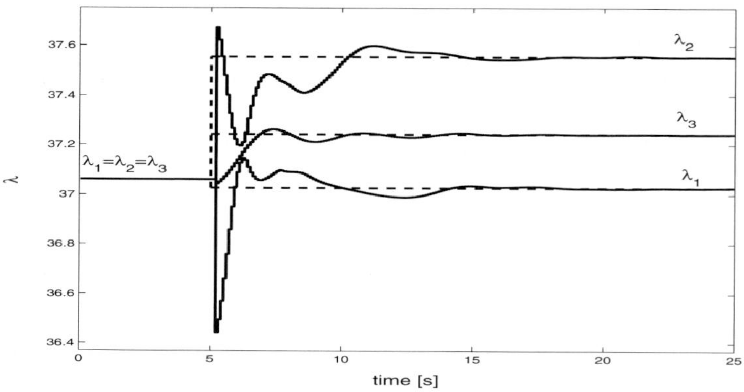

Fig. 3. Nodal prices λ as sum of λ^{EX} and λ^{MPC}. Dashed lines represent steady-state optimal nodal prices.

the values of the nodal prices λ as a function of time (solid lines). These trajectories correspond to the summation of the outputs from the explicit controller and the MPC controller. In the same figure, dashed lines represent the off-line calculated values for the steady-state optimal nodal prices, which depend on the value of $\hat{p}_2(t)$. The MPC control actions are presented in Figure 4, and the simulated trajectories of the node frequency deviations are presented in Figure 5. Note that, before the change in the load $\hat{p}_2$, the line is not congested and all nodal prices are equal for the steady-state optimal operating point. After the step increase in the load, in the optimal steady-state operating point the line is congested and, as a consequence, the optimal nodal prices have different values. The simulation results clearly illustrate the efficiency of the proposed hybrid control scheme for both steady-state and transient behavior of the closed-loop system.

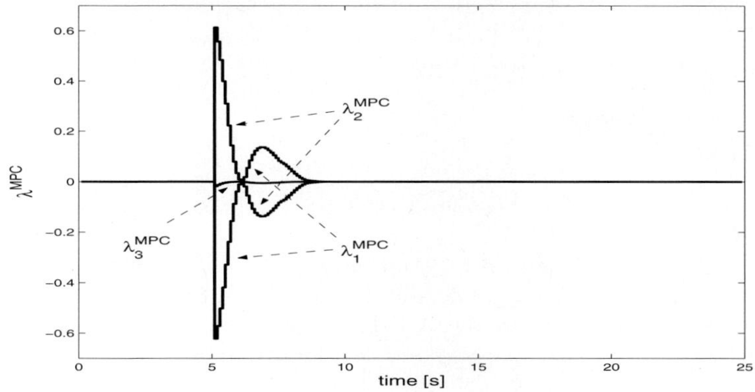

Fig. 4. The MPC control actions

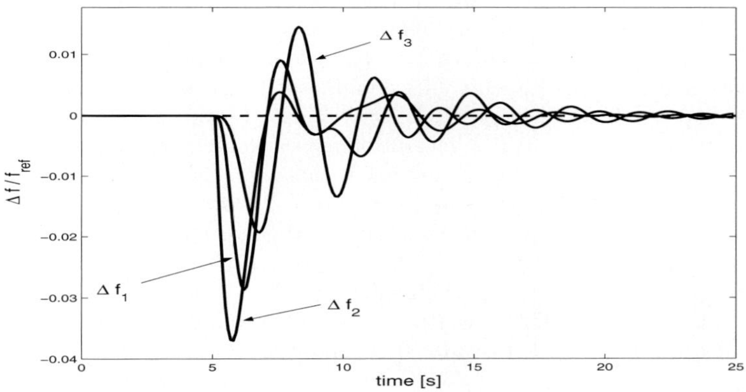

Fig. 5. Relative frequency deviations $\frac{\Delta f}{f_{\rm ref}}$

5 Conclusion

This paper presented a novel hybrid control scheme for achieving optimal power balancing and congestion control in electrical energy transmission networks. We developed an explicit dynamic controller that guarantees economically optimal steady-state operation while respecting all line flow constraints in steady-state. Due to these constraints, the resulting optimal control law has a piecewise affine structure. To improve the response of the system and to effectively and efficiently handle constraints during the transient period, the explicit optimal controller was complemented with a hybrid MPC controller. An example illustrated the proposed hybrid control scheme.

A scalable approach for stability analysis and distributed implementation of the proposed hybrid control scheme will be considered in future work.

Acknowledgments. This work is part of the research program "Intelligent Power Systems" that is supported financially by SenterNovem. SenterNovem is an agency of the Dutch Ministry of Economic Affairs.

References

1. Stoft, S.: Power System Economics: Designing Markets for Electricity. Kluwer Academic Publishers (2002)
2. Christie, R.D., Wollenberg, B.F., Wangensteen, I.: Transmission management in the deregulated enviroment. Proceedings of the IEEE **88**(2) (2000) 170–195
3. Alvarado, F.L., Meng, J., DeMarco, C.L., Mota, W.S.: Stability analysis of interconnected power systems coupled with market dynamics. IEEE Transactions on Power Systems **16**(4) (2001) 695–701
4. Alvarado, F.: The stability of power system markets. IEEE Transactions on Power Systems **14**(2) (1999) 505–511
5. Kundur, P.: Power System Stability and Control. McGraw-Hill (1994)
6. Geyer, T., Larsson, M., Morari, M.: Hybrid emergency voltage control in power systems. In: European Control Conference, Cambridge, UK (2003)
7. Beccuti, A.G., Geyer, T., Morari, M.: A hybrid system approach to power systems voltage control. In: IEEE Conference on Decision and Control, Seville, Spain (2005)
8. Ferrari-Trecate, G., Gallestey, E., Letizia, P., Spedicato, M., Morari, M., Antoine, M.: Modeling and control of co-generation power plants: A hybrid system approach. IEEE Transactions on Control Systems Technology **12**(5) (2004) 694–705
9. Bemporad, A., Muñoz de la Peña, D., Piazzesi, P.: Optimal control of investments for quality of supply improvement in electrical energy distribution networks. Automatica **42** (2006)
10. Saadat, H.: Power System Analysis. McGraw-Hill (1999)
11. Lazar, M.: Model predictive control of hybrid systems: Stability and robustness. PhD thesis, Eindhoven University of Technology, The Netherlands (2006)

A Appendix

This appendix concerns Remark 2 in Section 2. The controller modification includes replacing (8a) and (8b) with (14a) and (14b), respectively, where:

$$\begin{pmatrix} \dot{x}_{\lambda_0} \\ \dot{x}_{\Delta\lambda} \\ \dot{x}_{\mu} \end{pmatrix} = \begin{pmatrix} 0 & 0 & 0 \\ 0 & -K_\Delta B_\Delta & -K_\Delta L_\Delta^\top \\ 0 & 0 & 0 \end{pmatrix} \begin{pmatrix} x_{\lambda_0} \\ x_{\Delta\lambda} \\ x_{\mu} \end{pmatrix} + \begin{pmatrix} -k_f \mathbf{1}_n^\top & 0 \\ 0 & 0 \\ 0 & \Gamma \end{pmatrix} \begin{pmatrix} \Delta f \\ \Delta p_L \end{pmatrix}, \tag{14a}$$

$$\lambda = \left(\begin{pmatrix} 1 \\ \mathbf{1}_{n-1} \end{pmatrix} \begin{pmatrix} 0 \\ I_{n-1} \end{pmatrix} \begin{pmatrix} 0 \\ 0 \end{pmatrix} \right) \mathrm{col}(x_{\lambda_0}, x_{\Delta\lambda}, x_{\mu}). \tag{14b}$$

The diagonal and positive definite matrix K_Δ ($K_\Delta \in \mathbb{R}^{(n-1)\times(n-1)}$) and the positive scalar k_f represent controller gains. B_Δ is a submatrix of B obtained by removing the first column and the first row of B, and L_Δ is a submatrix of L, obtained by removing the first column of L. The dynamics of the scalar valued controller state x_{λ_0} acts as an integral control action on frequency deviations Δf, while the dynamics of the controller state $x_{\Delta\lambda}$ acts as an integral control action on violations of the optimality condition (7f).

Robust Test Generation and Coverage for Hybrid Systems

A. Agung Julius[1], Georgios E. Fainekos[2], Madhukar Anand[2], Insup Lee[2], and George J. Pappas[1]

[1] University of Pennsylvania, Department of Electrical and Systems Engineering, 200 South 33rd Street, Philadelphia PA-19104, USA
{agung,pappasg}@seas.upenn.edu

[2] University of Pennsylvania, Department of Computer and Information Sciences, 200 South 33rd Street, Philadelphia PA-19104, USA
{fainekos,anandm,lee}@seas.upenn.edu

Abstract. Testing is an important tool for validation of the system design and its implementation. Model-based test generation allows to systematically ascertain whether the system meets its design requirements, particularly the safety and correctness requirements of the system. In this paper, we develop a framework for generating tests from hybrid systems' models. The core idea of the framework is to develop a notion of robust test, where one nominal test can be guaranteed to yield the same qualitative behavior with any other test that is close to it. Our approach offers three distinct advantages. 1) It allows for computing and formally quantifying the robustness of some properties, 2) it establishes a method to quantify the test coverage for every test case, and 3) the procedure is parallelizable and therefore, very scalable. We demonstrate our framework by generating tests for a navigation benchmark application.

1 Introduction

As engineering systems gain more functionality and complexity, there is a need for sound discipline in their design, development and deployment. In particular, ensuring the safety and correctness of these large and complex systems is becoming increasingly hard. In recent years, a slew of model-based design efforts have been developed to address these problems. The promise of the model-based design paradigm is to develop design models and subject them to analysis, simulation, and validation prior to their implementation. Performing analysis early in the development cycle allows one to detect and fix design problems sooner and at a lower cost. There has been a lot of work [1,2,3,4,5,6,7,8] in the hybrid systems community toward analysis, validation and verification of systems developed from hybrid control models. The list [1,2,3,4,5,6,7,8] is by no means exhaustive. However, it does capture a broad spectrum of techniques that have been developed in the community to answer the reachability and verification problems.

Testing has been used in practice to check the conformance of an implementation to its specification. Although testing cannot provide formal guarantees on

A. Bemporad, A. Bicchi, and G. Buttazzo (Eds.): HSCC 2007, LNCS 4416, pp. 329–342, 2007.

correctness and reachability as it is possible with verification, disciplined use of testing, coupled with coverage criteria can be a great aid to system verification and validation.

Testing amounts to running or simulating the operation of the system for a finite period of time. It is comparable to taking a snapshot of the operation of the system. As we are interested in gaining some information about the system, testing is done repetitively with varying *testing parameters*, so as to simulate as many scenarios of operation as possible. By testing parameters, we mean the parameters that characterize the run of a test. For example, if we have an autonomous system whereof we can only influence the initial condition, then the testing parameter is the initial condition. If we have more degrees of freedom in influencing the execution of the system, for example, if we can also adjust some parameters in the system, then these parameters can be regarded as testing parameters as well. The ultimate goal of testing is to cover the entirety of the set of testing parameters.

When the set of testing parameters is an infinite set, it is obvious that we cannot exhaustively test each of the testing parameters. However, it is possible that one testing parameter is representative of many others. A testing parameter is said to be *robust* if a slight (quantifiable) perturbation of the parameter is guaranteed to result in a test with the same qualitative properties (for example, safety and correctness). It is obvious that robustness can lead to a significant reduction in the set of testing parameters. In fact, ideally, we would like to be able to reduce an infinite set of testing parameters into a finite set, and quantify the coverage by the performed tests. In this paper, we develop a framework where the robustness of a test can be formally quantified and computed. The framework is then applied to test a navigation benchmark problem [9].

Prior work on generating tests from hybrid systems' models has mainly focused on randomized testing or monitoring to check whether an implementation conforms to its model. Esposito [10] and Branicky et. al. [11] use Rapidly exploring Random Trees (RRT) to generate test cases from hybrid systems models. Another paper by van Osch [12] describes testing for input-output conformance by providing inputs to the implementation and comparing its outputs to those of its model. In [13], the author presents a case-study that identifies a minimal set of test scenarios required to determine, with some confidence interval, if the system meets the specification by casting the test generation problem as an optimal control problem. This approach suffers from the drawback that it is only applicable in scenarios where the optimal control problem can be solved efficiently. Other publications in this area include [14], where the authors present an integrated framework to test and monitor code generated from hybrid models, and [15], where test generation from Extended Finite State Machines (EFSM) is developed in order to test temporal logic properties.

2 Problem Formulation

In this paper, we consider a standard model of a hybrid automaton [16], $\mathcal{H} = (\mathcal{X}, \mathcal{L}, E, Inv, F)$, where $\mathcal{X}$ is the continuous state space of the system, $\mathcal{L}$ is the finite set of discrete states (locations), E is the set of transitions, $Inv : \mathcal{L} \to 2^{\mathcal{X}}$ is the invariant set of each location, and $F : \mathcal{X} \times \mathcal{L} \to \mathcal{X}$ is the vector field that defines the continuous dynamics in each location.

A transition $e \in E$ is a 4-tuple (l, l', g, r), where $l \in \mathcal{L}$ is the origin of the transition, $l' \in \mathcal{L}$ is the target of the transition, $g \subset \partial Inv(l)$ is the guard of the transition, which is a subset of the boundary of the invariant set of location l, and $r : g \to Inv(l')$ is the reset map that resets the continuous state at the new location. Here, we assume that the reset map r is continuous.

In this paper, we shall assume that the following statements hold. The state space is $\mathbb{R}^n$ and the invariant sets are closed. We denote the open interior of an invariant set as $\underline{Inv}(l)$ and we assume that the differential equation

$$\frac{dx}{dt} = F(x(t), l),$$

admits a unique solution for every location $l \in \mathcal{L}$, i.e. it satisfies the Lipschitz conditions. The transitions are deterministic[1] in the sense that the guards are forcing and all outgoing transitions from a location have disjoint guards. Finally, the system does not deadlock or possess Zeno behavior.

In analyzing the safety of the system, we assume that there is a subset `Unsafe` $\subset \mathcal{X} \times \mathcal{L}$ unsafe states. A trajectory of the hybrid system corresponds to an unsafe execution if it intersects with the unsafe set.

Example 1 (Navigation Benchmark [9]). As a case study in this paper, we consider a slightly modified version of the navigation benchmark proposed by Fehnker and Ivancic [9]. The benchmark studies a hybrid automaton $\mathcal{H}$ with 3×3 discrete locations and 4 continuous variables x_1, x_2, v_1, v_2 that form the state vector $x = [x_1\ x_2\ v_1\ v_2]^T$. We refer to the vectors $[x_1\ x_2]^T$ and $[v_1\ v_2]^T$ as the position and the velocity of the system, respectively. The structure of the hybrid automaton can be better visualised in Fig. 1. The invariant set of every (i, j) location is an 1×1 box that constraints the position of the system, while the velocity can flow unconstrained. The guards in each location are the edges and the vertices that are common among the neighboring locations.

Each location has affine constant dynamics with drift. In detail, in each location (i, j) of the hybrid automaton, the system evolves under the differential equation $\dot{x} = Ax - Bu(i, j)$ where the matrices A and B are

$$A = \begin{bmatrix} 0 & 0 & 1 & 0 \\ 0 & 0 & 0 & 1 \\ 0 & 0 & -1.2 & 0.1 \\ 0 & 0 & 0.1 & -1.2 \end{bmatrix} \quad \text{and} \quad B = \begin{bmatrix} 0 & 0 \\ 0 & 0 \\ -1.2 & 0.1 \\ 0.1 & -1.2 \end{bmatrix}$$

and the input in each location is $u(i, j) = [\sin(\pi C(i, j)/4)\ \cos(\pi C(i, j)/4)]^T$. The array C is one of the two parameters of the hybrid automaton that the user can control and it defines the input vector in each discrete location. Here, we consider the following input arrays:

$$C_1 = \begin{bmatrix} U & 2 & 4 \\ 4 & 3 & 4 \\ 2 & 2 & G \end{bmatrix} \quad C_2 = \begin{bmatrix} 2 & 3 & 6 \\ 3 & 3 & G \\ 2 & 2 & U \end{bmatrix} \quad C_3 = \begin{bmatrix} U & 2 & 4 \\ 2 & 2 & 4 \\ 1 & 1 & G \end{bmatrix}$$

[1] We limit the discussion in this paper to deterministic guards. However, the framework presented here is also applicable to nondeterministic guards.

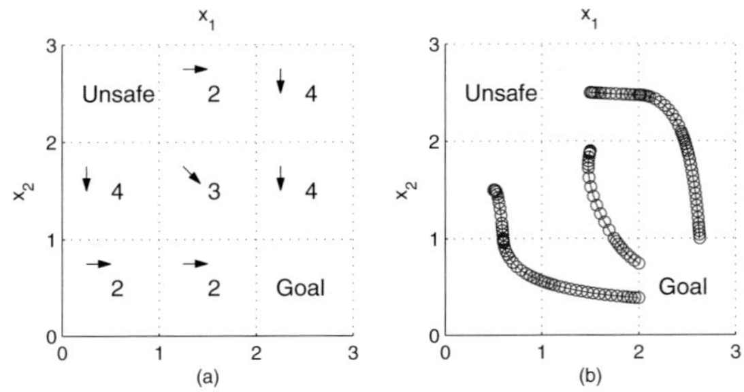

Fig. 1. A graphical representation of the benchmark hybrid automaton. The upper left box is the invariant set for the location (1, 1). (a) The constant input vector in each location. (b) Sample trajectories for different initial conditions.

where U denotes the unsafe set and G the goal set. The other user-input parameter is the set of initial conditions $\mathcal{X}_0 \times \mathcal{L}_0 \subseteq \mathcal{X} \times \mathcal{L}$. The requirement for $\mathcal{H}$ is that all of its trajectories starting in $\mathcal{X}_0 \times \mathcal{L}_0$ should avoid the unsafe set and eventually reach the goal set. Sample trajectories of the system appear in 1.(b).

Example 1 describes a typical verification problem for hybrid systems. The goal of exhaustive verification algorithms is to prove that there cannot exist a trajectory that falsifies the hybrid automaton assumptions, i.e. safety and reachability. In this paper, we try to solve a different problem in an attempt to overcome the theoretical and practical difficulties of exhaustive verification. Here, the target is not complete coverage of the set of initial conditions, but the computation of a (possibly) quick estimate of which part of the initial conditions is safe and/or unsafe for a bounded horizon using only a small number of tests. One of the most important aspects of such a testing methodology is that it should be completely transparent to the user with no (or very few) parameters to tune.

Problem 1 (Testing the benchmark example). Given the hybrid automaton $\mathcal{H}$ of Example 1 with a set of initial conditions $\mathcal{X}_0 \times \mathcal{L}_0$, a bounded horizon $T > 0$ and an unsafe `Unsafe` and/or `Goal` set, develop a strategy for picking test points in order to cover the set of initial conditions.

As mentioned in the previous section, we want to cover the whole set of initial conditions with finitely many test points. This requires the construction of robust neighborhoods around the test points. Each such neighborhood contains a set of points which initiate trajectories that have the same qualitative properties as the trajectory generated by the actual test point. By qualitative properties, we mean the sequence of locations that are visited and the safety property.

3 Robust Testing for Hybrid Systems

In this section, we discuss the computation of robust neighborhoods of initial conditions. First, we are going to review the theory of bisimulation functions for dynamical systems [17]. The concept of bisimulation functions introduced in [17] is more general than what we are going to review here since we only consider systems without input.

Let $\phi : \mathcal{X} \times \mathcal{X} \to \mathbb{R}_+$ be a bisimulation function between a dynamical system

$$\Sigma : \tfrac{dx}{dt} = F(x), \ x \in \mathcal{X} \tag{1}$$

and itself. Such a function ϕ satisfies the following requirements:

$$\phi(x_1, x_2) \geq 0, \tag{2}$$

$$\tfrac{\partial \phi(x_1,x_2)}{\partial x_1} f(x_1) + \tfrac{\partial \phi(x_1,x_2)}{\partial x_2} f(x_2) \leq 0, \tag{3}$$

for every $x_1, x_2 \in \mathcal{X}$.

Denote the continuous flow of the dynamical system Σ as $\xi : \mathbb{R}_+ \times \mathcal{X} \to \mathcal{X}$, that is, $\xi(t, x_0)$ satisfies the differential equation

$$\tfrac{\partial \xi(t,x_0)}{\partial t} = f(\xi(t, x_0)), \ \xi(0, x_0) = x_0. \tag{4}$$

Note that the bisimulation function is nonincreasing with respect to the flow.

Proposition 1 ([17]). *For any $x_1^0, x_2^0 \in \mathcal{X}$, the bisimulation function evaluated along the flows of the initial conditions x_1^0 and x_2^0 is nonincreasing, i.e. for any $t_2 \geq t_1 \geq 0$ it is $\phi(\xi(t_1, x_1^0), \xi(t_1, x_2^0)) \geq \phi(\xi(t_2, x_1^0), \xi(t_2, x_2^0))$.*

We denote the ε-neighborhood (or ε-ball) of $x \in \mathcal{X}$ with respect to a bisimulation function ϕ as $B_\phi(x, \varepsilon)$, i.e. $B_\phi(x, \varepsilon) = \{y \in \mathcal{X} \mid \phi(x, y) \leq \varepsilon\}$. The following corollary is a direct consequence of Proposition 1.

Corollary 1. *For any $x, y \in \mathcal{X}$, if $y \in B_\phi(x, \varepsilon)$ for some $\varepsilon > 0$, then for every $t \geq 0$ it is $\xi(t, y) \in B_\phi(\xi(t, x), \varepsilon)$.*

Thus, the ε-neighborhood defined by the bisimulation function ϕ is invariant with respect to the flow of the dynamical system. If we define the (directed) metric $d_\phi(x(\cdot), y(\cdot))$ between different state trajectories of the system Σ with respect to the bisimulation function ϕ as

$$d_\phi(x(\cdot), y(\cdot)) := \sup_{t \geq 0} \phi(x(t), y(t)),$$

then the corollary above is equivalent to $d_\phi(\xi(\cdot, x), \xi(\cdot, y)) \leq \phi(x, y)$ for any $x, y \in \mathcal{X}$. Hereunder, we shall assume that bisimulation functions are symmetric, that is, $\phi(x, y) = \phi(y, x)$. A bisimulation function that is symmetric and forms a metric on the space $\mathcal{X}$ is called a *contraction metric*. Such functions are used in contraction analysis in relation to the stability of a system [18,19].

When the dynamics are affine,

$$F(x) = Ax + b \text{ for } x \in \mathbb{R}^n, A \in \mathbb{R}^{n \times n}, b \in \mathbb{R}^{n \times 1},$$

we can propose that the bisimulation function assumes the form

$$\phi(x_1, x_2) = (x_1 - x_2)^T M (x_1 - x_2),$$

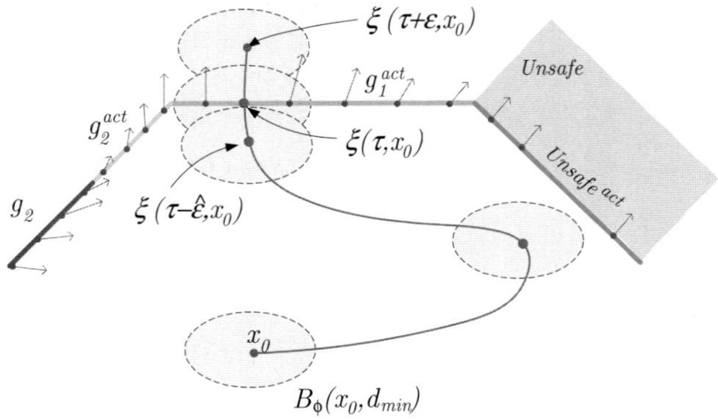

Fig. 2. An illustration for Definition 1 and Proposition 2 for $j = 1$. The line on the left is the guard g_2. Its part with lighter shade, g_2^{act} is the active part, where the vector field points outward. The guard g_1 is active everywhere, as the vector field there points outward. The boundary of the unsafe set in this picture is active since the vector field points into the unsafe set.

where M is a positive semidefinite matrix. Thus, the bisimulation function defines a Euclidean metric in a (linearly) transformed space. It can be shown that such a bisimulation function is essentially a Lyapunov function, and it exists if and only if the system is stable [20].

In the following, we are going to construct robust testing neighborhoods using the level sets of a bisimulation function. For that, we need a few definitions.

Definition 1. *For any location $l \in \mathcal{L}$ we define the set of outgoing transitions from l as $Out(l) \subseteq E$. For any transition $e = (l, l', g, r) \in Out(l)$, we denote by g^{act} the active part of the guard g, which is the part $g^{act} \subset g$ of the guard that can be reached from inside $Inv(l)$, i.e. we exclude from g the points where the vector field $F(\cdot, l)$ points inward. Similarly, we define $\texttt{Unsafe}^{act}$ to be the portion of the boundary of the unsafe set that is reachable from the safe portion of the state space.*

See Fig. 2 for an illustration of the definition of the active guard and of the proposition below.

Proposition 2. *Let $x_0 \in Inv(l)$ for some location $l \in \mathcal{L}$, and assume that the state trajectory $\xi(t, x_0)$ lies entirely in $Inv(l)\backslash\texttt{Unsafe}$ for $t \leq \tau$. Suppose that $Out(l) = \{e_1, e_2, \cdots, e_n\}$ and that g_i is the guard of e_i, $i = 1, \ldots, n$. Let τ be the time when the state trajectory hits a guard g_j, which is the guard of the transition e_j for some $j \in \{1, \ldots, n\}$. Suppose that we have a bisimulation function ϕ for the continuous dynamics in location l. We also assume that there is a positive time lag $\varepsilon > 0$ such that $\xi(\tau + \varepsilon, x_0) \notin Inv(l)$. We define*

$$
\begin{aligned}
d_{out} &:= \inf_{y \in g_j} \phi(\xi(\tau + \varepsilon, x_0), y), \\
d_i &:= \inf_{0 \le t \le \tau + \varepsilon} \inf_{y \in g_i^{act}} \phi(\xi(t, x_0), y), \ i \in \{1, \ldots, n\}/\{j\}, \\
d_{unsafe} &:= \inf_{0 \le t \le \tau + \varepsilon} \inf_{y \in Inv(l) \cap \mathtt{Unsafe}^{act}} \phi(\xi(t, x_0), y), \\
d_{\min} &:= \min\{d_{out}, d_{unsafe}, d_1, \ldots, d_{j-1}, d_{j+1}, \ldots, d_n\}, \\
\hat{\varepsilon} &:= \inf \{\delta > 0 \mid B_\phi(\xi(\tau - \delta, x_0), d_{\min}) \subset \underline{Inv}(l)\}.
\end{aligned}
$$

The following statement holds. For any $x_0' \in B_\phi(x_0, d_{\min}) \cap Inv(l)$*, the state trajectory* $\xi(t, x_0')$ *exits* $Inv(l)$ *through transition* e_j *at time* $t \in [\tau - \hat{\varepsilon}, \tau + \varepsilon]$ *and is safe at least until it exits location* l*.*

Proof. See Fig. 2 for an illustration. By construction of $d_{\min}$, we can infer that for any $t \in [0, \tau + \varepsilon]$ and $i \in \{1, \ldots, n\}/\{j\}$, $B_\phi(\xi(t, x_0), d_{\min}) \cap g_i^{act} = \emptyset$, and $B_\phi(\xi(t, x_0), d_{\min}) \cap \mathtt{Unsafe}^{act} \cap Inv(l) = \emptyset$.

We then invoke Corollary 1 and infer that any state trajectory originating in $B_\phi(x_0, d_{\min})$ will not be unsafe nor touch the active guards g_i^{act}, $i \in \{1, \ldots, n\}/\{j\}$, within the time interval $[0, \tau + \varepsilon]$. We also know that the neighborhood $B_\phi(\xi(\tau + \varepsilon, x_0), d_{\min})$ lies entirely outside of $Inv(l)$, beyond g_j. This implies that any trajectory starting in $B_\phi(x_0, d_{\min}) \cap Inv(l)$ crosses g_j before $t = \tau + \varepsilon$. Finally, since the neighborhood $B_\phi(\xi(t, x_0), d_{\min})$ does not touch any active guard, for $t \in [0, \tau - \hat{\varepsilon})$, we also know that the trajectories will not touch any active guard before time $t = \tau - \hat{\varepsilon}$.

Proposition 2 provides us with a way to compute a neighborhood around the initial state x_0, which consists of initial states that have the same qualitative behavior as x_0. Namely, they lead to a trajectory that exits location l by taking the same transition and which is safe at least until it performs that transition. In addition to that, we obtain a timing guarantee in the form of a time interval where the transition is guaranteed to occur if the initial state belongs to the computed neighborhood. The next step is to design an algorithm that uses Proposition 2 repetitively in order to deal with trajectories that take multiple transitions.

Given a hybrid automaton $\mathcal{H} = (\mathcal{X}, \mathcal{L}, E, Inv, F)$. We denote the continuous flow at every location $l \in L$ as $\xi_l(\cdot, \cdot)$, and we assume that we have a bisimulation function for the dynamics in location $l \in L$, which is $\phi_l(\cdot, \cdot)$. A testing trajectory is a sequence $(x_i, l_i, e_i, \tau_i)_{i=0,\cdots,N}$ such that:

- $l_i \in L, x_i \in Inv(l_i), e_i \in Out(l_i), \tau_i > 0$, for every $i \in \{0, 1, \ldots, N\}$,
- If we define $e_i = (l_i, l_{i+1}, g_i, r_i)$, then $\xi_{l_i}(\tau_i, x_i) \in g_i$, $x_{i+1} = r_i(\xi_{l_i}(\tau_i, x_i))$, $\xi_{l_i}(t, x_i) \in \underline{Inv}(l_i)$ for all $t \in [0, \tau_i)$, for every $i \in \{0, 1, \ldots, N-1\}$,

We define $T := \sum_{i=0}^{N-1} \tau_i$, which is the time where the trajectory enters the final state. The length of the test is $T + \tau_N$. Given a testing trajectory, the algorithm for constructing a robust tube around a nominal trajectory is given as follows.

Algorithm 1. *The following are the steps:*

1. *Define the avoided set as the union of the unsafe set and active parts of all the outgoing guards from* l_N*, i.e.*[2]

$$D_N := \mathtt{Unsafe}^{act} \cup_{g \in Out(l_N)} g^{act}. \tag{5}$$

2. *Compute (or obtain a lower bound on)*

$$d_{\min}^N := \inf_{t \leq \tau_N} \inf_{y \in D_N} \phi_{l_N}(\xi_{l_N}(t, x_N), y). \tag{6}$$

3. *Define the allowed guard*

$$A_{N-1} := r_{N-1}^{-1}(r_{N-1}(g_{N-1}) \cap B_{\phi_{l_N}}(x_N, d_{\min}^N)). \tag{7}$$

This is the set of states on the guard of the transition between l_{N-1} *and* l_N *that is reset into the* $d_{\min}^N-$ *neighborhood of* x_N *(with respect to the bisimulation function* ϕ_{l_N}*).*

4. *Define the avoided set*

$$D_{N-1} := (\mathtt{Unsafe}^{act} \cup_{g \in Out(l_{N-1})} g^{act}) \backslash A_{N-1}. \tag{8}$$

5. *Pick a time lag* $\varepsilon_{N-1} > 0$ *such that*

$$\xi_{l_{N-1}}(\tau_{N-1} + \varepsilon_{N-1}, x_{N-1}) \notin Inv(l_{N-1}).$$

We present an algorithm for picking a good time lag later in this paper.

6. *Compute (or obtain a lower bound on)*

$$d_{\min}^{N-1} := \min \left(\inf_{y \in g_{N-1}} \phi_{l_{N-1}}(\xi_{l_{N-1}}(\tau_{N-1} + \varepsilon_{N-1}, x_{N-1}), y), \right.$$
$$\left. \inf_{t \leq \tau_{N-1} + \varepsilon_{N-1}} \inf_{y \in D_{N-1}} \phi_{l_{N-1}}(\xi_{l_{N-1}}(t, x_{N-1}), y) \right).$$

7. *Define*

$$\hat{\varepsilon}_{N-1} := \inf \left\{ \delta > 0 \mid B_{\phi_{N-1}}(\xi_{l_{N-1}}(\tau_{N-1} - \delta, x_{N-1}), d_{\min}^{N-1}) \subset \underline{Inv}(l_{N-1}) \right\}.$$

8. *Repeat steps 3 - 7 to obtain* $A_i, D_i, \varepsilon_i, d_{\min}^i, \hat{\varepsilon}_i$, $i = 0, 1, \ldots, N-2$.

A property of the result of this iteration is presented in the following theorem, whose proof can essentially be constructed by repeated application of Proposition 2.

[2] Notice that for simplicity, we abuse the notation and associate the transition with its guard.

Theorem 2. *Given a testing trajectory of a hybrid system* $(x_i, l_i, e_i, \tau_i)_{i=0,\cdots,N}$, *let* $d^0_{\min}$, ε_i, $\hat{\varepsilon}_i$, $i = 0, 1, \ldots, N-1$ *be obtained from the iteration in Algorithm 1. Define*

$$\varepsilon := \sum_{i=0}^{N-1} \varepsilon_i, \ \hat{\varepsilon} := \sum_{i=0}^{N-1} \hat{\varepsilon}_i.$$

Any testing trajectory that starts in $B_{l_0}(x_0, d^0_{\min})$ *has the following properties.*
(i) It follows the same sequence of locations, $(l_i)_{i=0,\ldots,N}$ *and it enters the final location* l_N *at* $t \in [T - \hat{\varepsilon}, T + \varepsilon]$,
(ii) The trajectory is safe at least until τ_N *time unit after it enters* l_N.

An essential part of Algorithm 1 is the generation of the time lags ε_i (see Step 5). First of all, notice that a small ε_i is more desirable than a larger one. This is because ε_i is a measure in the slackness in the timing when the trajectories in the tube hit the desired guard (see Theorem 2). The idea is to construct ε_i as small as possible, but large enough so that by introducing this time lag, we are sure that all the trajectories in the constructed tube hit the desired guard within the time interval $[\tau_i, \tau_i + \varepsilon_i]$. In order to do this, we can replace Steps 5 and 6 in Algorithm 1 with the following steps.

Step 5'. Compute

$$\hat{d}^{N-1}_{\min} := \inf_{t \le \tau_{N-1}} \inf_{y \in D_{N-1}} \phi_{l_{N-1}}(\xi_{l_{N-1}}(t, x_{N-1}), y).$$

Step 5". Compute

$$\varepsilon_{N-1} = \min\Big(\inf\Big\{e \mid \inf_{y \in g_{N-1}} \phi_{l_{N-1}}(\xi_{l_{N-1}}(\tau_{N-1} + e, x_{N-1}), y) \ge \hat{d}^{N-1}_{\min},$$
$$\xi_{l_{N-1}}(\tau_{N-1} + e, x_{N-1}) \notin Inv(l_{N-1})\Big\}, \varepsilon_{\max}\Big).$$

Step 6'. If $\varepsilon_{N-1} < \varepsilon_{\max}$ then $d^{N-1}_{\min} = \hat{d}^{N-1}_{\min}$, otherwise

$$d^{N-1}_{\min} = \sup_{0 \le e \le \varepsilon_{\max}} \inf_{y \in g_{N-1}} \phi_{l_{N-1}}(\xi_{l_{N-1}}(\tau_{N-1} + e, x_{N-1}), y).$$

In Step 5' we compute the largest level set that fits within the allowed set. In Step 5", we want to find the minimum time lag such that the computed level set lies entirely beyond the desired guard (and hence outside of the invariant set $Inv(l_{N-1})$). See Fig. 2 for an illustration. Because such time lag might not exist, or is too large, we can establish a maximum allowed value for the time lag, $\varepsilon_{\max}$. If such time lag is found and is smaller than $\varepsilon_{\max}$, then this value is used. If it is not found, then we compute the largest level set that can be fit outside of the invariant set. This is done in Step 6'.

4 Test Generation and Coverage Strategies

In the benchmark problem that we are working on, our goal is to cover the given set of initial states with robust neighborhoods. In the previous section, we have presented an algorithm for computing the robust neighborhood around a given initial state. What needs to be done next is to select subsequent initial states from the given set, so as to (eventually) cover the whole set and/or to provide a quantitative measure of coverage based on the executed tests. The strategy for selecting the test points is called the *test generation*.

An important issue in test generation is the notion of coverage, which qualitatively characterizes the number and the type of tests generated. There are a number of coverage criteria based on the test requirement, which can be categorized into two classes: initial state coverage and structural coverage. The first type of coverage criteria is concerned with covering the set of initial states and characterizing each test case that has been generated. The second class of coverage criteria is concerned with analyzing the structural coverage of a test trajectory, such as location coverage and transition coverage. This notion of coverage can capture more aspects of the execution than just coverage of the initial states. The main challenge here is how to generate tests so as to meet particular coverage criteria. In this paper, we are only concerned with the coverage of the set of initial states and leave the prospect of using our framework to analyze structural coverage as future work.

There are a number of strategies for initial state coverage:

Randomized Strategy: The first strategy for covering the set of initial states is to pick points randomly. Consequently, it is hardly possible to guarantee efficient coverage. However, a randomized strategy might be an attractive option because of its simplicity.

Greedy Strategy: Under this strategy, we first pick a point and run the testing algorithm with it. Then, we subtract the computed robust ball around the initial point from the set of initial states and pick the center of the maximum ball that can be fitted into the remaining space as the next test point.

Tessellation-based Strategy: Picking points at random may not ensure uniform coverage. One possible strategy to ensure uniform coverage is to use tessellation of the initial state space based on an appropriate metric. This strategy does not scale well as the dimension of the state space increases.

Minimal Dispersal-based Strategy: Picking points so as to minimize the *dispersion* [21] of the points in the set of initial states. This strategy involves the generation of weighted Voronoi diagrams in the set. The goal is to pick the points incrementally so as to maximize the radius of a non-overlapping ball that can be inserted in the set. We use this method to analyze the benchmark problem (see the following section).

5 Numerical Results and Discussion

In this section, we present some numerical results using our prototype MATLAB implementation of the robust testing algorithm. The experimental results will helps us discuss the strengths and weaknesses of our approach.

One of the advantages of using robust testing methodologies is that we can obtain an estimate of the degree of coverage of initial conditions that we have

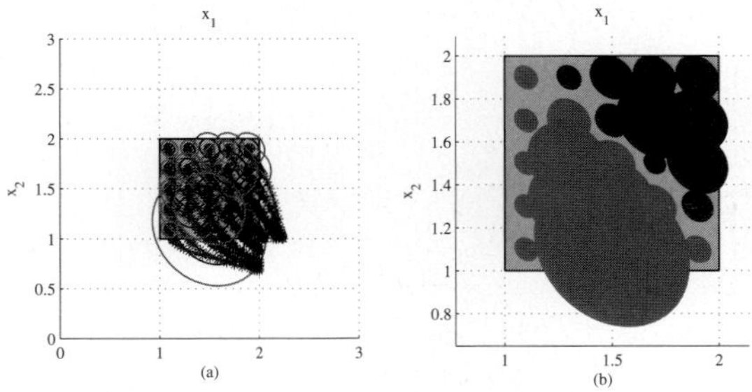

Fig. 3. Result after 25 simulations for the problem instance of Example 2

achieved. Theoretically, this can be done by computing the volume of the intersection of the robustness ellipsoid $\mathcal{E}$ with the polytope that defines the set of initial conditions $\mathcal{X}_0$. Nonetheless, this is not feasible computation-wise when the robustness ellipsoid $\mathcal{E}$ is not contained inside the set of initial conditions. Therefore, we compute the maximum ellipsoid that fits inside the intersection of $\mathcal{E}$ and $\mathcal{X}_0$. This can lead to a significant under-approximation of the actual covered space (see Example 3).

The following testing problem provides some inside on the principles behind our testing algorithm. The planar choice of initial conditions help us visualize the coverage of initial conditions, since the same is not possible when testing a 4D set of initial conditions.

Example 2. The first case that we consider is testing the navigation benchmark for the input array C_1 and the set of initial conditions $\mathcal{X}_0 = [1,2] \times [1,2] \times \{-0.2\} \times \{0\}$ with $\mathcal{L}_0 = \{(2,2)\}$ (*light gray* region in Fig. 3). Here instead of using the Minimal Dispersal-based Strategy, we create a grid of 25 points which serve as initial conditions for each simulation. The resulting simulations appear in Fig. 3.(a). The ellipsoids centered at the initial conditions denote the projections of the 4D ellipsoids on the position plane $x_1 - x_2$. In Fig. 3.(b), we present the covered space of initial conditions after 25 simulations. Here, the ellipsoids are the intersection of the corresponding 4D ellipsoids with the position plane. The *gray* and *black* ellipsoids denote covered initial conditions whose corresponding trajectories followed different discrete paths. Note that there exists a clear partition of $\mathcal{X}_0$ into two subsets of initial conditions that initiate trajectories that traverse different discrete paths. In this case, our proposed under-approximation algorithm for coverage computed 48% of covered initial conditions.

The next example indicates that when the set of initial conditions is thin and the system is robust with respect to the specifications (unsafe and/or goal set), the testing problem becomes easier.

Example 3. Consider again C_1, but now with the following set of initial conditions $\mathcal{X}_0 = [2.2, 2.8] \times [1.2, 1.8] \times \{-0.2\} \times \{0\}$ with $\mathcal{L}_0 = \{(2,3)\}$. This set of initial conditions has been verified to be safe with respect to the unsafe set in [9]. Using the testing algorithm we can cover the set of initial conditions with only 9

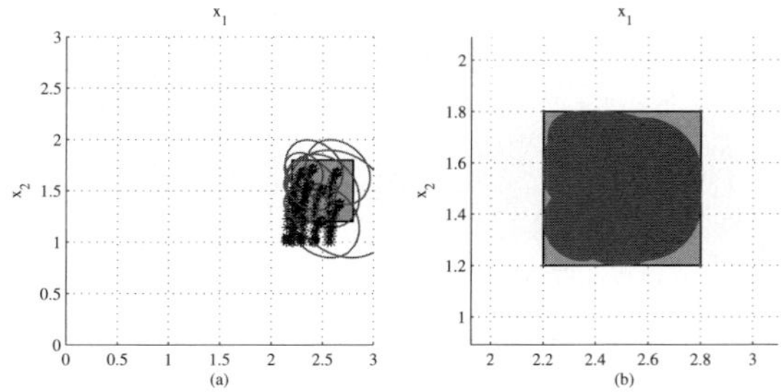

Fig. 4. Result after 9 simulations for the problem instance of Example 3

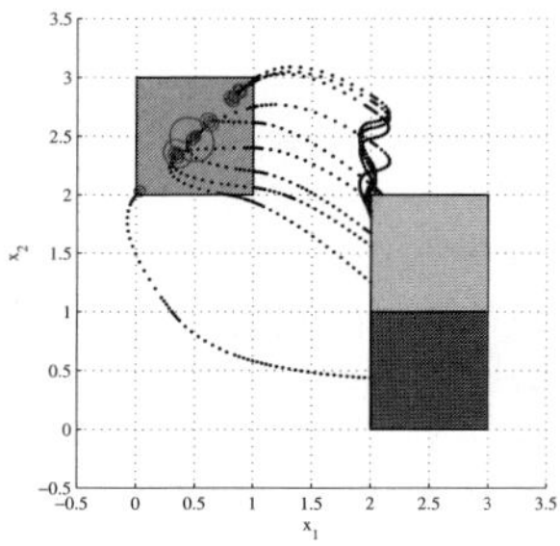

Fig. 5. The testing framework can potentially detect the unsafety of the system with just one test. Legend: *upper left square* - initial conditions, *lower right square* - unsafe set, *upper right square* - goal set

simulations (Fig. 4.(a)). In Fig. 4.(a) the ellipsoids represent the intersection of the corresponding 4D ellipsoids with the position plane, while in Fig. 4.(b) we present the under-approximation of the aforementioned ellipsoids with ellipsoids that fit inside $\mathcal{X}_0$. Numerically, we compute a coverage estimate of 72%.

The previous example also shows that even though by visual inspection we can verify that we have tested all the set of initial conditions, numerically we do not have an accurate way to confirm that. Next, we show the main strength of the testing framework, i.e. easy detection of robustly unsafe systems.

Example 4. Consider the input array C_2 with initial conditions $\mathcal{X}_0 = [0,1] \times [2,3] \times [-1,1] \times [-1,1]$ and $\mathcal{L}_0 = \{(1,1)\}$. This was proven to be unsafe with just 10 simulations (see Fig. 5). Notice the complicated hybrid dynamics.

Finally, we apply our framework to a more demanding example.

Example 5. Here, we use input array C_3 with initial conditions $\mathcal{X}_0 = [0,1] \times [0,1] \times [-0.1, 0.5] \times [-0.05, 0.25]$ and $\mathcal{L}_0 = \{(3,1)\}$. This example was proven to be safe in [22] using the verification toolbox PHAVer [7]. Our testing algorithm was able to cover 7% of the initial conditions after 300 simulations.

On-going research is focused on obtaining better estimates of the covered set of initial conditions. Finally, one of the main advantages of our robust testing framework is that it can be effectively parallelized by simply assigning a different test trajectory to each CPU.

6 Concluding Remarks

In this paper, we presented an algorithm for test generation for hybrid systems. The algorithm is based on a computational method for robust testing. We implemented the algorithm to verify a navigation benchmark problem [9]. One advantage of our algorithm, compared to some other tools, is that we do not need to tune any parameter beforehand.

As future research agenda, we identify a number of potential directions. For example, we are going to develop a framework for robust testing of linear temporal logic properties [23], and develop a probabilistic notion of robust testing by using the idea of stochastic bisimulation function [24]. The algorithm that we presented in this paper is also able to provide a timing guarantee for the occurrence of the transitions. Although this feature is not exploited in the example that we presented in this paper, it can potentially be applied in automatic translation of hybrid automata into timed automata. Such a translation is useful for example in verification and observer design for hybrid systems [25].

Acknowledgments. This research is partially supported by the NSF PECASE 0132716, NSF EHS 0311123, NSF ITR 0121431, ARO MURI Grant DAAD 19-02-01-0383, NSF CNS 0410662, NSF CNS 0509327, NSF CNS 0509143, and the ARO W911NF-05-1-0182 research grants.

References

1. Dang, T., Maler, O.: Reachability analysis via face lifting. In: Hybrid Systems: Computation and Control. Volume 1386 of LNCS., Springer Verlag (1998) 96–109
2. Kurzhanski, A.B., Varaiya, P.: Ellipsoidal technique for reachability analysis. In: Hybrid Systems: Computation and Control. Volume 1790 of LNCS., Springer Verlag (2000) 202–214
3. Mitchell, I., Tomlin, C.J.: Level set methods in for computation in hybrid systems. In: Hybrid Systems: Computation and Control. Volume 1790 of LNCS., Springer Verlag (2000) 310–323
4. Alur, R., Dang, T., Ivancic, F.: Reachability analysis of hybrid systems via predicate abstraction. In: Hybrid Systems: Computation and Control. Volume 2289 of LNCS., Springer Verlag (2002) 35–48
5. Han, Z., Krogh, B.H.: Reachability analysis of hybrid control systems using reduced-order. In: Proc. American Control Conference. (2004) 1183–1189
6. Prajna, S., Jadbabaie, A.: Safety verification of hybrid systems using barrier certificates. In: Hybrid Systems: Computation and Control. Volume 2993 of LNCS., Springer Verlag (2004) 477–492

7. Frehse, G.: PHAVer: Algorithmic verification of hybrid systems past HyTech. In: Hybrid Systems: Computation and Control. Volume 3414 of LNCS., Springer Verlag (2005) 258–273
8. Girard, A.: Reachability of uncertain linear systems using zonotopes. In: Hybrid Systems: Computation and Control. Volume 3414 of LNCS., Springer Verlag (2005) 291–305
9. Fehnker, A., Ivancic, F.: Benchmarks for hybrid systems verification. In: Hybrid Systems: Computation and Control. Volume 2993 of LNCS., Springer Verlag (2004) 326–341
10. Esposito, J.M.: Randomized test case generation for hybrid systems: metric selection. In: Proc. 36th Southeastern Symposium of System Theory. (2004)
11. Branicky, M.S., Curtiss, M.M., Levine, J., Morgan, S.: RRTs for nonlinear, discrete, and hybrid planning and control. In: Proc. IEEE Conf. Decision and Control, Hawaii, USA (2003)
12. van Osch, M.: Automated model-based testing of χ simulation models with TorX. In: Quality of Software Architectures and Software Quality. Volume 3712 of LNCS., Springer Verlag (2005) 227–241
13. Esposito, J.M.: Automated test trajectory for hybrid systems. In: Proc. 35th Southeastern Symposium of System Theory. (2003)
14. Tan, L., Kim, J., Lee, I.: Testing and monitoring model-based generated program. Electronic Notes in Theoretical Computer Science **89**(2) (2003) http://www.elsevier.nl/locate/lntcs/volume89.html.
15. Hong, H.S., Lee, I., Sokolsky, O., Ural, H.: A temporal logic based theory of test coverage and generation. In: TACAS '02: Proceedings of the 8th International Conference on Tools and Algorithms for the Construction and Analysis of Systems, London, UK, Springer-Verlag (2002) 327–341
16. Alur *et. al.*, R.: The algorithmic analysis of hybrid systems. Theoretical Computer Science **138** (1995) 3–34
17. Girard, A., Pappas, G.J.: Approximation metrics for discrete and continuous systems. accepted for publication on *IEEE Trans. Automatic Control* (March 2005)
18. Lohmiller, W., Slotine, J.J.E.: On contraction analysis for nonlinear systems. Automatica **34**(6) (1998) 683–696
19. Aylward, E., Parillo, P.A., Slotine, J.J.E.: Algorithmic search for contraction metrics via sos programming. In: Proc. American Control Conference, Minneapolis, USA (2006)
20. Brogan, W.L.: Modern control theory. Prentice Hall International, New Jersey (1991)
21. LaValle, S.M.: Planning algorithms. Cambridge University Press (2006)
22. Makhlouf, I.B., Kowalewski, S.: An evaluation of two recent reachability analysis tools for hybrid systems. In: Proc. IFAC Conf. Analysis and Design of Hybrid Systems, Alghero, Italy, IFAC (2006) 377–382
23. Fainekos, G.E., Girard, A., Pappas, G.J.: Temporal logic verification using simulation. In: Proceedings of FORMATS. Volume 4202 of LNCS., Springer (2006) 171–186
24. Julius, A.A.: Approximate abstraction of stochastic hybrid automata. In: Hybrid Systems: Computation and Control. Volume 3927 of LNCS., Springer Verlag (2006) 318–332
25. D'Innocenzo, A., Di Benedetto, M.D., Di Gennaro, S.: Observability of hybrid automata by abstraction. In: Hybrid Systems: Computation and Control. Volume 3927 of LNCS., Springer Verlag (2006) 169–183

Minimalilty of Finite Automata Representation in Hybrid Systems Control

Koichi Kobayashi and Jun-ichi Imura

Tokyo Institute of Technology, Tokyo 152-8552, Japan
Tel.:+81.3.5734.2646; Fax:+81.3.5734.2646
{k-kobayashi,imura}@cyb.mei.titech.ac.jp

Abstract. This paper discusses a new approach to representing a finite automaton as a combination of a linear state equation with a smaller set of free binary variables (i.e., input variables) and binary inequalities, in order to reduce the computational time for solving the model predictive control problem of a class of hybrid systems. In particular, this paper is devoted to proving that a system representation derived by our proposed method is minimal in the sense that the number of its binary input variables is minimal among system models over all linear equivalence transformations that preserve the binary property of free (input) variables.

1 Introduction

As is well known, the optimal/model predictive (MPC) control problem of hybrid systems is in general reduced to a mixed integer quadratic programming (MIQP) problem. However, the MIQP problem has crucial weakness for real-time control; i.e., the computational amount exponentially grows with the number of binary variables. So it will be desirable that hybrid systems are represented in terms of a smaller set of binary variables to reduce the number of binary variables in the MIQP problem. In particular, it is very important to discuss how to express the discrete dynamics such as finite automata in a concise way.

To our knowledge, however, few results from the above points of view have been obtained in the previous literatures. Indeed the binary-inequalities based representation is well known as a method for expressing finite automata [2,11], but it will not be desirable for the branch and bound method, which is one of the standard methods for solving the MIQP problem, because the relaxed problem in question does not effectively work for the binary inequalities expressing a finite automaton within the framework of the MPC problem (see [4] for more details).

The authors thus have proposed in [5] a new method of representing a finite automaton as *a linear state equation* with a relatively small number of free binary variables (called here binary input variables) and binary inequalities, based on the implicit system representation [1,8], to produce an MIQP problem with a smaller number of binary variables. It has also been shown that our method is

A. Bemporad, A. Bicchi, and G. Buttazzo (Eds.): HSCC 2007, LNCS 4416, pp. 343–356, 2007.

very effective, by numerical examples of the finite-time optimal control problem of switched systems [5] and piecewise affine systems including discrete transition constraints specified by finite automata [6]. Furthermore, in [7], we have discussed the minimality of system representation in the sense that the number of the binary input variables of our model [5] is smallest over *the binary field* $GF(2)$ [9]. More rigorously, it has been proven there that, for a finite automaton expressed by a linear implicit system model on $GF(2)$, a linear state equation on the real number field $\mathbf{R}$ derived via our method has the smallest number of binary input variables among the system models obtained under *all linear equivalence transformations on* $GF(2)$.

However, it is natural to discuss the minimality of system representation under *all linear equivalence transformations on* $\mathbf{R}$, because the linear system model used when reducing the control problem to the MIQP problem has to be given on $\mathbf{R}$. One of the main difficulties in such discussion on $\mathbf{R}$ is that the coordinates transformation on $\mathbf{R}$ does not in general preserve the binary property in the original model. Thus at the first step, we have considered in [7] only the coordinate transformations on $GF(2)$, which preserve the binary property. It is also remarked that the relation between the operations on $GF(2)$ and $\mathbf{R}$ is nonlinear. So the linear equivalence transformations on $GF(2)$ do not in general imply those on $\mathbf{R}$; that is, the minimality on $GF(2)$ is not equivalent to that on $\mathbf{R}$.

This paper continues upon a series of our research [5,6,7], and proves that, for a finite automaton expressed by a linear implicit system model on $\mathbf{R}$, a system representation on $\mathbf{R}$ derived via the proposed method, which is a more sophisticated version of our previous method [5,7], has the smallest number of binary input variables among the system models over *all linear equivalence transformations on* $\mathbf{R}$ *that preserve the binary property of free variables.*

First, the modeling method in [5] is roughly explained. Next, after the concept of a "minimal representation" used in this paper is defined, the problem of finding a minimal representation of a finite automaton is formulated. Third, a characterization of the linear equivalence transformations on $\mathbf{R}$ to be studied here are given, while mathematical properties of the system model derived via our modeling method are derived. These results are crucial keys for proving the minimality on $\mathbf{R}$. Based on these several results, the proof on the minimality is finally completed.

Notation: Let $\{0,1\}^{m\times n}$ express the set of $m \times n$ matrices, which consists of elements 0 and 1, and also let $\{-1,0,1\}^{m\times n}$ express the set of $m \times n$ matrices, which consists of elements -1, 0 and 1. Let I_n, $0_{m\times n}$ and e_n express the $n \times n$ identity matrix, the $m \times n$ zero matrix and the $n \times 1$ vector whose elements are all one, respectively. For a finite set of vectors made by the product of a matrix $T \in \mathbf{R}^{m\times n}$ and a vector in the finite set $\{0,1\}^n$, we use the notation $T\{0,1\}^n := \{\, Ta \,|\, a \in \{0,1\}^n \,\}$. For simplicity of notation, we sometimes use the symbol 0 instead of $0_{m\times n}$, and the symbol I instead of I_n.

2 Optimal Control of Hybrid Systems

2.1 MLD Model Based Approach

In the optimal/model predictive control problem of hybrid systems, one of the powerful methods is an approach based on the mixed logical dynamical (MLD) model [2] given by

$$x(k+1) = Ax(k) + Bv(k) \quad (1)$$
$$Cx(k) + Dv(k) \leq H \quad (2)$$

where $x(k) \in \mathbf{R}^{n_c} \times \{0,1\}^{n_d}$ is the state, $v(k)$ is given by $v(k) = [\ u^T(k)\ \ z^T(k)\ \ \delta^T(k)\ ]^T$, $u(k) \in \mathbf{R}^{m_{1c}} \times \{0,1\}^{m_{1d}}$ is the control input, and $z(k) \in \mathbf{R}^{m_2}$, $\delta(k) \in \{0,1\}^{m_3}$ are auxiliary continuous and binary variables. For this model, the following finite-time optimal/model predictive control problem is considered.

Problem 1. Suppose that the initial state $x(0) = x_0$ is given. Then for the system of (1), (2), find $v^*(k),\ k = 0,\ 1,\ \cdots,\ N-1$, minimizing the cost function

$$J = \sum_{i=0}^{N-1} \{x^T(i)Qx(i) + v^T(i)Rv(i)\} + x^T(N)Q_f x(N)$$

where $Q \geq 0$, $R > 0$, and $Q_f \geq 0$.

As is well-known, Problem 1 is reduced to the MIQP problem:

$$\min_{\bar{v} \in \mathcal{V}} \ \bar{v}^T M_1 \bar{v} + \bar{v}^T (M_2 x_0 + M_3)$$
$$\text{subject to} \ \ L_1 \bar{v} \leq L_2 x_0 + L_3$$

where the input set $\mathcal{V}$ is a set of $((\mathbf{R}^{m_{1c}} \times \{0,1\}^{m_{1d}}) \times \mathbf{R}^{m_2} \times \{0,1\}^{m_3})^N$ and $\bar{v} := [\ v^T(0)\ \ v^T(1)\ \ \cdots\ \ v^T(N-1)\ ]^T$. Then we can see that $(m_{1d} + m_3)N$ corresponds to the number of the binary variables in the MIQP problem. Thus we will discuss here how the number m_{1d} (or m_3) can be decreased when a finite automaton is expressed as an MLD model in some way. It is also stressed that the number of the binary variables in the MIQP problem does not depend on the dimension of the binary state variables, n_d.

2.2 Proposed Method and Conventional Method

Before describing the minimality problem of system models of representing an finite automaton, let us explain "what is our approach" in a simple way, which has been basically proposed in [5]. So consider an example of a finite automaton in Fig. 1, where each number in the node of Fig. 1 denotes the mode (the discrete state) of the system.

Our approach based on the implicit system representation is as follows (see [5] for the details). In our approach, a binary variable is assigned to the arc (directed edge), not to the node. So a binary variable δ_{ij} is assigned to the arc

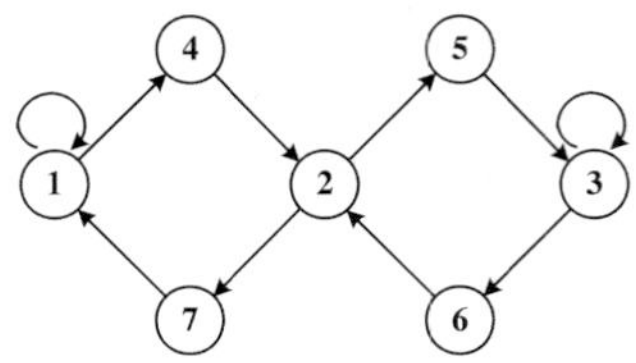

Fig. 1. Example of finite automaton

from node i to node j, by which we thus can express the input-output relation at each node. For example, the input-output relation at node 1 is expressed by the equation

$$\delta_{14}(k+1) + \delta_{11}(k+1) = \delta_{11}(k) + \delta_{71}(k).$$

By expressing a similar relation for every node, the automaton in Fig. 1 is expressed as the following discrete-time implicit system with an equality constraint

$$E\xi(k+1) = F\xi(k), \quad e_{10}^T\, \xi(0) = 1 \tag{3}$$

where $\xi = [\ \delta_{11}\ \ \delta_{14}\ \ \delta_{25}\ \ \delta_{27}\ \ \delta_{33}\ \ \delta_{36}\ \ \delta_{42}\ \ \delta_{53}\ \ \delta_{62}\ \ \delta_{71}\ ]^T$,

$$E = \begin{bmatrix} 1\,1\,0\,0\,0\,0\,0\,0\,0\,0 \\ 0\,0\,1\,1\,0\,0\,0\,0\,0\,0 \\ 0\,0\,0\,0\,1\,1\,0\,0\,0\,0 \\ 0\,0\,0\,0\,0\,0\,1\,0\,0\,0 \\ 0\,0\,0\,0\,0\,0\,0\,1\,0\,0 \\ 0\,0\,0\,0\,0\,0\,0\,0\,1\,0 \\ 0\,0\,0\,0\,0\,0\,0\,0\,0\,1 \end{bmatrix}, \quad F = \begin{bmatrix} 1\,0\,0\,0\,0\,0\,0\,0\,0\,1 \\ 0\,0\,0\,0\,0\,0\,1\,0\,1\,0 \\ 0\,0\,0\,0\,1\,0\,0\,1\,0\,0 \\ 0\,1\,0\,0\,0\,0\,0\,0\,0\,0 \\ 0\,0\,1\,0\,0\,0\,0\,0\,0\,0 \\ 0\,0\,0\,0\,0\,1\,0\,0\,0\,0 \\ 0\,0\,0\,1\,0\,0\,0\,0\,0\,0 \end{bmatrix}.$$

Then we can show that by a certain linear transformation of coordinates $\xi(k) = V\,[\ x_d^T(k)\ \ u_d^T(k)\ ]^T$ for $V \in \mathbf{R}^{10\times 10}$, the implicit system (3) can be expressed by the following state equation:

$$\begin{cases} x_d(k+1) = A_d x_d(k) + B_d u_d(k), \\ 0_{7\times 1} \le x_d(k) \le e_7, \quad e_7^T x_d(0) = 1 \end{cases} \tag{4}$$

where $u_d(k) \in \{0,1\}^3$, $x_d(k) \in \mathbf{R}^7$ (in fact, $x_d(k) \in \{0,1\}^7$ is guaranteed thanks to the inequalities and the initial condition in (4), and $u_d(k) \in \{0,1\}^3$). The derivation procedure from the implicit system (3) to the state equation (4) will be explained in Section 4.2. The point of our approach is that, based on the property that $\xi(k+1)$ is not uniquely determined by (3) and the value of $\xi(k)$, i.e., that the automaton in Fig. 1 is nondeterministic, the free variable $u_d(k)$ corresponding to the input variable and the state variables $x_d(k)$ uniquely determined by the state equation are derived from $\xi(k)$ of the implicit system. This model provides $n_d = 7$, $m_{1d} = 3$, $m_3 = 0$ for binary variables in the MLD model (1), (2).

On the other hand, the model via the standard method [2,11] is given as follows. Suppose that a binary variable δ_i is assigned to node i (mode i), and that $\delta_i(k) = 1$ and $\delta_j(k) = 0$ for all $j \neq i$ hold when the mode at time k is i, which implies that the equality constraint $\sum_{i=1}^{7} \delta_i(k) = 1$. Note here that the relation between δ_{ij} and a pair (δ_i, δ_j) is given by $[\delta_{ij}(k) = 1] \leftrightarrow [\delta_i(k-1) = 1] \wedge [\delta_j(k) = 1]$, i.e., $\delta_{ij}(k) = \delta_i(k-1)\delta_j(k)$. Then, based on the equivalence relations between propositional logic and linear-type inequalities, the automaton in Fig. 1 can be expressed by the following binary linear-type inequalities:

$$\begin{cases} \delta_1(k) - \delta_1(k+1) - \delta_4(k+1) \leq 0, \\ \delta_2(k) - \delta_5(k+1) - \delta_7(k+1) \leq 0, \\ \delta_3(k) - \delta_3(k+1) - \delta_6(k+1) \leq 0, \\ \delta_4(k) - \delta_2(k+1) \leq 0, \\ \delta_5(k) - \delta_3(k+1) \leq 0, \\ \delta_6(k) - \delta_2(k+1) \leq 0, \\ \delta_7(k) - \delta_1(k+1) \leq 0, \\ \sum_{i=1}^{7} \delta_i(k) \leq 1, \quad \sum_{i=1}^{7} \delta_i(k) \geq 1. \end{cases} \tag{5}$$

So letting the state equation $\delta_i(k+1) = u_i(k)$ for a new binary variable $u_i(k)$, we can express the binary linear-type inequalities in (5) as the form of (2). Note also that $\delta_i(k+1) = u_i(k)$ is embedded in (1). Then the number of binary variables in the MLD model is given by $n_d = m_{1d} = 7$.

In this case, we see that the number of binary variables $(m_{1d} + m_3)N$ in the MIQP problem produced by our model is smaller than that of the standard method. In [5], we have also shown by numerical examples that the computational time in our method is dramatically reduced compared with that in the standard method. One of the open, important questions at the next step is "for an implicit system expressing a finite automaton, does our modeling method provide a system model that has the smallest number of binary input variables among system models under all equivalence transformations on **R**?". It will not so easy to find an answer to this question. So we will address here "for what class of transformations on **R** our model has the smallest number of binary input variables?".

3 Definition of Minimal Representation

In this section, the problem of finding a minimal representation for an implicit system expressing a given finite automaton is formulated.

Consider the following linear implicit system expressing a finite automaton:

$$\Sigma_I : \begin{cases} E\xi(k+1) = F\xi(k), \\ \xi(k) \in \{0,1\}^n, \quad \xi(0) = \xi_0 \in \Xi_0 \end{cases} \tag{6}$$

where $E,\ F \in \mathbf{R}^{m \times n}$ and $\Xi_0 \subseteq \{0,1\}^n$. Furthermore, the following assumptions are made for this system:

Assumption A1. $\mathrm{rank}\,(F - zE) = m,\ \forall z \in \mathbf{C},$ where $\mathbf{C}$ denotes the complex number field.

Assumption A2. The system Σ_I of (6) has no noncausal behavior.

Assumption A1 implies that there exist no redundant equations in (6), in other words, (6) has m independent equations. For example, if (6) has two same equations $\xi_1(k+1)+\xi_2(k+1)=\xi_1(k)+\xi_2(k)$, then these are dependent (see [1,8]). The noncausal behavior in assumption A2 corresponds to the impulsive behavior in continuous-time systems.

In this paper, we consider a finite automaton that satisfies $M \le L$, where M and L denote the number of the nodes and that of the arcs, respectively. We often denote such an automaton by an automaton ($M \le L$). Such an automaton can be expressed as the above implicit system. In fact, this is proven as follows.

Lemma 1. *For every finite automaton ($M \le L$), there exists an implicit system model of the form (6) satisfying assumptions A1 and A2.*

Proof (sketch of proof). This is proven by construction. Suppose that a finite automaton is given. Then let $\mathcal{I}_a=\{i_1,\ i_2,\ \cdots,\ i_L\}$ denote the set of combinations of (i,j) for which the arc from node i to node j exists, and assign a binary variable $\delta_{i_l}\in\{0,1\}$, $l=1,\ 2,\ \cdots,\ L$ to each arc. Then by expressing the relation between inputs and outputs at every node, we obtain the following implicit system with:

$$\begin{cases} E\xi(k+1)=F\xi(k), \quad \xi(k)\in\{0,1\}^L, \\ \xi(0)=\xi_0\in\Xi_0:=\{\ \eta\in\{0,1\}^L \mid e_L^T\eta=1\ \} \end{cases} \tag{7}$$

where $\xi(k):=[\ \delta_{i_1}(k)\ \ \delta_{i_2}(k)\ \ \cdots\ \ \delta_{i_L}(k)\ ]^T\in\{0,1\}^L$ and $E,\ F\in\{0,1\}^{M\times L}$. Furthermore, let $E^{(l)}\in\{0,1\}^{1\times L}$, $l=1,\ 2,\ \cdots,\ M$, denote each row vector of E, and let $F^{(l)}\in\{0,1\}^{1\times L}$, $l=1,\ 2,\ \cdots,\ M$, denote each row vector of F. Then the following condition holds:

$$\sum_{i=1}^{M} E^{(i)}=e_L^T, \qquad \sum_{i=1}^{M} F^{(i)}=e_L^T. \tag{8}$$

From (8), $e_L^T\xi(k)=1$ holds under the initial condition $\xi(0)=\xi_0\in\Xi_0$. So the implicit system (7) expresses mode (node) transition constraints of a given finite automaton.

Finally, it will be proven that the implicit system (7) satisfies assumptions A1 and A2. since $\xi(k+1)$ depends on only $\xi(k)$ and is independent of $\xi(\bar{k})$, $\bar{k}>k+1$, the implicit system (7) satisfies assumption A2. On the other hand, from (8), we can show that each row of the system of (7) is independent. Therefore, since $M\le L$ holds, the implicit system (7) satisfies assumption A1. This completes the proof. □

Thus, for the implicit system (6) expressing a given automaton, this paper will consider transforming it into the following type of system, which can be considered as a part of the MLD model (1), (2):

$$\Sigma_{\text{target}} : \begin{cases} x(k+1) = Ax(k) + Bu(k), \quad Cx(k) + Du(k) \leq E, \\ x(k) \in \mathbf{R}^p, \quad u(k) \in \{0,1\}^r, \\ x(0) = x_0 \in X_0, \quad u(0) = u_0 \in U_0 \end{cases} \tag{9}$$

where $x(k)$ is the state, X_0 is an initial-state set, $u(k)$ is a free binary variable, called "input variable" hereafter, and U_0 is an initial-input set.

Let us define the class of the above transformations. So consider the following class of linear transformations for the implicit system (6).

Definition 1. *For nonsingular matrices U and V, the implicit system*

$$\bar{\Sigma}_I : \begin{cases} \bar{E}\bar{\xi}(k+1) = \bar{F}\bar{\xi}(k), \\ \bar{\xi}(k) \in V^{-1}\{0,1\}^n, \quad \bar{\xi}(0) = \bar{\xi}_0 \in V^{-1}\Xi_0 \end{cases} \tag{10}$$

where $\bar{E} := UEV$, $\bar{F} := UFV$, $\bar{\xi}(k) := V^{-1}\xi(k)$ is said to be equivalent to the system Σ_I of (6), and the transformation of (6) into an equivalent implicit system (10) is called a (U,V)-equivalence transformation.

It is remarked that (6) and (10) are in one-to-one correspondence under the (U,V)-equivalence transformation, since U and V are nonsingular.

Among all (U,V)-equivalence transformations, some class of them may transform the implicit system Σ_I into a combination of a linear state equation and an additional algebraic equation. The set of all such $(U,V) \in \mathbf{R}^{m\times m} \times \mathbf{R}^{n\times n}$ is denoted by $\mathcal{T}$, and the (U,V)-equivalence transformation for $(U,V) \in \mathcal{T}$ is simply called the *$\mathcal{T}$-equivalence transformation.*

Then the following result is straightforwardly obtained from the result in [10].

Lemma 2. *Suppose that an implicit system Σ_I of (6) satisfying assumptions A1 and A2 is given. Then there exist nonsingular matrices U and V such that the following relation holds:*

$$UEV = \begin{array}{c}\left[\begin{array}{cccc} I & 0 & 0 & 0 \\ 0 & 0 & 0 & 0 \end{array}\right] \\ \underbrace{\quad}_{\lambda}\underbrace{\quad}_{m-\lambda}\underbrace{\quad}_{\alpha}\underbrace{\quad}_{\beta}\end{array}\begin{array}{l} \}\ \lambda \\ \}\ m-\lambda \\ \end{array}, \qquad UFV = \begin{array}{c}\left[\begin{array}{cccc} A & B_0 & B & 0 \\ C & I & 0 & 0 \end{array}\right] \\ \underbrace{\quad}_{\lambda}\underbrace{\quad}_{m-\lambda}\underbrace{\quad}_{\alpha}\underbrace{\quad}_{\beta}\end{array}\begin{array}{l} \}\ \lambda \\ \}\ m-\lambda \\ \end{array}.$$

where $\lambda := \text{rank}E$, $\alpha := \text{rank}B$, and $\beta := n - m - \text{rank}B$. Furthermore, by denoting

$$\left[\ x^T(k)\ \ w^T(k)\ \ u^T(k)\ \ u_e^T(k)\ \right]^T := V^{-1}\xi(k)$$

i.e., $x := T_x\xi$, $w := T_w\xi$, $u := T_u\xi$, $u_e := T_{u_e}\xi$ and $\left[\ T_x^T\ \ T_w^T\ \ T_u^T\ \ T_{u_e}^T\ \right]^T := V^{-1}$. the implicit system (6) can be expressed as

$$\Sigma_w : \begin{cases} x(k+1) = Ax(k) + B_0w(k) + Bu(k), \quad Cx(k) + w(k) = 0, \\ x(k) \in T_x\{0,1\}^n, \quad w(k) \in T_w\{0,1\}^n, \quad u(k) \in T_u\{0,1\}^n, \\ x(0) = x_0 \in T_x\Xi_0, \quad w(0) = w_0 \in T_w\Xi_0, \quad u(0) = u_0 \in T_u\Xi_0 \end{cases} \tag{11}$$

where A, B_0, B, C are some matrices.

The proof will be omitted due to the limited space. From Lemma 2, we see that there exists a $\mathcal{T}$-equivalence transformation for every Σ_I of (6) satisfying assumptions A1 and A2. It is also remarked that by the $\mathcal{T}$-equivalence transformation, the vector ξ of the implicit system Σ_I is decomposed into the state $x(k)$, the dependant variable $w(k)$, the independent variable $u(k)$, and the redundant variable $u_e(k)$.

Since we are interested in the number of the *independent binary* variables in this paper, we will further restrict the class of the $\mathcal{T}$-equivalence transformations as follows. If a $\mathcal{T}$-equivalence transformation transforms the system Σ_I of (6) into the system Σ_w of (11) with $u(k) \in \{0,1\}^\alpha$, $u(0) = u_0 \in \{0,1\}^\alpha$, then it is called here the *$\mathcal{T}_\mathcal{B}$-equivalence transformation*, and let $\Sigma_\mathcal{B}$ denote the $\mathcal{T}_\mathcal{B}$-equivalence transformed system. Note that the binary property for the free (input) variable u is preserved in this system.

Furthermore, we will consider to rewrite $x(k) \in T_x\{0,1\}^n$, $w(k) \in T_w\{0,1\}^n$ into some inequality conditions, to transform $\Sigma_\mathcal{B}$ finally into Σ_{target}. In this case, note that, under such a transformation, the number of the binary free variables is greater than or equal to α in $\Sigma_\mathcal{B}$.

Then the minimal representation of a finite automaton is defined as follows.

Definition 2. *Suppose that an implicit system Σ_I of (6) satisfying assumptions A1 and A2, which expresses a finite automata ($M \le L$), is given. Then if among all $\mathcal{T}_\mathcal{B}$-transformations of Σ_I, the system Σ_{target} generated throughout $\Sigma_\mathcal{B}$ has the binary input vector $u(k)$ with minimal dimension, it is called a minimal representation of Σ_I, denoted by $\Sigma_{\text{target}}^{\min}$. In addition, the corresponding minimal dimension of the input vector $u(k)$ is denoted by α^*.*

It is remarked that the concept of the above minimal representation is different from that of the standard minimal realization of linear state equations at the point that the minimality of the dimension of the input variable is considered.

Finally, the problem to be addressed in the next sections is given as follows.

Problem 2. Suppose that an implicit system Σ_I of (6) satisfying assumptions A1 and A2, which expresses a finite automata ($M \le L$), is given. Then find a minimal representation $\Sigma_{\text{target}}^{\min}$ of Σ_I.

4 Derivation of Minimal Representation

The story of solving Problem 2 is as follows. First, a characterization of $\mathcal{T}$-equivalence transformations is given. Next, a procedure for deriving a system model Σ_{target} expressing a finite automaton is proposed, and several properties on its system model are proven. Finally, based on the above results, it is proven that the system model Σ_{target} derived by the proposed procedure is a minimal representation of Σ_I.

4.1 Characterization of $\mathcal{T}$-Equivalence Transformation

In this subsection, we characterize a class of $\mathcal{T}$-equivalence transformations preserving the matrix form in (11).

Lemma 3. *Suppose that matrices* $A_0 \in \mathbf{R}^{n_1 \times n_1}$*,* $B_0 \in \mathbf{R}^{n_1 \times m_1}$*,* $B \in \mathbf{R}^{n_1 \times m_2}$ *and* $C \in \mathbf{R}^{n_2 \times n_1}$ *is given. Then for nonsingular matrices* $U \in \mathbf{R}^{n \times n}$*,* $V \in \mathbf{R}^{m \times m}$*, where* $n = n_1 + n_2$*,* $m = n_1 + m_1 + m_2$*, there exists matrices* $\bar{A}_0$*,* $\bar{B}_0$*,* $\bar{B}$ *and* $\bar{C}$ *such that*

$$U \begin{bmatrix} I & 0 & 0 \\ 0 & 0 & 0 \end{bmatrix} V = \begin{bmatrix} I & 0 & 0 \\ 0 & 0 & 0 \end{bmatrix}, \quad U \begin{bmatrix} A_0 & B_0 & B \\ C & I & 0 \end{bmatrix} V = \begin{bmatrix} \bar{A}_0 & \bar{B}_0 & \bar{B} \\ \bar{C} & I & 0 \end{bmatrix} \tag{12}$$

hold if and only if $V_{11} = U_{11}^{-1}$*,* $V_{22} = U_{22}^{-1}$*,* $U_{21} = 0$*,* $V_{12} = 0$*,* $V_{13} = 0$*,* $V_{23} = 0$ *hold, and* V_{33} *is a nonsingular matrix, where*

$$U = \begin{bmatrix} \underbrace{U_{11}}_{n_1} & \underbrace{U_{12}}_{n_2} \\ U_{21} & U_{22} \end{bmatrix} \begin{matrix} \} \, n_1 \\ \} \, n_2 \end{matrix}, \quad V = \begin{bmatrix} \underbrace{V_{11}}_{n_1} & \underbrace{V_{12}}_{m_1} & \underbrace{V_{13}}_{m_2} \\ V_{21} & V_{22} & V_{23} \\ V_{31} & V_{32} & V_{33} \end{bmatrix} \begin{matrix} \} \, n_1 \\ \} \, m_1 \\ \} \, m_2 \end{matrix}. \tag{13}$$

Furthermore, coefficient matrices of (12) are given by $\bar{A}_0 = \Theta U_{11}^{-1} + (U_{11}B_0 + U_{12})V_{21} + U_{11}BV_{31}$*,* $\bar{B}_0 = \Theta U_{22}^{-1} + U_{11}BV_{32}$*,* $\bar{B} = U_{11}BV_{33}$*,* $\bar{C}_0 = U_{22}(C_0U_{11}^{-1} + V_{21})$*,* $\bar{a}_1 = U_{11}a_1 + U_{12}a_2$ *and* $\bar{a}_2 = U_{22}a_2$*, where* $\Theta = U_{11}A_0 + U_{12}C_0$*.*

Proof. By substituting (13) into (12) and then comparing both hand sides, the proof is completed. □

Lemma 3 gives a characterization of every $\mathcal{T}$-equivalence transformation that preserves (12). From the conditions that $\bar{B} = U_{11}BV_{33}$ and that U_{11} and V_{33} are nonsingular, we have $\mathrm{rank}B = \mathrm{rank}\bar{B}$. So we can conclude that the dimension of independent (input) vector u is invariant for all $\mathcal{T}_\mathcal{B}$-equivalence transformations. We thus have $\alpha^* = \mathrm{rank}B$. In the next subsection, hence, we will give one of $\mathcal{T}_\mathcal{B}$-equivalence transformations in an explicit way.

4.2 Derivation Procedure of Minimal Representation

This subsection proposes a procedure for deriving a minimal representation $\Sigma_{\mathrm{target}}^{\mathrm{min}}$ of Σ_I. The procedure is a more sophisticated version of our approach [5,7], which includes new lemmas.

Procedure for deriving a Σ_{target}
Step 1: For a given finite automaton ($M \le L$), let $\mathcal{I}_a = \{i_1, \ i_2, \ \cdots, \ i_L\}$ denote the set of combinations of (i, j) such that the arc from node i to node j exists, and assign a binary variable δ_{i_k} to the arc i_k. Furthermore, set $\xi(k) := [\ \delta_{i_1}(k) \ \ \delta_{i_2}(k) \ \ \cdots \ \ \delta_{i_L}(k)\ ]^T \in \{0,1\}^L$. Then the input-output relation of $\delta_{i_k}(k)$ on each node gives the implicit system of

$$\begin{cases} E\xi(k+1) = F\xi(k), \ \ \xi(k) \in \{0,1\}^L, \\ \xi(0) = \xi_0 \in \Xi_0 := \{\ \eta \in \{0,1\}^L \mid e_L^T\eta = 1\ \} \end{cases} \tag{14}$$

where $E, \ F \in \{0,1\}^{M \times L}$.

Step 2: Derive a permutation matrix P satisfying $EP = [\ I_M \quad \tilde{E}\ ]$, where $\tilde{E} \in \{0,1\}^{M\times(L-M)}$ is some matrix. Then by using $\hat{U} = I_{M+1}$ and

$$\hat{V} = P \begin{bmatrix} I_M & -\tilde{E} \\ 0_{(L-M)\times M} & I_{L-M} \end{bmatrix}, \tag{15}$$

we obtain $\hat{U}E\hat{V} = [\ I_M \quad 0_{M\times(L-M)}\ ]$ and $\hat{U}F\hat{V} = [\ \hat{A} \quad \hat{B}\ ]$ (This follows from the property of $\sum_{j=1}^{M} E^{(j)} = e_L^T$, where $E^{(j)}$ denotes the j-th row of E). Thus letting $[\ x^T(k) \quad \hat{u}^T(k)\ ]^T := \hat{V}^{-1}\xi(k)$, $\hat{V}^{-1} := [\ \hat{T}_x^T \quad \hat{T}_u^T\ ]^T$, the state equation is obtained as

$$\begin{cases} x(k+1) = \hat{A}x(k) + \hat{B}\hat{u}(k), \quad x(k) \in \hat{T}_x\{0,1\}^L, \quad \hat{u}(k) \in \hat{T}_u\{0,1\}^L, \\ x(0) = x_0 \in X_0 := \left\{\ \zeta \in \hat{T}_x\{0,1\}^M \mid e_M^T\zeta = 1\ \right\}, \\ \hat{u}(0) = \hat{u}_0 \in \hat{T}_u\{0,1\}^L. \end{cases} \tag{16}$$

Note here that in the proposed procedure, an initial-input set does not depend on Ξ_0 (see Lemma 5).
Step 3: Reduce the matrix $\hat{B}$ to

$$\hat{B} = S_B \begin{bmatrix} I_{\hat{\alpha}} & 0 \\ 0 & 0 \end{bmatrix} T_B \tag{17}$$

where $\hat{\alpha} := \mathrm{rank}\hat{B}$, and S_B, T_B are nonsingular matrices. Furthermore, by defining

$$\left[\ \tilde{u}^T(k) \quad \tilde{u}_e^T(k)\ \right] := \left[\ T_{B1}^T \quad T_{B2}^T\ \right]^T \hat{u}(k), \quad \left[\ T_{B1}^T \quad T_{B2}^T\ \right]^T := T_B$$

where $\tilde{u}(k) \in \{-1,0,+1\}^{\hat{\alpha}}$, $\tilde{u}_e(k) \in \{-1,0,+1\}^{\hat{\beta}}$, $\hat{\beta} := L - M - \hat{m}$ (see Lemma 6), and using $B = S_B\,[\ I_{\hat{\alpha}} \quad 0\ ]^T$, we obtain from (16)

$$\begin{cases} x(k+1) = \hat{A}x(k) + B\tilde{u}(k), \\ x(k) \in \hat{T}_x\{0,1\}^L, \quad \tilde{u}(k) \in T_{B1}\hat{T}_u\{0,1\}^L, \\ x(0) = x_0 \in \hat{T}_x\Xi_0, \quad \tilde{u}(0) = \tilde{u}_0 \in T_{B1}\hat{T}_u\{0,1\}^L. \end{cases} \tag{18}$$

Step 4: If input variables $\tilde{u}(k)$ are three-valued variables, which consist of elements -1, 0 and $+1$, then transform $\tilde{u}(k)$ into binary variables by using the relation between state/input variables and ξ as follows. First, let $\tilde{u}^{(s)}(k)$, $s = 1,\ 2,\ \cdots,\ \hat{\alpha}$ denote each element of $\tilde{u}$. Then $\tilde{u}^{(s)}(k)$ is expressed as

$$\tilde{u}^{(s)}(k) = \sum_{i\in\mathcal{I}} \delta_i(k) - \sum_{j\in\mathcal{J}} \delta_j(k) \in \{-1,0,+1\} \tag{19}$$

where $\mathcal{I},\ \mathcal{J} \in \mathcal{I}_a$ and $\mathcal{I} \bigcap \mathcal{J} = \emptyset$. Next, elements $x^{(p)}(k)$, $p = 1,\ 2,\ \cdots,\ M$, of the state variable x are expressed as $x^{(p)}(k) = \sum_{l\in\mathcal{L}_p} \delta_l(k)$, where $\mathcal{L}_p$ is the index set of nodes that have an arc from the node p (see the proof of Lemma 7).

Applying $x^{(p)}(k) = \sum_{l \in \mathcal{L}_p} \delta_l(k)$ to $\sum_{j \in \mathcal{J}} \delta_j(k)$, we obtain

$$\begin{aligned} \tilde{u}^{(k)}(k) &= \sum_{i \in \mathcal{I}} \delta_i(k) - \left(\sum_{q \in \mathcal{Q}} x^{(q)}(k) - \sum_{r \in \mathcal{R}} \delta_r(k) \right) \\ &= -\sum_{q \in \mathcal{Q}} x^{(q)}(k) + \left(\sum_{i \in \mathcal{I}} \delta_i(k) + \sum_{r \in \mathcal{R}} \delta_r(k) \right) \end{aligned} \tag{20}$$

where $\mathcal{Q}$ is an appropriate integer set, $\mathcal{R} \subseteq \mathcal{I}_a$, and $u^{(s)}(k) := \sum_{i \in \mathcal{I}} \delta_i(k) + \sum_{r \in \mathcal{R}} \delta_r(k)$ of (20) are binary variables. By executing the above operation for $s = 1, 2, \cdots, \hat{m}$, we obtain

$$\tilde{u}(k) = \hat{A}_u x(k) + u(k) \tag{21}$$

where $u(k) \in \{0,1\}^{\hat{\alpha}}$, and $\hat{A}_u$ is some matrix (see Lemma 7).
Step 5: In (18), replace $x(k) \in \hat{T}_x\{0,1\}^L$ by $x(k) \in \{0,1\}^M$ (see Lemma 4), $x(k) \in \{0,1\}^M$ by $x(k) \in \mathbf{R}^M$ and $0_{M\times 1} \le x(k) \le e_M$ (see Lemma 8). Then from Step 4 we obtain

$$\begin{cases} x(k+1) = Ax(k) + Bu(k), \ \ 0_{M\times 1} \le x(k) \le e_M, \\ x(k) \in \mathbf{R}^M, \ \ u(k) \in \{0,1\}^{\hat{\alpha}}, \\ x(0) = x_0 \in X_0, \ \ u(0) = u_0 \in \{0,1\}^{\hat{\alpha}} \end{cases} \tag{22}$$

where $A := \hat{A} + B\hat{A}_u$.

Note here that the computation cost of the proposed procedure is very small, since there does not exist iteration in all steps of the proposed procedure.

Next, several facts that are given in the procedures will be proven.

First, it is proven that a linear state equation obtained in Step 2 has binary state and input variables.

Lemma 4. *Consider a linear state equation (18) obtained in Step 2. Then the state variable $x(k)$ and the input variable $\hat{u}(k)$ are binary variables.*

Proof. From (15) in Step 2, the following relation holds:

$$\begin{aligned} \hat{V}^{-1}\xi(k) &= \begin{bmatrix} I_M & \tilde{E} \\ 0_{(L-M)\times M} & I_{L-M} \end{bmatrix} P^{-1}\xi(k) \\ &= \begin{bmatrix} E \\ \left[0_{(L-M)\times M} \ I_{L-M} \right] P^{-1} \end{bmatrix} \xi(k) = \begin{bmatrix} x(k) \\ \hat{u}(k) \end{bmatrix}. \end{aligned}$$

The input variable $\hat{u}(k)$ is the binary variable, because P^{-1} is also the permutation matrix. Furthermore, from $E \in \{0,1\}^{M\times L}$ and $e_L^T\xi(k) = 1$ (see the proof of Lemma 1), the state variable $x(k) = E\xi(k)$ is a binary variable. □

Next, it is shown that in Step 2 an initial-input set does not depend on Ξ_0.

Lemma 5. *$\hat{V}^{-1}\Xi_0$ transformed from Ξ_0 is equivalent to a combination of X_0 of (16) and $\hat{u}(0) = \hat{u}_0 \in \hat{T}_u\{0,1\}^L$.*

Proof. Noting here that $e_L^T\hat{V} = [\ e_M^T \quad 0_{1\times(L-M)}\ ]$ holds, the following relation holds:

$$\hat{V}^{-1}\Xi_0 = \left\{\ \hat{V}^{-1}\eta \in \hat{V}^{-1}\{0,1\}^L \mid e_L^T\hat{V}\hat{V}^{-1}\eta = 1\ \right\}$$
$$= \left\{\ [\ \zeta^T \quad \hat{u}_0^T\ ]^T \in \hat{V}^{-1}\{0,1\}^L \mid e_M^T\zeta = 1\ \right\}.$$

Therefore, by using $\hat{V}$ of (15), the equality constraint $e_L^T\eta = 1$ is transformed into the initial-state constraint $e_M^T\zeta = 1$, which does not depend on $\hat{u}_0$. □

Furthermore, it is shown that $T_B\hat{u}(k)$ in Step 3 is the three-valued variable.

Lemma 6. *$T_B\hat{u}(k)$ $(= [\ \tilde{u}(k)^T \quad \tilde{u}_e(k)^T\ ]^T)$ in Step 3 is the three-valued variable, which consists of elements -1, 0 and $+1$.*

Proof. Consider the matrix pencil $-F + zE$. From Step 1, in the case of $z = 1$, the matrix $-F + E$ is an incidence matrix. So in the case of $z = 1$, the matrix pencil $[\ zI - \hat{A} \quad -\hat{B}\ ]$ obtained via the transformation of Step 2, is also the incidence matrix of some directed graph, which one-to-one corresponds to a given finite automaton, because $x(k)$ and $\hat{u}(k)$ are binary variables. Then the matrix $-\hat{B}$ does not depend on $z \in \mathbf{C}$, and has the structure of an incidence matrix, i.e., elements $\hat{B}^{(i,j)}$, $j = 1, 2, \cdots, L - M$ of each column of $\hat{B} \in \{-1, 0, +1\}^{M\times(L-M)}$ are $\hat{B}^{(\hat{I}_j,j)} = -1$, $\hat{B}^{(\hat{J}_j,j)} = 1$ and $\hat{B}^{(i,j)} = 0$, $i \neq \hat{I}_j$, $\hat{J}_j$, where $\hat{I}_j$, $\hat{J}_j$ are some nonnegative integers. From the structure of $\hat{B}$, (17) can be derived by using the addition and the subtraction of column vectors or row vector, and appropriate permutation. Here $S_B \in \{-1, 0, +1\}^{M\times M}$, $T_B \in \{-1, 0, +1\}^{(L-M)\times(L-M)}$ hold. Finally, noting that $e_{L-M}^T\hat{u}(k) \leq 1$ holds from $e_L^T\xi(k) = 1$ and $\hat{u}(k) = [\ 0 \quad I\ ]P^{-1}\xi(k) \in \{0,1\}^{L-M}$, we obtain $T_B\hat{u}(k) \in \{-1, 0, +1\}^{L-M}$. □

The following proves that $\tilde{u}(k) \in \{-1, 0, +1\}^{\hat{m}}$ can be transformed into the binary variable.

Lemma 7. *Using some transformation of coordinate, $\tilde{u}(k) \in \{-1, 0, +1\}^{\hat{m}}$ can be expressed as (21).*

Proof. From Lemma 6, $\tilde{u}^{(s)}(k)$ is expressed as (19). Using $x(k) = E\xi(k)$, we obtain $x^{(p)}(k) = \sum_{l\in\mathcal{L}_p}\delta_l(k)$. Applying $x^{(p)}(k) = \sum_{l\in\mathcal{L}_p}\delta_l(k)$ to $\sum_{j\in\mathcal{J}}\delta_j(k)$, we obtain (20). Using $\mathcal{I}\cap\mathcal{J} = \phi$ and the structure of $\hat{B}$, we can show that $u^{(s)}(k) := \sum_{i\in\mathcal{I}}\delta_i(k) + \sum_{r\in\mathcal{R}}\delta_r(k)$ of (20) are binary variables. By executing the above operation for $s = 1, 2, \cdots, \hat{m}$, we obtain (21). □

Step 5 is proven as follows. First, by Step 4 and Lemma 4, (18) can be rewritten as

$$\begin{cases} x(k+1) = Ax(k) + Bu(k), \\ x(k) \in \{0,1\}^M, \quad u(k) \in \{0,1\}^{\hat{\alpha}}, \\ x(0) = x_0 \in X_0, \quad u(0) = u_0 \in \{0,1\}^{\hat{\alpha}}. \end{cases} \tag{23}$$

Next, it is proven that $x(k) \in \{0,1\}^M$ can be replaced by $x(k) \in \mathbf{R}^M$, $0_{M\times 1} \leq x(k) \leq e_M$, and $x(0) = x_0 \in X_0$, using following lemma.

Lemma 8. *Suppose that α_i, $i = 1, 2, \cdots, d$, are intergers. Then the following statements are equivalent:*

$$\text{(i)} \quad \delta_0 = \sum_{i=1}^{d} \alpha_i \delta_i, \quad \delta_i \in \{0,1\}, \ i \in \{0, \ 1, \ \cdots, \ d\},$$

$$\text{(ii)} \quad \delta_0 = \sum_{i=1}^{d} \alpha_i \delta_i, \quad \delta_i \in \{0,1\}, \ i \in \{1, \ 2, \ \cdots, \ d\}, \quad 0 \leq \delta_0 \leq 1.$$

Proof. Noting that each δ_i is the binary variable, $\delta_0 \in \{0,1\}$ follows from the inequality constraint $0 \leq \delta_\gamma \leq 1$. □

In Step 5, δ_0 corresponds to each element of $x(k+1)$, and δ_i corresponds to each element of $x(k)$, $u(k)$. Therefore, by applying Lemma 8 to each row of the state equation in (23), we conclude that, under $u(k) \in \{0,1\}^{\hat{\alpha}}$, $x(k) \in \{0,1\}^M$ is equivalent to $x(k) \in \mathbf{R}^M$, $0_{M\times 1} \leq x(k) \leq e_M$, and $x(0) = x_0 \in X_0$. From these results, (22) holds.

Finally, we arrive at the following main theorem.

Theorem 1. *Suppose that an implicit system Σ_I of (6) satisfying assumptions A1 and A2, which expresses a finite automaton ($M \leq L$), is given. Then a system model (22) derived by the proposed procedure is a minimal representation of Σ_I, i.e., $\Sigma_{\text{target}}^{\min}$.*

Proof. By Lemma 3, we only have to prove that Step 2 $\sim$ Step 4 of the proposed procedure corresponds to a $\mathcal{T}_\mathcal{B}$-equivalence transformation. The transformation in Step 2 is the $\mathcal{T}$-equivalence transformation using $\hat{U} = I_M$ and $\hat{V}$ of (15). The transformation in Step 3 is the $\mathcal{T}$-equivalence transformation from (16) to the state equation

$$x(k+1) = \hat{A}x(k) + \begin{bmatrix} B & 0_{M\times\hat{\beta}} \end{bmatrix} \begin{bmatrix} \tilde{u}(k) \\ \tilde{u}_e(k) \end{bmatrix} \tag{24}$$

which includes the redundant variable $\tilde{u}_e(k)$, by using

$$U_3 := I_M, \quad V_3 := \begin{bmatrix} I_M & 0 \\ 0 & T_B^{-1} \end{bmatrix}.$$

The transformation of (21) in Step 4 is the $\mathcal{T}$-equivalence transformation for (24), by using

$$U_4 := I_M, \quad V_4 := \begin{bmatrix} I_M & 0 & 0 \\ \hat{A}_u & I_{\hat{\alpha}} & 0 \\ 0 & 0 & W \end{bmatrix}$$

where $W \in \mathbf{R}^{\hat{\beta}\times\hat{\beta}}$ is some nonsingular matrix. Furthermore, since $u(k)$ is the binary variables, it is also the $\mathcal{T}_\mathcal{B}$-equivalence transformation. So the whole transformation in Step 2 $\sim$ Step 4 of the proposed procedure corresponds to a $\mathcal{T}_{\mathcal{B}}$-equivalence transformation, and we obtain $\Sigma_{\mathcal{B}}$. Therefore, Step 5 produces Σ_{target}, whose dimension of the binary input vector is the same as that of $\Sigma_{\mathcal{B}}$. This implies that it is $\Sigma_{\mathrm{target}}^{\mathrm{min}}$. □

5 Conclusions

This paper has proven that for a finite automaton expressed by an implicit system model, our method gives a minimal representation in the sense that it has the smallest number of binary input variables over all linear equivalence transformations on **R** that preserve the input binary property. What we would like to stress in this paper is that the obtained result provides a *systematic* modeling method of finite automata for control of hybrid/discrete systems, which can *guarantee* the minimum number of the binary variables representing finite automata in the above sense. As far as we know, no such results have been obtained so far. Also, note that the obtained minimal representation may be also useful for automata theory. However, there are many open problems in this line. In this paper, we restrict the class of transformations. So one of the significant future topics is to find a modeling method for realizing the minimality of the binary variables in a more rigorous way.

References

1. J. D. Aplevich. *Implicit Linear Systems. Lecture Notes in Control and Information Sciences*, 152, Springer-Verlag, 1991.
2. A. Bemporad and M. Morari. Control of systems integrating logic, dynamics, and constraints. *Automatica*, Vol. 35, pp. 407–427, 1999.
3. N. Biggs. *Algebraic Graph Theory.* Cambridge University Press, 1973.
4. M. Ishikawa, T. Seki, J. Imura and S. Hara. An efficient algorithm for optimal control of hybrid dynamical systems utilizing mode transition rule. *Proc. of the 43rd IEEE Conf. Decision and Control*, pp. 1866–1871, 2004.
5. K. Kobayashi and J. Imura. Modeling of Discrete Dynamics for Computational Time Reduction of Model Predictive Control. *Proc. of the 17th Intl. Symp. on Mathematical Theory of Networks and Systems*, pp. 628–633, 2006.
6. K. Kobayashi and J. Imura. Efficient Modeling of Piecewise Affine Systems in Model Predictive Control. *Proc. of the SICE-ICASE International Joint Conference*, pp. 2041–2044, 2006.
7. K. Kobayashi and J. Imura. Minimal Representation of Finite Automata for Hybrid Systems Control. *Proc. of the 45th IEEE Conf. Decision and Control*, pp. 930–935, 2006.
8. M. Kuijper. *First-order Representations of Linear Systems.* Birkhäuser, 1994.
9. R. Lidl and H. Niederreiter. *Introduction to finite fields and their applications.* Cambridge University Press, 1994.
10. K. Takaba. Linear Quadratic Optimal Control for Linear Implicit System. *Proc. of the 38th IEEE Conf. Decision and Control*, pp. 4074–4079, 1999.
11. H. P. Williams. *Model building in mathematical programming.* 4th ed., John Willey & Sons, Ltd, 1999.

Hybrid Control and Verification of a Pulsed Welding Process

Jesper A. Larsen, Rafael Wisniewski, and Roozbeh Izadi-Zamanabadi

Aalborg University, Department of Electronic Systems, Fredrik Bajers Vej 7C
9220 Aalborg Ø, Denmark
{jal,raf,riz}@es.aau.dk
http://www.control.aau.dk/hybrid

Abstract. Currently systems, which are desired to control, are becoming more and more complex and classical control theory objectives, such as stability or sensitivity, are often not sufficient to cover the control objectives of the systems.

In this paper it is shown how the dynamics of a pulsed welding process can be reformulated into a timed automaton hybrid setting and subsequently properties such as reachability and deadlock absence is verified by the simulation and verification tool UPPAAL.

1 Introduction

The lack of analytical methods for design of hybrid control systems can often result in excessive testing and validation, which is time consuming and even then might not guarantee that the system will meet the control objectives under all operating conditions. To overcome the design and implementation problems which may result from the deficient use of an analytical approach, a notation of hybrid automaton has been introduced in [1].

Most algorithmic verification and synthesis tools for hybrid systems today are limited to systems exhibiting simple continuous dynamics, such as piecewise-affine hybrid systems[2,3] or timed automata[4,5,6]. One of the main objective in [7] was to enlarge this class of systems to all linear controllable systems, which is continued in this paper by showing how this theory apply in practice to the Gas Metal Arc Welding (GMAW) process. By restricting the observations for the system to a finite set of partitions, enables a bisimulation of the system to be modeled using simple timed automata.

With a bisimilar model of the system built with the use of timed automata, it is possible to use a verification tools such as UPPAAL to simulate and verify different system properties. Especially questions such as reachability, liveliness and possibilities of deadlocks are new questions, which are of great interest to the designer of the supervisory system and which previously needed to be guessed by simulations or ad-hoc methods.

1.1 Gas Metal Arc Welding

In the GMAW process the electrode is consumable and is fed continuously at a certain rate by the pistol to the welding pool. The weld is protected from

A. Bemporad, A. Bicchi, and G. Buttazzo (Eds.): HSCC 2007, LNCS 4416, pp. 357–370, 2007.

the surrounding air by a gas which is also fed by the pistol. Normally argon or argon/CO_2 is used as shielding gas. The current between the workpiece (cathode) and the welding pistol (anode) causes an arc and an electromagnetic field. The strong current makes the electrode melt and drop into the welding pool.

The GMAW process can be divided into three modes; short arc mode, spray mode and a mixed mode of the two, of which only the spray mode will be considered in this paper. In spray mode the electrode should never touch the workpiece in order to obtain the best weld quality.

The melting process can be described by two contributions, *anode heating* and *ohmic heating*. When the current rises, the temperature of the arc rises and the tip of the electrode is heated. The energy from the arc, which contributes to melting the electrode, is known as *anode heating*. The second contribution to the melting process is the *ohmic heating*, which is the heat energy resulting from the ohmic resistance in the electrode. The high current also creates a higher electromagnetic field which contributes, together with the gravitational force, to detachment of the drop.

While the tip is melting, a liquid drop of metal is formed. This drop is detached from the tip of the electrode when the surface tension on the drop, is too small to resist the gravitational- and the electromagnetic forces. Also the aerodynamic drag force from the shielding gas, contributes to the detachment of the drop. After detachment, a small liquid drop is left at the tip of the electrode and the process repeats.

A submode of the spray mode is the pulsed GMAW method, which is similar to spray mode, but in addition to the steady current between the cathode and the anode, current pulses causes the drop to detach in intervals. The advantage of using pulsed GMAW is a lower heat development in the weld pool. Furthermore the current pulses makes it possible to control the drop detachment [8].

1.2 Weld Quality Criteria

As described in the introduction, one of the objectives of this paper is to integrate a control structure for the GMAW process into a hybrid framework. The nature of the GMAW process makes classic control theory specifications, such as stability, inadequate. Instead control objectives focusing on obtaining the best weld quality is desirable. The quality of a weld depends on several factors, which will be discussed in the following.

Basically a high-quality weld is characterized by a good penetration, which is essential for a strong weld, as it allows a larger area of the workpiece edges to join.

A good penetration is a necessary, but not a sufficient, condition for a good weld. If the work piece becomes too hot and cools down too quickly, the material can loose some of its characterizing properties, e.g. heat-treated metals or metal alloys, such as stainless steel, can loose its characterizing properties. [9, ch. 5] The facts described in the latter are related to the weld pool and are the minimum criteria, which must be fulfilled to obtain a high-quality weld and is defined as *direct weld quality influencing factors*. More indirectly an additional number of factors influences the quality of a weld. The following quality influencing factors will be referred to as *indirect quality influencing factors*. Specific for pulsed GMAW welding, the quality of the weld is influenced by the control of the drop detachment, meaning that the current pulses should ideally detach one drop per pulse to obtain the best weld possible. It is also desirable to obtain a uniform drop size, in order to achieve a homogeneous weld. An additional control objective is to keep a short arc length, since it is easier for the operator to work with. Moreover the energy input into the workpiece should be minimized.

The indirect quality influencing factors are related to the control of the electrode and the arc.

1.3 Delimitation of Control Tasks

As described in the latter the control can be separated into control of the weld pool, and control of the arc and electrode. As it is only hand held welding which will be the focus on this paper, the weld pool control is handled by the operator. Figure 1 describes the control structure [8]. The outer control loop is handled by the operator and the inner loop is handled by the welding machine. The rest of

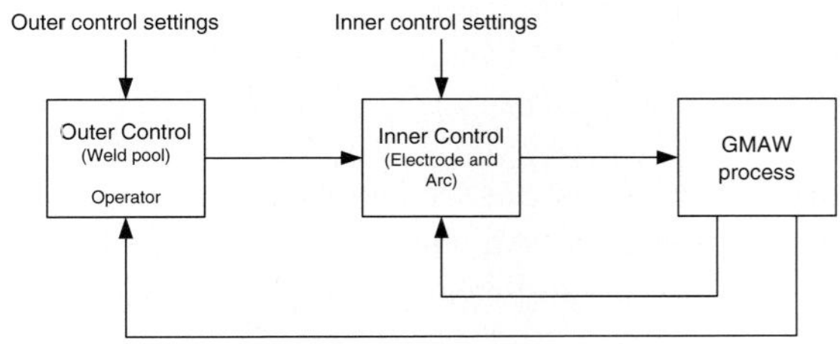

Fig. 1. Control structure for the GMAW process

this paper will concentrate on controlling the indirect quality influencing factors in the inner loop.

2 System Dynamics

The pulsed GMAW process is governed by the pulsing current, which is seen in figure 2(a). In order to control the pulsing, the base period, which is the time interval in which the electrode is melted, is variable, thus it becomes possible to control the amount of melt detached in each pulse. If the arc length between the work piece and the electrode becomes too big or too small, as shown in figure 2(b), then it is likewise possible to adjust the arc length, i.e. if the arc has become too small then by decreasing the base period, thus increasing the amount of electrode consumed the arc length will become longer. This is however done on the cost of a smaller drop size and is only possible within a small distance, the main part is still controlled manually.

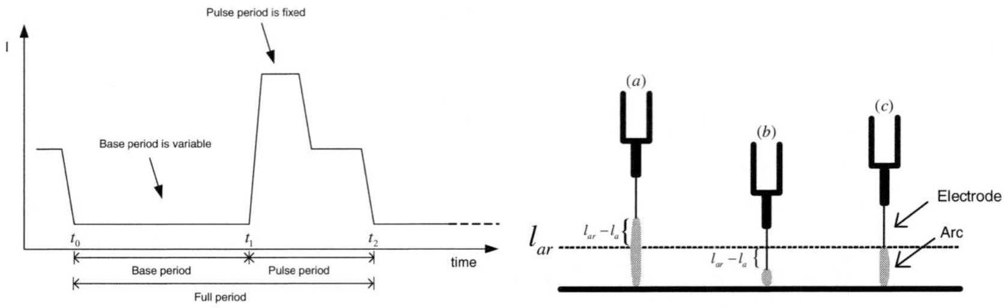

(a) Pulse by pulse method. (b) Arc length control scenarios.

Fig. 2. Left figure: The pulse by pulse method - the base current is fixed but the base period is variable. Right figure: The different arc length control scenarios - (a) the arc is too long (b) the arc is too short (c) the arc has the desired length l_{ar}.

The pulse condition can then be described as; a pulse should occur if the arc length is below the reference and the drop size is above minimum or if the arc length is longer than the reference and the drop size is bigger than the maximum, which can be written as:

$$\text{Pulse if:} \quad (l_a < l_{ar} \wedge x_{mb} \geq x_{mb\_min}) \bigvee (l_a > l_{ar} \wedge x_{mb} \geq x_{mb\_max}) \tag{1}$$

Where l_a is the arc length, l_{ar} is the arc length reference and x_{mb} is the current drop size with the indices *min* and *max* providing the bound on the desired drop size. The values of the bound can be regarded as weighting parameters for the controller design.

The weld process controller can thus be depicted as in figure 3, where an additional mode; *Short Circuit Handling*, is shown, which will not be discussed further in this paper. In the following the overall control strategy and dynamics of the underlying process will be presented. The model used in this paper, is derived in [8].

Drop Dynamics. The drop dynamics, i.e. the drop growth, can be expressed as the length of melted electrode, which is a function of the welding current and the electrode length:

$$x_m = \int_{t_0}^{t_1} v_m(I, l_s)dt \tag{2}$$

where v_m is the velocity of melted electrode given by

$$v_m = k_1 I + k_2 I^2 l_s \tag{3}$$

where $l_s = 0.0115$, $k_1 = 3.6733 \cdot 10^{-4}$ and $k_2 = 6.6463 \cdot 10^{-4}$ for the considered GMAW welding application.

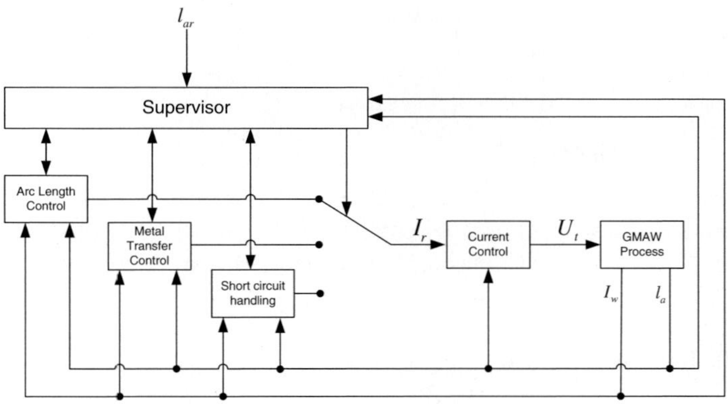

Fig. 3. Supervisory system for the GMAW process

Arc Length Dynamics. The governing equation for the arc length dynamics can be seen in (4).

$$\dot{l}_a = k_1 I + k_2 I^2 \cdot (l_c - l_a) - v_e \tag{4}$$

where k_1 and k_2 are constants, l_c is the length from the contact tip to the workpiece, l_a is the length of the arc ($l_s = l_c - l_a$) and v_e is the velocity of the electrode.

Equation (5) shows the current dynamics.

$$\dot{I} = -\frac{1}{\tau_i} I + \frac{1}{\tau_i} I_r \tag{5}$$

where $\tau_i = 66.7\mu s$ is a constant that characterizes the dynamics. I is the welding current and I_r is the current reference.

3 Hybrid System Modeling

The GMAW system, as described in the previous two sections, can be formulated as the following hybrid automaton using a commonly used formalism for hybrid systems, as presented in [1,10], with the dynamics in each state as described in the previous section. All transitions have a label, which is used for synchronization and a reset map, which in the "Drop detachment" case is the amount of melted wire which is set to $x_m := 0$, and in the "Pulse" and "Pulse done" case it is the current, which is set to the pulse and base current respectively.

As previously mentioned, the goal of this paper is to reformulate the hybrid system into a network of timed automata in order to expand the possibilities of verifying the system properties using an automated verification tool, such as UPPAAL[11]. This is essentially done because even though the system is exhibiting a nice and stable performance in each state, then it is possible by the right combination of switching to render the system unstable, of which a classical example can be seen in [12].

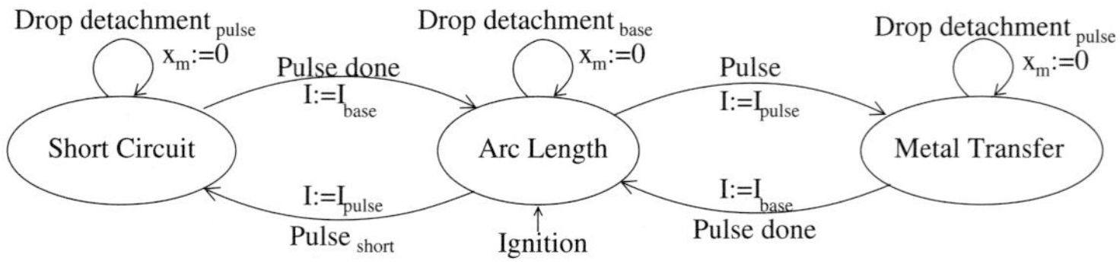

Fig. 4. Hybrid automaton for the controlled GMAW process, divided into the three control modes: Arc length, Metal transfer and Short circuit

3.1 Shift Register Form

In order to rewrite the dynamics of the system into shift register form, it is first put on Brunovsky normal form[7], for which a controllable linearized form of the system is needed.

To linearize equation (4) it is first rewritten into an operating point, $[\bar{I}, \bar{l}_a]$, and deviations from this, $[\hat{I}(t), \hat{l}_a(t)]$, where the cubed current term is split into a varying part and an operating mode part given by I_{op}:

$$\dot{l}_a = k_1\left(\bar{I} + \hat{I}(t)\right) + k_2\left(I_{op}\cdot\left(\bar{I} + \hat{I}(t)\right)\cdot\left(l_c - \left(\bar{l}_a + \hat{l}_a(t)\right)\right)\right) \tag{6}$$

The linearization is done around the point where v_e is equal to v_m thus v_e can be omitted from this equation. After multiplying (6) out and neglecting the product of time varying terms, an expression of the constant terms can be found as

$$k_1\bar{I} + k_2\cdot\left(I_{op}\cdot\left(\bar{I}\cdot\left(l_c - \bar{l}_a\right)\right)\right) \tag{7}$$

which, subtracted from (6), gives

$$\dot{l}_a = \hat{I}(t)\cdot\left(k_1 + k_2 I_{op}\bar{I}\cdot\left(l_c - \bar{l}_a\right)\right) - \hat{l}_a(t)\left(k_2 I_{op}\bar{I}\right) \tag{8}$$

The linearized system can now be written in state space form as

$$\begin{bmatrix}\dot{l}_a\\ \dot{I}\end{bmatrix} = \begin{bmatrix}-k_2 I_{op}\bar{I} & k_1 + k_2 I_{op}\bar{I}(l_c - l_a)\\ 0 & -\frac{1}{\tau_i}\end{bmatrix}\begin{bmatrix}l_a\\ I\end{bmatrix} + \begin{bmatrix}0\\ \frac{1}{\tau_i}\end{bmatrix}u \tag{9}$$

Although there are several operating points for the system only a single point will be used throughout this paper. This is done in order to simplify the presentation and is sufficient to prove the concept of the method.

The operating point used for the system is: $I_{op} = \bar{I} = 175A$, $l_c = 15mm, \bar{l}_a = 3.5mm$, which inserted into (9) and discretized using ZOH and a time step of $0.1s$ gives (10).

$$\begin{bmatrix}\dot{l}_a\\ \dot{I}\end{bmatrix} = \begin{bmatrix}0.1306 & 0.0020\\ 0 & 0\end{bmatrix}\begin{bmatrix}l_a\\ I\end{bmatrix} + \begin{bmatrix}9.5612\\ 1.000\end{bmatrix}u \tag{10}$$

The controllability matrix for this system has full rank, which shows that the system is controllable, thus the condition for getting the system into Brunovsky normal form is satisfied.

Following the method described in [13] the system is transformed into the normal form shown in (11) through the state transformation, $x = Tz$:

$$\begin{aligned} z(t+1) &= T^{-1}ATz(t) + T^{-1}Bu(t) \Leftrightarrow \\ z(t+1) &= A_z z(t) + B_z u_z(t) \end{aligned} \tag{11}$$

where

$$A_z = \begin{bmatrix} 0 & 1 \\ \alpha_1 & \alpha_2 \end{bmatrix} = \begin{bmatrix} 0 & 1 \\ 0 & 0.3659 \end{bmatrix}, \; B_z = \begin{bmatrix} 0 \\ 1 \end{bmatrix}, \; T = \begin{bmatrix} 0.392 & 0.143 \\ -2.73 & 0.001 \end{bmatrix}$$

The final step for getting the system into shift register form is to get the bottom row of the A_z matrix to be zeros, which is done through the following feedback transformation

$$u_z = u + \alpha_1 z_1 + \alpha_2 z_2 = \begin{bmatrix} F \\ 1 \end{bmatrix}^T \begin{bmatrix} z_1 \\ z_2 \\ u \end{bmatrix} \tag{12}$$

which gives the final system on shift register form as

$$z(t+1) = \tilde{A}_z z(t) + \tilde{B}_z u_z(t) \tag{13}$$

where

$$\tilde{A}_z = \begin{bmatrix} 0 & 1 \\ 0 & 0 \end{bmatrix}, \; \tilde{B}_z = \begin{bmatrix} 0 \\ 1 \end{bmatrix}, \; u_z = \begin{bmatrix} F \\ 1 \end{bmatrix}^T \begin{bmatrix} z \\ u \end{bmatrix}, \; F = \begin{bmatrix} 0 \\ 0.366 \end{bmatrix}^T, \; T = \begin{bmatrix} 0.392 & 0.143 \\ -2.73 & 0 \end{bmatrix}$$

The principle of the above computations is shown in figure 5.

3.2 State Space Partitioning

A discrete state space $\mathbb{Z}^2$ of $\mathbb{R}^2$ is now introduced, as described in [7], in order to form the space in which the shift-register form system operates. The 3 domains in which the system operates is, with reference to figure 4, the Arc Length Control (q_1), the Metal Transfer Control (q_2) and the Short Circuit Control (q_3).

- Domains

$$\begin{aligned} Dom(q_1) &= \{(l_a, I) \in \mathbb{R}^2 | \quad 0 \leq l_a \leq 0.01 \quad \wedge \quad 40 \leq I \leq 60\} \\ Dom(q_2) &= \{(l_a, I) \in \mathbb{R}^2 | \quad 0 \leq l_a \leq 0.01 \quad \wedge \quad 290 \leq I \leq 310\} \\ Dom(q_3) &= \{(l_a, I) \in \mathbb{R}^2 | \quad 0 \leq l_a \leq 0.01 \quad \wedge \quad 290 \leq I \leq 310\} \end{aligned}$$

where l_a [m] and I [A]. The values are specified from the normal operation of a GMAW welding machine.

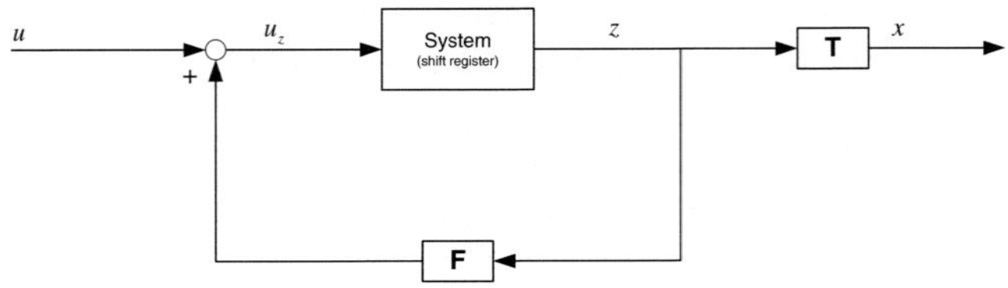

Fig. 5. Block diagram of the shift register transformation

These domains are then transformed into shift register form by $[z_1\ z_2]^T = \mathbf{T}\,[l_a\ I]^T$ which gives the new domains

$$
\begin{aligned}
Dom(q_1) &= \{(z_1, z_2) \in \mathbb{R}^2 |\, 0 \le z_1 \le 8596 \wedge 0 \le z_2 \le 0.0643\} \\
Dom(q_2) &= \{(z_1, z_2) \in \mathbb{R}^2 |\, 4.18 \cdot 10^4 \le z_1 \le 4.44 \cdot 10^4 \wedge 0.307 \le z_2 \le 0.328\} \\
Dom(q_3) &= \{(z_1, z_2) \in \mathbb{R}^2 |\, 4.18 \cdot 10^4 \le z_1 \le 4.44 \cdot 10^4 \wedge 0.307 \le z_2 \le 0.328\}
\end{aligned}
$$

The relation between the new domain space and the original one can be seen from figure 6, where the two regions of interests are marked, one being to the left in $I \in [40 - 60]$, which is the base period, and the region to the right, $I \in [290 - 310]$, being the pulse period. As it is seen from the figure then the domains of interest are no longer square. This deficiency is however remedied by relaxing the arc length constraint, which again makes the spaces of interest squares.

As described in [7] the partitioning needs to be equidistant, which would seem rather cumbersome for these domains due to the large ratio between z_1 and z_2, thus a scaling transformation is introduced, S_i, which transform each domains

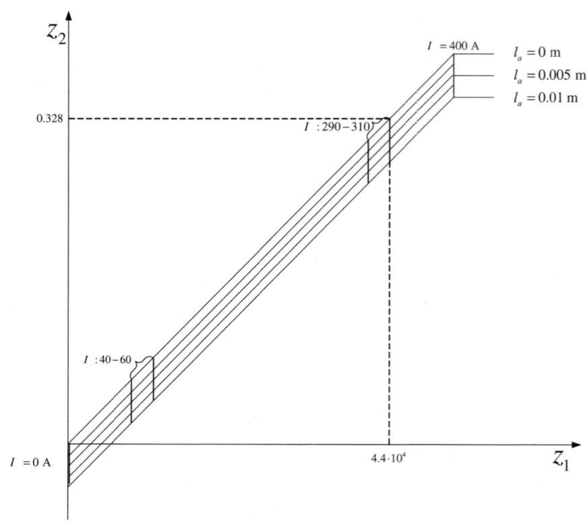

Fig. 6. Plot of state transformation: $[l_a\ I]^T \mapsto z$

into a sufficiently equiproportional domain. In this case it is only desirable to divide the spaces into a 3 by 3 grid to prove the concept, thus a transformation that scales the 3 domains into squares are used. Following this the drop forming dynamics is modelled as a 5 state timed automaton as shown in figure 7. The shifting time between the drop sizes dependents only on the current. Estimated shifting values for different current intervals are shown in table 1. The drop formation always starts in state 1 and will propagate through the states over time. State 3 is the reference state, i.e. the state in which it is desirable to do a drop detachment.

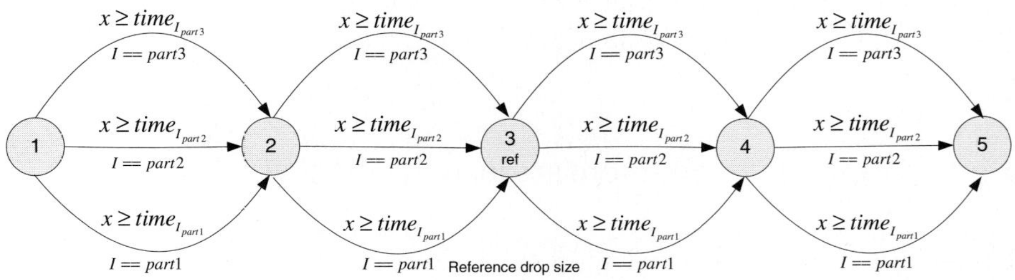

Fig. 7. The drop dynamics timed automaton. The automaton structure for the drop dynamics is the same in each domain q_1 and q_2.

Table 1. Estimated time between drop size partitions in $dom(q_1)$ to the left and $dom(q_2)$ and $dom(q_3)$ to the right

Current Partition	Current Interval [A]	Time Between Partitions [s]	Current Interval [A]	Time Between Partitions [s]
1	40.0 - 46.6	9.9 10^{-3}	290 - 296.6	3.9 10^{-4}
2	46.6 - 53.3	8.0 10^{-3}	296.6 - 303.3	3.7 10^{-4}
3	53.3 - 60.0	6.6 10^{-3}	303.3 - 310	3.5 10^{-4}

3.3 Control System Imposed on $\mathbb{Z}^2$

In section 3.2 a new state space $\mathbb{Z}^2$ was introduced. Utilizing that the system is in shift register form, insures a well defined controlled dynamics between the partition blocks. This means that under appropriate inputs the blocks will move into other partitions of equal division. To insure such *appropriate* inputs, a control law is needed.

The control law is constructed as described in [7] by starting with (11) and realizing that from a given position $(z_1, z_2){=}(p, q)$ the reachable set in one step is $(z_1, z_2){=}(q, r)$, where $r \in \mathbb{Z}$ is dependent on the input, thus it can be seen that the control law only has to ensure that z_2 will be within a control section of height δ, which is ensured by the control law:

$$u_z(k) = z_2(k) + \delta y(k) \ , \quad y \in \mathbb{Z} \tag{14}$$

which inserted into (11) results in the following system:

$$\begin{aligned} z_1(k+1) &= z_2(k) \\ z_2(k+1) &= z_2(k) + \delta y(k) \end{aligned} \tag{15}$$

which is not on shift register form any longer. This is however easily remedied by introducing the control law:

$$\epsilon(k) = z_2(k) + \delta y(k) \tag{16}$$

Which results in the system given by

$$\begin{bmatrix} z_{\varepsilon_1}(k+1) \\ z_{\varepsilon_2}(k+1) \end{bmatrix} = \begin{bmatrix} 0\,1 \\ 0\,0 \end{bmatrix} z(k) + \begin{bmatrix} 0 \\ 1 \end{bmatrix} \varepsilon(k) \tag{17}$$

4 Example of Implementation in UPPAAL

UPPAAL is a validation- and verification tool[11,14]. The tool consists of two main parts: A graphical user interface and a model checker engine. The idea in this paper is to model a system using timed automata, simulate it and then verify the system properties on it.

A system consists of a network of automata which are running in parallel. It is possible to step through the system, in order to check if the system behaves as intended and the system can be checked by the verifier to verify that it satisfies certain temporal specifications, such as if a certain state is reachable or if there is any deadlocks in the system. More generally speaking, the verifier can check all possible dynamical behaviors of a system[14].

4.1 The Controlled GMAW Process in UPPAAL

An overview of the implemented system is shown in figure 8, where the supervisor automaton controls the underlying automata, the drop dynamics automata and the GMAW dynamics automata. The supervisor decides by its two transitions which control mode the GMAW process should be in by a parallel composition with a shared label space in the sense of Milner[15], which is illustrated in figure 8. As pointed out in the previous section then the GMAW dynamics is only partitioned into 9 parts. This leads to a timed automaton for the arc length dynamics as shown in the middle of figure 8 with some of the possible transitions displayed. The automaton consists of 9 states, where each state represents a partition of the state space. The transitions between the states are decided by the shift register form, which is derived in the previous section.

In order to include disturbances into the model a disturbance automaton is included as shown in figure 9(a). It is designed to give a disturbance in the arc length in the base period. If the disturbance automation enables a disturbance (increase/decrease the arc length), it will affect the GMAW dynamics automaton as shown in figure 9(b).

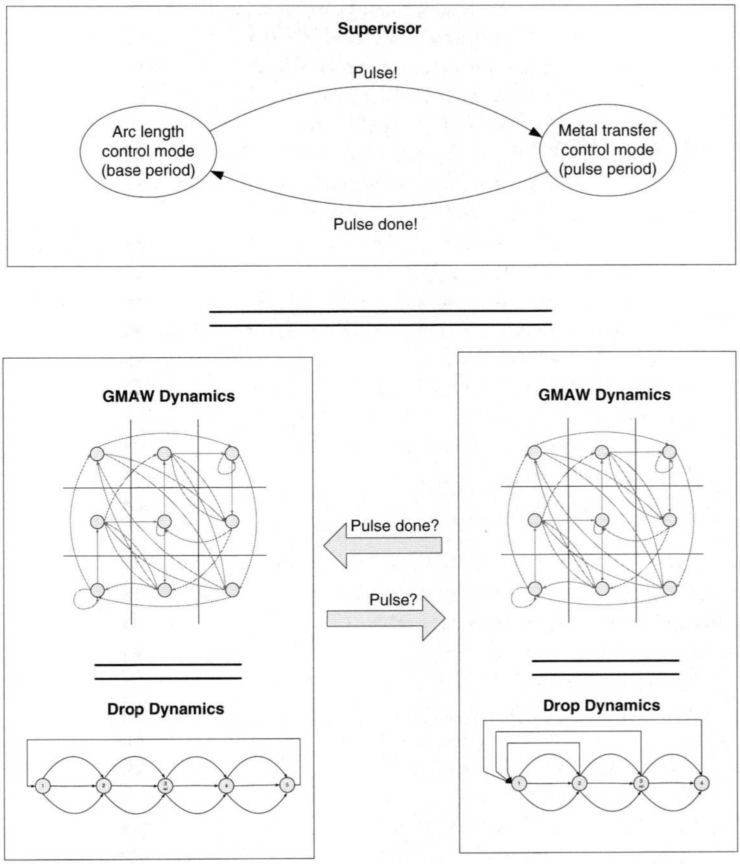

Fig. 8. The figure illustrates that it is possible to be in two different control modes; arc length control mode and metal transfer control mode. The supervisor controls which of the two modes to be in. In each control mode the processes are running in parallel.

4.2 Model Checking

It is possible in UPPAAL to use the model checker to get answers on specific questions, e.g. to check if there are deadlocks in the system. The deadlock check can be seen as a basic check of the systems behavior. By checking reachability and liveness properties the performance of a supervisor or controller can be analyzed. In UPPAAL the query language used is a simplified version of Computation Tree Logic (CTL)[16].

In the following specific questions regarding the GMAW process will be discussed.

Do deadlocks exists in the system?

Query:

```
A[] not deadlock
```

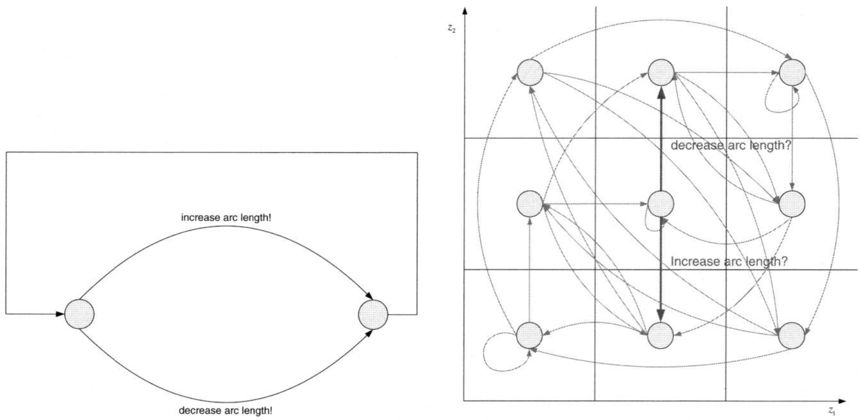

(a) Disturbance automaton (b) Effect of a disturbance

Fig. 9. (a) The disturbance automaton. (b) The two thick arrows shows the effect of a disturbance from the disturbance automaton to the GMAW dynamics.

Numerous factors can result in a deadlock in the system; A supervisor design flaw, faulty implementation etc.

Answer:

The property is satisfied.

Do the supervisor continuously cycle between the base period and the pulse period?

Query:

```
Supervisor.Arc_length --> Supervisor.Metal_transfer
Supervisor.Metal_transfer --> Supervisor.Arc_length
```

To guarantee the basic operation of the supervisor, a continuously cycle between the base period and the pulse period should take place. The first expression checks if the path between the states *Supervisor.Arc_length* and *Supervisor.Metal_transfer* will eventually be taken. The second expression checks if the path back from the state *Supervisor.Metal_transfer* to the state *Supervisor.Arc_length* will eventually be taken.

Answer:

The question is satisfied

Is the duration of the pulse period as specified?

Query:

```
A[] Supervisor.Metal_transfer imply x<=600
```

The duration in the pulse period is set to 600 clock cycles. This question checks if it is possible for the supervisor to jump from metal transfer control to arc length control before the specified time.

Answer:

The question is satisfied

The first question checks if there is some states from which the system cannot switch away from, which it is found that there are not. Secondly the liveliness of the supervisor is tested. This test can be seen as a check of the supervisor shown in figure 3. It is further interesting to verify if the system is staying too long in the different states, which is tested in the third query, where the time spend in the metal transfer state is tested. Similarly to the third query it could be tested if the supervisor is switching too fast between the different states, which will reveal if there is a possibility for Zeno behavior in the system.

5 Discussion

The objective of this paper was to show that it is possible to apply the theories developed in [7] to a given process, in this case the Pulsed GMAW process.

As seen from section 3 it is possible to formulate the GMAW welding process as a network of timed automata, which can be directly implemented in the simulation and verification tool, UPPAAL, thus giving the possibility of posing such questions as; is state A always reachable from state B or is it possible to end up in a deadlock - Questions, which is impossible to answer with classical control theory.

Acknowledgements

We would like to thank Jesper Sandberg Thomsen, Christian Krogh Nielsen and Jakob Brøchner Vilhelmsen for their insightful discussions and help with the first simulations.

References

1. Henzinger, T.A.: The theory of hybrid automata. In: Proceedings of the 11th Annual IEEE Symposium on Logic in Compute Science. (1996) 278–292
2. Bemporad, A., Ferrari-Trecate, G., Morari, M.: Observability and controllability of piecewise affine and hybrid systems. IEEE Transactions on Automatic Control **45**(10) (2000) 1864–1876
3. Collins, P., van Schuppen, J.H.: Observability of piecewise-affine hybrid systems. In: Hybrid Systems: Computation and Control. Volume 2993., Springer Berlin / Heidelberg (2004) 265–279
4. Feng, G.: Stability analysis of piecewise discrete-time linear systems. IEEE Transactions on Automatic Control **47**(7) (2002) 1108–1112
5. Alur, R., Dill, D.L.: A theory of timed automata. Theor. Comput. Sci. **126**(2) (1994) 183–235
6. Alur, R., Madhusudan, P.: Decision problems for timed automata: A survey. In: Formal Methods for the Design of Real-Time Systems. Volume 3185., Springer Berlin / Heidelberg (2004) 1–24
7. Tabuada, P., Pappas, G.J.: Model checking LTL over controllable linear systems is decidable. In: Hybrid Systems: Computation and Control, volume 2623 of Lecture Notes in Computer Sience, Springer-Verlag (2003)
8. Thomsen, J.S.: Advanced Control Methods for Optimization of Arc Welding. PhD thesis, Aalborg University (2004)

9. Storer, J.: The Haynes Manual on Welding. Haynes Group (2004)
10. Asarin, E., Bournez, O., Dang, T., Maler, O., Pnueli, A.: Effective synthesis of switching controllers for linear systems. Proceedings of the IEEE **88**(7) (2000) 1011–1025
11. Behrmann, G., David, A., Larsen, K.G., Möller, O., Pettersson, P., Yi, W.: Uppaal - present and future. In: Proc. of 40*th* IEEE Conference on Decision and Control, IEEE Computer Society Press (2001)
12. Decarlo, R., Branicky, M., Pettersson, S., Lennartson, B.: Perspectives and results on the stability and stabilizability of hybrid systems. Proceedings of the IEEE **88**(7) (2000) 1069–1082
13. Tabuada, P., Pappas, G.J.: Finite bisimulations of controllable linear systems. In: Proceedings of the 42nd IEEE Conference on Decision and Control. (2003)
14. Larsen, K.G., Pettersson, P., Yi, W.: Uppaal in a nutshell. International Journal on Software Tools for Technology Transfer (STTT) **1**(1-2) (1997) 134–152
15. Milner, R.: Communication and Concurrency. Prentice Hall (1989)
16. Huth, M., Ryan, M.: Logic in Computer Science: Modelling and Reasoning about Systems. 2 edn. Cambridge Univesity Press (2004)

On Self-triggered Full-Information H-Infinity Controllers

Michael Lemmon[1,⋆], Thidapat Chantem[2], Xiaobo Sharon Hu[2], and Matthew Zyskowski[1]

[1] University of Notre Dame, Department of Electrical Engineering, Notre Dame, IN 46556, USA
{lemmon,mzyskows}@nd.edu

[2] University of Notre Dame, Department of Computer Science and Engineering, Notre Dame, IN 46556, USA
{Sharon.Hu.8,tchantem}@nd.edu

Abstract. A self-triggered control task is one in which the task determines its next release time. It has been conjectured that self-triggering can relax the requirements on a real-time scheduler while maintaining application (i.e. control system) performance. This paper presents preliminary results supporting that conjecture for a self-triggered real-time system implementing full-information $\mathcal{H}_\infty$ controllers. Release times are selected to enforce upper bounds on the induced $\mathcal{L}_2$ gain of a linear feedback control system. These release times are treated as requests by the system scheduler, which then assigns actual release times using Buttazzo's elastic scheduling algorithm. Preliminary experimental results from a Matlab stateflow simulink model demonstrated a remarkable robustness to scheduling delays induced by real-time schedulers. These results show that self-triggered controllers are indeed able to maintain acceptable levels of application performance during prolonged periods of processor overloading.

1 Introduction

Computer controlled systems are often implemented using periodic real-time tasks. This approach can lead to significant over-provisioning of the real-time system since task periods are determined by the worst case time interval assuring closed loop system stability. In recent years, a number of researchers have proposed **aperiodic** task models in which tasks are either *event-triggered* [1] or *self-triggered* [2] controllers. Event-triggered control systems are systems whose control tasks are triggered by some asynchronous "event" within the control loop. These events are usually generated when an error signal crosses a specified threshold. The notion of event-triggered feedback [1] has appeared under a variety of names, such as interrupt-based feedback [3], Lebesgue sampling [4], asynchronous sampling [5], or state-triggered feedback [6].

⋆ The authors gratefully acknowledge the partial finanical support of the National Science Foundation (NSF-CNS-0410771).

A. Bemporad, A. Bicchi, and G. Buttazzo (Eds.): HSCC 2007, LNCS 4416, pp. 371–384, 2007.

Except for relay or pulse-width modulated feedback, event-triggered feedback can be impractical since it requires integrating an analog event detector into the physical plant. A more pragmatic approach for implementing aperiodic feedback is found in the **self-triggered** task model of Velasco et al. [2]. In self-triggered systems, the control task determines its next release time based on samples of the state gathered at the current release time. Self-triggered task models, therefore, can be implemented in existing computer controlled system without the need for any special analog event-detectors.

This paper presents experimental results examining the performance of a self-triggered control system. Our system's control tasks select sampling periods in a way that guarantees the closed-loop system's induced $\mathcal{L}_2$ gain satisfies a specified bound. We then consider a real-time system that schedules multiple self-triggered control tasks using traditional earliest-deadline-first (EDF) scheduling and Buttazzo's elastic scheduling algorithm [7]. Our implementation of Buttazzo's scheduler relies on a utilization constraint similar to that originally suggested by Chantem et al. [8]. Preliminary simulation results for a Simulink/Stateflow model of a real-time system controlling three inverted pendulums showed that the control system's performance under self-triggering was remarkably insensitive to the type of scheduler used by the real-time system. While preliminary, these results strongly suggest that self-triggering can provide a valuable way of ensuring control system performance in cases where the scheduler is unable to provide hard real-time guarantees on job completion.

2 Prior Work on Sample Period Selection

This section briefly reviews some of the prior work on sample period selection. Sample period selection for aperiodic real-time systems requires a detailed analysis of the system's intersample behavior. This usually involves studying a candidate Lyapunov function as was done in Zheng et al. [9] for a class of nonlinear sampled-data systems. Nesic et al. [10] used input-to-state stability (ISS) techniques to bound the intersample behavior of nonlinear systems [10]. This approach was used by Tabuada et al. [6] to estimate sampling periods for a class of nonlinear event-triggered control systems.

All of the aforecited works selected sampling periods to preserve some measure of the control system's stability, whether this is asymptotic stability or input-to-state stability. Applications, however, also need to ensure some minimum level of control system performance. Early work concerning the co-design of control systems and real-time systems viewed this as a schedulability problem in which sampling periods were selected to solve the following optimization problem,

$$\begin{array}{ll} \text{minimize:} & \text{Penalty on Control Performance} \\ \text{with respect to:} & \text{Sampling Periods} \\ \text{subject to:} & \text{closed loop stability} \\ & \text{task set schedulability} \end{array} \tag{1}$$

Early statements of this problem may be found in Seto et al. [11] with more recent studies in [12] and [13]. The penalty function used in the above problem is often a performance index for an infinite-horizon optimal control. The problem we face here, however, is that such performance indices [5] are rarely monotone functions of the sampling period. So it can be very difficult to identify "optimal" sampling periods for the above problem.

This paper uses Lyapunov techniques to select sampling periods that bound the intersample behavior of the system. This is similar to the approaches in [10], [9], and [6]. But rather than simply assuring closed loop stability, we select sampling periods to adjust the induced $\mathcal{L}_2$ gain of the system. This approach allows us to make the system responsive to variations in the intensity of the input disturbances driving the system. In the presence of high-intensity disturbances, for example, the system can reduce its gain to keep its output signal below some specified threshold. In the presence of low-intensity disturbances, it can then relax that gain and still ensure that the output remains below the same specified threshold. The variations in disturbance intensity are therefore mirrored in variations of the system gain which in turn result in large variations in the sampling period. This means that during periods of low disturbance intensity, the average sampling period can be much longer than during periods of high disturbance intensity. The bounds on intersample behavior ensuring a specified $\mathcal{L}_2$ gain are discussed in section 4. Because our controllers enforce a specified induced $\mathcal{L}_2$ gain, we confine our attention to linear full-information $\mathcal{H}_\infty$ controllers for which we can generate tight bounds on the system's intersample behavior.

3 System Model

The real-time system considered in this paper consists of N dynamical systems (called plants) that are controlled by N tasks running on a single processor. Each task samples (S) the system state, computes a state feedback control, and outputs that control to the plant through a zero-order hold (H). The state $x_i : \Re \rightarrow \Re^n$, of the ith plant satisfies the initial value problem,

$$\begin{aligned} \dot{x}_i(t) &= A_i x_i(t) + B_{1i} u_i(t) + B_{2i} w_i(t) \\ x_i(0) &= x_{i0} \end{aligned} \tag{2}$$

for $t \geq 0$ and $i \in \mathcal{N} = \{1, \ldots, N\}$. The function $w_i : \Re \rightarrow \Re^n$ is an uncontrolled and bounded disturbance. The function $u_i : \Re \rightarrow \Re^m$ is the control input generated by the ith real-time control task. A_i, B_{1i}, and B_{2i} are appropriately dimensioned matrices.

The ith plant's control, u_i, is generated by task $i \in \mathcal{N}$. Task i is associated with a sequence of **release times**, $\{r_i[j]\}_{j=1}^{\infty}$. The time $r_i[j] \in \Re$ is that time when the jth **job** of task i is available for execution. The task set is said to be **synchronous** if $r_i[0]$ is the same for all $i \in \mathcal{N}$. The **period** for the jth job of task i is denoted as

$$T_i[j] = r_i[j+1] - r_i[j] \tag{3}$$

If $T_i[j]$ is constant for all $j = 1, \ldots, \infty$, then the task is said to be periodic. A task that is not periodic is said to be **sporadic**.

The task set is associated with a **scheduling function**, $\sigma : \Re \to \mathcal{N}$. This function takes the value $\sigma(t) = i \in \mathcal{N}$ at time t when task i is executing at that time. The **finishing time** for job j of task i is denoted as $f_i[j] \in \Re$ and is formally defined as

$$f_i[j] = \max \left\{ t \in \Re \ : \ \begin{array}{c} i = \sigma(t^+), i \neq \sigma(t^-) \\ r_i[j] \leq t \leq r_i[j+1] \end{array} \right\} \tag{4}$$

where $\sigma(t^+) = \lim_{\tau \uparrow t} \sigma(\tau)$ and $\sigma(t^-) = \lim_{\tau \downarrow t} \sigma(\tau)$ are the left and right hand limits of σ at t, respectively.

The ith task's **worst-case execution time** (WCET) is denoted as $C_i \in \Re$. The task's **relative deadline** is denoted as $D_i \in \Re$. A task set is said to be **schedulable** if there exist sequences of release times $\{r_i[j]\}_{j=1}^{\infty}$ and a scheduling function σ such that

$$D_i \geq f_i[j] - r_i[j] \geq C_i \tag{5}$$

for all $i \in \mathcal{N}$ and $j = 1, \ldots, \infty$.

The ith task computes the control u_i for the ith plant. This control is assumed to be a state feedback control law of the form

$$u_i(t) = -k^T x(r_i[j]) \tag{6}$$

for $t \in [f_i[j], f_i[j+1])$ where $j = 1, \ldots, \infty$. Note that the control output is constant between finishing times and the value of that constant is determined by the system state at the job's release time, $r_i[j]$.

4 Sample Period Selection for Induced $\mathcal{L}_2$ Gain

This section states the paper's main result concerning sample period selection enforcing a specified bound on the closed-loop system's induced $\mathcal{L}_2$ gain. We confine our attention to the behavior of a single plant between consecutive release times. We therefore drop the task index, i, without a loss of generality.

We assume, for the purpose of analytic simplicity, that the control $u_i(t)$ satisfies equation 6 for all t between consecutive *release times* rather than consecutive *finishing times*. This simplification is done for analytical convenience at the expense of some loss in generality.

The following theorem provides conditions which guarantee that the induced $\mathcal{L}_2$ gain from the plant's disturbance w to its state be less than a specified positive number, γ.

Theorem 1. *Let G denote the sampled-data control system given by equations 2 and 6. Assume the control gain is $k^T = -B_1^T P$ where P is a positive symmetric matrix that satisfies the algebraic Riccati equation,*

$$0 = A^T P + PA + I - P\left(B_1 B_1^T - \frac{1}{\gamma^2} B_2 B_2^T\right) P \tag{7}$$

for some $\gamma > 0$.

Let x_r denote the system's state at release time $r[j]$. If the system state $x(t)$ satisfies

$$\begin{bmatrix} x(t) \\ x_r \end{bmatrix}^T \begin{bmatrix} -I + PB_1B_1^TP & -PB_1B_1^TP \\ -PB_1B_1^TP & 0 \end{bmatrix} \begin{bmatrix} x(t) \\ x_r \end{bmatrix} \leq -\|x(t)\|^2 \tag{8}$$

for all $t \in [r[j], r[j+1])$ and $j = 1, \ldots, \infty$, then the induced $\mathcal{L}_2$ gain of G is less than γ.

Proof. The directional derivative of $V = x^TPx$ is

$$\dot{V} = \frac{\partial V}{\partial x}\left(Ax - B_1B_1^TPx_r + B_2w\right)$$

Using the standard completing the square argument and the Riccati equation 7, we rewrite the above equation as

$$\begin{aligned} \dot{V} &= \overline{x}^TX\overline{x} - \frac{1}{\gamma^2}x^TPB_2B_2^TPx + 2w^TB_2^TPx \\ &= \overline{x}^TX\overline{x} - \left\|\gamma w - \frac{1}{\gamma}B_2^TPx\right\|^2 + \gamma^2\|w\|^2 \\ &\leq \overline{x}^TX\overline{x} + \gamma^2\|w\|^2 \end{aligned}$$

where

$$X = \begin{bmatrix} -I + PB_1B_1^TP & -PB_1B_1^TP \\ -PB_1B_1^TP & 0 \end{bmatrix}, \qquad \overline{x} = \begin{bmatrix} x \\ x_r \end{bmatrix} \tag{9}$$

Note that if

$$\overline{x}^TX\overline{x} \leq -\|x\|^2$$

then the above inequalities imply that

$$\dot{V}(x) \leq -\|x\|^2 + \gamma^2\|w\|^2$$

which is sufficient to ensure that the induced $\mathcal{L}_2$ gain of G is less than γ ◇

Remark: The matrix X in equation 9 can be viewed as a collection of rank-one perturbations of the block diagonal matrix diag$(-I, 0)$. We can use this observation to show that $\overline{x}^TX\overline{x} \leq \lambda_1(k^T(x(t) - x_r))^2$ where λ_1 is the largest eigenvalue of X. So if we can ensure that $\lambda_1(k^T(x(t) - x_r))^2 < \|x(t)\|^2$, then we can again guarantee that the conditions in theorem 1 are satisfied. This is a more conservative condition than the one in theorem 1 and it is similar to the switching condition used in [6]. Therefore by using an analysis similar to that in [6] we can show that the "sampling period" is bounded below by a positive constant. In general, however, this lower bound can be an extremely conservative estimate of the sampling period.

After the task's release, we're interested in approximating the interval over which we can guarantee the condition in theorem 1. This time interval can be taken as an estimate of the task's next release time, T^r, which we treat as the task period. A reasonable approximation for this period is obtained by integrating the differential equation

$$\begin{aligned} \dot{x}(t) &= Ax - B_1 B_1^T P x_r \\ x(r) &= x_r \end{aligned} \tag{10}$$

Let e^{At} be the transition matrix for A, so we can easily see that

$$x(t) = \left[e^{At} \left(I + \int_0^t e^{-As} B_1^T ds \right) \right] x_r = \Phi(t) x_r \tag{11}$$

Because A and B_1 are known, we can evaluate the matrix function $\Phi(t)$.

5 Schedulability with Deadlines Less Than Periods

The preceding section suggests that if our task retriggers itself so equation 8 is always satisfied, then our sampled-data system can guarantee the system's closed loop gain is less than γ. This may only happen, however, if the released tasks can meet their real-time deadlines. Earliest deadline first (EDF) schedulers are frequently used in periodically triggered real-time control systems. In the self-triggered system, however, release times will vary depending upon the system's current state. As a result we need to consider a scheduler that can adjust its task periods while assuring EDF schedulability and the "minimum" task period required by the application.

The **elastic task model** of Buttazzo et al., [7], is a popular method of adjusting task periods. The elastic task model uses a mechanical analogy to develop an algorithm for adjusting task periods. This analogy views tasks as being interconnected by "springs". The length of the spring represents the task's utilization and the spring constant represents that task's resistance to changing its utilization. Buttazzo's elastic task model was extended by Caccamo et al. [14] to handle uncertainties in computation time. A later paper [15] showed how to modify Buttazzo's algorithm to handle additional resource constraints. Hu et al [16] showed that Buttazzo's elastic scheduling algorithm can be viewed as minimizing a task set's summed squared utilization subject to the Liu-Layland EDF schedulability condition [17].

In our system, task deadlines will always be significantly less than task periods. This is needed to ensure a short time delay between the state sampling and the release of the control signal. Such task sets are schedulable under EDF if and only if

$$\sum_{i=1}^{N} \left(\left\lfloor \frac{L - D_i}{T_i} \right\rfloor + 1 \right) C_i \leq L \tag{12}$$

for all $L \in \mathcal{D}$ where

$$\mathcal{D} = \left\{ d_{i,k} \ : \ \begin{array}{l} d_{i,k} = kT_i + D_i \\ i \in \mathcal{N}, k \geq 0 \end{array} \right\}$$

This condition is proven by Baruah et al. [18] using a processor demand analysis.

Processor demand analysis requires that the total processor demand of all released tasks in an interval is less than or equal to the total processing power available in that interval. For task sets in which the deadline is less than the period, the ith task's processor demand over time interval $[0, L]$ is

$$C_i(0, L) = \left(\left\lfloor \frac{L - D_i}{T_i} \right\rfloor + 1 \right) C_i.$$

The condition in equation 12 is simply the sum of all demands, $C_i(0, L)$, over i which must be checked for all possible future releases of the tasks.

In our application, release times are recomputed each time the task is called and so the task set is really not periodic. As a result we only need to check equation 12 over a subset of $\mathcal{D}$; namely those times that are less than or equal to the next release time. This idea was used in Chantem et al. [8] to propose a heuristic generalization of Buttazzo's elastic scheduling algorithm. The following theorem introduces a schedulability condition similar to that used in [8] which can be directly used with Buttazzo's algorithm.

Theorem 2. *Consider a task set in which all tasks are released at time* 0. *Assume that the task set is sorted in order of non-decreasing relative deadlines* ($D_i \leq D_{i+1}$) *and let* $\{\underline{T}_j\}_{j=1}^N$ *be a set of bounds on the task period that are generated recursively from*

$$\left\lfloor \frac{D_2 - D_1}{\underline{T}_1} \right\rfloor = \left\lfloor \frac{D_2}{C_1} - \sum_{i=1}^{2} \frac{C_i}{C_1} \right\rfloor \tag{13}$$

$$\left\lfloor \frac{D_{j+1} - D_j}{\underline{T}_j} \right\rfloor = \left\lfloor \frac{D_{j+1}}{C_j} - \sum_{i=1}^{j+1} \frac{C_i}{C_j} - \sum_{i=1}^{j-1} \left\lfloor \frac{D_{j+1} - D_i}{\underline{T}_i} \right\rfloor \frac{C_i}{C_j} \right\rfloor \tag{14}$$

for $j = 1, \ldots, N$.

Let $i^* = \arg\min_i \{D_i + T_i\}$ *then the task set will miss no deadlines over the interval from* $[0, D_{i^*} + T_{i^*}]$ *if* $T_j \geq \underline{T}_j$ *for all* $j = 1, \ldots, N$ *and*

$$\sum_{i=1}^{N} U_i \leq 1 - \frac{1}{\underline{T}_{i^*}} \sum_{i=1}^{N} C_i \tag{15}$$

Proof. To prove this theorem we need to demonstrate that the processor demand satisfies

$$\sum_{i=1}^{N} \left(\left\lfloor \frac{L - D_i}{T_i} \right\rfloor + 1 \right) C_i \leq L$$

over intervals

$$L \in \{D_1, \cdots, D_N, \min_i \{T_i + D_i\}\}$$

Essentially this means that the processor demand is satisfied between the consecutive release times.

When $L = D_i$, the processor demand can be written as a triangular system of algebraic equations

$$0 \leq D_1 - C_1 \tag{16}$$

$$\left\lfloor \frac{D_2 - D_1}{T_1} \right\rfloor C_1 \leq D_2 - \sum_{i=1}^{2} C_i \tag{17}$$

$$\left\lfloor \frac{D_3 - D_1}{T_1} \right\rfloor C_1 + \left\lfloor \frac{D_3 - D_2}{T_2} \right\rfloor C_2 \leq D_3 - \sum_{i=1}^{3} C_i \tag{18}$$

$$\cdots \leq \cdots \tag{19}$$

in which the jth term has the form,

$$\sum_{i=1}^{j-1} \left\lfloor \frac{D_j - D_i}{T_i} \right\rfloor C_i \leq D_j - \sum_{i=1}^{j} C_i$$

This is a triangular system of equations that we can solve recursively for $\underline{T}_i$. In particular the second equation (eqn. 17) yields equation 13. Applying this in a recursive manner yields equation 14. So if these equations are satisified then we can guarantee the processor demand is sufficient for time intervals equal to the task deadlines.

Now consider $L = T_j + D_j$ for any arbitrary j and assume that $T_i \geq \underline{T}_i$ for all i. Then clearly,

$$\begin{aligned} D_j &\geq \sum_{i=1}^{N} \left(\left\lfloor \frac{D_j - D_i}{T_i} \right\rfloor + 1 \right) C_i \\ &\geq \sum_{i=1}^{N} \left(\left\lfloor \frac{T_j + D_j - D_i}{T_i} \right\rfloor - \left\lfloor \frac{T_j}{T_i} \right\rfloor - 1 + 1 \right) C_i \end{aligned}$$

where we used the fact that $\lfloor x + y \rfloor \leq \lfloor x \rfloor + \lfloor y \rfloor + 1$. We can now rearrange our last inequality to obtain

$$D_j + \sum_{i=1}^{N} C_i + \sum_{i=1}^{N} \left\lfloor \frac{T_j}{T_i} \right\rfloor C_i \geq \sum_{i=1}^{N} \left(\left\lfloor \frac{T_j + D_j - D_i}{T_i} \right\rfloor + 1 \right) C_i$$

The lefthand side of the above inequality can be bounded as

$$D_j + \sum_{i=1}^{N} C_i + \sum_{i=1}^{N} \left\lfloor \frac{T_j}{T_i} \right\rfloor C_i \leq D_j + \sum_{i=1}^{N} C_i + T_j \sum_{i=1}^{N} \frac{C_i}{T_i}$$

By the assumption in equation 15 we can see that

$$D_j + \sum_{i=1}^{N} C_i + \sum_{i=1}^{N} \left\lfloor \frac{T_j}{T_i} \right\rfloor C_i \leq D_j + \sum_{i=1}^{N} C_i - \sum_{i=1}^{N} C_i + T_j = D_j + T_j$$

So if the above condition holds we can ensure that

$$\sum_{i=1}^{N} \left(\left\lfloor \frac{T_j + D_j - D_i}{T_i} \right\rfloor + 1 \right) C_i \leq D_j + T_j$$

which is the inequality required to ensure that the processor demand is satisfied prior to the given release time. We choose $j = i^*$ to complete the proof. ♢

6 Self-triggered Real-Time $\mathcal{H}_\infty$ Controllers

This section discusses how theorems 1 and 2 are used to elastically schedule self-triggered control tasks. Upon release, the ith task numerically integrates equation 10 forward to determine T_i^r. T_i^r serves as the task's desired next release time. The scheduler handles T_i^r as a request for the specified task period. If the scheduler simply grants the task's requested period, we say it is a **rigid** task scheduler. If the scheduler adjusts the requested period as is done in Buttazzo's algorithm, then we say the task scheduler is **elastic**.

Let $U_i^r = C_i/T_i^r$ denote the utilization requested by the ith task. For the simulations in section 7, our elastic task scheduler assigns task utilization, $U_i = C_i/T_i$, in a manner that solves the following optimization problem.

$$\begin{array}{rl} \text{minimize:} & \sum_{i=1}^{N} (U_i - U_i^r)^2 \\ \text{subject to:} & \underline{U}_i \leq U_i \leq \min(U_i^r, C_i/\underline{T}_i) \\ & \sum_{i=1}^{N} U_i \leq 1 - \overline{U} \end{array} \tag{20}$$

where $\underline{U}_i$ is the minimum individual task utilization for closed loop stability, $\underline{T}_i$ is the minimum task period required in theorem 2, $U_i^r = C_i/T_i^r$ is the task's requested utilization and

$$\overline{U} = \frac{1}{\min\{\underline{T}_i\}} \sum_{i=1}^{N} C_i$$

is the upper bound on the total utilization given in equation 15 of theorem 2. This optimization problem seeks to minimize the squared difference between a task's desired utilization, U_i^r, and its actual utilization, U_i. The allocation is done subject to constraints in theorem 2. These constraints require that the allocated utilizations assure closed loop stability while remaining schedulable under EDF.

It is important to note that the optimization problem in equation 20 is precisely the problem considered in Hu et al. [16] and Chantem et al. [8]. In those papers it was shown that Buttazzo's elastic scheduling algorithm [15] actually assigns task periods in a way that always satisfies the optimization problem in equation 20. So in our proposed self-triggered system, we can simply use Buttazzo's algorithm directly to allocate task utilization.

7 Simulation Results

This section presents preliminary simulation results for the real-time control of three identical inverted pendulums that are controlled by three control tasks running on the same processor. The controllers are full-information $\mathcal{H}_\infty$ controllers. The plants are all identical and the linearized state equation for the ith plant is

$$\dot{x}_i(t) = \begin{bmatrix} 0 & 1 & 0 & 0 \\ 0 & 0 & -(mg/M) & 0 \\ 0 & 0 & 0 & 1 \\ 0 & 0 & g/\ell & 0 \end{bmatrix} x_i(t) + \begin{bmatrix} 0 \\ 1/M \\ 0 \\ -1/M\ell \end{bmatrix} u_i(t) + \begin{bmatrix} 1 \\ 0 \\ 1 \\ 0 \end{bmatrix} w_i(t)$$

where M is the cart mass, m is the mass of the pendulum bob, ℓ is the length of the pendulum and g is gravitational accleration. The initial state is zero. For these simulations we let $M = 10$, $\ell = 3$, and $g = 10$. The system state $x = \begin{bmatrix} y \; \dot{y} \; \theta \; \dot{\theta} \end{bmatrix}$ where y is the cart's position and θ is the pendulum bob's angle with respect to the vertical.

The external disturbance, w_i, is time-varying and takes the form,

$$w_i(t) = \nu(t) + W\text{Rect}(t - \tau_i)$$

where ν is band-limited white noise with a noise power of 100 and sampling period of .001 sec. Rect$(t-\tau_i)$ is a unit rectangle function of duration 0.25 seconds starting at time τ_i. W represents the strength of the rectangular disturbance. In these simulations, all three plants are hit with the same rectangular disturbance at the same time ($\tau_i = 2$) and the disturbance level, W, was set to 1000.

In these simulations we required the state magnitude to lie below a specified level, M. In other words, we require $\|x_i(t)\|_2 \leq M$ for some specified $M \in \Re$. Immediately after the rectangular pulse, the system state changes and we assume that the next released task can measure this change. Once the sampled system state exceeds M, the task reduces its gain to $\gamma = 100$ to more aggressively reject the rectangular pulse. M was set to 5 for these simulations. Once the system state is sufficiently small, the gain is reset to a large value ($\gamma = 500$) consistent with a less aggressive disturbance rejection objective. By adjusting the gain in this manner we kept the system output relatively small without having to use the more aggressive gain throughout the entire system's history.

We simulated the real-time system using a Matlab stateflow/simulink model. The model was built to accurately capture task timing that might be seen in actual real-time systems. The real-time computer was modeled as a stateflow chart that consisted of the parallel composition of three control tasks, six interrupt handlers, and two processes for the scheduler. The control tasks sample the plant state upon their release, compute the requested next release time T^r. Upon finishing their execution, these tasks output the control signal. This simulation, therefore, forces the control, $u_i(t)$, to be constant between consecutive finishing times rather than consecutive release times as assumed in theorem 1. The scheduler was implemented as two processes. One process assigned priorities to the released control tasks and the other process used the requested sampling period to compute the actual release times.

We simulated three different cases. The first "baseline" case simulated the response of a single inverted pendulum by an "idealized" real-time system in which scheduler performance was not an issue. The second case simulated the response of a self-triggered real-time system under "rigid" and "elastic" scheduling. The third case simulated the response of the periodically-triggered real-time system under two different task periods. These three cases are discussed below.

Baseline Case: The baseline case simulated the response of a single inverted pendulum in which the control, $u_i(t)$, was constant between consecutive release times. Tasks were periodically released every 0.25 seconds. The controller's state feedback gain was chosen to ensure the closed loop system's $\mathcal{H}_\infty$ gain was less than 100. This case is therefore an "idealized" real-time system in which missed deadlines and processor contention are abstracted away. The lefthand side of figure 1 shows the time histories for the four system states: cart position, cart velocity, pendulum bob angle and angular velocity. This plot serves as a baseline against which the other cases will be compared. The plot shows that when the rectangular disturbances hits the system, the cart moves quickly to ensure the pendulum bob angle, θ, remains small. This corrective action results in a large displacement, x, of the cart that takes about 10 seconds to return to the home position.

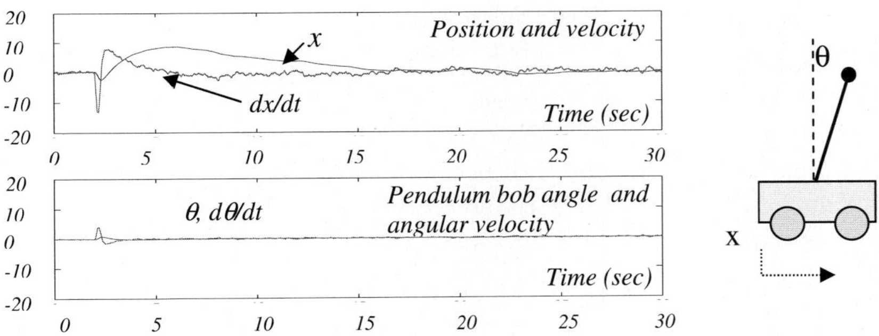

Fig. 1. Transient Response of Baseline System

Self-Triggered Case: The self-triggered case consisted of two simulations. One simulation used rigidly scheduled and the other used elastically scheduled task sets. The system clock ran at 0.001 seconds. All control tasks had identical computation times, $C_i = 50$ clock ticks, and deadlines, $D_i = 100$ clock ticks. Task periods were selected based on the results in theorem 1. In general, this resulted in a probabilistic distribution of task periods that were dependent on the system state at the release time. For a system gain $\gamma = 100$ and 500 the requested period averaged 200 and 500 clock ticks, respectively. In the "rigidly" self-triggered system, these requested task periods were granted by the scheduler. The "elastically" self-triggered system had the requested task periods adjusted by the Buttazzo algorithm using the schedulability condition of theorem 2.

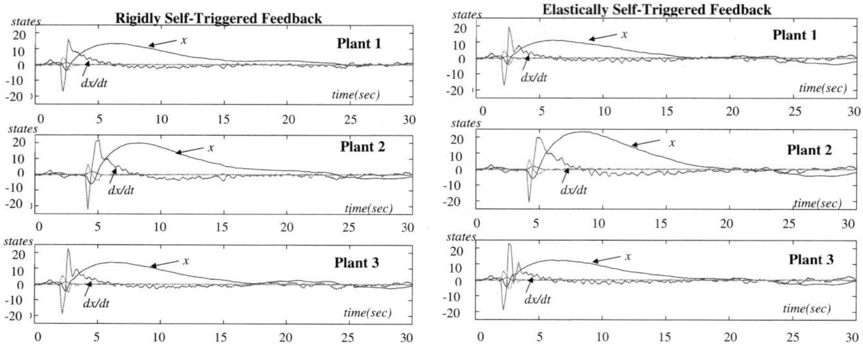

Fig. 2. State histories for rigidly (left) and elastically (right) self-triggered controllers

The state histories for the rigid and elastic self-triggered systems are shown in figure 2. The lefthand graphs are state histories for the three plants in the rigidly self-triggered system and the righthand graphs are for the elastically self-triggered system. What is perhaps most interesting here is that all systems have the same transient behavior regardless of whether task periods were assigned in an elastic or rigid manner. Comparing the state trajectories in figure 2 against the baseline trajectories in figure 1, we see little difference; thereby suggesting that the self-triggered system was maintaining the baseline transient behavior regardless of which scheduling scheme was used.

Periodically-Triggered Case: As a point of comparison we simulated a periodically triggered system. The task computation times and deadlines were 50 and 100 respectively for all tasks with a clock tick of 0.001 seconds. For one set of simulations, we set the task period to 250. The state trajectories for the three plants are shown in the lefthand plots of figure 3. A task period of 250 was the average task period for the rigidly scheduled self-triggered simulations. The lefthand plots in figure 3 are therefore comparable to the "rigidly scheduled" simulations in figure 2. In the other set of simulations, we set the task period to 500. This simulation's state histories are shown in the righthand plots of figure 3. A task period of 500 was the average task period for the "elastically scheduled" self-triggered simulations. The righthand plots are therefore comparable to the "elastically scheduled" simulations in figure 2. These simulations show that at the shorter task period (lefthand side of figure 3), the systems appears to have a transient response similar to that for the baseline system. At the longer task period (righthand side of figure 3), some of the plants become extremely oscillatory with one of the systems actually becoming unstable.

8 Final Remarks

In comparing the simulation results between the baseline, periodically triggered, and self-triggered systems it should be apparent that the self-triggered system was able to maintain an acceptable level of transient performance regardless of

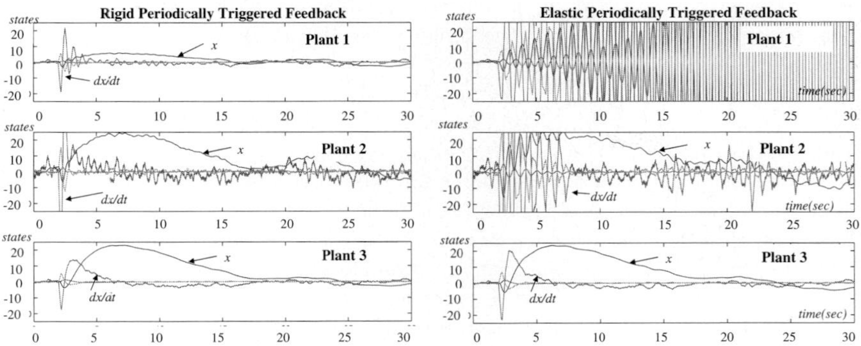

Fig. 3. State trajectories for periodically triggered system. (Left) Task period = 250, (Right) Task period = 500.

whether a rigid or elastic scheduler was used. This was somewhat surprising at first glance. But in hindsight it was conjectured that this might be due to the inherent feedback nature of self-triggering. In selecting the next release based on the current state, a self-triggered system is using state feedback to adjust task periods in a way that assures overall system "performance".

The feedback nature of this interaction suggests that the performance of self-triggered systems should be very robust to processor overloads and late (delayed) jobs. Our simulations showed that the rigidly scheduled system was overloaded whereas the elastically scheduled system was not overloaded. Inspite of the fact that the rigidly scheduled system was overloaded, figure 2 clearly shows that the transient response is very similar to the baseline response. The task periods in the elastically scheduled self-triggered system were long enough to destabilize the periodically triggered system (figure 3). The results in figure 2 clearly show that even with this longer task period, the self-triggered system was able to preserve control performance. In other words, self-triggered systems appear to maintain acceptable levels of application performance in the face of significant processor overloading.

This paper's results therefore demonstrate that self-triggering can maintain acceptable levels of system performance regardless of whether or not we schedule in an elastic manner. In particular, these results seem to suggest a practical way for relaxing the need for "hard" real-time support in computer controlled systems. These results, however, are only preliminary and future work will need to more rigorously verify them.

References

1. Arzen, K.: A simple event-based pid controller. In: Proceedings of the 14th IFAC World Congress. (1999)
2. Velasco, M., Marti, P., Fuertes, J.: The self triggered task model for real-time control systems. In: Work-in-Progress Session of the 24th IEEE Real-Time Systems Symposium (RTSS03). (2003)

3. Hristu-Varsakelis, D., Kumar, P.: Interrupt-based feedback control over a shared communication medium. In: Proceedings of the IEEE Conference on Decision and Control. (2002)
4. Astrom, K., Bernhardsson, B.: Comparison of riemann and lebesgue sampling for first order stochastic systems. In: Proceedings of the IEEE Conference on Decision and Control. (1999)
5. Voulgaris, P.: Control of asynchronous sampled data systems. IEEE Transactions on Automatic Control **39**(7) (1994) 1451–1455
6. Tabuada, P., Wang, X.: Preliminary results on state-triggered scheduling of stabilizing control tasks. In: IEEE Conference on Decision and Control. (2006)
7. Buttazzo, G., Lipari, G., Abeni, L.: Elastic task model for adaptive rate control. In: IEEE Real-Time Systems Symposium (RTSS). (1998)
8. Chantem, T., Hu, X., Lemmon, M.: Generalized elastic scheduling. In: Real-Time Systems Symposium (RTSS). (2006)
9. Zheng, Y., Owens, D., Billings, S.: Fast sampling and stability of nonlinear sampled-data systems: Part 2. sampling rate estimations. IMA Journal of Mathematical Control and Information **7** (1990) 13–33
10. Nesic, D., Teel, A., Sontag, E.: Formulas relating kl stability estimates of discrete-time and sampled-data nonlinear systems. Systems and Control Letters **38** (1999) 49–60
11. Seto, D., Lehoczky, J., Sha, L., Shin, K.: On task schedulability in real-time control systems. In: IEEE Real-time Technology and Applications Symposium. (1996) 13–21
12. Cervin, A., Eker, J., Bernhardsson, B., Arzen, K.E.: Feedback-feedforward scheduling of control tasks. Real-time Systems **23(1-2)** (2002) 25–53
13. Marti, P., Lin, C., Brandt, S., Velasco, M., Fuertes, J.: Optimal state feedback resource allocation for resource-constrained control tasks. In: IEEE Real-Time Systems Symposium (RTSS 2004). (2004) 161172
14. Caccamo, M., Buttazzo, G., Sha, L.: Elastic feedback control. In: IEEE Euromicro Conference on Real-Time Systems (ECRTS). (2000)
15. Buttazzo, G., Lipari, G., Caccamo, M., Abeni, L.: Elastic scheduling for flexible workload management. IEEE Transactions on Computers **51(3)** (2002) 289–302
16. Hu, X., Chantem, T., Lemmon, M.: Optimal elastic scheduling. In: IEEE Real-time and embedded technology and applications symposium - works in progress track. (2006)
17. Liu, C., Layland, J.: Scheduling for multiprogramming in a hard-real-time environment. Journal of the Association for Computing Machinery **20**(1) (1973) 46–61
18. Baruah, S., Rosier, L., Howell, R.: Algorithms and complexity concerning the preemptive scheduling of periodic , real-time tasks on one processor. Journal of Real-Time Systems **2** (1990) 301–324

Impulse Differential Inclusions Driven by Discrete Measures

John Lygeros[1,⋆], Marc Quincampoix[2], and Tadeusz Rzezuchowski[3]

[1] Automatic Control Laboratory, ETH Zurich
lygeros@control.ee.ethz.ch
[2] Laboratoire de Mathematiques, Université de Bretagne Occidentale
Marc.Quincampoix@univ-brest.fr
[3] Faculty of Mathematics & Information Science, Warsaw University of Technology
tarz@mini.pw.edu.pl

Abstract. We consider systems modeled by a differential inclusion subject to impulsive, set valued state resets. We study existence of solutions for this class of systems and derive conditions for a set of states to be viable. From the point of view of hybrid systems, of central interest is the fact that the class of systems and the solution concept considered allow any finite number of left and right accumulation points of the impulse times; in other words, very complex Zeno type behaviors. The results are demonstrated on simple examples that exhibit such behaviors.

1 Introduction

In this paper we consider a class of systems that comprise continuous dynamics, modeled by a differential inclusion

$$\dot{x} \in F(x)$$

and impulses that cause discrete jumps in the state. The magnitude of the jumps depends on the magnitude of the impulse as well as the value of the state before the jump. More specifically, the effect of an impulse of magnitude α is a jump whose magnitude takes values in a set $\alpha S(x, \alpha)$, or, in hybrid systems terminology, a set valued state reset of the form

$$x \longmapsto x + \alpha S(x, \alpha).$$

Systems of this type can be viewed as controlled systems, where the inputs can intervene in the continuous evolution (think of the differential inclusion as $F(x) = \{f(x, u) \mid u \in U(x)\}$) or by "hitting" the system with impulses. For related problems in game theory (in the presence of two antagonistic input signals) see [1].

This class of systems is related to the so-called measure driven differential inclusions studied in [2,3]. These are systems of the form

$$dx \in F(x)dt + G(x)\mu(dt)$$

⋆ Corresponding author.

A. Bemporad, A. Bicchi, and G. Buttazzo (Eds.): HSCC 2007, LNCS 4416, pp. 385–398, 2007.

where μ is a non-negative, Borel measure and $G(\cdot)$ and $F(\cdot)$ are suitable set valued maps. [2,3] provide an interpretation for this type of systems, leading to the notion of the so-called robust solutions. Roughly speaking, the idea is to generate an ordinary differential inclusion from the measure driven differential inclusion, by appropriately "stretching out" impulsive points of μ. The solutions of the resulting differential inclusion then correspond exactly to the robust solutions of the original measure driven differential inclusion (which thereby inherit desirable properties such as compactness of the trajectory set). Based on these results, [4,5] formulate optimal control problems for this class of systems. The main difference in this paper is that we take a hybrid approach and abstract the measure driven part of the differential inclusion $G(x)\mu(dt)$ by a discrete transition map. This simplifies the study of the existence results and allows us to treat the impulsive measure itself as a control signal (to study, for example, questions of viability).

From the point of view of hybrid systems, an interesting property of the class of systems considered here is that they place very few restrictions on the timing of the impulses that cause the discrete transitions. The solutions of the system can be defined, even if the set of impulse times contains left and right accumulation points, as long as the total number of impulses is countable and the number of accumulation points and total impulsive measure in the time interval of interest are finite. In other words, the class of systems considered here allow one to capture a very rich set of Zeno type behaviors.

The Zeno phenomenon has been studied extensively in the hybrid systems literature for a number of years. Many results on stability and optimal control of hybrid system indeed rely on an underlying assumption that the solutions of the system are Zeno free, even though it is known that it is impossible to check this assumption in general [6]. Most of the hybrid system solution concepts [7,8,9,10,11,12] allow one to define solutions that contain at most one point to which the discrete transition times converge from the left; in most cases the solution can be defined up to (and not including) this point [8,9,12], even though some authors are able to exploit special structure in the dynamics (e.g. linear complementarity) to define solutions beyond this point [7,10,11]. For more general classes of systems, a number of approaches have been proposed to augment the original model and extend solutions beyond the Zeno point. One such approach uses relaxation and averaging techniques [13] while another exploits physical intuition wherever this is possible (e.g. in mechanical systems) to extend the solutions [14,15].

Types of behavior related to the Zeno phenomenon have also been considered in the literature of discontinuous ordinary differential equations (ODE) [16]. Discontinuous differential equations can be viewed as switched systems whose state trajectory is continuous, and follows a differential equation that switches depending on the current state. Classical relaxed solution concepts for discontinuous ODE can deal with situations where the switch times accumulate from the left or from the right, exploiting the continuity of the solutions. However, most cases where this type of behavior is considered in the literature are provided as a warning of things (e.g. loss of uniqueness) that can go wrong in this case. [17] provides an overview of different solution concepts found in the hybrid systems and discontinuous ODE literature and compares (among other things) their treatment of the Zeno phenomenon.

The remainder of the paper is organized in five sections. Section 2 provides the formal definitions of the class of systems considered and their solutions, as well as a brief comparison with more familiar classes of systems found in the literature. Section 3 establishes conditions for the existence of solutions. Section 4 presents a simple example of a system that can be studied using the tools developed in this paper. The solutions of the example system are discontinuous and their transition times contain a left and right accumulation point, putting the system beyond the reach of most (if not all) of the hybrid system solution concepts known in the literature. Section 5 presents an initial treatment of reachability problems for this class of systems. Concluding remarks and possible extensions are given in Section 6.

2 Measure Driven Impulse Differential Inclusions

The dynamics we consider are defined using two set valued maps, $F(\cdot) : \mathbb{R}^d \rightarrow 2^{\mathbb{R}^d}$ and $S(\cdot,\cdot) : \mathbb{R}^d \times [0,1] \rightarrow 2^{\mathbb{R}^d}$. We consider also a scalar valued measure $\mu(\cdot)$ on the Borel subsets of $[0,\infty)$ and assume that $\mu(\{t\}) \in [0,1]$ for all $t \geq 0$[1]. To ensure that the impulsive system is well posed and solutions exist one needs to impose a number of assumptions on F, S and μ.

Assumption 1. *F, S and μ satisfy the following conditions:*

1. *F is Lipschitz and has convex, compact values.*
2. *S has compact values and is bounded, specifically for all $(x,\alpha) \in \mathbb{R}^d \times [0,1]$, $S(x,\alpha) \subseteq B(0,M)$. As usual, $B(x,r)$ denotes the closed ball in $\mathbb{R}^d$ with radius $r \geq 0$ centered at the point $x \in \mathbb{R}^d$.*
3. *The set valued map $S(\cdot,\alpha) : \mathbb{R}^d \rightarrow 2^{\mathbb{R}^d}$ is Lipschitz, uniformly in $\alpha \in [0,1]$.*
4. *There exists a countable index set I, a set of times $\{t_i\}_{i \in I}$ with $t_i \geq 0$ and a set of numbers $\{\alpha_i\}_{i \in I}$ with $\alpha_i \in [0,1]$ such that*

$$\mu(t) = \sum_{i \in I} \alpha_i \delta_{t_i}(t),$$

   where $\delta_{t_i}(\cdot)$ denotes the standard Dirac measure centered at t_i.
5. *There exists a $T > 0$ such that $\mu([0,T)) < \infty$.*

Most of the conditions imposed by Assumption 1 are relatively standard, typically imposed in the literature to ensure the existence of solutions to differential inclusions and/or impulsive systems. The conditions on the measure μ deserve

[1] For simplicity, the interval $[0,1]$ is used throughout as a bound on μ. The argument holds, however, for μ taking values in any compact interval of $\mathbb{R}_+$.

closer scrutiny. Assumption 1.4 requires that μ consists of a train of pulses, possibly infinite. Part 1.5 requires that the total measure on any compact set up to some time T is finite. Note, however, that the measure of the set $[0,\infty)$ may still be infinite. For example, a train of pulses with magnitude 1, one for each integer is acceptable as the measure μ. Moreover, the set $\{t_i \mid i \in I\}$ may also contain left and right accumulation points. In the hybrid systems terminology, the class of systems considered here allow multiple Zeno points, where discrete transitions accumulate from the left and/or from the right. Note that for the time being we have not specified how the impulse times t_i and magnitudes α_i are to be determined. In general we view these as control actions that can be used (together with the selection of a solution of $\dot{x} \in F(x)$) to steer the system.

Definition 1 (Solution of impulsive system). *Consider an initial time $t_0 \geq 0$, a horizon $T > t_0$, and an initial condition $x_0 \in \mathbb{R}^d$. We will say that a function $x(\cdot) : [t_0, T) \to \mathbb{R}^d$ is a solution to the impulsive system (F, S, μ) over the interval $[t_0, T)$ with initial condition x_0 if and only if $x(t_0) = x_0$ and there exist a Lebesgue integrable function $\Phi_1(\cdot) : [t_0, T) \to \mathbb{R}^d$ and a $\mu(\cdot)$ integrable function $\Phi_2(\cdot) : [t_0, T) \to \mathbb{R}^d$ such that for all $t \in [t_0, T)$,*

$$x(t) = x_0 + \int_{t_0}^{t} \Phi_1(s)ds + \int_{t_0}^{t} \Phi_2(s)\mu(ds), \text{ and}$$

1. *$\Phi_1(s) \in F(x(s))$ for Lebesgue-almost all $s \in [t_0, T)$.*
2. *$\Phi_2(s) \in S(x(s^-), \mu(\{s\}))$ for μ-almost all $s \in [t_0, T)$.*

If the measure μ is not fixed, we will say that $x(\cdot) : [t_0, T) \to \mathbb{R}^d$ is a solution to (F, S) if there exists a measure μ such that $x(\cdot)$ is a solution to (F, S, μ). Without loss of generality we will take $t_0 = 0$ from now on.

Sufficient conditions for the existence of solutions will be presented in Section 3. For the remainder of this section we highlight relations of the class of impulsive systems defined above with other classes considered in the literature, ignoring for the most part technical issues, such as assumptions on the model components.

It is easy to see that if we set $S(x, \alpha) = \{0\}$ for all $(x, \alpha) \in \mathbb{R}^d \times [0, 1]$ the impulsive system considered above reduces to the standard differential inclusion

$$\dot{x} \in F(x)$$

on $\mathbb{R}^d$ and the definition of the solution to the standard definition of solution for differential inclusions. Likewise, if we set $F(x) = \{0\}$ and let $\mu(\cdot)$ be a regularly spaced sequence of pulses

$$\mu(t) = \sum_{i=1}^{\infty} \alpha_i \delta_i(t),$$

then the system reduces to a standard difference inclusion

$$x(k+1) \in R(x(k)) = \{x(k) + \alpha S(x(k), \alpha) \mid \alpha \in [0, 1]\}.$$

If we assume that S is single valued and that any compact subset of $[0, \infty)$ contains only finitely many t_i, then the definition of solution given above reduces to the more familiar definition for impulsive systems,

$$x(t) = x_0 + \int_{t_0}^{t} \Phi(s)ds + \sum_{i \in I(t)} \alpha_i S(x(t_i^-), \alpha_i),$$

where $I(t) = \{i \in I \mid t_i \leq t\}$.

Finally, [2,3] consider systems of the form

$$dx(t) = F(x(t))dt + G(x(t))\mu(dt), \tag{1}$$

where $F(\cdot)$ and $G(\cdot)$ are set valued maps and μ a non-negative, scalar valued Borel measure on $[0, \infty)$. Systems of the type (1) are clearly more general than the impulsive systems considered here, since the measure μ is not restricted to be discrete (i.e. satisfy Assumption 1.4). In the special case where it is, however, systems of the type (1) can be reduced to our framework. To see this, let

$$\mathcal{A}_G(x, \alpha) = \{y(\alpha) \mid y(0) = x \text{ and } y(\cdot) \text{ solves } \dot{y} \in G(y)\} \subseteq \mathbb{R}^d.$$

denote the attainable set at time α of the differential inclusion $\dot{y} \in G(y)$ starting at $y(0) = x$. If we define

$$S(x, \alpha) = \begin{cases} G(x) & \alpha = 0 \\ \frac{1}{\alpha}\left(\mathcal{A}_G(x, \alpha) - x\right) & \alpha > 0, \end{cases}$$

it is easy to see that the robust solutions of (1) are equivalent to the solutions of the impulsive system (F, S) defined as above. Therefore, for the special case of discrete measures, the class of systems considered here appears to subsume systems of the type (1) (modulo technical assumptions).

3 Existence of Solutions

Theorem 1 (Existence). *Under Assumption 1 for any $x_0 \in \mathbb{R}^d$ there exists a solution $x(\cdot)$ of (F, S, μ) over the interval $[0, T)$ starting at x_0.*

Theorem 1 shows that under Assumptions 1 solutions in the sense of Definition 1 can be defined, at least locally, up to time T. A simple extension (Corollary 1 below) can then be used to establish additional conditions needed to ensure global existence of solutions. The argument used to prove Theorem 1 is somewhat involved notationally. To help the reader follow it we will first introduce a simple example, that will be used throughout the proof to illustrate the various steps.

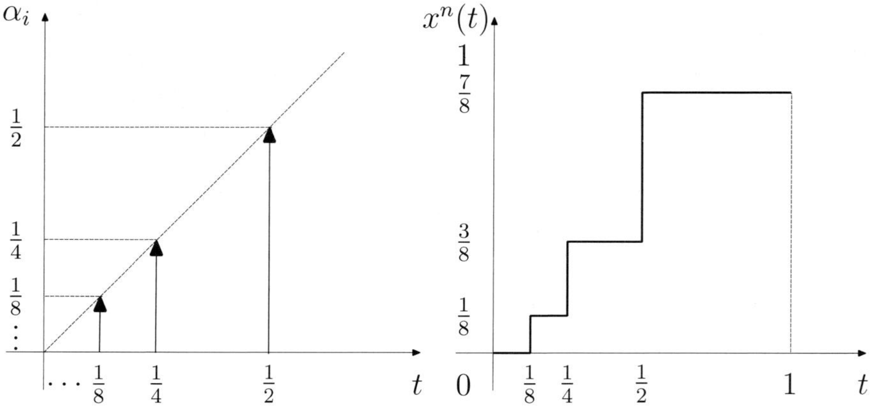

Fig. 1. The measure μ used in the example, and the approximating solution for $n = 3$ and $\gamma = \frac{1}{\sqrt{2}}$

Example: Consider an impulsive system with $d = 1$ (i.e. $x \in \mathbb{R}$), $F(x) = \{0\}$ (i.e. $\dot{x} = 0$) and $S(x, \alpha) = \alpha$ for all $x \in \mathbb{R}$ and $\alpha \in [0, 1]$. Consider also a measure μ with

$$I = \{1, 2, 3, \ldots\}, T = 1, t_i = \frac{1}{2^i}, \text{ and } \alpha_i = \gamma^i,$$

for some $\gamma \in (0, 1)$. Figure 1 illustrates the measure μ for the case $\gamma = 0.5$. Note that the times of the Dirac pulses converge to zero from the right. This type of behavior is already outside the scope of most traditional hybrid solution concepts. For simplicity, we assume that the initial state is $x_0 = 0$.

Proof of Theorem 1: Let $I(t) = \{i \in I \mid t_i \leq t\}$. If $I(T)$ is finite the system reduces to a usual impulsive system and the proof is standard. If $I(T)$ is infinite then assume, without loss of generality, that $I = \mathbb{N}$. Fix $n \in \mathbb{N}$ and consider the first n elements, t_1, t_2, ..., t_n of $\{t_i \mid i \in I(T)\}$. Rearrange these into an increasing sequence

$$\tau_1^n \leq \tau_2^n \leq \ldots \leq \tau_n^n \tag{2}$$

and add (if necessary) a first element $\tau_0^n = 0$ and a last element $\tau_{n+1}^n = T$; for simplicity we will assume that $0 \notin \{t_i \mid i \in I(T)\}$, so both additions are necessary.

In our case, $t_1 = \frac{1}{2}$, $t_2 = \frac{1}{4}$, ..., $t_n = \frac{1}{2^n}$ and $\tau_0^n = 0$, $\tau_1^n = \frac{1}{2^n}$, ..., $\tau_n^n = \frac{1}{2}, \tau_{n+1}^n = 1$.

Define an approximating trajectory $x^n(\cdot)$ over $[0, T)$ starting at x_0 by the algorithm of Table 1. The algorithm constructs a function $x^n(t)$ for $t \in [0, T)$ as well as a sequence of vectors $\{\zeta_i^n\}_{i=1}^n$ in $\mathbb{R}^d$ used to determine the destinations

Table 1. Algorithm for the generation of approximating trajectories

initialization:
$i = 0$, $x^n(\tau_0^n) = x_0$
while $i \leq n$
For $t \in [\tau_i^n, \tau_{i+1}^n)$ set $x^n(t)$ equal to a solution of $\dot{x} \in F(x)$
starting at $x^n(\tau_i^n)$
if $i < n$
Select $\zeta_{i+1}^n \in S(x^n((\tau_{i+1}^n)^-), \mu(\{\tau_{i+1}^n\}))$
Set $x^n(\tau_{i+1}^n) = x^n((\tau_{i+1}^n)^-) + \mu(\{\tau_{i+1}^n\})\zeta_{i+1}^n$
$i = i + 1$
end if
end while

of the discrete transitions. Let $\Phi_1^n(\cdot) : [0, T) \to \mathbb{R}^d$ denote the function that for $t \in [0, t)$ satisfies

$$x^n(t) = x_0 + \int_0^t \Phi_1^n(s)ds + \sum_{\{i \leq n \mid \tau_i^n \leq t\}} \mu(\tau_i^n)\zeta_i^n \tag{3}$$

Roughly speaking, the function $\Phi_1^n(\cdot)$ is constructed by concatenating the derivatives of the solutions to $\dot{x} \in F(x)$ generated by the algorithm. Notice that, by Assumption 1.1, $\Phi_1^n(\cdot)$ is bounded in $L^1([0,T])$. If we define $\xi_i^n = \zeta_j^n$ for the j for which $\tau_i^n = t_i^n$ (in other words, undo the reordering of (2)) equation (3) can equivalently be written as

$$x^n(t) = x_0 + \int_0^t \Phi_1^n(s)ds + \sum_{\{i \leq n \mid i \in I(t)\}} \mu(t_i)\xi_i^n. \tag{4}$$

In our case, at the first iteration of the while loop the algorithms solves $\dot{x} = 0$ over the interval $[0, \frac{1}{2^n})$ starting at $x^n(0) = 0$. Therefore, $x^n(t) = 0$ for $t \in [0, \frac{1}{2^n})$. The algorithm then selects

$$\zeta_1^n \in S(x^n((\tau_1^n)^-), \mu(\{\tau_1^n\})) = S(0, \mu\{\frac{1}{2^n}\}) = \gamma^n$$

and sets $x^n(\frac{1}{2^n}) = 0 + \mu\{\frac{1}{2^n}\}\zeta_1^n = \gamma^{2n}$. The while loop is then repeated, $\dot{x} = 0$ is solved over the interval $[\frac{1}{2^n}, \frac{1}{2^{n-1}})$ starting at $x^n(\frac{1}{2^n}) = \gamma^{2n}$, etc. At the final interval,

$$x^n(t) = \sum_{i=1}^n \gamma^{2i}, \text{ for } t \in [\frac{1}{2}, 1).$$

The trajectory constructed in this way, for $n = 3$ and $\gamma = \frac{1}{\sqrt{2}}$ is shown in Figure 1. Clearly $\Phi_1^n(t) = 0$ for all $t \in [0, 1)$, $\zeta_i^n = \gamma^{n+1-i}$ and $\xi_i^n = \gamma^i$.

Let now n go to infinity. Thanks to Assumption 1.1 the sequence Φ_1^n is bounded in both $L^1([0,T])$ and $L^\infty([0,T])$. Consequently, one can extract a subsequence which converges in the weak topology of $L^1([0,T])$ to some $\Phi_1 \in L^1([0,T])$.

$$\Phi_1^n(\cdot) \longrightarrow \Phi_1(\cdot) \in L^1([0,T]) \tag{5}$$

(For the sake of simplicity we denote the subsequence in the same way as the sequence.) Moreover, since by Assumption 1.2 $S(x,\alpha)$ is bounded, by the Cantor diagonal process, we can find a subsequence of $\{\xi_i^n\}_{i=1}^n$ (denoted the same way for simplicity) whose elements converge for all $i \in I(T)$. In other words, for all $i \in I(T)$ there exist ξ_i such that

$$\lim_{n\to\infty} \xi_i^n = \xi_i. \tag{6}$$

Moreover, it can be shown that as $n \to \infty$

$$\sum_{\{i\le n \mid i\in I(t)\}} \mu(t_i)\xi_i^n \longrightarrow \sum_{i\in I(t)} \mu(t_i)\xi_i \tag{7}$$

uniformly in $t \in [0,T)$.

Equations (5), (6) and (7) show that $x^n(\cdot)$ converges in $L^\infty([0,T))$ (up to a subsequence) to

$$x(t) = x_0 + \int_0^t \Phi_1(s)ds + \sum_{i\in I(t)} \mu(t_i)\xi_i. \tag{8}$$

We show that this $x(t)$ is a solution of (F,S,μ) over the interval $[0,T)$ starting at x_0. Let

$$\Phi_2(s) = \sum_{i\in I(T)} 1_{\{t_i\}}(s)\xi_i$$

(where, as usual, $1_{\{t_i\}}(s)$ denotes the indicator function of the set $\{t_i\}$). We will show that $x(\cdot)$, $\Phi_1(\cdot)$ and $\Phi_2(\cdot)$ satisfy the conditions of Definition 1.

Clearly $\Phi_1(\cdot)$ and $\Phi_2(\cdot)$ satisfy the desired integrability properties. Fix $s \in [0,T)$. Note that $\mu([0,T)\setminus\{t_i \mid i\in I(T)\}) = 0$. Therefore, unless $s \in \{t_i \mid i \in I(T)\}$, condition 2 of Definition 1 is automatically satisfied. If $s = t_i$ for some $i \in I(T)$, note that for all $t \in [0,T)$

$$x(t^-) = x_0 + \int_0^t \Phi_1(s)ds + \sum_{\{i\in I(t) \mid t_i<t\}} \mu(t_i)\xi_i. \tag{9}$$

and therefore, by (4), $\lim_{n\to\infty} x^n(t_i^-) = x(t_i^-)$. We know that

$$\xi_i^n \in S(x^n(t_i^-), \mu(\{t_i\}))$$

and that $\mu(t_i)\xi_i^n \to \mu(t_i)\xi_i$. Therefore, because $S(x,\alpha)$ is Lipschitz in x uniformly in α (Assumption 1.2), $\xi_i \in S(x(t_i^-), \mu(\{t_i\}))$ (Assumption 1.3). Hence

$$\Phi_2(t_i) = \xi_i \in S(x(t_i^-), \mu(\{t_i\})) \tag{10}$$

as required by Definition 1.

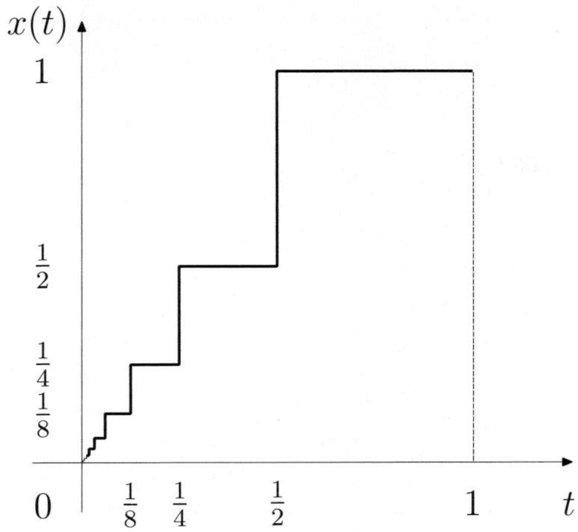

Fig. 2. Approximating solution for the example, with $n = 3$ and $\gamma = \frac{1}{\sqrt{2}}$

Finally, we know that $\Phi_1^n(s) \in F(x^n(s))$ almost everywhere in $[0, T)$. Since $x^n(\cdot)$ converges to $x(\cdot)$ in $L^\infty([0, T))$ and because F is Lipschitz (Assumption 1.3), by the Mazur Theorem (cf. Theorem 2.2.4, p.67 of [18]) we obtain that

$$\Phi_1(s) \in F(x(s)) \tag{11}$$

for almost all $s \in [0, T)$. Equations (10) and (11) show that $x(\cdot)$ is indeed a solution of the impulsive system, over the interval $[0, T)$ starting at x_0. ∎

Figure 2 shows the limit trajectory $x(t)$ of equation (8) for the example considered here. This trajectory is a solution to (F, S, μ) over the interval $[0, 1)$ starting at $x(0) = 0$.

Under some additional assumptions, the above construction can be extended (using a standard argument involving Zorn's Lemma) to establish the existence of solutions over the infinite interval $[0, \infty)$. For the sake of completeness, we state this observation as a corollary.

Corollary 1. *If, in addition to Assumption 1, F has linear growth and for all $T > 0$, $\mu([0, T]) < \infty$, then for every $x_0 \in \mathbb{R}^d$ there exists a solution $x(\cdot)$ of the impulsive system (F, S, μ) starting at x_0 and defined over the interval $[0, \infty)$.*

4 Example: Ball and Paddle

To demonstrate the type of complex Zeno phenomena that the class of systems considered here can encompass, we introduce one more simple example. It involves someone trying to keep a table tennis ball in the air by hitting it with the paddle. Our player is arbitrarily fast and elects to do this with a Zeno strategy

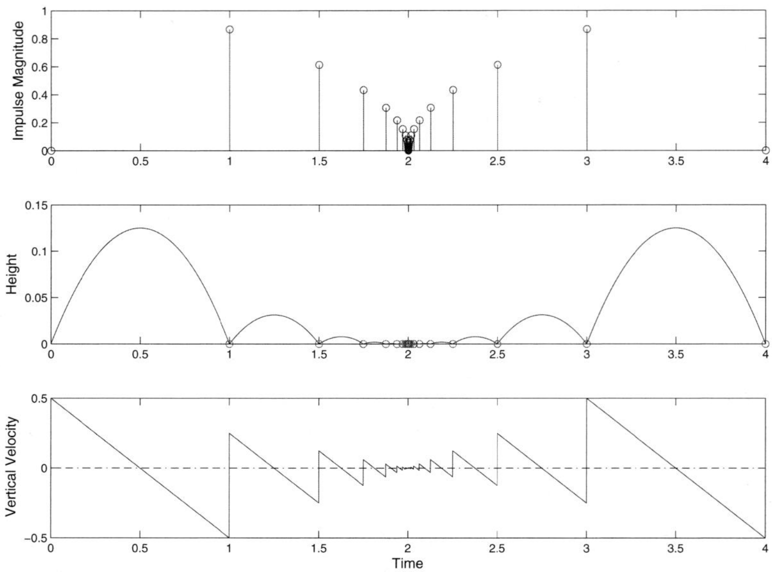

Fig. 3. Impulses and state trajectory for the ping-pong ball example

that consists of hitting the ball lighter and lighter until it comes to rest on the paddle and then starting it up again.

The continuous dynamics of the ball are determined by the standard equations of motion. If we let x_1 denote the height of the ball and x_2 its vertical velocity and normalize the mass and gravitational acceleration, we can simply write

$$F(x) = \begin{bmatrix} x_2 \\ -1 \end{bmatrix}.$$

The impulse magnitude, α, determines how much energy the player gives to the ball when he/she hits it with paddle. For the reset map we set

$$S(x, \alpha) = \begin{bmatrix} 0 \\ \alpha \end{bmatrix}.$$

In other words, an impulse of magnitude α resets the state (x_1, x_2) to $(x_1, x_2 + \alpha^2)$.

For simplicity we assume that the ball starts at $x = (0, 1/2)$ and that the player hits it whenever it gets back to $x_1 = 0$. We consider the interval of time $[0, 4]$ (the process can then be repeated) and assume that the impulses are placed symmetrically around $t = 2$ converging to it from the left and from the right,

$$i = 0, 1, 2, \ldots, \text{ with } t_{2i} = 2 - \frac{1}{2^i} \text{ and } t_{2i+1} = 2 + \frac{1}{2^i}.$$

It is easy to see that at $t_0 = 1$ (the first impulse time) the state of the ball is $(0, -1/2)$. Hitting the ball with an impulse of magnitude $\alpha_0 = \sqrt{3}/2$ at this time implies that the ball will leave the paddle flying upwards at a speed 1/4, half that with which it hit it coming down. The process is then repeated: at time $t_2 = 1.5$ the ball will come back to $(0, -1/4)$, we can then hit it with an impulse $\alpha_2 = \sqrt{3/2}/2$, it will leave the paddle flying at speed 1/8, etc. After the ball comes to rest at time $t = 2$, the process is reversed: the impulses get bigger and bigger until at time $t = 3$ the ball regains all its energy reaching state $(0, 1/2)$. In summary, the magnitudes of the impulses used are

$$i = 0, 1, 2, \ldots, \text{ with } \alpha_{2i} = \alpha_{2i+1} = \frac{1}{2}\sqrt{\frac{3}{2^i}}.$$

Figure 3 shows the impulse times and the solution generated in this way.

5 Viability

We now turn our attention to viability, in other words the question of whether solutions to the impulsive system that stay in a given set exist. For the class of systems considered here, one can envision two "controls" that can be used to keep the solution in the set: The direction in the differential inclusion followed by the continuous solution and the discrete intervention via the measure μ. To make this notion more precise, we first give a local definition.

Definition 2 (Viable set). *A set $K \subseteq \mathbb{R}^d$ is called viable under the system (F, S) if for all $x_0 \in K$ there exists a positive time $T > 0$, a measure μ satisfying Assumptions 1.4 and 1.5, and a solution $x(\cdot)$ of (F, S, μ) starting at x_0 and defined over $[0, T)$ such that $x(t) \in K$ for all $t \in [0, T)$.*

We will say that K is globally viable if we can take $T = \infty$ in the above definition (cf. Corollary 1 above).

Assumption 2. *For all $x \in \mathbb{R}^d$ the following conditions hold:*

1. *For all $0 < \alpha < \beta < 1$ with $(\beta - \alpha)S(x + \alpha S(x, \alpha), \beta - \alpha) \subseteq \beta S(x, \beta)$.*
2. *There exists $\alpha_x \in (0, 1]$ such that for all $\alpha \in [0, \alpha_x)$, $S(x, \alpha) = \{0\}$. We define $\alpha_x = \inf\{\alpha \mid S(x, [0, \alpha]) \neq \{0\}\}$.*
3. *With α_x defined as above, the map $R(x) = \{x + \alpha S(x, \alpha) \mid \alpha \geq \alpha_x\}$ is upper semi-continuous with closed values.*

Part 1 of the assumption implies that, in terms of reachability, there is nothing to be gained by taking many small jumps instead of one big one. Part 2 dictates that there is a minimum amount of "effort", α_x, needed to take a jump from state x. Finally, note that the last part of the assumption is satisfied in particular if $S(\cdot, \cdot)$ is continuous (both upper- and lower-semicontinuous).

For a closed subset, $K \subseteq \mathbb{R}^d$, and a point $x \in K$, we use $T_K(x)$ to denote the *contingent cone* to K at x, i.e. the set of $v \in \mathbb{R}^d$ such that there exists a sequence of real numbers $h_n > 0$ converging to 0 and a sequence of $v_n \in \mathbb{R}^d$ converging to v satisfying

$$\forall\, n \geq 0, \;\; x + h_n v_n \;\in\; K.$$

Theorem 2 (Viability). *Consider a system (F, S) satisfying Assumptions 1 and 2. A closed set $K \subseteq \mathbb{R}^d$ is viable under (F, S) if and only if for all $x \in K \setminus R^{-1}(K)$, $F(x) \cap T_K(x) \neq 0$.*

Proof: Necessity: Assume that K is viable. For $x_0 \in K$ consider $\mu(t) = \sum_{i \in I} \delta_{t_i}(t)$, $T > 0$ and a solution $x(\cdot)$ of (F, S, μ) over $[0, T)$ starting at x_0 such that $x(t) \in K$ for all $t \in [0, T)$. If $x_0 \in K \setminus R^{-1}(K)$, then there exists $\eta > 0$ such that $B(x_0, \eta) \cap R^{-1}(K) = \emptyset$ (note that $R^{-1}(K)$ is closed, since R is upper semi-continuous and K is closed). Since $x(t) \in K$ for all $t \in [0, T)$ and $R(x(t)) \cap K = \emptyset$ if $x(t) \notin R^{-1}(K)$, there exists $\epsilon < T$ such that $x(\cdot)$ solves $\dot{x} \in F(x)$ over $t \in [0, \epsilon)$. By Assumption 1.1 and the definition of $T_K(x_0)$, we have (cf. Proposition 3.4.1, p. 93 of [18])

$$\limsup_{t \to 0+} \frac{x(t) - x_0}{t} \in F(x_0) \cap T_K(x_0).$$

Hence $F(x_0) \cap T_K(x_0) \neq \emptyset$.

Sufficiency: Assume that for all $x \in K \setminus R^{-1}(K)$, $F(x) \cap T_K(x) \neq \emptyset$. For $x \in K \cap R^{-1}(K)$, let

$$\beta_x = \sup\{\alpha > \alpha_x \mid \frac{K - x}{\alpha} \cap S(x, \alpha)\}.$$

Fix $x_0 \in K$. We will construct a solution of the system that stays in K. We distinguish two cases.
Case 1: $x_0 \in R^{-1}(K) \cap K$. Note that since, by definition, $\beta_{x_0} > \alpha_{x_0}$, then

$$(x_0 + \beta_{x_0} S(x_0, \beta_{x_0})) \cap K \neq \emptyset.$$

Let $t_0 = 0$ and select $\mu(\{t_0\}) = \beta_{x_0}$ and

$$\Phi_2(t_0) \in \frac{K - x_0}{\beta_{x_0}} \cap S(x_0, \beta_{x_0}).$$

Note that the trajectory defined in this way starts with a discrete jump from x_0 to $x_0 + \Phi_2(t_0)\beta_{x_0} \in K$. Moreover, by Assumption 2.1 and the definition of β_x,

$$x_0 + \beta_{x_0} S(x_0, \beta_{x_0}) \cap K \cap R^{-1}(K) = \emptyset.$$

Therefore, after this initial jump we end up in Case 2, which we will treat now.

Case 2: $x_0 \notin K \setminus R^{-1}(K)$. Then by the viability with target theorem of [19] (see also the non-Lipschitz extension of [8]) there exists a solution of $\dot{x} \in F(x)$ starting at $x(0) \in x_0$ that either stays in K for ever, or stays in K until it reaches $R^{-1}(K)$. In the former case the proof is complete. In the latter, we have defined a solution $x(\cdot)$ of $\dot{x} \in F(x)$ (and hence of (F, S)) starting at x_0 over the interval $[0, \theta_{R^{-1}(K)}(x(\cdot)))$ with

$$\theta_{R^{-1}(K)}(x(\cdot)) = \inf\{t \geq 0 \mid x(t) \in R^{-1}(K)\}.$$

The process can then be iterated, leading to a (finite or infinite) sequence of impulses $\beta_{\hat{x}_i}\delta_{t_i}(t)$ (defined by case 1 above) and a sequence of solutions of $\dot{x}_i(t) \in F(x_i(t))$ (defined by case 2 above) such that for $i > 0$, $t_i = t_{i-1} + \theta_{R^{-1}(K)}(x_i(\cdot))$ and $\hat{x}_i = x_i(\theta_{R^{-1}(K)}(x_i(\cdot)))$. By construction the solution is viable in K and defined over the interval $\left[0, \sum_i \theta_{R^{-1}(K)}(x_i(\cdot))\right)$. ■

The result can be extended to global viability by introducing assumptions to ensure that the system is non-Zeno (e.g. compactness of $R(K)$ and $R^{-1}(K)$).

6 Concluding Remarks

A class of hybrid systems comprising differential inclusions driven by impulsive measures was introduced and basic properties (such as conditions for existence of solutions) were studied. An interesting feature of this class of systems is that one can define solutions even in the presence of very complex Zeno type behavior.

A number of extensions of the results discussed here are currently being pursued. An immediate improvement is the derivation of more powerful viability conditions. For example, the viability conditions given above implicitly require solutions to be non-Zeno to ensure global viability. One would like to relax these conditions and establish viability (perhaps with a bound on the impulsive measure used) in the presence of Zeno phenomena. An interesting related question is whether systems exist that are viable with Zeno type impulsive controls, but not viable if Zeno controls are disallowed. Very interesting (but also very challenging technically) is the extension of the results to more general, non-impulsive measures μ. In this case one could conceivably also capture behaviors such as the ball coming to rest on the paddle (in the example of Section 4) which is impossible to do if we restrict our attention to impulsive measures.

Acknowledgment. The work was supported by the European Commission under the project HYGEIA, FP6-NEST-4995, and the Network of Excellence HYCON, FP6-IST-511368. The authors would like to thank K. Koutroumpas for his help with the simulation of the ping-pong ball.

References

1. Bernhard, P., El Farouq, N., Thiery, S.: An impusive game arising in finance with interesting singularities. In A. Haurie, S. Muto, L. Petrosjan, T.E.S. Raghavan, eds.: Advances in Dynamic Games: Applications to Economics, Management Science, Engineering, and Environmental Management. Number 8 in Annals of the International Society of Dynamic Games. Birkhäuser, Boston, MA (2006) 335–364
2. Dal Maso, G., Rampazzo, F.: On systems of ordinary differential equations with measures as controls. Differential and Integral Equations **4** (1991) 739–765
3. Silva, G., Vinter, R.: Measure driven differential inclusions. Journal of Mathematical Analysis and Applications **202** (1996) 727–746
4. Mota, M., Rampazzo, F.: Dynamic programming for nonlinear systems driven by ordinary and impulsive controls. SIAM Journal on Control and Optimization **34**(1) (1996) 199–225

5. Silva, G., Vinter, R.: Necessary conditions for optimal impulsive control problems. SIAM Journal on Control and Optimization **35**(6) (1997) 1829–1846
6. Miller, J.: Decidability and complexity results for timed automata and semi-linear hybrid automata. In Lynch, N., Krogh, B.H., eds.: Hybrid Systems: Computation and Control. Number 1790 in LNCS. Springer-Verlag, Berlin (2000) 296–309
7. Imura, J., van der Schaft, A.J.: Characterization of well-posedness of piecewise linear systems. IEEE Transactions on Automatic Control **45**(9) (2000) 1600–1619
8. Aubin, J.P., Lygeros, J., Quincampoix, M., Sastry, S., Seube, N.: Impulse differential inclusions: A viability approach to hybrid systems. IEEE Transactions on Automatic Control **47**(1) (2002) 2–20
9. Lygeros, J., Johansson, K., Simić, S., Zhang, J., Sastry, S.: Dynamical properties of hybrid automata. IEEE Transactions on Automatic Control **48**(1) (2003) 2–17
10. Imura, J.: Well-posedness analysis of switch-driven piecewise affine systems. IEEE Transactions on Automatic Control **48**(11) (2003) 1926–1935
11. A.Yu. Pogromsky, Heemels, W., Nijmeijer, H.: On solution concepts and well-posedness of linear relay systems. Automatica **39** (2003) 2139–2147
12. Goebel, R., Hespanha, J., Teel, A., Cai, C., Sanfelice, R.: Hybrid systems: Generalized solutions and robust stability. In: IFAC Symposium on Nonlinear Control Systems. (2004)
13. Johansson, K., Egerstedt, M., Lygeros, J., Sastry, S.: On the regularization of Zeno hybrid automata. Systems and Control Letters **38**(3) (1999) 141–150
14. Zheng, H., Lee, E., Ames, A.: Beyond Zeno: Get on with it! In Hespanha, J., Tiwari, A., eds.: Hybrid Systems: Computation and Control. Number 3927 in LNCS. Springer-Verlag, Berlin (2006) 568–582
15. Ames, A., Zheng, H., Gregg, R., Sastry, S.: Is there life after Zeno? Taking executions past the breaking (Zeno) point. In: American Control Conference. (2006)
16. Filippov, A.F.: Differential equations with discontinuous right-hand sides. Kluwer Academic Publishers (1988)
17. Camlibel, M., Heemels, W., van der Schaft, A., Schumacher, J.: Solution concepts for hybrid dynamical systems. In: IFAC World Congress. (2002)
18. Aubin, J.P.: Viability Theory. Birkhäuser, Boston, MA (1991)
19. Quincampoix, M., Veliov, V.: Viability with target: Theory and applications. In Cheshankov, B., Todorov, M., eds.: Applications of Mathematics in Engineering. Heron Press, Sofia (1998) 47–54

CEGAR Based Bounded Model Checking of Discrete Time Hybrid Systems

Federico Mari and Enrico Tronci*

Dipartimento di Informatica, Università di Roma "La Sapienza",
Via Salaria 113, 00198 Roma, Italy
Tel.: +39 06 4991 8361; Fax: +39 06 8541 842
{mari,tronci}@di.uniroma1.it

Abstract. Many hybrid systems can be conveniently modeled as *Piecewise Affine Discrete Time Hybrid Systems* PA-DTHS. As well known *Bounded Model Checking* (BMC) for such systems comes down to solve a *Mixed Integer Linear Programming* (MILP) feasibility problem.

We present a SAT based BMC algorithm for automatic verification of PA-DTHSs. Using *Counterexample Guided Abstraction Refinement* (CEGAR) our algorithm *gradually* transforms a PA-DTHS verification problem into larger and larger SAT problems.

Our experimental results show that our approach can handle PA-DTHSs that are more then 50 times larger than those that can be handled using a MILP solver.

1 Introduction

Automatic analysis of *Hybrid Systems* poses formidable challenges both from a modeling as well as from a verification point of view. In fact the simultaneous presence of continuous and discrete variables may soon lead to *state explosion*, thus preventing completion of the verification process.

Many verification tools (*model checkers*) are available for automatic verification of hybrid systems. Here are a few examples. *Linear Hybrid Systems* (LHS) can be verified using HyTech [18,2,1]. If we restrict ourselves to LHSs in which all continuous variables have time derivative equal to 1 (clocks) then the UPPAAL [20,29] model checker can be used.

If we use a discrete model for time then we have *Discrete Time Hybrid Systems* (DTHSs). Many systems can be modeled or approximated using DTHSs. A model checker for (possibly nonlinear) DTHS is CMurphi [25,13,7].

Tools originally designed for hardware verification have also been used for hybrid systems verification. For example, in [28] SMV [23,26] has been used for verification of chemical processing systems.

By restricting our attention to linear DTHS we can design more efficient verification algorithms. *Piecewise Affine Discrete Time Hybrid Systems* (PA-DTHS) are an important subclass of DTHS. PA-DTHSs are DTHS whose behaviour can

* Corresponding author.

A. Bemporad, A. Bicchi, and G. Buttazzo (Eds.): HSCC 2007, LNCS 4416, pp. 399–412, 2007.

be defined using linear constraints involving real as well as discrete variables. An important subclass of PA-DTHSs are *Discrete Hybrid Automata* (DHA).

For DHA quite efficient *Mixed Integer Linear Programming* (MILP) based verification algorithms have been designed [4] and implemented within the symbolic DHA model checker HYSDEL [27,19].

SAT based *Bounded Model Checking* (BMC) [5,14] has turned out to be quite effective on hardware (e.g. see [21,31]), as well as on software systems (e.g. see [10,9]). Thus trying to use BMC in a DTHS setting is a quite natural step. For HYSDEL DHA this has been studied in [15].

Another interesting class of PA-DTHS is the one that can be handled by MathSAT [3,22]. SAT based *Counterexample Guided Abstraction Refinement* (CEGAR) [11] has also turned out to be a quite effective enhancement to BMC. This is the case for hardware verification (e.g. see [21]) as well as for software verification (e.g. see [17]). Our main contributions can be summarized as follows.

In Section 3 we define a quite large class of *Piecewise Affine* DTHS. Our class of DTHSs strictly contains those that can be handled by, e.g., UPPAAL, HYSDEL or MathSAT (Section 4).

In Section 5 we show how the BMC problem for a DTHS $\mathcal{H}$ in our class can be cast as an MILP feasibility problem $P_{\mathcal{H}}$. Of course, following [4], we could solve such MILP problem using a solver (e.g. GLPK [16], or CPLEX [12]). However for feasibility problems having many discrete variables (our target here) MILP solvers tend to be rather inefficient. For this reason our approach will be that of transforming our BMC problem for DTHSs into a SAT problem.

In Section 6 we give a *sound* and *complete* (up to ϵ) algorithm to transform our MILP problem $P_{\mathcal{H}}$ into a *Boolean Linear Programming* (BLP) problem $F^{\epsilon}_{\mathcal{H}}$. Our ϵ approximation of $P_{\mathcal{H}}$ only discretizes the continuous variables of $P_{\mathcal{H}}$. Effectiveness of our transformation rests on the fact that we *do not* discretize the real coefficients of the constraints in $P_{\mathcal{H}}$. Instead, for each contraint we generate a compact CNF (*conjunctive normal form*) representation of the set of assignments that falsify it.

In Section 7 we show how our BLP problem $F^{\epsilon}_{\mathcal{H}}$ can be effectively transformed into an equivalent SAT problem $B^{\epsilon}_{\mathcal{H}}$.

In Section 8, building on the transformation defined in Sections 7, we present a CEGAR based algorithm to solve our BLP problem $F^{\epsilon}_{\mathcal{H}}$ using a SAT solver. This yields a SAT based CEGAR BMC algorithm for our class of DTHSs. To the best of our knowledge for the class of systems we consider here no CEGAR BMC algorithms have been proposed in the literature. For example, our class of systems cannot be handled by the BMC algorithm proposed in [15].

In Section 9 we present experimental results showing effectiveness of the proposed approach. Given a DTHS $\mathcal{H}$ in our class we have essentially two ways of carrying out BMC: use an MILP solver to check feasibility of $P_{\mathcal{H}}$ *or* use our SAT based CEGAR approach. We present experimental results showing that using our SAT based CEGAR approach we can handle systems that are more than 50 times larger than those that can handled by an MILP solver.

2 Background

Notation 1. Let $X = [x_1, \ldots x_n]$, $Y = [y_1, \ldots y_n]$, be finite sequences (lists) of variables. By abuse of language we may regard sequences as set and we use $\cup$ to denote list concatenation. We denote with $f(X := Y)$ the expression $f([x_i := y_i | i = 1 \ldots n])$, that is the simultaneous substitution of the variables in X with those in Y. Moreover if X is clear form the context we just write $f(Y)$ for $f(X := Y)$.

Let $x \in X$, we denote with $\mathcal{D}_x$ the domain of x, that is the set on which x ranges. A *valuation* (over a list of variables X) is a function that maps each variable $x \in X$ to a value in $\mathcal{D}_x$. That is, a valuation is a point in $\times_{x \in X} \mathcal{D}_x$.

A *linear expression* over a list of variables X is a linear combination of variables in X with real coefficients. A *constraint* (over a list of variables X) is an expression of the form $\alpha \leq b$ where: α is a linear expression and b is a real constant. A *convex predicate* (over a list of variables X) is a finite conjunction of constraints. A *predicate* is defined as follows. A convex predicate is a predicate. If A and B are predicates then $(A \wedge B)$ is a predicate and $(A \vee B)$ is a predicate. As a *syntactic sugar*, if y is a discrete variable we will write $y < b$ for $y \leq b - 1$, $y > b$ for $y \geq b + 1$, $y \neq b$ for $((y \leq b - 1) \wedge (y \geq b + 1))$.

Classically a *Mixed Integer Linear Programming* (MILP) problem [8] is a linear optimization problem. However, in our context, we are only interested in finding *feasible* solutions. For this reason our definition of MILP does not include an objective function to be minimized.

Definition 1. *Let $M(X)$ be a convex predicate. The* Mixed Integer Linear Programming *(MILP) problem for $M(X)$ consists in finding a valuation V s.t.: 1. $M(V)$ holds 2. for all $x \in X$, $V(x) \in \mathcal{D}_x$. In other words we are looking for a satisfying assignment for the variables of M. An* Integer [Boolean] Linear Programming *(ILP [BLP]) problem is an MILP problem with only integer [boolean] variables.*

Usually an MILP problem P is represented as a list of constraints. That is $P = \{\sum_{j=1}^{n} a_{ij}x_j \leq b_j \mid i = 1, \ldots m$ and for $j = 1, \ldots n,\ x_j \in \mathcal{D}_{x_j}\}$. Using standard MILP techniques (e.g. [8]) it is possible to prove the following propositions.

Proposition 1. *Given a predicate $P(X)$ there exist a list Q of boolean variables and a convex predicate $L(Q, X)$ s.t. $\forall X\ [P(X)$ iff $\exists Q L(Q, X)]$.*

Proposition 2. *Given an MILP problem P there exists an MILP problem Align(P) s.t. 1. P is feasible iff Align(P) is feasible; 2. All variables in Align(P) are nonnegative.*

Proposition 3. *Given an ILP problem P there exists a BLP problem Q s.t. 1. P is feasible iff Q is feasible; 2. All variables in Q are boolean.*

It can be easily shown [8] that, in general, MILP feasibility is an NP-complete problem. However there are quite effective *solvers* (e.g. CPLEX [12], GLPK [16]) that can handle non trivial MILP optimization problems.

Remark 1. Note however that MILP solvers are designed to solve optimization problems rather than feasibility problems. In particular the *branch-and-bound* heuristics typically implemented in MILP solvers are not effective on feasibility problems since there is no objective function for computing *the bound* in the branch-and-bound process.

For the above reason, state-of-the-art commercial MILP solvers like CPLEX perform as poorly as state-of-the-art open source MILP solvers like GLPK on MILP feasibility problems with many discrete variables (our case here). In fact, feasibility problems do not have an objective function and thus CPLEX sophisticated heuristics tend to be quite ineffective.

3 Piecewise Affine Discrete Time Hybrid Systems

In this section we introduce a class of *Piecewise Affine Discrete Time Hybrid Systems* (DTHSs for short).

Many classes of piecewise affine hybrid systems have been studied in the literature, e.g. [2,20]. The same holds true for piecewise affine discrete time hybrid systems, e.g. [4,27,15,3]. The class of systems we are considering is essentially the one used in [30].

Definition 2. *A* Discrete Time Hybrid System *(DTHS for short) is a 6-tuple* $\mathcal{H} = (Q, X, Init, Inv, r, R)$ *where:*

- $Q = [q_1, \ldots q_k]$ *is a finite sequence of* discrete *variables. Each variable* $q \in Q$ *ranges on a finite subset* $[l_q, u_q]$ *of the integers* $\mathbb{Z}$. *Thus* $\mathcal{D}_q = [l_q, u_q]$.
- $X = [x_1, \ldots x_n]$ *is a finite sequence of* real-valued *variables. Each variable* $x \in X$ *ranges on a bounded interval* $[l_x, u_x]$ *of the reals* $\mathbb{R}$. *Thus* $\mathcal{D}_x = [l_x, u_x]$. *Of course* Q *and* X *are disjoint lists.*
- $Init(Q, X)$ *is a* predicate *over* $Q \cup X$.
- $Inv(Q, X)$ *is a* predicate *over* $Q \cup X$.
- $r(Q, X, X')$ *is a* predicate *over* $Q \cup X \cup X'$, *where* $X' = [x'_1, \ldots x'_n]$.
- $R(Q, X, Q', X')$ *is a* predicate *over* $Q \cup X \cup Q' \cup X'$, *where* $Q' = [q'_1, \ldots q'_k]$.

As usual primed variables denote "next state" values. Usually, when modeling a DTHS, R is used to define *reset* transitions, that is $R(q, x, q', x')$ implies $q \neq q'$. This is also our modeling style. However, from a formal point of view, Definition 2 only requires that R defines next state values for discrete states.

The list of *state variables* S for the DTHS $\mathcal{H} = (Q, X, \text{Init}, \text{Inv}, r, R)$ is $S = Q \cup X$. A *state* for $\mathcal{H}$ is a valuation $s = (q, x)$ of S, where q is a valuation of Q and x is a valuation of X.

A *run* for the DTHS $\mathcal{H}$ is a sequence $(q(0), x(0))$, $(q(1), x(1))$, $\ldots$ of states of $\mathcal{H}$ such that the following conditions are satisfied: 1. $\text{Init}(q(0), x(0)) \wedge \text{Inv}(q(0), x(0))$; 2. For all $k \geq 0$, $(R(q(k), x(k), q(k+1), x(k+1)) \vee (r(q(k), x(k), x(k+1)) \wedge q(k+1) = q(k))) \wedge \text{Inv}(q(k+1), x(k+1))$.

If $\pi = (q(0), x(0)), (q(1), x(1)), \ldots$ is a run of $\mathcal{H}$ we denote with $\pi(k)$ the k-th element of π. That is $\pi(k) = (q(k), x(k))$.

A state (q, x) of $\mathcal{H} = (Q, X, \text{Init}, \text{Inv}, r, R)$ is *k-reachable* if there exists a run π of $\mathcal{H}$ and there exists a $t \leq k$ s.t. $\pi(t) = (q, x)$.

In this paper we focus on bounded model checking of safety properties. That is, our goal is to check that for system $\mathcal{H}$ no *error state* is k-reachable. To this end we need to define the set of error states. This can be easily done using a predicate. The above considerations lead to the following definition.

Definition 3. *Let $\mathcal{H} = (Q, X, Init, Inv, r, R)$ be a DHTS, $E(Q, X)$ be a predicate over $(Q \cup X)$ and k be a natural number. We say that the triple $(\mathcal{H}, E, k)$ is* safe *(or that $\mathcal{H}$ is k-*safe *w.r.t. E) iff there is no run π of $\mathcal{H}$ for which there exists a $t \leq k$ s.t. $\pi(t) = (q, x)$ and $E(q, x)$ holds (i.e. $E(q, x) = 1$). In other words, no k-reachable state of $\mathcal{H}$ satisfies E.*

4 An Example of DTHS

We give an example of DTHS that will be useful to clarify the class of systems we are targeting. Consider a system consisting of k water pumps and $n > k$ tanks. Pump i $(i = 1, \ldots k)$ has (discrete) position $p_i \in \{1, \ldots n\}$, where $p_i = j$ means that pump i is above tank j. Each pump moves forward $(w_i = 1)$ and backward $(w_i = 0)$ between positions 1 and n. Pump i starts from position i. Each pump must stay in a position for at least α time units and must leave the position after at most $\beta \geq \alpha$ time units.

To compute the amount of water in a tank it is useful to introduce a boolean variable $z_{i,j}$ s.t. $z_{i,j} = 1$ iff water tank i is getting water from pump j, that is $z_{i,j} = 1$ iff $p_j = i$. This can be defined with the predicate $Z_{i,j} = (z_{i,j} = 0 \vee p_j = i) \wedge (z_{i,j} = 1 \vee p_j \neq i)$.

Each tank may get water from any of the pumps. Water flows out from tank i from a sink that can be open $(u_i = 1)$ or closed $(u_i = 0)$. The dynamics of the water level x_i of tank i satisfies the following constraint: $P_i = (x'_i \leq x_i - \gamma u_i + \sum_{j=1}^{k} \eta z_{i,j}) \wedge (x'_i \geq x_i - \mu u_i + \sum_{j=1}^{k} \theta z_{i,j})$, where: γ, μ model, respectively, min and max flow of water out of tank i and η, θ model, respectively, max and min flow of water out of pump j.

Water demand is modeled as follows. A tank sink can stay open for at most ζ time units and can stay closed for at most ξ time units. Moreover, the number of open sinks is at most $\Lambda \leq n$ and at least $\Gamma \leq \Lambda \leq n$. That is, $\sum_{i=1}^{n} u_i \geq \Gamma$ and $\sum_{i=1}^{n} u_i \leq \Lambda$. From the above description we can define a DTHS $\mathcal{H}$.

We expect that each tank i, has *enough* water $(x_i \geq m)$, but not *too much* $(x_i \leq M)$. Thus the predicate E representing our *error condition* can be defined as follows: $\bigvee_{i=1}^{n}((x_i < m) \vee (x_i > M))$.

Given an horizon k, our goal is to check that the triple $(\mathcal{H}, E, k)$ is *safe*.

Remark 2. Our class of DTHSs cannot be handled using the UPPAAL model checker, since we are not restricted to clock variables (e.g. see [20]). For example, tank water levels (x_i) cannot be modeled as UPPAAL clocks.

Remark 3. Our class of DTHS cannot be handled with HYSDEL since in our invariant we may have constraints consisting of discrete state variables. Such constraints are not handled by DHA (e.g. see [15]). For example our invariant constraints $\sum_{i=1}^{n} u_i \geq \Gamma$, $\sum_{i=1}^{n} u_i \leq \Lambda$ cannot be handled using HYSDEL. Moreover we can handle nondeterminism in the discrete time dynamics which cannot be modeled with the DHA in [15].

Remark 4. Our class of DTHS cannot be handled using the MathSAT tool since we have constraints involving continuous as well as discrete variables, whereas MathSAT only handles constraints built out of one type of variables, i.e. only continuous variables or only discrete variables (e.g. see [6]). For example x_i's constraints in $x'_i \leq x_i - \gamma u_i + \sum_{j=1}^{k} \eta z_{i,j}$, $x'_i \geq x_i - \mu u_i + \sum_{j=1}^{k} \theta z_{i,j}$ cannot be modeled using MathSat since they involve continuous (x_i) as well as discrete ($z_{i,j}$) variables.

5 BMC of DTHS as a MILP Problem

The BMC problem for DTHSs can be cast as an MILP problem. For DHA this has been shown in [4]. Along the same line we can show that the same holds for DTHS.

Theorem 1. *Let $\mathcal{H}$ = (Q, X, Init, Inv, r, R) be a DTHS, $E(Q, X)$ be an error condition for $\mathcal{H}$ and k be a natural number. Then there exists a convex predicate $M(Y, Q_0, X_0, \ldots Q_k, X_k)$ s.t.*

- *All variables in Y are booleans;*
- *($\mathcal{H}$, E, k) is safe iff the MILP problem $M(Y, Q_0, X_0, \ldots Q_k, X_k)$ does not have a solution*
- *Let $y, (q(0), x(0)), \ldots (q(k), x(k))$ be a solution to the MILP problem $M(Y, Q_0, X_0, \ldots Q_k, X_k)$. Then $\phi = (q(0), x(0)), \ldots (q(k), x(k))$ is a path in $\mathcal{H}$ containing an error state. That is, there is a $t \leq k$ s.t. $E(q(t), x(t))$.*

6 From MILP to BLP

A BMC problem for DTHSs can be transformed into an MILP problem (Theorem 1). In order to transform a BMC problem for DTHSs into a SAT problem, here we show how an MILP problem can be transformed into a *Boolean Linear Programming* (BLP) problem. We do this in three steps. First we transform our MILP problem P into a problem P_1 in which all variables are nonnegative (Proposition 2). Second, given $\epsilon > 0$, we transform problem P_1 into an *Integer Linear Programming* (ILP) problem P_2^ϵ (Proposition 4). Finally, we transform P_2^ϵ into a BLP problem F^ϵ (Proposition 3).

Approximating continuous variables with discrete ones generates discretization errors. To account for such errors we relax P constraints. Let P be an MILP problem and $\epsilon > 0$. We define the ϵ-relaxation P_ϵ of P which is obtained by replacing the rhs b of each constraint in P with $(b + \epsilon)$. Thus ϵ defines the relaxation we are willing to accept on P constraints.

The *size of a linear constraint* is the number of variables occurring in it. The *size of an MILP* problem P ($|P|$) is the sum of the sizes of its constraints. Because of lack of space we omit the proof of the following proposition.

Proposition 4. *Given $\epsilon > 0$ and an MILP problem P in which all variables are nonnegative there exists a linear time (in $|P|$) algorithm $\mathcal{Q}_\epsilon$ s.t. 1. $\mathcal{Q}_\epsilon(P)$ is an ILP problem. 2. If P is feasible then $\mathcal{Q}_\epsilon(P)$ is feasible. 3. If $\mathcal{Q}_\epsilon(P)$ is feasible then P_ϵ is feasible.*

Assembling the transformations in Propositions 2, 4, 3 we can constructively prove the following theorem.

Theorem 2. *Let P be an MILP problem and let $\epsilon > 0$. Then there exists a linear time (in $|P|$) algorithm $\mathcal{L}_\epsilon$ s.t. 1. $\mathcal{L}_\epsilon(P)$ is a BLP problem. 2.* (Soundness) *If P is feasible then $\mathcal{L}_\epsilon(P)$ is feasible. 3.* (Completeness) *If $\mathcal{L}_\epsilon(P)$ is feasible then P_ϵ is feasible. 4. If $\mathcal{L}_\epsilon(P)$ is feasible [infeasible] then $\forall \epsilon' \geq \epsilon$ $[\forall 0 \leq \epsilon' \leq \epsilon]$ $\mathcal{L}_{\epsilon'}(P)$ is feasible [infeasible].*

Theorem 2 says that by using $\mathcal{L}_\epsilon(P)$ instead of P we never have false negatives. That is we never declare safe an unsafe system. Moreover we never have false positives using $\mathcal{L}_\epsilon(P)$ instead of P_ϵ. On the other hand, $\mathcal{L}_\epsilon(P)$ may be feasible and P may be infeasible. That is we may declare unsafe a safe system. Note however that, as usual, by making ϵ *small enough* we can make the difference between P and P_ϵ arbitrarily small, at the price of making $|\mathcal{L}_\epsilon(P)|$ grow. Infact, it can be shown that the number of boolean variables of $\mathcal{L}_\epsilon(P)$ coming from the discretization of continuous variables is proportional to $\log_2(\epsilon^{-1})$.

Theorem 3, which proof we omit because of lack of space, guarantees that we never run into an infinite sequence of false positives. That is by taking smaller and smaller values for ϵ eventually we have that $\mathcal{L}_\epsilon(P)$ is infeasible (and so is P by Theorem 2) or that $\mathcal{L}_\epsilon(P)$ is feasible and so is P.

Theorem 3. *If P is infeasible then there exists $\epsilon > 0$ s.t. $\mathcal{L}_\epsilon(P)$ is infeasible.*

7 From BLP to SAT

Using Theorems 1 and 2, given a *tolerance* ϵ we can transform a BMC problem for DTHS $\mathcal{H}$ into a BLP problem M. In this Section we show how M can be transformed into a SAT problem.

Of course we may get a similar result by discretizing the real-valued coefficients in the linear inequalities in M, implementing floating point arithmetic and then translating the all problem into SAT. This is the approach followed, e.g., in the CBMC model checker [9]. However, if we follow such an approach even small systems will result in huge *Conjunctive Normal Forms* (CNFs), e.g. see the CBMC manual in [9]. Hence here we follow a different approach.

We represent a constraint P in M with its non-satisfying assignments. This yields an often compact CNF representation for P. The details follow.

Let $P = \sum_{i=n}^{n} a_i x_i \leq b$ be a constraint in M. We are looking for a CNF F s.t. $F(X) = 1$ iff $P(X) = 1$. Let V be the set of assignments that make P

```
int blp2sat(a, b) {m1 = 0; m2 = 0;
k = index of first undefined value in x;
if (k > n)
{if (∑_{i=1}^{n} a[i]x[i] > b) {F = (F ∧ clause(x)); return 1; }
 else return 0;}
if (min(a, x) > b) {F = (F ∧ clause(x)); return 1;}
if (a[k] > 0) {x[k] = 1} else {x[k] = 0;}
m1 = blp2sat(a, b);
if (m1 > 0) { x[k] = 1 - x[k]; m2 = blp2sat(a, b); }
return (m1 + m2); }
```

Fig. 1. Sketch of algorithm BLP2SAT

false. That is, $V = \{x \mid P(x) = 0\}$. The characteristic function for V is: $\chi_V = \bigvee_{(v_1,\ldots v_n)\in V} \bigwedge_{i=1}^{n} x_i^{v_i}$, where $x_i^{v_i}$ is $(x_i = v_i)$. Thus $x_i^1 = x_i$, $x_i^0 = \bar{x_i}$.

Let $F(X) = \neg\chi_V(X)$. Then $P(X) = 1$ iff $F(X) = 1$. In fact $P(X) = 1$ iff $\chi_V(X) = 0$ iff $\neg\chi_V(X) = 1$. Now $F(X) = \neg\chi_V(X) = \wedge_{(v_1,\ldots v_n)\in V} \vee_{i=1}^{n} x_i^{\bar{v_i}}$. Thus $F(X)$ is a CNF s.t. $F(X) = 1$ iff $P(X) = 1$.

Using the above procedure for all constraints in M we can build a CNF formula W s.t. $M(X) = 1$ iff $W(X) = 1$.

Example 1. As an example, let $X = [x_1, x_2, x_3]$ and $P(X) = 3x_1+2x_2+4x_3 \leq 5$. Then $V = \{(0, 1, 1), (1, 0, 1), (1, 1, 1)\}$. Thus $F(X) = (x_1 \vee \bar{x_2} \vee \bar{x_3}) \wedge (\bar{x_1} \vee x_2 \vee \bar{x_3}) \wedge (\bar{x_1} \vee \bar{x_2} \vee \bar{x_3})$.

Note that we do not need to discretize the real valued coefficients in the linear inequalities of M. However, to make the above approach interesting we need to generate F in a time efficient way and in such a way that F is not *too large.*

Our idea to produce a *small* CNF F in a *fast* way is the following. Let $P = \sum_{i=n}^{n} a_i x_i \leq b$. Let $y = (y_1, \ldots y_n)$ be an assignment s.t. $P(y) = 0$. If $a_i > 0$ and $y_i = 0$ then also for $z = (y_1, \ldots y_{i-1}, 1, y_{i+1}, y_n)$ we have $P(z) = 0$. Let F be the CNF representing P. Then $F = \ldots \wedge (x_1^{y_1} \vee \ldots \vee x_i^0 \vee \ldots x_n^{y_n}) \wedge (x_1^{y_1} \vee \ldots \vee x_i^1 \vee \ldots x_n^{y_n}) \wedge \ldots = \ldots \wedge (x_1^{y_1} \vee \ldots \vee x_{i-1}^{y_{i-1}} \vee x_{i+1}^{y_{i+1}} \vee \ldots x_n^{y_n})$.

That is, the value of y_i matters only if it is 1. Analogously, if $a_i < 0$ the value of y_i matters only if it is 0.

More formally, we say that variable y_i is *relevant* in P for the assignment y iff $(a_i > 0 \wedge y_i = 1) \vee (a_i < 0 \wedge y_i = 0)$. We denote with $\Gamma(P, y)$ the list of variables relevant in P for assignment y.

With the above considerations we can write the CNF for a constraint P as follows: $F(X) = \bigwedge_{v:P(v)=0} \bigvee_{x\in\Gamma(P,v)} x^{\overline{v(x)}}$.

Example 2. A CNF for P as in Example 1 is: $F = (\bar{x_2} \vee \bar{x_3}) \wedge (\bar{x_1} \vee \bar{x_3})$.

The above considerations suggest that, if our goal is to find a non-satisfying assignment, then we should first try setting to 1 the variables with large positive coefficients and to 0 the variables with large (in modulo) negative coefficients.

To generate F in a fast way, as a first step we *reorder* variables $x_1, \ldots x_n$ so that x_i precedes x_j in the ordering iff one of the following conditions hold: 1.

$a_i < 0$ and $a_j < 0$ and $a_i < a_j$; 2. $a_i > 0$ and $a_j > 0$ and $a_i > a_j$; 3. $a_i < 0$ and $a_j > 0$.

Let v be a partial assignment for X. That is $\forall x \in X\ v \in \{0, 1, \perp\}$. Function clause($v$) returns the clause $\bigvee_{x:v(x)\in\{0,1\}} x^{\overline{v(x)}}$.

Putting all the above considerations together leads to the *search-and-prune* algorithm sketched in the algorithm `blp2sat()` in Figure 1. As we shall see, our experimental results show that the search-and-prune heuristic used in function `blp2sat()` often allows us to quickly generate compact CNF representations for a linear inequality $\sum_{i=1}^{n} a_i x_i \leq b$.

In the following we describe the algorithm `blp2sat()` in Figure 1. First of all, `blp2sat()` assumes that variables have already been ordered as explained above. Let $P = \sum_{i=n}^{n} a_i x_i \leq b$ our constraint. Function `blp2sat()` has two arguments: the array a s.t. $a[i] = a_i$ and the rhs of P, b. The global variable x in `blp2sat()` stores the partial assignments found during the computation.

Line 2 of `blp2sat()` in Figure 1 computes in variable k the index of the first undefined assignment. If (line 3) this index is greater than n (number of variables in P) then all variables have been assigned a value and we go to line 4. Since all variables are defined we can evaluate the lhs $\sum_{i=1}^{n} a[i]x[i]$ of P. If $\sum_{i=1}^{n} a[i]x[i] > b$ (line 4) then constraint P is not satisfied and x contains a non-satisfying assignment for P. In such a case we add to F the clause clause(x) (line 4) and return 1, since we added one clause to F. If $\sum_{i=1}^{n} a[i]x[i] \leq b$ then x is a satisfying assignment for P and thus we generate no clause and return 0 (line 5).

If $k \leq n$ then $x[k]$ has not been assigned a value. Line 6 checks if there is hope to complete x to a satisfying assignment. If this is not the case then we add clause(x) to F (line 6) and return 1. If x may be completed to a satisfying assignment we go to line 7 and choose the value of $x[k]$ so as to make the lhs of P as large as possible in an attempt to find a non-satisfying assignment. After that in line 8 we recursively call `blp2sat()` and get in $m1$ the number of clauses produced by the setting for $x[k]$ chosen in line 7.

If the choice of $x[k]$ in line 7 has not produced non-satisfying assignments ($m1 = 0$) then, a fortiori, also the other possible assignment in line 7 for $x[k]$ will not produce any non-satisfying assignment. In other words, if $m1 > 0$ we consider the assignment (1 - $x[k]$), line 9, else we do not need to. Finally, line 10, we return the total number of non-satisfying assignments produced. Note how the test in line 7 allows us to prune the search tree for non-satisfying assignments.

8 Solving an MILP Problem with a SAT Based CEGAR

Rather than transforming an MILP problem into a SAT problem in one big step we can use a *Counterexample Guided Abstraction Refinement* (CEGAR) approach to *gradually* transform an MILP problem into a SAT problem.

First of all, given a linear constraint P, we can modify `blp2sat()` in Figure 1 so that each time it is called it generates at most `MaxClause` new clauses for the linear constraint P. The idea is that we can often prove correctness (UNSAT) or find an error (SAT) without generating all clauses for each constraint. This can be done by storing the *state* σ of the computation of `blp2sat()`. State σ summarizes all the information we need to save to safely *stop* and, above all, *resume*, the computation of `blp2sat()`. Initially the state is empty (meaning `blp2sat()` is at its start point). We call `cegar-blp2sat()` the version of `blp2sat()` thus modified.

In the following we describe the algorithm `cegar_milp2sat()` that *gradually* transforms an MILP problem into a SAT problem using a CEGAR based approach. A sketch of `cegar_milp2sat()` is in Figure 2.

First, for each constraint $(A[i], b[i])$ (that is $\sum_{j=1}^{n} A[i]_j x_j \leq b[i]$) we denote with σ_i the state of the computation of `blp2sat(`$A[i], b[i]$`)`. Initially σ_i is empty for $i = 1, \ldots n$. In the following lines refer to function `cegar_milp2sat()` in Figure 2.

The argument of `cegar_milp2sat()` is an MILP problem P, i.e. the pair (A, b).

Line 2 replaces P with Align(P) (see Proposition 2). Line 3, given the tolerance ϵ and using Proposition 4, computes the number of bits needed for x in order to achieve tolerance ϵ. Line 4 replaces in the MILP problem all continuous variables with discrete ones (Proposition 4). Line 5 transforms all discrete variables into boolean ones (Proposition 3). Line 6 initializes the state of the computation (i.e. clause generation) for each constraint.

Line 7, for each constraint i initializes the *history* $\gamma[i]$ of i. The history of constraint i records when i turned out to be false under a candidate solution ρ. More specifically. Assume we are at iteration t of the while loop starting at line 9. The history $\gamma[i]$ is a bitvector of size m that has a 1 in position $\gamma[i][j]$ iff at iteration $(t - (j - m + 1))$ constraint i was false under assignment ρ. Thus to update $\gamma[i]$ we simply shift it of 1 bit to the right and set $\gamma[i][m - 1]$ to 1 if the constraint is not satisfied by ρ, to 0 else. In our implementation we have set $m = 8$.

Line 8 initializes to 1 (empty set) the generated CNF F. Line 9 begins the main loop of `cegar_milp2sat()`. Line 10 begins the CEGAR for loop. Lines 10-12 generate an approximation of our MILP problem. Namely, for each constraint i, in line 11 we order constraint variables as described in Section 7 and in line 12 we generate at most `MaxClause` clauses for constraint i with lhs coefficients $A[i]$ and rhs coefficient $b[i]$ (function `cegar_blp2sat()`). Line 13 calls the SAT solver on CNF F. We use ZChaff [24,32] here as a SAT solver. The result may be UNSAT or SAT with an assignment ρ.

If we get UNSAT we are done since the original problem is then also UNSAT (line 14). However if we get SAT (line 15) the assignment ρ may not be a satisfying assignment for MILP (A, b) since we only generated at most `MaxClause` clauses for each constraint. In this case we go to line 16.

In lines 17-21, for each constraint i we update its *history* $\gamma[i]$. This is done by shifting $\gamma[i]$ one bit to the right and by writing a 0 (1) in the MSB (*Most Significant Bit*) of $\gamma[i]$ if ρ does (does not) satisfy constraint i.

If at the end of the loop in lines 17-21 no constraint has been found to be false then ρ is a *real* counterexample and we return SAT (line 22).

Lines 23-25 of `cegar_milp2sat()` for each constraint i compute in w_i the number of clauses to be generated by `cegar_blp2sat(`$A[i]$`, `$b[i]$`)`. This number

```
1  void cegar_milp2sat(A, b) {
2  Align variables to Zero;
3  Compute number of bits for continuous variables;
4  Discretize Continuous variables;
5  Transform discrete variables into boolean variables;
6  Initialize computation state of cegar_blp2sat();
7  Initialize history γ to all 0s;
8  F = 1;
9  while (1) {
10   foreach i {
11       Order variables accordingly to coefficients;
12       F = F ∧ cegar_blp2sat(A[i], b[i]); }
13   sol = call_SAT_solver(F);
14   if (sol == UNSAT) {return (UNSAT);}
15   else {ρ = decode_SAT_assignment();}
16   oksat = 1;
17   for each constraint i {
18         shift γ[i] 1 bit to the right;
19         if (ρ does not satisfy constraint i)
20            {oksat = 0; γ[i][m − 1] = 1;}
21         else {γ[i][m − 1] = 0;} }
22   if (oksat == 1) { return (SAT);}
23   for each constraint i {
24        compute clauses to be generated w_i using γ[i];
25        cegar_blp2sat(A[i], b[i]); }
26   let ρ′ be the assignment containing only the decision variables
      in ρ;
27   add to F the clause ¬ρ. }}
```

Fig. 2. Sketch of algorithm `cegar_milp2sat()`

$\Gamma = 1$	$\Lambda = 1$	$\gamma = 1$	$\mu = 2$	$\theta = 1$	$\eta = 2$	$\alpha = 2$	$\beta = 4$	$\zeta = 3$	$\xi = 2$	$m = 6$	$M = 24$

Fig. 3. Parameters for the water-tanks system in Section 4

is 0 if the constraint was true under the last found assignment ρ. Otherwise it is greater than 0 and depends on the history of constraint i. Intuitively, the more often *and* the more recently constraint i has turned out to be false, the larger the values of w_i. In any case $w_i \leq$ `MaxClause`. After computing w_i we call `cegar_blp2sat(`$A[i]$`, `$b[i]$`)`.

In line 26, from the SAT solver data structures we compute the *decision variables* of ρ. Let ρ' be the assignment containing only the decision variables in ρ. In line 27 we add to F the clause $\neg\rho'$.

Eventually we get UNSAT or a real (not spurious) satisfying assignment.

9 Experimental Results

To assess effectiveness of our approach we have implemented the proposed algorithms and compared their performance w.r.t. MILP based BMC verification.

Let $(\mathcal{H}, E, k)$ be a BMC problem. By using Theorem 1 we can transform such BMC problem into an MILP problem P. We then have two choices: use an MILP solver to check feasibility of P or use our SAT-CEGAR approach (Section 8) to check feasibility of P. In this section we compare these two approaches.

		GLPK		SAT				SAT-CEGAR			
k	h	Output	Time (secs)	Output	Time (secs)	CL	Mem (MB)	Output	Time (secs)	CL	Mem (MB)
1	7	SAT	88.09	SAT	3.12	334279	30.59	SAT	4.63	330318	30.56
2	4	SAT	4.95	SAT	5.66	600433	60.64	SAT	6.28	482930	47.74
3	4	UNSAT	7256.42	UNSAT	13.63	1.55e+06	123.96	UNSAT	3.09	370931	30.937
3	23		OOT	UNSAT	1902.32	8.93e+06	837.253	UNSAT	21.43	2.13e+06	240.83
3	98		OOT				OOM	UNSAT	153.35	9.08e+06	839.641
4	9		OOT	UNSAT	125.28	8.53e+06	833.473	UNSAT	9.79	1.1e+06	120.756
4	71		OOT				OOM	UNSAT	144.17	8.65e+06	836.938
5	3		OOT	UNSAT	62.44	6.75e+06	627.688	UNSAT	4.07	473834	47.8307
5	54		OOT				OOM	UNSAT	139.58	8.52e+06	836.518
6	1		OOT	UNSAT	46.99	5.21e+06	487.896	UNSAT	171.903	3.87e+06	381.539
6	48		OOT				OOM	UNSAT	157.57	8.2e+06	643.253

Fig. 4. Experimental results for the system in Section 4 with parameters in Figure 3

As a SAT solver we use ZChaff [24,32] a well know open source SAT solver. As an MILP solver for comparison we use GLPK [16]. In view of Remark 1 using GLPK rather than CPLEX [12] for our feasibility problems does not change meaningfully the results.

We use the example in Section 4 to assess effectiveness of our approach. This is a parametric example with most of the features that make *life hard* for reachability analysis.

Let k be the number of pumps. The number of tanks n is set to $2k$ and the system parameters are those in Figure 3.

Figure 4 shows our results when using GLPK, SAT without CEGAR (i.e. with `MaxClause` set to ∞) and our SAT-CEGAR (Section 8) with `MaxClause` set to 100. We set $\epsilon = 0.1$, that is we accept a relaxation of 0.1 on P constraints (Theorem 2). Note that (Theorem 2) if we get UNSAT then the original problem is UNSAT and for each $0 \leq \epsilon' \leq \epsilon$ we would get the same result (UNSAT).

The meaning of the columns in Figure 4 is the following.

Column k gives the number of pumps. The number of tanks is then $n = 2k$. *Column* h gives the BMC horizon. *Column Output* gives the outcome of the verification process. Namely, SAT if an error has been found within horizon h, UNSAT else. *Column Time* gives the CPU time in seconds. We have set a time limit of 180 minutes (10800 seconds) for the verification process. If a process does not complete by such time limit we report *Out Of Time* (OOT) in column Time. *Column Mem* gives the RAM used by the process. We report OOM if a process runs out of memory (1GB of RAM in our case). *Column CL* gives the number of clauses generated.

In our experiments we used a Mac Mini (CPU PowerPC G4 1.5 GHz; L2 Cache 512 KB; RAM 1GB).

For each k in Figure 4 we show the first horizon h to which we find an error or the last horizon that we were able to handle before going out of time or out of RAM. From Figure 4 we can clearly see how SAT behaves better than an MILP solver (GLPK) and how our SAT-CEGAR approach behaves better than SAT.

Figure 6 gives some detail about the SAT-CEGAR computation time (y axes) as a function of the horizon (x axes) with $k = 3$ pumps. We see that for SAT-CEGAR computation times are almost a linear function of the horizon.

Figure 5 gives some information about the MILP problem generated from our BMC problems.

		MILP Problem				
k	h	Rows	Real Vars	Non-Bool Int Vars	Bool Vars	Non-zeros
1	7	965	40	8	219	1826
2	4	1413	50	10	406	2764
3	4	2751	75	15	927	5550
3	98	64979	1485	297	21795	133014
3	99	65641	1500	300	22017	134370
4	71	82087	1440	288	30732	172856
4	72	83241	1460	292	31164	175288
5	54	100759	1375	275	40585	217330
5	55	102621	1400	280	41335	221350
6	48	136309	1470	294	57714	299844
6	49	139143	1500	300	58914	306084

Fig. 5. MILP problems generated for the water-tanks system in Section 4 with parameters as in Figure 3

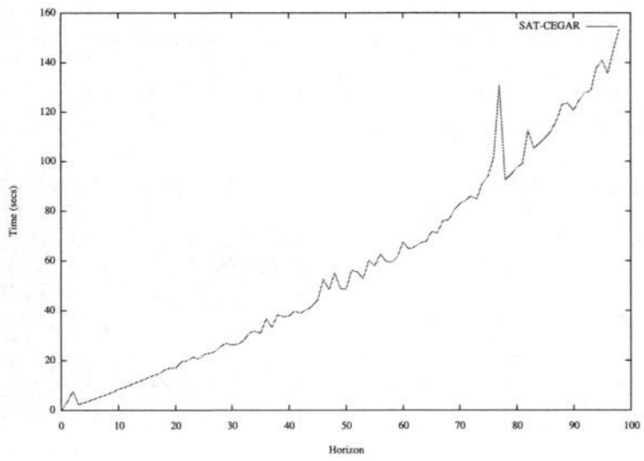

Fig. 6. SAT-CEGAR. Times for $k = 3$ and $n = 6$

Summing up, our experimental results show that our SAT-CEGAR approach can solve problems that are about 50 times larger than those that can be handled with an MILP solver. Namely, from Figure 4 we see that SAT-CEGAR can solve problems with $k = 6$ and $h = 48$ (i.e. 136309 linear constraints from Figure 5) whereas GLPK stops at $k = 3$, $h = 4$ (i.e. 2751 linear constraints from Figure 5).

10 Conclusions

We have presented a SAT based BMC algorithm for automatic verification of DTHSs. Using *Counterexample Guided Abstraction Refinement* (CEGAR) our algorithms *gradually* transforms a DTHS verification problem into larger and larger SAT problems.

Our experimental results show that our approach can handle DTHSs that are more then 50 times larger than those that can be handled using an MILP solver.

Acknowledgements. We are very grateful to HSCC'07 referees for their helpful comments on the submitted version of this paper.

References

1. url: `http://www.eecs.berkeley.edu/~tah/HyTech`.
2. R. Alur, T.A. Henzinger, and P.-H. Ho. Automatic symbolic verification of embedded systems. *IEEE Trans. on Software Engineering*, 22, 1996.
3. G. Audermand, M. Bozzano, A. Cimatti, and R. Sebastiani. Verifying industrial hybrid systems with mathsat. In *Proc. of the 2nd Int. Workshop on Bounded Model Checking*, 2004.
4. A. Bemporad and M. Morari. Verification of hybrid systems via mathematical programming. In *Proc. of: HSCC*, volume 1569 of *LNCS*. Springer, 1999.
5. A Biere, A. Cimatti, E. Clarke, and Y. Zhu. Symbolic model checking without bdds. In *Proc. of TACAS*, volume 1579 of *LNCS*. Springer, 1999.
6. M. Bozzano, R. Bruttomesso, A. Cimatti, T. Juntilla, S. Ranise, P. van Rossum, and R. Sebastiani. Efficient satisfiability modulo theories via delayed theory combination. In *Proc. of CAV*, volume 3576 of *LNCS*, 2005.

7. url: http://www.dsi.uniroma1.it/~tronci/cached.murphi.html.
8. M. W. Carter and C. C. Price. *Operations Research - A Practical Introduction.* CRC Press, 2001.
9. url: http://www.cs.cmu.edu/~modelcheck/cbmc/.
10. Edmund Clarke and Daniel Kroening. Hardware verification using ANSI-C programs as a reference. In *Proc. of ASP-DAC.* IEEE Computer Society Press, 2003.
11. Edmund M. Clarke, Orna Grumberg, Somesh Jha, Yuan Lu, and Helmut Veith. Counterexample-guided abstraction refinement. In *Proc. of CAV*, LNCS. Springer, 2000.
12. url: http://www.ilog.com/products/cplex/.
13. G. Della Penna, B. Intrigila, I. Melatti, E. Tronci, and M. Venturini Zilli. Exploiting transition locality in automatic verification of finite state concurrent systems. *International Journal of Software Tools for Technology Transfer (STTT)*, 6(4), 2004.
14. R. Raimi E. Clarke, A. Biere and Y. Zhu. Bounded model checking using satisfiability solving. *Formal Methods in system Design*, 19:7–34, July 2001.
15. N. Giorgetti, G. J. Pappas, and A. Bemporad. Bounded model checking of hybrid dynamical systems. In *Proc. of 44th IEEE Int Conf. CDC*, 2005.
16. url: http://www.gnu.org/software/glpk/glpk.html.
17. Anubhav Gupta and Ofer Strichman. Abstraction refinement for bounded model checking. In *CAV*, volume 3576 of *LNCS*, pages 112–124, 2005.
18. T.A. Henzinger, P.-H. Ho, and H. Wong-Toi. Hytech: A model checker for hybrid systems. *Software Tools for Technology Transfer*, 1, 1997.
19. url: http://control.ee.ethz.ch/~hybrid/hysdel.
20. Kim G. Larsen, Paul Pettersson, and Wang Yi. UPPAAL: Status and developments. In *CAV97*, number 1254 in LNCS. Springer–Verlag, 1997.
21. B. Li, C. Wang, and F. Somenzi. Abstraction refinement in symbolic model checking using satisfiability as the only decision procedure. *Software Tools for technology Transfer (STTT)*, 7(2):143–155, 2005.
22. url: http://mathsat.itc.it/.
23. K. L. McMillan. *Symbolic Model Checking: An Approach to the State Explosion Problem.* Kluwer Academic Publishers, 1993.
24. M. W. Moskewicz, C. F. Madigan, Y. Zhao, L. Zhang, and S. Malik. Chaff: Engineering an efficient sat solver. In *39th DAC*, 2001.
25. G. Della Penna, B. Intrigila, I. Melatti, M. Minichino, E. Ciancamerla, A. Parisse, E. Tronci, and M. V. Zilli. Automatic verification of a turbogas control system with the murphi verifier. In *Proc. of HSCC*, LNCS. Springer, 2003.
26. url: http://www.cs.cmu.edu/~modelcheck/.
27. F. D. Torrisi and A. Bemporad. Hysdel - a tool for generating computational hybrid models. *IEEE Trans. on Control Systems Technology*, 12(2):235–249, March 2004.
28. A. L. Turk, S. T. Probst, and G. J. Powers. Verification of real-time chemical processing systems. In *HRTS*, number 1201 in LNCS. Springer, 1997.
29. url: http://www.docs.uu.se/docs/rtmv/uppaal/.
30. R. Vidal, S. Schaffert, O. Shakernia, J. Lygeros, and S. Sastry. Decidable and semi-decidable controller synthesis for classes of discrete time hybrid systems. In *In Proc. of 40th IEEE CDC*, 2001.
31. url: http://vlsi.colorado.edu/~vis.
32. url: http://www.princeton.edu/~chaff/zchaff.html.

Solving Coverage Problems with Embedded Graph Grammars

John-Michael McNew[1], Eric Klavins[1], and Magnus Egerstedt[2]

[1] Univ. of Washington
{jmmcnew,klavins}@ee.washington.edu
[2] Georgia Institute of Technology
magnus@ece.gatech.edu

Abstract. We show how Embedded Graph Grammars (EGGs) are used to specify local interaction rules between mobile robots in a natural manner. This formalism allows us to treat local network topologies, geometric transition conditions, and individual robot dynamics and control modes in a unified framework. An example EGG is demonstrated that achieves sensor coverage in a provably stable and correct manner. The algorithm results in a global network with a lattice-like triangulation.

1 Introduction

The overarching scientific question facing the area of networked robot systems is how global behaviors arise from local rules in a well-defined and predictable manner. In a dynamic task, as agents move in and out of each others sensory or communication ranges, the network changes, resulting in inherently hybrid dynamics. Hybrid dynamics also result from the fact that individual agents can assume different roles depending on local network adjacency or on the geometrical characteristics of the local environment. Networked robot systems require methods that control both the network dynamics and the individual robot behaviors. This suggests a handful of potential formalisms [1,2]. Unfortunately, the available languages are not appealing for networked systems because the local network topology associated with each robot is not cleanly represented. Hybrid automata [3,4], while certainly expressive enough, are cumbersome in this setting. They additionally suffer from the state explosion problem since a distinct state typically has to be enumerated for every possible combination of role assignments and network topologies.

Thus, we take as our starting point *graph grammars* which provide a compact representation of subgraph transitions arising from local interactions. Although graph grammars have been successfully used to control robot systems [5], they best describe changing networks, not, for example, low-level robot motion. To address this, we augment the notion of a graph grammar, introducing the *Embedded Graph Grammar*(EGG) model and associated analysis methods.

To demonstrate and challenge the approach, we extend the geometric problem of *coverage* (see also [6]) to include specifications on network topology.

A. Bemporad, A. Bicchi, and G. Buttazzo (Eds.): HSCC 2007, LNCS 4416, pp. 413–427, 2007.

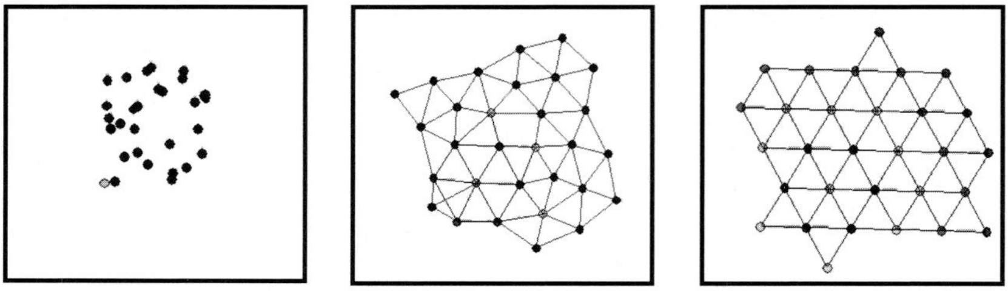

Fig. 1. Simulated formation of a δ-triangulation using an embedded graph grammar

In particular, we suppose that three communicating robots forming an equilateral triangle of side-length δ *cover* the convex hull of their locations. The goal (described more formally in Section 3) is to produce a triangulation of an entire region in which any three nearby robots form a δ-triangle and there are no holes in the resulting mesh (see Figure 1). Similar lattice-like geometries are generated by virtual potential leaders in [7] where the system shares a global coordinate system. What is novel here is (1) the EGG formalism that encodes the system dynamics, local network topology, guard conditions for switching between control modes, and inter-robot communication rules in a unified manner; (2) the lack of a common global reference frame among the robots; and (3) the superposition and successful coordination of several complementary control objectives.

2 Embedded Graph Grammar Definitions

2.1 Graphs and Other Notation

If Σ is a set of labels, a labeled graph is the quadruple (V, E, l, e) where V is a set of vertices, E is a set of edges, l is a vertex labeling function that maps vertices into Σ, and e is an edge labeling function that maps edges into Σ. We denote by $\mathcal{G}$ the space of labeled graphs. Sometimes the label set Σ is a cartesian product of atomic label spaces we refer to as *fields*. For example the system introduced in this paper has a node label space where the fields are named $(mode, dist)$. We use dot notation to indicate the values of label fields for specific robots. For instance $i.dist = 10$ indicates that robot i has the value 10 in its $dist$ field.

If S is a set of vertices, $G[S]$ denotes the subgraph of G induced by S. We denote by V_G and E_G the edges of a graph G and when there is no danger of confusion we write V to indicate the vertex set of the system under consideration. If i and j are vertices in a graph, we denote their graph distance by $d(i, j)$.

If f is a function defined on a domain A, we denote the restriction of the function to $B \subset A$ by $f_{|B}$. A function f is *well-defined* with respect to an equivalence relation $\sim$ if $x \sim y$ implies $f(x) = f(y)$.

2.2 Embedded Graphs

Consider a system of N communicating robots with identical state space X.

Definition 1. *An embedded graph γ is a pair $\gamma = (G, x)$ where G is a labeled graph and $x : V \rightarrow X$ is a* realization function. *The space of all embedded graphs is denoted by Γ.*

A vertex $i \in V$ indicates the index of the *ith* robot. The presence of an edge $ij \in E$ corresponds to a physical and/or *maintained* communication link between robots i and j. The vertex label indicates the operational mode of robot i and (along with the edge label $e(ij)$) keeps track of local information. The function x assigns to each robot a continuous state or *realization* in its state space X.

We write $G_\gamma, x_\gamma, V_\gamma$, and E_γ to denote the labeled graph, continuous state, vertices, and edges associated with an embedded graph γ. If $S \subseteq V_\gamma$, then the the embedded graph induced by S, $\gamma[S]$, is given by the pair $(G[S], x_{|S})$. We define the distance between two embedded graphs, γ and ρ, by

$$d(\gamma, \rho) \triangleq \begin{cases} \infty & \text{if } G_\gamma \neq G_\rho \\ \|x_\gamma - x_\rho\| & \textit{otherwise.} \end{cases}$$

The distance between an embedded graph γ and a set of embedded graphs T is denoted by $d(\gamma, T) = \min_{\rho \in T} d(\gamma, \rho)$.

2.3 Embedded Graph Transition Systems

A trajectory of an *embedded graph transition system* describes how the network topology and continuous states of a group of robots changes over time.

Definition 2. *An* embedded graph transition relation *is a relation $\mathcal{A} \subseteq \Gamma \times \Gamma$ such that $(\gamma_1, \gamma_2) \in \mathcal{A}$ implies $x_{\gamma_1} = x_{\gamma_2}$.*

Definition 3. *An* embedded graph transition system *is a triple $(\gamma_0, \mathcal{A}, u)$ where γ_0 is an initial embedded graph and $u : V \times \Gamma \rightarrow TX$ is the vector field describing the continuous flow.*

Definition 4. *A* trajectory *is a map $\sigma : \mathbb{R}^{\geq 0} \rightarrow \Gamma$ such that there exists a sequence $\tau_0, \tau_1, \tau_2, \ldots$ where*

1. *$x_{\sigma(t)}$ is continuous.*
2. *$\tau_k \leq \tau_{k+1}$ and if the sequence has any finite length N, $\tau_N \triangleq \infty$.*
3. *For all $t, t' \in [\tau_k, \tau_{k+1})$, $G_{\sigma(t)} = G_{\sigma(t')}$.*
4. *$\tau_i \neq \infty$ and $i > 0$ if and only if there exists a transition$((G, x^*), (H, x^*)) \in \mathcal{A}$ such that $(G, x^*) = \lim_{t \rightarrow \tau_i} \sigma(t)$ and $\sigma(\tau_i) = (H, x^*)$.*
5. *For all $i \in V$ and $t \in [\tau_i, \tau_i + 1)$, $\frac{d}{dt} x_{\sigma(t)}(i) = u(i, \sigma(t))$.*

We denote the set of nondeterministic trajectories of a system by $\mathcal{T}(\gamma_0, \mathcal{A}, u)$. Clearly an embedded graph transition system is a *globally* defined nondeterministic hybrid automata where the discrete states are labeled graphs.

2.4 Locality and Embedded Graph Grammars

Our interest is in modeling and implementing embedded graph transition systems in a *local* and *distributed* fashion. In other work [8,9], we model notions of "local" that include geometric restrictions on sensing and communications. In this paper we focus exclusively on the notion of local graph neighborhoods and we develop a model where:

1. Graph matching involves only a small subset of vertices.
2. The flow and transition relations use discrete information from the graph neighborhood.
3. The flow and transition relation are permutation-invariant.

We refer to the neighborhood of $i \in V_\gamma$ as the *friends* of i and denote this set by $F(i)$. In keeping with existing graph literature, we write $F[i]$ to mean $F(i) \cup i$ and denote the closed out-neighborhood by $F^+[i]$.

Definition 5. *Consider the pairs (A, γ) and (B, ρ) where γ and ρ are embedded graphs, $A \subset V_\gamma$, and $B \subset V_\rho$. $(A, \gamma) \sim (B, \rho)$ if there exists a bijective map ν between V_γ and V_ρ such that*

1. *For all $i \in A$, $\nu(i) \in B$,*
2. *For all $k \in V_\gamma$, $x_\gamma(k) = x_\rho(\nu(k))$, and*
3. *For all $i \in A$, ν is a label preserving isomorphism between $G_\gamma[F[i]]$ and $G_\rho[F[\nu(i)]]$.*

If $(A, \gamma) \sim (B, \rho)$ we say γ from the point of view of A is equivalent to ρ from the point of view of B .

Definition 6. *A control law $u : V \times \Gamma \to TX$ is* locally implementable *if u is well defined with respect to the point of view equivalence relation $\sim$.*

This definition captures the requirement that robots use discrete information from their local graph neighborhood in a permutation invariant manner. Robots form new links and update their labels on a *local* scale by using *guarded rules.*

Definition 7. *A* guard *g is a function $g : \mathcal{P}(V) \times \Gamma \to \{true, false\}$. A guard is* locally checkable *if it is well-defined with respect to the point of view equivalence relation on sets, $\sim$.*

Definition 8. *A* guarded rule *(or just rule), $r = (g, L, R)$, is a pair of labeled graphs over some* small vertex set *$V_L = V_R$ and a locally checkable guard g.*

Figure 4 shows an example of a guarded rule. Each breakout box contains a dummy variable on the left used to identify specific vertices in a rule. The right hand side of the box contains the labeling of that vertex. When the topology and labeling of a small group of robots matches that of the left hand graph of the rule and the robots are in a "safe" configuration (as defined by the guard), then they can update their state to match the right hand graph in the rule.

More formally suppose γ represents a possible state of a system and h is a label preserving subgraph isomorphism from V_L into G_γ such that $g(h(V_L), \gamma)$ is true. We call h a witness and the pair (r, h) an action. A rule r is *applicable* if a witness h can be found. The application of an action (r, h) on an embedded graph $\gamma = (G, x)$ produces a new embedded graph $\gamma' = ((V, E', l', e'), x)$ defined by

$$E' = (E - \{h(i)h(j) | ij \in E_L\}) \cup \{h(i)h(j) \mid ij \in E_R\}$$
$$l'(i) = \begin{cases} l(i) \text{ if } i \notin h(V_L) \\ l_R \circ h^{-1}(i) \text{ otherwise.} \end{cases} \quad e'(ij) = \begin{cases} e(ij) \text{ if } i \notin h(V_L) \text{ or } j \notin h(V_L) \\ e_R \circ h^{-1}(i)h^{-1}(j) \text{ otherwise.} \end{cases}$$

That is, we replace $h(L)$ (which is a copy of L) with $h(R)$ in the graph G_γ. We write $\gamma \xrightarrow{r,h} \gamma'$ to denote that we obtain $G'_{\gamma'}$ from G_γ by applying action (r, h).

Definition 9. *An embedded graph transition* $((G, x), (H, x))$ *is* consistent *with a rule* r *if there exists a witness* h *such that* (r, h) *is applicable to* (G, x) *and* $G \xrightarrow{r,h} H$. *We denote by* $\mathcal{A}(r)$ *the set of transitions consistent with rule* r. *If* Φ *is a set of rules,* $\mathcal{A}(\Phi) = \cup_{r\in\Phi}\mathcal{A}(r)$.

Definition 10. *An* embedded graph grammar system (EGG) *is a triple* (γ_0, Φ, u) *where* γ_0 *is an embedded graph representing the initial state,* Φ *is a set of rules, and* u *is a locally implementable controller.*

Proposition 1. *The set of trajectories of a local embedded graph grammar,* $\mathcal{T}(\gamma_0, \Phi, u)$ *are equivalent to the trajectories of an embedded graph transition system under the transition relation consistent with* Φ, $\mathcal{A}(\Phi)$, *and the vector field described by the locally implementable controller* u, *i.e.*

$$\mathcal{T}(\gamma_0, \Phi, u) = \mathcal{T}(\gamma_0, \mathcal{A}(\Phi), u).$$

Figure 2 shows an embedded graph grammar trajectory. Note that discrete transitions involve small sets of vertices and that concurrent application of rules is possible.

3 Coverage Via δ-Triangulation

3.1 Preliminaries

By a *triangle*, we mean any subgraph composed of three fully connected vertices (say i, j, k). A *δ-triangle* is a triangle where $||x_i - x_j|| = ||x_j - x_k|| = ||x_i - x_k|| = \delta$. We say that a region of the plane $A \subset \mathbb{R}^2$ is *covered* if A lies within the convex hull of the positions of three robots in a δ-triangle.

A graph G is *planar* if there exists an embedding $x : V \to \mathbb{R}^2$ such that when the edges are drawn as straight lines in the plane, no edges intersect except at vertices. An embedded graph $\gamma = (G, x)$ is a *plane graph* when G is planar via the realization function x. The regions of a plane graph bounded by the edges are called *faces* and every finite planar graph has exactly one unbounded face called the *outer face.*

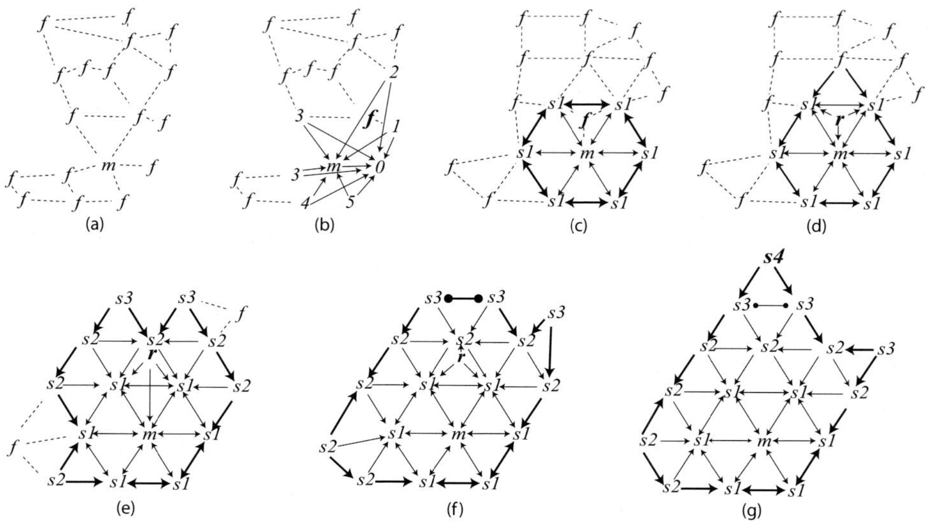

Fig. 2. A Sample Trajectory. (a) Initial state all `free` robots, f, are running the Gabriel graph pre-sorting controllers (indicated by dashed edges). (b) Hexagonal seed formation. (c) After the hexagonal seed is formed by R_2, the robot in bold labeled **f** is trapped inside the hexagon. (d) To correct this "error", rule R_6 changes the robot to `repulse` mode, **r**. (e)-(f) Utilizing the `repulsing` controller, the robot in bold moves away from the center. Also Rules R_5 and R_6 of the crystal growth process are applied. (g) By repeatedly applying the error correction rules, the robot in bold works its way out of the crystal where it can be absorbed as the final element in the structure.

Definition 11. *An plane graph γ is a* near-triangulation *if the outer face is a cycle and all inner faces are triangles.*

Definition 12. *A near-triangulation γ is a δ-*triangulation *if every inner face corresponds to a δ-triangle. We say a δ-triangulation γ is* maximal *if for every graph H created by adding an edge to G_γ, there does not exist a realization x such that (H, x) is a δ-triangulation.*

We denote by T the set of all δ-triangulations and by T_{max} the set of all maximal δ-triangulations.

3.2 General Algorithm

Consider N robots moving in the plane without a common reference frame and obeying integrator dynamics given by $\dot{x}_i = u$. Suppose Γ_0 is the class of initial embedded graphs where: (1) $E_{\gamma_0} = \varnothing$, (2) there is a unique vertex 0 labeled $l(0).mode =$ `master` and (3) all other vertices are labeled by $l(i).mode =$ `free`.
Task. *Design an embedded graph grammar, (γ_0, Φ, u) such that for all trajectories $\sigma \in \mathcal{T}(\gamma_0, \Phi, u)$*

$$\lim_{t\to\infty} \sigma(t) \in T_{max}.$$

Our method of constructing δ-triangulations is similar to that of growing "crystals" where placing a "seed crystal" in a "solution" causes the solution to replicate the seed and crystalize. We often view EGGs as collections of small behaviors interacting locally, a fact highlighted by our δ-triangulation grammar, Φ, which describes the four concurrent processes shown in Figure 2:

1. *Pre-sorting*–The robots not involved in the other three processes try (without communication) to organize into a mesh that is *close* to a δ-triangulation.
2. *Hexagon Formation*–The `master` chooses six robots and creates a hexagonal "seed" formation, labeling all edges on the boundary by *ij.boundary* = `outer`.
3. *Crystal Growth*–The δ-triangulation or "crystal" is grown by waiting until robots in a boundary edge are near settling, then a robot on the exterior of the crystal "attaches" to these robots to form a new triangle or two robots already in the crystal add an edge to enclose a δ-triangle.
4. *Error Correction*–Occasionally robots executing the pre-sorting algorithm become trapped in the interior of the crystal structure. The error correction controller routes these robots to the exterior using local label information.

Table 1. Control Law u for the δ-triangulation solution grammar. Robots execute the control law corresponding to their label.

Purpose	Label	Control
Pre-Sort	f	$-\sum_{ij\in Gabriel(i)} \nabla_{x_i} U_{ij}$
Hexagon Formation	m	0
	0	$-\sum_{ij\in F+(i)} \nabla_{x_i} U_{ij}$
	$1-5$	$- \nabla_{x_i}(\|\|x_i - P(i)\|\|)^2$
Crystal Growth	s	$-\sum_{ij\in F+(i)\cap Con(i)} \nabla_{x_i} U_{ij}$
Error Correction	r	$- \nabla_{x_i}(\|\|x_i - p(i)\|\|)^2$

Suppose Φ_{HEX} is the grammar for hexagon formation (Figure 3), Φ_{CG} is the grammar for crystal growth (Figure 4),Φ_{EC} is the grammar for error correction (Figure 6), and the solution grammar is $\Phi = \Phi_{HEX} \cup \Phi_{CG} \cup \Phi_{EC}$. The following notation is useful in describing the solution system (γ_0, Φ, u). Fix a trajectory σ and define $\gamma(t) = \gamma_{\sigma(t)}, x(t) = x_{\gamma(t)}, G(t) = G_{\gamma(t)}$. By $r(t)$ and $h(t)$ we denote the rule and witness applied at time t. The grammar Φ uses two vertex label fields, *mode* and *dist*(the graph distance from the `master`) and two edge label fields, *boundary* and *control*. The mode labels are: `master`,0, 1,..., 5, `slave`, `repulse`, and `free`. Figures refer to mode labels by their first letter.

The function $Con(V)$ takes any set of vertices V and returns another set of vertices V' where V' is the set of vertices reachable from V via paths where every edge is labeled as a `control` edge. The *crystal* function $C(t)$ is defined as the restriction of the embedded graph $\gamma(t)$ to the set of vertices labeled *`slave`* or *`master`* and defines the growing crystalline structure. We define the *interior* function $I(t)$ as the closed union of the faces in the crystal $C(t)$. We denote by

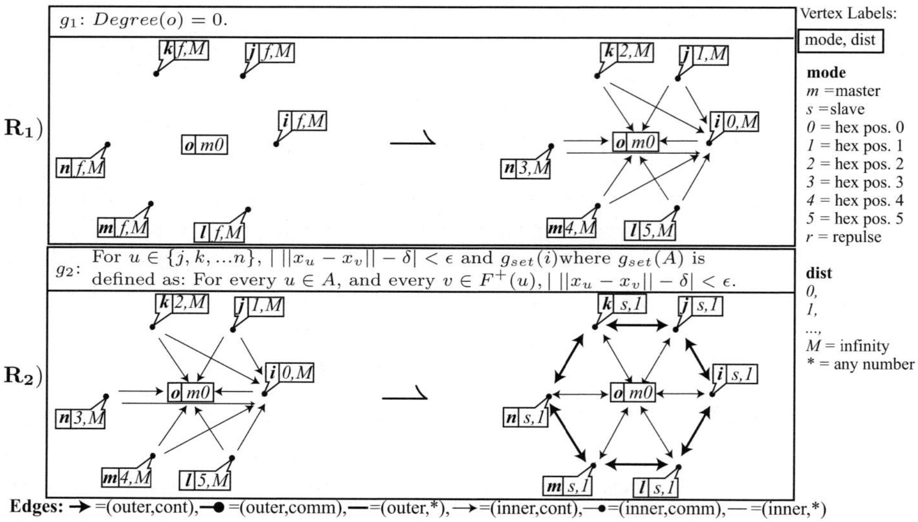

Fig. 3. Rule set Φ_{HEX} for hexagonal seed formation. Rule R_1 establishes the topology used to move the robots into a hexagonal configuration. Once the robots are in the attracting region, rule R_2 switches to the desired topology and control modes.

$d(i, l_{jk})$ the distance of x_i from the line l_{jk} through x_j and x_k. If l_{jk} defines a half plane in $\mathbb{R}^2$, we denote by $H_l^i(j,k)$ the region of state space where i lies in the half plane opposite l.

3.3 Pre-sorting Using Gabriel Graphs

We use a geometric switching algorithm by Schucker, Murphey and Bennet [10] to produce *near* δ-triangulations. The algorithm uses the edges of the Gabriel graph as a switching criterion and does not require explicit communication.

Definition 13. *Given a realization x, we denote the Gabriel graph of x by $Gabriel(x)$, where $ij \in E_{Gabriel(x)}$ if and only if for all $k \neq i$ or j,*

$$||x_i - x_j||^2 < ||x_i - x_k||^2 + ||x_j - x_k||^2.$$

For an edge ij we define an edge based potential $U_{ij} = (||x_i - x_j|| - \delta)^2$. As seen in Table 1, `free` robots follow the negative gradient of U_{ij} for Gabriel graph edges ij, trying to make the distance between those robots δ. The first few panels of Figure 2 show the Gabriel graph controller pre-sorting the robots.

3.4 Hexagon Formation

By a *hexagonal seed graph*, we mean any embedded graph μ where $G_\mu = W_7$ is a wheel graph and x_μ is a hexagonal configuration with edge lengths of δ (i.e. $\mu \in T_{max}$). Forming the hexagonal seed is challenging because the robots lack of

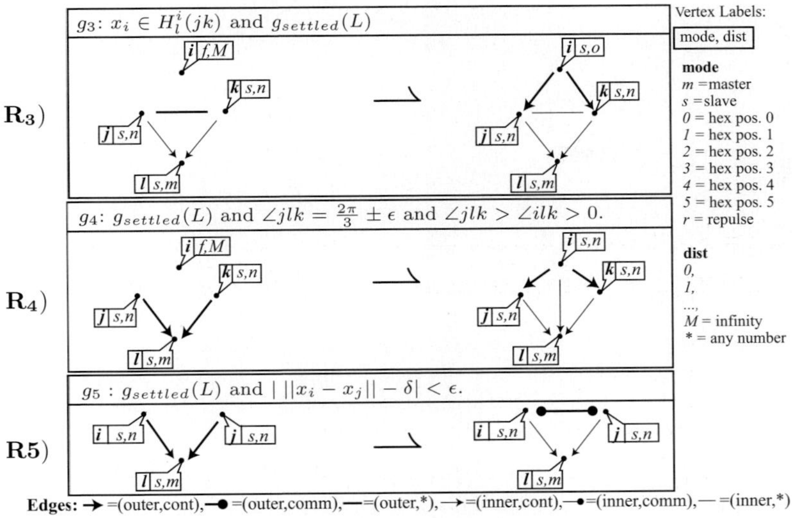

Fig. 4. Rule set Φ_{CG} for Crystal Growth. Rules R_3 and R_4 add robots to the crystal. Rule R_5 encloses δ-triangulations.

a common coordinate system. When rule R_1 is applied, the robot corresponding to vertex i in Figure 3 changes to $i.mode = \mathtt{0}$. The `master` robot corresponds to vertex o. The controller $\dot{x}_i = \nabla_i U_{io}$ limits the motion of i to a linear manifold defined by the configuration of i and o when the rule is applied. Since the vector $\mathbf{v}_{oi}$ is constant, the other robots (labeled by $mode \in \{1, 2, ...5\}$) form edges to robots 0 and i and use $\mathbf{v}_{oi}$ as the basis of a shared local coordinate system. Then each vertex v constructs a sink point $P(v)$ defined by their mode labels where

$$P(v) = x_o + \delta(cos(\frac{\pi}{3}v.mode), sin(\frac{\pi}{3}v.mode))^T.$$

Using the simple potential controllers shown in Table 1, the robots converge towards the hexagonal configuration defined by P. Once the robots' geometry is close enough to hexagonal, the robots apply rule R_2 to switch to the δ-triangulation topology. Proposition 2 states a hexagonal seed is always formed and is proven (along with other supporting propositions) in Section 4.2. Figure 2(a)-(c)shows hexagon formation in a partial trajectory.

Proposition 2. *For any ϵ, if $\gamma(t_0) = \gamma_0$, then there exists $t_2 > t_0$ and a δ-triangulation $\mu \in T$ where $G_\mu = W_7$ such that for all $t \geq t_2$, $d(C(t)[V_\mu], \mu) < \epsilon$.*

3.5 Crystal Growth

Consider any crystal $C(t)$ and an edge labeled $jk.boundary = \mathtt{outer}$ (we call these edges "boundary" edges). The controllers grow the crystal by attaching a "`free`" robot from the exterior (say i) to j and k and updating the boundary labeling. The rules in Figure 4 address two different geometries. In rule R_3,

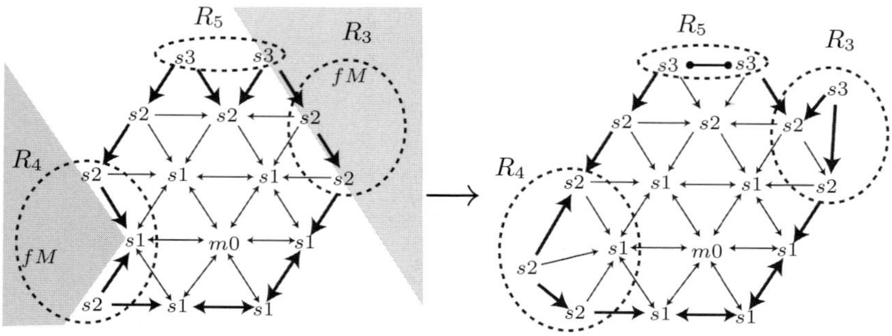

Fig. 5. Applications of Crystal Growth Rules. The gray areas indicate the regions in which the guards g_3 and g_4 are true. R_3 and R_4 add vertices to the crystal. R_5 closes a δ-triangulation. Note that after the application of the rules, the new labeling of the outer boundary remains consistent with the geometry.

if $j.dist = k.dist$, then i forms edges with j and k , marking the new edges *outer* and labeling $i.dist = j.dist + 1$ and $i.mode =$ `slave`. If applied in g_3, the controller for the `slave` label shown in Table 1 moves i to a point where $||x_i - x_j|| = ||x_i - x_k|| = \delta$ corresponding to a δ-triangulation. Figure 5 shows the guards of R_3 and the other crystal growth rules and their applications. The unidirectional information flow from the hexagonal seed outward guarantees that the crystal does not deform when a new robot is added.

Rule R_4 is applied at the corners of the hexagon where there are two boundary edges jl and kl with $j.dist = k.dist = l.dist + 1$. When $\angle jlk \approx \frac{2\pi}{3}$, R_4 adds i between j and k and updates the distance by $i.dist = j.dist$.

Proposition 3. *For all $\nu > \epsilon$ and for $\rho \in T$, if $d(C(t_1), \rho) = \nu$, then there exists $t_2 > t_1$ such that either $r(t_2) \in \{R_3, R_4\}$ or $d(C(t_2), \rho) = \epsilon$.*

Finally, if there are two robots, j and k where $j.dist = k.dist$, jl and $kl \in E$ and $\angle jlk \approx \frac{\pi}{3}$, then the grammar encloses a triangle by adding an edge jk via rule R_5. This guarantees that the resulting δ-triangulation is maximal. The edge is labeled $e(jk) = (outer, comm)$, indicating it is not used as a control input.

Proposition 4. *If $d(C(t_1), \rho) < \epsilon$, then there exists $t_2 > t_1$ such that either*

i. $r(t_2) = R_3$ *or* $r(t_2) = R_4$ *or*

ii. *There exists $\eta \in T_{max}$ such that (1) $d(C(t_2), \eta) < \epsilon$ and (2) if $i.mode = free$ and $x_i \notin I(t)$, R_3 or R_4 are applicable to i.*

3.6 Error Correction

Rules R_6-R_9 (Figure 6) and the controller (Table 1) associated with the label `repulse` remove robots trapped in the crystal structure. Suppose three robots $\{j, k, l\}$ define a face f_1 and a robot i (with $i.mode =$ `free`) lies in the closure of f_1. Then applying rule R_6 changes $i.mode$ to `repulse`. The `repulsing` robot calculates a sink point $p(i)$ such that when $l.dist = k.dist < j.dist$ then $p(i) =$

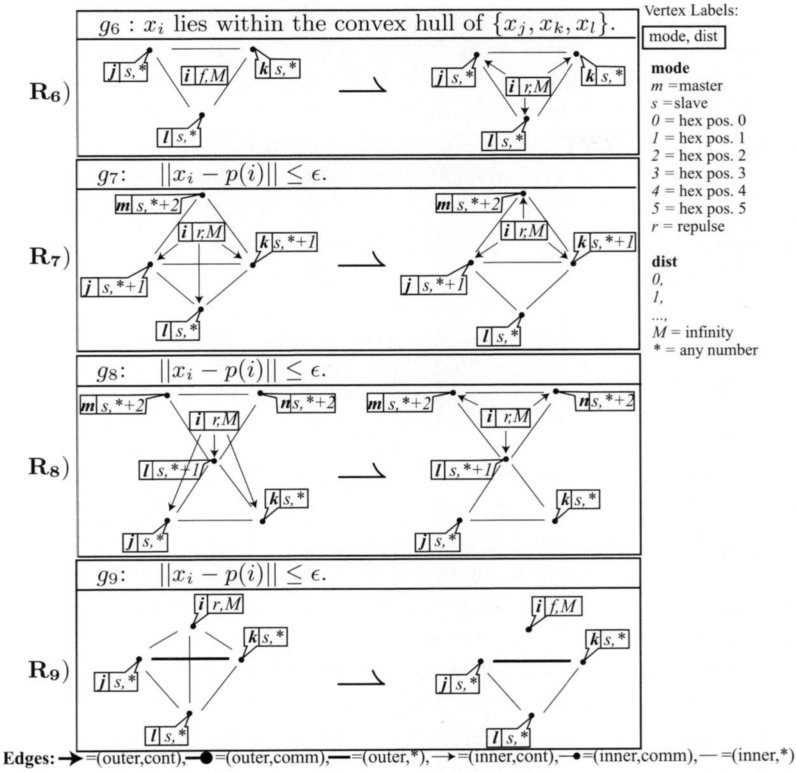

Fig. 6. Error Correction rules, Φ_{EC}. Rule R_6 initiates the error correction process, R_7 and R_8 move vertices towards the outside of the crystal and R_9 terminates the error correction process when the robot is sufficiently far from the crystal.

$2\delta(x_j - \frac{1}{2}(x_k + x_l))/(||x_j - \frac{1}{2}(x_k + x_l)||)$ and when $j.dist < k.dist = j.dist$, then $p(i) = 2\delta(\frac{1}{2}(x_k + x_l) - x_j)(||x_j - \frac{1}{2}(x_k + x_l)||)$. The `repulsing` robot then moves to p via the simple potential controller defined in Table 1. If x_i is near p and p is inside another face farther away from the center, either rule R_7 or R_8 are applied to continue moving away from the `master`. Or if $p \notin I(t)$, rule R_9 returns robot i to `free` mode. Applications of these rules can be seen in Figure 2(c)-(f).

Proposition 5. *For any $\rho \in T$, if $i \in I(t_1)$ and $d(C(t_1), \rho) < \epsilon$, then there exists $t_2 > t_1$ such that $i \notin I(t)$ or $r(t_2) = R_3$ or $r(t_3) = R_4$.*

Corollary 1. *There exists ϵ such that if $d(C(t_1), \rho) < \epsilon$ and R_9 is applicable, then there exists $t_2 \geq t_1$ such that $r(t_2) \in \{R_3, R_4\}$.*

3.7 Process Interaction

Theorem 1. *For all $M \in \{6, 7, ..., |V| - 1\}$, if $|C(t_1)| = M$, then there exists $t_2 \geq t_1$ such that $r(t_2) \in \{R_3, R_4\}$ and $|C(t_2^+)| = M + 1$.*

Proof. Assume to the contrary that $|C(t_1)| = M < V$ but neither rule R_3 nor R_4 are applied at any time $t' \geq t_1$. Since at time t_1, there is some $\rho \in T$ such that $d(C(t_1), \rho) < \infty$, then by Proposition 3 if R_3 or R_4 are not applied the system will flow to a state where $d(C(t_1), \rho) < \epsilon$. By Proposition 4, we know that if $i \notin I(t)$ then eventually R_3 or R_4 are applicable. Thus eventually it must be the case that $i \in I(t)$. However, by proposition 5 and Corollary 1, $i \in I(t)$ leads to R_3 or R_4 being applied. Thus it must be the case one is applied and for some i, $i.mode$ changes to `slave`, therefore $|C(t_2^+)| = M + 1$. □

Proposition 6. *For all $t_1 < t_2$, $|C(t_1)| \leq |C(t_2)|$.*

Proof. No rule in Φ changes the vertex labels `master` or `slave`. □

Theorem 2. *For all trajectories, $\gamma(t_0) = \gamma_0$ implies there exists a time t such that for all $t' \geq t$, $|C(t')| = |V|$.*

Proof. Proposition 2 implies eventually $|C| = 6$. Theorem 2 follows by induction on Propositions 1 and 6 and the finiteness of the initial graph.

Theorem 3. *For all trajectories $\gamma(t) \in \mathcal{T}(\gamma_0, \Phi, u)$,*

$$\lim_{t \to \infty} \gamma(t) \in T_{max}.$$

Proof. By Theorem 2 and Proposition 4 we have that eventually $G_\gamma(t) = G_\rho$ for some $\rho \in T_{max}$. By Proposition 3 we know the positions converge.

The embedded graph grammar system (γ_0, Φ, u) was simulated for a variety of initial conditions with graph sizes ranging from 12 to 100. All simulations resulted in δ-triangulations. Figure 1 shows snapshots from one such simulation. Although the error correction rules are central to the proofs above, in simulation error correction rules are rarely applied. Note that the final shape is nondeterministic and highly dependent on the initial conditions. Nonetheless, once an initial δ-triangulation is established, simple rules can be constructed to form almost any global shape from local control.

4 Proof of Supporting Propositions

4.1 Auxiliary Properties

Lemma 1. *Let $A \subset V$ be any set such that $Con(A) = A$, then the equilibrium and dynamics of A are independent of $V - A$.*

Proposition 7. *If $\gamma(t_0) = \gamma_0$, then there exists a time t_2 such that $r(t_2) = R_2$.*

Proof. $r(\tau_1) = R_1$ since only R_1 is applicable to G_{γ_0}. $Im(h_1)$ is the set of vertices to which R_1 is applied. R_2 is the only rule applicable to vertices in h_1. Since Lemma 1 holds for $Im(h_1)$ we construct a Lyapunov function for $x_{|_{Im(h_1)}}$ by

$$V(t) = \sum_{i \in Im(h_1) - \{0,j\}} ||x_i - P(i)||^2 + (||x_j - x_0|| - \delta)^2$$

If we treat x_0 as a constant, then $V \geq 0$, $\dot{V} = -\nabla V^T \nabla V \leq 0$. Furthermore, if $\dot{V} = 0$, $||x_i - P_i|| < \epsilon$, so we can satisfy the guard g_2. □

Proposition 8. *Suppose $C(t_k^-) = B$, $r(t_k) = R_3$, and $C(t_k^+) = D$. If for some ϵ, $d(B,T) < \epsilon$, then* $\lim_{t\to\infty} C(t)[V_D] \in T$

Proof. Suppose robot i is added to $C(t_k^-)$ by forming `control` edges ij and ik to robots j and k to form D. Since for any u, v the rules never allow an edge uv where $u.dist < v.dist$, Lemma 1 holds and we may consider the dynamics of $C(t)[V_D]$ without reference to the vertices in $V - V_D$. We choose ϵ in $g_{settled}$ small enough so that if for R_3, $g_{settled}(L)$ is satisfied for j and k, their motion relative to the coordinate frames of the robots to which they have directed control edges is zero. Since every time a robot is added to the structure, $g_{settled}(L)$ must be satisfied, by induction we may treat x_j and x_k as constants.

Let $V = (||x_i - x_j|| - \delta)^2 + (||x_i - x_k|| - \delta)^2$. $\dot{V} = \nabla V^T \frac{dx}{dt} = -\nabla_{x_i} V^T \nabla_{x_i} V \leq 0$. V is a Lyapunov function with stable fixed points where x_i^* satisfies $||x_i^* - x_j|| - \delta = ||x_i^* - x_k|| - \delta = 0$. We must show that the region of attraction of one of these fixed points is the open half plane defined by line l_{jk} containing x_i^*.

We have that $\dot{x}_i \propto (||v_{ij}|| - \delta)\hat{v}_{ij} + (||v_{ik}|| - \delta)\hat{v}_{ik}) = \alpha\hat{v}_{ij} + \beta\hat{v}_{ik}$. We show that there is no path from $x_i \notin l_{jk}$ to $x_i \in l_{jk}$. Suppose we pick x_i near l_{jk} such that $||v_{ij}|| < ||v_{ik}||$ and $\beta > \alpha > 0$. Then $\angle\hat{v}_{ij} < \angle\dot{x}_i < \angle\hat{v}_{ik}$. This remains true (near l_{jk}) until $||v_{ij}|| = \delta$. Now $\alpha \leq 0$and $\angle\hat{v}_{ik} \leq \angle\dot{x}_i < \pi + \angle\hat{v}_{ik}$. This suggests if the trajectory intersects l_{jk} it must do so between x_j and x_k. However, in this region, $\alpha < 0$ and $\beta < 0$ which means there is a component of $\dot{x}_i$ away from the line l_{jk}. Thus there is no path leading from $x_i \notin l_{jk}$ to $x_i \in l_{jk}$.

Proposition 9. *Suppose $C(t_k^-) = B$, $r(t_k) = R_4$, and $C(t_k^+) = D$. If for some ϵ, $d(B,T) < \epsilon$, then* $\lim_{t\to\infty} C(t)[V_D] \in T$. (See proof above).

4.2 Supporting Proposition Proofs

Proof of Proposition 2. By Proposition 7, rule R_2 is executed. We propose the Lyapunov function $V = \sum_{i \in Im(h_2)} \sum_{j \in F^+(i)} 1/2 U_{ij}$. $\dot{V} = \nabla V_x \dot{x} = -\nabla V_x^T \nabla V_x \leq 0$. In addition to fixed points corresponding to δ-triangulations, the `slave` controllers operating under the topology created by R_2 have local minima at some positions where $x_i = x_j$. However, by choosing ϵ in g_2 small enough ($18\epsilon^2 < \delta^2$), then for all V in the guard region, $V \leq 18\epsilon^2 < \delta^2$ where δ^2 is the contribution to the Lyapunov function for a single edge where $x_i = x_j$. Thus $\lim_{t\to\infty} \gamma[Im(h_2)](t) \in T$. □

Proof of Proposition 3. By propositions 8 and 9 and the fact that edges added by R_5 are not used by the controllers, every vertex that is not settled has its own decreasing Lyapunov function parameterized by the edge lengths. □

Proof of Proposition 4. Since Proposition 3 holds and since a small neighborhood near $\rho \in T$ satisfies guard g_5, every application of R_5 must be applied. Furthermore, since the intersection of the half-planes and cones defined by the boundary edges is a superset of the exterior of the crystal, it must be the case that $x_i \notin I(t)$ implying rule R_3 or R_4 are applicable. □

Proof of Proposition 5. When $d(C(t), \rho) < \epsilon$ for very small ϵ, it is clear that either (1) $p(i)$ lies in a face where the sum of the distances is greater or (2) $p(i) \notin I(t)$. In case (1), as $x_i \to p(i)$ either R_7 or R_8 becomes applicable. Since the graph is finite, by induction on rules R_7 and R_8 eventually $p(i) \notin I(t)$ (i.e case (2)). By the convergence property of the controllers, eventually $x_i \notin I(t)$. □

Proof of Corollary 1. If $d(C(t), \rho) < \epsilon$ and if $\delta + \epsilon \ll \sqrt{3}/2\delta - \varepsilon$, there are no Gabriel graph edges between i and vertices on the interior of the crystal. This implies that if i executes R_9 and is labeled `free`, while $||x_i - x_j|| < \delta$ for $j \in C(t)$, the motion is away from the crystal. Thus i remains in g_3 or g_4. □

5 Conclusions and Future Work

Embedded graph grammars are a unique combination of concurrency and hybrid systems in which we can model networked robotic systems. Our solution to the δ-triangulation coverage problem is meant to demonstrate how embedded graph grammars can be used to specify and reason about the correctness of complex, multi-mode coordination problems. We note that issues such as robustness or performance under sensing and communication limitations can be addressed by constructing more complicated grammars. However, these questions lie outside the scope of this paper.

We believe that as systems incorporate more complex combinations of reactive tasks, the need for the unified modeling of communication protocols and the ensuing hybrid dynamics becomes pronounced. Unfortunately, proving that such systems are correct is unwieldy as the proofs in this paper suggest. We plan to explore real time temporal logics, compositional methods, and automated verification techniques, thereby completing the formalism introduced here.

Acknowledgement

Eric Klavins and John-Michael McNew are partially supported by the AFOSR via the 2006 MURI award *Design, Specification and Verification of Distributed Embedded Systems*. The work by Magnus Egerstedt is supported under a grant from the NASA, Earth Science Technology Office, AIST Program.

References

1. Egerstedt, M., Brockett, R.: Feedback can reduce the specification complexity of motor programs. IEEE Transactions on Automatic Control (2003)
2. V. Manikonda, P.S.K., Hendler., J.: Languages, Behaviors, Hybrid Architectures and Motion Control. In: Mathematical Control Theory. (1998)
3. Lygeros, J., Johansson, K.H., Simic, S., Zhang, J., Sastry, S.: Dynamical properties of hybrid automata. IEEE Transactions on Automatic Control (2003)

4. Henzinger, T.: The theory of hybrid automata, 11th Annual Symposium on Logic in Computer Science (LICS), IEEE Computer Society Press (1996)
5. Klavins, E., Burden, S., Napp, N.: Optimal rules for programmed stochastic self-assembly. In: Robotics: Science and Systems, Philadelphia, PA (2006)
6. Cortes, J., Martinez, S., Bullo, F.: Robust rendezvous for mobile autonomous agents via proximity graphs in arbitrary dimensions. IEEE Transactions on Automatic Control (2006)
7. Leonard, N., Fiorelli, E.: Virtual leaders, artificial potentials and coordinated control of groups. Proc. of the IEEE Conference on Decision and Control (2001)
8. McNew, J.M., Klavins, E.: Locally interacting hybrid sytems using embedded graph grammars. In: Proc. of the Conference on Decision and Control. (2006)
9. Smith, B., McNew, J., Egerstedt, M., Klavins, E., Howard, A.: Embedded graph grammars for multi-robot coordination. In: Proc. of the International Conference on Robotics and Automation. (2007) submitted.
10. B. Shucker, T. Murphey, J.: A method of cooperative control using occasional non-local interactions. In: Proc. of the American Control Conference. (2006)

Comparing Forward and Backward Reachability as Tools for Safety Analysis

Ian M. Mitchell

Department of Computer Science, University of British Columbia
2366 Main Mall, Vancouver, BC, Canada V6T 1Z4
mitchell@cs.ubc.ca
http://www.cs.ubc.ca/~mitchell

Abstract. Using only the existence and uniqueness of trajectories for a generic dynamic system with inputs, we define and examine eight types of forward and backward reachability constructs. If the input is treated in a worst-case fashion, any forward or backward reach set or tube can be used for safety analysis, but if the input is treated in a best-case fashion only the backward reach tube always provides the correct results. Fortunately, forward and backward algorithms can be exchanged if well-posed reverse time trajectories can be defined. Unfortunately, backward reachability constructs are more likely to suffer from numerical stability issues, especially in systems with significant contraction—the very systems where forward simulation and reachability are most effective.

1 Introduction

Except for the simplest of examples, analytic verification of safety properties for continuous and hybrid systems is rarely possible. With the goal of broadening the applicability and automating the process, numerical methods for verifying or validating such properties have been the subject of much study. The approximation of reachable sets is one major category of such numerical methods. There are two fundamental types of reachability: forward and backward. Many algorithms have been proposed to compute one of these reachable sets (see Section 3), and some type of equivalence is often informally mentioned when a problem statement requires computation of the other set. The contribution of this paper is a detailed examination of the distinctions between these two sets. We make rather strong assumptions about the existence and uniqueness of trajectories, so it is the negative conclusions that hold the most significance.

Section 2 informally discusses the relationship between reachability and safety and defines some of the terminology, while Section 3 covers previous work. The body of the paper beings in Section 4 by examining the question of when various forms of forward and/or backward reachability can be used to prove system safety: in some cases any form will do, but in some cases only one type of backward reachability gives the correct result. Section 5 then demonstrates that the formulation of the reachability problem and the algorithm used to solve

A. Bemporad, A. Bicchi, and G. Buttazzo (Eds.): HSCC 2007, LNCS 4416, pp. 428–443, 2007.

it need not work in the same temporal direction, since forward and backward algorithms can be interchanged for systems which are reversible.

Unfortunately, these algorithms find only approximations. In Section 6, trajectory sensitivity analysis [1] is extended to examine the way in which numerical error may grow as these algorithms are run. Even though the backward reachability formulation may be applicable to more problems, we conclude that it is also more likely to experience numerical stability problems, regardless of whether it is implemented by a forward or backward algorithm.

2 Reachability and Safety Analysis

Safety analysis of a given system seeks to discover whether the system—or more accurately, the mathematical model representing the system—can enter a specified set of unsafe states. Since many systems operate correctly only when started correctly, a set of initial states is also often specified. Mathematically, we will specify a safety analysis problem by a tuple $\mathsf{S} = (\mathsf{H}, I, T)$ where H is a system model, I is the initial set, and T is the unsafe set or target.

We define the concepts more formally in Section 4, but informally *reachability analysis* seeks to determine whether trajectories of H can reach T from I. There are two types of analysis. *Forward reachability* starts with states in I and follows trajectories forward in time. If any of these trajectories intersect with T the system is unsafe. *Backward reachability* starts with states in T and follows trajectories backwards in time. If any of these backwards trajectories intersects I the system is unsafe.

Under these definitions it sounds like reachability can be determined by simulating individual trajectories of H, and simulation is in fact the typical method by which safety is disproved. Proof of safety, however, requires a guarantee that all possible trajectories have been investigated; a challenging task in continuous and hybrid systems where the number of states is infinite. Consequently, the term *reachability algorithm* is usually reserved for techniques that determine the set of states traversed by all trajectories emanating from a given set.

While the terms are not used consistently in the literature, we will in this paper distinguish two different objects that a reachability algorithm might generate: the *reach set* is the set of states occupied by trajectories at exactly some specified time, and the *reach tube* is the set of states traversed by those same trajectories over all times prior to and including the specified time. Thus, the reach tube always contains the reach set. Forward and backward versions of both reach sets and tubes can be specified.

While we examine their properties and appropriateness in terms of the fully specified safety analysis problem S, forward and backward reachable sets and tubes may be more or less appropriate for other tasks; for example, backward reach tubes for finding the set of states which achieves a target set despite the unknown but bounded disturbance of exogenous inputs, or forward reach sets for demonstrating system liveness.

3 Related Work

There are two main classes of *direct* reachability algorithms, those that work directly with continuous representations. *Lagrangian* approaches represent the set or tube with information that moves with the flow of the underlying dynamics, and are typically described in terms of forward reachability. A few are designed for systems without inputs [2], many permit inputs which expand the size of the reach set [3,4,5,6] and some permit inputs which shrink the reach set [7]. The theory is often based on linear continuous dynamics, although most schemes have demonstrated computational extensions to handle the nonlinear case. These schemes have also shown the best scalability; for example, results for systems with hundreds of dimensions have been reported in [6,2].

Eulerian approaches work with a discretization that is not moving with the dynamics (although it may be refined during computation), and are typically described in terms of backward reachability [8,9,10]. All schemes can support systems with inputs which expand the size of the reachable set, and most handle those that shrink it as well. The theory works directly with nonlinear systems, although scalability much beyond four dimensions has not been demonstrated.

The results in Section 6 are derived by a sensitivity analysis of trajectories. Lagrangian reachability algorithms that depend on numerical integration of these (or related) trajectories are clearly affected by such sensitivity. Despite the fact that they do not directly integrate the dynamics, Eulerian schemes will also be subject to similar numerical stability problems since the approximations that they use are based on the evolution of the underlying system.

In addition to the classes of direct algorithms, there are at least two other classes of *indirect* algorithms related to reachability for continuous and/or hybrid systems. Discretization of the state space and dynamics can yield a system on which discrete reachability algorithms can be run; for example [11,12]. Alternatively, automated Lyapunov type methods can be used to prove invariance properties, such as [13,14]. How the sensitivity results might apply to these algorithms has not yet been investigated.

The conclusions of Sections 4 and 5 apply to discrete systems as well; in fact, forward and backward reachability have been combined to verify some discrete systems (see [15] and the citations within). However, the nature of the approximation errors (if any) in discrete algorithms is different enough that Section 6 may not apply.

4 Comparing Forward and Backward Reachability

In this section we compare properties of forward and backward reachability for a very generically defined dynamic system H. Trajectories of H will be denoted by

$$\xi_{\mathsf{H}}(s; z, t, u(\cdot)) : \mathbb{T} \to \mathbb{Z},$$

where $\mathbb{T} = [-\mathcal{T}, +\mathcal{T}] \subset \mathbb{R}$ is the time interval over which the trajectory exists. We employ the semicolon to distinguish between the argument s of ξ_{H} and the trajectory parameters: initial state $z \in \mathbb{Z}$, initial time $t \in \mathbb{T}$ and input signal $u(\cdot) \in \mathbb{U}$. For systems lacking an input signal, we omit it and denote trajectories as $\xi_{\mathsf{H}}(s; z, t)$.

Existence and uniqueness of trajectories ξ_{H} for various types of dynamic systems is a challenging subject by itself; for example, see [16,17] and the citations within. To maintain the focus of this paper, we make the following rather idealized assumption.

Assumption 1. *For given initial state z, time t, and input signal $u(\cdot)$ drawn from an appropriate class, there exists a unique trajectory $\xi_{\mathsf{H}}(s; z, t, u(\cdot))$ for $s \in \mathbb{T}$.*

By making this strong but generic assumption, many of the results in the next two sections will apply to a broad group of dynamic systems, although we focus on continuous and hybrid systems. It is the negative conclusions that we draw that have the most relevance to future research—if a technique or formulation fails under such a strong but generic assumption, there is little point in pursuing its concrete implementation.

In *continuous systems*, the dynamics are given by an ordinary differential equation (ODE) of the form $\dot{z}(t) = f(z(t), u(t))$, where the state z is continuous. Typically $\mathbb{Z} \subseteq \mathbb{R}^d$, although some state variables may use other domains; for example, angles are often drawn from the periodic set $[0, 2\pi[$. If $f : \mathbb{Z} \times U \to \mathrm{T}\mathbb{Z}$ is uniformly continuous, bounded and Lipschitz continuous in z for fixed u, then Assumption 1 is satisfied [18] for fixed $u(\cdot) \in \mathbb{U}$, where

$$\mathbb{U} \triangleq \{\phi : \mathbb{T} \to U \mid \phi(\cdot) \text{ is measurable}\} \tag{1}$$

and $U \subset \mathbb{R}^{d_u}$ is convex and compact. Consequently, we can specify a continuous system as a tuple $\mathsf{H_C} = (\mathbb{Z}, f, U)$.

The generalization to *hybrid systems* involves a form of hybrid automaton (HA) adapted from [10]: we simplify to a single input, but that input may affect the guards and domains. The state of a hybrid system is $z = (q, x) \in \mathbb{Q} \times \mathbb{X} = \mathbb{Z}$, where q is the discrete state and x is the continuous state. The full HA is given by the tuple $\mathsf{H_H} = (\mathbb{Q}, \mathbb{X}, f, D, G, r, U)$, where

$$\begin{array}{ll}
\mathbb{Q} & \text{discrete states;} \\
\mathbb{X} & \text{continuous states;} \\
f : \mathbb{Q} \times \mathbb{X} \times U_C \to \mathrm{T}\mathbb{X} & \text{continuous dynamics (vector field);} \\
D : \mathbb{Q} \times U_D \to P(\mathbb{X}) & \text{domain of continuous evolution;} \\
G : \mathbb{Q} \times \mathbb{Q} \times U_D \to P(\mathbb{X}) & \text{guard conditions for discrete evolution;} \\
r : \mathbb{Q} \times \mathbb{Q} \times \mathbb{X} \times U \to \mathbb{X} & \text{reset function;} \\
U = (U_C, U_D) & \text{continuous and discrete input sets;}
\end{array} \tag{2}$$

where $P(\mathbb{X})$ is the power set (set of all subsets) of $\mathbb{X}$. As in [10], we assume that the discrete inputs are constant during continuous evolution. We will call the boundaries of the domains and guards the *switching surfaces*. Formal mathematical conditions under which Assumption 1 holds are available for some subclasses of this hybrid automata [16,17]. At a minimum, Assumption 1 will require that H_H be non-Zeno and non-blocking, and that f statisfies conditions to ensure existence of the continuous components of the trajectories.

The proofs for many of the propositions in this section were omitted due to space limitations, but can be found in [19].

4.1 Maximal Reachability

When performing safety analysis with forward reachability, the single input's authority is used to make the reach set and tube as large as possible. We will use the subscript "1+" to denote a single input used to maximize the size of the reachable set and tube and call these constructs *maximal.*

$$F_{1+}(\mathsf{H}, S, t) \triangleq \{\hat{z} \in \mathbb{Z} \mid \exists u(\cdot) \in \mathbb{U}, \exists z \in S, \xi_\mathsf{H}(t; z, 0, u(\cdot)) = \hat{z}\}, \tag{3}$$

$$F_{1+}(\mathsf{H}, S, [0,t]) \triangleq \{\hat{z} \in \mathbb{Z} \mid \exists u(\cdot) \in \mathbb{U}, \exists z \in S, \exists s \in [0,t], \xi_\mathsf{H}(s; z, 0, u(\cdot)) = \hat{z}\}. \tag{4}$$

In the corresponding backward reachability problems, the input is used to drive as many states as possible towards the target set. The result is that the size of the reachable set and tube are again maximized.

$$B_{1+}(\mathsf{H}, S, t) \triangleq \{z \in \mathbb{Z} \mid \exists u(\cdot) \in \mathbb{U}, \exists \hat{z} \in S, \xi_\mathsf{H}(0; z, -t, u(\cdot)) = \hat{z}\}, \tag{5}$$

$$B_{1+}(\mathsf{H}, S, [0,t]) \triangleq \{z \in \mathbb{Z} \mid \exists u(\cdot) \in \mathbb{U}, \exists \hat{z} \in S, \exists s \in [0,t], \xi_\mathsf{H}(0; z, -s, u(\cdot)) = \hat{z}\}. \tag{6}$$

The relationships between these four sets is easy to establish and should not be surprising.

Proposition 1

$$F_{1+}(\mathsf{H}, S, [0,t]) = \bigcup_{\hat{t} \in [0,t]} F_{1+}(\mathsf{H}, S, \hat{t}) \qquad B_{1+}(\mathsf{H}, S, [0,t]) = \bigcup_{\hat{t} \in [0,t]} B_{1+}(\mathsf{H}, S, \hat{t})$$

Reachability for zero input systems is a special case of maximal reachability; for example, the forward reach set is given by

$$F_0(\mathsf{H}, S, t) \triangleq \{\hat{z} \in \mathbb{Z} \mid \exists z \in S, \xi_\mathsf{H}(t; z, 0) = \hat{z}\}.$$

4.2 Minimal Reachability

Instead of the sets defined above, we can choose to seek only those states that trajectories are forced to reach no matter what input is chosen. Consequently, the reachable sets and tubes are as small as possible, we use the "1−" notation, and call these constructs *minimal.*

$$F_{1-}(\mathsf{H}, S, t) \triangleq \{\hat{z} \in \mathbb{Z} \mid \forall u(\cdot) \in \mathbb{U}, \exists z \in S, \xi_{\mathsf{H}}(t; z, 0, u(\cdot)) = \hat{z}\}, \tag{7}$$

$$F_{1-}(\mathsf{H}, S, [0,t]) \triangleq \{\hat{z} \in \mathbb{Z} \mid \forall u(\cdot) \in \mathbb{U}, \exists z \in S, \exists s \in [0,t], \xi_{\mathsf{H}}(s; z, 0, u(\cdot)) = \hat{z}\}, \tag{8}$$

$$B_{1-}(\mathsf{H}, S, t) \triangleq \{z \in \mathbb{Z} \mid \forall u(\cdot) \in \mathbb{U}, \exists \hat{z} \in S, \xi_{\mathsf{H}}(0; z, -t, u(\cdot)) = \hat{z}\}, \tag{9}$$

$$B_{1-}(\mathsf{H}, S, [0,t]) \triangleq \{z \in \mathbb{Z} \mid \forall u(\cdot) \in \mathbb{U}, \exists \hat{z} \in S, \exists s \in [0,t], \xi_{\mathsf{H}}(0; z, -s, u(\cdot)) = \hat{z}\}. \tag{10}$$

Unfortunately, the properties that hold in the purely existential maximal case above no longer apply.

Proposition 2

$$\bigcup_{\hat{t}\in[0,t]} F_{1-}(\mathsf{H}, S, \hat{t}) \subseteq F_{1-}(\mathsf{H}, S, [0,t]) \qquad \bigcup_{\hat{t}\in[0,t]} B_{1-}(\mathsf{H}, S, \hat{t}) \subseteq B_{1-}(\mathsf{H}, S, [0,t])$$

The problem arises because the choice of t in the reach set definitions is fixed before any other variable is quantified, while the choice of $s \in [0,t]$ in the reach tube definition occurs after all other variables are quantified. For maximal reachability all the quantifiers are existential, so their ordering does not matter. However, once the input's quantifier is changed to be universal, the order in which the trajectory's time interval is chosen matters a great deal.

We close by mentioning that systems with competing inputs (such as control and disturbance) are subject to the same negative results as the minimal reachability constructs (such as Propositions 2, 4 and 5); for more details see [19].

4.3 Application to Safety Analysis

Having defined the maximal and minimal forward and backward reach sets and tubes, we examine which can be used to solve the safety problem $\mathsf{S} = (\mathsf{H}, I, T)$ under various assumptions about the input's behaviour. Throughout this section we assume that H satisfies Assumption 1.

Proposition 3. *The following properties are equivalent.*

1. H *is safe over horizon* $t \leq T$ *for all possible inputs* $u(\cdot) \in \mathbb{U}$.
2. $F_{1+}(\mathsf{H}, I, s) \cap T = \emptyset$ *for all* $s \in [0,t]$.
3. $F_{1+}(\mathsf{H}, I, [0,t]) \cap T = \emptyset$.
4. $B_{1+}(\mathsf{H}, T, s) \cap I = \emptyset$ *for all* $s \in [0,t]$.
5. $B_{1+}(\mathsf{H}, T, [0,t]) \cap I = \emptyset$.

Based on this proposition, we can use any of the reach sets or tubes to demonstrate the safety of systems despite the actions of bounded exogenous inputs, or of systems without any inputs. The situation is not quite so favourable for proving the existence of an input which guarantees safety.

Proposition 4. *Given horizon* $t \leq T$, *there exists an input* $u(\cdot) \in \mathbb{U}$ *(which may depend on initial state) that keeps* H *safe if and only if* $B_{1-}(\mathsf{H}, T, [0,t]) \cap I = \emptyset$. *Such an input may exist only if* $B_{1-}(\mathsf{H}, T, s) \cap I = \emptyset$ *for all* $s \leq t$, *but the converse is not necessarily true.*

Proof. We first prove the claims for the reach tube. Let $S = B_{1-}(\mathsf{H}, T, [0,t]) \cap I$.

H safe $\Longrightarrow$ $(S = \emptyset)$: Assume $z \in S$ but that H is safe for input $u(\cdot) \in \mathbb{U}$ and derive a contradiction. By (10), there exists $\hat{z} \in T$ and $s \in [0,t]$ such that $\xi_{\mathsf{H}}(0; z, -s, u(\cdot)) = \hat{z}$. But then this trajectory reaches from I to T under input $u(\cdot)$, which is a contradiction that H is safe for input $u(\cdot)$

H safe $\Longleftarrow$ $(S = \emptyset)$: Assume that $S = \emptyset$. Then for all $z \in I$, z is in the complement of $B_{1-}(\mathsf{H}, T, [0,t])$. Negating (10), there exists $u(\cdot) \in \mathbb{U}$ such that for all $\hat{z} \in T$ and $s \in [0,t]$, $\xi_{\mathsf{H}}(0; z, -s, u(\cdot)) \neq \hat{z}$; in other words, for any initial state in I, there is an input which gives rise to a trajectory which does not reach T during the interval $[0,t]$. Hence, there is an input $u(\cdot)$ which makes H is safe during this interval.

The "only if" claim for the reach set is a simple outcome of combining Proposition 2 and the proof for $\Longrightarrow$ above. The converse is not necessarily true because for the reach set the input is chosen after the time t, and for larger t the input may drive trajectories right through the unsafe set T and out the other side [9]. An example can be found in [19]. □

Based on this proposition, we can use the minimal backwards reach tube to prove the existence of a safe input for any state in the initial set. Unfortunately, the same cannot be said of the minimal forward reachability constructs.

Proposition 5. *The forward minimal reach set and tube provide no information about whether there exists an input $u(\cdot) \in \mathbb{U}$ that makes H safe.*

Proof. Consider first the forward reach tube. Let $S = F_{1-}(\mathsf{H}, I, [0,t]) \cap T$. We show that any combination of S empty or nonempty with H safe or unsafe is possible. The two easy cases are the ones that should hold. For $S \neq \emptyset$ and H unsafe, take $I \cap T \neq \emptyset$. For $S = \emptyset$ and H safe, take $T = \emptyset$.

Now consider $S = \emptyset$. Then for all $\hat{z} \in T$, $\hat{z}$ is in the complement of $F_{1-}(\mathsf{H}, I, [0,t])$. Negating (8), there exists $u(\cdot) \in \mathbb{U}$ such that for all $z \in I$ and $s \in [0,t]$, $\xi_{\mathsf{H}}(s; z, 0, u(\cdot)) \neq \hat{z}$; in other words, for any unsafe state $\hat{z}$ in T, there is an input such that no trajectory emanating from the initial set I arrives at $\hat{z}$ during the interval $[0,t]$—so far, so good. Unfortunately, this proof only applies once $\hat{z} \in T$ is selected; there is nothing in this proof to stop the chosen input from driving all those trajectories into some other part of T, thus rendering the system unsafe.

Finally, consider $\hat{z} \in S$. By (8), for all $u(\cdot) \in \mathbb{U}$ there exists $z \in I$ and $s \in [0,t]$ such that $\xi_{\mathsf{H}}(s; z, 0, u(\cdot)) = \hat{z}$; in other words, for all inputs there exists a trajectory starting from somewhere in I that will arrive at $\hat{z}$ at or before time t. However, this is not the safety question that we sought to answer. For all these $z \in I$, there may still exist some other $\hat{u}(\cdot) \in \mathbb{U}$ that ensures $\xi_{\mathsf{H}}(s; z, 0, \hat{u}(\cdot)) \notin T$ for all $s \in [0,t]$, and hence that H is safe.

The forward reach set can fail for safety verification in either of the ways that the forward reach tube or the backward reach set fails. □

The essential problem with minimal forward reachability is that the state lying in the initial set is chosen after the input while the state lying in the target set is chosen before, rather than the other way around.

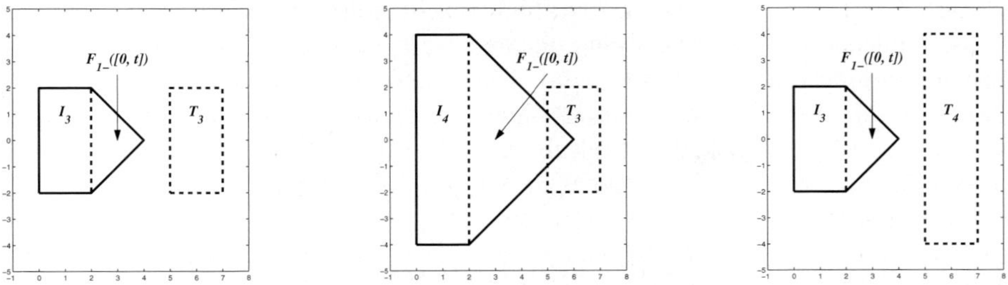

Fig. 1. Fixed points of the forward minimal reach tubes. The two cases on the left are actually safe, while the case on the right is unsafe. The forward reach tube demonstrates that it is inappropriate for existential safety verification in the two cases on the right.

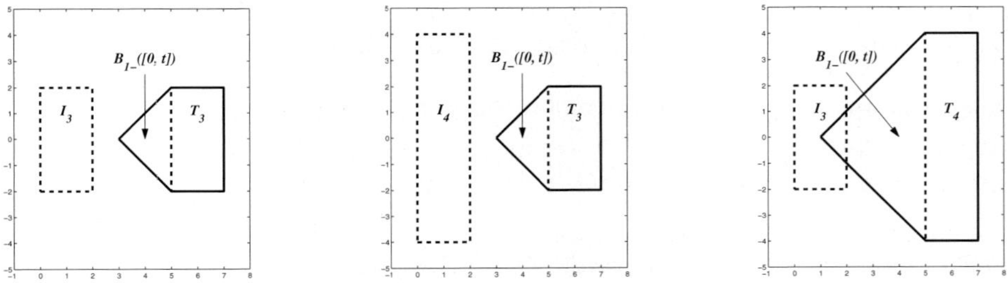

Fig. 2. Fixed points of the backward minimal reach tubes. Safety is correctly determined for the two cases on the left, and a lack of safety for the case on the right.

4.4 Examples of Forward and Backward Reachability for Safety

In this section we examine the various reachability constructs in terms of the purely continuous system H_2 for $x \in \mathbb{R}^2$.

$$\dot{x} = \begin{bmatrix} +1 \\ u \end{bmatrix}, \qquad \text{where } |u| \leq 1. \tag{11}$$

The motion of H_2 is easy to visualize: translation to the right at unit speed, and the choice of input determines vertical motion at unit speed.

Examples demonstrating Proposition 3 and the reach set components of Propositions 4 and 5 can be found in [19]. For the reach tube components of these latter two propositions, we choose two initial and two target sets.

$$\begin{aligned} I_3 &= [0, +2] \times [-2, +2] \quad T_3 = [+5, +7] \times [-2, +2] \\ I_4 &= [0, +2] \times [-4, +4] \quad T_4 = [+5, +7] \times [-4, +4] \end{aligned}$$

These initial and target sets are horizontally aligned, so for any initial state with $x_2 \geq 0$, choose $u(t) = +1$ and for any initial state with $x_2 \leq 0$, choose $u(t) = -1$. With these input signals it is easy to see that either initial set with T_3 is safe, while either initial set with T_4 is unsafe.

Figures 1 and 2 show the two minimal reach tubes for three of the combinations of initial and target sets (the unsafe case I_4 and T_4 is not shown but is an easy extrapolation from those given). Both tubes reach a fixed point at $t = 2$ (for I_3 or T_3) or $t = 4$ (for I_4 or T_4), and it is that fixed point which is shown. The failure of the forward reach tube to correctly distinguish safe and unsafe situations can be seen in the two right subplots of Figure 1.

5 Exchanging Forward and Backward Reachability

Despite the negative conclusions regarding the minimal forward reach tube $F_{1-}(\mathsf{H}, S, [0,t])$, algorithms for its computation may still be useful if they can be used to compute backward reach tubes. In order to establish the situations under which forward and backward reachability may be interchanged, we must be able to reverse the direction of time in our dynamic system. Under Assumption 1, the following assumption will be relatively easily satisfied.

Assumption 2. *For a given dynamic system* H*, there exists a* backward dynamic system $\overleftarrow{\mathsf{H}}$ *such that for all* $t, s \in \mathbb{T}$

$$\xi_{\mathsf{H}}(s; z, t, u(\cdot)) = \hat{z} \quad \Longleftrightarrow \quad \xi_{\overleftarrow{\mathsf{H}}}(s; \hat{z}, t, u(\cdot)) = z.$$

Furthermore, $\xi_{\overleftarrow{\mathsf{H}}}$ *satisfies the conditions of Assumption 1.*

For the continuous $\mathsf{H_C}$, $\xi_{\overleftarrow{\mathsf{H_C}}}$ satisfies the ODE

$$\dot{z}(t) = \overleftarrow{f}(z(t), u(t)) \triangleq -f(z(t), u(t)) \tag{12}$$

and $\overleftarrow{\mathsf{H_C}} = (\mathbb{Z}, -f, U)$. If f satisfies the sufficient conditions mentioned above for $\xi_{\mathsf{H_C}}$ to satisfy Assumption 1, then so will $\xi_{\overleftarrow{\mathsf{H_C}}}$.

The case for the HA $\mathsf{H_H}$ is considerably more complex. In addition to the reversed continuous evolutions satisfying (12), there must exist reversed versions $\overleftarrow{G}$ and $\overleftarrow{r}$ of the guards and reset which satisfy

$$\begin{aligned} x \in G(q, \hat{q}, u_D) \quad &\Longleftrightarrow \quad \hat{x} \in \overleftarrow{G}(\hat{q}, q, u_D), \\ r(q, \hat{q}, x, u) = \hat{x} \quad &\Longleftrightarrow \quad \overleftarrow{r}(\hat{q}, q, \hat{x}, u) = x. \end{aligned} \tag{13}$$

With these definitions, $\overleftarrow{\mathsf{H_H}} = (\mathbb{Q}, \mathbb{X}, -f, D, \overleftarrow{G}, \overleftarrow{r}, U)$. Conditions under which $\xi_{\overleftarrow{\mathsf{H_H}}}$ would satisfy Assumption 1 are even more challenging to come by, although there has been some work [20]. However, if we can find a well posed $\overleftarrow{\mathsf{H}}$, then the temporal direction of our favourite reachability algorithm is irrelevant.

Proposition 6. *If* H *satisfies the conditions of Assumptions 1 and 2, then*

$$\begin{aligned} F_{1+}(\mathsf{H}, S, [0,t]) &= B_{1+}(\overleftarrow{\mathsf{H}}, S, [0,t]) \quad & F_{1+}(\mathsf{H}, S, t) &= B_{1+}(\overleftarrow{\mathsf{H}}, S, t) \\ F_{1-}(\mathsf{H}, S, [0,t]) &= B_{1-}(\overleftarrow{\mathsf{H}}, S, [0,t]) \quad & F_{1-}(\mathsf{H}, S, t) &= B_{1-}(\overleftarrow{\mathsf{H}}, S, t) \end{aligned}$$

Proof. We prove the claim for the minimal reach tubes; the proofs for the remaining claims are similar. Assume $\hat{z} \in F_{1-}(\mathsf{H}, S, [0,t])$. By (7), for all $u(\cdot) \in \mathbb{U}$ there exists $z \in S$ and $s \in [0,t]$ such that $\xi_{\mathsf{H}}(t, z, 0, u(\cdot)) = \hat{z}$. Under Assumption 2, $\xi_{\overleftarrow{\mathsf{H}}}(t; \hat{z}, 0, u(\cdot)) = z$. Because $\overleftarrow{\mathsf{H}}$ is time independent, we can shift the time variable to get $\xi_{\overleftarrow{\mathsf{H}}}(0; \hat{z}, -t, u(\cdot)) = z$, which by (9) implies $z \in B_{1-}(\overleftarrow{\mathsf{H}}, S, [0,t])$. The proof in the converse direction is similar. □

6 Reachable Set Sensitivity

Section 4.3 demonstrated that the backward reach tube is the most generally applicable of the reachability operators to verification tasks. However, reach sets and tubes can rarely be determined analytically, so they must be approximated numerically. In this section we examine equations for the sensitivity of trajectories with respect to initial conditions. From these equations we can draw the conclusion that for some types of systems accurate numerical approximation of backwards reachability may not be possible.

The sensitivity analysis techniques used in this section force us to abandon the very general dynamic system definition used in the previous sections. Furthermore, we will assume that the number of states in discrete systems or the discrete component of hybrid systems is small enough that the discrete component of the reachable sets or tubes can be represented exactly. Therefore, we will focus our attention on continuous systems and the continuous component of hybrid systems. Since the former are a subset of the latter, we perform the analysis for hybrid systems and except where noted assume that $\mathsf{H} = \mathsf{H}_{\mathsf{H}}$ and that Assumptions 1 and 2 hold.

For the purposes of this analysis the domains D and guards G are specified by implicit surface functions

$$
\begin{aligned}
D(q, u_D) &= \{x \in \mathbb{X} \mid \psi_D(q, x, u_D) \leq 0\} \\
G(q, \hat{q}, u_D) &= \{x \in \mathbb{X} \mid \psi_G(q, \hat{q}, x, u_D) \leq 0\}
\end{aligned}
$$

for all $q, \hat{q} \in \mathbb{Q}$ and $u_D \in U_D$. The switching surfaces are then given by the zero level sets of these functions, and the normals of those switching surfaces by the local gradients. In order to study perturbations, we make the following assumption about the components of the hybrid system; the assumption also ensures that the switching surfaces and their normals are well defined.

Assumption 3. *The vector field f, reset r and implicit surface functions of the domains ψ_D and guards ψ_G are differentiable with respect to their continuous parameter x when all other parameters are held fixed.*

Sensitivity equations for a class of hybrid systems called differential-algebraic-discrete were derived in [1]. Here we adapt these results to HA of the form (2) by ignoring sensitivity with respect to parameter or discrete state, removing the algebraic component and adding a continuous reset. Details are omitted because the derivation follows directly from [1]. Sensitivity with respect to (constant) problem parameters can be derived in a similar manner. We do not consider sensitivity with respect to the input, and hence assume throughout that $u(\cdot) \in \mathbb{U}$ is fixed.

For convenience, define the matrices

$$\mathbf{F}(q, x, u) \triangleq \frac{\partial f(q, x, u)}{\partial x} \qquad \mathbf{R}(q, \hat{q}, x, u) \triangleq \frac{\partial r(q, \hat{q}, x, u)}{\partial x}$$

6.1 Trajectory Sensitivity Analysis

In this section we examine the effects on a trajectory's position due to small perturbations of the continuous portion of its initial state.

$$\xi_{\mathsf{H}}(t; z_0 + \delta x, 0, u(\cdot)) = \xi_{\mathsf{H}}(t; z_0, 0, u(\cdot)) + \Xi_{\mathsf{H}}(t; \xi_{\mathsf{H}}(\cdot))\delta x + \mathcal{O}(\delta x^2), \tag{14}$$

where the initial state is $z_0 = (q_0, x_0)$, the perturbation is purely continuous $z_0 + \delta x = (q_0, x_0 + \delta x)$, $\xi_{\mathsf{H}}(\cdot) = \xi_{\mathsf{H}}(\cdot; z_0, 0, u(\cdot))$, and the *sensitivity matrix* is defined as

$$\Xi_{\mathsf{H}}(t; \xi_{\mathsf{H}}(\cdot)) \triangleq \frac{\partial \xi_{\mathsf{H}}(t; z_0, 0, u(\cdot))}{\partial x_0}.$$

The continuous evolution of the HA is governed by an ODE, and sensitivity analysis of ODEs is well established; for example, see [21, section 4.6 and exercise 6.4]. Using what is essentially a Taylor series expansion, it can be shown that the sensitivity matrix solves the ODE

$$\tfrac{d}{dt}\Xi_{\mathsf{H}}(t) = \mathbf{F}(q, x, u)\Xi_{\mathsf{H}}(t), \tag{15}$$

where $z = (q, x) = \xi_{\mathsf{H}}(t; z_0, 0, u(\cdot))$ and $u = u(t)$. The initial condition for (15) is $\Xi_{\mathsf{H}}(0) = \mathbf{I}$, where $\mathbf{I}$ is the identity matrix of appropriate size.

To treat the discrete jumps that occur in hybrid systems, let t^- and t^+ indicate values just before and just after the instantaneous jump respectively, $z^- = (q^-, x^-) = \xi_{\mathsf{H}}(t^-; z_0, 0, u(\cdot))$ be the state just before the jump, and q^+ be the discrete state just after the jump (so $x^- \in G(q^-, q^+, u)$). For jumps that occur on switching surfaces the difference in post-jump state for two neighboring trajectories depends both on the reset and the difference in time when the jump is enabled (for guard switching surfaces) or forced (for domain switching surfaces). Let $t(z_0)$ be the time of the jump as a function of initial state and τ be its sensitivity. Then

$$\tau = \frac{\partial t(z_0)}{\partial z_0} = -\frac{\nabla \psi(x^-)^T\, \Xi_{\mathsf{H}}(t^-)}{\nabla \psi(x^-)^T f(q^-, x^-, u)} \tag{16}$$

where $\psi(x^-)$ is $\psi_D(q^-, x^-, u_D)$ for domain switching surfaces and $\psi_G(q^-, q^+, x^-, u_D)$ for guard switching surfaces. This equation is only valid if the vector field satisfies a transversality condition such that $\nabla \psi(x^-)^T f(q^-, x^-, u) \neq 0$ [1]. During this period, one trajectory is subject to the old vector field and one to the new vector field, so

$$\Xi_{\mathsf{H}}(t^+) = \mathbf{R}(q^-, q^+, x^-, u)\left(\Xi_{\mathsf{H}}(t^-) + f(q^-, x^-, u)\tau\right) - f(q^+, x^+, u)\tau, \tag{17}$$

where $x^+ = r(q^-, q^+, x^-, u)$ and τ is given in (16). Away from switching surfaces trajectories in a neighborhood can all jump at the same time, so $\tau = 0$.

6.2 Implications for Approximating Reach Sets and Tubes

Given a nominal system trajectory ξ_{H}, the sensitivity evolution equations (15) and (17) can be solved as if they were a dynamic system to provide quantitative estimates of the form (14) for the effects of small perturbations on the initial conditions. Here, though, we will use them to ascertain conditions under which we cannot expect accurate results from approximate reachability algorithms. Most such algorithms use floating point instead of exact arithmetic, and hence make small errors throughout computation. Taking δx as a small numerical error incurred, for example, by a single floating point operation at time t and state z, algebraic manipulation of (14) arrives at a bound for the error at another time s (ignoring the $\mathcal{O}(\delta x^2)$ terms)

$$\|\xi_{\mathsf{H}}(s; z + \delta x, t, u(\cdot)) - \xi_{\mathsf{H}}(s; z, t, u(\cdot))\| \leq \|\Xi_{\mathsf{H}}(s; \xi_{\mathsf{H}}(\cdot))\| \|\delta x\|. \qquad (18)$$

This trajectory-based sensitivity analysis is relevant to direct reachability algorithms because they either track trajectories explicitly (for Lagrangian approaches) or implicitly (for Eulerian); consequently, errors in locating a trajectory translate directly into errors in the approximation of the boundary of the reachable set or tube. It should be noted that Assumption 3 and the vector field transversality condition ensure that the two trajectories in (18) follow the same sequence of discrete states, so we need only consider the difference in their continuous states.

The multiplicative factor $\|\Xi_{\mathsf{H}}(s; \xi_{\mathsf{H}}(\cdot))\|$ in (18) depends on the trajectory $\xi_{\mathsf{H}}(\cdot)$, but there are three ways in which it might grow large.

$$\mathrm{Real}[\lambda(\mathbf{F})] \gg 0 \qquad \text{continuous evolution,} \qquad (19)$$

$$|\lambda(\mathbf{R})| \gg 1 \qquad \text{discrete jumps,} \qquad (20)$$

$$\nabla\psi^T f^- \approx 0 \qquad \text{grazing contact with switching surface,} \qquad (21)$$

where $\lambda(\mathbf{A})$ are the eigenvalues of matrix $\mathbf{A}$ and $f^-(x) = f(q^-, x, u)$. Because $\mathbf{F}$, $\mathbf{R}$, ψ and f depend on state (and potentially input), checking these conditions explicitly will usually be impractical. However, systems satisfying any of the conditions (19)–(21) are inherently unpredictable; consequently, deterministic models of the form studied here are rarely constructed for such systems. With the notable exception of chaotic systems, conditions (19)–(21) are unlikely to occur in practice when computing forward reachability.

Unfortunately, the same cannot be said of backward reachability. It may be defined in terms of the forward dynamics, but computational approximations will begin with the target set and work backwards along trajectories of the time reversed system. Therefore, let us consider the form of conditions (19)–(21) for $\overleftarrow{\mathsf{H}}$ in terms of the elements of a given H. From (12),

$$\overleftarrow{f} = -f \implies \overleftarrow{\mathbf{F}} = -\mathbf{F} \implies \lambda(\overleftarrow{\mathbf{F}}) = -\lambda(\mathbf{F}).$$

From (13), $r(q, \hat{q}, \overleftarrow{r}(\hat{q}, q, x, u), u) = x$. Taking the derivative with respect to x

$$\mathbf{R}\overleftarrow{\mathbf{R}} = \mathbf{I} \implies \overleftarrow{\mathbf{R}} = \mathbf{R}^{-1} \implies \lambda(\overleftarrow{\mathbf{R}}) = \lambda(\mathbf{R})^{-1}.$$

The equivalent of (21) is a little more difficult to deduce, but as explained in [20] the concern is that the flow field after a forward time jump (before the reverse time jump) is nearly parallel to the switching surface which triggered the jump. To summarize, we restate conditions (19)–(21) for $\overleftarrow{\mathsf{H}}$ in terms of the parameters of H

$$\text{Real}[\lambda(\mathbf{F})] \ll 0 \quad \text{backward continuous evolution,} \tag{22}$$

$$|\lambda(\mathbf{R})| \ll 1 \quad \text{backward discrete jumps,} \tag{23}$$

$$\nabla\psi^T f^+ \approx 0 \quad \text{backward grazing contact with switching surface,} \tag{24}$$

where $f^+(x) = f(q^+, r(q^-, q^+, x, u), u)$ includes the action of the reset. As demonstrated in the next section, these conditions can easily occur for systems whose forward simulations are very well behaved. From these conditions, we draw the following conclusion about the challenges of using numerically approximated backwards reachability.

Remark 1. *Systems which display large amounts of contraction in forward time (ie nearby trajectories get closer together) in either their continuous evolution (of the form (22)) or discrete evolution (of the form (23)) are likely to be numerically ill-conditioned for backwards reachability. Poorly conditioned switching events (of the form (24)) are also more likely to be overlooked when working backward, because the relevant switching surfaces and vector fields are in different discrete modes.*

As a final comment, we note that this ill-conditioning of backwards reachability depends on the set being sought, and not the manner in which it is calculated. Consequently there are unlikely to be issues of ill-conditioning when using a backward algorithm and Proposition 6 to compute a forward reach set—this process involves reversing the dynamics twice and ends up back with forward dynamics. On the other hand, using a forward algorithm and Proposition 6 to determine the backward reach tube may run into ill-conditioning because the dynamics are reversed before the algorithm is applied.

6.3 Continuous System Sensitivity Example

To illustrate how sensitivity of the continuous evolution can be a major issue in computing reachability for real systems, we examine the toggle circuit [22] whose schematic and typical trace are shown in Figure 3. The model H_3 is based on a simple, short channel transistor model with velocity saturation [23, pp. 62–63]. All capacitances are to ground and are of fixed value, and interconnect capacitance is ignored. To emulate the effect of connecting toggle elements together, the output node z is given an additional capacitative load equivalent to that seen by input ϕ.

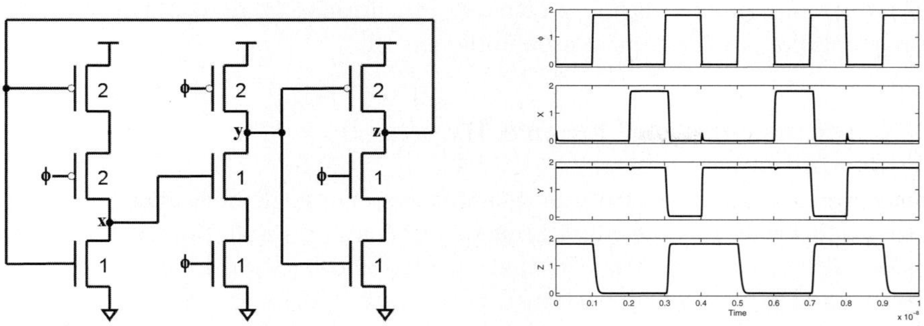

Fig. 3. Left: Yuan's and Svensson's toggle circuit [22]. The numbers next to the transistors are the relative sizing used in the simulations. Right: Simulation of the toggle model H_3 for a typical input signal ϕ.

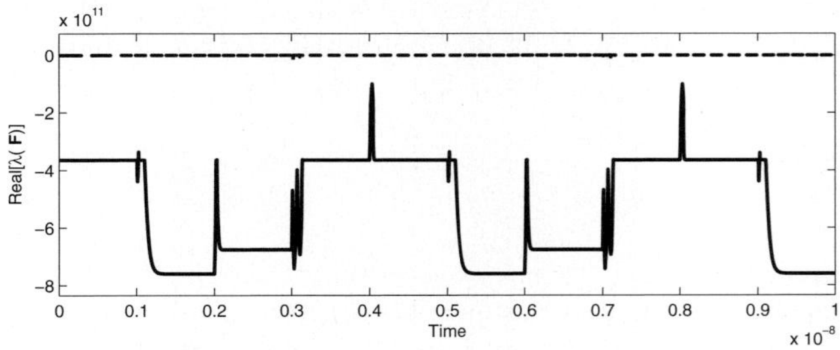

Fig. 4. Upper and lower bounds on the real components of the eigenvalues of the Jacobian $\mathbf{F}$ of the dynamics of H_3 during the simulation in Figure 3

The circuit is correctly operating if the period of the output z is twice the period of the input signal ϕ. Forward reachability has been used to demonstrate that under suitable constraints on the input, the output has twice the period of the input and satisfies the same constraints as the input; consequently, a chain of toggle circuits can be used to form a counter [24].

Unfortunately, a similarly successful analysis using backward reachability would be unlikely to succeed. Figure 4 shows the maximum and minimum real components of the eigenvalues of the Jacobian $\mathbf{F}$ of the dynamics for H_3 over the course of the simulation in Figure 3. Even after scaling by 10^{-8} to account for the very short time intervals typical of VLSI circuits, the minimum real component of the eigenvalues of $\mathbf{F}$ is $-(10^3)$ or less, which indicates a highly contractive dynamic system. Such systems are great for forward reachability calculations, since overapproximation errors will be rapidly contracted to the point of being negligible. But from (22) we see that backward reachability calculations are unlikely to maintain any accuracy for circuits of this type, since they face expansion factors of the same magnitude. In this case, error in backward reachability could grow by a factor of e^{1000} or more on time intervals as short as those in Figure 3.

An example demonstrating sensitivity of the forms (23) and (24), and its effect on reachability calculations can be found in [19].

7 Conclusions and Future Research

Using a very general definition of dynamic system, we demonstrated that backward reach tubes are the most broadly applicable formulation of reachability for demonstrating system safety; that forward and backward algorithms can be interchanged if well-posed backward trajectories can be defined; and that the backward reachability formulation is more likely to suffer from numerical stability problems, particularly for systems displaying significant contraction. We intend to continue studying the sensitivity of reachability algorithms to problem parameters such as inputs, initial and target sets.

Acknowledgments. The author would like to thank Professor Mark Greenstreet, Chao Yan and Suwen Yang for the model, code and help with the toggle example.

References

1. Hiskens, I.A., Pai, M.A.: Trajectory sensitivity analysis of hybrid systems. IEEE Transactions on Circuits and Systems **47**(2) (2000) 204–220
2. Han, Z., Krogh, B.H.: Reachability analysis of large-scale affine systems using low-dimensional polytopes. In Hespanha, J., Tiwari, A., eds.: Hybrid Systems: Computation and Control. Number 3927 in Lecture Notes in Computer Science. Springer Verlag (2006) 287–301
3. Henzinger, T.A., Ho, P.H., Wong-Toi, H.: Algorithmic analysis of nonlinear hybrid systems. IEEE Transactions on Automatic Control **43**(4) (1998) 540–554
4. Greenstreet, M., Mitchell, I.: Reachability analysis using polygonal projections. In Vaandrager, F., van Schuppen, J., eds.: Hybrid Systems: Computation and Control. Number 1569 in Lecture Notes in Computer Science. Springer Verlag (1999) 103–116
5. Bemporad, A., Torrisi, F.D., Morari, M.: Optimization-based verification and stability characterization of piecewise affine and hybrid systems. In Krogh, B., Lynch, N., eds.: Hybrid Systems: Computation and Control. Number 1790 in Lecture Notes in Computer Science. Springer Verlag (2000) 45–59
6. Girard, A., Guernic, C.L., Maler, O.: Efficient computation of reachable sets of linear time-invariant systems with inputs. In Hespanha, J., Tiwari, A., eds.: Hybrid Systems: Computation and Control. Number 3927 in Lecture Notes in Computer Science. Springer Verlag (2006) 257–271
7. Kurzhanski, A.B., Varaiya, P.: Reachability analysis for uncertain systems—the ellipsoidal technique. Dynamics of Continuous, Discrete and Impulsive Systems Series B: Applications and Algorithms **9**(3) (2002) 347–367
8. Saint-Pierre, P.: Hybrid kernels and capture basins for impulse constrained systems. In Tomlin, C.J., Greenstreet, M.R., eds.: Hybrid Systems: Computation and Control. Number 2289 in Lecture Notes in Computer Science. Springer Verlag (2002) 378–392

9. Mitchell, I.M., Bayen, A.M., Tomlin, C.J.: A time-dependent Hamilton-Jacobi formulation of reachable sets for continuous dynamic games. IEEE Transactions on Automatic Control **50**(7) (2005) 947–957
10. Gao, Y., Lygeros, J., Quincampoix, M.: The reachability problem for uncertain hybrid systems revisited: The viability theory perspective. In Hespanha, J., Tiwari, A., eds.: Hybrid Systems: Computation and Control. Number 3927 in Lecture Notes in Computer Science. Springer Verlag (2006) 242–256
11. Tiwari, A., Khanna, G.: Series of abstractions for hybrid automata. In Tomlin, C.J., Greenstreet, M.R., eds.: Hybrid Systems: Computation and Control. Number 2289 in Lecture Notes in Computer Science. Springer Verlag (2002) 465–478
12. Kloetzer, M., Belta, C.: Reachability analysis of multi-affine systems. In Hespanha, J., Tiwari, A., eds.: Hybrid Systems: Computation and Control. Number 3927 in Lecture Notes in Computer Science. Springer Verlag (2006) 348–362
13. Johansson, M., Rantzer, A.: Computation of piecewise quadratic Lyapunov functions for hybrid systems. IEEE Transactions on Automatic Control **43**(4) (1998) 555–559
14. Prajna, S., Jadbabaie, A.: Safety verification of hybrid systems using barrier certificates. In Alur, R., Pappas, G.J., eds.: Hybrid Systems: Computation and Control. Number 2993 in Lecture Notes in Computer Science. Springer Verlag (2004) 477–492
15. Stangier, C., Sidle, T.: Invariant checking combining forward and backward traversal. In Hu, A.J., Martin, A.K., eds.: Formal Methods in Computer-Aided Design. Number 3312 in Lecture Notes in Computer Science. Springer Verlag (2004) 414–429
16. Broucke, M., Arapostathis, A.: Continuous selections of trajectories of hybrid systems. Systems and Control Letters **47** (2002) 149–157
17. Lygeros, J., Johansson, K.H., Simic, S.N., Zhang, J., Sastry, S.: Dynamical properties of hybrid automata. IEEE Transactions on Automatic Control **48**(1) (2003) 2–17
18. Evans, L.C., Souganidis, P.E.: Differential games and representation formulas for solutions of Hamilton-Jacobi-Isaacs equations. Indiana University Mathematics Journal **33**(5) (1984) 773–797
19. Mitchell, I.M.: Comparing forward and backward reachability as tools for safety analysis. Technical Report TR-2006-23, Department of Computer Science, University of British Columbia, Vancouver, BC, Canada (2006)
20. Hiskens, I.A.: Non-uniqueness in reverse time of hybrid system trajectories. In Morari, M., Thiele, L., eds.: Hybrid Systems: Computation and Control. Number 3414 in Lecture Notes in Computer Science. Springer Verlag (2005) 339–353
21. Ascher, U.M., Petzold, L.R.: Computer Methods for Ordinary Differential Equations and Differential-Algebraic Equations. Society for Industrial and Applied Mathematics, Philadelphia (1998)
22. Yuan, J., Svensson, C.: High-speed CMOS circuit technique. IEEE Journal of Solid-State Circuits **24**(1) (1989) 62–70
23. Hodges, D.A., Jackson, H.G., Saleh, R.A.: Analysis and Design of Digital Integrated Circuits in Deep Submicron Technology. Third edn. McGraw Hill, New York (2004)
24. Greenstreet, M.R.: Verifying safety properties of differential equations. In Alur, R., Henzinger, T.A., eds.: Computer Aided Verification. Number 1102 in Lecture Notes in Computer Science. Springer Verlag (1996) 277–287

Approximation of the Joint Spectral Radius of a Set of Matrices Using Sum of Squares

Pablo A. Parrilo[1] and Ali Jadbabaie[2]

[1] Laboratory for Information and Decision Systems
Massachusetts Institute of Technology, Cambridge, MA 02139
parrilo@mit.edu
[2] Dept. of Electrical and Systems Engineering and GRASP laboratory
University of Pennsylvania, Philadelphia, PA 19104
jadbabai@seas.upenn.edu

Abstract. We provide an asymptotically tight, computationally efficient approximation of the joint spectral radius of a set of matrices using sum of squares (SOS) programming. The approach is based on a search for a SOS polynomial that proves simultaneous contractibility of a finite set of matrices. We provide a bound on the quality of the approximation that unifies several earlier results and is independent of the number of matrices. Additionally, we present a comparison between our approximation scheme and a recent technique due to Blondel and Nesterov, based on lifting of matrices. Theoretical results and numerical investigations show that our approach yields tighter approximations.

1 Introduction

Stability of discrete linear inclusions has been a topic of major research over the past two decades. Such systems can be represented as a switched linear system of the form $x(k+1) = A_{\sigma(k)}x(k)$, where σ is a mapping from the integers to a given set of indices. The above model has been studied extensively across multiple disciplines, ranging from control theory, theory of non-negative matrices and Markov chains, wavelets, dynamical systems, etc. The fundamental question of interest is to determine whether $x(k)$ converges to a limit, or equivalently, whether the infinite matrix products chosen from the set of matrices converge [1,2,3]. The research on convergence of infinite products of matrices spans across four decades. A majority of results in this area has been provided in the special case of non-negative and/or stochastic matrices. A non-exhaustive list of related research providing several necessary and sufficient conditions for convergence of infinite products includes [4,3,5]. Despite the wealth of research in this area, finding algorithms that can determine the convergence remains elusive. Much of the difficulty of this problem stems from the hardness in computation or efficient approximation of the joint spectral radius of a finite set of matrices [6]. This is defined as

$$\rho(A_1, \ldots, A_m) := \limsup_{k \to +\infty} \max_{\sigma \in \{1,\ldots,m\}^k} ||A_{\sigma_k} \cdots A_{\sigma_2} A_{\sigma_1}||^{1/k}, \quad (1)$$

A. Bemporad, A. Bicchi, and G. Buttazzo (Eds.): HSCC 2007, LNCS 4416, pp. 444–458, 2007.

and represents the maximum growth (or decay) rate that can be obtained by taking arbitrary products of the matrices A_i, and its value is independent of the norm chosen. Daubechies and Lagarias [3] conjectured that the joint spectral radius is equal to a related quantity, the *generalized spectral radius*, which is defined in a similar way except for the fact that the norm of the product is replaced by the spectral radius. Later Berger and Wang [1] proved this conjecture to be true for finite set of matrices. It is well known that computing ρ is hard from a computational viewpoint, and even approximating it is difficult [7,8,9]. For rational matrices, the joint spectral radius is not a semialgebraic function of the data, thus ruling out a very large class of methods for its exact computation. When the matrices are non-negative and stochastic, the problem is decidable but PSPACE complete [10], a complexity class widely believed to be worse than NP.

It turns out that a necessary and sufficient condition for stability of a linear difference inclusion is for the corresponding matrices to have a subunit join spectral radius [4], i.e., $\rho(A_1, \ldots, A_m) < 1$. This, however, is a condition that is impossible to verify in general. A subunit joint spectral radius is on the other hand equivalent to existence of a common norm with respect of which all matrices in the set are contractive [11,12,13]. Unfortunately, the proof of this result is not constructive, as knowledge of the joint spectral radius is needed to construct such an *extremal norm* [13]. In fact a similar result, due to Dayawansa and Martin [14], holds for nonlinear systems that undergo switching. A common approach in trying to approximate the joint spectral radius or to prove that it is indeed subunit, has been to try to prove simultaneous contractibility (i.e., existence of a common norm with respect to which matrices are contractive), by searching for a common ellipsoidal norm, or equivalently, searching for a common quadratic Lyapunov function. The benefit of this approach is due to the fact that the search for a common ellipsoidal norm can be posed as a semidefinite program and solved efficiently using interior point techniques. However, it is not too difficult to generate examples where the discrete inclusion is *absolutely asymptotically stable*, i.e. asymptotically stable for all switching sequences, but a common quadratic Lyapunov function, (or equivalently a common ellipsoidal norm) does not exist.

Ando and Shih describe in [15] a constructive procedure for generating a set of m matrices for which the joint spectral radius is $\frac{1}{\sqrt{m}}$, but no quadratic Lyapunov function exists. They prove that the interval $[0, \frac{1}{\sqrt{m}})$ is effectively the "optimal" range for the joint spectral radius necessary to guarantee simultaneous contractibility under an ellipsoidal norm for a finite collection of m matrices. The range is denoted as optimal since it is the largest subset of $[0, 1)$ for which if the joint spectral radius is in this subset the collection of matrices is simultaneously contractible. Furthermore, they show that the optimal joint spectral radius range for a *bounded* set of $n \times n$ matrices is the interval $[0, \frac{1}{\sqrt{n}})$. The proof of this fact is based on John's ellipsoid theorem [15]. Roughly speaking, John's ellipsoid theorem implies that every convex body in the n dimensional Euclidean space that is symmetric around the origin can be approximated by an ellipsoid (from the inside and outside) up to a factor of $\frac{1}{\sqrt{n}}$. A major consequence of this result is that finding a common Lyapunov function becomes increasingly hard as the dimension goes up.

Recently, Blondel, Nesterov and Theys [16] showed a similar result (also based on John's ellipsoid theorem), that the best ellipsoidal norm approximation of the joint spectral radius provides a lower bound and an upper bound on the actual value. Given a set of matrices $\mathcal{M}$ with joint spectral radius ρ, and best ellipsoidal norm approximation $\hat{\rho}$, it is shown there that

$$\frac{1}{\sqrt{m}}\hat{\rho}(\mathcal{M}) \leq \rho(\mathcal{M}) \leq \hat{\rho}(\mathcal{M}) \tag{2}$$

Furthermore, in [17], Blondel and Nesterov proposed a scheme to approximate the joint spectral radius, by "lifting" the matrices using Kronecker products to provide better approximations. A common feature of these approaches is the appearance of convexity-based methods to provide certificates of the desired system properties.

In this paper, we develop a sum of squares (SOS) based scheme for the approximation of the joint spectral radius, and prove several results on the resulting quality of approximation. For this, we use two different techniques, one inspired by recent results of Barvinok [18] on approximation of norms by polynomials, and the other one based on a convergent iteration similar to that used for Lyapunov inequalities. Our results provide a simple and unified derivation of most of the available bounds, including some new ones. As a consequence, we can use SOS polynomials to approximate the extremal norm that is equal to the joint spectral radius. We also show that this approximation is always tighter than the one provided by Blondel and Nesterov.

A description of the paper follows. In Section 2 we present a class of bounds on the joint spectral radius based on simultaneous contractivity with respect to a norm, followed by a sum of squares-based relaxation, and the corresponding suboptimality properties. In Section 3 we present some background material in multilinear algebra, necessary for our developments, and a derivation of a bound of the quality of the SOS relaxation. An alternative development is presented in Section 4, where a different bound on the performance of the SOS relaxation is given in terms of a very natural Lyapunov iteration, similar to the classical case. In Section 5 we make a comparison with earlier techniques and analyze a numerical example. Finally, in Section 6 we present our conclusions.

2 SOS Norms

A natural way of bounding the joint spectral radius is to find a common norm under for which we can guarantee certain contractiveness properties for all the matrices. In this section, we first revisit this characterization, and introduce our method of using SOS relaxations to approximate this common norm.

2.1 Norms and the Joint Spectral Radius

As we mentioned, there exists an intimate relationship between the spectral radius and the existence of a vector norm under which all the matrices are simultaneously contractive. This is summarized in the following theorem, a special case of Proposition 1 in [6] by Rota and Strang.

Theorem 1 ([6]). *Consider a finite set of matrices $\mathcal{A} = \{A_1, \ldots, A_m\}$. For any $\epsilon > 0$, there exists a norm $\|\cdot\|$ in $\mathbb{R}^n$ (denoted as JSR norm hereafter) such that*

$$||A_i x|| \leq (\rho(\mathcal{A}) + \epsilon)||x||, \quad \forall x \in \mathbb{R}^n, \quad i = 1, \ldots, m.$$

The theorem appears in this form, for instance, in Proposition 4 of [16]. The main idea in our approach is to replace the JSR norm that approximates the joint spectral radius with a homogeneous SOS polynomial $p(x)$ of degree $2d$. As we will see in the next sections, we can produce arbitrarily tight SOS approximations, while still being able to prove a bound on the resulting estimate.

2.2 Joint Spectral Radius and Polynomials

As the results presented above indicate, the joint spectral radius can be characterized by finding a common norm under which all the maps are simultaneously contractive. As opposed to the unit ball of a norm, the level sets of a homogeneous polynomial are not necessarily convex (see for instance Figure 1). Nevertheless, as the following theorem suggests, we can still obtain upper bounds on the joint spectral radius by replacing norms with homogeneous polynomials.

Theorem 2. *Let $p(x)$ be a strictly positive homogeneous polynomial of degree $2d$ that satisfies*

$$p(A_i x) \leq \gamma^{2d} p(x), \qquad \forall x \in \mathbb{R}^n \quad i = 1, \ldots, m.$$

Then, $\rho(A_1, \ldots, A_m) \leq \gamma$.

Proof. If $p(x)$ is strictly positive, then by compactness of the unit ball in $\mathbb{R}^n$ and continuity of $p(x)$, there exist constants $0 < \alpha \leq \beta$, such that

$$\alpha \, ||x||^{2d} \leq p(x) \leq \beta \, ||x||^{2d} \qquad \forall x \in \mathbb{R}^n.$$

Then,

$$\begin{aligned} ||A_{\sigma_k} \ldots A_{\sigma_1}|| &\leq \max_x \frac{||A_{\sigma_k} \ldots A_{\sigma_1} x||}{||x||} \\ &\leq \left(\frac{\beta}{\alpha}\right)^{\frac{1}{2d}} \max_x \frac{p(A_{\sigma_k} \ldots A_{\sigma_1} x)^{\frac{1}{2d}}}{p(x)^{\frac{1}{2d}}} \\ &\leq \left(\frac{\beta}{\alpha}\right)^{\frac{1}{2d}} \gamma^k. \end{aligned}$$

From the definition of the joint spectral radius in equation (1), by taking kth roots and the limit $k \to \infty$ we immediately have the upper bound $\rho(A_1, \ldots, A_m) \leq \gamma$.

2.3 Sums of Squares Programming

The condition in Theorem 2 involves positive polynomials, which are computationally hard to characterize. A useful scheme, introduced in [19,20] and relatively well-known by now, relaxes the nonnegativity constraints to a much more tractable *sum of squares* condition, where $p(x)$ is required to have a decomposition as $p(x) = \sum_i p_i(x)^2$. The SOS condition can be equivalently expressed in terms of a semidefinite programming constraint, hence its tractability. In what follows, we briefly describe the basic SOS construction.

Consider a given multivariate polynomial for which we want to decide whether a sum of squares decomposition exists. This question is equivalent to a semidefinite programming (SDP) problem, because of the following result:

Theorem 3. *A homogeneous multivariate polynomial $p(x)$ of degree $2d$ is a sum of squares if and only if*

$$p(x) = (x^{[d]})^T Q x^{[d]}, \tag{3}$$

where $x^{[d]}$ is a vector whose entries are (possibly scaled) monomials of degree d in the x_i variables, and Q is a symmetric positive semidefinite matrix.

Since in general the entries of $x^{[d]}$ are not algebraically independent, the matrix Q in the representation (3) *is not unique.* In fact, there is an affine subspace of matrices Q that satisfy the equality, as can be easily seen by expanding the right-hand side and equating term by term. To obtain an SOS representation, we need to find a positive semidefinite matrix in this affine subspace. Therefore, the problem of checking if a polynomial can be decomposed as a sum of squares is *equivalent* to verifying whether a certain affine matrix subspace intersects the cone of positive definite matrices, and hence an SDP feasibility problem.

Example 1. Consider the quartic homogeneous polynomial in two variables described below, and define the vector of monomials as $[x^2, y^2, xy]^T$.

$$\begin{aligned} p(x,y) &= 2x^4 + 2x^3y - x^2y^2 + 5y^4 \\ &= \begin{bmatrix} x^2 \\ y^2 \\ xy \end{bmatrix}^T \begin{bmatrix} q_{11} & q_{12} & q_{13} \\ q_{12} & q_{22} & q_{23} \\ q_{13} & q_{23} & q_{33} \end{bmatrix} \begin{bmatrix} x^2 \\ y^2 \\ xy \end{bmatrix} \\ &= q_{11}x^4 + q_{22}y^4 + (q_{33} + 2q_{12})x^2y^2 + 2q_{13}x^3y + 2q_{23}xy^3 \end{aligned}$$

For the left- and right-hand sides to be identical, the following linear equations should hold:

$$q_{11} = 2, \quad q_{22} = 5, \quad q_{33} + 2q_{12} = -1, \quad 2q_{13} = 2, \quad 2q_{23} = 0. \tag{4}$$

A positive semidefinite Q that satisfies the linear equalities can then be found using SDP. A particular solution is given by:

$$Q = \begin{bmatrix} 2 & -3 & 1 \\ -3 & 5 & 0 \\ 1 & 0 & 5 \end{bmatrix} = L^T L, \quad L = \frac{1}{\sqrt{2}} \begin{bmatrix} 2 & -3 & 1 \\ 0 & 1 & 3 \end{bmatrix},$$

and therefore we have the sum of squares decomposition:

$$p(x,y) = \frac{1}{2}(2x^2 - 3y^2 + xy)^2 + \frac{1}{2}(y^2 + 3xy)^2. \qquad \square$$

2.4 Norms and SOS Polynomials

The procedure described in the previous subsection can be easily modified to the case where the polynomial $p(x)$ is not fixed, but instead we search for an SOS polynomial in a given affine family (for instance, all polynomials of a given degree).

This line of thought immediately suggests the following SOS relaxation of the conditions in Theorem 2:

$$\rho_{SOS,2d} := \inf_{p(x)\in\mathbb{R}_{2d}[x],\gamma} \gamma \qquad \text{s.t.} \begin{cases} p(x) \text{ is SOS} \\ \gamma^{2d}p(x) - p(A_i x) \text{ is SOS} \end{cases} \tag{5}$$

where $\mathbb{R}_{2d}[x]$ is the set of homogeneous polynomials of degree $2d$.

For any fixed γ, the constraints in this problem are all of SOS type, and thus equivalent to semidefinite programming. Therefore, the computation of $\rho_{SOS,2d}$ is a quasiconvex problem, and can be easily solved with a standard SDP solver via bisection methods. By Theorem 2, the solution of this relaxation gives the bound

$$\rho(A_1, \ldots, A_m) \leq \rho_{SOS,2d}, \tag{6}$$

where $2d$ is the degree of the approximating polynomial.

2.5 Quality of Approximation

What can be said about the quality of the bounds produced by the SOS relaxation? We present next some results to answer this question; a more complete characterization is developed in Section 3.1. A very helpful result in this direction is the following theorem of Barvinok, that quantifies how tightly SOS polynomials can approximate norms:

Theorem 4 ([18], p. 221). *Let $||\cdot||$ be a norm in $\mathbb{R}^n$. For any integer $d \geq 1$ there exists a homogeneous polynomial $p(x)$ in n variables of degree $2d$ such that*

1. *The polynomial $p(x)$ is a sum of squares.*
2. *For all $x \in \mathbb{R}^n$,*
$$p(x)^{\frac{1}{2d}} \leq ||x|| \leq k(n,d)\, p(x)^{\frac{1}{2d}},$$
where $k(n,d) := \binom{n+d-1}{d}^{\frac{1}{2d}}$.

For fixed state dimension n, by increasing the degree d of the approximating polynomials, the factor in the upper bound can be made arbitrarily close to one. In fact, for large d, we have the approximation

$$k(n,d) \approx 1 + \frac{n-1}{2}\frac{\log d}{d}.$$

With these preliminaries, we can now present the main result of this paper that proves a bound on the quality of SOS approximations to the joint spectral radius:

To apply these results to our problem, consider the following. If $\rho(A_1, \ldots, A_m) < \gamma$, by Theorem 1 (and sharper results in [11,12,13]) there exists a norm $\|\cdot\|$ such that

$$||A_i x|| \leq \gamma ||x||, \quad \forall x \in \mathbb{R}^n, i = 1, \ldots, m.$$

By Theorem 4, we can therefore approximate this norm with a homogeneous SOS polynomial $p(x)$ of degree $2d$ that will then satisfy

$$p(A_i x)^{\frac{1}{2d}} \leq ||A_i x|| \leq \gamma ||x|| \leq \gamma\, k(n,d)\, p(x)^{\frac{1}{2d}},$$

and thus we know that there exists a feasible solution of

$$\begin{cases} p(x) \text{ is SOS} & \\ \alpha^{2d} p(x) - p(A_i x) \geq 0 & i = 1, \ldots, m, \end{cases}$$

for $\alpha = k(n,d)\rho(A_1, \ldots, A_m)$.

Despite these appealing results, notice that in general we cannot directly conclude from this that the proposed SOS relaxation will always obtain a solution that is within $k(n,d)^{-1}$ from the true spectral radius. The reason is that even though we can prove the existence of a $p(x)$ that is SOS and for which $\alpha^{2d} p(x) - p(A_i x)$ are nonnegative for all i, it is unclear whether the last m expressions are actually SOS. We will show later in the paper that this is indeed the case. Before doing this, we concentrate first on two important cases of interest, where the approach described guarantees a good quality of approximation.

Planar systems. The first case corresponds to two-dimensional (planar) systems, i.e., when $n = 2$. In this case, it always holds that nonnegative homogeneous bivariate polynomials are SOS (e.g., [21]). Thus, we have the following result:

Theorem 5. *Let $\{A_1, \ldots, A_m\} \subset \mathbb{R}^{2\times 2}$. Then, the SOS relaxation (5) always produces a solution satisfying:*

$$\tfrac{1}{2}\rho_{SOS,2d} \leq (d+1)^{-\frac{1}{2d}}\, \rho_{SOS,2d} \leq \rho(A_1, \ldots, A_m) \leq \rho_{SOS,2d}.$$

This result is independent *of the number m of matrices.*

Quadratic Lyapunov functions. For the quadratic case (i.e., $2d = 2$), it is also true that nonnegative quadratic forms are sums of squares. Since

$$\binom{n+d-1}{d}^{\frac{1}{2d}} = \binom{n}{1}^{\frac{1}{2}} = \sqrt{n},$$

the inequality

$$\frac{1}{\sqrt{n}}\, \rho_{SOS,2} \leq \rho(A_1, \ldots, A_m) \leq \rho_{SOS,2} \tag{7}$$

follows. This bound exactly coincides with the results of Ando and Shih [15] or Blondel, Nesterov and Theys [17]. This is perhaps not surprising, since in this case both Ando and Shih's proof [15] and Barvinok's theorem rely on the use of John's ellipsoid to approximate the same underlying convex set.

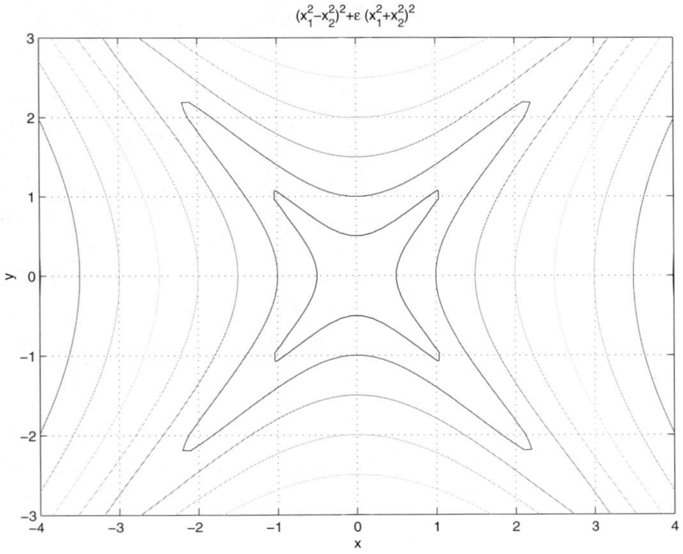

Fig. 1. Level sets of the quartic homogeneous polynomial $V(x_1, x_2)$. These define a Lyapunov function, under which both A_1 and A_2 are $(1+\epsilon)$-contractive. The value of ϵ is here equal to 0.01.

Level sets and convexity. Unlike the norms that appear in Theorem 1, an appealing feature of the SOS-based method is that we are not constrained to use polynomials with convex level sets. This enables in some cases much better bounds than what is promised by the theorems above, as illustrated in the following example.

Example 2. This is based on a construction by Ando and Shih [15]. Consider the problem of proving a bound on the joint spectral radius of the following matrices:

$$A_1 = \begin{bmatrix} 1 & 0 \\ 1 & 0 \end{bmatrix}, \qquad A_2 = \begin{bmatrix} 0 & 1 \\ 0 & -1 \end{bmatrix}.$$

For these matrices, it can be easily shown that $\rho(A_1, A_2) = 1$. Using a common quadratic Lyapunov function (i.e., the case $d = 2$), the upper bound on the joint spectral radius is equal to $\sqrt{2}$. However, even a simple quartic SOS Lyapunov function is enough to prove an upper bound of $1 + \epsilon$ for every $\epsilon > 0$, since the SOS polynomial

$$V(x) = (x_1^2 - x_2^2)^2 + \epsilon(x_1^2 + x_2^2)^2$$

satisfies

$$(1+\epsilon)V(x) - V(A_1 x) = (x_2^2 - x_1^2 + \epsilon(x_1^2 + x_2^2))^2$$
$$(1+\epsilon)V(x) - V(A_2 x) = (x_1^2 - x_2^2 + \epsilon(x_1^2 + x_2^2))^2.$$

The corresponding level sets of $V(x)$ are plotted in Figure 1, and are clearly non-convex.

3 Symmetric Algebra and Induced Matrices

In the upcoming sections, we present some further bounds on the quality of the SOS relaxation (5), either by a more refined analysis of the SOS polynomials in Barvinok's theorem or by explicitly producing an SOS Lyapunov function of guaranteed suboptimality properties. These constructions are quite natural, and parallel some lifting ideas as well as the classical iteration used in the solution of discrete-time Lyapunov inequalities.

Before proceeding further, we need to briefly revisit first some classical notions from multilinear algebra, namely the *symmetric algebra* of a vector space.

Consider a vector $x \in \mathbb{R}^n$, and an integer $d \geq 1$. We define its d-lift $x^{[d]}$ as a vector in $\mathbb{R}^N$, where $N := \binom{n+d-1}{d}$, with components $\{\sqrt{\binom{d}{\alpha}}x^\alpha\}_\alpha$, where $\alpha = (\alpha_1, \ldots, \alpha_n)$, $|\alpha| := \sum_i \alpha_i = d$, and $\binom{d}{\alpha}$ denotes the multinomial coefficient $\binom{d}{\alpha_1, \alpha_2, \ldots, \alpha_n} = \frac{d!}{\alpha_1! \alpha_2! \ldots \alpha_n!}$. In other words, the components of the lifted vector are the monomials of degree d, scaled by the square root of the corresponding multinomial coefficients.

Example 3. Let $n = 2$, and $x = [u, v]^T$. Then, we have

$$\begin{bmatrix} u \\ v \end{bmatrix}^{[1]} = \begin{bmatrix} u \\ v \end{bmatrix}, \qquad \begin{bmatrix} u \\ v \end{bmatrix}^{[2]} = \begin{bmatrix} u^2 \\ \sqrt{2}uv \\ v^2 \end{bmatrix}, \qquad \begin{bmatrix} u \\ v \end{bmatrix}^{[3]} = \begin{bmatrix} u^3 \\ \sqrt{3}u^2v \\ \sqrt{3}uv^2 \\ v^3 \end{bmatrix}.$$

The main motivation for this specific scaling of the components, is to ensure that the lifting preserves (some of) the norm properties. In particular, if $||\cdot||$ denotes the standard Euclidean norm, it can be easily verified that $||x^{[d]}|| = ||x||^d$. Thus, the lifting operation provides a norm-preserving (up to power) embedding of $\mathbb{R}^n$ into $\mathbb{R}^N$ (in the projective case, this is the so-called *Veronese* embedding).

This concept can be directly extended from vectors to linear transformations. Consider a linear map in $\mathbb{R}^n$, and the associated $n \times n$ matrix A. Then, the lifting described above naturally induces an associated map in $\mathbb{R}^N$, that makes the corresponding diagram commute. The matrix representing this linear transformation is the *d-th induced matrix* of A, denoted by $A^{[d]}$, which is the unique $N \times N$ matrix that satisfies

$$A^{[d]}x^{[d]} = (Ax)^{[d]}$$

In systems and control, these classical constructions of multilinear algebra have been used under different names in several works, among them [22,23] and (implicitly) [17]. Although not mentioned in the Control literature, there exists a simple explicit formula for the entries of these induced matrices; see [24,25]. The d-th induced matrix $A^{[d]}$ has dimensions $N \times N$. Its entries are given by

$$A^{[d]}_{\alpha\beta} = \frac{\sqrt{\alpha!\beta!}}{d!} \operatorname{per} A_{\tilde{\alpha}\tilde{\beta}}, \tag{8}$$

where the indices α, β are multisets of $\{1, \ldots, n\}$ of cardinality d, and per indicates the *permanent*[1] of a square matrix. It can be shown that these operations define an algebra homomorphism, i.e., one that respects the structure of matrix multiplication. In particular, for any matrices A, B of compatible dimensions, the following identities hold:

$$(AB)^{[d]} = A^{[d]} B^{[d]}, \qquad (A^{-1})^{[d]} = (A^{[d]})^{-1}.$$

Furthermore, there is a simple and appealing relationship between the eigenvalues of $A^{[d]}$ and those of A. Concretely, if $\lambda_1, \ldots, \lambda_n$ are the eigenvalues of A, then the eigenvalues of $A^{[d]}$ are given by $\prod_{j \in S} \lambda_j$ where $S \subseteq \{1, \ldots, n\}, |S| = d$; there are exactly $\binom{n+d-1}{d}$ such multisets. A similar relationship holds for the corresponding eigenvectors. Essentially, as explained below in more detail, the induced matrices are the symmetry-reduced version of the d-fold Kronecker product.

As mentioned, the symmetric algebra and associated induced matrices are classical objects of multilinear algebra. Induced matrices, as defined above, as well as the more usual *compound matrices*, correspond to two specific isotypic components of the decomposition of the d-fold tensor product under the action of the symmetric group S^d (i.e., the *symmetric* and *skew-symmetric* algebras). Compound matrices are associated with the alternating character (hence their relationship with determinants), while induced matrices correspond instead to the trivial character, thus the connection with permanents. Similar constructions can be given for any other character of the symmetric group, by replacing the permanent in (8) with the suitable immanants; see [24] for additional details.

3.1 Bounds on the Quality of $\rho_{SOS,2d}$

In this section we directly prove a bound on the approximation properties of the SOS approximation. As we will see, the techniques based on the lifting described will exactly yield the factor $k(n,d)^{-1}$ suggested by the application of Barvinok's theorem.

We first prove a preliminary result of the behavior of the joint spectral radius under lifting. The scaling properties described in the previous section can be applied to obtain the following:

Lemma 1. *Given matrices* $\{A_1, \ldots, A_m\} \subset \mathbb{R}^{n \times n}$ *and an integer* $d \geq 1$, *the following identity holds:*

$$\rho(A_1^{[d]}, \ldots, A_m^{[d]}) = \rho(A_1, \ldots, A_m)^d.$$

The proof follows directly from the definition (1) and the two properties $(AB)^{[d]} = A^{[d]} B^{[d]}$, $||x^{[d]}|| = ||x||^d$, and it is thus omitted.

Combining all these inequalities, we obtain the main result of this paper:

[1] The permanent of a matrix $A \in \mathbb{R}^{n \times n}$ is defined as $\mathrm{per}(A) := \sum_{\sigma \in \Pi_n} \prod_{i=1}^n a_{i,\sigma(i)}$, where Π_n is the set of all permutations in n elements.

Theorem 6. *The SOS relaxation (5) satisfies:*

$$\binom{n+d-1}{d}^{-\frac{1}{2d}} \rho_{SOS,2d} \leq \rho(A_1, \ldots, A_m) \leq \rho_{SOS,2d}. \tag{9}$$

Proof. From Lemma 1 and inequality (7), we have:

$$\binom{n+d-1}{d}^{-\frac{1}{2}} \rho_{SOS,2}(A_1^{[d]}, \ldots, A_m^{[d]}) \leq \rho(A_1, \ldots, A_m)^d \leq \rho_{SOS,2}(A_1^{[d]}, \ldots, A_m^{[d]}).$$

Combining this with the inequality (proven later in Theorem 9),

$$\rho_{SOS,2d}(A_1, \ldots, A_m)^d \leq \rho_{SOS,2}(A_1^{[d]}, \ldots, A_m^{[d]}),$$

the result follows.

4 Sum of Squares Lyapunov Iteration

We describe next an alternative approach to obtain bounds on the quality of the SOS approximation. As opposed to the results in the previous section, the bounds now explicitly depend on the number of matrices, but will usually be better in the case of small m.

Consider the iteration defined by

$$V_0(x) = 0, \qquad V_{k+1}(x) = Q(x) + \frac{1}{\beta} \sum_{i=1}^{m} V_k(A_i x), \tag{10}$$

where $Q(x)$ is a fixed n-variate homogeneous polynomial of degree $2d$ and $\beta > 0$. The iteration defines an affine map in the space of homogeneous polynomials of degree $2d$. As usual, the iteration will converge under certain assumptions on the spectral radius of this linear operator.

Theorem 7. *The iteration defined in (10) converges for arbitrary* $Q(x)$ *if* $\rho(A_1^{[2d]} + \cdots + A_m^{[2d]}) < \beta$.

Proof. Since the vector space of homogenous polynomials $\mathbb{R}_{2d}[x_1, \ldots, x_n]$ is isomorphic to the space of linear functionals on $(\mathbb{R}^n)^{[2d]}$, we can write $V_k(x) = \langle v_k, x^{[2d]} \rangle$, where $v_k \in \mathbb{R}^{\binom{n+2d-1}{2d}}$ is the vector of (scaled) coefficients of $V_k(x)$. Then, the iteration (10) can be simply expressed as:

$$v_{k+1} = q + \frac{1}{\beta} \sum_{i=1}^{m} A_i^{[2d]} v_k,$$

and it is well known that an affine iteration converges if the spectral radius of the linear term is less than one.

Theorem 8. *The following inequality holds:*

$$\rho_{SOS,2d} \leq \rho(A_1^{[2d]} + \cdots + A_m^{[2d]})^{\frac{1}{2d}}.$$

Proof. Choose a $Q(x)$ that is SOS, e.g., $Q(x) := (\sum_{i=1}^{n} x_i^2)^d$, and let $\beta = \rho(A_1^{[2d]} + \cdots + A_m^{[2d]}) + \epsilon$. The iteration (10) guarantees that V_{k+1} is SOS if V_k is. By induction, all the iterates V_k are SOS. By the choice of β and Theorem 7, the V_k converge to some homogeneous polynomial $V_\infty(x)$. By the closedness of the cone of SOS polynomials, the limit V_∞ is also SOS. Furthermore, we have

$$\beta V_\infty(x) - V_\infty(A_i x) = \beta Q(x) + \sum_{j \neq i} V_\infty(A_j x)$$

and therefore the expression on the right-hand side is SOS. This implies that $p(x) := V_\infty(x)$ is a feasible solution of the SOS relaxation (5). Taking $\epsilon \to 0$, the result follows.

Corollary 1. *For the SOS relaxation in (5), we always have*

$$\frac{1}{\sqrt[2d]{m}} \rho_{SOS,2d} \leq \rho(A_1, \ldots, A_m) \leq \rho_{SOS,2d}.$$

Proof. This follows directly from Theorem 8 and the inequalities (5.1) and (5.2) in [17].

The iteration (10) is the natural generalization of the Lyapunov recursion for the single matrix case, and of the construction by Ando and Shih in [15] for the quadratic case. Notice also that this corresponds exactly to the condition introduced in [17]. As a consequence, the SOS-based approach will *always* produce an upper bound at least as good as the one given by the Blondel-Nesterov procedure.

5 Comparison with Earlier Techniques

In this section we compare the $\rho_{SOS,2d}$ approach with some earlier bounds from the literature. We show that our bound is never weaker than those obtained by all the other procedures.

5.1 The Blondel-Nesterov Technique

In [17], Blondel and Nesterov develop a method based on the calculation of the spectral radius of "lifted" matrices. They in fact present two different lifting procedures (the "Kronecker" and "semidefinite" lifting), and in Section 5 of their paper, they describe a family of bounds obtained by arbitrary combinations of these two liftings.

In contrast, the procedure presented in our paper relies on a single canonically defined lifting, that always dominates the Blondel-Nesterov contruction. Furthermore, it can be shown that our construction exactly corresponds to a fully symmetry-reduced version of their procedure, thus yielding the same (or better) bounds, but at a much smaller computational cost since the corresponding matrices are exponentially smaller.

5.2 Common Quadratic Lyapunov Functions

This method corresponds to finding a common quadratic Lyapunov function, either directly for the matrices A_i, or for the lifted matrices $A_i^{[d]}$. Specifically, let

$$\rho_{CQLF,2d} := \inf\{\gamma \mid \gamma^{2d} P - (A_i^{[d]})^T P A_i^{[d]} \succeq 0, \quad P \succ 0\}$$

This is essentially equivalent to what is discussed in Corollary 3 of [17], except that the matrices involved in our approach are exponentially smaller (of size $\binom{n+d-1}{d}$ rather than n^d), as all the symmetries have been taken out[2]. Notice also that, as a consequence of their definitions, we have

$$\rho_{CQLF,2d}(A_1, \dots, A_m)^d = \rho_{SOS,2}(A_1^{[d]}, \dots, A_m^{[d]}).$$

We can then collect most of these results in a single theorem:

Theorem 9. *The following inequalities between all the bounds hold:*

$$\rho(A_1, \dots, A_m) \leq \rho_{SOS,2d} \leq \rho_{CQLF,2d} \leq \rho\left(\sum_{i=1}^{m} A_i^{[2d]}\right)^{\frac{1}{2d}}.$$

Proof. The left-most inequality is (6). The right-most inequality follows from a similar (but stronger) argument to the one given in Theorem 8 above, since the spectral radius condition $\rho(A_1^{[2d]} + \cdots + A_m^{[2d]}) < \beta$ actually implies the convergence of the matrix iteration in $\mathcal{S}^N$ given by

$$P_{k+1} = Q + \frac{1}{\beta} \sum_{i=1}^{m} (A_i^{[d]})^T P_k A_i^{[d]}, \qquad P_0 = I.$$

For the middle inequality, let $p(x) := (x^{[d]})^T P x^{[d]}$. Since $P \succ 0$, it follows that $p(x)$ is SOS. From $\gamma^{2d} P - (A_i^{[d]})^T P A_i^{[d]} \succeq 0$, left- and right-multiplying by $x^{[d]}$, we have that $\gamma^{2d} p(x) - p(A_i x)$ is also SOS, and thus $p(x)$ is a feasible solution of (5), from where the result directly follows.

Remark 1. We always have $\rho_{SOS,2} = \rho_{CQLF,2}$, since both correspond to the case of a common quadratic Lyapunov function for the matrices A_i.

5.3 Example

We present next a numerical example that compares the described techniques. In particular, we show that the bounds in Theorem 9 can all be strict.

---

[2] There seems to be a typo in equation (7.4) of [17], as all the terms A_i^k should likely read $A_i^{\otimes k}$.

Table 1. Comparison of the different approximations for Example 4

d	1	2	3
$\dim A_i^{[d]}$	4	10	20
$\dim A_i^{[2d]}$	10	35	84
$\rho_{SOS,2d}$	9.761	8.92	-
$\rho_{CQLF,2d}$	9.761	9.02	-
$\rho(\sum_i A_i^{[2d]})^{\frac{1}{2d}}$	12.519	9.887	9.3133

Example 4. Consider the three 4×4 matrices

$$A_1 = \begin{bmatrix} 0 & 1 & 7 & 4 \\ 1 & 6 & -2 & -3 \\ -1 & -1 & -2 & -6 \\ 3 & 0 & 9 & 1 \end{bmatrix}, \quad A_2 = \begin{bmatrix} -3 & 3 & 0 & -2 \\ -2 & 1 & 4 & 9 \\ 4 & -3 & 1 & 1 \\ 1 & -5 & -1 & -2 \end{bmatrix}, \quad A_3 = \begin{bmatrix} 1 & 4 & 5 & 10 \\ 0 & 5 & 1 & -4 \\ 0 & -1 & 4 & 6 \\ -1 & 5 & 0 & 1 \end{bmatrix}.$$

The value of the different approximations are presented in Table 1. A lower bound is $\rho(A_1A_3)^{\frac{1}{2}} \approx 8.9149$, which is extremely close (and perhaps exactly equal) to the upper bound $\rho_{SOS,4}$.

6 Conclusions

We provided an asymptotically tight, computationally efficient scheme for approximation of the joint spectral radius of a set of matrices using sum of squares programming. Utilizing the classical result of Rota and Strang on approximation of the joint spectral radius with a norm in conjunction with Barvinok's [18] result on approximation of any norm with a sum of squares polynomial, we provided an asymptotically tight estimate for the joint spectral radius which is independent of the number of matrices. Furthermore, we constructed an iterative scheme based on the generalization of a Lyapunov iteration which provides a bound on the joint spectral radius dependent on the number of matrices. Our results can be alternatively interpreted in a simpler way as providing a trajectory-preserving lifting to a polynomially-sized higher dimensional space, and proving contractiveness with respect to an ellipsoidal norm in that space. The results generalize earlier work of Ando and Shih [15], Blondel, Nesterov and Theys [16], and the lifting procedure of Blondel and Nesterov [17]. The good performance of our procedure was also verified using numerical examples.

References

1. Berger, M., Wang, Y.: Bounded semigroups of matrices. Linear Algebra and Applications **166** (1992) 21–27
2. Daubechies, I., Lagarias, J.C.: Sets of matrices all infinite products of which converge. Linear Algebra and Applications **161** (1992) 227–263

3. Daubechies, I., Lagarias, J.C.: Corrigendum/addendum to "Sets of matrices all infinite products of which converge". Linear Algebra and Applications **327** (2001) 69–83
4. Shih, M., Wu, J., Pang, C.T.: Asymptotic stability and generalized Gelfand spectral radius formula. Linear Algebra and Applications **251** (1997) 61–70
5. Leizarowitz, A.: On infinite products of stochastic matrices. Linear Algebra and Applications **168** (1992) 189–219
6. Rota, G.C., Strang, W.G.: A note on the joint spectral radius. Indag. Math. **22** (1960) 379–381
7. Blondel, V.D., Tsitsiklis, J.N.: A survey of computational complexity results in systems and control. Automatica **36** (2000) 1249–1274
8. Blondel, V.D., Tsitsiklis, J.N.: The boundedness of all products of a pair of matrices is undecidable. Systems and Control Letters **41** (2000) 135–140
9. Tsitsiklis, J.N., Blondel, V.: The Lyapunov exponent and joint spectral radius of pairs of matrices are hard- when not impossible- to compute and to approximate. Mathematics of Control, Signals, and Systems **10** (1997) 31–40
10. Hernek, D.: Random walks on colored graphs: Analysis and applications. PhD thesis, EECS Department, University of California at Berkeley (1995)
11. Barabanov, N.E.: Lyapunov indicators of discrete linear inclusions, parts I, II, and III. Translation from Avtomat. e. Telemekh. **2, 3 and 5** (1988) 40–46, 24–29, 17–44
12. Kozyakin, V.A.: Algebraic unsolvability of problem of absolute stability of desynchronized systems. Automation and Remote Control **51** (1990) 754–759
13. Wirth, F.: Joint spectral radius and extremal norms. Linear Algebra and Applications **251** (2002) 61–70
14. Dayawansa, W.P., Martin, C.F.: A converse Lyapunov theorem for a class of dynamical systems that undergo switching. IEEE Transactions on Automatic Control **44** (1999) 751–760
15. Ando, T., Shih, M.H.: Simultaneous contractibility. SIAM Journal on Matrix Analysis and Applications **19** (1998) 487–498
16. Blondel, V.D., Nesterov, Y., Theys, J.: On the accuracy of the ellipsoidal norm approximation of the joint spectral radius. Linear Algebra and Applications **394** (2005) 91–107
17. Blondel, V.D., Nesterov, Y.: Computationally efficient approximations of the joint spectral radius. SIAM J. Matrix Anal. Appl. **27**(1) (2005) 256–272
18. Barvinok, A.: A course in convexity. American Mathematical Society (2002)
19. Parrilo, P.A.: Structured semidefinite programs and semialgebraic geometry methods in robustness and optimization. PhD thesis, California Institute of Technology (2000) Available at `http://resolver.caltech.edu/CaltechETD:etd-05062004-055516`.
20. Parrilo, P.A.: Semidefinite programming relaxations for semialgebraic problems. Math. Prog. **96**(2, Ser. B) (2003) 293–320
21. Reznick, B.: Some concrete aspects of Hilbert's 17th problem. In: Contemporary Mathematics. Volume 253. American Mathematical Society (2000) 251–272
22. Brockett, R.: Lie algebras and Lie groups in control theory. In Mayne, D., Brockett, R., eds.: Geometric Methods in Systems Theory. D. Reidel Pub. Co. (1974) 17–56
23. Zelentsovsky, A.L.: Nonquadratic Lyapunov functions for robust stability analysis of linear uncertain systems. IEEE Trans. Automat. Control **39**(1) (1994) 135–138
24. Marcus, M.: Finite dimensional multilinear algebra. M. Dekker, New York (1973)
25. Marcus, M., Minc, H.: A survey of matrix theory and matrix inequalities. Dover Publications Inc., New York (1992) Reprint of the 1969 edition.

Metrics and Topology for Nonlinear and Hybrid Systems

Mihály Petreczky and René Vidal

Center for Imaging Science, Johns Hopkins University, Baltimore MD 21218, USA
{mihaly,rvidal}@cis.jhu.edu

Abstract. This paper presents an approach to defining distances between nonlinear and hybrid dynamical systems based on formal power series theory. The main idea is that the input-output behavior of a wide range of dynamical systems can be encoded by rational formal power series. Hence, a natural distance between dynamical systems is the distance between the formal power series encoding their input-output behavior. The paper proposes several computable distances for rational formal power series and discusses the application of such distances to various classes of nonlinear and hybrid systems. In particular, the paper presents a detailed discussion of distances for stochastic jump-linear systems.

1 Introduction

In this paper, we present several possible definitions of a computable distance for rational formal power series and their representations. The main motivation for studying distances between rational formal power series is that the input-output behavior of various classes of dynamical systems can be encoded by rational formal power series. For example, linear and bilinear systems, switched linear and bilinear systems, finite state hidden Markov models, and linear and bilinear hybrid systems [1,2,3,4,5]. Therefore, one can define a distance between two dynamical systems as the distance between the formal power series that encode the input-output behavior of the systems. By construction, the proposed distances will be invariant under any transformation that preserves the input-output behavior of the systems. In addition, restricting attention to *rational* formal power series will enable us to compute the distances by using the rational representations of the formal power series. In general, such rational representations can easily be computed from the dynamical system. Another advantage of the proposed approach is that it connects well with identification and realization theory, because several identification methods are based on realization theory and hence on finding an appropriate rational representation of a family of formal power series.

Endowing the space of dynamical systems with a topology and a metric not only is an interesting theoretical exercise, but also has several interesting applications. A classical application is in system identification, more precisely, in finding a continuous parameterization and suitable canonical forms of dynamical systems [6,7,8,9]. Another important application comes from the field of computer vision, where one is interested in automatically recognizing different types of motions in a video sequence. That is, given a sequence of images depicting moving objects and people at different time instances, we would like to determine automatically the object class, the person identify, and the type of motion we see in the video sequence. For instance, we would like to

A. Bemporad, A. Bicchi, and G. Buttazzo (Eds.): HSCC 2007, LNCS 4416, pp. 459–472, 2007.

determine whether the video sequence depicts a running person or a galloping deer. One of the traditional mathematical tools for recognition and classification is machine learning. However, many of the classical machine learning techniques require a metric on the observation space. Since our observations are video sequences depicting multiple motions, it is rather natural to model such videos as the output of one or more dynamical systems, where each dynamical system describes a particular motion. Our observations are then outputs of dynamical systems, or, after an identification procedure, dynamical systems themselves. Therefore, in order to apply machine learning algorithms for recognizing motions in video sequences, one needs to define a suitable metric and topology on the space of dynamical systems. The study of topological and metric properties of dynamical systems from this point of view is a relatively recent development, see [10,11,12,13].

The outline of the paper is as follows. Section 2 presents the background material on the theory of rational formal power series. Section 3 presents the definition of several possible distances for rational formal power series and their rational representations. Section 4 discusses the relationship between formal power series and various classes of dynamical systems. In particular, it presents a detailed description of this relationship as well as a distance for stochastic jump-linear systems. Section 5 discusses the relationship between the results of the current paper and earlier results in the literature, as well as issues concerning the practical computability of the defined distances.

2 Rational Power Series

This section presents several results on formal power series that will be used throughout the rest of the paper. The material in Subsections 2.1 and 2.2 can be found in [1]. The results in Subsection 2.3 are, to the best of our knowledge, new. For more details on the classical theory of rational formal power series, the reader is referred to [14,3,15].

2.1 Definition and Basic Theory

Let X be a finite set. We will refer to X as the *alphabet*. The elements of X will be called *letters*, and every finite sequence of letters will be called a *word* or *string* over X. Denote by X^* the set of all finite words from elements in X. An element $w \in X^*$ of length $|w| = k \geq 0$ is a finite sequence $w = w_1 w_2 \cdots w_k$ with $w_1, \ldots, w_k \in X$. The empty word is denoted by ϵ and its length is zero, i.e. $|\epsilon| = 0$. The concatenation of two words $v = v_1 \cdots v_k$ and $w = w_1 \cdots w_m \in X^*$ is the word $vw = v_1 \cdots v_k w_1 \cdots w_m$.

For any two sets J and A, an *indexed subset* of A with the *index set* J is simply a map $Z : J \to A$, denoted by $Z = \{a_j \in A \mid j \in J\}$, where $a_j = Z(j)$ for all $j \in J$. Notice that we do not require the elements a_j to be all different.

A *formal power series* S with coefficients in $\mathbb{R}^p$ is a map $S : X^* \to \mathbb{R}^p$. We will call the values $S(w) \in \mathbb{R}^p$, $w \in X^*$, the *coefficients* of S. We denote by $\mathbb{R}^p \ll X^* \gg$ the set of all formal power series with coefficients in $\mathbb{R}^p$. Consider the indexed set of formal power series $\Psi = \{S_j \in \mathbb{R}^p \ll X^* \gg \mid j \in J\}$ with an arbitrary (not necessarily finite) index set J. We will call such an indexed set of formal power series a *family of formal power series*. A family of formal power series Ψ is called *rational* if there exists an integer $n \in \mathbb{N}$, a matrix $C \in \mathbb{R}^{p \times n}$, a collection of matrices $A_\sigma \in \mathbb{R}^{n \times n}$ where $\sigma \in X$ runs through all elements of X, and an indexed set $B = \{B_j \in \mathbb{R}^n \mid j \in J\}$ of vectors in $\mathbb{R}^n$, such that for each index $j \in J$ and for all sequences $\sigma_1, \ldots, \sigma_k \in X$, $k \geq 0$,

$$S_j(\sigma_1 \sigma_2 \cdots \sigma_k) = C A_{\sigma_k} A_{\sigma_{k-1}} \cdots A_{\sigma_1} B_j. \tag{1}$$

The 4-tuple $R = (\mathbb{R}^n, \{A_\sigma\}_{\sigma \in X}, B, C)$ is called a *representation* of Ψ, and the number $n = \dim R$ is called the *dimension* of the representation R. If $S \in \mathbb{R}^p \ll X^* \gg$ is a single power series, then S will be called *rational* if the singleton set $\{S\}$ is rational, and by a representation of S we will mean a representation of $\{S\}$. A representation R_{min} of Ψ is called *minimal* if all representations R of Ψ satisfy $\dim R_{min} \leq \dim R$. Two representations of Ψ, $R = (\mathbb{R}^n, \{A_\sigma\}_{\sigma \in X}, B, C)$ and $\widetilde{R} = (\mathbb{R}^n, \{\widetilde{A}_\sigma\}_{\sigma \in X}, \widetilde{B}, \widetilde{C})$, are called *isomorphic*, if there exists a nonsingular matrix $T \in \mathbb{R}^{n \times n}$ such that $T\widetilde{A}_\sigma = A_\sigma T$ for all $\sigma \in X$, $T\widetilde{B}_j = B_j$ for all $j \in J$, and $\widetilde{C} = CT$.

Let $R = (\mathbb{R}^n, \{A_\sigma\}_{\sigma \in X}, B, C)$ be a representation of Ψ. In the sequel, we will use the following short-hand notation $A_w \doteq A_{w_k} A_{w_{k-1}} \cdots A_{w_1}$ for $w = w_1 \cdots w_k \in X^*$. A_ϵ will be identified with the identity map. The representation R is called *observable* if $O_R = \{0\}$ and *reachable* if $\dim R = \dim W_R$, where W_R and O_R are the following subspaces of $\mathbb{R}^n$

$$W_R = \mathrm{Span}\{A_w B_j \mid w \in X^*, |w| \leq n-1, j \in J\} \text{ and } O_R = \bigcap_{w \in X^*, |w| \leq n-1} \ker C A_w. \tag{2}$$

Observability and reachability of representations can be checked numerically. One can formulate an algorithm for transforming any representation to a minimal representation of the same family of formal power series (see [1] and the references therein for details).

Let $\Psi = \{S_j \in \mathbb{R}^p \ll X^* \gg \mid j \in J\}$ be a family of formal power series, and define the Hankel-matrix H_Ψ of Ψ as the matrix $H_\Psi \in \mathbb{R}^{(X^* \times I) \times (X^* \times J)}$, where $I = \{1, 2, \ldots, p\}$ and $(H_\Psi)_{(u,i)(v,j)} = (S_j(vu))_i$. That is, the rows of H_Ψ are indexed by pairs (u, i) where u is a word over X and i is and integer in the range $1, \ldots, p$. The columns of H_Ψ are indexed by pairs (v, j) where v is a word over X and j is an element of the index set J. The element of H_Ψ whose row index is (u, i) and whose column index is (v, j) is simply the ith row of the vector $S_j(vu) \in \mathbb{R}^p$. The following result on realization of formal power series can be found in [3,15,1].

Theorem 1 (Realization of formal power series). *Let $\Psi = \{S_j \in \mathbb{R}^p \ll X^* \gg \mid j \in J\}$ be a set of formal power series indexed by J. Then the following holds.*

(i) *Ψ is rational* $\iff$ $\mathrm{rank}\, H_\Psi < +\infty$.
(ii) *R is a minimal representation of Ψ* $\iff$ *R is reachable and observable* $\iff$ $\dim R = \mathrm{rank}\, H_\Psi$.
(iii) *All minimal representations of Ψ are isomorphic.*
(iv) *If the rank of the Hankel matrix H_Ψ is finite, i.e. $n = \mathrm{rank}\, H_\Psi < +\infty$, then one can construct a representation $R = (\mathbb{R}^n, \{A_\sigma\}_{\sigma \in X}, B, C)$ of Ψ using the columns of H_Ψ (see [1] for details).*

2.2 Realization Algorithm

In this subsection, we present an algorithm for computing a minimal representation of a family of formal power series Ψ from finite data, more precisely, from a finite left-upper block of the infinite Hankel matrix H_Ψ. The theorem guaranteeing the correctness of the algorithm, Theorem 2, will also enable us to define a distance between families of rational formal power series.

Let $\Psi = \{S_j \in \mathbb{R}^p \ll X^* \gg \mid j \in J\}$ be a family of formal power series indexed by a *finite* set J. Let $H_{\Psi,N,M} \in \mathbb{R}^{I_M \times J_N}$ be a finite upper-left block of the infinite Hankel matrix H_Ψ obtained by taking all columns of H_Ψ indexed by words over X of length at most N, and all the rows of H_Ψ indexed by words of length at most M. More specifically, $H_{\Psi,N,M} \in \mathbb{R}^{I_M \times J_N}$ is the matrix whose rows are indexed by elements of the set $I_M = \{(u,i) \mid u \in X^*, |u| \leq M, i = 1,\dots,p\}$, whose columns are indexed by elements of the set $J_N = \{(v,j) \mid j \in J, v \in X^*, |v| \leq N\}$, and whose entries are defined by $(H_{\Psi,N,M})_{(u,i),(v,j)} = (S_j(vu))_i$.

The following algorithm computes a representation of Ψ from $H_{\Psi,N+1,N}$.

Algorithm 1. [16] $(\mathbb{R}^r, \{A_\sigma\}_{\sigma \in X}, B, C) =$ *ComputeRepresentation*$(H_{\Psi,N+1,N})$

1: Let $r = \operatorname{rank} H_{\Psi,N,N}$ and choose $j_1,\dots,j_r \in J$, $i_1,\dots,i_r \in \{1,\dots,p\}$, $v_1,\dots,v_r, u_1,\dots,u_r \in X^*$ such that for all $l = 1,\dots,r$, $|v_l| \leq N$ and $|u_l| \leq N$, and the minor $T = ((S_{j_k}(v_k u_l))_{i_l}))_{l,k=1,\dots,r} \in \mathbb{R}^{r \times r}$ of $H_{\Psi,N,N}$ is of rank r.

2: For each symbol $\sigma \in X$ let $A_\sigma \in \mathbb{R}^{r \times r}$ be such that

$$A_\sigma T = Z_\sigma \text{ where } Z_\sigma = ((S_{j_k}(v_k \sigma u_l))_{i_l})_{l,k=1,\dots,r}.$$

Let $B = \{B_j \mid j \in J\}$, where for each index $j \in J$, the vector $B_j \in \mathbb{R}^r$ is given by

$$B_j = T^{-1}((S_j(u_1)_{i_1}), (S_j(u_2))_{i_2}, \dots (S_j(u_r))_{i_r})^T.$$

Let $C \in \mathbb{R}^{p \times r}$ be given by $C = \begin{bmatrix} C_1 & \cdots & C_r \end{bmatrix}$, where $C_l = S_{j_l}(v_l)$, for $l = 1,\dots,r$.

Theorem 2 ([1,3,16]). *If* $\operatorname{rank} H_{\Psi,N,N} = \operatorname{rank} H_\Psi$*, then the representation* $\widetilde{R}_N$ *of* Ψ *returned by* ComputeRepresentation *is minimal. Furthermore, if* $\operatorname{rank} H_\Psi \leq N$*, or, equivalently, there exists a representation* R *of* Ψ*, such that* $\dim R \leq N$*, then* $\operatorname{rank} H_\Psi = \operatorname{rank} H_{\Psi,N,N}$*, hence* $\widetilde{R}_N$ *is a minimal representation of* Ψ.

From a computational point of view, algorithm *ComputeRepresentation* may not be the best way to compute a representation of Ψ. However, we have chosen to present it, because it makes theoretical reasoning easier. The algorithm is essentially a reformulation of the construction presented in [16]. An alternative algorithm, which uses the factorization of the finite Hankel-matrix $H_{\Psi,N,N+1}$ can be found in [1].

2.3 A Notion of Stability for Formal Power Series

Since our ultimate goal is to compare formal power series, we might want to restrict our attention to formal power series that are stable in some sense. In this subsection, we consider the notion of square summability for formal power series, and translate the requirement of square summability into algebraic properties of their representations.

Consider a formal power series $S \in \mathbb{R}^p \ll X^* \gg$, and denote by $||\cdot||_2$ the Euclidean norm in $\mathbb{R}^p$. Consider the following sequence,

$$L_n = \sum_{k=0}^{n} \sum_{\sigma_1 \in X} \cdots \sum_{\sigma_k \in X} ||S(\sigma_1 \sigma_2 \cdots \sigma_k)||_2^2. \tag{3}$$

The series S will be called *square summable*, if the limit $\lim_{n \to +\infty} L_n$ exists and is finite. We will call the family $\Psi = \{S_j \in \mathbb{R}^p \ll X^* \gg \mid j \in J\}$ *square summable*, if for each $j \in J$, the formal power series S_j is square summable.

We now characterize square summability of a family of formal power series in terms of the stability of its representation. Let $R = (\mathbb{R}^n, \{A_\sigma\}_{\sigma \in X}, B, C)$ be an arbitrary representation of $\Psi = \{S_j \in \mathbb{R}^p \ll X^* \gg \mid j \in J\}$. Assume that $X = \{\sigma_1, \ldots, \sigma_d\}$, where d is the number of elements of X, and consider the matrix

$$\widetilde{A}_R = \begin{bmatrix} A_{\sigma_1} \otimes A_{\sigma_1} & A_{\sigma_2} \otimes A_{\sigma_2} & \cdots & A_{\sigma_d} \otimes A_{\sigma_d} \\ A_{\sigma_1} \otimes A_{\sigma_1} & A_{\sigma_2} \otimes A_{\sigma_2} & \cdots & A_{\sigma_d} \otimes A_{\sigma_d} \\ \vdots & \vdots & \vdots & \vdots \\ A_{\sigma_1} \otimes A_{\sigma_1} & A_{\sigma_2} \otimes A_{\sigma_2} & \cdots & A_{\sigma_d} \otimes A_{\sigma_d} \end{bmatrix} \in \mathbb{R}^{n^2 d \times n^2 d}, \tag{4}$$

where $\otimes$ denotes the Kronecker product. We will call R *stable*, if the matrix $\widetilde{A}_R$ is stable, i.e. if all its eigenvalues λ lie inside the unit disk ($|\lambda| < 1$). We have the following.

Theorem 3. *A rational family of formal power series is square summable if and only if all minimal representations are stable.*

Notice the analogy with the case of linear systems, where the minimal realization of a stable transfer matrix is also stable.

3 Distances for Rational Power Series

The goal of this section is to present a notion of distance for families of rational formal power series, or equivalently, a distance between their minimal representations. The choice of a distance is by no means unique, in fact, we will suggest several different distances. The common feature of all these distances is that they all can be computed either from a minimal representation of the family, or from a big enough but finite set of values of the formal power series constituting the families.

Through the section, we will fix the space of coefficients $\mathbb{R}^p$ and the alphabet X. Also, we will fix a *finite* index set J and consider the space of all rational families of formal power series indexed by J, i.e. $\mathcal{P}_J = \{\Psi = \{S_j \in \mathbb{R}^p \ll X^* \gg \mid j \in J\} \mid \Psi \text{ is rational }\}$. Define the subset $\mathcal{P}_{J,N} = \{\Psi \in \mathcal{P}_J \mid \text{rank } H_\Psi \leq N\}$ of all rational families of formal power series whose Hankel-matrix is of rank at most N. Then it is easy to see that $\mathcal{P}_{J,N} \subseteq \mathcal{P}_{J,K}$ for all $N \leq K$, and $\mathcal{P}_J = \bigcup_{N=0}^{+\infty} \mathcal{P}_{J,N}$.

3.1 Distances Based on Truncation

We will first consider distances based on truncation, that is distances that compare finitely many values of formal power series.

Fix a natural number N, and denote by $m = card(J)$ the cardinality of J. Since J is finite, without loss of generality, we can assume that $J = \{1, \dots, m\}$. Assume that the alphabet X is of the form $X = \{z_1, \dots, z_d\}$ where d is the number of elements of X. For each $N \geq 1$, let $F_{N,J} : \mathbb{R}^{mpN} \times \mathbb{R}^{mpN} \to \mathbb{R}$ be a distance on $\mathbb{R}^{mpN}$ and denote by $F = \{F_{N,J} \mid N \in \mathbb{N}\}$ the family of distances.

We now define a pseudo-metric on $\mathcal{P}_J$ using the family of distances F. The main idea is the following. If $S \in \mathbb{R}^p \ll X^* \gg$ is a formal power series, then it can be viewed as a map $S : X^* \to \mathbb{R}^p$ on words over X. There are $M(N) = \sum_{j=0}^{2N+1} d^j$ words of length at most $2N+1$ over the alphabet X, if X has d elements. Hence, we can view the restriction of the map S to the set of all words of length at most $2N+1$ as a vector in $\mathbb{R}^{M(N)p}$. We can then define the distance $d_{F,N,J}(\Psi_1, \Psi_2)$ between two families of formal power series indexed by J, Ψ_1 and Ψ_2, as the distance $F_{M(N),J}(\phi_1, \phi_2)$ between the vectors ϕ_1 and ϕ_2 in $\mathbb{R}^{mpM(N)}$ representing the restriction of the elements of Ψ_1 and Ψ_2, respectively, to the set of words of length at most $2N+1$. More formally,

1. Define an enumeration of all the words over the alphabet X as the bijective map $\psi : X^* \to \mathbb{N}$ defined as follows. For the empty word ϵ, let $\psi(\epsilon) = 0$ and for each letter z_i, $i = 1, \dots, d$, let $\psi(z_i) = i$. Then, for each word of the form $w = vz_j$, $j = 1, \dots, d$, $v \in X^*$ define $\psi(w)$ recursively as $\psi(w) = d \cdot \psi(v) + j$.
2. Denote by $X^{\leq 2N+1} = \{w \in X^* \mid |w| \leq 2N+1\}$ the set of all words on X of length at most $2N+1$. Notice that the restriction of ψ to the set $X^{\leq 2N+1}$ yields a bijective map with the range $[0, M(N)-1]$.
3. For each $S \in \mathbb{R}^p \ll X^* \gg$ define $\pi_N(S)$ as the vector $(Z_0^T, Z_1^T, \dots, Z_{M(N)-1}^T)^T$ in $\mathbb{R}^{pM(N)}$, where $Z_i = S(\psi^{-1}(i)) \in \mathbb{R}^p$. Since the integer i goes through all the values $[0, M(N)-1]$, $\psi^{-1}(i)$ goes through all possible words of length at most $2N+1$. Hence $\pi_N(S)$ is just the vector of all values $S(w)$ where $|w| \leq 2N+1$.
4. For each rational family of formal power series $\Psi = \{S_j \in \mathbb{R}^p \ll X^* \gg \mid j \in J\}$ define the vector $\pi_{J,N}(\Psi) = (\pi_N(S_1)^T, \dots, \pi_N(S_m)^T)^T \in \mathbb{R}^{mpM(N)}$. That is, $\pi_{J,N}(\Psi)$ is obtained by stacking up the vectors $\pi_N(S_1), \dots, \pi_N(S_m)$ representing the values of $S_1, \dots, S_m$ on words of length at most $2N+1$.
5. For each $\Psi_1, \Psi_2 \in \mathcal{P}_J$, define the functions $d_{F,N,J} : \mathcal{P}_J \times \mathcal{P}_J \to \mathbb{R}$ by

$$d_{F,N,J}(\Psi_1, \Psi_2) = F_{M(N),J}(\pi_{J,N}(\Psi_1), \pi_{J,N}(\Psi_2)) \tag{5}$$

We then have the following result.

Lemma 1 (Properties of $d_{F,N,J}$). *$d_{F,N,J}$ is a pseudo-distance in $\mathcal{P}_J$. That is, for each $\Psi_1, \Psi_2, \Psi_3 \in \mathcal{P}_J$*

$$\begin{aligned} d_{F,N,J}(\Psi_1, \Psi_2) &= d_{F,N,J}(\Psi_2, \Psi_1) \geq 0, \\ d_{F,N,J}(\Psi_1, \Psi_2) &\leq d_{F,N,J}(\Psi_1, \Psi_3) + d_{F,N,J}(\Psi_2, \Psi_3) \quad \textit{and} \\ \Psi_1 = \Psi_2 &\implies d_{F,N,J}(\Psi_1, \Psi_2) = 0. \end{aligned} \tag{6}$$

The following theorem formulates an important property of $d_{F,N,J}$ and it relies on the partial realization result of Theorem 2.

Theorem 4 (Distance for Rational Formal Power Series). *The restriction of* $d_{F,N,J}$ *to* $\mathcal{P}_{N,J}$ *is a distance. That is, in addition to the properties listed in Lemma 1, the following holds.*

$$\forall \Psi_1,\ \Psi_2 \in \mathcal{P}_{N,J}: \quad \Psi_1 = \Psi_2 \iff d_{F,N,J}(\Psi_1, \Psi_2) = 0 \tag{7}$$

Proof. If Ψ_1 and Ψ_2 belong to $\mathcal{P}_{N,J}$, then by Theorem 2 $\text{rank}\, H_{\Psi_i,N,N} = \text{rank}\, H_{\Psi_i}$ holds for $i = 1, 2$. It is easy to see that $d_{F,N,J}(\Psi_1, \Psi_2) = 0$ if and only if $\pi_{N,J}(\Psi_1) = \pi_{N,J}(\Psi_2)$, i.e. the values of the elements of Ψ_1 and Ψ_2 coincide for all the words of length at most $2N+1$. Hence, $H_{\Psi_1,N+1,N} = H_{\Psi_2,N+1,N}$. Therefore, by Theorem 2, the representation $\widetilde{R}_N$ produced by **Algorithm 1** with the input $H_{\Psi_1,N+1,N}$ is a minimal representation of both Ψ_1 and Ψ_2, which implies that $\Psi_1 = \Psi_2$.

3.2 The Hilbert Space of Square Summable Families of Formal Power Series

In what follows we will define a scalar product on the space of square summable rational families of formal power series. With this scalar product the space of square summable rational families becomes a Hilbert space, and the corresponding distance will take all values of the formal power series into account.

Consider the set $\mathcal{P}_{s,J} = \{\Psi \in \mathcal{P}_J \mid \Psi \text{ is square summable }\}$ of *square summable rational families of formal power series*. Assume that J is finite. It is clear that $\mathcal{P}_{s,J}$ is a vector space, if we define addition and multiplication by a scalar as follows. Let $\Psi_1 = \{S_j \in \mathbb{R}^p \ll X^* \gg \mid j \in J\}$ and $\Psi_2 = \{T_j \in \mathbb{R}^p \ll X^* \gg \mid j \in J\}$ be two square summable rational families of formal power series. Then for each $\alpha, \beta \in \mathbb{R}$, let $\alpha\Psi_1 + \beta\Psi_2 = \{\alpha T_j + \beta S_j \in \mathbb{R}^p \ll X^* \gg \mid j \in J\}$, where $\alpha S_j + \beta T_j$ is just the usual point-wise linear combination on formal power series ([14]), i.e. for all $w \in X^*$, $(\alpha S_j + \beta T_j)(w) = \alpha S_j(w) + \beta T_j(w)$. Now, for Ψ_1 and Ψ_2 defined as before, define the bilinear map $< \cdot, \cdot >_J$: $\mathcal{P}_{s,J} \times \mathcal{P}_{s,J} \to \mathbb{R}$ as

$$< \Psi_1, \Psi_2 >_J \doteq < \cdot, \cdot >_J (\Psi_1, \Psi_2) = \sum_{j \in J} \sum_{w \in X^*} S_j(w)^T T_j(w) \tag{8}$$

Since J is finite and S_j, T_j are both square summable, the infinite sum in (8) is well defined and finite. The following lemma characterizes some properties of $< \cdot, \cdot >_J$.

Lemma 2. *The map* $< \cdot, \cdot >_J$ *is a scalar product and the space* $\mathcal{P}_{s,J}$ *with the scalar product* $< \cdot, \cdot >_J$ *is a Hilbert space.*

As a consequence, we can view $\mathcal{P}_{s,J}$ as a normed space with the norm $||.||_J$ induced by $< \cdot, \cdot >_J$, i.e.

$$||\Psi||_J \doteq \sqrt{\sum_{j \in J} \sum_{w \in X^*} ||S_j(w)||_2^2} = \sqrt{< \Psi, \Psi >_J}. \tag{9}$$

The following theorem gives a formula for computing $< \Psi_1, \Psi_2 >_J$ for all $\Psi_1, \Psi_2 \in \mathcal{P}_{s,J}$, provided that the representations of Ψ_1 and Ψ_2 are available.

Theorem 5. *For $i = 1, 2$, assume that $R_i = (\mathbb{R}^{n_i}, \{A_{i,\sigma}\}_{\sigma \in X}, C_i, B_i)$ is a stable representation of $\Psi_i \in \mathcal{P}_{s,J}$ and that $B_i = \{B_{i,j} \in \mathbb{R}^{n_i} \mid j \in J\}$. Then there exists a unique solution $P \in \mathbb{R}^{n_1 \times n_2}$ to the Sylvester equation*

$$P = \sum_{\sigma \in X} A_{1,\sigma}^T P A_{2,\sigma} + C_1^T C_2 \tag{10}$$

and the scalar product $< \Psi_1, \Psi_2 >_J$ can be written explicitly

$$< \Psi_1, \Psi_2 >_J = \sum_{j \in J} B_{1,j}^T P B_{2,j}. \tag{11}$$

Notice from Theorem 3 that if R_1 and R_2 are minimal representations of Ψ_1 and Ψ_2 in $\mathcal{P}_{s,J}$, respectively, then the condition of Theorem 5 holds. Hence we can use any minimal representation to compute $< \Psi_1, \Psi_2 >_J$. From this we may compute the distance between Ψ_1 and Ψ_2 as $||\Psi_1 - \Psi_2||_J^2 = < \Psi_1, \Psi_1 >_J - 2 < \Psi_1, \Psi_2 >_J + < \Psi_2, \Psi_2 >_J$.

4 Rational Power Series and Input-Output Behavior of Dynamical Systems

The main motivation for introducing the framework of rational formal power series is that it provides a common algebraic framework for realization theory and system identification of a wide-variety of input/output systems. The classes of systems whose behaviors can be described in terms of rational formal power series include linear systems [17,9], bilinear systems, [15,3,2,18], multidimensional systems [4], finite state hidden Markov models [5], continuous-time linear and bilinear switched systems [1] and continuous-time linear and bilinear hybrid systems [1].

Hence, if we pick two dynamical systems Σ_1 and Σ_2 from any of the classes mentioned above we can compare them as follows. We can construct the families of formal power series Ψ_1 and Ψ_2 corresponding to the input-output behaviors of Σ_1 and Σ_2, respectively. Then, we can compare Ψ_1 and Ψ_2 using one of the distances defined in Section 3. Note that the choice of the families Ψ_i, $i = 1, 2$ is unique if Σ_i belongs to one of the classes of systems described in the previous paragraph. Alternatively, we can construct the rational representations R_{Σ_1} and R_{Σ_2} of the families Ψ_1 and Ψ_2 respectively. In general, the representations R_{Σ_1} and R_{Σ_2} can be easily computed from the parameters of Σ_1 and Σ_2. Then, we can use R_{Σ_1} and R_{Σ_2} to compute the distance between Ψ_1 and Ψ_2. This approach is particularly appealing if Ψ_1 and Ψ_2 are square summable and one wants to use the norm (9). Notice, that even if Ψ_1 and Ψ_2 are square summable, R_{Σ_1} or R_{Σ_2} might fail to be stable. In this case we have to minimize R_{Σ_1} and R_{Σ_2} first, and use the resulting stable minimal representations (see Theorem 3 and Theorem 5) for computing the distance. Algorithms for minimizing representations can be found in [1]. Notice also that, in general, there need not be any connection between square summability of Ψ_i, $i = 1, 2$ and stability of the dynamical systems Σ_i, $i = 1, 2$.

In what follows, we will demonstrate the use of rational formal power series for stochastic discrete-time jump-linear systems. This class of hybrid systems has a wide variety of applications including computer vision. To the best of our knowledge, the relationship between stochastic jump-linear systems and formal power series presented here is new, though there are some similarities between our approach and that in [18].

4.1 Stochastic Jump-Linear Systems

The terminology and notation used in this section is based on the conventions adopted in literature, see [19,9]. A *stochastic jump-linear systems* [19] is a discrete-time stochastic system described by the equations

$$\Sigma : \begin{cases} \mathbf{x}(k+1) = A_{\boldsymbol{\theta}(k)}\mathbf{x}(k) + B_{\boldsymbol{\theta}(k)}\mathbf{v}(k) \\ \quad\mathbf{y}(k) = C_{\boldsymbol{\theta}(k)}\mathbf{x}(k) \text{ and } \mathbf{o}(k) = \lambda(\boldsymbol{\theta}(k)) \end{cases} . \tag{12}$$

Here, $\mathbf{x}$, $\boldsymbol{\theta}$, $\mathbf{y}$, $\mathbf{o}$ and $\mathbf{v}$ are stochastic processes of the following form. The process $\mathbf{x}$ is called the *continuous state process* and takes values in the *continuous-state space* $\mathbb{R}^n$. The process $\boldsymbol{\theta}$ is called the *discrete state process* and takes values in the *set of discrete states* $Q = \{1, 2, \ldots, d\}$. The process $\mathbf{y}$ is the *continuous output process* and takes values in the *set of continuous outputs* $\mathbb{R}^p$. The process $\mathbf{o}$ is the *discrete output process* and takes values in the *set of discrete outputs* $O = \{1, 2, \ldots, l\}$. Finally, the process $\mathbf{v}$ is the *continuous noise* and takes values in $\mathbb{R}^m$. The matrices A_q, B_q, C_q, $q \in Q$, are of the form $A_q \in \mathbb{R}^{n\times n}$, $B_q \in \mathbb{R}^{n\times m}$, and $C_q \in \mathbb{R}^{p\times n}$. The map $\lambda : Q \to O$ is called the *readout map* and it assigns a discrete output to each discrete state. We will assume that $E[\mathbf{v}(k)\mathbf{v}(l)] = \delta_{k,l}I$ and $E[\mathbf{v}(k)] = 0$, for all $k, l \in \mathbb{N}$, that is $\mathbf{v}$ is a zero mean process and $\mathbf{v}(k)$, $k \in \mathbb{N}$ are uncorrelated. Furthermore, we will assume that for each $k, l \in \mathbb{N}$, $\mathbf{x}(0), \mathbf{v}(k), \boldsymbol{\theta}(l)$ are mutually independent random variables. We will also assume that the state-transition of the Markov process $\boldsymbol{\theta}$ is governed by the transition probabilities $p_{q_1,q_2}, q_1, q_2 \in Q$, where p_{q_1,q_2} is the probability that $\boldsymbol{\theta}$ changes its value from q_2 to q_1, i.e. $p_{q_1,q_2} = \text{Prob}(\boldsymbol{\theta}_{k+1} = q_1 \mid \boldsymbol{\theta}_k = q_2)$. In addition, we will assume that the initial probability distribution of $\boldsymbol{\theta}$ is given by the vector $\pi = (\pi_1, \ldots, \pi_d)^T \in \mathbb{R}^d$, where $\pi_q = \text{Prob}(\boldsymbol{\theta}(0) = q)$ denotes the probability that the process $\boldsymbol{\theta}$ is in state q at time 0. The evolution of system (12) is as follows. At each time instant k, the continuous state $\mathbf{x}$ and the continuous output $\mathbf{y}$ change according to the discrete-time stochastic linear system $(A_{\boldsymbol{\theta}(k)}, B_{\boldsymbol{\theta}(k)}, C_{\boldsymbol{\theta}(k)})$. The discrete state process $\boldsymbol{\theta}$, together with the discrete output process $\mathbf{o}$, form a finite state hidden Markov model [5].

In the next subsection, we study the concept of realization for stochastic jump-linear systems. To that end, we will assume that the stochastic processes $\mathbf{x}$ and $\mathbf{y}$ are wide-sense stationary and zero mean, which is guaranteed under the following assumptions.

Assumption 1. *The Markov process $\boldsymbol{\theta}$ is stationary and ergodic, hence for all $q \in Q$, $\sum_{s\in Q} p_{q,s}\pi_s = \pi_q$.*

Assumption 2. *There exists $n \times n$ matrices P_q, $q \in Q$, such that for each $q \in Q$*

$$P_q = \sum_{s\in Q} p_{q,s}A_sP_sA_s^T + B_sB_s^T p_{q,s}\pi_s, \tag{13}$$

$E[\mathbf{x}(0)] = 0$, and $E[\mathbf{x}(0)\mathbf{x}(0)^T\chi(\boldsymbol{\theta}(0) = q)] = P_q$, where χ denotes the indicator function, i.e. $\chi(A) = 1$ if the event A is true, and $\chi(A) = 0$ otherwise.

These assumptions are not particularly strong. For instance, under suitable conditions [19], there is a unique collection of positive semi-definite matrices P_q such that (13) holds.

4.2 Realization of Stochastic Jump-Linear Systems

Recall the notion of *weak realization* for linear stochastic systems [9]. In this subsection, we will formulate a similar concept for stochastic jump-linear systems.

Consider a stationary process $\widetilde{\mathbf{o}}$ taking values in the finite output space O, and a wide-sense stationary, zero-mean stochastic process $\widetilde{\mathbf{y}}$ taking values in the continuous output space $\mathbb{R}^p$. Let O^+ be the set of all nonempty words in O, i.e. $O^+ = O^* \setminus \{\epsilon\}$. For all $o_0, o_1, \ldots, o_k \in O$, $k \geq 0$, define the maps $\mathcal{P}_{\widetilde{\mathbf{o}}} : O^+ \to \mathbb{R}$ and $\mathcal{C}_{\widetilde{\mathbf{o}},\widetilde{\mathbf{y}}} : O^+ \to \mathbb{R}^{p \times p}$

$$\begin{aligned} \mathcal{P}_{\widetilde{o}}(o_0 o_1 \cdots o_k) &= \mathrm{Prob}(\widetilde{\mathbf{o}}(i) = o_i, i = 0, \ldots, k) \\ \mathcal{C}_{\widetilde{\mathbf{o}},\widetilde{\mathbf{y}}}(o_0 o_1 \cdots o_k) &= E[\widetilde{\mathbf{y}}(k)\widetilde{\mathbf{y}}(0)^T \chi(\widetilde{\mathbf{o}}(i) = o_i, i = 0, \ldots, k)]. \end{aligned} \tag{14}$$

Notice that the map $\mathcal{P}_{\widetilde{o}}$ gives the probability distribution of the stochastic process $\widetilde{\mathbf{o}}$, while the map $\mathcal{C}_{\widetilde{\mathbf{o}},\widetilde{\mathbf{y}}}$ gives the covariance of $\widetilde{\mathbf{y}}(k)$ and $\widetilde{\mathbf{y}}(0)$, provided that the proces $\widetilde{\mathbf{o}}$ takes values $o_0, \ldots, o_k$ in the first $k+1$ time instances. That is, $\mathcal{C}_{\widetilde{\mathbf{o}},\widetilde{\mathbf{y}}}$ collects information on the second-order moments of $\widetilde{\mathbf{y}}$.[1]

Consider now a jump-linear system Σ of the form (12) and recall the definition of the processes $\mathbf{y}$ and $\mathbf{o}$. If **Assumption 1** and **Assumption 2** hold, then $\mathbf{y}$ is wide-sense stationary and zero-mean and $\mathbf{o}$ is stationary. Hence, $\mathcal{C}_{\mathbf{o},\mathbf{y}}$ and $\mathcal{P}_{\mathbf{o}}$ are well-defined. In fact, they depend only on the matrices A_q, B_q, C_q, $q \in Q$, the discrete state-transition probabilities p_{q_1,q_2}, $q_1, q_2 \in Q$, the probability distribution of the initial discrete-state π, and the readout map λ. To emphasize that $\mathcal{C}_{\mathbf{o},\mathbf{y}}$ and $\mathcal{P}_{\mathbf{o}}$ depend only on the parameters of Σ, we will denote $\mathcal{C}_{\mathbf{o},\mathbf{y}}$ by $\mathcal{C}_{\Sigma}$ and $\mathcal{P}_{\mathbf{o}}$ by $\mathcal{P}_{\Sigma}$. These maps are important, because they contain information about the probability distribution of the output processes generated by Σ. In fact, the following is true.

Proposition 1. *If $\mathbf{x}(0)$ and $\mathbf{v}$ are Gaussian, $Q = O$, and $\lambda = id$, i.e. the discrete state is fully observed, then the map $\mathcal{C}_{\Sigma}$ uniquely determines the distribution of $\mathbf{y}$.*

The assumption that $Q = O$ is critical here. Intuitively, the more information about the discrete state is preserved by the discrete output, i.e. the closer $\mathbf{o}$ is to $\boldsymbol{\theta}$, the better estimate of the probability distribution of $\mathbf{y}$ is provided by $\mathcal{C}_{\Sigma}$.

With the notation above, we are now ready to define a notion of weak realization for stochastic jump-linear systems. Let $\widetilde{\mathbf{y}}$ be a wide-sense stationary, zero mean $\mathbb{R}^p$-valued process, and let $\widetilde{\mathbf{o}}$ be a stationary O-valued process. A stochastic-jump linear system Σ is said to be a *weak stochastic realization of* $(\widetilde{\mathbf{y}}, \widetilde{\mathbf{o}})$, if $\mathcal{P}_{\widetilde{\mathbf{o}}} = \mathcal{P}_{\Sigma}$ and $\mathcal{C}_{\widetilde{\mathbf{o}},\widetilde{\mathbf{y}}} = \mathcal{C}_{\Sigma}$.

Clearly, the fact that Σ is a weak-realization of the processes $(\widetilde{\mathbf{y}}, \widetilde{\mathbf{o}})$ imposes some constraints on the probability distribution of the processes $\widetilde{\mathbf{o}}$ and $\widetilde{\mathbf{y}}$. We will now show that $(\widetilde{\mathbf{y}}, \widetilde{\mathbf{o}})$ has a weak realization by a stochastic jump-linear system, only if certain families of formal power series are rational. We will construct two families of formal power series $\Psi_{\widetilde{\mathbf{o}},\widetilde{\mathbf{y}}}$ and $S_{\widetilde{\mathbf{o}}}$ based on the maps $\mathcal{C}_{\widetilde{\mathbf{o}},\widetilde{\mathbf{y}}}$ and $\mathcal{P}_{\widetilde{\mathbf{o}}}$, respectively, as follows.

Let $X = O = \{1, 2, \ldots, l\}$ and $J_{\widetilde{\mathbf{o}},\widetilde{\mathbf{y}}} = \{1, \ldots, p\} \times O$ be, respectively, the alphabet and the index set over which the formal power series will be defined. For each integer $i = 1, \ldots, p$, letter $o \in O$, and word $w \in O^*$, let $\mathcal{C}_{\widetilde{\mathbf{o}},\widetilde{\mathbf{y}},(i,o)}(w) \in \mathbb{R}^{lp}$ be the ith column of the matrix

$$\left[\mathcal{C}^T_{\widetilde{\mathbf{o}},\widetilde{\mathbf{y}}}(ow1), \mathcal{C}^T_{\widetilde{\mathbf{o}},\widetilde{\mathbf{y}}}(ow2), \cdots, \mathcal{C}^T_{\widetilde{\mathbf{o}},\widetilde{\mathbf{y}}}(owl)\right]^T \in \mathbb{R}^{lp \times p}. \tag{15}$$

[1] Notice the similarity between $\mathcal{C}_{\widetilde{\mathbf{o}},\widetilde{\mathbf{y}}}$ and the generalized covariances in [18].

We define the family of formal power series $\Psi_{\widetilde{\mathbf{o}},\widetilde{\mathbf{y}}}$ associated with $\mathcal{C}_{\widetilde{\mathbf{o}},\widetilde{\mathbf{y}}}$ as

$$\Psi_{\widetilde{\mathbf{o}},\widetilde{\mathbf{y}}} = \{\mathcal{C}_{\widetilde{\mathbf{o}},\widetilde{\mathbf{y}},(i,o)} \in \mathbb{R}^{pl} \ll O^* \gg \mid (i,o) \in J_{\widetilde{\mathbf{o}},\widetilde{\mathbf{y}}}\} \tag{16}$$

The construction of $S_{\widetilde{\mathbf{o}}}$ is much simpler. We can simply identify the map $\mathcal{P}_{\widetilde{\mathbf{o}}}$ with a formal power series $S_{\widetilde{\mathbf{o}}} \in \mathbb{R} \ll O^* \gg$ by defining $S_{\widetilde{\mathbf{o}}}(\epsilon) = 1$ for the empty word and $S_{\widetilde{\mathbf{o}}}(w) = \mathcal{P}_{\widetilde{\mathbf{o}}}(w)$ for all $w \in O^+$. By abuse of notation we will denote $S_{\widetilde{\mathbf{o}}}$ simply by $\mathcal{P}_{\widetilde{\mathbf{o}}}$. We will call $(\Psi_{\widetilde{\mathbf{o}},\widetilde{\mathbf{y}}}, \mathcal{P}_{\widetilde{\mathbf{o}}})$ *the pair of formal power series associated with* $(\widetilde{\mathbf{y}}, \widetilde{\mathbf{o}})$.

The next step is to define a *pair of representations* $(R_{\Sigma,C}, R_{\Sigma,D})$ *associated with a jump-linear system* Σ *of the form (12).*

We define the representation $R_{\Sigma,C}$ as follows. For each $o \in O$ and $q, q_1, q_2 \in Q$, let $C_q^o = C_q\chi(\lambda(q) = o) \in \mathbb{R}^{p\times n}$, $A^o_{q_1,q_2} = p_{q_1,q_2}\chi(\lambda(q_2) = o)A_{q_2} \in \mathbb{R}^{n\times n}$, and $B_q^o = P_q(C_q^o)^T \in \mathbb{R}^{n\times p}$, where $P_q \in \mathbb{R}^{n\times n}$ is the matrix defined in (13). Using this notation, define the matrices $\widetilde{A}_o \in \mathbb{R}^{nd\times nd}$, $\widetilde{C} \in \mathbb{R}^{lp\times nd}$ and $\widetilde{B}_o \in \mathbb{R}^{nd\times p}$, $o \in O$, as

$$\widetilde{A}_o = \begin{bmatrix} A^o_{1,1} & A^o_{1,2} & \cdots & A^o_{1,d} \\ A^o_{2,1} & A^o_{2,2} & \cdots & A^o_{2,d} \\ \vdots & \vdots & \vdots & \vdots \\ A^o_{d,1} & A^o_{d,2} & \cdots & A^o_{d,d} \end{bmatrix}, \quad \widetilde{C} = \begin{bmatrix} C^1_1 & C^1_2 & \cdots & C^1_d \\ C^2_1 & C^2_2 & \cdots & C^2_d \\ \vdots & \vdots & \vdots & \vdots \\ C^l_1 & C^l_2 & \cdots & C^l_d \end{bmatrix}, \text{ and } \widetilde{B}_o = \begin{bmatrix} B^o_1 \\ B^o_2 \\ \vdots \\ B^o_d \end{bmatrix}.$$

Then, for each $o \in O$, and $i = 1, \ldots, p$ let $\widetilde{B}_{(i,o)} \in \mathbb{R}^{nd}$ be the ith column of $\widetilde{B}_o$ and define the set $\widetilde{B} = \{\widetilde{B}_{(i,o)} \in \mathbb{R}^{nd} \mid (i,o) \in J_{o,y}\}$ indexed by $J_{o,y} = \{1, \ldots, p\} \times O$. We define the representation $R_{\Sigma,C}$ as

$$R_{\Sigma,C} = (\mathbb{R}^{nd}, \{\widetilde{A}_o\}_{o\in O}, \widetilde{B}, \widetilde{C}). \tag{17}$$

As per the representation $R_{\Sigma,D}$, we define it as

$$R_{\Sigma,D} = (\mathbb{R}^d, \{M_o\}_{o\in O}, \{\pi\}, e), \tag{18}$$

where $e = (1, 1, \ldots, 1) \in \mathbb{R}^{1\times d}$, and for each $o \in O$ and $q_1, q_2 \in Q$, the (q_1, q_2) entry of the matrix $M_o \in \mathbb{R}^{d\times d}$ is defined from the transition probabilities of the process $\boldsymbol{\theta}$ as $p_{q_1,q_2}\chi(\lambda(q_2) = o)$. Notice the similarity between the definition of $R_{\Sigma,D}$ and the definition of a quasi-realization for the finite state hidden Markov model formed by $(\boldsymbol{\theta}, \mathbf{o})$ given in [5].

We these definitions, we have the following result.

Theorem 6 (Weak Realization). *A jump-linear system* Σ *of the form (12) is a weak realization of* $(\widetilde{\mathbf{y}}, \widetilde{\mathbf{o}})$ *if and only if* $R_{\Sigma,C}$ *is a representation of* $\Psi_{\widetilde{\mathbf{o}},\widetilde{\mathbf{y}}}$ *and* $R_{\Sigma,D}$ *is a representation of* $\mathcal{P}_{\widetilde{\mathbf{o}}}$. *Hence,* $(\widetilde{\mathbf{y}}, \widetilde{\mathbf{o}})$ *admits a weak stochastic realization by a stochastic jump-linear system, only if* $\Psi_{\widetilde{\mathbf{o}},\widetilde{\mathbf{y}}}$ *is a rational family of formal power series and* $\mathcal{P}_{\widetilde{\mathbf{o}}}$ *is a rational formal power series.*

An important implication of the theorem above is the following. If we know that the processes $(\widetilde{\mathbf{y}}, \widetilde{\mathbf{o}})$ admit a weak stochastic realization by a stochastic jump-linear system, then we can find representations of $\Psi_{\widetilde{\mathbf{o}},\widetilde{\mathbf{y}}}$ and $\mathcal{P}_{\widetilde{\mathbf{o}}}$ from finite data. More precisely, if $\operatorname{rank} H_{\Psi_{\widetilde{\mathbf{o}},\widetilde{\mathbf{y}}}} \leq N$ and $\operatorname{rank} H_{\mathcal{P}_{\widetilde{\mathbf{o}}}} \leq N$, then from $\mathcal{C}_{\widetilde{\mathbf{o}},\widetilde{\mathbf{y}}}(o_0 \cdots o_k)$, $\mathcal{P}_{\widetilde{\mathbf{o}}}(o_0 \cdots o_k)$

$k \leq 2N+1$, $o_0, \ldots, o_k \in O$, we can construct the Hankel matrices $H_{\Psi_{\tilde{\mathbf{o}},\tilde{\mathbf{y}}},N+1,N}$ and $H_{\mathcal{P}_{\tilde{\mathbf{o}}},N+1,N}$ and compute a representation $R_{\tilde{\mathbf{o}},\tilde{\mathbf{y}}}$ of $\Psi_{\tilde{\mathbf{o}},\tilde{\mathbf{y}}}$ and $R_{\tilde{\mathbf{o}}}$ of $\mathcal{P}_{\tilde{\mathbf{o}}}$ respectively. Note that if we know that $(\tilde{\mathbf{y}}, \tilde{\mathbf{o}})$ has a weak stochastic realization by a jump-linear system Σ with a continuous state-space of dimension n and a discrete state-space of cardinality d, then we can take $N \geq nd > 0$. Finally, recall that the problem of estimating $\mathcal{C}_{\tilde{\mathbf{o}},\tilde{\mathbf{y}}}(o_0 \cdots o_k)$ and $\mathcal{P}_{\tilde{\mathbf{o}}}(o_0 \cdots o_k)$ is a classical statistical problem. In particular, if $\tilde{\mathbf{y}}$ and $\tilde{\mathbf{o}}$ are ergodic, then these quantities can easily be estimated from a long enough sequence of measurements.

4.3 Distances Between Stochastic Jump-Linear Systems

Imagine we would like to compare the probability distributions of the output processes $(\tilde{\mathbf{o}}_1, \tilde{\mathbf{y}}_1)$, and $(\tilde{\mathbf{o}}_2, \tilde{\mathbf{y}}_2)$ of two stochastic jump-linear systems Σ_1 and Σ_2, respectively. We can do that by using one of the distances defined in Section 3 to compare their associated pairs of families of formal power series: $\Psi_{\tilde{\mathbf{o}}_1,\tilde{\mathbf{y}}_1}$ with $\Psi_{\tilde{\mathbf{o}}_2,\tilde{\mathbf{y}}_2}$ and $\mathcal{P}_{\tilde{\mathbf{o}}_1}$ with $\mathcal{P}_{\tilde{\mathbf{o}}_2}$. When Σ_1 and Σ_2 are known, we can construct the representations $R_{\Sigma_i,C}$ and $R_{\Sigma_i,D}$, $i = 1, 2$. Then, we can use $R_{\Sigma_i,C}$, $i = 1, 2$ to compute the distance between $\Psi_{\tilde{\mathbf{o}}_1,\tilde{\mathbf{y}}_1}$ and $\Psi_{\tilde{\mathbf{o}}_2,\tilde{\mathbf{y}}_2}$. Likewise, we can use $R_{\Sigma_i,D}$, $i = 1, 2$ to compute the distance between $\mathcal{P}_{\tilde{\mathbf{o}}_1}$ and $\mathcal{P}_{\tilde{\mathbf{o}}_2}$. The advantage of using distances on formal power series is even more apparent if Σ_1 and Σ_2 are unknown, because the identification of stochastic jump-linear systems is poorly developed.[2] Instead, one could use the estimates of finitely many values of $\mathcal{C}_{\tilde{\mathbf{o}}_i,\tilde{\mathbf{y}}_i}$ and $\mathcal{P}_{\tilde{\mathbf{o}}_i}$, $i = 1, 2$ to compute the minimal representations $R_{C,i}$ of $\Psi_{\tilde{\mathbf{o}}_i,\tilde{\mathbf{y}}_i}$, $i = 1, 2$ and $R_{D,i}$ of $\mathcal{P}_{\tilde{\mathbf{o}}_i}$, $i = 1, 2$, and use the computed representations to compare the behavior of the two systems. The procedure for computing such representations from their Hankel matrices is known [1,15,16] and it is likely to be computationally less costly than identifying the original jump-linear systems.

5 Discussion and Conclusion

In this paper several definitions of distances for rational formal power series and rational representations were presented. It was argued that the results can be used to define metrics and topology on the space of a wide variety of dynamical systems. The key argument is that for many classes of dynamical systems there is a correspondence between the input-output behaviors of the systems and rational formal power series. In particular, this is the case for a number of hybrid systems and some nonlinear systems.

To the best of our knowledge, the problem of distances between hybrid systems had not been addressed so far. In the case of nonlinear systems, there are some results on the topological and geometric structure of the space of bilinear systems, see for example [16], where the algebraic variety structure of that space was described. In contrast, there is a fair amount of literature on distances between linear systems and on the topological and geometric structure of the space of linear systems. Note the relationship between input-output maps and output processes of linear systems and families of formal power series over the one letter alphabet $X = \{z\}$. Because of this correspondence, any distance on rational families of formal power series will give us a distance between linear

[2] Even in the linear case, the full identification procedure for linear stochastic systems is computationally costly.

systems. Spaces of equivalence classes of minimal linear systems were already studied before, for both the stochastic and deterministic settings. In fact, it was shown in [7] that, for each N, the set of all equivalence classes $M_N^{m,p}$ of minimal linear systems of dimension N with m inputs and p outputs forms both an analytic manifold and an algebraic variety and admits a natural topology. Here two minimal linear systems belong to the same equivalence class if they are algebraically similar. Denote by $M_N^{m,p,a}$ the set of equivalence classes of stable minimal linear systems. Then it was shown in [8,6] that $M_N^{m,p,a}$ and $M_N^{m,p}$ are diffeomorphic as analytic manifolds and the topology of $M_N^{m,p,a}$ as an analytic manifold can be obtained by the metric induced by the H_2 norm. It is easy to see that the distance induced by the H_2 norm is a particular case of the distance induced by the norm (9), if we identify equivalence classes of minimal linear systems with equivalence classes of minimal rational representations (two minimal representations are equivalent, if they are isomorphic). More recent papers on distances between stochastic linear systems can be found in [20,21,12]. In particular, [12] introduces the trace distance between linear systems and gives a formula to compute it. Surprisingly, the distance induced by the norm (9) is closely related to the trace distance.

In practical situations, the families of formal power series are likely to encode the external behaviors of some dynamical systems. In such cases, the available information is either a rational representation of each family of formal power series, or a finite collection of values of the formal power series.

If we are given a rational representation R_Ψ of each family Ψ, then any of the distances described in Section 3 can easily be computed. Notice that if Ψ is square summable, then we can minimize R_Ψ and the obtained minimal representation of $R_{m,\Psi}$ will be stable, due to Theorem 3. Hence, using Theorem 5 we can compute the distance induced by $< \cdot, \cdot >_J$ by solving the corresponding Sylvester equation from Theorem 5. Hence, except for the computational complexity, there are few restrictions on using any of the distances. Therefore, one may choose different distances depending on the particular application and the computational costs. Note that the issue of computational complexity is still open for the distances we presented. However, the computational cost of computing distances of type $d_{F,N,J}$ between formal power series is exponential in N, if the underlying alphabet X has more than one element.

If only the finite collection of values is available then the task of choosing the right distance is more complex. Assume that we know the values of the elements of the families for all words of length at most N. Then there are two cases to be considered. If $N \geq 2M + 1$ and $M = \text{rank}\, H_\Psi$ for all families Ψ involved, then we may apply the algorithm described in Theorem 2 to compute a minimal representation R_Ψ of Ψ. If Ψ is a square summable family, then the resulting representation R_Ψ is stable. Even in this ideal case when N is big enough several issues require attention. First of all, computing R_Ψ might be computationally expensive. If we want to compute one of the distances $d_{F,M,J}$, $M \leq N$, then we might do better by using the data directly, rather than computing the representations first and then computing the distance from the representations. However, computing the representations R_Ψ might be a good idea if we want to use the distances induced by the norm (9). Moreover, if the data are noisy, we do not know whether our algorithm will still produce a stable representation, which is a prerequisite for the existence of the solution of the Sylvester equation. If we cannot ensure that N is big enough, then the algorithm from Theorem 2 might fail to produce a representation of Ψ.

Acknowledgements. This work was supported by grants NSF EHS-05-09101, NSF CAREER IIS-04-47739, and ONR N00014-05-1083.

References

1. Petreczky, M.: Realization Theory of Hybrid Systems. PhD thesis, Vrije Universiteit, Amsterdam (2006) available online: `http://www.cis.jhu.edu/~mihaly`.
2. Isidori, A.: Nonlinear Control Systems. Springer Verlag (1989)
3. Sontag, E.D.: Realization theory of discrete-time nonlinear systems: Part I – the bounded case. IEEE Transaction on Circuits and Systems **26**(4) (1979)
4. Ball, J.A., Groenewald, G., Malakorn, T.: Structured noncommutative multidimensional linear systems. In: Proceedings Sixteenth International Symposium on Mathematical Theory of Networks and Systems. (2004)
5. Anderson, B.D.O.: The realization problem for hidden Markov models. Mathematical Control of Signals and Systems **12** (1999) 80–120
6. Peeters, R.: System Identification Based on Riemannian Geometry: Theory and Algorithms. PhD thesis, Free University, Amsterdam (1994)
7. Hazewinkel, M.: Moduli and canonical forms for linear dynamical systems II: The topological case. Mathematical Systems Theory **10** (1977) 363–385
8. Hanzon, B.: On the differentiable manifold of fixed order stable linear systems. Systems and Control Letters **13** (1989) 345–352
9. Caines, P.: Linear Stochastic Systems. John Wiley and Sons, New-York (1988)
10. Bissacco, A., Chiuso, A., Ma, Y., Soatto, S.: Recognition of human gaits. In: IEEE Conference on Computer Vision and Pattern Recognition. Volume 2. (2001) 52–58
11. Doretto, G., Chiuso, A., Wu, Y., Soatto, S.: Dynamic textures. International Journal of Computer Vision **51**(2) (2003) 91–109
12. Vishwanathan, S., Smola, A., Vidal, R.: Binet-Cauchy kernels on dynamical systems and its application to the analysis of dynamic scenes. International Journal of Computer Vision (2006)
13. Bissacco, A., Chiuso, A., Soatto, S.: Classification and recognition of dynamical models. Technical Report CSD-TR 060020, UCLA (2006)
14. Berstel, J., Reutenauer, C.: Rational series and their languages. EATCS Monographs on Theoretical Computer Science, Springer-Verlag (1984)
15. Sontag, E.D.: Polynomial Response Maps. Volume 13 of Lecture Notes in Control and Information Sciences. Springer Verlag (1979)
16. Sontag, D.E.: A remark on bilinear systems and moduli spaces of instantons. Systems and Control Letters **9**(5) (1987) 361–367
17. Callier, M.F., Desoer, A.C.: Linear System Theory. Springer-Verlag (1991)
18. Desai, U.: Realization of bilinear stochastic systems. IEEE Transactions on Automatic Control **31**(2) (1986)
19. Costa, O., Fragoso, M., Marques, R.: Discrete-Time Markov Jump Linear Systems. Springer-Verlag, London (2005)
20. Martin, A.: A metric for ARMA processes. IEEE Trans. on Signal Processing **48**(4) (2000) 1164–1170
21. Cock, K.D., Moor, B.D.: Subspace angles and distances between ARMA models. System and Control Letters **46**(4) (2002) 265–270

The Image Computation Problem in Hybrid Systems Model Checking*

André Platzer[1] and Edmund M. Clarke[2]

[1] University of Oldenburg, Department of Computing Science, Germany
platzer@informatik.uni-oldenburg.de
[2] Carnegie Mellon University, Computer Science Department, Pittsburgh, PA
emc@cs.cmu.edu

Abstract. In this paper, we analyze limits of approximation techniques for (non-linear) continuous image computation in model checking hybrid systems. In particular, we show that even a single step of continuous image computation is not semidecidable numerically even for a very restricted class of functions. Moreover, we show that symbolic insight about derivative bounds provides sufficient additional information for approximation refinement model checking. Finally, we prove that purely numerical algorithms can perform continuous image computation with arbitrarily high probability. Using these results, we analyze the prerequisites for a safe operation of the *roundabout maneuver* in air traffic collision avoidance.

Keywords: model checking, hybrid systems, image computation.

1 Introduction

The fundamental operation in model checking [1] is *image computation*, i.e., determining the set of states reachable from some (initial) set of states by following all transitions of the system. Verifying safety amounts to checking whether a bad state can be reached by repeating image computation from the initial states until convergence or a bound is reached. Today, the primary challenge for verification of industrial hybrid systems is to improve (a) scalability by building model checkers that are able to deal with higher-dimensional continuous state-spaces, and (b) modeling capabilities by providing verification techniques for systems having richer continuous dynamics. In this paper, we focus on (b) using approximation techniques and delineate the borderline of decidability of the image computation

* This research was sponsored by a fellowship of the German Academic Exchange Service (DAAD), by the German Research Council (DFG) under grant SFB/TR 14 AVACS, the National Science Foundation under grant nos. CNS-0411152, CCF-0429120, CCR-0121547, and CCR-0098072, the US Army Research Office under grant no. DAAD19-01-1-0485, and the Office of Naval Research under grant no. N00014-01-1-0796. The views and conclusions contained in this document are those of the author and should not be interpreted as representing the official policies, either expressed or implied, of any sponsoring institution, government or other entity.

A. Bemporad, A. Bicchi, and G. Buttazzo (Eds.): HSCC 2007, LNCS 4416, pp. 473–486, 2007.

problem for hybrid systems model checking. In particular, we show that even a *single step* of continuous image computation is not semidecidable.

In this paper we analyze techniques for approximating image computation for hybrid systems having non-linear continuous flows. First-order real arithmetic is among the most expressive theories for continuous values that is known to be decidable [2]. It is used successfully in hybrid system verification [3,4,5,6]. We thus investigate approximations of system flows using real arithmetic, including polynomial and spline approximations. For verification, we argue that uniform approximations, i.e., approximations with a uniform global error bound, are crucial, since verification is about making sure that the system is well-behaved even in worst-case scenarios. For this, we analyze the conditions required to guarantee that uniform approximations of flows can be constructed computationally. In addition, we study approximation-based model checking for hybrid systems with flows that are given implicitly as numerical solutions of differential equations.

Throughout the paper, we observe that numerical algorithms need additional knowledge about the system behavior to be successful in model checking. We show a strong undecidability result about the purely numerical treatment of even the basic operation of image computation in hybrid systems to support this observation.

The distinguishing feature of *numerical algorithms* in this context is that they compute their output with specific real values or rational approximations like 1.421. In contrast, *symbolic algorithms* are capable of computing with symbolic terms like $x^2 + 2xy$ that involve variable symbols to obtain results that are valid for *all* instantiations of x and y with real values. However, all terms that occur during the symbolic computation need to have a common representation that is effective. Further, numerical computations are generally more scalable to higher dimensions. See [7] for details on machine models for numerical computations; see [8] for symbolic computation and symbolic representations.

Model checking depends on image computation of *sets* of states. As they operate on concrete values, numerical algorithms can only compute images at a finite number of individual points in bounded time. Thus, the primary challenge in using numerical methods for verification is caused by the need for such a finite mesh of points on which solutions are computed numerically. This imposes two primary causes for errors: (a) there is only limited knowledge about the behavior in between the finite mesh, and (b) the numerical computations themselves introduce errors. For proper verification, these errors have to be controlled computationally to make sure the system is safe under *all* circumstances.

While the certainty required for verification is impossible to obtain by numerical means alone, we additionally show that numerical methods can provide a stochastic understanding of system safety. The probability of a wrong verification result can be made arbitrarily small under fairly mild assumptions.

We use our techniques to obtain results about *roundabout maneuvers for collision avoidance* in air traffic management (ATM) [9,10]. We show that a classical collision avoidance maneuver is unsafe for more realistic model assumptions. To overcome this limitation, we propose a modified roundabout maneuver that uses adaptive flight paths following a tangential geometric construction. Since the image computation techniques presented in this paper are suitable for automation, they have impact on improving verification tools like HyTech [11], CheckMate [12], or PHAVer [13] to cover more complicated dynamics. Supporting more general dynamics is important for verifying hybrid systems, for instance, in ATM [9, 10, 14] and for systems biology [6].

Structure of this Paper. After giving the basics of model checking in Sect. 2, we present the roundabout maneuver in Sect. 3. In Sect. 4, we present the framework for approximation refinement model checking. We analyze flow approximation techniques in Sect. 5. In addition, we cover flows that are specified implicitly as solutions of differential equations in Sect. 6. Experimental results of our preliminary model checker for *roundabout maneuvers* are presented in Sect. 7. Related work is discussed in Sect. 8.

2 Preliminaries

For model checking to be effective, both representing sets of states and computing images of sets of states under transitions have to be computable. Hybrid systems have two kinds of transitions: discrete jumps in the state space caused by mode switches, and continuous evolution along flows within a mode; see [6, 11].

Definition 1 (Hybrid Automata). *A* hybrid automaton *A consists of*

- *a continuous state space* $\mathbf{R}^n$*;*
- *a directed graph with vertices* Q *(as modes) and edges* E *(control switches);*
- flows φ_v*, where* $\varphi_v(t; x) \in \mathbf{R}^n$ *is the state reached after staying in mode* v *for time* $t \geq 0$ *when continuous evolution starts in state* $x \in \mathbf{R}^n$*;*
- *invariant conditions* $inv_v \subseteq \mathbf{R}^n$ *for* $v \in Q$*;*
- *jump relations* $jump_e \subseteq \mathbf{R}^n \times \mathbf{R}^n$ *for edges* $e \in E$*;*

where $jump_e$ *and* inv_v *are definable in first-order real arithmetic [2]. Typically, the jump relation* $jump_e$ *contains transition guards and variable resets as in [6].*

To simplify the formal machinery, we define the semantics of hybrid automata in terms of image computation (see, e.g. [6, 15, 11] for details on the relationship to trace semantics). Numerical algorithms typically work within a compact domain. For simplicity, we assume that all flows share the same domain of relevance $D \subseteq \mathbf{R} \times \mathbf{R}^n$, which comprises all relevant states and observation times. In (1) of Fig. 1, the *post-image* for automaton A is defined in terms of its discrete and continuous transitions: $Post_A(Y)$ is the set of states reachable from $Y \subseteq Q \times \mathbf{R}^n$ in one step. The post-image under the continuous flow φ_v restricted to D is defined in (2). For discrete jumps along edge $e \in E$ from $v \in Q$ to $w \in Q$, the post-image is defined in (3). Reachability in an arbitrary number of steps is defined by the least fixpoint equation (4). The *pre-image* $Pre_A(Y)$ is defined accordingly. Model checking reachability of bad states $B \subseteq Q \times \mathbf{R}^n$ from the initial set of states $I \subseteq Q \times \mathbf{R}^n$ amounts to checking emptiness of $Post_A^*(I) \cap B$.

$$\begin{aligned}
Post_A(Y) &:= \bigcup_{v\in Q} Post_{\varphi_v|_D}(Y) \cup \bigcup_{e\in E} Post_{jump_e}(Y) & (1)\\
Post_{\varphi_v|_D}(Y) &:= \{(v,\varphi_v(t;x)) \in Q\times\mathbf{R}^n : (v,x)\in Y, (t,x)\in D \text{ for some } t\geq 0 \\
&\quad \text{and } \varphi_v(t';x)\in inv_v \text{ for all } 0\leq t'\leq t\} & (2)\\
Post_{jump_e}(Y) &:= \{(w,y)\in Q\times\mathbf{R}^n \ :\ (x,y)\in jump_e \text{ for some } (v,x)\in Y \\
&\quad \text{and } y\in inv_w \text{ where } e=(v,w)\} & (3)\\
Post^*_A(Y) &:= \mu Z.(Y\cup Z\cup Post_A(Z)) & (4)
\end{aligned}$$

Fig. 1. Image computation semantics of hybrid automata

3 Air Traffic Management

Tomlin et al. [9] presented conflict resolution protocols for air traffic management, which direct two airplanes flying too close to each other to perform collision avoidance maneuvers. Assuming, for simplicity, aircraft remain at the same altitude, a configuration can be described in the special Euclidean group of $\mathbf{R}^2$ [9] and relative coordinates can be used to reduce the state-space dimension. The relative position of aircraft 2 with aircraft 1 at the origin is represented by its (planar) position x, y and orientation ϕ; see Fig. 2a. With linear velocities v_i and angular velocities ω_i (in radians per time unit) of the respective aircraft i, the in-flight dynamics in relative coordinates are as follows (see [9] for details):

$$\dot{x} = -v_1 + v_2\cos\phi + \omega_1 y \qquad \dot{y} = v_2\sin\phi - \omega_1 x \qquad \dot{\phi} = \omega_2 - \omega_1 \ . \tag{5}$$

A configuration is unsafe if there is another aircraft within a 5mi-radius protected zone, i.e., $x^2 + y^2 < 5^2$.

Straight line protocols [9, 10] for collision avoidance are unrealistic. Between straight lines, they assume instant turns, which are impossible in mid-flight. As a more realistic model, we investigate roundabout maneuvers [9], which also contain proper flight curves with $\omega_i \neq 0$, see Fig. 2b. The roundabout maneuver refines several instant turns to realistic curves with more complicated dynamics. For this refinement, we show that the standard maneuvers are unsafe.

Fig. 2c contains the hybrid automaton for *roundabout collision avoidance*, which generalizes the protocols in [9, 10, 14]. This protocol initiates evasive actions when the distance drops to α. The clock c determines when it is safe to turn back into the original direction after a half turn of duration $\frac{\pi}{\omega}$. For a concise presentation, (5)$[\omega_i := s]$ is an abbreviation for the dynamics of equation (5), with ω_1 and ω_2 replaced by s. Further, $\mathsf{rot}[\theta_1, \theta_2]$ denotes the action of the first aircraft turning by θ_1 and the second by θ_2, simultaneously. Typically, the θ_i are chosen as fixed values like $\theta = \pi/2$ [14]. We use $[-r, r]^2 \times [0, 2\pi]$ with $r = \alpha + 8$ as the relevant domain for states (x, y, ϕ) and choose observation times in $[0, 400]$. By continuity, other safety-relevant trajectories trespass a point in D.

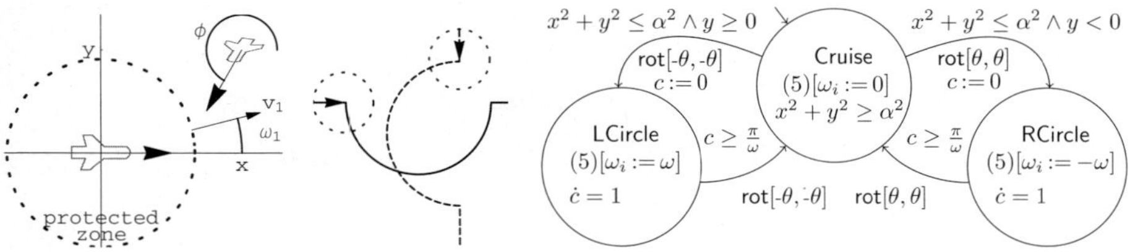

(a) Configuration. (b) Maneuver. (c) Roundabout maneuver automaton.

Fig. 2. Roundabout collision avoidance maneuver

4 Approximation in Hybrid Systems Model Checking

In this section, we provide the theoretical foundations for flow approximation in model checking hybrid systems and outline the approximation refinement model checking algorithm. Model checking depends on image computation of *sets* of states, which is particularly crucial for infinite-state systems. Yet, computing the image of a set under complicated flows is not possible in general. Hence, our guiding principle is to first approximate complicated dynamics using simpler flows (Sect. 4.2) and then compute images of sets under simple flows (Sect. 4.3).

4.1 Approximation Refinement Model Checking

For approximate set operations, we define the distance between sets $X \subseteq \mathbf{R}^n$ and $Y \subseteq \mathbf{R}^n$ as $d(X,Y) := \inf_{x\in X, y\in Y} \|x-y\|$, and $d(x,Y) := d(\{x\},Y)$ for a point $x \in \mathbf{R}^n$. Further, for an $\epsilon > 0$, let $\mathcal{U}_\epsilon(Y) := \{x \in \mathbf{R}^n : d(x,Y) < \epsilon\}$ be the ϵ-*neighborhood* of $Y \subseteq \mathbf{R}^n$. For convenience, we define $\mathcal{U}_0(Y) := Y$. Finally, let $S[x,y] \subseteq \mathbf{R}^n$ be the *line segment* connecting $x \in \mathbf{R}^n$ and $y \in \mathbf{R}^n$.

Image computation for discrete transitions is as usual in model checking [1]. Hence, we focus on a treatment of continuous evolutions that combines well with techniques for handling discrete image computation. Since the mode does not change during a continuous flow, we drop modes from $Post_{\varphi|_D}(Y)$.

We handle complicated dynamics by approximating flows and we conservatively over-approximate the resulting images. For an approximation of error $\le\epsilon$, safety proofs require that all states reachable in this approximation have a distance $>\epsilon$ to B. This is captured formally in the following decision problem.

Problem 1 (Approximate reachability in image computation). Given an *arbitrarily effective* function $\varphi \in C^k(D \subseteq \mathbf{R}^n, \mathbf{R}^m)$, i.e., for rational input x, the value $\varphi(x)$ can be computed up to arbitrary precision, and given effective representations of $B \subseteq \mathbf{R}^m$ and of a compact closure D of an open set, decide the following problem with tolerance $\epsilon \ge 0$: "$\mathcal{U}_\epsilon(Post_{\varphi|_D}(Y)) \cap B = \emptyset$?"

Exact image computation is retained with $\epsilon = 0$. Extensions to $Post^*_A(Y)$ are defined inductively using approximate flow images $\mathcal{U}_\epsilon(Post_{\varphi|_D}(Y))$ in Fig. 1.

Safety of an approximation with tolerance ϵ implies safety of the actual system by monotonicity of image computation. If the over-approximation is unsafe,

```
choose initial δ > 0
while true do
    Ã := approx(δ,A); ϵ := errorbound(Ã)
    reachable := check(U_ϵ(Post*_Ã(I)) ∩ B ≠ ∅)
    if not reachable then
        return 'A is safe'
    else if ϵ ≪ 1
        return 'A is unsafe with fragility ϵ'
    else δ := δ/2
```

Fig. 3. Approximation Refinement Model Checking (AMC)

however, counterexamples can be *spurious*. This happens if the approximation is too coarse because the current guaranteed error bound, ϵ, is still too large and permits behavior that is impossible in reality. Hence, refining the approximation tolerance is necessary [16] until the system is (a) proven safe after closer analysis, or (b) the system is considered *fragile* [3,14] because it is unsafe for a sufficiently small value of ϵ (below the stability advised by general engineering principles). An approximation refinement algorithm (AMC) exploiting those circumstances for Problem 1 is depicted in Fig. 3. It is parametric in a procedure approx for approximating the flows of the hybrid automaton A with a means to determine a uniform error bound. Techniques for this will be examined in Sect. 5–6 using the theory in Sect. 4.2. AMC further depends on the ability to check reachability by image computation in the approximation $\tilde{A}$, which we investigate in Sect. 4.3.

In order to support approximations with posterior error bound reporting, our algorithm distinguishes the refinement tolerance δ from the resulting error bound ϵ. The required assumption to ensure convergence is that ϵ decreases with δ and converges to zero when δ does. Modes can be split into modes that apply for different subregions by partitioning D (using the techniques in [13]) to keep refinements of δ local to smaller parts of the state space. As a further improvement, it is simple to extend AMC to stop if a counterexample has been found that reaches a bad state with a distance $>\epsilon$ to good states (beyond the approximation error). In that case, the concrete system is unsafe without fragility.

4.2 Image Approximation

As a theoretical framework for flow approximations in approx to solve Problem 1 with AMC, we present the following result. It shows that continuous flows support uniform approximation of images with polynomials on compact domains.

Proposition 1 (Weierstraßian flows). *Let $\varphi \in C(D, \mathbf{R}^n)$ on a compact closure $D \subset \mathbf{R}\times\mathbf{R}^n$ of an open set. Then, $\forall \epsilon > 0\ \exists p \in \mathbf{R}[t, x_1, \ldots, x_n]^n\ \forall Y \subseteq \mathbf{R}^n$*

$$Post_{\varphi|_D}(Y) \subseteq \mathcal{U}_\epsilon(Post_{p|_D}(Y)) \tag{6}$$

$$Pre_{\varphi|_D}(Y) \subseteq Pre_{p|_D}(\mathcal{U}_\epsilon(Y)) \ . \tag{7}$$

Proof. For any $\epsilon > 0$, let p be a vector of polynomials approximating φ on D with uniform error $<\epsilon$ according to the generalized Weierstraß theorem [17]. Equation (6) is a consequence of the following representation (case (7) is similar):

$$\mathcal{U}_\epsilon(Post_{p|_D}(Y)) = \{z \in \mathbf{R}^n \ : \ \exists x \in Y \ \exists t \ (t,x) \in D\,,\ \|z - p(t,x)\| < \epsilon\} \ .$$

Let $z \in Post_{\varphi|_D}(Y)$, i.e., let $(t,x) \in D, x \in Y$ with $z = \varphi(t;x)$. The Weierstraß theorem implies $z \in \mathcal{U}_\epsilon(Post_{p|_D}(Y))$, as $\|z - p(t,x)\| = \|\varphi(t;x) - p(t,x)\| < \epsilon$.

This result shows that image computation can be split into approx, i.e., finding a uniform approximation p of φ that satisfies (7), and check, i.e., computing the right-hand side of (7). Further, it proves the existence of an approximation p.

4.3 Polynomial Image Computation and Beyond

In this section, we present classes of flows that support exact image computation of sets of states for the procedure check. These are adequate choices for functions with which approx can approximate more complicated dynamics. Beyond polynomial flows, we generalize exact image computation to piecewise polynomials and rational functions—in particular to multivariate rational splines.

Proposition 2 (Decidability of polynomial image computation). *Given definable Y and D, the right-hand sides of* (6) *and* (7) *in Proposition 1 are definable in first-order real arithmetic, hence decidable by Tarski's theorem [2].*

Proof. Let F_D and F_Y define D and Y, respectively. Then, $z \in \mathcal{U}_\epsilon(Post_{p|_D}(Y))$ is definable by: $\exists x \, \exists t \, (F_Y(x) \wedge F_D(t,x) \wedge \|z - p(t,x)\| < \epsilon)$. As the square function increases strictly monotonically on $[0, \infty)$ and $\epsilon \geq 0$, the Euclidean norm can in turn be defined by: $\|z\| < \epsilon \equiv \sum_{i=1}^{n} z_i^2 < \epsilon^2$. With this, we can implement check.

Proposition 3. *Piecewise polynomials are definable in first-order arithmetic.*

Proof. Let $s : D \to \mathbf{R}$ be a function consisting of polynomial pieces $P_i : D_i \to \mathbf{R}$ for disjoint domains D_i with $D = D_1 \cup \cdots \cup D_n$ that are definable in first-order real arithmetic. Then, the following equivalence defines the piecewise function s:

$$s(x) = t \quad \equiv \quad \bigvee_{i=1}^{n} (x \in D_i \wedge p_i(x) = t) \ .$$

The image computation corresponding to (6) follows from the decomposition

$$Post_{s|_D}(Y) \ = \ \bigcup_{i=1}^{n} Post_{p_i|_{D_i}}(Y) \quad \text{and} \quad \mathcal{U}_\epsilon(X \cup Y) \ = \ \mathcal{U}_\epsilon(X) \cup \mathcal{U}_\epsilon(Y) \ . \tag{8}$$

Due to their piecewise definitions, splines provide a better approximation with lower degree than polynomials do. Hence, we propose to use splines for image computation, and solve a multitude of simpler polynomial problems as opposed to using a single high-degree polynomial problem. For this, splines in (8) split into a disjoint set of polynomial reachability problems of lower degree. For a result on uniform approximation with multivariate splines, we refer to [18, 19]. Even rational approximations can be used, but AMC does not yet apply them:

Proposition 4. *Tarski's theorem [2] can be extended from semialgebraic sets formed with polynomials over real-closed fields to rational functions.*

Proof. In first-order formulas of real arithmetic with rational expressions, the following equivalences reduce rational (in-)equalities to polynomial formulas:

$$\begin{aligned} p(x)/q(x) = 0 &\equiv p(x) = 0 \land q(x) \neq 0 \\ p(x)/q(x) > 0 &\equiv (p(x) > 0 \land q(x) > 0) \lor (p(x) < 0 \land q(x) < 0) \ . \end{aligned}$$

By using the fact that the field of fractions of $\mathbf{Q}[X_1, \ldots, X_n]$ is a field, all atomic formulas can be reduced to one of the above forms.

5 Flow Approximation

In this section, we analyze which flows can be approximated effectively. In addition to giving an approximation result for bounded flows, we identify the limits of numerical methods for approximating hybrid systems with the certainty that is needed for verification. Further, we show that numerical methods can give sufficient justification of verification in stochastic terms up to arbitrary probability. Throughout the section we assume φ is a flow of a mode of a hybrid system.

Using the results presented so far, we can reduce Problem 1 to the following problem for approx, for which Proposition 1 guarantees the existence of solutions.

Problem 2 (Uniform approximation). Given an arbitrarily effective continuous function $\varphi \in C(D, \mathbf{R}^n)$ on a compact closure $D \subset \mathbf{R} \times \mathbf{R}^n$ of an open set, with an effective representation of D, find an approximation of φ with multivariate splines of uniform error $<\epsilon$.

5.1 Bounded Flow Approximation

In order to turn the theoretical existence result of Proposition 1 into an algorithm approx that solves Problem 2, we need an effective form of Weierstraß approximation. The following result shows that solutions of Problem 2 can be computed effectively when derivatives $\dot{\varphi}$ are continuous and have a known bound.

Proposition 5 (Effective Weierstraß approximation). *If* $\varphi \in C^1(D, \mathbf{R}^n)$ *and* $b := \max_{x \in D} \|\dot{\varphi}(x)\|$ *are given, then Problem 2 is computable.*

Proof. Using component-wise approximation and norm properties, we can assume the range of φ is in $\mathbf{R}^1$ rather than $\mathbf{R}^n$. Let $\epsilon > 0, x \in D$. Further, we can assume D is connected (otherwise the problem can be treated separately on each connected component). By premise, φ is arbitrarily effective, i.e., for each $\delta_c > 0$ there is an effective function f_{δ_c} such that for all $y \in D$: $\|\varphi(y) - f_{\delta_c}(y)\| < \delta_c$. Let x_i be a point on a δ_g-grid with distance $\|x - x_i\| < \delta_g$. We assume that $x_i \in D$ and D is convex on the grid cell around x_i. Due to convexity, the mean-value theorem applies and yields a $\xi \in S[x, x_i]$ such that

$$\|\varphi(x) - \varphi(x_i)\| = \|\dot{\varphi}(\xi)(x - x_i)\| = \|\dot{\varphi}(\xi)\| \cdot \|(x - x_i)\| < b\delta_g \ .$$

As φ is arbitrarily effective at the grid point x_i, this inequality implies

$$\|\varphi(x) - f_{\delta_c}(x_i)\| \leq \|\varphi(x) - \varphi(x_i)\| + \|\varphi(x_i) - f_{\delta_c}(x_i)\| < b\delta_g + \delta_c \ .$$

Thus, φ can be approximated by step functions up to precision $b\delta_g + \delta_c$, which can be chosen $<\epsilon$. Such step functions are defined as $f_{\delta_c}(x_i)$ on the $\pm\delta_g/2$ hypercube around x_i (or sufficiently close rational approximations thereof). As step functions are piecewise polynomials there is no need to prove that step functions can be approximated by polynomials (cf. Proposition 3).

5.2 Continuous Image Computation

In this section, we demonstrate a fundamental limitation of numerical approaches to verification of hybrid systems. Despite the fact that Proposition 1 guarantees the existence of a uniform polynomial approximation, effectively constructing such an approximation using numerical computations is impossible without additional symbolic techniques. More generally, we show that even a single step of continuous image computation is not semidecidable using numerical evaluations.

As they require concrete values, numerical algorithms can only evaluate the input function φ *at individual points* but do not have access to its symbolic representation. Even evaluating derivatives of φ at points is not sufficient to obtain decidability:

Proposition 6 (Undecidability of image computation). *Problem 1 is not semidecidable using numerical evaluation of derivatives $\varphi^{(j)}(x)$ for $j \geq 0$ at individual points, even for arbitrarily large tolerable errors $\epsilon > 0$ and arbitrary high degrees of derivatives. This remains true even for smooth functions where all derivatives are effectively known, and when functions are restricted to one-dimensional (effective) smooth polynomial functions with rational coefficients.*

Proof. In Problem 1, choose $n = m = 1$, $D = [0, 1]$, $B = [\epsilon, \infty)$ for the tolerable error $\epsilon > 0$. Assume there is an algorithm $\mathcal{A}$, which solves Problem 1 for this case. Choose a function φ with $\varphi(D) \cap B = \emptyset$, say $\varphi = 0$. Running $\mathcal{A}$ with input φ yields correct output "$=\emptyset$", since $\varphi(x) = 0 < \epsilon$. Tracing the run identifies the set of all points x_i at which $\mathcal{A}$ evaluates at least one of the $\varphi^{(j)}$. Although the set of all x_i is unbounded, it is finite after termination, since $\mathcal{A}$ can only make a finite number of computation steps in a bounded interval of time. After termination, the maximum j where a $\varphi^{(j)}(x_i)$ has been evaluated by $\mathcal{A}$ is finite as well.

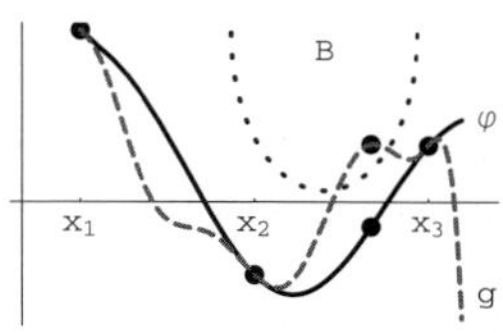

Fig. 4. Indistinguishable

Now let $0 < \delta < \min_{i \neq j} \|x_i - x_j\|$, and assume x_2 is not the right-most point, hence $x_2 + \delta \in D$ (otherwise reorder). However, by Hermite interpolation, there is an (effective) polynomial function $g \in C^k(D, \mathbf{R})$ with $g^{(j)}(x_i) = \varphi^{(j)}(x_i) = 0$ and $g(x_2 + \delta) = 2\epsilon > \epsilon$ but $g(D) \cap B \neq \emptyset$. Since φ and g are indistinguishable by the $\varphi^{(j)}(x_i)$ that $\mathcal{A}$ asked about φ, the hypothetical algorithm $\mathcal{A}$ would yield the same output for φ

and g, one of which is wrong. Fig. 4 depicts this situation with a more general choice of φ and B that gives better graphics. Moreover, Turing machines choose $x_i \in \mathbf{Q}$, from which $g \in \mathbf{Q}[X]$ can be concluded.

The proof principles of Proposition 6 are highly general and apply for all machine models that only allow a finite number of evaluations of input function φ (and derivatives) at individual points in bounded time. This includes the generalization of "numerical" Turing machines for real values by Blum et al. [7].

As a simple corollary, the same undecidability results apply for Problem 2 using the reduction in Sect. 4. In particular, this shows that the mere presence of a bound is not sufficient if the bound b is not known for Problem 2.

5.3 Probabilistic Model Checking

While Proposition 6 shows that image computation is not semidecidable even in quite robust scenarios with large tolerable errors, increasing the number of points x_i where φ (or its derivatives) are evaluated increases the constraints on the counterexample g, hence—intuitively speaking—increases the likelihood of the reachability problem being answered correctly (when assuming a non-degenerate probability distribution P). If arbitrarily large derivatives are unlikely by system design, model checking algorithms based on purely numerical information can provide stochastic certainty of verification. In that case, the following result shows that such algorithms can perform image computation with arbitrarily high probability by evaluating φ on a sufficiently dense grid.

Proposition 7 (Stochastic model checking). *If $P(\|\dot{\varphi}\|_\infty > b) \to 0$ when the bound $b \to \infty$, and if D is an open set, then any evaluation of φ on a finite set of points $G \subseteq D$ obtains sufficient information to decide Problem 1 correctly with probability $p \to 1$ as $\|d(\cdot, G)\|_\infty \to 0$.*[1]

Proof. Let (φ, D, B) be a problem instance with tolerance $\epsilon > 0$. Let $G \subseteq D$ be the set of points where φ is evaluated and $\nu := \|d(\cdot, G)\|_\infty$. If $\varphi(x_i) \in \mathcal{U}_\epsilon(B)$ for some $x_i \in G$, the output "$\neq\emptyset$" is correct with tolerance ϵ. Otherwise, we show that the probability of the output "$=\emptyset$" being wrong converges to zero for $\nu \to 0$. Suppose there is an $x \in D$ with $\varphi(x) \in B$. Let $x_i \in G$ have smallest distance to x. Then we can assume $S[x, x_i] \subseteq D$ (otherwise use a $\nu > 0$ such that $\mathcal{U}_\nu(x) \subseteq D$, which exists since D is open). Thus, by mean-value theorem, there is a $\xi \in S[x, x_i]$ such that

$$\epsilon \leq \|\varphi(x) - \varphi(x_i)\| = \|\dot{\varphi}(\xi)(x - x_i)\| = \|\dot{\varphi}(\xi)\| \cdot \|x - x_i\| \ . \tag{9}$$

The first inequality holds since $\varphi(x) \in B$ but $\varphi(x_i) \notin \mathcal{U}_\epsilon(B)$. Yet, $\nu \geq \|x - x_i\|$. Thus, dividing (9) by $\nu > 0$ leads to $\frac{\epsilon}{\nu} \leq \|\dot{\varphi}(\xi)\| \leq \|\dot{\varphi}\|_\infty$. But this becomes arbitrarily improbable when refining ν, because $P(\|\dot{\varphi}\|_\infty \geq \frac{\epsilon}{\nu}) \to 0$ for $\nu \to 0$ by premise, as ϵ is a constant independent of ν and $\frac{\epsilon}{\nu} \to \infty$ as $\nu \to 0$.

[1] This result also applies for a compact D by working (separately) on a finite open subcover. $\|d(\cdot, G)\|_\infty = \max_{x \in D} d(x, G)$ corresponds to the "density" of G in D.

6 Differential Flow Approximation

In this section, we investigate how the results of the previous sections can be extended when the flow φ of a mode in a hybrid system is not given to the model checker, but implicitly generated as a numerical solution of a differential equation. For verification, we have to control several sources of errors: (a) initial conditions between the points of the numerical mesh can lead to different behavior of the solutions, (b) observation times t off the mesh lead to interpolation errors, and (c) numerical computations introduce errors. Proposition 6 shows that we have to assume additional knowledge, e.g., a Lipschitz-constant. Note that the undecidability proof of Proposition 6 shows that it is *not* sufficient to assume Lipschitz-continuity without knowledge of the actual Lipschitz-constant.

Proposition 8. *Let $f \in C([a,b] \times \mathbf{R}^n, \mathbf{R}^n)$ be ℓ-Lipschitz-continuous in x, i.e., $\|f(t,x_1) - f(t,x_2)\| \leq \ell \|x_1 - x_2\|$ for all t, x_1, x_2. Then there is a computable set of points sufficient for solving Problem 1 numerically, where φ is a solution of the differential equation $\dot{x}(t) = f(t,x)$.*

Proof. Let $\epsilon > 0$. For t, x_0 let t_2, x_2 be the closest points on a mesh. Then the solution flow $\varphi(t;x_0)$ after time t, with initial value $\varphi(t_0;x_0) = x_0$, is arbitrarily close to the mesh values $\varphi(t_2;x_2)$, which can be approximated numerically:

$$\begin{aligned} \|\varphi(t;x_0) - \varphi(t_2;x_2)\| &\leq \|\varphi(t;x_0) - \varphi(t;x_2)\| + \|\varphi(t;x_2) - \varphi(t_2;x_2)\| \\ &\leq e^{\ell|t-t_0|}\|x_0 - x_2\| + \|\dot{\varphi}(\xi;x_2)\| \cdot |t - t_2| \\ &= e^{\ell|t-t_0|}\|x_0 - x_2\| + \|f(\xi, \varphi(\xi;x_2))\| \cdot |t - t_2| \end{aligned} \tag{10}$$

by a consequence of Picard-Lindelöf [20, theorem 7.1.4] and mean-value theorem with a ξ between t and t_2. Further, (10) can be bounded by any $\frac{\epsilon}{2} > 0$ by refining the mesh such that $\|x_0 - x_2\|$ and $|t - t_2|$ are sufficiently small, since the remaining factors are bounded on a compact domain in bounded time and f is Lipschitz-continuous. Moreover, by [20, theorem 7.2.2.3] there are "Lipschitz-continuous one-step methods of order p" (see [20]) that approximate the mesh quantity $\varphi(t_2;x_2)$ with a global discretization error that is bounded by $\frac{\epsilon}{2}$ when refining the mesh. The rate of convergence can be computed from the Lipschitz-constants and p (see [20] for details). Hence, the overall error is bounded by ϵ.

The most crucial influence on the error bound analysis comes from the exponential term in the proof of Proposition 8. Yet, this bound is tight in general: $\dot{x} = \ell x$ is ℓ-Lipschitz-continuous with unique global solution $\varphi(t;x_0) = x_0 e^{\ell t}$ for $t_0 = 0$, hence $\varphi(t;x_0) - \varphi(t;x_2) = e^{\ell t}(x_0 - x_2)$.

7 Experimental Results

Using the results presented in this paper, we have implemented a preliminary approximation refinement model checker for a class of hybrid systems. For a reasonable range of parameter choices (in particular for α, ω, θ), it always produces

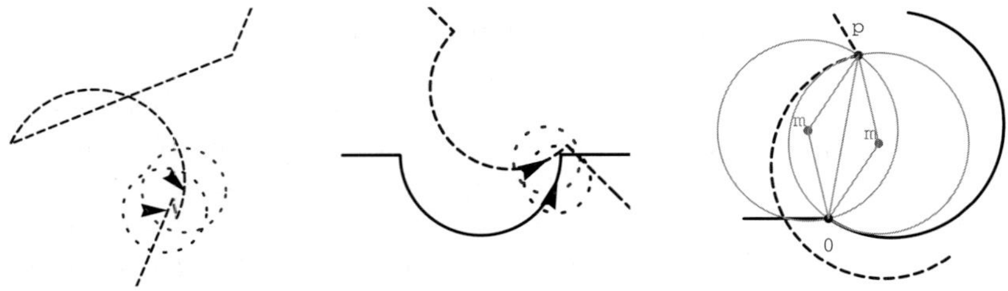

(a) Relative CEX. (b) Absolute CEX. (c) Tangential ATM.

Fig. 5. Counterexample flight and adaptive tangential construction

a counterexample to the safety property in Sect. 3 (with distances of ≈0.0016mi, the first after 3 mesh refinements). Fig. 5a contains a counterexample flight in relative coordinates with aircraft 1 fixed at the origin, 5b in absolute coordinates. This counterexample shows that the verification results in [14,10] for roundabout maneuvers starting from orthogonal flight paths do not extend to non-orthogonal initial flight paths. To maintain safe operation for general free flight, we propose the following modified roundabout maneuver with *adaptive tangential rotation.*

Instead of choosing a fixed rotation angle θ as in Sect. 3, we choose rotation angles θ_i for the individual aircraft depending on the current relative position $p = (x, y)$. Let m be the center of any circle of radius α through the plane positions 0 and p (cf. construction in Fig. 5c). Those (gray) circles correspond to worst-case evasive flight curves at maximum angular speed $\omega = v/\alpha$. Actual evasive actions use smaller ω (dark curves). Let γ_1 and γ_2 be the angles of the plane positions 0 and p, respectively, to m according to the following equation system:

$$\alpha^2 = \|m - 0\|^2 \qquad \alpha^2 = \|m - p\|^2 \qquad \gamma_1 = \angle(m - 0) \qquad \gamma_2 = \angle(m - p) \ . \qquad (11)$$

We define the angle $\angle(u)$ as the argument of the complex number $u_1 + u_2\imath$. Then, we choose rotation angles (θ_1, θ_2) as $(\gamma_1 - \pi/2, \gamma_2 - \pi/2)$ or $(\gamma_1 + \pi/2, \gamma_2 + \pi/2)$ with all solutions of (11) for γ_j. This rotates the aircraft tangentially to the gray circles such that the aircraft follow the dark curves in circle mode. Among the resulting choices for θ_j, we choose minimal turning angles. Thus, the primary change for the automaton in Fig. 2c is a position-dependent rotation $\mathsf{rot}[\theta_1, \theta_2]$ due to our construction (which happens before the check on $y \geq 0$).

8 Related Work

Several approaches [3,4,5,6] emphasize the importance of quantifier elimination in first-order real arithmetic for hybrid system verification. Our results thus use first-order real arithmetic for approximating more general dynamics.

Piazza et al. [6] propose Taylor series approximation of known flows. For applications in systems biology, they do not handle approximation errors.

Lanotte and Tini [15] propose a syntactic Taylor approximation of hybrid automata with known flows, modified by the maximum error. They use a complicated computation of error bounds from given Lipschitz-constants. Taylor approximations, though, have a non-uniform and more complicated error distribution, which makes them less useful for verification.

Tomlin et al. [9] derive results for the straight line ATM scenario using Hamilton-Jacobi-Isaacs partial differential equations. Our techniques avoid complicated PDEs and are thus more suitable for automatic model checking.

Massink and Francesco [10] investigate ATM using purely discrete linearized or untimed models. They primarily focus on the straight line protocol but also use coarse over-relaxations to investigate the roundabout maneuver. Massink and Francesco do not investigate the resulting error bounds.

Damm et al. [14] investigate model checking of LTL properties for discrete time robust hybrid systems using interval-constraint solving. They emphasize the importance of robustness in safety-critical control applications and show safety only for a *discrete* roundabout maneuver with orthogonal trajectories.

Asarin et al. [21] approximate non-linear differential equations by piecewise linear differential equations using interpolation. We propose non-linear polynomial and spline approximations of flows and investigate hybrid dynamics.

9 Conclusions and Future Work

We analyzed the image computation problem in hybrid systems model checking with a focus on approximation techniques for continuous dynamics. We presented a model checking algorithm that successively refines flow approximations. It approximates complicated dynamics using simpler flows (approx), and then computes images of sets of states under simple flows (check) taking into account error bounds. Flow approximations are refined when counterexamples are spurious.

Uniform polynomial approximations always exist for continuous functions on compact domains. Despite that, we have shown that the image computation problem for continuous flows is not semidecidable with numerical evaluations even for very restricted dynamics. With a priori knowledge about the system behavior, uniform approximation is effective. We have illustrated that such additional knowledge can either be obtained from information on bounds of flows or differential equations, or from stochastic information about likely system behavior. Definitely, numerical computations are invaluable for verification speed-up. Yet, for the mathematical rigor and certainty that is required in verification, they always have to be accompanied by symbolic analysis.

Additionally, we gave results for the *roundabout maneuver* in air traffic management using our preliminary model checker implementation. For free flight, we show that a classical maneuver is unsafe and propose a solution.

Future work includes improvements of our model checker. For the roundabout maneuver, we want to analyze situations arising from discrepancies in relative position recording of the aircraft, and extend our collision avoidance protocol to full curve dynamics using compositional verification. Finally, we want to investigate the impact of rational spline approximations for hybrid system verification.

References

1. Clarke, E.M., Grumberg, O., Peled, D.A.: Model Checking. MIT Press (1999)
2. Tarski, A.: A Decision Method for Elementary Algebra and Geometry. 2nd edn. University of California Press, Berkeley (1951)
3. Fränzle, M.: Analysis of hybrid systems. In Flum, J., Rodríguez-Artalejo, M., eds.: CSL. Volume 1683 of LNCS., Springer (1999) 126–140
4. Lafferriere, G., Pappas, G.J., Yovine, S.: A new class of decidable hybrid systems. In Vaandrager, F.W., van Schuppen, J.H., eds.: HSCC. Volume 1569 of LNCS., Springer (1999) 137–151
5. Anai, H., Weispfenning, V.: Reach set computations using real quantifier elimination. In Benedetto, M.D.D., Sangiovanni-Vincentelli, A.L., eds.: HSCC. Volume 2034 of LNCS., Springer (2001) 63–76
6. Piazza, C., Antoniotti, M., Mysore, V., Policriti, A., Winkler, F., Mishra, B.: Algorithmic algebraic model checking I: Challenges from systems biology. In Etessami, K., Rajamani, S.K., eds.: CAV. Volume 3576 of LNCS., Springer (2005)
7. Blum, L., Cucker, F., Shub, M., Smale, S.: Complexity and real computation. Springer New York, Inc., Secaucus, NJ, USA (1998)
8. Mora, T.: Solving Polynomial Equation Systems II. Cambridge Univ. Press (2005)
9. Tomlin, C., Pappas, G.J., Sastry, S.: Conflict resolution for air traffic management. IEEE Transactions on Automatic Control **43**(4) (1998) 509–521
10. Massink, M., Francesco, N.D.: Modelling free flight with collision avoidance. In: ICECCS, IEEE Computer Society (2001) 270–280
11. Alur, R., Henzinger, T.A., Ho, P.H.: Automatic symbolic verification of embedded systems. IEEE Trans. Software Eng. **22**(3) (1996) 181–201
12. Silva, B.I., Richeson, K., Krogh, B.H., Chutinan, A.: Modeling and verification of hybrid dynamical system using CheckMate. In: ADPM. (2000)
13. Frehse, G.: PHAVer: Algorithmic verification of hybrid systems past HyTech. [22]
14. Damm, W., Pinto, G., Ratschan, S.: Guaranteed termination in the verification of LTL properties of non-linear robust discrete time hybrid systems. In Peled, D., Tsay, Y.K., eds.: ATVA. Volume 3707 of LNCS., Springer (2005)
15. Lanotte, R., Tini, S.: Taylor approximation for hybrid systems. [22] 402–416
16. Clarke, E.M., Grumberg, O., Jha, S., Lu, Y., Veith, H.: Counterexample-guided abstraction refinement. In Emerson, E.A., Sistla, A.P., eds.: CAV. Volume 1855 of LNCS., Springer (2000) 154–169
17. Stone, M.H.: The generalised Weierstrass approximation theorem. Math Mag **21** (1948) 167–184 and 237–254
18. Bejancu, A.: The uniform convergence of multivariate natural splines. Technical Report NA1997/07, Applied Mathematics, Cambridge, UK (1997)
19. Wang, R.H.: Multivariate Spline Functions and Their Applications. Kluwer (2001)
20. Stoer, J., Bulirsch, R.: Introduction to Numerical Analysis. Springer, NY (2002)
21. Asarin, E., Dang, T., Girard, A.: Reachability analysis of nonlinear systems using conservative approximation. In Maler, O., Pnueli, A., eds.: HSCC. Volume 2623 of LNCS., Springer (2003) 20–35
22. Morari, M., Thiele, L., eds.: HSCC. In Morari, M., Thiele, L., eds.: HSCC. Volume 3414 of LNCS., Springer (2005)

A New Hybrid State Estimator for Systems with Limited Mode Changes*

Kaushik Roy[1] and Claire J. Tomlin[2]

[1] Stanford University
[2] Stanford University; University of California at Berkeley

Abstract. A new algorithm for hybrid state estimation, the K-Limited Mode-Change (KLMC) algorithm, is presented. Given noisy measurements, this algorithm estimates the continuous and discrete state histories for a class of hybrid systems that exhibit limited mode changes over time. The KLMC algorithm is compared to an existing hybrid state estimator, the Interacting Multiple Model (IMM), using a newly developed performance metric based on the concept of *probability of error*. Monte Carlo methods are used to obtain numerical estimates of the performance metric for simple hybrid system models. Simulation results show that KLMC outperforms IMM in terms of the estimate-error metric but requires larger storage and computational resource consumption.

Keywords: hybrid systems, hybrid state estimation, Monte Carlo.

1 Introduction

Stochastic hybrid system models consist of discrete modes, each with continuous states governed by a set of continuous dynamics. Various fields, including target tracking, fault detection, and systems biology, involve systems characterized by a set of modes that partition the continuous behavior of the system. Although continuous models for each discrete mode are often well known, discrete mode evolution is often dependent on the particular instance of the system. For example, for commercial aircraft tracking, continuous dynamics for flight modes, such as straight-and-level flight or constant-yaw-rate turns, are known to relatively high accuracy [1,2]. However, without a flight plan and real-time air traffic control updates, it is difficult to determine the flight mode of an aircraft [3].

Indeed, estimation and control of hybrid trajectories is highly dependent on accurate tracking of the discrete mode over time. Unfortunately, for a model with N modes and T time steps, there are N^T possible discrete trajectories, each with a different continuous state model. A variety of techniques have been developed to combat the intractability of exact modeling of discrete trajectories. The Interacting Multiple Model (IMM) [4,5,6], combined with the ideas of Identity-Mass-Flow [7] and Joint Probability Data Association [8,9], aggregates

* This research was supported by ONR under the CoMotion MURI contract N00014-02-1-0720, by NASA JUP under grant NAG2-1564, and by NASA grant NCC2-5536.

A. Bemporad, A. Bicchi, and G. Buttazzo (Eds.): HSCC 2007, LNCS 4416, pp. 487–500, 2007.

all discrete trajectories that pass through a given discrete mode as mass proportional to the likelihood of being in that mode. Continuous state estimation is then carried out using a bank of estimators, one for each discrete mode. This approach reduces the number of discrete trajectories to N. Other approaches, based on Multiple Hypothesis Testing [10] or extensions of IMM [11], maintain a set of trajectories that are aggregated or pruned according to Bayesian techniques. Previous work has also explored the comparison of different estimators to understand the effectiveness of any particular methodology [12,13].

An overview of the vast number of stochastic hybrid models is found in [14]. In this paper, a subset of these systems which exhibit limited mode changes over time is considered; this restriction allows for the possibility of tracking all possible discrete trajectories without pruning or aggregation. The K-Limited Mode-Change (KLMC) algorithm is a hybrid state estimator for such systems, tracking only those discrete trajectories possible for a system with K mode changes over a given time horizon. For each discrete trajectory, the continuous state evolution model is known and an estimator such as a Kalman filter is used to estimate the continuous state and a likelihood of this estimate, given a noisy measurement. These likelihoods are then used to weight the likelihood of the discrete trajectories. The process is repeated at each time step, building an estimate of the most likely hybrid trajectory. Though KLMC requires more data storage and longer run-times than IMM, the number of trajectories stored is polynomial of order K in the number of discrete modes. The upshot is that computing resources required by KLMC are feasible. In this paper, a metric based on the concept of probability of error is developed to quantify the relative performance of the two algorithms. Monte Carlo simulation is used to show the superior performance of the KLMC algorithm in several scenarios.

The paper is organized as follows. In the next section, hybrid systems and the problem of hybrid state estimation are introduced, and the KLMC algorithm is presented. Section 3 discusses a metric to measure the performance of a hybrid state estimation algorithm, as well as analytic results for simple systems. In Section 4, realistic scenarios from air traffic and other systems are simulated to show that the new algorithm outperforms existing algorithms such as IMM. Finally, conclusions and future work directions are presented in Section 5.

2 K-Limited Mode-Change (KLMC) Algorithm

In this section, hybrid system models and the problem of hybrid state estimation are first introduced. Next, an outline of the K-Limited Mode-Change algorithm is given, with a proof that the algorithm requires only polynomial storage. Finally, an example of a simple hybrid system is used to show the efficacy of the algorithm in capturing all possible behaviors of a system.

2.1 Hybrid System Description

Hybrid systems consist of a set of discrete states, such that for each discrete state, there exists a model description for the evolution of a set of continuous states. We restrict our analysis to systems with a finite set of N discrete states, denoted x_d. Without loss of generality, $x_d \in \{0, 1, ..., N-1\}$. Also, models are restricted to discrete-time systems. Similar definitions and arguments can be presented for continuous-time systems but are not covered in this paper. Finite-state, discrete-time hybrid systems are defined more formally as follows.

Definition 1. *Finite-state Discrete-time Hybrid System (FDHS). Consider a discrete-time system consisting of both discrete $x_d(k)$ and continuous states $x_c(k)$, where the time-variable $k \in \mathbb{Z}+$. There are assumed to be a finite set of N possible discrete states, or modes; that is, $x_d(k) \in \{0, 1, ..., N-1\}$, for all times k. The continuous state is assumed to be finite-dimensional ($x_c(k) \in \mathbb{R}^n$ for n finite). Continuous state evolution is dependent on the discrete state of the system; that is, there are N distinct continuous models that correspond to the N discrete modes. F_j is the continuous model corresponding to mode j, and therefore, $x_c(k+1) = F_j(x_c(k))$ for $x_d(k) = j$. The discrete state evolves according to a stochastic transition matrix $H(k)$, where $H(k)_{ij}$ represents the probability that $x_d(k+1) = i$ given that $x_d(k) = j$.*

The history of states is referred to as the *trajectory* of the system. That is, the discrete trajectory to time κ is denoted $X_d(\kappa) = (x_d(0), x_d(1), ..., x_d(\kappa))$. Similarly, the continuous trajectory is $X_c(\kappa) = (x_c(0), x_c(1), ..., x_c(\kappa))$. The hybrid trajectory is denoted $X(\kappa) = (X_d(\kappa); X_c(\kappa))$. Note that the models F_j may be stochastic or deterministic, linear or nonlinear, and include fixed or unknown parameters, but it is assumed that the structures of the models are known *a priori*. Inputs to the system can also be captured in the F_j models. For example, the F_j can be linear systems with additive inputs and Gaussian noise. Also, though not explicitly represented, the transition matrix H can depend on parameters other than time, including the continuous or discrete state. The hybrid system is useful in modeling a variety of scientific and natural phenomena, as described in Section 1. The estimation problem for these systems is known as hybrid state estimation and is introduced in the next subsection.

2.2 Hybrid State Estimation

The estimation problem is key to many applications. For hybrid system models, the estimation problem is twofold: estimation of the discrete mode as well as of the continuous states of the system [15,16,17,18]. The hybrid state estimation problem, in discrete-time, consists of an FDHS and *observations* of this system. The term *measurement* is used synonymously with observation.

Definition 2. *Observations of an FDHS. Given an FDHS, an observation is defined as a (possibly noisy) measurement of the continuous state. That is, the observation of the system at time k, $z(k)$, is a function of the continuous state, $z(k) = C(x_c(k))$. A set of observations to time κ is denoted $Z(\kappa) = (z(0), z(1), ..., z(\kappa))$.*

For any observation of an FDHS, the discrete state is not measured in any direct sense. However, the discrete mode has an effect on the continuous state and therefore on the observation. The hybrid state estimation problem is one of estimating the hybrid trajectory of an FDHS, $X(\kappa)$, given the observation history $Z(\kappa)$. Variations include scenarios in which some observations may be missing or in which predictions are to be made for times greater than κ.

The estimated hybrid trajectory is denoted $\widehat{X}(\kappa)$; for any quantity q, the corresponding estimated quantity is $\hat{q}$. Also, when a specific time horizon is not relevant, it is dropped from the notation: X_d and X refer to the discrete and hybrid trajectory for an unspecified time horizon. The quality of this estimated hybrid trajectory will be discussed further in Section 3.

2.3 Hybrid Systems with Limited Mode Changes

A common restriction of the discrete state of a hybrid system is that mode changes are limited in frequency. Hybrid state estimation analysis often depends on the idea of a dwell time τ, which represents the minimum time the system must spend in a given mode before switching to another mode. This quantity is useful in representing events that take a minimum required time, such as an aircraft turn. Thus, it is reasonable to expect that for a limited time horizon, the discrete modes are unlikely to change a large number of times. For a variety of applications, including air traffic target tracking, discrete mode changes are limited to a handful over the course of hundreds of observations [19]. For example, regression models used for estimating arrival time split the continuous estimation problem into discrete segments corresponding to mode changes [20,21]. Thus, we focus our attention on a class of hybrid systems in which mode changes occur infrequently, and thus discrete trajectories can be maintained exactly. Formally, the restricted systems exhibit K or fewer mode changes over a time horizon of T time steps, where K is small ($K << T$).

2.4 K-Limited Mode-Change (KLMC) Hybrid State Estimation Algorithm

The K-Limited Mode-Change (KLMC) algorithm is a new approach to the hybrid state estimation problem posed for systems with limited mode changes. At each time step, the algorithm generates a hybrid trajectory estimate based on the conditions from the previous time step and the measurement available. Consider the process at time κ. Available to the estimator are m possible discrete trajectories denoted $X_d^i(\kappa-1), i \in \{1, 2, ..., m\}$ and a quantity $\mu(k-1)$ such that $\mu_i(k-1) = P(X_d^i$ is true). For each of these discrete trajectories there is an estimated continuous state trajectory $X_c^i(\kappa-1)$. Note that at time 1, the inputs to the algorithm are just the initial conditions of the system. For each i, the algorithm first determines all discrete states $x_d^{i,j}(\kappa)$ that are possible given discrete trajectory $X_d^i(\kappa-1)$, based on the transition matrix H. Without loss of generality, assume there are n possible $x_d^{i,j}(\kappa)$. These correspond to n discrete trajectories up to time κ: $X_d^{i,j}(\kappa) = (X_d^i(\kappa-1), x_d^{i,j}(\kappa))$, $j \in \{1, ..., n\}$. For each of these n trajectories, a continuous state estimator is used to estimate $x_c^{i,j}(\kappa)$ given $x_c^i(\kappa-1)$ and $z(\kappa)$. This estimator must also provide an estimate of its relative accuracy, such as a covariance matrix or a likelihood function $\mathcal{L}_{i,j}(k)$. This likelihood function is then used to weight the relative probability of all mn possible discrete trajectories up to time κ: $\mu_{i,j}(k) = c\mathcal{L}_{i,j}(k)H_{i,j}(k)\mu_i(k-1)$, where c is a normalization constant. Trajectories that have probability below a preset tolerance ϵ (e.g. $\epsilon = 10^{-6}$) are discarded. The result is a set of hybrid trajectories up to time κ, $X^i(\kappa)$, with associated probabilities of likelihood, $\mu(k)$, and are the input to the estimator at the next time step. The hybrid state at time κ of the hybrid trajectory with highest likelihood is the output of the algorithm at time κ: $\hat{x}^\kappa = x^{i^*}(\kappa)$ such that $\mu_{i^*}(\kappa) > \mu_i(\kappa), i \neq i^*$.

Proof of Polynomial Storage and Computation Requirements. The KLMC algorithm has to maintain a large set of possible discrete trajectories and associated continuous state trajectories. For constant time horizon T, the space required for any single discrete trajectory is constant, because at most T integers are required to store the discrete trajectory and nT real numbers for the continuous trajectory (since $x_c(k) \in \mathbb{R}^n$). Run-time is also directly proportional to the number of possible discrete trajectories, because each trajectory must be updated at each time step.

Claim. The growth of possible discrete trajectories for a hybrid system with N discrete states and limited to a fixed number of mode changes K is polynomial in N.

To prove this claim, we count the number of possible discrete trajectories that have exactly j transitions, for $j \in \{0, 1, ..., K\}$. For $j = 0$, there are obviously N possible trajectories in which the discrete mode is one of the N possible discrete states during the entire time period. For $j = 1$, the problem is equivalent to asking what the initial mode is (N possibilities), what the final mode is ($N-1$ possibilities), and where the mode change occurs ($T-1$ possibilities), for a total of $(N^2 - N)(T-1)$ possible discrete trajectories. For general j, the problem is that of choosing (1) the initial mode of the hybrid system, (2) the next j modes that the system exhibits, and (3) the locations of the j mode changes. (1) has N possible choices, (2) has $(N-1)^j$ choices, and (3) has $\binom{T}{j}$ choices. Thus, the number of possible discrete trajectories with exactly j mode changes is

$$\frac{N(N-1)^j T!}{j!(T-j)!}, \tag{1}$$

which is $O(N^{j+1})$. The number with at most K mode changes is then

$$\sum_{j=0}^{K} \frac{N(N-1)^j T!}{j!(T-j)!}, \tag{2}$$

which is $\sum_{j=0}^{K} O(N^{j+1})$. This is simply $O(N^{K+1})$, which proves the claim.

It is noted that systems with minimum dwell times have far fewer possible mode changes, and therefore far fewer discrete trajectories are obtained. Also, the KLMC algorithm ignores trajectories with probability less than a preset tolerance, which greatly reduces the storage requirements. However, in all cases, more discrete trajectories have to be maintained than for other hybrid state estimators such as the Interacting Multiple Model (IMM). This implies longer run times as well, since more trajectories have to be updated at each time step. However, simulation results in Section 4 show that both algorithms can be used in a real-time setting (in terms of run-time and computing requirements); because KLMC outperforms IMM, it is more effective than IMM in certain conditions.

2.5 Stochastic Hybrid Linear System Example

A common instance of an FDHS is a Discrete-Time Stochastic Hybrid Linear System (DTSHLS) defined as follows [19].

Definition 3 (Discrete-time Stochastic Hybrid Linear System DTSHLS). *Consider N linear systems with discrete-time, continuous-state dynamics:*

$$\begin{aligned} x_c(k+1) &= A_j x_c(k) + B_j w_j(k) \\ z_j(k) &= C_j x_c(k) + D_j v_j(k), \end{aligned} \tag{3}$$

$j \in \{0, ..., N-1\}$, *where* $x_c \in \mathbb{R}^n$ *and* $z \in \mathbb{R}^p$ *are the continuous state and output, respectively. For mode* j, *the terms* $w_j \in \mathbb{R}^m$ *and* $v_j \in \mathbb{R}^r$ *are the uncorrelated, white Gaussian process noise and measurement noise with zero means and covariances* Q_j *and* R_j, *respectively. In each mode, evolution of the state and observation of the state are defined by system matrices* $A_j \in \mathbb{R}^{n\times n}$, $B_j \in \mathbb{R}^{n\times m}$, $C_j \in \mathbb{R}^{p\times n}$, *and* $D_j \in \mathbb{R}^{p\times r}$ *respectively. The discrete mode of the hybrid system, denoted* $x_d(k)$, $x_d(k) \in \{0, ..., N-1\}$, *determines which linear system model is used to update the state and output at time* k. *Evolution of the discrete state is described by the mode transition matrix, denoted* H. *Entries* H_{ij} *represent the probability that the system will transition from mode* j *to* i.

For stochastic linear systems and measurements with Gaussian noise, Kalman filters are optimal for minimum mean square error [4,22]. Given a linear system model and a continuous state trajectory to time $\kappa - 1$ and a noisy measurement at time κ, a Kalman filter can be used to optimally estimate the continuous state at time κ. The Kalman filter also provides a likelihood $\mathcal{L}(\kappa)$ of the estimate, used to weight the probability of possible discrete trajectories. Thus, for a DTSHLS, for discrete mode j, the associated continuous model F_j has an appropriate estimator for generating state and likelihood estimates, as required by KLMC.

Consider a numerical example for a stochastic hybrid system with two modes (0 and 1), and continuous states that are scalars. The mode transition matrix is $H = [.9\ .1;\ .1\ .9]$, with the additional stipulation that each mode has a minimum dwell time of 10 time units, which is equivalent to forcing $H = [1\ 0;\ 0\ 1]$ during the minimum dwell time. For a time horizon of 50, this implies at most 5 mode changes. In mode i, the system model F_i at time k is

$$\begin{aligned} x_c(k+1) &= A_i x_c(k) + w(k) \\ z(k) &= x_c(k) + v(k), \end{aligned} \tag{4}$$

where $A_0 = -1$, $A_1 = 1$, and $w(k)$ and $v(k)$ are zero-mean, unit-variance Gaussians.

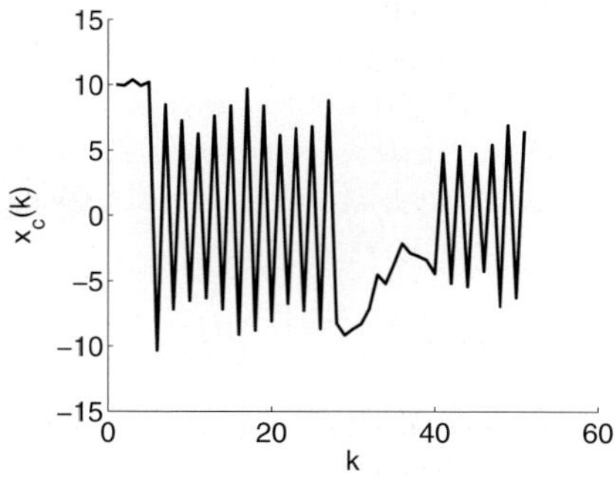

Fig. 1. Continuous state trajectory for simulated hybrid system with two discrete modes: $x_c(k+1) = x_c(k)$ and $x_c(k+1) = -x_c(k)$

For initial conditions $x_d(0) = 1$ and $x_c(0) = 10$, simulations are used to generate several possible hybrid trajectories. The continuous state trajectory for one simulation is plotted in Figure 1. Visually, the discrete state can be seen easily: when the continuous state alternates between positive and negative, the system is in mode 0 ($x_c(k+1) = -x_c(k)+w(k)$), and when the continuous state is only affected by additive noise, the system is in mode 1 ($x_c(k+1) = x_c(k)+w(k)$). Mode changes occur at times 5, 28, and 40.

The KLMC algorithm is used to determine hybrid state estimates using only noisy measurements of the continuous trajectory. In Figure 2(a), the actual, measured, and estimated continuous trajectory are shown, while Figure 2(b) displays the actual and estimated discrete trajectory. While the estimated discrete

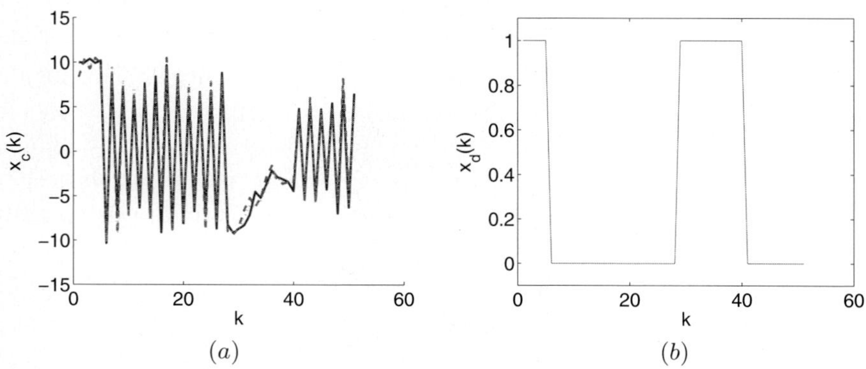

Fig. 2. Results from application of KLMC algorithm to trajectory from Figure 1. In (a), actual (solid line), measured (dashed line), and estimated (dotted line) continuous state trajectories are shown. Because the actual and estimated trajectories are almost identical, they are indistinguishable on the plots. In (b), actual and estimated discrete trajectories are shown (trajectories are coincident).

trajectory is exactly the same as the actual, there are necessarily errors in the continuous estimates (due to measurement noise). Such an estimate may be useful if discrete mode estimation is the essential problem, while in other situations it may be more important to estimate continuous states even if discrete mode estimation is inaccurate. These ideas are explored further in the next section.

3 Performance Metric for Estimate-Error

In this section, the hybrid state estimate quality is explored further and a quantitative metric is presented. For linear systems, the covariance of the continuous state estimation, or simply the error bar, is often used to compare different estimators. The concept of *probability of error* is used extensively in fields such as communications to describe the efficacy of various information coding schemes. In this section, this concept is adapted to a metric for hybrid state estimators. A numerical scheme for comparing estimators is presented.

3.1 Probability of Error

Probability of error is defined for a decision-making process under a set of system conditions [23]. In communications, for example, a quantity of interest may be the probability of an error if a bit is sent using a specific coding scheme and power level. A related quantity is the probability of error given a stochastic description of the transmitted information. As an example, consider the scheme shown in Figure 3, where a 0 corresponds to -1V and a 1 corresponds to +1V, and the received signal has additive white Gaussian noise with variance σ^2. The estimator chooses 0 when the received voltage is negative and 1 when positive. In terms of the quantities defined above, the probability of error given a 0 is sent is the probability of receiving a positive voltage when 0 is sent, which is $\frac{1}{2}\text{erfc}(\frac{1}{\sigma\sqrt{2}})$, where erfc is the complementary error function defined as $\text{erfc}(x) = \frac{2}{\sqrt{\pi}}\int_x^\infty \exp -t^2 dt$. The associated probability of error for the scheme as a whole is $P_e = P(0 \text{ sent}) * P(\text{positive received}|0 \text{ sent}) + P(1 \text{ sent}) * P(\text{negative received}|1 \text{ sent})$. If 0 is sent with probability α, then this reduces to $\frac{\alpha}{2}\text{erfc}(\frac{1}{\sigma\sqrt{2}}) + \frac{1-\alpha}{2}\text{erfc}(\frac{1}{\sigma\sqrt{2}})$, or $\frac{1}{2}\text{erfc}(\frac{1}{\sigma\sqrt{2}})$. For any transmission that satisfies this stochastic description of the channel noise and the data, this quantity describes the probability of making an error for a random bit.

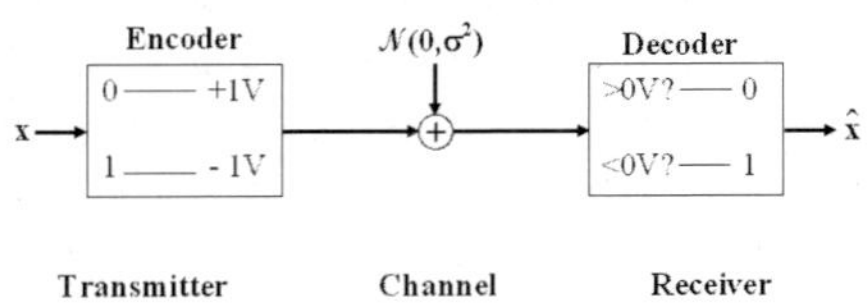

Fig. 3. Transmission scheme with encoder, additive white Gaussian noise channel, and decoder

An associated quantity is the expected cost of an error. While probability of error is the expectation of making an error, expected cost of an error is the expectation of making an error weighted by the cost of this error. For the above example, consider a scenario where estimating 1 when 0 is sent costs e_0 units, but estimating 0 when 1 is sent costs e_1 units. Then, the expected cost of error is $\frac{e_0\alpha+e_1(1-\alpha)}{2}\text{erfc}(\frac{1}{\sigma\sqrt{2}})$.

In relation to hybrid systems, these concepts are applied to the discrete mode estimation part of hybrid state estimation. That is, the probability of error at time k is the probability that the wrong discrete mode is chosen ($\hat{x}_d(k) \neq x_d(k)$). More specifically, there is a probability of estimating $\hat{x}_d(k) = j$ when the actual mode is $x_d(k) = i$. It is assumed there is an associated cost function $\mathcal{C}$ such that $\mathcal{C}(i,j)$ represents the cost of estimating mode j when the actual mode is i. Then the expected cost of an error in discrete mode estimation at a given time is

$$\sum_{i,j} \mathcal{C}(i,j) * P(\hat{x}_d(k) = j | x_d(k) = i) P(x_d(k) = i), \tag{5}$$

where it is assumed that the actual discrete mode probability $P(x_d(k))$ is a known parameter of the system. Thus, there is a quantity to describe the cost of incorrectly estimating the discrete mode for a given description of the hybrid system. Combining this quantity with the cost of error in the continuous estimate gives an error metric, which is introduced in the next subsection.

3.2 Development of Performance Metric

A method for assigning cost of discrete mode errors is given above. Continuous state error can be quantified according to traditional residual norms. For a given scenario and measurements, the error of the hybrid estimate is then defined as

$$\Gamma(X_d, X_c, Z) = \sum_{i=0}^{T} \lambda_i \mathcal{C}(x_d(i), \hat{x}_d(i)) + \alpha \sum_{i=0}^{T} \nu_i \|x_c(i) - \hat{x}_c(i)\|^2, \tag{6}$$

where $\mathcal{C}(m,n)$ denotes the cost of error associated with estimating mode n when the real mode is m, and λ_i, ν_i, and α are weights that depend on the quantity of interest. For example, if only the estimate at time κ is of interest, $\lambda_i = \nu_i = 0$ for $i \neq \kappa$. If the estimation problem is concerned more with the discrete mode (e.g. - fault tolerance), α is small, while if the estimation problem is concerned with the continuous state estimates (e.g. - target tracking), α is large. For many systems, both discrete mode transitions and the continuous state are relevant to the problem and an intermediate α is chosen.

The estimate error function $\Gamma(X_d, X_c, Z)$ is a function of the specific discrete trajectory, continuous trajectory, and measurements, respectively. The larger goal of this metric is to quantify the expected error of a hybrid estimate for a set of scenarios rather than a specific instance of the problem. First, consider the expected estimate error for all possible measurements Z for a given hybrid trajectory (X_d, X_c). The set of possible measurements is assumed to have probability distribution function $f(Z|X_d, X_c)$ as described by the observation functions C_j. Then the expected error estimate is denoted $\Gamma(X_d, X_c)$ and is defined as

$$\Gamma(X_d, X_c) = \int f(Z|X_d, X_c)\Gamma(X_d, X_c, Z)dZ. \tag{7}$$

The quantity in (7) represents the expected estimated error for a given trajectory. By integrating over all possible hybrid trajectories, one can obtain the overall expected estimated error for a hybrid state estimator for a given hybrid system:

$$\Gamma = \int f(X_d, X_c)\Gamma(X_d, X_c)dX_d dX_c. \tag{8}$$

Note that Γ without arguments is the expectation of $\Gamma(X_d, X_c, Z)$ over the three arguments. Analytic results for the evaluation of Γ are limited due to the integration involved. In the next subsection, a numerical approach to evaluating Γ is proposed, and results of simulation are presented in Section 4.

3.3 Numerical Methods for Comparing Estimators

While $\Gamma(X_d, X_c, Z)$ can be determined for a specific instance of a hybrid system, it is difficult to determine the expected error metric Γ for more general scenarios. The problem can be posed as a multidimensional integration problem, but is usually not solvable analytically. Rather, numerical methods such as Monte Carlo simulation can be used as an approximation. By calculating $\Gamma(X_d, X_c, Z)$ for specific instances drawn from the full set of instances according to the stochastic description of X_d, X_c, and Z, one can generate an estimate of Γ. This numerical estimate of the error metric by Monte Carlo methods can be made arbitrarily accurate by taking a large enough sampling.

4 Comparison of IMM and KLMC Using Monte Carlo Simulation

In this section, two hybrid state estimators, the KLMC algorithm and the IMM algorithm, are compared by the metric developed in Section 3 and by computing resources used. Performance is estimated via Monte Carlo simulation; 1000 simulations are run for each parameter set. For each run, various random quantities are sampled: the initial distribution of the hybrid state, the discrete state trajectory using the transition matrix H, and the additive measurement noise. Two scenarios are presented in this section: a simple system with two modes and scalar continuous state, and a more realistic target tracking problem.

4.1 Simple Hybrid System Example

Consider the hybrid system from Section 2. The two modes are stochastic linear systems as expressed in (4), with $A_0 = -1$, $A_1 = 1$, and $w(k)$ and $v(k)$ zero-mean, unit-variance Gaussians. For this system, the metric weights, α, λ_i, and

μ_i, $i \in \{0, 1, ..., T\}$, are set to unity and the error cost matrix ($\mathcal{C}$) is zero on the diagonal and unity for all other entries. The time horizon T is 50, and the minimum dwell time τ for both modes is 10. Two parameter sets are considered. In the first set, the initial discrete mode has uniform distribution and the initial continuous state is 10. In the second set, the only change is that the initial continuous state is 0.1.

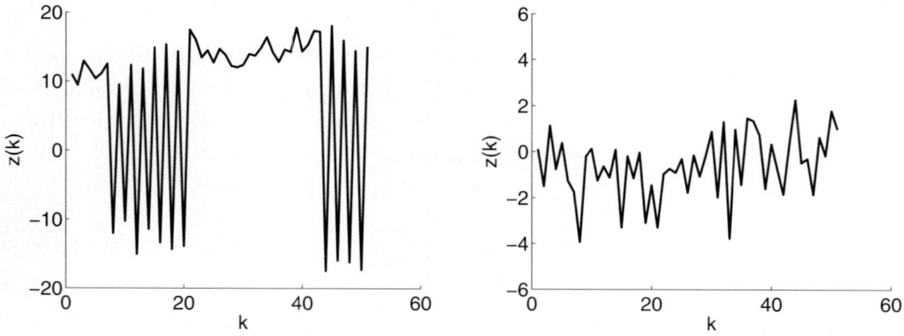

(a) (b)

Fig. 4. Sample measurement trajectories for stochastic hybrid linear system with two modes and initial state: (a) $(x_d(0), x_c(0)) = (1, 10)$ and (b) $(x_d(0), x_c(0)) = (1, 0.1)$

Table 1. Table summarizing performance of and computing resources used by hybrid state estimators KLMC and IMM for the scenario illustrated in Fig. 4(a)

Algorithm	**KLMC**	**IMM**
Average Performance Γ	32.1	33.0
Average Run-Time/Time Step (ms)	97.4	24.1
Average Maximum Storage (kB)	409.8	20.9

For the first parameter set, one set of measurements is shown in Figure 4(a). These are obtained from the continuous state trajectory with noise added. Because the initial continuous state is large compared to the noise, it is easier to distinguish modes. Table 1 shows the performance of the IMM and KLMC algorithms, averaged over 1000 runs. The two algorithms perform almost identically, though the KLMC algorithm is slightly better. The table also shows the average run-time and average maximum number of discrete trajectories of the two algorithms. IMM outperforms KLMC, because it stores far fewer trajectories.

In Figure 4(b), measurements for a sample trajectory are shown for the second parameter set. Because the noise is relatively large compared to the continuous state, modes are hard to distinguish. Results for this parameter set are shown in Table 2. IMM continues to outperform KLMC in run-time and storage requirements, but KLMC is now better according to the estimate-error metric. This result is because KLMC maintains all possible discrete trajectories for a system

Table 2. Table summarizing performance of and computing resources used by hybrid state estimators KLMC and IMM for the scenario illustrated in Fig. 4(b)

Algorithm	**KLMC**	**IMM**
Average Performance Γ	36.1	41.9
Average Run-Time/Time Step (ms)	115.7	26.3
Average Maximum Storage (kB)	490.4	21.1

with limited mode changes, while IMM maintains only the total probability of either mode at any given time. The overall performance of both algorithms is significantly decreased from the first parameter set, showing that this estimation problem is more difficult for both estimators.

Table 3. Table summarizing performance of and computing resources used by hybrid state estimators KLMC and IMM for aircraft target-tracking scenario

Algorithm	**KLMC**	**IMM**
Average Performance Γ	51.3	63.2
Average Run-Time/Time Step (ms)	280.4	53.1
Average Maximum Storage (kB)	2143	120.2

4.2 Target Tracking Example

The second example is taken from an air traffic application: the tracking of an aircraft as it alternates between straight flight at constant velocity and turning at a constant yaw rate of 3 degrees per second. Both modes involve six continuous state variables (position, velocity, and acceleration in each of the x- and y-directions), and measurements are taken of the position and velocity components of the continuous state. For further details on the hybrid system model, see [24]. The initial discrete mode is 0 (straight flight) with probability 0.5. Also, the minimum dwell time for straight flight mode is 20 time units, but only 10 time units for turns (turns take less time to complete in general than straight segments). Again, 1000 runs are completed for both estimators to obtain numerical estimates of their estimation performance and use of computing resources. The results are compiled in Table 3. It is noted that the KLMC algorithm does significantly better than the IMM algorithm in estimate performance, but at a significant cost in run-time and storage. This performance gain corresponds to an average decrease in RMS position error on the order of 1m or velocity error of several m/s. Also, both run-time and storage are at levels that are sufficient for real-time use with a tracking system running at 1 Hz on a standard desktop computer, for example. Thus, it can be argued that the KLMC outperforms the IMM algorithm without overburdening the computing resources available for the problem. Testing on more complicated systems and scenarios remains necessary to adequately compare the two algorithms.

5 Conclusions and Future Work

In this paper, the K-Limited Mode-Change Algorithm, a hybrid state estimator for systems with limited mode changes, is introduced. The algorithm maintains a set of possible discrete trajectories that is polynomially bounded with order K. A metric based on probability of error is developed for hybrid state estimators. It is used to show that KLMC outperforms other existing algorithms such as IMM for two case study problems, especially in more difficult scenarios. This performance gain requires longer run-times and storage space, but not more than available on a standard desktop.

Future work directions are twofold: developing estimators for specific problems and improving the analysis of the hybrid state estimation metric. Critical issues include the appropriate handling of stochastic constraints and systems with unknown parameters. The ability to manage more complex hybrid models (e.g. nonlinear hybrid systems) is also needed. The refinement of the metric requires further analysis for systems with complicated noise characteristics. Subclasses of hybrid systems that have exploitable structure must be explored to generate more analytic results.

References

1. Li, X., Bar-Shalom, Y.: Design of an Interacting Multiple Model Algorithm for Air Traffic Control tracking. IEEE Transactions on Control Systems Technology **1**(3) (1993) 186–194
2. Hwang, I., Balakrishnan, H., Roy, K., Shin, J., Guibas, L., Tomlin, C.: Multiple-target Tracking and Identity Management algorithm in clutter. In: Proceedings of the AACC American Control Conference, Boston, MA (2004)
3. Bayen, A.M., Tomlin, C.J.: Real-time discrete control law synthesis for hybrid systems using MILP: applications to congested airspaces. In: Proceedings of the 2003 American Control Conference, Denver, CO (2003)
4. Bar-Shalom, Y., Fortmann, T.: Tracking and Data Association. Academic Press (1988)
5. Blom, H., Bar-Sharlom, Y.: The Interacting Multiple Model algorithm for systems with Markovian switching coefficients. IEEE Trans. on Automatic Control **33**(8) (1988) 780–783
6. Mazor, E., Averbuch, A., Bar-Shalom, Y., Dayan, J.: Interacting multiple model methods in tracking: A survey. IEEE Transactions on Aerospace and Electronic Systems **34**(1) (1998) 103–123
7. Shin, J., Guibas, L., Zhao, F.: A distributed algorithm for managing multi-target identities in wireless ad-hoc sensor networks. In Zhao, F., Guibas, L., eds.: Information Processing in Sensor Networks. Lecture Notes in Computer Science 2654, Palo Alto, CA (2003) 223–238
8. Zhou, B., Bose, N.: Multitarget tracking in clutter: Fast algorithms for data association. IEEE Trans. on Aerospace and Electronic Systems **29**(2) (1993) 352–362
9. Hadzagic, M., Michalska, H., Jouan, A.: IMM-JVC and IMM-JPDA for closely maneuvering targets. In: Conference Record of the Thirty-Fifth Asilomar Conference on Signals, Systems and Computers. Volume 2. (2001) 1278 –1282

10. Danchick, R., Newnam, G.: A fast method for finding the exact N-best hypotheses for multitarget tracking. IEEE Trans. on Aerospace and Electronic Systems **29**(2) (1993) 555–560
11. Boers, Y., Driessen, H.: A multiple model multiple hypothesis filter for Markovian switching systems. Automatica **41**(4) (2005) 709–716
12. Egerstedt, M., Wardi, Y., Axelsson, H.: Transition-time optimization for switched systems. IEEE Transactions on Automatic Control (**51**(1))
13. Girard, A., Pappas, G.J.: Verification using simulation. In: Hybrid Systems: Computation and Control. Volume 3927 of Lecture Notes in Computer Science., Santa Barbara, CA, Springer (2006)
14. Pola, G., Bujorianu, M., Lygeros, J., Di Benedetto, M.: Stochastic hybrid models: An overview with applications to air traffic management. In: IFAC Conference on Analysis and Design of Hybrid Systems, Saint-Malo, France (2003)
15. Hofbaur, M., Williams, B.: Mode estimation of probabilistic hybrid systems. In Tomlin, C., Greenstreet, M.R., eds.: Hybrid Systems: Computation and Control. Volume 2289 of Lecture Notes in Computer Science., Stanford, CA, Springer (2002)
16. Baballi, M., Egerstedt, M., Kamen, E.W.: A direct algebraic approach to observer design under switched measurement equations. IEEE Transactions on Automatic Control (**49**(11))
17. Morari, M., Bemporad, A., Mignone, D.: A Framework for Control, State Estimation, Fault Detection, and Verification of Hybrid Systems. Scientific Computing in Chemical Engineering II **2** (1999) 46–61
18. Bar-Shalom, Y., Li, X.: Estimation and Tracking: Principles, Techniques and Software. Artech House, Boston, MA (1993)
19. Roy, K., Levy, B., Tomlin, C.: Target tracking and estimated time of arrival (ETA) prediction for arrival aircraft. In: Proceedings of AIAA Guidance, Navigation, and Control Conference, Keystone, CO (2006)
20. Levy, B., Bedada, S.: A real-time ETA-to-threshold prediction tool. In: 25th Digital Avionics Systems Conference, Portland, OR (2006)
21. Idris, H., Anagnostakis, I., Delcaire, B., Clarke, J.P., Hansman, R., Feron, E., Odoni, A.: Observations of departure processes at Logan Airport to support the development of departure planning tools. Air Traffic Control Quarterly **7**(4) (1999) 229–257
22. Blom, A., Hogendoorn, R., van Doorn, B. In: Design of a multisensor tracking system for advanced air traffic control. Volume 2 of Multitarget-Multisensor Tracking: Application and Advances. Artech House (1990) 31–63
23. Proakis, J., Salehi, M.: Communication Systems Engineering. Second edn. Prentice Hall (2001)
24. Hwang, I., Balakrishnan, H., Roy, K., Shin, J., Guibas, L., Tomlin, C.: Multiple-target Tracking and Identity Management algorithm for Air Traffic Control. In: Proceedings of the Second IEEE International Conference on Sensors, Toronto, Canada (2003)

Ant Colony and Genetic Algorithm for Constrained Predictive Control of Power Systems

Guillaume Sandou and Sorin Olaru

Supélec, Automatic Control Department
3, rue Joliot Curie, F-91192 Gif-sur-Yvette, France
{Guillaume.Sandou,Sorin.Olaru}@supelec.fr

Abstract. In this paper, a cooperative metaheuristic based on ant colony optimization and genetic algorithm is developed for constrained predictive control of power systems. The classical Unit Commitment solution is an open loop control for power systems which cannot be applied to real system, since it is affected by important uncertainties, a typical source being the consumer load. Predictive control offers an efficient way to use optimization results in a closed loop framework, implying the online solution of successive constrained mixed optimization problems. The algorithm proposed here is able to explicitly deal with constraints, and to quickly find high quality suboptimal solutions for computationally involving predictive control schemes. Simulation results show the efficiency of the developed method, even for Unit Commitment problems with underestimated consumer demand.

1 Introduction

The control of hybrid systems necessitates pertinent answers to several challenging problems, like the switching between different operating regimes, the interaction of continuous-time and discrete event subsystems and the overall satisfaction of operational constraints [1]. An inherent question is whether the classical automatic control methods (for example frequency domain shaping or optimal control) can be adapted in a systematic manner to the design of control laws for hybrid systems. It turns out that the answer is generally negative (due to the structural disparity) with a notable exception, that of predictive control law [2]. Indeed this technique, which is based on finite time optimal control problems over a receding horizon, has the important advantage of being a time-domain design procedure and thus being effective in the hybrid systems framework, too.

Optimizing the evolution of the hybrid system over a receding horizon generally leads to mixed integer quadratic/linear programs (MIQP/MILP) which are computationally involving (NP-complete problems) due to the presence of logical optimisation arguments. The exact solvers have a worst case combinatorial complexity as it is the case for example with the classical branch and bound

A. Bemporad, A. Bicchi, and G. Buttazzo (Eds.): HSCC 2007, LNCS 4416, pp. 501–514, 2007.

routines, even if they are tailored to the problem specificity [3]. The real-time implementation constraints might force the optimization routines to act on a relative short time interval and to offer at least a feasible suboptimal solution. This can be obtained in a slightly different manner by renouncing to the exhaustive search and the use of evolutionary methods.

Figure 1 presents a classification for the MPC implementations. In a first stage, a cautious control design would pass by the tuning of the prediction horizon such that the attainability demands to be met and the optimization problem to be simplified. However, the real-time implementation would be resumed by a MIQP/MILP or MINP if nonlinearities are considered.

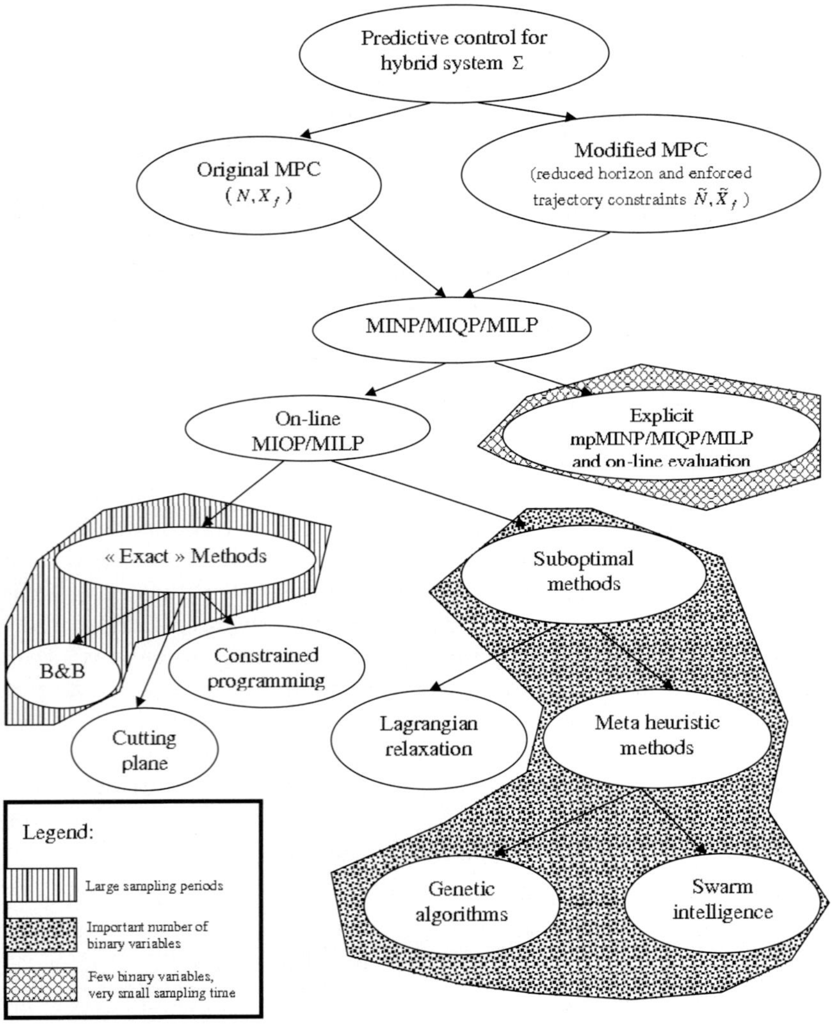

Fig. 1. A classification of predictive control possible implementations

The same figure presents a demarcation between the implementations based on on-line optimisation and those where an explicit analytical dependence of the optimum on the current state is available [4]. In the second case, there is no need for on-line optimisation, the computational load being reduced to the evaluation of the analytical function which gives the exact solution. Unfortunately explicit solution can be practically constructed only for control problems with few states and small prediction horizon due to the curse of dimensionality.

Returning to the classification, on the on-line implementation part a separation can be made between the exact methods and the routines which allow from the design stage a certain degree of suboptimality as it is the case for example with genetic algorithms [5] or the Lagrangian relaxation where the quality of the solution is improved, but it can not go beyond the duality gap.

In the present paper the attention will be given to cooperative metaheuristics based on ant colony optimization and genetic algorithm situated on the class of suboptimal methods. From the standpoint of computational complexity, finding out if an MIQP model has a feasible solution is essentially as hard as actually finding the optimum. Branch and Bound procedures, for example, have to explore the entire search tree to prove this. The developed algorithm proves to have the ability of proposing (high quality) feasible solutions due to the contraints handling mechanism.

For the sake of illustration, a popular problem in power systems will be used, the Unit Commitment. Indeed, power systems can be casted in the class of hybrid systems since they have to be controlled both by integer variables (on/off variables) and real variables (produced powers when switched on). Benefiting from this practical aspects, the details of the algorithm will be described in a practical manner.

The paper is organized as follows. The predictive control of power systems is presented in section 2. This control law is based on a scheduling algorithm. This scheduling algorithm has to be highly tractable even for such large scale mixed integer optimization problems. Furthermore it has to take into account all technical constraints of the system. That is why a cooperative stochastic method using ant colony optimization and genetic algorithm has been developed and is depicted in section 3. Numerical results are given in section 4, showing the efficiency of the proposed control method even in the case of underestimated consumer demand. Finally, conclusions and forthcoming works are drawn in section 5.

2 Predictive Control of Power Systems

2.1 Open Loop Control and Unit Commitment

Unit Commitment is a classical large scale mixed integer problem in power systems, which aims to compute the optimal scheduling of several production units while satisfying consumer demand and technical constraints:

$$\min_{\{u_n^k, Q_n^k\}} \sum_{n=m}^{m+N-1} \left(\sum_{k=1}^{K} \left(c_{prod}^k(Q_n^k, u_n^k) + c_{on/off}^k(u_n^k, u_{n-1}^k) \right) \right). \tag{1}$$

N is the length of time horizon, K the number of production unit, u_n^k (resp. Q_n^k) the on/off status (resp. produced power) of production unit k during time interval n. Production costs and start up and shut down costs are defined by:

$$\begin{cases} c_{prod}^k(Q_n^k, u_n^k) = \alpha_1^k Q_n^k + \alpha_0^k u_n^k \\ c_{on/off}^k(u_n^k, u_{n-1}^k) = c_{on}^k u_n^k (1 - u_{n-1}^k) + c_{off}^k u_{n-1}^k (1 - u_n^k) \end{cases} . \tag{2}$$

Integer variables are on/off status of production units, and real variables are produced powers. The usual restriction comes from:

– capacity constraints

$$Q_{\min}^k u_n^k \leq Q_n^k \leq Q_{\max}^k u_n^k, \forall n \in \{m, \ldots, m+N-1\}, \forall k \in \{1, \ldots, K\}, \tag{3}$$

– consumer demand satisfaction

$$\sum_{k=1}^{K} Q_n^k \geq \hat{Q}_n^{dem}, \forall n \in \{m, \ldots, m+N-1\}, \tag{4}$$

– time up and time down constraints

$$\begin{cases} \left(u_{n-1}^k = 0, u_n^k = 1\right) \Rightarrow \left(u_{n+1}^k = 1, u_{n+2}^k = 1, \cdots, u_{n+T_{up}^k-1}^k = 1\right) \\ \left(u_{n-1}^k = 1, u_n^k = 0\right) \Rightarrow \left(u_{n+1}^k = 0, u_{n+2}^k = 0, \cdots, u_{n+T_{down}^k-1}^k = 0\right) \end{cases}, \tag{5}$$

– ramp constraints

$$\left|Q_n^k - Q_{n-1}^k\right| \leq \Delta Q^k, \quad \forall n \in \{m, \ldots, m+N-1\}, \forall k \in \{1, \ldots, K\}. \tag{6}$$

Discrete dynamics on the system are expressed through logical equations (5) and continuous dynamics are stated as power increment limitations (6).

2.2 Closed Loop Control

For optimisation, consumer demand is supposed to be perfectly known over the whole time horizon: the computation of the optimal scheduling is a reference trajectory for integer and real control inputs. However, prediction errors of the consumer load may lead to a deficient behavior. Thus, a closed loop control is required: the problem refers to the control of a hybrid system. A convenient way to extend optimisation results in a closed loop framework is the receding horizon. The closed loop structure is presented in figure 2.

The idea is to compute the optimal scheduling on time interval $[m, m+N-1]$, considering predicted consumer load $\hat{Q}_n^{dem}$. The first values of integer scheduling are applied to the system. Simultaneously, real variables are slightly updated so as to fulfill, if possible, the real demand at time m, Q_m^{dem}. Production unit local regulations are assumed to be perfect: if production unit k has produced Q_{m-1}^k during time interval $m-1$, it can produce, if still switched on, whatever power in the range $[\max(Q_{m-1}^k - \Delta Q^k, Q_{\min}^k), \min(Q_{m-1}^k + \Delta Q^k, Q_{\max}^k]$ during time interval m. The scheduling algorithm will now be detailed.

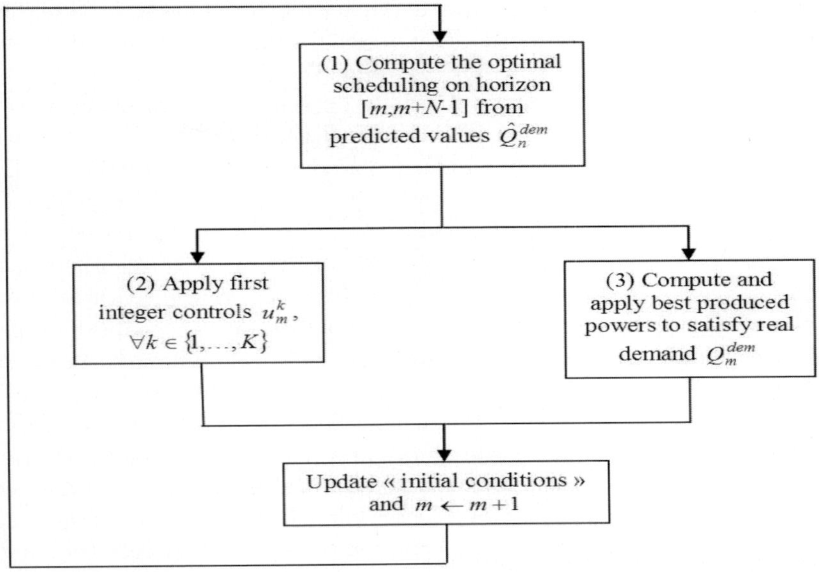

Fig. 2. Closed loop control synopsis

3 Optimization Procedure

3.1 Optimization Methods Analysis and Proposition of a Well Suited Algorithm

Numerous methods have been applied to solve Unit Commitment and related problems such as facility location. They are listed for instance in [6] and are here briefly depicted. Exact solution methods (exhaustive enumeration, Branch and Bound [7], dynamic programming [8]) have been tested. These methods suffer from combinatorial complexity: an efficient approximated method is requested. Deterministic approximated methods can be used (priority lists in [9]). But, due to numerous constraints, they are often strongly suboptimal. Constraints are explicitly considered using Lagrangian relaxation [10]. Coupling constraints are relaxed, and the problem is divided into several optimisation problems (one per production unit). However, no guarantee can be given on the actual optimality. Further, an iterative procedure has to be performed: solution of the optimisation problems and updating of Lagrange multipliers. The update can be made with genetic algorithms [11] or subgradient methods [12]. For large scale cases, metaheuristics are interesting methods: simulated annealing in [13], tabu search in [14] and genetic algorithms in [15]. No guarantee can be given on the actual optimality of the solution, but an often suitable solution with low computation times can be found. One of the problems of these methods is the handling of constraints. The algorithm "moves" randomly in the search space, and so, there is no guarantee that the final solution is in the feasible set. This is particularly the case for Unit Commitment, as the feasible set is much smaller than the search space.

Considering these arguments, ant colony appears to be an efficient way to solve this kind of problems, as it is able to find near optimal solutions with an explicit handling of all constraints. Indeed, ant colony is a constructive stochastic algorithm and solutions are explicitly built as feasible ones (see section 3.2). From this initial population of "medium quality solutions" quickly computed by ant colony, a feasibility criterion is defined (see section 3.4). Genetic algorithm is then used to intensively explore the search space (see section 3.5), with an implicit management of problem constraints. Indeed, due to a positive feedback in the ant colony algorithm formulation, it may converge to a local minimum. An intensive exploration of the search space is thus required to circumvent this issue, and genetic algorithm, which is an efficient stochastic algorithm for unconstrained problems is used. Note that the algorithm supposes that real variables are quickly computed: the developed cooperative algorithm is also hybridized with an exact solution algorithm for real variables (see section 3.3). Finally, this method allows simultaneously using the interesting properties of ant colony (explicitly handling of constraints) and of genetic algorithm (deep exploration of the search space, and so high quality of the solution). The general synopsis of the method is depicted on figure 3.

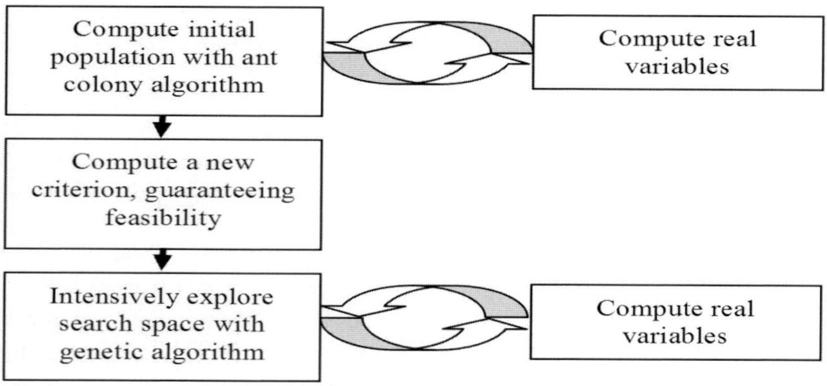

Fig. 3. Cooperative algorithm synopsis

3.2 Ant Colony Optimization

Real world ants. Ant colony optimization was firstly introduced by Marco Dorigo [16] and [17]. It is based on the way ants are looking for food.

Suppose (see figure 4a), that ants have managed to find food. Each particular ant does not know where to go. It only chooses its path depending on the pheromone trail which has been laid on the ground by previous ants. If the path of pheromone is broken because of an obstacle, see figure 4b, first ants randomly

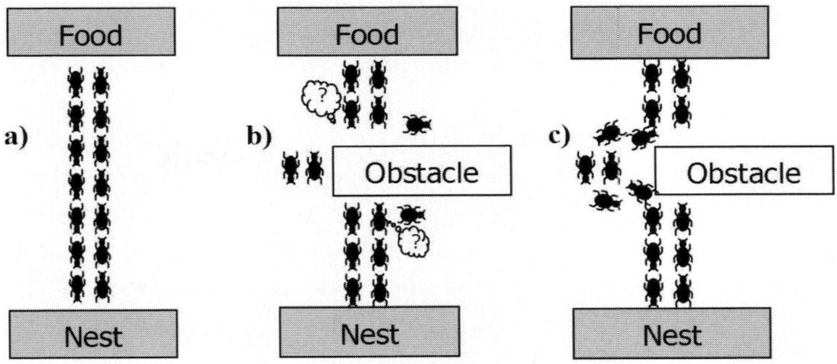

Fig. 4. Ants looking for food

choose their path. But, the ants which have chosen the shortest path will arrive first: the trail of pheromone in the shortest path is increasing faster than in the longest path. The positive feedback structure makes all ants finally choose the shortest path at the end of the experience (see figure 4c).

Ant colony optimization for the Unit Commitment problem. The Unit Commitment problem can be formulated as a graph exploration problem as shown in figure 5, as in previous work [18].

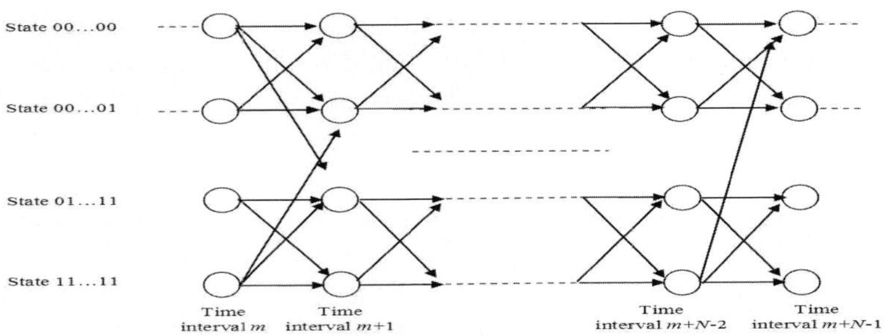

Fig. 5. Graph structure for Unit Commitment

The nodes of the graph represent all the possible states of production system, for all time intervals: $(u_n^1, \ldots, u_n^K)$. The aim is to go from one of the possible states at time m, to one of the possible states at time $m+N-1$, while satisfying all the constraints and minimizing global costs defined in equation (1). For each edge $(u_n^1, \ldots, u_n^K) \rightarrow (u_{n+1}^1, \ldots, u_{n+1}^K)$ of the graph, start-up and shut-down costs are added. Production costs are also associated to nodes.

During iteration t of the algorithm, F ants walk on this graph. If an ant f has reached state $i = (u_n^1, \ldots, u_n^K)$, the probability that it chooses the next state $j = (u_{n+1}^1, \ldots, u_{n+1}^K)$ is defined by the probabilistic law:

$$p_i^{(f)}(j) = \frac{\eta_{ij}^{\alpha} \tau_{ij}(t)^{\beta}}{\sum\limits_{m \in J_f(i)} \eta_{im}^{\alpha} \tau_{im}(t)^{\beta}}. \tag{7}$$

- $\tau_{ij}(t)$ is the pheromone trail on edge $i = (u_n^1, \ldots, u_n^K) \rightarrow j = (u_{n+1}^1, \ldots, u_{n+1}^K)$ during iteration t. Its value depends on the results of previous ants.
- η_{ij} is the attractiveness. It refers to the " local choice ". When next node has to be chosen, the best local candidate is the node for which the gap between the maximum produced power and the predicted demand is the smallest. It is not sure that this is the best "global" choice, as the security margin is quite low. For more details, see previous work [18].
- α and β are weighting factors.
- $J_f(i)$ is the feasible set. This feasible set contains a priori all 2^K states. But, those states which do not satisfy time up and time down constraints, and those states which do not satisfy consumers' demands, are to be removed. Note that, even if produced powers are not known yet, it is possible to check the possibility of consumer's demand satisfaction with the equation:

$$\sum_{k=1}^{K} \left(\begin{pmatrix} Q_{\min}^k (1 - u_n^k) + \\ (\min(Q_{\max}^k, Q_n^k + \Delta Q^k)) \, u_n^k \end{pmatrix} u_{n+1}^k \right) \geq \hat{Q}_{n+1}^{dem}. \tag{8}$$

$J_f(i)$ sets are recursively constructed for each ant, and lead to the guarantee of the feasibility of solutions. After the ant has completed its path, it is possible to evaluate the solution by solving the real optimisation problem defined in equation (1), with fixed binary variables. Due to the positive feedback of the algorithm the past mistakes have to be forgotten to avoid premature convergence. This is done by the pheromone evaporation. The pheromone trail is updated:

$$\tau_{ij}(t+1) = (1 - \rho)\tau_{ij}(t) + \Delta\tau_{ij}(t) \tag{9}$$

ρ is the evaporation coefficient. This coefficient is viewed as an analogy with natural evaporation. $\Delta\tau_{ij}$ is the updating coefficient, depending on the results of ants in iteration t. An elitism algorithm is used: only the best ant is allowed to lay some pheromone on each edge it has used. The ant evaluation supposes the computation of real variables, which will now be depicted.

3.3 Computation of Real Variables

Binary variables are computed by ant colony. For each feasible sequence u_n^k the corresponding real variables Q_n^k are computed as the solution of:

$$\begin{aligned}
&\arg\min_{\{Q_n^k\}} \sum_{n=m}^{m+N-1} \left(\sum_{k=1}^{K} \left(c_{prod}^k \left(Q_n^k, u_n^k \right) + c_{on/off}^k \left(u_n^k, u_{n-1}^k \right) \right) \right) \\
&= \arg\min_{\{Q_n^k\}} \left(\sum_{n=m}^{m+N-1} \sum_{k=1}^{K} \left(c_{prod}^k \left(Q_n^k, u_n^k \right) \right) \right) \\
&= \arg\min_{\{Q_n^k\}} \left(\sum_{n=m}^{m+N-1} \sum_{k=1}^{K} \alpha_1^k Q_n^k u_n^k + \alpha_0^k u_n^k \right) = \arg\min_{\{Q_n^k\}} \left(\sum_{n=m}^{m+N-1} \sum_{k=1}^{K} \alpha_1^k Q_n^k u_n^k \right)
\end{aligned} \tag{10}$$

As there are no temporally coupling constraints anymore (they have been guaranteed by the constructive ant colony algorithm), the problem can be divided into N successive optimisation problems:

$$\min_{\{Q_n^k, k=1,\ldots,K\}} \left(\sum_{k=1}^{K} \alpha_1^k Q_n^k u_n^k \right)$$
$$\text{subject to} \begin{cases} \sum_{k=1}^{K} Q_n^k \geq \hat{Q}_n^{dem} \\ Q_{\min}^k u_n^k \leq Q_n^k \leq \underbrace{\begin{pmatrix} Q_{\min}^k (1 - u_{n-1}^k) + \\ (\min(Q_{\max}^k, Q_{n-1}^k + \Delta Q^k))\, u_{n-1}^k \end{pmatrix}}_{=Q_{\max}^k(n)} u_n^k \end{cases} \tag{11}$$

Without loss of generality, consider that $\alpha_1^1 \leq \alpha_1^2 \leq \ldots \leq \alpha_1^K$. Then, the optimal solution of problem (10) is to produce as much as possible with low-cost units, while satisfying capacity constraints. Then, the following recursive algorithm is performed:

$$\begin{cases} Q_n^1 = \min\left(\max\left(\hat{Q}_n^{dem} - \sum_{i=2}^{K} Q_{\min}^i u_n^i, Q_{\min}^1 \right), Q_{\max}^1(n) \right) u_n^1 \\ \vdots \\ Q_n^k = \min\left(\max\left(\hat{Q}_n^{dem} - \sum_{i=1}^{k-1} Q_n^i - \sum_{i=k+1}^{K} Q_{\min}^i u_n^i, Q_{\min}^k \right), Q_{\max}^k(n) \right) u_n^k \\ \vdots \\ Q_n^K = \min\left(\max\left(\hat{Q}_n^{dem} - \sum_{i=1}^{K-1} Q_n^i, Q_{\min}^K \right), Q_{\max}^K(n) \right) u_n^K \end{cases}$$
$$Q_{\max}^k(n+1) = Q_{\min}^k (1 - u_n^k) + (\min(Q_{\max}^k, Q_n^k + \Delta Q^k))\, u_n^k \tag{12}$$

3.4 Feasibility Criterion

To compute a feasibility criterion one has just to know a feasible solution. If the cost of this feasible known solution is c^f, the feasibility criterion can be:

$$\min_{\substack{\{u_n^k, Q_n^k\} \\ n=1,\cdots,N \\ k=1,\ldots,K}} \left(\begin{array}{l} \sum_{n=1}^{N} \sum_{k=1}^{K} \left(c_{prod}^k (Q_n^k, u_n^k) + c_{on/off}^k (u_n^k, u_{n-1}^k) \right) \\ \quad + \left((1+\varepsilon) c^f + h\left(\{u_n^k, Q_n^k\}\right) \right) B\left(\{u_n^k, Q_n^k\}\right) \end{array} \right) \tag{13}$$

where one can distinguish:

- ϵ is a small positive real,
- $h\left(\{u_n^k, Q_n^k\}\right)$ is a penalty function for non feasible solutions $\{u_n^k, Q_n^k\}$,
- $B\left(\{u_n^k, Q_n^k\}\right)$ is a boolean function with value 1 for non feasible solutions and 0 for feasible ones.

With this criterion, all infeasible solutions will have a higher cost than the cost of the feasible known solution. For feasible solutions, the penalty is null, and the initial cost function is just considered. Finally, any unconstrained optimisation algorithm can be used to solve the problem, the constraints being implicitly taken into account. In this study, the known feasible solution is the best solution found by ant colony optimization.

3.5 Knowledge Based Genetic Algorithm

Classical genetic algorithm. A knowledge based genetic algorithm, similar to the one developed in current work [19] is used. The general flow chart of a genetic algorithm is called up in figure 6.

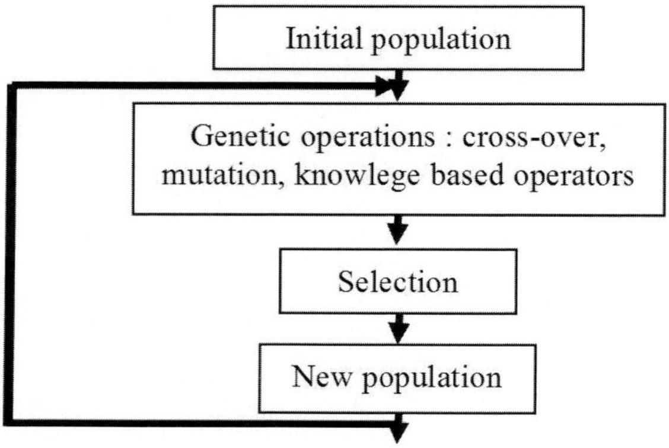

Fig. 6. Flow chart of a genetic algorithm

The main idea is to make a population of potential solutions evolves to create new solutions by using stochastic (or "genetic") operators. In this cooperative method, the initial population is made of all feasible solutions computed by ant colony optimization. Classical operators are crossing-over operator and mutation operator (see figure 7). For the crossing over operation, two potential solutions are randomly chosen in the population. They randomly merge their variables (or "genes") to create two new solutions. The mutation operation is the random selection of a potential solution and of one of its genes. This gene is changed to another. The aim of this operator is to keep the population genetic diversity.

The selection operator is an operator which aims to choose a new population from parents and children. This operation is made using the roulette wheel selection. After having computed the fitness value of each individual in the population, the probability of selection is proportional to the quality of individuals.

Knowledge based operator. Classical genetic operators may not be well suited to Unit Commitment problem. Consider for instance the situation of figure 8. Because of time down and time up constraints, a random mutation will very often lead to an infeasible solution. Switching times are particular points of

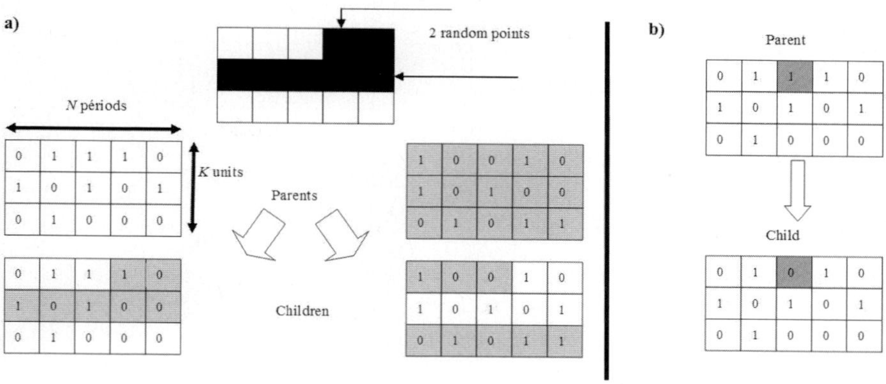

Fig. 7. Classical crossing-over (a) and mutation (b) operators

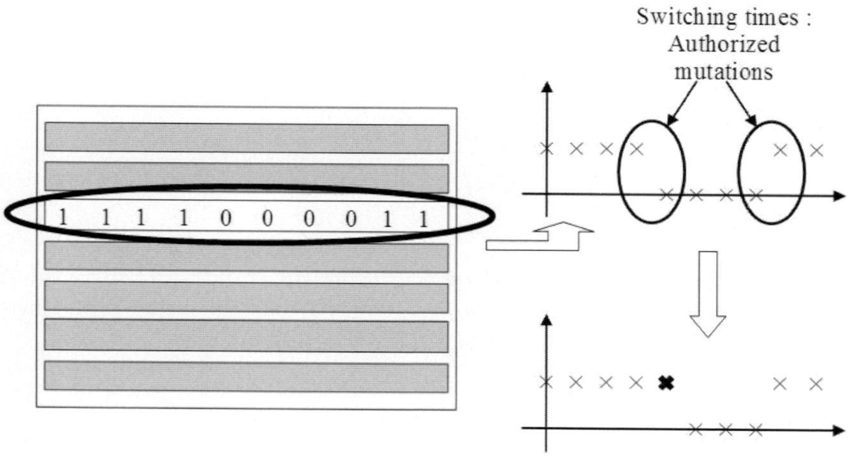

Fig. 8. Selective mutation operator

the solution where a mutation has a higher probability to create a new feasible solution. Thus, a selective mutation operator has been designed. Switching times are detected, and mutations are allowed only for these genes. No guarantee on the feasibility is achieved, but the "probability of feasibility" is higher.

4 Simulation Results

4.1 Real Time Updating of Produced Powers

As already explained, the "optimal" scheduling of the power system is computed from predicted values $\hat{Q}_n^{dem}$ on time interval $[m, m+N-1]$. The first values of integer scheduling u_m^k are applied to the system. Real variables are updated

Table 1. Characteristics of the benchmark example

Unit	Q_{min} (MW)	Q_{max} (MW)	ΔQ (MW)	α_0 (€)	α_1 (€/MWh)	c_{on} (€)	c_{off} (€)	T_{down} (h)	T_{up} (h)
1	10	40	10	25	2.6	10	2	2	4
2	10	40	10	25	7.9	10	2	2	4
3	10	40	10	25	13.1	10	2	3	4
4	10	40	10	25	18.3	10	2	3	4

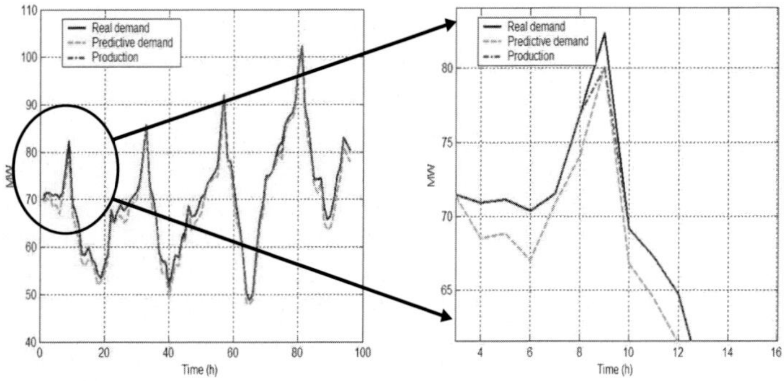

Fig. 9. Simulation results

at time m, when real value Q_m^{dem} is known, using the equation (12) considering Q_m^{dem} instead of $\hat{Q}_n^{dem}$. These values are given to the scheduling procedure at time m+1, for instance for the satisfaction of ramp constraints.

4.2 Case Study

To test the algorithm, a "four unit" academic case is considered. The characteristics of this case are given in table 1. A worst case is considered: consumer load is always underestimated. The prediction error is a random value in the range $[-5\%, 0\%]$. The time horizon is $N = 24$ hours. Consumer load has a daily oscillation. Thus, the dynamic of the system is about 24 hours and the time horizon has to be greater: a high value has to be given to N. The simulation is performed on a 4 day total horizon. Results, obtained with Matlab 6.5 on a PIV 2GHz, are given on figure 9.

Results show that the production is very close to the real demand, except for some peaks which have been underestimated. The optimization of the 96 binary variable problem is performed in just 25 seconds with the developed ant colony/genetic algorithm method. Due to the computation of successive economical near optimal solutions and real time slight updates, global costs are very close to global optimal costs. More precisely, a comparison has been performed with classical MILP solver, with a "Branch and Bound" method. It shows that the developed metaheuristic optimisation algorithm can compute a sub-optimal solution with a mean slight increase of 0.4% compared with the optimal solution. Furthermore, the MILP solver is very sensitive to the problem characteristics such as start up and shut down costs, leading to varying computation times going from 10 seconds to 10 hours. Finally, the algorithm has been tested for 10-unit case, leading to satisfying results with 10 minute computation times, and for which MILP solver is untractable.

5 Conclusions and Perspectives

In this paper a control structure has been defined for power systems. A meta-heuristic algorithm, based on ant colony and genetic algorithm, is used to perform the solution of Unit Commitment. As this kind of algorithm is highly tractable and explicitly manage all the constraints of the problem, it can be used to extend the optimization results in a closed loop real time framework.

This study is an experimentation of the use of stochastic algorithms for predictive control of hybrid systems. Of course, the main drawback of the proposed method is that no guarantee can be given on the actual optimality of solutions. Further, the problem of stability of the closed loop predictive control structure is a very tough task. However, the use of ant colony for the constrained predictive control of hybrid systems is promising. The proposed algorithm is tractable even for large scale systems and could be applied to various kinds of systems such as non linear or non analytical systems. Further, numerous constraints can be explicitly taken into account as the proposed method is based on a constructive algorithm. Finally, it allows the use of large receding horizons. Forthcoming works deal with the use of such algorithms in a more general framework for control of generic non linear constrained hybrid systems.

References

1. Branicky M.S., Borkar V.S., Mitter S.K.: A unified framework for hybrid control: model and optimal control theory, IEEE Transaction on Automatic Control, **43**, (1998), 31-45.
2. Mayne D.Q., Rawlings J.B., Rao C.V., Scockaert P.O.M.: Constrained model predictive control: Stability and optimality, Automatica, **36**, (2000),789–814.
3. Fletcher R., Leyffer S., "Numerical experience with lower bounds for MIQP branch and bound", *Technical report*, Dept. of Mathematics, University of Dundee, Scotland. *http://www.mcs.-dundee.ac.uk:8080/sleyffer/miqp_art.ps.z*, (1995).
4. Bemporad A., Borelli F., Morari M.: Piecewise Linear Optimal Controllers for Hybrid Systems, IEEE American Control Conference, (2000).
5. Olaru S., Thomas J., Dumur D., Buisson J.: Genetic Algorithm based Model Predictive Control for Hybrid Systems under a Modified MLD Form, International Journal of Hybrid Systems, **4**, (2004), 113-132.
6. Sen S., Kothari D. P.: Optimal Thermal Generating Unit Commitment: a Review. Electrical Power and Energy Systems, **20(7)**, (1998), 443-451.
7. Chen C.-L, Wang S.-C.: Branch and Bound scheduling for thermal generating units. IEEE Transactions on Energy Conversion, **8**, (1993), 184-189.

8. Ouyang Z., Shahidehpour S. M.: An intelligent dynamic programming for unit commitment application. IEEE Transactions on Power Systems, **6**, (1991), 1203-1209.
9. Senjyu T., Shimabukuro, K., Uezato K., Funabashi T.: A fast technique for Unit Commitment problem by extended priority list. IEEE Transactions on Power Systems, **19**, (2004), 2119-2120.
10. Zhai Q; Guan X.: Unit Commitment with identical units: successive subproblems solving method based on Lagrangian relaxation. IEEE Transactions on Power Systems, **17**, (2002), 1250-1257.
11. Cheng C.-P., Liu C.-W., Liu C.-C.: Unit Commitment by Lagrangian Relaxation and Genetic Algorithms. IEEE Transactions on Power Systems, **15**, (2000), 707-714.
12. Dotzauer E., Holmstrm K., Ravn H. F.: Optimal Unit Commitment and Economic Dispatch of Cogeneration Systems with a Storage. 13th Power Systems Computation Conference, Trondheim, Norway (1999), 738-744.
13. Yin Wa Wong S.: An Enhanced Simulated Annealing Approach to Unit Commitment. Electrical Power & Energy Systems, **20**, (1998), 359-368.
14. Rajan C. C. A, Mohan M. R.: An evolutionary programming-based tabu search method for solving the unit commitment problem. IEEE Transactions on Power Systems, **19**, (2004), 577-585.
15. Swarup K . S., Yamashiro: Unit commitment solution methodology using genetic algorithm. IEEE Transactions on Power Systems, **17**, (2002), 87-91.
16. Dorigo M., Maniezzo V., Colorni A.: The Ant System: Optimization by a Colony of Cooperating Agents. IEEE Transactions on Systems, Man and Cybernetics-Part B, **26**, (1996), 1-13.
17. Dorigo M., Gambardella, L. M., Ant Colony System : a Cooperative Learning Approach to the Traveling Salesman Problem, IEEE Transactions on Evolutionary Computation, **1** (1997), 53-66.
18. Sandou G., Font S., Tebbani S., Hiret A., Mondon C.: Optimisation par colonies de fourmis d'un site de génération d'énergie. Journal Européen des Systèmes Automatisés, Numéro spécial: Métaheuristiques pour l'optimisation difficile, **38**, (2004), 1097-1119.
19. Sandou G., Font S., Tebbani S., Hiret A., Mondon C.: Enhanced genetic algorithm with gurantee of feasibility for the Unit Commitment problem, submitted to IEEE Transactions on Power Systems.

Stabilization of Limit Cycles of Discretely Controlled Continuous Systems by Controlling Switching Surfaces

Axel Schild and Jan Lunze

Institute for Automation and Computer Control, Ruhr-Universitaet Bochum,
Universitaetsstrasse 150, 44780 Bochum, Germany
{Schild,Lunze}@atp.rub.de

Abstract. Discretely controlled continuous systems (DCCS) represent an important class of hybrid systems, in which a continuous process is regulated by a discrete controller. The paper considers the problem of stabilizing the periodic operation of such systems under event-driven switching. The key idea presented is to adjust switching surface parameters on-line to the current behavior of the continuous system. To this end, only measurements at switching instants are necessary. The presented approach extends the well-known OGY-method for controlling unstable periodic orbits in smooth chaotic systems to this class of hybrid systems.

1 Introduction

Discretely controlled continuous systems (DCCS) are composed of a continuous plant being arranged in a feedback loop with a discrete-event controller (Fig. 1(a)). The control task is to appropriately switch the mode of operation of the continuous process to meet objectives, which are specified in terms of the continuous variables at stationary operation. Internally, the controller consists of a switching logic and an event generator (Fig. 1(b)). The described system structure is found in various application domains, such as process engineering, manufacturing systems and power electronics. An important feature of DCCS is that no common equilibrium exists, which is shared by all the continuous mode dynamics. Moreover the system's function requires a perpetual switching among the modes of operation. Thus even for simple continuous mode dynamics, the resulting stationary behavior of the overall system is complex, namely either periodic or chaotic. In many applications a stable periodic stationary behavior is preferred.

The occurrence and the stability of periodic behavior of DCCS has been thoroughly investigated as for instance in [1,2] and many others. This paper however concerns the problem of designing a controller that locally stabilizes a given limit cycle of the closed-loop system. The focus is put onto the design of the event generator for a given switching logic. The novel idea presented is to introduce a switching surface controller (SSC), which adapts the location of the switching surfaces in the state space in response to the behavior of the continuous

A. Bemporad, A. Bicchi, and G. Buttazzo (Eds.): HSCC 2007, LNCS 4416, pp. 515–528, 2007.

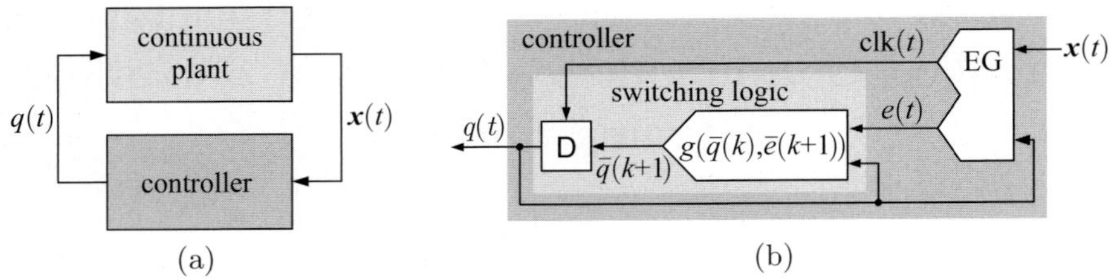

Fig. 1. Structure of a DCCS: (a) control loop, (b) discrete controller

system. The adjustments are *computed and executed on-line*. To this end, only measurements at the switching instants are necessary.

The approach adopts and extends methods from literature on controlling chaos in smooth chaotic systems like [3] to DCCS. Its central aspect concerns the analytic derivation of a suitable model of the closed-loop dynamics of a DCCS, which explicitly describes the influence of variations in the switching surfaces parameters on the system's behavior (Sect. 3.3). Based on this description, the original problem can be cast into a simple periodic control problem (Sect. 3.4). The performance of this method is demonstrated experimentally (Sect. 4).

Literature. Over the past decades a large effort has been spent on deriving state-feedback switching laws for many classes of hybrid systems. For the stabilization of common equilibria of switched linear systems, methods relying on the computation of common or piecewise Lyapunov functions have been elaborated [4]. They result in a conic state space partition, with the cones' borders constituting fixed switching surfaces. [5] presents a strategy for the practical stabilization of DCCS in an ϵ-environment around a reference. There, the motion of the continuous state inside this environment is arbitrary. Recently, optimal control was applied by [6] for determining switching surface parameters off-line, which minimize a suitable performance criterion for one particular initial state. An extension concerning the adjustment of switching times in the case of executions starting in different initial conditions, with potential for on-line implementation, was outlined in [7]. In contrast to the above, the stabilization of a particular periodic behavior of a DCCS considered here, has not attracted much attention yet and remains an open problem.

On the other hand, stabilizing unstable periodic orbits of smooth chaotic systems has been a hot topic over the past two decades. Many different solutions to this problem have been proposed as summarized in [8]. The first and most prominent solution is the OGY-method [3]. It relies on the knowledge of a *controlled Poincaré map* $\boldsymbol{P}_x$, in which the system parameters $\boldsymbol{p}$ appear as explicit arguments. Deriving a linear approximation of $\boldsymbol{P}_x$ around the fix-point $\bar{\boldsymbol{x}}^\star = \boldsymbol{P}_x(\bar{\boldsymbol{x}}^\star, \boldsymbol{p}_0)$, where $\bar{\boldsymbol{x}}^\star$ represents an unstable limit cycle for nominal parameter values $\boldsymbol{p}_0$, the initial problem can be cast into a classical linear time-invariant control problem. Several improvements and extensions of the original method have been developed [9,10], which mainly differ in the applied control laws, the number of samplings per period (multi-step methods) and the extension of the region around $\bar{\boldsymbol{x}}^\star$, in which the control signal is issued. Successful experimental application underline the feasibility of the approach.

For the design of a DCCS presented in this paper, the parameters of the switching surfaces represent a natural choice of externally accessible parameters as a means to delay switching to the successor mode or to execute it earlier. This influence can be employed for attaining stability and improving the convergence rate towards a limit cycle.

2 Models for Discretely Controlled Continuous Systems

2.1 Hybrid Model

A suitable hybrid model explicitly reflecting the components of a DCCS shown in Fig. 1 was presented in [11]. This model can be considered as a differential automaton [1] with an explicit realization of the discrete-event dynamics. The governing equation of the continuous plant is

$$\dot{\boldsymbol{x}}(t) = \boldsymbol{f}(\boldsymbol{x}(t), \boldsymbol{d}(t), q(t)), \ \boldsymbol{x}(0) = \boldsymbol{x}_0, \tag{1}$$

where $\boldsymbol{x} \in \mathrm{I\!R}^n$, $\boldsymbol{d} \in \mathrm{I\!R}^{n_\mathrm{d}}$ denote the measurable continuous state and the disturbance input, respectively. $q(t) \in \mathcal{Q}$ denotes the *mode of operation* of the overall system, where $\mathcal{Q}$ is a finite set. It provides the sole controllable input to the plant. For each mode q, the function $\boldsymbol{f}$ defines a Lipschitz-continuous n-dimensional vector field ensuring uniqueness of solutions to (1) between consecutive switching instants.

As depicted in Fig. 1(b), the controller can be separated into two components: an event generator (EG) and a switching logic. For event-driven switching considered in the sequel, the event generator is described by a set of *event functions*

$$\begin{aligned} &\Phi : \mathrm{I\!R}^n \times \mathcal{Q} \times \mathcal{E} \to \mathrm{I\!R} \\ &\Phi(\boldsymbol{x}, q, e) = \begin{cases} \Phi_{e_1}(\boldsymbol{x}, q), & \text{if } e = e_1 \\ \quad\vdots & \\ \Phi_{e_E}(\boldsymbol{x}, q), & \text{if } e = e_E, \end{cases} \end{aligned} \tag{2}$$

with the set of possible events $\mathcal{E}$. These functions implicitly define switching surfaces

$$\mathcal{S}(q, e) = \{\boldsymbol{x} \, : \, \Phi(\boldsymbol{x}, q, e) = 0\} \tag{3}$$

in $\mathrm{I\!R}^n$. The event generator outputs two signals: a trigger signal

$$\mathrm{clk}(t) = \begin{pmatrix} 0, & \text{if } |\Phi(\boldsymbol{x}(t), q(t), e)| > 0, \ \forall e \in \mathcal{E} \\ 1, & \text{if } \exists e \in \mathcal{E} \text{ s.t. } \Phi(\boldsymbol{x}(t), q(t), e) = 0, \end{pmatrix} \tag{4}$$

which is fed to a positive edge triggered memory, and an event signal

$$e(t) = \arg\min_{e \in \mathcal{E}} |\Phi(\boldsymbol{x}(t), q(t), e)| \,. \tag{5}$$

At the *switching time*

$$\bar{t}(k+1) = \min_{t \in \mathbb{R}} t \,|\mathrm{clk}\,(t) = 1 \wedge t > \bar{t}(k)\,, \tag{6}$$

the memory block (D) transfers the successor mode

$$\begin{aligned}\bar{q}(k+1) = q\big(\bar{t}(k+1)^{+}\big) &= g(q(\bar{t}(k+1))\,,\, e(\bar{t}(k+1))) \\ &= g(\bar{q}(k)\,,\, \bar{e}(k+1))\,,\ \bar{q}\,(0) = q_0,\end{aligned} \tag{7}$$

which is determined by the deterministic transition function g of the switching logic, to its output. Here $\bar{t}(k)^{+}$ denotes the limit from above. It is assumed, that $\bar{e}(k+1)$ is unique for all pairs $(\boldsymbol{x}, q)$, which can be ensured by establishing a priority order among the elements of $\mathcal{E}$. In the following, signal values at switching instants are defined as

$$\bar{\boldsymbol{x}}(k) = \boldsymbol{x}(\bar{t}(k))\,,\ \bar{q}(k) = q\big(\bar{t}(k)^{+}\big)\,,\ \bar{e}(k) = e(\bar{t}(k)) = e\big(\bar{t}(k)^{-}\big)$$

and are indicated by a bar and a discrete counter k. The distance of two consecutive switching times

$$\bar{\tau}(k) = \bar{t}(k+1) - \bar{t}(k) > 0$$

is called the *activation duration* of mode $\bar{q}(k)$ and τ denotes the elapsed time since the last switching. The *execution* of a DCCS is defined by the triplet

$$\chi\big(\boldsymbol{x}_0^h, t_0\big) = (\boldsymbol{x}\,(t)\,,\, q\,(t)\,,\, \mathcal{T})\,, \quad \boldsymbol{x}_0^h = \big(\boldsymbol{x}_0 \;\; q_0\big)^{\mathrm{T}}$$

with $\boldsymbol{x}\,(t)$ denoting the piecewise smooth continuous state trajectory, $q\,(t)$ denoting the piecewise constant, left-continuous discrete state trajectory and $\mathcal{T} = (([t_0, \bar{t}(0)]\,,\, (\bar{t}(0), \bar{t}(1)]\,,\, ...)$ being the infinitely long sequence of activation intervals. Furthermore the continuous evolution of the system during the activation of the particular mode q is denoted by $\boldsymbol{x}(\tau, \boldsymbol{x}_0, q)$.

2.2 Sampled Data Modeling of DCCS

If no disturbances occur, sampling the trajectories at switching instants

$$\bar{\chi}\big(\boldsymbol{x}_0^h, t_0\big) = \big(\big(\,\bar{\boldsymbol{x}}(0)\;\bar{q}(0)\;\bar{\tau}(0)\,\big)^{\mathrm{T}}, \big(\,\bar{\boldsymbol{x}}(1)\;\bar{q}(1)\;\bar{\tau}(1)\,\big)^{\mathrm{T}}, ...\big) \tag{8}$$

provides sufficient information to uniquely reconstruct the continuous behavior by means of (1). The sequence (8) is called the *sampled execution* and the vector $(\bar{\boldsymbol{x}}(k)\,, \bar{q}(k)\,, \bar{\tau}(k))^{\mathrm{T}}$ is the k-th *hybrid switch point.* To model this deterministic discrete-event evolution of a DCCS, the *embedded map* is defined:

Definition 1. *The* embedded map *of a DCCS is a function*

$$\begin{aligned}&\boldsymbol{H} : \mathbb{R}^n \times \mathcal{Q} \times \mathbb{R} \rightarrow \mathbb{R}^n \times \mathcal{Q} \times \mathbb{R} \\ &\big(\,\bar{\boldsymbol{x}}(k+1)\;\bar{q}(k+1)\;\bar{\tau}(k+1)\,\big)^{\mathrm{T}} = \boldsymbol{H}(\bar{\boldsymbol{x}}(k)\,,\, \bar{q}(k)\,,\, \bar{\tau}(k)),\end{aligned} \tag{9}$$

which determines the $(k+1)$-st hybrid switch point from the k-th one.

Unfortunately, an analytic expression for $\boldsymbol{H}$ can rarely be stated. However it can always be computed numerically for any DCCS. For notational convenience the map (9) can be decomposed into three parts, one for each hybrid switch point component:

$$\boldsymbol{H} = (\boldsymbol{H}_x \; \boldsymbol{H}_q \; \boldsymbol{H}_\tau)^{\mathrm{T}}.$$

3 Stabilization of Limit Cycles by Feedback Control

3.1 Control Objective of DCCS

A typical objective of DCCS is to stabilize a given limit cycle, which in analogy to [1,12] is defined as follows:

Definition 2. *An execution* $\chi\big(\boldsymbol{x}_0^h, \bar{t}(0)\big)$ *of a DCCS with* $\boldsymbol{x}_0^h = \big(\bar{\boldsymbol{x}}(0) \;\; \bar{q}(0)\big)^{\mathrm{T}}$ *is called* periodic *of* order p, *if it repeats itself after every p-th switching, i.e.*

$$\bar{q}(k+lp) = \bar{q}(k)\,,\; k,l \in \mathbb{N}\;,$$
$$\boldsymbol{x}(\tau, \bar{\boldsymbol{x}}(k+lp)\,, \bar{q}(k+lp)) = \boldsymbol{x}(\tau, \bar{\boldsymbol{x}}(k)\,, \bar{q}(k))\,,\; \forall \tau \in (0,\, \bar{\tau}(k)]\,,\; k,l \in \mathbb{N}$$

with $\bar{\boldsymbol{x}}(k)$, $\bar{q}(k)$ *and* $\bar{\tau}(k)$ *obtained from the corresponding sampled execution* (8).

Definition 3. *An isolated periodic execution* $\chi\big(\boldsymbol{x}_0^h, \bar{t}(0)\big)$ *of order p is is called a* limit cycle *of order p. Its limit set*

$$\mathcal{L}_{\mathrm{LC}} = \{(\boldsymbol{x}, q) \mid \exists k \in \{0, ..., p-1\}\,, \exists \tau \in (0, \bar{\tau}(k)] \;\; s.t.$$
$$\bar{q}(k) = q \in \mathcal{Q}_{\mathrm{LC}} \;\wedge\; \boldsymbol{x} = \boldsymbol{x}(\tau, \bar{\boldsymbol{x}}(k)\,, \bar{q}(k))\} \qquad (10)$$

corresponds to the closed hybrid orbit *traced out by* $\chi\big(\boldsymbol{x}_0^h, \bar{t}(0)\big)$ *over one switching cycle* $Q_{\mathrm{LC}} = (\bar{q}(0),\, \bar{q}(1),\, ...,\, \bar{q}(p-1))$. $\mathcal{Q}_{\mathrm{LC}} \subseteq \mathcal{Q}$ *is the set of modes activated during* Q_{LC}.

With a suitable mode definition, all elements of $\mathcal{Q}_{\mathrm{LC}}$ are unique. Therefore denote $\bar{q}(k+lp) = \bar{q}_k^\star$, $\bar{e}(k+lp) = \bar{e}_k^\star$, $k,l \in \mathbb{N}$ and refer to the associated switch points and activation durations of mode $\bar{q}_k^\star$ as $\bar{\boldsymbol{x}}(k+lp) = \bar{\boldsymbol{x}}^\star(\bar{q}_k^\star)$ and $\bar{\tau}(k+lp) = \bar{\tau}^\star(\bar{q}_k^\star)$. A sample limit cycle (thick dashed grey line) of order four is shown in Figure 2. It intersects the switching surfaces (dashed, black lines) in four switch points $\bar{\boldsymbol{x}}^\star(\bar{q}_k^\star)\,,\, k = 0, ..., 3$ (black diamonds). It is well known that local stability of a limit cycle can be determined from its *characteristic multipliers*

$$m_i = \lambda_i \left(\frac{\mathrm{d}\boldsymbol{H}_x^{(p)}}{\mathrm{d}\boldsymbol{x}} (\bar{\boldsymbol{x}}^\star(\bar{q}_k^\star)\,,\, \bar{q}_k^\star,\, \bar{\tau}^\star(\bar{q}_k^\star)) \right),\; i = 1, ..., n, \qquad (11)$$

which coincide with the eigenvalues λ_i of the Jacobian of the p-th iterate $H_x^{(p)}$ of (9) (return map) around its fix-point $\big(\bar{\boldsymbol{x}}^\star(\bar{q}_k^\star)\; \bar{q}_k^\star \; \bar{\tau}^\star(\bar{q}_k^\star)\big)^{\mathrm{T}}$ [12]. These multipliers characterize the dynamics of local deviations over one period. If $|m_i < 1|$ holds for all multipliers, all trajectories starting in a local environment $\mathcal{U}$ of the limit

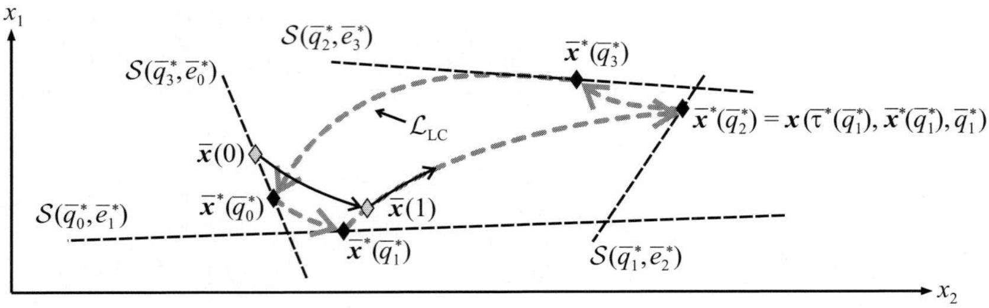

Fig. 2. Limit cycle executed by a DCCS

cycle converge to it asymptotically. Small $|m_i|$ imply a fast convergence and a large stability degree. Moreover they ensure robustness against bifurcations.

The situation investigated in this paper concerns the local stabilization of a limit cycle. Hence, it is feasible to assume that all executions starting in $\mathcal{U}$ generate the same cyclic mode sequence Q_{LC}. For the design task to be solved here, this assumption allows to reduce the event function (2) to only those switching conditions, which trigger one of the transitions $\bar{q}_k^\star \to \bar{q}_{k+1}^\star$.

Problem 1. Consider a limit cycle $\mathcal{L}_{\text{LC}}$, a DCCS with $\boldsymbol{x} \in \mathbb{R}^n$, $\mathcal{Q} = \mathcal{Q}_{\text{LC}}$ and predetermined dynamics (1), (7). Design the event generator, such that the stability degree $\Delta\lambda = 1 - \max|\lambda_i|$ of the limit cycle is maximized.

For DCCS with $\boldsymbol{x} \in \mathbb{R}^2$, where the dimensions of the switching surfaces and the continuous trajectory are equal, it is easy to determine event functions, which provide a solution to this problem. As illustrated in Fig. 2 by the sample trajectory starting in $\bar{\boldsymbol{x}}(0)$ (solid line), single-step convergence to $\mathcal{L}_{\text{LC}}$ at the transition $\bar{q}_0^\star \to \bar{q}_1^\star$ could be enforced, if the next switching was executed on the limit cycle at $\bar{\boldsymbol{x}}(1)$. Obviously, single-step convergence is ensured for all executions starting in $\mathcal{U}$, if each switching surfaces $\mathcal{S}\left(\bar{q}_k^\star, \bar{e}_{k+1}^\star\right)$ is identical to the orbit section of $\mathcal{L}_{\text{LC}}$ traced out during the activation of the successor mode $\bar{q}_{k+1}^\star$. Concerning the controller implementation, this knowledge is of no direct use, because firstly, analytic expression for $\Phi(\boldsymbol{x}, q, e)$ generally does not exist and secondly, an equivalent result for DCCS with $\boldsymbol{x} \in \mathbb{R}^n$ is not available. Consequently an event generator is sought, which locally reconstructs the optimal switching surfaces on-line. Such a method is described in the following.

3.2 Outline of Solution

The main idea is to parameterize the switching surfaces $\mathcal{S}\left(\bar{q}_k^\star, \bar{e}_{k+1}^\star\right)$ and to adapt them dynamically. Let $\mathcal{S}\left(\bar{q}_k^\star, \bar{e}_{k+1}^\star\right)$ be switching hyperplanes, i.e. (2) becomes

$$\Phi\big(\boldsymbol{x},\, \bar{q}_k^\star,\, \bar{e}_{k+1}^\star\big) = \boldsymbol{c}^{\mathrm{T}}\left(\bar{q}_k^\star, \bar{e}_{k+1}^\star\right) \boldsymbol{x} - d\big(\bar{q}_k^\star, \bar{e}_{k+1}^\star\big)\,. \tag{12}$$

In these event functions $\boldsymbol{c}^{\mathrm{T}}\left(\bar{q}_k^\star, \bar{e}_{k+1}^\star\right)$ denotes the plane's normal and $d\big(\bar{q}_k^\star, \bar{e}_{k+1}^\star\big)$ its distances to the origin.

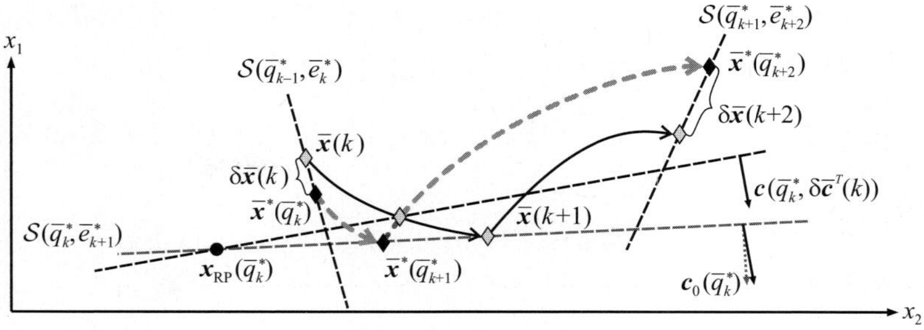

Fig. 3. Influence of the adjustment of a switching plane on subsequent switch points

First design step: Initialize the parameters of (12) with *nominal values* $\boldsymbol{c}_0^{\mathrm{T}}(\bar{q}_k^\star, \bar{e}_{k+1}^\star)$, $d_0(\bar{q}_k^\star, \bar{e}_{k+1}^\star)$, such that for all $\bar{q}_k^\star \in \mathcal{Q}_{\mathrm{LC}}$

$$\Phi(\boldsymbol{x}(\tau, \bar{\boldsymbol{x}}^\star(\bar{q}_k^\star), \bar{q}_k^\star), \bar{q}_k^\star, \bar{e}_k^\star) > 0, \forall \tau \in (0, \bar{\tau}^\star(\bar{q}_k^\star)], \tag{13}$$

$$\Phi(\bar{\boldsymbol{x}}^\star(\bar{q}_{k+1}^\star), \bar{q}_k^\star, \bar{e}_{k+1}^\star) = 0, \tag{14}$$

$$\boldsymbol{c}_0^{\mathrm{T}}(\bar{q}_k^\star, \bar{e}_{k+1}^\star)\, \boldsymbol{f}(\bar{\boldsymbol{x}}^\star(\bar{q}_{k+1}^\star), \bar{q}_k^\star) > 0 \text{ (transversal intersection)} \tag{15}$$

hold. As a further requirement for the choice of nominal parameters, $\bar{\boldsymbol{x}}^\star(\bar{q}_{k+1}^\star)$ must vary continuously under small perturbations of $\boldsymbol{c}_0^{\mathrm{T}}(\bar{q}_k^\star, \bar{e}_{k+1}^\star)$, $d_0(\bar{q}_k^\star, \bar{e}_{k+1}^\star)$.

With these nominal switching planes, $\mathcal{L}_{\mathrm{LC}}$ need not be locally stable. However it can be stabilized by perturbing the planes' positions (parameters) appropriately, which is illustrated in Fig. 3 for $\bar{\boldsymbol{x}} \in \mathbb{R}^2$.

As depicted, the deviation $\delta\bar{\boldsymbol{x}}(k) = \bar{\boldsymbol{x}}(k) - \bar{\boldsymbol{x}}^\star(\bar{q}_k^\star)$ of the current execution (solid line, grey diamonds) from the limit cycle (grey dashed line, black diamonds) can be eliminated after two switchings, if $\mathcal{S}(\bar{q}_k^\star, \bar{e}_{k+1}^\star)$ is adjusted, such that its first intersection with the execution coincides with the intersection of the execution with the limit cycle. Without this adjustment, the deviation will increase over the next two switchings, instead. For $\boldsymbol{x} \in \mathbb{R}^n$, $n > 2$, eliminating $\delta\bar{\boldsymbol{x}}(k)$ will generally require the adaptation of more than one switching plane, but is often possible over a full switching cycle.

Second design step: Determine a switching surface controller, which executes the described cyclic action, namely

1. At $\bar{t}(k)$, determine the deviation $\delta\bar{\boldsymbol{x}}(k)$ in the departure plane $\mathcal{S}(\bar{q}_{k-1}^\star, \bar{e}_k^\star)$
2. Compute the adjustments $\delta\bar{\boldsymbol{c}}^{\mathrm{T}}(k)$, $\delta\bar{d}(k)$ of the arrival plane $\mathcal{S}(\bar{q}_k^\star, \bar{e}_{k+1}^\star)$
3. Actuate $\mathcal{S}(\bar{q}_k^\star, \bar{e}_{k+1}^\star)$ by adjusting the corresponding parameters of (12)
4. After the occurrence of *two more* switchings, repeat the procedure,

such that small deviations $\delta\bar{\boldsymbol{x}}(k)$ decrease over one switching cycle. As will be shown, this two-step design procedure of the event generator solves the posed problem. In the next section, a model is derived, which describes the influence of $\delta\bar{\boldsymbol{c}}^{\mathrm{T}}(k)$, $\delta\bar{d}(k)$ onto $\bar{\boldsymbol{x}}(k+2)$. This model provides the basis for the systematic design of the switching surfaces controller. As for each $\bar{q}_k^\star$ the next event $\bar{e}_{k+1}^\star$

is unique, in the following nominal parameters $\boldsymbol{c}_0^{\mathrm{T}}(\bar{q}_k^\star, \bar{e}_{k+1}^\star)$ and $d_0(\bar{q}_k^\star, \bar{e}_{k+1}^\star)$ will be referred to as $\boldsymbol{c}_0^{\mathrm{T}}(\bar{q}_k^\star)$ and $d_0(\bar{q}_k^\star)$.

3.3 Analytical Expressions of the Linearized Controlled Embedded Map

The design of the switching surface controller requires an analytical expression of the consecutively applied *controlled embedded map*

$$\bar{\boldsymbol{x}}(k+2) = \boldsymbol{H}_x^{(2)}\left(\bar{\boldsymbol{x}}(k)\,,\,\bar{q}(k)\,,\,\bar{\tau}(k)\,,\,\mathbf{c}^{\mathrm{T}}(\bar{q}(k))\,,\,d(\bar{q}(k))\right), \tag{16}$$

which, compared to (9), explicitly depends on the switching condition parameters. As will be shown, it is possible to derive its linear approximation for each $\bar{q}_k^\star \in Q_{\mathrm{LC}}$ around $\bar{\boldsymbol{x}}^\star(\bar{q}_k^\star)$

$$\begin{aligned}
\delta\bar{\boldsymbol{x}}(k+2) &= \frac{\mathrm{d}\boldsymbol{H}_x^{(2)}}{\mathrm{d}\boldsymbol{x}}\left(\bar{\boldsymbol{x}}^\star(\bar{q}_k^\star)\,,\,\bar{q}_k^\star,\,\bar{\tau}^\star(\bar{q}_k^\star)\,,\,\boldsymbol{c}_0^{\mathrm{T}}(\bar{q}_k^\star)\,,\,d_0(\bar{q}_k^\star)\right)\,\delta\bar{\boldsymbol{x}}(k)\\
&+\frac{\mathrm{d}\boldsymbol{H}_x^{(2)}}{\mathrm{d}\boldsymbol{c}}\left(\bar{\boldsymbol{x}}^\star(\bar{q}_k^\star)\,,\,\bar{q}_k^\star,\,\bar{\tau}^\star(\bar{q}_k^\star)\,,\,\boldsymbol{c}_0^{\mathrm{T}}(\bar{q}_k^\star)\,,\,d_0(\bar{q}_k^\star)\right)\,\delta\bar{\boldsymbol{c}}(k)\\
&+\frac{\mathrm{d}\boldsymbol{H}_x^{(2)}}{\mathrm{d}d}\left(\bar{\boldsymbol{x}}^\star(\bar{q}_k^\star)\,,\,\bar{q}_k^\star,\,\bar{\tau}^\star(\bar{q}_k^\star)\,,\,\boldsymbol{c}_0^{\mathrm{T}}(\bar{q}_k^\star)\,,\,d_0(\bar{q}_k^\star)\right)\,\delta\bar{d}(k)\\
&= \boldsymbol{A}_{\mathrm{d}}(\bar{q}_k^\star)\,\delta\bar{\boldsymbol{x}}(k) + \underbrace{\left(\boldsymbol{b}_{\mathrm{d},1}(\bar{q}_k^\star)\;\boldsymbol{b}_{\mathrm{d},2}(\bar{q}_k^\star)\right)}_{\boldsymbol{B}_{\mathrm{d}}(\bar{q}_k^\star)}\,\underbrace{\left(\delta\bar{\boldsymbol{c}}^{\mathrm{T}}(k)\;\;\delta\bar{d}(k)\right)^{\mathrm{T}}}_{\boldsymbol{u}(k)}.
\end{aligned} \tag{17}$$

The matrices $\boldsymbol{A}_{\mathrm{d}}(\bar{q}_k^\star)$ and $\boldsymbol{B}_{\mathrm{d}}(\bar{q}_k^\star)$ are abbreviations of the Jacobians appearing in the first lines. As one possible parameterization of the parameters in (12) take

$$\mathbf{c}^{\mathrm{T}}\left(\bar{q}_k^\star, \bar{e}_{k+1}^\star, \delta\boldsymbol{c}^{\mathrm{T}}\right) = \boldsymbol{c}_0^{\mathrm{T}}(\bar{q}_k^\star) + \delta\boldsymbol{c}^{\mathrm{T}} \tag{18}$$

$$d(\bar{q}_k^\star, \bar{e}_{k+1}^\star, \delta\boldsymbol{c}^{\mathrm{T}}, \delta d) = d_0(\bar{q}_k^\star) + \left(\boldsymbol{c}_0^{\mathrm{T}}(\bar{q}_k^\star) + \delta\boldsymbol{c}^{\mathrm{T}}\right)\boldsymbol{c}_0(\bar{q}_k^\star)\,\delta d + \delta\boldsymbol{c}^{\mathrm{T}}\,\boldsymbol{x}_{\mathrm{RP}}(\bar{q}_k^\star)\;, \tag{19}$$

with $\boldsymbol{x}_{\mathrm{RP}}(\bar{q}_k^\star)$ denoting a suitable point of rotation of the switching plane $\mathcal{S}\left(\bar{q}_k^\star, \bar{e}_{k+1}^\star\right)$ (see Fig. 3). From the application of the chain rule, all total derivatives of (16) appearing in (17) can be expressed by a product of partial derivatives. To obtain these, consider the dependencies

$$\bar{\boldsymbol{x}}(k+1) = \boldsymbol{x}\left(\bar{\tau}(k)\,,\bar{\boldsymbol{x}}(k)\,,\bar{q}(k)\right), \quad \bar{\tau}(k) = \tau\left(\bar{\boldsymbol{x}}(k)\,,\bar{q}(k)\right)$$

and additional fundamental relations, which are presented next. The sensitivity of a trajectory for a fixed mode q with respect to its initial conditions is described by the solution to the variational equation

$$\frac{\mathrm{d}}{\mathrm{d}\tau}\left(\frac{\partial\boldsymbol{x}(\tau,\boldsymbol{x}_0,q)}{\partial\boldsymbol{x}_0}\right) = \frac{\partial\boldsymbol{f}}{\partial\boldsymbol{x}}\left(\boldsymbol{x}(\tau,\boldsymbol{x}_0,q)\,,q\right)\,\frac{\partial\boldsymbol{x}(\tau,\boldsymbol{x}_0,q)}{\partial\boldsymbol{x}_0},\quad \frac{\partial\boldsymbol{x}(0,\boldsymbol{x}_0,q)}{\partial\boldsymbol{x}_0} = \boldsymbol{I} \tag{20}$$

$$\frac{\mathrm{d}\boldsymbol{x}(\tau,\boldsymbol{x}_0,q)}{\mathrm{d}\tau} = \boldsymbol{f}\left(\boldsymbol{x}(\tau,\boldsymbol{x}_0,q)\,,q\right),\quad \boldsymbol{x}(0,\boldsymbol{x}_0,q) = \boldsymbol{x}_0. \tag{21}$$

With (20)–(21), a switch point's sensitivity with respect to its predecessor is

$$\frac{\partial \bar{\boldsymbol{x}}(k+1)}{\partial \bar{\boldsymbol{x}}(k)} = \frac{\partial \boldsymbol{x}(\bar{\tau}(k)\,,\bar{\boldsymbol{x}}(k)\,,\bar{q}(k))}{\partial \bar{\boldsymbol{x}}(k)}\,. \tag{22}$$

Furthermore the derivative of the switch point with respect to the activation duration is

$$\frac{\partial \bar{\boldsymbol{x}}(k+1)}{\partial \bar{\tau}(k)} = \boldsymbol{f}(\bar{\boldsymbol{x}}(k+1)\,,\bar{q}(k)) \tag{23}$$

The dependence of the activation duration on small perturbations in the departure switch point can be obtained by differentiating (12) with respect to $\boldsymbol{x}$ under consideration of (18)–(19) and by subsequently reorganizing the expression [12]:

$$\frac{\mathrm{d}\bar{\tau}(k)}{\mathrm{d}\bar{\boldsymbol{x}}(k)} = -\frac{\boldsymbol{c}_0^{\mathrm{T}}(\bar{q}(k))\;\frac{\partial \bar{\boldsymbol{x}}(k+1)}{\partial \bar{\boldsymbol{x}}(k)}}{\boldsymbol{c}_0^{\mathrm{T}}(\bar{q}(k))\;\boldsymbol{f}(\bar{\boldsymbol{x}}(k+1)\,,\bar{q}(k))}\;. \tag{24}$$

In the same way, the derivatives of the activation duration with respect to variations $\delta\boldsymbol{c}^{\mathrm{T}}$, δd of a switching surface parameterized by (18)–(19) can be obtained:

$$\frac{\mathrm{d}\bar{\tau}(k)}{\mathrm{d}\boldsymbol{c}} = -\frac{\bar{\boldsymbol{x}}(k+1)^{\mathrm{T}} - \boldsymbol{x}_{\mathrm{RP}}(\bar{q}(k))^{\mathrm{T}}}{\boldsymbol{c}_0^{\mathrm{T}}(\bar{q}(k))\;\boldsymbol{f}(\bar{\boldsymbol{x}}(k+1)\,,\bar{q}(k))} \tag{25}$$

$$\frac{\mathrm{d}\bar{\tau}(k)}{\mathrm{d}d} = \frac{\boldsymbol{c}_0^{\mathrm{T}}(\bar{q}(k))\,\boldsymbol{c}_0(\bar{q}(k))\;\delta d}{\boldsymbol{c}_0^{\mathrm{T}}(\bar{q}(k))\;\boldsymbol{f}(\bar{\boldsymbol{x}}(k+1)\,,\bar{q}(k))}\;. \tag{26}$$

With these fundamental relations the total derivatives appearing in (17) can be pieced together, as stated in the following theorem.

Theorem 1. *A linear approximation of the consecutively applied controlled embedded map* (16) *can be computed for any DCCS with linear event functions. The expressions of the total derivatives in* (17) *follow from* (22)–(26) *and yield*

$$\begin{aligned}\boldsymbol{A}_{\mathrm{d}}(\bar{q}_k^\star) &= \frac{\mathrm{d}\boldsymbol{H}_x^{(2)}}{\mathrm{d}\boldsymbol{x}}(\bar{\boldsymbol{x}}^\star(\bar{q}_k^\star)\,,\,\bar{q}_k^\star,\,\bar{\tau}^\star(\bar{q}_k^\star)\,,\,\boldsymbol{c}_0^{\mathrm{T}}(\bar{q}_k^\star)\,,\,d_0(\bar{q}_k^\star)) = \\ &\left(\frac{\partial \bar{\boldsymbol{x}}^\star(\bar{q}_{k+2}^\star)}{\partial \bar{\boldsymbol{x}}^\star(\bar{q}_{k+1}^\star)} + \frac{\partial \bar{\boldsymbol{x}}^\star(\bar{q}_{k+2}^\star)}{\partial \bar{\tau}^\star(\bar{q}_{k+1}^\star)}\frac{\mathrm{d}\bar{\tau}^\star(\bar{q}_{k+1}^\star)}{\mathrm{d}\bar{\boldsymbol{x}}^\star(\bar{q}_{k+1}^\star)}\right)\left(\frac{\partial \bar{\boldsymbol{x}}^\star(\bar{q}_{k+1}^\star)}{\partial \bar{\boldsymbol{x}}^\star(\bar{q}_k^\star)} + \frac{\partial \bar{\boldsymbol{x}}^\star(\bar{q}_{k+1}^\star)}{\partial \bar{\tau}^\star(\bar{q}_k^\star)}\frac{\mathrm{d}\bar{\tau}^\star(\bar{q}_k^\star)}{\mathrm{d}\bar{\boldsymbol{x}}^\star(\bar{q}_k^\star)}\right)\end{aligned} \tag{27}$$

$$\begin{aligned}\boldsymbol{b}_{\mathrm{d},1}(\bar{q}_k^\star) &= \frac{\mathrm{d}\boldsymbol{H}_x^{(2)}}{\mathrm{d}\boldsymbol{c}}(\bar{\boldsymbol{x}}^\star(\bar{q}_k^\star)\,,\,\bar{q}_k^\star,\,\bar{\tau}^\star(\bar{q}_k^\star)\,,\,\boldsymbol{c}_0^{\mathrm{T}}(\bar{q}_k^\star)\,,\,d_0(\bar{q}_k^\star)) = \\ &\left(\frac{\partial \bar{\boldsymbol{x}}^\star(\bar{q}_{k+2}^\star)}{\partial \bar{\boldsymbol{x}}^\star(\bar{q}_{k+1}^\star)} + \frac{\partial \bar{\boldsymbol{x}}^\star(\bar{q}_{k+2}^\star)}{\partial \bar{\tau}^\star(\bar{q}_{k+1}^\star)}\frac{\mathrm{d}\bar{\tau}^\star(\bar{q}_{k+1}^\star)}{\mathrm{d}\bar{\boldsymbol{x}}^\star(\bar{q}_{k+1}^\star)}\right)\left(\frac{\partial \bar{\boldsymbol{x}}^\star(\bar{q}_{k+1}^\star)}{\partial \bar{\tau}^\star(\bar{q}_k^\star)}\frac{\mathrm{d}\bar{\tau}^\star(\bar{q}_k^\star)}{\mathrm{d}\boldsymbol{c}}\right)\end{aligned} \tag{28}$$

$$\begin{aligned}\boldsymbol{b}_{\mathrm{d},2}(\bar{q}_k^\star) &= \frac{\mathrm{d}\boldsymbol{H}_x^{(2)}}{\mathrm{d}d}(\bar{\boldsymbol{x}}^\star(\bar{q}_k^\star)\,,\,\bar{q}_k^\star,\,\bar{\tau}^\star(\bar{q}_k^\star)\,,\,\boldsymbol{c}_0^{\mathrm{T}}(\bar{q}_k^\star)\,,\,d_0(\bar{q}_k^\star)) = \\ &\left(\frac{\partial \bar{\boldsymbol{x}}^\star(\bar{q}_{k+2}^\star)}{\partial \bar{\boldsymbol{x}}^\star(\bar{q}_{k+1}^\star)} + \frac{\partial \bar{\boldsymbol{x}}^\star(\bar{q}_{k+2}^\star)}{\partial \bar{\tau}^\star(\bar{q}_{k+1}^\star)}\frac{\mathrm{d}\bar{\tau}^\star(\bar{q}_{k+1}^\star)}{\mathrm{d}\bar{\boldsymbol{x}}^\star(\bar{q}_{k+1}^\star)}\right)\left(\frac{\partial \bar{\boldsymbol{x}}^\star(\bar{q}_{k+1}^\star)}{\partial \bar{\tau}^\star(\bar{q}_k^\star)}\frac{\mathrm{d}\bar{\tau}^\star(\bar{q}_k^\star)}{\mathrm{d}d}\right)\,.\end{aligned} \tag{29}$$

The proof follows directly from the rules of differentiation. Except for the matrix (22), all expressions are in analytic form.

3.4 Switching Surface Controller Design

To describe the dynamics of the deviation $\delta\bar{\boldsymbol{x}}(k)$ in one particular switching plane $\mathcal{S}\left(\bar{q}^\star_{p-1}, \bar{e}^\star_p\right)$, linear approximations (17) of several controlled embedded maps must be concatenated. This concatenation results in a discrete-time periodic linear system with a period of $(p(1+p \bmod 2)/2)$:

$$\delta\bar{\boldsymbol{x}}(2k+2) = \boldsymbol{A}_\mathrm{d}(\bar{q}^\star_{2k})\ \delta\bar{\boldsymbol{x}}(2k) + \boldsymbol{B}_\mathrm{d}(\bar{q}^\star_{2k})\ \boldsymbol{u}(2k)\,,\ \delta\bar{\boldsymbol{x}}(0) = \delta\bar{\boldsymbol{x}}_0 \text{ and} \tag{30}$$

$$\boldsymbol{A}_\mathrm{d}(\bar{q}^\star_k) = \boldsymbol{A}_\mathrm{d}(\bar{q}^\star_{k+p}),\ \boldsymbol{B}_\mathrm{d}(\bar{q}^\star_k) = \boldsymbol{B}_\mathrm{d}(\bar{q}^\star_{k+p}),\ \boldsymbol{c}^\mathrm{T}\left(\bar{q}^\star_{p-1}, \bar{e}^\star_p\right)\ \delta\bar{\boldsymbol{x}}(0) = 0. \tag{31}$$

From (11) and (17), it follows, that the eigenvalues $\tilde{\lambda}_i\left(\boldsymbol{\Psi}_{A_\mathrm{d}}\right)$ of the monodromy matrix of this periodic system (30)–(31)

$$\boldsymbol{\Psi}_{A_\mathrm{d}} = \boldsymbol{A}_\mathrm{d}\left(\bar{q}^\star_{(p(1+p \bmod 2)-2)}\right)\ \boldsymbol{A}_\mathrm{d}\left(\bar{q}^\star_{(p(1+p \bmod 2)-4)}\right)\cdots\boldsymbol{A}_\mathrm{d}(\bar{q}^\star_0) \tag{32}$$

and the characteristic multipliers of the limit cycle are identical, if p is even. If p is odd, then $\tilde{\lambda}_i = m_i^2$ holds. A similar relation can be shown for the characteristic multipliers and the eigenvalues of the closed-loop dynamics of (30)–(31) under

$$\bar{\boldsymbol{u}}(2k) = -\underbrace{\left(\boldsymbol{K}^\mathrm{T}_{\delta c}(\bar{q}(2k))\ \ \boldsymbol{K}^\mathrm{T}_{\delta d}(\bar{q}(2k))\right)^\mathrm{T}}_{\boldsymbol{K}(\bar{q}(2k))}\underbrace{\left(\bar{\boldsymbol{x}}(2k) - \bar{\boldsymbol{x}}^\star(\bar{q}(2k))\right)}_{\delta\bar{\boldsymbol{x}}(2k)}. \tag{33}$$

By considering the controlled event functions

$$\begin{aligned}&\Phi\big(\boldsymbol{x},\, \bar{q}(k)\,,\, \bar{e}(k+1)\,,\, \delta\bar{\boldsymbol{c}}^\mathrm{T}(k)\,,\, \delta\bar{d}(k)\big) = \\ &\boldsymbol{c}^\mathrm{T}\left(\bar{q}(k)\,, \bar{e}(k+1)\,,\, \delta\bar{\boldsymbol{c}}^\mathrm{T}(k)\right)\ \boldsymbol{x} - d\big(\bar{q}(k)\,, \bar{e}(k+1)\,, \delta\bar{\boldsymbol{c}}^\mathrm{T}(k)\,,\, \delta\bar{d}(k)\big)\,, \text{ with}\end{aligned} \tag{34}$$

$$\delta\bar{\boldsymbol{c}}^\mathrm{T}(k) = \begin{cases}\delta\bar{\boldsymbol{x}}^\mathrm{T}(k)\ \boldsymbol{K}^\mathrm{T}_{\delta c}(\bar{q}(k)) \\ \boldsymbol{0}\end{cases},\ \delta\bar{d}(k) = \begin{cases}\boldsymbol{K}^\mathrm{T}_{\delta d}(\bar{q}(k))\ \delta\bar{\boldsymbol{x}}(k)\,, & k=0,2,.. \\ \boldsymbol{0}, & k=1,3,..\end{cases} \tag{35}$$

parameterized by (18)–(19) and the control action of the SSC as described in Sect. 3.2, the following can be shown:

Lemma 1. *A limit cycle of a DCCS with controlled switching planes defined by event functions* (34), (35), *which are actuated at every second switching, possesses characteristic multipliers m_i, which are identical to the eigenvalues $\tilde{\lambda}_i\left(\boldsymbol{\Psi}_{(A_\mathrm{d}-B_\mathrm{d}\,K)}\right)$ of the monodromy matrix of the closed-loop periodic system* (30), (31), (33), *if p is even. Otherwise $\tilde{\lambda}_i = m_i^2$ holds.*

Therefore, designing a stabilizing event generator, which solves the postulated Problem 1, boils down to solving a dead-beat control problem of the periodic linear system (30)–(31), for which a well developed theory exists.

Theorem 2. *A switching surface controller, which locally stabilizes the limit cycle $\mathcal{L}_\mathrm{LC}$ and maximizes the stability degree $\Delta\lambda$, exists, if a periodic deadbeat-control law of the form* (33) *can be designed for the equivalent periodic linear system* (30)–(31), *which places the eigenvalues of $\boldsymbol{\Psi}_{(A_\mathrm{d}-B_\mathrm{d}\,K)}$ at zero.*

Proof. If a deadbeat controller $\boldsymbol{K}(\bar{q}(k))$ for the periodic system exists, Lemma 1 implies that $\Delta\lambda$ is maximal. Since the deviation $\delta\bar{\boldsymbol{x}}(k)$ disappears for $t \to \infty$, the control action becomes zero, leaving the nominal switching surfaces in effect. Thus condition (14) ensures that the stationary execution intersects the nominal switching planes exactly in the switch points $\bar{\boldsymbol{x}}^\star(\bar{q}_k^\star)$ of the limit cycle $\mathcal{L}_{\mathrm{LC}}$, which guarantees that the limit cycle is actually executed. □

If the periodic system is completely reachable or stabilizable with all uncontrollable eigenvalues at zero, the control law (33) can be determined according to different approaches as presented in [13,14]. Depending on the order p and the mode dynamics, this controller is generally not unique. As (17) is a first order approximation of (16), the periodic gains should be chosen balanced and as small as possible to avoid large control actions. In the case of $\boldsymbol{x} \in \mathbb{R}^2$ and $\boldsymbol{f}\left(\bar{\boldsymbol{x}}^\star\left(\bar{q}_{k+1}^\star\right), \bar{q}_{k+1}^\star\right) \neq \boldsymbol{f}\left(\bar{\boldsymbol{x}}^\star\left(\bar{q}_{k+1}^\star\right), \bar{q}_k^\star\right)$ for all $\bar{q}_k^\star \in \mathcal{Q}_{\mathrm{LC}}$, the feedback gains

$$\boldsymbol{K}(\bar{q}_k^\star) = \boldsymbol{B}_{\mathrm{d}}^{+}(\bar{q}_k^\star)\ \boldsymbol{A}_{\mathrm{d}}(\bar{q}_k^\star) \tag{36}$$

can be determined independently for each $\bar{q}_k^\star$. Here $\boldsymbol{B}_{\mathrm{d}}^{+}$ denotes the pseudo-inverse. With these SSC gains, single-step convergence towards $\mathcal{L}_{\mathrm{LC}}$ is achieved.

The extension of the discrete controller (Fig. 1(b)) by the switching surface controller is shown in Fig. 4. The function of the block $\mathcal{L}_{\mathrm{LC}}$ is to output the switch point $\bar{\boldsymbol{x}}^\star(\bar{q}(k))$ of the limit cycle corresponding to the currently activated mode. The currently measured deviation from this desired orbit is amplified by the gain $\boldsymbol{K}(\bar{q}(k))$ and issued from the memory block to the event generator every other switching. At all other switchings the parameters are left unchanged, which is achieved by the two-port switch and the source block.

4 Experimental Validation

The presented approach for stabilizing limit cycles was applied to the discretely controlled 2-Tank system, depicted in Figure 5(a). Like the one considered in [2], it consists of two coupled tanks with a controllable number of inlets and outlets.

Tank 1 is fed by a pump, whose output can be switched among two set-points by a binary control signal u_1. Complementarily, Tank 2 possesses two sinks: a permanent outflow and a supplementary outflow, which is opened or closed by

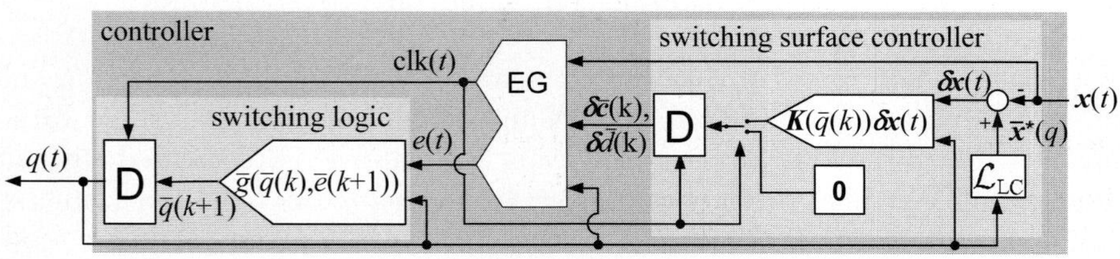

Fig. 4. Extended controller of DCCS

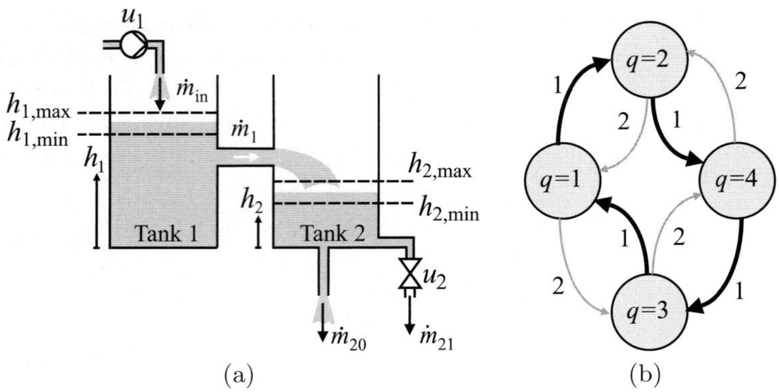

Fig. 5. Discretely controlled 2-Tank: (a) System setup, (b) Switching logic

an electromagnetic valve. The valve is actuated by the binary signal u_2. Because u_1, u_2 can be set independently, the DCCS possesses four operation modes:

$$q \leftrightarrow (u_1, u_2) \ : \ 1 \leftrightarrow (0,0),\ 2 \leftrightarrow (1,0),\ 3 \leftrightarrow (0,1),\ 4 \leftrightarrow (1,1) \ .$$

The implemented switching logic is depicted in Fig. 5(b) Furthermore, the nominal switching planes are defined by the event functions

$$\Phi\big((h_1 \ \ h_2)^{\mathrm{T}}, q, e\big) = \begin{cases} \ \ h_1 - h_{1,\min}, & \text{if } q=1, e=1;\ q=3, e=2 \\ -h_2 + h_{2,\max}, & \text{if } q=1, e=2;\ q=2, e=1 \\ -h_1 + h_{1,\max}, & \text{if } q=2, e=2;\ q=4, e=1 \\ \ \ h_2 - h_{2,\min}, & \text{if } q=3, e=1;\ q=4, e=2 \end{cases} . \qquad (37)$$

These switching planes stabilize a desired stationary periodic behavior for nominal system parameters. In Fig. 5(b) the transitions executed during the stationary switching sequence Q_{LC} are highlighted by thick arrows. The dynamics in the vicinity of the chosen operating point are accurately described by four affine mode dynamics $\delta\dot{\boldsymbol{h}} = \boldsymbol{A}_q\,\delta\boldsymbol{h} + \boldsymbol{b}_q, q = 1, ..., 4$ with

$$\boldsymbol{A}_1 = \boldsymbol{A}_2 = \begin{pmatrix} -0.0206 & 0 \\ 0.0206 & -0.0148 \end{pmatrix}, \boldsymbol{A}_3 = \boldsymbol{A}_4 = \begin{pmatrix} -0.0206 & 0 \\ 0.0206 & -0.0258 \end{pmatrix},$$
$$\boldsymbol{b}_1 = \big(-0.0026 \ \ 0.0015\big)^{\mathrm{T}},\ \boldsymbol{b}_2 = \big(-0.0017 \ \ 0.0015\big)^{\mathrm{T}},$$
$$\boldsymbol{b}_3 = \big(-0.0026 \ \ 0.0014\big)^{\mathrm{T}},\ \boldsymbol{b}_4 = \big(-0.0017 \ \ 0.0014\big)^{\mathrm{T}}.$$

Experimental data revealed that the flow constants of the interconnection and the outflow pipe vary considerably. Although small in amplitude, these variations destabilize the limit cycle, if the event generator is implemented without the proposed switching surfaces controller. According to Sect. 3.1 the switching planes, which trigger the transitions of the limit cycle ($e=1$ in (37)), are "controlled" in an experiment, whereas all others ($e=2$ in (37)) are left unchanged at nominal position. Deriving the four linearized embedded maps from (17) with

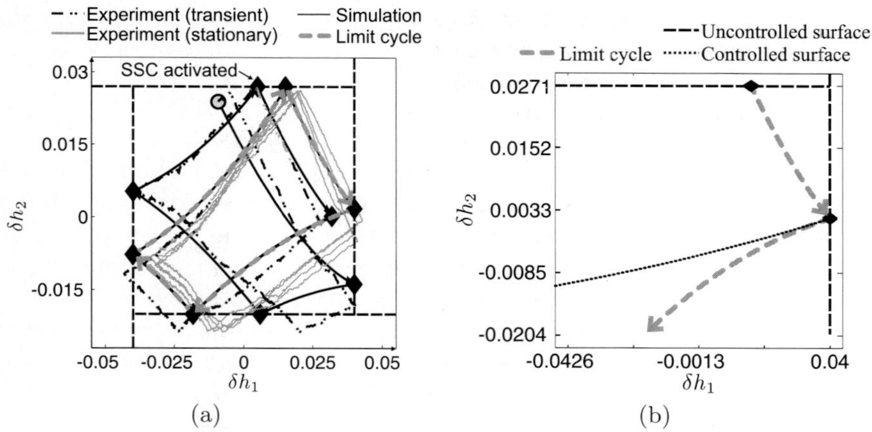

Fig. 6. Experimental data obtained from the 2-Tank: (a) Phase plot of simulated and measured executions, (b) Shape of a "controlled" switching surface

$\boldsymbol{x}_{\text{RP}}(\bar{q}_k^\star) = \bar{\boldsymbol{x}}^\star(\bar{q}_k^\star)$ and subsequently designing the SSC according to (36), the following switching surfaces controller gains are obtained:

$$\boldsymbol{K}(1) = \begin{pmatrix} 0 & 0 \\ 0 & 0 \\ 0.7031 & 1.0820 \end{pmatrix}, \quad \boldsymbol{K}(2) = \begin{pmatrix} 0 & 0 \\ 0 & 0 \\ 0.1190 & -0.3656 \end{pmatrix},$$

$$\boldsymbol{K}(3) = \begin{pmatrix} 0 & 0 \\ 0 & 0 \\ -0.0599 & 0.3113 \end{pmatrix}, \quad \boldsymbol{K}(4) = \begin{pmatrix} 0 & 0 \\ 0 & 0 \\ -0.8011 & 0.6270 \end{pmatrix}.$$

Figure 6(a) depicts a measured (grey solid line) and a simulated execution (black solid line) starting in the same initial state (circle). The dashed black paraxial lines represent the nominal switching surfaces and the grey diamonds correspond to the switch points $\bar{\boldsymbol{x}}(k)$ of the simulated trajectory. The limit cycle to be stabilized is indicated by the bold grey dashed line, which forms a closed parallelogram-like curve. Its switch points are highlighted by black diamonds.

At the beginning of the depicted execution, the switching surfaces controller is deactivated. As shown, both the simulated and the measured continuous state evolve far away from the desired motion and convergence towards $\mathcal{L}_{\text{LC}}$ is slow. Right after the activation of the SSC (white diamond), the simulated execution is redirected onto the limit cycle at the next mode transition. For all subsequent switchings, the control action vanishes, such that the switching planes take on their nominal position. As shown, the measured trajectory is also directed onto the limit cycle at the subsequent switching by the SSC. However, as a consequence of model uncertainties, neglected nonlinearities and disturbances, the measured trajectory constantly deviates from $\mathcal{L}_{\text{LC}}$ again. This is compensated by the SSC at every second switching and hence periodic operation is maintained. The same cannot be achieved, if the planes are fixed at their nominal position.

An analysis of the shape of the controlled switching planes reveals that this SSC succeeds in the local reconstruction of the optimal surfaces (Fig. 6(b)).

5 Conclusion

The paper presented an approach for locally stabilizing limit cycles in discretely controlled continuous systems by adapting switching surface parameters on-line. Given predetermined nominal switching hyperplanes, it was shown how to obtain a linear approximation of the controlled embedded map. This approximation yields time-invariant, discrete-time linear models describing the effect of variations in a switching hyperplane's location on the switch point two switchings ahead. By concatenating these models over a complete switching cycle, a discrete-time linear periodic system was obtained. With this, the design of a switching surface controller was reformulated as a deadbeat control problem of a periodic linear system and solved according to classical control theory. The application of the method to a switched 2-Tank system demonstrated an excellent performance even under considerable parameter uncertainties and disturbances.

References

1. Matveev, A., Savkin, A.: Qualitative Theory of Hybrid Dynamical Systems. Birkhauser (2000)
2. Rubensson, M., Lennartson, B., Pettersson, S.: Convergence to limit cycles in hybrid systems: An example. In: Proc. IFAC-LSS'98, Rio Patras (1998)
3. Ott, E., Grebogi, C., Yorke, J.: Controlling chaos. Physical Review Letters **64**(11) (1990) 1196–1199
4. Pettersson, S.: Synthesis of switched linear systems handling sliding motions. In: Proc. of IEEE International Symposium on Intelligent control, Limassol (2005)
5. Xu, X., Zhai, G.: On practical stability and stabilization of hybrid and switched systems. In: Hybrid Systems: Computation and Control, Springer Verlag (2004)
6. Boccadoro, M., Wardi, Y., Egerstedt, M., Verriest, E.: Optimal control of switching surfaces in hybrid systems. Discrete Event Dynamic Systems **15**(4) (2005) 433–448
7. Boccadoro, M., Egerstedt, M., Valigi, P., Wardi, Y.: Beyond the construction of optimal switching surfaces for autonomous hybrid systems. In: Proc. of ADHS, Alghero, Italy (2006) 101–105
8. Fradkov, A., Evans, R.: Control of chaos: survey 1997-2000. In: Proc. of 15th IFAC World Congress on Automatic Control., Barcelona (2002) 143–154
9. Romeiras, F., Grebogi, C., Ott, E., Dayawansa, W.: Controlling chaotic dynamical systems. Physica D **58** (1992) 165–192
10. Huebinger, B., Doerner, R., Martienssen, W.: Local control of chaotic motion. Zeitschrift fuer Physik B **90** (1993) 103–106
11. Krupar, J., Schild, A., Schwarz, W., Lunze, J.: Modeling and analysis of a class of hybrid systems by return maps. In: Proc. of NOLTA, Bologna, Italy (2006) 59–62
12. Parker, T.S., Chua, L.O.: Practical Numerical Algorithms for Chaotic Systems. Springer Verlag (1989)
13. Sreedhar, J., Dooren, P.V.: Pole placement via the periodic schur decomposition. In: Proc. American Control Conference, San Fransisco (1993)
14. Varga, A.: Robust minimum norm pole assignment with periodic state feedback. IEEE Trans. on Automatic Control **45**(5) (2000) 1017–1022

Approximate Simulation Relations and Finite Abstractions of Quantized Control Systems*

Paulo Tabuada

Electrical Engineering Department
University of California at Los Angeles
Los Angeles, CA 90095-1594
tabuada@ee.ucla.edu
http://www.ee.ucla.edu/~tabuada

Abstract. In this paper we revisit the construction of quantized models of control systems. Based on an approximate notion of simulation relation and under a stabilizability assumption we show how we can *force* a lattice structure on the reachable space of a quantized control system for any finite input quantization. When we are only interested in a compact subset of the state space, as is the case in concrete applications, our results immediately provide a finite model for the quantized control system.

Finite abstractions of continuous and hybrid systems are one of the most frequently used techniques to reduce the *verification* of these systems to the verification of finite-state systems. Recently, finite abstractions have also been used for the *synthesis* of correct-by-design embedded control software. This is accomplished by reformulating the synthesis of control software as the construction of a supervisory controller acting on the discrete abstraction. This finite-state controller is then refined to an hybrid system model of the control software enforcing the specification. The key enabling step in the whole synthesis process is the construction of the finite abstraction which is also the central theme of this paper.

The existing techniques for the construction of finite abstractions can be divided in two different (and dual) approaches: quotients and sub-systems. Techniques based on quotients construct a finite partition of the state space by identifying points having similar reachability properties. Examples include finite models for linear control systems in discrete-time [Tab07, TP06] and multi-affine control systems in continuous-time [KB06]. Finite abstractions based on subsystems are obtained by restricting the behaviors described by a control system to a subset having desirable properties. This restriction is achieved in the setting of quantized control systems [BMP02, BMP06] by selecting a subset of all the admissible input trajectories. Elements of this subset are termed control quanta and its choice can be seen as a quantization of the space of admissible input

* This research was partially supported by the National Science Foundation CAREER award 0446716.

A. Bemporad, A. Bicchi, and G. Buttazzo (Eds.): HSCC 2007, LNCS 4416, pp. 529–542, 2007.

trajectories. For certain classes of control systems, such as drift-free systems in chained form, the reachable space of the quantized system admits the structure of a lattice which can be rendered as fine as desired by properly choosing the control quanta. This structure can then be exploited to obtain efficient motion planning algorithms for these systems [PLPB02].

The work described in this paper is a contribution to the construction of finite abstractions based on the sub-system approach. By resorting to a notion of approximate simulation relation inspired by the work of Girard and Pappas [GP05b] we show how to impose a lattice structure on the reachable set of a quantized control system provided that: the resulting model is related to the original (unquantized) control system not by a simulation relation but by an approximate simulation relation; a certain stabilizability assumption holds. Moreover, this lattice structure is independent of the chosen input quantization which makes our results useful even if the quantized control system admits a lattice structure on the reachable set *for some but not for all* input quantizations. We also briefly discuss how this abstraction can be computed by resorting to numerical methods. As mentioned before, these finite abstractions are an essential ingredient for the correct-by-design synthesis of embedded control software. For reasons of space we cannot, in this paper, give details on the whole synthesis process and refer the interested reader to [Tab06] where he can also find several examples.

1 Definitions, Control Systems and Stability Notions

1.1 Definitions

The following definitions and notations will be used throughout the paper. Given a map $f : A \rightarrow B$ we denote by $\Gamma(f)$ the graph of f, that is, the set $\Gamma(f) = \big\{(a,b) \in A \times B \mid b = f(a)\big\}$. If A is a subset of B we denote by $\imath_A : A \hookrightarrow B$ or simply by $\imath$ the natural inclusion map taking any $a \in A$ to $\imath(a) = a \in B$. The identity map on a set A is denoted by 1_A. For $x \in \mathbb{R}^n$ we denote by x_i the ith element of the vector x. Let now $A \subseteq \mathbb{R}^n$ and $\mu \in \mathbb{R}$. We will use the notation $[A]_\mu$ to denote the subset of A defined by all the vectors whose elements are integer multiples of μ or equivalently $[A]_\mu = \big\{a \in A \mid a_i = k_i\mu \text{ for some } k_i \in \mathbb{Z} \text{ and } i = 1, \ldots, n\big\}$. The set $[A]_\mu$ is thus a subset of the lattice $[\mathbb{R}^n]_\mu$. When $x \in \mathbb{R}^n$, $\lceil x \rceil$ will denote the smallest integer $n \in \mathbb{N}$ such that $x \leq n$. We will say that $a \in \mathbb{R}$ integrally divides $b \in \mathbb{R}$ when $b/a \in \mathbb{Z}$. The standard Euclidean norm of $x \in \mathbb{R}^n$ is denoted by $\|x\|$ while $\|x\|_S$ denotes the usual point to set distance defined by:

$$\|x\|_S = \inf_{s \in S} \|x - s\|$$

We can thus recover $\|x\|$ as $\|x\|_{\{0\}}$. The closed ball centered at $x \in \mathbb{R}$ with radius ε is denoted by $\mathcal{B}_\varepsilon(x)$ or equivalently:

$$\mathcal{B}_\varepsilon(x) = \big\{y \in \mathbb{R}^n \mid \|x - y\| \leq \varepsilon\big\}$$

A continuous function $\gamma : \mathbb{R}_0^+ \to \mathbb{R}_0^+$, is said to belong to class $\mathcal{K}_\infty$ if it is strictly increasing, $\gamma(0) = 0$ and $\gamma(r) \to \infty$ as $r \to \infty$. A continuous function $\beta : \mathbb{R}_0^+ \times \mathbb{R}_0^+ \to \mathbb{R}_0^+$ is said to belong to class $\mathcal{KL}$ if, for each fixed s, the map $\beta(r, s)$ belongs to class $\mathcal{K}_\infty$ with respect to r and, for each fixed r, the map $\beta(r, s)$ is decreasing with respect to s and $\beta(r, s) \to 0$ as $s \to \infty$.

We now review some formal language concepts. Given a set S we denote by S^* the set of all finite strings obtained by concatenating elements in S. An element s of S^* is therefore given by $s = s_1 s_2 \dots s_n$ with $s_i \in S$ for $i = 1, \dots, n$. Given a string s belonging to S^* we denote by $s(i)$ the ith element of s. The length of a string $s \in S^*$ is denoted by $|s|$ and a subset of S^* is called a language. Given a map $f : A \to B$ we shall use the same letter to denote the extension of f to $f : A^* \to B^*$ defined by:

$$f\big(s(1)s(2)\dots s(n)\big) = f\big(s(1)\big)f\big(s(2)\big)\dots f\big(s(n)\big)$$

1.2 Control Systems

One the main objects of study in this paper are control systems defined as follows:

Definition 1. *A control system is a quadruple $\Sigma = (\mathbb{R}^n, U \subset \mathbb{R}^m, \mathcal{U}, f)$ where:*

- *U is a compact subset of $\mathbb{R}^m$ containing the origin;*
- *$\mathcal{U}$ is a subset of the set of all measurable functions from intervals of the form $]a, b[\subseteq \mathbb{R}$ to U with $a < 0$ and $b > 0$;*
- *$f : \mathbb{R}^n \times U \to \mathbb{R}^n$ is a continuous map satisfying the following Lipschitz assumption: for every compact set $K \subset \mathbb{R}^n$, there exists a constant $L > 0$ such that $\|f(x, u) - f(y, u)\| \leq L\|x - y\|$ for all $x, y \in K$ and all $u \in U$.*

An absolutely continuous curve $\mathbf{x} :]a, b[\to \mathbb{R}^n$ is said to be a trajectory of Σ if there exists $\mathbf{u} \in \mathcal{U}$ satisfying:

$$\dot{\mathbf{x}}(t) = f(\mathbf{x}(t), \mathbf{u}(t)) \tag{1}$$

for almost all $t \in]a, b[$. Control system Σ is said to be forward complete if every trajectory is defined on an interval of the form $]a, \infty[$.

Although we have defined trajectories over open domains, we shall refer to trajectories $\mathbf{x} : [0, \tau] \to \mathbb{R}^n$ defined on closed domains $[0, \tau]$, $\tau \in \mathbb{R}^+$ with the understanding of the existence of a trajectory $\mathbf{x}' :]a, b[\to \mathbb{R}^n$ such that $\mathbf{x} = \mathbf{x}'|_{[0,\tau]}$. We will also write $\mathbf{x}(\tau, x, \mathbf{u})$ to denote the point reached at time τ under the input $\mathbf{u}$ from initial condition x. This point is uniquely determined since the assumptions on f ensure existence and uniqueness of trajectories. For certain results we will need to assume that Σ is control affine meaning that $f(x, u)$ can be written as:

$$f(x, u) = f_0(x) + \sum_{i=1}^{m} f_i(x)u_i$$

where the f_i satisfy the same regularity conditions as f and $(u_1, \dots, u_m) \in U$.

1.3 Stability Notions

The results presented in this paper will assume certain stabilizability assumptions that we now recall. We will say that a set $S \subseteq \mathbb{R}^n$ is invariant under a control system Σ if for any trajectory $\mathbf{x}$ of Σ, $\mathbf{x}(0) \in S$ implies $\mathbf{x}(t) \in S$ for all $0 \leq t < b$. We will also need to refer to the diagonal set on $\mathbb{R}^{2n}$, denoted by Δ, and defined by $\Delta = \{(x, y) \in \mathbb{R}^n \times \mathbb{R}^n \mid x = y\}$.

Definition 2. *A control system $\Sigma = (\mathbb{R}^n, U \subset \mathbb{R}^m, \mathcal{U}, f)$ is uniformly globally asymptotically stable with respect to a closed invariant set S if it is forward complete and there exists a class $\mathcal{KL}$ function β such that the following estimate holds for all $x \in \mathbb{R}^n$, $\mathbf{u} \in \mathcal{U}$ and $t \geq 0$:*

$$\|\mathbf{x}(t, x, \mathbf{u})\|_S \leq \beta(\|x\|_S, t) \tag{2}$$

Definition 3 (Stabilizability Assumption). *A control system $\Sigma = (\mathbb{R}^n, U \subset \mathbb{R}^m, \mathcal{U}, f)$ is said to satisfy the Stabilizability Assumption (SA) if there exists a function $k : \mathbb{R}^n \times \mathbb{R}^n \times U \to U$ satisfying:*

1. *k is continuously differentiable on $\mathbb{R}^{2n} \backslash \Delta$;*
2. *$k(y, x, u) = u$ for $(x, y) \in \Delta$,*

and rendering control system $(\mathbb{R}^n \times \mathbb{R}^n, U \subset \mathbb{R}^m, \mathcal{U}, f \times_k f)$ with $f \times_k f$ defined by:

$$(f \times_k f)((x, y), u) = \big(f(x, u), f(y, k(y, x, u))\big) \tag{3}$$

uniformly globally asymptotically stable with respect to Δ, that is, enforcing the following estimate for all $x, y \in \mathbb{R}^n$, $\mathbf{u} \in \mathcal{U}$ and $t \geq 0$:

$$\|\mathbf{x}(t, x, \mathbf{u}) - \mathbf{y}(t, y, k(\mathbf{y}, \mathbf{x}, \mathbf{u}))\| \leq \beta(\|x - y\|, t) \tag{4}$$

The possible lack of regularity of k on Δ does not pose a problem with respect to existence and uniqueness of trajectories. On the open set $\mathbb{R}^{2n} \backslash \Delta$ existence and uniqueness of trajectories is guaranteed by the regularity assumptions on k and f. On the set Δ, the requirement $k(y, x, u) = u$ ensures that Δ is an invariant set since $f \times_k f$ degenerates into $(f(x, u), f(x, u))$ which guarantees existence and uniqueness of trajectories.

A control system satisfying the SA is able to track its own trajectories since for any trajectory $\mathbf{x}$ defined by an input curve $\mathbf{u}$, the feedback controller k will guarantee that the trajectory $\mathbf{y}$ starting at any initial condition and defined by the input curve $k(\mathbf{y}, \mathbf{x}, \mathbf{u})$ will asymptotically converge to $\mathbf{x}$.

A sufficient condition for the stabilizability assumption introduced in Definition 3 can be obtained through the concept of control Lyapunov function. Following the ideas initially presented in [Art83] and later extended by many authors, the existence of a control Lyapunov function allows one to recover the controller k. To simplify the presentation let us consider new coordinates given by:

$$z = x - y \qquad w = x + y \tag{5}$$

Proposition 1. *Let* $\Sigma = (\mathbb{R}^n \times \mathbb{R}^n, U \subset \mathbb{R}^m, \mathcal{U}, f)$ *be a control affine system with* $U = \{u \in \mathbb{R}^m \mid u_1^2 + u_2^2 + \ldots + u_m^2 \leq 1\}$ *and assume the existence of a continuously differentiable function* $V : \mathbb{R}^n \times \mathbb{R}^n \to \mathbb{R}_0^+$ *and class* $\mathcal{K}_\infty$ *functions* $\underline{\alpha}, \overline{\alpha}, \alpha$ *for which the following inequalities hold:*

1. $\forall z, w \in \mathbb{R}^n \qquad \underline{\alpha}(\|z\|) \leq V(z,w) \leq \overline{\alpha}(\|z\|)$;
2. $\forall z \in \mathbb{R}\, \forall u \in U\, \exists v \in U\, \forall w \in \mathbb{R}^n \quad \frac{\partial V}{\partial z} f((z,w),u) + \frac{\partial V}{\partial w} f((z,w),v) \leq -\alpha(\|z\|)$.

Then, control system Σ *satisfies the SA.*

Proof. The result follows, for example, from the formulas given in [LS95]. In this reference only asymptotic stability towards a compact closed set is considered. However, condition (2) guarantees that the resulting controller is a function of z and u alone thus guaranteeing global uniform asymptotic stability. This is not the case when instead of (2) we use the usual condition $\inf_{v \in V} \frac{\partial V}{\partial z} f((z,w),u) + \frac{\partial V}{\partial w} f((z,w),v) \leq -\alpha(\|z\|)$ which corresponds to (note the change in the quantification order):

$$\forall z \in \mathbb{R}\ \forall w \in \mathbb{R}^n\ \forall u \in U\ \exists v \in U \qquad \frac{\partial V}{\partial z} f((z,w),u) + \frac{\partial V}{\partial w} f((z,w),v) \leq -\alpha(\|z\|)$$

The previous result provides a more efficient way to determine if the SA is satisfied by searching for a single scalar function V instead of having to search for a controller k.

2 Approximate Simulations

In this section we introduce the notion of approximate simulation upon which all the results in this paper rely. Approximate simulations relate transition systems that will be used in this paper as abstract models for control systems.

Definition 4. *A transition system* T *is a quintuple* $(Q, L, \longrightarrow, O, H)$ *consisting of:*

- *A set of states* Q;
- *A set of labels* L;
- *A transition relation* $\longrightarrow\ \subseteq Q \times L \times Q$;
- *An output set* O;
- *An output function* $H : Q \to O$.

A metric transition system is a transition system $(Q, L, \longrightarrow, O, H)$ *in which the output set* O *is equipped with a metric* $\mathbf{d} : O \times O \to \mathbb{R}_0^+$.

We will follow standard practice and denote an element $(p, l, q) \in\ \longrightarrow$ by $p \xrightarrow{l} q$. We will also use the notation $p \xrightarrow{l} q$ when $l = l_1 l_2 \ldots l_n \in L^*$ is a string of elements in L. In this case $p \xrightarrow{l} q$ denotes the existence of a sequence of transitions $p \xrightarrow{l_1} p_1 \xrightarrow{l_2} p_2 \xrightarrow{l_3} \ldots \xrightarrow{l_n} q$. We shall say that a transition system T is finite when Q is finite. Transition systems capture dynamics through the transition relation. For any states $p, q \in Q$, $p \xrightarrow{l} q$ simply means that it is possible to evolve or jump from state p to state q under the action labeled by l. Note that we cannot model $\longrightarrow$ as a function since, in general, there may be several states $q_1, q_2 \in Q$ such that $p \xrightarrow{l} q_1$ and $p \xrightarrow{l} q_2$.

We will use transition systems as an abstract representation of control systems. There are several different ways in which we can transform control systems into transition systems. We now describe one of these which has the property of capturing all the information contained in a control system Σ:

Definition 5. *Let $\Sigma = (\mathbb{R}^n, U, \mathcal{U}, f)$ be a control system. The transition system $T(\Sigma) = (Q, L, \longrightarrow, O, H)$ associated with Σ is defined by:*

- *$Q = \mathbb{R}^n$;*
- *$E = \mathcal{U}$;*
- *$p \xrightarrow{\mathbf{u}} q$ if there exists a trajectory $\mathbf{x} : [0, \tau] \to \mathbb{R}^n$ of Σ satisfying $\mathbf{x}(\tau, p, \mathbf{u}) = q$ for some $\tau \in \mathbb{R}^+$;*
- *$O = \mathbb{R}^n$;*
- *$H = 1_{\mathbb{R}^n}$.*

Note that $T(\Sigma)$ is a metric transition system when we regard $O = \mathbb{R}^n$ as being equipped with the metric $d(p, q) = \|p - q\|$.

Definition 6. *A run of a transition system $T = (Q, L, \longrightarrow, O, H)$ is a string $r \in Q^*$ for which there exists $l \in L^*$ satisfying $r(i) \xrightarrow{l(i)} r(i+1)$ for $i = 1, \ldots, |r| - 1$. A string $s \in O^*$ is said to be an output run of T if there exists a run r of T such that $H(r) = s$. The language of T, denoted by $L(T)$, is the set of all output runs of T.*

Simulation and bisimulation relations are standard mechanisms to relate the properties of transition systems [CGP99]. Intuitively, a simulation relation from a transition system T_1 to a transition system T_2 is a relation between the corresponding state sets explaining how a run r of T_1 can be transformed into a run s of T_2. While typical simulation relations require that runs r and s are observationally indistinguishable, that is, $H_1(r) = H_2(s)$, we shall relax this by requiring $H_1(r)$ to simply be close to $H_2(s)$ where closeness is measured with respect to the metric on the output set:

Definition 7. *Let $T_1 = (Q_1, L_1, \underset{1}{\longrightarrow}, O, H_1)$ and $T_2 = (Q_2, L_2, \underset{2}{\longrightarrow}, O, H_2)$ be metric transition systems with the same output space and let $\varepsilon, \delta \in \mathbb{R}^+$. A relation $R \subseteq Q_1 \times Q_2$ is said to be a (ε, δ)-approximate simulation relation from T_1 to T_2 if:*

1. *$(q_1, q_2) \in R$ implies $\mathbf{d}(H(q_1), H(q_2)) \le \varepsilon$;*
2. *$\mathbf{d}(H(q_1), H(q_2)) \le \delta$ implies $(q_1, q_2) \in R$;*
3. *$(q_1, q_2) \in R$ and $q_1 \underset{1}{\xrightarrow{l_1}} q_1'$ imply the existence of $q_2 \in Q_2$ such that $q_2 \underset{2}{\xrightarrow{l_2}} q_2'$ with $(q_1', q_2') \in R$.*

A different notion of approximate simulation appeared[1] in the work of Girard and Pappas [GP05a] where it was termed δ-approximate simulation relation. Such a notion is essentially the same as a (ε, δ)-approximate simulation relation except for requirement (2) which is not present in [GP05a]. The need for this requirement and for two parameters, namely ε and δ, will become apparent in Section 3 where we provide a characterization in terms of the stabilizability concepts reviewed in Section 1.

While the existence of a simulation relation between two transition systems implies language containment, the existence of an approximate simulation only implies a weaker version of this containment:

Proposition 2. *If there exists a (ε, δ)-approximate simulation relation from T_1 to T_2 satisfying $R(Q_1) = Q_2$, then $L(T_1) \subseteq \mathcal{B}_\varepsilon\big(L(T_2)\big)$ where $\mathcal{B}_\varepsilon\big(L(T_2)\big)$ denotes the language:*

$$\big\{s \in O^* \mid \mathbf{d}(s(i), r(i)) \leq \varepsilon \text{ for some } r \in L(T_2) \text{ with } |s| = |r| \text{ and } i = 1, \ldots, |s|\big\}$$

Proof. For any $s \in L(T_1)$ there exist strings $r \in Q^*$ and $l \in L^*$ such that:

$$r(1) \xrightarrow{l(1)} r(2) \xrightarrow{l(2)} \ldots \xrightarrow{l(|s|-1)} r(|s|)$$

and $H(r) = s$. Let now $q_2 \in Q_2$ satisfy $(r(1), q_2) \in R$ and note that q_2 exists since $R(Q_1) = Q_2$. By the definition of approximate simulation relation we have $u(1) = q_2 \xrightarrow{m(1)} u(2)$ for some $m(1) \in L_2$ and $(r(2), u(2)) \in R$. Invoking (1) in the definition of approximate simulation we conclude that $\mathbf{d}(r(2), u(2)) \leq \varepsilon$. Extending this argument by induction on the length of s we conclude the existence of $u \in Q_2^*$, $m \in L_2^*$ with $|u| = |r|$, $(r(i), u(i)) \in R$ and thus $\mathbf{d}(H(r(i)), H(u(i))) \leq \varepsilon$ for $i = 1, \ldots, |s|$ or $H(r) \in \mathcal{B}_\varepsilon(L(T_2))$. □

The notion of sub-transition system formalizes the idea of constructing a new transition system by isolating certain states and certain transitions of an existing transition system:

Definition 8. *Transition system $T_1 = (Q_1, L_1, \underset{1}{\longrightarrow}, O, H_1)$ is said to be a sub-transition system of $T_2 = (Q_2, L_2, \underset{2}{\longrightarrow}, O, H_2)$ if $Q_1 \subseteq Q_2$, $H_1 = H_2|_{Q_1}$, and the graph $\Gamma(\imath)$ of the natural inclusion $\imath : Q_1 \hookrightarrow Q_2$ is a relation satisfying requirement (3) in Definition 7.*

In the remaining paper we will work with sub-transition systems of $T(\Sigma)$ obtained by selecting those transitions from $T(\Sigma)$ describing trajectories of duration τ for some chosen $\tau \in \mathbb{R}^+$. This can be seen as a time discretization or sampling process.

Definition 9. *Let Σ be a control system and $T(\Sigma)$ its associated transition system. For any $\tau \in \mathbb{R}^+$, the sub-transition system $T_\tau(\Sigma) = (Q, L, \longrightarrow, O, H)$ of $T(\Sigma)$ is defined by:*

[1] The authors of [GP05b] only discuss approximate bisimulations but one can easily derive the corresponding notion of approximate simulation.

1. $Q = \mathbb{R}^n$;
2. $L = \{\mathbf{u} \in \mathcal{U} \mid$ *the domain of* $\mathbf{u}$ *is* $[0, \tau]\}$;
3. $p \xrightarrow{\mathbf{u}} q$ *if there exists a trajectory* $\mathbf{x}$ *of* Σ *satisfying* $\mathbf{x}(\tau, p, \mathbf{u}) = q$;
4. $O = \mathbb{R}^n$;
5. $H = 1_{\mathbb{R}^n}$.

3 Existence of Approximate Simulations

The adequacy of the notion of approximate simulation relation introduced in the previous section will be justified in this paper with two arguments: its characterization in terms of known stabilizability concepts and its essential role in the existence and computation of finite abstractions. In this section we provide the first argument by relating existence of approximate simulation relations with the SA:

Theorem 1. *Let* Σ *be a control system satisfying the SA. Then, for any* $\varepsilon \in \mathbb{R}^+$ *there exists a* $\delta \in \mathbb{R}^+$ *such that for all* $\tau \in \mathbb{R}^+$ *there exists a* (ε, δ)*-approximate simulation relation from* $T_\tau(\Sigma)$ *to* $T(\Sigma)$.

The proof of this result can be found in [Tab06] where a converse result is also presented under an assumption weaker than the SA. In concrete applications we shall not work with $T_\tau(\Sigma)$ but with a finite sub-transition system of $T_\tau(\Sigma)$. In this case we can still guarantee existence of a (ε, δ)-approximate simulation relation:

Corollary 1. *Let* Σ *be a control system satisfying the SA and consider a finite sub-transition system* $T = (Q, L, \longrightarrow, O, H)$ *of* $T_\tau(\Sigma)$. *Then, there exists a contractible compact set* $S \subset \mathbb{R}^n$ *containing* Q *and a* (ε, δ)*-approximate simulation relation* R *from* T *to* $T_\tau(\Sigma)$ *satisfying* $R(Q) = S$.

Existence of approximate simulations from arbitrary sub-transition systems of $T_\tau(\Sigma)$ to $T_\tau(\Sigma)$ is thus guaranteed by the SA which, according to Proposition 1, can be checked by resorting to a control Lyapunov function. The correct-by-design methodology that is being introduced in this paper thus leverages on the extensive work that has been done by the control community on stability, stabilization and its Lyapunov characterizations.

4 Computation of Finite Sub-transition Systems

We consider the computation of finite sub-transition systems in the framework of quantized control systems [BMP02, PLPB02, BMP06] where one restricts attention to a denumerable subset of $\mathcal{U}$ whose elements are termed control quanta. In this paper, control quanta are defined by constant input curves assuming values in a finite set $\mathtt{U} \subset U$. Although this restriction on the class of input curves may appear to be quite drastic, there are several reasons to consider it. In many man-made systems, input signals are physically implemented as piece-wise constant signals. Our assumptions are then in consonance with real physical constraints. Moreover, input quantization can be seen as a very powerful complexity reduction mechanism simplifying several control synthesis problems [BMP02, PLPB02, BMP06].

From Corollary 1 we know that under the SA we can construct a (ε, δ)-approximate simulation relation from any finite sub-transition system T of $T_\tau(\Sigma)$ to $T_\tau(\Sigma)$. The question we address in this section is:

How do we compute such finite sub-transition systems?

We assume that parameters τ and ε, describing the desired sampling time and state accuracy, respectively, are given along with a finite set $\mathtt{U} \subset U$ of inputs and compact subset $S \subset \mathbb{R}^n$ of the state space. The finite set $\mathtt{U}$ describes the input quantization while the set S represents the working region that is of interest and which will be compact (at least bounded) in concrete applications. The naive approach to obtain a sub-transition system based on the given data would be to construct the transition relation by rounds. The first round would compute all the transitions $p \xrightarrow{u} q$ with $u \in \mathtt{U}$, $p \in [S]_\eta$, (where $\eta \in \mathbb{R}^+$ is chosen so that any $x \in S$ belongs to $\mathcal{B}_\varepsilon(p')$ for some $p' \in [S]_\eta$) and for which there exists a trajectory $x : [0, \tau] \to \mathbb{R}^n$ of Σ satisfying $\mathbf{x}(\tau, p, u) = q$. The second round would repeat the same construction, enlarging the transition relation with transitions starting at the states q obtained in the first round. The sub-transition system T could then be seen as the limit of this process. One immediate difficulty with this naive approach is to determine at which round to terminate the construction of T since this process is not guaranteed to terminate. But there are also other difficulties that we now illustrate with the following linear control system:

$$\begin{bmatrix} \dot{x}_1 \\ \dot{x}_2 \end{bmatrix} = \begin{bmatrix} 0 & 1 \\ -1 & 2 \end{bmatrix} \begin{bmatrix} x_1 \\ x_2 \end{bmatrix} + \begin{bmatrix} 0 \\ u \end{bmatrix} \tag{6}$$

with states $(x_1, x_2) \in [-5, 5] \times [-5, 5] = S$ and input $u \in \{-1, 0, 1\} = \mathtt{U}$. The outcome of the naive approach to the construction of a finite sub-transition system of $T_{0.5}(\Sigma)$ is displayed in Figure 1.

The first observation is that terminating the process after some predetermined number n of rounds may lead to a sub-transition system that is only guaranteed to be nonblocking during the first n steps. Since many control tasks require the system to run for an arbitrarily long sequence of steps this is a serious drawback of the naive approach. Moreover, the states of the constructed sub-transition system are not evenly distributed across the state space thus implying that we have a better description of the dynamics in some areas than in others. These difficulties can be avoided when a (ε, δ)-approximate sub-transition system can be found:

Definition 10. *Let Σ be a control system and let $\tau, \varepsilon \in \mathbb{R}^+$ and a finite $\mathtt{U} \subset U$ be given. A transition system T is said to be a (ε, δ)-approximate sub-transition system if:*

1. *there exists a (ε, δ)-approximate simulation relation from T to $T_\tau(\Sigma)$;*
2. *$Q \subseteq [\mathbb{R}^n]_\chi$ for some $\chi \in \mathbb{R}^+$;*
3. *for every $p \in Q$ and $l \in \mathtt{U}$ there exists a $q \in Q$ such that $p \xrightarrow{l} q$ in T.*

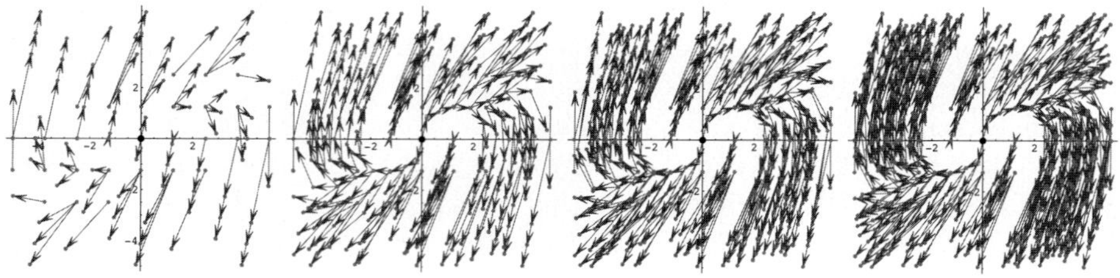

Fig. 1. Finite sub-transition system of the linear system (6) obtained through the naive method. States are represented by red dots while black dots represent states for which there exists a self transition. Transitions are represented by blue arrows. From left to right we have the result of the first, third, fifth and tenth simulation rounds.

Transition system T is equipped with a (ε, δ)-approximate simulation relation to $T_\tau(\Sigma)$ and solves the difficulties illustrated by the previous example by guaranteeing that its state set is a subset of a lattice. When working on a compact subset of the state space, usually the case in most applications, T is in fact finite and the construction of T is guaranteed to terminate. Furthermore, T is nontrivial in the sense that for every state p of T all the transitions labeled by inputs in $\mathtt{U}$ are captured in T and lead to states of T. Note that when one restricts attention to a compact subset S of the state space, some states may fail to have transitions defined for every element of $\mathtt{U}$. However, this is the case only when these transitions would lead to states outside S. Existence of (ε, δ)-approximate sub-transition systems is guaranteed by the SA:

Theorem 2. *For any control system Σ satisfying the SA, for any $\varepsilon \in \mathbb{R}^+$, for any finite $\mathtt{U} \subset U$ and for any $\tau \in \mathbb{R}^+$ such that $\beta(\varepsilon, \tau) < \varepsilon$ there exists a (ε, δ)-approximate sub-transition system T of $T_\tau(\Sigma)$. Furthermore, χ can be chosen to be any positive real number integrally dividing $2\varepsilon/\sqrt{n}$ and satisfying:*

$$0 < \chi \leq \frac{2}{\sqrt{n}}\big(\varepsilon - \beta(\varepsilon, \tau)\big), \tag{7}$$

and the (ε, δ)-approximate simulation relation R from T to $T_\tau(\Sigma)$ satisfies $R(Q) = \mathbb{R}^n$ where Q is the state set of T.

Note that the condition $\beta(\varepsilon, \tau) < \varepsilon$ can always be satisfied by choosing a sufficiently large τ since β is a decreasing function of τ.

Proof. We start by constructing T. Let $\xi = 2\varepsilon/\sqrt{n}$, assume that χ integrally divides ξ and that it satisfies inequality (7). Let now $F : \mathbb{R}^n \to 2^{[\mathbb{R}^n]_\chi}$ be the function defined by $q \in F(p)$ if $p \in \mathcal{B}_{\varepsilon-\beta(\varepsilon,\tau)}(q)$. The set Q of states of T is the smallest set satisfying:

1. $[\mathbb{R}^n]_\xi \subseteq Q$;
2. $p \in Q$, $u \in \mathtt{U}$ and $q \in F\big(\mathbf{x}(\tau, p, \mathbf{u})\big)$ for some trajectory $\mathbf{x} : [0, \tau] \to \mathbb{R}^n$ of Σ with $\mathbf{u}(t) = u$ for $0 \leq t \leq \tau$ imply $q \in Q$.

The transition relation is defined by $p \xrightarrow{u} q$ if $p, q \in Q$, $u \in \mathtt{U}$ and there exists a trajectory $\mathbf{x} : [0, \tau] \to \mathbb{R}^n$ of Σ satisfying $q \in F\big(\mathbf{x}(\tau, p, \mathbf{u})\big)$ with $\mathbf{u}(t) = u$ for $0 \leq t \leq \tau$. Transition system T is thus defined by $T = (Q, \mathtt{U}, \longrightarrow, \mathbb{R}^n, \imath : Q \hookrightarrow \mathbb{R}^n)$. The approximate simulation relation is given by $(q, x) \in R$ iff $\|q - x\| \leq \varepsilon$. Note that requirements (1) and (2) in Definition 7 are satisfied by construction if we take $\delta = \varepsilon$. By noting that any point $x \in \mathbb{R}^n$ belongs to $\mathcal{B}_\varepsilon(q)$ for some $q \in [\mathbb{R}^n]_\xi$ we conclude that $R(Q) = \mathbb{R}^n$. We now show that R also satisfies requirement (3) in Definition 7. Let $(p, y) \in R$ and assume that $p \xrightarrow{\mathbf{u}} q$ with $\mathbf{u}(t) = u \in \mathtt{U}$ for $0 \leq t \leq \tau$. This implies the existence of a trajectory $\mathbf{x}$ of Σ satisfying $q \in F\big(\mathbf{x}(\tau, p, \mathbf{u})\big)$ or equivalently $\|q - \mathbf{x}(\tau, p, \mathbf{u})\| \leq \varepsilon - \beta(\varepsilon, \tau)$. Since $(p, y) \in R$ implies $\|p - y\| \leq \varepsilon$ we have, by the SA, $\|\mathbf{x}(\tau, p, \mathbf{u}) - \mathbf{y}(\tau, y, k(\mathbf{y}, \mathbf{x}, \mathbf{u}))\| \leq \beta(\|p - y\|, \tau) = \beta(\varepsilon, \tau)$. It then follows:

$$\begin{aligned}\|q - \mathbf{y}(\tau, y, k(\mathbf{y}, \mathbf{x}, \mathbf{u}))\| &\leq \|q - \mathbf{x}(\tau, p, \mathbf{u}) + \mathbf{x}(\tau, p, \mathbf{u}) - \mathbf{y}(\tau, y, k(\mathbf{y}, \mathbf{x}, \mathbf{u}))\| \\ &\leq \|q - \mathbf{x}(\tau, p, \mathbf{u})\| + \|\mathbf{x}(\tau, p, \mathbf{u}) - \mathbf{y}(\tau, y, k(\mathbf{y}, \mathbf{x}, \mathbf{u}))\| \\ &\leq \varepsilon - \beta(\varepsilon, \tau) + \beta(\varepsilon, \tau) = \varepsilon\end{aligned}$$

thus showing $y \xrightarrow{k(\mathbf{y},\mathbf{x},\mathbf{u})} \mathbf{y}(\tau, y, k(\mathbf{y}, \mathbf{x}, \mathbf{u}))$ in $T_\tau(\Sigma)$ with $(q, \mathbf{y}(\tau, y, k(\mathbf{y}, \mathbf{x}, \mathbf{u}))) \in R$ which concludes the proof. □

The proof of Theorem 2 is constructive since it defines how to construct the (ε, δ)-approximate sub-transition system T and the (ε, δ)-approximate simulation relation. Intuitively, the construction of T proceeds as follows. We use $[\mathbb{R}^n]_{\frac{2\varepsilon}{\sqrt{n}}}$ as the initial state set Q of T. This state set has the property that for every $x \in \mathbb{R}^n$ there exists a $q \in Q$ such that $x \in \mathcal{B}_\varepsilon(q)$. Starting from this initial state set we construct all the transitions $q \xrightarrow{u} p$ with $q \in Q$ and $u \in \mathcal{U}$. However, instead of declaring $q \xrightarrow{u} p$ to be a transition in T we declare that all the transitions $q \xrightarrow{u} p'$, with $p' \in [\mathbb{R}^n]_\chi$ and $p \in \mathcal{B}_{\chi\frac{\sqrt{n}}{2}}(p')$, are transitions of T. It thus follows by construction that $Q \subset [\mathbb{R}^n]_\chi$ since χ integrally divides χ. Moreover, when working on a compact subset S of $\mathbb{R}^n$, there are only finitely many points of the lattice $[\mathbb{R}^n]_\chi$ which are contained in S and this guarantees termination of the construction of T in a finite number of steps. Note also that the resulting transition system T is nondeterministic since for every p there exist, in general, several p' satisfying the conditions $p' \in [\mathbb{R}^n]_\chi$ and $p \in \mathcal{B}_{\chi\frac{\sqrt{n}}{2}}(p')$. We thus see that the construction of T consists in approximating p by several points $p' \in [\mathbb{R}^n]_\chi$. This is only possible since we only ask for the existence of a (ε, δ)-approximate simulation relation from T to $T_\tau(\Sigma)$ and since we can use the feedback controller k to correct for the introduced errors when replacing p with p'. It can also be seen from the construction of T that if S is of the form $S = [-s/2, s/2]^n$ for some $s \in \mathbb{R}$ then T will have at most $\lceil s/\chi \rceil^n$ states and $\lceil s/\chi \rceil^n |\mathtt{U}|$ transitions. This exponential dependence on n is unavoidable if we want to keep the resolution ε constant when n increases. We shall further comment on this fact in Section 5. The (ε, δ)-approximate simulation relation R from T to $T_\tau(\Sigma)$ is simply given by $(q, x) \in R$ if $\|q - x\| \leq \varepsilon$ for any state $q \in Q$ of T and $x \in \mathbb{R}^n$ of $T_\tau(\Sigma)$. Note that in this case we have $\delta = \varepsilon$.

We now return to the linear example to illustrate Theorem 2. One possible stabilizing controller for (6) is given by $u = Kx = -80x_1 - 20x_2$ which places the eigenvalues of the closed loop system at -9 and -9, respectively (the open loop eigenvalues are 1 and 1). Since the dynamics of $x - y$ is given by:

$$\dot{x} - \dot{y} = Ax - Ay + Bu - Bv = A(x - y) + B(u - v)$$

we see that the controller $v = u - K(x - y)$ can be used to enforce the SA. In order to obtain a (ε, δ)-approximate sub-transition system for $\varepsilon = 1$ we solve for the flow of the closed loop system and obtain $\|\mathbf{x}(0.5) - \mathbf{y}(0.5)\| \leq 0.46$ for all initial conditions satisfying $\|x - y\| \leq 1$. Using 0.46 as our estimate for $\beta(1, 0.5)$ we pick $\chi = \frac{2}{\sqrt{2}}0.5$. The resulting (ε, δ)-approximate sub-transition system is displayed in Figure 2. It has 439 states while the naive approach produced a transition system with 4437 states after ten rounds.

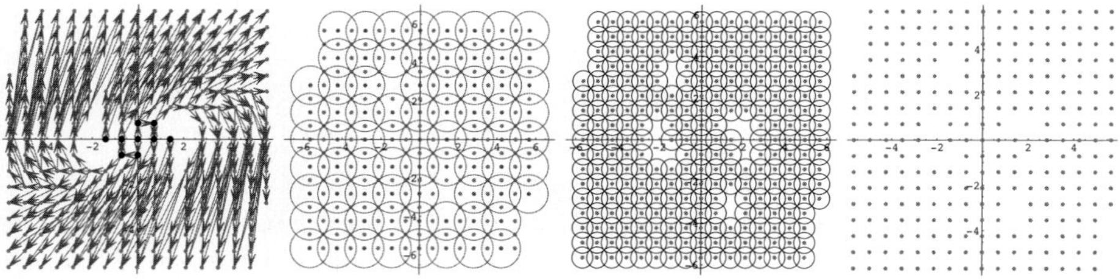

Fig. 2. (ε, δ)-approximate sub-transition system associated with the linear system (6) for $\varepsilon = 1$ and $\chi = 0.5(2/\sqrt{2})$. From left to right we have: the (ε, δ)-approximate sub-transition system; the states of the sub-transition system belonging to $[S]_{1\frac{2}{\sqrt{2}}}$ represented in blue and enclosed in a circle of radius 1 while the remaining states are represented in red; all the states of the sub-transition system, enclosed in a circle of radius $0.5 = \chi\frac{\sqrt{2}}{2}$; all the states of the sub-transition system. The sates belonging to $[S]_\chi$ which are not displayed are states whose transitions lead to points outside S and which have no incoming transitions.

5 Discussion

1. Instead of working with the Euclidean norm we could have constructed T and defined the (ε, δ)-simulation relation by directly using the level sets of the Lyapunov function whose existence is implied by the SA. Since given an equation of the form $\dot{V} \leq -\alpha(V)$ we can always transform V into another Lyapunov function U satisfying $\dot{U} \leq -U$, we can sidestep the need to estimate β. However, working directly with level sets of U increases the complexity of the computations since U is a general nonlinear function.
2. The global nature of the SA upon which Theorem 2 relies was assumed for simplicity of presentation and can be relaxed. Since we have a (ε, δ)-approximate simulation relation from T to $T_\tau(\Sigma)$, states $x \in \mathbb{R}^n$ of $T(\Sigma)$ related to states $q \in Q$ of T will necessarily satisfy $\|q - x\| \leq \varepsilon$. We can thus

relax the SA by requiring that it holds only for initial conditions $(x, y) \in \mathbb{R}^n \times \mathbb{R}^n$ satisfying $\|x - y\| \leq \varepsilon$.

3. The SA was defined in terms of the existence of a single controller k. In many situations, however, we have not only one but several controllers $\{k_i\}_{i \in I}$, each designed to track a familly of trajectories. It is clear that the conclusions of Theorem 2 still hold in this case provided that we use for β a $\mathcal{KL}$ function satisfying $\beta_i(r, s) \leq \beta(r, s)$ for all $i \in I$ and where β_i is the $\mathcal{KL}$ function associated with controller k_i.
4. Although for linear systems we can explicitly compute the flow for each of the inputs in $\mathtt{U}$ the same is no longer true in the nonlinear case. We are thus forced to resort to numerical simulation methods in order to construct T. Theorem 2 is still of value in this case since given a bound $\eta > 0$ on the simulation error, that is, $\|\mathbf{x}(\tau) - \tilde{\mathbf{x}}(\tau)\| \leq \eta$ where $\tilde{\mathbf{x}}$ is the simulated value, the conclusions of Theorem 2 still hold provided that we choose τ such that $\beta(\varepsilon, \tau) < \varepsilon + \eta$. Note that such τ always exists since $\beta(r, s)$ is a decreasing function of s. In this case, χ can be any positive real number integrally dividing $2\varepsilon/\sqrt{n}$ and satisfying:
$$0 < \chi \leq \frac{2}{\sqrt{n}}\big(\varepsilon + \eta - \beta(\varepsilon, \tau)\big)$$
5. The proposed methodology enforces a constant accuracy ε on the state set of a (ε, δ)-approximate sub-transition system T by guaranteeing that it is a subset of a lattice $[\mathbb{R}^n]_\chi$. Although this guarantees a spatially uniform description of the dynamics of Σ, it also forces the size of T to grow exponentially with n. Since the specification may not require a spatially uniform resolution, we can instead construct specification dependent multi-resolution finite abstractions. This kind of finite abstractions is currently being investigated by the author as a lower complexity alternative to the (ε, δ)-approximate sub-transition systems introduced in this paper.

References

[Art83] Z. Artstein. Stabilization with relaxed controls. *Nonlinear Analysis, Theory, Methods, and Applications*, 7:1163–1173, 1983.

[BMP02] A. Bicchi, A. Marigo, and B. Piccoli. On the reachability of quantized control systems. *IEEE Transaction on Automatic Control*, 47(4):546–563, April 2002.

[BMP06] A. Bicchi, A. Marigo, and B. Piccoli. Feedback encoding for efficient symbolic control of dynamical systems. *IEEE Transaction on Automatic Control*, 51(6):987– 1002, June 2006.

[CGP99] E. M. Clarke, O. Grumberg, and D. Peled. *Model Checking.* MIT Press, 1999.

[GP05a] A. Girard and G. Pappas. Approximate bisimulations for constrained linear systems. In *Proceedings of the 44th IEEE Conference on Decision and Control*, Seville, Spain, 2005.

[GP05b] A. Girard and G. J. Pappas. Approximate bisimulations for nonlinear dynamical systems. In *Proceedings of the 44th IEEE Conference on Decision and Control*, Seville, Spain, 2005.

[KB06] Marius Kloetzer and Calin Belta. Reachability analysis of multi-affine systems. In Joao P. Hespanha and Ashish Tiwari, editors, *Hybrid Systems: Computation and Control 2006*, volume 3927 of *Lecture Notes in Computer Science*, pages 348–362. Springer-Verlag, Santa Barbara, CA, USA, 2006.

[LS95] Y. Lin and E.D. Sontag. Control-Lyapunov universal formulae for restricted inputs. *Control: Theory and Advanced Technology*, 10:1981–2004, 1995.

[PLPB02] Stefania Pancanti, Laura Leonardi, Lucia Pallottino, and Antonio Bicchi. Optimal control of quantized linear systems. In Claire Tomlin and Mark R. Greenstreet, editors, *Hybrid Systems: Computation and Control*, Lecture Notes in Computer Sience, pages 351–363. Springer-Verlag, 2002.

[Tab06] P. Tabuada. An approximate simulation approach to symbolic control. 2006. Submitted for publication. Available at `http://www.ee.ucla.edu/~tabuada`.

[Tab07] Paulo Tabuada. Symbolic models for control systems. *Acta Informatica*, 2007. Accepted for publication. Available at `http://www.ee.ucla.edu/~tabuada`.

[TP06] P. Tabuada and G. J. Pappas. Linear Time Logic control of discrete-time linear systems. *IEEE Transactions on Automatic Control*, 51(12):1862–1877, 2006.

Finite State Controllers for Stabilizing Switched Systems with Binary Sensors*

Danielle C. Tarraf[1], Alexandre Megretski[1,2], and Munther A. Dahleh[1,2]

[1] Laboratory for Information and Decisions Systems
[2] Department of Electrical Engineering and Computer Science
Massachusetts Institute of Technology
Cambridge, MA 02139
dtarraf@alum.mit.edu, ameg@mit.edu, dahleh@mit.edu

Abstract. This paper presents a systematic, semi-automated, constructive procedure for designing switching controllers to stabilize pairs of unstable, homogeneous second order systems based on binary sensor measurements. The plant is first approximated by a finite state machine, and a bound on the quality of approximation is established. A control law is then designed to robustly stabilize the nominal finite state machine model in the presence of admissible approximation uncertainty. The resulting controller thus consists of a finite state machine observer and a corresponding full state feedback switching control law. The design procedure is demonstrated on a simple benchmark example.

1 Introduction

Switched systems are a special class of hybrid systems consisting of a family of plants and a switching law among them. Switched control systems arise in many engineering applications [1,2,3]. In particular, the problem of finding a stabilizing switching law for a family of unstable linear systems has received much attention in the past decade [4,5,6,7,8]. An overview of recent results in this area can be found in [9]. Typically, the controllers implementing the switching strategy are either assumed to have full access to the state or to have access to a sensor output that is a linear function of the state. However, there may be a practical need to base the design and implementation of the switching controller on coarse, discrete sensing. In particular, coarse sensors are increasingly common in autonomous vehicles and sensor networks, in order to keep operating cost, weight, and power consumption low. The problem of stabilizing a family of unstable systems where only very coarse sensing is available has not been as extensively studied to date. Problems involving pairs of second order LTI systems were considered in [10] and [11], where it was shown that stabilization based on coarse sensing is in fact possible. In this paper, a new constructive approach for designing finite state machine switching controllers for pairs of homogeneous, discrete-time second order systems is presented. Binary, but noiseless, sensor information is assumed to be available at each time step.

* Research supported by Air Force Aerospace Research-OSR Grant FA9550-04-1-0052.

A. Bemporad, A. Bicchi, and G. Buttazzo (Eds.): HSCC 2007, LNCS 4416, pp. 543–556, 2007.

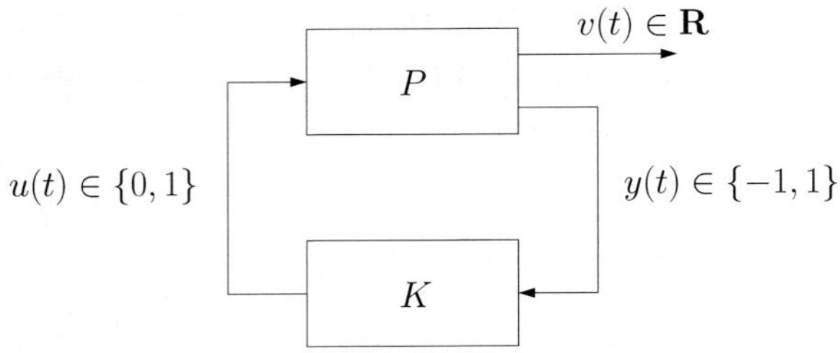

Fig. 1. Closed Loop System

The paper is organized as follows: A formal statement of the control design problem and an overview of the controller design procedure are presented in Section 2. The algorithms for constructing a finite state machine approximation of the plant, and for computing an a-posteriori bound on the resulting approximation error are presented in Section 3. Design of the robust stabilizing switching law, and the structure of the resulting switching controller, are described in Section 4. An illustrative example is presented in Section 5. The paper concludes in Section 6 where recommendations for future work are given.

2 Problem Statement and Design Procedure

Consider a discrete-time plant P described by:

$$x(t+1) = f_{u(t)}(x(t)) \tag{1}$$

$$y(t) = sgn\big(Cx(t)\big) \tag{2}$$

$$v(t) = log\Big(\frac{\|x(t+1)\|_2}{\|x(t)\|_2}\Big) \tag{3}$$

where the time index $t \in \mathbf{Z}_+$ (the set of non-negative integers), $\|\cdot\|_2$ denotes the Euclidean norm, state $x(t) \in \mathbf{R}^2$, control input $u(t) \in \mathcal{U} = \{0,1\}$, performance output $v(t) \in \mathbf{R}$ and sensor output $y(t) \in \mathcal{Y} = \{-1,1\}$. $f_{0,1} : \mathbf{R}^2 \to \mathbf{R}^2$ and $C \in \mathbf{R}^{1\times 2}$ are given, with the following conditions assumed to hold for all $u \in \mathcal{U}$:

1. Function f_u is continuous and homogeneous with degree 1. [1]
2. Function $\frac{\|f_u(x)\|_2}{\|x\|_2}$ is bounded above.

The objective is to design a controller $K \subset \mathcal{Y}^{\mathbf{Z}_+} \times \mathcal{U}^{\mathbf{Z}_+}$ such that the closed loop system (P, K) with output v (Figure 1) satisfies the following performance objective for some $R > 0$:

$$\sup_{T\geq 0} \sum_{t=0}^{T} (v(t) + R) < \infty \tag{4}$$

[1] That is, $f_u(cx) = cf_u(x)$ for all $c \in \mathbf{R}$, $x \in \mathbf{R}^2$.

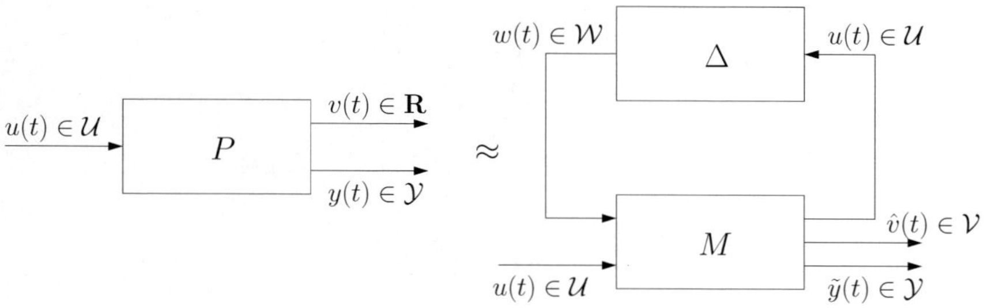

Fig. 2. Plant P and its finite state machine approximation M

Satisfaction of this objective guarantees that the state of the closed loop system exponentially converges to the origin, on average, at a rate not less than R.

The procedure for designing a stabilizing controller is iterative: First, P is approximated by the feedback interconnection of a finite state machine M and a system Δ representing the approximation error (Figure 2). Next, a gain bound is established for the error system Δ. Finally, an attempt is made to synthesize a full state feedback switching law for the nominal finite state model M, that is robust to the approximation error. If synthesis is successful, the resulting controller is guaranteed to stabilize the system at a rate of exponential convergence not less than R. Otherwise, a better approximation (meaning a finite state machine with a larger number of states) is constructed and the above process is repeated.

3 A Finite State Machine Approximation of the Plant

3.1 Finite State Machine Models and Notions of Approximation

A deterministic finite state machine (DFM) is understood to mean a discrete-time system with a finite number of states, and with finite input and output alphabets (i.e. the inputs and outputs take their values in finite sets). Given a plant P with binary control input u, binary sensor output y, and bounded real performance output v. Consider system M with the internal structure shown in Figure 3, where $\hat{M}$ is some deterministic finite state machine. The disturbance input to M, w, is assumed to take binary values in the set $\mathcal{W} = \{0, 1\}$. Block "ϕ" is a memoryless map defined by $\phi(\hat{y}) = \hat{y}$ if $w = 0$ and $\phi(\hat{y}) = -\hat{y}$ if $w = 1$. System M is thus also a finite state machine. Consider also the corresponding system Δ shown in Figure 4. Block "β" is a memoryless map defined by $\beta(y, \hat{y}) = 0$ if $y = \hat{y}$ and $\beta(y, \hat{y}) = 1$ otherwise. Note that Δ is not a finite state machine in general since P is not one.

Consider the feedback interconnection, as in Figure 2, of M and Δ with these particular structures, and suppose that the two copies of $\hat{M}$ are *identically initialized*[2]. It can be seen by direct inspection that for arbitrary initial conditions

[2] A fair assumption since we have full access to $\hat{M}$.

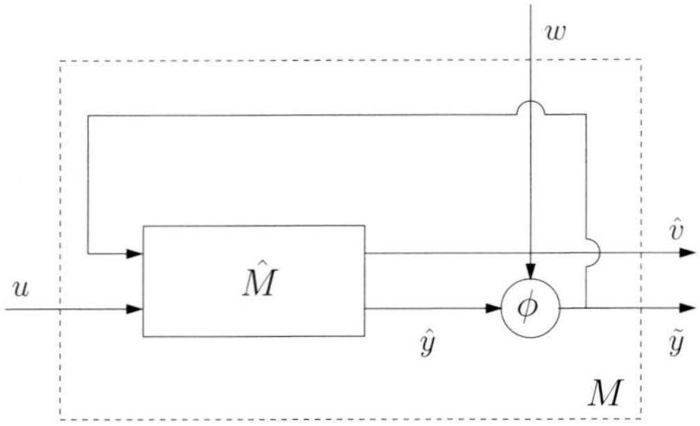

Fig. 3. Structure of M, the finite state machine approximation of P

of P and for any input $\mathbf{u} \in \mathcal{U}^{\mathbf{Z}_+}$, the corresponding outputs $\mathbf{y} \in \mathcal{Y}^{\mathbf{Z}_+}$ of P and $\tilde{\mathbf{y}} \in \mathcal{Y}^{\mathbf{Z}_+}$ of the interconnection (M, Δ) are identical for any choice of $\hat{M}$. Now suppose that we can construct a deterministic finite state machine $\hat{M}$ such that, for any input $\mathbf{u} \in \mathcal{U}^{\mathbf{Z}_+}$ and any initial condition of P, we have:

1. Outputs $\hat{\mathbf{v}} \in \mathcal{V}^{\mathbf{Z}_+}$ of (M, Δ) and $\mathbf{v} \in \mathbf{R}^{\mathbf{Z}_+}$ of P satisfy:
$$\hat{v}(t) \geq v(t), \text{ for all } t \in \mathbf{Z}_+ \tag{5}$$
2. Every input/output pair $(\mathbf{u}, \mathbf{w}) \in \mathcal{U}^{\mathbf{Z}_+} \times \mathcal{W}^{\mathbf{Z}_+}$ of Δ satisfies:
$$\inf_{T \geq 0} \sum_{t=0}^{T} \gamma\rho(u(t)) - \mu(w(t)) > -\infty \tag{6}$$
for some positive scalar γ, some function $\rho : \mathcal{U} \to [1, 2]$, and function $\mu : \mathcal{W} \to \mathbf{R}$ defined by $\mu(w) = w$.

The resulting machine M is then said to be a *finite state machine approximation* of P, and the resulting system Δ is said to be the *approximation error.*

Remark 1. Traditional model order reduction deals with stable systems: a lower order model is considered a good approximation of the original system if the outputs of the two systems are close when driven side by side by the same input. The internal structure proposed here for Δ, where the output of the plant P is fed back to $\hat{M}$ (essentially an observer setup), is needed because the original system is not stable (i.e. remembers its initial condition forever). Two copies of P initialized differently and driven side by side may end up with different outputs at every step. Thus, there is a need to explicitly 'estimate' the initial condition of P, otherwise there is no hope for satisfying gain condition (6) for any choice of $\hat{M}$.

Remark 2. If $\rho(0) = \rho(1) = 1$, the smallest value of γ for which (6) holds (the 'gain' of Δ) represents the fraction of time that the outputs of P and $\hat{M}$ are mismatched in the worst-case scenario. A smaller value of γ is hence desirable and indicative of a better approximation. Allowing a choice of $\rho(1) > 1$ serves to penalize the corresponding control input.

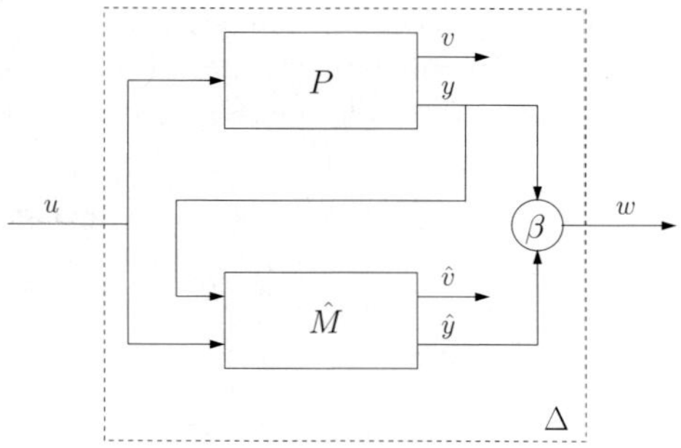

Fig. 4. The error system Δ of M and P

Remark 3. The smaller the gap between v and $\hat{v}$ in (5), the better the approximation and the less conservative the resulting design procedure.

In the following sections a systematic procedure for constructing finite state machine $\hat{M}$, and an algorithm for computing an a-posteriori upper bound on the gain of the resulting error system Δ, are proposed.

3.2 Construction of the Nominal Model

While there are several potential approaches for constructing a nominal model of plant P (see [12] for an overview), the approach presented here takes into account the dynamical properties specific to homogeneous systems, evident after a coordinate transformation. The dynamics of system P described in (1), (2) and (3) in a polar coordinate system are given by:

$$r(t+1) = r(t)\|f_{u(t)}\big(\beta(\theta(t))\big)\|_2$$

$$\theta(t+1) = \tan^{-1}\left(\frac{f_{2,u(t)}\big(\beta(\theta(t))\big)}{f_{1,u(t)}\big(\beta(\theta(t))\big)}\right) \tag{7}$$

$$y(t) = sgn\Big(C\beta(\theta(t))\Big) \tag{8}$$

$$v(t) = log\Big(\|f_{u(t)}\big(\beta(\theta(t))\big)\|_2\Big) \tag{9}$$

where

$$\beta(\theta) = \begin{bmatrix} cos(\theta) \\ sin(\theta) \end{bmatrix}, \quad f_u = \begin{bmatrix} f_{1,u} \\ f_{2,u} \end{bmatrix}$$

Thus, the evolution of the angular coordinate and both outputs of system P are independent of the radial coordinate. The state of P relevant to the stabilization problem at hand effectively evolves on a compact set, the unit circle. $\hat{M}$ is constructed by quantizing[3] the unit circle and defining the states of $\hat{M}$ to be the quantization intervals and unions of *adjacent* quantization intervals[4].

Consider a partition of the unit circle consisting of intervals $I_0, \ldots, I_{n-1}$ defined by $I_i = [\alpha_i, \alpha_{i+1})$ for some even integer n (a design parameter) and some sequence of angles, $\alpha_0 < \ldots < \alpha_n$, satisfying:

$$\alpha_0 = tan^{-1}\Big(-\frac{c_1}{c_2}\Big) \qquad \alpha_{n/2} = \alpha_0 + \pi \qquad \alpha_n = \alpha_0 + 2\pi$$

Construct a set S of intervals as follows:

$$S = \bigcup_{k=0}^{n-1}\Big(\bigcup_{i=0}^{n-1}\{I_{<i>_n} \cup \ldots \cup I_{<i+k>_n}\}\Big)$$

where $< a >_n$ denotes 'a modulo n'. Note that S can be partitioned into:

$$S_1 = \Big\{I \in S | C\beta(\alpha) \leq 0, \forall \alpha \in I\Big\} \qquad S_2 = \Big\{I \in S | C\beta(\alpha) \geq 0, \forall \alpha \in I\Big\}$$

$$S_3 = \Big\{I \in S | \exists \alpha_1, \alpha_2 \in I, C\beta(\alpha_1) < 0, C\beta(\alpha_2) > 0\Big\}$$

The set $\mathcal{Q}$ of states of $\hat{M}$ is then $\mathcal{Q} \doteq \{q | I_q \in S\}$, with $\mathcal{Q}_i$ denoting the subset of $\mathcal{Q}$ corresponding to intervals in S_i. The dynamics of $\hat{M}$ are described by:

$$\begin{aligned} q(t+1) &= f(q(t), u(t), \tilde{y}(t)) \\ \hat{y}(t) &= g(q(t)) \\ \hat{v}(t) &= h(q(t), u(t)) \end{aligned} \tag{10}$$

with state transition function $f : \mathcal{Q} \times \mathcal{U} \times \mathcal{Y} \rightarrow \mathcal{Q}$ and output functions $g : \mathcal{Q} \rightarrow \mathcal{Y}$ and $h : \mathcal{Q} \times \mathcal{U} \rightarrow \mathcal{V}$ defined by:

$$f(q, u, \tilde{y}) \doteq \begin{cases} \delta(q, u) & \text{if } q \in \mathcal{Q}_1 \cup \mathcal{Q}_2 \\ \delta(P_1(q), u) & \text{if } q \in \mathcal{Q}_3, \tilde{y} = -1 \\ \delta(P_2(q), u) & \text{if } q \in \mathcal{Q}_3, \tilde{y} = 1 \end{cases}$$

$$g(q) \doteq \begin{cases} -1 & \text{if } q \in \mathcal{Q}_1 \text{ or } q \in \mathcal{Q}_3,\ |I_{P_1(q)}| \geq |I_{P_2(q)}| \\ 1 & \text{if } q \in \mathcal{Q}_2 \text{ or } q \in \mathcal{Q}_3,\ |I_{P_1(q)}| < |I_{P_2(q)}| \end{cases}$$

$$h(q, u) \doteq \sup_{\theta \in I_q} log\Big(f_u\big(\beta(\theta(t))\big)\Big)$$

where function $\phi : (\mathcal{Q}_1 \cup \mathcal{Q}_2) \times \mathcal{U} \rightarrow \mathcal{Q}$ is defined by:

$$\delta(q, u) = \arg\min_{q \in \mathcal{A}_u^q} |I_q|$$

with $|I_q|$ denoting the length of interval $I_q = [a_q, b_q)$, that is $|I_q| \doteq |b_q - a_q|$, and with set $\mathcal{A}_u^q$ defined as:

$$\mathcal{A}_u^q = \Big\{q \in \mathcal{Q} | I_q \supset [\tan^{-1}\Big(\frac{f_{2,u}(\beta(a_q))}{f_{1,u}(\beta(a_q))}\Big), \tan^{-1}\Big(\frac{f_{2,u}(\beta(b_q))}{f_{1,u}(\beta(b_q))}\Big)\Big\}$$

[3] The quantization need not be uniform in general.

[4] Continuity of f_u rules out non-adjacent quantization intervals as potential states.

and where $P_1 : \mathcal{Q}_3 \rightarrow \mathcal{Q}_1$ and $P_2 : \mathcal{Q}_3 \rightarrow \mathcal{Q}_2$ are defined by $P_i(q) = q_i$, where (q_1, q_2) are the unique pair of elements in $\mathcal{Q}_1 \times \mathcal{Q}_2$ with the property that $I_q = I_{q_1} \cup I_{q_2}$. In particular, let q_o denote the state corresponding to $I_{q_o} = [\alpha_0, \alpha_n)$.

Proposition 1. *Consider a plant P and let $\hat{M}$ be the corresponding finite state machine defined by (10) for some choice of parameter n. If $q(0) = q_o$, then for any input $\mathbf{u}$ and initial condition of P, the outputs $\hat{\mathbf{v}}$ of (M, Δ) and $\mathbf{v}$ of P satisfy (5).*

Proof. Let $\theta(t)$ and $q(t)$ be the states of P and $\hat{M}$, respectively, at time t. It follows from the construction of $\hat{M}$ that:

1. $\theta(t) \in I_{q(t)} \Rightarrow \theta(t+1) \in I_{q(t+1)}$
2. $\theta(t) \in I_{q(t)} \Rightarrow \hat{v}(t) \geq v(t)$

When $q(0) = q_o$, $\theta(0) \in I_{q_o}$. Hence, the statement follows by induction on t. □

Remark 4. The size of the finite state approximation grows polynomially with the number of quantization intervals, with the total number of states N of $\hat{M}$ satisfying: $n + 1 \leq N \leq n(n-1) + 1$.

3.3 Description of the Approximation Error

The next step is to compute an upper bound for the gain of the corresponding error system Δ, as in (6), for a choice of ρ specific to the problem at hand.

Proposition 2. *If there exists a function $V : \mathcal{Q} \rightarrow \mathbf{R}$ and a $\gamma > 0$ such that:*

$$V(f(q, u, y)) - V(q) \leq \gamma\rho(u) - d(q) \tag{11}$$

holds for all $q \in \mathcal{Q}$, $u \in \mathcal{U}$ and $y \in \mathcal{Y}$, where $d : \mathcal{Q} \rightarrow \{0, 1\}$ defined by:

$$d(q) = \begin{cases} 0 & \text{for } q \in \mathcal{Q}_1 \cup \mathcal{Q}_2 \\ 1 & \text{for } q \in \mathcal{Q}_3 \end{cases}$$

then the error system Δ satisfies (6).

Proof. By summing up (11) along any state trajectory of $\hat{M}$ from $t = 0$ to $t = T$, we get:

$$\sum_{t=0}^{T} \gamma\rho(u(t)) - d(q(t)) \geq V(q(T)) - V(q(0)) \geq \min_{q_1, q_2} V(q_1) - V(q_2)$$

Hence, we have:

$$\inf_{T \geq 0} \sum_{t=0}^{T} \gamma\rho(u(t)) - d(q(t)) \geq \min_{q_1, q_2} V(q_1) - V(q_2) > -\infty$$

It follows from Proposition 1 that when $\hat{M}$ is initialized to $q(0) = q_o$, we have $\theta(t) \in I_{q(t)}$ for all t, where θ and q are the states of P and $\hat{M}$ respectively. Thus, when $q(t) \in \mathcal{Q}_1 \cup \mathcal{Q}_2$, $y(t) = \hat{y}(t)$ and $w(t) = 0$, whereas when $q(t) \in \mathcal{Q}_3$, $w(t) \in \{0, 1\}$. Hence by definition, $\mu(w(t)) = w(t) \leq d(q(t))$, for all t. Consequently, all feasible input/output signal pairs of Δ satisfy (6). □

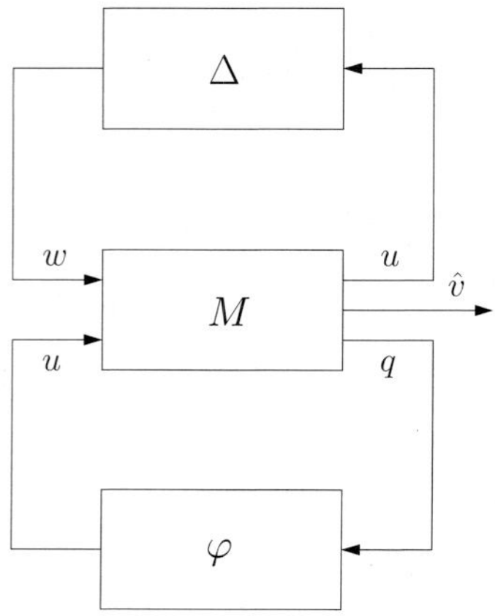

Fig. 5. Robust control setup

An upper bound for the gain of Δ can thus be computed by simply solving a linear program in which we minimize γ such that (11) holds for all $q \in \mathcal{Q}$, $u \in \mathcal{U}$ and $y \in \mathcal{Y}$. If $N = card(\mathcal{Q})$, the linear program has $N+1$ decision variables (γ, $V(q_0)$,..., $V(q_N)$) and $4N$ inequality constraints. An alternative algorithm, which grows as $\mathcal{O}(N^2)$, can be found in [13].

The relevance of Proposition 2 is that it allows us to compute an upper bound for the gain of complex system Δ by looking at the much simpler system $\hat{M}$. However, the computed gain bound is conservative for two reasons: first, we are assuming that an an error occurs at every opportunity, and second, we are assuming that any pair $(\mathbf{u}, \mathbf{y}) \in \mathcal{U}^{\mathbf{Z}_+} \times \mathcal{Y}^{\mathbf{Z}_+}$ is a valid input for $\hat{M}$, which is not the case since $\mathbf{y}$ is an output of P corresponding to $\mathbf{u}$.

4 Controller Design

4.1 Design of a Robust Switching Law

The problem considered in this section is that of designing a switching law $\varphi : \mathcal{Q} \to \mathcal{U}$ such that the closed loop system consisting of the interconnection (M, Δ) in feedback with φ (Figure 5) satisfies the following performance objective:

$$\sup_{T \geq 0} \sum_{t=0}^{T} \hat{v}(t) + R < \infty \tag{12}$$

for some $R > 0$ for all admissible uncertainty (i.e. for all systems Δ satisfying (6)). The largest value of R for which (12) holds is then the *guaranteed* rate of exponential convergence; the actual rate of convergence is generally better.

The switching law φ will be designed based on the nominal model M using a small gain argument. Consider the feedback interconnection of two systems

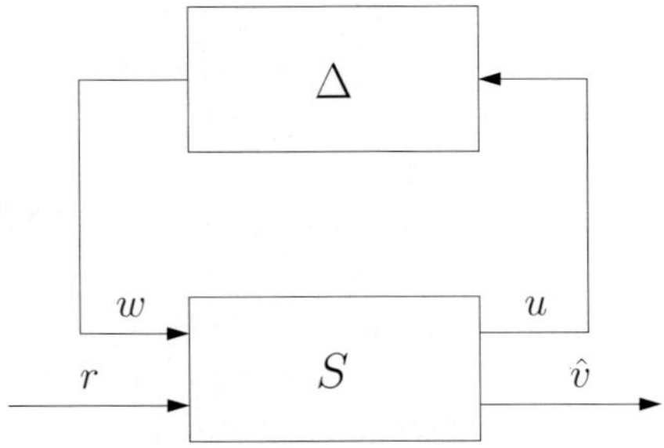

Fig. 6. Setup for the small gain theorem

S and Δ as in Figure 6. The following 'Small Gain' Theorem, presented here without proof, is adapted from Theorem 1 in [13]. It characterizes the 'stability' of an interconnection of two 'stable' systems.

Theorem 1. *(A 'Small Gain' Theorem) Suppose that S satisfies:*

$$\inf_{T\geq 0}\sum_{t=0}^{T}\rho_S(r(t),w(t))-\mu_S(\hat{v}(t),u(t))>-\infty \tag{13}$$

for some $\rho_S:\mathcal{R}\times\mathcal{W}\to\mathbf{R}$ and $\mu_S:\mathcal{V}\times\mathcal{U}\to\mathbf{R}$, and that Δ satisfies:

$$\inf_{T\geq 0}\sum_{t=0}^{T}\rho_\Delta(u(t))-\mu_\Delta(w(t))>-\infty \tag{14}$$

for some $\rho_\Delta:\mathcal{U}\to\mathbf{R}$ and $\mu_\Delta:\mathcal{W}\to\mathbf{R}$, where $\mathcal{R}$, $\mathcal{V}$, $\mathcal{W}$ and $\mathcal{U}$ are finite sets. Then, the interconnected system (S,Δ) with input r and output $\hat{v}$ satisfies:

$$\inf_{T\geq 0}\sum_{t=0}^{T}\rho(r(t))-\mu(\hat{v}(t))>-\infty \tag{15}$$

for $\rho:\mathcal{R}\to\mathbf{R}$ and $\mu:\mathcal{V}\to\mathbf{R}$ defined by:

$$\rho(r)\doteq\max_{w\in\mathcal{W}}\{\rho_S(r,w)-\mu_\Delta(w)\}\qquad\qquad \mu(\hat{v})\doteq\min_{u\in\mathcal{U}}\{\mu_S(\hat{v},u)-\rho_\Delta(u)\}$$

Corollary 1. *The interconnected system satisfies (15) for $\rho:\mathcal{R}\to\mathbf{R}$ and $\mu:\mathcal{V}\to\mathbf{R}$ defined by:*

$$\rho(r)\doteq\max_{w\in\mathcal{W}}\{\rho_S(r,w)-\tau_d\mu_\Delta(w)\}\qquad\qquad \mu(v)\doteq\min_{u\in\mathcal{U}}\{\mu_S(\hat{v},u)-\tau_d\rho_\Delta(u)\}$$

for any strictly positive scalar parameter τ_d.

The auxiliary performance objective in (12) can thus be achieved by designing a switching law $\varphi:\mathcal{Q}\to\mathcal{U}$ for which there exists a $\tau_d>0$ such that the feedback interconnection $S = (M, \varphi)$, with input[5] w and outputs $\hat{v}$ and u, satisfies (13) with:

$$\rho_S(r, w) \doteq R + \tau_d \mu(w) \qquad \mu_S(\hat{v}, u) \doteq \hat{v} + \tau_d \gamma \rho(u) \tag{16}$$

for some $R > 0$, for ρ and μ defined as in (6), and for the corresponding upper bound γ on the gain of Δ computed based on Proposition 2.

The design of the full state feedback switching law can be carried out using standard Dynamic Programming techniques [14]. Value Iteration (described in statement (c) of Theorem 2) is used to solve a Bellman inequality (18) for the cost-to-go function J; the desired switching law is then simply the optimizing argument. The following Theorem states this rigorously; a proof can be found in [12].

Theorem 2. *Consider a deterministic finite state machine M with state set $\mathcal{Q}$ and inputs u and w in $\mathcal{U}$ and $\mathcal{W}$ respectively, and defined by state transition equation:*

$$q(t+1) = f(q(t), u(t), w(t))$$

Let $\sigma : \mathcal{Q} \times \mathcal{U} \times \mathcal{W} \to \mathbf{R}$ be a given map. The following statements are equivalent:

(a) There exists a $\varphi : \mathcal{Q} \to \mathcal{U}$ such that the closed loop system (M, φ) satisfies:

$$\inf_{T \geq 0} \sum_{t=0}^{T} \sigma(q(t), u(t), w(t)) > -\infty \tag{17}$$

(b) There exists a function $J : \mathcal{Q} \to \mathbf{R}$ such that for any $q \in \mathcal{Q}$, the following inequality holds:

$$J(q) \geq \min_{u \in \mathcal{U}} \max_{w \in \mathcal{W}} \{-\sigma(q, u, w) + J(f(q, u, w))\} \tag{18}$$

(c) The sequence of functions $J_k : \mathcal{Q} \to \mathbf{R}$, $k \in \mathbf{Z}_+$, defined recursively by:

$$\begin{aligned} J_0 &= 0 \\ J_{k+1} &= \max\{0, T(J_k)\} \end{aligned} \tag{19}$$

where $T : \mathbf{R}^{\mathcal{Q}} \to \mathbf{R}^{\mathcal{Q}}$ is defined by:

$$T(J(q)) \doteq \min_{u \in \mathcal{U}} \max_{w \in \mathcal{W}} \{-\sigma(q, u, w) + J(f(q, u, w))\} \tag{20}$$

converges.

For the particular control problem of interest, the per-stage cost function σ depends on two parameters, R and the scale τ_d. In reality, the goal is to maximize $R > 0$ for which there exists a $J : \mathcal{Q} \to \mathbf{R}$ and a $\tau_d > 0$ such that (18) holds. However, it is not possible to directly compute the optimal value R^*. Instead, a search is carried out resulting in a suboptimal value of R: first, the range of values of τ_d for which 'stability' with $R = 0$ is possible is computed. Then, different values of τ_d are sampled in this range, and the largest value of R is computed for each sampled value τ_d, with the largest of those being a suboptimal guaranteed rate of convergence.

[5] Alphabet set $\mathcal{R}$ can be thought of as a singleton here.

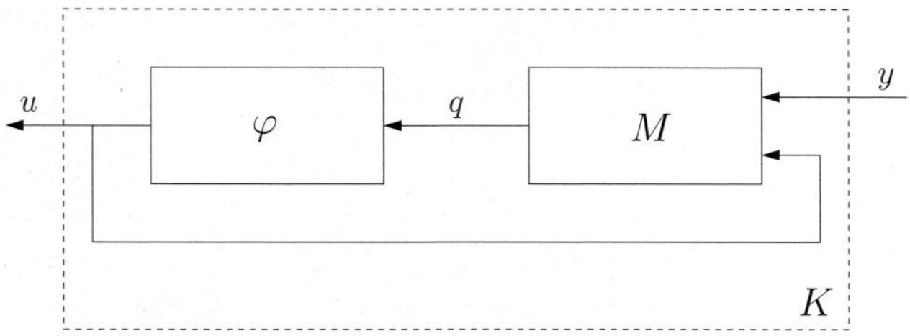

Fig. 7. The finite state stabilizing controller K

4.2 The Stabilizing Controller K

The final step is to ensure that the actual closed loop system satisfies the desired performance objective, described in (4), for the suboptimal rate R computed. By construction, the condition in equation (5) holds whenever the same input $\mathbf{u}$ drives the plant P and the interconnection (M, Δ). Thus, to ensure that (4) holds whenever (12) holds, it is sufficient to ensure that the controller K connected in feedback with plant P is identical to the subsystem with input y and output u in the interconnection (M, Δ, φ). The structure of the resulting controller K is shown in Figure 7. Controller K thus consists of a finite state machine observer for the plant and a corresponding full state feedback control law.

5 Illustrative Example

Consider a harmonic oscillator described by:

$$\dot{x}_1 = x_2$$
$$\dot{x}_2 = kx_1$$

The corresponding equations in polar coordinates are: $\dot{r} = (1+k)r\sin(\theta)\cos(\theta)$ and $\dot{\theta} = (1+k)\cos^2(\theta) - 1$. For $k = -1$, $\dot{r} = 0$ and the state trajectories are concentric circles centered at the origin. For any $k \neq -1$, $\dot{r} < 0$ in exactly two orthants. Thus, when the switching controller has full access to the state and when switching can occur at any time, it is always possible to stabilize the analog system by appropriately switching between gains -1 and k, for any choice of $k \in \mathbf{R} \backslash \{-1\}$. Now suppose that the only sensor information available for use is the sign of the position measurement (Figure 8). The stabilization problem becomes more difficult because of the non-trivial state estimation problem that arises due to binary sensing. The approach described in this paper will be used to design a stabilizing controller.

The sampled system is described by:

$$\begin{aligned} x(t+1) &= A_T(u(t))x(t) \\ y(t) &= sgn(x_1(t)) \\ v(t) &= log\Big(\frac{\|x(t+1)\|_2}{\|x(t)\|_2}\Big) \end{aligned} \tag{21}$$

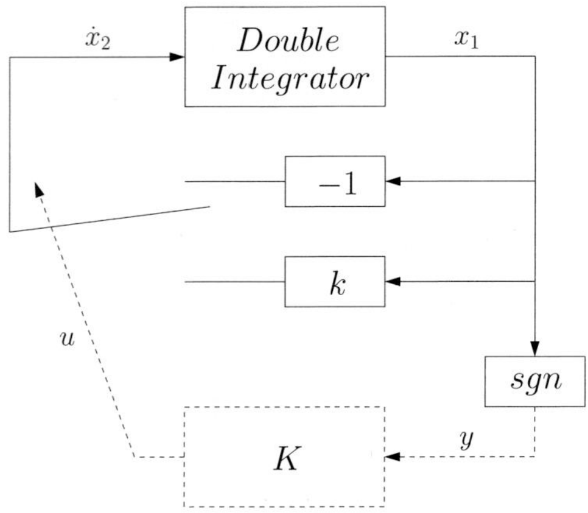

Fig. 8. Double Integrator with Switched Static Feedback and Binary Position Sensor

$A_T(0)$ and $A_T(1) \in \mathbf{R}^{2\times 2}$ are of the form $A_T(u) = e^{\overline{A}(u)T}$ where:

$$\overline{A}(u) = \begin{bmatrix} 0 & 1 \\ k(u) & 0 \end{bmatrix}$$

with $k(0) = -1$, $k(1) = k$ for some given value k in $\mathbf{R}\backslash\{-1\}$. The sampling rate T is assumed to be a design parameter in this case. Satisfaction of the performance objective in (4) guarantees that the state of the actual (continuous-time) closed loop system exponentially converges to the origin, on average, at a rate not less than $R/(T+1)$. The particular structure of the problem can be used to counteract the conservatism introduced in the computation of an upper bound for the gain of Δ. Note that for $k = -1$, the state transition equation (7) reduces to $\theta(t+1) = \theta(t) - T$. The unit circle is thus uniformly quantized into n quantization intervals, and the sampling rate is matched to the quantization rate; that is, $T = \frac{2\pi}{n}$.

For instance, suppose that the system can switch between gains $k(0) = -1$ (passive control) and gain $k(1) = -3$ (aggressive control), and where $\rho(0) = 1$, $\rho(1) = 2$. The smallest value of design parameter n for which we can guarantee convergence in this case is $n = 10$. The properties of the design procedure for several choices of n are shown in Table 1. n is the number of quantization

Table 1. Data for the example

n	T	γ	R	$R/(T+1)$	p	Plot color
10	0.6283	0.5	0.016	0.0098	13	Green
12	0.5236	0.4286	0.024	0.0158	16	Red
18	0.3491	0.4545	0.0275	0.0204	24	Blue

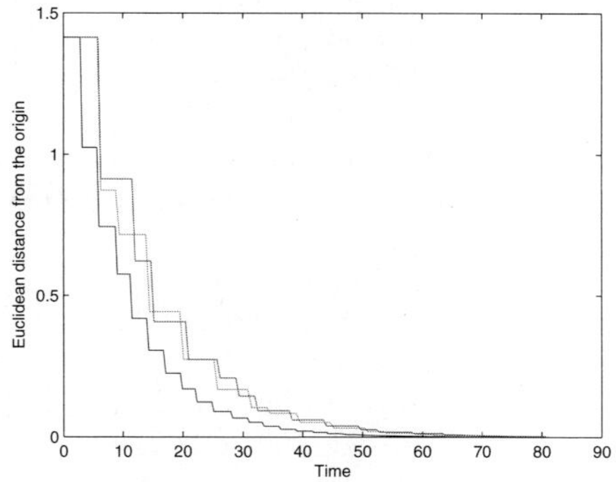

Fig. 9. Implementation of the DFM Controller in the Illustrative Example

intervals, T is the corresponding sampling time, R is the guaranteed rate of exponential convergence of the sampled system, $R/(T+1)$ is the corresponding guaranteed rate for the analog system, and p is the number of iterations needed for convergence of the Value Iteration Algorithm. Simulated implementations of the resulting controllers are shown in Figure 9.

6 Conclusions and Future Works

A new constructive approach for designing switching controllers to stabilize a pair of homogeneous second order systems, based only on binary sensing, was presented. The approach is inspired from classical robust control: the hybrid plant is approximated by a model from the class of finite state machines models, and an upper bound on the gain of the approximation error is established. A controller is then synthesized based on the nominal model such as to allow us to guarantee closed loop stability of the original system using a small gain argument.

While this approach can be generalized in principle to higher order systems (for the purpose of stabilization, homogeneous systems of order $n+1$ effectively evolve on the n-sphere with unity radius), impractically large finite state machine models may be required (the number of states grows exponentially with the order of the original system). Future work will focus on addressing scalability issues, as well as developing approaches for computing tighter bounds on Δ leading to less conservative design. Another direction of future work will be quantifying the robustness of the controllers designed using this approach to sensor noise.

References

1. Shorten, R.N., Leith, D.J., Wirth, F., Kellett, C.: Switched linear systems and internet congestion control. In: Proceedings of the 13th Yale Workshop on Adaptive and Learning Systems, Yale University, New Haven CT (2005)

2. Savaresi, S.M., Silani, E., Bittanti, S., Porciani, N.: On performance evaluation methods and control strategies for semi-active suspension systems. In: Proceedings of the 42nd IEEE Conference on Decision and Control, Maui, Hawaii (2003) 2264–2269
3. Oishi, M., Tomlin, C.: Switching in non-minimum phase systems: Applications to a VSTOL aircraft. In: Proceedings of the American Control Conference, Chicago, IL (2000) 487–491
4. Liberzon, D., Morse, A.S.: Basic problems in stability and design of switched systems. IEEE Control Systems Magazine **19**(5) (1999) 59–70
5. Santarelli, K.R., Megretski, A., Dahleh, M.: On the stabilizability of two-dimensional linear systems via switched output feedback. In: Proceedings of the American Control Conference, Portland, OR (2005) 3778–3783
6. Hespanha, J.P., Morse, A.S.: Stabilization of nonholonomic integrators via logic-based switching. Automatica **35** (1999) 385–393
7. Xu, X., Antsaklis, P.: Stabilization of second-order LTI switched systems. International Journal of Control **73**(14) (2000) 1261–1279
8. Wicks, M., Peleties, P., DeCarlo, R.: Switched controller synthesis for the quadratic stabilization of a pair of unstable linear systems. European Journal of Control **4** (1998) 140–147
9. Lin, H., Antsaklis, P.J.: Stability and stabilizability of switched linear systems: A short survey of recent results. In: Proceedings of the IEEE International Symposium on Intelligent Control, Limassol, Cyprus (2005) 25–29
10. Megretski, A.: Robustness of finite state automata. In Giarre, L., Bamieh, B., eds.: Multidisciplinary Research in Control: The Mohammed Dahleh Symposium 2002. Number 289 in Lecture Notes in Control and Information Sciences. Springer, Berlin; New York (2003) 147–160
11. Santarelli, K.R., Megretski, A., Dahleh, M.A.: Input-to-state stability of nonlinear discrete time systems via R-cycle. In: Proceedings of the 45th IEEE Conference on Decision and Control, San Diego, CA (2006) 333–337
12. Tarraf, D.C.: A Finite State Machine Framework for Robust Analysis and Control of Hybrid Systems. PhD dissertation, Massachusetts Institute of Technology, Department of Mechanical Engineering (2006)
13. Tarraf, D.C., Megretski, A., Dahleh, M.A.: A framework for robust stability of systems over finite alphabets. (IEEE Transactions on Automatic Control) (To appear). First version appeared as LIDS Technical Report LIDS-P-2680, Cambridge, MA 02139, February 2006.
14. Bertsekas, D.P.: Dynamic Programming and Optimal Control. Second edn. Volume 1. Athena Scientific, Belmont, Massachusetts (2000)

Safety Verification of an Aircraft Landing Protocol: A Refinement Approach*

Shinya Umeno and Nancy Lynch

CSAIL, Massachusetts Institute of Technology, Cambridge MA, USA
{umeno,lynch}@theory.csail.mit.edu

Abstract. In this paper, we propose a new approach for formal verification of hybrid systems. To do so, we present a new refinement proof technique, *a weak refinement using step invariants*. As a case study of the approach, we conduct formal verification of the safety properties of NASA's *Small Aircraft Transportation System* (SATS) landing protocol. A new model is presented using the *timed I/O automata* (TIOA) framework [1], and key safety properties are verified. Using the new refinement technique presented in the paper, we first carry over the safety verification results from the previous discrete model studied in [2] to the new model. We also present properties specific to the new model, such as lower bounds on the spacing of aircraft in specific areas of the airspace.

1 Introduction

Hybrid systems are complex. In order to obtain a manageable mathematical model of a real hybrid system, a certain level of abstraction needs to be taken. A high-level abstraction of a system gives us a *discrete state-transition model*, where timing-dependent and continuous behavior of a real system are abstractly represented as discrete transitions. This high-level abstraction is particularly useful for a system that has algorithmically complex behavior. For instance, in [3], the initial start-up algorithm for the Time-Triggered Architecture [4] is formally verified using such a high-level abstraction. An important question here is whether the properties proved for the discrete abstraction hold for a real system, or for a refined, more realistic model.

In this paper, we propose a new approach to formally verify a given hybrid system. Basic concept of this approach is to use two levels of abstraction to verify a given hybrid system. The low-level *continuous model* includes descriptions of timing-dependent and continuous behavior, whereas in the high-level *discrete model*, timing-dependent and continuous behavior are abstracted away. Verification for these two models is done in the following steps.

1. First, the formal verification of the discrete model is conducted. This can be done either by the *invariant-proof technique*, or a *model-checking*.
2. Next, to carry over verification results from the discrete model to the continuous model, we prove a *refinement* from the continuous model to the discrete model.

* This project is supported by Air Force Contract FA9550-04-C-0084.

A. Bemporad, A. Bicchi, and G. Buttazzo (Eds.): HSCC 2007, LNCS 4416, pp. 557–572, 2007.

3. Finally, by using invariants carried over from the discrete model, we prove safety properties in the continuous model. Some of these properties immediately follow from the invariants carried over. On the other hand, some other properties can be expressed only in the continuous model, since they involve time-dependent or continuous behavior.

We technically contribute to the second step in the above stated approach. We often need some invariants of both the discrete model and the continuous model to prove a refinement. To make use of the invariants of the discrete model to larger extent than the existing techniques, we introduce a new refinement technique, called a *weak refinement using step invariants*. This technique differs from the existing techniques in that, by using this, we can use *invariants of the discrete model* in order to prove *invariants of the continuous model* needed for a refinement proof. Since we can assert the fact that invariants of the discrete model also hold for the continuous model *only after* proving a refinement between them, using the existing techniques causes a circular reasoning. Our new technique, a *weak refinement using step invariants*, resolves this problem.

As a case study of an application of the newly presented approach and refinement technique, we conduct a safety verification of the aircraft landing protocol that is part of NASA's *Small Aircraft Transportation System* (SATS) concept of operation [5]. Some formal verification studies for this protocol have been conducted so far. In [6], Dowek, Muñoz, and Carreño presented a state-transition model of the protocol. This model was a discrete model in that the airspace of airport is divided into several logical zones, and movements of aircraft are represented as discrete transitions Using this discrete model, safety verification of the model was done in [6], using a model-checking. The safety properties the authors model-checked were key upper bounds on the number of aircraft in the specific divided zones. In [7], Muñoz and Dowek extended their previous work [6] by presenting a *hybrid model* of the protocol, in which aircraft in a specific portion of the airspace of the airport exhibit continuous behavior, but movements of aircraft in the remaining portion are still discretized. Using this model, in [7], the authors verified key spacing properties of aircraft in the continuous portion of the hybrid model, using *symbolic model-checking* technique. We previously presented in [2] *invariant-proof-style verification* of the discrete model presented in [6]. In doing so, we first re-constructed the discrete model of [6] using an untimed *I/O automata* (IOA) framework, and verified key safety properties model-checked in [6] by using the invariant-proof technique. The proof for this case study has been mechanically checked using the PVS theorem prover [8].

In this paper, we present a new model of the protocol, *ContSATS*, which represents the continuous model in our new approach for this case study. This model more realistically reflects the dynamics of aircraft movement in a real system than the previous models presented in [6] and [7]. In contrast to the previous models, our new model captures continuous movements of aircraft in the entire airspace of the airport. The model is constructed using the *timed I/O automata* (TIOA) framework [1]. This framework and the *hybrid I/O automata* (HIOA) framework [9][1] have been used successfully to model several hybrid systems, such as a helicopter controller [10], the Traffic Alert and Collision Avoidance System [11] and a Lego car [12]. We first carry over the result from the discrete model to *ContSATS* by proving a refinement, and then prove key spacing properties of aircraft in *ContSATS*, which can be expressed by *ContSATS*, but not by the discrete model.

This paper is organized as follows: In Section 2, we briefly explain the TIOA framework, and introduce a new refinement technique. In Section 3, we quickly review the discrete model of [6], and present key invariants of the model proved in [2]. In Section 4, we introduce the new model *ContSATS*. Section 5 is devoted to proving a refinement from *ContSATS* to the discrete model. In Section 6, we present lower bounds on the spacing of aircraft in *ContSATS*. These are obtained by using the results carried over from the discrete model by a refinement. Finally, in Section 7, we summarize the results, and discuss some future work.

2 Timed I/O Automata Framework

In this section, we explain some basics of the timed I/O automata (TIOA) framework [1]. In Section 2.2, we introduce a new refinement technique, a *weak refinement using step invariants.* We also present a theorem that states that this refinement from automaton A to B implies that invariants of B also hold in A in some specific sense.

2.1 Timed I/O Automata

A timed I/O automaton (TIOA) is a state transition machine with an extension of continuous behavior. Every discrete transition is defined in a precondition-effect style, and continuous behavior is defined using *trajectories.* A trajectory is a partial function from a time to the current values of the state components of the automaton. The domain of a trajectory must be some interval in the time domain, and the size of the domain represents the duration that elapses by that trajectory. A trajectory can be a *point trajectory*, whose domain is a point $[t, t]$, for some time t. The trajectories of an automata are specified by the **evolve** and the **stop when** statement in the trajectory definition. In the **evolve** statement, we state the rate of the value change of a real-time variable x by differential equations or inequalities in terms of $\mathsf{d}(x)$, the first derivative of x. Informally, the **stop when** statement specifies the time when we want the model to perform some discrete transition. An *execution fragment* of an automata is a (possibly infinite) alternating sequence of trajectories and discrete transitions $\tau_0 a_1 \tau_1 a_2 \tau_2 ...$ that satisfies the following three conditions: 1). Each trajectory satisfies the constraints defined by the **evolve** and the **stop when** statements;

[1] The latest version of the TIOA framework presented in [1] is the restricted version of the HIOA framework in that external analog variables cannot be used for TIOA. Since we do not have any analog variables in automata for our case study, the two frameworks are intrinsically same in this study.

2). a_{i+1} is enabled in $\tau_i.lstate$ (the last state of trajectory τ_i); 3). a_{i+1} represents the transition from $\tau_i.lstate$ to $\tau_{i+1}.fstate$ (the first state of trajectory τ_{i+1}). We call an execution fragment an *execution* if it starts with one of the designated *start states*. We say that state s is a reachable state if there is an execution α such that $\alpha.lstate = s$. Informally, the *trace* of an execution is the externally visible part of the execution. More formally, it is the alternating sequence of the duration that elapses by the trajectory and the *external* transitions, such that each duration matches up the duration of the corresponding trajectory in the execution, and all *internal* transitions are hidden.

Let A be a TIOA. Q_A denotes the set of the states of A. Θ_A denotes the set of the start states of A. $reachable(A)$ denotes the set of the reachable states of A. $traces_A$ denotes the set of the the traces of A. An *invariant* of automaton A is a predicate over Q_A that is satisfied for any $s \in reachable(A)$. A *step of A starting with state s* is an execution fragment of A starting with s that consists of either one discrete transition surrounded by two point trajectories, or one closed trajectory with no discrete transition.

2.2 Weak Refinement Using Step Invariants

A *refinement* is a proof technique that has been used to show trace inclusion between two automata A and B ($traces_A \subseteq traces_B$). Informally, the above stated trace inclusion tells us that the external behavior of A does not go beyond what we expect from B. In some cases, we want to use invariants of automata in a proof of a refinement. A *weak refinement*[2] has been used for such cases. These refinement techniques, and simulation relations (more general version of refinements) are well studies in the computer science community, and several kinds of such simulation techniques for TIOA are summarized in [13].

In some cases (as we will see in Section 5), we actually need *invariants of B* in order to prove some *invariants of A* needed in the proof of a refinement from A to B. Since we can assert the fact that invariants of B also hold for A *only after* proving a refinement from A to B, we end up with circular reasoning if we use an existing refinement technique. This is why we need our new technique, a *weak refinement using step invariants*. Informally, our solution to this problem is to prove only the inductive case of the invariant proof for such invariants of A, assuming some additional conditions. In the following, we present a new definition of invariants that captures the above informal discussion.

Definition 1. Let A be a TIOA. Let P_1 and P_2 be predicates over Q_A. We say that P_1 is a *step invariant of A using* P_2, or simply *a step invariant using* P_2 when A is obvious from the context, if, for any reachable state s of A and any step α of A starting with s, the following condition holds.

$$P_1(\alpha.fstate) \wedge P_2(\alpha.fstate) \Rightarrow P_1(\alpha.lstate)$$

[2] This usage of the term "weak" here comes from [13]. We use this term since we have more assumptions (namely, invariants of automata) in some conditions of the definition of this refinement, than an ordinary refinement.

That is, to show that P_1 is a step invariant using P_2, we prove only the step condition of the invariant proof for P_1, assuming the additional condition P_2. The following lemma easily follows from the definition of a step invariant.

Lemma 2. *$P_1 \wedge P_2 \wedge ... \wedge P_n$ is a step invariant for automaton A using condition Q if P_1 is a step invariant of A using Q, and P_i, $2 \leq i \leq n$, is a step invariant of A using $Q \wedge P_1 \wedge ... \wedge P_{i-1}$.*

Now we define the new refinement. The main difference from the definition of an ordinary weak refinement[3] is that we assume an additional predicate P^* over Q_A in the step condition (Conditions 2) of the refinement. This P^* must be a step invariant using $\lambda s.P_B(r(s))$[4], where P_B is an invariant of B. This captures the above informal discussion: since we need invariant P_B of B in order to prove that P^* is an invariant of A, we just require P^* to be a step invariant using $\lambda s.P_B(r(s))$, invariant P_B "adapted" to A using mapping r.

Definition 3. Let A and B be TIOA. Let P_A be an invariant of A, and P_B be an invariant of B. Let r be a partial function from Q_A to Q_B. Let P^* be a step invariant of A using $\lambda s.P_B(r(s))$.

We say that r is a *weak refinement using P_A, P_B, and P^** if it satisfies the following two conditions for all states x_A and x_B of A and B, respectively.

1. If $x_A \in \Theta_A$ then $x_A \in dom(r)$, $r(x_A) \in \Theta_B$, and $P^*(x_A)$ hold.
2. If α is a step of A, and $\alpha.fstate \in dom(r)$, and

$$P_A(\alpha.fstate) \wedge P_B(r(\alpha.fstate)) \wedge P^*(\alpha.fstate)$$

holds, then $\alpha.lstate \in dom(r)$ and B has a closed execution fragment β with $\beta.fstate = r(\alpha.fstate)$, $trace(\beta) = trace(\alpha)$, and $\beta.lstate = r(\alpha.lstate)$.

We can prove the following soundness theorem for this new refinement technique. A proof appears in the full version of this paper [14].

Theorem 4. *Let A and B be TIOA and r be a weak refinement from A to B, using P_A, P_B, and P^*. Then $traces_A \subseteq traces_B$.*

The existence of a refinement from A to B actually implies more than just trace inclusion. Due to space limitation, we cannot present general theorems about this close correspondence (they appear in [14]). Here we present one theorem regarding invariants of automata.

Theorem 5. *Let A and B be TIOA. Let r be a refinement, a weak refinement, or a weak refinement using step invariants, from A to B. Let P_B be an invariant of B. Then, the predicate $\lambda s.P_B(r(s))$ is an invariant of A.*

[3] Due to space limitation, we cannot give the definition of an ordinary refinement or that of a weak refinement in this paper. The definition appears in the full version of this paper [14].

[4] $\lambda s.P_B(r(s))$ is the function that, given $s_1 \in Q_A$, returns $P_B(r(s_1))$.

Theorem 5 is used in Section 5 to carry over the invariants of the discrete model that have been proved in [2] to our new continuous model presented in Section 4.

Related works: The new refinement introduced in this section has a flavor of *assume-guarantee* reasoning, which has also been applied to hybrid systems [15,16]. Assume-guarantee reasoning is used for compositional verification of a system. When we verify a composed system $S_1||S_2$, instead of verifying S_1 and S_2 separately, we sometimes want to assume some properties of the system to be composed with. For example, to prove that S_1 works correctly, we may have to assume that S_2 "well behaves" in some particular sense. Assume-guarantee techniques allows us to have deduction rules that if S_1 is correct assuming S_2 well behaves and S_2 is correct assuming S_1 well behaves, then, the composed final system $S_1||S_2$ is correct. In contrast to the existing assume-guarantee techniques, with our new technique, we can assume that the high-level abstraction behaves correctly in order to prove that the low-level abstraction has invariants needed to prove the refinement. To our best knowledge, we have not seen any other technique that uses assume-guarantee reasoning in the above sense.

3 Discrete Model

A discrete state-transition model of the SATS landing protocol is presented in [6]. In this model, the airspace of the airport is discretized, and every movement of the aircraft is represented as a transition of the model. In [2], we reconstructed the model using the I/O automata framework. Due to space limitation, we cannot present a formal description of the discrete model. However, we present a formal description of our new model in Section 4, and also discuss differences between the discrete model and the new model in the same section.

Aircraft: An aircraft is defined as a tuple that has two attributes: the mahf assignment, mahf, which will be explained shortly, of type Side (an enumeration of left and right); and a unique ID, id.

Logical zones: In the discrete model, the airspace of the airport is logically divided into 13 zones (see Fig. 1). Each zone is modeled as a first-in first-out queue of aircraft. A movement of aircraft is represented by moving an aircraft from the head of one queue to the end of another queue. We refer to the T-shaped area consists of base(right), base(left), intermediate, and final as *the approach area.* This area is where aircraft perform the final approach to the ground.

Landing sequence: When an aircraft enters the system, the system assigns its *leader* aircraft, or the aircraft it has to follow. This leader relation is used in the protocol as a guard that delays the aircraft's final approach initiation for safe landings: an aircraft cannot enter the approach area until its leader has done so. In our discrete model, we encode this notion of the leader aircraft as an explicit queue of aircraft, called the *landing sequence.* When an aircraft enters the logical zones, it is also added to the end of the landing sequence, and is removed when it finishes landing. We define the leader of aircraft a as the aircraft just in front of a in the landing sequence.

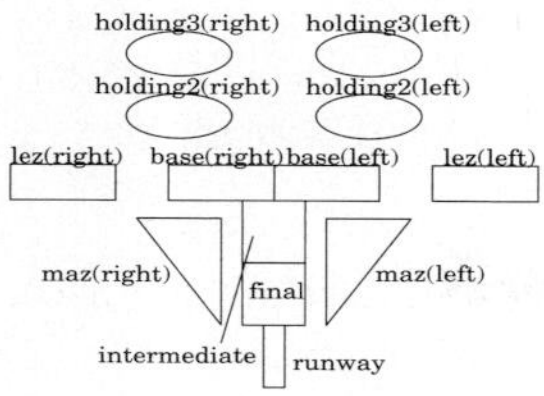

Fig. 1. 13 logical zones in the discrete model

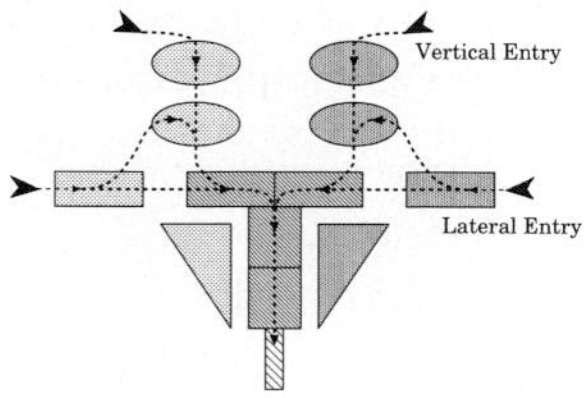

Fig. 2. Paths of aircraft

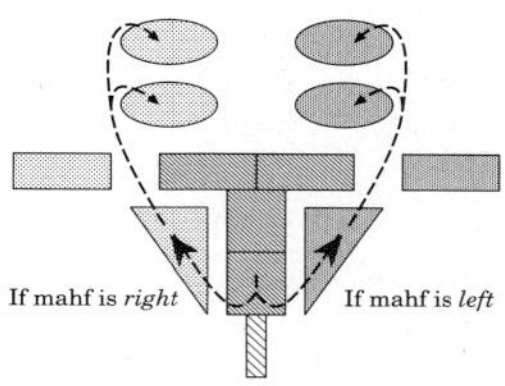

Fig. 3. Paths of aircraft that have missed the approach

Paths of aircraft: Here we present a high level picture of aircraft movements in the logical zones. All movements are represented by transitions, which are described in the precondition-effect style. A transition moves one aircraft from one zone to another in a way that it satisfies the rules specified in the protocol. The paths of aircraft are depicted in Fig. 2. An aircraft may miss the approach to the ground at the final zone. In such a case, it goes back to a holding fix (either holding3 or holding2), and makes the next try to land. An aircraft needs to determine the side of the holding fixes to which it goes in case it misses the approach. For this purpose, the assignment of the side, called the *missed approach holding fix (mahf)* is given to an aircraft when it enters the system. These paths of missed aircraft are depicted in Fig. 3.

Properties: In [6], some interesting properties of the discrete model that express safe separation of aircraft are presented and are exhaustively checked using an exhaustive exploration technique. In [2], using the invariant-proof technique, we proved key safety properties presented in [6]. Here we review some of the properties proved in [2]. The following condition Φ is defined as the conjunction of the listed seven conditions. An auxiliary predicate on_approach_qn(σ) checks if there is some aircraft assigned σ as its mahf in the approach area. In the rest of the paper, we refer to the first condition of Φ by Φ.1, the second condition by Φ.2, and so on. In Section 5.1, we present auxiliary invariants of the new model that is needed to prove a refinement as a step invariant using this Φ. It is worth to note here that, Conditions 3, 4, and 5 cannot be derived from the main safety properties taken from [6], but are derived from auxiliary lemmas to prove the main properties. Since we need these three conditions in Φ to prove a refinement in Section 5.2, this indicates that, by proving these auxiliary invariants, the assertional-style techniques give us more insight to how the system works, than an exhaustive exploration.

Condition Φ

1. $\forall\sigma : \text{side},\ \text{length}(\mathsf{holding3}(\sigma)) \le 1 \wedge \text{length}(\mathsf{holding2}(\sigma)) \le 1$
2. $\forall\sigma : \text{side},\ \neg\text{empty\_qn}(\mathsf{lez}(\sigma)) \Rightarrow$
   $\text{empty\_qn}(\mathsf{holding2}(\sigma)) \wedge \text{empty\_qn}(\mathsf{holding3}(\sigma)) \wedge \text{empty\_qn}(\mathsf{maz}(\sigma))$
3. $\text{first}(\mathsf{final}) = \text{first}(\mathsf{landing\_seq})$
4. $\forall\sigma : \text{side},\ (\text{on\_approach\_qn}(\sigma) \wedge \neg\text{empty\_qn}(\mathsf{maz}(\sigma))) \Rightarrow \text{empty\_qn}(\mathsf{holding3}(\sigma))$

5. $\forall\sigma$: side, on_approach_qn(σ) $\Rightarrow$ length($\mathsf{holding2}(\sigma)$) + length($\mathsf{holding3}(\sigma)$) ≤ 1
6. $\forall\sigma$: side, $length(\mathsf{maz}(\sigma)) \geq 2 \Rightarrow$ empty_qn($\mathsf{holding2}(\sigma)$) $\wedge$ empty_qn($\mathsf{holding3}(\sigma)$)
7. $\forall\sigma$: side, $\neg$empty_qn($\mathsf{maz}(\sigma)$)) $\Rightarrow$ length($\mathsf{holding2}(\sigma)$) + length($\mathsf{holding3}(\sigma)$) ≤ 1

4 Our New Continuous Model

In this section, we present our new continuous model, *ContSATS*, which more realistically reflects the dynamics of the aircraft movement in a real system than the discrete model or the *hybrid model* presented in [7]. In the hybrid model of [7], the movement of the aircraft in the approach area and the missed approach zones is modeled as continuous behavior. These areas are modeled as abstract lines[5] representing paths of aircraft on which aircraft continuously move according to their velocity vectors. Now a discrete transition for aircraft in the approach area and the maz zones is performed when an aircraft reaches the intersection points of the lines, in order to reassign the line on which that aircraft move.

To describe continuous dynamics of aircraft in the entire airspace of the airport, we use the same strategy as used for the hybrid model of [7]: in *ContSATS*, we model the paths of aircraft predetermined by the protocol as a collection of lines, with aircraft moving on them according to their velocity. (see Fig. 4, and compare it with Fig. 2 and 3). In the new model, analogous to the hybrid model of [7], we use transitions to re-assign the line on which aircraft move. The pre-determined paths in *ContSATS* include holding points (holding3hold and holding2hold in Fig. 4), where aircraft hover until the condition for the next procedure (transition) is satisfied.

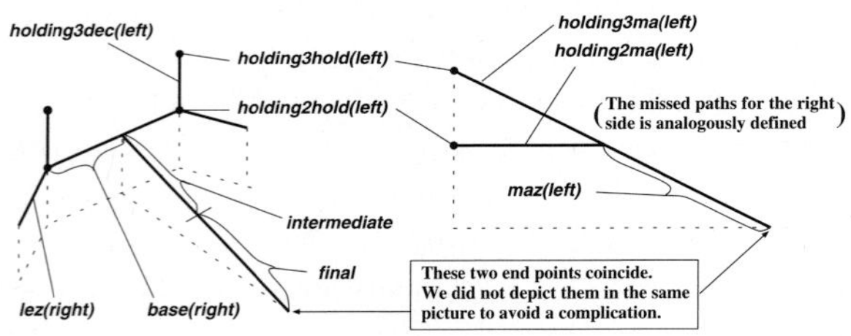

Fig. 4. Our new continuous model: *ContSATS*

4.1 Formal Specification for *ContSATS*

In this subsection, we present formal code for *ContSATS*, written in the TIOA specification language [17]. We explain auxiliary constants and functions first.

The line on which a specific aircraft currently moves is specified by a new attribute of aircraft, line. We use the prefix "LINE_" for the line names; for example, the final zone as a line is represented as LINE_final. The position of

[5] These lines forms "trajectories" of aircraft flying on the pre-determined paths. However, we avoid using the term "trajectories", and instead use "lines", since the term is also used in the TIOA framework and thus two usages may confuse the reader.

a specific aircraft in the line is specified by another new attribute of aircraft, pos. Using both the line value and pos value of a particular, we can uniquely determine on which line, and at what position in that line that aircraft is now.

Another new attribute of aircraft is t, which is used to express a time bound for some specific transitions to be performed. When one of the designated transitions becomes enabled, the aircraft a corresponding to that transition (the aircraft that will move by the transition) has its t value set to the current value of now. By the **stop when** clause in the trajectory definition, *ContSATS* is guaranteed to fire the transition corresponding to aircraft a either before or at the time the value of now $- a$.t reaches the pre-determined time bound for that transition. T_3, T_2, and T_{Tax} represents the time bounds for StartDescending, VerticalApproachInitiation, and Taxiing, respectively. We use function T that maps the name of a zone to the above specified time bounds for aircraft in that zone. We set t of aircraft outside of the holding zones or of the runway to -1, indicating that a timer is not set for those aircraft.

For simplicity, we assume that the lines are exactly symmetric on the right and left sides of the airport. L_{3dec}, L_{3ma}, L_B, L_I, L_F, and L_M respectively represents the lengths of holding3dec, holding3ma, base, intermediate, final, and maz. We use the function L to represent the length of the line for a given line. We denote by L_T the length aircraft fly in the entire approach area, that is, $L_B + L_I + L_F$. We use the function D to represent the distance a specific aircraft has flown in the approach area, and then in the missed approach zone; for example, if aircraft a is in final, $D(a) = L_B + L_I + a.pos$, and if a is in maz(σ), $D(a) = L_B + L_I + + L_F + a.pos$. If an aircraft is not in the approach area nor in the missed approach zones, the D function returns 0.

The velocity of the aircraft is bounded by some constants. This constraint is specified in the **evolve** statement in the trajectory definition.

We present formal code for *ContSATS* in the following. Due to space limitation, we only show the definitions of three transitions (VerticalApproachInitiation, MissedApproach, and LowestAvailableAltitude), and the trajectory definition. The full specification appears in [14]. The above three transitions are chosen because of the following three reasons. 1: VerticalApproachInitiation is one of the most interesting transitions, which represents an initiation of the aircraft's final approach to the ground. The precondition of the transition represents the guard so that an aircraft cannot initiate its approach until its leader has done so and the separation between the aircraft and its leader becomes at least S_0. 2: MissedApproach is also an interesting transition, which represents missed approaches of aircraft. As we can see from the precondition, this transition is preformed nondeterministically whenever an aircraft reaches the end point of the final line (a.pos $= L_F$). 3: In addition to the extra structure needed to represent the continuous behavior (such as now, pos, or the trajectory definition), we also modified three transitions inherited from the discrete model, in order to more faithfully represent a real system (how we modified them is explained in [14]). These are LowestAvailableAltitude, Landing, and HoldingPatternDescend. Due to this modification, we need some nontrivial auxiliary invariants of *ContSATS* to prove a refinement from *ContSATS* to the discrete model. As we will see in Section 5.1, these invariants are proved as step invariants using Φ (Corollary 7). In this paper, we focus on LowestAvailableAltitude among the three transitions.

We use three effects set_pos, set_line, and set_t to re-assign the pos, line, and t attributes of aircraft, respectively. The code of the automaton imports a *vocabulary*, ContSatsVocab, where auxiliary functions used in *ContSATS* are defined. We do not have a space to explain all these functions (it appears in [14]), but will explain those we need for the lemma statement and the proof. leader(a, landing_seq) represents the leader of aircraft a in the landing sequence. The predicate on_approach_qn(a) where a is an aircraft checks if a is in the approach area. The predicate on_approach_qn(σ) where σ is a side checks if there is some aircraft assigned to σ as its mahf in the approach area. The predicate on_zone_qn(z, a) checks if aircraft a is in zone z.

```
automaton ContSATS
  imports ContSatsVocab

%% All original discrete transitions are considered as the output transitions.
%% We added four new internal transitions, as well as the trajectory definition.
signature
  output
    VerticalEntry(ac:Aircraft, id:ID, side:Side),  LateralEntry(ac:Aircraft, id:ID, side:Side),
    HoldingPatternDescend(ac:Aircraft,side:Side),  VerticalApproachInitiation(ac:Aircraft,side:Side),
    LateralApproachInitiation(ac:Aircraft,side:Side),  Merging(ac:Aircraft,side:Side),
    Exit(ac:Aircraft),  FinalSegment(ac:Aircraft),  Landing(ac:Aircraft),  Taxiing(ac:Aircraft),
    MissedApproach(ac:Aircraft),  LowestAvailableAltitude(ac:Aircraft,side:Side),
  internal
    StartHolding2(ac:Aircraft,side:Side),  StartHolding3(ac:Aircraft,side:Side),
    StartDescending(ac:Aircraft,side:Side),  SetTime
states
  zones : zone_map, % mapping from a zone name to a zone
  nextmahf : Side, % Next missed approach holding fix
  landing_seq : Zone % landing sequence is defined as a queue
  now : AugumentedReal % the time elapsed from the initial state
  initially
    zones = initialZones ∧ nextmahf = right ∧ landing_seq = empty ∧ now = 0

%% Definitions of auxiliary functions are not shown in this code due to space limitation.

transitions
 output VerticalEntry(a, id, side)

 output LateralEntry(a, id, side)

 internal StartDescending(a, side)

 output HoldingPatternDescend(a, side)

 output VerticalApproachInitiation(a, side)
  pre ¬(empty_qn(holding2(side))) ∧
    a = first(holding2(side)) ∧
    length(base(opposite(side))) ≤ 1 ∧
    (first_in_seq_qn(a) ∨
      (on_approach_qn(leader(a,landing_seq)) ∧
      D(leader(a,landing_seq)) ≥ S0))
  eff set_line(a, AC_base(side)); set_pos(a, 0);
    set_t(a, -1);
    zones := move(holding2(side),base(side),zones)

 output LateralApproachInitiation(a, side)

 internal SetTime

 output Merging(a, side)

 output Exit(a)

 output FinalSegment(a)

 output Landing(a)

 output Taxiing(a)

 output MissedApproach(a)
  pre ¬(empty_qn(final)) ∧ ¬(empty_qn(landing_seq))
    ∧ a = first(final) ∧ a.pos = LF
  eff set_line(a, AC_maz(a.mahf)); set_pos(a, 0)
    zones:= assign(zones, final, rest(final));
    zones:= assign(zones, maz(a.mahf),
      add(maz(a.mahf),reassign(a)));
   landing_seq := add(rest(landing_seq),reassign(a));
    nextmahf := opposite(reassign(a).mahf);
```

```
output LowestAvailableAltitude(a, side)
 pre ¬(empty_qn(maz(side))) ∧
   a = first(maz(side)) ∧ a.pos = L_M;
 eff IF empty_qn(holding3(side)) ∧
     empty_qn(holding2(side))
   THEN set_line(a, AC_holding2ma(side));
      set_pos(a,0);
     zones := move(maz(side),holding2(side),zones);
   ELSE set_line(a, AC_holding3ma(side));
      set_pos(a,0)
     zones := move(maz(side),holding3(side),zones);
   FI

internal StartHolding3(a, side)

internal StartHolding2(a, side)
```

```
trajectories
  stop when
   (∃ a:Aircraft,
     (∃ z:Zone, on_zone_qn(z, a)) ∧
     a.pos ≥ L(a.line))
  ∨  (∃ a:Aircraft,
     (∃ z:Zone, on_zone_qn(z, a)) ∧
     a.t ≠ −1 ∧ now − a.t ≥ T(a.line))
  ∨  (∃ a:Aircraft,
     (∃ z:Zone, on_zone_qn(z, a)) ∧
     ((a.line = holding2L ∨ a.line = holding2R)∧
      a.t = −1 ∧ ¬first_in_seq_qn(a) ∧
        on_approach_qn(leader(a,landing_seq)) ∧
        D(leader(a,landing_seq)) = S_0))
  evolve
   d(now) = 1
   ∀ a: Aircraft
    IF (a.line=holding3decL ∨ a.line=holding3decR)
     THEN (V_d_min ≤ d(a.pos) ≤ V_d_max)
     ELSE (V_min ≤ d(a.pos) ≤ V_max) FI
```

In order to obtain a refinement, we have to assume the following condition: $(\frac{\mathsf{L}_{3ma}}{\mathsf{V}_{min}} + \mathsf{T}_3 + \frac{\mathsf{L}_{3dec}}{V_{d\_min}})\mathsf{V}_{max} < \mathsf{L}_\mathsf{T} + \mathsf{L}_\mathsf{M}$. This is used in the refinement proof in the case of the LowestAvailableAltitude transition.

5 Carrying over the Results from the Discrete Model Using a Refinement

In [2], we formally verified the safe separation of aircraft in the discrete model, by proving bounds on the number of aircraft in the logical zones. If we can carry over these results to *ContSATS*, the properties carried over tell us important spacing properties in *ContSATS*. For example, from the property that there is at most one aircraft in one holding3(σ), we can guarantee that two aircraft would never get close in the holding3 line in *ContSATS*. On the other hand, we cannot guarantee spacing properties of two aircraft on two adjacent lines from the properties of the discrete model. Some of these properties are actually proved as auxiliary lemmas for the refinement. We also examine several spacing properties in Section 6.

To make the discrete model (an ordinary IOA) comparable to *ContSATS* (a TIOA), we first construct *ExtSATS*, a natural extension of the discrete model to a TIOA. This extension can be done in the following generic way: Given an ordinary IOA A, we construct A' that is an timed extension (TIOA version) of A. First, in A', we add a new now state component to A which evolves at rate 1 (d(now) = 1). There is no **stop when** statement for A', and all discrete transitions are exactly the same as before the extension. From this straightforward extension, it is easy to see that all invariants of A are also invariants of A'. From Theorem 5, if we prove a refinement from *ContSATS* to *ExtSATS*, any invariant of *ExtSATS* is guaranteed to be an invariant of *ContSATS*.

One straightforward refinement mapping to consider (and actually the one we use for the refinement proof) is the following mapping r from a state of $ContSATS$ to a state of $ExtSATS$: for all $s \in Q_{ContSATS}$, $r(s) = t$ such that

$zones\_equal(s.\mathsf{zones}, t.\mathsf{zones}) \wedge s.\mathsf{nextmahf} = t.\mathsf{nextmahf} \wedge$
$queue\_equal(s.\mathsf{landing\_seq}, t.\mathsf{landing\_seq}) \wedge t.\mathsf{now} = s.\mathsf{now}$,

where $zones\_equal$ and $queue\_equal$ represent the equalities for two zone maps and two aircraft queues, respectively, defined by ignoring the new attributes of aircraft in $ContSATS$, such as pos (formal definitions appear in [14]). This mapping r maps a state of $ContSATS$ to a state of $ExtSATS$ so that every component of the state of $ContSATS$ matches the corresponding component of the state of $ExtSATS$. Note that such a state $r(s)$ in $ExtSATS$ is uniquely determined for every state s of $ContSATS$, since the above conditions specify all components of $ExtSATS$.

It turns out that we have to use a weak refinement using step invariants introduced in Section 2.2 for this mapping r. This is because in order to prove some invariants of $ContSATS$ needed to prove a refinement, we actually need some invariants of $ExtSATS$ that have been verified.

5.1 Auxiliary Invariants

In this subsection, we present the auxiliary invariants needed for the refinement proof. Due to space limitation, we cannot present a proof for these auxiliary invariants (it appears in [14]). We use Condition Φ defined in Section 3 as a state proposition of $ContSATS$.

Lemma 6. *Consider the following conditions A_1, A_2, B, C_1, and C_2.*

$(\mathbf{A_1})$: $\forall a, b$: Aircraft, $\forall \sigma$: side, on_approach_qn$(a) \wedge$
$a.\mathsf{mahf} = \sigma \wedge$ on_zone_qn$(\mathsf{holding3}(\sigma), b) \Rightarrow (1) \wedge (2) \wedge (3)$

(1) $b.\mathsf{line} = \mathsf{LINE\_holding3ma}(\sigma) \Rightarrow \mathsf{D}(a) \leq \frac{b.\mathsf{pos}}{\mathsf{V_{min}}} \cdot \mathsf{V_{max}}$.
(2) $b.\mathsf{line} = \mathsf{LINE\_holding3hold}(\sigma) \Rightarrow \mathsf{D}(a) \leq (\frac{\mathsf{L_{3ma}}}{\mathsf{V_{min}}} + (\mathsf{now} - b.\mathsf{t})) \cdot \mathsf{V_{max}}$.
(3) $b.\mathsf{line} = \mathsf{LINE\_holding3dec}(\sigma) \Rightarrow \mathsf{D}(a) \leq (\frac{\mathsf{L_{3ma}}}{\mathsf{V_{min}}} + \mathsf{T_3} + \frac{b.\mathsf{pos}}{\mathsf{V_{d\_min}}}) \cdot \mathsf{V_{max}}$.

$(\mathbf{A_2})$: $\forall a, b$: Aircraft, $\forall \sigma$: side, on_zone_qn$(\mathsf{maz}(\sigma), a) \wedge$
on_zone_qn$(\mathsf{holding3}(\sigma), b) \Rightarrow (1) \wedge (2) \wedge (3)$

(1) $b.\mathsf{line} = \mathsf{LINE\_holding3ma}(\sigma) \Rightarrow \mathsf{D}(a) \leq \frac{b.\mathsf{pos}}{\mathsf{V_{min}}} \cdot \mathsf{V_{max}}$.
(2) $b.\mathsf{line} = \mathsf{LINE\_holding3hold}(\sigma) \Rightarrow \mathsf{D}(a) \leq (\frac{\mathsf{L_{3ma}}}{\mathsf{V_{min}}} + (\mathsf{now} - b.\mathsf{t})) \cdot \mathsf{V_{max}}$.
(3) $b.\mathsf{line} = \mathsf{LINE\_holding3dec}(\sigma) \Rightarrow \mathsf{D}(a) \leq (\frac{\mathsf{L_{3ma}}}{\mathsf{V_{min}}} + \mathsf{T_3} + \frac{b.\mathsf{pos}}{\mathsf{V_{d\_min}}}) \cdot \mathsf{V_{max}}$.

$(\mathbf{B})$: $\forall a$: Aircraft, $\forall \sigma$: side,
(on_zone_qn$(\mathsf{holding3}(\sigma)) \wedge a.\mathsf{line} = \mathsf{LINE_holding3dec}(\sigma)) \Rightarrow$ empty_qn$(\mathsf{holding2}(\sigma))$.

$(\mathbf{C_1})$: $\forall a$: Aircraft, (on_approach_qn$(a) \wedge \neg$first_in_seq_qn$(a)) \Rightarrow$
$\mathsf{D}(\mathrm{leader}(a, \mathsf{landing\_seq})) - \mathsf{D}(a) \geq \mathsf{S_0} - \frac{\mathsf{D}(\mathrm{leader}(a, \mathsf{landing\_seq})) - \mathsf{S_0}}{\mathsf{V_{min}}} (\mathsf{V_{max}} - \mathsf{V_{min}})$.

$(\mathbf{C_2})$: $\forall a, b$: Aircraft, (on_zone_qn$(\mathsf{runway}, a) \wedge$ on_approach_qn$(b)) \Rightarrow$
$\mathsf{now} - a.\mathsf{t} \geq \frac{\mathsf{D}(b) - (\mathsf{L_T} - \mathsf{S_T})}{\mathsf{V_{max}}}$

The following conditions hold:

1. *A_1, B, and C_1 are step invariants using Φ.*
2. *A_2 is a step invariant using Φ and A_1.*
3. *C_2 is a step invariant using Φ and C_1.*

From Lemmas 2 and 6, we have the following corollary.

Corollary 7. *The conjunction $A_1 \wedge A_2 \wedge B \wedge C_1 \wedge C_2$ forms a step invariant of $ContSATS$ using Φ.*

Conditions A_2, B, and C_2 are used in the refinement proof (Theorem 8) for transitions LowestAvailableAltitude, HoldingPatternDescend, and Taxiing, respectively. Recall that these three transitions are modified from those in the original discrete model, so that $ContSATS$ more realistically models a real system. This is why we need these nontrivial conditions A_2, B, and C_2 in the refinement proof, in order to show that the modified transitions of $ContSATS$ matches with the original transitions of the discrete model. In the proof sketch of Theorem 8, we demonstrate how A_2 is used in the case of LowestAvailableAltitude in the refinement proof.

5.2 Refinement Proof

Now we prove a refinement from $ContSATS$ to $ExtSATS$. We use the mapping r defined in the beginning of Section 5. We use Inv_{Cont}, some auxiliary invariants of $ContSATS$ proved in [14], and Inv_{Ext}, invariants of the discrete model (and thus of $ExtSATS$) proved in [2]. We use $A_1 \wedge A_2 \wedge B \wedge C_1 \wedge C_2$ as a step invariant using Inv_{Ext} (since Inv_{Ext} implies Φ).

Theorem 8. *The function r is a weak refinement from $ContSATS$ to $ExtSATS$ using Inv_{Cont}, Inv_{Ext}, and $A_1 \wedge A_2 \wedge B \wedge C_1 \wedge C_2$.*

Proof sketch: Condition 1 is easy to prove.
Condition 2: Suppose α is a step of A. We refer to $\alpha.fstate$ as s and $\alpha.lstate$ as s' in the following. It is easy to see that $s' \in dom(r)$ since r is a total function. We also assume invariants of $ContSATS$, Conditions Φ, and $A_1 \wedge A_2 \wedge B \wedge C_1 \wedge C_2$ hold in s. We demonstrate how a proof goes for Condition 2 by proving the case of the LowestAvailableAltitude(σ) transition. We use Condition A_2 for this case.

Suppose α consists of one LowestAvailableAltitude(σ) transition. From the precondition of the transition, there is at least one aircraft in maz(σ) in s, and thus also in $r(s)$. It follows that LowestAvailableAltitude(σ) is enabled in $r(s)$, and thus an execution fragment β of $ExtSATS$ starting with $r(s)$ that consists of one LowestAvailableAltitude(σ) is a valid execution fragment of $ExtSATS$. It is easy to see $trace(\alpha) = trace(\beta)$. Now we prove $\beta.lstate = r(s')$. If holding3(σ) is empty in s, LowestAvailableAltitude(σ) actually has the exact same effects in $ContSATS$ and $ExtSATS$ (see [14]). Hence it is sufficient to prove that holding3(σ) is empty in s. From the precondition, there is an aircraft a such that

$\mathsf{a.pos} = \mathsf{L_M}$, and $\mathsf{a.line} = \mathsf{LINE\_maz}(\sigma)$. From Condition A_2 and an invariant of *ContSATS*: $\forall b : \mathsf{Aircraft}, b.\mathsf{pos} \leq \mathsf{L}(b.\mathsf{line})$ (this can be easily proved by induction), if $\mathsf{holding3}(\sigma)$ is not empty, then $\mathsf{a.x} = \mathsf{L_M} \leq (\frac{\mathsf{L_{3ma}}}{\mathsf{V_{min}}} + \mathsf{T_3} + \frac{\mathsf{L_{3dec}}}{V_{d\_min}})\mathsf{V_{max}} - \mathsf{L_T}$. This contradicts the assumption that $(\frac{\mathsf{L_{3ma}}}{\mathsf{V_{min}}} + \mathsf{T_3} + \frac{\mathsf{L_{3dec}}}{V_{d\_min}})\mathsf{V_{max}} < \mathsf{L_T} + \mathsf{L_M}$. □

From Theorems 8 and 5, we have the following corollary.

Corollary 9. *Let P be an invariant of ExtSATS. Then $\lambda s.P(r(s))$ is an invariant of ContSATS.*

6 Spacing Properties of Aircraft in *ContSATS*

In the previous section, by using a refinement technique, we proved as Corollary 9 that all invariants of the discrete model of SATS that have been proved in [2] are also invariants of *ContSATS*. For example, from Φ.1 (the number of aircraft in each vertical fix is at most one), we can guarantee two aircraft would never get close in $\mathsf{holding2}$ and $\mathsf{holding3}$ zones in *ContSATS*. This kind of spacing properties of *ContSATS* are derivable from the invariants of the discrete model, and they express the safe separation of aircraft in one specific zone (represented by a line in *ContSATS*). However, one might be interested in the safe separation of aircraft in two consecutive zones. In this section, we conclude the analysis of safe separation properties for *ContSATS* in this paper, by presenting such spacing properties for all pairs of consecutive zones in *ContSATS*. The *spacing* between two aircraft is defined as the distance of the two aircraft with respect to the pre-determined paths of *ContSATS*.

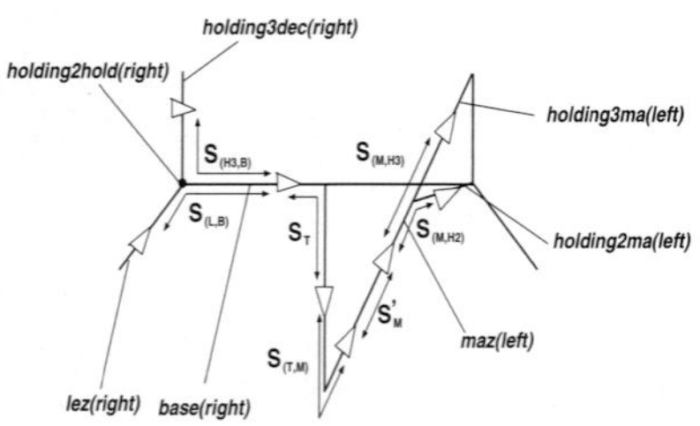

$\mathsf{S_{(H3,B)}}$	$\mathsf{L_{3dec}} - \frac{\mathsf{L_{3dec}}}{\mathsf{V_{maz}}}(\mathsf{V_{max}} - \mathsf{V_{min}})$
$\mathsf{S_{(L,B)}}$	$\mathsf{L_l} - \frac{\mathsf{L_l}}{\mathsf{V_{maz}}}(\mathsf{V_{max}} - \mathsf{V_{min}})$
$\mathsf{S_T}$	$\mathsf{S_0} - \frac{\mathsf{L_T} - \mathsf{S_0}}{\mathsf{V_{min}}}(\mathsf{V_{max}} - \mathsf{V_{min}})$
$\mathsf{S_{(T,M)}}$	$\mathsf{S_0} - \frac{\mathsf{L_T}}{\mathsf{V_{max}}}(\mathsf{V_{max}} - \mathsf{V_{min}})$
$\mathsf{S'_M}$	$2\mathsf{S_0} - (\mathsf{L_T} + \mathsf{L_M} - \mathsf{S_0})\Delta$
$\mathsf{S_{(M,H2)}}$	$(1 + \frac{\mathsf{V_{min}}}{\mathsf{V_{max}}})\mathsf{S_0} - \frac{\mathsf{V_{max}} - \mathsf{V_{min}}}{\mathsf{V_{max}}}(\mathsf{L_T} + \mathsf{L_M})$
$\mathsf{S_{(M,H3)}}$	$\mathsf{L_M} + \mathsf{L_T} - \mathsf{L_{3ma}}\Delta.$

Fig. 5. Lower bounds on the spacing of aircraft in two consecutive zones in *ContSATS*

An overview of the spacing properties of aircraft in two consecutive zones that we have proved in [14] is depicted in Figure 5. Each bi-directional arrow in the picture represents a lower bound on the spacing of aircraft. We have proved these properties by induction over the length of the execution of *ContSATS*. To do so, we used invariants carried over from the discrete model to *ContSATS*, by Corollary 9. Among these spacing properties, $\mathsf{S_T}$ and $\mathsf{S'_M}$ are the ones model-checked in [7] (we actually obtained a better bound for $\mathsf{S'_M}$ than [7], using some reasonable assumption stated in [14]).

7 Conclusion

In this paper, we presented a new approach to verify a given hybrid system. For the new approach, we introduced a new refinement proof technique, a *weak refinement using step invariants.* To demonstrate how the approach can be used, we conduct formal verification of NASA's SATS aircraft landing protocol. We believe that this approach is highly applicable to other hybrid systems as well. Proving the soundness of the abstraction used in [3] for a start-up algorithm of TTA by this approach appears one possible interesting future work.

Acknowledgment. we thank anonymous reviewers for their fruitful comments on an earlier version of this paper.

References

1. Kaynar, D.K., Lynch, N., Segala, R., Vaandrager, F.: The Theory of Timed I/O Automata. Synthesis Lectures on Computer Science. Morgan & Claypool Publishers (2006)
2. Umeno, S., Lynch, N.: Proving safety properties of an aircraft landing protocol using I/O automata and the PVS theorem prover: a case study. In: FM 2006: Formal Methods. Volume 4084 of Lecture Notes in Computer Science., Hamilton, Ontario Canada (2006) 64 – 80
3. Steiner, W., Rushby, J., Sorea, M., Pfeifer, H.: Model Checking a Fault-Tolerant Startup Algorithm: From Design Exploration To Exhaustive Fault Simulation. In: Proc. of the 2004 International Conference on Dependable Systems and Networks, Florence, Italy, IEEE Computer Society (2004) 189–198
4. Kopetz, H., Bauer, G.: The time-triggered architecture. Proceedings of The IEEE **91; PART 1** (2003) 112–126
5. T.Abbott, Jones, K., Consiglio, M., Williams, D., Adams, C.: Small Aircraft Transportation System, High Volume Operation concept: Normal operations. Technical Report NASA/TM-2004-213022, NASA Langley Research Center, NASA LaRC,Hampton VA 23681-2199, USA (2004)
6. Dowek, G., Muñoz, C., Carreño, V.: Abstract model of the SATS concept of operations: Initial results and recommendations. Technical Report NASA/TM-2004-213006, NASA Langley Research Center, NASA LaRC,Hampton VA 23681-2199, USA (2004)
7. Muñoz, C., Dowek, G.: Hybrid verification of an air traffic operational concept. In: Proceedings of IEEE ISoLA Workshop on Leveraging Applications of Formal Methods, Verification, and Validation, Columbia, Maryland (2005)
8. Owre, S., Rushby, J.M., Shankar, N.: PVS: A prototype verification system. In Kapur, D., ed.: 11th International Conference on Automated Deduction (CADE). Volume 607 of Lecture Notes in Computer Science., Saratoga, NY (1992) 748 – 752
9. Lynch, N., Segala, R., Vaandraager, F.: Hybrid I/O automata. Information and Computation **185(1)** (2003) 105–157
10. Mitra, S., Wang, Y., Lynch, N., Feron, E.: Safety verification of model helicopter controller using hybrid Input/Output automata. In: HSCC'03, Hybrid System: Computation and Control, Prague, the Czech Republic (2003)

11. Livadas, C., Lygeros, J., Lynch, N.A.: High-Level Modeling and Analysis of the Traffic Alert and Collision Avoidance System (TCAS). Proceedings of the IEEE, Special Issue on Hybrid Systems: Theory & Applications **88**(7) (2000) 926–948
12. Fehnker, A., Zhang, M., Vaandrager, F.: Modeling and verifying a lego car using hybrid I/O automata. In: Third International Conference on Quality Software (QSIC 2003), Dallas, Texas, USA, IEEE Computer Society Press (2003)
13. Lynch, N., Vaandrager, F.: Forward and backward simulations – part II: Timing-based systems. Information and Computation **128(1)** (1996) 1 – 25
14. Umeno, S.: Proving safety properties of an aircraft landing protocol using timed and untimed I/O automata: a case study. Master's thesis, Massachusetts Institute of Technology, Cambridge, MA (2006)
15. Henzinger, T.A., Minea, M., Prabhu, V.: Assume-guarantee reasoning for hierarchical hybrid systems. In: Proc. of HSCC'01, Hybrid Systems: Computation and Control. Volume 2034 of Lecture Notes in Computer Science. (2001) 275 – 290
16. Frehse, G., Han, Z., Krogh, B.: Assume-guarantee reasoning for hybrid I/O-automata by over-approximation of continuous interaction. In: CDC 2004: IEEE Conference on Decision and Control. (2004)
17. Garland, S.: TIOA User Guide and Reference Manual. (2005)

Rate Admission Control for Hard Real-Time Task Scheduling

Vladimiro Vacca, Francesco Vasca, and Luigi Iannelli

Department of Engineering, University of Sannio,
Piazza Roma 21, 82100 Benevento, Italy
vasca@unisannio.it
http://www.grace.ing.unisannio.it

Abstract. Hard real-time scheduling problem of periodic tasks in the presence of aperiodic tasks occurrence is analyzed. A hybrid model of the tasks and of the Earliest Deadline First scheduling policy is proposed and its translation into a Mixed Logical Dynamic model is presented. The model is used to design an optimal admission controller that selects the best feasible period for the periodic tasks based on the prediction of the processor utilization.

1 Introduction

In many real-time applications, such as in space avionics [1], periodic activities represent the major computational demand in the system. Periodic tasks typically arise from sensory data acquisition, low level servoing, control loops, action planning, and system monitoring. Such activities need to be cyclically executed at specific rates, which can be derived from the application requirements. When an application consists of several concurrent periodic tasks with individual timing constraints, the real-time operating system has to guarantee that each periodic instance is regularly activated at its proper rate and is completed within its deadline [2]. Such constraints can be hard (as in most control systems) or soft (as in multimedia systems). In hard real-time systems the scheduling problem is eventually complicated by the occurrence of external events, related for instance to emergency or fault operating conditions that imply the activation of aperiodic tasks.

Classical real-time scheduling algorithms are usually based on *a priori* knowledge of resource requirements, precedence constraints, resource contention, and future arrival times [3]. When the new task activations are not known a priori, a dynamic scheduling technique should be adopted. Earliest Deadline First (EDF) is an efficient dynamic scheduling algorithm in resource sufficient environments [4], however its performance degrades rapidly in overload situations. The Spring scheduling algorithm [5] can dynamically guarantee incoming tasks via on-line admission control and planning and thus it is applicable in resource insufficient environments. Roughly speaking the main idea consists of selecting the period of the task to be admitted into the scheduler queue so that schedulability and some desired performance (i.e. Quality of Service, QoS) are achieved.

A. Bemporad, A. Bicchi, and G. Buttazzo (Eds.): HSCC 2007, LNCS 4416, pp. 573–586, 2007.

In the presence of resource insufficient environment or unpredictable events, the admission control can be much improved by using a feedback of the actual state of the scheduler queue and processor [6]-[8]. Dynamic feedback with an elastic scheduling model and period adjustment algorithms have been proposed for soft real-time application with flexible workload such as multimedia systems [9]-[11]. In multimedia applications, when a reservation-based CPU scheduling policy is used, the issue of dynamically adjusting the bandwidth for a set of periodic tasks has been also solved by proposing a QoS control strategy [12]-[13]. A feedback scheduler for digital control systems has been presented in [14]. In [15] another feedback scheduling scheme based on predictive control has been presented in order to dynamically adjust the reserved processor time for each task, and thus to optimize the performance. In [16] a feedback-based admission controller was designed to maintain desired utilization of an Apache Web server. A feedback control real-time scheduling framework was proposed in [17] to provide performance guarantees for real-time systems with unknown task execution times.

In this paper we investigate the hard real-time periodic tasks scheduling scenario in the presence of aperiodic tasks. An interesting approach that uses timed automata for modeling and simulation analysis of dynamic task scheduling problem is proposed in [18]-[19]. Here we deal with the problem by using a different modeling framework: a hybrid model of the scheduling process represented into a Mixed Logical Dynamic (MLD) form is proposed. Moreover, by assuming an EDF scheduling policy, an admission controller, based on the feedback of the CPU queue state and on the prediction of the processor utilization, selects the "best" periods of the tasks to be admitted into the CPU queue. The controller selects the shortest tasks periods (associated to the highest QoSs) so that schedulability is ensured also in the presence of the aperiodic task. The optimal admission controller is designed by using mixed integer linear programming based on the MLD model. The performance of the proposed solution are evaluated through simulations based on a realistic C-written scheduling code.

2 Hybrid Model for Task Scheduling

At each time unit (or step) k the scheduler has to dispatch a single task execution time unit to the processor by considering a task sorting policy (scheduling policy) into the ready task queue and the periodic activations of the tasks (see Fig. 1).

Let us consider a set of periodic tasks τ_i, $i \in I \equiv \{1, ..., N\}$. Without loss of generality, for the sake of simplicity in the sequel we assume zero initial phase and relative deadline equal to the task period. Thus each ith task is identifiable by the triple of Worst Case Execution Time (WCET) C_i, task period T_i and absolute deadline d_i. Since C_i and (for the moment) T_i are fixed task parameters in a general periodic task scheduler, the task scheduling modeling can be only focused on the task absolute deadline d_i, which also plays an important role in the EDF preemption. The task absolute deadline updating can be modeled as follows:

$$d_i(k+1) = \begin{cases} d_i(k) & if\ x_c(k) + 1 < d_i(k) \\ d_i(k) + T_i & if\ x_c(k) + 1 = d_i(k) \end{cases} \quad (1)$$

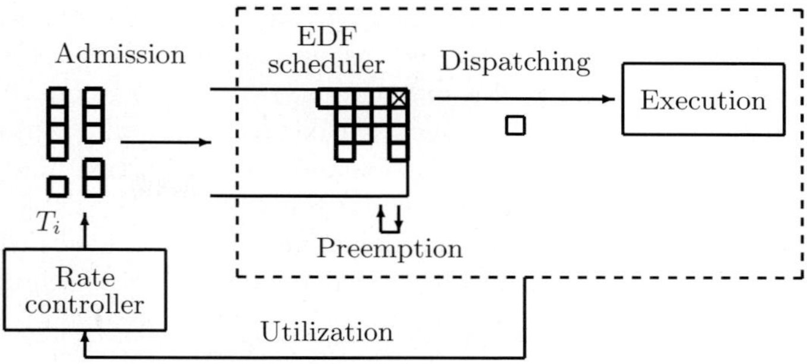

Fig. 1. Rate admission control scheme with EDF scheduler. Note that tasks preemption is also allowed.

with $d_i(0) = T_i$ and where $x_c(k)$ is a discrete time counter variable defined as

$$x_c(k+1) = x_c(k) + 1 \tag{2}$$

with $x_c(0) = 0$. By indicating with $S(k)$ the task dispatched at time instant k, i.e. $S(k) = i$ means that a unit of the ith task is in execution between k and $k+1$, the EDF scheduling function can be described as follows:

$$S(k) = \arg \min_{i \in \Gamma(k)} \{d_i(k)\} \tag{3}$$

where $\Gamma(k)$ is the subset of tasks included in the ready task queue at time step k. Thus (1)–(3) can represent an hybrid model of EDF task scheduler. In order to estimate the processor utilization it is necessary to define a more detailed task model, which is able to dynamically describe also the variable task execution time and the deadline keeping of each task instance. For such purpose the task execution time $E_i(k)$ is introduced, which depicts the time units of the task already executed. The corresponding dynamics are modeled as

$$E_i(k+1) = \begin{cases} E_i(k) & if\ S(k) \neq i \\ E_i(k) + 1 & if\ S(k) = i \end{cases} \tag{4}$$

with $E_i(0) = 0$. A reset condition for the execution time when a time instant multiple of the task period has been reached must be used. Therefore (4) can be replaced by

$$E_i(k+1) = \begin{cases} E_i(k) & if\ S(k) \neq i,\ k \neq nT_i \\ E_i(k) + 1 & if\ S(k) = i,\ k \neq nT_i \\ 0 & if\ S(k) \neq i,\ k = nT_i \\ 1 & if\ S(k) = i,\ k = nT_i \end{cases} \tag{5}$$

where $n = 0, 1, 2, \ldots$, and we assumed that a task can be dispatched at the same time instant in which it has been released by the admission controller. Note that the deadline keeping condition can be expressed as $E_i(nT_i) = C_i$.

The task execution time variable allows to model more precisely the EDF scheduler. Indeed, in our framework the scheduling is based on a ready-task-queue in which a task dispatched could enter again into the queue because of a preemption policy, until it is fully executed. In order to model the queue state during EDF scheduling policy it is necessary to introduce two more variables: a binary variable $INQ_i(k)$ for checking the full execution of the ith task and its consequent exit from queue, and an integer variable $DINQ_i(k)$ which masks with a large integer, say $M \gg \max_{i \in I} d_i(k)$, the deadline of a task fully executed and not already rescheduled until the next release time. We consider $DINQ_i(k)$ as a virtual task deadline defined in order to model the EDF preemption by not excluding the tasks out of the task ready queue at time instant k. The variables $INQ_i(k)$ and $DINQ_i(k)$ can be modeled as follows:

$$INQ_i(k) = \begin{cases} 0 & if\ E_i(k) = C_i \\ 1 & if\ E_i(k) < C_i \end{cases} \tag{6}$$

$$DINQ_i(k) = \begin{cases} d_i(k) & if\ INQ_i(k) = 1 \\ M & if\ INQ_i(k) = 0. \end{cases} \tag{7}$$

Since the EDF policy is based on dispatching the task with minimum absolute deadline and we consider all tasks (i.e. $i \in I$ including also the tasks which are out of the task ready queue), the scheduling function (3) can be replaced by

$$S(k) = \arg\min_{i \in I} \{DINQ_i(k)\}. \tag{8}$$

Typical aperiodic tasks in real-time control systems are linked to fault conditions or they are reaction tasks to unpredictable events. These tasks, which will be indicated with the $\bar{i}^{th}$ index, can be characterized by a known WCET $C_{\bar{i}}$ and by a very short deadline for a fast reactivity to the faults in order to carry out some recovery operations. Since the aperiodic task must be executed without preemption, its task execution updating can be modeled as

$$E_{\bar{i}}(k+1) = \begin{cases} 0 & if\ S(k) \neq \bar{i} \\ E_{\bar{i}}(k) + 1 & if\ S(k) = \bar{i}\,. \end{cases} \tag{9}$$

In this scenario we can associate to the occurring of a fault or an unpredictable event a binary variable $\delta_{\bar{i}}(k)$ which is 1 for one step when a fault or an event happens. By considering this binary variable, the inclusion of the aperiodic task into the ready queue can be modeled as follows:

$$INQ_{\bar{i}}(k) = \begin{cases} 0 & if\ (\delta_{\bar{i}}(k) = 0) \wedge (E_{\bar{i}}(k) = C_{\bar{i}}) \\ 1 & if\ (\delta_{\bar{i}}(k) = 1) \vee (0 < E_{\bar{i}}(k) < C_{\bar{i}}) \end{cases} \tag{10}$$

$$DINQ_{\bar{i}}(k) = \begin{cases} 0 & if\ INQ_{\bar{i}}(k) = 1 \\ M & if\ INQ_{\bar{i}}(k) = 0 \end{cases} \tag{11}$$

such that when a fault happens the aperiodic task is included into the ready queue and immediately it is dispatched because of its zero virtual deadline. Thus (1)-(2) together with (5)-(9) represent the scheduling model of a mixed task set.

3 Mixed Logical Dynamical Task Scheduling Model

In order to design a control strategy for the tasks periods selection it is useful to transform the previous scheduling model into a MLD form [20]-[21].

3.1 Task MLD Model

The hybrid dynamics represented by (1) are ruled by the quantity $x_c(k) + 1 - d_i(k)$. We can define a binary variable $\delta_{d_{3i}}(k) \in \{0, 1\}$ such that

$$[\delta_{d_{3i}}(k) = 1] \Leftrightarrow [x_c(k) + 1 - d_i(k) = 0]. \tag{12}$$

By using the condition $x_c(k) + 1 \leq d_i(k)$, which is valid $\forall k$ and $\forall i$, we can rewrite (1) as

$$d_i(k + 1) = d_i(k) + \delta_{d_{3i}}(k)T_i. \tag{13}$$

Since the MLD framework deals with linear inequalities, we have to represent (12) as a set of inequalities. The first step is to rewrite the equivalence (12) by introducing two further binary variables $\delta_{d_{1i}}(k) \in \{0, 1\}$ and $\delta_{d_{2i}}(k) \in \{0, 1\}$ as

$$[\delta_{d_{1i}}(k) = 1] \Leftrightarrow [x_c(k) + 1 - d_i(k) \leq 0] \tag{14}$$

$$[\delta_{d_{2i}}(k) = 1] \Leftrightarrow [x_c(k) + 1 - d_i(k) \geq 0] \tag{15}$$

such that $\delta_{d_{3i}}(k) = \delta_{d_{1i}}(k) \wedge \delta_{d_{2i}}(k)$. The previous logical condition can be transformed by considering the equivalence

$$[\delta_{d_{3i}}(k) = \delta_{d_{1i}}(k) \wedge \delta_{d_{2i}}(k)] \Leftrightarrow \begin{cases} -\delta_{d_{1i}}(k) + \delta_{d_{3i}}(k) \leq 0 \\ -\delta_{d_{2i}}(k) + \delta_{d_{3i}}(k) \leq 0 \\ \delta_{d_{1i}}(k) + \delta_{d_{2i}}(k) - \delta_{d_{3i}}(k) \leq 1 \end{cases} \tag{16}$$

that is trivial to prove. In order to obtain a set of inequalities which replace (14)-(15), we use the BigM approach and transformations shown in [20]. Indeed the right hand side of (14) can be expressed as a set of inequalities by considering the following equivalence [22]

$$[x_c(k) + 1 - d_i(k) \leq 0] \Leftrightarrow \begin{cases} x_c(k) + 1 - d_i(k) \leq M_{d_i}(1 - \delta_{d_{1i}}(k)) \\ x_c(k) + 1 - d_i(k) \geq \varepsilon + (m_{d_i} - \varepsilon)\delta_{d_{1i}}(k) \end{cases} \tag{17}$$

where $m_{d_i}(M_{d_i})$ is an under(over)-estimate of the minimum (maximum) of $x_c(k) + 1 - d_i(k)$ and ε is a small positive scalar. Moreover it is simple to prove also the following equivalence:

$$[x_c(k) + 1 - d_i(k) \geq 0] \Leftrightarrow \begin{cases} x_c(k) + 1 - d_i(k) \geq -M_{d_i}(1 - \delta_{d_{2i}}(k)) \\ x_c(k) + 1 - d_i(k) \leq -\varepsilon + (-m_{d_i} + \varepsilon)\delta_{d_{2i}}(k) \end{cases} \tag{18}$$

Thus, the absolute deadline evolution of each periodic task can be represented in the form (13) subject to the linear inequality constraints reported in (16), (17) and (18).

In order to represent the task execution time evolution (4) in MLD form, we use the same approach presented for the conversion of (1). By introducing the binary variables $\delta_{S_{1i}}(k)$, $\delta_{S_{2i}}(k)$, $\delta_{S_{3i}}(k)$, the updating logic of (4) can be represented by the condition

$$[\delta_{S_{3i}}(k) = 1] \Leftrightarrow [S(k) - i = 0]. \tag{19}$$

The model (4) can now be rewritten in the following MLD form:

$$E_i(k+1) = E_i(k) + \delta_{S_{3i}}(k) \tag{20}$$

subject to linear inequality constraints similar to (16)-(18) with δ_S playing the role of δ_d and $S(k)-i$ replacing $x_c(k)+1-d_i(k)$. For the reset condition reported in (5) one can substitute (20) with

$$E_i(k+1) = E_i(k) + \delta_{S_{3i}}(k) - \delta_{d_{3i}}(k) \cdot E_i(k). \tag{21}$$

In order to recast (21) in a mixed integer linear programming problem, we introduce a new auxiliary integer variable $Z_{E_i}(k) = \delta_{d_{3i}}(k) \cdot E_i(k)$ by considering the linear constraints

$$\begin{aligned} Z_{E_i}(k) &\le M_{E_i}\delta_{d_{3i}}(k) \\ Z_{E_i}(k) &\ge m_{E_i}\delta_{d_{3i}}(k) \\ Z_{E_i}(k) &\le E_i(k) - m_{E_i}(1-\delta_{d_{3i}}(k)) \\ Z_{E_i}(k) &\ge E_i(k) - M_{E_i}(1-\delta_{d_{3i}}(k)). \end{aligned} \tag{22}$$

Thus we can express (5) in MLD form as follows:

$$E_i(k+1) = E_i(k) + \delta_{S_{3i}}(k) - Z_{E_i}(k) \tag{23}$$

subject to the linear inequality constraints on $\delta_{S_i}(k)$, $Z_{E_i}(k)$ and $S(k) - i$.

It is interesting to note that the variable $Z_{E_i}(k)$ can be also used to model, i.e. to impose, the condition that deadlines must not be violated. Indeed one can write

$$Z_{E_i}(k) = \delta_{d_{3i}}(k) \cdot (C_i - \delta_{S_{3i}}(k)) \tag{24}$$

which means that if next step is a deadline for the ith task, i. e. $\delta_{d_{3i}}(k) = 1$, then it must be $E_i(k) = C_i - 1$ if $S(k) = i$ or $E_i(k) = C_i$ if $S(k) \neq i$. The condition (24) can be simply rewritten in MLD form.

Then, the MLD periodic task model can be expressed by using the dynamic equations (13) and (23) with linear inequality constraints. In [22] are reported the details showing that by using a similar procedure also the aperiodic task model can be reformulated in the MLD formalism. In order to complete the model we now need to transform in MLD form the scheduling function (8) that determines the task to be executed.

3.2 Scheduler MLD Model

The scheduler model can be obtained by considering the variables $INQ_i(k)$, $DINQ_i(k)$ and $S(k)$. The static model (6) is ruled by the relation between the values of $E_i(k)$ and C_i, to which we can associate a binary variable $\delta_{E_{3i}}(k) \in \{0,1\}$ such that

$$[\delta_{E_{3i}}(k) = 1] \Leftrightarrow [E_i(k) < C_i] \tag{25}$$

Then, we can rewrite (6) as

$$INQ_i(k) = \delta_{E_{3i}}(k). \tag{26}$$

Since by definition $E_i(k)$ is never greater than C_i, the equivalence (25) can be rewritten as

$$[\delta_{E_{3i}}(k) = 1] \Leftrightarrow [E_i(k) \neq C_i] \tag{27}$$

In order to replace (27) by a set of inequalities we can introduce two new binary variables $\delta_{E_{1i}}(k)$ and $\delta_{E_{2i}}(k)$ such that

$$[\delta_{E_{1i}}(k) = 1] \Leftrightarrow [E_i(k) - C_i \leq 0] \tag{28}$$

$$[\delta_{E_{2i}}(k) = 1] \Leftrightarrow [E_i(k) - C_i \geq 0] \tag{29}$$

and we define the following relation:

$$\delta_{E_{3i}}(k) = \overline{\delta_{E_{1i}}(k) \wedge \delta_{E_{2i}}(k)} = 1 - (\delta_{E_{1i}}(k) \wedge \delta_{E_{2i}}(k)) \tag{30}$$

Relations (28) and (29) can be replaced by a set of inequalities with the BigM approach. Thus, the model (6) can be rewritten in the MLD form (26) subject to the following linear constraints:

$$\begin{array}{l} -\delta_{E_{1i}}(k) - \delta_{E_{3i}}(k) \leq -1 \\ -\delta_{E_{2i}}(k) - \delta_{E_{3i}}(k) \leq -1 \\ \delta_{E_{1i}}(k) + \delta_{E_{2i}}(k) + \delta_{E_{3i}}(k) \leq 2 \end{array} \tag{31}$$

and other constraints similar to (17)-(18) after replacing $x_c(k) + 1 - d_i(k)$ by $E_i(k) - C_i$ and δ_d by δ_E.

We now show how it is possible to write in the MLD form the static model (7) representing the deadlines into the queue. By defining an integer auxiliary variable $Z_{d_i}(k) = INQ_i(k) \cdot d_i(k)$ in order to avoid a product between the binary variable $INQ_i(k)$ and the integer variable $d_i(k)$, the model (7) can be represented in the following MLD form:

$$DINQ_i(k) = Z_{d_i}(k) + M(1 - INQ_i(k)) \tag{32}$$

subject to a suitable set of linear constraints on $Z_{d_i}(k)$.

Equation (8) can be now transformed into a set of inequalities in order to complete the MLD scheduler model. We tackle this problem by splitting it into two subproblems: the *min* modeling and the *arg* modeling. The first subproblem can be solved by considering the equivalence between the classes of hybrid dynamical systems [23]. Using Proposition 6 in [23] the *min* function

$Z_{dmin}(k) = \min_{i \in I}\{DINQ_i(k)\}$ can be expressed in the following Extended Linear Complementary (ELC) form:

$$\begin{cases} DINQ_i(k) - Z_{dmin}(k) \geq 0, \quad \forall i \\ \prod_{i \in I}(DINQ_i(k) - Z_{dmin}(k)) = 0 \end{cases} \tag{33}$$

where the last equality is called ELC condition. Thus, for Proposition 8 in [23] if an upper bound M_{DQ_i} exists for each component $DINQ_i(k) - Z_{dmin}(k)$ we can replace the ELC complementary condition by the following set of inequalities:

$$\begin{cases} DINQ_i(k) - Z_{dmin}(k) \leq M_{DQ_i}\delta_{LD_i}(k) \\ \sum_{i \in I} \delta_{LD_i}(k) \leq N - 1 \end{cases} \tag{34}$$

where N is the cardinality of I and $\delta_{LD_i}(k)$ are N binary variables. Therefore we can express the initial *min* problem in the following MLD form:

$$\begin{cases} DINQ_i(k) - Z_{dmin}(k) \geq 0 \\ DINQ_i(k) - Z_{dmin}(k) \leq M_{DQ_i}\delta_{LD_i}(k) \\ \sum_{i \in I} \delta_{LD_i}(k) = N - 1. \end{cases} \tag{35}$$

Note that in (35) the last relation is now fixed to be an equality so that only one among the possible multiple solutions of the problem, corresponding to different tasks with the same deadline, is selected.

The second subproblem relative to the *arg* function can be represented as

$$S(k) = \sum_{i \in I} i \cdot (1 - \delta_{LD_i}(k)). \tag{36}$$

Equation (36) subject to (35) is a MLD representation of the scheduler (3).

Assume that at a certain time unit the queue is empty, all tasks have been executed and all tasks deadlines are later than next time step. Under such situation $DINQ_i(k) = M \; \forall i \in I$, but no one of the tasks should be selected by (8). To solve this problem we introduce the "idle task", identified by $i = 0$, which is characterized by the only variable $DINQ_0(k) = M - 1 \; \forall k$, so that we can extend (35)-(36) also to the idle task letting $i \in I + \{0\}$.

3.3 State-Space Model and Constraints

We are now able to show that all MLD models formulated above can be represented in a state space form together with linear inequality constraints. In fact, (2), (13), (23) and (9) can be rewritten as

$$x(k+1) = A\,x(k) + B\delta(k) + CZ(k) + b_x \tag{37}$$

where $x(k)$ is the state vector whose components are the time variable $x_c(k)$, the deadline $d_i(k)$ and execution time $E_i(k)$ for all periodic tasks and the execution times $E_{\bar{i}}(k)$ for the aperiodic tasks; $\delta(k)$ is the vector of all binary variables defined above; $Z(k)$ the vector of the integer variables and A, B, C and b_x are suitable matrices. Moreover it is simple to show that all inequality constraints can be written in the compact form

$$Gx(k) + H\delta(k) + KZ(k) + b_c \geq 0 \tag{38}$$

where G, H, K and b_c are suitbale matrices.

4 Rate Control

In this section we use the MLD model to design a rate admission controller based on the prediction of the processor utilization. Given a set of N periodic tasks, the processor utilization factor U is defined by

$$U = \sum_{i \in I} \frac{C_i}{T_i}, \tag{39}$$

where C_i/T_i is the fraction of the processor time spent in executing the ith task (called also *task density*). The processor utilization provides a measure of the computational load on the CPU due to the periodic tasks set. The CPU utilization can be increased by increasing tasks' computation or by decreasing their periods. In what follows we assume that for each task, C_i is fixed and T_i can be chosen within a set T_{i1}, T_{i2}, ..., T_{iL}. The easiest scheduling is achieved by choosing the largest periods however in that case a lower number of instances of the task will be processed in a given time interval (the utilization is small and the QoS is low). In such sense we can associate to each task different QoS levels proportional to the inverse of the periods. Since the task period is not fixed a priori, the utilization factor should be redefined as

$$U = \frac{1}{T_S} C_{T_S} = \frac{1}{T_S} \sum_{i \in I} C_i N_i^{T_s} \tag{40}$$

where C_{T_s} is the total computation load associated to the task set I and $N_i^{T_S}$ is the number of activations of ith task in T_S time units. If the task period is fixed, then $N_i^{T_S} = T_S/T_i$ and (40) becomes equal to (39).

4.1 Static Optimization Problem

Let us consider a set I of only periodic tasks and the goal of achieving a desired processor utilization reference U^{ref}. Since we assume that each task has a fixed WCET C_i, a possible QoS selection coincides with a task invocation period selection, similarly to what is done for the elastic task algorithm [10]. By considering the possible selection among L invocation periods T_{ij}, $j \in J = \{1, \ldots, L\}$ for each ith task and by assuming that T_S is multiple of any task period, we can rewrite (40) as follows:

$$U = \frac{1}{T_S} \sum_{i \in I} C_i T_S (\sum_{j \in J} \delta_{u_{ij}} \frac{1}{T_{ij}}) = \sum_{i \in I} C_i (\sum_{j \in J} \delta_{u_{ij}} \frac{1}{T_{ij}}) \tag{41}$$

where $\delta_{u_{ij}}$ is a binary variable associated for each ith task to the jth possible selectable invocation period T_{ij} subject to the following constraint:

$$\sum_{j \in J} \delta_{u_{ij}} = 1 \tag{42}$$

such that a unique selection is guaranteed. Hence, in order to achieve a processor utilization U^{ref} we can select the invocation period of the task set I by solving the following *static* optimization problem (Mixed Integer Linear Programming problem):

$$\min_{\delta_u}(U^{ref} - U(\delta_u)) \tag{43}$$

subject to the constraint (42) and to the classical EDF schedulability constraint

$$U(\delta_u) \leq 1. \tag{44}$$

The optimization problem (43)-(44) can be also used in order to select the task invocation periods for a time varying processor utilization reference by splitting the problem in several static optimization problems similar to (43)-(44). However, in this case one must neglect the eventually non empty state of the queue.

4.2 MLD Model for Dynamic Optimization

In the previous optimization problem we have assumed an empty queue and the presence of only periodic tasks. In this case the problem is static and, because of the schedulability condition (44), the dynamic model (37)-(38) is not needed. The tasks periods can be fixed once forever and do not need to be changed during time. However, if during T_S time units there is an arrival of an aperiodic task into the queue of the scheduler with a known WCET and a deadline "as soon as possible", condition (44) does not ensure schedulability. The motivation for that can be argued by considering that the definition (39) of the utilization is based on a density concept which is not consistent with the strict and absolute processing time requested by the aperiodic task. To solve this problem we can try to adapt the period of the not yet activated instances of the periodic tasks so that schedulability and some desired performance are achieved. To this aim we need to reformulate the utilization definition. Since the scheduler queue can be assumed to be non empty, together with the aperiodic task we need to consider also the WCETs of the already admitted instances of the periodic tasks. Therefore the prediction of the utilization over T_S time units can be written as (see Fig. 2)

$$\begin{aligned} U(\delta_u(k), x(k)) = & \frac{1}{T_S} \sum_{i \in I} C_i (\sum_{j \in J} \delta_{u_{ij}}(k) \frac{\Delta_i^{T_S}(k)}{T_{ij}}) \\ & + \frac{1}{T_S} \sum_{i \in I} (C_i N_i(k) - E_i(k)) + \frac{1}{T_S} C_{ap}^{T_S}(k), \end{aligned} \tag{45}$$

where $\Delta_i^{T_S} = T_S - \bar{k}_i$ is the fraction of T_S still to be allocated, $\bar{k}_i$ is the sum of the periods of the already activated instances, N_i is the number of instances of the ith task into the queue at k (fixed by the activation patterns of the tasks) and $C_{ap}^{T_S}$ is the WCET associated to the aperiodic task.

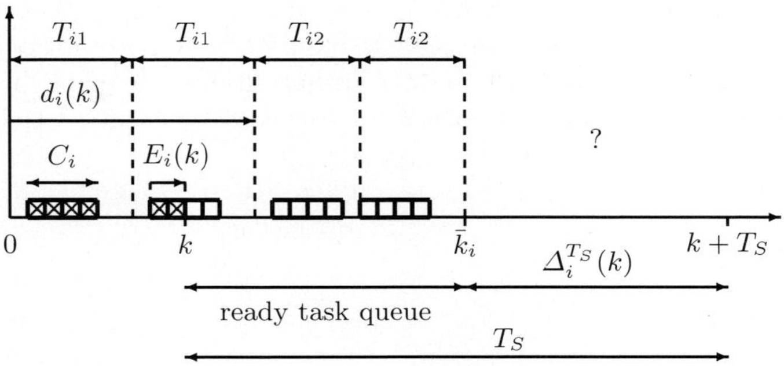

Fig. 2. Time allocation corresponding to the proposed rate admission control in the presence of a non empty task ready queue

Before formulating the optimization problem we represent the period selection in MLD form. In particular, we can define the period selection as follows:

$$P_i(k) = \sum_{j \in J} \delta_{u_{ij}}(k) T_{ij} \tag{46}$$

together with the constraints

$$\sum_{j \in J} \delta_{u_{ij}}(k) \leq 1, \quad \sum_{j \in J} \delta_{u_{ij}}(k) \geq 1 \tag{47}$$

where the new integer variable $P_i(k)$ represents the time-variant invocation period associated to one selectable period (in such particular case also QoS level) at k time instant for each ith task. The introduction of $P_i(k)$ requires some modifications in (13) in order to avoid the product term $\delta_{d_{3i}}(k) \cdot P_i(k)$ [22]. Thus, we can formulate the optimization problem for the rate admission control of periodic and aperiodic tasks as follows:

$$\min_{\delta_u} \sum_{k=1}^{T_S} (U^{ref} - U(\delta_u(k), x(k))) \tag{48}$$

subject to (37)-(38). Then at each step k the optimization problem is solved and for each task the best period that also ensures schedulability is selected and virtually assigned to that task for the entire "free" time interval $\Delta_i^{T_S}$. The admission controller decides the number of instances of the tasks to be admitted into the CPU queue that determines $N_i(k+1)$ and $\Delta_i^{T_S}(k+1)$. Moreover, if $x_c(k) + 1 - d_i(k) = 0$ then $N_i(k+1)$ must be also decremented by one. At step $k+1$ the updated optimization problem (48) is solved, and so on. For each task the proposed controller considers the processor queue and operates only on the instances still to be activated. In fact the period(s) of the task instances already admitted into the queue are not modified and are taken into account in the second term of (45).

Different optimization problems, i.e. cost functions, could be formulated. For instance, one can define the cost function by incorporating application performance measures so that the admission is regulated, e. g. by means of $N_i(k)$, according to the application dynamics. Another alternative could be to weight properly some of the tasks parameters in the cost function so that resource allocation management is taken into account [24].

5 Simulation Results

The proposed rate admission control strategy has been tested by simulating the kernel of a real-time operating system based on the EDF scheduling policy. A detailed simulation study is out the scope of this paper and for that goal one can refer to dedicated tools already available from the literature [19]. Here the EDF kernel simulator has been developed in the Simulink environment. In particular, the task ready queue managing has been implemented by exploiting the typical algorithms of the linked-list class and the corresponding C++ code has been run in Simulink through a S-function block. This real-time kernel simulator periodically receives the tasks as arrays in which are indicated all task features such as the triple C_i, T_i, d_i.

For comparative purposes also the more classical feedback real-time scheduling strategy reported in [17] has been implemented. Let us consider three periodic tasks with $C_1 = 1$, $T_1 \in \{20, 15, 10\}$, $C_2 = 2$, $T_2 \in \{25, 21, 8\}$, $C_3 = 3$, $T_3 \in \{23, 20, 10\}$, and an aperiodic task with $C_{ap} = 2$ which is admitted suddenly into ready task queue at $k = 130$, 180 and 210. Our objective is to control a step change of the utilization processor (from 58% to 40% at $k = 100$). By considering the dispatching of the tasks it simple to verify that when the aperiodic task is

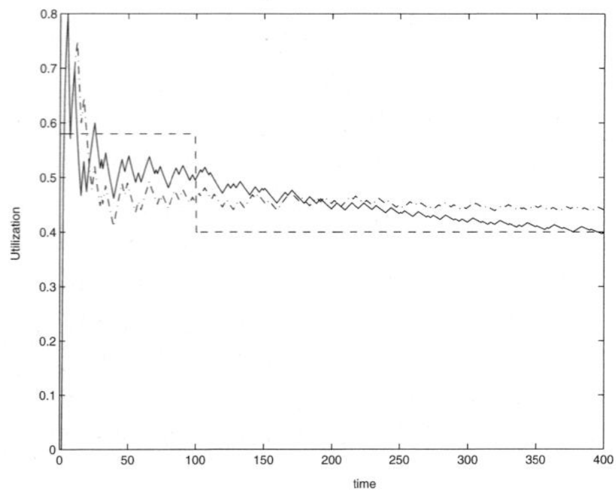

Fig. 3. Average utilization over T_S: reference (dashed line), classical control strategy (dash-dotted line), proposed QoS control (solid line)

admitted the proposed controller ensures the deadlines keeping, which is not the case for the other controller. In Fig. 3 is reported the average time relative to T_S spent by the CPU in executing the tasks. The numerical test shows that the proposed control strategy provides good regulation to the desired utilization.

6 Conclusions

A mixed logical dynamical model of hard real-time periodic tasks EDF scheduling in the presence of aperiodic tasks has been proposed. The model is used to predict the processor utilization and to design an optimal period selection for the tasks to be admitted into the CPU queue. If the aperiodic tasks must be executed in specific time slots, the condition on the utilization to be less than one is not enough to ensure schedulability and the optimization problem must be constrained by the proposed dynamic model. The rate admission control guarantees schedulability and good performance so as shown by simulating the kernel of a real-time operating system. Future analysis will deal with the estimation of the minimum prediction time T_S that ensures schedulability. On the other hand one should say that, since avoiding deadline violation can be explicitly included into the model as a constraint, one could make the estimation of the minimum T_S looking at the lowest T_S that includes the aperiodic task and that ensures the optimization problem to be feasible. Future work will also deal with implementation issues of the proposed technique, in particular as regards the computational load, and with the use of other cost functions.

References

1. Elia, M., Sanner, A. R. M.: QoS Tradeoffs for Guidance, Navigation, and Control. Proc. IEEE Aerospace Conference, Big Sky, MT **7** (2002) 3333–3341
2. Buttazzo, G. C.: Hard Real-Time Computing Systems. Kluwer, Boston (1997)
3. Klein, M., Ralya, T., Pollak, B., Obenza, R., Harbour, M. G.: A Practitioner 's Handbook for Real-Time Analysis. Guide to Rate Monotonic Analysis for Real-Time Systems. Kluwer, Boston (1993)
4. Stankovic, J. A., Spuri, M., Ramamritham, K., Buttazzo, G. C.: Deadline Scheduling for Real-Time Systems, EDF and Related Algorithms. Kluwer, Boston (1998)
5. Zhao, W., Ramamritham, K., Stankovic, J. A.: Preemptive Scheduling Under Time and Resource Constraints. IEEE Transactions on Computers **36**(8) (1987) 949–960
6. Abdelzaher, T. F., Stankovic, J. A., Lu, C., Zhang, R., Lu, Y.: Feedback Performance Control in Software Services. IEEE Control Systems **23**(3) (2003) 74–90
7. Steere, D. C., Goel, A., Gruenberg, J., McNamee, D., Pu, C., Walpole, J.: A Feedback Driven Proportion Allocator for Real-Rate Scheduling. Proc. Symposium on Operating Systems Design and Implementation. New Orleans, LA (1999) 145–158
8. Abeni, L., Palopoli, L., Lipari, G., Walpole, J.: Analysis of a Reservation-Based Feedback Scheduler. Proc. IEEE Real-Time Systems Symposium. Austin, TX (2002) 71–80
9. Chen, Y., Dai, Q.: Research on Dynamic Feedback and Elastic Scheduling Model and Algorithm for Flexible Workload. Proc. International Conference on Computer Networks and Mobile Computing. Shangai, China (2003) 283–290

10. Buttazzo, G. C., Lipari, G., Caccamo, M.: Elastic Scheduling for Flexible Workload Management. IEEE Transaction on Computers **51**(3) (2002) 289–302
11. Buttazzo, G. C.: Achieving Scalability in Real-Time Systems. IEEE Computer Magazine **39**(5) (2006) 54–59
12. Palopoli, L., Cucinotta, T., Bicchi, A.: Quality of Service Control in Soft Real-Time Applications. Proc. IEEE Conference on Decision and Control. Maui, Hawaii (2003) 664–669
13. Cucinotta, T., Palopoli, L., Marzario, L.: Stochastic Feedback-Based Control of QoS in Soft Real-Time Systems. Proc. IEEE Conference on Decision and Control. Paradise Island, Bahamas (2004) 3533–3538
14. Cervin, A., Eker, J., Bernhardsson, B., Arzen, K. E.: Feedback-Feedforward Scheduling of Control Tasks. Journal of Real-Time Systems **23**(1-2) (2002) 25–53
15. Zhou P., Xie, J.: Feedback Scheduling for Resource-Constrained Real-Time Control Systems. Proc. IEEE International Conference on Computer and Information Technology. Dhaka, Bangladesh (2005) 800–804
16. Abdelzaher, T. F., Shin, K. G., Bhatti, N.: Performance Guarantees for Web Server End-Systems: A Control Theoretical Approach. IEEE Transactions on Parallel and Distributed Systems **13**(1) (2002) 80–96
17. Lu, C., Stankovic, J. A., Tao, G., Son, S. H.: Feedback Control Real-Time Scheduling: Framework, Modeling, and Algorithms. Journal of Real-Time Systems. **23**(1-2) (2002) 85–126
18. Norstrom, C., Wall, A., Yi, W.: Timed Automata as Task Models for Event-Driven Systems. Proc. IEEE International Conference on Real-Time Computing Systems and Applications. Hong Kong, China (1999) 182–189
19. Behrmann, G., Larsen, K. G., Moller, O., David, A., Pettersson, P., Yi, W.: UPPAAL - Present and Future. Proc. IEEE Conference on Decision and Control. Orlando, FL (2001) 2881–2886
20. Bemporad A., Morari, M.: Control of Systems Integrating Logic, Dynamics, and Constraints. Automatica **35**(3) (1999) 407–427
21. Maciejowski, J. M.: Predictive Control with Constraints. Prentice Hall (2002)
22. Vacca, V.: Quality of Service Control for Real-Time Task Scheduling in Space Avionics. PhD Thesis at Department of Engineering, University of Sannio, Benevento, Italy. Avalaible at www.grace.ing.unisannio.it (2006)
23. Heemels, W. P. M. H., De Schutter, B., Bemporad, A.: Equivalence of Hybrid Dynamical Models. Automatica. **37**(7) (2001) 1085–1091
24. Hansen, J. P., Lehoczky, J., Rajkumar, R.: Optimization of Quality of Service in Dynamic Systems. Proc. IEEE International Parallel and Distributed Processing Symposium, San Francisco, CA (2001) 1001–1008

Foundations of a Compositional Interchange Format for Hybrid Systems*

D.A. van Beek, M.A. Reniers, R.R.H. Schiffelers, and J.E. Rooda

Eindhoven University of Technology (TU/e)
P.O. Box 513, NL-5600 MB Eindhoven, The Netherlands
{D.A.v.Beek,M.A.Reniers,R.R.H.Schiffelers,J.E.Rooda}@tue.nl

Abstract. A compositional interchange format for hybrid systems is defined in terms of an interchange automaton, allowing arbitrary differential algebraic equations, including fully implicit or switched DAEs, discrete, continuous and algebraic variables, that can be internal or external, urgency conditions, and operators for parallel composition, action hiding, variable hiding and urgent actions. Its compositional semantics is formally defined in terms of a hybrid transition system. This allows development of transformations to and from other formalisms that can be proven to preserve essential properties, and it allows a clear separation between the mathematical meaning of a model and implementation aspects such as algorithms used for solving differential algebraic equations.

1 Introduction

Our intention is to establish inter-operability of a wide range of tools by means of model transformations to and from a compositional interchange format that is defined in terms of an *interchange automaton*. The domain of the interchange automaton format consists of languages and tools from computer science and from dynamics and control for modeling, simulation, analysis, controller synthesis, and verification in the area of hybrid and timed systems. The purpose of an interchange format is to avoid the implementation of many bi-lateral translators between specific formalisms. Instead, the translation from a formalism A to a formalism B is divided in two steps: first, the model in formalism A is translated into a representation (model) in the interchange automaton format, then, this representation is translated into a model in formalism B [1].

Our main requirements for the interchange format are summarized below. A more detailed discussion of these requirements follows in Sections 2 and 3.

1. It should have a formal and compositional semantics, based on (hybrid) transition systems, and allow property preserving model transformations.

* Work partially done in the framework of the HYCON Network of Excellence, contract number FP6-IST-511368.

A. Bemporad, A. Bicchi, and G. Buttazzo (Eds.): HSCC 2007, LNCS 4416, pp. 587–600, 2007.

2. Its concepts should be based on mathematics, and independent of implementation aspects such as equation sorting, and numerical equation solving algorithms.
3. It should support arbitrary differential algebraic equations (DAEs), including fully implicit equations, higher index systems, algebraic loops, steady state initialization, switched systems such as piecewise affine systems, and DAEs with discontinuous right hand sides.
4. It should support a wide range of concepts originating from hybrid automata, including different kinds of urgency, such as 'urgency predicates', 'deadline predicates', 'triggering guard semantics', and 'urgent actions'.
5. It should support parallel composition with synchronization by means of shared variables and shared actions.
6. It should support hierarchy and modularity to allow the definition of parallel modules and modules that can contain other modules (hierarchy), and to allow the definition of variables and actions as being local to a module, or shared between modules.

Other work on interchange formats for hybrid systems has been carried out in different projects: in the MoBIES project, the Hybrid System Interchange Format (HSIF) [2] is defined; in [3] an 'abstract semantics' of an interchange format based on the Metropolis meta model is defined (this work is a continuation of the COLUMBUS project [4]); and in the HYCON NoE [5], an interchange format for switched linear systems [6] in the form of piecewise affine systems (PWAs) is defined.

In HSIF, a network of hybrid automata is used for model representation. The network behaves as a parallel composition of its automata, without hierarchy or modules. Variables can be shared or local, and the communication mechanism is based on broadcasting of boolean 'signals', where signals are partitioned in input and output signals. Each signal is required to be either a global input to the network or to be modified by exactly one automaton. The semantics is defined only for 'acyclic dependency graphs' with respect to the use of signals. The time dependent behavior is specified by means of ordinary differential equations (ODEs), together with algebraic relations of the form $x = f(x_1, \ldots, x_n)$, and invariants. The equation $\dot{x} = 0$ is assumed for each shared variable. Circular dependencies of the algebraic equations, i.e. algebraic loops, are not allowed, and urgency predicates or urgent actions are not available [2]. The interchange automaton format defined in this article aims to be more general than HSIF, and does not incorporate tool limitations, such as restrictions on circular dependencies, or restrictions on shared variables or algebraic loops, in its compositional formal semantics.

The 'abstract semantics' presented in [3], takes implementation considerations into account, such as equation sorting, iterations that may be required for state-event detection, and iterations for reaching a fixed-point in case of algebraic loops. The semantics is defined in terms of functions and algorithms such as init, markchange, and solve. This is different from the compositional formal semantics as defined in Section 5, which aims at defining the *mathematical* meaning of interchange automata, independently of implementation aspects such as equation sorting or state-event detection. For example, the semantics defines the mathematical meaning of a switched system of equations, such as a PWA system, but an implementation may choose to implement such switching behavior with or without state-event detection.

A transformation from the PWA-based interchange format [6] to the interchange automaton format will be developed. Based on this transformation, several tools, based on among others PWA, HYSDEL, MLD (see [7] for an overview relating these languages) can then be connected to the interchange automaton format.

The remainder of this article is organized as follows: Section 2 discusses the importance of a compositional formal semantics, Section 3 discusses the concepts present in the interchange automaton format, Sections 4 and 5 define the syntax and semantics of interchange automata, respectively, and Section 6 presents concluding remarks.

2 Importance of a Compositional Formal Semantics

To use the interchange automaton format for verification purposes, translations from models to the interchange format, and vice versa, should preserve essential properties. I.e., verification results obtained for a derived model should also be valid for the original model specified in another language.

The different languages may have different features and semantics, complicating translations between them. To keep the translations between the interchange automaton format and the other languages manageable, in terms of complexity, it is important that transformations between parts of specifications within the interchange format itself can be defined. To allow such transformations, it is essential that the semantics of the interchange automaton format is *compositional*. I.e., that the notion of equivalence is a congruence for all operators of the interchange automaton format, see [8]. Parts of a model can then be replaced by equivalent parts without changing the meaning of the model.

Consider, for instance, a transformation of a model in a simulation language, such as Modelica [9] or EcosimPro [10], to a verification tool, such as PHAVER [17] or HYTECH [16]. The simulation languages use a triggering guard semantics (see Section 3.4), whereas the verification tools use invariants to force switching to a different location. Defining direct translations from the simulation languages to the verification tools and reasoning about the correctness of the translations would be difficult, since the simulation languages do not have a formal semantics and the urgent guards in two languages would need to be transformed into invariants in combination with non-urgent guards in two other languages. By using the compositional interchange format as an intermediate, the complicated direct translations can be replaced by more straightforward translations from the simulation languages to the interchange format, using urgent guards, and from the interchange format to the verification tools, using invariants; in combination with a transformation in the interchange format from urgent guards to invariants.

3 Concepts in the Interchange Automaton Format

3.1 Differential Algebraic Equations

Modeling of physical systems, such as mechanical or chemical systems frequently leads to DAEs. Algebraic constraints can also be the result of stateless components such as proportional controllers. DAEs can be modeled and simulated using languages such as Modelica and EcosimPro. DAEs can be specified in the invariants of an interchange automaton, since such invariants are predicates over all variables, including the dotted variables. Flow clauses are supported for reasons of compatibility with existing hybrid automata. The reason for not enforcing a separation between invariants (over non-dotted variables) and flow clauses (over dotted variables), as in existing hybrid automata, is that such a separation is absent in the mathematical theory of dynamical systems, including control theory. In many cases, fully implicit DAEs, such as $\mathbf{f}(\dot{\mathbf{x}}, \mathbf{x}, \mathbf{y}, t) = \mathbf{0}$, cannot even be rewritten to a form where the algebraic constraints and the differential constraints are separated, such as the semi-explicit form $\dot{\mathbf{x}} = \mathbf{g}(\mathbf{x}, \mathbf{y}, t)$, $\mathbf{h}(\mathbf{x}, \mathbf{y}, t) = \mathbf{0}$, where $\mathbf{x}$ and $\mathbf{y}$ are the continuous and algebraic variables, respectively. The generalized invariant allows us to consider the four expressions $x = 1 \wedge x = 2$, $\dot{x} = 1 \wedge \dot{x} = 2$, $\dot{x} = y \wedge \dot{x} = 2y \wedge y = 1$ and 'false' to be equivalent (bisimilar): no behavior is possible.

The initialization clause of the interchange automaton is also defined as a predicate over all variables, including the dotted variables. This allows more general initializations than usually allowed in hybrid automata. In particular, steady state initialization, as available in Modelica and EcosimPro, is supported. E.g. by defining $\dot{\mathbf{x}} = \mathbf{0}$ as initialization predicate for a location with invariant $\mathbf{f}(\dot{\mathbf{x}}, \mathbf{x}, \mathbf{y}, t) = \mathbf{0}$, the initial state is defined as the 'steady state', that is the solution of the set of DAEs such that all derivatives are zero: $\mathbf{f}(\mathbf{0}, \mathbf{x}, \mathbf{y}, t) = \mathbf{0}$.

3.2 Discrete, Continuous and Algebraic Variables

The interchange automaton defines three classes of variables: the discrete and continuous variables, and in addition the algebraic variables. The differences are as follows: Continuous variables are the only variables for which dotted variables (derivatives) can be used in models. The values of discrete variables remain constant when model time progresses, the values of continuous variables may change according to a continuous function of time when model time progresses, and the values of algebraic variables may change according to a discontinuous function of time. Finally, there is a difference between the different classes of variables with respect to how the resulting values of the variables in a transition relate to the starting values of the variables in the next transition. The resulting value of a discrete or continuous variable in a transition always equals its starting value in the next transition. For algebraic variables there is no such relation, because algebraic variables are not part of the state.

The state of an interchange automaton consists of, among others, the interchange automaton itself, and a valuation of the discrete and continuous variables (see Section 5 for a more precise definition of the state). The values of the dotted variables and the algebraic variables are not contained in the state. The reason for this is that the state of an interchange automaton represents all information needed to determine future behavior, i.e., the state of a system makes the system's history irrelevant. The dotted and algebraic variables are not needed in the state, because their values are determined completely by the interchange automaton: in particular by the initial conditions, the flow conditions, the invariants and the jump predicates as defined in Section 4.

In most languages that allow (implicit) DAEs, such as Modelica [9], EcosimPro [10], and Simulink [11], the distinction between continuous and algebraic variables is implicitly made by considering all continuous variables that do not occur differentiated as algebraic.

3.3 Automata Related Concepts

Many different hybrid automaton definitions exist. Some definitions require solutions for the continuous variables to be differentiable functions, e.g. in [12,13]. Other definitions allow the more general case of piecewise differentiable or piecewise continuous functions, e.g. in [14]. Such restrictions can be realized in the interchange automaton format by means of the parameters F and G as defined in Section 5. In [15], for each variable a 'dynamic type' can be defined. However, since we did not find such expressivity in tools, the interchange automaton format allows the definition of the dynamic type for the algebraic and continuous variable classes, not for each individual variable.

With respect to the meaning of jump predicates, that define the behavior of the variables in action transitions, differences also occur: in [12] the variables can in principle perform arbitrary jumps unless restricted by the jump predicate, in [16], variables in principle remain unchanged unless changes are enforced by the jump predicate by means of primed variables. The first behavior is obtained by an interchange automaton that defines the set of jumping variables W (see Sections 4 and 5.1) at each edge to be equal to the set of all variables. The second kind of behavior is obtained by defining the set W as the union of all primed variables of the jump predicate. The specification of a set of jumping variables and a jump predicate for each edge of an interchange automaton is based on [13].

The interchange automaton format is expressive enough to deal with verification tools such as PHAVer [17] and HyTech [16]. The behavior of the algebraic variables from the interchange automaton is related to the external variables from the semantical hybrid I/O automaton defined in [15]. In this I/O automaton, the external variables are also not part of the state, and they can have a dynamic type that allows discontinuous trajectories. The state is defined by the values of the internal variables, and discrete transitions (action transitions) are defined only on internal variables. The interchange automaton format can express this as a special case, since the different classes of variables, action transitions, and hiding/abstraction are orthogonal concepts in the interchange automaton format.

The basic elements of hierarchy and modularity that are supported by the interchange format are parallel composition, and hiding of variables and actions. Interchange automata can be grouped by means of parallel composition, and variables and/or actions that are meant to be local to that group can be hidden from the environment of the group. The concrete interchange format, that will be developed, will define modularity in terms of the basis elements of the abstract format.

3.4 Urgency

The concept of urgency allows the passing of time up to a certain point. There are essentially two different kinds of urgency:

1. Urgency that is defined for an atomic automaton by means of one or more predicates. Such predicates can be associated to a location, or to outgoing edges of the location.
2. Urgency that is defined as an operation on a composition of one or more automata. Such an operation defines a set of actions as urgent for the composition. The operation allows the passing of time up to the point when one or more of the urgent actions can be executed.

The first kind of urgency is defined in many different forms. The `tcp` (*time can progress*) predicate [18], is a predicate over the variables of the automaton and time. The predicate is associated to a location. It allows passing of time in a location for as long as the predicate is true. Related to the `tcp` predicate is the *stopping condition* [19], which is a predicate on the variables of the automaton, also associated to a location, and which allows passing of time in a location for as long as the stopping condition is false, or in other words, until the time-point when the stopping condition is true. *Deadline predicates* [20] and *urgency predicates* [19] are associated to the edges of an automaton. Deadline predicates allow passing of time in a location until the time-point that one or more deadline predicates of the outgoing edges of the location become true. Whenever a deadline predicate of an edge becomes true, the guard associated to that edge must also be true: the deadline predicate must imply the guard. Urgency predicates are similar to deadline predicates; the only difference is that they do not have the restriction that the urgency predicate should imply the guard. Urgency predicates allow passing of time in a location until the point of time that for one or more of the outgoing edges, the guard and the urgency predicate are both true.

Restricting a `tcp` predicate as a predicate over the variables of an automaton makes it equal to the negation of a stopping condition. Deadline predicates and urgency predicates are less expressive. They can both be expressed in terms of stopping conditions, see [19], or as `tcp` predicates. E.g. the stopping condition of a location corresponds to the disjunction of all deadline predicates of the outgoing edges of the location. Note that a flow condition which is false in a hybrid automaton is equivalent to a stopping condition that is true, or a `tcp` predicate that is false. The interchange automaton format adopts stopping conditions, which we refer to as *urgency conditions*.

In simulation languages, such as Modelica, EcosimPro, and HyVisual, usually a triggering or urgent guard semantics is used, meaning that the passing of time in a location is allowed until the time-point that any of the guards of the outgoing edges becomes true. This is equivalent to a stopping condition associated to the location that is the disjunction of the guards of all outgoing edges.

The second kind of urgency, as for example defined in [21] and [22], is often available with restrictions only. E.g. in HyTech, edges can be defined as urgent. The composition of urgent actions is required to be well-formed: 'whenever two components synchronize on a label, if one transition is urgent then the other must either be urgent, or have a jump condition expressible as a guarded command with its guard being either the predicate true or the predicate false' [16]. A second restriction is that 'if there exists an urgent transition from a location v to a location v', then for all valuations satisfying the invariant of v, an urgent transition to v' should exist'.

The interchange automaton format defines the second kind of urgency by means of the urgent action operator. Note that defining this kind of urgency by means of labeling certain edges or actions as urgent may lead to bisimulation not being a congruence for parallel composition, as described in [23]. Another example is the ASAP flag that can be attached to an edge in HyTech. In such cases, replacing a part of a specification by another part with the same behavior may lead to different behavior of the complete system. Straightforward translations of such languages to and from the interchange format is in principle possible if each action (or edge with a certain action) is either always urgent or never urgent in a model. If the same action (or edge with a certain action) occurs both as urgent and not urgent, it may be necessary to eliminate parallel composition, as for example described in [8], before translation to the interchange format is possible.

4 Syntax of Interchange Automata

Notation 1. *The following notations are defined:*

- *A set $\mathcal{X}$ of variables and a set of action labels $\mathcal{L}$, which does not include the predefined non-synchronizing action τ, are assumed. The set $\mathcal{L}_\tau$ denotes the set $\mathcal{L} \cup \{\tau\}$.*
- *For a set of variables $S \subseteq \mathcal{X}$, $\dot{S} = \{\dot{x} \mid x \in S\}$ denotes the set of dotted variables.*
- *For a set of variables $S \subseteq \mathcal{X}$, $\mathrm{Pred}(S)$ denotes the set of all predicates over variables from S.*
- *$f : A \mapsto B$ and $g : A \to B$ define a partial function f and a total function g, both with domain A and range B.*

Definition 1 (Atomic Interchange Automaton). *An* atomic interchange automaton *is a tuple* $(X, X_\mathrm{i}, \mathrm{dtype}, V, v_0, \mathrm{init}, \mathrm{flow}, \mathrm{inv}, \mathrm{urgent}, L, E)$ *where*

- *$X \subseteq \mathcal{X}$ is a finite set of variables, $X_\mathrm{i} \subseteq X$ is the set of* internal *variables, and $X_\mathrm{e} = X \setminus X_\mathrm{i}$ is the set of* external *variables.*

- dtype : $X \to \{\text{disc}, \text{cont}, \text{alg}\}$ *is a function that associates to each variable a dynamic type:* discrete, continuous *or* algebraic. *The sets* $X_{\text{disc}}, X_{\text{cont}}, X_{\text{alg}}$ *are defined as* $X_t = \{x \in X \mid \text{dtype}(x) = t\}$ *for* $t \in \{\text{disc}, \text{cont}, \text{alg}\}$, *and* $X_{\text{state}} = X_{\text{disc}} \cup X_{\text{cont}}$ *is the set of* state *variables.*
- V *is a finite non-empty set of* vertices, *called* locations, *and* $v_0 \in V$ *is the initial location.*
- init $\in \text{Pred}(\tilde{X})$ *is the initial condition. For* $Y \subseteq X$, $\tilde{Y} = Y \cup \{\dot{y} \mid y \in Y \cap X_{\text{cont}}\}$ *is the extension of* Y *with the dotted versions of the continuous variables in* Y.
- flow, inv, urgent : $V \to \text{Pred}(\tilde{X})$, *are functions that each associate to each location* $v \in V$ *a predicate describing the* flow condition, *the* invariant, *and the* urgency condition, *respectively.*
- $L \subseteq \mathcal{L}$ *is a finite set of action labels.*
- $E = V \times \text{Pred}(\tilde{X}) \times (L \cup \{\tau\}) \times (\mathcal{P}(\tilde{X}) \times \text{Pred}(\tilde{X} \cup \tilde{X}^{-})) \times V$ *is a finite set of* edges, *such that for each element* $(v, g, \ell, (W, r), v') \in E$, v *and* v' *are the* source *and* target *locations, respectively,* g *is the* guard, ℓ *is the* action label, $W \subseteq \tilde{X}$ *is a set of jumping variables (the value of which may change as a result of an action transition), and* r *is the* jump predicate, *also called* reset map. *For any* $Y \subseteq \tilde{X}$, $Y^{-} = \{y^{-} \mid y \in Y\}$ *denotes the set of minus superscripted variables that represent the values of variables before an action transition.*

Note that the *dynamic* type of a variable gives information about its time dependent behavior. E.g. the value of a discrete variable remains constant when time passes, whereas the value of a continuous variable changes as a continuous function of time. The *static* type, such as real, integer or boolean, mainly gives information about the domain in which the variable takes values.

The interchange automaton format consists of automata, and operators for parallel composition, for hiding of action labels and variables, and for the definition of urgent actions. The automata and operators can be freely combined:

Definition 2 (Interchange automaton). *The set of interchange automata* A *is defined by the following grammar for the interchange automata* $\alpha \in A$:

$\alpha ::= \alpha_{\text{atom}}$	*atomic interchange automaton*
$\mid \alpha \parallel \alpha$	*parallel composition*
$\mid \text{hide}_{\text{act}}(L_{\text{h}}, \alpha)$	*action hiding operator*
$\mid \text{hide}_{\text{var}}(X_{\text{h}}, \alpha, \sigma_{\text{h}})$	*variable hiding operator*
$\mid \text{urgent}(L_{\text{u}}, \alpha)$	*urgent action operator,*

where

- α_{atom} *denotes an arbitrary atomic interchange automaton;*
- $L_{\text{h}} \subseteq \mathcal{L}$ *denotes a set of actions to hide;*
- $X_{\text{h}} \subseteq \mathcal{X}$ *denotes a set of variables to hide and* $\sigma_{\text{h}} : X_{\text{h}} \mapsto \Lambda$ *denotes a (partial) valuation for the hidden state variables of interchange automaton* α;
- $L_{\text{u}} \subseteq \mathcal{L}$ *denotes a set of urgent actions.*

In the next sections, two auxiliary functions on interchange automata are used. These are defined below. The functions $\mathrm{var}_{\mathrm{e,st}}$ and act extract the sets of external state variables and external (non-hidden) action labels[1], respectively, from an interchange automaton:

Definition 3. *For* $\alpha_{\mathrm{a}} = (X, X_{\mathrm{i}}, \mathrm{dtype}, V, v, \mathrm{init}, \mathrm{flow}, \mathrm{inv}, \mathrm{urgent}, L, E)$*, and* α*,* α_1*,* $\alpha_2 \in A$ *we define the functions* $\mathrm{var}_{\mathrm{e,st}} : A \to \mathcal{P}(\mathcal{X})$ *and* $\mathrm{act} : A \to \mathcal{P}(\mathcal{L})$ *as follows:*

$\mathrm{var}_{\mathrm{e,st}}(\alpha_{\mathrm{a}}) = X_{\mathrm{state}} \cap X_{\mathrm{e}}$	$\mathrm{act}(\alpha_{\mathrm{a}}) = L$
$\mathrm{var}_{\mathrm{e,st}}(\alpha_1 \parallel \alpha_2) = \mathrm{var}_{\mathrm{e,st}}(\alpha_1) \cup \mathrm{var}_{\mathrm{e,st}}(\alpha_2)$	$\mathrm{act}(\alpha_1 \parallel \alpha_2) = \mathrm{act}(\alpha_1) \cup \mathrm{act}(\alpha_2)$
$\mathrm{var}_{\mathrm{e,st}}(\mathrm{hide}_{\mathrm{act}}(L_{\mathrm{h}}, \alpha)) = \mathrm{var}_{\mathrm{e,st}}(\alpha)$	$\mathrm{act}(\mathrm{hide}_{\mathrm{act}}(L_{\mathrm{h}}, \alpha)) = \mathrm{act}(\alpha) \setminus L_{\mathrm{h}}$
$\mathrm{var}_{\mathrm{e,st}}(\mathrm{hide}_{\mathrm{var}}(X_{\mathrm{h}}, \alpha, \sigma_{\mathrm{h}})) = \mathrm{var}_{\mathrm{e,st}}(\alpha) \setminus X_{\mathrm{h}}$	$\mathrm{act}(\mathrm{hide}_{\mathrm{var}}(X_{\mathrm{h}}, \alpha, \sigma_{\mathrm{h}})) = \mathrm{act}(\alpha)$
$\mathrm{var}_{\mathrm{e,st}}(\mathrm{urgent}(L_{\mathrm{u}}, \alpha)) = \mathrm{var}_{\mathrm{e,st}}(\alpha)$	$\mathrm{act}(\mathrm{urgent}(L_{\mathrm{u}}, \alpha)) = \mathrm{act}(\alpha)$

5 Semantics of Interchange Automata

The formal semantics associates to each interchange automaton an action transition relation, a time transition relation, and a consistency predicate on states. A different way of looking at such a semantics is as a labeled transition system with two types of transitions and a predicate. The states S of the labeled transition system associated to an interchange automaton consist of an interchange automaton, a valuation of the external state variables of that automaton, and a set of jumping external state variables: i.e., $S = A \times \mathit{Val} \times \mathcal{P}(\mathcal{X})$, where $\mathit{Val} = \mathcal{X} \cup \dot{\mathcal{X}} \mapsto \Lambda$ is the set of all partial mappings from $\mathcal{X} \cup \dot{\mathcal{X}}$ to the set of values Λ. The set of jumping variables J is defined by other automata executing in the environment of (in parallel to) automaton α. The valuation σ of a state of a transition system defines values for precisely the externally visible state variables, i.e., $\mathrm{dom}(\sigma) = \mathrm{var}_{\mathrm{e,st}}(\alpha)$ for all $(\alpha, \sigma, J) \in S$.

The intuition of an action transition $(\alpha, \sigma, J) \xrightarrow{\xi, \ell, W, \xi'} (\alpha', \sigma', J)$ is that the state (α, σ, J) executes a discrete action (with action label) ℓ with visible valuations ξ, ξ', before and after execution of the action, respectively, and thereby transforms into the state (α', σ', J), where σ' denotes the accompanying valuation of the automaton α', after the discrete action ℓ is executed. The set W represents the external state variables that are allowed to change (jump) in this action transition. They need to be visible for synchronization in a parallel composition of interchange automata.

The intuition of a time transition $(\alpha, \sigma, J) \overset{t, \rho}{\mapsto} (\alpha', \sigma', J)$ is that model time passes for t time units, and the valuation at each time-point $s \in [0, t]$ is given by $\rho(s)$ for the externally visible variables. At the end-point t, the resulting state is (α', σ', J).

The intuition of the consistency predicate $(\alpha, \sigma, J) \overset{\xi}{\rightsquigarrow}$ is that the interchange automaton α is consistent with extended valuation ξ, which means that the invariants of all active locations of α are satisfied in ξ.

[1] This does not mean that these actions are actually used. It is allowed to specify the set of actions much broader than the actions that appear on transitions.

Notation 2. *In this section, some notations and operators are used. These are defined as follows:*

- $\restriction$ *is the restriction operator on functions. If* f *is a function, and* S *is a set,* $f \restriction S$ *denotes the restriction of* f *to* S*, that is, the function* g *with* $\mathrm{dom}(g) = \mathrm{dom}(f) \cap S$*, such that* $g(c) = f(c)$ *for each* $c \in \mathrm{dom}(g)$*.*
- $\downarrow$ *is the projection operator on functions, which is used here on trajectories. For* $\rho : T \mapsto (Y \to \Lambda)$*,* $S \subseteq Y$ *and* $x \in Y$*,* $\rho \downarrow S$ *denotes the function* $\rho' : T \mapsto (S \to \Lambda)$ *such that* $\rho'(t) = \rho(t) \restriction S$ *for each* $t \in T$*; and* $\rho \downarrow x$ *denotes the function* $f : T \mapsto \Lambda$ *such that* $f(t) = \rho(t)(x)$ *for each* $t \in \mathrm{dom}(\rho)$*.*
- *For atomic interchange automaton* α_{atom} *given by* $(X, X_{\mathrm{i}}, \mathrm{dtype}, V, v, \mathrm{init}, \mathrm{flow}, \mathrm{inv}, \mathrm{urgent}, L, E)$*,* $\alpha_{\mathrm{atom}}[v', \mathrm{init}'/v, \mathrm{init}]$ *denotes the atomic interchange automaton obtained from atomic interchange automaton* α_{atom} *by replacing* v *by* v' *and* init *by* init$'$*, and* $\alpha_{\mathrm{atom}}[v'/v] = \alpha_{\mathrm{atom}}[v', \mathrm{init}/v, \mathrm{init}]$*.*

5.1 Semantics of Atomic Interchange Automata

Definition 4 (Action transitions). *Consider an atomic interchange automaton* $\alpha = (X, X_{\mathrm{i}}, \mathrm{dtype}, V, v, \mathrm{init}, \mathrm{flow}, \mathrm{inv}, \mathrm{urgent}, L, E)$*. The* action transition relation $\_ \xrightarrow{\_} \_ \subseteq S \times (\mathit{Val} \times \mathcal{L}_\tau \times \mathcal{P}(\mathcal{X}) \times \mathit{Val}) \times S$ *is for* $(\alpha, \sigma, J), (\alpha', \sigma', J) \in S$*,* $\xi_{\mathrm{e}}, \xi'_{\mathrm{e}} \in \mathit{Val}$*,* $\ell \in \mathcal{L}_\tau$*, and* $W_{\mathrm{e}} \subseteq \mathcal{X}$*, defined as follows:* $(\alpha, \sigma, J) \xrightarrow{\xi_{\mathrm{e}}, \ell, W_{\mathrm{e}}, \xi'_{\mathrm{e}}} (\alpha', \sigma', J)$*, if and only if there exist an edge* $(v, g, \ell, (W, r), v') \in E$ *and* $\xi, \xi' \in \mathit{Val}$ *with* $\mathrm{dom}(\xi) = \mathrm{dom}(\xi') = \tilde{X}$ *such that*

- $\xi \restriction \tilde{X}_{\mathrm{e}} = \xi_{\mathrm{e}}$ *and* $\xi' \restriction \tilde{X}_{\mathrm{e}} = \xi'_{\mathrm{e}}$*;*
- $\xi_{\mathrm{e}} \restriction X_{\mathrm{state}} = \sigma$ *and* $\xi'_{\mathrm{e}} \restriction X_{\mathrm{state}} = \sigma'$*;*
- $\xi \models \mathrm{init}$ *and* $\xi \models g$*;*
- $\xi \models \mathrm{inv}(v)$ *and* $\xi' \models \mathrm{inv}(v')$*;*
- $\xi' \cup \xi^- \models r$*;*
- $\xi \restriction X_{\mathrm{nonjmp}} = \xi' \restriction X_{\mathrm{nonjmp}}$*, where* $X_{\mathrm{nonjmp}} = X_{\mathrm{state}} \setminus (W \cup (J \cap X_{\mathrm{e}}))$*;*
- $W_{\mathrm{e}} = W \cap X_{\mathrm{state}} \cap X_{\mathrm{e}}$*;*
- $\alpha' = \alpha[v', \mathrm{init}'/v, \mathrm{init}]$*,* $\mathrm{init}' = \left(\bigwedge_{x \in X_{\mathrm{state}} \cap X_{\mathrm{i}}} x = c_x \right)$*, and* $c_x \in \Lambda$ *is given by* $c_x = \xi'(x)$*.*

Here, *val* $\models$ *pred*, where *val* is a valuation and *pred* a predicate, means that *pred* is satisfied when all variables occurring in it are substituted by their values as defined in *val*. Minus superscripted variables, such as x^-, occurring in r are evaluated in ξ^-, which is defined as $\mathrm{dom}(\xi^-) = \{x^- \mid x \in \mathrm{dom}(\xi)\}$, and $\xi^-(x^-) = \xi(x)$. The 'non-jumping' variables in the set $X_{\mathrm{nonjmp}} = X_{\mathrm{state}} \setminus (W \cup (J \cap X_{\mathrm{e}}))$ are the variables the values of which are not allowed to change in an action transition. These variables are the discrete and continuous variables apart from two sets of variables: the variables from set W and the externally visible variables from set J ($J \cap X_{\mathrm{e}}$). The jumping variables in set J are the result of changes in external variables of synchronizing automata, as defined in the semantics of parallel composition in [8]. The updated initial condition init$'$ acts as a local valuation which ensures that for each local state variable x, its starting value for the next transition equals its resulting value (here: $\xi'(x)$, in Definition 5: $\rho(t)(x)$) for the current transition. Note that we do not combine the valuation of the internal variables with the valuation of the external variables in σ. Having the valuations of the local variables in σ would lead to restrictions on the parallel composition of two automata, namely that the sets of internal variables of two automata in a parallel composition would need to be disjoint. Otherwise, local variables with the same names in parallel automata would become shared. See for example [15], where the local variables and local actions of automata in a parallel composition are required to be disjoint.

Definition 5 (Time transitions). *Consider an atomic interchange automaton* $\alpha = (X,\ X_{\mathrm{i}},\ \mathrm{dtype},\ V,\ v,\ \mathrm{init},\ \mathrm{flow},\ \mathrm{inv},\ \mathrm{urgent},\ L,\ E)$. *The* time transition *relation* $\_ \overset{\_}{\longmapsto} \_ \subseteq S \times (T \times (T \mapsto \mathit{Val})) \times S$ *is for* $(\alpha, \sigma, J), (\alpha', \sigma', J) \in S$, $t \in T$, *and* $\rho_{\mathrm{e}} : [0,t] \to \mathit{Val}$, *defined as follows:* $(\alpha, \sigma, J) \overset{t,\rho_{\mathrm{e}}}{\longmapsto} (\alpha', \sigma', J)$, *if and only if there exists a* $\rho : [0,t] \to \mathit{Val}$ *with* $\mathrm{dom}(\rho(s)) = \tilde{X}$ *for all* $s \in [0,t]$ *such that*

- $\rho \downarrow \tilde{X}_{\mathrm{e}} = \rho_{\mathrm{e}}$;
- $\rho_{\mathrm{e}}(0) \upharpoonright X_{\mathrm{state}} = \sigma$ *and* $\rho_{\mathrm{e}}(t) \upharpoonright X_{\mathrm{state}} = \sigma'$;
- $\rho(0) \models \mathrm{init}$;
- $\rho \downarrow x$ *is a constant function for all* $x \in X_{\mathrm{disc}}$;
- $(\rho \downarrow x) \in F$ *for all* $x \in X_{\mathrm{alg}}$;
- $\rho \downarrow \dot{x}$ *is an integrable function in the Lebesgue sense for all* $x \in X_{\mathrm{cont}}$;
- $\rho(s) \models \mathrm{flow}(v)$ *and* $\rho(s) \models \mathrm{inv}(v)$ *for all* $s \in [0,t]$;
- $(\rho \downarrow x)(s) = (\rho \downarrow x)(0) + \int_0^s (\rho \downarrow \dot{x})(s')ds'$ *for all* $x \in X_{\mathrm{cont}}$ *and* $s \in [0,t]$;
- $(\rho \downarrow x, \rho \downarrow \dot{x}) \in G$ *for all* $x \in X_{\mathrm{cont}}$;
- $\rho(s) \models \neg\mathrm{urgent}(v)$ *for all* $s \in \{0\} \cup [0,t)$;
- $\alpha' = \alpha[\mathrm{init}'/\mathrm{init}]$, $\mathrm{init}' = \left(\bigwedge_{x \in X_{\mathrm{state}} \cap X_{\mathrm{i}}} x = c_x\right)$, *and* $c_x \in \Lambda$ *is given by* $c_x = \rho(t)(x)$.

Item $(\rho \downarrow x) \in F$ for all $x \in X_{\mathrm{alg}}$, requires the trajectories of the algebraic variables to be functions of type F. This set of functions is a global parameter of the solution concept of an interchange automaton specification.

The relation between the trajectory of a continuous variable x and the trajectory of its 'derivative' $\dot{x}$ is given by the Caratheodory solution concept [24]: $(\rho \downarrow x)(s) = (\rho \downarrow x)(0) + \int_0^s (\rho \downarrow \dot{x})(s')ds'$. This integral relation can hold only for those continuous variables for which $\rho \downarrow x$ is an absolutely continuous function, but it does allow a non-smooth trajectory for a continuous variable in the case that the trajectory of its 'derivative' $\rho \downarrow \dot{x}$ is non-smooth or even discontinuous, as in, for example, in the solution of the invariant $\dot{y} = \mathrm{step}(t-1) \wedge \dot{t} = 1$, where t and y are continuous variable with initial value of 0, and $\mathrm{step}(x)$ equals 0 for $x \le 0$ and 1 for $x > 0$.

In hybrid automata, the solution concept usually defines the function $\rho \downarrow \dot{x}$ to be the derivative function of $\rho \downarrow x$ for continuous variables $x \in X_{\mathrm{cont}}$. This can be realized for the interchange automaton format semantics by restricting the set G, which is used in the requirement $(\rho \downarrow x, \rho \downarrow \dot{x}) \in G$ for all $x \in X_{\mathrm{cont}}$, as $G = \{(f, f') \mid f$ is differentiable, and f' is the derivative function of $f\}$. In this way, the semantics of the interchange automaton format corresponds to the usual semantics of hybrid automata.

Definition 6 (Consistency predicates). *Consider an atomic interchange automaton* $\alpha = (X, X_{\mathrm{i}}, \mathrm{dtype}, V, v, \mathrm{init}, \mathrm{flow}, \mathrm{inv}, \mathrm{urgent}, L, E)$. *The* consistency predicate $\_ \overset{\_}{\rightsquigarrow} \subseteq S \times \mathit{Val}$ *is for* $(\alpha, \sigma, J) \in S$ *and* $\xi_{\mathrm{e}} \in \mathit{Val}$*, defined as follows:* $(\alpha, \sigma, J) \overset{\xi_{\mathrm{e}}}{\rightsquigarrow}$*, if and only if there exists a valuation* $\xi \in \mathit{Val}$ *with* $\mathrm{dom}(\xi) = \tilde{X}$ *such that* $\xi \upharpoonright \tilde{X}_{\mathrm{e}} = \xi_{\mathrm{e}}$*,* $\xi_{\mathrm{e}} \upharpoonright X_{\mathrm{state}} = \sigma$*,* $\xi \models \mathrm{init}$*, and* $\xi \models \mathrm{inv}(v)$*.*

5.2 Semantics of the Operators

The informal semantics of the operators is defined below. The formal semantics of the operators is defined in a structured operational semantics in [8].

Parallel composition. The most common operator for composing hybrid automata is parallel composition. There are no compatibility requirements for the parallel composition of interchange automata: any pair of interchange automata can be composed by the parallel composition operator. The parallel composition operator synchronizes on all external actions that the arguments share and allows interleaving of any other actions (under the condition that they maintain the consistency of the other automaton). Time transitions must be synchronized, and consistency is established only if both automata agree on it. The external state variables that are shared by the argument automata need to have the same values (all the time).

Hiding. The action hiding operator applied to an automaton, $\mathrm{hide}_{\mathrm{act}}(L_{\mathrm{h}}, \alpha)$, hides (abstracts from) the actions from set L_{h} by replacing them by the internal action τ. This only affects the action behavior of α; its delay behavior and consistency remain unchanged.

The variable hiding operator applied to an automaton, $\mathrm{hide}_{\mathrm{var}}(X_{\mathrm{h}}, \alpha, \sigma_{\mathrm{h}})$, hides the variables from set X_{h} by removing information about them from the action and time transitions of α. The values of the hidden state variables are stored in valuation σ_{h}.

Urgent action operator. The urgent action operator applied to an automaton, $\mathrm{urgent}(L_{\mathrm{u}}, \alpha)$, gives actions from the set L_{u} priority over time passing. The action behavior and consistency of α are not affected by the urgent action operator. Time transitions are allowed only if at the current state, and at each intermediate state while delaying, no actions from set L_{u} are possible.

6 Concluding Remarks

The proposed interchange automaton format integrates formalisms rooted in computer science with those rooted in dynamics and control. It is indeed compositional, since bisimilarity is proved to be a congruence for all operators of the interchange format in [8]. Future work entails, among others, adding the notion of input/output variables and input/output actions, adding channels as communication mechanism between interchange automata in a parallel composition, adding additional operators, such as sequential composition, and possibly extending the interchange format with stochastic model primitives. The development of translations and simulator implementations will be done by different partners in Work Package 3 of the HYCON NoE [5]. The translations that are to be developed should also specify incompatibilities, if present, or subsets for which property preserving translations are possible.

The atomic interchange automata as introduced syntactically in Definition 1 and for which the semantics has been introduced in Section 5 are very expressive. For any application of an action or variable hiding operator on an atomic interchange automaton, it is possible to obtain an equivalent atomic interchange automaton. Also, the parallel composition of any two atomic interchange automata for which the shared variables have compatible types can be replaced by an equivalent atomic interchange automaton. The only operator that cannot be eliminated in all relevant cases is the urgent action operator. Further substantiation of these claims and ideas can be found in [8].

References

1. Remelhe, M.A.P., Beek, D.A.v.: Requirements for an interchange format for non-linear hybrid systems. Technical Report D 3.6.1, HYCON NoE (2006)
2. MoBIES team: HSIF semantics. Technical report, University of Pennsylvania (2002) internal document.
3. Pinto, A., Carloni, L.P., Passerone, R., Sangiovanni-Vincentelli, A.L.: Interchange format for hybrid systems: Abstract semantics. In Hespanha, J.P., Tiwari, A., eds.: Hybrid Systems: Computation and Control, 9th International Workshop. Volume 3927 of Lecture Notes in Computer Science., Santa Barbara, Springer-Verlag (2006) 491–506
4. Columbus IST project: http://www.columbus.gr (2006)
5. HYCON NoE: http://www.ist-hycon.org/ (2005)
6. Cairano, S.D., Bemporad, A., Kvasnica, M.: An architecture for data interchange of switched linear systems. Technical Report D 3.3.1, HYCON NoE (2006)
7. Heemels, W.P.M.H., Schutter, B.d., Bemporad, A.: Equivalence of hybrid dynamical models. Automatica **37**(7) (2001) 1085–1091
8. Beek, D.A.v., Reniers, M.A., Schiffelers, R.R.H., Rooda, J.E.: A compositional interchange format for hybrid systems. Technical Report SE-Report 2006-05, Eindhoven University of Technology, Department of Mechanical Engineering, The Netherlands (2006) http://se.wtb.tue.nl/sereports/.
9. Tiller, M.: Introduction to Physical Modeling with Modelica. Volume 615 of The International Series in Engineering and Computer Science. Springer-Verlag (2001)

10. EcosimPro: http://www.ecosimpro.com (2006)
11. The MathWorks, Inc: Using Simulink, version 6, http://www.mathworks.com. (2005)
12. Henzinger, T.A.: The theory of hybrid automata. In Inan, M., Kurshan, R., eds.: Verification of Digital and Hybrid Systems. Volume 170 of NATO ASI Series F: Computer and Systems Science. Springer-Verlag, New York (2000) 265–292
13. Alur, R., Henzinger, T.A., Ho, P.H.: Automatic symbolic verification of embedded systems. IEEE Transactions on Software Engineering **22**(3) (1996) 181–201
14. Schaft, A.J.v.d., Schumacher, J.M.: An Introduction to Hybrid Dynamical Systems. Volume 251 of Lecture Notes in Control and Information Sciences. Springer-Verlag (2000)
15. Lynch, N., Segala, R., Vaandrager, F.: Hybrid I/O automata. Information and Computation **185**(1) (2003) 105–157
16. Henzinger, T.A., Ho, P.H., Wong-Toi, H.: A user guide to HyTech. In Brinksma, E., Cleaveland, R., Larsen, K.G., Margaria, T., Steffen, B., eds.: Tools and Algorithms for Construction and Analysis of Systems, First International Workshop. Volume 1019 of Lecture Notes in Computer Science., Aarhus, Denmark, Springer-Verlag (1995) 41–71
17. Frehse, G.: PHAVer: Algorithmic verification of hybrid systems past HyTech. In Morari, M., Thiele, L., eds.: Hybrid Systems: Computation and Control, 8th International Workshop. Volume 3414 of Lecture Notes in Computer Science. Springer-Verlag (2005) 258–273
18. Nicollin, X., Olivero, A., Sifakis, J., Yovine, S.: An approach to the description and analysis of hybrid systems. In Grossman, R.L., Nerode, A., Ravn, A.P., Rischel, H., eds.: Hybrid Systems. Volume 736 of Lecture Notes in Computer Science., Springer (1993) 149–178
19. Gebremichael, B., Vaandrager, F.: Specifying urgency in timed i/o automata. In: Proc. Third IEEE Conference on Sofware Engineering and Formal Methods. (2005) 64–74
20. Bornot, S., Sifakis, J.: An algebraic framework for urgency. Information and Computation **163**(1) (2000) 172–202
21. Bolognesi, T., Lucidi, F.: Timed process algebras with urgent interactions and a unique powerful binary operator. In de Bakker, J.W., Huizing, C., de Roever, W.P., Rozenberg, G., eds.: Real-Time: Theory in Practice, REX Workshop. Volume 600 of Lecture Notes in Computer Science., Mook, The Netherlands, Springer-Verlag (1991) 124–148
22. Beek, D.A.v., Man, K.L., Reniers, M.A., Rooda, J.E., Schiffelers, R.R.H.: Syntax and consistent equation semantics of hybrid Chi. Journal of Logic and Algebraic Programming **68**(1-2) (2006) 129–210
23. Bohnenkamp, H.C., D'Argenio, P.R., Hermanns, H., Katoen, J.P.: Modest: A compositional modeling formalism for real-time and stochastic systems. Technical Report Technical Report TR-CTIT-04-46, University of Twente, Centre for Telematics and Information Technology, The Netherlands (2004)
24. Filippov, A.F.: Differential Equations with Discontinuous Right Hand Sides. Kluwer Academic Publishers, Dordrecht (1988)

Automata Based Interfaces for Control and Scheduling

Gera Weiss and Rajeev Alur

University of Pennsylvania
{gera,alur}@seas.upenn.edu

Abstract. We propose the use of formal languages of infinite words over the alphabet of task identifiers as an interface between control designs and software implementations. We argue that this approach is more flexible than the classical real-time scheduling framework based on periodic tasks, and allows composition of interfaces by language-theoretic operations. We show that finite automata over infinite words offer analyzable representation and can capture many interesting interface specifications such as exponential stability of switched linear systems.

1 Introduction

Modern software engineering heavily relies on clearly specified interfaces for separation of concerns among designers implementing components and programmers using those components. The interface of a component describes the functionality and constraints on the correct usage in a succinct manner. For example, the interface of a Java class describes all the methods it supports, along with the types of input and output parameters for each method, and client code is written based on this interface without much understanding of the implementation of these methods. The need for interfaces is evident for assembling complex systems from components, but more so in control applications where the components are designed by control engineers using mathematical modeling tools and invoked by software executing on digital computers. The notion of an interface for a control component must incorporate some information about timing, and standard programming languages do not provide a way of capturing such resource requirements (cf. [1,2]).

In current practice, typically the real-time aspect of the interface of a control component is captured by a period, sometimes along with a deadline, which specifies the frequency at which the component must execute. The control engineer makes sure that the control objectives will be met as long as the component is executed consistent with its period. The software (for example, the real-time operating system) performs a worst-case execution time analysis on all the components, followed by schedulability analysis to check whether all the timing requirements can be met (cf. [3,4]). Using period as an interface specification of a control component is simple and intuitive, but has some key deficiencies. First, a specification such as "execute the component every 5ms" does not say

A. Bemporad, A. Bicchi, and G. Buttazzo (Eds.): HSCC 2007, LNCS 4416, pp. 601–613, 2007.

whether the scheduler should or should not execute it more frequently if enough computing resources are available. Second, such specifications do not compose in the sense that a system composed of two control components cannot be specified by a single period. These deficiencies create problems for integrating control components, and this has led researchers to explore many variations of the basic real-time scheduling framework [4, 3, 5].

In this paper, we propose to use formal languages as interfaces for control components. Our approach can be best explained using an example. Consider a control component with two tasks 1 and 2. Assuming that there is a single processor that is allocated in discrete slots of some fixed duration, a schedule with respect to this component can be described by an infinite word over the alphabet $\{0, 1, 2\}$, where 0,1,2, respectively, denote that the processor executes neither, first, second task of this component. The control engineer can express the interface of the component as an ω-language (see [6] for an introduction to theory of languages of infinite words) that contains all acceptable schedules. The software must ensure, then, that the runtime allocation is in this language. The main benefit of this approach is composability: conjoining specifications of two components corresponds to a simple language-theoretic operation on interfaces (for instance, renaming of alphabet symbols and intersection). Schedulability analysis corresponds to checking the emptiness of the language of acceptable schedules. Another benefit of this approach is predictability. A mathematical formulation of sets of schedules allows for an analysis of the type used for static (fixed) schedules. This may allow dynamic scheduling for safety critical control systems [7].

More specifically, we focus on discrete-time switched linear systems, and explore the use of finite Büchi automata as interfaces. Using automata as specifications fits nicely with current trends in type systems and static analysis tools. While control engineers may be less familiar with automata, we show that a variety of scheduling constraints can be expressed as automata. A sample specification expressible by finite automata is the language of all schedules such that any subsequence of length ℓ is contracting at least by ϵ, where ℓ and ϵ are parameters. Periodic schedules can be expressed using automata, and can also be composed (for example, the set of all schedules such that 1 appears at least once every five slots and 2 appears at least once every 6 slots). The properties studied in this paper are, so called, safety properties. Safety properties are properties of systems such that every violation of a property occurs after a finite execution. Such properties can be used for online detection of violation of constraints.

While automata-based specifications are flexible, composable, and analyzable, they can express only regular languages. We show, for example, that set of all schedules that ensure stability is not regular. This fact does not necessarily means that regular languages are not suitable for the purpose of specifying scheduling constraints for control tasks. For example, the set of schedules that ensures stability is not regular because the definition of stability allows unlimited excursion from the control objective. This, of course, is not desirable in practical applications. We show that some common stronger stability requirements, such as periodic and exponential stability, are regular.

Related Work

The use of automata as formal specification is common. Frameworks such as I/O automata [8] and interface automata [9] are focussed on capturing functionality and interaction of individual components, while timed automata have been used for schedulability analysis [10]. As far as we know, the idea of using formal languages and Büchi automata as an interface to capture the set of acceptable schedules over the alphbet of task identifiers, does not appear in the literature.

There are many approaches to scheduling safety-critical control systems [7, 11, 12, 13]. Most dynamic scheduling approaches are based on priority; that is, a task is dynamically chosen according to a priority order. In this case, the analysis of the effect of scheduling on control performance becomes difficult and inaccurate. The other popular approach is static scheduling - tasks are executed in a predetermined order. Static schedules can be analyzed to verify the effect of scheduling on control performance but they restrict the scheduling of sporadic tasks to fixed slots [14]. The approach presented in this paper is a way in the middle: it allows to model a set of schedules for which the effect of scheduling is verified.

Our work also relates to the research on stability of switched systems [15,16,17, 18]. Most of the stability results for switched systems are about stability under all switching signals. This point of view regards switching as an uncontrolled signal. We, instead, identify those switchings that render the system stable. This point of view is appropriate for scheduling, where we have some control over switching but need the smallest set of constraints.

The only attempt, we know of, to use automata to describe stability related criteria for switched systems is given in [19]. For polyhedral norms, it is shown that if a switched system satisfies $\|A_i\| \leq 1$ for all $i \in I$ then the set $\{\sigma \in I^* : \|A_\sigma\| < 1\}$ is regular. It is also shown that the language $\{\sigma \in I^* : \|A_\sigma\| < \epsilon\}$ may not be context free for $\epsilon < 1$.

2 Problem Formulation and Main Results

A discrete-time switched linear system is modeled by a finite set of real $n \times n$ matrices $\Sigma = \{A_i\}_{i \in I}$, $|I| < \infty$. Infinite words are used to describe schedules. Given a schedule $\sigma \in I^\omega$ and an initial state $x_0 \in \mathbb{R}^n$, the dynamics of the system is given by

$$x_{k+1} = A_{\sigma_k} x_k. \tag{1}$$

One interpretation of this model is as an abstraction of control mode scheduling. In this interpretation, one assumes that the choice of a control mode induces a linear transformation over the state variables. A schedule induces a sequence of transformations that gives rise to the dynamics formulated in equation (1). Given a finite word $\sigma = \sigma_1 \cdots \sigma_l$, we will use $A_\sigma = A_{\sigma_l} \cdots A_{\sigma_1}$ to denote the transformation induced by the finite schedule σ.

In the following propositions we identify some types of constraints that are expressible by Büchi automata. A Büchi automaton is a tuple $A = (Q, I, \delta, q_0, F)$, where Q is a finite set of states with $q_0 \in Q$ an initial state, I is an alphabet, $\delta : Q \times I \to 2^Q$ is a transition function and $F \subseteq Q$ is a set of accepting states. A run of the automaton A on an infinite word $\sigma \in I^\omega$ is an infinite sequence $q_0, q_1, \ldots$ of states such that $q_{k+1} \in \delta(q_k, \sigma_k)$ for every $k \in \mathbb{N}$. The run $q_0, q_1, \ldots$ is accepting if $q_k \in F$ for infinitely many k's. An infinite word $\sigma \in I^\omega$ is accepted by A if there exists an accepting run for σ. The automaton is called *deterministic* if $|\delta(q, i)| = 1$ for all $q \in Q$ and $i \in I$. Deterministic automata give a unique run for every input word. In the setting of this paper, the alphabet is the set of task identifiers.

A safety automaton is such that there is no transition from a non-accepting state to an accepting state. Such automata specify safety properties, namely, those properties where a violation of the property can be detected after only a finite execution of the system.

Liveness properties, on the other hand, are conditions that will eventually hold. A typical example of a liveness property is: "task x is executed infinitely often". One may use such a requirement, for schedulability analysis, in early design phases when an explicit bound is not provided. Other examples of liveness properties involve interaction with the environment. Since this paper is focused on scheduling, we restricted attention to automata whose alphabet is the set of task identifier. To allow finer specifications, one may add symbols that model observations. Then, a specification such as: "if an event is detected infinitely often, eventually some process is invoked" may be useful.

In diagrams representing Büchi automata, we adopt the following graphical convention: if a symbol does not decorate any transition going out of a state then an implicit transition to a non accepting sink (a state with self loop for all the symbols) is assumed for this symbol.

The next two propositions show that many natural requirements that appear in control applications can be expressed using safety automata.

Proposition 1. *Given a switched system, the following languages can be recognized by a deterministic safety Büchi automaton:*

1. *Exponential stability: for $0 < \epsilon < 1$ and $l \in \mathbb{N}$; the language of all schedules such that any interval of length l is contracting by at least ϵ*

$$\mathit{ExpStab}(l, \epsilon) = \{\sigma \in I^\omega : \|A_{\sigma_{k+l}} \cdots A_{\sigma_{k+1}}\| < \epsilon \text{ for every } k \in \mathbb{N}\}.$$

2. *Directional growth: for $c \in \mathbb{R}^n$, $\delta > 1$ and $l \in \mathbb{N}$; the language of all schedules such that the projection on c grows exponentially fast,*

$$\mathit{DirG}(c, \delta, l) =$$
$$\{\sigma \in I^\omega : |\langle c, A_{\sigma_{k+l}} \cdots A_{\sigma_{k+1}} x\rangle| > \delta |\langle c, x\rangle| \text{ for every } k \in \mathbb{N} \text{ and } x \in \mathbb{R}^n\}.$$

3. *Cost function: for positive definite matrices $Q_1, \ldots, Q_m$, horizon $h \in \mathbb{R}$ and maximal cost $J \in \mathbb{R}$;*

$$\mathit{Cost}(Q_1, \ldots, Q_m, h, J) =$$
$$\{\sigma \in I^\omega : \sum_{k=k_0}^{k_0+h} \sum_{i=1}^{m} x_k^T Q_i x_k < J \text{ for every } k_0 \in \mathbb{N} \text{ and } x_{k_0} \in \mathbb{R}^n\}.$$

The above languages are examples of control performance type of constraints. These constraints refer to the model of the system. In the next proposition, we give some languages that represent constraints that are not directly related to the model of the plant as a switched system. These languages represent external considerations such as sensor or actuator usage constraints. They can also represent scheduling constraints that come from the implementation (e.g. if the implementation can only handle periodic schedules).

Proposition 2. *The following languages can be recognized by a deterministic safety Büchi automaton:*

1. *Minimal separation: for $i, j \in I$ and $m_{i,j} \in \mathbb{N}$; the language of all schedules that separate the schedule of mode i from mode j by at least $m_{i,j}$ slots,*

$$MinSep(i, j, m_{i,j}) = \{\sigma \in I^\omega : \sigma_k = i \implies j \notin \{\sigma_{k+1}, \ldots, \sigma_{k+m_{i,j}}\}\}.$$

2. *Maximal separation: for $i, j \in I$ and $M_{i,j} \in \mathbb{N}$; the language of all schedules that separate the schedule of mode i from mode j by at most $M_{i,j}$ slots,*

$$MaxSep(i, j, M_{i,j}) = \{\sigma \in I^\omega : \sigma_k = i \implies j \in \{\sigma_{k+1}, \ldots, \sigma_{k+M_{i,j}}\}\}.$$

3. *Periodic execution: for $i \in I$ and $P_i \in \mathbb{N}$; the language of all schedules that execute mode i every P_i slots,*

$$Per(i, P_i) = \\ \{\sigma \in I^\omega : \sigma_k = i \implies \sigma_{k+P_i} = i \text{ and } \sigma_{k+j} \neq i \text{ for all } j = 1, \ldots, P_i - 1\}.$$

4. *Dependency: for a relation $R \subset I \times I$; the language of all schedules such that mode i executes after mode j only if $(i, j) \in R$,*

$$Dep(R) = \{\sigma \in I^\omega : (\sigma_k, \sigma_{k+1}) \in R \text{ for every } k \in \mathbb{N}\}.$$

5. *Sequential execution: for an ordered tuple $(i_1, \ldots, i_l) \subseteq I^l$; the language of all schedules such that modes $i_1, \ldots, i_l$ are executed periodically in a sequence (other modes may get in between),*

$$Seq(i_1, \ldots, i_l) = \{\sigma \in I^\omega : \pi(\sigma, \{i_1, \ldots, i_l\}) \text{ is a prefix of } (i_1 \cdots i_l)^\omega\}.$$

where $\pi(\sigma, S)$ denotes the projection of the word σ over the set S, i.e., the word obtained by deleting from σ all the letters not in S.

6. *Cyclic schedules: for $P \in \mathbb{N}$; the set of cyclic schedules with cycle length C*

$$Cyc(C) = \{\sigma \in I^\omega : \sigma_{k+C} = \sigma_k \text{ for all } k \in \mathbb{N}\}$$

Since the set of languages expressible by deterministic safety Büchi automata is closed under finite intersections and unions, disjunctions and conjunctions of the above constraints are also expressible. For example, we can express that modes i and j should interlace every 10 execution slots by the formula $Per(i, 20) \cap$

$Per(j, 20) \cap MinSep(i, j, 10) \cap MaxSep(i, j, 10)$. Of course, finite automata can model many other languages. The above list is only a collection of examples, put together in order to convince the reader that many typical specifications are expressible using automata.

The above propositions, together with the succeeding example, show that control engineers do not have to construct automata manually. We envision a tool that accepts specifications in a language tailored to allow easy scheduling specifications for control systems. The tool may take specifications of the form exhibited in the proposition, allowing to combine them using logical operators. The output of this tool can be a finite automaton representing all acceptable schedules for the system.

From the software engineering perspective, a representation of scheduling constraints by finite automata lets a scheduling mechanism use all the flexibility allowed by the control system. This gives the flexibility of dynamic scheduling with the predictability of static scheduling. Assume, for example, that we want to control several systems by the same computer. Each system poses some scheduling constraints. If these constraints are expressed by automata, the intersection can be easily computed. If the intersection is not empty, the system is schedulable. In this setting, each subsystem has a distinct matrix A_0 that models the behavior of the system when the processor is assigned to control another subsystem. In other settings, it can be that controlling one system has an affect on the other systems. Such dependencies can be easily incorporated into the automata composition algorithm using an adequate renaming of the alphabet. This approach also allows to schedule non-control tasks more efficiently. The automaton that represents the intersection can also be used as an admission control mechanism (if we want to allow online registration of new subsystems).

For completeness, we exhibit some limitations of using automata to express scheduling constraints for control systems. In the following proposition, we identify a type of constraint not expressible using automata. A language of finite words is called *regular* if it can be accepted by a finite automaton.

Proposition 3. *For a switched system, if there exits $i \in I$ such that A_i is not stable (one of its eigenvalues is not in the complex unit disc) then the language*

$$L = \{\sigma \in I^* : \|A_\sigma\| < 1\}$$

is either empty or not regular.

This result implies that we cannot model the set of all finite schedules, $\sigma \in I^*$, such that $\|A_\sigma x\| < \|x\|$ for every $x \in \mathbb{R}^n$. This language corresponds to the set of periodic schedules for which every period is contracting.

Intuitively, the reason this language is not regular is because a contracting word may begin with an arbitrarily long expanding prefix. The only way to check that an expansion is compensated is to count arbitrarily high (a formal proof is given in Section 5 below). We argue that the key property that makes this language irregular, also makes it inadequate for scheduling specifications of control modes. Requiring that the transformation be contracting does look natural at the first glance, since it means that perturbations are steered to equilibrium. However, allowing an arbitrarily long expansion is not usually acceptable because it means that perturbations may explode before they are cleared. To fix this problem we offer constraints of the type given in Proposition 1, where an explicit bound on the maximal length of an excursion is given. Such requirements capture practical control constraints and are describable by finite automata.

3 Compositional Schedulability Analysis

Consider a control system composed of l subsystems of dimensions $n_1, \ldots, n_l$ (the state space of the ith subsystem is $\mathbb{R}^{n_i}$, $i = 1, \ldots, l$, and the state space of the whole system is $\mathbb{R}^n = \mathbb{R}^{n_1} \times \cdots \times \mathbb{R}^{n_l}$). Assume that the system is controlled by a single computer and that the control engineer designed m control tasks.

In each computation slot, the states of the subsystems are transformed, depending on the task scheduled for execution in this slot. Assuming that the transformations are linear and time invariant, we can model this dependency by a map

$$A : \{1, ..., m\} \rightarrow \mathbb{R}^{n_1 \times n_1} \times \cdots \times \mathbb{R}^{n_l \times n_l}$$

that takes a task identifier to a list of matrices. The ith matrix in $A(j)$ models the transformation of the state of the ith subsystem when task j is scheduled for execution.

Given a requirements specification for each subsystem, based on the propositions in Section 2, we can compute a Büchi automaton for every subsystem. This automaton specifies the sequences of task executions that will not violate the specification of the subsystem.

Since there are efficient algorithms to find an automaton recognizing the intersection of the languages recognized by a finite set of automata, there is an efficient way to get the set of schedules that keep all subsystems within their requirements. An empty intersection means that the system is not schedulable. A nonempty intersection can be used for dynamic or static scheduling mechanisms, or serve for further schedulability analysis (by intersecting it with another automaton).

When a new subsystem is added, we do not need to repeat all the computation. Once the automaton specifying the requirements for the new subsystem is computed, it can be intersected with the product of the other subsystems.

Assume that we want to add a control task $m+1$ such that the ith matrix in the list $A(m+1)$ is the same as the ith matrix in $A(j)$ for some $j \in \{1, \ldots, m\}$. This happens when the effect of task $m+1$ on the ith subsystem is the same as the effect of the task j on that subsystem. In that case, the automaton for the requirements of the ith subsystem can be adjusted incrementally by adding an $m+1$ transition whenever a j transition exists.

One practical example, is when each task is designed to regulate a specific subsystem. In this case, each subsystem is assigned with a list of matrices. The list $L_i = (A_0, \ldots, A_{m_i})$ describes the possible transformations of the state of the ith system. The first element in this list, A_0, models the response of the system when the controller attends another subsystem. Then, task identifiers are pairs (i, j) and the global map is

$$A(i,j) = (L_1(0), \ldots, L_i(j), \ldots, L_l(0)).$$

Namely, when the jth transformation of the ith subsystem is scheduled for execution, all other subsystems are transformed according to the first entry in their list. In this case, schedulability analysis is completely compositional: when a new subsystem is defined, schedulability analysis reduces to computing the product of the automaton for the new subsystem with the automaton for the rest of the system (in both automata, the missing alphabet symbols are identified with the zero symbol).

4 Examples

Consider the system $\Sigma = \{A_1, A_2\}$ where $A_1 = \begin{pmatrix} 2 & -\frac{7}{4} \\ 2 & -2 \end{pmatrix}$ and $A_2 = \begin{pmatrix} \frac{1}{4} & \frac{7}{4} \\ \frac{1}{4} & -\frac{1}{4} \end{pmatrix}$. Both A_1 and A_2 are stable (their eigenvalues lie in the unit ball of the complex plain), but their product is not stable. In particular, the schedule $\sigma = 1212\ldots$ steers the initial state $(1,1)^T$ arbitrarily far from the origin and so does the schedule $\sigma = 2121\ldots$ to the vector $(1,0)^T$.

Figure 1 depicts an automaton that models the set of schedules that satisfy $ExpStab(4,1) = \{\sigma \in \{1,2\}^\omega : \|A_{\sigma_{k+4}} \cdots A_{\sigma_{k+1}}\| < 1 \text{ for every } k \in \mathbb{N}\}$. This automaton was constructed by first computing the set

$$\{1112, 1121, 1211, 1212, 1222, 2111, 2121, 2122, 2212, 2221\}$$

of the words of length four that are not contracting. Then, an automaton that rejects any infinite word that contains one of these four letter words as a subword was derived.

Assume that, in addition to the above constraint, we are not allowed to apply the first mode consecutively more than two times. Such a constraint may arise when A_1 models the use of some sensor or actuator that needs a time to re-calibrate after two consecutive operations. This is captured by the language $MaxCon(1,2)$ whose automaton is given in Figure 2. The language that captures both constraints together is $ExpStab(4,1) \cap MaxCon(1,2)$ whose automaton is given in Figure 3.

5 Proofs

In this section we give the proofs of the propositions brought in Section 2. The proofs are constructive. In particular the proofs of Proposition 1 and Proposition 2 suggest algorithms to compute the automata representations. The constructions provided in the proofs will not always give the smallest possible automata. For practical use, an automata minimization algorithm may be needed (cf. [20]).

Since this paper is intended for both control theoretic and computer science audience, we include some details that may seem trivial to some readers.

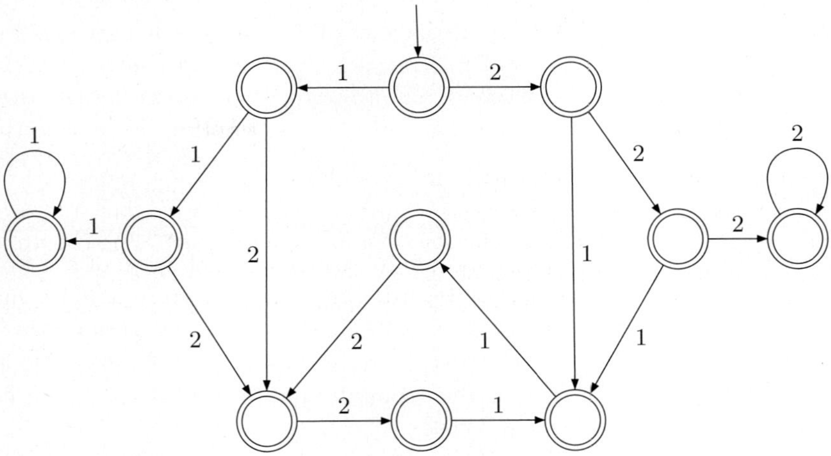

Fig. 1. A deterministic Büchi automaton for $ExpStab(4, 1)$

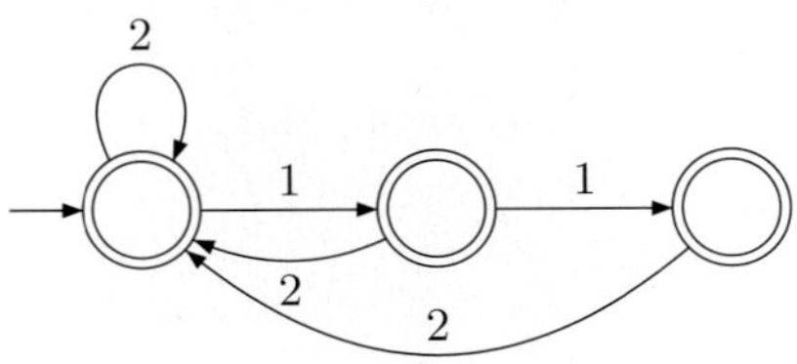

Fig. 2. A deterministic Büchi automaton for $MaxCon(1, 2)$

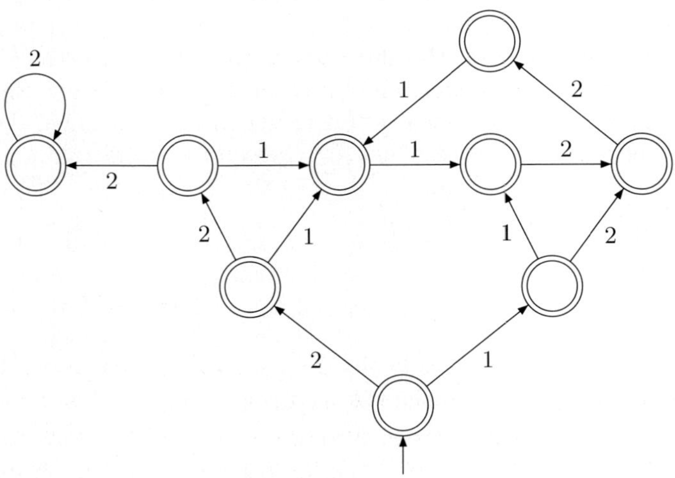

Fig. 3. A deterministic Büchi automaton for $ExpStab(4, 1) \cap MaxCon(1, 2)$

Proof (of Proposition 1)
The language $ExpStab(l, \epsilon)$ is the language of all infinite words that avoid any word $\sigma \in I^l$ such that $\|A_\sigma\| \geq \epsilon$. For each such word we can construct the automaton that describes the language of infinite words that avoid it as a sub-word (see Figure 4 for an example), and then take the intersection. Note that there are only finite number of such words. Similarly, the language $DirG(c, \delta, l)$ is the language of of all words such $|\langle c, A_\sigma x\rangle| > \delta|\langle c, x\rangle|$ for every subword σ of length l and any $x \in \mathbb{R}^n$. For each $\sigma \in I^l$, we can compute $B_\sigma := \min_{x\in\mathbb{R}^n} \frac{|\langle c, A_\sigma x\rangle|}{|\langle c, x\rangle|}$. Then, $DirG(c, \delta, l)$ is the language of infinite words in which none of the words $\{\sigma \in I^l : B_\sigma < \delta\}$ appear as a subword. Again,there is only a finite number of such words. For the language $Cost(Q_1, \ldots, Q_m, h, J)$, consider the set $S = \{\sigma \in I^l : \max_{x\in\mathbb{R}^n} \sum_{i=1}^{l} \sum_{j=1}^{m} (A_{\sigma_i} \cdots A_{\sigma_1} x)^T Q_j (A_{\sigma_i} \cdots A_{\sigma_1} x) \geq J\}$. The language $Cost(Q_1, \ldots, Q_m, h, J)$ is the set of all infinite words that do not contain any of the words in S as a subword. □

Proof (of Proposition 2)
Figure 5 depicts an automaton that accepts $MinSep(i, j, m_{i,j})$. The automaton counts the letters different from j after each occurrence of i and moves to a non-accepting state when a j is too close to an i. Since we do not need to count more than $m_{i,j}$, the automaton is finite. The automaton for $MaxSep(i, j, M_{i,j})$ is similar. We need to carry the same counting but accept only words for which the counter does not exceed the limit. See Figure 6 for an automaton that accepts $Per(i, 3)$. Generalization to arbitrary period is straightforward. Figure 7 depicts an automaton that accepts only words in which only mode j is allowed to follow mode i. If we are given a relation, R, as a set of ordered pairs, we can construct an automaton as in Figure 7 for each pair and take the union of the languages. See Figure 8 for an automaton that accepts the language $Seq(1, 2, 3)$. This example can be easily extended to longer sequences and other indices. For the language $Cyc(C)$, consider the automaton depicted in Figure 9. Let $L(i, m, C)$ be the language of all infinite words such that, starting with the mth letter, Cth letter is again i. This language can be accepted by an automaton similar to the one given in Figure 9 (possibly with different initial and cycle counts). Then, the language $Cyc(C)$ is given as the union of $L(i, m, C)$ over all $i \in I$ and $m = 1, \ldots, C$. □

Proof (of Proposition 3)
Assume towards contradiction that the language is regular but not empty. By the pumping lemma, there is $p > 0$ such that for every $w \in L$ there are words $x, y, z \in I^*$ such that $w = xyz$, $|xy| \leq p$, $|y| > 0$ and $xy^k z \in L$ for every $k \geq 0$. Since L is not empty there is a word $u \in L$. In particular A_u is contracting. There must be an $m \in \mathbb{N}$ such that $\|A_u^m A_i^p\| < 1$ because $\|A_u^m A_i^p\| \leq \|A_u\|^m \|A_i\|^p$ which is smaller than one for every m larger than $\log_{\|A_u\|}(\|A_i\|^{-p})$. Consider the word $w = i^p u^m$. This word is in L because $\|A_u^m A_i^p\| < 1$. By the pumping lemma, the words $i^l u^m$ must also belong to L for every $l \in \mathbb{N}$. But this leads to a contradiction because we allow arbitrarily large expansion before a fixed contraction which will eventually give an expanding product. □

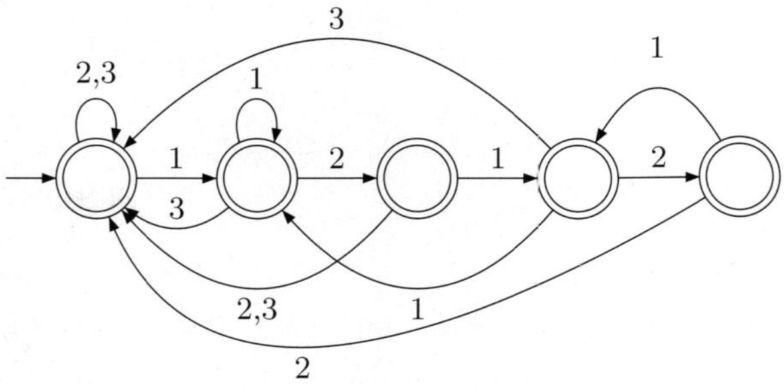

Fig. 4. A deterministic Büchi automaton that accepts all infinite words that avoid the subword 12123

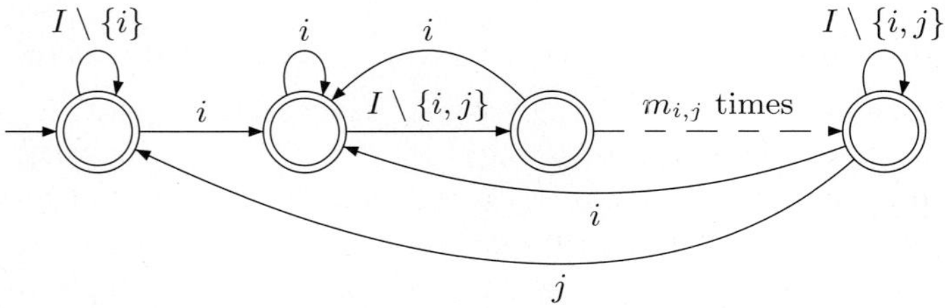

Fig. 5. A deterministic Büchi automaton that accepts $MinSep(i, j, m_{i,j})$

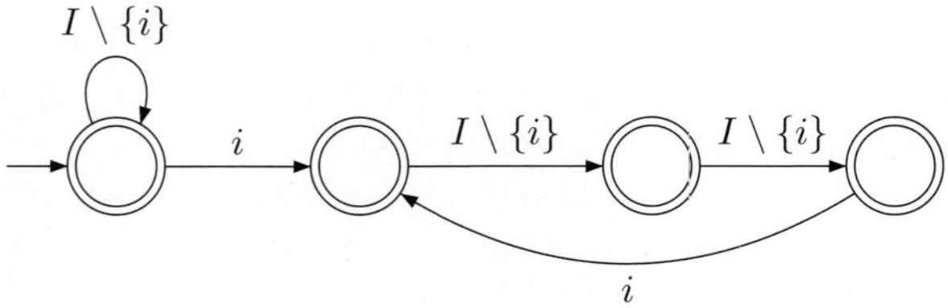

Fig. 6. A deterministic Büchi automaton that accepts $Per(i, 3)$

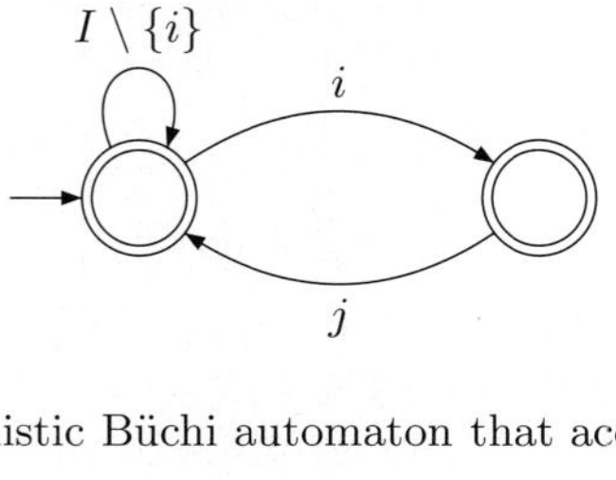

Fig. 7. A deterministic Büchi automaton that accepts $Dep(\{(i, j)\})$

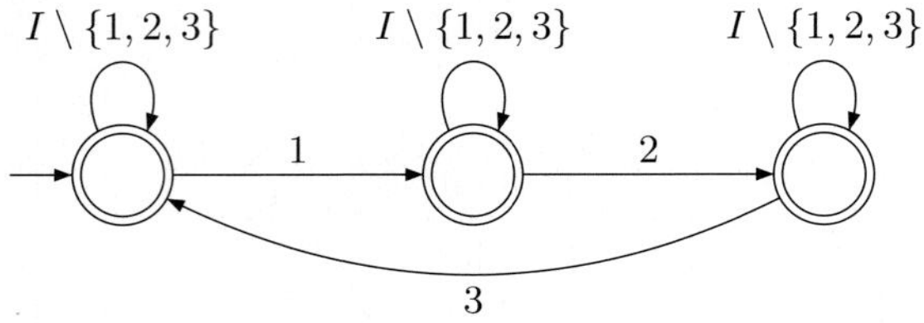

Fig. 8. A deterministic Büchi automaton that accepts $Seq(1, 2, 3)$

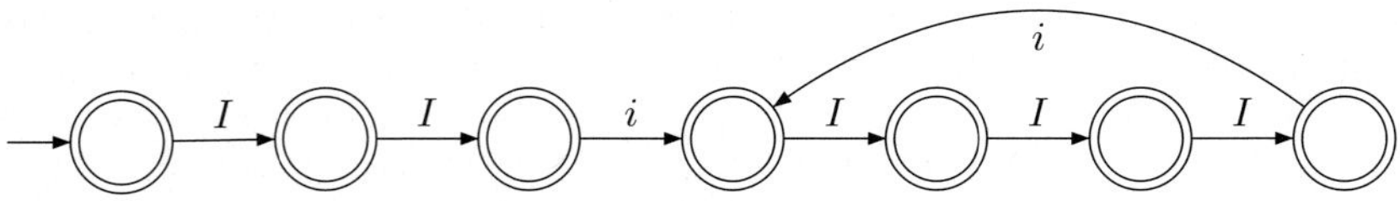

Fig. 9. A deterministic Büchi automaton that accepts words with i as the third letter and then every fourth letters

Acknowledgments

We thank George Pappas, Oded Maler and Truong Nghiem for fruitful discussions. This research was supported by NSF grants CCR 0410662 and CSR-EHS 0509143.

References

1. Sastry, S., Sztipanovits, J., Bajcsy, R., Gill, H.: Modeling and design of embedded software. Proceedings of the IEEE **91**(1) (2003)
2. Lee, E.: What's ahead for embedded software. IEEE Computer (2000) 18–26
3. Kopetz, H.: Real-Time Systems: Design Principles for Distributed Embedded Applications. Kluwer Academic Publishers (2000)
4. Buttazo, G.: Hard real-time computing systems: Predictable scheduling algorithms and applications. Kluwer Academic Publishers (1997)
5. Shin, I., Lee, I.: Compsitional real-time scheduling framework. In: Proceedings of the 25th IEEE Real-Time Systems Symposium. (2004)
6. Thomas, W.: Automata on infinite objects. In van Leeuwen, J., ed.: Handbook of Theoretical Computer Science. Volume B. Elsevier Science Publishers (1990) 133–191
7. Balarin, F., Lavagno, L., Murthy, P., Sangiovanni-vincentelli, A.: Scheduling for embedded real-time systems. IEEE Design and Test of Computers **15**(1) (1998) 71–82
8. Lynch, N., Segala, R., Vaandrager, F.: Hybrid I/O automata. Information and Computation **185**(1) (2003) 105–157
9. de Alfaro, L., Henzinger, T.: Interface theories for component-based design. In: Embedded Software, First International Workshop. LNCS 2211, Springer (2001) 148–165

10. Abdeddaïm, Y., Maler, O.: Job-shop scheduling using timed automata. In: CAV 01: Proc. of 13th Conf. on Computer Aided Verification. LNCS 2102, Springer (2001) 478–492
11. Balbastre, P., Ripoll, I., Crespo, A.: Control tasks delay reduction under static and dynamic scheduling policies. rtcsa **00** (2000) 522
12. Bate, I., Burns, A.: A framework for scheduling in safety-critical embedded control systems. In: RTCSA '99: Proceedings of the Sixth International Conference on Real-Time Computing Systems and Applications, Washington, DC, USA, IEEE Computer Society (1999) 46
13. Cervin, A.: Improved scheduling of control tasks. In: Proceedings of the 11th Euromicro Conference on Real-Time Systems, York, UK (1999) 4–10
14. Audsley, N., Tindell, K., Burns, A.: The end of the line for static cyclic scheduling? In: Real-Time Systems, 1993. Proceedings., Fifth Euromicro Workshop on on Real-Time Systems. (1993)
15. Blondel, V.D., Tsitsiklis, J.N.: The boundedness of all products of a pair of matrices is undecidable. Systems Control Lett. **41**(2) (2000) 135–140
16. Hespanha, J., Morse, A.: Stability of switched systems with average dwell-time (1999)
17. Liberzon, D.: Switching in systems and control. Systems & Control: Foundations & Applications. Birkhäuser Boston Inc., Boston, MA (2003)
18. Schultz, P.: Research Problems: Mortality of 2×2 Matrices. Amer. Math. Monthly **84**(6) (1977) 463–464
19. Gurvits, L.: Stability of discrete linear inclusion. Linear Algebra Appl. **231** (1995) 47–85
20. Etessami, K., Holzmann, G.J.: Optimizing Büchi automata. Lecture Notes in Computer Science **1877** (2000) 153+

Modeling and Optimal Control of Hybrid Rigidbody Mechanical Systems

Kerim Yunt and Christoph Glocker

Swiss Federal Institute of Technology, Center of Mechanics
Tannenstr. 3, 8092 Zurich, Switzerland

Abstract. A measure differential inclusion (MDI) based modeling approach for rigidbody mechanical systems will be introduced, that can exhibit autonomous or controlled mode transitions, accompanied by discontinuities on velocity and acceleration level. The hybrid optimal control necessitates the consideration of an uncommon concept of control, namely, controls of unbounded, impulsive and set-valued type. Examples to manipulators and wheeled robots are presented.

Keywords: Impulsive Optimal Control, Impactive Systems, non-smooth analysis, hybrid, mechanical systems.

1 Introduction

A measure differential inclusion (MDI) based modeling approach for rigidbody mechanical systems will be introduced, that can exhibit autonomous or controlled mode transitions, accompanied by discontinuities on velocity and acceleration level. The introduced framework will have the ability to model and control hybrid mechanical systems with discontinuous transitions among different system modes. Modeling of rigidbody Lagrangian systems as Linear Complementarity Problem (LCP) will be presented, and the associated optimal control problem of hybrid Lagrangian systems will be stated. In recent years, several works have been presented in order to establish the relations between complementarity dynamical systems and hybrid systems. There are general results in literature that investigate the relation between different representations of dynamical systems and hybrid systems such as [8], [13]. The properties of the optimal control problem derive from the underlying modeling approach. Optimal Control of hybrid systems is addressed in several publications such as [4], [5], [23] and [25] on various field of applications, in which the modeling is based on approaches to the ones similar as in [3], [6]. The treatment of discontinuous transitions and the combinatorial nature of mode sequencing are partially treated in these publications about optimal control, due to the modeling approach chosen. In this work, the MDI modeling of mechanical hybrid systems will be proposed and the suitability from the viewpoints above presented. There are some works devoted to the complementarity modeling and optimal control of mechanical systems as given in references [26], [27], [28], [29]. The main issue in the optimal control of hybrid mechanical systems has been the blending of impact mechanics with

A. Bemporad, A. Bicchi, and G. Buttazzo (Eds.): HSCC 2007, LNCS 4416, pp. 614–627, 2007.

impulsive optimal control. Indeed, the optimal control of such systems entails unavoidably impulsive control. References [18] and [2] consider impulsive systems and impulsive differential inclusions from the viewpoint of hybrid systems. There has been much interest in the research of modeling discontinuities and nonlinearities in multibody systems for which a compact overview is provided in [7]. The discontinuities arising from impacts and stick-slip transitions are primarily contact phenomena, which concurre temporally and spatially. The spatial concurrence of discontinuity is due to the fact that discontinuities on velocity level (e.g. collisions) can occur along with discontinuities on acceleration level (e.g.stick-slip transitions). Temporal concurrence is caused by collision, shock and impact phenomena occuring at multiple locations of the system at the same time as well as stick-slip transitions. Recent research showed that such rigid-body systems can best be described by variational inequalities which lead to nonlinear and linear complementarity type of systems to be solved in order to obtain the accelerations/velocities and forces. In the modeling considered in this work, impulsive forces can arise autonomously, due to effects such as collisions or controlled/nonautonomously, due to actions such as blocking some DOF. The hybrid optimal control requires the consideration of an uncommon concept of control, namely, controls of unbounded, impulsive and set-valued type. The existence of force and impulsive/discrete type of controls will through the solution of the complementarity problem take influence on the course of system trajectories. The presence of impulsive forces require to solve impact equations and constitutive laws that relate post- and pre-impact velocities of the system. The topic of impacts with and without friction is investigated in references [15], [16].

2 The Measure Differential Inclusion Representation of Hybrid Mechanical Systems

One of the cornerstones of non-smooth modeling of mechanical systems has been the introduction of the MDI concept that considers the equations of motion subject to variational inequalities as a balance of measures. The concept of MDI and its applications to mechanics stems from J. J. Moreau, and related works of him are given in [20], [21]. The application of the measure-differential inclusion approach to rigid-body mechanics can be overviewed in [14].

Let $\mathbf{q}$, $\dot{\mathbf{q}}$, $\ddot{\mathbf{q}}$ represent the position, velocity and acceleration in the generalized coordinates of a scleronomic rigid body mechanical system with n degrees of freedom (DOF), respectively. The equations of motion (EOM) are obtained by using the well-known Lagrange II formalism for the smooth dynamics :

$$\frac{d}{dt}\left(\frac{\partial T}{\partial \dot{\mathbf{q}}}\right) - \left(\frac{\partial T}{\partial \mathbf{q}}\right) + \left(\frac{\partial V}{\partial \mathbf{q}}\right) - \mathbf{f} = \mathbf{0} \ . \tag{1}$$

Here T denotes the total kinetic, and V the total potential energy of the system. The unilateral forces, which are non-potential in the classical sense, are incorporated by the appropriate generalized force directions in the generalized force vector $\mathbf{f}$. The controls will also be introduced into the equations of motion by means of the structure of $\mathbf{f}$.

The tangential and normal local kinematics need to be defined in order to relate the contact distance to the set-valued force element. For the detection of the closing of a contact let the vector $\mathbf{g}_\mathrm{u}(\mathbf{q})$ entail the normal contact distances between the rigid bodies in the system which are always non-negative. The normal and tangential contact velocities $\boldsymbol{\gamma}_\mathrm{u}$ and $\boldsymbol{\gamma}_\mathrm{s}$ are defined as:

$$\boldsymbol{\gamma}_\mathrm{u} = \mathbf{W}_\mathrm{u}^T\dot{\mathbf{q}}, \quad \boldsymbol{\gamma}_\mathrm{s} = \mathbf{W}_\mathrm{s}^T\dot{\mathbf{q}}, \tag{2}$$

respectively, and $\boldsymbol{\gamma}_\mathrm{u}$ is obtained as the total time derivative of $\mathbf{g}_\mathrm{u}(\mathbf{q})$. The normal and tangential contact accelerations are given by the following equations:

$$\dot{\boldsymbol{\gamma}}_\mathrm{u} = \mathbf{W}_\mathrm{u}^T\ddot{\mathbf{q}} + \boldsymbol{\omega}_\mathrm{u}, \quad \dot{\boldsymbol{\gamma}}_\mathrm{s} = \mathbf{W}_\mathrm{s}^T\ddot{\mathbf{q}} + \boldsymbol{\omega}_\mathrm{s} \ . \tag{3}$$

Here $\mathbf{W}_\mathrm{u}(\mathbf{q})$ and $\mathbf{W}_\mathrm{s}(\mathbf{q})$ represent the generalized force directions of normal and tangential contact forces. In order to formulate the contact situations properly following index sets will be defined:

$$\mathcal{I}_\mathrm{G} = \{1, 2, ..n_\mathrm{G}\}, \mathcal{I}_\mathrm{S} = \{i \in \mathcal{I}_\mathrm{G} | \, g_{\mathrm{u}_i} = 0\}, \mathcal{I}_\mathrm{N} = \{i \in \mathcal{I}_\mathrm{S} | \, \gamma_{\mathrm{u}_i} = 0\} \ . \tag{4}$$

$\mathcal{I}_\mathrm{G}$ denotes the set of all contacts that can occur on position level of the non-smooth mechanical system. $\mathcal{I}_\mathrm{S}$ denotes the set of all contacts that are closed on position level. $\mathcal{I}_\mathrm{N}$ denotes the set of all contacts such that normal contact velocity and normal contact distance equal to zero. Further, the definition of following index sets are necessary:

$$\mathcal{C}_{\mathrm{u}_i} = \{\lambda_{\mathrm{u}_i} \, | \, \lambda_{\mathrm{u}_i} \geq 0, \, \forall i \in \mathcal{I}_\mathrm{G}\}, \quad \mathcal{C}_{\mathrm{s}_i} = \{\lambda_{\mathrm{s}_i} \, | \, |\lambda_{\mathrm{s}_i}| \leq \mu_i \, \lambda_{\mathrm{u}_i}, \, \forall i \in \mathcal{I}_\mathrm{S}\}, \tag{5}$$

where the vectors $\boldsymbol{\lambda}_\mathrm{s}$, $\boldsymbol{\lambda}_\mathrm{u}$ are the tangential and the normal contact forces, respectively and μ_i denotes the friction coefficient at contact i. The differential inclusion of a general non-autonomous mechanical system $\mathcal{S}_\mathrm{a}$ subject to spatial Coulomb friction and unilateral contact forces in the absence of impacts can then be stated as:

$$\begin{aligned} &\mathbf{M}\,\dot{\mathbf{u}} - \mathbf{h} - \mathbf{W}_\mathrm{s}\boldsymbol{\lambda}_\mathrm{s} - \mathbf{W}_\mathrm{u}\boldsymbol{\lambda}_\mathrm{u} - \mathbf{B}\,\boldsymbol{\tau} = \mathbf{0}, \\ &\dot{\gamma}_{\mathrm{u}_i} \in -\mathcal{N}_{\mathcal{C}_{\mathrm{u}_i}}(\lambda_{\mathrm{u}_i}), \quad \forall i \in \mathcal{I}_\mathrm{N}, \dot{\boldsymbol{\gamma}}_{\mathrm{s}_i} \in -\mathcal{N}_{\mathcal{C}_{\mathrm{s}_i}}(\boldsymbol{\lambda}_{\mathrm{s}_i}), \quad \forall i \in \mathcal{I}_\mathrm{N} \ . \end{aligned} \tag{6}$$

$\mathbf{M}(\mathbf{q})$ is the symmetric positive-definite (PD) mass matrix and $\mathbf{h}(\mathbf{q}, \, \mathbf{u})$ represents the vector with gyroscopic and coriolis accelerations along with smooth potential forces such as gravity. $\mathbf{B}(\mathbf{q})$ entails columnwise the generalized force directions of controls. The vector $\boldsymbol{\tau} \in \mathbb{R}^d$ denotes the vector of controls. The normal cone to a set $\mathcal{C}$ at the point $\mathbf{x} \in \mathcal{C}$ is given by $\mathcal{N}_\mathcal{C}(\mathbf{x})$ in the sense of convex analysis [24]. The normal cone representation in contact mechanics and friction problems are first treated in [21] and [1], respectively. In order to treat impacts in this framework constitutive laws will have to be introduced. In [17] a representation of Moreau's impact law in local contact coordinates has been derived, showing that:

$$0 \leq \Lambda_i \quad \perp \quad (\gamma_i^+ + \epsilon_i \, \gamma_i^-) \geq 0, \quad \forall i \in \mathcal{I}_\mathrm{S} \ . \tag{7}$$

Here ϵ_i denotes the restitution coefficient. These concepts can be extended to the tangential and normal impact by introducing following variables ξ_{s_i} and ξ_{u_i}, which vectorially are given by the following expressions if the Newton-like impact law is used:

$$\boldsymbol{\xi}_s = \boldsymbol{\gamma}_s^+ + \boldsymbol{\epsilon}_s \boldsymbol{\gamma}_s^- , \quad \boldsymbol{\xi}_u = \boldsymbol{\gamma}_u^+ + \boldsymbol{\epsilon}_u \boldsymbol{\gamma}_u^- \ . \tag{8}$$

The diagonal matrices $\boldsymbol{\epsilon}_s$ and $\boldsymbol{\epsilon}_u$ have as entries the tangential and normal restitution coefficients. The dynamics of a mechanical System $\mathcal{S}_v$ can be formulated on the measure-differential level:

$$\begin{aligned} &\mathbf{M}\, d\mathbf{u} - \mathbf{h}\, dt - \mathbf{W}_s d\boldsymbol{\Lambda}_s - \mathbf{W}_u d\boldsymbol{\Lambda}_u - \mathbf{B}\, d\boldsymbol{\Gamma} = \mathbf{0}, \\ &\xi_{u_i} \in -\mathcal{N}_{\mathcal{C}_{u_i}}(d\Lambda_{u_i}), \quad \forall i \in \mathcal{I}_S, \boldsymbol{\xi}_{s_i} \in -\mathcal{N}_{\mathcal{C}_{s_i}}(d\boldsymbol{\Lambda}_{s_i}), \quad \forall i \in \mathcal{I}_S \ . \end{aligned} \tag{9}$$

Here $d\boldsymbol{\Lambda}_s$ and $d\boldsymbol{\Lambda}_u$ are the differential measures of the tangential and normal contact forces, respectively. The vector $d\boldsymbol{\Gamma}$ denotes the differential measure of controls, which is unbounded, and can therefore induce impulsions. The differential measure of the generalized velocity is given by $d\mathbf{u}$.

2.1 The Linear Complementarity (LCP) Representation of MDI

If the normal cones in the force laws are finitely generated, then the determination of accelerations and forces can be represented in a linear complementarity form. This is the case, when the line of action of the friction forces are known, which is not the case for spatial friction. In the sequel between two types of frictional contacts is distinguished, because of their differing roles in the LCP. The set $\mathcal{I}_{NA}$ denotes the set of contacts at which the normal force, that induces the friction force is known a priori, whereas $\mathcal{I}_{NC}$ denote the set of contacts, where the value of the normal contact force depends on the friction force value at a given moment. The definition of following index sets are made:

$$\mathcal{C}_{u_i} = \{\lambda_{u_i} \,|\, \lambda_{u_i} \geq 0, \quad \forall i \in \mathcal{I}_G\}, \tag{10}$$
$$\mathcal{C}_{s_i} = \{\lambda_{sc_i} \,| \quad |\lambda_{sc_i}| \leq \mu_i \lambda_{u_i}, \quad \forall i \in \mathcal{I}_{NC} \subset \mathcal{I}_N\}, \tag{11}$$
$$\mathcal{C}_{a_i} = \{\lambda_{sa_i} \,| \quad |\lambda_{sa_i}| \leq a_{u_i}, \quad \forall i \in \mathcal{I}_{NA} \subset \mathcal{I}_N\}, \tag{12}$$

such that $\mathcal{I}_{NA} \bigcap \mathcal{I}_{NC} = \emptyset$, $\mathcal{I}_{NA} \bigcup \mathcal{I}_{NC} = \mathcal{I}_N$, $N(\mathcal{I}_{NA}) = w$, $N(\mathcal{I}_{NC}) = v$, $N(\mathcal{I}_N) = m$ and $N(\mathcal{I}_S) = k$. Here, the vectors $\boldsymbol{\lambda}_{sa}$ and $\boldsymbol{\lambda}_{sc}$, denote tangential contact forces of Tresca-type and Coulomb-type, respectively. The entity a_{u_i} denotes the apriori known sliding contact force at contact i. Further, for both type frictional contacts between sticking and sliding contacts will be distinguished, which gives rise to the definition of new index sets such that $\mathcal{I}_{NA} = {}_H\mathcal{I}_{NA} \bigcup {}_G\mathcal{I}_{NA}$ and $\mathcal{I}_{NC} = {}_H\mathcal{I}_{NC} \bigcup {}_G\mathcal{I}_{NC}$ are valid. The number of elements of these sets are related to each other by

$$N(\mathcal{I}_{NA}) = w = N({}_H\mathcal{I}_{NA}) + N({}_G\mathcal{I}_{NA}) = p + s, \tag{13}$$

$$N(\mathcal{I}_{NC}) = v = N({}_H\mathcal{I}_{NC}) + N({}_G\mathcal{I}_{NC}) = r + t \ . \tag{14}$$

Here subscript H refers to sticking and G refers to sliding. Based on this classification, the vector $\boldsymbol{\gamma}_{\mathrm{s}}$ is decomposed as $\boldsymbol{\gamma}_{\mathrm{s}} = \left[\boldsymbol{\gamma}_{\mathrm{sa}}\ \boldsymbol{\gamma}_{\mathrm{sc}}\right]^T$ and the related relative contact accelerations are given by:

$$\dot{\boldsymbol{\gamma}}_{\mathrm{sa}} = \mathbf{W}_{\mathrm{sa}}^T \ddot{\mathbf{q}} + \boldsymbol{\omega}_{\mathrm{sa}}, \quad \dot{\boldsymbol{\gamma}}_{\mathrm{sc}} = \mathbf{W}_{\mathrm{sc}}^T \ddot{\mathbf{q}} + \boldsymbol{\omega}_{\mathrm{sc}} \ . \tag{15}$$

The force $\boldsymbol{\lambda}_{\mathrm{sc}}$ can be decomposed into sliding contact forces ${}_{\mathrm{G}}\boldsymbol{\lambda}_{\mathrm{sc}}$ and sticking contact forces ${}_{\mathrm{H}}\boldsymbol{\lambda}_{\mathrm{sc}}$ at a given instant uniquely. The tangential contact forces and relative tangential contact velocities of contacts with Tresca-type friction can be decomposed analogously. The differential inclusion of a general non-autonomous mechanical system $\mathcal{S}_{\mathrm{a}}$ subject to planar friction (known line of action) and unilateral contact forces is stated as:

$$\begin{aligned}
&\mathbf{M}\,\dot{\mathbf{u}} - \bar{\mathbf{h}} - {}_{\mathrm{H}}\mathbf{W}_{\mathrm{sc}}\,{}_{\mathrm{H}}\boldsymbol{\lambda}_{\mathrm{sc}} - {}_{\mathrm{H}}\mathbf{W}_{\mathrm{sa}}\,{}_{\mathrm{H}}\boldsymbol{\lambda}_{\mathrm{sa}} - \mathbf{W}_{\mathrm{u}}\boldsymbol{\lambda}_{\mathrm{u}} - \mathbf{B}\,\boldsymbol{\tau} = \mathbf{0}, \\
&\lambda_{\mathrm{u}_i} \in -\mathrm{Upr}(\dot{\gamma}_{\mathrm{u}_i}),\ \forall i \in \mathcal{I}_{\mathrm{N}}, \\
&\lambda_{sc_j} \in -\mu_j\,\lambda_{\mathrm{u}_j}\mathrm{Sgn}\left(\dot{\gamma}_{sc_j}\right),\ \forall j \in {}_{\mathrm{H}}\mathcal{I}_{\mathrm{NC}}, \\
&\lambda_{sa_k} \in -a_{\mathrm{u}_k}\mathrm{Sgn}\left(\dot{\gamma}_{sa_k}\right),\ \forall k \in {}_{\mathrm{H}}\mathcal{I}_{\mathrm{NA}} \ .
\end{aligned} \tag{16}$$

Defining $\bar{\mathbf{h}} = \mathbf{h} + {}_{\mathrm{G}}\mathbf{W}_{\mathrm{sc}}\,{}_{\mathrm{G}}\boldsymbol{\lambda}_{\mathrm{sc}} + {}_{\mathrm{G}}\mathbf{W}_{\mathrm{sa}}\,{}_{\mathrm{G}}\boldsymbol{\lambda}_{\mathrm{sa}}$, enables the incorporation of all sliding contact forces in the vector $\bar{\mathbf{h}}$. In the sequel for ease of notation the entities ${}_{\mathrm{H}}\mathbf{W}_{\mathrm{sc}} \in \mathbb{R}^{n\times r}$, ${}_{\mathrm{H}}\boldsymbol{\lambda}_{\mathrm{sc}} \in \mathbb{R}^{r}$, ${}_{\mathrm{H}}\mathbf{W}_{\mathrm{sa}} \in \mathbb{R}^{n\times p}$, ${}_{\mathrm{H}}\boldsymbol{\lambda}_{\mathrm{sa}} \in \mathbb{R}^{p}$ will be represented by $\mathbf{W}_{\mathrm{sc}}$, $\boldsymbol{\lambda}_{\mathrm{sc}}$, $\mathbf{W}_{\mathrm{sa}}$, $\boldsymbol{\lambda}_{\mathrm{sa}}$, respectively. Here, the matrix $\mathbf{W}_{\mathrm{u}}$ has dimensions $\mathrm{n} \times \mathrm{m}$. In the sequel, the linear complementarity problem will be constructed. The set-valued signum type friction force characteristics can be decomposed into two unilateral force laws by introducing new variables as depicted in Fig. (1) and given below:

$$\dot{\boldsymbol{\gamma}}_{\mathrm{sc}} = \dot{\boldsymbol{\gamma}}_{\mathrm{rc}} - \dot{\boldsymbol{\gamma}}_{\mathrm{lc}}, \quad \boldsymbol{\lambda}_{\mathrm{lc}} = \boldsymbol{\mu}\,\boldsymbol{\lambda}_{\mathrm{u}} - \boldsymbol{\lambda}_{\mathrm{sc}}, \quad \boldsymbol{\lambda}_{\mathrm{rc}} = \boldsymbol{\mu}\,\boldsymbol{\lambda}_{\mathrm{u}} + \boldsymbol{\lambda}_{\mathrm{sc}}, \tag{17}$$

$$\dot{\boldsymbol{\gamma}}_{\mathrm{sa}} = \dot{\boldsymbol{\gamma}}_{\mathrm{ra}} - \dot{\boldsymbol{\gamma}}_{\mathrm{la}}, \quad \boldsymbol{\lambda}_{\mathrm{la}} = \mathbf{a} - \boldsymbol{\lambda}_{\mathrm{sa}}, \quad \boldsymbol{\lambda}_{\mathrm{ra}} = \mathbf{a} + \boldsymbol{\lambda}_{\mathrm{sa}}, \tag{18}$$

along with following nonnegativity and complementarity conditions:

$$\dot{\boldsymbol{\gamma}}_{\mathrm{u}} \geq \mathbf{0}, \quad \dot{\boldsymbol{\gamma}}_{\mathrm{rc}} \geq \mathbf{0}, \quad \dot{\boldsymbol{\gamma}}_{\mathrm{lc}} \geq \mathbf{0}, \quad \dot{\boldsymbol{\gamma}}_{\mathrm{ra}} \geq \mathbf{0}, \quad \dot{\boldsymbol{\gamma}}_{\mathrm{la}} \geq \mathbf{0}, \tag{19}$$

$$\boldsymbol{\lambda}_{\mathrm{u}} \geq \mathbf{0}, \quad \boldsymbol{\lambda}_{\mathrm{rc}} \geq \mathbf{0}, \quad \boldsymbol{\lambda}_{\mathrm{lc}} \geq \mathbf{0}, \quad \boldsymbol{\lambda}_{\mathrm{ra}} \geq \mathbf{0}, \quad \boldsymbol{\lambda}_{\mathrm{la}} \geq \mathbf{0}, \tag{20}$$

$$\dot{\boldsymbol{\gamma}}_{\mathrm{u}}\boldsymbol{\lambda}_{\mathrm{u}} = \mathbf{0},\ \dot{\boldsymbol{\gamma}}_{\mathrm{rc}}\boldsymbol{\lambda}_{\mathrm{rc}} = \mathbf{0},\ \dot{\boldsymbol{\gamma}}_{\mathrm{lc}}\boldsymbol{\lambda}_{\mathrm{lc}} = \mathbf{0},\ \dot{\boldsymbol{\gamma}}_{\mathrm{ra}}\boldsymbol{\lambda}_{\mathrm{ra}} = \mathbf{0},\ \dot{\boldsymbol{\gamma}}_{\mathrm{la}}\boldsymbol{\lambda}_{\mathrm{la}} = \mathbf{0} \ . \tag{21}$$

Here $\boldsymbol{\mu} \in \mathbb{R}^{r\times r}$ is a diagonal matrix with friction coefficients. The vector $\mathbf{a} \in \mathbb{R}^{\mathrm{p}}$ denotes the normal force vector for frictional contacts of Tresca-type. Further $\boldsymbol{\lambda}_{\mathrm{rc}}, \boldsymbol{\lambda}_{\mathrm{lc}} \in \mathbb{R}^{r}$; $\boldsymbol{\lambda}_{\mathrm{ra}}, \boldsymbol{\lambda}_{\mathrm{la}} \in \mathbb{R}^{\mathrm{p}}$; $\boldsymbol{\gamma}_{\mathrm{rc}}, \boldsymbol{\gamma}_{\mathrm{lc}} \in \mathbb{R}^{r}$ and $\boldsymbol{\gamma}_{\mathrm{ra}}, \boldsymbol{\gamma}_{\mathrm{la}} \in \mathbb{R}^{\mathrm{p}}$ and are related entities to sticking contacts. The generalized accelerations of the system S_{a} are given by:

$$\dot{\mathbf{u}} = \mathbf{M}^{-1}\left(\bar{\mathbf{h}}(\mathbf{q},\mathbf{u}) + \mathbf{W}_{\mathrm{sc}}(\mathbf{q})\boldsymbol{\lambda}_{\mathrm{sc}} + \mathbf{W}_{\mathrm{sa}}(\mathbf{q})\boldsymbol{\lambda}_{\mathrm{sa}} + \mathbf{W}_{\mathrm{u}}(\mathbf{q})\boldsymbol{\lambda}_{\mathrm{u}} + \mathbf{B}(\mathbf{q})\,\boldsymbol{\tau}\right) \ . \tag{22}$$

Insertion of (22) in the expressions for relative contact accelerations given in equations (3) and (15) reveals following set of equations:

$$\dot{\boldsymbol{\gamma}}_{\mathrm{u}} = \mathbf{W}_{\mathrm{u}}^T\mathbf{M}^{-1}\left(\bar{\mathbf{h}} + \mathbf{W}_{\mathrm{sc}}\boldsymbol{\lambda}_{\mathrm{sc}} + \mathbf{W}_{\mathrm{sa}}\boldsymbol{\lambda}_{\mathrm{sa}} + \mathbf{W}_{\mathrm{u}}\boldsymbol{\lambda}_{\mathrm{u}} + \mathbf{B}\,\boldsymbol{\tau}\right) + \boldsymbol{\omega}_{\mathrm{u}}, \tag{23}$$

$$\dot{\gamma}_{\rm sc} = \mathbf{W}_{\rm sc}^T \mathbf{M}^{-1} \left(\bar{\mathbf{h}} + \mathbf{W}_{\rm sc}\boldsymbol{\lambda}_{\rm sc} + \mathbf{W}_{\rm sa}\boldsymbol{\lambda}_{\rm sa} + \mathbf{W}_{\rm u}\boldsymbol{\lambda}_{\rm u} + \mathbf{B}\,\boldsymbol{\tau}\right) + \boldsymbol{\omega}_{\rm sc}, \quad (24)$$

$$\dot{\gamma}_{\rm sa} = \mathbf{W}_{\rm sa}^T \mathbf{M}^{-1} \left(\bar{\mathbf{h}} + \mathbf{W}_{\rm sc}\boldsymbol{\lambda}_{\rm sc} + \mathbf{W}_{\rm sa}\boldsymbol{\lambda}_{\rm sa} + \mathbf{W}_{\rm u}\boldsymbol{\lambda}_{\rm u} + \mathbf{B}\,\boldsymbol{\tau}\right) + \boldsymbol{\omega}_{\rm sa}, \quad (25)$$

which can be arranged in the form of a linear complementarity problem [10], [22]:

$$\mathbf{y} = \mathbf{A}\,\mathbf{x} + \mathbf{b}, \quad \mathbf{y} \succeq \mathbf{0}, \quad \mathbf{x} \succeq \mathbf{0}, \quad \mathbf{x}\,\mathbf{y} = \mathbf{0}\ . \quad (26)$$

Where the complementarity vectors $\mathbf{x}$ and $\mathbf{y}$ are identified as:

$$\mathbf{x} = \left[\boldsymbol{\lambda}_{\rm u}\ \boldsymbol{\lambda}_{\rm rc}\ \boldsymbol{\lambda}_{\rm ra}\ \dot{\gamma}_{\rm lc}\ \dot{\gamma}_{\rm la}\right]^T, \mathbf{y} = \left[\dot{\gamma}_{\rm u}\ \dot{\gamma}_{\rm rc}\ \dot{\gamma}_{\rm ra}\ \boldsymbol{\lambda}_{\rm lc}\ \boldsymbol{\lambda}_{\rm la}\right]^T, \quad (27)$$

respectively. The matrix $A \in \mathbb{R}^{(m+2p+2r)\times(m+2p+2r)}$ and vector $\mathbf{b} \in \mathbb{R}^{(m+2p+2r)}$ are given by:

$$\mathbf{A} = \begin{bmatrix} \mathbf{W}_{\rm u}^T\mathbf{M}^{-1}(\mathbf{W}_{\rm u} - \mathbf{W}_{\rm sc}\boldsymbol{\mu}) & \mathbf{W}_{\rm u}^T\mathbf{M}^{-1}\mathbf{W}_{\rm sc} & \mathbf{W}_{\rm u}^T\mathbf{M}^{-1}\mathbf{W}_{\rm sa} & \mathbf{0}_{\rm m\times r} & \mathbf{0}_{\rm m\times p} \\ \mathbf{W}_{\rm sc}^T\mathbf{M}^{-1}(\mathbf{W}_{\rm u} - \mathbf{W}_{\rm sc}\boldsymbol{\mu}) & \mathbf{W}_{\rm sc}^T\mathbf{M}^{-1}\mathbf{W}_{\rm sc} & \mathbf{W}_{\rm sc}^T\mathbf{M}^{-1}\mathbf{W}_{\rm sa} & \mathbf{I}_{\rm r\times r} & \mathbf{0}_{\rm r\times p} \\ \mathbf{W}_{\rm sa}^T\mathbf{M}^{-1}(\mathbf{W}_{\rm u} - \mathbf{W}_{\rm sc}\boldsymbol{\mu}) & \mathbf{W}_{\rm sa}^T\mathbf{M}^{-1}\mathbf{W}_{\rm sc} & \mathbf{W}_{\rm sa}^T\mathbf{M}^{-1}\mathbf{W}_{\rm sa} & \mathbf{0}_{\rm p\times r} & \mathbf{I}_{\rm p\times p} \\ 2\boldsymbol{\mu} & -\mathbf{I}_{\rm r\times r} & \mathbf{0}_{\rm r\times p} & \mathbf{0}_{\rm r\times r} & \mathbf{0}_{\rm r\times p} \\ \mathbf{0}_{\rm p\times m} & \mathbf{0}_{\rm p\times r} & -\mathbf{I}_{\rm p\times p} & \mathbf{0}_{\rm p\times r} & \mathbf{0}_{\rm p\times p} \end{bmatrix} \quad (28)$$

$$\mathbf{b} = \left[\mathbf{W}_{\rm u}^T\,\mathbf{f}_{\ddot{\rm q}} + \boldsymbol{\omega}_{\rm u} \quad \mathbf{W}_{\rm sc}^T\,\mathbf{f}_{\ddot{\rm q}} + \boldsymbol{\omega}_{\rm sc} \quad \mathbf{W}_{\rm sa}^T\,\mathbf{f}_{\ddot{\rm q}} + \boldsymbol{\omega}_{\rm sa} \quad \mathbf{0}_{r\times 1} \quad 2\mathbf{a}\right]^T, \quad (29)$$

where $\mathbf{f}_{\ddot{\rm q}}$ is given by $\mathbf{f}_{\ddot{\rm q}} = \mathbf{M}^{-1}(\bar{\mathbf{h}} + \mathbf{B}\,\boldsymbol{\tau})$ and $\mathbf{I}$ is an identity matrix of appropriate size. Making use of the formulation given in (7), these concepts can be extended

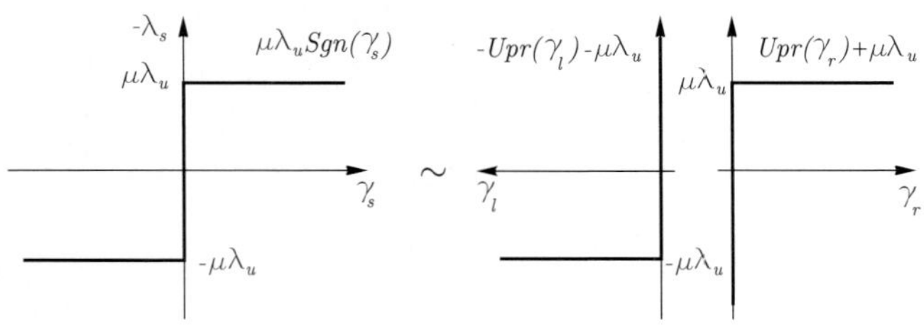

Fig. 1. Decomposition of the set-valued Sgn relation into two Unilateral Primitives

to the tangential and normal impact by introducing following variables $\xi_{{\rm sc}_i}$, $\xi_{{\rm sa}_i}$ and $\xi_{{\rm u}_i}$, which vectorially are given by the following expressions if the Newton-like impact law is used:

$$\boldsymbol{\xi}_{\rm sa} = \gamma_{\rm sa}^+, \quad \boldsymbol{\xi}_{\rm sc} = \gamma_{\rm sc}^+ + \boldsymbol{\epsilon}_{\rm sc}\gamma_{\rm sc}^-, \quad \boldsymbol{\xi}_{\rm u} = \gamma_{\rm u}^+ + \boldsymbol{\epsilon}_{\rm u}\,\gamma_{\rm u}^-\ . \quad (30)$$

The matrices $\boldsymbol{\epsilon}_{\rm u}$ and $\boldsymbol{\epsilon}_{\rm sc}$ are diagonal matrices with normal and tangential restitution coefficients, respectively. The tangential impact in contacts with Coulomb type friction is induced by the dependence on the normal impactive force at the relevant contact. The dynamics of a mechanical System $\mathcal{S}_{\mathrm{v}}$ can be formulated on the measure-differential level:

$$\begin{aligned}
&\mathbf{M}\,d\mathbf{u} - \mathbf{h}\,dt - \mathbf{W}_{\mathrm{sa}}d\boldsymbol{\Lambda}_{\mathrm{sa}} - \mathbf{W}_{\mathrm{sc}}d\boldsymbol{\Lambda}_{\mathrm{sc}} - \mathbf{W}_{\mathrm{u}}d\boldsymbol{\Lambda}_{\mathrm{u}} - \mathbf{B}\,d\boldsymbol{\Gamma} = \mathbf{0}, \qquad (31)\\
&d\Lambda_{\mathrm{u}_i} \in -\mathrm{Upr}(\xi_{\mathrm{u}_i}), \quad \forall i \in \mathcal{I}_{\mathrm{S}},\\
&d\Lambda_{\mathrm{sa}_i} \in -dA_i\,\mathrm{Sgn}(\xi_{\mathrm{sa}_i}), \quad \forall i \in \mathcal{I}_{\mathrm{SA}},\\
&d\Lambda_{\mathrm{sc}_i} \in -\mu_i\,d\Lambda_{\mathrm{u}_i}\,\mathrm{Sgn}(\xi_{\mathrm{sc}_i}), \quad \forall i \in \mathcal{I}_{\mathrm{SC}}\ .
\end{aligned}$$

such that $\mathcal{I}_{\mathrm{SA}} \bigcap \mathcal{I}_{\mathrm{SC}} = \emptyset$, $\mathcal{I}_{\mathrm{SA}} \bigcup \mathcal{I}_{\mathrm{SC}} = \mathcal{I}_{\mathrm{S}}$ and $N(\mathcal{I}_{\mathrm{SA}}) = l$, $N(\mathcal{I}_{\mathrm{SC}}) = z$. As a consequence of the changing force laws in (31), the dimensionality of matrices also change. In equation (31) , the matrix $\mathbf{W}_{\mathrm{sa}} \in \mathbb{R}^{\mathrm{n}\times \mathrm{l}}$ is given by $\mathbf{W}_{\mathrm{sa}} = \mathrm{col}\{_{\mathrm{H}}\mathbf{W}_{\mathrm{sa}}\} \bigcup \mathrm{col}\{_{\mathrm{G}}\mathbf{W}_{\mathrm{sa}}\}$, $\mathrm{col}\{\cdot\}$ denotes the column set of the matrix in argument. Analogously, the matrix $\mathbf{W}_{\mathrm{sc}} \in \mathbb{R}^{\mathrm{n}\times \mathrm{z}}$ is given by $\mathbf{W}_{\mathrm{sc}} = \mathrm{col}\{_{\mathrm{H}}\mathbf{W}_{\mathrm{sc}}\} \bigcup \mathrm{col}\{_{\mathrm{G}}\mathbf{W}_{\mathrm{sc}}\}$. A linear complementarity problem on the measure-differential level can be formulated in the form of (26). The set-valued signum relations in the MDI representation (31) will again be decomposed in analogy to Fig. (1). After introduction of slack variables, the complementarity vectors $\mathbf{x} \in \mathbb{R}^{\mathrm{k}+2\mathrm{l}+2\mathrm{z}}$ and $\mathbf{y} \in \mathbb{R}^{\mathrm{k}+2\mathrm{l}+2\mathrm{z}}$ are identified as:

$$\mathbf{x} = \left[d\boldsymbol{\Lambda}_{\mathrm{u}}\; d\boldsymbol{\Lambda}_{\mathrm{rc}}\; d\boldsymbol{\Lambda}_{\mathrm{ra}}\; \boldsymbol{\xi}_{\mathrm{lc}}\; \boldsymbol{\xi}_{\mathrm{la}}\right]^T, \mathbf{y} = \left[\boldsymbol{\xi}_{\mathrm{u}}\; \boldsymbol{\xi}_{\mathrm{rc}}\; \boldsymbol{\xi}_{\mathrm{ra}}\; d\boldsymbol{\Lambda}_{\mathrm{lc}}\; d\boldsymbol{\Lambda}_{\mathrm{la}}\right]^T, \qquad (32)$$

respectively. The vector $\mathbf{b} \in \mathbb{R}^{\mathrm{k}+2\mathrm{l}+2\mathrm{z}}$ of the linear complementarity problem is given by:

$$\mathbf{b} = \begin{bmatrix} (\mathbf{I} + \boldsymbol{\epsilon}_{\mathrm{u}})\mathbf{W}_{\mathrm{u}}^T\mathbf{u}^- + \mathbf{W}_{\mathrm{u}}^T\mathbf{M}^{-1}(\mathbf{h}\,dt + \mathbf{B}\,d\boldsymbol{\Gamma})\\ (\mathbf{I} + \boldsymbol{\epsilon}_{\mathrm{sc}})\mathbf{W}_{\mathrm{sc}}^T\mathbf{u}^- + \mathbf{W}_{\mathrm{sc}}^T\mathbf{M}^{-1}(\mathbf{h}\,dt + \mathbf{B}\,d\boldsymbol{\Gamma})\\ \mathbf{W}_{\mathrm{sa}}^T\mathbf{u}^- + \mathbf{W}_{\mathrm{sa}}^T\mathbf{M}^{-1}(\mathbf{h}\,dt + \mathbf{B}\,d\boldsymbol{\Gamma})\\ \mathbf{0}_{z\times 1}\\ 2\mathbf{dA} \end{bmatrix}. \qquad (33)$$

The matrix $\mathbf{A}$ remains the same in the structure as in (28) but the dimensionality of the associated elements vary since the contact force laws are modified as in equation (31). As illustrated, for contacts which are closed in normal direction, the MDI representation removes the distinction between sliding and sticking, and considers all contacts closed on position level, which reduce the burden of management of index sets.

2.2 Statement of the Hybrid Optimal Control Problem

The non-smooth optimal control problem subject to a mechanical dynamical system described as a measure-differential inclusion can be stated as follows:

$$\begin{aligned}
\min \mathcal{J}(\boldsymbol{\tau}, d\boldsymbol{\Gamma}, t_{\mathrm{f}}) &= \Phi(\mathbf{q}(t_{\mathrm{f}}), \mathbf{u}^+(t_{\mathrm{f}}), t_{\mathrm{f}}) + \int_{t_0}^{t_{\mathrm{f}}} g(\mathbf{u}^+, \mathbf{q}, \boldsymbol{\tau})\,dt, \qquad (34)\\
d\mathbf{u} &= \mathbf{f}(\mathbf{u}^+, \mathbf{q}, \boldsymbol{\tau}, t)\,dt + \mathbf{p}_{\Lambda}(\mathbf{q}, t)\,d\boldsymbol{\Lambda} + \mathbf{p}_{\Gamma}(\mathbf{q}, t)\,d\boldsymbol{\Gamma},\\
(d\boldsymbol{\Lambda}, d\boldsymbol{\Gamma}) &\in \boldsymbol{\Upsilon}(d\boldsymbol{\Lambda}, d\boldsymbol{\Gamma}, \mathbf{q}, \mathbf{u}^+),
\end{aligned}$$

$$\boldsymbol{\Pi}(\mathbf{u}^+, \mathbf{q}, \boldsymbol{\tau}) \leq \mathbf{0},$$
$$\boldsymbol{\Psi}(\mathbf{q}(t_0), \mathbf{u}^-(t_0), \mathbf{q}(t_\mathrm{f}), \mathbf{u}^+(t_\mathrm{f})) = \mathbf{0},$$
$$t_0 \quad \text{fixed}, t_\mathrm{f} \quad \text{free}, \quad t \in [t_0, t_\mathrm{f}],$$

with absolutely continuous positions $\mathbf{q} \in \mathbb{R}^\mathrm{n}$, right continuous bounded variation (RCBV) generalized velocities $\mathbf{u} \in \mathbb{R}^\mathrm{n}$, control variables $\boldsymbol{\tau} \in \mathbb{R}^\mathrm{m}$, unilateral force differential measures $d\boldsymbol{\Lambda} \in \mathbb{R}^\mathrm{p}$, impulsive and set-valued control differential measures $d\boldsymbol{\Gamma} \in \mathbb{R}^r$. Further, the Lebesgue measurable system dynamics is given by $\mathbf{f} : \mathbb{R}^\mathrm{n} \times \mathbb{R}^\mathrm{n} \times \mathbb{R}^\mathrm{m} \times \mathbb{R} \rightarrow \mathbb{R}^\mathrm{n}$. The set-valued variational constraints on the measure variables $\boldsymbol{\Upsilon} : \mathbb{R}^\mathrm{p} \times \mathbb{R}^r \times \mathbb{R}^\mathrm{n} \times \mathbb{R}^\mathrm{n} \rightarrow \mathbb{R}^k$, influence matrix of contact force differential measures $\mathbf{p}_\Lambda \in \mathbb{R}^{\mathrm{n}\times\mathrm{p}}$, influence matrix of control differential measures $\mathbf{p}_\Gamma \in \mathbb{R}^{n\times r}$, state and control constraints $\boldsymbol{\Pi} : \mathbb{R}^\mathrm{n} \times \mathbb{R}^\mathrm{n} \times \mathbb{R}^\mathrm{m} \rightarrow \mathbb{R}^\mathrm{l}$ and boundary constraints $\boldsymbol{\Psi} : \mathbb{R}^\mathrm{n} \times \mathbb{R}^\mathrm{n} \times \mathbb{R}^\mathrm{n} \times \mathbb{R}^\mathrm{n} \rightarrow \mathbb{R}^\mathrm{q}$ are incorporated in the optimal control problem. The end state cost $\Phi : \mathbb{R}^\mathrm{n} \times \mathbb{R}^\mathrm{n} \times \mathbb{R} \rightarrow \mathbb{R}$, integrand of the cost functional $g : \mathbb{R}^\mathrm{n} \times \mathbb{R}^\mathrm{n} \times \mathbb{R}^\mathrm{m} \rightarrow \mathbb{R}$ constitute the goal function to be minimized.

3 Examples

3.1 Example: Two-DOF Underactuated Planar Manipulator with Impactively Blockable DOF

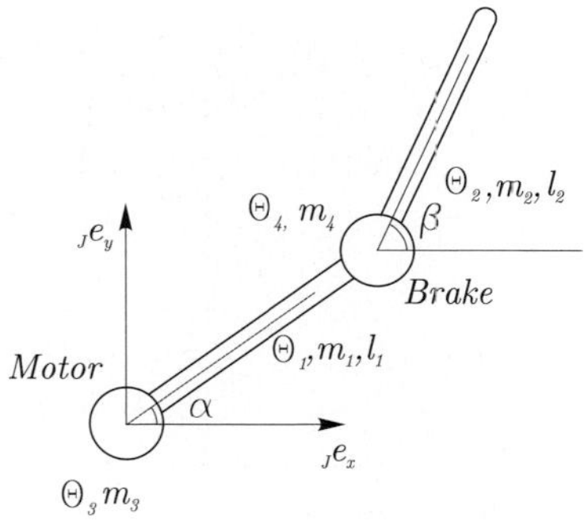

Fig. 2. The 2-DOF planar manipulator with 1 blockable DOF

The first application will deal with underactuated manipulators with blockable degrees of freedom (DOF). The example mechanical system can be seen in Fig. (2) and is treated in detail in [28]. The robot has two rotational degrees of freedom denoted by α and β. The DOF α is controlled continuously in a single valued manner, whereas DOF β can be blocked meaning that any given position the relative angular velocity $\dot{\beta} - \dot{\alpha}$ can be reduced to zero immediately. In the presented maneuver in Fig. (3), it is supposed to start at $\alpha_0 = 0\,\mathrm{rad}$ and $\beta_0 = 0\,\mathrm{rad}$ from standstill, and to reach the final position $\alpha_\mathrm{f} = \pi\,\mathrm{rad}$ and $\beta_\mathrm{f} = \pi\,\mathrm{rad}$, time-optimally. The DOF β is blocked at times $t_1 = 0\,\mathrm{s}$, $t_3 = 1.07\,\mathrm{s}$, $t_5 = 2.39\,\mathrm{s}$

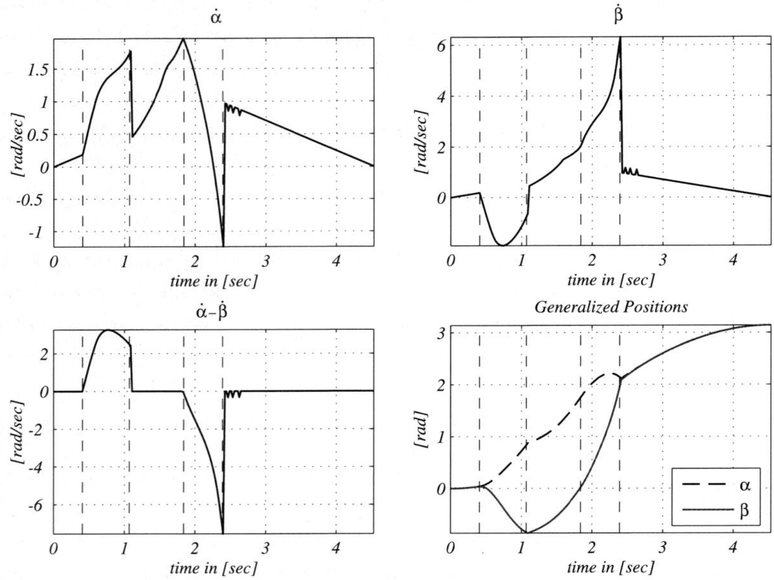

Fig. 3. The generalized and relative velocities, generalized positions of the optimal maneuver

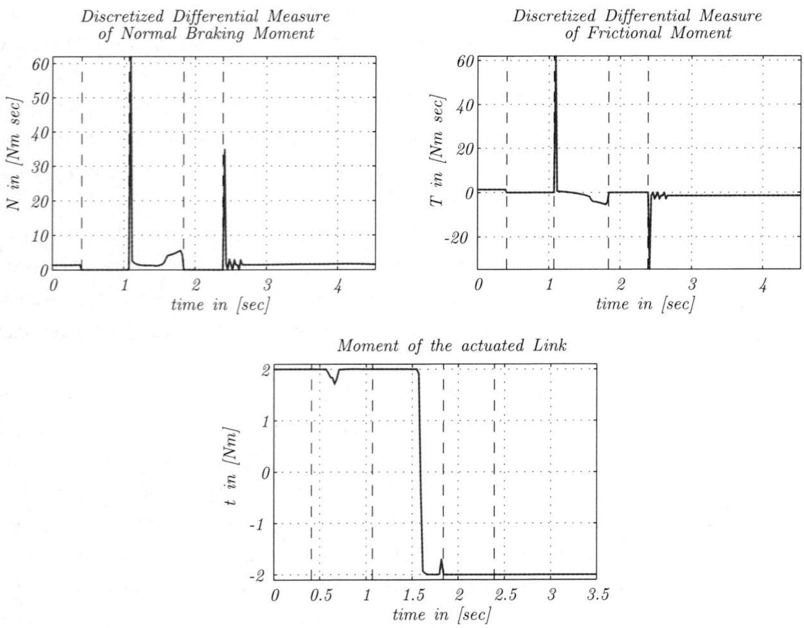

Fig. 4. The differential measures of the controls

and is released at times $t_2 = 0.41\,\text{s}$, $t_4 = 1.84\,\text{s}$, whereas the total maneuver takes 4.53 s. When the link is blocked, the whole system possesses one mechanical DOF, whereas when released it has two DOF. The transitions at times t_3, t_5 are impactive as can be seen in Fig. (3). The control history has a time-optimal bang-bang character as can be seen in Fig. (4). The blocking control is set-valued because its value is within a set in a phase of blocking, unbounded; because it has to bring the link immediately to zero relative velocity and therefore impulsive as can be seen in Fig. (4).

3.2 Example: Three-Wheeled Differential Drive Robot

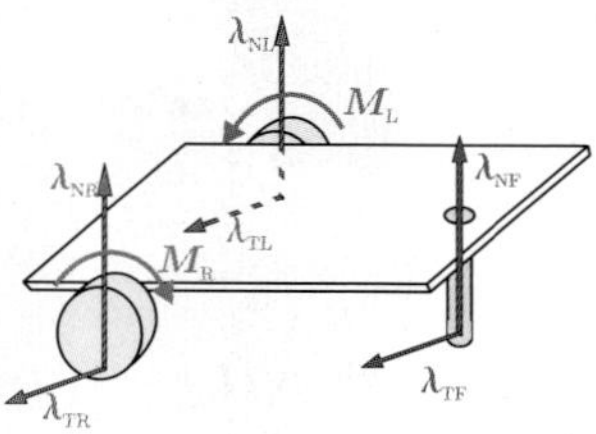

Fig. 5. Contact forces and motor moments on the simplified model

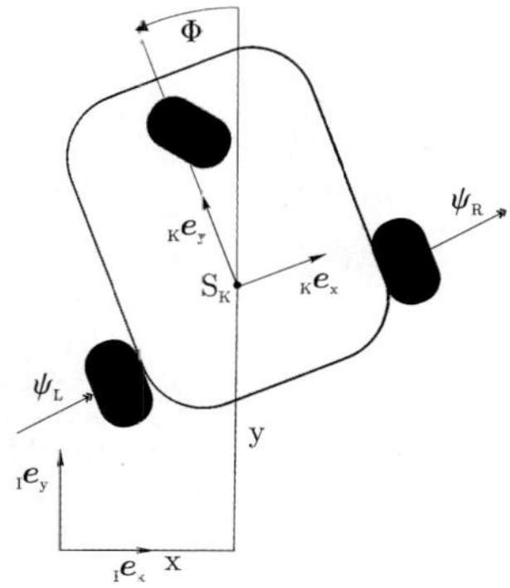

Fig. 6. The generalized coordinates of the wheeled robot

The second application will be about a three-wheeled robot with stick-slip transitions at the wheel contacts. The differential-drive robot is a three-wheeled actuated robot of which the rear wheels are actuated and controlled separately contrary to the front wheel which is neither actuated nor steered. The properties of the optimal control of this nonholonomically constrained system are studied in detail in [27] and [29]. A rigid-body mechanical model is used, in which the friction between wheels and ground is modeled as isotropic spatial Coulomb friction. The non-steered unactuated front wheel is replaced by a stick as a simplification, removing two degrees of freedom (DOF) to be modeled. The rotational inertia of the total actuation consisting of the components of motor rotors and transmissions are added to the rotational inertias of the wheels. Under the given assumptions there are five mechanical DOF necessary in order to model the mechanical system depicted in Figs. (5) and (6). The following set of generalized coordinates and velocities are used to describe the system:

$$\mathbf{q}^T = \left[x,\, y,\, \phi,\, \psi_\mathrm{L},\, \psi_\mathrm{R}\right], \quad \dot{\mathbf{q}}^T = \left[\dot{x},\, \dot{y},\, \dot{\phi},\, \dot{\psi}_\mathrm{L},\, \dot{\psi}_\mathrm{R}\right] \; . \tag{35}$$

These are the planar translational coordinates of center of mass (CM) of the chassis x and y as well as the planar orientation of the chassis ϕ, the angular

Table 1. Relation of different modes to the contact state and relative contact velocities

Modes	${}_{K}\gamma_{Rx}$	${}_{K}\gamma_{Ry}$	${}_{K}\gamma_{Lx}$	${}_{K}\gamma_{Ly}$
5-DOF mode	$\neq 0$	$\neq 0$	$\neq 0$	$\neq 0$
3R-DOF mode	$= 0$	$\neq 0$	$= 0$	$= 0$
3L-DOF mode	$= 0$	$= 0$	$= 0$	$\neq 0$
2-DOF mode	$= 0$	$= 0$	$= 0$	$= 0$

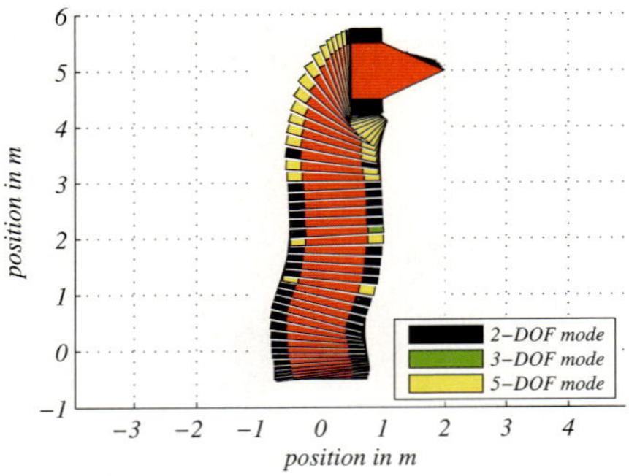

Fig. 7. Maneuver A: Number of DOF during the maneuver

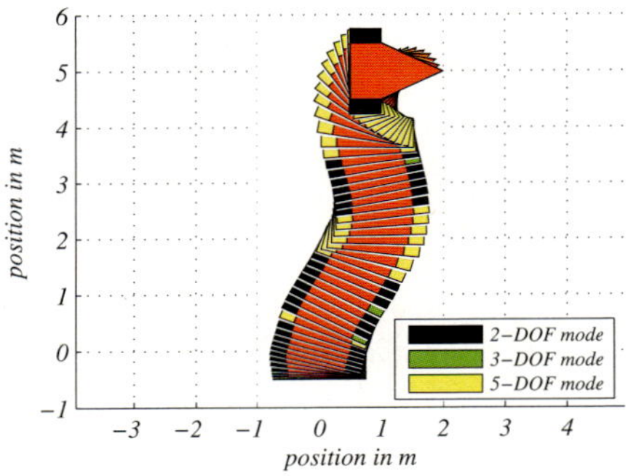

Fig. 8. Maneuver B: Number of DOF during the maneuver

positions ψ_{R} and ψ_{L} of the wheels with respect to the chassis frame. There are two coordinate systems, namely, the inertial coordinate system (subscript $_{\mathrm{I}}$), a body attached coordinate system for chassis (subscript $_{\mathrm{K}}$). The K-system of the chassis has its origin in the CM S_{K}. When the axial slip constraints and rolling constraints are fulfilled, the non-actuated mechanical system possesses two DOF. If both wheels slide it is a mechanical system with five DOF. In the three-DOF mode one wheel contact sticks and the other wheel slides, meaning that the axial slip constraints are fulfilled but one wheel does not fulfill the rolling condition. If the actuated mechanical system is considered then one observes that the system is fully actuated when it moves in the two-DOF mode. In the other modes it is an underactuated system with less actuators than mechanical degrees of freedom. The transitions among all four operating modes are possible. The modes are classified according to relative contact velocities at the wheel contacts as represented in the chassis fixed frame in table 1. In Fig. (9) the contact velocities and forces for maneuver A are shown. In maneuver A the task is to reach the following end-point control-effort optimally. The sum of squares of the actuating torques is being minimised. The desired end state to be reached is

$$\big(x_{\mathrm{f}},\, y_{\mathrm{f}},\, \phi_{\mathrm{f}},\, \psi_{\mathrm{Lf}},\, \psi_{\mathrm{Rf}}\big) = \big(2,\, 5,\, -\tfrac{\pi}{2},\, \text{free},\, \text{free}\big) \ . \qquad (36)$$

The robot accomplishes this task in 4.05 seconds. In Maneuver B the robot is supposed to reach the same final end-state time-optimally. Maneuver B is characterized by a high dynamical activity in the orientation of the chassis.

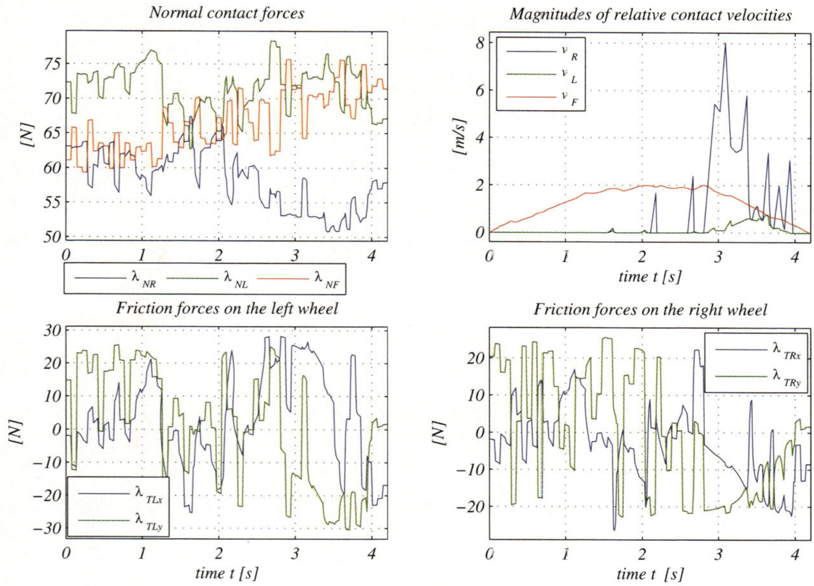

Fig. 9. Maneuver A: Contact forces and contact relative velocities

The robot accomplishes this task in 4.03 seconds. These maneuvers are depicted in Figs. (7) and (8).

4 Discussion and Conclusion

The necessity to represent multibody systems as MDI's emanates from several facts. The optimal control of hybrid rigidbody mechanical systems benefits from several aspects of the MDI approach, which are summarized below:

a. The index sets that are used to take account of the behaviour of contacts on different levels such as position, velocity and acceleration for stick-slip transitions etc. is not managable for large systems with many contacts. The index-set reduction can be seen in going from the acceleration level representation in (6) to the measure-differential MDI representation in (9).

b. The impacts, that may occur with or without collisions e.g. Painleve Paradox, velocity jumps due to C^0 constraints are a strong incentive to describe the mechanical systems as MDI.

c. Systems which are zeno (e.g jumping ball on the ground) are problematic for event-driven schemes whereas MDI approach can handle them properly and realistically.

d. The hybrid optimal control requires the consideration of an uncommon concept of control, namely, controls of unbounded, impulsive and set-valued type for non-autonomous impulsive transitions, such as sudden blocking of DOF, which can in a natural way be added to the MDI structure.

e. As a novel property, the location and time of phase transitions where the system changes DOF is not prespecified but is determined as an outcome of the optimization. Though the underlying system might undergo structure-variant phase changes such as impactive phase transitions a mixed integer approach is not necessary.

f. Numerical methods can be deviced that calculate the costate dynamics which itself is described as a MDI. The costate dynamics in the sense of optimal control is discontinuous and nonsmooth as well, as presented in [28] in the sense of optimal control.

References

1. Alart, P., Curnier, A.: A mixed formulation of frictional contact problems prone to newton-like solution methods. Computer Methods in Applied Mechanics and Engineering, **92** (1991) 353–375
2. Aubin, J.-P. , Lygeros, J., Quincampoix, M., Sastry, S.: Impulse Differential Inclusions: A Viability Approach to Hybrid Systems, IEEE Trans. on Automatic Control, **47** (2002) 2–20
3. Bemporad, A., Morari, M.: Control of systems integrating logic, dynamic, and constraints. Automatica. **35** (1999) 407–427
4. Bengea, S. C., DeCarlo, R. A.: Optimal Control of switching systems. Automatica. **41** (2005) 11–27
5. Borelli, F., Baotić, M., Bemporad, A., Morari, M., Dynamic programming for constrained optimal control of discrete-time linear hybrid systems, Automatica. **41** (2005) 1709–1721
6. Branicky, M. S., Borkar V. S., Mitter S. M.: A unified framework for hybrid control: Model and optimal theory, IEEE Transactions on Automatic Control, **43** (1998) 31–45
7. Brogliato, B.: Non-smooth Impact Mechanics, Lecture Notes in Control and Information Sciences Springer Verlag (1996)
8. Brogliato, B., Daniilidis, A., Lemaréchal, C., Acary, V.: On the equivalence between complementarity systems, projected systems and differential inclusions. Systems and Control Letters. **55** (2006) 45–51
9. Clarke, F. H.: Optimization and Nonsmooth Analysis. SIAM Classics in Applied Mathematics Wiley New York (1983)
10. Cottle, R. W., Pang, J.-S., Stone, R. E.: The Linear Complementarity Problem. Academic Press Boston (1992)
11. De Schutter, B., Heemels, W. P. M. H, Bemporad, A.: On the equivalence of linear complementarity problems. Operations Research Letters. **30** (2002) 211–222
12. Ferrari-Trecate, G., Cuzzola, F. A., Mignone, D., Morari, M.: Analysis of discrete-time piecewise affine and hybrid systems. Automatica. **38** (2002) 2139–2146
13. Heemels, W. P. M. H, De Schutter, B., Bemporad, A.: Equivalence of hybrid dynamical models. Automatica. **37** (2001) 1085–1091
14. Glocker, Ch.: Set-Valued Force Laws, Dynamics of Non-Smooth Systems. Lecture Notes in Applied Mechanics. **1** (2001) Springer-Verlag Berlin
15. Glocker, Ch.: On Frictionless Impact Models in Rigid-Body Systems, Phil. Trans. Royal Soc. Lond. **A359** (2001) 2385–2404
16. Glocker, Ch., Pfeiffer, F.: Multiple Impacts with Friction in Rigid Multibody Systems. Nonlinear Dynamics **7** 471–497 (1995)

17. Glocker, Ch.: The Geometry of Newtonian Impacts with Global Dissipation Index for Moving Sets. Proc. of the Int. Conf. on Nonsmooth/ Nonconvex Mechanics, Thessaloniki (2002) 283–290
18. Miller, B. M., Bentsman, J.: Optimal control problems in hybrid systems with active singularities. Nonlinear Analysis. **65** (2006) 999–1017
19. Moreau, J. J.: Quadratic programming in mechanics: Dynamics of one-sided constraints. SIAM Journal of Control. **4** (1966) 153–158
20. Moreau, J. J.: Bounded Variations in time. In: Topics in Non-smooth Mechanics, Edts: J. J. Moreau, P. D. Panagiotopoulos, G. Strang: 1–74 (1988), Birkhäuser, Basel
21. Moreau, J. J.: Unilateral Contact and Dry Friction in Finite Freedom Dynamics. Non-smooth Mechanics and Applications, CISM Courses and Lectures. **302** Springer Verlag Wien (1988)
22. Murty, K. G.: Linear Complementarity, Linear and Nonlinear Programming Helderman-Verlag (1988)
23. Potočnik, B., Bemporad, A., Torrisi, F. D., Mušič, G., Zupančič, B.: Hybrid Modelling and optimal control of a multiproduct Batch plant. Control Engineering Practice, **12** 1127–1137 (2004)
24. Rockafellar R. T.: Convex Analysis, Princeton Landmarks in Mathematics. Princeton University Press. (1970)
25. Shahid Shaikh, M., Caines P. E., On the optimal control of hybrid systems: Optimization of trajectories, switching times, and location schedules, In O. Maler, A. Pnueli (Eds.), HSCC (2003), Lecture Notes in Computer Science. Berlin Springer Vol: 2623 (2003) 466–481
26. Yunt, K., Glocker, Ch.: Trajectory Optimization of Hybrid Mechanical Systems using SUMT. IEEE Proc. of Advanced Motion Control. Istanbul 665–671 (2006)
27. Yunt, K., Glocker, Ch.: Time-Optimal Trajectories of a Differential-Drive Robot, Proceedings of the 2005 ENOC Conference, Eindhoven, Netherlands.
28. Yunt, K., Glocker, Ch.: A combined continuation and penalty method for the determination of optimal hybrid mechanical trajectories. IUTAM Symposium on Dynamics and Control of Nonlinear Systems with Uncertainty (2006), Nanjing, China, Springer (to appear)
29. Yunt, K.: Trajectory Optimization of structure-variant mechanical Systems, Proc. on of Int. Workshop on Variable Structure Systems Alghero Italy (2006) 298–303

The Concept of Deadlock and Livelock in Hybrid Control Systems

Alessandro Abate[1], Alessandro D'Innocenzo[2], Giordano Pola[2,3], Maria Domenica Di Benedetto[2], and Shankar Sastry[1]

[1] Department of Electrical Engineering and Computer Sciences, University of California, at Berkeley - Berkeley, USA
{aabate,sastry}@eecs.berkeley.edu
[2] Department of Electrical Engineering and Computer Science, Center of Excellence DEWS, University of L'Aquila - L'Aquila, Italy
{adinnoce,pola,dibenede}@ing.univaq.it
[3] Department of Electrical Engineering, University of California, at Los Angeles - Los Angeles, USA
pola@ee.ucla.edu

Abstract. This short paper qualitatively introduces the definition of the concepts of Deadlock and Livelock for a general class of Hybrid Control Systems (HCS). Such a characterization hinges on three important aspects: firstly, the concept of composition of HCS; secondly, the general concept of specifications and their composition for HCS; finally, the dynamical structure and behaviors of HCS. The first aspect is introduced in a novel manner, including ideas from the literature of discrete transition systems and accounting for concepts such as that of dynamical feedback interconnection. The second point includes general properties that are of interest from a systems and control theory perspective. The third part categorizes the diverse and possibly pathological behaviors that are distinctive of HCS. A first look at the problem of Deadlock and Livelock Verification concludes the manuscript.

1 Introduction

The concept of deadlock and its close relative, that of livelock, have been widely investigated in the literature of various branches of computer science. Deadlock, in particular, has often been regarded as a pathology and associated with the deficiency of a liveness specification, that of forward progress [6]. Much interesting work has been focused on verifying the presence of deadlock situations in algorithms or programs, or on ensuring its absence upon their composition [3][4].

Hybrid Systems are rather general mathematical models that connect between discrete, logical, synchronous systems and continuous, real-time, asynchronous ones [2]. It has often been observed that they present behaviors or are endowed with properties that are "at the limit" between classical transition systems and dynamical models [2].

Motivated by a number of case studies, this work aims at "exporting" the notions of deadlock and livelock to the Hybrid Control Systems (HCS) case. More

A. Bemporad, A. Bicchi, and G. Buttazzo (Eds.): HSCC 2007, LNCS 4416, pp. 628–632, 2007.

precisely, the objective has been that of first introducing a mathematically rigorous definition of the phenomena and providing a clear characterization of them. We stress that the introduced concepts naturally tailor back to the corresponding ones in the literature of, respectively, discrete and continuous systems.

2 Deterministic Hybrid Control Systems

The model for HCS is a melange between the classic hybrid automaton [2] and the HIOA [4]. In particular, it adheres to a *denotational* definition at the internal, state-space level, while it is inspired by an *operational* characterization at the external, input/output level. More precisely, an HCS is characterized by a finite collection of modes, each of which is associated with a domain and a control-dependent vector field. The set of transition relations is composed of a collection of edges (ordered pairs of modes), guards (subsets of the domains, possibly control-dependent), and deterministic reset functions. The control space, which contains real time-dependent control functions, is assumed to be bounded. Finally, the HCS is endowed with an observation space: the output functions will be obtained from the hybrid executions via a static output map. The set of initial conditions is a subset of the hybrid state space. To introduce the concept of executions of the HCS, it is first necessary to define the *hybrid time set*, a rather classical notion in the literature, as an ordered sequence of time intervals that represent the "dwelling times" of the continuous evolution within a mode. The *hybrid execution* is a hybrid trajectory (a pair of discrete and continuous evolutions of the flow) which is defined on the hybrid time set and abides by the flowing and switching within a HCS and is thus characteristic of its internal structure. It is possible to raise some rather general assumptions to enforce the determinism of the model.

The output of the hybrid system is, for each execution, a function from the hybrid time set to the output space. Since our purpose is to set up a notion of input-output interconnection, in the spirit of [4], we suppose that the *interconnectible* output of hybrid systems considered is instead a set of physical signals, function of the real time, obtained by a simple operation on the output of the HCS. This assumption is motivated by the need to give an asynchronous notion of interconnection.

3 Hybrid Systems Composition

Abstractly, the concept of systems *composition* may be introduced in many ways, depending on the characteristics and properties of the systems that are considered, the structure of the operation, and the particular properties that we may want to check for. In this work we consider an operation that may be interpreted as a form of *parallel composition*. Unlike previous work though, which simply performed parallel compositions as crude variables "sharing", inspired here by a more control theoretical perspective we allow the connections between inputs and outputs of the systems to depend on general functions endowed with some properties. Doing so, we naturally introduce an *output feedback* framework. Notice that the introduction of a model structure with internal and external components, similar to that in [4], allows to conceive the system at the level of its hidden/internal variables (the hybrid state space with its vector fields and transition relations) as a black box and only focus on the external components when performing the interconnection.

Proper "compatibility" conditions on two general HCS need to be raised before composing them. The actual HCS, result of the composition, is defined as follows: the "internal" structure of the composed system is basically the cartesian product of the two original hybrid automata. Two interconnecting static maps turn a transformation of the original output space of one of the two systems into part of the original input space of the other system, and vice versa. The new output space is simply the cartesian product of the original two, while the input space of the composition is, intuitively, the set of "unused inputs" of the composition. In the extreme case, the composition may be purely dynamical.

Asynchronism is preserved in the composition. The semantics of the composed model allow to not care about the presence of "cyclic constraints". The composition does not exclude the presence of pathological events (Zeno or blocking, for instance), which arises at an internal level. A rather slack condition on the continuity of the interconnecting maps allows to preserve determinism in the composition. Furthermore, the commutativity and associativity properties hold.

4 Composing Hybrid Systems Specifications

In this section we consider rather general specifications defined on hybrid trajectories in the observation space. They may be defined, for instance, via temporal logic formulae for real-time systems. Furthermore, we shall also introduce an explicit dependence on the control signals: this would allow to express specifications that are general enough to cover the most important problems in control theory. Instances of such specifications are that of *reachability*, *invariance*, *viability*, *attractivity*. Safety, liveness and forward progress can be reinterpreted through the above properties, as well as verification and control synthesis tasks.

We look for the set of trajectories, that is the behaviors, that verify a particular specification. Because of the deterministic hypothesis for the model, it is possible to associate this set of trajectories to a certain collection of initial conditions.

Given two HCS, two corresponding specifications and a composition procedure, the composed specification is defined as the *conjunction* of the two original specifications, modulo proper *variables substitutions* according to the interconnection maps associated with the composition procedure.

Consider the cartesian product of the sets of initial conditions of the single systems associated with trajectories that verify the corresponding property. Within this set, it is particularly interesting to look at the set of initial conditions in the composed system, that originates trajectories that do not verify the composed specification. These initial conditions are associated to "pathological" executions. It is indeed among the trajectories in this set that we shall categorize those associated with deadlock and livelock.

5 Definition of Deadlock and Livelock for Hybrid Control Systems

From a dynamical standpoint, the concepts of deadlock and livelock are intrinsically related to the idea of a trajectory being "constrained" or "stalled" somewhere in the state space. This locking condition is then further specified with regards to the presence or absence of indefinite motion within the region.

The fundamental concepts of this paper are then qualitatively introduced as follows. The "pathological" trajectories singled out above can be of two kinds: those that end up in a *hybrid invariant set*, and those that do not. The executions that do enter in an invariant set are either *deadlock* or *livelock*: the first are characterized by the *absence of motion* in finite time ("stalling" situations). The second are instead characterized by *endless motion*, either in their continuous or discrete component.

Notice that the definition above hinges on a purely *dynamical* level. This represents the last point, after that of *composition* and that of *specification*, which is regarded as necessary to introduce the notions of deadlock and livelock in the framework of HCS. Special instances of the above behaviors that are "notorious" for HCS are, in the case of deadlock situations, blocking conditions, stable equilibria in finite time, chattering and genuine Zeno. For the case of livelock, examples are represented by stable equilibria in infinite time and limit cycles.

6 Conclusions and Future Work

This extended abstract only qualitatively introduces the concept of deadlock and livelock for HCS. A number of fundamental details have been skipped for the sake of space. Also, interesting interpretations of the above concepts in a number of application instances have not been reported in this work. An extended and detailed manuscript can be found in the form of a technical report [1].

From the above discussions, it comes at no surprise that the next obligatory step after the definition and characterization of the notion of deadlock and livelock for HCS is that of looking at ways to *detect* it. Deadlock and Livelock *prevention* and *resolution* are other topics that do not find space in the present paper. The authors are also working on other extensions of the presented results. The concept of composition is prone to be generalized, and the issue of "deep composition", i.e. of a composition procedure preserving certain properties through its structure, clearly connects with the above ideas when the absence of deadlock or livelock is the specification to be exported.

References

1. Alessandro Abate, Alessandro D'Innocenzo, Giordano Pola, Maria Domenica Di Benedetto, Shankar Sastry : *The Concept of Deadlock and Livelock in Hybrid Control Systems.* Technical Report, UCB/EECS-2006-181, `http://www.eecs.berkeley.edu/Pubs`, Dec 2006.
2. John Lygeros, Karl Henrik Johansson, Slobodan N. Simic, Jun Zhang, Shankar Sastry : *Dynamical Properties of Hybrid Automata.* IEEE Transactions on Automatic Control, vol. 48, no. 1, Jan 2003.
3. Rajeev Alur, Thomas Henzinger: *Modularity for Timed and Hybrid Systems.* Proceedings of the 8th International Conference on Concurrency Theory (CONCUR 97), LNCS 1243, pp. 74-88, 1997.
4. Nancy Lynch, Roberto Segala, Frits Vaandrager: *Hybrid I/O Automata.* Information and Computation, 185(1):105-157, 2003.
5. Martin Abadi, Leslie Lamport: *Composing Specifications.* REX Workshop on Stepwise Refinement of Distributed Systems, Mook, NL, May 1989.
6. Bowen Alpern and Fred Schneider: *Defining Liveness.* Information Processing Letters, vol. 21, pp. 181-185, 1985.

Reachability Algorithm for Biological Piecewise-Affine Hybrid Systems

Anil Aswani and Claire Tomlin

University of California at Berkeley
Dept. of Electrical Engineering and Computer Sciences
{aaswani,tomlin}@eecs.berkeley.edu

Abstract. We consider a class of qualitative biological models which describe species (e.g. protein) interactions in terms of promotion or inhibition, from which a piecewise-affine (PWA) hybrid system model can be generated. These models have a special structure under which negative feedback is a necessary condition for the presence of limit cycles, centers, and foci. We describe modifications to reachability algorithms to take advantage of this special structure, and we give conditions on the qualitative model for termination of the algorithm. An example of applying the algorithm to a simple biological system is given.

1 Trajectory Cycles

Consider a general hybrid system $\mathcal{H}$ described by [1], such that the union of the invariants is positively invariant. Suppose there exists a non-blocking execution $(\tau, q[\cdot], \boldsymbol{x}(\cdot))$ of the hybrid system. If there exists a discrete state $s \in Q$ such that $q[i] = s$ for infinitely many i, then the execution is called a *trajectory cycle*; note that this encompasses Zeno behavior. Trajectory cycles are important because of the following lemma:

Lemma 1. *If a PWA system has no trajectory cycles, then every trajectory in the system converges to an equilibrium point that lies inside an invariant.*

This tell us that trajectory cycles are a necessary condition for the presence of limit cycles, centers, and foci. The absence of trajectory cycles rules out any complicated behaviors from occurring.

2 Promotion-Inhibition Networks

A *promotion-inhibition network* is a collection $\mathcal{N} = (S, R, T)$ where S is a set of nodes, R is a set of directed edges, and $T : R \to \{+, -\}$ designates a label for each edge. It is a directed graph with two types of edges: promotion (labeled $+$) and inhibition (labeled $-$). Such networks are a common qualitative model for biological systems, with nodes used to mark biological species and directed edges to indicate qualitative relationships between species quantities.

A. Bemporad, A. Bicchi, and G. Buttazzo (Eds.): HSCC 2007, LNCS 4416, pp. 633–636, 2007.

Table 1. PWA Model of Subsystem of Segment Polarity Network

q	A_q	b_q	Σ_q	q	A_q	b_q	Σ_q
1	$\begin{pmatrix} -10 & 0 & 0 \\ 0 & -0.1 & 0 \\ 0 & 0 & -2 \end{pmatrix}$	$\begin{pmatrix} 0 \\ 0 \\ 0 \end{pmatrix}$	$0.0 < x^1 < 0.5$ $0.0 < x^2 < 0.5$ $0.0 < x^3 < 0.5$	5	$\begin{pmatrix} -10 & 0 & 0 \\ 0 & -0.1 & 0 \\ 0 & 0 & -2 \end{pmatrix}$	$\begin{pmatrix} 0 \\ 0.1 \\ 2 \end{pmatrix}$	$0.5 < x^1 < 1.0$ $0.5 < x^2 < 1.0$ $0.0 < x^3 < 0.5$
2	$\begin{pmatrix} -10 & 0 & 0 \\ 0 & -0.1 & 0 \\ 0 & 0 & -2 \end{pmatrix}$	$\begin{pmatrix} 0 \\ 0.1 \\ 0 \end{pmatrix}$	$0.5 < x^1 < 1.0$ $0.0 < x^2 < 0.5$ $0.0 < x^3 < 0.5$	6	$\begin{pmatrix} -10 & 0 & 0 \\ 0 & -0.1 & 0 \\ 0 & 0 & -2 \end{pmatrix}$	$\begin{pmatrix} 0 \\ 0.1 \\ 0 \end{pmatrix}$	$0.5 < x^1 < 1.0$ $0.0 < x^2 < 0.5$ $0.5 < x^3 < 1.0$
3	$\begin{pmatrix} -10 & 0 & 0 \\ 0 & -0.1 & 0 \\ 0 & 0 & -2 \end{pmatrix}$	$\begin{pmatrix} 0 \\ 0 \\ 2 \end{pmatrix}$	$0.0 < x^1 < 0.5$ $0.5 < x^2 < 1.0$ $0.0 < x^3 < 0.5$	7	$\begin{pmatrix} -10 & 0 & 0 \\ 0 & -0.1 & 0 \\ 0 & 0 & -2 \end{pmatrix}$	$\begin{pmatrix} 0 \\ 0.1 \\ 2 \end{pmatrix}$	$0.0 < x^1 < 0.5$ $0.5 < x^2 < 1.0$ $0.5 < x^3 < 1.0$
4	$\begin{pmatrix} -10 & 0 & 0 \\ 0 & -0.1 & 0 \\ 0 & 0 & -2 \end{pmatrix}$	$\begin{pmatrix} 0 \\ 0.1 \\ 0 \end{pmatrix}$	$0.0 < x^1 < 0.5$ $0.0 < x^2 < 0.5$ $0.5 < x^3 < 1.0$	8	$\begin{pmatrix} -10 & 0 & 0 \\ 0 & -0.1 & 0 \\ 0 & 0 & -2 \end{pmatrix}$	$\begin{pmatrix} 0 \\ 0.1 \\ 2 \end{pmatrix}$	$0.5 < x^1 < 1.0$ $0.5 < x^2 < 1.0$ $0.5 < x^3 < 1.0$

A promotion-inhibition network $\mathcal{N}$ can be turned into a dynamical model by generating a PWA hybrid system $\mathcal{H}$ that maintains the qualitative relationships; see, for example, [2,3,4]. A subsystem in a biological model from [5] is shown in Fig. 1, and the generated PWA system is shown in Table 1, where Σ_q denotes an invariant and the accompanying vector field is $\dot{x} = A_q x + b_q$.

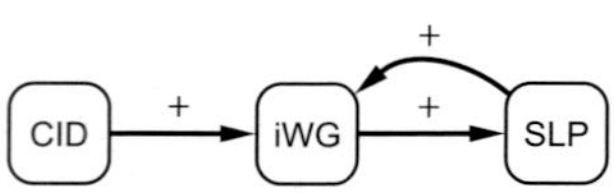

Fig. 1. Subsystem of Segment Polarity Network

These generated PWA models display rich behavior, including limit cycles and foci. It is difficult to intuit the global behavior of the system based on the promotion-inhibition interactions alone. However, an interesting result can be shown relating the local properties of promotion-inhibition to the global properties of the system trajectories:

Theorem 1. *If for each edge $e_i \in E$ of $\mathcal{N}$ that connects nodes in a cycle we have $T(e_i) = +$, then the continuous states in $\mathcal{H}$ are not involved in a trajectory cycle.*

The biological interpretation of this if the species are connected such that there is no negative feedback, that is no cycles with edges of type $-$, then all trajectories converge to a node equilibrium point. Furthermore, negative feedback is a necessary condition for the presence of limit cycles, centers, and foci.

3 Reachability Algorithm

Our reachability algorithm is essentially the bisimulation algorithm originally described by [6] and implemented using different approximation schemes by

[4,7,8,9]. For biological systems, reach sets can be used to determine system behavior by telling us which parameter instantiations lead to which steady states. Our approximation method for computing backwards reach sets is better suited than [4,7,8,9] for this purpose.

3.1 Backwards Reach Set Calculation

Our distinguishing approach is that: (1) we store the guards and reach sets as exact, symbolic, parameterized equations and (2) make approximations on the ranges of the parameterized variables. This approximation is better suited for biological systems because it allows the algorithm to compute reach sets that are within some tolerance of the actual reach sets, which gives a clearer description of system behavior than possible with over- or under-approximations.

Suppose that we have a guard that has been parameterized by $(n-1)$-variables: $\boldsymbol{y}(\boldsymbol{u})$ for $\boldsymbol{u} \in \mathcal{R}$. We populate the guard with sample points, and for each sample point we determine: (1) when the backwards trajectory starting from this point hits a boundary of the invariant and (2) which boundary it hits. This can be performed by doing computations of the form:

$$t^*(\boldsymbol{y}(\boldsymbol{u})) = \max_i \left\{ \frac{1}{\lambda^i} \ln \left(\frac{\theta^i + \frac{b^i}{\lambda^i}}{y^i(\boldsymbol{u}) + \frac{b^i}{\lambda^i}} \right) \right\} \tag{1}$$

where θ^i is a boundary of the invariant. This approach differs from previous algorithms which use numerical integration to do the backwards trajectory calculations [7,8,9].

A portion of the guard will lie on the boundary of an invariant. The equation of this guard on a particular boundary is given by:

$$z^j(\boldsymbol{u}) = \left(y^j(\boldsymbol{u}) + \frac{b^j}{\lambda^j} \right) \left(\frac{\theta + \frac{b^i}{\lambda^i}}{y^i(\boldsymbol{u}) + \frac{b^i}{\lambda^i}} \right)^{\lambda^j/\lambda^i} - \frac{b^j}{\lambda^j} . \tag{2}$$

The range of the parameterization $\boldsymbol{u}$ is given by those sample points that were determined to hit this particular boundary.

3.2 Algorithm Termination

Our version of the reachability algorithm can be proven to terminate under certain conditions on the system structure or behavior. The abbreviated conditions are:

Theorem 2. *If a PWA system has no trajectory cycles, then our reachability algorithm will terminate.*

Corollary 1. *If for each edge $e_i \in E$ of $\mathcal{N}$ that connects nodes in a cycle we have $T(e_i) = +$, then our reachability algorithm will terminate.*

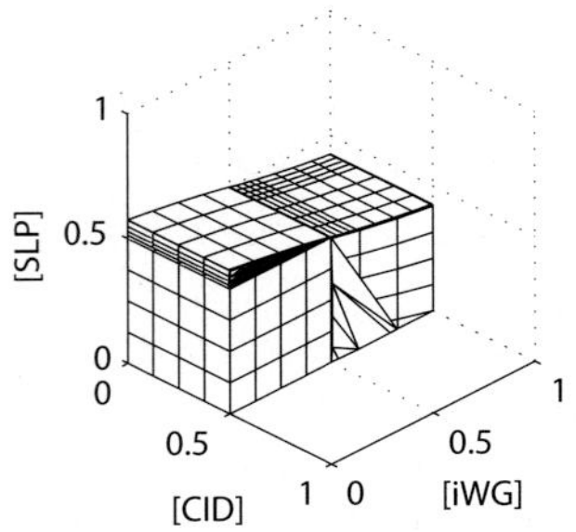

Fig. 2. Reach Set for Subsystem of Segment Polarity Network

3.3 Example: Subsystem of Segment Polarity Network (cont.)

Applying the full reachability algorithm to the system in Fig. 1 and Table 1 results in the output shown in Fig. 2. Other algorithms use numerical integration to calculate the reach set; we use the analytical solution. Thus, we get a more accurate approximation of the reach set.

The shaded area converges to the equilibrium point $([\text{CID}], [\text{iWG}], [\text{SLP}]) = (0, 0, 0)$ and the unshaded area converges to the equilibrium point $(0, 1, 1)$. These reach sets emphasize the bistability of the system, which is caused by the positive feedback that maintains these two equilibrium points.

References

1. Lygeros, J., Tomlin, C., Sastry, S.: Controllers for reachability specifications for hybrid systems. Automatica **35**(3) (1999) 349–370
2. Gouzé, J.L., Sari, T.: A class of piecewise linear differential equations arising in biological models. Dynamical Systems **17**(4) (2002) 299–316
3. de Jong, H., Gouzé, J.L., Hernandez, C., Page, M., Sari, T., Geiselmann, J.: Qualitative simulation of genetic regulatory networks using piecewise-linear models. Bulletin of Mathematical Biology **66**(2) (2004) 301–340
4. Ghosh, R., Tomlin, C.: Symbolic reachable set computation of piecewise affine hybrid automata and its application to biological modelling: Delta-Notch protein signalling. Systems Biology **1**(1) (2005) 170–183
5. von Dassow, G., Odell, G.M.: Design and constraints of the *Drosophila* segment polarity module. Journal of Experimental Zoology **294** (2002) 179–215
6. Bouajjani, A., Fernandez, J.C., Halbwachs, N.: Minimal model generation. In: CAV '90: Proceedings of the 2nd International Workshop on Computer Aided Verification, London, UK, Springer-Verlag (1991) 197–203
7. Belta, C., Esposito, J., Kim, J., Kumar, V.: Computational techniques for analysis of genetic network dynamics. International Journal of Robotics Research **24**(2-3) (2005)
8. Frehse, G.: PHAVer: Algorithmic verification of hybrid systems past HyTech. In: HSCC. Volume 3414 of LNCS., Springer (2005) 258–273
9. Girard, A., Le Guernic, C., Maler, O.: Efficient computation of reachable sets of linear time-invariant systems with inputs. In: HSCC. Volume 3927 of LNCS., Springer (2006) 257–271

Necessary Optimality Conditions for a Class of Hybrid Optimal Control Problems

Vadim Azhmyakov[1], Sid Ahmed Attia[1,2], Dmitry Gromov[1], and Jörg Raisch[1,2]

[1] Fachgebiet Regelungssysteme, Technische Universität Berlin, Einsteinufer 17, D-10587 Berlin, Germany
{azhmyakov,attia,gromov,raisch}@control.tu-berlin.de
http://www.control.tu-berlin.de

[2] Systems and Control Theory Group, MPI for Dynamics of Complex Technical Systems, Sandtorstr. 1, D-39106 Magdeburg, Germany

Abstract. In this paper we study a class of Mayer-type hybrid optimal control problems. Using Lagrange techniques, we formulate a version of the Hybrid Maximum Principle for optimal control problems governed by hybrid systems with autonomous location transitions in the presence of additional target constraints.

1 Introduction and Problem Formulation

Over the last few years hybrid optimal control theory has been formalized as a natural generalization of classical optimal control theory. For hybrid optimal control problems, the main tool towards the construction of optimal control signals and optimal trajectories is the Hybrid Maximum Principle, which generalizes the classical Pontryagin Maximum Principle (see, e.g., [4]). Several variants of the Hybrid Maximum Principle were developed in [6,7,9,10,3,8]. It is well-known that the standard proof of the Pontryagin Maximum Principle is based on the technique of "needle variations" [4]. In this paper we derive necessary optimality conditions for a class of hybrid optimal control problems without recourse to the technique of needle variations. Instead, we apply a Lagrange based approach. This approach allows us to obtain necessary conditions for a *weak minimum* as opposed to the standard Maximum Principle which gives necessary conditions for a *strong minimum.*

In this paper, we consider a hybrid control system $\mathcal{H}$ with autonomous location transitions (see, e.g.,[8]). Its state at time $t \in [0, t_f]$ is the pair $(q, x(t))$, where $q \in \mathcal{Q}$, $x(t) \in \mathbb{R}^n$, and $\mathcal{Q}$ is a finite set of locations. While in location q, the temporal evolution of x is determined by $\dot{x} = f_q(x, u)$, where the functions f_q are assumed to be continuously differentiable and bounded, the admissible control sets U_q are compact and convex, and

$$\mathcal{U}_q := \{u(\cdot) \in \mathbb{L}^m_\infty(0, t_f) \ : \ u(t) \in U_q, \ \text{a.e. on}[0, t_f]\}$$

represent the sets of admissible control signals. Switchings between locations are determined by smooth functions $m_q : \mathbb{R}^n \to \mathbb{R}$, $q \in \mathcal{Q}$ with nonzero gradients

A. Bemporad, A. Bicchi, and G. Buttazzo (Eds.): HSCC 2007, LNCS 4416, pp. 637–640, 2007.

such that the hypersurfaces ("switching sets") $M_q := \{x \in \mathbb{R}^n : m_q(x) = 0\}$ are pairwise disjoint and divide $\mathbb{R}^n$ into open sets.

Let us now introduce the following concept (see [9,3]).

Definition 1. *A* **hybrid trajectory** *of $\mathcal{H}$ is a triple $\mathcal{X} = (x, \{q_i\}, \tau)$, where $x(\cdot) : [0, t_f] \to \mathbb{R}^n$, $\{q_i\}_{i=1,\dots,r}$ is a finite sequence of locations and τ is the corresponding sequence of switching times $0 = t_0 < \dots < t_i < \dots < t_r = t_f$ such that for each $i = 0, \dots, r$ there exists $u_i(\cdot) \in \mathcal{U}_i$ such that:*

- *$x(0) = x_0 \notin \bigcup_{q\in\mathcal{Q}} M_q$ and $x_i(\cdot) = x(\cdot)|_{(t_{i-1},t_i)}$ is an absolutely continuous function in (t_{i-1}, t_i) continuously prolongable to $[t_{i-1}, t_i]$, $i = 1, \dots, r$;*
- *$\dot{x}_i(t) = f_{q_i}(x_i(t), u_i(t))$ for almost all $t \in [t_{i-1}, t_i]$, $i = 1, \dots, r$;*
- *the switching condition $(x_i(t_i), x_{i+1}(t_i)) \in M_{q_i}$ holds for each $i = 1, \dots, r-1$.*

Note that the evolution equation for the continuous trajectory $x(\cdot)$ of a given $\mathcal{H}$ can be represented as follows

$$\dot{x}(t) = \sum_{i=1}^{r} \beta_{(t_{i-1},t_i]}(t) f_{q_i}(x(t), u(t)), \text{a.e. on } [0, t_f],\ i = 1, \dots, r, \tag{1}$$

where $\beta_{(t_{i-1},t_i]}(\cdot)$ is a characteristic function of the interval $(t_{i-1}, t_i]$. Under the above assumptions for the family of vector fields $\{f_q(x, u)\}_{q\in\mathcal{Q}}$, the right-hand side of equation (1) satisfies the Caratheodory conditions (see e.g., [4]). Next we consider $x(\cdot)$ as an element of the Sobolev space $x(\cdot) \in \mathbb{W}^n_{1,\infty}([0, t_f])$, which contains all absolutely continuous functions with essentially bounded derivatives. Let $\phi : \mathbb{R}^n \to \mathbb{R}$ and $g : \mathbb{R}^n \to \mathbb{R}$ be continuously differentiable functions. We consider an additional target manifold given by the equation $g(x) = 0$ and introduce the notation $M_{q_r}(x) := \{x \in \mathbb{R}^n : m_r(x) = 0\}$, where $m_r(x) = g(x)$. Given a hybrid system $\mathcal{H}$ we formulate the following Mayer-type hybrid optimal control problem $(\mathcal{HOCP})$:

$$\begin{aligned} &\text{minimize } \phi(x(t_f)) \\ &\text{over all trajectories } \mathcal{X} \text{ of } \mathcal{H} \text{ such that } g(x(t_f)) = 0. \end{aligned} \tag{2}$$

2 A Variant of the Hybrid Maximum Principle

We will study necessary optimality conditions for (2) with the help of the general Lagrange multiplier rule for optimization problems in Banach spaces [4,5]:

Theorem 1. *Let Y and Z be real* Banach *spaces, $\psi : Y \to \mathbb{R}$ be a cost functional and $h : Y \to Z$ be a mapping. Let $\mathcal{F}$ be a convex subset of Y with a nonempty interior and y^0 be a solution of the following optimization problem*

$$\begin{aligned} &\text{minimize } \psi(y) \\ &\text{subject to } h(y) = \mathbf{0}_Z,\ y \in \mathcal{F}, \end{aligned} \tag{3}$$

where $\mathbf{0}_Z$ is the zero element of Z. Assume that ψ and h are Fréchet *differentiable in a neighborhood of y^0, the derivative $h'(\cdot)$ is continuous at the point y^0 and* *the image set* $h'(y^0)(Y)$ *is closed. Then there are a real number* $\mu \geq 0$ *and a continuous linear functional* $\ell \in Z^*$ *(the topological dual space to* Z*) with* $(\mu, \ell) \neq (0, \mathbf{0}_{Z^*})$ *such that*

$$\mathcal{L}'_y(y^0, \mu, \ell)(y - y^0) = \left(\mu\psi'(y^0) + \ell \circ h'(y^0)\right)(y - y^0) \geq 0 \quad \forall y \in \mathcal{F}, \qquad (4)$$

where $\mathcal{L}(y, \mu, \ell) := \mu\psi(y) + \ell \circ h(y)$ *is the* Lagrange *function for* (3) *and* $\circ$ *is referred to as the duality pairing* [5], *which maps a pair from* $\mathbb{Z}^* \times \mathbb{Z}$ *to* $\mathbb{R}$.

Note that in the case of a regular problem (3) one can put $\mu = 1$.

We now introduce the mapping

$$P(v(\cdot), \xi(\cdot)) := \begin{pmatrix} \xi(\cdot) - x_0 - \int_0^{\cdot} \sum_{i=1}^{r} \beta_{(t_{i-1}, t_i]}(t) f_{q_i}(\xi(t), v(t))dt \\ \left[m_{q_i}(\xi(t_i))\right]_{i=1,\ldots,r} \end{pmatrix}.$$

where $(v(\cdot), \xi(\cdot)) \in \mathbb{L}^m_\infty([0, t_f]) \times \mathbb{W}^n_{1,\infty}([0, t_f])$. The first element of P is an operator of differential equation (1) in integral form whereas the second one determines switching times t_i from a sequence τ according to the following switching rule: $t_i = \inf\{t \,|\, m_{q_i}(x(t)) = 0\}$, where $i = 1, ..., r - 1$. The operator equation $P(v(\cdot), \xi(\cdot)) = 0$ determines the evolution of the hybrid control system $\mathcal{H}$ as a function of control $v \in \mathcal{U}$. Note that the mapping P is specified in the same manner as the mapping h from Theorem 1. Here $Y := \mathbb{L}^m_\infty([0, t_f]) \times \mathbb{W}^n_{1,\infty}([0, t_f])$ and $Z := \mathbb{W}^n_{1,\infty}([0, t_f]) \times \mathbb{R}^r$. We now rewrite the $\mathcal{HOCP}$ (2) in the form (3) as

$$\begin{aligned} &\text{minimize } J(u(\cdot), x(\cdot)) = \phi(x(t_f)) \\ &\text{subject to } P(u(\cdot), x(\cdot)) = \mathbf{0}_{\mathbb{W}^n_{1,\infty}([0,t_f]) \times \mathbb{R}^r}, \ (u(\cdot), x(\cdot)) \in \mathcal{F}. \end{aligned} \qquad (5)$$

where $\mathcal{F} := \{(v(\cdot), \xi(\cdot)) \in \mathbb{L}^m_\infty([0, t_f]) \times \mathbb{W}^n_{1,\infty}([0, t_f]) \,:\, v_i(t) \in U_{q_i}, \, i = 1, \ldots, r\}$, and $v_i(\cdot)$ is a restriction of the function $v(\cdot)$ on the time interval $[t_{i-1}, t_i]$. Note that $\mathcal{F}$ is a convex set. A solution of (5) is denoted by $(x^0(\cdot), u^0(\cdot))$. For (5) we introduce the Lagrange function

$$\mathcal{L}((u(\cdot), x(\cdot)), \mu, l) := \mu J(u(\cdot), x(\cdot)) + (\ell, a) \circ P(u(\cdot), x(\cdot)),$$

where $\mu \in \mathbb{R}_{\geq 0}$, $a \in \mathbb{R}^r$ and the continuous linear functional ℓ belongs to the (topological) dual space of $\mathbb{W}^n_{1,\infty}([0, t_f])$. We now are in the position to state our main result.

Theorem 2. *Let functions* ϕ, f_q, m_q, g *be continuously differentiable and the optimization problem (5) be regular. Then there exist a function* $p(\cdot)$ *from* $\mathbb{W}^n_{1,\infty}([0, t_f])$ *and a vector* $a \in \mathbb{R}^r$ *such that*

$$\begin{aligned} \dot{p}_i(t) &= -\frac{\partial H_{q_i}(x_i^0(t), u_i^0(t), p(t))}{\partial x} \text{ a. e. on } (t_{i-1}^0, t_i^0), \ i = 1, ..., r, \\ p_r(t_f) &= -\left(a_r \frac{\partial g(x^0(t_f))}{\partial x} + \frac{\partial \phi(x^0(t_f))}{\partial x}\right), \\ p_i(t_i^0) &= p_{i+1}(t_i^0) + \left(a_i, \frac{\partial m_{q_i}(x^0(t_i^0))}{\partial x}\right), i = 1, ..., r - 1, \end{aligned} \qquad (6)$$

where $p(t) = \sum_{i=1}^{r} \beta^0_{(t^0_{i-1},t^0_i]}(t)p_i(t)$. *Moreover, for every admissible control* $u(\cdot)$ *the following inequalities are satisfied*

$$\left(\frac{\partial H_{q_i}(x^0(t), u^0(t), p(t))}{\partial u}, (u(t) - u^0(t)) \right) \leq 0 \text{ a. e. on } [t^0_{i-1}, t^0_i], \qquad (7)$$

where $i = 1, ..., r$ *and* $H_{q_i}(x, u, p) := (p, f_{q_i}(x, u))$ *is a "partial" Hamiltonian for the location* $q_i \in \mathcal{Q}$ *and* $(\cdot, \cdot)$ *denotes the corresponding scalar product.*

3 Discussion

In this contribution we have proposed a version of the Hybrid Maximum Principle based on the Lagrange multiplier rule. This implies that the necessary optimality conditions for a weak minimum are obtained. However, in many practical optimal control problems weak minima coinside with strong minima. Note that the suggested proof-technique can also be applied to hybrid systems with state jumps and to some classes of hybrid systems with controlled location transitions.

The Hamilton minimization conditions from Theorem 2 are presented in the form of variational inequalities. This form is closely related to the Weierstraß conditions for a strong minimum (see, e.g.,[4]) and to the gradient-based computational approach studied in [1]. Finally note that condition (7) make it possible to take into consideration some effective methods for numerical treatment of variational inequalities.

References

1. V. Azhmyakov and J. Raisch, A gradient-based approach to a class of hybrid optimal control prolems. In: *Proceedings of the 2nd IFAC Conference on Analysis and Design of Hyrid Systems*, Alghero, 2006, pp. 89 – 94.
2. V. Azhmyakov, S.A. Attia, D. Gromov and J. Raisch, Optimal dymanics of hybrid and switched systems, Technical Report, Fachgebiet Regelungstechnik, TU Berlin, 2007.
3. M. Garavello and B. Piccoli, Hybrid necessary priniple, *SIAM J. Control Optim.*, **43** (2005), pp. 1867 – 1887.
4. A.D. Ioffe and V.M. Tichomirov, *Theory of Extremal Problems*, North Holland, Amsterdam, 1979.
5. A.J. Kurdila and M. Zabarankin, *Convex Functional Analysis*, Birkhäuser, 2005
6. B. Piccoli, Hybrid systems and optimal control. In: *Proceedings of the 37th IEEE CDC*, Tampa, 1998, pp. 13 – 18.
7. B. Piccoli, Necessary conditions for hybrid optimization. In: *Proceedings of the 38th IEEE CDC*, Phoenix, 1999, pp. 410 – 415.
8. M.S. Shaikh and P.E. Caines, On the hybrid optimal control problem: The hybrid maximum principle and dynamic programming theorem. *IEEE Trans. Automat. Contr.* submitted (2004).
9. H. Sussmann, A maximum principle for hybrid optimal control problems. In: *Proceedings of the 38th IEEE CDC*, Phoenix, 1999, pp. 425 – 430.
10. H. Sussmann, A nonsmooth hybrid maximum priniple, *Stability and Stabilization of Nonlinear Systems (Ghent, 1999)*, LNCIS, **246**, pp. 325 – 354, Springer, 1999.

Optimal Switches in Multi–inventory Systems

Dario Bauso

Dipartimento di Ingegneria Informatica Università di Palermo,
Viale Delle Scienze, 90128 Palermo, Italy*
dario.bauso@unipa.it
http://www.unipa.it/ dario.bauso

Abstract. Given a switched multi–inventory system we wish to find the optimal schedule of the resets to maintain the system in a safe operating interval, while minimizing a function related to the cost of the resets. We discuss a family of instances that can be solved in polynomial time by linear programming. We do this by introducing a *set-covering* formulation with a *totally unimodular* constraint matrix.

1 Problem Description

Consider the family of continuous time linear multi–inventory system

$$\dot{x}(t) = B_i u_c(t) - d(t), \quad i \in \{1,2\} \tag{1}$$

where $x(t) \in \mathbb{R}^n$ is a vector whose components are the buffer levels, $u_c(t) \in \mathbb{R}^m$ is the controlled flow vector, $B_i \in \mathbb{Q}^{n \times m}$ is the controlled process matrix and $d(t) \in \mathbb{R}^n$ is the unknown demand. To model backlog $x(t)$ may be less than zero. Controls and demands are bounded within polytopes, i.e.,

$$u_c(t) \in \mathcal{U}_c = \{u \in \mathbb{R}^m : u^- \leq u \leq u^+\}$$
$$d(t) \in \mathcal{D} = \{d \in \mathbb{R}^n : d^- \leq d \leq d^+\},$$

where u_c^-, u_c^+, d^-, and d^+ are assigned vectors. We also assume that matrix B_i is a "fat matrix" and has full row rank. A point of interest is the existence of *feedback stabilizing strategy* [1], that is, strategies able to drive the state within a neighborhood of x_{ref} of radius ϵ in finite time. Henceforth assume $x_{\text{ref}} = 0$.

Theorem 1. *([1]) For the generic system $\dot{x}(t) = B_i u_c(t) - d(t)$ there exists a feedback stabilizing strategy if and only if*

$$\mathcal{D} \subseteq int\{B_i \mathcal{U}\}. \tag{2}$$

Assume that only B_2 satisfies the above condition, namely, $\mathcal{D} \subseteq int\{B_2 \mathcal{U}\}$ and $\mathcal{D} \nsubseteq int\{B_1 \mathcal{U}\}$. Henceforth we refer to systems B_1 and B_2 as the unstable and stable mode respectively.

* This work was supported by PRIN "Analysis, optimization, and coordination of logistic and production systems".

A. Bemporad, A. Bicchi, and G. Buttazzo (Eds.): HSCC 2007, LNCS 4416, pp. 641–644, 2007.

After introducing the stable and unstable modes, the switched system is alternatively in one of the two modes as described by the following dynamics

$$\begin{array}{c}\dot{x} = \alpha_1(B_1 u_c(t) - d(t)) + \alpha_2(B_2 u_c(t) - d(t)) \\ \alpha_1 + \alpha_2 = 1, \quad \text{binary.}\end{array} \tag{3}$$

Transitions from the unstable mode B_1 to the stable mode B_2 are controlled by a continuous time reset signal $r(t) \in \{0, 1\}$ and occur anytime the reset signal is set to one (the value $\alpha_1[1 - r(t)]$ is equal to zero). Once in B_2 the system remains in this mode for the time necessary to drive the state x to zero (until $\alpha_2 x$ is equal to zero) before switching back to B_1. To model transitions we can use the following discontinuity function

$$g(\alpha, r, x) = \alpha_1[1 - r(t)] + \alpha_2 x, \tag{4}$$

and say that transitions occur at the earliest time τ at which $g(\alpha, r, x)$ becomes zero.

Given an n-tuple of times $\bar{t}_1, \dots, \bar{t}_N$ at which resets may occur, the problem of interest consists in finding the optimal schedule of the resets to maintain the system in a safe operating region, while minimizing a function related to the cost of the resets. The decision variables are thus binary (whether to reset the state at a given time instant or not). In formulas,

$$\min_{\mathbf{r} \in \{0,1\}^N} \quad \mathbf{c}^T \mathbf{r} \tag{5}$$

$$\text{s.t. dynamics (3)-(4),}$$

$$\|x(t)\| \leq \gamma \tag{6}$$

where $\mathbf{r} = [r(\bar{t}_1), \dots, r(\bar{t}_N)]^T$ are the resets, $\mathbf{c} = [c(1), \dots, c(N)]^T$ the costs of resets. Our aim is to show that the above problem leads to tractable solutions, though, in general, hybrid optimal control problems are difficult to solve [3].

2 Set-Covering Reformulation

A general recipe is to take the continuous-time hybrid optimal control problem (5)-(6), discretize it, and reformulate it thus to apply discrete optimization techniques [2]. Given an n-tuple of times $\bar{t}_1, \dots, \bar{t}_N$ at which resets may occur, let $k = 1, 2, \dots$ discrete sample times such that

$$t(k) = \bar{t}_k + T r(\bar{t}_k),$$

the latter meaning that the kth sample is at time $t(k)$ where $t(k)$ coincides with $\bar{t}_k$ if $r(\bar{t}_k) = 0$ (no reset at $\bar{t}_k$), or with $\bar{t}_k + T$ if $r(\bar{t}_k) = 1$ (reset at $\bar{t}_k$).

Let us introduce a discrete time reset $u(k)$ related to the continuous time reset as follows

$$u(k) = r(\bar{t}_k).$$

Assume no switches between $t(k)$ and $t(k+1)$, let $B_{\alpha(t(k))}$ be the active mode and define the disturbance

$$w(k, u_c, d) = \int_{t(k)}^{t(k+1)} (B_{\alpha(t(k))} u_c(t) - d(t))dt. \quad (7)$$

If we sample state dynamics (3)-(4) and constraints (6) at discrete times $k = 1, 2, \ldots$, problem (5)-(6) becomes

$$\min_{\mathbf{u} \in \{0,1\}^N} \quad \mathbf{c}^T \mathbf{u}, \quad (8)$$

$$\text{s.t. } x(k+1) = x(k) + w(k, ., .) - x(k)u(k),\ u(0) = 1, \quad (9)$$

$$\|x(k)\| \ \leq \gamma \quad \forall k = 1, \ldots, N+1, \quad (10)$$

where $\mathbf{u} = [u(1), \ldots, u(N)]^T$ are the vector of resets. Note that violation of constraints (6) for $t \neq t(k)$ may not be an issue when sample intervals are small. The constraint $u(0) = 1$ means that the initial state $x(0)$ is reset to zero.

Now, we show that (8)-(10), can be reformulated as a set-covering problem with a totally unimodular constraint matrix. Let $\mathcal{U}$, be the set of feasible discrete controls. To transcribe $\mathcal{U}$ in the space of controls, we must rewrite constraints on the state (10) in terms of controls $u(1), \ldots, u(N)$. To do this, we need to introduce the cover inequalities. A *cover* is any subset $C = \{\bar{k}, \bar{k}+1, \ldots, \tilde{k}-1, \tilde{k}\}$ of consecutive time instants such that $\min_{u_c \in \mathcal{U}_c} \max_{d \in \mathcal{D}} \| \sum_{k=\bar{k}}^{\tilde{k}} w(k, u_c, d)\| > \gamma$ and $\min_{u_c \in \mathcal{U}_c} \max_{d \in \mathcal{D}} \| \sum_{k=\bar{k}}^{\tilde{k}-1} w(k, u_c, d)\| \leq \gamma$. A cover defines the minimal time interval within which, for a solution to be feasible under the worst demand, we must reset at least one time.

Lemma 1. *(Cover Inequality) For any cover $C = \{\bar{k}, \bar{k}+1, \ldots, \tilde{k}-1, \tilde{k}\}$, the constraint*

$$\sum_{k=\bar{k}+1}^{\tilde{k}} u(k) \geq 1 \quad (11)$$

is a valid inequality for the feasible solution set $\mathcal{U}$.

Proof. For a given time $\tilde{k} < N+1$, let $\hat{k} < \tilde{k}$ the time of last reset. Proving (11) corresponds to proving that $\hat{k} \geq \bar{k}+1$. Assume, by contradiction, that it holds $\hat{k} < \bar{k}+1$. Then, dynamics between $\hat{k}$ and $\tilde{k}$ yields $\|x(\tilde{k}+1)\| = \| \sum_{k=\hat{k}}^{\tilde{k}} w(k)\| \geq \| \sum_{k=\bar{k}}^{\tilde{k}} w(k)\| > \gamma$, where the latter inequality derives from the very definition of cover. □

Given the above lemma, for each time k, we can find, if exists, the associated cover, and transcribe the set of feasible controls $\mathcal{U}$ by replacing constraints on the state (10) with the corresponding inequalities (11). For sake of simplicity we can assume $\gamma \geq \max_k \|w(k)\|$ in order to disregard special covers that are singletone $C = \{\tilde{k}\}$. Such covers would make the problem always unfeasible.

We are now ready to introduce the set-covering reformulation.

$$\min_{\mathbf{u}\in\{0,1\}^N} \quad \mathbf{c}^T\mathbf{u} \tag{12}$$

$$\text{s.t. } \sum_{\bar{k}+1}^{\tilde{k}} u(k) \geq 1, \text{ for all covers } C = \{\bar{k}, \ldots, \tilde{k}\}. \tag{13}$$

We can rewrite the above valid inequalities (11) by using a matrix $A \in \{0,1\}^{m\times N}$, with only entries 0 and 1, one row for each cover inequality (assume the cover inequalities are m), one column for each time k. Such a matrix is an *interval matrix*, i.e., it has 0-1 entries and each row is of the form

$$(0,\ldots,0\ \underbrace{1,\ldots\ldots\ldots,1}_{\text{consecutive 1's}}\ 0,\ldots,0).$$

It is well known from the literature that each interval matrix is totally unimodular (the determinant of any square sub-matrix is equal to -1, 0 or 1). This means that the polyhedron obtained from $\mathcal{U}$ by replacing the integrity constraints $u(k) \in \{0,1\}$ with the linear constraint $0 \leq u(k) \leq 1$ is an integral polyhedron. As a consequence we have the following result.

Theorem 2. *Solving the set-covering reformulation (12)-(13) is equivalent to solving the linear programming problem*

$$\min_{\mathbf{u}} \quad c^T\mathbf{u} \tag{14}$$

$$s.t. \quad A\mathbf{u} \geq 1 \tag{15}$$

$$0 \leq \mathbf{u} \leq 1. \tag{16}$$

Proof. Since constraints (15)-(16) define an integral polyhedron, then the linear programming problem (14)-(16) has as optimal solution an integer solution. □

From the above theorem we can conclude that solving the linear program (14)-(16) corresponds to solving problem (5)-(6).

References

1. Bauso, D., Blanchini, F., Pesenti, R.: Robust control strategies for multiinventory systems with average flow constraints. Automatica, **42** (2006) 1255–1266
2. Bemporad, A., Morari, M.: Control of systems integrating logic, dynamics, and constraints. Automatica **35** (1999) 407-427
3. Branicky, M. S., Borkar, V. S., Mitter, S. K.: A Unified Framework for Hybrid Control: Model and Optimal Control Theory. IEEE Trans. on Automatic Control, **43** (1998) 31–45

Viability-Based Computations of Solutions to the Hamilton-Jacobi-Bellman Equation

Alexandre M. Bayen[1], Christian Claudel[2], and Patrick Saint-Pierre[3]

[1] University of California at Berkeley, Civil and Environmental Engineering, Davis Hall 711, Berkeley, CA 94720-1710
bayen@berkeley.edu
http://www.berkeley.edu

[2] LICIT Laboratory, ENTPE/INRETS, Rue Maurice Audin, 69518 Vaulx-en-velin Cedex, France
christian.claudel@gmail.com
http://www.inrets.fr

[3] Department of Mathematics, Université Paris-Dauphine, Place du Maréchal de Lattre de Tassigny, 75775 Paris cedex 16, France
patrick.saint.pierre@gmail.com
http://www.dauphine.fr

Abstract. This article proposes a new capture basin algorithm for computing the numerical solution of a class of *Hamilton-Jacobi-Bellman* (HJB) *partial differential equations* (PDEs) [3], based on a Lax-Hopf formula [2]. The capture basin algorithm is derived and implemented to perform numerical computations. Its performance is measured with highway data obtained for interstate I80 in California.

Assumptions. We posit the following assumptions

1. A concave function $\psi : X \mapsto \mathbb{R}$ on $[0,\omega]$, which vanishes at 0 and ω, equal to $\psi'(0)v$ for $v \leq 0$ and $\psi'(\omega)(\omega - v)$ for $v \geq \omega$.
2. A bounded continuous function $v : \mathbb{R}_+ \mapsto \mathrm{Dom}(\psi)$,
3. An upper semicontinuous initial datum $\mathbf{N}_0 : X \mapsto \mathbb{R}_+$. We set $\mathbf{N}_0(0,x) := \mathbf{N}_0(x)$ and $\mathbf{N}_0(t,x) := -\infty$ if $t > 0$.
4. A Lipschitz function $\mathbf{b} : \mathbb{R}_+ \times X \mapsto \mathbb{R} \cup \{-\infty\}$ setting the upper constraint.

We set $\forall\, x \in \partial K,\ \ \forall\, t \geq 0,\ \ \gamma(t,\xi) \ := \ 0$ and $\forall\, x > \xi,\ \ \gamma(t,x) \ = \ -\infty$. This is required to satisfy consistency assumptions

$$\begin{cases} (i)\ \ \forall\, t \geq 0,\ \forall\, x \in K,\ \ \max\left(\mathbf{N}_0(t,x), \gamma(t,x)\right) \ \leq \ \mathbf{b}(t,x) \\ (ii)\ \forall x \in K,\ \ \mathbf{N}_0(x) \ \leq \ \inf\limits_{s \geq 0} \left(\dfrac{x-\xi}{s} \displaystyle\int_0^s v(\tau)d\tau \right) \end{cases} \tag{1}$$

When the function $v(\cdot) \equiv v$ is constant, condition (1)(ii) boil down to

$$\forall\, x \geq \xi,\ \ \mathbf{N}_0(x) \ \leq \ v(x-\xi)$$

Problem statement. Under the above mentioned assumptions, that are assumed all along this paper, we shall solve the existence of a solution to the *non-homogenous HJB PDE:*

A. Bemporad, A. Bicchi, and G. Buttazzo (Eds.): HSCC 2007, LNCS 4416, pp. 645–649, 2007.

$$\forall\, t > 0,\ \ x \in \mathrm{Int}(K),\ \ \frac{\partial \mathbf{N}(t,x)}{\partial t} + \psi\left(\frac{\partial \mathbf{N}(t,x)}{\partial x}\right) = \psi(v(t)) \tag{2}$$

satisfying the *initial and Dirichlet conditions*

$$\begin{cases} (i)\ \ \forall\, x \in K,\ \ \mathbf{N}(0,x) = \mathbf{N}_0(x) \text{ (initial condition)} \\ (ii)\ \forall\, t \geq 0,\ \mathbf{N}(t,\xi) = 0 \text{ (Dirichlet boundary condition)} \end{cases} \tag{3}$$

and the user defined *viability constraints.*

$$\forall\, t \geq 0,\ x \in K,\ \ \mathbf{N}(t,x)\ \leq\ \mathbf{b}(t,x) \text{ (upper inequality constraint)} \tag{4}$$

Flux functions. The assumption that the flux function ψ is concave and upper semicontinuous plays a crucial role for defining the viability hyposolution. Indeed, since ψ is concave, the function $\varphi(p) := -\psi(p)$ is convex and its Fenchel transform is defined by:

$$\varphi^*(u)\ :=\ \sup_{p\in \mathrm{Dom}(\varphi)} [p\cdot u - \varphi(p)]\ =\ \sup_{p\in \mathrm{Dom}(\psi)} [p\cdot u + \psi(p)] \tag{5}$$

Recall that the fundamental theorem of convex analysis states that $\varphi = \varphi^{**}$ if and only if φ is convex, lower semicontinuous, and non trivial (i.e. $\mathrm{Dom}(\varphi) := \{p \mid \varphi(p) < +\infty\} \neq \emptyset$). Therefore we can recover the function ψ from φ^* by

$$\psi(p)\ :=\ \inf_{u\in \mathrm{Dom}(\varphi^*)} [\varphi^*(u) - p\cdot u] \tag{6}$$

Proposition 1. *Let us consider a concave flux function ψ_0 defined on a neighborhood of the interval $[0,\omega]$ and satisfying $\psi_0(0) = \psi_0(\omega) = 0$. We assume for simplicity that ψ is differentiable at 0 and ω, and we set $\nu^\flat = \psi'(0) \geq 0$ and $\nu^\sharp = -\psi'(\omega) \geq 0$ We associate with it the continuous concave function ψ:*

$$\psi(p)\ =\ \begin{cases} \nu^\flat p & \text{if } p \leq 0 \\ \psi_0(p) & \text{if } p \in [0,\omega] \\ \nu^\sharp(\omega - p) & \text{if } p \geq \omega \end{cases}$$

Then the Fenchel transform φ^* is bounded above, and its domain $Dom(\varphi^*) = [-\nu^\flat, +\nu^\sharp]$ is bounded:

$$\varphi^*(u)\ =\ \begin{cases} \varphi_0^*(u) & \text{if } u \in [-\nu^\flat, +\nu^\sharp] \\ +\infty & \text{if } u \notin [-\nu^\flat, +\nu^\sharp] \end{cases}$$

Viability hyposolution of the HJB equation [1]. We define a target $\mathcal{C} := \mathcal{H}yp(\mathbf{c})$ as the subset of triples $(T,x,y) \subset \mathbb{R}_+ \times X \times \mathbb{R}$ such that $y \leq \mathbf{c}(T,x)$ (which is the *hypograph* of the function $\mathbf{c}$), where the function $\mathbf{c}(t,x)$ is defined (here) by:

$$\mathbf{c}(t,x)\ :=\ \begin{cases} -\infty & \text{if } t > 0 \text{ and } x > \xi \\ \mathbf{N}_0(x) & \text{if } t = 0 \text{ and } x \geq \xi \\ 0 & \text{if } t \geq 0 \text{ and } x = \xi \end{cases}$$

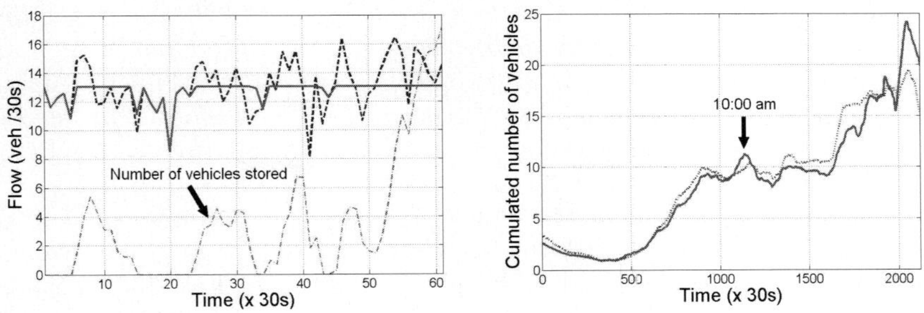

Fig. 1. Left: Flow-time curves. The (actual) measured inflow on the boundary $x = \xi$ is represented by the dashed curve. The continuous curve shows the simulated inflow through boundary $x = \xi$, taking into account the highway capacity. The number of corresponding (remaining) vehicles stored at the $x = \xi$ boundary is shown on the dash dotted curve. **Right:** Comparison between experimental values and simulated values for the cumulated vehicle number $\mathbf{N}(t, \xi + L)$ between ξ and L.

The environment $\mathcal{K} := \mathcal{H}yp(\mathbf{b})$ is the subset of triples $(T, x, y) \subset \mathbb{R}_+ \times X \times \mathbb{R}$ such that $y \leq \mathbf{b}(T, x)$, which is a user-defined function (this is the *hypograph* of the function $\mathbf{b}$). We define the auxiliary control system :

$$\begin{cases} \tau'(t) &= -1 \\ x'(t) &= u(t) \\ y'(t) &= \varphi^*(u(t)) - \psi(v(\tau(t))) \end{cases} \qquad \text{where } u(t) \in [-\nu^\flat, +\nu^\sharp] \tag{7}$$

where φ^* is the Fenchel conjugate function of ψ, as defined previously. To be rigorous, we have to mention *once and for all* that the controls $u(\cdot)$ are measurable integrable functions with values in $\mathrm{Dom}(\varphi^*)$, and thus, ranging $L^1(0, \infty; \mathrm{Dom}(\varphi^*))$, and that the above system of differential equations is valid for almost all $t \geq 0$.

Definition 1. ***The Viability Hyposolution.*** *The capture basin* $\mathrm{Capt}_{(7)}(\mathcal{K}, \mathcal{C})$ *of a target* $\mathcal{C}$ *viable in the environment* $\mathcal{K}$ *under control system (7) is the subset of initial states* (t, x, y) *such that there exists a measurable control* $u(\cdot)$ *such that the associated solution*

$$s \mapsto \left(t - s, x + \int_0^s u(\tau)d\tau, y + \int_0^s (\varphi^*(u(\tau)) - \psi(v(t - \tau)))d\tau\right)$$

to system (7) is viable in $\mathcal{K} = \mathcal{H}yp(\mathbf{b})$ *until it reaches the target* $\mathcal{C} = \mathcal{H}yp(\mathbf{c})$. *The* viability hyposolution $\mathbf{N}$ *is defined by*

$$\mathbf{N}(t, x) := \sup_{(t,x,y) \in \mathrm{Capt}_{(7)}(\mathcal{K}, \mathcal{C})} y \tag{8}$$

Theorem 1. Non-homogenous Dirichlet/Initialvalue Problem with inequality contraints. *The viability hyposolution* $\mathbf{N}$ *defined by (8) is the* largest *upper semicontinuous solution to Hamilton-Jacobi equation (2) satisfying initial and Dirichlet conditions (3) and inequality constraints (4). If the functions ψ, φ^* and v are furthermore Lipschitz, then the viability hyposolution* $\mathbf{N}$ *is its* unique *upper semicontinuous solution in both the contingent Frankowska sense and in the Barron-Jensen/Frankowska sense.*

Other numerical illustrations of the article [2], such as a sup-morphism property, have been carried out using the viability algorithm. The results can be compared with explicit analytical solutions obtained from Lax-Hopf formulas extended to the case of boundary-value problems. The viability kernel algorithm [4] is adapted to the case in which the target $\mathcal{C}$ and the environment $\mathcal{K}$ are hypographs, which allows us to take some specificities of the problem into account. An example of a solution computed with the algorithm is provided in Figure 2. In this Figure, one can see two computations: one of an unconstrained solution (thick line), and one of a constrained solution (thin lines). A cap of $\mathbf{b}(t, x) = 40$ is imposed on the constrained solution, as can be seen in this Figure. From Figure 2, one can observe that the solution with constraints is not the supremum of the solution without constraints and the function $\mathbf{b}$.

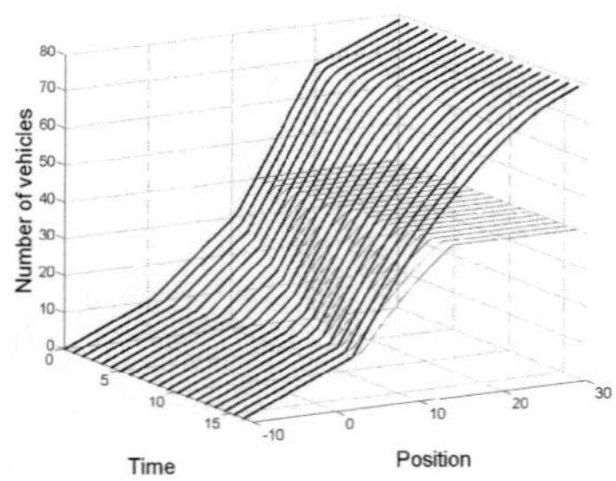

Fig. 2. Plots of $\mathbf{N}(t, x)$ versus t and x. The unconstrained solution is represented by a thick line. The thin line represents the constrained solution of the same system (we set $\mathbf{b}(t, x) = 40$ in this example). Result obtained using a Greenshields [2] flux function.

We use the same experimental set up as in earlier work [5] to assess the performance of the algorithm with highway traffic data: we use three loop detectors in interstate I80 at Emeryville. When the measured inflow (upstream) exceeds the modeled highway capacity (because of noise in the measurements or model inaccuracy), we "store" the corresponding vehicles at $x = \xi$ until they can be released into the highway. The resulting curves are shown in Figure 1 (left): the cutoff happens at $\psi(v(t)) = \delta$ above which the vehicles have to be stored until the highway capacity allows them to enter at $x = \xi$. In this Figure, all numbers of vehicles are per lane. The corresponding number of "stored" vehicles is shown in the same subfigure, and the corresponding $\mathbf{N}(t, \xi + L)$ curves are shown in the right subfigures, where L represents the length of the corresponding highway stretch. The evolution $\mathbf{N}(t, \xi + L)$ thus represents the evolution of the cumulated number of vehicles between ξ and $\xi + L$ as a function of time. Both simulated and measured curves are represented on this plot and show remarkable agreement. The differences between simulation and theory are mainly linked with uncertainties on the numerical values of the parameters of the model.

References

1. J.-P. Aubin. *Viability Theory*. Birkhäuser, Boston, MA, 1991.
2. J.-P. Aubin, A. M. Bayen, and P. Saint-Pierre. Dirichlet problems for some Hamilton-Jacobi equations with inequality constraints. Technical report, Scuola Normale Superiore, Pisa, Italy, May 2006. Preprint di Matematica, no. 4.
3. H. Frankowska. Lower semicontinuous solutions of Hamilton-Jacobi-Bellman equations. *SIAM Journal on Control and Optimization*, 31:257–272, 1993.
4. P. Saint-Pierre. Approximation of the viability kernel. *Applied Mathematics and Optimization*, 29:187–209, 1994.
5. I. S. Strub and A. M. Bayen. Weak formulation of boundary conditions for scalar conservation laws. *International Journal on Nonlinear and Robust Control*, 16(16):733–748, Nov. 2006.

A Method for the Design of Optimal Switching Surfaces for Autonomous Hybrid Systems

Mauro Boccadoro[1], Paolo Valigi[1], and Yorai Wardi[2]

[1] DIEI, Università di Perugia, via Duranti, 93 I-06125 Italy
{mauro.boccadoro,paolo.valigi}@diei.unipg.it
[2] Georgia Institute of Technology, Atlanta, GA 30332
ywardi@ece.gatech.edu

Abstract. This paper discusses the design of optimal feedback control laws for hybrid systems with autonomous (continuous) modes. The solution of similar problems, when the dynamic modes are non linear, is, in general, dependent on the initial conditions. The same problem affects the "open loop" control based on the definition of the optimal switching times. In this work an algorithm is proposed, allowing to build switching surfaces which are optimal for any possible initial condition. It is based on the sensitivity analysis of the optimal switching times with respect to the initial conditions and on the identification of the set of initial conditions maximizing the information relevant to the design of the surface.

1 Introduction

Consider a switched system with autonomous continuous dynamics,

$$\dot{x}(t) = f_{q(t)}(x(t)), \tag{1}$$

$$q^{+}(t) = s(x(t), q(t)). \tag{2}$$

where (1) describes the continuous dynamics of the state variable $x \in \mathcal{X} \subseteq \mathbb{R}^n$ and (2) describes the discrete event dynamics of the system. Since the continuous modes are autonomous, the evolution of the system is determined by the active modes, according to (2). When the function s does not depend by the (continuous) state variable x, the switching instants are determined as exogenous inputs, and the system is controlled in open loop (timing control); when s is dependent only on the state variables, the switching law is given in a feedback form, and it may be defined by switching surfaces in the state space.

To formulate the problem we are interested in, consider a simple execution of (1,2), starting at $x(t_0) = x_0$ with mode 1, switching (exogenously) to mode 2 at time t_1, and terminating either at a *fixed final time* t_2 or in correspondence of a *terminal manifold*, e.g., defined by a function $g_2(x)$, so that t_2 satisfies $g_2(x(t_2)) = 0$. Denote such two sets of executions by χ_t and χ_g, respectively, and let the evolution under mode i be expressed by $x(t) = \varphi_i(t, s, x(s))$.

When an optimal control problem for this system is formulated, according to a cost function $J = \int_{t_0}^{t_2} L(x(t))dt$, for some continuously differentiable function

A. Bemporad, A. Bicchi, and G. Buttazzo (Eds.): HSCC 2007, LNCS 4416, pp. 650–655, 2007.

L, then, if $t_1 = t_1^*$, a (locally) optimal switching time, the following condition is satisfied, see e.g. [1]:

$$c(t_1^*) := p^T(t_1^*)[f_1(x(t_1^*)) - f_2(x(t_1^*))] = 0 \tag{3}$$

where p is the costate (adjoint) variable.

In this work a feedback solution to this problem is pursued through the determination of an (optimal) switching surface. Computational methods exist and are based on the optimization of parametrized switching surfaces [2]; however, the optimal parameter values depend on the particular trajectory chosen to run an optimization program, and thus, on the initial conditions. Ref. [3] addressed a timing optimization problem, and discovered the special structure of the solution for linear quadratic problems by identifying homogeneous regions in the state space which dictate the optimal switches, thus providing a feedback solution to a problem which is formulated in terms of an open loop strategy.

Here we explicitly investigate the relation existing between optimal switching times and initial conditions, studying how the condition of optimality (3) that switching times must satisfy, vary in dependence of the initial conditions. By this analysis, a procedure which builds the switching surface which is the optimal one for any initial condition, denoted G^*, can be set up once the optimal switching instant for one single execution is known.

2 Optimal Switching Times v/s Initial Conditions

It is well known that, under mild assumptions, executions of switched systems are continuous w.r.t. the initial conditions [4]. It is reasonable to expect that also the dependence of t_1^* on x_0 is continuous a.e. in the state space. In this hypothesis, the function $t_1^*(x_0)$, limited to that set of x_0's in which t_1^* is smooth, may be characterized by deriving (3) w.r.t. x_0 and setting this derivative to zero[1]. In fact, if starting from $\tilde{x}_0 = x_0 + \delta x_0$, it results $\tilde{t}_1^* = t_1^* + \delta t_1^*$; then, by continuity, $0 = c(\tilde{t}_1^*) = c(t_1^*) + \frac{dc}{dx_0}\delta x_0 + o(\delta x_0)$. Hence, setting $\frac{dc}{dx_0} = 0$, to satisfy optimality condition for $\tilde{t}_1^*$, allows to compute dt_1^*/dx_0.

The derivation of such formula, carried out in [7], for both types of executions χ_t and χ_s, relies on various results of sensitivity analysis for switched systems, given, among others, in [5,6]. The present case, however, deals with an optimal control problem, and hence needed further derivations relative also to costate variables, which evolve "backward" in time.

Now, the idea is to "build up" G^* by finding a set of optimal switching states, and to determine such states by the variations of the optimal switching times when initial conditions different than a nominal one, for which t_1^* is known, are considered. These variations can be computed, approximated to the first order, using dt_1^*/dx_0.

[1] For this to be valid, also c must be continuuous in t_1, as c depends on x_0 via t_1^*. The RHS of (3) is the sensitivity of a cost function w.r.t. a switching time. As such, c is intrinsically continuous in t_1^*.

Prior to set up this program, the conditions under which optimal switching times yield "well defined" optimal switching states are given.

Theorem 1. *Consider a nominal and a perturbed execution of the set* χ_g, $x(\cdot)$ *and* $y(\cdot)$, *respectively, the first starting at* x_0 *and the latter starting from a point* y_0 *such that it exists a duration* $\delta t_0 : y_0 = \varphi_1(t_0 + \delta t_0, t_0, x_0)$ *(if* $\delta t_0 > 0$, y_0 *lies on the nominal trajectory, and vice versa). Then, for all* $\delta t_0 < t_1^* - t_0$

$$t_1^*(y_0) = t_1^*(x_0) - \delta t_0 \tag{4}$$

This theorem, proved in [7], states that for the two evolutions described the optimal switching *state* is the same. In the same situation, but for evolutions of the type χ_t, the optimal switching state may vary because the perturbed trajectory, switching at $t_1^* - \delta t_0$, reaches the point $x(t_2)$ (of the nominal trajectory) at time instant $t_2 - \delta t_0$, thence "visits" additional states from $t_2 - \delta t_0$ to t_2 (in other words $y(\cdot)_{(t_2 - \delta t_0, t_2]}$ is a set of states not visited by $x(\cdot)$). Such remnants of the perturbed trajectory add further costs, so that two different trajectories, even if the starting point of one of them lies in the trajectory of the other, cannot be properly compared, in terms of optimal switching states.

Remark 1. This may be verified also through the formula dt_1^*/dx_0 which in this situation yields a first order approximation of the kind $\delta t_1^* = -\frac{b}{b+a}\delta t_0$, for suitable terms a and b. As the condition (4) is equivalent to $\delta t_1^* = -\delta t_0$, the additional term a prevents this to hold. It is possible to verify [7], however, that for a case similar to those considered in [3], where the the final dynamic mode is linear, stable and the terminal time tends to infinity, the term a is vanishing.

In force of Theorem 1 and Remark 1, the objective to characterize optimal switching surfaces *independent of the initial conditions* should be pursued considering evolutions ending at terminal manifolds or those evolutions of the family χ_t with the restrictions of Remark 1, since in such cases variations in the switching times define soundly optimal switching states as well (for ease of reference, these evolutions will be said to belong to the family χ_{ss}).

3 Drawing the Optimal Switching Surface

Summarizing the developments up to this point, condition (3) characterizes an optimal switching time in relation to some actual evolution of the system, which depends (only) on the initial condition, being the system autonomous. A switching state is optimal when $c(t_1, x(t_1)) = 0$; accordingly, the optimal switching states are parametrized by the variable t_1^* as $x_1^* = \varphi(t_1^*, x_0)$ (here t_0 is not considered in φ). Notice that the choice of x_0 does not affect the optimality of x_1 for those executions $x(\cdot) \in \chi_{ss}$. In this case, G^* can be drawn by finding the locus of those t_1-parametrized states satisfying (3), a task that can be accomplished, thanks to the sensitivity analysis carried out previously, as follows.

As a starting point, assume that for some initial condition x_0 the optimal switching time t_1^* is known (this may be even found by brute force). Choose a variated initial condition $\tilde{x}_0 = x_0 + \delta x_0$ and compute δt_1 by a linear approximation using dt_1^*/dx_0, for this evolution. The approximation to the optimal switching state for the evolution starting in $\tilde{x}_0$ is $\hat{x}_1 = \varphi(\hat{t}_1, \tilde{x}_0)$, with $\hat{t}_1 = t_1^* + \delta t_1$. To maximize the relevance of the switching state thus found, the variated initial condition should be chosen orthogonal, w.r.t. to x_0, to the flow $f_1(x_0)$. This choice is justified by Theorem 1, since the component of the variation δx_0 on x_0 which is tangent to the flow $f_1(x_0)$ yields no difference on the optimal switching *state*, hence giving no relevant information to the construction of G^*.

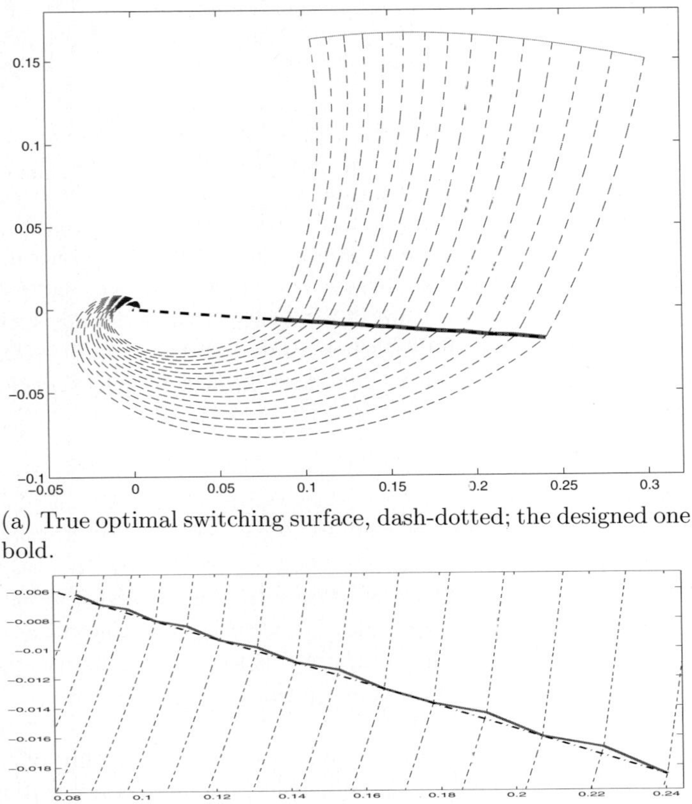

(a) True optimal switching surface, dash-dotted; the designed one, bold.

(b) Close up to the two switching surfaces.

Fig. 1. Executions of the system and the optimal switching surface

To set an iterative procedure, now assign $x_0 := \tilde{x}_0$ and $t_1^* := \hat{t}_1$. The collection of the approximated optimal switching states (possibly interpolated) computed by the above procedure yield the *designed* switching surface.

The main drawback of this scheme is that as the algorithm proceeds, the sensitivity dt_1^*/dx_0 is no longer computed with reference to a true optimal switching state (as it happens for the first step, only) but to an approximated one. For this reason, the surface drawn in this way is likely to "drift away" from the true one. Also, assuming that the *true* surface is piecewise smooth, then near a critical point of it (i.e., a discontinuity in the derivative of the functional description of the surface) a switching state found by the above described first order approximation, may lie quite far from the true surface.

It is possible to overcome this problem by checking, for every approximated switching state $\hat{x}_1$, the corresponding value of c (in Eq. 3), which is of no relevant computational cost, as the evaluation of this condition is a by-product of the computation of dt_1^*/dx_0. This allows to pick, instead of $t_1^* + \delta t_1$, a better, "corrected", value for $\hat{t}_1$, hence "correcting" also $\hat{x}_1$. One way to do this is, for example, to pick two sample switching times s_1 and s_2, both very close to the first choice $t_1^* + \delta t_1$, and to compute, by linear extrapolation, $\hat{t_1}$ according to the found values of $c(s_1)$ and $c(s_2)$ (thus assuming that $c(s, x(s))$ is linear in s close to a point where c is zero).

A numerical implementation of this algorithm was carried out for a system with $f_i(x) = A_i x$, where $A_1 = [-1\ 1; -2\ -1]$ and $A_2 = [-1\ 2; -1\ -1]$, considering $J = \int_0^{t_2} \|x(t)\|^2 dt$ and "very large" t_2 (thus, for an evolution of the type described in Remark 1). For the nominal $x_0 = [.3; .15]$ and the corresponding evolution $x(\cdot) = \varphi(\cdot, x_0)$ the optimal switching time is $t_1^* = .279$. The procedure was run for other initial conditions, which form a transversal surface, showed in figure 1(a), to the flow of f_1, starting from the first x_0. The switching surface found through the procedure described matches quite well the true one (for the case considered the optimal switching surfaces are known to be homogeneous, see [3]), as appears in figure 1.

4 Conclusion

This paper presents a method to determine optimal switching surfaces for hybrid systems with autonomous modes. The idea is to characterize the variations in the optimal switching times corresponding to variations in the initial conditions of the evolution of the system, and to apply this formula for transverse shifts in the initial conditions. An algorithm based on this concept was set up and showed good results.

The results given are relative to a simple evolution with only one switch, a setting which was useful to develop, and illustrate, the main concept. Future work will be devoted to extend the results to more general event driven dynamics.

Acknowledgements. The authors would like to thank Magnus Egerstedt for fruitful discussions.

This research was partially supported by MIUR under grant PRIN 2005092439 (Mauro Boccadoro and Paolo Valigi) and by the National Science Foundation under Grant Number 0509064 (Yorai Wardi).

References

1. M. Egerstedt, Y. Wardi, and F. Delmotte, "Optimal control of switching times in switched dynamical systems," in *42nd IEEE CDC*, (Maui, Hawaii, USA), Dec. 2003.
2. M. Boccadoro, Y. Wardi, M. Egerstedt, and E. Verriest, "Optimal control of switching surfaces in hybrid dynamical systems," *JDEDS*, vol. 15, pp. 433 – 448, Dec. 2005.
3. A. Giua, C. Seatzu, and C. Van Der Mee, "Optimal control of switched autonomous linear systems," in 40th *IEEE CDC*, (Orlando, FL, USA), pp. 2472 – 2477, Dec. 2001.
4. M. Broucke and A. Arapostathis, "Continuous selections of trajectories of hybrid systems," *Systems and Control Letters*, vol. 47, pp. 149–157, 2002.
5. D. Flieller, P. Riedinger, and J. Louis, "Computation and stability of limit cycles in hybrid systems," *Nonlinear Analysis*, vol. 64, p. 352 - 367, 2006.
6. A. Nordmark, "Discontinuity mappings for vector fields with higher order continuity," *International Journal Dynamical Systems*, vol. 17, p. 359 - 376, 2002.
7. M. Boccadoro, P. Valigi, and Y. Wardi, "A method for the design of optimal switching surfaces for autonomous hybrid systems," Tech. Rep. RT-006-06 in http://www.diei.unipg.it/rt/rt_frames.htm

A Hybrid Bellman Equation for Bimodal Systems

Peter Caines[1], Magnus Egerstedt[2], Roland Malhame[3], and Angela Schöllig[4]

[1] McGill University, Electrical and Computer Engineering,
Montreal, Quebec, Canada H3A-2A7
peterc@cim.mcgill.ca

[2] Georgia Institute of Technology, Electrical and Computer Engineering
Atlanta, GA 30323, USA
magnus@ece.gatech.edu

[3] Ecole Polytechnique Motreal, Electrical Engineering,
Montreal, Quebec, Canada H3C 3A7
roland.malhame@polymtl.ca

[4] Georgia Institute of Technology, Civil and Environmental Engineering
Atlanta, GA 30323, USA
Angela.Schoellig@web.de

Abstract. In this paper we present a dynamic programming formulation of a hybrid optimal control problem for bimodal systems with regional dynamics. In particular, based on optimality-zone computations, a framework is presented in which the resulting hybrid Bellman equation guides the design of optimal control programs with, at most, N discrete transitions.

1 Introduction

Optimal control of hybrid systems is certainly not a new topic. For example, the hybrid maximum principle has been well-studied [3,6], and the community now has a clear grasp of what constitutes necessary optimality conditions for very general classes of hybrid systems. Moreover, a number of results of a more computational flavor have complemented the work on the maximum principle, in which specialized classes of systems are considered. (See for example [1,4,7].)

The contribution in this paper fits squarely in between the hybrid maximum principle work and the more computationally flavored work, in that we produce a Bellman equation for hybrid systems, along the lines of [2,5], that can be easily solved once a so-called optimality zone computation has been performed to seed the computation.

2 The Bimodal Hybrid System

2.1 Geometric Framework

Given two open, connected, and simply connected regions D_1, D_2 such that $D_1 \cap D_2 = \emptyset$, forming a partition of the compact state space X in the sense that

A. Bemporad, A. Bicchi, and G. Buttazzo (Eds.): HSCC 2007, LNCS 4416, pp. 656–659, 2007.

$X = (D_1 \cup \partial D_1) \cup (D_2 \cup \partial D_2)$, where the boundaries $\partial D_1, \partial D_2$ are assumed to be finite unions of closed, smooth codimension one submanifolds of X.

The vector fields $f_i(x, u),\ i = 1, 2$ associated with each region are further assumed to satisfy a transversality condition in the sense that for $i = 1, 2$, (i) the vector field $f_i(x, u)$ is non-tangential to the boundary ∂D_i at any point in the relative interior of each component of D_i; (ii) at points x on ∂D_i at which smooth components intersect, the vector field $f_i(x, u)$ is non-tangential to each of the tangent spaces of the intersecting components, for all control values.

2.2 Specifications of Executions

The controlled continuous dynamics of the bimodal hybrid system in any of the two regions, on any bounded time interval, are given by:

$$\dot{x}(t) = \begin{cases} f_1(x(t), u(t)), \, x(t) \in D_1 \\ f_2(x(t), u(t)), \, x(t) \in D_2, \end{cases} \quad x(0) \in D_1 \cup D_2,$$

where f_i is continuously differentiable in x (for all u) on the closure of $D_i,\ i = 1, 2$ (and hence uniformly continuous and uniformly Lipschitz in x on the closure of D_i) for each i. The solutions are interpreted in the Caratheodory sense, and the initial condition $x(0)$ of an admissible execution satisfies $x(0) \in D_1 \cup D_2$.

3 Optimal Control Formulation

3.1 The Hybrid Bellman Equation

Given an initial condition $x(0) \in D_i$, the control input $u \in \mathcal{U}$ gives rise to a trajectory x^u passing through a sequence of N regions (regions 1 and 2 repeatedly). We let $i(x^u)$ denote this index, i.e. $i(x^u) = N$. Corresponding to this index there is an ordered set of half open intervals $\{[t_{i_k}, t_{i_{k+1}});\ 0 \leq k \leq N - 1\}$, such that $x^u(t) \in \overline{D}_i$ for $t \in [t_{i_k}, t_{i_{k+1}})$.

The hybrid optimization problem addressed in this paper is the following:

$$\mathcal{P}_N : \inf_{u \in \mathcal{U}} \int_0^T \ell(x(t), u(t)) dt$$

subject to the constraints that $\dot{x} = f_i(x, u)$ (when $x \in D_i$), $x(0) = x_0,\ x(T) = x_T$, and $|i(x^u)| \leq N$, for a given upper limit on the total number of switches $N \leq \infty$.

Given $x_1, x_2 \in \overline{D}_i$ we let $c_i(x_1, x_2, \Delta)$ denote the infimum of the costs associated with driving the system from x_1 to x_2 during a time horizon Δ without leaving D_i (except possibly at time 0 or Δ). Our ambition is to produce a hybrid Bellman equation describing the cost-to-go dynamics, and for this we define $V_i^M(x_1, \tau)$ as the cost of going from x_1 to x_T in time τ, using exactly M switches, starting with mode i. In other words, by defining the complementary indicator i^c as $i^c = 2$ if $i = 1$ and $i^c = 1$ if $i = 2$, we get

$$V_i^M(x_1, \tau) = \inf_{t \in [0, \tau], x_2 \in \partial \mathcal{D}} \left\{ c_i(x_1, x_2, t) + V_{i^c}^{M-1}(x_2, \tau - t) \right\}.$$

This relation holds as long as $M \geq 1$. If $M = 0$ then we get the following direct simplification

$$V_i^0(x_1, \tau) = c_i(x_1, x_T, \tau).$$

It should be noted already at this point that $V^M(x_1, \tau) = \infty$ for all τ if M is even and x_1 and x_T belong to different regions, or if M is odd and they belong to the same region.

Since we do not want to insist on an *a priori* given number of intersections of the switching surface $\partial\mathcal{D}$, we need to relate V^N to the original problem. If we let $x_0 \in D_i$ then the optimal cost associated with the original problem $W_i^N(x_0, T)$ is given by

$$W_i^N(x_0, T) = \min_{0 \leq k \leq N} V_i^k(x_0, T).$$

3.2 Optimality Zones

The Bellman equation from the previous section immediately allows an interpretation in terms of optimality zones [3]. In fact, it can be noted that except for the initial and final pieces of the trajectory, from x_0 to the first intersection of the switching surface and from the last intersection to x_T, the trajectory is simply given by a concatenation of trajectories from points on the switching surface. This observation leads to a computational framework in which a large computational burden is needed initially when preparing the so-called optimality zones, but once that price is paid, fast solutions are possible.

4 Examples

We first consider an example in which a planar system $x \in \mathbb{R}^2$ is driven between the boundary points $x(0) = (-1, 0)^T$, $x(T) = (1, 0)^T$, under the system dynamics

$$\dot{x} = \begin{cases} \begin{pmatrix} -0.3 & 0.05 \\ -0.5 & 0 \end{pmatrix} x + \begin{pmatrix} 0.1 \\ 1 \end{pmatrix} u, & (1\;1)\, x < 0 \\ \begin{pmatrix} 0.8 & 1 \\ -3 & -5 \end{pmatrix} x + \begin{pmatrix} -0.3 \\ 3 \end{pmatrix} u, & (1\;1)\, x > 0. \end{cases}$$

Moreover, the final time is $T = 1$, with the maximum number of intersection being given by $N = 20$.

The numerical solution is obtained by discretizing the area over which the optimality zone is computed, and we let the space-time domain be discretized into 50 temporal steps (over $(0, T)$) and 40 spatial steps (over each dimension of the state space.) The particular cost function under consideration is the control energy of the control signal (in the L_2-sense), and the resulting optimal solution is given in Figure 1a. In this case, the optimal solution was obtained when only one crossing of the switching surface took place, with the corresponding optimal cost being $W_1^{20}(x_0, 1) = V_1^1(x_0, 1) \approx 22.91$.

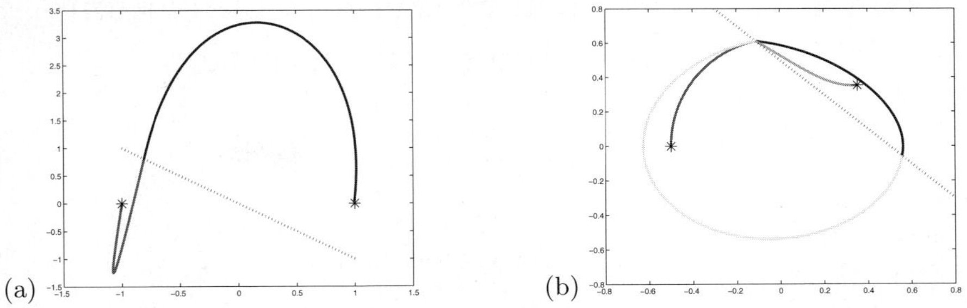

Fig. 1. One switch is optimal (a). Three switches are optimal (b).

In order to highlight the fact that multiple switches may be to prefer, we now consider another linear situation in which

$$\dot{x} = \begin{cases} \begin{pmatrix} \varepsilon_{11} & \omega_1 \\ -\omega_1 & \varepsilon_{12} \end{pmatrix} x + \begin{pmatrix} 0 \\ 1 \end{pmatrix} u, & \begin{pmatrix} 1 & 1 \end{pmatrix} x < 0 \\ \begin{pmatrix} \varepsilon_{21} & \omega_2 \\ -\omega_2 & \varepsilon_{22} \end{pmatrix} x + \begin{pmatrix} 0 \\ 1 \end{pmatrix} u, & \begin{pmatrix} 1 & 1 \end{pmatrix} x > 0. \end{cases}$$

In fact, by making ε_{ij} small, we have a (slightly disturbed) oscillation in the system and if we let $\omega_1 = \pi/4$, $\omega_2 = \pi/2$, $T = \pi/2\omega_1 + \pi/2\omega_2 + 3\pi/2\omega_1 + \pi/4\omega_2 = 9.5$, we get that, using the initial and final conditions $x_0 = (-1/2, 0)^T$, $x_T = (1/\sqrt{8}, 1/\sqrt{8})^T$, a zero control effort would result in a three-switch situation if $\varepsilon = 0$. Using exactly the same numerical parameters and costs as in the previous example, with small but non-zero ε_{ij}, we still get that the three switch-situation is optimal, as seen in Figure 1b, with $W_1^{20}(x_0, 9.5) = V_1^3(x_0, 9.5) \approx 0.014 < V_1^1(x_0, 9.5) \approx 0.043$.

References

1. A. Bemporad, F. Borrelli, and M. Morari. On the Optimal Control Law for Linear Discrete Time Hybrid Systems. *HSCC* 2002.
2. M. Branicky, V. Borkar, S. Mitter. A Unified Framework for Hybrid Control: Model and Optimal Control Theory. *IEEE Transactions on Automatic Control*, Vol. 43, No. 1, pp. 31-45, January, 1998.
3. P. Caines and S. Shaikh. Optimality Zone Algorithms for Hybrid Systems: Efficient Algorithms for Optimal Location and Control Computation. *HSCC* 2006.
4. M. Egerstedt, Y. Wardi, and H. Axelsson. Transition-Time Optimization for Switched Systems. *IEEE Transactions on Automatic Control*, Vol. 51, No. 1, pp. 110-115, Jan. 2006.
5. S. Hedlund and A. Rantzer. Optimal Control of Hybrid System. Proceedings of the *IEEE Conference on Decision and Control*, pp. 3972-3977, Phoenix, AZ, December 1999.
6. H.J. Sussmann. Set-Valued Differentials and the Hybrid Maximum Principle. *IEEE Conference on Decision and Control*, Vol. 1, pp. 558 -563, Dec. 2000.
7. X. Xu and P. Antsaklis. Optimal Control of Switched Autonomous Systems. *IEEE Conference on Decision and Control*, Las Vegas, NV, Dec. 2002.

Networks of Hybrid Systems: Connections Faults Modelling and Detection

Marta Capiluppi[1,2] and Manfred Morari[2]

[1] CASY - DEIS, University of Bologna, 40136 Bologna, Italy
mcapiluppi@deis.unibo.it
[2] Automatic Control Laboratory, ETH Zurich, 8092 Zurich, Switzerland
morari@control.ee.ethz.ch

Abstract. This work aims at presenting a new framework for detection of connection faults in networks of hybrid systems. The considered faults are losses of connections which are represented through automata. The hybrid systems which are the nodes of the network are modelled using a qualitative representation through stochastic automata. A distributed fault detection policy for the considered faults is proposed.

1 Networks of Hybrid Systems

Networks of hybrid systems are composed of hybrid systems cooperating through a network. The systems in the network can be components of the same hybrid system, but can also be complete hybrid systems communicating through the channels of the network. The subsystems composing the network are called **nodes**. In this work the hybrid systems composing the network are connected point-to-point (see fig. 1(a)), meaning also that each system is directly connected with all the others. The choice of this configuration is motivated by its usefulness for fault detection and reconfiguration. In fact if a connection is lost it will be possible to reach a node following a path which is different from the direct one. Another more general reason for this choice is that it is the more widely used configuration for large scale systems, while using, e.g., a broadcast model will increase the transmission time with the scale of the network (see [1] for details). Suppose, then, that the nodes communicate using a protocol similar to the so called *start-stop*. In this protocol the channel can transmit one message at a time in both directions, but not at the same time. The policy of the protocol is the following:

1. the sender sends a *request to send* (*req*) signal to the receiver;
2. the receiver sends a *receive ready* (*rr*) signal to the sender;
3. the sender transmits the message;
4. the receiver sends an *acknowledgement* (*ack*) signal to say it has received the message correctly;
5. the sender sends a *clear to send* (*cs*) signal to say it has finished transmitting messages;
6. both sender and receiver disconnect.

A. Bemporad, A. Bicchi, and G. Buttazzo (Eds.): HSCC 2007, LNCS 4416, pp. 660–663, 2007.

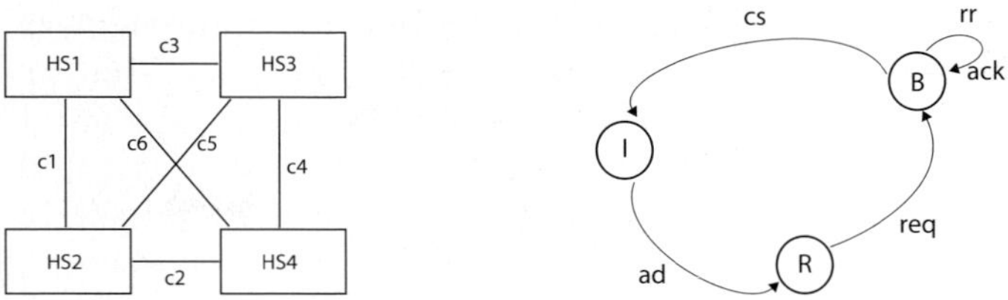

(a) Point-to-point HSN.

(b) Automaton of a connection.

Fig. 1. Hybrid Systems Network (HSN)

For the sake of simplicity in the following it will be considered that the systems always send only one message at a time. It is also necessary to establish an address for each node. Assume the use of local addresses: the nodes can be numbered 1 to n and in our representation we can think they are contained in some signals (ad) the sender node uses to activate the right channel. An automaton representing a connection is reported in fig. 1(b) where it is shown when the connection is considered busy or idle. The connection is normally in state **Idle** (I). The sender specifies the address of the receiver in the signal ad which is related and recognized by the link connecting the correct nodes. Then the connection moves to state **Ready** (R). When the req signal is sent by the sender, the connection moves to state **Busy** (B). With the rr and ack signals the connection remains in B. When the signal cs occurs the connection moves back to I.

A modelling framework which is of particular interest for fault detection is based on the Artificial Intelligence theory of Qualitative Reasoning. Qualitative models are abstractions of the system's behavior, they represent qualitative relations among physical systems. Building qualitative models does not require a mathematical knowledge of the systems and their behavior. This kind of modelling is frequently used for fault diagnosis, when we can be aware of the effects of the faults but not of their quantitative description. Then, in this work, the nodes of the network are modelled using the qualitative approach presented in [2], based on the quantisation of hybrid systems. A quantised system is a system in which all the continuous dynamics are converted into discrete dynamics. This means that in a quantised system only symbolic input/output information is available. Its behavior is then qualitative and non-deterministic, i.e. it is impossible to predict the qualitative output sequence of the quantised system unambiguously for given qualitative initial state and input. In any case it is always possible to know with which probability a qualitative output value $[y(k)]$ occurs (here $[y(k)]$ is the quantised output, and k is the time instant). Using the theory of stochastic automata (see [3]), it is possible to represent the quantised system with a stochastic automaton $S(N_x, N_u, N_y, L, P(z(0)))$, where N_x, N_u, N_y are the sets of states, inputs and outputs of the system respectively, L is the behavioural relation and $P(z(0))$ is the probability distribution of the initial state $z(0)$ of the automaton. The behavioural relation is defined as $L(z', w|z, v) = P([x(1)] = z', [y(0)] = w|[x(0)] = z, [u(0)] = v)$, where the notation $[\cdot]$ refers to quantised signals, z' is the new (meaning next at time 1) state of the automaton, w is the actual output, z the actual state and v the actual input. So the behavioural relation express the probability of the next state to be z' and the actual output to be w if the actual state is z and the actual input is v. In [4] the stochastic automata networks are introduced by extending the concept of stochastic automata with the definition of exogenous signals such as inputs, outputs and faults. In [5] a way to cast distributed quantised systems into stochastic automata networks is presented. The above presented modelling framework is adapted to the kind of systems we want to study. Each node is represented by a quantised system, modelled by a stochastic automaton. The connections are represented by systems themselves, as shown in fig. 1(b). Connections and nodes exchange data using their input/output signals which are supposed to be known. The signals of the protocol are divided into inputs and outputs as follows:

$$\text{Sender}\begin{cases} req, cs, ad \in N_y \\ rr, ack \in N_u \end{cases} ; \text{ Receiver}\begin{cases} req, cs, ad \in N_u \\ rr, ack \in N_y \end{cases} .$$

2 Faults Modelling and Fault Detection Issues

There are many different kinds of faults that can occur in a system: sensor, actuator, communication faults. The effects of these kinds of faults show up in different parts of the system. Sensor faults change the output dynamics, while the actuator faults change the state dynamics. In this work a special attention to communication faults is given. They affect the communication channel. Depending on the severity of the fault the channel can be unreliable or completely lost. Unreliable channels are usually handled by the protocol policy, i.e. if some messages are lost or some bits are changed, the protocol usually is robust enough to recall the message or to recover the wrong transmission. The above specified communication policy is weak in this sense, but some extensions exist such as the so called *go back to n* protocol (see [1]). In this protocol the sender has buffers of the sent and received messages because the receiver sends an acknowledgement signal each time it receives a message; if a message is lost the buffers allow to compare the sent and the received messages and go back to the (n^{th}) message lost. Then more severe faults such as the complete loss of a channel are interesting, because they cannot be managed by a robust protocol. The model of the connection with this fault is represented in fig. 2. Here the fault is called f and after its occurrence the connection moves to the faulty state F. The fault obviously depends on the transmission signals. In fact when one of these signals is expected but not received, a fault has possibly occurred. Note that faults on the nodes are not considered at this level, but a signal missing can also be due to a node internal fault, which should be distinguished from the connection fault. Note also that from the connection point of view it is not possible to relate the fault to one of the signal missing, because the connection is not aware of the nodes communication policy, while the protocol is managed by the nodes themselves. The connection faults for the presented systems can be detected using a

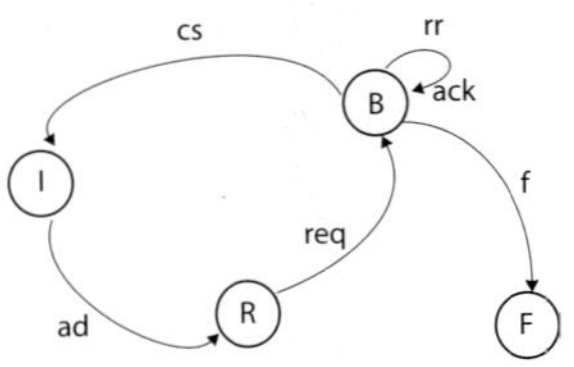

Fig. 2. Automaton of a connection with the faulty state

decentralized policy. This allows the fault detection and isolation (FDI) unit not to check all the connections and the nodes even for big complex systems. This can be realised giving a state estimator to each node and a fault detection unit to each pair of nodes. The state estimator aims at giving an estimate of the state of the system to avoid mistaking internal node faults for connection faults, while the fault detection unit supervises the pair of nodes that communicate to detect if a fault on the channel occurs. Practically the estimator checks if the state of the node is the nominal one or one of the possible faulty ones. If the estimators of both communicating nodes claim that they are in the nominal state but the fault detection unit claims that some transmission signal (related to the fault f) is missing, then the fault detection unit can establish that a fault on the connection has occurred.

3 Conclusions

In this work a framework for detection of connections breakdowns in networks of hybrid systems has been presented. With the use of qualitative modelling and estimation techniques presented in literature it is possible to achieve fault detection in a distributed way, by supervision of pairs of communicating nodes of the network. This method is applicable to large scale systems, where a plant has to communicate data to a remote station and when many connections are present because it is impossible to schedule a polling for all of them.

References

1. Tanenbaum, A.: Computer Networks. 4 edn. Prentice Hall (2003)
2. Lunze, J., Raisch, J.: Discrete models for hybrid systems. in Modelling, Analysis and Design of Hybrid Systems, LNCIS 279, Springer (2002)
3. Blanke, M., Kinnaert, M., Lunze, J., Staroswiecki, M.: Diagnosis and Fault Tolerant Control. Springer (2003)
4. Neidig, J., Lunze, J.: Decentralised diagnosis of automata networks. In: Proceedings of the 16th IFAC World Congress, Prague, CZ (2005)
5. Schröder, J.: Modelling, State Observation and Diagnosis of Quantised Systems. Lecture Notes in Control and Information Sciences. Springer (2003)

Switching-Based Lyapunov Function and the Stabilization of a Class of Non-holonomic Systems*

Daniele Casagrande[1], Alessandro Astolfi[2,3], and Thomas Parisini[1]

[1] Università degli Studi di Trieste, Dipartimento di Elettrotecnica, Elettronica ed Informatica, Via Valerio 10 - 34127 Trieste, Italy
dcasagrande@units.it, t.parisini@paperplaza.net

[2] Imperial College London, Department of Electrical and Electronic Engineering, Exhibition Road, London SW7 2AZ, UK
a.astolfi@ic.ac.uk

[3] Università degli Studi di Roma "Tor Vergata", Dipartimento di Informatica, Sistemi e Produzione, Via del Politecnico, 1 - 00133 Roma, Italy
astolfi@disp.uniroma2.it

Abstract. This paper describes a sufficient condition, based on a new definition of Lyapunov function for switched systems, for the existence of a time-varying switching control scheme which globally asymptotically stabilizes the zero equilibrium of a class of non-holonomic systems.

We present the new concept of *switching-based Lyapunov function* (SBLF) which can be used to analize the stabilizability of the equilibrium of non-holonomic systems (NHS's). Without providing a general overview on hybrid systems (HS's), for which we refer the reader to the extensive literature (see, for instance, [1,2,3]), we focus on autonomous *switched systems* (SS's); in particular, we consider a generic control system $\dot{\mathbf{x}}(t) = \mathbf{f}[\mathbf{x}(t), \mathbf{w}(t)]$ to which a switched feedback is applied, each control law of which, $\mathbf{w}_k$, depends on the state (i.e. $\mathbf{w}_k = \mathbf{w}_k(\mathbf{x})$) in such a way that the resulting closed loop system is an autonomous SS. Now, several theoretical tools are already available for the analysis of the stability of the equilibrium of HS's, of which SS's are a sub-class, such as the concepts of *common* or *multiple Lyapunov functions*. The new approach that we follow herein is based on the consideration that the main idea behind Lyapunov's results is that a function V of the state $\mathbf{x}$ can be used to characterized the stability of an equilibrium of the system, provided that V is non-negative for all $\mathbf{x}$ and non-increasing over time. Analogously, the concept that we want to formalize is the following: *for a dynamical system whose equilibrium state is* $\mathbf{x} = \mathbf{0}$, *it is possible to design a switching control scheme based on the value of a positive definite function* $V(\mathbf{x})$ *in such a way that* $V[\mathbf{x}(t)]$ *is non-increasing over time for*

* This work has been partially supported by the Italian Ministry of University and Research. D. Casagrande has been partially supported through a European Community Marie Curie Fellowship in the framework of the CTS, contract number: HPMT-CT-2001-00278. A. Astolfi is partially supported by the LEVERHULME TRUST.

A. Bemporad, A. Bicchi, and G. Buttazzo (Eds.): HSCC 2007, LNCS 4416, pp. 664–667, 2007.

the switched system. To this aim, not only the first-order time-derivative will be considered but also the time-derivatives of higher order. To see how this concept can lead to a stability theorem, we start with the following definitions.

Definition 1. *Consider an autonomous SS characterized by a continuous state* $\mathbf{x} \in \mathbb{R}^n$ *and by a discrete state* $q \in \mathcal{Q} = \{1, \ldots, N\}$*:*

$$\begin{cases} \dot{\mathbf{x}}(t) = \tilde{\mathbf{f}}_{q_k}[\mathbf{x}(t)]\,, \\ \tau_{k+1} = \tau_k + \sigma_\tau[q_k, \mathbf{x}(\tau_k)]\,, \\ q_{k+1} = \sigma_q[q_k, \mathbf{x}(\tau_{k+1})]\,, \end{cases} \tag{1}$$

where $\tilde{\mathbf{f}}_{q_k} : \mathbb{R}^n \to \mathbb{R}^n$ *is associated to* q_k*,* $\tilde{\mathbf{f}}_{q_k} \in \{\tilde{\mathbf{f}}_1, \tilde{\mathbf{f}}_2, \ldots, \tilde{\mathbf{f}}_N\}$ *and* $\tilde{\mathbf{f}}_q(\mathbf{0}) = \mathbf{0}$*, for all* $q \in \mathcal{Q}$*. Moreover, in (1), the functions* $\sigma_\tau : \mathcal{Q} \times \mathbb{R}^n \to \mathbb{R}^+$ *and* $\sigma_q : \mathcal{Q} \times \mathbb{R}^n \to \mathcal{Q}$ *are the updating rules of* τ *and* q*, respectively, and* $\mathbf{x}(\tau_k)$ *denotes the value of the continuous state at the beginning of the time-interval* $[\tau_k, \tau_{k+1})$*. A smooth function* $V : \mathbb{R}^n \to \mathbb{R}$ *is a* switching-based *(SB)* Lyapunov function *for system (1) if* $V(\mathbf{0}) = 0$*,* V *is radially unbounded and positive definite and for each pair of time-instants* $t_1 \in \mathbb{R}^+$ *and* $t_2 \in \mathbb{R}^+$*, with* $t_2 \geqslant t_1$*,* $V[\mathbf{x}(t_2)] \leqslant V[\mathbf{x}(t_1)]$*.*

Note that the function may assume a constant value for a whole interval $[\tau_k, \tau_{k+1}]$ and may be not differentiable w.r.t. time in some switching time-instants.

Definition 2. *The equilibrium in* $\mathbf{x} = \mathbf{0}$ *of the continuous dynamics of the HS (1) is* SB stable *if system (1) admits a SBLF.*

The notions of SBLF is useful to state the following results (for the proof see [4]).

Theorem 1. *Consider a system* $\Sigma : \dot{\mathbf{x}}(t) = \mathbf{f}[\mathbf{x}(t), \mathbf{w}(t)]$ *with* $\mathbf{f}(\mathbf{0}, \mathbf{0}) = \mathbf{0}$*. If there exist a set* $\mathcal{Q}^* \subset \mathbb{N}$*, whose cardinality is* N*, two functions* $\sigma_\tau^* : \mathcal{Q}^* \times \mathbb{R}^n \to \mathbb{R}^+$ *and* $\sigma_q^* : \mathcal{Q}^* \times \mathbb{R}^n \to \mathcal{Q}^*$ *and a set of* N *control laws* $\mathbf{w}_1^*[\mathbf{x}(t)], \ldots, \mathbf{w}_N^*[\mathbf{x}(t)]$ *such that setting* $\tilde{\mathbf{f}}_{q_k}[\mathbf{x}(t)] = \mathbf{f}\{\mathbf{x}(t), \mathbf{w}_{q_k}^*[\mathbf{x}(t)]\}$*, the equilibrium of the HS*

$$\begin{cases} \dot{\mathbf{x}}(t) = \tilde{\mathbf{f}}_{q_k}[\mathbf{x}(t)]\,, \\ \tau_{k+1} = \tau_k + \sigma_\tau^*[q_k, \mathbf{x}(\tau_k)]\,, \\ q_{k+1} = \sigma_q^*[q_k, \mathbf{x}(\tau_{k+1})]\,, \end{cases}$$

is SB stable, then, if $\mathbf{w}$ *is selected according to a proper switching strategy, the equilibrium in* $\mathbf{x} = \mathbf{0}$ *is locally stable in the sense of Lyapunov.* □

Theorem 2. *Suppose that, in addition to the hypotheses of Theorem 1,* V *is such that for all* $m \in \mathbb{Z}^+$ *there exists* L_m *such that* $\left|\frac{d^m V[\mathbf{x}(t)]}{dt^m}\right| < L_m$ *for all* $t > 0$*. Moreover, suppose that*[1] *there exists* $T_m > 0$ *such that* $\sigma_\tau^*[q_k, \mathbf{x}(\tau_k)] > T_m$ *for all* $k \in \mathbb{Z}^+$ *and that it is possible to associate to* $\mathcal{Q}^*$ *a set of* N *continuous negative semidefinite functions from* $\mathbb{R}^n$ *to* $\mathbb{R}$*,* $\eta_1, \ldots, \eta_N$*, such that for all* $\mathbf{x} \neq \mathbf{0}$ *there exists* $q \in \mathcal{Q}^*$ *such that* $\eta_q(\mathbf{x}) < 0$*. Suppose also that* $\mathcal{Q}^*$ *can be partitioned into* M *(disjoint) subsets,* $\mathcal{Q}_1^*$*,* $\mathcal{Q}_2^*, \ldots,$ $\mathcal{Q}_M^*$*, such that, defining, for* $i = 1, \ldots, M$*,*

$$\mathcal{X}_i \triangleq \{\mathbf{x} \in \mathbb{R}^n \text{ such that } \mathbf{x} \neq \mathbf{0} \text{ and } \eta_q(\mathbf{x}) = 0\,, \forall\, q \in \mathcal{Q}_i^*\}\,,$$

[1] Note that this requirement implies the absence of Zeno behaviour and chattering, conditions that a switched system should fulfill in practice.

and, for a given set $\mathbf{X} \subset \mathbb{R}^n$, $\mathcal{Q}^*_{\mathbf{X}} \triangleq \{q \in \mathcal{Q}^* \text{ such that } \eta_q(\mathbf{x}) = 0\,, \forall\, \mathbf{x} \in \mathbf{X}\}$, *the following three conditions hold*[2]*.*

(C1) For all $\mathbf{x}(\tau_k) \in \mathbb{R}^n$ *and for all* $m \in \mathcal{Q}^*_1$ *we have* $\lim_{t\to\tau_k^+} \left[\dot{V}(t)\Big|_{q=m} \right] = \eta_m[\mathbf{x}(\tau_k)]$.

(C2) For all $j = 2, \ldots, M$, *for all* $\mathbf{x}(\tau_k) \in \bigcap_{i=1}^{j-1} \mathcal{X}_i$ *and for all* $m \in \mathcal{Q}^*_j$ *we have*[3] $\lim_{t\to\tau_k^+} \left[\frac{d^j V(t)}{dt^j}\Big|_{q=m} \right] = \eta_m[\mathbf{x}(\tau_k)]$.

(C3) σ^*_q *is such that for all* $\mathbf{X} \subset \mathbb{R}^n \setminus \{\mathbf{0}\}$ *eiher*[4] $\lim_{k\to\infty} [\min_{\bar{\mathbf{x}}\in\mathbf{X}} \|\mathbf{x}(\tau_k) - \bar{\mathbf{x}}\|] \neq 0$ *or for all* $\bar{k}$ *there exists* $k > \bar{k}$ *such that* $q_k \notin \mathcal{Q}^*_{\mathbf{X}}$.

Then the zero-state equilibrium is globally asymptotically stable. □

The control system Σ appearing in the statements of the theorems above is very general. Nevertheless, we have developed the theory of SB stability motivated by the problem of the stability of the equilibrium of NHS's. These are systems of great interest as they model a number of mechanical systems such as multi-fingered robot hands and wheeled mobile robots (see, among several references, [5] and [6]). For these systems, the controllability problems have been widely investigated (see, e.g., [7] and the references cited therein) while the stability problem is still a challenging issue since they are not asymptotically stabilizable by means of smooth control laws (see [8] and the cited reference), motivating the use either of time-varying or of discontinuous feedback control laws. Dedicated solutions for particular NHS's can be found in the literature (see, for instance, [9,10,11]); a general treatment of the problem has also been addressed, mainly to NHS's in "chained" (or "power") form (see, for instance, [12,13]). The class of systems considered herein is even more general than the class of chained systems and is defined as follows. For a generic $p \in \mathbb{N}$, consider the system constituted by two linear equations (level L0) together with, for each $m = 1, \ldots, p$, a basis of the subspace of m-dimensional non-integrable forms (level Lm)

$$\begin{array}{cccc} \text{L0} & \text{L1} & \cdots & \text{L}p \\ \dot{x} = u\,, \dot{y} = v & \dot{z}_1 = xv & & \dot{\mathbf{z}}_p = \left(x^{p-1}, x^{p-2}y, \ldots, xy^{p-2}, y^{p-1}\right)^\top xv \end{array} \tag{2}$$

where u and v are the input variables. The above equations describe a class of systems, as p varies, each of which has dimension equal to $n = 2 + \sum_{i=1}^p i = \frac{p(p+1)}{2} + 2$. Note that the set of the two equations of level L0 together with the first equation of each level is equivalent to a NHS in power form.

We conclude the paper claiming that the equilibrium in $\mathbf{x} = \mathbf{0}$ of the above class can be rendered globally asymptotically stable by properly designing the

---

[2] For a generic function $F[\mathbf{s}(t)]$ we denote by $F(t)|_{q=i}$ the value of $F[\mathbf{s}(t)]$ when $\mathbf{s}(t)$ varies according to the dynamics associated to the discrete state $q = i$.

[3] Condition (C2) expresses the fact that when, for a particular vale of q, the trajectory approaches a state in which the first $j-1$ time-derivatives of V vanish there is at least another value of q for which the j-th derivative is strictly negative.

[4] The meaning of condition (C3) is to introduce a constraint that the switching strategy must fulfill to guarantee that the feedback scheme does not get stuck in a discrete state $\bar{q}$ while the trajectory $\mathbf{x}(t)$ tends to a state $\bar{\mathbf{x}} \neq \mathbf{0}$ such that $\eta_{\bar{q}}(\bar{\mathbf{x}}) = 0$.

switching control scheme [14]. Two earlier works [15,16] take into account the first two instances of the above class and show that one of the main advantages of the proposed method is that the control laws may be selected in different ways, in particular accomodating *saturations*. Unfortunately, the application of the method to real systems may yield a highly oscillatory behaviour [16] what will be the motivation for further research.

References

1. Antsaklis, P., ed.: Special Issue on Hybrid Systems. Volume 88. Proceedings of the IEEE (2000)
2. Morse, A., Pantelides, C., Sastry, S., Schumacher, J., eds.: Special issue on hybrid systems. Volume 35. Automatica (1999)
3. van der Schaft, A., Schumacher, H.: An Introduction to Hybrid Dynamical Systems. Volume 251 of Lecture Notes in Control and Information Sciences. Springer-Verlag, London (2000)
4. Casagrande, D., Astolfi, A., Parisini, T.: Achieving stability in non-holonomic systems by means of switched control laws. In F. Lamnabhi-Lagarrigue, A. Loria, E.P., Laghrouche, S., eds.: Taming Heterogeneity and Complexity of Embedded Control, (International Scientific and Technical Encyclopedia) to appear.
5. R.M. Murray, Li, Z., S.S. Sastry: A mathematical introduction to robotic manipulation. CRC Press (1994)
6. A.M. Bloch: Nonholonomic mechanics and control. Springer (2003)
7. Kolmanovsky, I., N.H. McClamroch: Developments in nonholonomic control problems. IEEE Control Systems **15** (1995) 20–36
8. E.P. Ryan: On Brockett's condition for smooth stabilizability and its necessity in a context of nonsmooth feedback. SIAM J. Contr. Optim. **32**(6) (1994) 1597–1604
9. Aicardi, M., Casalino, G., Bicchi, A., Balestrino, A.: Closed loop steering of unicycle-like vehicles via Lyapunov techniques. IEEE Robotics & Automation Magazine **2**(1) (1995) 27–35
10. J.P. Hespanha, A.S. Morse: Stabilization of nonholonomic integrators via logic-based switching. Automatica **35** (1999) 385–393
11. Oriolo, G., Vendittelli, M.: A framework for stabilization of general nonholonomic systems with an application to the ball-plate mechanism. IEEE Trans. Robot. **21**(2) (2005) 162–174
12. Astolfi, A.: Discontinuous control of nonholonomic systems. Systems & Control Letters **27**(1) (1996) 37–45
13. O.J. Sørdalen, Egeland, O.: Exponential stabilization of nonholonomic chained systems. IEEE Trans. Autom. Contr. **40**(1) (1995) 35–49
14. Casagrande, D.: Stabilizzazione di sistemi non lineari mediante leggi di controllo a commutazione. caso dei sistemi non lineari e generalizzazioni. PhD thesis, Università degli Studi di Trieste, ITALY. An english version can be found at http://control.units.it/dcasagrande/PhD_thesis.pdf (2006)
15. Casagrande, D., Astolfi, A., Parisini, T.: Control of nonholonomic systems: a simple stabilizing time-switching strategy. In: Proc. of the 16th IFAC World Congress. (2005)
16. Casagrande, D., Astolfi, A., Parisini, T.: A stabilizing time-switching control strategy for the rolling sphere. In: Proc. of the 44th CDC-ECC. (2005)

Composing Semi-algebraic O-Minimal Automata*

A. Casagrande[1,2], P. Corvaja[1], C. Piazza[1], and B. Mishra[3,4]

[1] DIMI, Università di Udine, Via delle Scienze, 206, 33100 Udine, Italy
[2] Istituto di Genomica Applicata, Via J.Linussio, 51, 33100 Udine, Italy
[3] Courant Institute of Mathematical Science, NYU, New York, U.S.A.
[4] NYU School of Medicine, 550 First Avenue, New York, 10016 U.S.A.
{casa,corvaja,piazza}@dimi.uniud.it, mishra@nyu.edu

Abstract. This paper addresses questions regarding the decidability of hybrid automata that may be constructed hierarchically and in a modular way, as is the case in many exemplar systems, be it natural or engineered. Since an important step in such constructions is a product operation, which constructs a new product hybrid automaton by combining two simpler component hybrid automata, an essential property that would be desired is that the reachability property of the product hybrid automaton be decidable, provided that the component hybrid automata belong to a suitably restricted family of automata. Somewhat surprisingly, the product operation does not assure a closure of decidability for the reachability problem. Nonetheless, this paper establishes the decidability of the reachability condition over automata which are obtained by composing two semi-algebraic o-minimal systems. The class of semi-algebraic o-minimal automata is not even closed under composition, i.e., the product of two automata of this class is not necessarily a semi-algebraic o-minimal automaton. However, we can prove our decidability result combining the decidability of both semi-algebraic formulæ over the reals and linear Diophantine equations. All the proofs of the results presented in this paper can be found in [1].

1 Semi-algebraic O-Minimal Automata and Composition

Hybrid automata are systems in which discrete and continuous evolutions are mixed. In particular, their discrete nature is usually modeled through *labeled directed graphs* (called graphs in the rest of this paper), i.e., directed graphs with labels on the edges. On this kind of graphs we define: a *path ph* as sequence of edges; a *cycle* as a path in which the first and the last edges coincide; a *simple cycle* as a cycle without other repetitions.

A *hybrid automaton* $H = (Z, Z', \mathcal{V}, \mathcal{E}, Inv, \mathcal{F}, Act, Res)$ of dimension k consists of the following components:

* This work is developed within HYCON, contract number FP6-IST-511368 and supported by the projects PRIN 2005 2005015491 and BIOCHECK. B.M. is supported by funding from two NSF ITR grants and one NSF EMT grant.

A. Bemporad, A. Bicchi, and G. Buttazzo (Eds.): HSCC 2007, LNCS 4416, pp. 668–671, 2007.

1. $Z = \langle Z_1, \ldots, Z_k\rangle$ and $Z' = \langle Z'_1, \ldots, Z'_k\rangle$ are two vectors of reals variables;
2. $\langle \mathcal{V}, \mathcal{E}\rangle$ is a labeled directed graph; the vertices, $\mathcal{V}$, are called *locations*;
3. Each vertex $v \in \mathcal{V}$ is labeled by the formulæ $Inv(v)[Z]$ and $Dyn(v)[Z, Z', T] \stackrel{\text{def}}{=} Z' = f_v(Z, T)$, where f_v is the solution of the continuous vector field $\mathcal{F}$;
4. Each edge $e \in \mathcal{E}$ is labeled by the two formulæ $Act(e)[Z]$ and $Res(e)[Z, Z']$.

A *state* q of H is a pair $\langle v, r\rangle$, where $v \in \mathcal{V}$ is a location and $r = \langle r_1, \ldots, r_k\rangle \in \mathbb{R}^k$ is an assignment of values for the variables of Z. A state $\langle v, r\rangle$ is said to be *admissible* if $Inv(v)[r]$ is true. The semantics of hybrid automata is given in terms of *continuous* $\xrightarrow{t}_C$ and *discrete* $\xrightarrow{e}_D$ transitions over asmissible states in the standard way [1]. We use the notation $q \to q'$ to denote that either $q \xrightarrow{t}_C q'$ or $q \xrightarrow{e}_D q'$. A *trace* $tr = q_0, q_1, \ldots, q_n$is a sequence of admissible states connected through transitions. The automaton H *reaches* a point $s \in \mathbb{R}^k$ (in time t) from a point $r \in \mathbb{R}^k$ if there exists a trace $tr = q_0, \ldots, q_n$ of H such that $q_0 = \langle v, r\rangle$ and $q_n = \langle u, s\rangle$, for some $v, u \in \mathcal{V}$ (and t is the sum of the continuous transitions elapsed times). Given a trace tr of H we can identify at least one path of $\langle \mathcal{V}, \mathcal{E}\rangle$ underlying tr. We call such paths *corresponding paths* of tr.

A well-known class of hybrid automata is the class of *o-minimal hybrid automata* [2], defined by using formulæ taken over an ambient o-minimal theory [3] and by imposing the constraints of *constant resets at discrete transitions*. In the case of o-minimal automata defined by a decidable theory, reachability can be decided through bisimulation [2]. A theory which is both o-minimal and decidable is the first-order theory of $(\mathbb{R}, 0, 1, +, *, <)$ [4], also known as the theory of semi-algebraic sets. In this paper we focus on *semi-algebraic o-minimal hybrid automata*, i.e., o-minimal hybrid automata built over the theory of $(\mathbb{R}, 0, 1, +, *, <)$.

Let $H_1 = (Z1, Z1', \mathcal{V}_1, \mathcal{E}_1, Inv_1, \mathcal{F}_1, Act_1, Res_1)$ and $H_2 = (Z2, Z2', \mathcal{V}_2, \mathcal{E}_2, Inv_2, \mathcal{F}_2, Act_2, Res_2)$ be hybrid automata over distinct variables and let ϵ be a label not occurring in $\mathcal{E}_1 \cup \mathcal{E}_2$. The *product* (see, e.g., [5,6]) of H_1 and H_2 is the hybrid automaton $H_1 \otimes H_2 = (Z, Z', \mathcal{V}, \mathcal{E}, Inv, \mathcal{F}, Act, Res)$, where:

1. Z (Z') is the concatenation of $Z1$ and $Z2$ ($Z1'$ and $Z2'$, respectively);
2. $\mathcal{V} = \mathcal{V}_1 \times \mathcal{V}_2$ and $\mathcal{E} = \mathcal{E}_x \cup \mathcal{E}^1 \cup \mathcal{E}^2$, where: $\mathcal{E}_x = \{e_{e_1,e_2} \,|\, e_1 \in \mathcal{E}_1 \text{ and } e_2 \in \mathcal{E}_2\}$, $\mathcal{E}^1 = \{e_{e,v} \,|\, e\mathcal{E}_1 \text{ and } v \in \mathcal{V}_2\}$, and $\mathcal{E}^2 = \{e_{v,e} \,|\, v \in \mathcal{V}_1 \text{ and } e \in \mathcal{E}_2\}$.
3. $Inv(\langle v_1, v_2\rangle)[Z] \stackrel{\text{def}}{=} Inv(v_1)[Z1] \wedge Inv(v_2)[Z2]$;
4. $Dyn(\langle v_1, v_2\rangle)[Z, Z', T] \stackrel{\text{def}}{=} Dyn(v_1)[Z1, Z1', T] \wedge Dyn(v_2)[Z2, Z2', T]$;
5. $Act(e_{a,b})[Z] \stackrel{\text{def}}{=} \begin{cases} Act(a)[Z1] \wedge Act(b)[Z2] & \text{if } e_{a,b} \in \mathcal{E}_x \\ Act(a)[Z1] & \text{if } e_{a,b} \in \mathcal{E}^1 \\ Act(b)[Z2] & \text{if } e_{a,b} \in \mathcal{E}^2 \end{cases}$
6. $Res(e_{a,b})[Z, Z'] \stackrel{\text{def}}{=} \begin{cases} Res(a)[Z1] \wedge Res(b)[Z2] & \text{if } e_{a,b} \in \mathcal{E}_x \\ Res(a)[Z1] \wedge Z2' = Z2 & \text{if } e_{a,b} \in \mathcal{E}^1 \\ Z1' = Z1 \wedge Res(b)[Z2] & \text{if } e_{a,b} \in \mathcal{E}^2 \end{cases}$

We study the reachability problem over $H_1 \otimes H_2$, where H_1 and H_2 are semi-algebraic o-minimal hybrid automata, considering sets of points of the form $I = I_1 \times I_2$ and $F = F_1 \times F_2$. As noticed in [6] the decidability of reachability is not always preserved under product operations, i.e., it is possible that reachability is decidable over two classes of automata, but not over the product class.

2 Our Results

A common approach in deciding reachability of hybrid automata is that of discretizing the automata using equivalence relations (see, e.g., [2]). A powerfull equivalence reduction preserving reachability is *time-abstract simulation.* Let H and $\overline{H}$ be two automata, a relation R between H and $\overline{H}$ states is a *time-abstract simulation* if and only if, for each pair of states q and $\tilde{q}$ of H and for each state q' of $\overline{H}$, if $(q, q') \in R$ then: for each edge e of H such that $q \xrightarrow{e}_D \tilde{q}$ in H there exist an edge e' and a state $\tilde{q}'$ such that $\text{Label}(e) = \text{Label}(e')$, $q' \xrightarrow{e'}_D \tilde{q}'$ in $\overline{H}$, and $(\tilde{q}, \tilde{q}') \in R$; if $q \to_C \tilde{q}$ in H, then there exists a state $\tilde{q}'$ such that $q' \to_C \tilde{q}'$ in $\overline{H}$ and $(\tilde{q}, \tilde{q}') \in R$. We cannot use time-abstract simulation to decide reachability.

Theorem 1. *There exist products of two semi-algebraic o-minimal automata, which possess an infinite simulation quotient.*

In order to study the reachability problem over the product of two semi-algebraic o-minimal automata we exploit a characterization of the reachability problem over hybrid automata based on first-order formulæ over the reals (see [1]): there exists a formula $Reach(H)(ph)[Z, Z', T]$ such that $r \in \mathbb{R}^k$ reaches $s \in \mathbb{R}^k$ in time t through a trace tr having ph as a corresponding path if and only if $Reach(H)(ph)[r, s, t]$ holds. We can also characterize through a first-order formula the set of time instants $Time(ph)$ in which a path ph can be covered starting and ending with discrete transitions. This means that $Time(ph)$ is a finite union of intervals and points. Moreover, we exploit the existence of a canonical path decomposition: given a semi-algebraic o-minimal automaton, from any path we can extract both an acyclic part and a set of simple cycles. In this case we say that the set of simple cycles is *augmentable* to the acyclic part. The global time necessary to cover the path is then equal to the sum of the time necessary to cover the acyclic part plus multiples of the times we can spend over the simple cycles. What is important is that in the case of o-minimal automata the time we can spend over a cycle does not depend on the starting and ending points.

Theorem 2. *Let H_1 and H_2 be o-minimal automata of dimensions k_1 and k_2, respectively, and $I_1, F_1 \subseteq \mathbb{R}^{k_1}$ and $I_2, F_2 \subseteq \mathbb{R}^{k_2}$ be characterized by the first-order formulæ $\mathcal{I}_1[Z1]$, $\mathcal{F}_1[Z1]$, $\mathcal{I}_2[Z2]$, and $\mathcal{F}_2[Z2]$. The automaton $H_1 \otimes H_2$ reaches $F_1 \times F_2$ from $I_1 \times I_2$ if and only if there exist two acyclic paths ph_1 and ph_2 and two sets of paths $\mathsf{PH}_1 = \{ph^1_1, \ldots, ph^1_{n_1}\}$ and $\mathsf{PH}_2 = \{ph^2_1, \ldots, ph^2_{n_2}\}$ augmentable to ph_1 and ph_2, respectively, such that for each $h \in \{1, 2\}$ it holds that there exists t_h satisfying $\exists Zh, Zh'(Reach(H_h)(ph_h)[Zh, Zh', T] \wedge \mathcal{I}_h[Zh] \wedge \mathcal{F}_h[Zh'])$ and for each ph^h_i there are two finite non empty sets $\{t^0_{(i,h)}, \ldots, t^{m(i,h)}_{(i,h)}\} \subseteq Time(ph^h_i)$ and $\{k^0_{(i,h)}, \ldots, k^{m(i,h)}_{(i,h)}\} \subseteq \mathbb{N}_{>0}$ such that*

$$\sum_{i=1}^{n_1} \sum_{j=0}^{m(i,1)} k^j_{(i,1)} * t^j_{(i,1)} + t_1 = \sum_{i=1}^{n_2} \sum_{j=0}^{m(i,2)} k^j_{(i,2)} * t^j_{(i,2)} + t_2$$

We say that $H_1 \otimes H_2$ *reaches* $F_1 \times F_2$ *from* $I_1 \times I_2$ *through* $ph_1, \mathsf{PH}_1, ph_2, \mathsf{PH}_2$ if the hypothesis of Theorem 2 are satisfied. Given a set PH of paths we say that PH is *time-empty* if either $\mathsf{PH} = \emptyset$ or for each $ph \in \mathsf{PH}$ it holds that $Time(ph) = \{0\}$.

We prove the decidability of $H_1 \otimes H_2$ reaches $F_1 \times F_2$ from $I_1 \times I_2$ through $ph_1, \mathsf{PH}_1, ph_2, \mathsf{PH}_2$ by the following case analysis: (0) both PH_1 and PH_2 are time-empty; (1) only PH_1 or PH_2 is not time-empty and there exists a simple cycle ph_i^h such that $Time(ph_i^h)$ contains an interval; (2) both PH_1 and PH_2 are not time-empty and there exists a simple cycle ph_i^h such that $Time(ph_i^h)$ contains an interval; (3) either PH_1 or PH_2 is not time-empty and for each simple cycle ph_i^h the set $Time(ph_i^h)$ consists of a finite number of points. In case (0) the decidability follows from Tarski's result [4]. In case (1) we map our problem into that of deciding a first-order formula with a bounded integer parameter, since, if $Time(ph_1^1)$, with $ph_1^1 \in \mathsf{PH}_1$, contains an interval (t_a, t_b) and PH_2 is time-empty, then either $t_a = 0$ or $t_a > 0$. In the former case H_1 can spend any wanted time t by cycling on ph_1^1. In the latter, the number of cycles elapsing a time $t \in \mathbb{R}$ is upper bounded. In case (2) the decidability is a consequence of the density of the time interval. In particular, if there exist two simple cycles $ph^1 \in \mathsf{PH}_1$ and $ph^2 \in \mathsf{PH}_2$ such that $Time(ph^1)$ contains an interval (t_a, t_b) and $t_2 \in Time(ph^2)$, with $t_2 > 0$, then there exist a number n_1 of iterations over ph^1 and a number n_2 of iterations over ph^2 such that H_1 and H_2 can elapse the same amount of time over ph^1 and ph^2, respectively. Case (3) requires the use of algorithms to solve membership problems over algebraic fields [7] and algorithms for solving systems of Diophantine equations.

Since graphs have a finite number of acyclic paths and simple cycles, it holds:

Corollary 1. *Let H_1 and H_2 be semi-algebraic o-minimal automata of dimensions k_1 and k_2, respectively. Let $I_1, F_1 \subseteq \mathbb{R}^{k_1}$ and $I_2, F_2 \subseteq \mathbb{R}^{k_2}$ be characterized by first-order semi-algebraic formulæ. Verifying that $H_1 \otimes H_2$ reaches $F_1 \times F_2$ from $I_1 \times I_2$ is decidable.*

References

1. Casagrande, A., Corvaja, P., Piazza, C., Mishra, B.: Synchronized Product of Semi-Algebraic O-Minimal Hybrid Automata. Technical report, University of Udine (2006) Available at http://fsv.dimi.uniud.it/downloads/synchro.pdf.
2. Lafferriere, G., Pappas, G.J., Sastry, S.: O-minimal Hybrid Systems. Mathematics of Control, Signals, and Systems **13** (2000) 1–21
3. Van den Dries, L.: Tame Topology and O-minimal Structures. Cambridge University Press (1998)
4. Tarski, A.: A Decision Method for Elementary Algebra and Geometry. Univ. California Press (1951)
5. Henzinger, T.A.: The Theory of Hybrid Automata. In: Proc. of IEEE Symp. on Logic in Computer Science (LICS'96), IEEE Computer Society Press (1996) 278–292
6. Miller, J.S.: Decidability and Complexity Results for Timed Automata and Semilinear Hybrid Automata. In: Proc. of Hybrid Systems: Computation and Control (HSCC'00). Volume 1790 of LNCS., Springer (2000) 296–309
7. Cohen, H.: A Course in Computational Algebraic Number Theory. Volume 138 of Graduate Texts in Mathematics. Springer (1993)

A Hybrid Model for Subliminal Air Traffic Control*

Eva Crück and John Lygeros

Automatic Control Laboratory, ETH Zurich, ETL Gebäude, Physikstrasse 3, CH-8092 Zurich, Switzerland
{cruck,lygeros}@control.ee.ethz.ch
http://www.control.ee.ethz.ch/

Abstract. We consider an automation problem in air-traffic management, in which small speed variations, unnoticed by the air-traffic controllers emulate a "*lucky traffic*". We formulate this control problem as a hybrid dynamical game in which the control has to minimize a cost representing the risk perceived by air-traffic controllers against uncertainty on the aircraft dynamics. We prove that this game can be solved using a reachability computation for an auxiliary hybrid system.

1 The Subliminal Control Problem

It has been shown [1,2] that in enroute airspace, small adjustments of speeds executed early enough can prevent a large percentage of conflicts between aircraft. In subliminal control, speed resets are small enough not to raise the air-traffic controller (ATCo)'s awareness. He/she is then delivered a *lucky traffic* and can handle more aircraft. We illustrate our approach for the computation of subliminal control on a simple example.

Let us consider two aircraft A_1 and A_2 cruising at level flight on straight trajectories. We denote by x_1 and x_2 their positions with respect to the crossing point of the paths. ATCo do not have a precise knowledge of the speeds, so we can assume that they estimate future positions using a 2D-model

$$(S_{ATCo}) \qquad \begin{cases} x_1'(t) \in (1+[-\alpha,\alpha])\,\widehat{V_1} \\ x_2'(t) \in (1+[-\alpha,\alpha])\,\widehat{V_2} \end{cases} \tag{1}$$

with $\widehat{V_1}$, $\widehat{V_2}$ estimated ground speeds and α the uncertainty margin.

The automated system is provided with data-links with both aircraft. It receives a measurement of positions and speeds and can send speed adjustment orders. Since the aircraft are cruising, we can use the model

$$\begin{aligned} &x_1'(t) \in V_1(t) + w_1(t), \quad w_1(t) \in W \\ &V_1'(t) = 0 \\ &x_2'(t) \in V_2(t) + w_2(t), \quad w_2(t) \in W \\ &V_2'(t) = 0 \end{aligned}$$

* Research supported by the European Commission under project ERASMUS, FP6-TREN-518276.

A. Bemporad, A. Bicchi, and G. Buttazzo (Eds.): HSCC 2007, LNCS 4416, pp. 672–675, 2007.

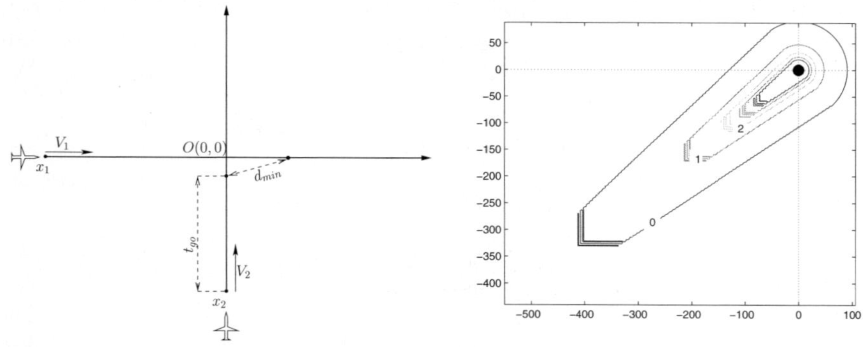

Fig. 1. Example. *left:* Conflict geometry. *right:* Evaluation of the risk perceived by ATCo.

where $W \subset \mathbb{R}$ represents the residual uncertainty. We assume that V_1 and V_2 can be instantaneously reset under the following constraints:

- for all t, $V_i(t) \in (1 + [-\alpha, \alpha]) \widehat{V_i}$, so that the control is *subliminal*;
- on any time interval $[t, t+T_{pil})$, there can be only one speed reset per aircraft;
- a speed reset norm has to belong to $[\delta_V, \overline{p}]$, where δ_V is the minimal speed variation that can be implemented by the flight management system, and $\overline{p}$ is the maximal admissible speed variation from the airliners point of view.

We introduce a cost function $V_{ATCo} : \mathbb{R}^2 \longrightarrow [0, +\infty)$ which describes the risk perceived by ATCo in the following sense: for aircraft at positions (x_1, x_2), the higher the perceived risk of collision in the next future, the higher $V_{ATCo}(x_1, x_2)$. Then the control objective is to minimize $V_{ATCo}(x_1(t), x_2(t))$ against all possible perturbations. The computation of V_{ATCo} is ongoing work which is out of the scope of this paper. We provide an example of such a function in Fig. 1-*(right)* for straight crossing.

2 Game Formulation of the Subliminal Control Problem

We use a receding horizon approach, with control horizon denoted by T_{sub}. The determination of optimal control for the next step is made upon arrival of measurement which resets uncertainty on positions to 0. We describe below the optimization procedure for time interval $[t_0, t_0 + T_{sub}]$.

For aircraft A_i, $i = 1, 2$, let us introduce variables ξ_i, and y_i such that $x_i = \xi_i + y_i$, with $\xi_i' = V_i$. Then for initial position $x_i(0)$ with $y_i(0) = 0$, $y_i(t)$ represent the contribution of uncertainty to the position $x_i(t)$. We also introduce an auxiliary variable τ_i to keep track of time since the last command was issued to aircraft i in the following way: we set $\tau_i' = -1$, and it is reset to T_{pil} at each speed reset. Then the next speed reset can be sent only when $\tau_i(t) \leq 0$.

If we set $z = (\xi_1, V_1, \tau_1, \xi_2, V_2, \tau_2)$ and $H : z \rightarrow \{(V_1, 0, -1, V_2, 0, -1)\}$, the dynamics of z is described by an impulse differential inclusion (see for instance [3]):

between speed reset orders, $z' \in H(z)$, and resets are allowed only if $\tau_1 \leq 0$ and/or $\tau_2 \leq 0$. We denote by Q the set valued map describing the possible resets, and by $S_{H,Q}(z_0)$ the set of possible trajectories such that $z(0) = z_0$. We define the cost functional

$$J(t, y(\cdot), z(\cdot)) = \begin{cases} +\infty & \text{if } \exists \tau \leq (T_{Sub} - t), \text{ such that } z(t) \notin K_z, \\ \sup_{\tau \leq (T_{sub}-t)} V_{ATCo}(\xi_1(\tau) + y_1(\tau), \xi_2(\tau) + y_2(\tau)) & \text{otherwise,} \end{cases}$$

with $K_z = \mathbb{R} \times (\widehat{V}_1 + [-\alpha, +\alpha]\widehat{V}_1 - W) \times \mathbb{R} \times \mathbb{R} \times (\widehat{V}_2 + [-\alpha, +\alpha]\widehat{V}_2 - W) \times \mathbb{R}$. The control has to minimize this cost against the worst possible $y(\cdot)$.

We consider the following game setting: At initial condition (y_0, z_0) the disturbance chooses a trajectory $y(\cdot) \in S_G(y_0)$, and the control has to determine a trajectory $z(\cdot) \in S_{H,Q}(z_0)$ without *a priori* knowledge of $y(\cdot)$. We denote by $\mathbb{B}(y_0, z_0)$ the set of admissible control strategies (see [3]).

Proposition 1. *The optimal reset strategy can be written in feedback form. Specifically, there exists a closed set $\Psi \subset \mathbb{R}^+ \times \mathbb{R}^2 \times K_z$ and a map $\pi : \Psi \longrightarrow K_z$ such that a reset order must be sent when $(t, y(t), z(t))$ enters Ψ; the value of the reset is given by $\pi(t, y(t), z(t))$.*

Indeed, we can define the value function

$$V(t_0, y_0, z_0) = \inf_{\mathbf{B} \in \mathbb{B}(y_0, z_0)} \sup_{y(\cdot) \in S_G(y_0)} J(t_0, y(\cdot), B(y(\cdot))).$$

A dynamic programming approach provides the optimal strategy as a feedback.

Theorem 1. *The epigraph of V, ($\mathcal{E}$pi$(V) := \{(a,b) : V(a) \leq b\}$) is the victory domain of the control in an auxiliary game in which the disturbance plays with the trajectories of differential inclusion $(\theta', y') \in \{1\} \times G(y)$ and the control plays with the impulse differential inclusion*

$$\begin{cases} z' \in H(z) \\ \eta' = 0 \end{cases} \quad \text{and} \quad \begin{cases} z^+ \in Q(z) \\ \eta^+ = \eta^- \end{cases}$$

We set $\widetilde{V}_{ATCo}((y_1, y_2), (\xi_1, V_1, \tau_1, \xi_2, V_2, \tau_2)) = V_{ATCo}(\xi_1 + y_1, \xi_2 + y_2)$. The objective of the control is to keep the state in $K = [0, T_{sub}] \times (\mathbb{R}^2 \times K_z \times \mathbb{R}) \cap \mathcal{E}\mathrm{pi}(\widetilde{V}_{ATCo})$ until it can drive it to $C = \{T_{sub}\} \times \mathcal{E}\mathrm{pi}(\widetilde{V}_{ATCo})$ whatever the uncertainty.

The computation of V can be derived from this definition or from geometrical conditions of the victory domain of the control [3]. The formalism is simpler than in [4] since control and disturbance play with separated dynamics.

3 Implementation of Subliminal Control

Let us denote by $\{t_j\}_{j \in \mathbb{N}}$ a sequence of measurement time. According to our model, the speeds V_1 and V_2 are known precisely and are constant except when a reset order is sent, and the uncertainty on positions is reset to 0 whenever a new measurement occurs. Therefore, $\xi_i(t_j^+) = x_i(t_j^+)$ and $y_i(t_j^+) = 0$. Hence, the value of $y_i(t_j^-) = 0$ can be deduced from the measurement. Since τ_1 and τ_2 are internal variables of the controller, they are known precisely. Therefore, at time t_j, the controller knows the state $(y(t_j^-), z(t_j^-))$. This leads to the following control procedure:

Algorithm 1

j =0 and $\theta(t_0) = 0$
while TRUE
  Instantiate S_{ATCo} and S_{sub} and compute the optimal strategy (Ψ, π)
  while model S_{sub} is valid **do**
    Measure $(x_1(t_j), x_2(t_j))$ and deduce $y_i(t_j)$ and $\xi_i(t_j)$
    if $(\theta(t_j), z(t_j), y(t_j)) \in \Psi$,
      send speed reset order $\pi(\theta(t_j), z(t_j), y(t_j))$
      set $V_1^+ = V_1(t_j^+)$, $V_2^+ = V_2(t_j^+)$
    end if
    Set $\theta^+ = 0$, $\xi_1^+ = x_1(t_j)$, $\xi_2^+ = x_2(t_j)$ and $y_1^+ = y_2^+ = 0$
    j = j+1
  end while
end while

We have illustrated our approach on a very simple traffic situation. If more than two aircraft are involved, the computation of the optimal control strategy has to be performed by the central subliminal control unit for all aircraft simultaneously. Moreover, in order to model accurately aircraft dynamics and to take into account the flight plans, a hybrid model such as the one in [5] may be used. Our formalism can cope with this as long as uncertainty is additive. However, the dimension of the state space will then increase drastically. Therefore, the development of algorithmic solutions to reduce the computation load is crucial for the application of our approach.

References

1. J. Villiers, "Automatisation du contrôle de la circulation aérienne - Projet "ERASMUS" : une voie originale pour mieux utiliser l'espace aérien," ITA, Tech. Rep. Volume 58, 2004.
2. J.-M. Alliot, N. Durand, and G. Granger, "A statistical analysis of the influence of vertical and ground speed errors on conflict probe," in *4th Air Traffic Management Research & Development Seminar, Santa Fe (USA)*.
3. E. Crück, M. Quincampoix, and P. Saint-Pierre, *Pursuit-Evasion games with impulsive dynamics*. Advances in Dynamical Games. Birkhäuser, 2006, vol. 9, ch. 11, pp. 223–247.
4. Y. Gao, J. Lygeros, and M. Quincampoix, "The reachability problem for uncertain hybrid systems revisited: A viability theory perspective," in *Hybrid Systems: Computation and Control*. LNCS, J. Hespanha and A. Tiwari, Eds. Berlin: Springer-Verlag, 2006, no. 3927, pp. 242–256.
5. A. Lecchini, W. Glover, J. Lygeros, and J. Maciejowski, "Monte carlo optimization for conflict resolution in air traffic control," *IEEE Transactions on Intelligent Transportation Systems*, vol. 7, no. 4, pp. 470–482, December 2006.

On Bicontinuous Bisimulation and the Preservation of Stability

P.J.L. Cuijpers

Technische Universiteit Eindhoven, Department of Mathematics and Computer Science, P.O.Box 513, 5600 MB Eindhoven, Netherlands
p.j.l.cuijpers@tue.nl
http://www.win.tue.nl/~pcuijper

1 Introduction

Since Pappas et all. transferred the notion of *bisimulation* from computer science to control theory, it has attracted quite some attention in the hybrid systems community[1]. Notably, bisimulation relations are used to reduce the complexity of dynamic systems, while preserving reachability notions [1]. In [2], we argued that bisimulation needs to be strengthened with continuity conditions, in order to preserve other control science notions as well. This idea was independently explored in [3,4,5] where modal and temporal logics are extended with topological operators, to be able to reason about robustness of a control strategy for embedded systems.

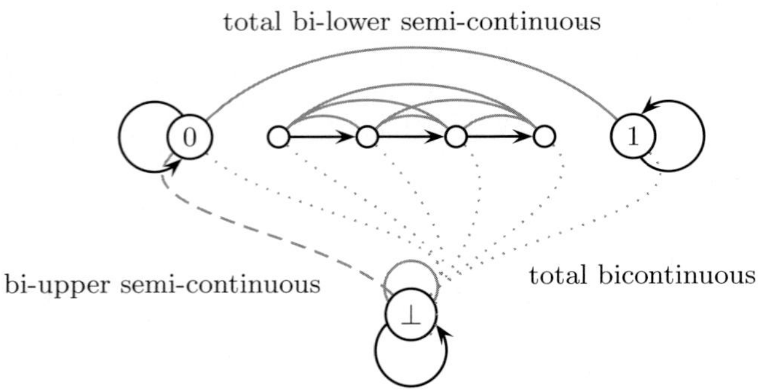

Fig. 1. Bisimilar systems $\bot, 1$ (stable) and 0 (unstable)

In this short paper, we explore the concept of stability for regions or sets of states [6], and show that bicontinuous bisimulation preserves this notion to some extent. Using Fig. 1 above, we also show that weaker notions of continuity do not suffice, and as a rather unexpected side result it turns out that stability is not expressible as a formula in the logic of [4].

2 Mathematical Preliminaries

Our formalism of preference for describing system behavior is using two mechanisms: a *transition relation* [7] describing how a system may evolve from one

A. Bemporad, A. Bicchi, and G. Buttazzo (Eds.): HSCC 2007, LNCS 4416, pp. 676–679, 2007.

state to another, and a *topology* [8] describing how states are positioned with respect to one another.

Definition 1 (Topological transition system). *A* topological transition system *is a tuple* $\langle X, A, \rightarrow, \mathcal{T} \rangle$ *where* X *is a* set of states, A *is a set of* actions, $\rightarrow \subseteq X \times A \times X$ *is a* transition relation, *and* $\mathcal{T} \subseteq 2^X$ *is a* topology[1] *on* X.

Analysis of a system may be done by comparing it to a simpler system. Without loss off generality, we may assume that both systems are part of a common topological transition system, each represented within the transition relation. Comparison of systems then becomes comparison of state transitions.

Traditionally, transition relations are compared using so-called *simulation relations* [7], while topologies are compared by *continuous relations* [9] or *continuous functions* [8]. Topological transition systems are compared by combinations of those. We use $\mathcal{R}^{-1}$ to denote the inverse relation $\{(y, x) \mid (x, y) \in \mathcal{R}\}$, $\mathcal{R}^l(S)$ to denote the lower (or existential) image $\{y \in X \mid \exists_{x\mathcal{R}y}\ x \in S\}$ and $\mathcal{R}^u(S)$ to denote the upper (or universal) image $\{y \in X \mid \forall_{x\mathcal{R}y}\ x \in S\}$. Furthermore, $\mathcal{R}^{-l}(S) = (\mathcal{R}^{-1})^l(S)$ and $\mathcal{R}^{-u}(S) = (\mathcal{R}^{-1})^u(S)$.

Definition 2 (Simulation Relation). *A relation* $\mathcal{R} \subseteq X \times X$ *is a* simulation relation *iff for all* $x, y \in X$ *we find that* $x\mathcal{R}y$ *and* $x \xrightarrow{a} x'$ *implies there exists* $y' \in X$ *such that* $y \xrightarrow{a} y'$ *and* $x'\mathcal{R}y'$. *If* $\mathcal{R}$ *and* $\mathcal{R}^{-1}$ *are simulations, then* $\mathcal{R}$ *is called a bisimulation.*

Definition 3 (Continuous Relation). *A relation* $\mathcal{R} \subseteq X \times X$ *is* lower semi-continuous *iff* $U \in \mathcal{T}$ *implies* $\mathcal{R}^{-l}(U) \in \mathcal{T}$, *i.e. every open set has an open (lower) inverse to which it is related. It is* upper semi-continuous *iff* $U \in \mathcal{T}$ *implies* $\mathcal{R}^{-u}(U) \in \mathcal{T}$, *i.e. every open set has an open (upper) inverse outside which it is not related. Finally,* $\mathcal{R}$ *is* continuous *iff we have both, and if* $\mathcal{R}$ *and* $\mathcal{R}^{-1}$ *are (upper/lower semi-)continuous, then* $\mathcal{R}$ *is called bi-(upper/lower semi-)continuous.*

Definition 4 (Functionality). *A relation* $\mathcal{R} \subseteq X \times X$ *is* functional *iff for each* $x, y, z \in X$ *we have that* $x\mathcal{R}y$ *and* $x\mathcal{R}z$ *implies* $y = z$. *If both* $\mathcal{R}$ *and* $\mathcal{R}^{-1}$ *are functional, then* $\mathcal{R}$ *is a bifunctional relation (i.e. a partial bijection).*

Assuming functionality, the notions of upper and lower semi-continuity coincide to the familiar notion of a continuous function.

3 Stability

Stability is a fundamental notion from control science, stating that small deviations in an initial set of states will not trigger large variations in behavior. The formal definition of stability [6] requires a way to reason about the consecutive transitions in a topological transition system, a reachability relation.

[1] A topology on X is a set of (open) subsets, closed under finite intersection and arbitrary union, and including both X and $\emptyset$.

Definition 5 (Reachability relation). *Given a transition relation* $\rightarrow \subseteq X \times A \times X$*, a* reachability relation $\twoheadrightarrow \subseteq X \times X$ *is the smallest relation such that for all* $x, y, z \in X$ *and* $a \in A$*:* $x \xrightarrow{a} y$ *implies* $x \twoheadrightarrow y$*, and* $x \twoheadrightarrow y$ *and* $y \twoheadrightarrow z$ *implies* $x \twoheadrightarrow z$*.*

Definition 6 (Stable set). *A closed set* $S \subseteq X$ *of states is* stable *if for every open set* $U \in \mathcal{T}$ *with* $S \subseteq U$*, there exists an open set* $V \in \mathcal{T}$ *such that* $S \subseteq V$ *and* $\twoheadrightarrow^l (V) \subseteq U$*.*

According to [6], stability is preserved under so-called *conjugacies*, i.e. under bicontinuous bifunctional bisimulations.

Theorem 1. *Let* $\langle X, A, \rightarrow, \mathcal{T}\rangle$ *be a topological transition system, and let* $\mathcal{R} \subseteq X \times X$ *be a bicontinuous bifunctional bisimulation. Then, a closed set* $S \subseteq X$ *is stable if and only if its lower image* $\mathcal{R}^l(S)$ *is closed and stable.*

In [2] we have proven that these conditions can be relaxed as follows.

Theorem 2. *Let* $\langle X, A, \rightarrow, \mathcal{T}\rangle$ *be a topological transition system, and let* $\mathcal{R} \subseteq X \times X$ *be a relation such that* $\mathcal{R}$ *is upper semi-continuous,* $\mathcal{R}^{-1}$ *is lower semi-continuous and* $\mathcal{R}^{-1}$ *is a simulation. Then, if a closed set* $S \subseteq X$ *is stable, its lower image* $\mathcal{R}^l(S)$ *is also closed and stable.*

Corollary 1. *Let* $\langle X, A, \rightarrow, \mathcal{T}\rangle$ *be a topological transition system, and let* $\mathcal{R} \subseteq X \times X$ *be a relation such that* $\mathcal{R}$ *is a bicontinuous bisimulation. Furthermore, let* $S \subseteq X$ *be a closed set such that* $\mathcal{R}^{-l}(\mathcal{R}^l(S)) = S$*, then this set is stable if and only if its lower image* $\mathcal{R}^l(S)$ *is closed and stable.*

In order to show that none of the conditions in the corollary is redundant, we now give a number of counterexamples (see figure 1). Consider a topological transition system with $X = [0,1] \cup \{\bot\}$, $A = \{\surd\}$, and $\mathcal{T}$ the union of the Euclidean topology on $[0,1]$ and the singleton topology on $\{\bot\}$. This means U is open in $\mathcal{T}$ iff U is open in $[0,1]$ or else $U = U' \cup \{\bot\}$ where U' is open in $[0,1]$. The transition relation $\rightarrow$ is defined by $x \xrightarrow{\surd} \sqrt{x}$, for all $x \in [0,1]$ and $\bot \xrightarrow{\surd} \bot$. Clearly, the set $\{0\}$ is not stable in this system, while the sets $\{1\}$ and $\{\bot\}$ are stable. Now, observe the following:

- $\mathcal{R}_{\rm bi} = X \times \{\bot\}$ (dotted in Fig. 1), is a bicontinuous bisimulation, mapping $\{0\}$ to $\{\bot\}$, but ${\mathcal{R}_{\rm bi}}^{-l}({\mathcal{R}_{\rm bi}}^l(\{0\})) = {\mathcal{R}_{\rm bi}}^{-l}(\{\bot\}) = X \neq \{0\}$.
- $\mathcal{R}_{\rm usc} = \{(0,\bot)\}$ (dashed in Fig. 1), is a bi-upper, but not bi-lower, semi-continuous bisimulation, mapping $\{0\}$ to $\{\bot\}$, with $\mathcal{R}^{-l}_{\rm usc}(\mathcal{R}^l_{\rm usc}(\{0\})) = \{0\}$.
- $\mathcal{R}_{\rm lsc} = (J \times J) \cup \{(0,1),(1,0),(\bot,\bot)\}$ with J the open interval $(0,1)$ (solid in Fig. 1), is a bi-lower, but not bi-upper, semi-continuous bisimulation, mapping $\{0\}$ to $\{1\}$, with $\mathcal{R}^{-l}_{\rm lsc}(\mathcal{R}^l_{\rm lsc}(\{0\})) = \{0\}$, and $\mathcal{R}$ and $\mathcal{R}^{-1}$ both total.

4 Discussion

We have proven that bicontinuous bisimulation is sufficient to preserve stability of a set, under the condition that the set itself is preserved by the bisimulation relation. Also, we have shown that none of the upper- and lower- semicontinuity conditions can be dropped without compromising stability in some way. Finally, by our construction of the bi-total bi-lower semi-continuous bisimulation $\mathcal{R}_{\mathrm{lsc}}$ that violates stability, we have indirectly also proven that stability cannot be expressed as a formula in the logic of [4], since the truth of formulas in this logic is preserved under bi-total bi-lower semi-continuous bisimulation relations.

This last observation justifies further research from a control/hybrid systems point of view, focusing on the limits of expressibility within topological modal and temporal logics. A possible remedy might be to introduce operators with the power to express *compactness* of a set, for example a kind of compact closure. Indeed, using similar counter examples, we can prove that the basic topological notion of compactness is not preserved by total lower semi-continuity either, and can therefore not be expressed as a formula in the logic of [4]. It remains to be seen, however, whether such a closure operator can be found.

Acknowledgements. Many thanks to Jen Davoren, for putting me on the trail of this counterexample, and for straightening out several misconceptions. Importantly, in our email correspondence she noted that although concepts such as dynamic stability and compactness may not be expressible by single formulas in the relatively simple logic of [4], there may be further means to express them using formula schemes. Each of the semi-continuity properties of relations are expressible in this way, for example.

References

1. Haghverdi, E., Tabuada, P., Pappas, G.: Bisimulation relations for dynamical, control, and hybrid systems. Theoretical Computer Science **342** (2005) 229–261
2. Cuijpers, P.: Hybrid Process Algebra. PhD thesis, Technische Universiteit Eindhoven (TU/e), Eindhoven, Netherlands (2004)
3. Moor, T., Davoren, J.: Robust controller systhesis for hybrid systems using modal logic. In Benedetta, M.D., Sangiovanni-Vincentelli, A., eds.: Hybrid Systems: Computation and Control (HSCC'01). Volume 2034 of Lecture Notes in Computer Science., Springer-Verlag (2001) 433–446
4. Davoren, J.: Topological semantics and bisimulations for intuitionistic modal logics and their classical companion logics. Manuscript (2006)
5. Fainekos, G., Pappas, G.: Robustness of temporal logic specifications. In: Formal Approaches to Testing and Runtime Verification, Seatle, WA (2006)
6. Akin, E.: The General Topology of Dynamical Systems. Volume 1 of Graduate Studies in Mathematics. American Mathematical Society (1993)
7. Bergstra, J., Ponse, A., Smolka, S., eds.: Handbook of Process Algebra. Elsevier Science B.V., Amsterdam (2001)
8. Munkres, J.: Topology. 2nd edn. Prentice Hall inc. (2000)
9. Berge, C.: Topological Spaces: Including a treatment of multi-valued functions, vectors spaces and convexity. Oliver and Boyd Ltd., London (1963)

Efficient Simulation of Component-Based Hybrid Models Represented as Hybrid Bond Graphs

Matthew Daigle, Indranil Roychoudhury, Gautam Biswas, and Xenofon Koutsoukos

EECS Department/ISIS, Vanderbilt University
Nashville, TN 37235, USA
matthew.j.daigle@vanderbilt.edu

1 Introduction

Modern engineering systems consist of a large number of interacting components with nonlinear, hybrid behaviors. Building accurate and computationally efficient simulation models for these systems is a challenging task. Researchers have adopted component- [1] and actor-oriented [2] frameworks for modeling large hybrid systems. Mathematical models specify individual component behaviors and formal models of computation define component interactions in these frameworks, and they provide the basis for developing efficient schemes for simulating the hybrid system behavior.

In our work, we adopt the Hybrid Bond Graph (HBG) paradigm [3], an extension of the Bond Graph (BG) modeling language [4], for component-based modeling of embedded systems. HBGs are a domain-independent topological modeling language that capture interactions among the physical and logical processes that constitute a system. The parametric component-based modeling of hybrid systems and the inherent topological structure offer significant advantages for analyzing system behavior and model-based fault diagnosis [5].

In this paper, we address the challenge of translating HBG models to computationally efficient simulation models exploiting causal information that is derived from the topological structure. Mode changes in HBG models, represented as discrete switching events, cause dynamic changes in the topological structure, and, therefore, the computational model during execution. We develop efficient simulation algorithms by converting the HBG models to reconfigurable block diagram structures, using the *Hybrid Sequential Causal Assignment Procedure* to dynamically update the causal information. We demonstrate the technique by deriving the block diagram model of an electrical power system, and running simulation experiments in MATLAB® Simulink® [7].

2 Translating Hybrid Bond Graphs to Block Diagrams

BGs are domain-independent, topological, lumped-parameter models that capture the energy exchange mechanisms in physical processes [4]. The nodes of a bond graph model energy storage, dissipation, transformation, and input-output

A. Bemporad, A. Bicchi, and G. Buttazzo (Eds.): HSCC 2007, LNCS 4416, pp. 680–683, 2007.

elements. Connections in the system are idealized, and modeled by two additional nodes: 0- (or parallel) and 1- (or series) junctions. The connecting edges, called *bonds*, define energy pathways between elements. Parameters of nonlinear BG elements are defined by algebraic *modulating functions*, whose parameters are system variables and external input signals [8]. HBGs introduce discrete configuration changes in continuous BG models by allowing junctions to be turned on and off [3]. A two state (*on* and *off*) finite state machine implements the junction *control specification* with the transition guards expressed as boolean functions of system variables and inputs. When a controlled junction is on, it behaves like a conventional junction. When off, all bonds incident on the junction are deactivated. The system mode at any time is determined by composing states of the individual switched junctions. Details of the language are presented in [3].

There are two primary challenges in deriving simulation models from HBGs. First is to *avoid pre-enumeration of model configurations.* A HBG model with m components, each with n_i switching junctions, defines $2^{\sum_{i=1}^{m} n_i}$ different system modes (or model configurations), where $i = 1, 2, \ldots m$. When large, it is infeasible to pre-enumerate all the model configurations. Therefore, model reconfiguration at mode changes must be executed at run-time. Second is to *avoid algebraic loops.* Component-based modeling of hybrid systems produces an underlying mathematical model, which is a set of differential-algebraic equations (DAEs) that may include algebraic loops. Generating fixed-point solutions for DAEs with algebraic loops becomes computationally expensive when the fixed-point method has to iterate to converge to a solution.

Causality Assignment. The *Sequential Causal Assignment Procedure* (`SCAP`) [4] applied to well-formed BG models assigns causal directions to all bonds in the model. Causality defines the input-output relations between the associated effort and flow variables. This provides the basis for a graphical block diagram (BD) representation, which captures the complete computational model of the system (This is equivalent to the DAE model of the system). The causally derived BD model will also have the minimum number of algebraic loops [4].

Given causal assignments, there is a one-to-one mapping from the BG to the BD model. For HBGs, however, the causal assignments may change when junctions switch state. To avoid the costly pre-enumeration of system modes, we implement an efficient BD reconfiguration scheme that recomputes the causal assignments incrementally, starting from the junctions that switch state, and propagating causal assignment changes till a new consistent assignment is derived. Corresponding changes are made only to those blocks that have changes in the causal assignments of their incident bonds.

The *Hybrid Sequential Causal Assignment Procedure* (`Hybrid SCAP`) performs the causality assignment dynamically when mode changes occur in the system. We assume that the states of all junctions are available before `Hybrid SCAP` is applied. The algorithm starts with a queue of switched junctions. It picks one junction off the queue, makes all the forced causal assignments, and propagates effects of these assignments, making all the consequent forced changes till none remain. Junctions with incomplete causal assignments to their bonds are added to a second queue. When the first queue is empty, the algorithm picks elements off the second queue, makes a valid causal assignment to an unforced element, and propagates its effects to make any forced changes that result from the chosen assignment. This process continues until all bonds have been assigned causality. The propagation is local, so only a subset of the bonds change causal assignments. Details of the algorithm can be found in [6].

Implementation. In MATLAB Simulink implementations, we have explored two approaches. *Implicit switching* uses conditional statements to model the variable input-output relations for block elements whose incident bond(s) can change causality. The switching of the data flow between blocks is, therefore, implicit in the model. The models generated are compact because mode descriptions are expressed concisely as code. However, this approach results in more algebraic loops in the Simulink model, because the input-output directional structure gets buried in the code. During simulation, Simulink invokes fixed point solvers, and the computational overhead affects the simulation efficiency.

Explicit switching uses switching elements to enumerate the data flow paths and the corresponding computational structure for each configuration. At run time, the appropriate switches are triggered to produce the changed block diagram structure. The models created by this approach have many more atomic blocks than the implicit models because multiple BD expansions are enumerated for each element. Since the data flow paths are made explicit for each configuration, no additional algebraic loops are created, however, the switching elements incur overhead associated with zero-crossing detection.

3 Case Study: Electrical Power System

We applied our modeling and simulation framework to the Advanced Diagnostics and Prognostics Testbed (ADAPT) system deployed at NASA Ames. The system consists of *power generation* (solar panel and battery chargers), *power storage* (three sets of lead-acid batteries), and *power distribution* (a number of DC to AC converters and AC and DC loads) subsystems. Relays are used to configure the system in different modes of operation, e.g., charge and/or discharge modes of the batteries, as well as different power supply and load configurations. Because of the large number of possible configurations involving different components, it is infeasible to pre-enumerate all possible modes of operation. We have developed HBG models for all of the components in the ADAPT testbed and used our approach to simulate the system in different configurations [6].

We present the simulation results for a battery supplying power to two DC loads in parallel, with relays that enable the loads to be switched on and off (see Fig. 1). The Simulink model was run for 7,000 seconds of simulation time with the battery discharging through different load configurations. For this experiment, the explicit switching implementation executed about 20% faster than the implicit switching implementation. This difference in the simulation efficiency was consistent with other configurations of the system. In future work, we will

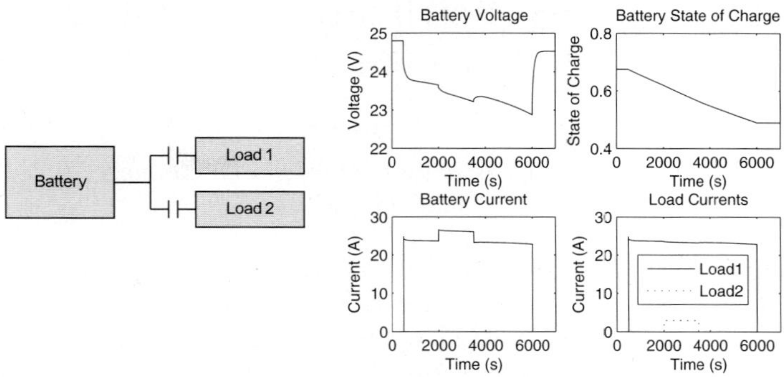

Fig. 1. Example model and simulation results

formalize the computational modeling framework, and further study the computational efficiency for different systems.

Acknowledgments. This work was supported in part by grants NSF CNS-0347440 and NSF CNS-0615214, and a NASA USRA grant from Ames Research Center. We gratefully acknowledge the help from Scott Poll and Ann Patterson-Hine in building the ADAPT system models.

References

1. Liu, J., Lee, E.A.: A component-based approach to modeling and simulating mixed-signal and hybrid systems. ACM Trans. Model. Comput. Simul. **12**(4) (2002) 343–368
2. Lee, E.A., Neuendorffer, S., Wirthlin, M.J.: Actor-oriented design of embedded hardware and software systems. Journal of Circuits, Systems, and Computers **12**(3) (2003) 231–260
3. Mosterman, P.J., Biswas, G.: A theory of discontinuities in physical system models. J Franklin Institute **335B**(3) (1998) 401–439
4. Karnopp, D.C., Margolis, D.L., Rosenberg, R.C.: Systems Dynamics: Modeling and Simulation of Mechatronic Systems. Third edn. John Wiley & Sons, Inc., New York (2000)
5. Narasimhan, S., Biswas, G.: Model-based Diagnosis of Hybrid systems. IEEE Transactions on Systems, Man and Cybernetics, Part A, to appear.
6. Daigle, M., Roychoudhury, I., Biswas, G., Koutsoukos, X.: Efficient simulation of component-based hybrid models represented as hybrid bond graphs. Technical Report ISIS-06-712, Institute for Software Integrated Systems, Vanderbilt University, Nashville, TN, USA (2006).
7. MATLAB/Simulink: (http://www.mathworks.com/products/simulink/)
8. Manders, E.J., Biswas, G., Mahadevan, N., Karsai, G.: Component-oriented modeling of hybrid dynamic systems using the Generic Modeling Environment. In: Proc of the 4th Workshop on Model-Based Development of Computer Based Systems, Potsdam, Germany, (2006).

Diagnosability Verification for Hybrid Automata*

Maria Domenica Di Benedetto, Stefano Di Gennaro, and Alessandro D'Innocenzo

Department of Electrical Engineering and Computer Science, Center of Excellence DEWS
University of L'Aquila, Italy
{dibenede,digennar,adinnoce}@ing.univaq.it

Abstract. We define a notion of diagnosability for hybrid automata, which generalizes the notion of observability. We propose a procedure to check diagnosability on a given hybrid automaton, and show that the complexity of the verification problem is in PTIME. We apply our procedure to an electromagnetic valve system for camless engines. This paper, because of space limitations, only summarizes the obtained results. An extended and detailed manuscript can be found in the form of a technical report on line [1].

1 Introduction

Diagnosability corresponds to failure detection in finite time. Given a plant, a system is diagnosable if it is possible to detect, within a finite time bound and only using the observable outputs of the plant, if a fault has occurred. We say that a system has a fault if its trajectory crosses a given *faulty* subset of the internal state space. Diagnosability generalizes the concept of observability [2].

Diagnosability has many applications in several fields, e.g. the detection of an error in an Air Traffic Management procedure [3,4], of a failure in an automotive system [5], in a component of an industrial plant, or in a communication system [6]. Given a plant and a set of faulty states, an important problem often addressed in the literature is to verify automatically whether the system is diagnosable. For the class of discrete event systems (DESs), the diagnosability verification problem has been treated in several papers [7,8,9,10,11]. The time complexity was collocated in PTIME. Since the concept of time is not present in DESs, the diagnosability definition given in [11] is associated to a finite number of steps, rather than to a time bound: namely, a plant is diagnosable if it is possible to detect a failure after a finite number of transitions since the fault has occurred. For the class of timed automata, a definition of δ-diagnosability has been proposed in [12]: a plant is δ-diagnosable if it is possible to detect a failure after a time delay bounded by $\delta \in \mathbb{N}$ since the fault has occurred. The diagnosability verification problem for timed automata was demonstrated to be

* This work was partially supported by European Commission under Project IST NoE HyCON contract n. 511368 and the National Science Foundation ITR 0121431.

A. Bemporad, A. Bicchi, and G. Buttazzo (Eds.): HSCC 2007, LNCS 4416, pp. 684–687, 2007.

in PSPACE. Diagnosability of hybrid systems was considered in [5], where a notion of diagnosability is proposed for input-output automata, diagnosability conditions are stated, but no complexity analysis is performed. In [13], a hybrid diagnosis problem was formulated, and qualitative techniques for diagnosis of continuous systems was proposed. The computational analysis of diagnosability for hybrid systems is an interesting and challenging issue and, to the best of our knowledge, only few results are available in the literature. This is the main topic of this paper.

2 Diagnosability Definition and Verification

The two main contributions are the following. First, *we propose a procedure to verify diagnosability of a hybrid automaton.* We use here the same hybrid automata model as in [2], where the continuous state evolves following deterministic autonomous dynamics, and the discrete state evolution depends only on the continuous state according to guards, possibly with non deterministic transitions. The observable output is given by the discrete output (possibly unobservable) of the system, that is associated to each edge. We assume that the system is non-blocking, while it might be Zeno. The non-determinism of our model is due to the non-singleton set of initial states, to a possible intersection of two or more guards that simultaneously enable more than one discrete transition, to the "enabling nature" of the guards, that introduces non-determinism in the time instant of a discrete transition, and to the reset map.

Then, we propose a definition of δ-diagnosability where $\delta \in \mathbb{R}^+$, and a failure is modeled as a faulty set, that is a subset of the discrete state space. Our δ-diagnosability definition implies that it is *always* possible to detect by the output string, and after a delay upper bounded by δ, that the associated execution has visited the faulty set. This is equivalent to say that, for any output string of the system, it is always possible to determine whether the associated execution is $\bar{\delta}$-faulty with $\bar{\delta} \geq \delta$ or not. We also prove that δ-diagnosability generalizes observability as defined in [2].

Since hybrid models are complex, we address the diagnosability verification problem by constructing a conservative abstraction and studying diagnosability on this abstraction. We consider here the procedure proposed in [2], that translates a hybrid automaton in a *durational graph (DG)*, a special case of timed automaton, and we prove that this procedure preserves diagnosability. DGs are similar to the durational transition graphs defined in [14]: their expressive power is set between DES (where diagnosability verification is in PTIME) and timed automata (where diagnosability verification is in PSPACE). Our second contribution is to prove that the complexity of the diagnosability verification problem for DGs is in PTIME.

3 Electromagnetic Valve System for Camless Engines

Camless electromagnetic valves are devices recently considered to decouple the camshaft and the valve lift dynamics, namely to command the opening and closing phases of the intake and exhaust valves. The main advantage that these devices should bring is the possibility of obtaining the optimal engine efficiency in every operating condition. One of the main open problems is the control of the impact velocities between the valve and the constraints (typically the valve seat), which should be sufficiently low in order to eliminate acoustic noises and avoid damages of the mechanical components. The problem is complicated by the short time (typically 3×10^{-5} s) available at high engine speed to make a transition between the two valve's terminal positions, and the constraint in terms of actuator cost and space limitations. These last aspects imply that one typical request is the absence of the valve position sensor.

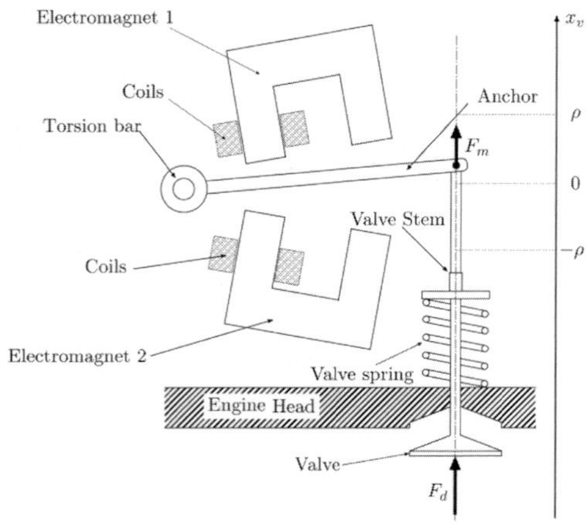

Fig. 1. Scheme of an Electromagnetic Valve System

Referring to [15,16] and references therein for details, we consider here a simplified model of the electromagnetic valve, represented in Figure 1. We define a faulty set of "defective" parameters of the system's dynamics, meaning that they correspond to a unsatisfactory behavior of the system: e.g. the anchor could hit an electromagnet with a velocity higher then a critical value, and this could damage the mechanical system. We assume that the system parameters might change abruptly to a faulty value, and we model it as an unobservable transition to a discrete state associated with "faulty dynamics". Then, we apply our abstraction procedure and verify diagnosability of the system.

4 Conclusions

We proposed a novel verification procedure for checking diagnosability on a given hybrid automaton whose output is a timed string on a finite alphabet, and analyzed the computational complexity of the verification problem. Theoretical results were applied to a case study in automotive.

References

1. Di Benedetto, M.D., Di Gennaro, S., D'Innocenzo, A.: Diagnosability verification for hybrid automata. Technical report, Department of Electrical and Computer Science, University of L'Aquila (2006) http://www.diel.univaq.it/research/.
2. D'Innocenzo, A., Di Benedetto, M.D., Di Gennaro, S.: Observability of hybrid automata by abstraction. In Hespanha, J., Tiwari, A., eds.: Hybrid Systems: Computation and Control. Volume 3927 of Lecture Notes in Computer Science. Springer Verlag (2006) 169–183
3. Di Benedetto, M.D., Di Gennaro, S., D'Innocenzo, A.: Error detection within a specific time horizon and application to air traffic management. In: Proceedings of the Joint 44th IEEE Conference on Decision and Control and European Control Conference (CDC–ECC'05), Seville, Spain. (2005) 7472–7477
4. Di Benedetto, M.D., Di Gennaro, S., D'Innocenzo, A.: Critical states detection with bounded probability of false alarm and application to air traffic management. In: Proceedings of the 2nd IFAC Conference on Analysis and Design of Hybrid Systems (ADHS), Alghero, Sardinia, Italy. (2006)
5. Fourlas, G.K., Kyriakopoulos, K.J., Krikelis, N.J.: Diagnosability of hybrid systems. In: Proceedings of the 10th Mediterranean Conference on Control and Automation - MED2002, Lisbon, Portugal. (2002) 3994–3999
6. Sheth, A., Hartung, C., Han, R.: A decentralized fault diagnosis system for wireless sensor networks. In: Proceedings of the 2nd IEEE International Conference on Mobile Ad-Hoc and Sensor Systems (MASS) 2005. (1999) 192–194
7. Lin, F.: Diagnosability of discrete event systems and its applications. Journal of Discrete Event Dynamic Systems **4(2)** (2005) 197–212
8. Frank, P.: Fault diagnosis in dynamic systems using analytical and knowledge–based redundancy – a survey and some new results. Automatica **26(3)** (1990) 459–474
9. Paoli, A., Lafortune, S.: Safe diagnosability for fault tolerant supervision of discrete event systems. Automatica **41(8)** (2005)
10. Sampath, M., Sengupta, R., Lafortune, S., Sinnamohideen, K., Teneketzis, D.: Diagnosability of discrete-event systems. IEEE Transactions on Automatic Control **40(9)** (1995) 1555–1575
11. Yoo, T., Lafortune, S.: Polynomial-time verification of diagnosability of partially-observed discrete-event systems. IEEE Transactions on automatic control **47(9)** (2002) 1491–1495
12. Tripakis, S.: Fault diagnosis for timed automata. Lecture Notes in Computer Science, W. Damm and E.R. Olderog Eds., Springer Verlag **2469** (2002) 205–221
13. McIlraith, S., Biswas, G., Clancy, D., Gupta, V.: Hybrid systems diagnosis. In Lynch, N., Krogh, B., eds.: Hybrid Systems: Computation and Control. Volume 1790 of Lecture Notes in Computer Science. Springer (2000) 282–295
14. Laroussinie, F., Markey, N., Schnoebelen, P.: Efficient timed model checking for discrete-time systems. Theoretical Computer Science **353**(1-3) (2006) 249–271
15. Di Gennaro, S., Toledo, B.C., Di Benedetto, M.D.: Nonlinear regulation of electromagnetic valves for camless engines. In: Proceedings of the 45th IEEE Conference on Decision and Control, San Diego, CA, USA. (2006)
16. Acosta Lua, C., Castillo Toledo, B., M. D. Di Benedetto, S. Di Gennaro: Nonlinear control of electromagnetic valves for camless engines. Technical report, Department of Electrical and Computer Science, University of L'Aquila (2006) http://www.diel.univaq.it/research/.

Piecewise Constant Feedback Control of Piecewise Affine Gene Network Models

Etienne Farcot and Jean-Luc Gouzé

COMORE INRIA, UR Sophia Antipolis
2004 route des Lucioles
BP 93, 06902 Sophia Antipolis, France
{etienne.farcot,gouze}@sophia.inria.fr

Abstract. The use of hybrid dynamical systems to model gene regulation is impelled by the switch-like behaviour of the latter. Piecewise affine differential equations is one of the most extensively studied among such kind of models. We propose an extension of this class, introducing some input variables. A special focus is given to degradation and production rates being affine functions of the inputs. Some generic control problems are proposed, formulated in terms of an underlying discrete structure. Piecewise constant feedback laws that solve these problems are characterized in terms of affine inequalities. These general feedback laws are then applied to a well-known two dimensional example: the *toggle switch*. It is shown how to control this system toward various behaviours, especially bistability and bisimilarity with a discrete quotient.

1 Introduction

This work deals with control theoretic aspect of a class of piecewise-affine differential equations, introduced in the 1970's [5] to model gene networks, and since widely studied, both theoretically [5], and to model concrete biological systems [2,9]. Furthermore, recent advances have shown that gene networks may be synthesized in labs [4,8]. This fact strongly motivates the elaboration of a control theory for such systems [6,8]. This work is an attempt in this direction: input variables are added to the original models to represent some physico-chemical influence on production and degradation rates. Among concrete realizations, one may use specific inhibitors or activators, introduced in a chosen quantity. Other techniques, such as directed mutagenesis, the use of interfering RNA [7], would also fit within the present framework. Mathematically, similar approaches have been proposed for multiaffine dynamical systems on rectangles [1], which would apply here, but these more general techniques are much less efficient than the specific methods introduced here. More details and references can be found in [3].

2 Piecewise Affine Models

The general form of models considered here is:

$$\frac{dx}{dt} = \kappa(x, u) - \Gamma(x, u)x. \tag{1}$$

A. Bemporad, A. Bicchi, and G. Buttazzo (Eds.): HSCC 2007, LNCS 4416, pp. 688–692, 2007.

The state vector x represents (mRNA or protein) concentrations: x_i is the expression level of the ith gene among n. Then, $u \in \mathcal{U} \subset \mathbb{R}_+^p$ is an input variable, meaning that additional biochemical species can be introduced in the system, or that some physical parameter (e.g. light intensity or temperature) is modified. In any case, input variables are bounded: $\mathcal{U} = \prod_{j=1}^{p} [0, U_j]$.

$\kappa(x, u) \in \mathbb{R}_+^n$ is a production term, and $\Gamma(x, u)$ a diagonal matrix, with positive diagonal entries γ_i representing degradation rates. For a fixed u, both are piecewise constant functions of x with a rectangular underlying partition, due to the switch-like nature of gene regulation.

Succinctly, the continuous dynamics is as follows: in any rectangle $\mathcal{D}$ in phase space, all trajectories are explicitly known, and tend toward a so-called *focal point* $\phi(\mathcal{D}, u)$. If $\phi(\mathcal{D}, u) \in \mathcal{D}$, one has a stable steady state and the system stays in $\mathcal{D}$ forever. Otherwise, the system leaves $\mathcal{D}$ in finite time, reaching another rectangular region. Repeating this provides well defined trajectories. This naturally leads to a discrete quotient of the dynamics, often called *state transition graph*, denoted $\mathsf{TG}(u)$: rectangular regions are the states of this discrete system, and its transitions are defined as those pairs $(\mathcal{D}, \mathcal{D}')$ such that at least one continuous trajectory crosses $\mathcal{D}$ and $\mathcal{D}'$ successively.

3 Control Problems

We focus on piecewise constant feedback control laws: $u = u(x)$, and the restriction $u|_{\mathcal{D}}$ is constant for each $\mathcal{D}$. This relies on the assumption that threshold crossings, or switchings, can be detected accurately. Then, typical control problems are as below, where $\mathcal{V}$ denotes the set of vertices of TG, i.e. of rectangular regions in state space:

Global Control Problem: Let $\mathsf{TG}^{\heartsuit}$ be a transition graph. Find a feedback law $u : \mathcal{V} \to \mathcal{U}$ such that $\mathsf{TG}(u) = \mathsf{TG}^{\heartsuit}$.

Locally, at a region $\mathcal{D}$, solving this problem is tantamount to finding an input $u(\mathcal{D})$ such that $\phi(\mathcal{D}, u(\mathcal{D})) \in \mathcal{D}'$, where $\mathcal{D}'$ is easily deduced from the arrows in $\mathsf{TG}(u)$ having $\mathcal{D}$ as initial vertex. Since the vector $u(\mathcal{D})$ can be chosen arbitrarily in $\mathcal{U}$, this yields a *controllable focal set*, which is the whole set in which focal points can be chosen: $\phi(\mathcal{D}, \mathcal{U})$. A transition $(\mathcal{D}, \mathcal{D}')$ is then said *controllable* if $\phi(\mathcal{D}, \mathcal{U}) \cap \mathcal{D}' \neq \varnothing$. This non-emptiness condition is equivalent to a system of affine inequalities, in the case when production and degradation terms are affine functions of u: $\Gamma(\mathcal{D}, u) = \operatorname{diag}\left(\Gamma(\mathcal{D})u + \gamma^0(\mathcal{D})\right)$, and $\kappa(\mathcal{D}, u) = \kappa(\mathcal{D})u + \kappa^0(\mathcal{D})$, where $\kappa(\mathcal{D}) \in \mathbb{R}^{n\times p}$, $\kappa^0(\mathcal{D}) \in \mathbb{R}_+^{n\times 1}$, $\Gamma(\mathcal{D}) \in \mathbb{R}^{n\times p}$ and $\gamma^0(\mathcal{D}) \in \mathbb{R}_+^{n\times 1}$.

A general form of local problem is then:

$$\exists u \in \mathcal{U}, \forall i \in \{1 \cdots n\},\ \theta_i^- < \phi_i(\mathcal{D}, u) < \theta_i^+ \qquad (P)$$

and its solution is known. Denoting $T^{\pm} = \operatorname{diag}(\theta_1^{\pm} \cdots \theta_n^{\pm}) \in \mathbb{R}^{n\times n}$, one gets

Proposition 1. *An input u solves problem (P) if and only if it satisfies:*

$$\begin{cases} \left(\kappa(\mathcal{D}) - T^-\Gamma(\mathcal{D})\right) u > T^-\,\gamma^0(\mathcal{D}) - \kappa^0(\mathcal{D}) \\ \left(\kappa(\mathcal{D}) - T^+\Gamma(\mathcal{D})\right) u < T^+\,\gamma^0(\mathcal{D}) - \kappa^0(\mathcal{D}) \end{cases} \qquad (\star)$$

Two particularly interesting forms of transitions are depicted in figure 1: if all transitions in $\mathsf{TG}(u)$ are of one of these two forms, this graph yields a deterministic transition system, in which every path represents a continuous trajectory of the original system.

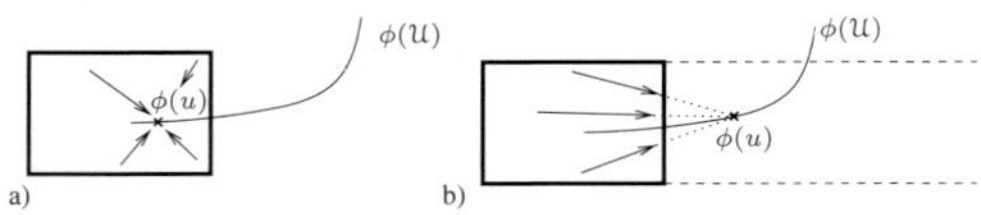

Fig. 1. Among all points in the focal set $\phi(\mathcal{U})$, one has to chose a particular u. Case a) is to make a region invariant, while in b), one has to find an input u such that $\phi(u)$ is situated 'behind' a single facet of the box under consideration.

4 The *Toggle Switch* Example

One considers two genes inhibiting each other, often called *toggle switch*, and notably used as a building block of larger biological circuits [4,6,8]. Its biological function is that of a switch between two steady states, each being a long-term response to some transient induction. Here we suppose that the autonomous system is not bistable, and that a scalar input can affect the decay rates:

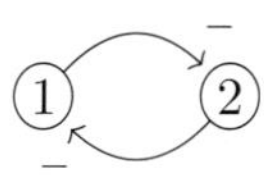

$$\begin{cases} \dfrac{dx_1}{dt} = \kappa_1^1 \mathsf{s}^-(x_2, \theta_2^1) + \kappa_1^0 - (\gamma_1^1 u + \gamma_1^0) x_1 \\ \dfrac{dx_2}{dt} = \kappa_2^1 \mathsf{s}^-(x_1, \theta_1^1) + \kappa_2^0 - (\gamma_2^1 u + \gamma_2^0) x_2 \end{cases}, \qquad (2)$$

where $\mathsf{s}^-(x, \theta)$ is the decreasing Heaviside (or step) function.

Solving a system of the form ($\star$) at each of the 4 rectangular regions in this system, yields all the possible graphs $\mathsf{TG}(u)$. Below are those obtained for a particular set of parameters, see [3] for details. The disposition of vertices reflects the geometry of state space, and ■ represent regions containing a steady state.

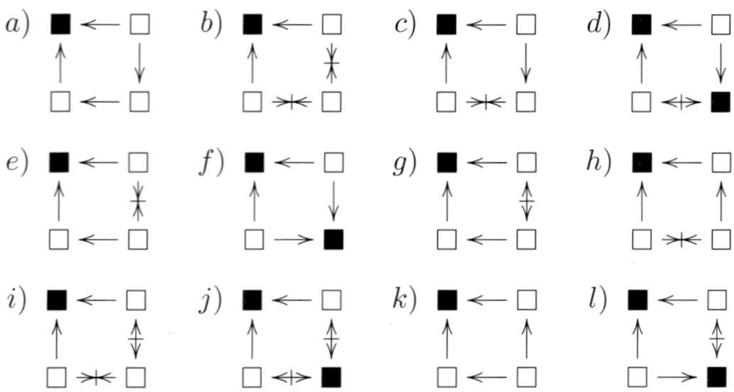

One can see that the upper-left vertex is always fixed, whatever the input value. The lower-right vertex, on the other hand, may be fixed or not. $a)$ is the autonomous case. It appears that bi-stability may be ensured for transition graphs:

d), f), j) and l). Another objective may be to require that the graph be deterministic, with transitions as in figure 1 only. Here, this may be achieved by inputs associated to the graphs g) and j). The first presents a single global equilibrium, while j) is bistable.

5 Conclusion

Among modern advances in cell biology, *synthetic biology* is one of the most striking and promising topic, which might involve control theoretic problems in the years to come. For gene regulatory networks modeled by piecewise affine differential equations, we have characterized the piecewise constant feedback laws that solve some qualitative control problems. An important follow-up of this work would be the analysis of more global control problems. In particular, instead of an explicit transition graph, one may aim at satisfying a formal property, like bi-stability or bisimilarity as exemplified in section 4. Systematizing this with the aid of tools from model checking might lead to efficient algorithms. Also, the input u may model uncertainties of the system. In particular, a property that holds for all u (or whose negation is not controllable) could be called robust with respect to uncertainties, like for example the existence of at least one steady state in section 4.

Acknowledgments. This work was partially supported by the European Commission, project Hygeia Nest-004995. The authors would like to thank D. Ropers and H. de Jong for their helpful advice during the elaboration of this paper, and the referees for their suggestions.

References

1. C. Belta, L. Habets, V. Kumar, *Control of multi-affine systems on rectangles with applications to hybrid biomolecular networks*, 41st IEEE Conference on Decision and Control, pp. 534-539 (2002).
2. G. Batt, D. Ropers, H. de Jong, J. Geiselmann, R. Mateescu, M. Page, D. Schneider, *Validation of qualitative models of genetic regulatory networks by model checking: Analysis of the nutritional stress response in Escherichia coli*, Bioinformatics, 21(Suppl 1):i19-i28, (2005).
3. E. Farcot, J.-L. Gouzé, *How to control a biological switch: a mathematical framework for the control of piecewise affine models of gene networks*, Research Report RR-5979, INRIA Sophia-Antipolis, `https://hal.inria.fr/inria-00094853` (2006).
4. T.S. Gardner, C.R. Cantor, J.J. Collins, *Construction of a genetic toggle switch in Escherichia Coli*, Nature, 403:33 9-342 (2000).
5. L. Glass, *Combinatorial and topological methods in nonlinear chemical kinetics*, J. Chem. Phys. 63:1325-1335 (1975).
6. Hasty, J., Isaacs, F., Dolnik, M., McMillen, D., Collins, JJ, *Designer gene networks: towards fundamental cellular control*, Chaos 11, 207-220 (2001).

7. F.J. Isaacs, D.J. Dwyer,J.J. Collins, *RNA synthetic biology*, Nature Biotechnology 24, 545-554 (2006)
8. Kobayashi H, Kaern M, Araki M, Chung K, Gardner TS, Cantor CR, Collins JJ, *Programmable cells: interfacing natural and engineered gene networks*, Proc. Natl. Acad. Sci. U.S.A., 101(22):8414-9 (2004).
9. D. Ropers, H. de Jong, M. Page, D. Schneider, J. Geiselmann, *Qualitative simulation of the carbon starvation response in Escherichia coli*, BioSystems, 84(2):124-152 (2006).

Hybrid Models for Gene Regulatory Networks: The Case of *lac* Operon in E. Coli

Marcello Farina and Maria Prandini

Dipartimento di Elettronica e Informazione, Politecnico di Milano, Milano, Italy
{farina,prandini}@elet.polimi.it

Abstract. In this paper, we study a hybrid automaton model of the *lac* operon regulating lactose metabolism in E. Coli. We identify two stationary conditions for the system corresponding to lactose metabolism being induced/uninduced, and characterize the extra-cellular concentrations of glucose and lactose that are compatible with them. We also investigate conditions for switching between the two identified stationary conditions.

1 Modeling Gene Regulatory Networks

Gene regulatory networks (GRNs) are a subject of great interest in biology. They involve the interaction of genes, RNA, proteins, and other chemicals in the cell through gene transcription (genes→ RNA, inhibited/enhanced by proteins), protein synthesis (RNA→ proteins), and catalytic reactions. Understanding the mechanisms responsible of the process gene → protein is fundamental. This unidirectional process represents the central dogma of molecular biology, [1], and it is the core of all phenomena occurring in living organisms since proteins are involved in almost all biological activities, structural and enzymatic.

Modeling and analysis methods provided by systems theory can help improving the level of understanding of biological phenomena, [2]. In particular, hybrid systems are becoming the reference modeling formalism for GRNs (see, *e.g.*, [3,4,5]): activation and inhibition of gene activity is intrinsically an on/off mechanism, and the dynamics governing proteins concentration are described by ordinary differential equations (ODEs), while the activation and the deactivation of these dynamics are triggered by discrete switches encoding protein concentration reaching some threshold. In this work, we study a hybrid system model for the *lac* operon regulatory network in the E. Coli bacterium.

The *lac operon* regulatory network: The *lac* operon is a group of nucleotides in the genome required for the transport/metabolism of the lactose l. It is expressed if l is present in the cell and glucose g is missing. It is affected by the extra-cellular concentrations of lactose and glucose, l_{ext} and g_{ext}. The *lac* operon encodes two enzymes: *β-galactosidase* and *lactose permease*, w and y. The main function of w is to enhance the conversion of l into g, while that of y is to pump l_{ext} into the cell through the membrane. If the operon transcription is active, the RNA z responsible of the synthesis of y and w is produced. An inhibitor h of the operon transcription is present in the cell.

A. Bemporad, A. Bicchi, and G. Buttazzo (Eds.): HSCC 2007, LNCS 4416, pp. 693–697, 2007.

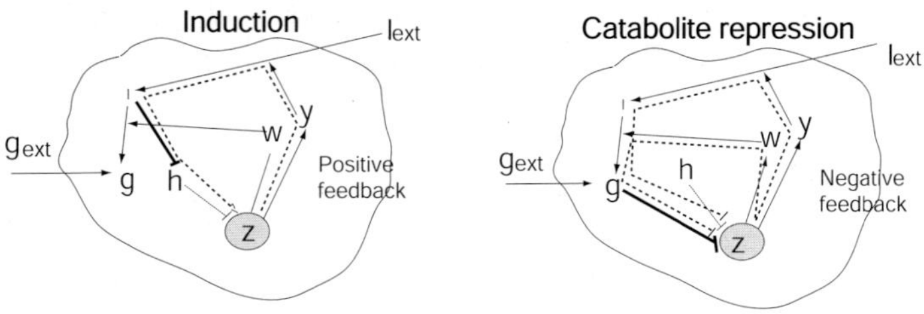

Fig. 1. Directed graph model of the lac operon

In the graph model of Figure 1, we describe the two main mechanisms involved in the *lac* operon regulatory network: *induction* and *catabolite repression.* Induction occurs when l is present and consists in the inhibition of the action of h by y and l (positive feedback). Catabolite repression depends on g. If g is missing the transcription of the operon is stimulated, thus enhancing the conversion of l into g (negative feedback). Other mechanisms are neglected here.

We now present a piecewise affine model of the *lac* operon regulatory network inspired by [6] and by Griffith's model in [7]. The state variables w, y, z, l, and g are affected by inputs l_{ext} and g_{ext}, and evolve according to:

$$\dot{w} = \alpha z - \nu_2 w \tag{1a}$$

$$\dot{y} = \beta z - \nu_3 y \tag{1b}$$

$$\dot{z} = \gamma_{RNA}(g, l) - \nu_1 z \tag{1c}$$

$$\dot{l} = \gamma_{cat}(l_{ext}, y) - \gamma_{cat}(l, w) \tag{1d}$$

$$\dot{g} = \gamma_{cat}(l, w) + \delta\, g_{ext} - \delta\, g \tag{1e}$$

where

$$\gamma_{RNA}(g,l) = \begin{cases} \varepsilon & l < \theta_l^{RNA} \\ \varepsilon + \gamma & l > \theta_l^{RNA} g > \theta_g^{RNA} \\ \varepsilon + K\gamma & l > \theta_l^{RNA} g < \theta_g^{RNA} \end{cases}$$

$$\gamma_{cat}(l_{ext}, y) = \begin{cases} v_1 & l_{ext} < \theta_{l_{ext}}^{cat} \\ v_1 + k_1\, y & l_{ext} > \theta_{l_{ext}}^{cat} y < y_{sat} \\ v_1 + k_1\, y_{sat} & l_{ext} > \theta_{l_{ext}}^{cat} y > y_{sat} \end{cases} \quad \gamma_{cat}(l, w) = \begin{cases} v_2 & l < \theta_l^{cat} \\ v_2 + k_2\, w & l > \theta_l^{cat} w < w_{sat} \\ v_2 + k_2\, w_{sat} & l > \theta_l^{cat} w > w_{sat} \end{cases}$$

with all the constants involved positive. The dynamics of w and y are characterized by **formation** and **degradation** linear terms, thus resulting in the linear ODEs (1a) and (1b). Two **catalytic reactions** take place in the network: $l + w \to g + w$ and $l_{ext} + y \to l + y$. The rate of these reactions can be expressed by the *Hill* formula, [7], which is sigmoidal as a function of the substrate concentration (l and l_{ext}, respectively) and can be approximated with a step function, thus originating the piecewise affine terms $\gamma_{cat}(l, w)$ and $\gamma_{cat}(l_{ext}, y)$ in (1d) and (1e). As for the activation/inhibition of **gene expression**, modeling with piecewise constant functions is quite common. Term $\gamma_{RNA}(g, l)$ in (1c) represents the operon transcription leading to the RNA synthesis. The rate of transport of g_{ext}

inside of the cell is constant, according to [6]. Furthermore, both g and z are affected by a degradation rate proportional to their concentration.

The open hybrid automaton model, [8], of the *lac* operon with (continuous) state $x := [w\,y\,z\,l\,g]'$ and input $u := [l_{ext}\,g_{ext}]'$ is given by $\mathcal{H} = \{Q, X, U, f, Inv, R\}$ where $X = \mathbb{R}^5_+$ is the continuous state space and $U = \mathcal{R}^2_+$ is the input space. Based on (1), we can distinguish 21 disjoint open regions with different affine dynamics for x. As shown in Figure 2, these regions can be identified through their projections onto the (y, l_{ext})-space and the (l, g, w)-space. We can then define the discrete state space Q as $Q = \{(i, j) : i = 1, 2, 3, j = 1, \ldots, 7\}$, where i (j) denotes the projection onto the (y, l_{ext})-space (the (l, g, w)-space). For each $q \in Q$, the set $\{(x, u) : (q, x, u) \in Inv\}$ of (x, u) values for which continuous evolution is allowed is the region in the $X \times U$ space used to define q, whereas the affine vector field $f(q, \cdot, \cdot) : X \times U \to X$ governing the dynamics of x within q can be derived from (1). The reset relation $R : Q \times X \times U \to 2^{Q \times X}$ is defined so that transitions are allowed only between discrete states corresponding to adjacent invariant sets, and the continuous state is maintained constant.

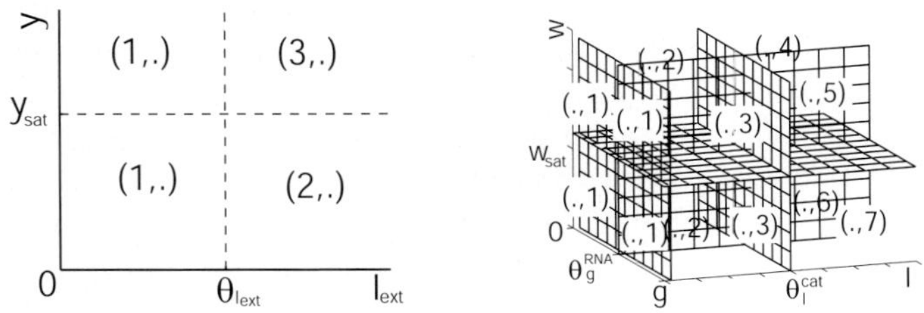

Fig. 2. Projections of the invariant sets onto the (y, l_{ext})-space and (l, g, w)-space

2 Analysis of the *lac* Operon Hybrid Automaton

In this section we illustrate some results on the analysis of the introduced *lac* operon hybrid automaton. Due to space limitations, details are omitted. The interested reader is referred to [9]. The analysis is based on decoupling the (w, y, z) component of the continuous state x from (l, g) by treating $\gamma_{RNA}(g, l)$ as an exogenous input to subsystem (1a)–(1c). We can associate to the three different values taken by $\gamma_{RNA}(g, l)$ a minimum value (w_m, y_m, z_m), an intermediate value (w_i, y_i, z_i), and a maximum value (w_M, y_M, z_M) for (w, y, z).

Assumption 1. *The parameters in* (1) *satisfy: 1.* $y_M < y_{sat}$ *and* $w_M < w_{sat}$*; 2.* $v_2 + k_2 w_M > v_1 + k_1 y_M$*; 3.* $y_i < \hat{y}$ *where* $\hat{y} := (v_2 - v_1)/k_1$*; 4.* $\theta_l^{RNA} < \theta_l^{cat}$.

Condition 1 can be justified based on Griffith's model, 2 means that the effect of reaction $l \to g$ dominates over the transport $l_{ext} \to l$, thus avoiding that l exceeds the θ_l^{cat} threshold and grows unbounded. In condition 3, $\hat{y}$ is a threshold value for y: if y is smaller than $\hat{y}$, then $\dot{l} < 0$ in equation (1d), irrespectively of the other quantities involved. Here, we require $\hat{y}$ to be sufficiently high.

For a constant input $u = [l_{ext}\, g_{ext}]'$, we can identify qualitatively different stationary behaviors for x, corresponding to the metabolism being *active/induced* (x_a) and *not-active/uninduced* (x_{na}): $x_a = [w_M\ y_M\ z_M\ \theta_l^{cat}\ g_{ext}+(v_2+k_2 w_M)/\delta]'$ and $x_{na} = [w_m\ y_m\ z_m\ 0\ g_{ext} + v_2/\delta]'$. In x_{na}, $l = 0$ and the concentrations of enzymes and RNA correspond to the basal level ε of gene expression.

We can show that the active stationary condition x_a is compatible with $l_{\text{ext}} > \theta_{l_{\text{ext}}}^{cat}$ and $g_{\text{ext}} < \theta_g^{RNA} - \frac{1}{\delta}(v_2 + k_2 w_M)$. If $l_{\text{ext}} < \theta_{l_{\text{ext}}}^{cat}$ or $g_{\text{ext}} > \theta_g^{RNA} - \frac{v_2}{\delta}$, then the system ends up in the not-active x_{na} state. Interestingly enough, the system cannot move from this state because the not-active condition is compatible with all input values. This means that if the cell ends up in the not-active state, it cannot be driven back to the active state, irrespectively of the values taken by the extra-cellular concentrations of lactose and glucose. The not-active state is then a sort of "shut-down" condition for the metabolism. This reveals the limitations of the proposed model.

Consider now the case when the system is in the stationary active state x_a, and, at some instant, say $t = 0$, the external supply of lactose l_{ext} drops under threshold: $l_{ext} < \theta_{l_{\text{ext}}}^{cat}$. Then, if l_{ext} keeps under threshold, x will tend to x_{na}, to remain there indefinitely. A question then naturally arises: how long can the system "survive", not reaching the shut-down x_{na} condition, when the lactose supply is missing? In [9], we determine analytically a time t_{re} such that if at time $t < t_{re}$ extra-cellular lactose is supplied, the system will move towards the activation condition, and a "breakpoint" time instant $t_{in} > t_{re}$ such that if extra-cellular lactose is supplied only at $t > t_{in}$, the system will move towards the inactivation stationary condition. Figure 3 shows a simulation obtained when the condition $l_{ext} > \theta_{l_{\text{ext}}}^{cat}$ is restored at $t < t_{re}$. The projections of the continuous state trajectory onto the (l, g, w)-space and (l_{ext}, y)-space are plotted.

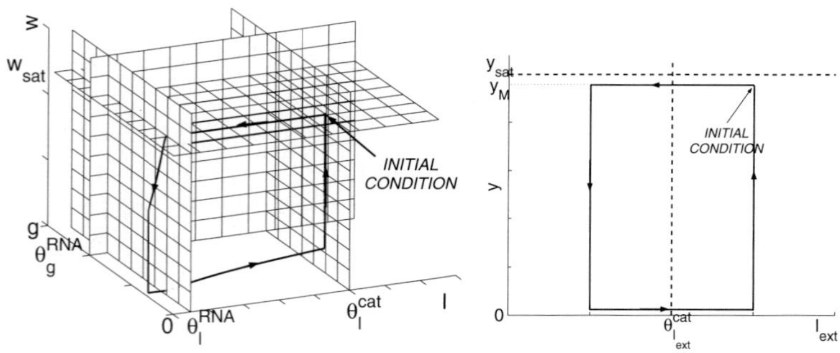

Fig. 3. System trajectory when $t < t_{re}$

References

1. Jacob, F., Monod, J.: Gene regulatory mechanisms in the synthesis of proteins. J. Mol. Biol. (3) (1961) 318–356
2. Sontag, E.D.: Molecular systems biology and control. European Journal of Control **11** (2005) 396–435

3. Ghosh, R., Tomlin, C.J.: Lateral inhibition through delta-notch signaling: A piecewise affine hybrid model. Proceedings of HSCC (2001)
4. Batt, G., Ropers, D., de Jong, H., Geiselmann, J., Page, M., Schneider, D.: Qualitative analysis and verification of hybrid models of genetic regulatory networks: Nutritional stress response in *Escherichia Coli.* Proceedings of HSCC (2005)
5. de Jong, H., Gouzè, J.L., Casey, R.: Piecewise linear models of genetic regulatory networks: Equilibria and their stability. J. Math. Biol. (2005)
6. Tournier, L., Farcot, E.: Hybrid model of gene regulatory networks, the case of *lac* operon. Rapport LMC-IMAG (2002)
7. Klipp, E., Herwig, R., Kowald, A., Wierling, C., Lehrach, H.: Systems Biology in Practice: Concepts, Implementation and Application. Wiley-VCH (2005)
8. Tomlin, C., Lygeros, J., Sastry, S.: A game theoretic approach to controller design for hybrid systems. Proceedings of the IEEE **88**(7) (2000) 949–970
9. Farina, M., Prandini, M.: Hybrid models for gene regulatory networks: The case of *lac* operon in E. Coli. Technical Report 2006.75, Politecnico di Milano (2006)

Reachability Analysis of a Switched Buffer Network

Goran Frehse and Oded Maler

Verimag
{goran.frehse,oded.maler}@imag.fr
http://www-verimag.imag.fr/~{frehse,maler}

1 Introduction

Many situation in various application domains can be formalized as *switched buffer networks*, that is, networks of containers in which quantities of some substances are stored and transported at various rates to other buffers. A *mode* of such a system is defined by the channels that are active at a given time, which determine the rates of change in the quantities of the substances in all the buffers. Switching occurs while opening or closing channels, starting or stopping a reaction, thus causing the system to move from one mode to another. Hybrid automata provide a natural modeling formalism for such systems, a model on which one can verify properties (lack of overflow or deadlock, arrival of products to certain buffers at pre-specified times and quantities) and even automatically synthesize switching controllers that achieve such goals in an efficient manner. Such verification and synthesis techniques can complement traditional analytic techniques that are harder to apply as the switching aspects become more dominant. Looking from the other side of the spectrum, reachability-based methods can be seen adding more rigor and coverage to simulation-based methodologies.

As a first step toward a computer-aided methodology for designing such systems, we present a simple class of such networks, motivated by chemical engineering applications. For this class of networks we define an automatic and compositional translation into "linear" hybrid automata (LHA), that is, automata where in each mode the derivatives of all continuous state variable are constant or bounded by some linear relation over constants. Once a translation is established, we use the tool PHAVer for verifying the correct functioning of the system [1].

The idea of modeling switched buffer systems using hybrid automata is quite natural and has been explored, to a certain extent, in the early days of hybrid systems research, see, for example, [2] or [3] or modeling approaches based on continuous Petri Nets [4]. Dynamic properties, such as stability, of switched buffer networks has been the object of study of many papers, e.g., [5,6], but as in other domains, methods based on hybrid automata are not restricted to the steady-state behavior of the network but can handle also transients. The contribution of our approach is in the combination of a general rigorous translation combined with the availability of a powerful tool like PHAVer that can handle automata derived from nontrivial networks and find subtle bugs in their controllers.

A. Bemporad, A. Bicchi, and G. Buttazzo (Eds.): HSCC 2007, LNCS 4416, pp. 698–701, 2007.

2 Switched Buffer Networks

In this paper we focus on a simple class of networks where substances are only *transported* between buffers without being subject to "reactions" that change their type. Most of the modeling problems and solutions are demonstrated already by this type of networks and their generalization is a topic of ongoing work. Let $\mathbb{I}^+$ be the set of positive closed intervals over $\mathbb{R}$.

Definition 1 (Switched Buffer Network). *A switched buffer network* $\mathcal{B} = (\mathcal{S}, \mathcal{C}, \sigma, \gamma)$ *consists of:*

- *A set of buffers* $\mathcal{S} = \{s_1, \ldots, s_n\}$*,*
- *a set of channels* $\mathcal{C} \subseteq S \times S$*,*
- *a storage capacity over* $\mathcal{S}$ *given by a function* $\sigma : \mathcal{S} \to \mathbb{R}_+$*, stating an upper bound on the quantity of material at each buffer, and*
- *a channel capacity over* $\mathcal{C}$ *given by a function* $\gamma : \mathcal{C} \to \mathbb{I}^+$*, determining the rates at which material can be transported through an active channel.*

The interval $\gamma(c) = [\underline{\gamma}(c), \overline{\gamma}(c)]$ reflects outside influences (disturbances) on the flow rate in an active channel. The decisions whether to open or close channels are taken by a controller that observes the state of the system. In this paper we restrict ourselves to memoryless controllers, that is, controllers that do not have a state of their own and can be expressed as a function of the form $u : P \times X \to P$ meaning that when the network is in state (p, x), it will switch immediately to $(p', x) = (u(p, x), x)$. Note that this restriction is just for notational convenience and our framework can accomodate for any controller specified as a LHA.

The continuous state x evolves in a discrete state p according to a derivative $\dot{x}$ that depends on the active channels. Due to the nondeterminism in the channel capacities, this derivative is only known up to an interval. However, we wish to impose conservation of material, that is, the material leaving a buffer via some channel must appear in the exact same quantity in the target buffer of the channel. So while the change is only known up to some bounds, there must be a pairwise match across channels that we describe with a function $v : \mathcal{C} \to \mathbb{R}^+$ called *throughput* and satisfying $v(s, s') \in \gamma(s, s')$ if $p(s, s') = 1$ and $v(s, s') = 0$ otherwise. The *inflow* and *outflow* of a buffer s are defined as $v_{in}(s) = \sum_{s'} v(s', s)$ and $v_{out}(s) = \sum_{s'} v(s, s')$. The derivative of each buffer is $\dot{x}(s) = v_{in}(s) - v_{out}(s)$. Note that due to the nondeterministic choice of throughput this is a differential *inclusion*.

When a buffer becomes empty in a discrete global state in which one or more of its outgoing channels is active, we need to fix the throughput of this channel. The case when a buffer becomes *full* when it has an active *incoming* channel is symmetric. When a buffer s is empty we relax the lower bounds on the throughput of its outgoing channels, allowing them to be as low as zero. The system can stay for a non-zero duration at a state where $x(s) = 0$ only if $v_{in}(s) = v_{out}(s)$, otherwise it moves immediately to a state where $x(s) > 0$. Likewise, when a buffer is full we relax the lower bounds on the rates of incoming channels.

Table 1. PHAVer performance*

Instance	Time [s]	Mem. [MB]	Depth[a]	Checks[b]	Reached Loc.	Poly.
BP8.1	120	267	173	279	130	279
BP8.2	139	267	173	422	131	450
BP8.3[c]	845	622	302	2669	143	2737
BP8.4[c]	1243	622	1071	4727	147	4772

* on Xeon 3.2GHz, 4GB RAM, Linux;
[a] lower bound on depth in breadth-first search;
[b] nr. of applied post-operators;
[c] computation stopped when violation of safety detected.

3 Multi-product Batch Plant

We demonstrate the approach by modeling and analyzing the multi-product batch plant from [7]. The plant has three levels: On the top level, three buffer tanks B11 to B13 contain the raw materials *yellow*, *red* and *white*. On the second level, there are three reactors R21 to R23 that can be filled from B11, B12, B13. Mixing Yellow and White in a reactor results in the product *Blue*, while Red and White become *Green*. From the reactors, the product is drained into either of two buffer tanks B31 and B32 on the third level, from which it is extracted by the consumer. We verify for a given strategy that it never results in overflow and that $B31$,$B32$ are never empty, i.e., the consumer demand is always met.

The plant is modeled as a switched buffer network with 8 buffers and 20 channels. We transform it into a product of hybrid automata, with one automaton for each buffer and each channel. Changes between stationary behavior, saturation, and starvation are modeled by transitions between locations, in each of which the corresponding constraints on the throughputs are imposed. The controller commands are modeled with transitions with ASAP semantics. At a first glance, the class of hybrid automata that we use seems richer than LHA, because the derivative of $x(s)$ is a function of the throughput variables of the forms $v(s,s')$ and $v(s',s)$ but since these variables have no "state" of their own, they can be projected away and the obtained automaton is a LHA with 266 locations and 823 transitions.

We verify a control strategy that uses R21 solely to produce Blue, and R22 to produce Green, while R23 is alternatingly used to produce either one, see Table 1. In case BP8.1, the consumer drains at a fixed rate of 1 batch/30s from B31 and B32. The throughputs of all channels are deterministic. Figure 1(a) shows a plot of the levels of B31 and B32 over time. B32 drops considerably below B31 because of an asymmetry in the strategy, but the specification holds. In case BP8.2, B11–B13 initially contain anywhere between zero and one, B31 and B32 between 5 and 6 batches. In case BP8.3, there is a single initial state, as in BP8.1, but the consumer demand varies by ±1s/batch. PHAVer aborts the computation as soon as it detects the violation. The set of error traces found is shown in Fig. 1(c). Case BP8.4 is similar to BP8.3, but with the consumer

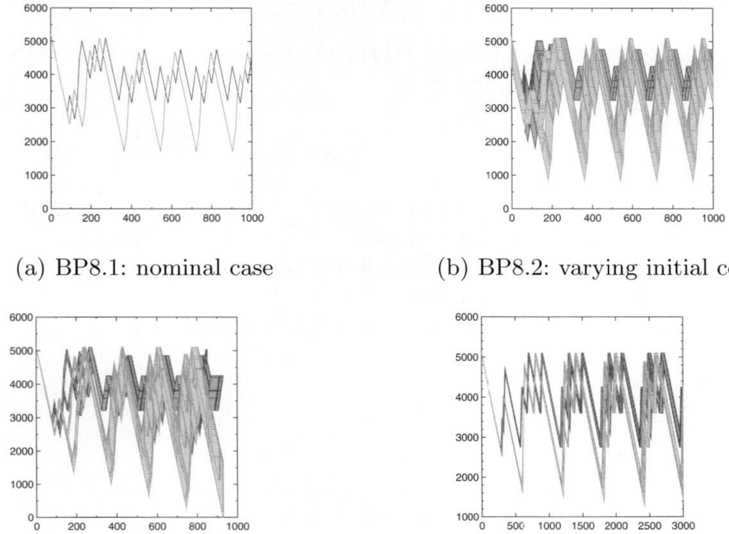

(a) BP8.1: nominal case

(b) BP8.2: varying initial cond.

(c) BP8.3: varying demand

(d) BP8.4: varying, slow demand

Fig. 1. Levels of product buffers B31 and B32 [ml] over time [s]

drawing only 1 batch/100s, i.e., much below production capacity. Nonetheless, the strategy fails, and the error traces found are shown.

References

1. Frehse, G.: PHAVer: Algorithmic verification of hybrid systems past HyTech. In Morari, M., Thiele, L., eds.: HSCC'05. Volume 2289 of LNCS., Springer (2005)
2. Tittus, M.: Control Synthesis for Batch Processes. PhD thesis, Chalmers University of Technology (1995)
3. Kowalewski, S., Engell, S., Preußig, J., Stursberg, O.: Verification of logic controllers for continuous plants using timed condition/event-system models. Automatica **35**(3) (1999) 505–518 Special Issue on Hybrid Systems.
4. David, R., Alla, H.: Discrete, Continuous and Hybrid Petri Nets. Springer (2005)
5. Horn, C., Ramadge, P.: Dynamics of switched arrival systems with thresholds. In: IEEE Conf. Decision and Control. Volume 1. (1993) 288–293
6. Kumar, P., Meyn, S.: Stability of queueing networks and scheduling policies. IEEE Transactions on Automatic Control **40**(2) (1995) 251–260
7. Bauer, N., Kowalewski, S., Sand, G., Löhl, T.: A case study: Multi product batch plant for the demonstration of control and scheduling problems. In Engell, S., Kowalewski, S., Zaytoon, J., eds.: ADPM 2000, Shaker (2000) 383–388

Composition of Dynamical Systems for Estimation of Human Body Dynamics

Sumitra Ganesh[1], Aaron D. Ames[2], and Ruzena Bajcsy[1]

[1] Department of Electrical Engineering and Computer Sciences
University of California, Berkeley, CA 94720
{sumitra,bajcsy}@eecs.berkeley.edu
[2] Control and Dynamical Systems
California Institute of Technology, Pasadena, CA 91125
ames@cds.caltech.edu

Abstract. This paper addresses the problem of estimating human body dynamics from 3-D visual data. That is, our goal is to estimate the state of the system, joint angle trajectories and velocities, and the control required to produce the observed motion from indirect noisy measurements of the joint angles. For a two-link chain in the human body, we show how two independent spherical pendulums can be *composed* to create a behaviorally equivalent double spherical pendulum. Therefore, the estimation problem can be solved in parallel for the low-dimensional spherical pendulum systems and the composition result can be used to arrive at estimates for the higher dimensional double spherical pendulum system. We demonstrate our methods on motion capture data of human arm motion.

1 Introduction and Related Work

The analysis of human motion is motivated by applications such as classification of motion, analysis of motion in activities such as sports and dance, animation and biologically inspired robotic design. Our goal is to extract, from noisy visual observations, a physically meaningful mid-level representation of motion—namely, the joint angle trajectories, angular velocities and joint torques for an assumed nonlinear model. Higher level descriptions, such as motion categories, can be constructed over this representation by using discrete state variables to represent the specific categories. A similar approach was taken in [1] and [2], where switching linear dynamical systems were used for action recognition. Since nonlinear state estimations methods usually scale poorly as the state dimension increases, we propose a more scalable approach to solving the estimation problem. Our approach is based on *composing* estimates for a set of low dimension systems to create a behaviorally equivalent higher dimension system of interest. The paper is organized as follows: Sec. 2 presents our approach and in Sec. 3 we present our results on human arm motion.

A. Bemporad, A. Bicchi, and G. Buttazzo (Eds.): HSCC 2007, LNCS 4416, pp. 702–705, 2007.

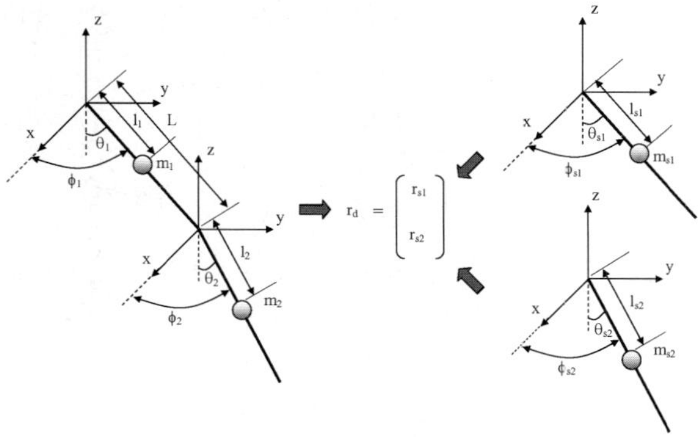

Fig. 1. Equivalent models for the human arm

2 Composition-Based Estimation

The human body can be modeled as a set of rigid links connected by joints. The motion of any open chain of links in the human body is described by a set of nonlinear differential equations [3] of the form

$$\underbrace{M(\boldsymbol{r},\boldsymbol{\lambda})}_{\text{mass matrix}}\ \ddot{\boldsymbol{r}} + \underbrace{C(\boldsymbol{r},\dot{\boldsymbol{r}},\boldsymbol{\lambda})}_{\text{coriolis matrix}}\ \dot{\boldsymbol{r}} + \underbrace{N(\boldsymbol{r},\boldsymbol{\lambda})}_{\text{gravity}} = \underbrace{\boldsymbol{\tau}(t)}_{\text{torques}}\ , \tag{1}$$

where $\boldsymbol{r}(t)$ is the vector of joint angles and $\boldsymbol{\lambda}$ denotes the model parameters. We assume that entire mass of a limb is concentrated at its center of mass and model any open chain of links as a series of connected spherical pendulums. The model parameters are designed using anthorpometric data available in [4]. Given observations of the form $\boldsymbol{y}(t) = \boldsymbol{r}(t) + \boldsymbol{\eta}(t)$, where the statistics of the additive noise $\boldsymbol{\eta}(t)$ are known, the goal is to estimate the joint angle trajectories $\boldsymbol{r}(t)$, angular velocities $\dot{\boldsymbol{r}}(t)$ and the required torques $\boldsymbol{\tau}(t)$.

Different dynamical models could produce a given set of joint angle tracjectories $\boldsymbol{r}(t)$. These dynamical models are equivalent with respect to the observations. We refer to the joint angle trajectories as the behavior of the system and use the notation $\mathcal{B}(\Psi|\,\boldsymbol{\lambda},\,\boldsymbol{\tau}\,) \triangleq \boldsymbol{r}$ to denote that the behavior of the system $\Psi = (M, C, N)$ (see (1)), with parameters $\boldsymbol{\lambda}$ and torques $\boldsymbol{\tau}(t)$ is $\boldsymbol{r}(t)$. For instance, consider the joint angle trajectories of a two-link chain such as the human arm—this data could be produced by two independent spherical pendulums actuated by appropriate torques or, equivalently, by an actuated double spherical pendulum.

We now present our composition result for the 2-link case.

Theorem 1. *Let $\Psi_{\rm s1}$ and $\Psi_{\rm s2}$ denote two spherical pendulum models and let $\Psi_{\rm d}$ denote a double spherical pendulum model. Then*

$$\mathcal{B}(\Psi_{\rm d}|\,\boldsymbol{\lambda}_{\rm d},\,\boldsymbol{\tau}_{\rm d}\,) \triangleq \boldsymbol{r}_{\rm d} = \begin{bmatrix}\boldsymbol{r}_{\rm s1}\\ \boldsymbol{r}_{\rm s2}\end{bmatrix} \triangleq \begin{bmatrix}\mathcal{B}(\Psi_{\rm s1}|\,\boldsymbol{\lambda}_{\rm s1},\,\boldsymbol{\tau}_{\rm s1}\,)\\ \mathcal{B}(\Psi_{\rm s2}|\,\boldsymbol{\lambda}_{\rm s2},\,\boldsymbol{\tau}_{\rm s2}\,)\end{bmatrix} \tag{2}$$

Table 1. The matrices M_{12}, C_{12} and C_{21}, with $\mu = m_2 l_2 L$, $s\phi = \sin(\phi_1 - \phi_2)$, $c\theta_1 = \cos\theta_1$, *et cetera*

$$C_{12} = \mu \begin{bmatrix} (c\theta_2\, s\theta_1 - c\phi\, c\theta_1\, s\theta_2)\dot{\theta}_2 + s\phi\, c\theta_1\, c\theta_2\, \dot{\phi}_2 & s\phi\, c\theta_1\, c\theta_2\, \dot{\theta}_2 - c\phi\, c\theta_1\, s\theta_2\, \dot{\phi}_2 \\ s\phi\, s\theta_1\, s\theta_2\, \dot{\theta}_2 + c\phi\, c\theta_2\, s\theta_1\, \dot{\phi}_2 & c\phi\, c\theta_2\, s\theta_1\, \dot{\theta}_2 + \mu\, s\phi\, s\theta_1\, s\theta_2\, \dot{\phi}_2 \end{bmatrix}$$

$$C_{21} = \mu \begin{bmatrix} (c\theta_1\, s\theta_2 - c\phi\, c\theta_2\, s\theta_1)\dot{\theta}_1 - s\phi\, c\theta_1\, c\theta_2\, \dot{\phi}_1 & -s\phi\, c\theta_1\, c\theta_2\, \dot{\theta}_1 - c\phi\, c\theta_2\, s\theta_1\, \dot{\phi}_1 \\ -s\phi\, s\theta_1\, s\theta_2\, \dot{\theta}_1 + c\phi\, c\theta_1\, s\theta_2\, \dot{\phi}_1 & c\phi\, c\theta_1\, s\theta_2\, \dot{\theta}_1 - s\phi\, s\theta_1\, s\theta_2\, \dot{\phi}_1 \end{bmatrix}$$

$$M_{12} = \mu \begin{bmatrix} c\phi\, c\theta_1\, c\theta_2 + s\theta_1\, s\theta_2 & s\phi\, c\theta_1\, s\theta_2 \\ -s\phi\, c\theta_2\, s\theta_1 & c\phi\, s\theta_2\, s\theta_1 \end{bmatrix}$$

for $\boldsymbol{\lambda}_{\mathrm{d}}$*,* $\boldsymbol{\lambda}_{\mathrm{s1}}$ *and* $\boldsymbol{\lambda}_{\mathrm{s2}}$ *satisfying:*

$$\begin{aligned} \boldsymbol{\lambda}_{\mathrm{d}} &= (m_1, l_1, L, m_2, l_2), \\ \boldsymbol{\lambda}_{\mathrm{s1}} &= \left(\frac{(m_1 l_1 + m_2 L)^2}{m_1 l_1^2 + m_2 L^2}, \frac{m_1 l_1^2 + m_2 L^2}{m_1 l_1 + m_2 L} \right), \quad \boldsymbol{\lambda}_{\mathrm{s2}} = (m_2, l_2), \end{aligned} \tag{3}$$

and for $\boldsymbol{\tau}_{\mathrm{d}}$*,* $\boldsymbol{\tau}_{\mathrm{s1}}$ *and* $\boldsymbol{\tau}_{\mathrm{s2}}$ *satisfying:*

$$\boldsymbol{\tau}_{\mathrm{d}} = \begin{bmatrix} \boldsymbol{\tau}_{\mathrm{s1}} + M_{12}(\boldsymbol{r}_{\mathrm{s1}}, \boldsymbol{r}_{\mathrm{s2}})\, \ddot{\boldsymbol{r}}_{\mathrm{s2}} + C_{12}(\boldsymbol{r}_{\mathrm{s1}}, \boldsymbol{r}_{\mathrm{s2}}, \dot{\boldsymbol{r}}_{\mathrm{s2}})\, \dot{\boldsymbol{r}}_{\mathrm{s2}} \\ \boldsymbol{\tau}_{\mathrm{s2}} + M_{12}^T(\boldsymbol{r}_{\mathrm{s1}}, \boldsymbol{r}_{\mathrm{s2}})\, \ddot{\boldsymbol{r}}_{\mathrm{s1}} + C_{21}(\boldsymbol{r}_{\mathrm{s2}}, \boldsymbol{r}_{\mathrm{s1}}, \dot{\boldsymbol{r}}_{\mathrm{s1}})\, \dot{\boldsymbol{r}}_{\mathrm{s1}} \end{bmatrix}, \tag{4}$$

with M_{12}, C_{12} *and* C_{21} *as given in Table 1.*

As a result of this theorem, spherical pendulum models with parameters $\boldsymbol{\lambda}_{\mathrm{s1}}$ and $\boldsymbol{\lambda}_{\mathrm{s2}}$ can be used to obtain estimates of upper and lower angle trajectories, velocities and the torques $\boldsymbol{\tau}_{\mathrm{s1}}(t)$ and $\boldsymbol{\tau}_{\mathrm{s2}}(t)$, respectively. The *composition* relation (4) can then be used to arrive at an estimate of the torques $\boldsymbol{\tau}_{\mathrm{d}}(t)$ required for an equivalent double spherical pendulum model. The estimation of the independent spherical pendulum models is done using an auxiliary particle filter [5]. The control utilized in the particle filter is designed to mimic a controller that feedback-linearizes and uses Linear-Quadratic optimal control [6] to track the observations. This structure also provides us with angular acceleration estimates which are used in the composition.

3 Results

We tested our approach on motion capture data from the Carnegie Mellon motion capture database. The motion capture data consisted of human arm motion sampled at 120 Hz. The joint angles extracted from the motion capture data were used as the ground truth reference. Gaussian noise was added to the data to simulate noisy observations. The standard deviation of the observation noise was fixed at 0.2σ, where σ is the standard deviation of the reference. An auxiliary particle filter with 1000 particles was used for the estimation. The results for one action are presented in Figs. 2-4.

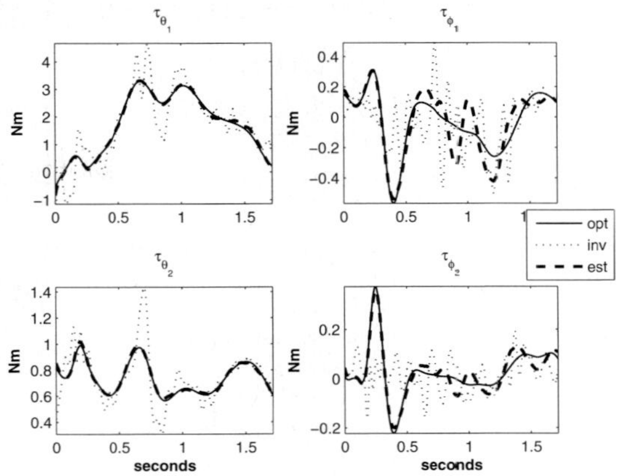

Fig. 2. Double Spherical Pendulum Torques $\boldsymbol{\tau}_{\mathrm{d}}$ for Arm. *est*-torques estimated by composition, *opt*-torques applied by an optimal controller that tracks the reference, *inv*-torques obtained by differentiating clean reference and plugging into equations of motion.

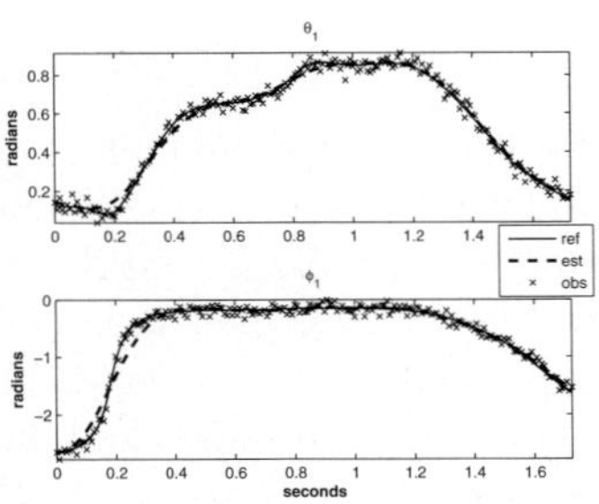

Fig. 3. Estimation of Upper Arm Angles

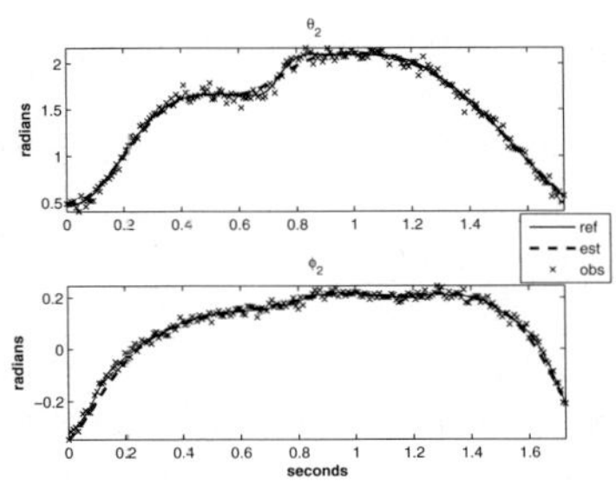

Fig. 4. Estimation of Lower Arm Angles

References

1. Bregler, C., Malik, J.: Learning and recognizing human dynamics in video sequences. In: IEEE Conference on Computer Vision and Pattern Recognition. (1997) pp 568–674
2. Del Vecchio, D., Murray, R., Perona, P.: Decomposition of human motion into dynamics based primitives with application to drawing tasks. Automatica **39(12)** (2003) 2085–2098
3. Murray, R., Sastry, S., Li, Z.: A Mathematical Introduction to Robotic Manipulation. CRC Press (1994)
4. Winter, D.A.: Biomechanics and Motor Control of Human Movement. Wiley (1990)
5. Pitt, M.K., Shephard, N.: Filtering via simulation: Auxiliary particle filters. Journal of the American Statistical Association **94**(446) (1999) 590–9
6. Lewis, F., Syrmos, V.: Optimal Control. Wiley-Interscience (1995)

Computation in One-Dimensional Piecewise Maps

Oleksiy Kurganskyy[1], Igor Potapov[2], and Fernando Sancho Caparrini[3]

[1] Institute of Applied Mathematics and Mechanics, NAS of Ukraine
[2] Computer Science Department, University of Liverpool
[3] Department of Computer Science and AI, University of Seville

Abstract. In this paper we show that the one-dimensional Piecewise Affine Maps (PAMs) are equivalent to planar Pseudo-Billiard Systems (PBSs) or so called "strange billiards". The reachability problem for PAMs is still open, however the more general model of rational one-dimensional maps is shown to be universal with undecidable reachability problem.

1 Introduction

In the present work we investigate a class of hybrid systems defined by one-dimensional piecewise maps. We mainly interested in a class of one-dimensional piecewise-affine maps (PAMS) for which reachability problem is still open. It was recently shown that PAM is equivalent to hierarchical piecewise constant derivatives system (HPCD)[1]. In this paper we show that PAM is equivalent to planar pseudo-billiard system (PBS). PBS is also referred as "strange billiards" model that is a well known object in bifurcation and chaos theory [3]. HPCD is a hybrid automaton where each state is defined by planar piecewise constant derivatives system (PCD). In contrast to HPCD, the model of PBS can also be seen as two dimensional linear hybrid automaton but with only one state. In the second part of this paper we are exploring the complexity of more general class of one-dimensional maps that includes a class of affine maps. We show that the one-dimensional piecewise rational map (PRM) is universal model of computation with undecidable reachability problem. Moreover it is possible to show that there is a particular map, that corresponds to the universal Minsky machine, for which the reachability problem is undecidable.

2 Equivalence Between PBS and PAM

The pseudo billiard model is already appeared in a different context and became an abstract framework for several practical problems. By the pseudo billiard we understand a number of segments with assigned to them vector fields. The computation in this system can be described by the dynamics of the particle, which initially moves with the constant velocity (in a particular direction) inside a given region (not necessarily a polyhedron) and changes it instantaneously at

A. Bemporad, A. Bicchi, and G. Buttazzo (Eds.): HSCC 2007, LNCS 4416, pp. 706–709, 2007.

the moment of a collision with the boundary to the velocity defined by a given vector field (not necessarily a constant one) on the boundary. We start with a more general definition for PBS's, where we have no constraints on distributing the segments around the space. In this case, a particle can touch the segments by both faces, and therefore it may cross them by the action of their projection vectors.

Definition 1. *A Pseudo Billiard System (PBS) is a pair* $(\mathcal{A}, \mathcal{V})$*, where* $\mathcal{A}$ *is a set of pairwise disjoint segments in* $\mathbb{R}^2$ *(closed, open or semi-open), and* $\mathcal{V} = \{\boldsymbol{v}_A\}_{A\in\mathcal{A}}$ *is a set of vectors in* $\mathbb{R}^2$ *(*$\boldsymbol{v}_A$ *is called the projection vector of* A*).*

The dynamic of a particle in PBS can be defined as follows. Let a particle P that is represented by a vector x and is located on a segment $A \in \mathcal{A}$, i.e. $x \in A$. The transition function that move P from x to a position x' can be defined as follows: $x' = x + \lambda \boldsymbol{v}_A$, where $x \in A$ and $\lambda = \min\{\delta > 0 \ : \ x + \delta \boldsymbol{v}_A \in \bigcup_{A' \in \mathcal{A}} A'\}$. In this case we say that x' is (directly) reachable from x and we denote it as $x \Rightarrow x'$. Since we have a set of pairwise disjoint segments it is clear that for any x there is a unique x' if $x \Rightarrow x'$ and $x \neq x'$. We also assume that minimum in λ always exists (particles do not go to infinity).

Definition 2. *A PBS is* reflecting, *if for every* $A \in \mathcal{A}$*, two sets of points* $Pre(A)$ *and* $Post(A)$ *are in the same half-plane determined by* A*, where* $Pre(A)$ *is a set of points from which points on* A *are directly reachable and* $Post(A)$ *is a set of points which are directly reachable from points on* A*.*

Definition 3. *We say that* $f : \mathbb{R} \to \mathbb{R}$ *is piecewise affine map (PAM) if there exists a partition of* $dom(f)$ *in a finite number of pairwise disjoint intervals of* $\mathbb{R}$ *(we allow the intervals to be closed, open or semi-open intervals),* $\mathcal{I}$*, and for every* $I \in \mathcal{I}$*, there exists* $a_I, b_I \in \mathbb{R}$ *such that:* $\forall x \in I, \ f(x) = a_I x + b_I$*.*

In this section we will study the equivalence between the models introduced above. We will say that two models are equivalent if for every system of one type there exists a system of another type that simulates it and vice versa. In particular, the equivalence of one-dimensional PAM, planar PBS and planar reflective PBS can be shown by several geometric constructions. Moreover using the result that model of hierarchical piecewise constant derivative systems (HPCDs) [1] is equivalent to one-dimensional PAMs we can state that planar PBS is equivalent to two-dimensional HPCDs. Hence the complexity that can be obtained with any of them is the same.

Theorem 1. *For every PBS,* $\{\mathcal{A}, \mathcal{V}\}$*, there exists a PAM that simulates it and the number of intervals in the PAM is bounded*[1] *by* $|\mathcal{A}|(|\mathcal{A}| + 2)$*.*

In the proof of Theorem 1 for every segment of the PBS, we construct all possible projections on the other segments that in their turn are bounded in size by $|\mathcal{A}|+2$.

[1] In case of reflecting PBSs, the bound can be reduced to $|\mathcal{A}| + 2$.

Theorem 2. *Let f be a PAM with N affine functions. Let R be the number of affine maps, f_i, with $a_i < 0$. Then, there is a reflecting PBS simulating f using, at most, $2N + R$ reflecting segments.*

Proof. Let $f : I \to I$ be a PAM expressed in such a way that $I = \bigcup_{i=1}^{n} I_i$ is union of pairwise disjoint intervals, and for every i, $f_{|I_i} = f_i$, where $f_i(x) = a_i x + b_i$ is an affine function.

The first step of the proof consists in assigning to every interval of the PAM a segment in $\mathbb{R}^2$ where we simulate the dynamic of the system. Since $f_i : I_i \to I$ is affine, and I_i is an interval, $f_i(I_i)$ must be an interval too. Hence, the image of every interval of our partition must be inside an union of intervals of our partition that constitutes a larger interval. To make more direct the proof, we will maintain the continuity among intervals of f by considering for every interval, $I_i \subseteq \mathbb{R}$, of f, the segment $A_i = I_i \times \{0\} \subseteq \mathbb{R}^2$.

Now, we will simulate the dynamic of each affine map separately. Because the segments A_i are in the same line, we can't go directly from one to another by using projections, therefore we will make use of auxiliary reflection segments to produce the same result as f produces. Depending on the coefficients of the affine map, there are three different cases:

Case 1: $a_i > 0$. In case 1 there is no flip from A_i to $f_i(A_i)$, so we will need only one reflecting auxiliary segment to simulate the application of f, B_i.
Case 2: $a_i < 0$. In case 2 there is a flip from A_i to $f_i(A_i)$, so we will need two reflecting auxiliary segments, B_i and B'_i, to simulate the function f.
Case 3: $a_i = 0$. In case 3 $f(A_i)$ is a point, and we will make use of only one reflecting auxiliary segment, B_i, to project to this point.

We can construct simultaneously all these segments with projection vectors on $\mathbb{R}^2$ without disturbing one to each other, obtaining a reflecting PBS for a complete construction for PAM. It is easy to see from the above construction that the resulting PBS simulates the given PAM. From above construction, we can obtain an upper bound to the number of segments we need in a reflecting PBS to simulate a PAM. The presented method of construction is not efficient in general, but it works for any possible PAM. In a number of PAM's, it is possible to reduce the number of elements of the PBS simulating the PAM.

3 Unpredictability in Rational Piecewise Maps

Now we consider the more general class of rational functions. We define it over $\mathbb{Q}$ to show that even in this case the predictability of its behaviour is an undecidable problem.

Definition 4. *A Piecewise Rational Map (PRM) is a function that is defined on a finite sequence of disjoint intervals $I_- = (-\infty, r_-]$, $I_+ = [l_+, +\infty)$, $I_i = [l_i, r_i]$ with $r_-, l_+, l_i, r_i \in \mathbb{Q}$, $i = 1..k$ and uses rational functions*[2] *for different parts of its domain $I = \{I_1 \cup \ldots \cup I_k\}$.*

[2] $f(x) = P(x)/Q(x)$ is a rational function, where P and Q are polynomials in x as indeterminate, and Q is not the zero polynomial.

Let A be a 2-counter machine with a set of states $S = \{1, 2, \ldots, n\}$. The configuration of A is a triple $[k, l, s]$ where k and l are values of two counters and s is a current state of A. Let us define the mapping $\phi : \mathbb{N} \times \mathbb{N} \times \mathbb{N} \to \mathbb{Q}$ that is an isomorphism between a configuration $[k, l, s]$ of A and a rational number $s + \frac{1}{2^{k+1}3^{l+1}}$ that is shifted to the interval (0,1):

$$\phi([k, l, s]) \to \frac{1}{10^H}(s + \frac{1}{2^{k+1}3^{l+1}}), H = \lceil \lg(|S|) \rceil.$$

Instead of classical Minsky machine with two counters c,d from now on we will consider the equivalent model of one counter machine where the counter holds an integer whose prime factorization is $2^c \cdot 3^d$. Increment (decrement) of the counters c and d can be done by multiplication (division) by 2 and 3 and zero testing corresponds to testing of divisibility by 2 and 3.

Let A be in a configuration $[k, l, s]$ that is mapped to x by ϕ. We can multiply/divide a virtual counter by 2 or/and 3 using the following expression

$$\frac{(10^H x - s)2^a 3^b + s}{10^H},$$

where a,b are integers. For example, to check the emptiness of the first counter we need to add an integer $2^k 3^{l+1}$ using the expression $\frac{1}{2(10^H x - s)} + x$. Then we can easily check whether a virtual counter is divisible by 2 iteratively applying $x - 2$ until the point x in the interval $[3, +\infty)$. Finally a point x should reach either the interval $[2, 3]$, which corresponds to $k \neq 0$, or the interval $[1, 2]$, which corresponds to $k = 0$.

In a similar way we can check divisibility by 3 from a state s using negative numbers. If $x \in [\frac{s}{10^H}, \frac{s+1}{10^H}]$ we apply $-(\frac{1}{3(10^H x - s)} + x)$ and then $x + 3$ for any point in the interval $(-\infty, -4]$. After that the number x should appear in the interval $[-4, -3]$, which corresponds to $l \neq 0$ or in the interval $[-3, -1]$, which corresponds to $l = 0$.

Theorem 3. *One-dimensional piecewise rational map with a finite number of intervals is the universal model of computation.*

Thus, the problem whether a point $x \in \mathbb{Q}$ can be mapped to $y \in \mathbb{Q}$ in a one-dimensional piecewise rational map is undecidable. In contrast to the work [2] we have shown that the more natural extension of affine functions in dimension one is universal. As a next step it would be reasonable to raise a question about the complexity of piecewise linear rational maps.

References

1. E. Asarin and G. Schneider. Widening the boundary between decidable and undecidable hybrid systems. CONCUR'2002,LNCS 2421 (2002) 193-208.
2. O.Kurganskyy, I.Potapov. Computation in One-Dimensional Piecewise Maps and Planar Pseudo-Billiard Systems. Unconventional Computation, LNCS 3699 (2005) 169-175.
3. K.Peters, U.Parlitz. Hybrid systems forming strange billiards. Int. J. of Bifurcations and Chaos, 19 (2003) 2575-2588.

Toward Flexible Scheduling of Real-Time Control Tasks: Reviewing Basic Control Models

Pau Martí and Manel Velasco

Technical University of Catalonia, Automatic Control Department, Pau Gargallo 5, 08028 Barcelona, Spain
pau.marti@upc.edu
http://webesaii.upc.edu

Abstract. We review state-space control models in order to identify timing properties that can favour flexible scheduling of real-time control tasks. First, from the state-space model of a linear time-invariant discrete-time control system with time delay, we derive a new model that involves computing the control signal with a predicted state vector at the actuation instant. Second, by allowing irregular sampling instants, we show that the new model only forces a single synchronization point (actuation instant) while having a feasible implementation. This augurs better schedulability for a set of real-time control tasks, and can provide robustness against scheduling induced jitters.

1 Introduction

The state space model for a linear time-invariant discrete-time control system with time delay [1] implies a synchronous timing at the sampling and actuation instants that can be perfectly met by a single periodic real-time control task. Assuming that sampling and actuation occurs at the beginning and at the end of each task execution [2], the timing of the control task execution corresponds to the timing of the model if task period and deadline are set equal to the sampling period and time delay of the model. However, in a multitasking real-time control system, the tight specification of these timing constraints for control tasks impairs schedulability in the general case [3].

Relaxing this specification by setting the deadline greater than the time delay introduces jitters in control tasks executions that violate the synchronous timing of the model, causing control performance degradation [2]. To maintain synchronism, abstract computational models that force two synchronization points (sampling and actuation) for each control task have been developed ([4] or [5]).

Such computational models introduce three main drawbacks. First, they impose an artificially longer time delay in the closed loop system. Second, they constrain system schedulability by requiring two synchronization points. Third, if state feedback control is used, they involve computing the control signal considering a state vector that becomes outdated at the actuation instant.

The contribution of this paper is to derive a novel state-space model for real-time control tasks aimed at solving the above identified deficiencies. First, we

A. Bemporad, A. Bicchi, and G. Buttazzo (Eds.): HSCC 2007, LNCS 4416, pp. 710–713, 2007.

derive a new model, *prediction-based model*, that involves computing the control signal using the updated (predicted) state vector at the actuation instant, eliminating the possibly unknown and varying time delay between sampling and actuation. Second, by allowing in the new model irregular sampling instants, we show that the new model only forces a single synchronization point (actuation instant) while having a feasible implementation. This offers flexibility for real-time control tasks scheduling, which can improve schedulability and can provide robustness against jitters.

2 Standard Model

Consider the state space model of a linear time-invariant continuous-time system with time delay τ

$$\begin{aligned} \frac{dx(t)}{dt} &= Ax(t) + Bu(t-\tau) \\ y(t) &= Cx(t). \end{aligned} \tag{1}$$

where $x(t)$ is the plant state, $u(t)$ and $y(t)$ are the input and output of the plant, and $A \in \mathbb{R}^{n\times n}$, $B \in \mathbb{R}^{n\times m}$ and $C \in \mathbb{R}^{p\times n}$ are the system, input and output matrices respectively. The time delay can model an input/output latency that appears due to the computation of the control algorithm or due to the insertion of a network within a control loop. For periodic sampling, with sampling period h, where $\tau \leq h$, we obtain the standard discrete time model

$$\begin{aligned} x_{k+1} &= \Phi(h)x_k + \Phi(h-\tau)\Gamma(\tau)u_{k-1} + \Gamma(h-\tau)u_k \\ y_k &= Cx_k, \end{aligned} \tag{2}$$

where matrices $\Phi(t)$, $\Gamma(t)$ are obtained using the following

$$\Phi(t) = e^{At}, \quad \Gamma(t) = \int_0^t e^{As}Bds. \tag{3}$$

Note that eq. (2) slightly differs from conventional notation. The purpose of the new notation is to explicitly indicate dependencies on h and τ. An state space model of the system (2) is given by

$$\begin{bmatrix} x_{k+1} \\ z_{k+1} \end{bmatrix} = \begin{bmatrix} \Phi(h) & Phi(h-\tau)\Gamma(\tau) \\ 0 & 0 \end{bmatrix} \begin{bmatrix} x_k \\ z_k \end{bmatrix} + \begin{bmatrix} \Gamma(h-\tau) \\ I \end{bmatrix} u_k \tag{4}$$

where $z_k \in \mathbb{R}^{m\times 1}$ are the values that represent the past control signals. For closed loop operation, the control signal will be

$$u_k = \begin{bmatrix} L_1 & L_2 \end{bmatrix} \begin{bmatrix} x_k \\ z_k \end{bmatrix} = L_1x_k + L_2z_k \quad \text{with} \quad L_1 \in \mathbb{R}^{1\times n}, \quad L_2 \in \mathbb{R}^{1\times m}. \tag{5}$$

Remark 1. The closed loop model given by (4) and (5) is based on two synchronization points, the sampling and actuation instants. At time t_k the kth sample[1] (x_k) is taken, and at time $t_{k+\tau}$ the kth control signal (u_k) is sent out. The sampling period h is defined from t_k to t_{k+1}, and the time delay τ from t_k to $t_{k+\tau}$.

[1] Sample is used to refer to the full state vector, regardless of whether it has been sampled or observed.

Remark 2. The closed loop model given by (4) and (5) involves computing the control signal to be sent out at time $t_{k+\tau}$ using a sample taken at t_k, τ time units before.

Remark 3. The closed loop model given by (4) and (5) holds the control signal u_k from $t_{k+\tau}$ to $t_{k+1+\tau}$.

3 Prediction-Based Model

A more consistent model could involve computing the control signal with the updated state vector at the actuation instant $t_{k+\tau}$. Instead of (5), we will have

$$u_k = Lx_{k+\tau}. \tag{6}$$

Remark 4. With (6), the sampling period h is the time elapsed from $t_{k+\tau}$ to $t_{k+1+\tau}$. Moreover, no delay is present. An u_k still is held from $t_{k+\tau}$ to $t_{k+1+\tau}$. Recall remarks 1, 2 and 3.

The new closed loop dynamics can be modeled by (6) and

$$x_{t+\tau+1} = \Phi(h)x_{k+\tau} + \Gamma(h)u_k. \tag{7}$$

Remark 5. Although u_k in (6) uses $x_{k+\tau}$, the sample is still taken at time t_k.

According to remark 5, $x_{k+\tau}$ has to be predicted from x_k, that is

$$x_{k+\tau} = \Phi(\tau)x_k + \Gamma(\tau)u_{k-1}. \tag{8}$$

Proposition 1. *All closed loop dynamics given by (6), (7) and (8) can be obtained by (4) and (5).*

Proof. Let L be the state feedback gain of (6). Substituting (8) into (6) we have

$$u_k = L\Phi(\tau)x_k + L\Gamma(\tau)u_{k-1}. \tag{9}$$

By comparing (9) to (5), we obtain that the state feedback gain of (5) is $L_1 = L\Phi(\tau)$ and $L_2 = L\Gamma(\tau)$ (when $z_k \in \mathbb{R}^{1\times 1}$). □

Proposition 2. *Not all closed loop dynamics given by (4) and (5) can be obtained by (6), (7) and (8).*

Proof. If $L_1 = L\Phi(\tau)$ and $L_2 = L\Gamma(\tau)$, by simple algebraic manipulations we obtain that $L_2 = L_1\Phi^{-1}(\tau)\Gamma(\tau)$. This is the family of controllers of model (4) and (5) that can be obtained by model (6), (7) and (8). □

Proposition 3. *All closed loop dynamics given by (4) and (5) can be obtained by (6), (7) and (8) if $m = n$.*

Proof. By adding $L_1 = L\Phi(\tau)$ and $L_2 = L\Gamma(\tau)$, and isolating L we obtain $L = (L_1 + L_2)(\Phi(\tau) + \Gamma(\tau))^{-1}$. This is only feasible if $m = n$. □

Remark 6. By proposition (1) and (2) we deduce that the model given by (4) and (5) is more general than the specified by (6), (7) and (8).

Remark 7. The closed loop system given by (6) and (7) is based on one synchronization point, $t_{k+\tau}$, that is, on the actuation instant.

Taking advantage of remark 7, we can relax the constraint imposed by remark 5. That is, the sample may be taken at times different than t_k.

Proposition 4. *For irregular sampling with $t_k \in (t_{k-(h+\tau)} \quad t_{k+\tau})$, model (4) and (5) can not be applied while model (6), (7) and (8) still holds.*

Proof. In terms of two consecutive irregular sampling instants, t_k and t_{k+1}, eq. (5) is given by

$$u_k = \begin{bmatrix} L_1(h_k) \; L_2(h_k) \end{bmatrix} \begin{bmatrix} x_k \\ z_k \end{bmatrix} \tag{10}$$

where $h_k = t_{k+1} - t_k$. Since at the actutaion instant $(t_{k+\tau})$, the next sampling instant (t_{k+1}) is not known, the feedback gain can not be correctly computed. However, in (6) L depends on h, which is constant since it goes from actuation to actuation instant (recall remark 4). The irregular sample only influences eq. (8), where $\tau = t_{k+\tau} - t_k$. Since t_k can be known at the actutaion instant $(t_{k+\tau})$, the new model accepts irregular sampling. □

4 Discussion and Conclusions

The discrete-time closed loop model given by (6), (7) and (8), a) eliminates the input/output delay, b) is based on a single synchronization point, c) uses an updated (predicted) state at the actuation instants, and more important, d) can absorb irregular sampling. From the real-time perspective, accepting irregular sampling has two main benefits. It can eliminate the degradation introduced by jitters, and it can improve schedulability (since task releases can be performed earlier). Future work will focus on a) analysing the impact of the new model on the schedulability of a set of real-time control tasks, and b) evaluating in a multitasking real-time control system the control performance achieved by a set of tasks implementing control laws based on the new model with respect to the standard one or to those using existing abstract computational models.

References

1. Åström, K.J., Wittenmark, B.: Computer controlled systems. Prentice Hall (1997)
2. Årzén, K.-E., Cervin, A., Eker, J., Sha, L.: An introduction to control and scheduling co-design. In Proc. of the 39th IEEE Conference on Decision and Control. (2000)
3. Martí, P., Fuertes, J.M., Villà, R., Fohler, G: On real-time control tasks schedulability. In European Control Conference. (2001)
4. Cervin, A., Eker, J.: The Control Server: a computational model for real-time control tasks. In Proc. of the 15th Euromicro Conference on Real-Time Systems. (2003)
5. Henzinger, T.A., Horowitz, B., and Kirsch, C.M.: Giotto: a time-triggered language for embedded programming. In Proc. of the First International Workshop on Embedded Software, LNCS **2211** (2001) 166–184

Invertibility and Flatness of Switched Linear Discrete-Time Systems

Gilles Millerioux and Jamal Daafouz

Research Center on Automatic Control of Nancy, Nancy-University-CNRS
gilles.millerioux@esstin.uhp-nancy.fr, jamal.daafouz@ensem.inpl-nancy.fr

1 Introduction

The problem of inversion and flat output characterization for switched linear discrete-time systems is tackled through a unified framework. For this class of hybrid systems, we first derive algebraic conditions to conclude on invertibility, which extends the previous results of Sain and Massey [1] dealing with linear time-invariant systems. Then, the structure of the switched inverse system is provided. Finally, based on the connection with the inversion problem and the related notion of inverse dynamics, we derive conditions which enable to check whether a given output is flat. Due to space limitation, the proofs are not provided but can be found in [2].

Notation : For any integer l, $\mathbf{1}_l$ refers to the $l-$dimensional identity matrix and $\mathbf{O}_{l\times l'}$ stands for the $l \times l'$ zero matrix. If irrelevant, the dimension of the zero matrix will be omitted and we shall merely write $\mathbf{O}$. For a matrix X, $X^\dagger$ stands for the Moore-Penrose generalized inverse of X.

2 Problem Statement and Definitions

We examine switching linear discrete-time systems of the form

$$\begin{cases} x_{k+1} &= A_{\sigma(k)}x_k + B_{\sigma(k)}u_k \\ y_k &= C_{\sigma(k)}x_k + D_{\sigma(k)}u_k \end{cases} \tag{1}$$

where $x_k \in \mathbb{R}^n$, $u_k \in \mathbb{R}^m$ and $y_k \in \mathbb{R}^p$ are the states, the inputs and the measurements, respectively. All the matrices, namely $A \in \mathbb{R}^{n\times n}$, $B \in \mathbb{R}^{n\times m}$, $C \in \mathbb{R}^{p\times n}$ and $D \in \mathbb{R}^{p\times m}$ belong to the respective finite sets $(A_j)_{1\leq j\leq J}$, $(B_j)_{1\leq j\leq J}$, $(C_j)_{1\leq j\leq J}$ and $(D_j)_{1\leq j\leq J}$. The index j corresponds to the mode of the system at a given time k and results from a switching function $\sigma : k \in \mathbb{N} \mapsto j = \sigma(k) \in \{1, \dots, J\}$. The function $\sigma \in \Sigma$, where Σ is the set of all possible switching rules, orchestrates the switches which are triggered by exogenous events. No restriction on the time separation between switches ("dwell time") is imposed. For any time-dependent matrix $X_{\sigma(k)}$, $X_{\sigma(k_0)}^{\sigma(k_1)}$ stands for

$$\begin{aligned} X_{\sigma(k_0)}^{\sigma(k_1)} &= X_{\sigma(k_1)}X_{\sigma(k_1-1)}\dots X_{\sigma(k_0)} \text{ if } k_1 \geq k_0 \\ &= \mathbf{1}_n \text{ if } k_1 < k_0 \end{aligned}$$

A. Bemporad, A. Bicchi, and G. Buttazzo (Eds.): HSCC 2007, LNCS 4416, pp. 714–717, 2007.

Let $\mathcal{U}$ be the space of input sequences over $[0, \infty)$ and $\mathcal{Y}$ the corresponding output space. For each initial state $x_0 \in \mathbb{R}^n$, when the system (1) is driven by the input sequence $\{u\}_0^T = \{u_0, \ldots, u_T\} \subset \mathcal{U}$, for a given mode sequence $\{\sigma\}_0^T = \{\sigma(0), \ldots, \sigma(T)\}$ with $\sigma \in \Sigma$, $\{x(x_0, \sigma, u)\}_0^T$ refers to the solution of (1) in the interval of time $[0, T]$ and $\{y(x_0, \sigma, u)\}_0^T \subset \mathcal{Y}$ refers to the corresponding output in the same interval of time $[0, T]$.

Definition 1. *The system (1) is* invertible for a given switching rule $\sigma \in \Sigma$, *if there exists a positive integer $r < \infty$ such that, for two any input sequences $\{u\}_0^r, \{u'\}_0^r \subset \mathcal{U}$, the following implication applies:*

$$\{y(x_0, \sigma, u)\}_0^r = \{y(x_0, \sigma, u')\}_0^r \Rightarrow u_0 = u'_0 \quad \forall x_0 \tag{2}$$

The integer r will be called the inherent delay of (1).

Definition 2. *The system (1) is* invertible *if it is invertible for all $\sigma \in \Sigma$*

By *invertibility*, we thereby mean here the ability to infer the input u_0 from a finite number r of measurements y_i $(i = 0, \ldots, r)$, the state vector x_0 and the mode sequence $\{\sigma\}_0^r$ both associated to $\{y(x_0, \sigma, u)\}_0^r$ and $\{y(x_0, \sigma, u')\}_0^r$ being identical.

Remark: In the foregoing, the initial condition is considered at the special discrete time $k = 0$. However, we could take any discrete time k and then refer to the interval of time $[k, k + r]$ instead of $[0, r]$.

We are concerned with three closely related issues:

i) determining algebraic conditions, in terms of the state matrices description, under which the system (1) is invertible (invertibility)
ii) looking for a second system, which, when cascaded with the system (1) and so driven by its outputs, produces the same input under an identical initial condition and an identical mode sequence (inversion)
iii) providing some algebraic conditions, in terms of the state matrices description, to check whether a given output of (1) is flat (flatness)

We introduce the following matrices and vectors : $M^0_{\sigma(k)} = D_{\sigma(k)}$, $M^{i(>0)}_{\sigma(k)} =$

$$\begin{pmatrix} D_{\sigma(k)} & \mathbf{0}_{p\times m} & \cdots & \cdots & \cdots \\ C_{\sigma(k+1)}B_{\sigma(k)} & D_{\sigma(k+1)} & \mathbf{0}_{p\times m} & \cdots & \cdots \\ \vdots & \vdots & \ddots & \ddots & \ddots \\ \vdots & \vdots & \ddots & \ddots & \ddots \\ C_{\sigma(k+i)}A^{\sigma(k+i-1)}_{\sigma(k+1)}B_{\sigma(k)} & C_{\sigma(k+i)}A^{\sigma(k+i-1)}_{\sigma(k+2)}B_{\sigma(k+1)} & \cdots & C_{\sigma(k+i)}B_{\sigma(k+i-1)} & D_{\sigma(k+i)} \end{pmatrix}$$

$$\mathcal{O}^i_{\sigma(k)} = \begin{pmatrix} C_{\sigma(k)} \\ C_{\sigma(k+1)}A_{\sigma(k)} \\ \vdots \\ C_{\sigma(k+i)}A^{\sigma(k+i-1)}_{\sigma(k)} \end{pmatrix}, \quad \underline{y}^i_k = \begin{pmatrix} y_k \\ y_{k+1} \\ \vdots \\ y_{k+i} \end{pmatrix}, \quad \bar{I}_m = (\mathbf{1}_m \; \mathbf{0}_{m\times(m\cdot r)})$$

3 Main Results

3.1 Invertibility

Theorem 1. *The following statements are equivalent.*

i) The system (1) is invertible
ii) There exists a finite integer r such that the equation

$$Q^r_{\sigma(k)} M^r_{\sigma(k)} = \bar{I}_m \tag{3}$$

has a solution in $Q^r_{\sigma(k)}$ for all $\sigma \in \Sigma$
iii) There exists a finite integer r such that for all $\sigma \in \Sigma$

$$rank \begin{pmatrix} M^r_{\sigma(k)} \\ \bar{I}_m \end{pmatrix} - rank\ M^r_{\sigma(k)} = 0 \tag{4}$$

As a matter of fact, $\bar{I}_m$ can be replaced by any matrix of the form $(F\ \mathbf{0}_{m\times(m\cdot r)})$ where F is any rank m square matrix of size m. An explicit solution $Q^r_{\sigma(k)}$ of (3) is:

$$Q^r_{\sigma(k)} = \bar{I}_m M^{r\dagger}_{\sigma(k)} \tag{5}$$

3.2 Inversion

Definition 3. *A system is a r-delayed inverse for (1) if, under an identical initial condition x_0 and an identical switching rule σ, when driven by $\underline{y}^r_k$, its output fulfills $\hat{u}_{k+r} = u_k$ for all $k \geq 0$*

Theorem 2. *Assume that (1) is* invertible *with inherent delay r. The system*

$$\begin{cases} \hat{x}_{k+r+1} = P^r_{\sigma(k)}\hat{x}_{k+r} + B_{\sigma(k)}Q^r_{\sigma(k)}\underline{y}^r_k \\ \hat{u}_{k+r} \quad = -Q^r_{\sigma(k)}\mathcal{O}^r_{\sigma(k)}\hat{x}_{k+r} + Q^r_{\sigma(k)}\underline{y}^r_k \end{cases} \tag{6}$$

with

$$P^r_{\sigma(k)} = A_{\sigma(k)} - B_{\sigma(k)}Q^r_{\sigma(k)}\mathcal{O}^r_{\sigma(k)} \tag{7}$$

is a r-delayed inverse system for (1).

Actually, the system (6) enables to retrieve not only the input u_k but the state x_k of (1) as well. Thus, when x_0 is unknown, it can act as an unknown input observer by incorporating an extra gain.

3.3 Flatness

Let us recall that flatness was introduced by Fliess *and al.* [3] in 1995.

Definition 4. *A system with input u_k and state x_k, assumed to be square ($m = p$), is said to be* flat *if there exists a set of independent variables y_k, referred to* *as flat outputs, such that all system variables can be expressed as a function of the flat outputs and a finite number of its backward and/or forward iterates. In particular, there exist two functions $\mathcal{F}$ and $\mathcal{G}$, $\mathbb{Z}$-valued integers $k_{\mathcal{F}}$, $k'_{\mathcal{F}}$, $k_{\mathcal{G}}$ and $k'_{\mathcal{G}}$ such that*

$$\begin{cases} x_k = \mathcal{F}(y_{k+k_{\mathcal{F}}}, \ldots, y_{k+k'_{\mathcal{F}}}) \\ u_k = \mathcal{G}(y_{k+k_{\mathcal{G}}}, \ldots, y_{k+k'_{\mathcal{G}}}) \end{cases} \tag{8}$$

Theorem 3. *A componentwise independent output y_k of the system (1) assumed to be square ($m = p$), invertible for a given switching rule $\sigma \in \Sigma$ and with inherent delay r, is a* flat output *if there exists a positive integer $K < \infty$ such that the following equality applies for all $k \geq 0$*

$$P_{\sigma(k)}^{\sigma(k+K-1)} = \mathbf{0} \tag{9}$$

References

1. Sain M.K., Massey J.L.: Invertibility of linear time-invariant dynamical systems. IEEE Trans. Automatic Control **14** (1969) 141–149
2. Millerioux G., Daafouz J.: Invertibility, flatness and state reconstruction with unknown inputs for switched linear discrete-time systems. Technical report, CRAN-CNRS (2006) available at gilles.millerioux@esstin.uhp-nancy.fr.
3. Fliess M., Levine J., Martin P., Rouchon P.: Flatness and defect of non-linear systems: introductory theory and examples. Int. Jour. of Control **61**(6) (1995) 1327–1361

Trace-Based Semantics for Probabilistic Timed I/O Automata*

Sayan Mitra and Nancy Lynch

MIT Computer Science and AI Laboratory
{lynch,mitras}@csail.mit.edu

Abstract. We describe the main features of the Probabilistic Timed I/O Automata (PTIOA)—a framework for modeling and analyzing discretely communicating probabilistic hybrid systems. A PTIOA can choose the post-state of a discrete transition either nondeterministically or according to (possibly continuous) probability distributions. The framework supports modeling of large systems as compositions of concurrently executing PTIOAs, which interact through shared transition labels. We develop a trace-based semantics for PTIOAs and show that PTIOAs are compositional with respect a new notion of external behavior.

1 Introduction

Probabilistic automata with continuous state spaces provide a framework for studying computing systems that interact with unpredictable environments. In distributed systems, nondeterminism enables us to describe arbitrary interleaving of concurrently executing processes. For modeling and analyzing systems which have traits of both hybrid and distributed systems, such as sensor networks and mobile-robotic systems, we need frameworks that support continuous dynamics, probabilistic transitions, and nondeterminism. There are continuous state automaton frameworks which eschew *internal nondeterminism* in favor of fully probabilistic evolution such as [1,5,12], and those that support both nondeterministic and probabilistic transitions but restrict the state spaces and the probability distributions to be discrete [6,4,2,7]. In this note we describe the main features of the *Probabilistic Timed Input/Output Automaton (PTIOA)* framework (full version available as [10]) which generalizes both the *Timed I/O Automaton* [9] and the *Probabilistic I/O Automaton* [2] frameworks, and provides a basis for describing concurrent, continuous state-space systems, with both probabilistic and nondeterministic transitions.

Definition 1. *A* PTIOA *is a 6-tuple* $\mathcal{A} = ((X, \mathcal{F}_X), \bar{x}, A, \mathcal{R}, \mathcal{D}, \mathcal{T})$ *where: (1)* $(X, \mathcal{F}_X)$ *is a measurable space called the* state space. *(2)* $\bar{x} \in X$ *is the* start state. *(3)* A *is a countable set of* actions, *partitioned into* internal H, input I *and* output O *actions.* $L = O \cup H$ *is the set of* local actions *and* $E = O \cup I$ *is the set of* external actions. $\mathcal{A}$ *is said to be* closed *if* $I = \emptyset$. *(4)* $\mathcal{R}$ *is an equivalence*

* Supported by the MURI project:DARPA/AFOSR MURI F49620-02-1-0325 grant.

A. Bemporad, A. Bicchi, and G. Buttazzo (Eds.): HSCC 2007, LNCS 4416, pp. 718–722, 2007.

relation on L; the equivalence classes of $\mathcal{R}$ are called tasks. *A task T is called an* output task *if $T \subseteq O$. (5) $\mathcal{D} \subseteq X \times A \times \mathcal{P}(X, \mathcal{F}_X)$ is the set of* probabilistic transitions. If (x, a, μ) is an element of $\mathcal{D}$, we write $x \xrightarrow{a} \mu$ and action a is said to be *enabled* at x. *(6) $\mathcal{T}$ is a set of deterministic trajectories for X.* In addition, a PTIOA satisfies the following axioms:

M0 For all $B \subseteq A$, set of states in which at least one action from B is enabled is measurable. For measurable sets $R \subseteq \mathbb{R}_{\geq 0}$ and $Y \subseteq X$, the set of states from which some $r \in R$ amount of time can elapse and a state $y \in Y$ is reached (according to some trajectory in $\mathcal{T}$), is measurable.
D0 Input actions are enabled in all states.
D1 For any state x at most one of the following may exist: (1) a local action enabled at x (2) a non-point trajectory starting from x.
D2 For any state x, if there are actions a, b in the same task T and $x \xrightarrow{a} \mu_1$ and $x \xrightarrow{b} \mu_2$ then $a = b$ and $\mu_1 = \mu_2$.
D3 An execution of finite duration has at most finite number of internal actions.

The **M0** axiom ensures measurability of reasonable sets of executions. **D0** is a non-blocking axiom standard in I/O automata literature. Axiom **D1** allows resolution of nondeterminism in a structured manner; this axiom will be removed in Section **4. D1** prevents an action to remain enabled while time elapses. If time can elapse from x, then the state evolves according to the longest trajectory starting from x. If local actions are enabled at x then time cannot elapse and $\mathcal{A}$ nondeterministically chooses one action a from the set of enabled actions. This nondeterministic choice is resolved by a *task scheduler* (defined below). If a task T is specified then **D2** implies that at x there can be at most one enabled action in T, and at most one probabilistic transition corresponding to that action.

2 Distributions over Executions and Traces

An *execution fragment* of an PTIOA $\mathcal{A}$ is a sequence $\alpha = \tau_0 a_1 \tau_1 a_2 \ldots$, where each $\tau_i \in \mathcal{T}$, $a_i \in A$ and a_i is enabled at the last state of τ_{i-1}. An execution fragment α is an *execution* of $\mathcal{A}$ if the first trajectory starts from $\bar{x}$. The *trace* of an execution α represents its externally visible part, namely the external actions and time passage. It is obtained by removing internal actions, concatenating consecutive trajectories, and replacing all the trajectories with their lengths. We denote the set of executions, and the set of traces of $\mathcal{A}$ by $\mathsf{Execs}_{\mathcal{A}}$ and $\mathsf{Traces}_{\mathcal{A}}$.

In order to construct a probability measure over $\mathsf{Execs}_{\mathcal{A}}$ we have to first define a σ-algebra over $\mathsf{Execs}_{\mathcal{A}}$. Adapting a construction given in [3], we proceed as follows: A *base* is a finite sequence $\Lambda = X_0 R_0 X_1 A_1 R_1 \ldots X_m A_m R_m X_{m+1}$, where for every $i \in \{0, \ldots, m{+}1\}$, $X_i \in \mathcal{F}_X$, R_i is a measurable set in $\mathbb{R}_{\geq 0}$ and for every $i \in \{1, \ldots, m\}$, $A_i \subseteq A$. The *basic set* C_Λ corresponding to this base Λ is the set of all executions which have a prefix that matches the pattern of actions and trajectories in Λ. We show that collection $\mathscr{C}$ of all basic sets of $\mathcal{A}$ generates a σ-algebra over $\mathsf{Execs}_{\mathcal{A}}$, which we denote by $\mathcal{F}_{\mathsf{Execs}_{\mathcal{A}}}$. We define the measurable space of executions of $\mathcal{A}$ to be $(\mathsf{Execs}_{\mathcal{A}}, \mathcal{F}_{\mathsf{Execs}_{\mathcal{A}}})$. In a similar manner, we construct a measurable space $(\mathsf{Traces}_\mathcal{A}, \mathcal{F}_{\mathsf{Traces}_\mathcal{A}})$ of traces of $\mathcal{A}$ from *trace bases*, which are defined to be finite alternating sequences of the measurable sets in $\mathbb{R}_{\geq 0}$ and subsets of E.

We combine $\mathcal{A}$ with a *task scheduler* for resolving nondeterministic choice over enabled actions, which is is simply a finite or infinite sequence $\rho = T_1 T_2 \ldots$ of tasks in $\mathcal{R}$. A task scheduler [2] chooses the next action deterministically and independently of the information produced during an execution. In [10] we inductively define a function apply such that given any task schedule ρ for $\mathcal{A}$, $\mathsf{apply}(\delta_{\bar{x}}, \rho)$ gives a probability measure over $(\mathsf{Execs}_\mathcal{A}, \mathcal{F}_{\mathsf{Execs}_\mathcal{A}})$ by "applying" ρ to $\mathcal{A}$, one task at a time. This probability distribution over executions is called a *probabilistic execution* of $\mathcal{A}$.

For any probabilistic execution μ of a PTIOA we would like to have a single corresponding measure on $\mathsf{Traces}_\mathcal{A}$. This requires $trace : (\mathsf{Execs}_\mathcal{A}, \mathcal{F}_{\mathsf{Execs}_\mathcal{A}}) \rightarrow (\mathsf{Traces}_\mathcal{A}, \mathcal{F}_{\mathsf{Traces}_\mathcal{A}})$ to be a measurable function. A difficulty in proving this arises because the *trace* function concatenates trajectories that are separated by internal actions. Consider, for example, a simple trace base $[0, r]\{a\}$, where r is a positive real and a is an external action. Then, $\mathcal{E} = trace^{-1}(C_{[0,r]\{a\}})$ is the set of all finite executions of the form $\tau_1 h_1 \tau_2 h_2 \ldots h_{n-1} \tau_n a$, such that all the h_i's are internal actions and $\sum_{i=1}^{n} \tau_i.ltime \leq r$. For *trace* to be a measurable function $\mathcal{E}$ expressible as a countable union of basic sets. We prove the measurability of *trace* in [10] making use of several PTIOA properties including the **D3** axiom. With this result, we are able to prove a key theorem which asserts that each task schedule for a PTIOA $\mathcal{A}$ gives rise to a single distribution over the traces of $\mathcal{A}$. For each task schedule the corresponding distribution over traces is called a *trace distribution* and the set of all possible trace distributions of $\mathcal{A}$ is denoted by $tdists(\mathcal{A})$.

3 Composition and External Behavior

The composition operation allows a PTIOA representing a complex system to be constructed by composing PTIOAs representing smaller subsystems. In [10] we state the conditions under which two PTIOAs $\mathcal{A}_1$ and $\mathcal{A}_2$ can be composed, we define the *parallel composition* operator, and we prove a theorem asserting that a composition $\mathcal{A}_1 || \mathcal{A}_2$ is a valid PTIOA.

An *environment* for PTIOA $\mathcal{A}$ is a PTIOA $\mathcal{E}$ such that $\mathcal{A}$ and $\mathcal{E}$ can be composed and their composition $\mathcal{A}||\mathcal{E}$ is closed. The *external behavior* of a PTIOA $\mathcal{A}$, written as $extbeh_\mathcal{A}$, is defined as a function that maps each environment $\mathcal{E}$ for $\mathcal{A}$ to the set $tdists(\mathcal{A}||\mathcal{E})$. Two PTIOAs $\mathcal{A}_1$ and $\mathcal{A}_2$ are *comparable* if $E_1 = E_2$. If $\mathcal{A}_1$ and $\mathcal{A}_2$ are comparable then $\mathcal{A}_1$ is said to *implement* $\mathcal{A}_2$, written as $A_1 \leq A_2$ if, for every environment PTIOA $\mathcal{E}$ for both $\mathcal{A}_1$ and $\mathcal{A}_2$, $extbeh_{\mathcal{A}_1}(\mathcal{E}) \subseteq extbeh_{\mathcal{A}_2}(\mathcal{E})$. Viewing external behavior as a mapping from environments as opposed to a set of trace distributions is natural in many applications, including analysis of security protocols [2], and indeed, it lets us circumvent some of the difficulties that underlie compositionality in the probabilistic setting.

Theorem 1. *Suppose $\mathcal{A}_1$ and $\mathcal{A}_2$ are comparable PTIOAs and $\mathcal{A}_1 \leq \mathcal{A}_2$. If PTIOA $\mathcal{B}$ can be composed with both $\mathcal{A}_1$ and $\mathcal{A}_2$ then $\mathcal{A}_1||\mathcal{B} \leq \mathcal{A}_2||\mathcal{B}$.*

4 Generalized PTIOAs

In this section, we relax the deterministic assumptions (axiom **D1**) on PTIOAs. A *Generalized PTIOA* is a tuple $\mathcal{A} = ((X, \mathcal{F}_X), \bar{x}, A, \mathcal{R}, \mathcal{D}, \mathcal{T})$ as in Definition 1, but $\mathcal{A}$ does not necessarily satisfy **D1** and $\mathcal{T}$ is not necessarily deterministic. Thus, from a given state $x \in X$ of a generalized PTIOA $\mathcal{A}$ there may be non-deterministic choice of actions and also choice of distinct trajectories. A *local scheduler* for generalized PTIOA $\mathcal{A}$, is a PTIOA $S = ((X, \mathcal{F}_X), \bar{x}, A, \mathcal{R}, \mathcal{D}', \mathcal{T}')$ that is identical to $\mathcal{A}$ except that $\mathcal{D}' \subseteq D$ and $\mathcal{T}' \subseteq \mathcal{T}$. A *local scheduler* S satisfies **D1** and has deterministic trajectories.

A *probabilistic-system* is a pair $\mathcal{M} = (\mathcal{A}, \mathcal{S})$, where $\mathcal{A}$ is a generalized PTIOA and $\mathcal{S}$ is a set of local schedulers for $\mathcal{A}$. An *environment* for $\mathcal{M}$ is any PTIOA $\mathcal{E}$ such that $\mathcal{A}||\mathcal{E}$ is closed. A probabilistic execution for $\mathcal{M}$ is defined to be any probabilistic execution of S, for any $S \in \mathcal{S}$. The notion of trace distribution carries over naturally to generalized PTIOAs. For probabilistic system $\mathcal{M} = (\mathcal{A}, \mathcal{S})$, we define the *external behavior* of $\mathcal{M}$ to be the total function $extbeh_{\mathcal{M}}$ that maps each environment PTIOA $\mathcal{E}$ for $\mathcal{M}$ to the set $\cup_{S' \in \mathcal{S}} tdists(S'||\mathcal{E})$. Thus for each environment, we consider the set of trace distributions that arise from the choices of the local scheduler of $\mathcal{M}$ and the task scheduler ρ. This leads to a notion of implementation of probabilistic systems, similar to that of PTIOAs. Let $\mathcal{M}_1 = (\mathcal{A}_1, \mathcal{S}_1)$ and $\mathcal{M}_2 = (\mathcal{A}_2, \mathcal{S}_2)$ be probabilistic systems such that $\mathcal{A}_1$ and $\mathcal{A}_2$ are comparable generalized PTIOAs. Then, $\mathcal{M}_1$ *implements* $\mathcal{M}_2$ if for every environment $\mathcal{E}$ of $\mathcal{M}_1$ and $\mathcal{M}_2$, $extbeh_{\mathcal{M}_1}(\mathcal{E}) \subseteq extbeh_{\mathcal{M}_2}(\mathcal{E})$. Theorem 2 gives a sufficient condition for implementation of probabilistic systems:

Theorem 2. *If $\mathcal{M}_1 = (\mathcal{A}_1, \mathcal{S}_1)$, $\mathcal{M}_2 = (\mathcal{A}_2, \mathcal{S}_2)$ are comparable and there exists $f : \mathcal{S}_1 \to \mathcal{S}_2$, such that for all $S_1 \in \mathcal{S}_1$, S_1 implements $f(S_1)$, then $\mathcal{M}_1$ implements M_2.*

In the future we would like to extend PTIOAs to support shared variables and develop a suite of analysis techniques for proving probabilistic safety, stability and approximate implementation relations [11].

We thank Sanjoy Mitter for many valuable comments on this work.

References

1. M. Bujorianu and J. Lygeros. General stochastic hybrid systems: Modelling and optimal control. In *IEEE Conference on Decision and Control*, December 2004.
2. R. Canetti, L. Cheung, D. Kaynar, M. Liskov, N. Lynch, O. Pereira, and R. Segala. Task-structured probabilistic I/O automata. Tech. Report MIT-CSAIL-TR-2006-060, MIT, Cambridge, 2006.
3. S. Cattani, R. Segala, M. Z. Kwiatkowska, and G. Norman. Stochastic transition systems for continuous state spaces and non-determinism. In *FoSSaCS'05*, LNCS 3441, pages 125–139, 2005.

4. L. Cheung. *Reconciling nondeterministic and probabilistic choices.* PhD thesis, ICIS, Radboud University Nijmegen, The Netherlands, 2006.
5. V. Danos, J. Desharnais, F. Laviolette, and P. Panangaden. Bisimulation and co-congruence for probabilistic systems. *Information and Computation, Special issue for selected papers from CMCS04*, 2005.
6. L. de Alfaro. *Formal Verification of Probabilistic Systems.* PhD thesis, Stanford University, CA, 1997. Technical Report STAN-CS-TR-98-1601.
7. S. Smolka E. Stark, R. Cleaveland. A process-algebraic language for probabilistic I/O automata. In *Proc. CONCUR 03*, *LNCS* 2761:189–203, 2003.
8. H. Hermanns. *Interactive Markov Chains : The Quest for Quantified Quality.* Springer, 2002.
9. D. Kaynar, N. Lynch, R. Segala, and F. Vaandrager. *The Theory of Timed I/O Automata.* Synthesis Lectures on Computer Science. Morgan Claypool, 2005.
10. S. Mitra and N. Lynch. Trace-based Semantics for Probabilistic Timed I/O Automata Full version `http://theory.lcs.mit.edu/~mitras/research/PTIOA06-full.pdf`
11. S. Mitra and N. Lynch Approximate simulations for task-structured probabilistic I/O automata. In LICS Workshop on Probabilistic Automata and Logics, 2006.
12. F. van Breugel, M. W. Mislove, J. Ouaknine, and J. Worrell. Domain theory, testing and simulation for labelled markov processes. *Theoretical Computer Science*, 2005.

Asymptotic Stability of Switched Higher Order Laplacians

Abubakr Muhammad and Ali Jadbabaie

Department of Electrical and Systems Engineering
University of Pennsylvania, Philadelphia, PA 19104, USA
{abubakr,jadbabai}@seas.upenn.edu

1 Introduction

Recently, several properties in networked sensing and distributed systems have been modeled by various researchers [1,2,3,5,7,9,11,12] using topological spaces and their topological invariants. The unifying theme in these approaches has been that the local properties of a network, as dictated by local interactions among agents, can be captured by certain topological spaces. These spaces are mostly combinatorial in nature and are a generalization of the more familiar graphical models. Moreover, the global properties of the network characteristics correspond to certain topological invariants of these spaces such as genus, homology, homotopy, and the existence of simplicial maps. Examples of such modeling attempts include coverage problems for sensor networks [1,2,3,7]; consensus & concurrency modeling in asynchronous distributed systems [9]; and routing in networks without geographical information [5]. One notable characteristic of these studies has been that the topological abstractions preserve many global geometrical properties of the network while abstracting away the redundant geometrical details at small scales. This promises great simplification of algorithms as well as hardware, which is an important requirement for realizing large-scale robust networks.

To elaborate further on this point, consider a network of nodes in a plane, each capable of performing a sensing task within a radially symmetric neighborhood. Then, one can study the problem of *blanket coverage*, i.e. whether the union of the coverage discs about the nodes cover a certain area of interest? This problem has been studied recently using computational homology methods [1,2,3,7]. The homological methods permit the study of this problem for sensors which are remarkably minimal, having no means of measuring distance, orientation, or location in their environment.

In this paper, we report our study on dynamic coverage in sensor networks using homological methods by focussing on the mobility of the nodes and the effect of changes in the topology of the underlying network. More precisely, we want to determine if it is possible to provide *sweep coverage* for a given planar domain. By sweep coverage we mean that every point in the environment is required to be covered infinitely often (possibly with some frequency), so that no point is left undetected for long. Thus by deploying a combination of static and mobile sensors, one can produce exploratory patrol-like behavior. This problem

A. Bemporad, A. Bicchi, and G. Buttazzo (Eds.): HSCC 2007, LNCS 4416, pp. 723–726, 2007.

is also related to pursuit-evasion problems and to exploration problems in an unknown environment. This problem is well studied and many researchers have proposed different solutions [6]. The aim of our work is not to provide another solution but to provide an online robust *verification* method for any given solution, in order to provide safeguards against node failure and uncertainties in modeling. As explained later, this verification can be implemented with minimal complexity on a large network of homogeneous mobile agents.

2 Homology and Higher Order Laplacians

The above mentioned criteria for studying coverage are based on the concepts of *simplicial complexes* and their *homology groups* [8]. The simplicial complexes arising in sensor networks are derived from or related to the unit-disk connectivity graphs in networks. In addition to the vertices and edges in graphs, they are made up of higher dimensional objects such as 'triangles' and 'tetrahedrons' that capture tertiary, quaternary or even higher-order relations between basic entities. These objects are collectively known as *simplices*. The homology groups of a simplicial complex X, denoted by $H_k(X)$ for $k \geq 0$, are used to distinguish one simplicial complex from one another by identifying the number of 'holes' of various dimension, contained in them. Each non-trivial member of the homology group in a certain dimension helps identify a corresponding hole in that dimension. Crudely speaking, the dimension of $H_0(X)$ is the number of connected components of X. The dimension of $H_1(X)$ is the number of non-contractible cycles in X, where each cycle encircles a 'puncture' in space. $H_2(X)$ identifies the number of 3-dimensional voids in a space and so on. In sensor networks, the members of the homology groups of certain simplicial complexes help identify *coverage losses* or *network communication holes* [1,2,3,7].

Recent advances in computational algebraic topology has enabled the successful verification of these topological results in simulation. However, the algorithms available are not meant for a distributed implementation. Therefore, in order to properly utilize these tools for networked sensing and control, one needs a new approach towards computing topological invariants that is well suited for implementation on real networked systems, with manageable complexity and scalability. In [11,12], the first steps towards this goal have been taken.

The central objects of this approach are the so-called higher order Laplacian operators on simplicial complexes. It has been shown in [11] that the spectral decomposition of the higher order Laplacian operators is a way to compute homology groups. To understand this operator, we recall that the homology groups are defined as a quotient space $H_k(X) = \ker \partial_k / \mathrm{im}\, \partial_{k+1}$, where ∂_k are the so called *boundary operators*, that linearly map simplices of a particular dimension to simplices at one dimension less. As an example, the familiar incidence matrix in graph theory is the boundary operator ∂_1, mapping edges to vertices. The higher order combinatorial Laplacian $\mathcal{L}_k$ is an operator (matrix) between simplices of the same dimension, and is given by $\mathcal{L}_k := \partial_{k+1}\partial_{k+1}^T + \partial_k^T\partial_k$. Observe that since there are no simplices below dimension zero, $\partial_0 = 0$ and the zeroth order Laplacian is the familiar graph Laplacian $\mathcal{L}_0 = \partial_1 \partial_1^T$. The key observation [4] is that $\ker \mathcal{L}_k \cong H_k(X)$. Again, the readers familiar with algebraic graph theory can verify that the dimension of $\ker \mathcal{L}_0$ is the number of connected components of X, which was earlier identified with $H_0(X)$.

The flow of these Laplacians, roughly described by $\partial\omega/\partial t = -\mathcal{L}_k \omega$, has been used in [11] to detect the absence or presence of network holes. In light of the above discussion and the fact that $\mathcal{L}_k$ is positive definite, it is easy to see that $\omega(t) \to 0$ if and only if $H_k(X) \cong \ker \mathcal{L}_k = \{0\}$, i.e. no holes in dimension k. In order to distinguish between multiple holes, the flow alone does not help unless there is direct method of doing a decentralized spectral decomposition. Such an algorithm has been studied in another related work by the authors [12].

3 Mobility, Switching Topologies and Dynamic Coverage

Let us now consider the case when the nodes are mobile, giving rise to a switching connectivity graph structure, which induces a switching structure for the simplicial complexes as well. Thus we can study a switched k-Laplacian flow, described by

$$\frac{\partial \omega(t)}{\partial t} = -\mathcal{L}_k^{\sigma} \omega(t),$$

where σ indexes the appropriate simplicial complex whenever there is a change of topology via a *switching signal.* If we are ensured that each simplicial complex encountered during the evolution is hole-free (zero homology), then the switched linear system shows asymptotic stability. However, one can show that the system exhibits asymptotic stability under an even weaker condition, whereby the simplicial complexes encountered in bounded, non-overlapping time intervals are *jointly hole-free.* To understand this condition, let the simplicial complexes encountered in one such interval be $X^1, X^2, \ldots, X^m$. Then, they are said to be jointly hole-free in dimension k, if $H_k(\cup_{i=1}^m X^i) \cong \{0\}$. The proof of this result closely follows the presentation in [10], the details of which have been given in [13].

The consequences of this result for sweep coverage are now easy to explain. The property of being jointly hole-free guarantees that in contiguous bounded time-intervals, the union of the simplicial complexes has zero homology. If the simplicial complexes model coverage properties, where coverage gaps are modeled by holes, then this condition guarantees that the entire region is covered in a bounded time interval. Moreover, the existence of an infinite sequence of contiguous intervals guarantees that each point of the region is visited infinitely often, thus verifying sweep coverage by the mobile network.

We should mention that two issues have not been fully explained in this paper, and we refer the interested reader to a more detailed version of this work [13]. One is that to infer coverage gaps from simplicial complexes, some extra structure is needed. Either, the hole exhibits a certain robustness with respect to the connectivity radius of the underlying graph [2,3], or some knowledge about the nodes monitoring the boundary has to be assumed [1]. Secondly, we have presented our results without incorporating the effects of coverage losses at the boundaries. To fix this, one needs to compute homology groups for the complexes *relative* to the boundaries nodes. For these reasons, we only have a *verification* algorithm i.e. coverage can be maintained even when the joint-hole free condition is violated. Still, the verification is computationally inexpensive and can be implemented in a decentralized manner. The reason for this is that the k-Laplacians are essentially local averaging or mixing operations, and therefore work in the spirit of gossip algorithms [12].

The switched linear systems studied above are a natural generalization of the work on distributed consensus algorithms [10], based on the standard graph Laplacian from algebraic graph theory. The idea of *joint connectedness* of a set of graphs has been generalized to the idea of when a collection of simplicial complexes are jointly hole-free. Thus, this theme of research allows us to view algebraic graph theory as a special case of the spectral theory of simplicial complexes, which in turn has proven useful in the context of networked sensing and control.

References

1. V. de Silva and R. Ghrist, "Coordinate-free coverage in sensor networks with controlled boundaries via homology, to appear, Intl. J. Robotics Research.
2. V. de Silva and R. Ghrist, "Coverage in sensor networks via persistent homology," *Algebraic and Geometric Topology.* (To appear)
3. V. de Silva, R. Ghrist and A. Muhammad, "Blind Swarms for Coverage in 2-D," *Robotics: Science and Systems*, MIT, Cambridge, MA, June 8-11, 2005.
4. B. Eckmann, "Harmonische Funktionen und Randwertaufgaben in Einem Komplex," *Commentarii Math. Helvetici*, Vol. 17, pp. 240–245, 1945.
5. Q. Fang, J. Gao, L. Guibas, V. de Silva, and L. Zhang, "Glider: gradient landmark-based distributed routing for sensor networks," *Proc. IEEE Infocom*, 2005.
6. D. Gage, "Command control for many-robot systems," in the *Nineteenth Annual AUVS Technical Symposium*, pp. 22–24, Huntsville, Alabama, USA, 1992.
7. R. Ghrist and A. Muhammad, "Coverage and Hole-Detection in Sensor Networks via Homology," *The Fourth International Conference on Information Processing in Sensor Networks* (IPSN'05), UCLA, Los Angeles, CA, April 25-27, 2005.
8. A. Hatcher, *Algebraic Topology*, Cambridge University Press, 2002.
9. M. Herlihy and N. Shavit, "The Topological Structure of Asynchronous Computability," *Journal of the ACM*, vol. 46, no. 6, pp. 858–923, November 1999.
10. A. Jadbabaie, J. Lin, and A. Morse, "Coordination of groups of mobile autonomous agents using nearest neighbor rules," *IEEE Transactions on Automatic Control*, Vol. 48, No. 6, pp. 988-1001, 2003.
11. A. Muhammad and M. Egerstedt, "Control Using Higher Order Laplacians in Network Topologies," in *Proc. of 17th International Symposium on Mathematical Theory of Networks and Systems*, Kyoto, Japan , 2006.
12. A. Muhammad and A. Jadbabaie, "Decentralized Computation of Homology Groups in Networks by Gossip", American Controls Conference, 2007. (Submitted)
13. A. Muhammad and A. Jadbabaie, "From Consensus in Switching Graphs to Coverage in Switching Simplicial Complexes", University of Pennsylvania Technical Report, 2006.

Qualitative Analysis of Nonlinear Biochemical Networks with Piecewise-Affine Functions

M.W.J.M. Musters[1], H. de Jong[2], P.P.J. van den Bosch[1], and N.A.W. van Riel[1]

[1] Eindhoven University of Technology, 5600 MB Eindhoven, The Netherlands
{m.w.j.m.musters, p.p.j.v.d.bosch, n.a.w.v.riel}@tue.nl
[2] INRIA Rhône-Alpes, Montbonnot, 38334 Saint Ismier cedex, France
hidde.de-jong@inrialpes.fr

Abstract. Nonlinearities and the lack of accurate quantitative information considerably hamper modeling and system analysis of biochemical networks. Here we propose a procedure for qualitative mathematical analysis of piecewise-affine (PWA) approximations of these networks. First the biochemical model size was reduced with quasi-steady state approaches by taking *a priori* information into account. Second, a conversion of a nonlinear model into a PWA approximation and subsequent qualitative analysis of this model was performed. This resulted in different sets of transition graphs that depend on the parameter values, which enables reduction of the parameter search space.

1 Introduction

Understanding biological processes is essential and a challenging task in which computer modeling and analysis have become indispensable, but simulation and analysis are seriously hampered by a lack of accurate experimental data and the complexity of the mathematical models. Certain biochemical networks show switch-like behavior. For instance, genetic regulatory networks can be described as a collection of discrete switches that react on continuous fluctuations in protein concentration. An obvious choice would be to model these networks with a framework that can handle both discrete and continuous components, i.e. the hybrid approach. This hybrid systems paradigm is therefore appropriate and has been applied to various biological systems with distinct switches, e.g. [1,2,3]. However, the field of biology is much larger, covering other nonlinear biochemical networks that contain non-switch-like components, e.g. [4]. In this short paper, we use hybrid systems to analyze a more general class of nonlinear biochemical networks.

2 Systematic Approach for Biological System Analysis

We will present a methodology that consists of four consecutive steps: 1) model reduction by quasi-steady state approximation, 2) conversion of the nonlinear model into a hybrid system, in a piecewise-affine (PWA) form, 3) determination of the symbolic equilibrium points and transitions, and 4) transition graph

A. Bemporad, A. Bicchi, and G. Buttazzo (Eds.): HSCC 2007, LNCS 4416, pp. 727–730, 2007.

construction. The procedure will be illustrated with a biologically relevant biochemical network, the Transforming Growth Factor-β_1 (TGF-β_1) pathway. We remark explicitly that this method is not limited to this example, but is applicable to other biochemical networks with nonlinear terms as well.

2.1 Step I: Model Reduction

A nonlinear mathematical model can be formulated by means of standard kinetic modeling in biochemistry. Typical biological models are composed of many nonlinear differential equations. The analysis procedure presented here is not automated yet; for manual analysis, a reduced model is required. A suitable model reduction method is the quasi-steady-state approximation, which leads to a more condensed set of state equations by means of timescale separation. The original model of the example consists of a seventh order ODE and is reduced to a second order set:

$$\begin{aligned} \dot{x}_4 &= k_4 \frac{1 - x_4}{K_{\mathrm{m}_2}\left(1 + \frac{x_7}{K_\mathrm{I}}\right) + 1 - x_4} - k_6 x_4, \\ \dot{x}_7 &= k_8 \frac{x_4^r}{K_{\mathrm{m}_3}^r + x_4^r} - k_9 x_7, \end{aligned} \tag{1}$$

with k_4, K_{m_2}, K_I, k_6, k_8, K_{m_3}, r and k_9: eight parameters (positive values); x_4 and x_7: two states (not negative).

2.2 Step II: PWA Approximation of the Nonlinear Model

Subsequently, the nonlinear terms in Eq. (1) are approximated by PWA functions by applying the rules in [5], which have been extended to deal with multivariable functions. In Eq. (1), the PWA approximation results in a hybrid system f of maximally four regular modes $(q_1, \ldots, q_4)$ containing two states $x = [x_4; x_7]^T$:

$$f = \begin{cases} [-k_6 x_4 + \beta(1 - x_7); -k_9 x_7]^T & \text{if } x_4 - \alpha x_7 < 1 - \alpha \wedge x_4 < K_{\mathrm{m}3}, \\ [-k_6 x_4 + \frac{\beta}{\alpha}(1 - x_4); -k_9 x_7]^T & \text{if } x_4 - \alpha x_7 > 1 - \alpha \wedge x_4 < K_{\mathrm{m}3}, \\ [-k_6 x_4 + \beta(1 - x_7); \gamma - k_9 x_7]^T & \text{if } x_4 - \alpha x_7 < 1 - \alpha \wedge x_4 > K_{\mathrm{m}3}, \\ [-k_6 x_4 + \frac{\beta}{\alpha}(1 - x_4); \gamma - k_9 x_7]^T & \text{if } x_4 - \alpha x_7 > 1 - \alpha \wedge x_4 > K_{\mathrm{m}3}, \end{cases} \tag{2}$$

with α, β, and γ: parameters required for the PWA approximation, which could be linked to the parameters in the nonlinear model with Eq. (1).

2.3 Step III: Equilibrium Points

Analysis of system dynamics requires knowledge about the equilibrium points in the system. The equilibrium points have to satisfy the invariants and biochemical constraints, e.g. all concentrations are larger than zero. Symbolic expressions of the equilibrium points in each mode $(q_1, ..., q_4)$ are derived with $\dot{x} = 0$. For the example, all equilibrium points and their associated existence conditions are listed in Table 1. The equilibrium points are mutually exclusive, which implies that no multiple equilibrium points can occur. Furthermore, a common Lyapunov function can be found for all modes, which guarantees that the complete system is globally uniformly asymptotically stable (GUAS), regardless of the parameter values. Note that switching modes are also present in this model, but omitting these modes will not influence the conclusions of the analysis.

Table 1. Equilibrium points and corresponding existence conditions

	Equilibrium point	Existence condition
q_1 :	$\left(\frac{\beta}{k_6}, 0\right)$	$0 < \frac{\beta}{k_6} < \min\left(1-\alpha, K_{\mathrm{m}3}\right)$
q_2 :	$\left(\frac{\beta}{\alpha k_6+\beta}, 0\right)$	$1-\alpha < \frac{\beta}{k_6} < K_{\mathrm{m}3}$
q_3 :	$\left(\frac{\beta(k_9-\gamma)}{k_6 k_9}, \frac{\gamma}{k_9}\right)$	$K_{\mathrm{m}3}\frac{k_9}{k_9-\gamma} < \frac{\beta}{k_6} < \frac{k_9}{k_9-\gamma} - \alpha$
q_4 :	$\left(\frac{\beta}{\alpha k_6+\beta}, \frac{\gamma}{k_9}\right)$	$\max\left(K_{\mathrm{m}3}\frac{k_9}{k_9-\gamma}, \frac{k_9}{k_9-\gamma} - \alpha\right) < \frac{\beta}{k_6} < 1$

2.4 Step IV: Transition Graph Construction

Mode transitions occur if the inner product of the model equations f and the normal n of the guard condition, i.e. the inequality in Eq. (2), is larger than zero [5]:

$$f^T \cdot n > 0. \tag{3}$$

Eq. (3) is applied for all transitions on the guard conditions, for each equilibrium point. Since the equilibria are bounded by existence conditions (Table 1) and transitions have to comply with Eq. (3), a transition has to satisfy specific inequalities of parameter values in order to be feasible. For the example it has been shown that multiple transition graphs can exist. With experimental information from the literature, one can exclude specific system behavior and, consequently, impose additional constraints on the parameter values. However, due to limitations in space, the plots of these transition graphs and corresponding restrictions on the parameter values have been omitted.

3 Discussion and Conclusion

The qualitative analysis method described above has been tailored for processes that are typically observed in biochemical networks, e.g. Michaelis-Menten and Hill kinetics, but can be applied to other research fields outside biology as well. A model of a biochemical network was used to illustrate our method. Since the approach was partially based on model reduction, a relatively small model was constructed as a result, which simplifies qualitative analysis [5,6]. Such small models can of course be analyzed with standard phase plane techniques, but larger biochemical networks are currently under study. Automation of our qualitative hybrid approach with quantifier elimination [7] will be done in the near future, enabling the analysis of larger systems. Comparison of the transition graphs of the example with preliminary experimental data shows qualitative similarities: the system converges to a single, stable steady-state and contains neither a limit cycle nor exhibits multistability, exactly according to the model predictions. The next step is to link the parameters of the PWA model to their nonlinear counterparts. The hybrid analysis procedure could fulfill an important role in obtaining an initial estimate for nonlinear parameter estimation [8] and support the parameter estimator to select a more appropriate solution that satisfies both the limited amount of measurement data and the *a priori* qualitative information.

Acknowledgements

This research is financially supported by the Netherlands Organization for Scientific Research, grant R 61-594, and by the Dutch Ministry of Economic Affairs and Unilever Research and Development, Senter grant TSGE1028.

References

1. Alur, R., Belta, C., Kumar, V., Mintz, M., Pappas, G.J., Rubin, H., Schug, J.: Modeling and analyzing biomolecular networks. CiSE **4**(1) (2002) 20 – 31
2. Batt, G., Ropers, D., de Jong, H., Geiselmann, J., Page, M., Schneider, D.: Qualitative analysis and verification of hybrid models of genetic regulatory networks: Nutritional stress response in escherichia coli. In Morari, M., Thiele, L., eds.: Hybrid Systems: Computation and Control HSCC. Volume 3414 of LNCS. Springer-Verlag (2005) 134–150
3. Belta, C., Schug, J., Dang, T., Kumar, V., Pappas, G.J., Rubin, H.: Stability and reachability analysis of a hybrid model of luminescence in the marine bacterium vibrio fischeri. In: Proceedings of the 40th IEEE Conference on Decision and Control. (2001) 869–874
4. Hoffmann, A., Levchenko, A., Scott, M.L., Baltimore, D.: The IκB-NF-κB signaling module: temporal control and gene activation. Science **298**(5596) (2002) 1241–1245
5. Sacks, E.: Automatic qualitative analysis of dynamic systems using piecewise linear approximations. Artif Intell **41** (1990) 313–364
6. Nishida, T., Doshita, S.: Qualitative analysis of behavior of systems of piecewise linear differential equations with two state variables. Artif Intell **75** (1995) 3–29
7. Tiwari, A., Khanna, G.: Series abstractions for hybrid automata. In Tomlin, C.J., Greenstreet, M.R., eds.: Hybrid Systems: Computation and Control (HSCC 2002). Volume 2289 of LNCS. Springer-Verlag, Berlin (2002) 465–478
8. Musters, M.W.J.M., Lindenaar, D.J.W., Juloski, A.L., van Riel, N.A.W.: Hybrid identification of nonlinear biochemical processes. In: IFAC SYSID, Newcastle, Australia. (2006)

Guided Randomized Simulation

Tarik Nahhal and Thao Dang

VERIMAG, 2 avenue de Vignate
38610 Gières, France

Abstract. The paper presents a simulation-based method for the validation of continuous and hybrid systems, which is built upon the RRT algorithm (Rapidly-exploring Random Trees). We propose a coverage measure, defined as the star discrepancy of the explored points, and use it to guide the simulation towards the behaviors of interest.

1 Introduction

Recent results in formal verification of hybrid systems have been successfully applied to various interesting case studies. Nevertheless, its applicability is still limited to systems of small size. Simulation is another validation approach, which can be used for much larger systems and is a standard tool in industry, although simulation can only reveal an error but does not permit proving the correctness of the system. Therefore, a question of great interest is to bridge the gap between these two approaches by developing a light-weight method that can guarantee some level of confidence in the results. The goal of this paper is to develop a validation method for continuous and hybrid systems by combining numerical simulation with a guided search method. For simplicity, we only focus on continuous systems, but the approach we propose can be extended to hybrid systems. We consider a continuous system defined by the differential equation $\dot{x}(t) = f(x(t), u(t))$ where $x \in \mathcal{X}$ ($\mathcal{X}$ is the state space), and u denotes the input. Due to the infiniteness of the input space and of the state space, it is important to choose the inputs that lead to the scenarios that are interesting with respect to the property to check. The essential ideas of our approach can be summarized as follows. First, it is based on the RRT (Rapidly-exploring Random Tree) algorithm, a probabilistic motion planning technique. Second, we introduce a simulation coverage measure, defined as the star discrepancy of the visited points, and use it to guide the simulation process. In the rest of the paper, we first present our simulation coverage measure and then describe gRRT, a guided variant of RRT. We finally show some experimental results. Before continuing, we briefly summarize the RRT algorithm.

In path and motion planning, Rapidly-exploring Random Trees (RRTs) are used to find feasible robot trajectories in an environment with obstacles (see [6] for a survey). The reachable ponts are stored in a tree, the root of which corresponds to an initial condition $x(0)$. In each iteration, a goal point x_{goal} in $\mathcal{X}$ is sampled. Expanding the tree towards x_{goal} is done by determining an initial point x_{init} for the current integration step and then finding the input to take

A. Bemporad, A. Bicchi, and G. Buttazzo (Eds.): HSCC 2007, LNCS 4416, pp. 731–735, 2007.

the system from x_{init} as closest to x_{goal} as possible. In the 'classic' versions of RRT, the sampling distribution of x_{goal} is uniform over $\mathcal{X}$ and x_{init} is a nearest neighbor of x_{goal} (in some distance). Our goal is to build on top of the RRT algorithm a guiding tool, which biases the exploration in order to achieve a good coverage of the system's behaviors that we want to check.

2 Coverage Measure

Simulation coverage is a way to relate the number of simulations to carry out with the fraction of the system's behaviors effectively explored. The classic coverage notions mainly used in software testing, e.g. statement and path coverage are not appropriate for the trajectories of continuous and hybrid systems defined by differential equations. In this work, we are interested in *point coverage* and focus on a measure that describes how 'well' the points represent the reachable set of the system. This measure is defined using the notion of star discrepancy in statistics (see for example [5]). The state space $\mathcal{X}$ could be assumed to be a box $\mathcal{B} = [l_1, L_1] \times \ldots \times [l_n, L_n]$. Given a set P of k points in $\mathcal{B}$, the star discrepancy of P w.r.t. $\mathcal{B}$ is defined as: $D^*(P, \mathcal{B}) = sup_{J \in \Gamma} D(P, J)$ where Γ is the set of all subboxes $J = \prod_{i=1}^{n} [l_i, \beta_i]$ with $\beta_i \in [l_i, L_i]$. The local discrepancy $D(P, J)$ is defined as: $D(P, J) = |\frac{A(P,J)}{k} - \frac{\lambda(J)}{\lambda(\mathcal{B})}|$ where $A(P, J)$ is the number of points of P that are inside J, and $\lambda(J)$ is the volume of J. Note that $0 < D^*(P, \mathcal{B}) \leq 1$. Intuitively, the star discrepancy is a measure for the irregularity of a set of points. A large value $D^*(P, \mathcal{B})$ means that P are not much equidistributed over $\mathcal{B}$.

Simulation Coverage and Estimation. Let P be the set of all points explored by a simulation, and the simulation coverage is defined as $Cov(P) = 1 - D^*(P, \mathcal{B})$. The computation of the star discrepancy is not easy; in this work, we do not try to compute the star discrepancy but estimate a lower and upper bound, exploiting the results in [7] that we now briefly present. A box partition of $\mathcal{B}$ is a set of boxes $\Pi = \{\boldsymbol{b}^1, \ldots, \boldsymbol{b}^m\}$ such that $\cup_{i=1}^{m} \boldsymbol{b}^i = \mathcal{B}$ and the interiors of the boxes do not intersect. Given $\boldsymbol{b} = [\alpha_1, \beta_2] \times \ldots \times [\alpha_n, \beta_n] \in \Pi$, we define $\boldsymbol{b}^+ = [l_1, \beta_1] \times \ldots \times [l_n, \beta_n]$ and $\boldsymbol{b}^- = [l_1, \alpha_1] \times \ldots \times [l_n, \alpha_n]$. For any finite box partition Π of $\mathcal{B}$, the star discrepancy satisfies $C(P, \Pi) \leq D^*(P, \mathcal{B}) \leq B(P, \Pi)$ where the upper bound is: $B(P, \Pi) = \max_{\boldsymbol{b} \in \Pi} \max\{\frac{A(P, \boldsymbol{b}^+)}{k} - \frac{\lambda(\boldsymbol{b}^-)}{\lambda(\mathcal{B})}, \frac{\lambda(\boldsymbol{b}^+)}{\lambda(\mathcal{B})} - \frac{A(P, \boldsymbol{b}^-)}{k}\}$, and the lower bound is: $C(P, \Pi) = \max_{\boldsymbol{b} \in \Pi} \max\{|\frac{A(P, \boldsymbol{b}^-)}{k} - \frac{\lambda(\boldsymbol{b}^-)}{\lambda(\mathcal{B})}|, |\frac{A(P, \boldsymbol{b}^+)}{k} - \frac{\lambda(\boldsymbol{b}^+)}{\lambda(\mathcal{B})}|\}$.

3 The gRRT Algorithm

Discrepancy Guided Sampling. In the gRRT algorithm, the goal point sampling distribution is biased in order to improve the coverage in each iteration. Let Π be a finite box partition of $\mathcal{B}$ that is used to estimate the star discrepancy. The goal point sampling consists of two steps: first sample a box $\boldsymbol{b}$ in Π and then uniformly sample a point in $\boldsymbol{b}$. The box sampling distribution is defined as follows. Intuitively, we favor the selection of a box such that adding a new point p in this box results in a smaller star discrepancy of the new point set $P \cup \{p\}$. The strategy is thus to reduce both the lower and upper bounds. Due to space limitation, we present only the strategy to reduce the lower bound (see [1] for more details). Let $\tilde{P}$ be a set of k points in $\mathcal{B}$ such that for any box $\boldsymbol{b} \in \Pi$, we have $\frac{\lambda(\boldsymbol{b})}{\lambda(\mathcal{B})} = \frac{A(\tilde{P},\boldsymbol{b})}{k}$, where $A(\tilde{P}, \boldsymbol{b})$ is the number of points inside $\boldsymbol{b} \cap \tilde{P}$. We denote $\Delta_A(\boldsymbol{b}) = A(P, \boldsymbol{b}) - A(\tilde{P}, \boldsymbol{b})$ and $c(\boldsymbol{b}) = \max\{|\Delta_A(\boldsymbol{b}^+)|, |\Delta_A(\boldsymbol{b}^-)|\}$. Thus, the lower bound is $C(P, \Pi) = \frac{1}{k}\max_{\boldsymbol{b}\in\Pi}\{c(\boldsymbol{b})\}$. Note that in comparison with $\tilde{P}$, the negative (respectively positive) sign of $\Delta_A(\boldsymbol{b})$ indicates that in this box there is a lack (respectively an excess) of points; its absolute value indicates how significant the lack (or the excess) is. We observe that adding a point in $\boldsymbol{b}$ reduces $|\Delta_A(\boldsymbol{b}^+)|$ if $\Delta_A(\boldsymbol{b}^+) < 0$, and increases $|\Delta_A(\boldsymbol{b}^+)|$ otherwise. However, doing so does not affect $\Delta_A(\boldsymbol{b}^-)$ (see the definition of $\boldsymbol{b}^-$ and $\boldsymbol{b}^+$). Thus, for each box $\boldsymbol{b} \in \Pi$ we define a *potential function*: $\xi(\boldsymbol{b}) = \frac{1-\Delta_A(\boldsymbol{b}^+)/k}{1-\Delta_A(\boldsymbol{b}^-)/k}$, and its intepretation is as follows. If $\Delta_A(\boldsymbol{b}^+)$ is negative and its absolute value is large, the 'lack' of points in $\boldsymbol{b}^+$ is significant, and $\xi(\boldsymbol{b})$ is large, meaning that the selection of $\boldsymbol{b}$ is favored. If $\Delta_A(\boldsymbol{b}^-)$ is negative and its absolute value is large, then $\xi(\boldsymbol{b})$ is small, because it is preferable not to select $\boldsymbol{b}$ in order to increase the chance of adding new points in $\boldsymbol{b}^-$. The box sampling distribution can then be defined by combining the potential functions for the lower and upper bounds [1].

Implementation and Experimental Results. To store the explored points, we use a data structure similar to a k-d tree, which allows to efficiently perform the required operations (such as, adding a new point, updating the dicrepancy estimation, computing nearest neighbors). Indeed, for complexity reasons, in the gRRT, we only compute approximate nearest neighbors and we can prove that this approximation preserves the probabilistic completeness of the RRT algorithm (see [1]). The results reported here were obtained by running our prototype implementation of gRRT on a 1.4 GHz Pentium III. For illustration purposes, we first show the results on a simple nonlinear model of competing species: $\dot{x} = 2x(1 - x/2) - xy + u_1$, $\dot{y} = 3y(1 - y/3) - 2xy + u_2$. We let their dynamics be slightly perturbed by an input u. Figure 1 shows the simulation results after 50000 iterations, using the basic RRT (left) and the gRRT algorithm (right). The run times of the RRT and gRTT algorithms are respectively 1.2 minutes and 50 seconds. From Figure 1 we can see that the coverage of gRRT is better, which can be explained as follows. Due to the uniform sampling of goal points, the RRT exploration is biased by the Voronoi diagram of the explored points, that is, if the volume of the Voronoi cell of a point is large, it has a high probability of being selected to be the initial point. When the actual reachable set is a small fraction of the state space, the uniform sampling leads to a strong bias in selecting points on the boundary of the tree. In addition, for scalability evaluation, we tested gRRT on a set of randomly generated linear systems $\dot{x} = Ax + u$ in various dimensions up to 100 (but we did not exploit the dynamics

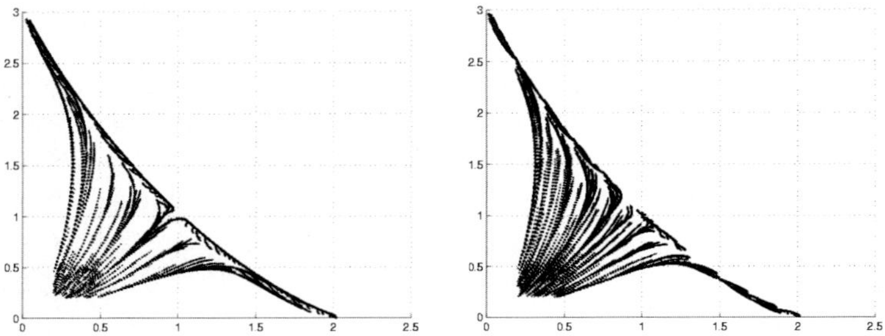

Fig. 1. Simulation results for the competing species model

linearity). The run times after 50000 iterations are: dimension $n = 5$, time 1 mn; $n = 10$, time 3.5mn; $n = 20$, time 7.3mn; $n = 50$, time 24mn, $n = 100$, time 71mn.

4 Conclusion

The RRT algorithm has been used to solve a variety of reachability-related problems (see for example [2,3,4]). Another coverage measure (using dispersion) was proposed in [3]. While in our work, the coverage measure is used to guide the simulation, in [3] it is used as a termination criterion. Our idea of guiding via sampling has some similarity with the sampling domain control in [8]. The main difference, which is also the novelty in our guiding method, is that we use the information about the current coverage to improve it.

To conclude, the main contribution of our paper is a way to guide the simulation by a coverage measure. The experimental results show the scalability of gRRT to high dimensional systems and an improvement in simulation coverage quality. A direction for future research is to extend the approach to hybrid systems. Convergence rate of the gRRT algorithm and trace coverage measures are also interesting theoretical problems.

References

1. T. Dang and T. Nahhal. Randomized simulation of hybrid systems. Technical report, Verimag, IMAG, May 2006.
2. M. S. Branicky, M. M. Curtiss, J. Levine, and Stuart Morgan. Sampling-based reachability algorithms for control and verification of complex systems. *Proc. Thirteenth Yale Workshop on Adaptive and Learning Systems*, New Haven, CT, 2005.
3. J.M. Esposito, J. Kim, and V. Kumar. Adaptive RRTs for validating hybrid robotic control systems. In *Int. Workshop on Algorithmic Foundations of Robotics* , 2004.
4. A. Bhatia and E. Frazzoli. Incremental Search Methods for Reachability Analysis of Continuous and Hybrid Systems. In HSCC, LNCS 2993, 142-156, Springer, 2004.
5. L. Kuipers and H. Niederreiter. Uniform Distribution of Sequences. John Wiley, New York, 1974.

6. S. M. LaValle and J. J. Kuffner. Rapidly-exploring random trees: Progress and prospects. In *Algorithmic and Computational Robotics: New Directions*, pages 293–308. A K Peters, Wellesley, MA, 2001.
7. E. Thiémard. An algorithm to compute bounds for the star discrepancy. J. Complexity 17(4):850, Dec 2001.
8. A. Yershova, L. Jaillet, T. Simeon, and S.M. LaValle. Dynamic-domain RRTs: Efficient exploration by controlling the sampling domain. In *Proc. IEEE Int. Conf. on Robotics and Automation*, 2005.

Controller Parameters Selection Through Bifurcation Analysis in a Piecewise-Smooth System

Eva M. Navarro-López[1,⋆] and Domingo Cortés[2]

[1] ETS de Ing. Industriales, Univ. de Castilla-La Mancha
Av. Camilo José Cela s/n, 13071 Ciudad Real, Spain
navarro.lopez@gmail.com

[2] Centro de Inv. y Estudios Avanzados del IPN, Dpto. de Ing. Eléctrica
Av. IPN, 2508, Col. San Pedro Zacatenco, 07360 Mexico, D.F., Mexico
domingo.cortes@gmail.com

Abstract. Classical linear and nonlinear control techniques applied to piecewise-smooth (PWS) systems can be ineffective if an additional bifurcation analysis is not made. Due to the presence of discontinuities, PWS systems present a wide variety of standard and non-standard bifurcations. Safe ranges of system and controller parameters can be established lest these bifurcations appear. This is studied by means of a simplified torsional model of an oilwell drillstring of three degrees of freedom (DOF). A PID-type controller is applied. The bifurcation analysis of the open-loop and the closed-loop systems is used to choose the controller parameters for which non-desired bit sticking situations are avoided. The control goal of driving the rotary velocities to a constant positive value is achieved despite the existence of a sliding motion.

Keywords: Discontinuous systems, sliding motions, bifurcation analysis, oilwell drillstrings, dry friction, stick-slip oscillations.

1 Introduction

Hybrid systems are dynamical systems consisting of continuous-time and discrete-event dynamics. They are characterized by discontinuous changes in system properties. PWS systems are an interesting subclass of hybrid systems. An example of PWS system is the model of a conventional vertical oilwell drillstring.

Oilwell drillstrings are systems exhibiting a wide variety of complex phenomena and non-desired oscillations. Stick-slip friction-induced oscillations and the permanent stuck-bit situation are particulary harmful [1]. Several approaches have appeared in order to model and control drillstring stick-slip oscillations. Most of them use lumped-parameter models of one DOF and two DOF (see [2] and references therein). In addition, in most of these works, no bifurcation

⋆ This work has been partially supported by the *Ministerio de Educación y Ciencia* (MEC) of Spain under *Ramón y Cajal* research contract.

A. Bemporad, A. Bicchi, and G. Buttazzo (Eds.): HSCC 2007, LNCS 4416, pp. 736–740, 2007.

analysis of system and controller parameters is made, and the influence of the weight on the bit (WOB) (a key drilling parameter) is not analysed.

Recently in [2], an analysis of bifurcations and transitions between several bit dynamics is reported for an n-DOF drillstring. Changes in drillstring dynamics are studied through variations in three parameters: 1) the WOB, 2) the rotary speed at the top-rotary system, 3) the torque given by the surface motor.

This paper follows [2] and sums up new results. On the one hand, the bit sticking problems are related to the existence of a sliding motion when the bit velocity is zero. On the other hand, the presence of multiple Hopf bifurcations in the vicinity of the standard equilibrium point when velocities are greater than zero is studied. A PID-type control which overcomes the existing sliding motion and drives the rotary velocities to a desired value is proposed. Controller parameters are chosen lest non-desired system transitions appear. These non-desired situations are permanent stuck bit, bit stick-slip oscillations and other periodic motions. A lumped-parameter torsional PWS model with three DOF including the bit-rock interaction is considered. This model is more general than the torsional lumped-parameter models of one and two DOF previously proposed.

2 A Torsional Discontinuous Model of the Drillstring

Three main parts of a conventional vertical oilwell drillstring are considered: 1) the rotating mechanism in the surface (inertia J_r), 2) the set of drill pipes which form a long pipeline (inertia J_p), 3) the bottom-hole assembly (BHA), which consists of the drill collars, the stabilizers, a heavy-weight drill pipe and the bit (inertia J_b). Hereinafter, the BHA will be also referred to as bit. A simplified torsional model is proposed in which the inertias are connected to each other by linear springs with torsional stiffness (k_t, k_{tb}) and torsional damping (c_t, c_{tb}).

The following system state vector is defined:

$$\mathbf{x} = (\dot{\varphi}_r, \varphi_r - \varphi_p, \dot{\varphi}_p, \varphi_p - \varphi_b, \dot{\varphi}_b)^T = (x_1, x_2, x_3, x_4, x_5)^T, \tag{1}$$

with φ_i, $\dot{\varphi}_i$ ($i \in \{r, p, b\}$) the angular displacements and angular velocities of drillstring elements, respectively. At the top-drive system, a viscous damping torque is considered ($T_{a_r} = c_r x_1$). $T_m = u$ is the torque coming from the electrical motor at the surface, with u the control input. $T_b(x_5) = T_{a_b}(x_5) + T_{f_b}(x_5)$ is the torque on the bit with $T_{a_b} = c_b x_5$ approximating the influence of the mud drilling on the bit behaviour. $T_{f_b}(x_5)$ is the friction modelling the bit-rock contact, and

$$T_{f_b}(x_5) = W_{ob} R_b \left[\mu_{c_b} + (\mu_{s_b} - \mu_{c_b}) \exp^{-\frac{\gamma_b}{v_f}|x_5|}\right] \mathrm{sgn}(x_5), \tag{2}$$

with $W_{ob} > 0$ the WOB, $R_b > 0$ the bit radius; μ_{s_b}, $\mu_{c_b} \in (0, 1)$ the static and Coulomb friction coefficients associated with J_b, $0 < \gamma_b < 1$ and $v_f > 0$. Finally, the drillstring behaviour is described by the following equations:

$$\dot{x}_1 = \frac{1}{J_r}\left[-(c_t + c_r)x_1 - k_t x_2 + c_t x_3 + u\right], \qquad \dot{x}_2 = x_1 - x_3,$$

$$\dot{x}_3 = \frac{1}{J_{\rm p}}\left[c_{\rm t}x_1 + k_{\rm t}x_2 - (c_{\rm t}+c_{\rm tb})x_3 - k_{\rm tb}\,x_4 + c_{\rm tb}\,x_5\right], \quad \dot{x}_4 = x_3 - x_5, \qquad (3)$$

$$\dot{x}_5 = \frac{1}{J_{\rm b}}\left[c_{\rm tb}\,x_3 + k_{\rm tb}\,x_4 - (c_{\rm tb}+c_{\rm b})x_5 - T_{\rm f_b}(x_5)\right],$$

or in a compact form: $\dot{\mathbf{x}}(t) = \mathbf{A}\mathbf{x}(t) + \mathbf{B}u + \mathbf{T}_{\rm f}(\mathbf{x}(t))$, where $\mathbf{A}$, $\mathbf{B}$ are constant matrices depending on system parameters and $\mathbf{T_f}$ is the torque on the bit.

3 Transitions and Bifurcations in the Open-Loop System

Two key dynamical characteristics of system (3) determine its behaviour: 1) the existence of a sliding regime which implies the existence of a sliding surface locally attractive; 2) the existence of a unique standard equilibrium point when velocities are greater than zero, which is locally asymptotically stable depending on the values of $W_{\rm ob}$, the torque u and the rotary velocity at the equilibrium.

The existence of a sliding motion (see [3]) in system (3) is directly related to different bit sticking phenomena, mainly, stick-slip motion and the permanent stuck-bit situation. System (3) is a PWS system of the form,

$$\dot{\mathbf{x}} = \begin{cases} \mathbf{f}^+(\mathbf{x}, W_{\rm ob}, u) = \mathbf{A}\mathbf{x} + \mathbf{B}u + \mathbf{T}_{\rm f}(\mathbf{x})|_{T_{\rm f_b}=T^+_{\rm f_b}} & \text{if } x_5 > 0, \\ \mathbf{f}^-(\mathbf{x}, W_{\rm ob}, u) = \mathbf{A}\mathbf{x} + \mathbf{B}u + \mathbf{T}_{\rm f}(\mathbf{x})|_{T_{\rm f_b}=T^-_{\rm f_b}} & \text{if } x_5 < 0, \end{cases} \qquad (4)$$

where $T^+_{\rm f_b}$ and $T^-_{\rm f_b}$ are $T_{\rm f_b}(x_5)$ for $x_5 > 0$ and $x_5 < 0$, respectively. Let $\Sigma := \{\mathbf{x} \in \mathrm{I\!R}^5 : x_5 = 0\}$ be the switching manifold, $\tilde{\Sigma} \subset \Sigma$ the sliding region with $\tilde{\Sigma} = \{\mathbf{x} \in \Sigma : |k_{\rm tb}\,x_4 + c_{\rm tb}\,x_3| < W_{\rm ob}R_{\rm b}\mu_{\rm s_b}\}$ and $\tilde{\mathbf{x}}$ the quasiequilibrium point existing on $\tilde{\Sigma}$, which can be shown to be asymptotically stable. If $x_5 > 0$ then the system is described by $\dot{\mathbf{x}} = \mathbf{f}^+(\mathbf{x}, W_{\rm ob}, u)$ and has a unique standard equilibrium point $\overline{\mathbf{x}}$ such that $\mathbf{f}^+(\overline{\mathbf{x}}, W_{\rm ob}, u) = 0$, which is the solution of the set of equations:

$$\overline{x}_1 = \overline{x}_3 = \overline{x}_5 > 0,\ u - (c_{\rm r}+c_{\rm b})\overline{x}_5 - T^+_{\rm f_b}(\overline{x}_5) = 0,\ \overline{x}_2 = \frac{h(\overline{x}_5, u)}{k_{\rm t}},\ \overline{x}_4 = \frac{h(\overline{x}_5, u)}{k_{\rm tb}},$$

with $h(\overline{x}_5, u) = \left[c_{\rm r}T^+_{\rm f_b}(\overline{x}_5) + c_{\rm b}u\right]/(c_{\rm r}+c_{\rm b})$ and $u > W_{\rm ob}R_{\rm b}\mu_{\rm s_b} > 0$.

Three main kinds of bit dynamical behaviours can be identified: 1) stick-slip at x_5, that is, the trajectory enters and leaves repeatedly the sliding mode; 2) permanent stuck bit, i.e., $\mathbf{x}(t) \in \tilde{\Sigma}$, $\forall t$; 3) the trajectory converges to the standard equilibrium $\overline{\mathbf{x}}$. The bit moves with a constant positive velocity (the third situation) when $\tilde{\mathbf{x}}$ is far away enough from the boundaries of $\tilde{\Sigma}$, which is accomplished when u is greater enough than $W_{\rm ob}R_{\rm b}\mu_{\rm s_b}$.

The local stability of $\overline{\mathbf{x}}$ must be also assured. Periodic orbits around it due to the presence of Hopf bifurcations (HB) can be avoided. Ranges of the WOB and velocities at equilibrium for which a HB may appear can be identified, they intersect the values for which stick-slip oscillations are present. The stability region of $\overline{\mathbf{x}}$ corresponds to low W_{ob} and high enough values of the rotary velocities.

4 Linear Control to Eliminate Non-desired Transitions

The control goal is to eliminate bit sticking problems and establish a stable standard equilibrium point in which the angular velocities are all equal to a desired positive velocity. This is accomplished by means of the following control:

$$u = K_1 x_6 + K_2(\Omega - x_1) + K_3(x_1 - x_5) + u^*, \tag{5}$$

with $x_6 = \int_0^t [\Omega - x_1(\tau)]\, d\tau$, $\dot{x}_6 = \Omega - x_1$, $\Omega > 0$ the desired rotary velocity, K_i positive constants and $u^* = W_\mathrm{ob} R_\mathrm{b} \mu_{\mathrm{s_b}} > 0$ the minimum value of u for the system trajectory to cross the boundary of $\tilde{\Sigma}$. This value of u^* prevents the bit from sticking when control (5) is initially switched on. The closed-loop system has a unique standard equilibrium point $\overline{\mathbf{x}}_\mathbf{c}$ with the form,

$$\overline{x}_{\mathrm{c},1} = \overline{x}_{\mathrm{c},3} = \overline{x}_{\mathrm{c},5} = \Omega,\ \overline{x}_{\mathrm{c},2} = \frac{h(\Omega)}{k_\mathrm{t}},\ \overline{x}_{\mathrm{c},4} = \frac{h(\Omega)}{k_\mathrm{tb}},\ \overline{x}_{\mathrm{c},6} = \frac{1}{K_1}\left[c_\mathrm{r}\Omega + h(\Omega) - u^*\right],$$

where $h(\Omega) = \left[c_\mathrm{b}\Omega + T^+_{\mathrm{f_b}}(\Omega)\right]$. The dynamical changes introduced by control (5) in the open-loop system (3) are mainly: 1) the standard equilibrium point has the angular velocities equal to Ω; 2) the sliding motion is maintained, however, there is no quasiequilibrium point on the switching manifold, and the permanent stuck-bit situtation is eliminated; 3) periodic orbits (including stick-slip oscillations) may still arise in the system.

Stick-slip oscillations appear in the closed-loop system due to the existence of a sliding region $\tilde{\Sigma}$ locally attractive and a standard equilibrium which can become unstable or whose domain of attraction can be reduced due to the variation of the controller parameters, Ω and W_ob. The analysis of the zero crossings of the real parts of the pairs of the complex eigenvalues of the Jacobian at $\overline{\mathbf{x}}_\mathrm{c}$ of the closed-loop system (referred to as ZC) is a good approach for studying the changing stability properties of $\overline{\mathbf{x}}_\mathrm{c}$. The ZC points might imply a Hopf bifurcation and be the origin of periodic orbits. The extension of the parameters region where no ZC point is present, is indicative of the domain of attraction of $\overline{\mathbf{x}}_\mathrm{c}$. When the domain of attraction of $\overline{\mathbf{x}}_\mathrm{c}$ increases, stick-slip motion may disappear.

Extensive simulations of the closed-loop system lead to establish safe ranges of controller parameters K_i in order to avoid bit sticking problems and the unstability of the equilibrium $\overline{\mathbf{x}}_\mathrm{c}$. The main conclusion is that for typical operation values of Ω (between 10 and 14 rad/s), stick-slip oscillations disappear in the closed-loop system and trajectories converge to $\overline{\mathbf{x}}_\mathrm{c}$ with low enough values of K_3 ($K_3 \leqq K_3^\mathrm{a}$) despite variations in W_ob. For $K_3 > K_3^\mathrm{a}$, stick-slip oscillations will disappear if $K_2 > K_3$ and $K_2 - K_3$ is high enough. In this case, the higher K_2 is, the bigger the (Ω, W_ob)-region without ZC points is. K_1 does not imply significant changes in the curves (Ω, W_ob) corresponding to ZC points which delimit safe parameters regions. The value of K_1 influences the transient system response: the higher K_1 is, the higher the overshooting in the velocities is.

Other non-standard bifurcations typical in PWS systems with sliding motion should be studied. The analysis shown could be successfully applied to other discontinuous mechanical systems exhibiting stick-slip oscillations and dry friction.

References

1. Brett, J.F.: The genesis of torsional drillstring vibrations. SPE Drilling Engineering **September** (1992) 168–174
2. Navarro-López, E.M., Cortés, D.: Avoiding harmful oscillations in a drillstring through dynamical analysis. Journal of Sound and Vibration (to appear)
3. Utkin, V.I. Sliding Modes in Control Optimization. Springer-Verlag, Berlin (1992)

Fully Automated Stability Verification for Piecewise Affine Systems*

Jens Oehlerking, Henning Burchardt, and Oliver Theel

Department of Computing Science, University of Oldenburg, Germany
{jens.oehlerking,henning.burchardt,
oliver.theel}@informatik.uni-oldenburg.de

1 Introduction

One of the most desired properties of a closed-loop control system is stability, as a stable loop is inherently resistant to outside disturbances. Of particular interest is the notion of *asymptotic stability*. An asymptotically stable system will always converge towards an equilibrium state, once the disturbances have ceased. For hybrid systems, however, there is no known method for proving asymptotic stability directly from the system model. Instead, a promising approach is the use of *Lyapunov functions*, which can be utilized to show stability indirectly. A Lyapunov function is a formalization of an abstract "energy function" of the system. If the "energy" of the system monotonically decreases over time, converging towards zero in the designated equilibrium state, then a system is asymptotically stable. The existence of such a Lyapunov function proves asymptotic stability, but finding such a function for a hybrid system is not a simple task.

For *piecewise affine systems*, methods for synthesizing piecewise quadratic Lyapunov functions in the continuous time domain have been proposed in [1] and [2] and methods for the discrete time domain in [3]. With these methods, it is possible to provide Lyapunov functions *semi*-automatically. Each of these methods requires a partitioning of the state space into regions, each region being the domain of one quadratic "component" of the piecewise quadratic Lyapunov function. With a given partitioning and knowledge of the possible transitions between these regions, the problem can then be reduced to a *linear matrix inequality* (LMI) problem [4]. This LMI problem can be solved automatically, leading to a piecewise quadratic Lyapunov function for the system.

However, the success of the method heavily depends on 1) the choice of partitioning and 2) a full understanding of the possible transitions between these regions. Unfortunately, up to now, these two steps have required a thorough understanding of the synthesis methods and could therefore not easily be solved without expert knowledge on the matter.

Our goal is to alleviate this problem by providing *full automation* of Lyapunov function synthesis for piecewise affine hybrid systems. This includes providing

* This work was partly supported by the German Research Foundation (DFG) as part of the Transregional Collaborative Research Center "Automatic Verification and Analysis of Complex Systems" (SFB/TR 14 AVACS).

A. Bemporad, A. Bicchi, and G. Buttazzo (Eds.): HSCC 2007, LNCS 4416, pp. 741–745, 2007.

fully automatable techniques for both 1) the choice of partitioning and 2) the transition computation. The result is a framework for proving stability fully automatically for piecewise affine hybrid systems.

2 Piecewise Quadratic Stability of Hybrid Systems

We consider piecewise affine hybrid systems with affine switch sets between the discrete modes of the system. They can be defined as follows:

Definition 1 (Piecewise Affine System). *A piecewise affine system is a hybrid system given by dynamics $\dot{x} = A_m(x) + b_m$ for each discrete mode $m \in \mathcal{M}$, with A_m being a $n \times n$-matrix of reals and b_m being a vector of reals. The transitions between the modes are governed by a set of affine switch planes $\mathcal{S}$. Each $s_i \in \mathcal{S}$ is given by an affine equation of the form $c_i^T x + d_i = 0, c_i \in \mathbb{R}^n, d_i \in \mathbb{R}$. The source and target modes for each switch plane are given by two mappings source : $\mathcal{S} \to \mathcal{M}$ and target : $\mathcal{S} \to \mathcal{M}$. A switch plane s_i is active for a point in time t if $c_i^T x(t) + d_i = 0$ and source$(s_i) = m(t)$. Whenever this happens, a switch to the mode given by target(s_i) takes place. Should several switch planes be active at the same time, then one switch plane is non-deterministically chosen. Furthermore, infinitely many switchings in finite time are excluded. Several consecutive switches at the same point in time are permissible, as long as the previous condition is not violated.*

To show asymptotic stability, LMIs [4] can be employed to find piecewise quadratic (PWQ) Lyapunov functions [1,2]. However, they need a partitioning of the hybrid state space into regions, such that the resulting PWQ Lyapunov function will be continuous within each region. The authors of [1,2] assumed that this partitioning is given as an input. Furthermore, the more flexible approach in [2] requires that the possible transitions of trajectories between regions are known. Under these conditions, it is possible to compute PWQ Lyapunov functions that prove asymptotic stability. In the context of an automatic tool, both of these issues need to be dealt with in a fully automated fashion.

3 Automated Search for Piecewise Quadratic Lyapunov Functions

Identifying a Suitable Partitioning. To obtain a PWQ Lyapunov function, a suitable partitioning of the state space needs to be identified, such that the associated LMI problem given in [2] has a solution. We partition the state space separately for each discrete mode – therefore, different modes can have different partitionings of the continuous state space. The partitioning for each mode is given by a set of hyperplanes that "cut up" the state space.

The search for such a partitioning can be done in an iterative manner: start with a simple partitioning and then refine it by splitting up regions in an appropriate way. As the initial partitioning, we use the partitioning given by the switch planes for each mode. We then try to solve the LMI problem given in [2]. If we are successful then the system is asymptotically stable. If we are not, then a refined partitioning might prove stability. To refine the partitioning, we use a two-step-approach.

First, we identify the region that is the "worst offender" with respect to the non-solvability of the LMI problem. This is done by converting the non-solvable original LMI problem into a solvable problem with the help of slack variables. One slack variable $\gamma_i \geq 0$ is inserted per region in such a way that the LMI problem is solvable and that $\forall i : \gamma_i = 0$ would result in the solvability of the original LMI problem. For example, consider a block of the LMI problem of the form $A_m^T P_i + P_i A_m + Q \leq 0$. Here P_i is the quadratic form describing the Lyapunov function in region i, and A_m is the vector field of the associated mode. This equation is changed to $A_m^T P_i + P_i A_m + Q - \gamma_i I \leq 0$, where I is the identity matrix and $\gamma_i \geq 0$. This is done for all regions simultaneously. If one solves this augmented LMI problem, minimizing the slack variables, one obtains a "best non-solution" of the original LMI problem, with respect to the metric given by the slack variables. The region with the highest corresponding value γ_i is then considered the "worst offender."

Second, this "worst offender" region is split in two. This is done by adding another hyperplane to the partitioning, such that the region is cut into two new regions of similar size. To achieve this, we take the average of two of the region's bounding hyperplanes, such that one angle of the polytopic regions is cut exactly in half. This yields a refined partitioning with increased chance of the existence of a PWQ Lyapunov function.

Then, after computing the transition relation as outlined in the next paragraph, we try to solve the original LMI problem for the refined partitioning, and, if necessary, repeat the entire procedure until a solution is found. If a "worst offender" region is bounded by more than two hyperplanes, several refined partitionings are possible, resulting in a tree of partitionings. This tree is traversed breadth-first, so precedence is given to partitionings that are made up of less regions.

Computing the region transition relation. The approach from [2] requires that the PWQ Lyapunov function does not increase when a transition from one region into another occurs. Therefore, knowledge on possible transitions between regions is desirable to reduce conservativeness. As a partitioning is given through hyperplanes in the continuous state space, the regions are all convex polytopes. An adequate transition relation can then be computed with the help of linear programming, by examining the vector field on the boundary between two regions. This has also been fully automated.

Example. In the following, an example system illustrating the functionality of the automatic partitioning procedure is presented. The system has two modes, namely m_1 and m_2, where m_1 is stable and m_2 is unstable. As there is an unstable mode, it is not possible to verify stability of the system without partitioning the state space.

The dynamics of the example system is given by $\dot{x}_1 = -x_1 + x_2$, $\dot{x}_2 = -5x_1 - x_2$ (mode m_1) and $\dot{x}_1 = 0.5x_1 + 5x_2$, $\dot{x}_2 = -x_1 + 0.5x_2$ (mode m_2). The system has two switch planes s_1 and s_2 represented by the equations $x_1 = x_2$ and $x_1 = -x_2$, respectively. The two mappings *source* and *target* are defined as *source*(s_1) = *target*$(s_2) = m_1$ and *source*(s_2) = *target*$(s_1) = m_2$.

The verification procedure proceeds as follows: In a first step, the state space is partitioned by means of the two hyperplanes representing the switch plane s_1 and s_2. As with this partitioning stability cannot be shown the state space is further partitioned. When terminating, there is a total of 16 regions – six regions for mode m_1 and ten regions for mode m_2. With the resulting partitioning, it can be verified that the system is globally asymptotically stable. Figure 1 shows the partitioning hyperplanes per step as the state space partitioning evolves: a dotted line denotes a hyperplane partitioning the state space for mode m_1, a dashed line a hyperplane that partitions the state space for mode m_2, and a solid line denotes a hyperplane partitioning it for both modes.

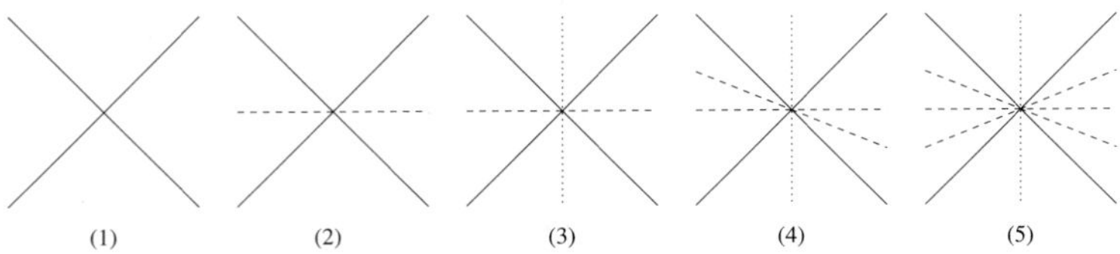

Fig. 1. Partitioning Steps

4 Conclusion and Future Work

We have described a procedure to fully automate the computation of PWQ Lyapunov functions for piecewise affine systems. Drawing from existing linear matrix inequality approaches, we gave algorithms for all additional steps needed for full automation. These algorithms include 1) the choice of a partitioning, which is implemented as an iterative algorithm that successively refines the initial partitioning by splitting, and 2) the computation of the region transition relation which is handled with the help of linear programming.

The algorithms have been implemented as a software tool which uses CSDP [5] and GLPK [6] to solve LMI problems and linear programs, respectively. It is planned to further refine the splitting procedure.

References

1. Johansson, M., Rantzer, A.: Computation of piecewise quadratic Lyapunov functions for hybrid systems. IEEE Transactions on Automatic Control **43** (1998)
2. Pettersson, S.: Analysis and Design of Hybrid Systems. PhD thesis, Chalmers University of Technology, Gothenburg (1999)

3. Feng, G.: Stability analysis of piecewise discrete-time linear systems. IEEE Transactions on Automatic Control **47**(7) (2002) 1108–1112
4. Boyd, S., Ghaoui, L.E., Feron, E., Balakrishnan, V.: Linear Matrix Inequalities in System and Control Theory. SIAM (1994)
5. Borchers, B.: CSDP, a C library for semidefinite programming. Optimization Methods and Software **10** (1999) 613–623
6. GLPK: GNU Linear Programming Kit – Reference Manual. (2005)

Differential Logic for Reasoning About Hybrid Systems*

André Platzer

Carnegie Mellon University, Computer Science Department, Pittsburgh, PA
University of Oldenburg, Department of Computing Science, Germany
platzer@informatik.uni-oldenburg.de

Abstract. We propose a first-order dynamic logic for reasoning about hybrid systems. As a uniform model for discrete and continuous evolutions in hybrid systems, we introduce hybrid programs with differential actions. Our logic can be used to specify and verify correctness statements about hybrid programs, which are suitable for symbolic processing by calculus rules. Using first-order variables, our logic supports systems with symbolic parameters. With dynamic modalities, it is prepared to handle multiple system components.

Keywords: dynamic logic, hybrid systems, parametric verification.

1 Introduction

A key idea for scalable verification of hybrid systems [1] is to decompose [2] reasoning into: (*a*) a closer investigation of the actual complex dynamics of a single system component; and (*b*) an integration of local correctness results into global system verification. Furthermore, both (*a*) and (*b*) need to handle parameters, which naturally arise from the degrees of freedom of how a single component can be instantiated in a system environment.

As first-order logic has widely proven its flexible power in handling symbolic parameters with logical variables, we extend it for reasoning about hybrid systems. Moreover, in order to be able to relate statements about a component and statements about its environment for compositional reasoning (*b*), we propose a dynamic logic in which such relations are naturally expressible [3]. Since hybrid systems are subject to continuous evolution along differential equations and discrete state change, we propose a first-order dynamic logic, $\mathsf{d}\mathcal{L}$, that provides both as fundamental system behaviour. Further, $\mathsf{d}\mathcal{L}$ can even be used for *parameter extraction*, i.e., automatic derivation of constraints for safety parameters.

Related work primarily uses propositional modal logic [4]. Unlike our first-order dynamic logic, propositional modal logic is restricted to handling abstract actions and does not support reasoning about concrete behaviour of hybrid systems like, for instance, continuous evolution along a differential equation $\ddot{z} = a$.

* This research was supported by a fellowship of the German Academic Exchange Service (DAAD) and by the German Research Council (DFG) as part of the Transregional Collaborative Research Center "Automatic Verification and Analysis of Complex Systems" (SFB/TR 14 AVACS, see www.avacs.org)

A. Bemporad, A. Bicchi, and G. Buttazzo (Eds.): HSCC 2007, LNCS 4416, pp. 746–749, 2007.

2 Differential Logic of Hybrid Programs

Dynamic Logic with Hybrid Programs. Dynamic logics (DL) [5] combine descriptions of system behaviour and correctness statements about the system state within a single specification language. By permitting arbitrary system operations α as actions of a labelled multi-modal logic, DL provides formulas of the form $[\alpha]\phi$ and $\langle\alpha\rangle\phi$. The formula $[\alpha]\phi$ expresses that all (terminating) runs of system α lead to states in which condition ϕ holds, whereas $\langle\alpha\rangle\phi$ expresses that there is at least one (terminating) run of α after which ϕ holds.

In this paper, we propose to extend DL to use hybrid systems for α. In particular, we propose a logic $\mathsf{d}\mathcal{L}$ that extends discrete DL [5] by *differential actions* such that α can display continuous evolution. Due to the symbolic nature of logic, it is beneficial to use simple system actions of an isolated effect in α. As a model for hybrid systems, we introduce hybrid programs, which are much more amenable to step-wise symbolic processing by calculus rules than graph structures of automata. Since hybrid automata [1] can be embedded, there is no loss of expressivity. Our differential logic $\mathsf{d}\mathcal{L}$ is a first-order dynamic logic with three basic characteristics to meet the requirements of hybrid systems:

Discrete jumps. Projections in state space are represented as instantaneous *assignments* of values to state variables. With this, mode switches like $\mathsf{mode} := 4$ or $\mathsf{signal} := 1$ can be expressed with discrete jumps, as well as resets $z := 0$ or discrete adjustments of control variables like $z := z - 2$.

Continuous evolution. Continuous variation in system dynamics is represented with differential equations as evolution constraints. For example, the evolution of a system with constant braking can be expressed with a *differential action* for the differential equation $\ddot{z} = -5$ with second time-derivative $\ddot{z}$ of z.

Regular combinations. Discrete and continuous evolutions can be combined to form *hybrid programs* using regular expression operators $(\cup, {}^*, ;)$ as structured behaviour of hybrid systems. For example, $\mathsf{mode} := 4 \cup \ddot{z} = -5$ describes a train controller that can choose by a nondeterministic choice $(\cup)$ to either switch its state to an alert mode (4) or initiate braking along the differential equation $\ddot{z} = -5$. In conjunction with other regular combinations, control constraints can be expressed using conditions like $z \geq 9?$ as guards for the system state.

Transition Semantics. There is a variety of slightly different semantics of hybrid system models. Since the interplay of discrete change with continuous evolution raises peculiar subtleties, we carefully motivate the advantages of our choice of semantics for $\mathsf{d}\mathcal{L}$ and hybrid programs. Consider the possible hybrid evolution with one system variable x over time t in Fig. 1. The semantics has to restrict the behaviour of the hybrid system during the continuous evolution phase, e.g., on the interval $[1, 2]$ to respect the differential equation $\dot{x} = f(x)$. Yet, the discrete jump at time 2 will necessarily lead to a discontinuity in the overall system trajectory.

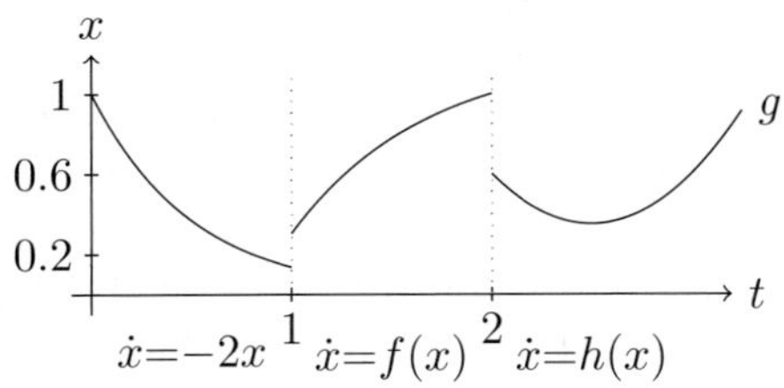

Fig. 1. Discontinuous hybrid trajectory

A global system trajectory function g (where $g(t)$ records the value of x at time t) can only assume a *single* value at time 2, say the value $g(2) = 0.6$. Hence, the continuous evolution—as visible in g—will only be continuous on the open interval $(1, 2)$. Still, the evolution along $\dot{x} = f(x)$ has to be constrained at time 2 to possess a *left*-continuous continuation towards a projected value of 1, although this value will never be assumed by g. This complicates the well-posed definition of semantics on the basis of an overall system trajectory. Note that leaving out this condition of left-continuity would lead to a total transition relation with *all* states being reachable, which, of course, would not reflect the proper system behaviour.

In contrast to this, the d$\mathcal{L}$ semantics inflates points in time with instantaneous discrete progression by associating an *individual* trajectory for each continuous evolution or instant jump phase, e.g. the phases $[0, 1], \{1\}, [1, 2], \{2\}, [2, 4]$. Hence, the trajectories remain continuous within each differential evolution phase and discontinuities are isolated purely in discrete jump transitions. Thereby, the d$\mathcal{L}$ semantics directly traces the succession of values assumed during the hybrid evolution, even if they belong to states which occur without model time passing in between. In addition to the fact that those so-called *super-dense* time effects naturally occur at mode switches between differential evolutions, they are necessary for joint mode switches of several system variables at once, like in $x := 3; y := 5$. We argue that the resulting d$\mathcal{L}$ semantics is much simpler to define than for approaches with a global overall system trajectory as, for example, in [2].

3 Parametric Verification of Train Control Systems

Symbolic parameters occurring in system dynamics raise a couple of challenges. Firstly, even very simple parametric flows and guards are *non-linear*: With parameter p, the flow constraint $2x + py \leq 5$ is an algebraic inequality but not linear. Thus, our logic needs to handle dynamics in full real arithmetic. Secondly, parameters often arise from system decomposition, e.g. in [2]. For this, safety statements about a parametric component typically have to take its interaction with the environment into account. In particular, local correctness statements need to reflect this interaction to obtain global correctness for every possible instantiation. Thus, the verification logic needs to support this interactive character with rely-guarantee reasoning; see, e.g. [2].

We argue that logic is the right level for handling the symbolic nature of parameters. All the more, the ability of dynamic logic to relate statements about multiple components is extremely valuable for compositional reasoning [3].

In the European Train Control System (ETCS) [6], the movement of trains is controlled by decentralised Radio Block Centres (RBC), which grant or deny movement authorities (MA) to trains by wireless communication. In case of an emergency, trains always have to stop within the MA issued by the RBC. In ETCS, the actual acceleration and braking behaviour is determined by the train and subject to MA limits, weather conditions, slope of track etc. For simplicity, assume that—depending on those conditions—the train motion control determines a safety envelope s around the train, within which it considers driving safe. When an MA has been granted up to the track position m and the train is currently located at position z then d$\mathcal{L}$ can analyse, for example, the following safety statement about the (simplified) acceleration system:

$$\psi \rightarrow [((m - z < s?; a := -b) \cup (m - z \geq 2s?; a := 0.1)); \ddot{z} = a]\; z < m \;. \quad (1)$$

It expresses that, under a condition ψ about parameters, trains always remain within their MA m. Further, it specifies that the train decelerates using engine brakes of force b if the safety envelope is underrun ($m - z < s$). It slowly accelerates if there is sufficient distance ($m - z \geq 2s$). To give a more concise program, we have omitted the case where the train keeps its current speed if there is no need to brake nor sufficient distance (i.e., $s \leq m - z < 2s$). The resulting transition structure for the hybrid program in (1) is depicted in Fig. 2. Formula (1) can be analysed successfully by our calculus for verifying d$\mathcal{L}$ formulas, which is similar to the one in [3]. With such an analysis, parameter constraints on the free variables of (1) can be discovered.

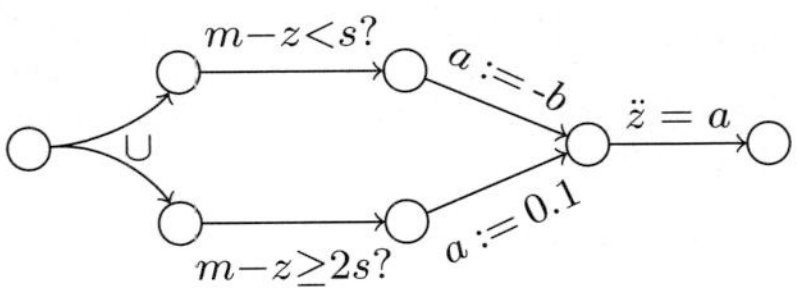

Fig. 2. Acceleration transitions

The behaviour of the program in (1) can be analysed in d$\mathcal{L}$, which is a first-order dynamic logic. In contrast, the hybrid program in (1) would collapse to a mere abstract shape $((\alpha_1; \alpha_2) \cup (\alpha_3; \alpha_4)); \alpha_5$ in propositional modal logics [4]. There, the truth of (1), which depends on the actual effects of the α_i, cannot be analysed, since the state changes induced by the abstract actions α_i are unknown in propositional programs. For this reason, d$\mathcal{L}$ is devised as a first-order logic.

References

1. Henzinger, T.A.: The theory of hybrid automata. In: LICS. (1996) 278–292
2. Damm, W., Hungar, H., Olderog, E.R.: Verification of cooperating travel agents. International Journal of Control **79**(5) (2006) 395–421
3. Platzer, A.: Towards a hybrid dynamic logic for hybrid dynamic systems. In Blackburn, P., Bolander, T., Braüner, T., de Paiva, V., Villadsen, J., eds.: Proc., LICS International Workshop on Hybrid Logic, Seattle, USA. ENTCS (2006)
4. Davoren, J.M., Nerode, A.: Logics for hybrid systems. Proceedings of the IEEE **88**(7) (2000) 985–1010
5. Harel, D., Kozen, D., Tiuryn, J.: Dynamic logic. MIT Press (2000)
6. Faber, J., Meyer, R.: Model checking data-dependent real-time properties of the European train control system. In: FMCAD, IEEE Computer Society Press (2006)

A Sound and Complete Proof Rule for Region Stability of Hybrid Systems

Andreas Podelski[1] and Silke Wagner[2]

[1] Universität Freiburg, Germany
[2] Max-Planck-Institut für Informatik, Saarbrücken, Germany*

Abstract. Region stability allows one to formalize hybrid systems whose trajectories may oscillate (within a given allowance) even after having 'stabilized'. Unfortunately, until today no proof rule (giving necessary and sufficient conditions for the purpose of verifying region stability) has been available. This paper fills the gap. Our (sound and complete) proof rule connects region stability with the finiteness of specific state sequences and thus with the emerging set of verification methods for program termination.

1 Introduction

We investigate the correctness of hybrid systems whose trajectories all eventually end up in a given region even though they may continue to oscillate (within the allowance prescribed by the region). A simple example is a heating control system, where the region is set by a thermostat. Given the premise (established in [17]) that *region stability* is a useful class of correctness properties for such hybrid systems, we need a proof rule, i.e. a set of conditions for the purpose of verification (conditions that together are necessary and sufficient for region stability).

Proof rules for many well-known notions of stability are based on Lyapunov functions [1,2,11,19]. These proof rules seem not applicable to region stability since they entail a form of convergence which is not present in region stability.

A major obstacle in the design of a necessary and sufficient set of conditions for region stability is given by an artefact of region stability. Namely, a trajectory stabilizing in the region may leave the region arbitrarily often and arbitrarily long before it eventually stabilizes, i.e. before it comes to a point when it lies in the region and from when on it never leaves the region again. The amount of time before stabilization, though finite in each trajectory, can in general not be bounded. The contribution of this paper is to overcome this obstacle.

The idea behind our result is based on two observations: there are exactly three basic situations that may be repeated finitely often in correct trajectories before stabilization, and it is possible to treat the three basic situations in a modular way. As a result, we can formulate three specific conditions and show that together they are necessary and sufficient for region stability. Each of the three conditions states the finiteness of sequences of states (each state being of one specific kind of snapshot of a trajectory of the hybrid system).

* This work was partly supported by the German Research Council (DFG) as part of the Transregional Collaborative Research Center "Automatic Verification and Analysis of Complex Systems" (SFB/TR 14 AVACS). See www.avacs.org for more information.

A. Bemporad, A. Bicchi, and G. Buttazzo (Eds.): HSCC 2007, LNCS 4416, pp. 750–753, 2007.

2 Preliminary Definitions

We follow the terminology and notation of [17]. The definitions of hybrid systems, states and trajectories can be found there.

Definition 1 (Region stability). *We call a hybrid system* stable *with respect to a region* φ *if for every trajectory* τ *there exists a point of time* t_0 *such that from then on, the trajectory is always in the region* φ.

$$\forall \tau \, \exists t_0 \, \forall t \geq t_0 \; : \; \tau(t) \in \varphi$$

Definition 2 (Sequence of snapshots). *Given a hybrid system A and a region* φ *a se-*quence of snapshots *is a sequence of states such that (i) all states of the sequence lie on the same trajectory* τ *of A, (ii) all states are not in the region* φ *and (iii) all pairs of consecutive states have a minimum time distance* δ*, where* δ *is an arbitrary but fixed constant greater than 0.*

3 Sufficient and Necessary Conditions

We will motivate informally the conditions for stability of hybrid system wrt. **interval regions**, i.e. regions φ that are given by

$$\varphi \quad \equiv \quad x \in [x_{min}, x_{max}] \, .$$

If we consider monotone trajectories of a stable system we see that they can never leave the region $\varphi \equiv x \in [x_{min}, x_{max}]$ again after they have reached it *once*. (Otherwise the trajectory could never return to the region again.) Or the other way round: if a monotone trajectory τ does not stabilize wrt. φ then (1) it either never reaches the region or (2) reaches the region but leaves it again for good.

In both cases exists an infinite computation of the system (starting either from the beginning, i.e. at $\tau(0)$, or from the time point t_0 when the trajectory has just left φ, i.e. at $\tau(t_0)$) such that all states of the computation are not in the region φ.

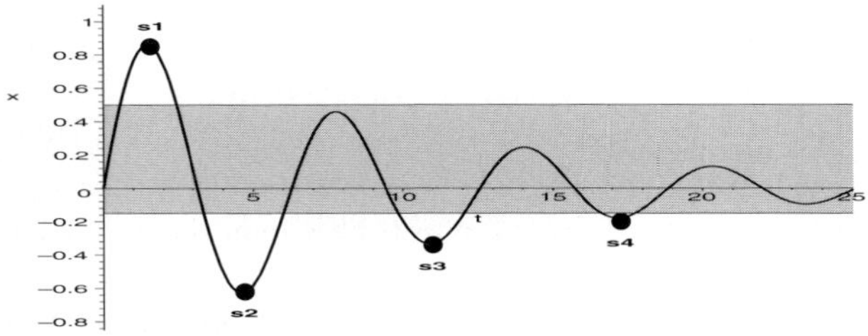

Fig. 1. Sample trajectory of a hybrid system that is stable wrt. the grey interval region $\varphi \equiv x \in [-0.15, 0.5]$. All monotone parts of the trajectory that lie outside φ are finite and the sequence of extremal-points is finite outside φ.

For general (non-monotone) trajectories we observe that they consist of (finitely or infinitely many) monotone parts. A trajectory can change its direction (i.e. its monotonicity behavior) either during a discrete jump or during continuous flows in one location. We call the states that occur just after a jump **entry-points**; states where the trajectory changes its x-direction during a continuous flow are called **extremal-points** wrt. x. If for a trajectory τ only finitely many extremal- or entry-points (i.e. states that split the trajectory into monotone parts) lie outside the region φ and if additionally all monotone parts of τ are finite outside φ then the trajectory τ must inevitably end up in the region φ.

Altogether we can reduce stability wrt. an interval region φ to the existence of infinite state sequences outside φ that (i) either lie on monotone parts of a trajectory τ, or (ii) are sequences of entry-points on τ, or (iii) are sequences of extremal-points on τ. To check whether or not such infinite computations exist we consider sequences of snapshots of the hybrid system.

Condition 1: There is no infinite sequence of snapshots such that
(i) no entry-point lies between two states of the sequence
(ii) no extremal-point lies between two states of the sequence.

Condition 2: There is no infinite sequence of snapshots such that
all states of the sequence are entry-points.

Condition 3: There is no infinite sequence of snapshots such that
all states of the sequence are extremal-points wrt. x.

Theorem 1. *Condition 1, Condition 2 and Condition 3 together are sufficient and necessary for region stability of a hybrid system A wrt. an interval region* φ. □

We can extend our previous result to n-dimensional region, i.e. to regions φ that can be expressed as cartesian products of intervals.

$$\varphi \equiv (x_1,\ldots,x_n) \in [x_1^{min}, x_1^{max}] \times \ldots \times [x_n^{min}, x_n^{max}],$$

We call such regions **box regions**.

Corollary 1. *A hybrid system A is stable wrt. an n-dimensional box region* φ *if and only if A is stable wrt. each interval region* φ_i. □

4 Conclusion

Region stability is a useful concept for the formal investigation of correctness properties of hybrid systems with certain characteristics. We have given a sound and complete proof rule, i.e. a set of conditions for the purpose of verification (conditions that together are necessary and sufficient for region stability). Before, no such proof rule has been available.

Our result is the first of two steps towards a set of verification methods for region stability. The first step has been to connect the verification of region stability with termination, a classical topic in program verification. The connection is achieved by our proof rule which uses the finiteness of certain sequences of states. (Such sequence are generated by a relation over states, the finiteness of all sequences means the well-foundedness of the relation, or the termination of the program whose state transitions are defined by the relation.) The second step consists of applying and extending the results of the research on program termination. Program termination is presently a very active research area in both theory and application. [3,7,5,4,9,14,15,16,8,18]

We believe that our work may trigger follow-up work in two directions, corresponding to two groups of researchers. The first group consists of experts in control theory who investigate a specific notion of stability, and who may carry over our techniques to their setting. The second group consists of experts in program verification who may develop specialized methods for the three termination conditions in our proof rule. The development of such methods and tools and their experimental evaluation for region stability is a research topic on its own.

References

1. M.S. Branicky. Stability of hybrid systems: State of the art. In CDC'97.
2. M.S. Branicky. Multiple Lyapunov functions and other analysis tools for switched and hybrid systems. In Trans. on Automatic Control, 1998.
3. A.R. Bradley, H. Sipma, Z. Manna. Termination of Polynomial Programs. In VMCAI'05.
4. P. Cousot. Proving Program Invariance and Termination by Parametric Abstraction, Lagrangian Relaxation and Semidefinite Programming. In VMCAI'05.
5. B. Cook, A. Podelski, A. Rybalchenko.Abstraction Refinement for Termination. In SAS'05.
6. B. Cook, A. Podelski, A. Rybalchenko. Termination proofs for systems code. In PLDI'06.
7. M. Colon, H. Sipma. Synthesis of Linear Ranking Functions. In TACAS'01.
8. M. Colon, H. Sipma. Practical Methods for Proving Program Termination. In CAV'02.
9. A. Gotsman, B. Cook, A. Podelski, A. Rybalchenko, M. Vardi. Proving that software eventually does something good. In POPL'07 (to appear).
10. T.A. Henzinger. The Theory of Hybrid Automata. In LICS'96.
11. D. Liberzon. Switching in Systems and Control. Birkhäuser, 2003.
12. V. Lakshmikantham, S. Leela, A.A. Martynyuk. Practical Stability of Nonlinear Systems. World Scientific Pub Co Inc, 1990.
13. S. Pettersson. Analysis and Design of Hybrid Systems. Ph.D. Thesis, Chalmers University of Technology, Göteborg, Sweden, 1999.
14. A. Podelski, A. Rybalchenko. Transition Invariants. In LICS, 2004.
15. A. Podelski, A. Rybalchenko. A Complete Method for the Synthesis of Linear Ranking Functions. In VMCAI'04.
16. A. Podelski, A. Rybalchenko. Transition Predicate Abstraction and Fair Termination. In POPL'05.
17. A. Podelski, S. Wagner. Model Checking of Hybrid Systems: From Reachability towards Stability. In HSCC'06.
18. A. Tiwari. Termination of linear programs. In CAV'04.
19. H. Ye, A.N. Michel, L. Hou. Stability Analysis of discontinuous Dynamical Systems with Applications. In IFAC'96.

Switch Detection in Genetic Regulatory Networks

Riccardo Porreca[1], Giancarlo Ferrari-Trecate[1], Daniela Chieppi[1], Lalo Magni[1], and Olivier Bernard[2]

[1] Dipartimento di Informatica e Sistemistica, Università degli Studi di Pavia,via Ferrata 1, 27100 Pavia, Italy
{riccardo.porreca,giancarlo.ferrari,daniela.chieppi,lalo.magni}@unipv.it
http://sisdin.unipv.it/lab

[2] INRIA, Sophia Antipolis, 2004 route des Lucioles, B.P. 93, 06902 Sophia Antipolis, France
olivier.bernard@inria.fr
http://www-sop.inria.fr/comore/main.html

Abstract. This paper considers piecewise affine models of genetic regulatory networks and focuses on the problem of detecting switches among different modes of operation in gene expression data. This task constitutes the first step of a procedure for the complete identification of the network. We propose two methods and illustrate the application to the reconstruction of switching times in data produced by a piecewise affine model of the network regulating the carbon starvation response in *Escherichia coli.*

1 Introduction

The reconstruction of biochemical networks from experimental data is nowadays recognized as one of the most important goals of systems biology. Research in this field has been promoted by the availability of experimental techniques for measuring the concentration of various molecules regulating the functioning of cells. As far as Genetic Regulatory Networks (GRN) are concerned, several measurements techniques have been developed for sampling gene expression, ranging from DNA microarrays [1] to RT-PCR and gene reporter systems [2].

We consider the problem of identifying the dynamics of GRNs using gene expression data collected with a sampling time that is sufficiently short with respect to the time constants of the network. Moreover, we restrict our attention to PieceWise Affine (PWA) models of GRNs [3,4] since they are attractive under many respects. First, they preserve the essential nonlinear character of the underlying biological process. Second, they usually involve a reduced number of parameters with respect to general nonlinear models of GRNs, a feature that is appealing from the identification viewpoint. Third, powerful techniques exist for analysis and qualitative simulation of PWA models of GRNs [4].

Recently, many different algorithms have been proposed for the identification of PWA input-output models [5], and in principle they could be used for the

A. Bemporad, A. Bicchi, and G. Buttazzo (Eds.): HSCC 2007, LNCS 4416, pp. 754–757, 2007.

data-based reconstruction of GRNs. However, PWA systems describing GRNs possess a specific structure that must be preserved in order to guarantee the biological interpretability of the identified model and all existing identification methods have a limited capability of incorporating such constraints.

The data-based reconstruction of GRNs can be conceptually split in the following tasks:

1. detection of the switches in single time series of gene expression data;
2. attribution of the data to distinct modes of operation of the whole GRN (classification problem);
3. reconstruction of thresholds on concentration variables characterizing the underlying PWA model (see [4] for details) and of all combinations of thresholds consistent with the data;
4. estimation of the kinetic parameters in each mode of operation for all models generated in point 3.

In this paper we focus on the first task in the whole identification procedure: the detection of switches in data generated by PWA input-output models of GRNs. In particular, our aim is to find switches between different modes of operation without assuming any knowledge of the model parameters. *Ad hoc* methods for task 2 are currently under study, while task 3 can be performed, under suitable assumptions, using the multicut algorithm proposed in [6]. Finally, task 4 can be easily carried out relying on the data classification produced in task 2.

2 Switch Detection in Genetic Regulatory Networks

In the sequel we consider the problem of detecting switches in the profile of a single molecule concentration.

Consider a network involving the interaction of n genes, each coding for a molecule (e.g. a protein), and denote with $\boldsymbol{x} = [x_1, \ldots, x_n]'$ the vector of molecule concentrations. Due to the PWA structure of the overall model, a molecule concentration x_i can evolve according to different modes of operation, each one characterized by an affine dynamics. Considering noisy measurements y_i of x_i, collected with a uniform sampling time, we introduce the PWA Output Error (PWA-OE) model

$$\begin{cases} x_i(k+1) = \kappa_i^j - \gamma_i^j x_i(k) \\ \quad\; y_i(k) = x_i(k) + n_i(k) \end{cases} , \quad \text{if } \boldsymbol{x}(k) \in M_i^j , \tag{1}$$

where $n_i(k) \sim WGN(0, \sigma_n^2)$. Therefore the j-th mode of operation is defined by the pair $\left(\kappa_i^j, \gamma_i^j\right)$ of positive rate parameters and the so-called *molecule domain* M_i^j. Molecule domains are unions of hyperrectangular regions in the state space of the overall network [7].

The first method to detect switches in a molecule concentration dynamics is based on the following *switching index*:

$$o(k) = \frac{y_i(k+1) - y_i(k)}{y_i(k) - y_i(k-1)} . \tag{2}$$

The index $o(k)$ enjoys the following property: in the noiseless case (i.e. $y_i(k) = x_i(k)$) when $y_i(k-1)$, $y_i(k)$ and $y_i(k+1)$ belong to the j-th mode, $o(k)$ takes the constant value $-\gamma_i^j$. The occurrence of a switch is therefore emphasized by time instants where $o(k)$ is not constant. A statistical analysis and suitable generalizations of $o(k)$ considering consecutive data on a time window of arbitrary length can be found in [7]. In particular, closed form expression of the $(1-\alpha)$-level confidence sets for $o(k)$ has been derived.

The algorithm based on $o(k)$ aims at aggregating data belonging to the same mode of operation. Given a set of consecutive data already aggregated, they are characterized by a switching index value and the associated confidence set I_M. Then a switching index involving the next measurement is considered, and its confidence set I_a is computed. As a decision rule, a switch is detected if $I_M \cap I_a = \emptyset$.

A different approach is to base the aggregation of data on the estimation of the concentration dynamics within a molecule domain. In particular, the output error model (1) is estimated over the already aggregated data, and an hypothesis test is performed to assess if the next data point belongs to the same mode of operation. Under the null hypothesis H_0 that $y_i(\bar{k}+1)$ is generated according to the model estimated over data up to $y_i(\bar{k})$, the confidence set I_a for the measurement at $\bar{k}+1$ is computed. Then the aggregation of the next data point is performed if $y_i(\bar{k}+1) \in I_a$, detecting a switch otherwise.

The algorithms based on the above-mentioned switch detection techniques are described in [7]. In particular, they turn out to be complementary in the following sense: on the one hand, the method based on switching indexes outperforms the method based on nonlinear estimation for low noise levels, on the other hand, the strategy based on nonlinear estimation is the most accurate for high noise levels. These properties have been observed in [7] on the basis of extensive simulations.

3 Switch Detection in a PWA Model of the Carbon Starvation Response of *E. coli*

In this section, we present the application of switch detection algorithms to data generated by the GRN regulating carbon starvation response in *E. coli*. For this study we considered the simplified GRN used by Drulhe et al. in [6].

The network involves interactions between genes *crp*, *fis*, *gyrAB* and their products (proteins CRP, Fis, GyrAB), regulating the synthesis of stable RNAs. In response to a carbon starvation signal, the regulatory mechanisms inhibit the synthesis of stable RNAs and then *E. coli* cells abandon their exponentially-growth state to enter a more resistent non-growth state called stationary phase.

Data in Fig. 1 represent the reentry into exponential phase following a carbon upshift and have been generated by a PWA-OE model where the chosen noise level and sampling time are comparable with the ones found in real experiments with gene reporter systems. In Fig. 1 the real switching times and those reconstructed by applying the switching detection technique based on nonlinear

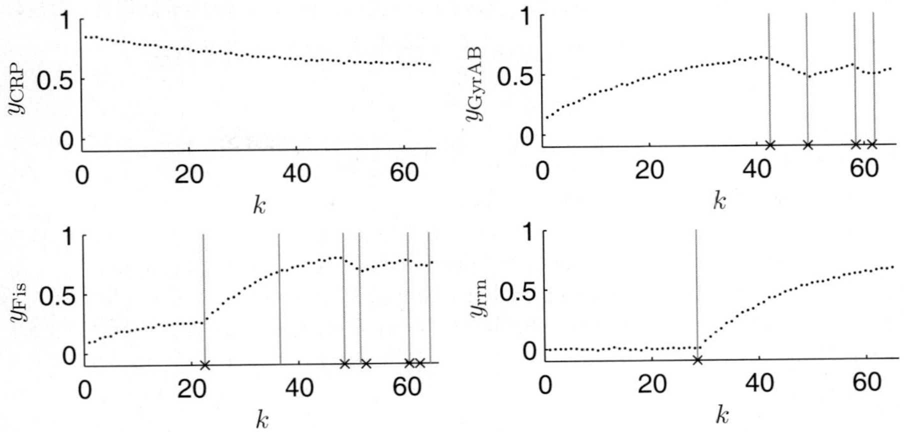

Fig. 1. Switch detection results on simulated data of the GRN regulating carbon starvation response in *E. coli*. Variables $y_{\rm CRP}$, $y_{\rm Fis}$, $y_{\rm GyrAB}$, and $y_{\rm rrn}$ denote the concentration measurements of proteins CRP, Fis, GyrAB, and stable RNAs. Vertical lines denote detected switches, while crosses correspond to real switching times.

estimation are also shown. One can observe that all switches have been identified with a satisfactory precision except for the concentration of protein Fis where a spurious switch appears.

References

1. Lockhart, D., Winzeler, E.: Genomics, gene expression and DNA arrays. Nature **405**(6788) (2000) 827–836
2. Ronen, M., Rosenberg, R., Shraiman, B., Alon, U.: Assigning numbers to the arrows: parameterizing a gene regulation network by using accurate expression kinetics. Proc. Natl. Acad. Sci. USA **99**(16) (2002) 10555–10560
3. Glass, L., Kauffman, S.: The logical analysis of continuous, non-linear biochemical control networks. J. Theor. Biol. **39**(1) (1973) 103–129
4. de Jong, H., Gouzé, J.L., Hernandez, C., Page, M., Sari, T., Geiselmann, J.: Qualitative simulation of genetic regulatory networks using piecewise-linear models. Bull. Math. Biol **66**(2) (2004) 301–340
5. Juloski, A., Heemels, W., Ferrari-Trecate, G., Vidal, R., Paoletti, S., Niessen, J.: Comparison of four procedures for the identification of hybrid systems. In Morari, M., Thiele, L., eds.: Proc. Hybrid Systems: Computation and Control (HSCC 2005). Volume 3414 of LNCS. Springer-Verlag, Berlin (2005) 354–369
6. Drulhe, S., Ferrari-Trecate, G., de Jong, H., Viari, A.: Reconstruction of switching thresholds in piecewise-affine models of genetic regulatory networks. In Hespanha, J., Tiwari, A., eds.: Proc. Hybrid Systems: Computation and Control (HSCC 2006). Volume 3927 of LNCS. Sringer-Verlag, Berlin (2006) 184–199
7. Porreca, R., Ferrari-Trecate, G., Chieppi, D., Magni, L., Bernard, O.: Switch detection in genetic regulatory networks. Technical Report 142/06, Università degli Studi di Pavia (2006) http://sisdin.unipv.it/lab/publications/TechRepSDGRN.pdf.

Safety Analysis of Sugar Cataract Development Using Stochastic Hybrid Systems

Derek Riley[1], Xenofon Koutsoukos[1], and Kasandra Riley[2]

[1] ISIS/EECS Vanderbilt University
Nashville, TN 37235, USA
[2] Mayo Foundation
Rochester, MN 55905, USA
{Derek.Riley,Xenofon.Koutsoukos}@vanderbilt.edu; riley.kasandra@mayo.edu

1 Introduction

Modeling and analysis of biochemical systems are important tasks because they can unlock insights into the complicated dynamics of systems which are difficult or expensive to test experimentally. A variety of techniques have been used to model biochemical systems, but the effectiveness of the analysis techniques is often limited by tradeoffs imposed by the modeling paradigms. Stochastic differential equations have been used to model biochemical reactions [5,2]; however, analysis of these models has mainly been limited to simulation. Hybrid systems have also been used to model biochemical systems [4]; however, verification methods based on deterministic hybrid systems fail to capture the probabilistic nature of some biochemical processes and therefore may not be able to correctly analyze certain systems. Stochastic Hybrid Systems (SHS) have been used to capture the stochastic nature of biochemical systems but have previously only been used for simulations [10] or analysis of systems with simplified continuous dynamics [6].

In this paper we model Sugar Cataract Development (SCD) as a SHS, and we present a probabilistic verification method for computing the probability of sugar cataract formation for different chemical concentrations. An accumulation of sorbitol in the eye is theorized to be the main factor in the SCD process. Understanding the exact conditions that lead to the development of sugar cataracts will help scientists better predict and prevent the condition [2]. The chemical reactions and kinetic constants for the system have been previously studied [8].

The stochastic dynamics for biochemical processes can be accurately modeled by the chemical master equation which, however, is impossible to solve for most practical systems [5]. The Stochastic Simulation Algorithm (SSA) is equivalent to solving the master equation based on a discrete model by simulating one reaction at a time, but if the number of molecules of any of the reactants is large, the SSA is not efficient [10]. For verification, it is computationally intractable to enumerate all possible states of the model employed by the SSA. Our approach suggests starting with the continuous stochastic dynamics and generating discrete approximations with coarser (and variable) resolution unlike the fixed, overly-fine resolution of the SSA. The discrete approximations can then be used for verification of reachability properties [7].

A. Bemporad, A. Bicchi, and G. Buttazzo (Eds.): HSCC 2007, LNCS 4416, pp. 758–761, 2007.

The proposed verification method employs dynamic programming based on a discretization of the state space and suffers from the curse of dimensionality. To verify the SCD process, we have developed a parallel dynamic programming implementation of the verification algorithm that can handle large systems. Although scalability is a limiting factor, this work demonstrates that the technique is feasible for realistic biochemical systems.

2 Modeling SCD Using SHS

A sugar cataract is a type of cataract which distorts the light passing through the lens of an eye by attracting water to the lens when an excess of sorbitol is present. The reactions involved in SCD are given in Table 1. The chemical species and concentration ranges for the SCD process are described in Table 2. The ranges are bounded and are estimated using realistic concentration values derived from experimental data and Michaelis-Menten constants (Km) defined as the rate of the reaction at half-maximal velocity [8].

Table 1. SCD reactions and kinetic constants

Reaction	Kinetic constant	Rate
$E + NADH \rightarrow E - NADH$	$k_1 = 6.2$	Fast
$E - NADH \rightarrow E + NADH$	$k_2 = 33$	Fast
$E - NADH + F \rightarrow E - NAD^+ + S$	$k_3 = 0.0022$	Fast
$E - NAD^+ + S \rightarrow E - NADH + F$	$k_4 = 0.0079$	Fast
$E - NAD^+ \rightarrow E + NAD^+$	$k_5 = 227$	Fast
$E + NAD^+ \rightarrow E - NAD^+$	$k_6 = .61$	Fast
$E \rightarrow Z$	$k_7 = 0.0019$	Slow

Following [10] we classify the reaction rates as either fast or slow. The slow reaction firing is described by a probabilistic rate function [10]. The dynamics of the fast reactions are described by the equation

$$dx_i = \sum_{j=1}^{M_{fast}} v_{ji} a_j(x(t))dt + \sum_{j=1}^{M_{fast}} v_{ji} \sqrt{a_j(x(t))} dW_j, \quad i = 1, ..., 7 \tag{1}$$

where x_i is the concentration of the ith chemical species, M_{fast} as the number of fast reactions, a_j as the reaction propensity of the jth reaction, and W as an M_{fast}-dimensional Wiener process. The stoichiometric matrix v is a (M_{fast} X N) matrix which holds values representing the concentration of chemical species lost or gained in each reaction.

To capture the discrete dynamics due to the slow chemical reaction, it is sufficient to consider a hybrid system with one discrete state and with a self-transition representing an occurrence of the slow reaction. As time progresses, the state $x_i, i = 1, ..., 7$ evolves according to (1). When the discrete transition occurs, the concentration of E (x_5) jumps instantaneously according to the assignment

Table 2. Chemical species properties for the SCD model

Reactant	Var.	[Min, Max] (μM)	Resolution (μM)
$NADH$	x_1	[0.0005, 10.0005]	1.0
$E - NADH$	x_2	[0.0005, 10.0005]	1.0
NAD^+	x_3	[0.0009, 10.0009]	1.0
$E - NAD^+$	x_4	[0.0009, 10.0009]	1.0
sorbitol dehydrogenase (E)	x_5	[0.0002, 1.0002]	0.1
fructose (F)	x_6	[0.2, 500.2]	20
sorbitol (S)	x_7	[0.2, 500.2]	20
Inactive form of E (Z)	-	[0.000002, 0.200002]	-

$x_5 := x_5 - d$ where d is a constant representing the Molar volume which the discrete transition consumes.

Biologists have determined that a ratio of sorbitol to fructose that is greater than one is correlated to the beginning stages of sugar cataract formation [2]. Therefore, we define the set of safe states as the set of all concentrations that satisfy $x_7 - x_6 < 1$, and we can then perform safety analysis on the system to determine the probability that a patient will develop a sugar cataract from any starting state.

3 Probabilistic Verification of SCD

In this section, we analyze the safety probability for the SCD model using the technique presented in [7]. The resolution parameters for the SCD system result in a discrete Markov Chain (MC) with approximately 800 million states. Storing the values at each state alone requires several gigabytes of memory, so we developed a parallel value iteration implementation to improve the performance of the algorithm. Assuming that the value at each state is updated periodically, the value iteration algorithm is guaranteed to converge in a parallel implementation [3]. The MC has a regular structure which improves the efficiency of the value iteration algorithm and allows us to implement a fairly straitforward partitioning technique for parallel implementation [9].

To visualize our results we plot projections of the data which show the safety probability for fixed values of the first five variables and the entire range for sorbitol and fructose. Figure 1 shows a projection of the value function along the safety boundary where $x_1 = 1.0, x_2 = 1.0, x_3 = 1.0, x_4 = 1.0, x_5 = 0.1$. This data could possibly be used to help better predict sugar cataracts by demonstrating where the safest and most unsafe concentrations exist. It could also give guidance for choosing the most effective of economical treatment to avoid the cataract development.

The SCD experiment took approximately 10 hours using 32 processors of the ACCRE computing cluster [1]. Currently, the bottlenecks of this approach are memory size and speed.

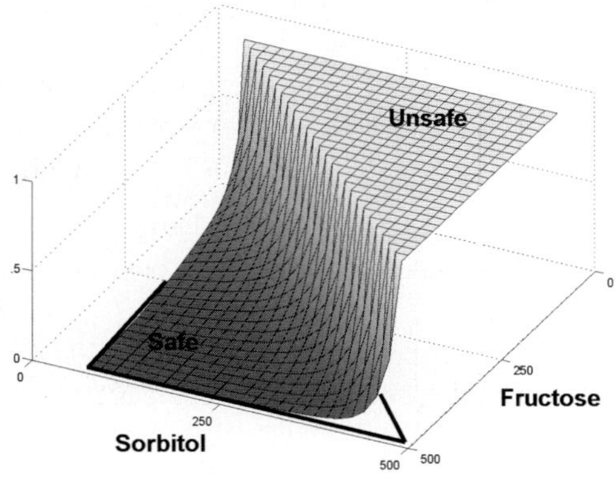

Fig. 1. Projection of the value function

Acknowledgements. This research is partially supported by the National Science Foundation (NSF) CAREER grant CNS-0347440. We would like to thank Howard Salis and Yiannis Kaznessis from the University of Minnesota for their help with this project.

References

1. ACCRE computing cluster, http://www.accre.vanderbilt.edu
2. R. Barbuti, A. Maggiolo-Schettini, P. Milazzo, A. Triona, An Alternative to Gillespie's Algorithm for Simulating Chemical Reactions, *Computational Methods in Systems Biology 2005*, Edinburgh, 2005.
3. D. Bertsekas, J. Tsitsiklis, Parallel and Distributed Computation: Numerical Methods, Prentice-Hall, 1989.
4. R. Ghosh, C. Tomlin, Symbolic Reachable Set Computation of Piecewise Affine Hybrid Automata and its Application to Biological Modeling: Delta-Notch Protein Signalling, *Systems Biology*, 1:170-183, 2004.
5. D. Gillespie, A General Method for Numerically Simulating the Stochastic Time Evolution of Coupled Chemical Reactions. *J Comp. Phys.*, 22:403-434, 1976.
6. J. Hespanha, A. Singh. Stochastic Models for Chemically Reacting Systems Using Polynomial Stochastic Hybrid Systems, *Int. J. on Robust Control, Special Issue on Control at Small Scales*, 15:669689, 2005.
7. X. Koutsoukos and D. Riley, Computational Methods for Reachability Analysis of Stochastic Hybrid Systems, *Hybrid Systems: Computation and Control 2006* LNCS 3927, pp. 377-391, 2006.
8. I. Marini, L. Bucchioni, P. Borella, A. Del Corso, U. Mura, Sorbitol Dehydrogenase from Bovine Lens: Purification and Properties, *Archives of Biochemistry and Biophysics*, 340:383-391, 1997.
9. D. Riley, X. Koutsoukos, K. Riley, Safety Analysis of Sugar Cataract Development Using Stochastic Hybrid Systems, ISIS Technical Report ISIS-06-708.
10. H. Salis and Y. Kaznessis, Accurate hybrid stochastic simulation of a system of coupled chemical or biochemical reactions, *The Journal of Chemical Physics*, v122, pp. 54-103, 2005.

Case Studies in Event-Driven Control*

J.H. Sandee[1], W.P.M.H. Heemels[1,2], and P.P.J. van den Bosch[1]

[1] Eindhoven Univ. of Techn., P.O. Box 513, 5600 MB Eindhoven, The Netherlands
[2] Embedded Systems Institute, P.O. Box 513, 5600 MB Eindhoven, The Netherlands
{j.h.sandee,m.heemels,p.p.j.v.d.bosch}@tue.nl

Abstract. The majority of research in control engineering considers periodic or time-triggered control systems with equidistant sample intervals. However, practical cases abound in which it is of interest to consider event-driven control systems, where the sampling is event-triggered. Although there are various benefits of using event-driven control like reducing resource usage (e.g. processor and communication load), their application in practice is hampered by the lack of a system theory for event-driven control systems. In this paper we present two types of event-driven controllers and show their potential via industrially relevant case studies and indicate initial theoretical results.

1 Introduction

The majority of research in digital control theory and engineering considers periodic or time-driven control systems in which continuous-time signals are represented by their sampled values at a fixed sample frequency. This leads to equidistant sampling intervals for which the analysis and synthesis problems can be coped with by the vast literature on sampled-data systems.

In most applications, these digital control algorithms are implemented in a real-time embedded software environment. As a consequence of the time-driven nature of controllers, control engineers pose strong, non-negotiable requirements on the real-time implementations of their algorithms as the required control performance can only be guaranteed in this manner. In the end, this leads to non-optimal solutions if the design problem is considered from a multi-disciplinary system perspective. As an example, time-driven controllers perform control calculations all the time at a fixed high rate, so also when nothing significant has happened in the process. This is clearly an unnecessary waste of resources like processor usage and communication bus load. As a consequence, a time-driven controller might not be optimal, when considered in a broader sense.

To reduce the severe real-time constraints imposed by the control engineer and the accompanying disadvantages, this paper proposes to drop the strict

* This work has been carried out as part of the Boderc project under the responsibility of the Embedded Systems Institute. This project is partially supported by the Dutch Ministry of Economic Affairs under the Senter TS program. This research was sponsored by the European 6th Framework Network of Excellence HYCON (contract number FP6-IST-511368).

A. Bemporad, A. Bicchi, and G. Buttazzo (Eds.): HSCC 2007, LNCS 4416, pp. 762–765, 2007.

requirement of equidistant sampling. The claim is that this enables better multi-disciplinary trade-off making to achieve a better overall system performance. This type of feedback controllers are called *event-driven controllers* as it is an event (e.g. the arrival of a new measurement), rather than the elapse of time, that triggers the controller to perform an update. As event-driven control loops typically deal with discrete events with strong interaction with the continuous-time dynamics of the plant, they can be considered as a specific class of hybrid systems.

In literature, only few examples of event-driven control have been presented and hardly any theory on control performance analysis can be found. Two good overviews can be found in [1] and [2]. To stimulate research in this direction, we consider in this paper two types of event-driven controllers and show their potential via case studies and indicate initial ideas for analyzing the resulting loops. Further details and related research can be found in [3].

2 Event-Driven Control for Reducing Resource Usage

We consider the system described by

$$\dot{x}(t) = A_c x(t) + B_c u(t) + E_c w(t) \tag{1a}$$

$$u(t) = F x(\tau_k), \qquad \text{for } t \in [\tau_k, \tau_{k+1}) \tag{1b}$$

where $x(t) \in \mathbb{R}^n$ is the state, $u(t) \in \mathbb{R}^m$ the control input and $w(t)$ the unknown disturbance, respectively, at time $t \in \mathbb{R}_+$. As a controller a discrete-time state-feedback controller with gain $F \in \mathbb{R}^{m \times n}$ is considered, i.e. $u_k = F x_k$, where $x_k = x(\tau_k)$, $u_k = u(\tau_k)$ using the zero-order hold $u(t) = u_k$ for all $t \in [\tau_k, \tau_{k+1})$.

The *control update times* τ_k are in conventional time-driven control related through $\tau_{k+1} = \tau_k + T_s$, where T_s is a fixed sample time, meaning that the control value is updated every T_s time units. To reduce the number of control calculations, we propose not to update the control value if the state $x(\tau_k)$ is contained in a set $\mathcal{B}$ around the origin. The control update times are now

$$\tau_{k+1} = \inf\{jT_s > \tau_k \mid j \in \mathbb{N},\ x(jT_s) \notin \mathcal{B}\}. \tag{2}$$

The control objective is a "practical stabilization problem" in the sense of controlling the state towards a region Ω close to the origin and keeping it there, as asymptotic stability cannot be obtained because the plant is operated in open-loop inside the set $\mathcal{B}$.

The performance of this novel control strategy is addressed in terms of ultimate boundedness and guaranteed speed of convergence. Depending on the particular event-triggering mechanism used for the control updates, properties like ultimate boundedness for the perturbed event-driven linear system can be derived either from a perturbed discrete-time linear system or from a perturbed discrete-time piecewise linear (PWL) system. Since results for ultimate boundedness are known for discrete-time linear systems and piecewise linear systems, these results can be carried over to event-driven controlled systems. In this way we can tune the parameters of the controller to obtain satisfactory control performance on one hand and low processor/communication load on the other. An initial experimental case study in paper flow control in a printer investigates the achievable reduction in the processor load by the particular type of event-driven controllers proposed here [3, Ch. 6]. In the typical case study, the processor load can be reduced with 50% without sacrificing the control performance significantly.

3 Sensor-Based Event-Driven Control

A second line of event-driven control is sensor-based control, which is related to the situation in which the measurement method is intrinsically event-based in nature. Examples are e.g. internal combustion engines that are sampled against engine speed; level sensors for measuring the height of a fluid in a tank; and transportation systems where the longitudinal position of a vehicle is only known when certain markers are passed. We will introduce sensor-based event-driven control via a typical and industrially relevant example of motor control, although the lines of reasoning in this section are more general.

In the case study, we use an (extremely) low resolution encoder to measure the angular position of a motor. The event-driven controller is designed such that actuation is performed at the detection of an encoder pulse. In this way, the controller can use the *exact* position measurement, and is not affected by the quantization errors of the encoder. Moreover, the controller can respond fast to measurement data. When the motor is not running at constant velocity, the updates are not equidistant in time. It is therefore not possible to use classical design methods which assume that updates are equally spaced in time. We can however apply variations on classical design methods if we define our models of the plant and the controller in the (angular) position domain instead of the time domain. This idea is based on the observation that the encoder pulses arrive equally spaced in the position domain. It is shown that, by applying this event-driven controller, we not only decrease the encoder resolution - and therefore the system cost price - but also the average processor load, compared to the originally applied controller in industry. This was accomplished without degrading the control performance. In the typical example of a motor controller applied to transport images through a printer we could accurately control the motor by means of an encoder with a resolution of only 1 pulse per revolution, with the controller running at an average sample frequency of 62 Hz. Compared with the originally applied controller, running at a constant sample frequency of 250 Hz in combination with an encoder resolution of 12 pulses per revolution, the processor load was reduced by a factor 5.

4 Conclusions

Although in many practical control problems it is natural and logical to use event-driven controllers, their application is scarce in both industry and academia.

A major reason why time-driven control still dominates is the absence of a system theory for event-driven control loops. To stimulate research in this direction, this paper presented two types of event-driven controllers with a clear industrial relevance. Given the potential benefits of such controllers as shown in the case studies, we believe it is worthwhile to invest research effort in this line of work and to develop a mature theory for event-driven control systems.

References

1. Årzén, K.E.: A simple event-based PID controller. In: 14th World Congress of IFAC. Volume 18., Beijing, P.R. China (1999) 423–428
2. Åström, K.J., Bernhardsson, B.M.: Comparison of Riemann and Lebesgue sampling for first order stochastic systems. In: 41st IEEE Conference on Decision and Control. (2002) 2011–2016
3. Sandee, J.H.: Event-driven control in theory and practice - trade-offs in software and control performance. Phd thesis, Eindhoven Univ. of Techn., Netherlands (2006)

Hybrid Estimation for Stochastic Piecewise Linear Systems

Chze Eng Seah and Inseok Hwang

Purdue University, West Lafayette, IN47907
{seah,ihwang}@purdue.edu

Abstract. We propose a hybrid estimation algorithm based on Bayesian inference approach for stochastic piecewise linear systems. We derive the discrete state (or mode) transition probabilities based on a system of inequalities that describe the probabilistic state space partition of the system. Simulation results are presented to demonstrate the performance of the algorithm.

1 Problem Formulation

We consider a stochastic piecewise linear system described by

$$q(k+1) = j \qquad \text{if} \qquad \boldsymbol{L}_{ij}\boldsymbol{x}(k) - \boldsymbol{\theta}_{ij} \geq 0 \tag{1}$$

$$\boldsymbol{x}(k+1) = \boldsymbol{A}_i\boldsymbol{x}(k) + \boldsymbol{B}_i\boldsymbol{u}(k) + \boldsymbol{w}_i(k) \tag{2}$$

$$\boldsymbol{z}(k) = \boldsymbol{C}_i\boldsymbol{x}(k) + \boldsymbol{v}_i(k) \tag{3}$$

where $q(k) \in \{1, 2, \ldots, r\}$, $\boldsymbol{x}(k) \in \mathbb{R}^n$, $\boldsymbol{u}(k) \in \mathbb{R}^p$ and $\boldsymbol{z}(k) \in \mathbb{R}^s$ are the discrete state, the (discrete-time) continuous state, the input and the measurement respectively; $\boldsymbol{A}_i, \boldsymbol{B}_i$ and $\boldsymbol{C}_i$ are system matrices corresponding to the discrete state $q(k) = i$; $\boldsymbol{w}_i(k)$ and $\boldsymbol{v}_i(k)$ are uncorrelated zero-mean Gaussian noise; $\boldsymbol{L}_{ij}$ are $m \times n$ constant matrices and $\boldsymbol{\theta}_{ij}$ are m-dimensional random vectors which represent uncertainties in the state space partitions. The vector $\boldsymbol{\theta}_{ij}$ is assumed to have a multivariate Gaussian probability density function (pdf), with mean $\boldsymbol{\mu}_{\theta_{ij}}$ and covariance $\boldsymbol{\Sigma}_{\theta_{ij}}$, denoted as

$$p[\boldsymbol{\theta}_{ij}] = \mathcal{N}_m(\boldsymbol{\theta}_{ij}; \boldsymbol{\mu}_{\theta_{ij}}, \boldsymbol{\Sigma}_{\theta_{ij}}) \tag{4}$$

Let the sequence $\boldsymbol{Z}^k = [\boldsymbol{z}(1), \boldsymbol{z}(2), \ldots, \boldsymbol{z}(k)]$ denotes the set of measurements up to time k. Using the Bayesian approach, we consider the problem of updating the pdfs $p[q(k+1)|\boldsymbol{Z}^{k+1}]$ and $p[\boldsymbol{x}(k+1)|\boldsymbol{Z}^{k+1}]$ at time $k+1$ using $p[q(k)|\boldsymbol{Z}^k]$, $p[\boldsymbol{x}(k)|\boldsymbol{Z}^k]$, and the new measurement $\boldsymbol{z}(k+1)$. Note that $p[\boldsymbol{x}(k)|\boldsymbol{Z}^k] = \sum_{i=1}^{r} p[\boldsymbol{x}(k)|q(k) = i, \boldsymbol{Z}^k]p[q(k) = i|\boldsymbol{Z}^k]$.

2 Algorithms

The algorithm consists of a bank of r mode-matched Kalman filters. At time k, we have, from each Kalman filter i, a posterior continuous state pdf

A. Bemporad, A. Bicchi, and G. Buttazzo (Eds.): HSCC 2007, LNCS 4416, pp. 766–770, 2007.

$$p\left[\boldsymbol{x}(k)=\boldsymbol{x}|q(k)=i,\boldsymbol{Z}^k\right]=\mathcal{N}_n\big(\boldsymbol{x};\hat{\boldsymbol{x}}_i(k|k),\boldsymbol{P}_i(k|k)\big) \tag{5}$$

where $\hat{\boldsymbol{x}}_i(k|k)$ and $\boldsymbol{P}_i(k|k)$ are the posterior mean and covariance respectively; and the discrete state pdf $\alpha_i(k|k) := p[q(k)|\boldsymbol{Z}^k]$. We then update the pdfs $p\left[\boldsymbol{x}(k+1)|q(k+1)=j,\boldsymbol{Z}^{k+1}\right]$ and $\alpha_j(k+1|k+1)$, for $j=1,\ldots,r$, as follows.

1. **Compute mode transition probability:** From (1), the mode transition probability is given by

$$\begin{aligned}\gamma_{ij}(k) &:= p[q(k+1)=j|q(k)=i,\boldsymbol{Z}^k]=p\left[\boldsymbol{L}_{ij}\boldsymbol{x}(k)-\boldsymbol{\theta}_{ij}\geq 0|q(k)=i,\boldsymbol{Z}^k\right]\\ &=\int_{\mathbb{R}^n}p\left[\boldsymbol{\theta}_{ij}\leq\boldsymbol{L}_{ij}\boldsymbol{x}|\boldsymbol{x}(k)=\boldsymbol{x},q(k)=i,\boldsymbol{Z}^k\right]p[\boldsymbol{x}(k)=\boldsymbol{x}|q(k)=i,\boldsymbol{Z}^k]d\boldsymbol{x}\end{aligned}$$

Definition 1. *Let* $\boldsymbol{y}=[y_1\ y_2\ \ldots\ y_m]^T\sim\mathcal{N}_m(\boldsymbol{y};\boldsymbol{0},\boldsymbol{\Sigma})$. *Let* $\boldsymbol{\alpha}=[\alpha_1\ \alpha_2\ \ldots\ \alpha_m]^T\in\mathbb{R}^m$. *We define an m-dimensional Normal cumulative density function as*

$$\Phi_m(\boldsymbol{\alpha};\boldsymbol{\Sigma}):=p[\boldsymbol{y}\leq\boldsymbol{\alpha}]=\int_{-\infty}^{\alpha_m}\int_{-\infty}^{\alpha_{m-1}}\cdots\int_{-\infty}^{\alpha_1}\mathcal{N}_m(\boldsymbol{y};\boldsymbol{0},\boldsymbol{\Sigma})dy_1dy_2\ldots dy_m$$

Using (4), (5) and Definition 1, it can be shown that [1]

$$\begin{aligned}\gamma_{ij}(k) &=\int_{\mathbb{R}^n}\Phi_m(\boldsymbol{L}_{ij}\boldsymbol{x}-\boldsymbol{\mu}_{\theta_{ij}};\boldsymbol{\Sigma}_{\theta_{ij}})\mathcal{N}_n\big(\boldsymbol{x};\hat{\boldsymbol{x}}_i(k|k),\boldsymbol{P}_i(k|k)\big)d\boldsymbol{x}\\ &=\Phi_m(\boldsymbol{L}_{ij}\hat{\boldsymbol{x}}_i(k|k)-\boldsymbol{\mu}_{\theta_{ij}};\boldsymbol{\Sigma}_{\theta_{ij}}+\boldsymbol{L}_{ij}\boldsymbol{P}_i(k|k)\boldsymbol{L}_{ij}^T)\end{aligned} \tag{6}$$

Numerical algorithms for computing the function $\Phi_m(\cdot)$, some with computational complexity that increases linearly with m, can be found in [1,2]

2. **Filtering:** Each Kalman filter j computes its own posteriors $\hat{\boldsymbol{x}}_j(k+1|k+1)$ and $\boldsymbol{P}_j(k+1|k+1)$ using the initial conditions, $\hat{\boldsymbol{x}}_{j0}$ and $\boldsymbol{P}_{j0}$, given by [3]

$$\hat{\boldsymbol{x}}_{j0}=\sum_{i=1}^{r}\hat{\boldsymbol{x}}_i(k|k)\bar{\alpha}_{ji}(k+1)$$

$$\boldsymbol{P}_{j0}=\sum_{i=1}^{r}\{\boldsymbol{P}_i(k|k)+[\hat{\boldsymbol{x}}_i(k|k)-\hat{\boldsymbol{x}}_{j0}][\hat{\boldsymbol{x}}_i(k|k)-\hat{\boldsymbol{x}}_{j0}]^T\}\bar{\alpha}_{ji}(k+1)$$

where $\bar{\alpha}_{ji}(k+1):=p[q(k)=i|q(k+1)=j,\boldsymbol{Z}^k]=\frac{1}{c_j}\gamma_{ij}(k+1)\alpha_i(k|k)$ and c_j is a normalizing constant.

---

[1] The proof, which is omitted due to space constraints, can be provided via correspondence with the authors.

3. **Mode probability update:** The posterior mode probability is given by

$$\alpha_j(k+1|k+1) = \frac{1}{\beta}\Lambda_j(k+1)\alpha_j(k+1|k)$$

$$\alpha_j(k+1|k) := p[q(k+1) = j|\boldsymbol{Z}^k] = \sum_{i=1}^{r}\gamma_{ij}(k+1)\alpha_i(k|k)$$

$$\Lambda_j(k+1) = \mathcal{N}_p(\boldsymbol{r}_j(k+1); 0, \boldsymbol{S}_j(k+1))$$

where β is a normalizing constant, $\boldsymbol{r}_j(k+1) = \boldsymbol{z}(k+1) - \boldsymbol{C}_j\hat{\boldsymbol{x}}_j(k+1|k)$ is the residual of Kalman filter j, and $\boldsymbol{S}_j(k+1)$ is its covariance.

3 Example

We consider the following dynamical system which models a rocket propulsion system shown in Fig. 1. Let $\boldsymbol{x} = [y\ \dot{y}\ P_t]^T$, where y is the lateral displacement and P_t is the pressure in the pneumatic tank. The system is described by

$$\boldsymbol{x}(k+1) = \boldsymbol{A}_i\boldsymbol{x}(k) + \boldsymbol{B}_i\boldsymbol{u}(k) + \boldsymbol{w}_i(k)$$
$$\boldsymbol{z}(k) = \boldsymbol{C}_i\boldsymbol{x}(k) + \boldsymbol{v}_i(k)$$

$$\boldsymbol{A}_1 = \begin{bmatrix} 1 & T_s & 0 \\ 0 & 1 & 0 \\ 0 & 0 & 1 \end{bmatrix} \qquad \boldsymbol{A}_2 = \begin{bmatrix} 1 & T_s & 0.005T_s^2 \\ 0 & 1 & 0.01T_s \\ 0 & 0 & 0.98 \end{bmatrix} \qquad \boldsymbol{A}_3 = \begin{bmatrix} 1 & T_s & 0.01T_s^2 \\ 0 & 1 & 0.02T_s \\ 0 & 0 & 0.95 \end{bmatrix}$$

$$\boldsymbol{B}_1 = \boldsymbol{B}_2 = \boldsymbol{B}_3 = \begin{bmatrix} 0 & 0 & T_s \end{bmatrix}^T \qquad \boldsymbol{C}_1 = \boldsymbol{C}_2 = \boldsymbol{C}_3 = \begin{bmatrix} 1 & 0 & 0 \\ 0 & 0 & 1 \end{bmatrix} \qquad T_s = 0.2\ sec$$

$$E[\boldsymbol{w}_i(k)\boldsymbol{w}_i^T(k)] = \begin{bmatrix} (3\frac{T_s^2}{2})^2 & 0 & 0 \\ 0 & (3T_s)^2 & 0 \\ 0 & 0 & (10T_s)^2 \end{bmatrix} \qquad E[\boldsymbol{v}_i(k)\boldsymbol{v}_i^T(k)] = \begin{bmatrix} 20^2 & 0 \\ 0 & 20^2 \end{bmatrix}$$

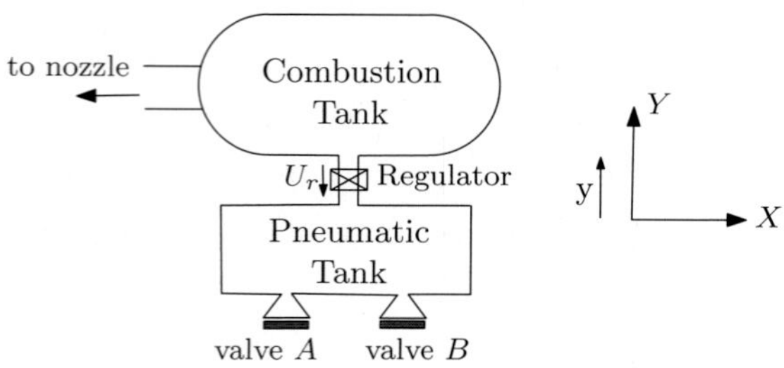

Fig. 1. Schematic of a rocket propulsion system

The discrete state $q(k) = 1, 2$ or 3 corresponds to (i) valves A and B closed, (ii) valve A opened and valve B closed, and (iii) valves A and B opened, respectively.

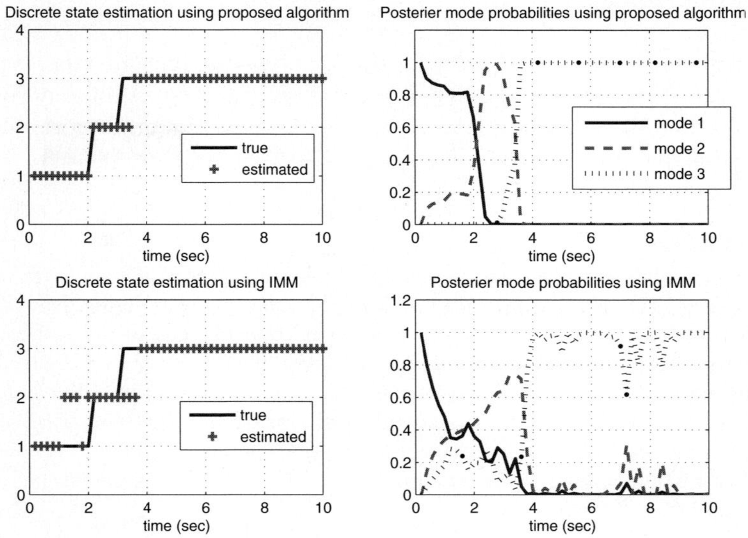

Fig. 2. Discrete state estimation and mode probabilities of a typical simulation run

Table 1. Root-mean-square errors of continuous state (for 50 simulation runs)

	Proposed algorithm	IMM algorithm	% Difference
y	6.9	9.2	25%
$\dot{y}$	3.3	5.4	39%
P_t	6.8	10.1	33%

The valves A and B will be opened when the pressure P_t exceeds P_A and P_B respectively, where P_A and P_B have uncertainties which are represented as

$$P_A \sim \mathcal{N}_1(P_A; 300, 25) \qquad P_B \sim \mathcal{N}_1(P_B; 600, 25)$$

The discrete state transitions can then be modeled using (1) and (4) with, for example, $\boldsymbol{L}_{12} = [0\ 0\ 1]$, $\boldsymbol{\theta_{12}} = P_A$, $\boldsymbol{\mu}_{\theta_{12}} = 300$, $\boldsymbol{\Sigma}_{\theta_{12}} = 25$.

We compare our results with that of the Interacting Multiple Model (IMM) algorithm [4] which models the mode transition probabilities as constants. Here, we use $[\gamma_{ij}]_{i,j=1,2,3} = \begin{bmatrix} 0.8 & 0.15 & 0.05 \\ 0.05 & 0.8 & 0.15 \\ 0.05 & 0.05 & 0.9 \end{bmatrix}$ for IMM. Figure 2 shows the discrete state (or mode) estimates and the posterior mode probabilities. In most of the time, the proposed algorithm gives the correct discrete state estimates with probability close to one. Table 1 compares the errors in the continuous state estimation for 50 Monte Carlo runs. The errors of the proposed algorithm are significantly smaller than those of the IMM algorithm. The total computational times of the 50 runs for the proposed algorithm and the IMM algorithm are 3.61 *sec* and 3.36 *sec* respectively.

4 Conclusions

A hybrid estimation algorithm for stochastic piecewise linear system is proposed. In the algorithm, the probability of a discrete state transition is derived based on the probabilistic state space partitions of the system. Performance of the proposed algorithm has been illustrated in simulations.

References

1. Y.L. Tong. *The Multivariate Normal Distribution.* Springer-Verlag, New York, 1990.
2. S. Kuriki T. Miwa, A.J. Hayter. The evaluation of general non-centred orthant probabilities. *J.R. Statist. Soc. B*, 65:223–234, 2003.
3. I. Hwang and C.E. Seah. An estimation algorithm for stochastic linear hybrid systems with continuous-state-dependent mode transitions. In *Proc. IEEE Conference on Decision and Control*, San Deigo, CA, USA, December 2006.
4. Y. Bar-Shalom and T.F. Fortmann. *Tracking and Data Association.* Academic Press, 1988.

On-Line Optimization of Switched-Mode Hybrid Dynamical Systems

Yorai Wardi, Xu Chu Ding, and Shun-ichi Azuma

Georgia Institute of Technology, Kyoto University
Atlanta, GA 30332 USA
Kyoto 611-0011 Japan
{ywardi,ding}@ece.gatech.edu,
sazuma@i.kyoto-u.ac.jp

Abstract. This paper concerns the optimal mode-switching problem in hybrid dynamical systems, where it is desired to compute the switching times between the modes in order to minimize a given cost functional defined on the state trajectory of the system. The state variable cannot be directly observed, and it has to be estimated by an observer. The paper proposes an on-line algorithm and presents bounds on its convergence rate.

1 Introduction

This paper concerns switched-mode hybrid dynamical systems having the following form. Given a time horizon $[0, T]$, $N+1$ differentiable functions $f_i : R^n \to R^n$, $i = 1, \ldots, N+1$, and a finite sequence of time points τ_i, $i = 1, \ldots, N$ satisfying the inequalities $0 \leq \tau_1 \leq, \ldots, \leq \tau_N \leq T$, consider the following differential equation

$$\dot{x} = \begin{cases} f_1(x), & \text{for all } t \in [0, \tau_1), \\ f_i(x), & \text{for all } t \in [\tau_{i-1}, \tau_i), \qquad i = 2, \ldots, N, \\ f_{N+1}(x), & \text{for all } t \in [\tau_N, T), \end{cases} \tag{1}$$

with a given initial condition $x_0 := x(0) \in R^n$. The variable $x(t) \in R^n$ is the state of the system, the functions f_i, $i = 1, \ldots, N+1$, are called the *modal functions*, and the times τ_i, $i = 1, \ldots, N$, are called the *switching times*. To simplify the notation in (1) we define $\tau_0 := 0$ and $\tau_{N+1} := T$, and we further define

$$F(x,t) := f_i(x) \quad \text{for all } t \in [\tau_{i-1}, \tau_i), \quad \text{for every } i = 1, \ldots, N+1, \tag{2}$$

so that Eq (1) has the following form,

$$\dot{x} = F(x,t). \tag{3}$$

Let $L : R^n \to R$ be a continuously-differentiable function, and consider the *cost functional* J, defined by

$$J := \int_0^T L(x) dt. \tag{4}$$

A. Bemporad, A. Bicchi, and G. Buttazzo (Eds.): HSCC 2007, LNCS 4416, pp. 771–774, 2007.

Defining the vector of switching times by $\bar{\tau} := (\tau_1, \ldots, \tau_N)^T \in R^N$, we observe that the cost functional J can be viewed as a function of $\bar{\tau}$ via Eq. (1), and hence it is denoted by $J := J(\bar{\tau})$. The *timing optimization problem* is the problem of minimizing $J(\bar{\tau})$ subject to the constraints

$$0 = \tau_0 \leq \tau_1 \leq \ldots, \leq \tau_N \leq \tau_{N+1} = T. \tag{5}$$

This problem arises in a number of applications, including situations where a control module has to switch its attention among a number of subsystems [9,12], or collect data sequentially from a number of sensory sources [2,5]. It has been addressed by several authors, and a number of gradient-descent algorithms have been developed for its solution [4,13,14,10,11,3]. Solutions to nonlinear optimal control problems generally depend explicitly on the initial state except under special circumstances. In practice, however, the state variable may not be measurable and it has to be estimated by a suitable observer. For this situation, Reference [1] has developed an on-line algorithm that tracks the optimal cost-to-go along the trajectory of the state observer. That algorithm has one major drawback: it must stay on the trajectory of the optimal cost-to-go at all times. This has motivated the development of the algorithm described in this paper, which does not have the above restriction.

2 On-Line Algorithm

Suppose that the system has a state observer, like a Luenberger observer if the system is linear, or and a Moraal-Grizzle observer if the system is nonlinear [6]. Let us denote by $\hat{x}(t) \in R^n$ the state estimator at time $t \in [0, T]$ as computed by the observer. The cost-to-go at time t is defined as follows. Given a switching-time vector $\bar{\tau}$, define $x(\xi)$, $\xi \in [t, T]$, as the solution of the equation

$$\dot{x}(\xi) \;=\; F(x, \xi), \tag{6}$$

with the boundary condition $x(t) = \hat{x}(t)$. The cost-to-go is defined by

$$J(t, \hat{x}(t), \bar{\tau}) \;:=\; \int_t^T L(x(\xi))d\xi, \tag{7}$$

where we note its explicit dependence on t and $\hat{x}(t)$. The problem of computing the switching-time vector that minimizes the cost-to-go will be denoted by $\pi_{\hat{x}(t),t}$.

Suppose for a moment that the switching-times vector $\bar{\tau}$ is computed for every time t by a suitable algorithm, and hence we denote it by $\bar{\tau}(t) : (\tau_1(t), \ldots, \tau_N(t))^T$. Obviously, if $\tau_i(t) < t$ for some $i \in \{1, \ldots, N\}$ then $\tau_i(t)$ cannot be changed at time t and it must retain its constant value. Therefore, only the future switching times can be modified by an on-line algorithm at time t. To simplify the exposition, we assume throughout this paper that $\tau_1 > t$ so that all of the switching times can be modified on line, whereas a treatment of the general case can be found in [15].

Now suppose that the functions f_i are three-times continuously differentiable, and hence (see, e.g., the appendix of [7]) the function $J(t, \hat{x}(t), \bar{\tau})$ is twice continuously differentiable in $\bar{\tau}$. Furthermore, we assume henceforth that the Hessian $\frac{\partial^2 J}{\partial \tau^2}(t, \hat{x}(t), \bar{\tau}(t))$ is positive definite. Consequently, the following process $\{\bar{\tau}(t)\}$ is well defined:

$$\bar{\tau}(t+dt) \;=\; \bar{\tau}(t) = \;-\; \left(\frac{\partial^2 J}{\partial \tau^2}(t, \hat{x}(t), \bar{\tau}(t)\right)^{-1} \frac{\partial J}{\partial \tau}(t+dt, \hat{x}(t+dt), \bar{\tau}(t)). \quad (8)$$

We implicitly assume throughout that the switching-time vector defined by (8) lies in the interior of the constraint set defined by (5). We point out that Reference [15] has proved that (8) preserves stationarity in the sense that, if $\bar{\tau}(t)$ is a stationary point for $\pi_{\hat{x}(t),t}$ then $\bar{\tau}(t+dt)$ is a stationary point for $\pi_{\hat{x}(t=dt),t+dt}$.

In order to turn Eq. (8) into an algorithm we have to replace the infinitesimal time-step dt by a finite-length time-step $\Delta t > 0$, which results in the following equation.

$$\bar{\tau}(t+\Delta t) \;=\; \bar{\tau}(t) \;-\; \left(\frac{\partial^2 J}{\partial \tau^2}(t, \hat{x}(t), \bar{\tau}(t)\right)^{-1} \frac{\partial J}{\partial \tau}(t+\Delta t, \hat{x}(t+\Delta t), \bar{\tau}(t)). \quad (9)$$

Now the algorithm consists of repeated computation of Eq. (9) with a given constant $\Delta t > 0$. To address its convergence properties it makes no sense to mention asymptotic convergence because the algorithm computes a finite number of iterations during the time-interval $0, T]$. Instead, we characterize convergence by its approach-rate to a stationary point. To clarify this issue, consider the Newton-Raphson algorithm for minimizing a function $f : R^n \rightarrow R$, and suppose that it computes a sequence of iteration points, $\{x_k\}_{k=1}^{\infty}$ convergent to a stationary point x^*, where $\frac{df}{dt^2}(x^*)$ is positive definite. Then (see [7]) there exist $\delta > 0$ and $K > 0$ such that, if $||x_k - x^*|| < \delta$, then $||x_{k+1} - x^*|| \leq K||x_k - x^*||^2$. Our algorithm has a similar property in the following sense.

Let $\bar{\tau}$ be a stationary point for the problem $\pi_{t,x(t)}$, and suppose that it lies in the interior of the constraint set, and the Hessian $\frac{\partial^2 J}{\partial \tau^2}(t, x(t), \bar{\tau})$ is positive definite. Define the error $e(t)$ by $e(t) := \hat{x}(t) - x(t)$.

Proposition 1. *There exist constants $\delta > 0$ and $K > 0$ such that, if $||\bar{\tau}(t) - \bar{\tau}|| < \delta$, $\Delta t < \delta$, and $||e(t)|| < \delta$ and $||e(t+\Delta t)|| < \delta$, then*

$$||\bar{\tau}(t+\Delta t) - \bar{\tau}|| \;\leq\; K\Big(||\bar{\tau}(t) - \bar{\tau}||^2 + \Delta t||\bar{\tau}(t) - \bar{\tau}|| + ||e(t+\Delta t)||\Big). \quad (10)$$

Proof. Please see [15].

The above equation is noteworthy in that it tells us something about convergence rates and explicitly incorporates terms stemming from the numerical accuracy, the rate at which the state estimate is converging, and the quadratic convergence rate of Newton's method.

Acknowledgments

The authors would like to thank Magnus Egerstedt for helpful discussions. The work of Wardi and Ding was supported in part by the National Science Foundation under Grant Number 0509064.

References

1. S. Azuma, M. Egerstedt, and Y. Wardi. Output-Based Optimal Timing Control of Switched Systems. *Hybrid Systems: Computation and Control*, Springer-Verlag, pp. 64-78, Santa Barbara, CA, March 2006.
2. R. Brockett. Stabilization of Motor Networks. *IEEE Conf. on Decision and Control*, pp. 1484–1488, 1995.
3. M. Egerstedt, Y. Wardi, and H. Axelsson. Transition-Time Optimization for Switched-Mode Dynamical Systems. *IEEE Trans. on Automatic Control*, Vol. AC-51, pp. 110-115, 2006.
4. A. Guia, C. Seatzu, and C. Van der Mee. Optimal Control of Switched Autonomous Linear Systems. In *IEEE Conf. on Decision and Control*, pp. 1816-1821, 1999.
5. D. Hristu-Varsakelis. Feedback Control Systems as Users of Shared Network: Communication Sequences that Guarantee Stability. *IEEE Conf. on Decision and Control*, pp. 3631–3631, 2001.
6. P. E. Moraal and J. W. Grizzle. Observer Design for Nonlinear Systems with Discrete-Time Measurements. *IEEE Trans. on Automatic Control*, vol. 40, pp. 395–404, 1995.
7. E. Polak. *Optimization Algorithms and Consistent Approximations.* Springer-Verlag, New York, New York, 1997.
8. A. Rantzer and M. Johansson. Piecewise Linear Quadratic Optimal Control. *IEEE Trans. on Automatic Control*, Vol. 54, pp. 629-637, 2000.
9. H. Rehbinder and M. Sanfirdson. Scheduling of a Limited Communication Channel for Optimal Control. *IEEE Conf. on Decision and Control*, pp. 1011–1016, 2000.
10. M.S. Shaikh and P. Caines. On Trajectory Optimization for Hybrid Systems: Theory and Algorithms for Fixed Schedules. *IEEE Conf. on Decision and Control*, pp. 1997–1998, 2002.
11. M.S. Shaikh and P.E. Caines. On the Optimal Control of Hybrid Systems: Optimization of Trajectories, Switching Times and Location Schedules. In *6th International Workshop on Hybrid Systems: Computation and Control*, 2003.
12. G. Walsh, H. Ye, and L. Bushnell. Stability Analysis of Networked Control Systems. *American Control Conf.*, pp. 2876–2880, 1999.
13. X. Xu and P. Antsaklis. Optimal Control of Switched Autonomous Systems. *IEEE Conf. on Decision and Control*, pp. 4401–4406, 2002.
14. X. Xu and P.J. Antsaklis. Optimal Control of Switched Systems via Nonlinear Optimization Based on Direct Differentiations of Value Functions. *Int. J. of Control*, Vol. 75, pp. 1406-1426, 2002.
15. Y. Wardi, X.C. Ding, and S. Azuma. On-Line Optimization of Switched-Mode Systems: Algorithms and Convergence Properties. Technical Memorandum, School of Electrical and Computer Engineering, Georgia Institute of Technology, 2006.

State Nullification of Switched Systems by Linear Output Feedback

Gera Weiss

University of Pennsylvania
gera@seas.upenn.edu

Abstract. We study the possibility to steer the state of a single-input single-output linear switched system to the origin in finite time by output feedback with finite memory. We show that if the system is in a controller canonical form, a causal state nullifying controller exists. We also show that, generically, only acausal feedback can achieve state nullification.

1 Introduction

We analyze the possibility to synthesize an output feedback controller that steers the state of a linear switched system to the origin in finite time. This control objective is called state nullification. We provide theoretical results and offer an algorithm for constructing a controller that nullifies a given system.

As time-invariant systems and time-varying systems are special cases of switched systems (can be viewed as switched systems with a single switching signal), the presented work generalizes some of the results presented in [1] and [2] where nullification of time-invariant and time-varying control systems is analyzed, respectively. In particular, the algorithm presented here has the algorithms displayed in [1] and [2] as special cases.

It turns out that considering a set of switching signals introduces new obstacles and challenges. We show how to overcome some of the obstacles and identify a class of switched systems that are nullifiable. We also show that unlike the case with a single switching signal, where nullification is generically possible, it is generically not possible to achieve nullification when all switching are allowed. Due to space limitations, proofs and examples are moved to the appendix. Formal proofs and examples can be found in the online version.

2 Theoretical Results

An n-dimensional single-input single-output switched system is described by a parametrized set of triplets $\{(A_s, b_s, c_s) : s \in \Sigma\}$ where, for every $s \in \Sigma$, the matrix A_s is $n \times n$-dimensional, b_s is n-dimensional column vector and c_s is n-dimensional row vector and Σ is a finite set.

Given a switching signal $\sigma : \mathbb{N} \to \Sigma$, the dynamics of the switched system are defined by

$$\begin{aligned} x(t+1) &= A_{\sigma(t)}x(t) + b_{\sigma(t)}u(t) \\ y(t) &= c_{\sigma(t)}x(t) \end{aligned} \quad (1)$$

A. Bemporad, A. Bicchi, and G. Buttazzo (Eds.): HSCC 2007, LNCS 4416, pp. 775–778, 2007.

where $x : \mathbb{N} \to \mathbb{R}^n$ represents the state evolution, $u : \mathbb{N} \to \mathbb{R}$ is a scalar input signal and $y : \mathbb{N} \to \mathbb{R}$ is a scalar output signal.

An output feedback controller is a map of the output signal and the switching signal to an input signal. We will study the possibility to design simple output feedback controllers to achieve state nullification, defined as follows.

Definition 1. *We say that a controller achieves state nullification for the system (1) if there exists $N \in \mathbb{N}$, called nullification time, such that for every $x(0) \in \mathbb{R}^n$ and any switching signal $\sigma : \mathbb{N} \to \Sigma$, the Nth state is the null vector ($x(N) = 0$).*

Note that we do not assume control over the switching signal. The controller needs to steer the state of the system to the origin, using the input signal, under all switching signals.

For the formulation of the first result we need to define the controller canonical form and n-step observability as follows. A triplet $(A, b, c) \in \mathbb{R}^{n\times n} \times \mathbb{R}^{n\times 1} \times \mathbb{R}^{1\times n}$ is in a *controller canonical form* if b equals $(0, \ldots, 0, 1)^T$ and A is in a companion form. The system (1) is called *n-step observable* if for all switching and control signals, the mapping of the initial state $x(0)$ to the sequence $y(0), \ldots, y(n-1)$ of observations is one-to-one.

The first main result is formulated as follows.

Theorem 1. *If the system (1) is n-step observable and, for every $s \in \Sigma$, the triplet (A_s, b_s, c_s) is in a controller canonical form and the first coordinate of c_s is not zero; then there exists a feedback of the form*

$$u(t) = K\big(\sigma(0), \ldots, \sigma(t)\big)y(t) \tag{2}$$

that achieves state nullification.

Note that a feedback of the form (2) is causal, i.e., the output $u(t)$ at time $t \in \mathbb{N}$ does not depend on the future. Note also that an efficient implementation of the controller requires only finite memory, since we only have to store the history of the switching signal, drawn from a finite set Σ, over a finite time (the nullification time).

Associate a system $(A, b, c) \in \mathbb{R}^{n\times n} \times \mathbb{R}^{n\times 1} \times \mathbb{R}^{1\times n}$ with the vector in $\mathbb{R}^{n^2+2n}$ obtained by flattening the matrix and the vectors to a single vector. With this association, the topology of $\mathbb{R}^{n^2+2n}$ induces a topology over linear systems. One can similarly define a topology over switched systems. We call a set of systems generic, if it is open and dense in that topology. A property is called generic if it is valid for a generic set of systems.

Note that the property that all the triplets are in a controller canonical form is not generic. This is consistent with the fact that nullification by causal output feedback is generically impossible, as stated in the following theorem.

Theorem 2. *There is a generic set of systems such that, for every system in the set, there exists no output feedback of the form (2) that achieves nullification.*

However, if we allow acausality (preview) it is generically possible to achieve nullification. This is consequence of a result reported in [2], as follows. If we allow an acausal controller of the form $u(t) = K(\sigma(0), \ldots, \sigma(t+N))y(t)$ where N is the nullification time, then there is no difference between time-varying systems and switched systems (because, for all we care, the switching signal is given in advance). In particular, since nullification is generically possible for time varying systems [3], it is also generically possible by an acasual feedback for switched systems. Note that there is a gap between the impossibility result that involves a feedback of the form (2) and this positive result, namely, we do not know if a feedback of the form $u(t) = K(\sigma(0), \ldots, \sigma(t+L))y(t)$ for $0 < L < N$ can achieve nullification.

3 Algorithm

The algorithm for the computation of a feedback mapping that achieves state nullification is composed of two subroutines labeled Algorithm 1 and Algorithm 2 below.

Algorithm 1. Synthesizes a feedback that steers a given initial state to the origin.

Require: a switched system, a state $x \in \mathbb{R}^n$ and a partial assignment to variables $K(\sigma), \sigma \in \cup_{t\in\mathbb{N}} \Sigma^t$.
Ensure: the coefficients $K(\sigma), \sigma \in \cup_{i=1}^{t+1} \Sigma^i$, when substituted in (2), assure $x(0) = x \implies x(t+1) = 0$.

$t \leftarrow 0$
inequalities $\leftarrow \emptyset$
equalities $\leftarrow \{x = 0\}$
while there is no solution to *equalities* $\cup$ *inequalities* **do**
  equalities $\leftarrow \emptyset$
  for all $\sigma \in \Sigma^{t+1}$ **do**
    $x^- \leftarrow x(\sigma(0), \ldots, \sigma(t-1))$
    if $K(\sigma(0), \ldots, \sigma(t))$ has a value K **then**
      $x^+ \leftarrow (A_{\sigma(t)} + K b_{\sigma(t)} c_{\sigma(t)}) x^-$
    else if $c_{\sigma(t)} x^- = 0$ **then**
      $x^+ \leftarrow A_{\sigma(t)} x^-$
      $K(\sigma(0), \ldots, \sigma(t)) \leftarrow 0$
    else
      $x^+ \leftarrow A_{\sigma(t)} x^- + b_{\sigma(t)} \delta_\sigma$
      inequalities $\leftarrow$ *inequalities* $\cup \{c_{\sigma(t)} x^- \neq 0\}$
      $K(\sigma(0), \ldots, \sigma(t)) \leftarrow \frac{\delta_\sigma}{c_{\sigma(t)} x^-}$
    end if
    equalities $\leftarrow$ *equalities* $\cup \{x^+ = 0\}$
    $x(\sigma(0), \ldots, \sigma(t)) \leftarrow x^+$
  end for
  $t \leftarrow t + 1$
end while
Choose a solution for *equalities* $\cup$ *inequalities*.
Use this solution to compute $K(\sigma)$ for all $\sigma \in \Sigma^{i+1}$, $i \in \{0, ..., t\}$.

Algorithm 1 computes a feedback that steers a single state to the origin. More specifically, given a switched system $(A_s, b_s, c_s)_{s\in\Sigma}$ and an initial state $x \in \mathbb{R}^n$; the coefficients $\{K(\sigma) : \sigma \in \Sigma^{i+1}, i = 0, \ldots, t\}$ are computed such that, when these parameters are plugged in equation (2) and the resulting control law is plugged in (1), the initial state $x(0) = x$ implies $x(t+1) = 0$.

Algorithm 2 is the main entry point for the synthesis algorithm. It invokes Algorithm 1 iteratively to compute coefficients that steer any initial state to the origin. More precisely, every iteration ends with a higher value of t and fixes the coefficients $K(\sigma), \sigma \in \cup_{i=1}^{t+1}\Sigma^i$. When Algorithm 2 halts, the coefficients are set such that $x(t+1) = 0$ independent of the switching signal and the initial state.

Algorithm 2. Synthesizes a feedback of the form (2) that steers any initial state to the origin, under all switching signals.

```
Require: a system of the form (1) that satisfies the conditions of Theorem 1.
Ensure: the feedback (2) steers any initial state to the origin in t steps, under any switching signal.
Choose a linear basis B that spans ℝ^n.
for all x ∈ B do
    Run Algorithm 1 with x as an initial state.
end for
```

Remark 1. The above algorithm can easily be adjusted to handle situations where not all switching signals are admissible. This is done by replacing line #1 of Algorithm 1. More specifically, σ should go over all admissible switching signals of length $t+1$. In particular, when the system is time-varying, we can think of it as a switched system with a single admissible signal. In that case, the resulting algorithm is similar to that proposed in [2]. We can also think of a time invariant system as a switched system with a single subsystem. In the case of time invariant systems, the above algorithm is similar to the algorithm proposed in [1]. The set of switching signals can be given in any form that allows the enumeration of all admissible signals of length $t+1$ (e.g. automata).

Remark 2. The proposed description of the algorithm does not depend on a representation of the system in a canonical form. The procedure applies to systems that are equivalent to a system in a controller canonical form, without the need to carry the actual transformation. In [1] and [2] it is shown that the pre-conditions of the algorithm can be checked without computing the canonical form representation.

Acknowledgment

I would like to acknowledge the help provided to me by Zvi Artstein. His insights and ideas where invaluable to the development of this work. An earlier version of this report appeared in my Ph.D thesis. This research was supported by NSF grants CCR 0410662 and CSR-EHS 0509143.

References

1. Artstein, Z., Weiss, G.: State nullification by memoryless output feedback. Math. Control Signals Systems **17**(1) (2005) 38–56
2. Weiss, G.: Memoryless output nullification and canonical forms, for time varying systems. International Journal of Control **78**(15) (2005) 1174–1181
3. Weiss, G.: A combinatorial game approach to state nullification by hybrid feedback. `http://www.wisdom.weizmann.ac.il/~gera/game.pdf` (2006) Draft.

Fault Accommodation for Hybrid Systems with Continuous and Discrete Faults*

Hao Yang[1,2], Bin Jiang[1], and Vincent Cocquempot[2,**]

[1] College of Automation Engineering
Nanjing University of Aeronautics and Astronautics
29 YuDao Street, Nanjing, 210016, P.R. China
binjiang@nuaa.edu.cn

[2] LAGIS-CNRS, UMR 8146, Université des Sciences et Technologies de Lille
59655 Villeneuve d'Ascq cedex, France
Tel.: +33 (0)3 20 43 62 43, Fax: +33 (0)3 20 33 71 89
vincent.cocquempot@univ-lille1.fr

Abstract. Fault accommodation (FA) problem is discussed for a class of hybrid systems with both continuous and discrete faults, and without full state measurements. The proposed observer-based FA strategy can maintain the stability of hybrid systems in spite of these two faults.

1 Preliminaries

Two kinds of faults may corrupt the behavior of a switched system [1]: *Continuous faults* that affect each continuous system mode. *Discrete faults* that affect the switching sequence. In this note, a FA framework is proposed for a class of switched nonlinear systems (SNS) with above two faults, and without full state measurements.

The contributions are in 3 aspects : 1) For the continuous faults in each mode, an adaptive observer technique is proposed to provide the rapid fault estimation, based on which the fault tolerant control law is designed. 2) For the discrete faults, a model-free sliding mode observer is developed to estimate the states at each switching instant, which together with a series of observers according to system modes, can identify the current mode quickly. 3) The above two strategies are combined with the average dwell time scheme, which can guarantee the input-to-state practical stability of overall SNS.

Consider a class of switched nonlinear systems:

$$\dot{x}(t) = A_\sigma x(t) + g_\sigma(t, x(t)) + B_\sigma u_\sigma(t) + E_\sigma f_\sigma^c(t) \tag{1}$$

$$y(t) = Cx(t) \tag{2}$$

where $x(t) \in \Re^n$ is the non measured state, $y(t) \in \Re^r$ is the output, $u_\sigma(t) \in \Re^m$ is the control. (A_σ, B_σ) is controllable. $g_\sigma(x(t), t)$ is a continuous Lipschitz

* This work is partially supported by NSF of China (60574083) and Key Laboratory of Process Industry Automation, State Education Ministry of China.

** Corresponding author.

A. Bemporad, A. Bicchi, and G. Buttazzo (Eds.): HSCC 2007, LNCS 4416, pp. 779–782, 2007.

function, i.e., $|g_\sigma(x_1,t) - g_\sigma(x_2,t)| \leq L_\sigma|x_1 - x_2|$ for $L_\sigma > 0$, where $|\cdot|$ is the Euclidean norm. Moreover, $g_\sigma(0,t) = 0$.

The *continuous actuator fault* is modelled by a "fault pattern" [2], which consists of the distribution matrix E_σ, and a "fault signal" $f_\sigma^c(t) \in \Re^q$. Assume that there exists two constants f_σ^0 and f_σ^1 such that $|f_\sigma^c| \leq f_\sigma^0$, $|\dot{f}_\sigma^c| \leq f_\sigma^1$.

Define $\mathcal{M} = \{1, 2, \ldots, N\}$, where N is the number of modes. $\sigma(t) : [t_0, \infty) \to \mathcal{M}$ denotes the *switching function*, which is assumed to be a piecewise constant function continuous from the right.

Denote t_j as the jth switching instant of the system (1)(2). At t_j, the system switches to mode k, where $k \in \mathcal{M}$, $j = 1, 2, ...$

The switching property is considered as in [3]: a) the switching sequence is fixed. b) there is a series of prescribed dwell periods between each switching. We also assume that the states do not jump at the switching instants.

The *discrete fault* is represented by the faulty switching function $\sigma_f(t)$, that forces the system to switch to a mode which is not the prescribed successor at the switching instant. Similarly, $\sigma_H(t)$ denotes the healthy switching function. If $\sigma(t) = \sigma_H(t)$, then there is no discrete fault in the current mode.

2 FA for Continuous Faults

Consider mode k of the system (1)(2) starting from $t = t_j$

Assumption 1. *There exists a matrix K_k such that $G_k(s) = C[sI - (A_k - K_kC)]^{-1}E_k$, is strictly positive real (SPR) :*

$$\forall \omega > 0 : Re(G_k(j\omega)) > 0 \tag{3}$$

Moreover

$$\min_{\omega \in R^+} \sigma_{min}(A_k - K_kC - j\omega I) > L_k \tag{4}$$

where $\sigma_{min}(M)$ is the smallest singular value of M.

The continuous fault diagnostic scheme for mode k is designed as

$$\dot{\hat{x}}(t) = A_k\hat{x}(t) + g_k(t, \hat{x}(t)) + B_ku_k(t) + E_k\hat{f}_k^c(t) + K_k(y(t) - \hat{y}(t)) \tag{5}$$

$$\dot{\hat{f}}_k^c(t) = \Gamma_kR_k^T(y(t) - \hat{y}(t)) - \vartheta_k\Gamma_k\hat{f}_k^c(t) \tag{6}$$

$$\hat{y}(t) = C\hat{x}(t) \tag{7}$$

where $\hat{x}(t), \hat{f}_k^c(t), \hat{y}(t)$ are the estimates of $x(t), f_k^c(t), y(t)$. The weighting matrix $\Gamma_k = \Gamma_k^T > 0$, and the constant $\vartheta_k > 0$ are chosen such that $\vartheta_k - \lambda_{max}(\Gamma_k^{-1}) > 0$. Denote $e_x(t) = x(t) - \hat{x}(t)$, $e_f(t) = f_k^c(t) - \hat{f}_k^c(t)$.

Recall that (A_k, B_k) is controllable. Let $W_k = W_k^T > 0$ be associated with a given symmetric positive definite matrix H_k by the Ricatti equation

$$A_k^TH_k + H_kA_k - 2H_kB_kB_k^TH_k + W_k = 0 \tag{8}$$

Assumption 2. *There exists a symmetric positive definite matrix H_k and a bounded function $\eta_k(x,t) > 0$ such that*

$$|H_k g_k(x,t)| \ \leq \eta_k(x,t)|x^T H_k B_k| \tag{9}$$

Assumption 3. $rank(B_k, E_k) = rank(B_k)$.
The fault-tolerant controller is constructed as $u_k(\hat{x}) = u_{k1}(\hat{x}) + u_{k2}(\hat{x})$ with

$$u_{k1}(\hat{x}) \triangleq -B_k^T H \hat{x} - B_k^* E_k \hat{f}_k^c, \tag{10}$$

$$u_{k2}(\hat{x}) \triangleq -\frac{\eta_k(\hat{x},t)|\hat{x}|}{|\phi_k(\hat{x})| + \epsilon/2}\phi_k(\hat{x}), \quad \phi_k(\hat{x}) \triangleq \eta_k(\hat{x},t) B_k^T H_k \hat{x} \tag{11}$$

where ϵ is a small positive scalar. B_k^* is such that $(I - B_k B_k^*)E_k = 0$.

We can prove that, *under assumptions 1-3, the fault diagnostic scheme (5)-(7) and the feedback control (10)-(11) guarantee*

- *(e_x, e_f) of mode k converges to a closed set which can be made arbitrarily small.*
- *e_x is input-to-state stable over $[t_j, t)$ w.r.t. e_f.*
- *mode k is input-to-state practically stable (ISpS) over $[t_j, t)$ w.r.t. e_x, e_f and a constant $\varsigma_k > 0$.*

Now consider the stability of overall SNS with each mode satisfying assumptions 1-3. At each switching instant, the fault diagnostic scheme (5)-(7) is switched according to the current mode. $\hat{x}$ of the current observer are chosen as the final states of the previous observer, and $\hat{f}_k^c$ is set to zero.

Let the switching function σ has an average dwell time τ_a. We can prove that, *if τ_a is large enough, then under the fault diagnostic scheme and the feedback control, the switched system is ISpS for $[0, T)$ w.r.t. e_x, e_f and a constant $\bar{\varsigma} > 0$, where $T > 0$ is an arbitrary time.*

3 FA for Discrete Faults

The main idea is to first identify the current mode at the beginning of each time interval $[t_j, t_{j+1})$ using a short time period $\Delta t_j \ll t_{j+1} - t_j$, and then control the identified mode in the rest of the time interval.

In each identifying period, the control signal is set to zero, which implies that there are no controller and continuous actuator fault present in each Δt_j.

The *observer-based identifier* consists of three parts:

1) A model free sliding observer.

$$\dot{\bar{x}}(t) = \bar{A}\bar{x}(t) + S(\bar{e}_x(t), \rho_j) + L(y(t) - \bar{y}(t)), \quad \bar{y}(t) = C\bar{x}(t) \tag{12}$$

where $\bar{A}$, $\bar{L}$ are chosen such that $(\bar{A}, C)$ is observable and $\bar{A} - \bar{L}C$ is stable. $\bar{e}_x \triangleq x - \bar{x}$, $S(\bar{e}_x, \rho_j) \triangleq \frac{\bar{P}^{-1}C^T C\bar{e}_x}{|C\bar{e}_x|}\rho_j$ with $\bar{P}$ a symmetric positive definite matrix. It can be proved that *there exists a $\rho_j > 0$ such that the observer (12) converges to the system without input, if x in Δt_j is bounded.*

2) A bank of observers

$$\text{Observer } k: \quad \begin{aligned} \dot{\hat{x}}_k(t) &= A_k\hat{x}_k(t) + g_k(t, \hat{x}_k(t)) + K_k(y(t) - \hat{y}_k(t)) \\ \hat{y}_k(t) &= C\hat{x}_k(t), \quad k \in \mathcal{M} \end{aligned} \tag{13}$$

which are the same as (5)-(7) without u_k and $\hat{f}_k^c$.

3) The identifier algorithm summarizing as: *The current mode is mode j, $j \in \mathcal{M}$, if $|\bar{x} - \hat{x}_j|$ is minimal at time instants $t_j + \Delta t_j$, where Δt_j can be made arbitrarily small.*

To avoid that the system states escape before a proper controller is invoked into action, Δt_j is chosen in the following set

$$\Omega_{\Delta t_j} \triangleq \{\Delta t_j | \Delta t_j < t_{j+1} - t_j \text{ and } |\bar{x}(t_j + \Delta t_j)| \leq \xi |\bar{x}(t_j)|\} \tag{14}$$

where $\xi > 1$. The selection of ξ depends on system dynamics. Δt_j is chosen such that the current mode can be detected in this period. The bound of x in Δt_j can be measured by $\bar{x}$.

4 FA Framework

Based on sections 2 and 3, the *FA framework* is proposed as follows:

1) At switching instant t_j, stop the fault diagnostic scheme, and set control signals and fault estimates to zero, invoke the identifier to identify the current mode.
2) At time instants $t_j + \Delta t_j$, stop the identifier and switch the fault diagnostic scheme and controller (10)-(11) into the system according to the current mode.
3) At switching instant t_{j+1}, go to 1).

The main result is described as: *If τ_a is large enough, then under assumptions 1-3, the proposed FA framework guarantees that the switched system is ISpS for $[0, T)$ w.r.t. $e_x(t)$, $e_f(t)$, $\bar{e}_x(t_j)$ and a constant $\bar{\varsigma}_2 > 0$, where $T > 0$ is an arbitrary time and $j = 1, 2....$*

References

1. V. Cocquempot, T. EI Mezyani, and M. Staroswiecki, Fault detection and isolation for hybrid systems using structured parity residuals, *Proc. of 5th Asian Control Conference,* 1204-1212, 2004.
2. B. Jiang, M. Staroswiecki, and V. Cocquempot, Fault accommodation for a class of nonlinear dynamic systems, *IEEE Trans. on Automatic Control,* 51 (9), 1578-1583, 2006.
3. H. Yang, B. Jiang, V. Cocquempot, and M. Staroswiecki, Adaptive fault tolerant strategy for a class of hybrid systems. *Proc. of IFAC Safeprocess 06',* 1021-1026, 2006.

A Heuristic Predictive Logic Controller Applied to Hybrid Solar Air Conditioning Plant

Darine Zambrano[1], Winston García-Gabín[1], and Eduardo F. Camacho[2]

[1] Universidad de Los Andes, Escuela de Ingeniería Eléctrica, 5101 Mérida, Venezuela
{darine,winston}@cartuja.us.es

[2] Universidad de Sevilla, Departamento de Ingenieria de Sistemas y Automática, 41092 Sevilla, España
eduardo@esi.us.es

Abstract. This paper shows the development of a heuristic predictive logic controller (HPLoC) applied to a solar air conditioning plant. The plant uses two energy sources, solar and gas, in order to warm up the water. The hot water feeds a single-effect absorption chiller. A hybrid controller using Model based Predictive Control (MPC) and heuristic logic conditions have been implemented in the real plant, the controller allows configuring the plant with the minimizing of the gas consumption.

1 Introduction

An evident use of the solar energy like a source of renewable energy is its application to air conditioning systems. A solar air conditioning plant is located at the University of Seville. The different operating modes of the process are defined by the components that provide thermal energy to the absorption machine. This process has to be modelled as a hybrid system, a more complete description of the plant and the model can be seen in [1]. Several control strategies have been tested on different solar power plants (see, for instance, [2,3,4]), but none of them take the hybrid nature of the process into account. A hybrid controller is needed in order to configure the plant according to the requirement of the control objectives. The configuration decisions depend of the environmental conditions.

2 Plant Description and Control Objectives

The main components of the plant are the following: the solar system, it is composed of a set of flat solar collectors; the accumulation system, it is composed of two tanks storing hot water; and the cooling machine. There also exist an auxiliary gas-fired heater that can supply energy in those situations where solar radiation is not enough, and a load simulator (a heat pump) that allows performing tests for different profiles of load. The hybrid nature of the plant comes from the use of two different energy sources (solar and gas), which can be combined or used independently. The plant can be re-configured on-line manipulating open/close valves and on/off pumps allowing to select the components

A. Bemporad, A. Bicchi, and G. Buttazzo (Eds.): HSCC 2007, LNCS 4416, pp. 783–786, 2007.

for energy supply. The plant evolves among several operating modes during its daily operation. The operating modes are selected using the discrete manipulated variables and are the following: 1⟩ Recirculation of water through the solar collectors, 2⟩ Loading the tanks with hot water from solar collectors, 3⟩ Using the water from solar collectors for feed the absorption machine, 4⟩ Using the water from the solar collectors and gas heater for feed the absorption machine generator, 5⟩ Using the gas heater for feed the absorption machine generator, 6⟩ Using the tanks and gas heater for feed the absorption machine generator, 7⟩ Using tanks for feed the absorption machine generator, 8⟩ Loading the tanks with water from solar collectors and using the gas heater for feed the absorption machine generator, 9⟩ Recirculation in solar system and using the gas heater for feed the absorption machine generator, 10⟩ Using the solar collectors for feed the absorption machine and loading tank, and, 11⟩ Using the solar collectors and gas heater to feed the absorption machine and loading the tanks.

The main control objective is to supply chilled water to the air distribution system according to the demanded temperature. The absorption machine must be fed with hot water between 75 $°C$ and 100 $°C$ for cold production. Consumption of the auxiliary energy must be minimized to diminish the environmental impact and save money.

3 Heuristic Predictive Logic Controller

An HPLoC has been designed based in the basic formulation of MPC and the operation knowledge. The controller is divided in two levels, the operating mode of the solar plant is obtained using an HPLoC, and the temperatures are controlled using a MPC for each operating mode. The HPLoC algorithm realizes the prediction of the solar collectors output temperature during a horizon N_p, Eq. (1), the input of the model is the collectors flow (u), and, the disturbances are the solar radiation (d_1) and the inlet temperature (d_2). A, B, B_{d1} and B_{d2} are polynomials in the back shift operator z^{-1}.

$$A(z^{-1})\hat{y}(t) = B(z^{-1})u(t-1) + B_{d1}(z^{-1})d_1(t-1) + B_{d2}(z^{-1})d_2(t-1) \quad (1)$$

The HPLoC algorithm is based in the minimization of a weighed function subject to constraints, Eq. (2), where x is a vector of the logic variables associated to the operating modes, and A_C and b_C matrix avoid that the operating modes are overlapping, and also it guarantee that the F vector is greater than the unit.

The F vector, Eq. (3), is defined for each operating mode in function of the factors associated to each equipment (k_j, $j = 1, \ldots, 3$) and of operating factors of the plant (k_{lm}, $m = 1, \ldots, 5$). F is the availability status of the operating modes according to the environmental conditions and output temperatures.

$$\min_{x \epsilon \{0,1\}} (F \cdot \omega_i)^T x \quad (2)$$

$$s.t. \qquad A_C x \leq b_C$$

$$F = [f_1(k_{l5}) \ f_2(k_{l5}) \ f_3(k_1) \ f_4(k_1,k_3) \ f_5(k_3) \ f_6(k_3,k_2,k_{l1}) \ldots \quad (3)$$
$$f_7(k_2) \ f_8(k_3) \ f_9(k_3) \ f_{10}(k_1) \ f_{11}(k_1,k_3)]$$

A factor is associated to each equipment that can provide hot water to the absorption machine. The factor of the solar collectors k_1 is expressed in function of the prediction output temperature, Eq. (1), and the solar radiation. The factor of the accumulators k_2 is defined in function of its output temperature. The factor of the gas heater k_3 is assumed constant because it is available in any time. Others logic conditions k_{li} associated to the operating of the solar plant are defined: k_{l1} is true if the absorption machine is turn off, k_{l2} is true if the solar collectors output temperature is inside of the working range of the absorption machine, k_{l3} is true if the accumulators output temperature is inside of the working range of the absorption machine, k_{l4} is true if the solar radiation is enough, and, k_{l5} is true if exist demand of cold.

The weight vector ω_i is calculated using the Analytic Hierarchy Process (AHP) [5]. This technique allows the development of weights for the criteria using pair wise comparisons, the pair wise comparison method for decision-maker is asked to give the relative importance to the criteria by comparing them two by two, in this case, they are established according to the control objectives. The expert criteria are given by the first row of the Saaty's matrix, (see Eq. 4).

The expert criteria a_i, Eq. (5), have been defined weighting the logic conditions, it is defined by each operating mode, the weighting g_i is according to the important of the operating modes.

$$A(1,1,\ldots,11) = [a_{1,1} \quad a_{1,2} \quad a_{1,3} \quad \ldots \quad a_{1,9} \quad a_{1,10} \quad a_{1,11}] \tag{4}$$

$$a_{1,i} = g_i m_{li} + h_i(\sim m_{li}) \qquad for \qquad i = 1,\ldots,11; \quad h_i \ll g_i \tag{5}$$

A m_{li} logic condition has been defined for each operating mode in function of the k_{li}, these conditions indicate when an operating mode can be used according to the actual conditions of the plant. Then m_{li} take a binary value, as the $a_{1,1}$ element must be equal to 1, then an h_i value is added for when the logic condition m_{li} is zero. When m_{li} is zero, the expert criteria takes a value less than when it is true, therefore h_i coefficient must be greater than g_i. Finally, x_i solution determines the operating mode and the configuration of the discrete manipulated variables.

4 Results and Conclusions

The heuristic predictive logic controller has been implemented on the solar air conditioning plant. Figure 1 shows the operating mode of the plant. The cold demand begins at 11 hour, the operating mode is 9, and then the generator input temperature is warmed by the gas heater. While the solar radiation is low, the configurations switch between modes 6 and 9. At 12:30 hour approximately, the sky is without clouds and the configuration is switch to the mode 4 that uses the solar energy combines with the auxiliary energy. After 15 hours the day is cleared, and then the absorption machine is fed by the flow from solar collectors only. The generator and evaporator temperature of the absorption machine are shown. Can be observed that the absorption machine is working right, the water is chilled and the demand of cold is satisfied.

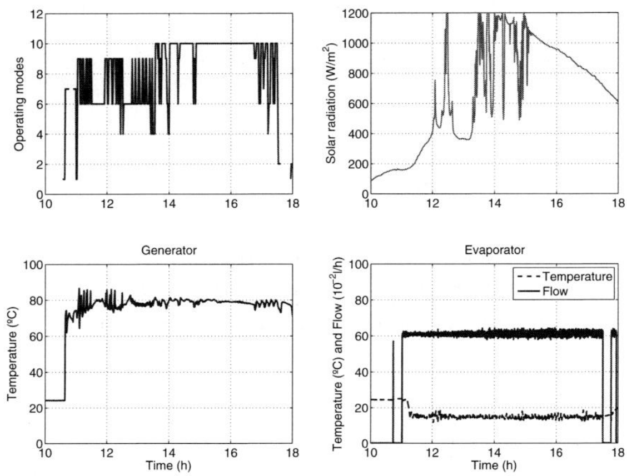

Fig. 1. Operating mode, solar radiation and other temperatures of the solar plant

Finally, the hybrid solar cooling plant requires efficient algorithms that allow optimizing the use of the energy, and other important characteristic, as the disturbance rejection and economic objectives. The hybrid controller include an MPC for the continuous variables, and, an integer optimization algorithm for the solar plant configuration, it has been design using the prediction concepts and logic expert criteria. The results shown as the demand of cold and the minimum gas consumption have been satisfied.

Acknowledgments

Supported by the Programme Alβan, the European Union Programme of High Level Scholarships for Latin America scholarship N° E05D053808VE, and, European Union under contract (FP6-511368) (HYCON).

References

1. Zambrano, D., Bordons, C., García-Gabín, W., Camacho, E.F.: A solar cooling plant: a benchmark for hybrid systems control. 2nd IFAC Conference on Analysis and Design of Hybrid Systems (2006)
2. Camacho, E., Berenguel, M., Rubio, F.: Advanced Control of Solar Plants. Springer, London (1997)
3. Lemos, J.M., Mosca, E., Silva, R.N., Shirley, P.O.: Industrial applications of predictive adaptive control based on multiple identifiers. 15th IFAC World Congress (2002)
4. Silva, R.N., Rato, L.M., Lemos, J.: Observer based non uniform sampling predictive controller for a solar plant. 15th IFAC World Congress (2002)
5. Saaty, R.: The analytic hierarchy process - what it is and how it is used. Mathematical Modelling **9**(3-5) (1987) 161–176

Distributed Hybrid Control for Multiple-Pursuer Multiple-Evader Games*

Michael M. Zavlanos and George J. Pappas

Department of Electrical and Systems Engineering
University of Pennsylvania
3330 Walnut Street, GRASP Laboratory, Philadelphia, PA 19104, USA
{zavlanos,pappasg}@grasp.upenn.edu

Abstract. Multiple-pursuer multiple-evader games raise fundamental and novel problems in control theory and robotics. In this paper, we propose a distributed solution to this problem that simultaneously addresses the discrete assignment of pursuers to evaders as well as the continuous control strategies for capturing individual evaders. The resulting hybrid control framework guarantees the mutual exclusion property of the final assignment for all initial conditions as well as capturing all evaders after exploring at most a polynomial number of assignments, dramatically reducing the combinatorial nature of purely discrete assignment problems.

1 Background and Problem Definition

Consider n identical pursuers in a p-dimensional space $\mathbb{R}^p$ and assume kinematic models for the pursuers with, in general, unbounded velocities. Consider, further, $m \geq n$ evaders with velocities that can be time varying but are assumed to be bounded for all time. We then say that pursuer i can *capture* evader k if there exists a finite time instant $T > 0$ such that pursuer i is in a sufficiently small neighborhood of evader k, for all time $t > T$. The time instant that every pursuer has captured a distinct evader corresponds to the *termination of the game* and the notion of capturing the evaders becomes equivalent to that of assigning evaders to pursuers.

Pursuit-evasion games can be classified in continuous and purely discrete games. Continuous games explicitly model the physical motion and constraints of the players [1], [2], [3] and often assume worst case motion for the evaders [1], [2]. Studying optimality of such strategies involves numerically solving the computationally challenging Hamilton-Jacobi-Isaacs partial differential equations. On the other hand, discrete games are either played in purely discrete environments such as graphs [4], [5], or in continuous environments disregarding though any physical dynamics of the players [6], [7]. Such models may result in discrete strategies which are dynamically infeasible. Recently, however, hybrid control

* This work is partially supported by ARO MURI SWARMS Grant W911NF-05-1-0219 and the NSF ITR Grant 0324977.

A. Bemporad, A. Bicchi, and G. Buttazzo (Eds.): HSCC 2007, LNCS 4416, pp. 787–789, 2007.

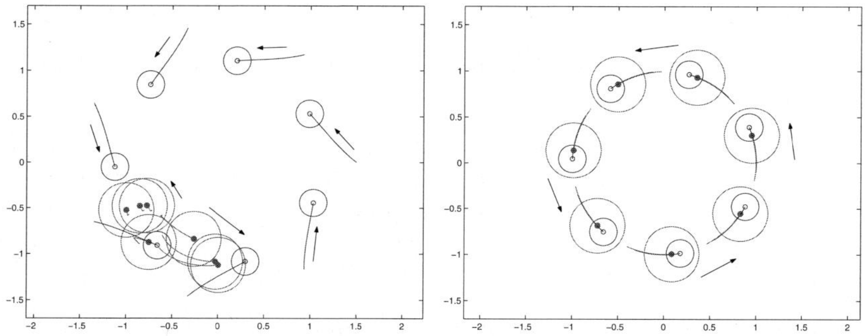

(a) Intermediate Configuration. (b) Final Configuration.

Fig. 1. Simulation for $n = 7$ pursuers and $m = 7$ evaders

has been used to bridge the gap between the discrete and continuous games and model, more realistic, pursuers with limited sensing and communication capabilities. Approaches involve visibility-based [8] and probabilistic [9], [10] games. Closely related to the topics discussed in this paper is also multiple-target tracking by sensor networks [11], [12] which, however, focuses more on the sensing and estimation problem of tracking rather than the actuation and control.

In this paper, we propose a novel distributed approach to the multiple-pursuer multiple-evader game, which is inspired by our previous work on dynamic assignment for stationary targets [13], [14]. Unlike *centralized* or *off-line* approaches [15], [16], [17], that decouple the assignment and navigation problems and focus on designing control laws for every pursuer to capture a preassigned evader, we *simultaneously* address the discrete assignment of pursuers to evaders as well as the continuous control strategies for tracking and capturing the individual evaders. The resulting hybrid controller for every pursuer consists of both local coordination protocols [14], [18] guaranteeing that distinct evaders are captured by distinct pursuers, and single-pursuer single-evader time-varying potential fields ensuring convergence of the tracking error to any neighborhood of zero. Composition of the hybrid controllers for all pursuers results in a highly efficient overall system that is illustrated in nontrivial multiple-pursuer multiple-evader games (Figure 1).

The main contribution of our approach is in the distributed coordination protocols that enable nontrivial multiple-pursuer multiple-evader behavior based on already available results for single-pursuer single-evader games. In particular, under the assumption that the pursuers have knowledge of the evaders' locations, the proposed nearest neighbor coordination scheme leads to a dynamically determined final assignment [14] which is shown to have the desired mutual exclusion property for all initial conditions. Furthermore, the overall coordination system is shown to have at most polynomial complexity, i.e., at most a polynomial number of transitions, despite the exponential growth of the number of assignments with respect to the number of pursuers.

References

1. Basar, T., Olsder, G.: Dynamic Noncooperative Game Theory. Number 23, SIAM, 2nd edition, 1999.
2. Isaacs,R.: Differential Games. Dover Publications, Mineola, New York, 1965.
3. Aguiar, A. P., Hespanha J.: Trajectory-Tracking and Path-Following of Underactuated Autonomous Vehicles with Parametric Modeling Uncertainty. IEEE Transactions on Automatic Control, 2006. (to appear)
4. Adler, M., Racke, H., Sivadasan, N., Sohler, C., Vocking, B.: Randomized Pursuit-Evasion in Graphs. Combinatorics, Probability and Computing, vol. 12(3), pp. 225 - 244, 2003.
5. Megiddo, N., Hakimi, S. L., Garey, M. R., Johnson, D. S., Papadimitriou, C. H.: The Complexity of Searching a Graph. Journal of the Association for Computing Machinery, vol. 35(1), pp. 18 44, 1988.
6. Guibas, L. J., Latombe, J. C., LaValle, S. M., Lin, D., Motwani, R.: A Visibility-Based Pursuit-Evasion Problem. International Journal of Computational Geometry and Applications, vol. 9(4/5), pp. 471, 1999.
7. LaValle, S. M., Hinrichsen, J. E.: Visibility-Based Pursuit-Evasion: The Case of Curved Environments. IEEE Transactions Robotics and Automation, vol. 17(2), pp. 196 - 202, April 2001.
8. Isler, V., Belta, C., Daniilidis, K., Pappas, G. J.: Hybrid Control for Visibility-Based Pursuit Evasion Games. Proceedings of the IEEE/RSJ International Conference on Intelligent Robots and Systems, Sendai, Japan, September 2004.
9. Vidal, R., Shakernia, O., Kim, H. J., Shim, D. H., Sastry, S.: Probabilistic Pursuit-Evasion Games: Theory, Implementation and Experimental Evaluation. IEEE Transactions on Robotics and Automation, vol. 18(5), pp. 662 - 669, Oct. 2002.
10. Prandini, M., Hespanha, J., Pappas, G. J.: Greedy Control for Hybrid Pursuit Games. Proceedings of the 2001 European Control Conference, Sep. 2001.
11. Hwang, I., Balakrishnan, H., Roy, K., Tomlin, C.: Multiple-Target Tracking and Identity Management in Clutter, with Application to Aircraft Tracking. Proceedings of the 2004 American Control Conference, Boston, MA, June 2005.
12. Oh, S., Hwang, I., Roy, K., Sastry, S.: A Fully Automated Distributed Multiple-Target Tracking and Identity Management Algorithm. Proceedings of the AIAA Guidance, Navigation, and Control Conference, San Francisco, CA, August 2005.
13. Zavlanos, M. M., Pappas, G. J.: Sensor-Based Dynamic Assignment in Distributed Motion Planning. IEEE International Conference on Robotics and Automation, Rome, Italy, April 2007. (submitted)
14. Zavlanos, M. M., Pappas, G. J.: Dynamic Assignment in Distributed Motion Planning with Local Coordination. American Control Conference, New York, NY, July 2007. (submitted)
15. Kloder, S., Hutchinson, S.: Path Planning for Permutation-Invariant Multirobot Formations. IEEE Transactions on Robotics, vol. 22(4), pp. 650 - 665, Aug. 2006.
16. Ji, M., Azuma, S., Egerstedt, M.: Role-Assignment in Multi-Agent Coordination. International Journal of Assistive Robotics and Mechatronics, vol. 7(1), pp. 32-40, March 2006.
17. Zavlanos, M. M., Pappas, G. J.: A Dynamical Systems Approach to Weighted Graph Matching. 45th IEEE Conference on Decision and Control, San Diego, December 2006. (to appear)
18. Pallottino, L., Scordio, V. G., Frazzoli, E., Bicchi, A.: Decentralized Cooperative Policy for Conflict Resolution in Multi-Vehicle Systems. IEEE Transactions on Robotics, 2006. (submitted)

A Controller Design Method Under Infrequent, Asynchronous Sensing

Fumin Zhang and Naomi Ehrich Leonard

Princeton University, Mechanical and Aerospace Engineering, Princeton, NJ 08544, USA
{fzhang,naomi}@princeton.edu
http://www.princeton.edu/~fzhang,~naomi

Abstract. We use discrete quadratic Lyapunov functions to design controllers for a class of systems where time intervals between state measurements are longer than time intervals between control actions and different components of the state vector are not measured at the same time. The discrete Lyapunov function is a discretization of a continuous Lyapunov function assumed to be known for the idealized system. With this framework, we determine the maximum time interval between measurements of each state variable to guarantee the non-increasing property for the discrete Lyapunov function.

1 Introduction

We propose a controller design method for systems where the time interval between state measurements is longer than the time interval between control actions. Furthermore, we consider asynchronous sensing which arises naturally from using multiple sensors with different clocks and time scales. We investigate systems where each component in the state vector is updated based on an independent clock, i.e. different state variables are measured at different times. In [11] and [2], asynchronous distributive algorithms and asynchronous multi-agent systems are discussed with a similar definition of asynchronicity. Aysnchronicity is also investigated in [1] and its references. There, however, the asynchronicity relates to non-uniform intervals at which the system is discretized, and state variables are all updated at the same time.

By introducing a formulation of asynchronicity in measurements of different components of the state vector, we establish a procedure that allows us to incorporate sensing asynchronicity in the discrete control Lyapunov function (DCLF) based controller design method. We suppose that for the idealized continuous system when both the control and sensing intervals are infinitely small, a (continuous) controller based on a quadratic control Lyapunov function (CLF) can be found. For the asynchronous system, we use the same CLF as a candidate for the DCLF and show how to change the continuous control law into a discrete law. From this we can estimate an upper bound on the maximum sensing interval for each component in the state. There have been ongoing efforts to generalize Lyapunov stability theory to systems with hybrid nature [4,5,6,7,8].

A. Bemporad, A. Bicchi, and G. Buttazzo (Eds.): HSCC 2007, LNCS 4416, pp. 790–794, 2007.

Our developments allow us to take advantage of the abundant tools available for the continuous design.

2 Controller Design

Consider a continuous system with control on $\mathbb{R}^n$ described by $\dot{\mathbf{x}} = f(\mathbf{x}) + g(\mathbf{x})u + F(\mathbf{x}, t)$ where $\mathbf{x} = (x_1, ..., x_n) \in \mathbb{R}^n$, $u \in \mathbb{R}^m$ is the control and $F(\mathbf{x}, t)$ represents the perturbations. Suppose such system has an equilibrium at $\mathbf{x} = 0$. It may be possible to find a control Lyapunov function $V(\mathbf{x})$ from which a control law u can be derived to stabilize the equilibrium following [10]. Suppose an estimated perturbation $\hat{F}$ is available. If there exist functions $Y(\mathbf{x})$ and $Z(\mathbf{x})$ such that $Y^T(\mathbf{x}) = \nabla V^T g(\mathbf{x})$ and $Y^T(\mathbf{x})Z(\mathbf{x}) = \nabla V^T f(\mathbf{x})$ for all $\mathbf{x}$, then $\dot{V} = Y^T(\mathbf{x})(\hat{F}(\mathbf{x}, t) + u + Z(\mathbf{x})) + \nabla V^T F(\mathbf{x}, t) - Y^T(\mathbf{x})\hat{F}(\mathbf{x}, t)$. The design of u is $u = -Z(\mathbf{x}) - K\,Y(\mathbf{x}) - \hat{F}(\mathbf{x}, t)$.

We assume that the control interval Δt, the time between two control actions, is fixed. We also assume each sensing interval ΔT_i, the time between measurements of x_i, is fixed and $\Delta T_i = P_i \Delta t$ where $P_i > 1$ is an integer for $i = 1, ..., n$. A simple recursive Euler predictor can be employed to estimate the states from measurements: $\mathbf{x}_{k+1}^- = \mathbf{x}_k^+ + f(\mathbf{x}_k^+)\Delta t + g(\mathbf{x}_k^+)u_k \Delta t + \hat{F}(\mathbf{x}_k^+, t_k)\Delta t$. The symbol $\mathbf{x}_k^+$ represents the updated state at step k that has incorporated available sensor information. The symbol $\mathbf{x}_{k+1}^-$ represents the predicted state at step $k+1$. If no measurements are taken at step k, then we let $\mathbf{x}_k^+ = \mathbf{x}_k^-$. Note that more sophisticated predictors can be applied [9].

Suppose we use a discretized control law: $u_k = -Z(\mathbf{x}_k^+) - K\,Y(\mathbf{x}_k^+) - \hat{F}(\mathbf{x}_k^+, t_k)$. Suppose further that the Lyapunov function V is quadratic. Then $V_{k+1}^- - V_k^+ = \nabla(V_k^+)^T(\mathbf{x}_{k+1}^- - \mathbf{x}_k^+) + \frac{1}{2}(\mathbf{x}_{k+1}^- - \mathbf{x}_k^+)^T H_k^+ (\mathbf{x}_{k+1}^- - \mathbf{x}_k^+)$ where ∇V_k^+ and H_k^+ represents the gradient and Hessian of function V at $\mathbf{x}_k^+$.

Suppose the estimate of the perturbation satisfies $Y^T\hat{F} - \nabla V^T F = 0$ when measurements are taken. Under the proposed control we have

$$V_{k+1}^- - V_k^+ = -K\,Y_k^T Y_k \Delta t + \left\| f_k + \hat{F}_k - g_k(Z_k + \hat{F}_k + KY_k) \right\|_{H_k^+}^2 \Delta t^2 \qquad (1)$$

where we let $\hat{F}_k = \hat{F}(\mathbf{x}_k^+, t_k)$ and g_k, Y_k, Z_k and f_k are defined similarly. Matrix H_k^+ defines a pseudo inner product $\langle \cdot, \cdot \rangle_{H_k^+}$ and its induced norm $\| \cdot \|_{H_k^+}$.

From (1) we can determine the value of K that achieves the largest decrease in V as $K_k = (2\langle A_k, g_k Y_k \rangle_{H_k^+} \Delta t + Y_k^T Y_k)/(2\, \| g_k Y_k \|_{H_k^+}^2 \Delta t)$ where $A_k = f_k + \hat{F}_k - g_k(Z_k + \hat{F}_k)$. The maximum decrease in V is

$$(V_{k+1}^- - V_k^+)^* = \| A_k \|_{H_k^+}^2 \Delta t^2 - \frac{\left(2\langle A_k, g_k Y_k \rangle_{H_k^+} \Delta t + Y_k^T Y_k\right)^2}{4\, \| g_k Y_k \|_{H_k^+}^2}. \qquad (2)$$

3 Asynchronous Sensing

If there exist rational numbers a_{il} such that $\Delta T_i = a_{il}\Delta T_l$ for all $i, l = 1, ..., n$, we say the asynchronous sensing is *resonant*. By assumptions made in Sec. 2 we develop results for the resonant case.

Let j_m and j_{m+1} be two time indices. We use the symbol $\Gamma_i(j_m, j_{m+1})$ to represent the set of time indices when measurements for x_i are available within the interval $[j_m, j_{m+1})$. In the resonant case, we can always find an (infinite) sequence $\{j_m\}$ such that: (S1) The interval $j_{m+1} - j_m$ is constant for all m. (S2) $\Gamma_i(j_m, j_{m+1})$ is not empty for all i. (S3) There exists i such that a measurement for x_i is available at each j_m.

We often write $\Gamma_i(j_m, j_{m+1})$ simply as Γ_i. Let N_i be the total number of elements in Γ_i and let $N = \sum_i N_i$. Since N is finite, we can construct an N dimensional vector $\mathbf{v}$ that contains all members of Γ_i for all i. We call $\mathbf{v}$ *the asynchronous index vector*. Note that N is constant for all $[j_m, j_{m+1})$.

Consider the DCLF $V(\mathbf{x})$ where $\mathbf{x} \in \mathbb{R}^n$. Let p assume all possible values for $p \in \Gamma_i$ and for $i = 1, 2, ..., n$. We use V_p^- and V_p^+ to denote the function value before and after the update. Let $\mathbf{x}_p^-$, $\mathbf{x}_p^+$, ∇V_p^- and H_p^- be the state vector before measurement, the state vector after measurement, the gradient vector and the Hessian at the time instant indexed by p.

Definition 1. *Consider the interval $[j_m, j_{m+1})$. On this interval, we define the* asynchronous state vector before update *as the N dimensional vector $\hat{\mathbf{x}}^{\mathrm{a}}$ whose r-th component satisfies $(\hat{\mathbf{x}}^{\mathrm{a}})_r = (\mathbf{x}_{v_r}^-)_i$ where v_r is the r-th component of the asynchronous index vector $\mathbf{v}$ and $v_r \in \Gamma_i$. We define the* asynchronous state vector after update *as the N dimensional vector $\mathbf{x}^{\mathrm{a}}$ whose r-th component satisfies $(\mathbf{x}^{\mathrm{a}})_r = (\mathbf{x}_{v_r}^+)_i$. We define the* asynchronous gradient *as the N dimensional vector $\nabla^{\mathrm{a}} V$ whose r-th component satisfies $(\nabla^{\mathrm{a}} V)_r = (\nabla V_{v_r}^-)_i$. We define the* asynchronous Hessian *as the $N \times N$ matrix H^{a} whose elements satisfy, for $r, l = 1, 2, ..., N$, $(H^{\mathrm{a}})_{r,l} = (H_{v_r}^-)_{i_1,i_1}$ if $r = l$, $(H^{\mathrm{a}})_{r,l} = (H_{v_r}^-)_{i_1,i_2}$ if $v_r = v_l$ but $i_1 \neq i_2$, and $(H^{\mathrm{a}})_{r,l} = 0$ for other cases, where indices i_1 and i_2 are such that $v_r \in \Gamma_{i_1}$ and $v_l \in \Gamma_{i_2}$.*

As a convention, we let $V_p^- = V_p^+$ if $p \notin \Gamma_i$ for any i. We compare the value of the candidate function V at two time instants j_m and j_{m+1} when the same part of the states have been updated. The difference between the function values $V_{j_m}^-$ and $V_{j_{m+1}}^-$ can be written as $V_{j_{m+1}}^- - V_{j_m}^- = \sum_{p=j_m}^{j_{m+1}-1} \left(V_p^+ - V_p^-\right) + \sum_{p=j_m}^{j_{m+1}-1} \left(V_{p+1}^- - V_p^+\right)$. Using the discretized controller u_k with adaptive gain K_k from Sec. 2, for $V_{j_{m+1}}^- - V_{j_m}^- \leq 0$ to hold, we need $\sum_{p=j_m}^{j_{m+1}-1} \left(V_p^+ - V_p^-\right) \leq \sum_{p=j_m}^{j_{m+1}-1} (V_p^+ - V_{p+1}^-)^*$. We introduce the notion of an asynchronously positive invariant set. This notion is closely related to Poisson stability c.f. [3].

Definition 2. *Consider an infinite sequence $\{j_m\}$ that satisfies assumptions (S1-S3). We say a set $M \subset \mathbb{R}^N$ is* asynchronously positive invariant *for $\{j_m\}$ if for $\mathbf{x}^a(j_0, j_1) \in M$, we have $\mathbf{x}^{\mathrm{a}}(j_m, j_{m+1}) \in M$ for all $m \geq 0$.*

We let the set $E_1 \subset \mathbb{R}^N$ be the set of all asynchronous states $\mathbf{x}^{\mathrm{a}}$ where $V^-_{j_{m+1}} - V^-_{j_m} \leq 0$ is satisfied for the sequence $\{j_m\}$. Let the set M_1 be the set of $\mathbf{x}^{\mathrm{a}}$ where $V(\Phi^-_{j_{m+1}}(\mathbf{x}^{\mathrm{a}}(j_m, j_{m+1}))) \leq c_1$ for some $c_1 \geq 0$ where $\Phi^-_{j_{m+1}}$ is the state transition map from $\mathbf{x}^{\mathrm{a}}(j_m, j_{m+1})$ to $\mathbf{x}^-_{j_{m+1}}$.

Proposition 1. *If $M_1 \subset E_1$, then M_1 is asynchronously positive invariant.*

Let $\mathbf{e}^{\mathrm{a}} = \mathbf{x}^{\mathrm{a}} - \hat{\mathbf{x}}^{\mathrm{a}}$. This $\mathbf{e}^{\mathrm{a}}$ contains all the corrections to the states when measurements are available. To estimate the upper bound for sensing intervals, some assumptions are needed regarding how the prediction error depends on time. If $x_i^+(t) = x_i^-(t)$, we assume that $|x_i^+(t+\tau) - x_i^-(t+\tau)| \leq L\tau^q$ for $i = 1, 2, ..., n$ where $L > 0$, $q \geq 1$ and for all $t, \tau \geq 0$. Under this assumption, the asynchronous error satisfies $\| \mathbf{e}^{\mathrm{a}} \| \leq L\, \Delta T_i^q \left(\sum_{l=1}^n N_l a_{li}^2\right)^{\frac{1}{2}}$ for any given i.

Proposition 2. $\sum_{p=j_m}^{j_{m+1}-1} \left(V_p^+ - V_p^-\right) = (\nabla^{\mathrm{a}} V)^T \mathbf{e}^{\mathrm{a}} + \frac{1}{2}(\mathbf{e}^{\mathrm{a}})^T H^{\mathrm{a}} \mathbf{e}^{\mathrm{a}}$.

We now know $| \sum_{p=j_m}^{j_{m+1}-1} \left(V_p^+ - V_p^-\right) | \leq \| \nabla V^{\mathrm{a}} \| \| \mathbf{e}^{\mathrm{a}} \| + \frac{1}{2} \| H^{\mathrm{a}} \| \| \mathbf{e}^{\mathrm{a}} \|^2$. Then a sufficient condition for $V^-_{j_{m+1}} - V^-_{j_m} \leq 0$ is $\| \nabla V^{\mathrm{a}} \| \| \mathbf{e}^{\mathrm{a}} \| + \frac{1}{2} \| H^{\mathrm{a}} \| \| \mathbf{e}^{\mathrm{a}} \|^2 \leq \Delta V^*$ where $\Delta V^* = \sum_{p=j_m}^{j_{m+1}-1} (V_p^+ - V_{p+1}^-)^*$. If this condition is satisfied for all $\mathbf{x}^a \in M_1$, then $M_1 \subset E_1$ and M_1 is asynchronously positive invariant by Prop. 1. We can estimate the maximum sensing interval ΔT_i for M_1 to be asynchronously positive invariant as

$$\Delta T_i^* = \inf_{\mathbf{x}^{\mathrm{a}} \in M_1} \left(\frac{- \| \nabla V^{\mathrm{a}} \| + \left(\| \nabla V^{\mathrm{a}} \|^2 + 2 \| H^{\mathrm{a}} \| \Delta V^* \right)^{\frac{1}{2}}}{L \left(\sum_{l=1}^n N_l a_{li}^2\right)^{\frac{1}{2}} \| H^{\mathrm{a}} \|} \right)^{\frac{1}{q}}. \quad (3)$$

References

1. A. Balluchi, P. Murrieri, and A. L. Sangiovanni-Vincentelli. Controller synthesis on non-uniform and uncertain discrete-time domains. In M. Morari and L. Thiele, editors, *HSCC 2005*, LNCS 3414:118–133. Springer-Verlag, 2005.
2. M. Cao, A. S. Morse, and B.D.O. Anderson. Agreeing asynchronously in continuous time, *IEEE Trans. Automat. Cont.*, submitted 2006.
3. P. E. Crouch. Spacecraft attitude control and stabilization: applications of geometric control theory to rigid body models. *IEEE Trans. Automat. Cont.*, 29(4): 321-331,1984.
4. J.W. Grizzle and J-M. Kang. Discrete-time control design with positive semi-definite Lyapunov functions. *Syst. & Cont. Letters*, 40(3):329–342, 1984.
5. T. Hu and Z. Lin. Absolute stability analysis of discrete-time systems with composite quadratic Lyapunov functions. *IEEE Trans. Automat. Cont.*, 50(6):781–797, 2005.
6. C.J.B. Macnab and G.M.T. D'Eleuterio. Discrete-time Lyapunov design for neuroadaptive control of elastic-joint robots. *Int. J. Robot. Res.*, 19(5):511–525, 2000.
7. K. M. Passino, A. N. Michel, and P. J. Antsaklis. Lyapunov stability of a class of discrete event systems. *IEEE Trans. Automat. Cont.*, 39(2):269–279, 1994.

8. M. S. Branicky. Multiple Lyapunov functions and other analysis tools for switched and hybrid systems. *IEEE Trans. Automat. Cont.*, 43:475–482, 1998.
9. B. Sinopoli, L. Schenato, M. Franceschetti, K. Poolla, M. Jordan and S. Sastry. Kalman filtering with intermittent observations *IEEE Trans. Automat. Cont.*, 49(9):1453-1464, 2004.
10. E. D. Sontag. *Mathematical Control Theory: Deterministic Finite Dimensional Systems 2nd ed.*, Springer, 1998.
11. J. N. Tsitsiklis and D. P. Bertsekas. Distributed asynchronous routing in data communication networks. *IEEE Trans. Automat. Cont.*, 31:325–332, 1986.

Author Index

Printing: Mercedes-Druck, Berlin
Binding: Stein + Lehmann, Berlin